Interaction of Nanomaterials With Living Cells

Faheem Arjamend Sheikh •
Shafquat Majeed • Mushtaq A. Beigh
Editors

Interaction of Nanomaterials With Living Cells

Springer

Editors
Faheem Arjamend Sheikh
Department of Nanotechnology
University of Kashmir
Srinagar, Jammu and Kashmir, India

Shafquat Majeed
Department of Nanotechnology
University of Kashmir
Srinagar, Jammu and Kashmir, India

Mushtaq A. Beigh
Department of Nanotechnology
University of Kashmir
Srinagar, Jammu and Kashmir, India

ISBN 978-981-99-2118-8 ISBN 978-981-99-2119-5 (eBook)
https://doi.org/10.1007/978-981-99-2119-5

This Springer imprint is published by the registered company Springer Nature Singapore Pte Ltd.
The registered company address is: 152 Beach Road, #21-01/04 Gateway East, Singapore 189721, Singapore

This book is dedicated to the Department of Nanotechnology, University of Kashmir.

Preface

As the scientific community encounters new and complex problems with each passing day and as a dynamically evolving field of science, a need to search for advanced biocompatible nanomaterials and the mode of their interaction with living systems needs to be explored for their use in human welfare.

This book entitled "*Interaction of Nanomaterials with Living Cells*" will further enrich the understanding of readers about various classes of nanoscale materials vis-à-vis their behavior and interaction with the cellular system.

This book comprises 31 uniquely drafted chapters critically analyzing various aspects of materials in nano dimensions that are currently being explored for biomedical applications. Chapter 1 discusses the role of synthetic and natural biomaterials in modulating the autoimmune response for better therapeutic outcomes, especially in countering autoimmune diseases. Chapter 2 introduces nano-biosensors to diagnose a range of conditions as well as their therapeutic significance. Chapter 3 sheds light on materials like calcium, phosphorous, bioglass, collagen, chitosan, hyaluronic acid, etc. that can be used efficiently for bone tissue engineering. Similarly, Chap. 4 covers the use of hydroxyapatite in bioimaging, controlled medication administration, gene treatments, and tissue engineering. Chapter 5 covers antiviral, antibacterial, antioxidant, anticancer, anti-inflammatory, and antiparasitic wound-healing activities properties of nanoparticles fabricated by green synthetic routes. Chapter 6 highlights carbon nanotube properties, characterization, functionalization, toxicity, and future prospects. Chapter 7 brings up the role of nanomaterials in improving animal health and production. Further, the toxicity of nanomaterials and considerations for animal health and safety are also discussed. Chapter 8 discusses the bioinspired materials that are inherited with antimicrobial properties that can be used for tissue engineering applications. Chapter 9 introduces 3D and 4D bioprinting technology for tissue engineering applications focusing on organ and tissue bioprinting.

Chapter 10 discusses the hemocompatibility of differently modified polymeric nanofibers and the testing of these nanofibers to determine hemolysis rate, platelet adhesion, plasma recalcification time, free hemoglobin estimation, attachment and release of red blood cells, and adsorption of plasma proteins. Chapter 11 discusses the role of polyurethane nanofibers fabricated by electrospinning used as drug carrier systems for treating cancers. Chapter 12 focuses on different types of nanoparticles used in cancer therapy and diagnosis. Chapter 13 introduces the applications of bioactive compounds and biomaterials in promoting cell differentiation, proliferation, and regenerating tissue. Chapter 14 will discuss the materials, for example, lipid, polymer, and peptide-based, for gene delivery. Chapter 15 introduces the natural hydrogels used as dressing for skin wound-healing applications. Chapter 16 offers a better understanding of nanomaterial-based cancer theranostics and recent innovations in this area. Chapter 17 describes nanocellulose as a sustainable nanomaterial for films and coating layers via spray-coating and applications. Chapter 18 describes a nanoparticles-based targeted drug delivery systems. Chapter 19 gives insight into the treatment of osteoarthritis using different materials. Chapter 20 emphasizes the potency of nanoencapsulation techniques using bioactive compounds. Chapter 21 reviews the green routes of synthesis, modification, characterization, properties, and applications of palladium-based nanoparticles in biomedical applications. Chapter 22 describes the different innovative nanomaterials with profound antibacterial action applied in biomedical sciences. Chapter 23 introduces musculoskeletal pains and diseases with novel insights in biomaterial based treatment. Chapter 24 describes the electrospun cellulose and derivative-based nanofibers loaded with bioactive agents that can be used in wound dressing and healing applications. Chapter 25 discusses the challenges and advantages of nanomedicine, especially the delivery routes, for example, using nanoparticles, dendrimers, micelle, drug conjugate, nanocapsules, nanoliposomes exosome, noisome, and nanogel. Chapter 26 mentions the silver nanoparticles incorporated in textile substrates for antimicrobial applications. Chapter 27 describes the biocompatibility and biodegradability of gelatin and its favorable physicochemical characteristics. Chapter 28 sheds light on biomedical applications of fused filament fabrication. Chapter 29 describes the role of stem cells in the delivery of essential pharmaceuticals. Chapter 30 describes various biomaterials used to treat autoimmune diseases (e.g., rheumatoid arthritis, multiple sclerosis, and type I diabetes). Finally, the regulatory and ethical issues associated with the use of various types of nanoscale materials are discussed in Chap. 31.

Srinagar, India

Faheem Arjamend Sheikh
Shafquat Majeed
Mushtaq A. Beigh

Acknowledgments

The Editors would like to express their heartfelt gratitude to all the contributing authors who, despite their busy schedules, contributed to this work at concise notice. However, at the same time, editing this book was a time-consuming process, and for that, we are thankful to our families for their patience and support during the completion of this work.

We also appreciate the kind efforts of Dr. Bhavik Sawhney (Editor, Biomedicine), Mahalakshmi Shankar (Production Editor) and Harshini Punithan, (Project Manager) at Springer Nature.

Contents

Editors and Contributors

About the Editors

Faheem Arjamend Sheikh is currently working as an Assistant Professor in the Department of Nanotechnology, University of Kashmir, India. He served as an Assistant Professor at the Department of Biotechnology at the Central University of Kashmir, India (2015–2016); Assistant Professor of Research at Myongji University, South Korea (2014–2015); Research Professor at Hallym University, South Korea (2012–2014); Post-Doc/Research fellow at the University of Texas Rio Grande Valley, Texas, United States (2010–2012) and Research Professor at Myongji University, South Korea (2010). His research mainly focuses on fabricating nanomaterials used in tissue engineering. Dr. Faheem did his Ph.D. at Jeonbuk National University, South Korea, Jeonju (2006–2010).

Currently, he is heading a lab that focusses on creating different nanomaterials that can be used in various biological applications. He has considerable expertise in the fabrication of polymeric and metal oxide nanofibers using electrospinning, as well as the production of porous materials by solvent casting, salt-leaching, 3D printing, gas forming, sol-gel synthesis, phase separation, freeze-drying, and particulate leaching for hard and soft tissue engineering. He has more than 17 years of research experience, focusing on tissue engineering and drug delivery. He has published more than 100 peer-reviewed articles, conference papers and has authored several books.

Shafquat Majeed is currently working as an Assistant Professor in the Department of Nanotechnology, University of Kashmir, India. Dr. Majeed did his Master's in Chemistry from the Department of Chemistry, University of Kashmir. He did his Ph.D. from the Indian Institute of Science Bangalore (2008–2015). He has postdoctoral research experience in rare-earth-based nanomaterials for biomedical applications and fabrication of Perovskite-based solar cells from the Centre for Nanoscience and Engineering, IISc Bangalore (2015–2016).

Presently, he heads a Multifunctional Nanomaterials Laboratory at the University of Kashmir where the research group works on the synthesis, characterization, optical and magnetic properties of nanomaterials prepared through various solution-based routes.

Mushtaq A. Beigh is currently working as an Assistant Professor at the Department of Nanotechnology, University of Kashmir, India. Dr. Beigh was a DST-Young scientist fellow and did his postdoctoral research in Cellular signaling and disease modeling (2013–2016). He completed his Ph.D. from the Laboratory of Signal transduction, Department of Biotechnology, University of Kashmir, India (2007–2012). He currently heads Cellular signaling and Nanotherapeutics laboratory at the University of Kashmir, where they are trying to understand the molecular mechanisms of cancer metastasis and bioengineer nanomaterial-based drug conjugates for targeted therapies in cancer. Dr. Beigh has qualified several national level exams like CSIR-NET JRF, GATE, SET, DBT-RA and to his credit has multiple international and national publications.

Contributors

Sabreena Aashaq Department of Immunology and Molecular Medicine, Sher-i-Kashmir Institute of Medical Sciences, Srinagar, Jammu and Kashmir, India
Cellular Signalling and Nanotherapeutics Laboratory, Department of Nanotechnology, University of Kashmir Hazratbal, Srinagar, Jammu and Kashmir, India

Ardeshir Abbasi Department of Immunology, Faculty of Medical Sciences, Tarbiat Modares University, Tehran, Iran

Sarbari Acharya Department of Biology, School of Applied Sciences, Kalinga Institute of Industrial Technology, Bhubaneswar, Odisha, India

Blessing A. Aderibigbe Department of Chemistry, University of Fort Hare, Alice, Eastern Cape, South Africa

Azadeh Izadyari Aghmiuni Department of Nanobiotechnology, Pasteur Institute of Iran, Tehran, Iran

Syed Mudasir Ahmad Faculty of Veterinary Sciences, Sher-e-Kashmir University of Agricultural Sciences and Technology, Srinagar, Jammu and Kashmir, India

Imran Ahmad Ganai Mountain Research Centre for Sheep and Goat, Sher-e-Kashmir University of Agricultural Sciences and Technology, Srinagar, Jammu and Kashmir, India

Akanksha Amity Institute of Nanotechnology, Amity University Uttar Pradesh, Noida, Uttar Pradesh, India

Abdullah A. A. Alghamdi Department of Biology, Faculty of Science, Al-Baha University, Albaha, Saudi Arabia

Mehvish Ali Cellular Signalling and Nanotherapeutics Laboratory, Department of Nanotechnology, University of Kashmir Hazratbal, Srinagar, Jammu and Kashmir, India

Rasha Alnefaie Department of Biology, Faculty of Science, Al-Baha University, Albaha, Saudi Arabia

Sibusiso Alven Department of Chemistry, University of Fort Hare, Alice, Eastern Cape, South Africa

Saurin Amin Aavis Pharmaceuticals, Hoschton, GA, USA

Touseef Amna Department of Biology, Faculty of Science, Al-Baha University, Albaha, Saudi Arabia

Abdul Salam Ansari Centre for Advanced Studies, Department of Zoology, University of Rajasthan, Jaipur, India

Mahdieh Asadi Department of Sport Physiology, Islamic Azad University, Sari Branch, Sari, Iran

Lubna Azmi CSIR-National Botanical Research Institute, Hazratgunj, Lucknow, Uttar Pradesh, India

Behnaz Bakhshandeh Department of Biotechnology, College of Science, University of Tehran, Tehran, Iran

Ramachandran Balaji Department of Biotechnology and Chemical Engineering, National Taipei University of Technology, Taipei, Taiwan

Aruna Kumar Barick Department of Chemistry, Veer Surendra Sai University of Technology, Burla, Odisha, India

Ananya Barui Center for Healthcare Science and Technology, Indian Institute of Engineering Science and Technology, Howrah, India

Howa Begam Center for Healthcare Science and Technology, Indian Institute of Engineering Science and Technology, Howrah, India

Mushtaq A. Beigh Cellular Signalling and Nanotherapeutics Laboratory, Department of Nanotechnology, University of Kashmir Hazratbal, Srinagar, Jammu and Kashmir, India
Department of Immunology and Molecular Medicine, Sher-i-Kashmir Institute of Medical Sciences, Srinagar, Jammu and Kashmir, India

Yasir Afzal Beigh Faculty of Veterinary Sciences, Sher-e-Kashmir University of Agricultural Sciences and Technology, Srinagar, Jammu and Kashmir, India

Irene Buj-Corral Department of Mechanical Engineering, Barcelona School of Industrial Engineering (ETSEIB), Universitat Politècnica de Catalunya (UPC), Barcelona, Spain

Narendhar Chandrasekar Department of BioNano Technology, Gachon University, Seongnam-si, Gyeonggi-do, Republic of Korea

Timanshi Chansoriya Centre for Advanced Studies, Department of Zoology, University of Rajasthan, Jaipur, India

Jaeyeop Choi Industry 4.0 Convergence Bionics Engineering, Department of Biomedical Engineering, Pukyong National University, Busan, Republic of Korea

V. R. Giri Dev Department of Textile Technology, Anna University, Chennai, Tamil Nadu, India

Mariana Díaz-Zaragoza Departamento de Ciencias de la Salud, Centro Universitario de los Valles (CUVALLES), Universidad de Guadalajara, Ameca, Jalisco, Mexico

Fahima Dilnawaz Department of Biotechnology, School of Engineering and Technology, Centurion University of Technology and Management, Bhubaneswar, India

Javaid Farooq Faculty of Veterinary Sciences, Sher-e-Kashmir University of Agricultural Sciences and Technology, Srinagar, Jammu and Kashmir, India

Gabriela Fletes-Vargas Departamento de Ciencias Clínicas, Centro Universitario de los Altos (CUALTOS), Universidad de Guadalajara, Tepatitlán de Morelos, Jalisco, Mexico
Tecnología de Alimentos, Centro de Investigación y Asistencia en Tecnología y Diseño del Estado de Jalisco A.C. (CIATEJ, A.C.), Zapopan, Jalisco, Mexico

Abdul Majeed Ganai Faculty of Veterinary Sciences, Sher-e-Kashmir University of Agricultural Sciences and Technology, Srinagar, Jammu and Kashmir, India

Hosahalli Veerabhadrappa Gangadharappa Department of Pharmaceutics, JSS College of Pharmacy, JSS Academy of Higher Education and Research, Mysuru, Karnataka, India

Arezoo Ghadi Department of Chemical Engineering, Islamic Azad University, Ayatollah Amoli Branch, Amol, Iran

Pon G. K. Gokul Department of Textile Technology, Anna University, Chennai, Tamil Nadu, India

M. P. Gowrav Department of Pharmaceutics, JSS College of Pharmacy, JSS Academy of Higher Education and Research, Mysuru, Karnataka, India

Gowher Gull Sheikh Faculty of Veterinary Sciences, Sher-e-Kashmir University of Agricultural Sciences and Technology, Srinagar, Jammu and Kashmir, India

Dorrin Mohtadi Haghighi Department of Pharmaceutics, Faculty of Pharmacy, Tehran University of Medical Sciences, Tehran, Iran

Henna Hamadani Faculty of Veterinary Sciences, Sher-e-Kashmir University of Agricultural Sciences and Technology, Srinagar, Jammu and Kashmir, India

M. Shamshi Hassan Department of Chemistry, Faculty of Science, Al-Baha University, Albaha, Saudi Arabia

Somia Abd Alla Mohammed El Hassan Department of Biology, Faculty of Science, Al-Baha University, Albaha, Saudi Arabia

T. Hemamalini Department of Textile Technology, Anna University, Chennai, Tamil Nadu, India

Carolina Herranz-Diez Unit of Biophysics and Bioengineering, Faculty of Medicine and Health Sciences, Universitat de Barcelona (UB), Barcelona, Spain

Zohreh Jahanafrooz Department of Biology, Faculty of Sciences, University of Maragheh, Maragheh, Iran

Bigyan Ranjan Jali Department of Chemistry, Veer Surendra Sai University of Technology, Burla, Odisha, India

Iqra Jan Cellular Signalling and Nanotherapeutics Laboratory, Department of Nanotechnology, University of Kashmir Hazratbal, Srinagar, Jammu and Kashmir, India

Asha Puthuvilayil Johnson Department of Pharmaceutics, JSS College of Pharmacy, JSS Academy of Higher Education and Research, Mysuru, Karnataka, India

Anwesha Kanungo Department of Biology, School of Applied Sciences, Kalinga Institute of Industrial Technology, Bhubaneswar, Odisha, India

Saeed Heidari Keshel Department of Tissue Engineering and Applied Cell Science, School of Advanced Technologies in Medicine, Shahid Beheshti University of Medical Sciences, Tehran, Iran

Azmat Alam Khan Faculty of Veterinary Sciences, Sher-e-Kashmir University of Agricultural Sciences and Technology, Srinagar, Jammu and Kashmir, India

Nisar Ahmad Khan Department of Pharmaceutical Sciences, University of Kashmir Hazratbal, Srinagar, Jammu and Kashmir, India

Rumysa Saleem Khan Nanostructured and Biomimetic Lab, Department of Nanotechnology, University of Kashmir Hazratbal, Srinagar, Jammu and Kashmir, India

Barkha Khilwani Centre for Advanced Studies, Department of Zoology, University of Rajasthan, Jaipur, India

Azim Akbarzadeh Khiyavi Department of Nanobiotechnology, Pasteur Institute of Iran, Tehran, Iran

Ekta Arjundas Kukreja Amity Institute of Nanotechnology, Amity University Uttar Pradesh, Noida, Uttar Pradesh, India

R. S. Mahalakshmi Department of Textile Technology, Anna University, Chennai, Tamil Nadu, India

Aasiya Majeed Department of Biochemistry, Sher-e-Kashmir University of Agricultural Sciences and Technology (Jammu), Srinagar, Jammu and Kashmir, India

Shafquat Majeed Laboratory for Multifunctional Nanomaterials, Department of Nanotechnology, University of Kashmir Hazratbal, Srinagar, Jammu and Kashmir, India

Laboratory for Multifunctional Nanomaterials, Department of Nanotechnology, University of Kashmir Hazratbal, Srinagar, Jammu and Kashmir, India

S. K. Manohar Department of Pharmaceutics, JSS College of Pharmacy, JSS Academy of Higher Education and Research, Mysuru, Karnataka, India

Saima Masood Department of Basic Sciences and Humanities, SKUAST-K, Srinagar, Jammu and Kashmir, India

Amarendra Narayan Misra School of Applied Sciences, Centurion University of Technology and Management, Bhubaneswar, India

Pooja Mohapatra Department of Chemistry, Veer Surendra Sai University of Technology, Burla, Odisha, India

Priyaranjan Mohapatra Department of Chemistry, Veer Surendra Sai University of Technology, Burla, Odisha, India

Sudip Mondal New-Senior Healthcare Innovation Center (BK21 Plus), Pukyong National University, Busan, Republic of Korea

Naveed Ul Mushtaq Department of Bioresources, School of Biological Sciences, University of Kashmir Hazratbal, Srinagar, India

Tobeka Naki Department of Chemistry, University of Fort Hare, Alice, Eastern Cape, South Africa

Samit Kumar Nandi Department of Veterinary Surgery and Radiology, West Bengal University of Animal & Fishery Sciences, Kolkata, India

Sadaf Nazir Multifunctional Nanomaterials Laboratory, Department of Nanotechnology, University of Kashmir Hazratbal, Srinagar, Jammu and Kashmir, India

Sindi P. Ndlovu Department of Chemistry, University of Fort Hare, Alice, South Africa

Junghwan Oh New-Senior Healthcare Innovation Center (BK21 Plus), Pukyong National University, Busan, Republic of Korea
Industry 4.0 Convergence Bionics Engineering, Department of Biomedical Engineering, Pukyong National University, Busan, Republic of Korea
Ohlabs Corp., Busan, Republic of Korea

Jordi Otero Unit of Biophysics and Bioengineering, Faculty of Medicine and Health Sciences, Universitat de Barcelona (UB), Barcelona, Spain
CIBER de Enfermedades Respiratorias, Madrid, Spain

Sajad Ahmed Padder Department of Bioresources, School of Biological Sciences, University of Kashmir Hazratbal, Srinagar, India

Sampita Pal Department of Natural Products Chemistry, National Institute of Pharmaceutical Education and Research, Kolkata, West Bengal, India

Fatma Nur Parın Polymer Materials Engineering Department, Faculty of Engineering and Natural Sciences, Bursa Technical University, Bursa, Turkey

Sumin Park Industry 4.0 Convergence Bionics Engineering, Department of Biomedical Engineering, Pukyong National University, Busan, Republic of Korea

Anita Patel Sankalchand Patel University, Visnagar, Gujarat, India

Jayvadan K. Patel Aavis Pharmaceuticals, Hoschton, GA, USA
Faculty of Pharmacy, Sankalchand Patel University, Visnagar, Gujarat, India

Nisarg C. Patel APMC College of Pharmaceutical Education and Research, Himatnagar, Gujarat, India

Sijongesonke Peter Department of Chemistry, University of Fort Hare, Alice, Eastern Cape, South Africa

Armaghan Pourramezanali Department of Microbiology, School of Biology, Faculty of Science, Tehran University, Tehran, Iran

Rakesh Kumar Pradhan Department of Chemistry, Veer Surendra Sai University of Technology, Burla, Odisha, India

Mingfeng Qiu School of Pharmacy, Shanghai Jiao Tong University, Shanghai, China

Asma Rafiq Cellular Signalling and Nanotherapeutics Laboratory, Department of Nanotechnology, University of Kashmir Hazratbal, Srinagar, Jammu and Kashmir, India

Muheeb Rafiq Nanostructured and Biomimetic Lab, Department of Nanotechnology, University of Kashmir Hazratbal, Srinagar, Jammu and Kashmir, India

Nasim Rahmani-Kukia Department of Biochemistry, School of Medicine, Shiraz University of Medical Sciences, Shiraz, Iran

R. Rajesh Department of Textile Technology, Anna University, Chennai, Tamil Nadu, India

Aadil Rasool Department of Bioresources, School of Biological Sciences, University of Kashmir Hazratbal, Srinagar, India

Anjum Hamid Rather Nanostructured and Biomimetic Lab, Department of Nanotechnology, University of Kashmir Hazratbal, Srinagar, Jammu and Kashmir, India

Faisal Raza School of Pharmacy, Shanghai Jiao Tong University, Shanghai, China

Reiaz Ul Rehman Department of Bioresources, School of Biological Sciences, University of Kashmir Hazratbal, Srinagar, India

Bilal A. Reshi Department of Biotechnology, University of Kashmir Hazratbal, Srinagar, Jammu and Kashmir, India

Sergio Yair Rodríguez-Preciado Departamento de Ciencias de la Salud, Centro Universitario de los Valles (CUVALLES), Universidad de Guadalajara, Ameca, Jalisco, Mexico

Rogelio Rodríguez-Rodríguez Departamento de Ciencias Naturales y Exactas, Centro Universitario de los Valles (CUVALLES), Universidad de Guadalajara, Ameca, Jalisco, Mexico

Seerat Saleem Department of Bioresources, School of Biological Sciences, University of Kashmir Hazratbal, Srinagar, India

Sayantika Sanyal Center for Healthcare Science and Technology, Indian Institute of Engineering Science and Technology, Howrah, India

Komal Iranna Savadatti Department of Pharmaceutics, JSS College of Pharmacy, JSS Academy of Higher Education and Research, Mysuru, Karnataka, India

Wasifa Hafiz Shah Department of Bioresources, School of Biological Sciences, University of Kashmir Hazratbal, Srinagar, India

Zaffar Amin Shah Department of Immunology and Molecular Medicine, Sher-i-Kashmir Institute of Medical Sciences, Srinagar, Jammu and Kashmir, India

Faamiya Shajar Department of Bioresources, School of Biological Sciences, University of Kashmir Hazratbal, Srinagar, India

Ke Shang College of Animal Science and Technology/College of Veterinary Medicine, Henan University of Science and Technology, Luoyang, China

Kirubanandan Shanmugam Saveetha School of Engineering, SIMATS, Chennai, Tamil Nadu, India

Faheem A. Sheikh Nanostructured and Biomimetic Lab, Department of Nanotechnology, University of Kashmir Hazratbal, Srinagar, Jammu and Kashmir, India

Neeraj Shrivastava Amity Institute of Microbial Technology, Amity University Uttar Pradesh, Noida, Uttar Pradesh, India

Lipsa Shubhadarshinee Department of Chemistry, Veer Surendra Sai University of Technology, Burla, Odisha, India

Alven Sibusiso Department of Chemistry, University of Fort Hare, Alice, South Africa

Hasham S. Sofi Nanostructured and Biomimetic Lab, Department of Nanotechnology, University of Kashmir Hazratbal, Srinagar, Jammu and Kashmir, India

Jing Su School of Pharmacy, Shanghai Jiao Tong University, Shanghai, China

Inayatullah Tahir Department of Bioresources, School of Biological Sciences, University of Kashmir Hazratbal, Srinagar, India

Aitor Tejo-Otero Department of Mechanical Engineering, Barcelona School of Industrial Engineering (ETSEIB), Universitat Politècnica de Catalunya (UPC), Barcelona, Spain

Ravi Mani Tripathi Amity Institute of Nanotechnology, Amity University Uttar Pradesh, Noida, Uttar Pradesh, India

Zulfqar ul Haq Faculty of Veterinary Sciences, Sher-e-Kashmir University of Agricultural Sciences and Technology, Srinagar, Jammu and Kashmir, India

Taha Umair Wani Nanostructured and Biomimetic Lab, Department of Nanotechnology, University of Kashmir Hazratbal, Srinagar, Jammu and Kashmir, India

Mohammad Zahraei Department of Orthopedics, Nikan Education & Research center (NERC), West Nikan Hospital, Tehran, Iran

Jun-Feng Zhang Department of Veterinary Infectious Diseases and Avian Diseases, College of Veterinary Medicine and Center for Poultry Diseases Control, Jeonbuk National University, Iksan, South Korea

Jian Zhong Xinhua Hospital, Shanghai Institute for Pediatric Research, Shanghai Key Laboratory of Gastroenterology and Nutrition, Shanghai Jiao Tong University School of Medicine, Shanghai, China

The Role of Synthetic and Natural Biomaterials in Modulating the Autoimmune Response

1

Sampita Pal, Hasham S. Sofi, Lubna Azmi, Nisar Ahmad Khan, Shafquat Majeed, Mushtaq A. Beigh, and Faheem A. Sheikh

S. Pal
Department of Natural Products Chemistry, National Institute of Pharmaceutical Education and Research, Kolkata, West Bengal, India

H. S. Sofi
Nanostructured and Biomimetic Lab, Department of Nanotechnology, University of Kashmir Hazratbal, Srinagar, Jammu and Kashmir, India

Department of Pharmaceutical Sciences, University of Kashmir, Hazratbal, Srinagar, Jammu and Kashmir, India
e-mail: hasham.scholar@kashniruniversity.net

L. Azmi
CSIR-National Botanical Research Institute, Hazratgunj, Lucknow, Uttar Pradesh, India

N. A. Khan
Department of Pharmaceutical Sciences, University of Kashmir, Hazratbal, Srinagar, Jammu and Kashmir, India

S. Majeed
Laboratory for Multifunctional Nanomaterials, Department of Nanotechnology, University of Kashmir Hazratbal, Srinagar, Jammu and Kashmir, India
e-mail: smshah@uok.edu.in

M. A. Beigh
Cellular Signalling and Nanotherapeutics Laboratory, Department of Nanotechnology, University of Kashmir Hazratbal, Srinagar, Jammu and Kashmir, India
e-mail: beighm@uok.edu.in

F. A. Sheikh (✉)
Nanostructured and Biomimetic Lab, Department of Nanotechnology, University of Kashmir Hazratbal, Srinagar, Jammu and Kashmir, India
e-mail: faheemnt@uok.edu.in

F. A. Sheikh et al. (eds.), *Interaction of Nanomaterials With Living Cells*,
https://doi.org/10.1007/978-981-99-2119-5_1

Abstract

The human immune system sketches an indispensable role in modulating disease advancement and maintaining tissue homeostasis. Conventionally, the immune system is recognized as a defensive system preventing bacterial and/or viral infections. Besides the immune system's role in preventing pathogenesis, it plays an essential role in countering allergies, cancer, autoimmune diseases, tissue repair mechanisms, and regeneration. Modulation of the immune system plays a vital role in eliminating many diseases and immune disorders. However, therapeutic modulation of the immune system may result in lifelong comorbidities and severe to fatal side effects. This chapter presents a brief outlook of the immune system and its role in pathogenesis and maintaining tissue homeostasis. Furthermore, this chapter highlights the role of different bioengineered materials in modulating the host immune system for better therapeutic outcomes, especially in countering autoimmune diseases.

Keywords

Novel scaffolds · Immune cell types · Autoimmune · Nanomaterials

1.1 Introduction

Autoimmunity-assisted disorders (ADs) can be defined as diseases resulting from the body's abnormal immune response, which recognizes the body's self-antigens as "foreign particles," thereby attacking and eventually destroying healthy tissue or organs (Davidson and Diamond 2001). ADs are often found to be idiopathic. The symptomatic outburst of such autoimmune diseases generally includes chronic inflammation of tissue and systemic circulation, severe pain in joints, fatigue, etc. Almost 5–7% of the global population is affected, with a higher prevalence in women (Hayter and Cook 2012). Some autoimmune diseases primarily affect women, such as rheumatoid arthritis (RA), which is three times higher than in men, and systemic lupus erythematosus, commonly known as lupus (Angum et al. 2020). About 90% of the patients who have lupus are women, and surprisingly, the susceptibility of developing the disease is ten times higher in young females during and after their adolescence (Angum et al. 2020). There are more than 80 different autoimmune diseases reported globally, among which most commonly occurring are multiple sclerosis, inflammatory bowel diseases (ulcerative colitis and Crohn's disease), type 2 diabetes, psoriasis, rheumatoid arthritis, Addison's disease, systemic lupus erythematous, Graves' disease, pernicious anemia, etc. Although most clinical manifestations and exact etiology are unknown, significant studies reveal that genetic alterations make people more susceptible to ADs. However, uncontrolled lifestyle and environmental factors may also trigger the breakdown of the typical immune system. For example, those living in an urban environment are more prone to develop an autoimmune disease, likely due to high exposure to various pollutants, chemicals, smoke, etc. (Javierre et al. 2011; Rook 2012; Rose and Mackay 2006).

ADs constitute a significant concern worldwide since the diagnosis and management of these frequently relapsing and chronic conditions leave a substantial socioeconomic burden. The diagnosis of ADs is often very challenging due to the scarcity of standard and clinically relevant biomarkers and the lack of early-stage clinical symptoms. For example, clinical diagnosis of inflammatory bowel diseases (IBD) involves endoscopic surveillance, such as colonoscopy, flexible sigmoidoscopy, and a complete profile of the small intestine (mainly duodenum), entire colon, and rectum. In this context, scoring systems, namely, Crohn's Disease Activity Index (CDAI) and Ulcerative Colitis Endoscopic Index of Severity (UCEIS), have been developed and validated in order to standardize the endoscopic assessment using a combination of factors. To visualize the entire bowel's more profound appearance, few imaging techniques like MRI (magnetic resonance imaging) and CT scan are often chosen as diagnostic tools. Other diagnoses include invasive tests instead of the scoring system. For instance, type 1 diabetes is typically diagnosed by a glycated hemoglobin test (A1C) indicating levels of 6.5 and above. Diagnosis of multiple sclerosis (MS) is more complicated, often based on evaluating cerebrospinal fluid.

1.2 Human Immune System: A Brief Outlook

The immune system constantly defends the body against invasion of pathogens and xenobiotics or tissue damage. In the host body, both the innate and adaptive immune responses actively participate in the detection and elimination of pathogenic invaders. Innate immunity encompasses immediate, nonspecific response, while adaptive or acquired immune strategy is pathogen- or antigen-specific, mediated by B and T cell activation that potentiates the production of antibodies and immunologic memory to combat the relapsing incidence of a disease. Adaptive immune cells may sometimes attack self-antigens by mistake, which leads to the loss of immune tolerance, thereby leading to autoimmune diseases.

When an exogenous threat is recognized, the immune system constitutes a complex series of mechanisms that aims to clear damaged cells, expel toxins, and restore tissue homeostasis (Medzhitov 2008). The first-line host defense is initiated with neutrophils that act nonspecifically to engulf the pathogens (phagocytosis), and simultaneously, a large amount of reactive oxygen species (ROS) is generated that is highly cytotoxic to potentially harmful invaders (Nathan 2006; Shi and Pamer 2011). Like neutrophils, circulating monocytes that mature into macrophages play a significant role in destroying pathogens by phagocytosis. Activated macrophages secrete pro-inflammatory factors that potentiate the killing response, followed by its conversion to an anti-inflammatory phenotype that ensures tissue repair after injury. In case of infection, the primary component of physiological immune response is antigen-representing cells (APCs). Dendritic cells (DCs) are a specialized subset of antigen-representing cells that can take up the invaded pathogens under the skin and mucous membrane via phagocytosis and pinocytosis. Furthermore, they migrate to

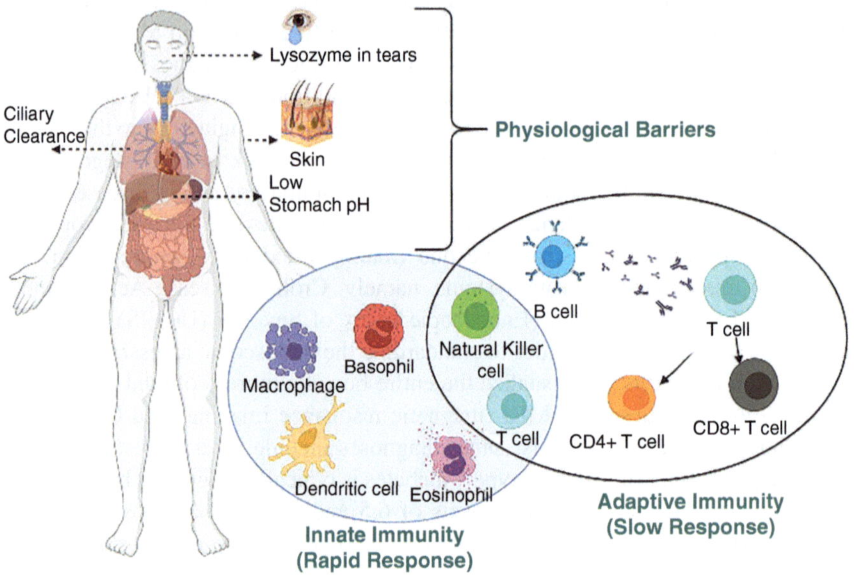

Fig. 1.1 Representation of different physiological barriers as a part of the human immune system along with important components of the innate and adaptive immunity that form part of rapid and slow response to an antigen. (Image created using BioRender)

the lymph nodes, where they become matured DC. Mature DCs then process the pathogens into short peptides to make them available on the cell surface, where they are displayed to B and T lymphocyte cells. Once those peptides are represented by the major histocompatibility complex (MHC) molecules, the cytotoxic T cells or $CD8^+$ T cells directly kill the pathogens. In contrast, helper T cells, also known as $CD4^+$ T cells, indirectly initiate the maturation of B-lymphocytes which, in turn, divide into plasma cells followed by secretion of immunoglobulins. Thus, the activated B cells mediate humoral or Th2 response by producing specific antibodies upon recognition of antigens (Fig. 1.1).

1.3 The Human Immune System Accounts for the Etiology and Pathogenesis of ADs

The term "autoimmune disorder" refers to the condition when the body's immune system fails to apprehend "self" from "nonself" and produces immune cells or antibodies (called autoantibodies) that target its own cells, tissues, and/or organs, leading to inflammation and significant tissue destruction. In this scenario, the body's defense serves as the body's enemy. Different organ targets and immune cells responsible for autoimmune disorders are shown in Fig. 1.2.

B cell and T cell play a vital role in triggering the onset of autoimmunity. The primary elements attributed to T cell-associated autoimmunity majorly include Type

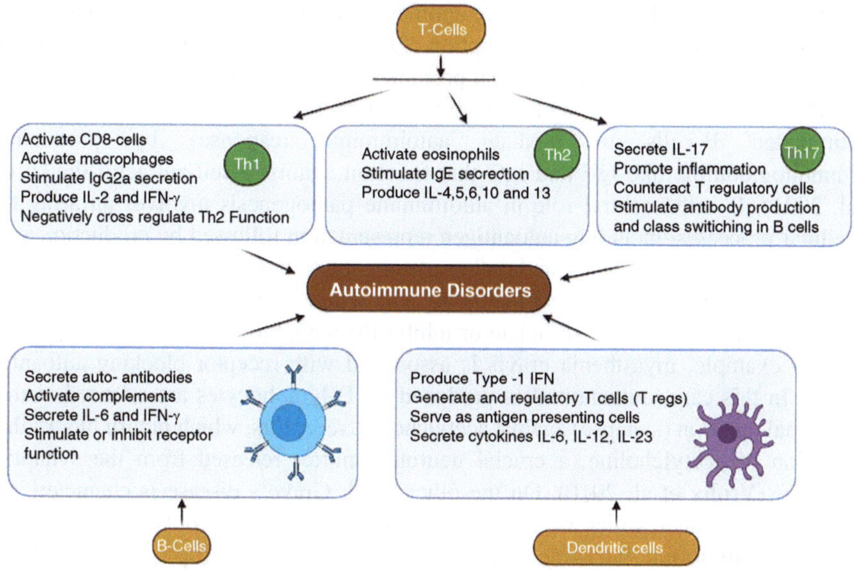

Fig. 1.2 Diagrammatic representation of the different mechanisms adapted by the T cells, B cells, and dendritic cells in various autoimmune diseases. The dendritic cells are usually activated by the antigens released during an active tissue injury, migrate to the lymph nodes and further functional B-cells and antigen-presenting cells. Activation of the B cells further leads to a cascade of events like the generation of effector T cells and memory B and T cells. Cytokine storms, for example, in response to an external aggravation of the immune system, damage the tissues. (Pictorial representation has been created using BioRender)

1 T helper cells (Th1), Type 2 T helper cells (Th2), and/or Th17 cells that secrete interleukin-17 (IL-17) (Lewis and Allen 2016). In the case of rheumatoid arthritis (RA) and type 1 diabetes, Th1 response significantly predominates, while in multiple sclerosis, destruction of nerve axons takes place (Aarvak et al. 2000). Pro-inflammatory cytokines such as interferon-γ (IFN-γ), IL-2, granulocyte-macrophage colony-stimulating factor (GMCSF), and tumor necrosis factor-α (TNF-α) are secreted by Th1 $CD4^+$ cells that efficiently activate the effector functions of macrophages (Charlton and Lafferty 1995).

Th17 (T helper 17) cells are widely considered as a potential pathogenic effector in autoimmune disorder development and progression. Th17 cells, identified first in the year 2005, are a novel subset of $CD4^+$ T cells which actuate the secretion of several interleukins (e.g., IL-21, IL-22, IL-17F, IL-17A) (Park et al. 2005). IL-17 is an efficient pro-inflammatory cytokine that induces the expression of IL-6, TNF-α, and IL-1β present in the endothelial and epithelial cell in conjunction with macrophage and fibroblasts cells. This ultimately results in increased inflammation (Waite and Skokos 2012). T helper 17 cells (Th17) and their cytokines give rise to several autoimmune disorders such as psoriasis, multiple sclerosis (MS), arthritis (RA), lupus (SLE), etc. (Maddur et al. 2012). For instance, a preclinical study showed that mice lacking in IL-17 failed to develop collagen-induced arthritis (Nakae et al.

2003). Regulatory T cells (Tregs) are another novel subset of $CD4^+$ T lymphocytes that provide self-tolerance. The disruption of the delicate balance between Treg cells and Th17 cells often acts as a critical promoter of autoimmune complications.

B lymphocytes are the significant source of immunoglobulins in our body and stimulated B-cells to mediate autoimmune responses by producing immunoglobulins like IgE and IgG that function against a self-antigen (Sheikh et al. 2021). B cells' central role in autoimmune pathogenesis involves an array of cellular processes, including autoantigen representation followed by production and secretion of autoantibodies and inflammatory cytokines (Christiane and Hampe 2012). Antibodies against self-antigens can cause severe systemic problems. Autoantibodies can either stimulate or inhibit the signaling receptors by binding to it. For example, myasthenia gravis is associated with receptor blocking autoantibody. In this case, autoantibodies produced by B-lymphocytes are released in the neuronal junction (synapse) against acetylcholine receptors, which in turn blocks the function of acetylcholine, a crucial neurotransmitter released from the synaptic boutons (Vrolix et al. 2010). On the other hand, Grave's disease is characterized by hyperthyroidism which is caused as a result of elevated thyroid level (thyroxine and triiodothyronine hormone) secreted by stimulated thyrotropin receptor (TSH receptor) upon binding to the thyroid-stimulating autoantibodies (Chen et al. 2003). In addition, activated B cells often serve as the potentiators for the secretion of pro-inflammatory cytokines, for example, IL-4, IL-6, TGF-β, IFN-γ, etc. (Duddy et al. 2004; Lund et al. 2005), which help in the pathogenesis of autoimmunity by orchestrating a cascade of events, including the activation of macrophages for phagocytosis, the migration of APCs (dendritic cells) to lymph nodes, upregulation of T cell effector functions, and activation of further B-lymphocytes by providing stimulatory signals (Matsumura et al. 2006).

1.4 Current Strategies and Therapeutics Against Autoimmune Diseases

Due to lack of adequate information regarding the etiology, effective treatment and cure for most autoimmune disorders have not been established, necessitating more detailed and more in-depth scientific investigation to understand the role of crucial inflammatory mediators present in ADs. In modern medicine, clinically relevant therapies can be broadly categorized as replacement therapy (symptomatic) and immunosuppressive or immune-modulation therapy (Chandrashekara 2012). Symptomatic therapy aims to manage disease symptoms in order to decrease the number of relapsing events. The best example of replacement therapy is autoimmune thyroid disease, predominantly treated either by reducing the thyroxin production or by replacing the hormone once the gland is damaged (Singer et al. 1995). Another example is Sjogren's syndrome, where patients undergo replacements of tear and saliva in addition to drugs used for alleviating with additional complications (Mavragani and Moutsopoulos 2014).

Similarly, management of type 1 diabetes requires constant monitoring of blood glucose levels besides regular administration of subcutaneous insulin or pressurized pumps to ensure normal blood glucose concentrations (Pickup 2012). In comparison, systemic diseases like systemic lupus erythematosus (SLE) are primarily treated with immunosuppressive agents to prevent organ damage. Previously, immunosuppressive drugs used to be nonspecific and toxic. However, the latest developments include more target-specific immunosuppressant drugs, with little to no distribution to the nontarget organs and a more profound immunosuppressive effect.

Therapeutic options for the treatment of RA are also sparse. Physiotherapy and medication can help slow down the disease's progression. To relieve pain and subdue inflammation, nonsteroidal anti-inflammatory drugs (NSAIDs) are used. Immune suppressants like corticosteroids such as prednisone reduce inflammation and pain and slow down joint damage. Disease-modifying antirheumatic drugs (DMARDs) are highly efficacious in RA; however, they need to be administered during the early stages of the disease (Aletaha and Smolen 2002). Commonly used DMARDs in clinical setting include sulfasalazine methotrexate, hydroxychloroquine, and leflunomide. However, as the joint damage and bone erosion become severe with disease progression (Heidari 2011), the use of DMARDs is not a viable option (Smolen et al. 2010). Furthermore, these therapies are associated with significant adverse effects such as bone marrow depletion, liver damage, and severe lung infections. Biological interventions (such as antibodies and proteins) have also been developed as alternatives for treating ADs. In IBDs, this therapeutic approach suppresses the expression of tumor necrosis factor-α (TNF-α), which is associated with abnormal expression of cytokine in IBDs (Sandborn et al. 2003; Sands and Kaplan 2007). Similarly, treatment with interferon-β (IFN-β) has shown promising impact in treating MS. This is the reason that injectable IFN-β products represent 50% of the pharmaceutical market for treating MS (Hegen et al. 2015). However, events of disease relapse are familiar with biological therapies (Targan 2006).

In the last few decades, immunotherapy has been conquered extensively in clinical research for the mitigation of a broad spectrum of immune disorders. Immunotherapy as a modulation to counter the immune disorders is a strategy that either interposes immune dysregulation or induces a specific immune tolerance to self-antigens. Immunotherapy was first launched in the early 1980s with a study demonstrating clinical evidence of polyclonal IgG immunoglobulin (IVIG) in improving autoimmune thrombocytopenia (Targan 2006). More than 70% of IVIG is currently clinically used to treat autoimmune diseases in the USA. With the advancement of molecular biology, novel immunotherapeutic approaches have emerged, broadly classified as regulatory T cell therapy, effector cell depletion (e.g., antithymocyte globulin [ATG]), and nonspecific molecular immunomodulation (e.g., IL-1 agonist).

1.5 Biomolecules: Promising Candidates for Autoimmune Therapy

The emerging incidence of autoimmune disorders has become a global threat, leading researchers to focus on developing more target-specific and nontoxic treatment strategies. In this context, biomolecules have been well explored for the detection and treatment of ADs. Biomaterials are selected or designed based on their ability to exert the desired effects concomitant with biocompatibility and most negligible adverse effects. To overcome biocompatibility issues of the in vivo devices required for external diagnostics, a broad range of biomaterials of synthetic and natural origin have been investigated. These include polymers, metal nanoparticles, hydrogels, quantum dots (QDs), and biomolecules such as antibodies and enzymes.

1.5.1 Polymeric Hydrogels

Polymeric hydrogels are among the most frequently utilized biomaterials due to their unique properties, such as high flexibility, biocompatibility, softness, high water content, resemblance to the living tissue, and degradation property. Polymeric hydrogels are of prime importance in the delivery of drugs and tissue engineering (Ashraf et al. 2018; Sofi et al. 2019). These can be either synthetic (prepared by the polymerization of functional monomers) or naturally derived through cross-linking, which offers excellent biocompatibility and degradation properties (Clegg et al. 2017). The hydrogels are often used as an artificial extracellular niche to support, nourish, and protect encapsulated cells (Chan and Neufeld 2010; Truong et al. 2015). These gel systems can also be prepared to preserve bioactive agents for targeted drug delivery (Chan and Neufeld 2010). These systems exhibit a wide range of responsive behaviors. For example, a cross-linked hydrogel composed of pH-responsive monomers such as diethylaminoethyl methacrylate or methacrylic acid shows pH-dependent swelling behavior.

Hydrogels may be charged or non-charged based on the nature of functional groups present in their structure. In the charged state, the hydrogel gets expanded or swollen, driven by electrostatic repulsion. A reverse phenomenon is observed when the gels are exposed to certain pH conditions that facilitate removal of the "interacting group" charges. This type of pH-responsive behavior can be controlled by incorporating comonomers that contain hydrophobes capable of altering the p*K*a of the ionizable groups to minimizing the electrostatic interactions (Liechty et al. 2013).

Thermoresponsive behavior is a characteristic feature of polymers like poly(*N*-isopropylacrylamide) or poly(*N*-vinylcaprolactam). These polymers experience a transition from hydrophilicity to hydrophobicity near body temperature capable of alteration by comonomers (Zhang et al. 2015a). However, pure synthetic polymers usually comprised of polyvinyl, polyacrylate, and polyacrylamide backbone which are typically nondegradable in nature. This has eventually potentiated the use of cross-linking agents containing disulfide bonds or biodegradable polymers such as poly(lactic-co-glycolic acid) (Lv et al. 2014; Tang et al. 2009). Polymeric hydrogels derived from natural sources, including collagen, alginates, chitosan, etc., have become very popular, especially in tissue engineering applications, because of their excellent biodegradability and biocompatibility (Lv et al. 2014).

1.5.2 Inorganic Biomaterials

The use of inorganic biomaterials, such as gold nanomaterials, iron oxide nanoparticles (IONPs), and semiconductor quantum dots (QDs), has promising applications in the field of disease detection, biomarker development, and drug delivery systems. Such materials' primary advantage is that they can quickly conjugate to bioactive molecules and act as a promising drug carrier for target-specific delivery. In autoimmune diseases, nanomaterials have been designed to modulate the antigen-representing cells (APCs) as well as downregulating the innate immune signals so as to strengthen adaptive autoimmune responses (Klippstein and Pozo 2010). For instance, in the treatment of experimental autoimmune encephalomyelitis (EAE), liposomes loaded with glucocorticoid were found to be effective at a lower dose as compared to usual glucocorticoid therapy by directly targeting the macrophages (Schweingruber et al. 2011).

Gold nanoparticles (GNPs) have specific advantages due to their size- and shape-related optoelectronic properties (Hu et al. 2006; Yeh et al. 2012), localization, biocompatibility, and low toxicity (Dykman and Khlebtsov 2012; Murphy et al. 2008). They are chemically stable and inert. Important physical properties of GNPs include surface plasmon resonance (SPR) and the ability to quench fluorescence. Moreover, the surface plasmon resonance is characterized by an absorption band that arises from the collective oscillation of the electrons excited by the photons of the incident light (Mock et al. 2002; Piliarik et al. 2011). The most common structures that have been explored for biomedical applications include nanorods, nanospheres, nanoshells, nanostars, nanocubes, and nanopyramids.

The QDs are semiconductor nanocrystals with a diameter in the range of 2–10 nm that can emit lights of distinct colors, predominantly depending upon their size, shape, and composition. QDs were first discovered in 1980 (Ekimov and Onushchenko 1981). QDs have been widely employed in biomedical imaging and biosensors due to the unique size that allows them to go anywhere in the body, improved brightness, and fluorescence property, making them more appealing over conventional organic dyes (Pinaud et al. 2006). QDs are usually conjugated with bioactive molecules or probes to improve their uptake during imaging or to facilitate the detection of biomarkers with their respective fluorescent antibodies (Knipe et al. 2013). However, the major drawback of QDs is the associated toxicity as seen in many standard formulations, such as CdTe, which can be known to release toxic cadmium ions (King-Heiden et al. 2009; Lovrić et al. 2005), although strategies like the encapsulation of QDs (Zhang et al. 2006) or preparing formulations with less poisonous metals (Pons et al. 2010) have been developed to reduce toxicity.

Superparamagnetic iron oxide nanoparticles (IONPs) with diameters between 1 and 100 nm serve as a potential alternative to photo-responsive nanomaterials. Magnetically responsive IONPs have gained much interest in biomedical applications since these are extensively used as contrast agents for MRI (Quan et al. 2011; Xie et al. 2010), drug carriers for target accumulation (Chertok et al. 2010), gene carriers for gene therapy, magnetic sensing probe for in vitro diagnostics (Li et al. 2011), and nanoadjuvants for vaccines and antibody production.

1.6 Role of Biosensors in the Detection of Autoimmune Disorders: Rationale and Application

Autoimmune diseases can be classified as organ-specific (e.g., thyroid, type 1 diabetes, Grave's disease) or systemic (like antiphospholipid syndrome, rheumatoid arthritis, and systemic lupus erythematosus). These are some of the major reasons for severe physical disability, organ failure, and morbidity; therefore, early diagnosis is essential for the augmentation of patients' health quality. Instead of invasive diagnostic tests, biomarkers found in the biofluids of patients with ADs have gained enormous interest in detecting the disease. Most of the relevant biomarkers identified in ADs are autoantibodies (Ghorbani et al. 2019), proteins (Derkus et al. 2017; Zasońska et al. 2018), chemokines (Vega et al. 2013), cytokines (La Belle et al. 2007), and ions (Tadi et al. 2017). Proteomic studies such as IIF (indirect immunofluorescence), ELISA (enzyme-linked immunosorbent assay), and Western blotting are commonly used to detect autoantibodies associated with autoimmune disorders (Campuzano et al. 2019; Zhang et al. 2017). However, these techniques are time-consuming and expensive, require many sophisticated instruments, and are even not highly sensitive in all cases. These disadvantages, therefore, render for designing more sensitive and novel techniques. One of the most promising and intriguing approaches is the development of biosensors for qualitative and quantitative detection of biomarkers ensuing proper disease monitoring. Table 1.1 provides a list of common biomarkers related to different autoimmune diseases.

As per IUPAC, *a biosensor* is a self-integrated device comprising of a biological recognition element in combination with a transducer element (both are in direct spatial contact) which is used to provide specific analytical information in qualitative or semiquantitative fashion by conversion and amplification of biological/chemical signals into measurable electrical signals. These are of emerging importance due to their excellent specificity, sensitivity, low cost, robustness, and real-time detection capability. Micro- and nanoscale biosensors have been explored for AD markers, especially when a sufficient quantity of sample (biofluid) is not available (e.g., patients with Sjogren's syndrome suffer ocular dryness; in such case, reduced amount of lacrimal fluid is available for biomarker detection and quantification). Biomaterials can be used to design biosensors for detecting immune diseases and, at the same instant, target these diseases with suitable drugs. The working principle is based on the hypothesis that a localized inflammatory response leads to the production of inflammatory cytokines, which can be detected using an antibody of the

Table 1.1 A list of biomarkers of different autoimmune diseases and their source for sampling

Disease	Biomarkers	Source of biomarkers	Reference
Sjögren's syndrome	Lysozyme, lipocalin, lactoferrin, cathepsin	Tears	Caffery et al. (2008), Danjo et al. (1994)
	Alpha-amylase, alpha-enolase, lysozyme C, cathepsin, carbonic anhydrase (VI), beta-2-microglobulin, lactoferrin	Saliva	Hu et al. (2007), Ryu et al. (2006)
Type 1 diabetes	Insulin, anti-GAD (anti-glutamic acid decarboxylase), glucose, IA2 (islet antigen-2)	Serum	Mathieu et al. (2018)
Multiple sclerosis	Fetuin-A, β 2-microglobulin, kallikrein-6, anti-Myelin Basic Protein (anti-MBP)	Cerebral spinal fluid	Harris et al. (2013)
Crohn's disease	Calprotectin, calgranulin C, anti-saccharomyces cerevisiae (ASCA), anti-neutrophil cytoplasmic antibodies, C-reactive protein	Serum	Chen et al. (2020)
Hashimoto thyroiditis	TSH, antithyroid peroxidase antibody (anti-TPO)	Serum	Santhoshkumar et al. (2021)
Graves' disease	Thyroid-stimulating immunoglobulin (TSI), anti-TPO	Serum	Longo and Higgins (2019)
Celiac disease	Anti-tissue transglutaminase (anti-tTG), antigliadin autoantibodies (AGA), anti-deamidated gliadin peptide (anti-DGP)	Serum	Vives-Pi et al. (2013)
	Gluten	Urine	Soler et al. (2016)
Lupus (SLE)	Anti-dsDNA, anti-ribonucleoprotein	Serum	Ahearn et al. (2012)
Rheumatoid arthritis	Calgranulins, CRP (C-reactive protein), alpha-2-plasmin inhibitor	Synovial fluid	Liao et al. (2004)
	CRP (C-reactive protein), anti-CCP (anticyclic citrullinated peptide), calgranulins	Serum	Otterness (1994)

released cytokine in a drug-laden biomaterial scaffold. In response to the cytokine release, drugs will be released from the scaffold. As the level of the cytokines will recede, the drug concentration will subsequently adjust to the receded levels, resulting in targeting and response-based drug release (Fig. 1.3).

Lipocalin 1 was detected by a sandwich immunoassay that exploited microparticles made up of polystyrene coated with a primary antibody in the presence of a secondary antibody attached to a quantum dot or a dye (Shodeinde et al. 2020). A biosensor was developed by Chen et al. for the detection of lysozyme, using gold nanoparticles coated with aptamer in a solution containing quantum dots (CdTe QDs). As a result of lysozyme-aptamer coupling, quenching of the cadmium-telluride QD was observed along with a significant change in the gold nanoparticles' absorbance (Chen et al. 2016).

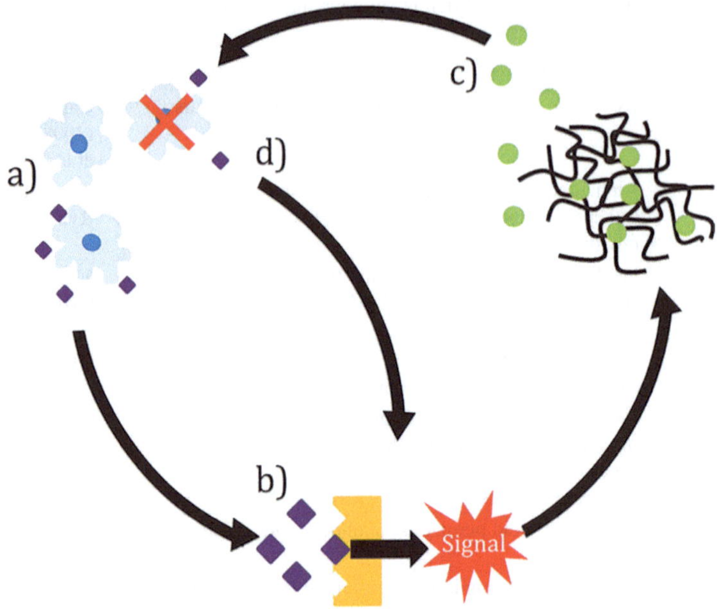

Fig. 1.3 A pictorial representation of the role of biomaterials in the detection and treatment of an autoimmune disease. (a) Localized inflammation and the production of inflammatory markers in autoimmune disease. (b) Detection of the inflammatory markers that are produced during autoimmune response leading to disease diagnosis. (c) Drugs are released from biomaterial scaffolds as per the signal received, and inflammatory response is mitigated. (d) Any alteration in the production of inflammatory markers inside the host is detected by biosensors that can directly be involved to monitor treatment efficacy and dosage adjustment as well (Shodeinde et al. 2020)

Based on the category of the signal transducer, biosensors are classified as electrochemical, optical, and piezoelectric sensors. The most frequently used biosensing device is electrochemical biosensors with high efficiency for biomarker detection. Fagúndez et al. fabricated an electrochemical immunosensor for the assay of autoantibodies (mainly anti-dsDNA) present in serum samples collected from lupus (SLE) patients (Fagúndez et al. 2018). Patients who have celiac disease produce an autoantibody, namely, anti-transglutaminase or anti-tTG, which can be detected by EIS (electrical impedance spectroscopy), a label-free subtype of electrochemical immunosensor (Wilson et al. 2015). Another highly sensitive, label-free immunosensor based on EIS technique was designed for the detection of a serum cytokine (interleukin-12) in order to diagnose multiple sclerosis, and this immunosensor was able to detect as low as 3.5 pg/mL sample (La Belle et al. 2007). Recently, a novel nanoimmunosensor has been reported based on differential pulse voltammetry (DPV) which is a nanocomposite comprised of poly (propyleneglycol) and graphene oxide. This nano-sensor exhibited a concurrent detection of Tau proteins and myelin basic protein (MBP) present in serum and cerebrospinal fluid samples collected from patients with MS (Derkus et al. 2017).

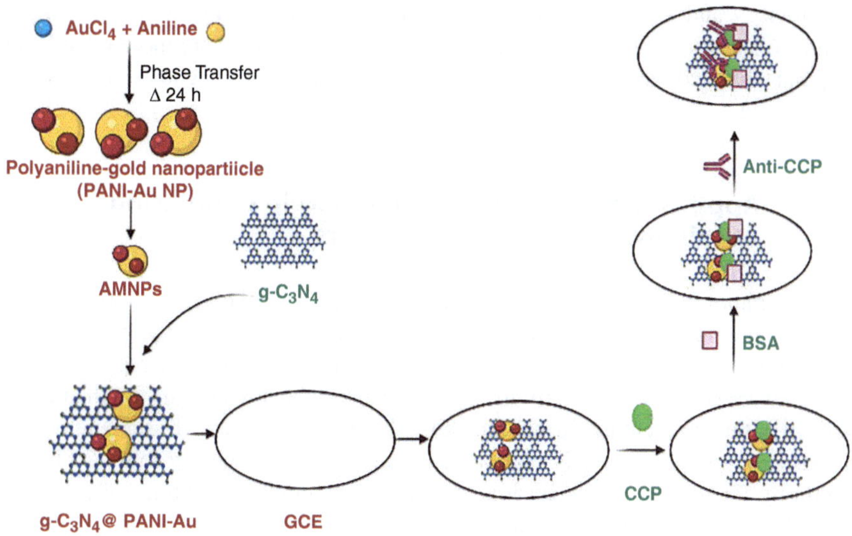

Fig. 1.4 A diagrammatic representation of an electro-chemiluminescence-based sensor for quantifying minute quantity (in picograms) of anticyclic citrullinated peptide (anti-CCP) autoantibody. This sensor had been employed for the diagnosis of early-stage rheumatoid arthritis. (Image created using Biorender software)

Optical biosensors are extremely significant for their label-free detection and rapid response (Perumal and Hashim 2014). Electro-chemiluminescence (ECL), surface plasmon resonance (SPR), and fluorescence and surface plasmon resonance imaging (SPRi) are the most popular optical biosensors that have been used for the detection of specific biomarkers associated with different ADs. An electro-chemiluminescence-based, label-free immunosensor was fabricated with asymmetric polyaniline-gold nanomaterial (PANI-Au), which encompassed highly sensitive estimation of anticyclic citrullinated peptide antibody (anti-CCP), responsible for early-stage diagnosis of RA (Fig. 1.4). Graphite-like carbon nitride (g-C3N4) was incorporated to improve the stability of the PANI-Au nanomaterial. This device was efficient enough to detect samples even in nanogram concentration (approx. 0–15 ng/mL) (Zhao et al. 2018).

Recently, fluorescence detection has significantly emerged for the development of biosensors. MicroRNA-145, a biomarker present in the plasma of MS patients, has been reported to be detected by a nanobiosensor composed of silver nanoclusters (AgNCs) tagged with fluorescent DNA, and the sample was amplified by hybridization chain reaction (HCR) using as low as 0.1 nM of microRNA-145 (Dong et al. 2014). Vega et al. demonstrated an SPR biosensor that detected CXCL12, a potentially relevant chemokine present in RA patients' urine samples, and, notably, a very minute quantity of sample (5–40 nM) was sensitive to this device (Vega et al. 2013). The crucial advantage of SPR biosensors is their efficiency in real-time, quantitative, and label-free detection in conjugation with their specificity and cost-effectiveness. Moreover, they can monitor binding characteristics to estimate kinetic parameters. Therefore, they are more reliable and popular than conventional laboratory tests in terms of diagnosis and disease monitoring.

1.7 Applications of Tissue Engineering in Countering Autoimmunity-Assisted Disorders

Most commonly available treatment strategies for ADs involve physiotherapy and drugs to provide symptomatic relief, but they do not cure the disease by addressing its root cause. Tissue engineering serves as a promising tool in the regeneration of defected cells (e.g., islet cells in type 1 diabetes, synovial cells in rheumatoid arthritis), thus preventing further attack by the immune system. Tissue engineering is based on the principle of isolation of cells from a tissue biopsy followed by cell culturing for further cell expansion. An in vitro assembly constitutes of a scaffold material seeded with cells to build a live tissue construct that is competent for the implantation in a host body. On the other hand, in vivo construct utilizes an empty scaffold inserted in the recipient's body, enabling the host cells to intrude on and cling to the scaffold material (Irvine et al. 2008). Scaffolds are commonly implanted by surgical intervention, while some scaffolds carrying encapsulated cells can be injected into the host body instead of implantation (Nicodemus and Bryant 2008).

Biomaterials serve as an integral and promising scaffold for regenerative medicine and advanced tissue engineering. Biomaterial scaffolds have been designed to successfully deliver cells or tissue constructs and bioactive compounds. Such scaffolds are 3D structures that can be synthesized or naturally obtained. The novel concept of "smart biomaterials" has emerged over the past few decades that can provide a specific immune microenvironment since they possess specific adaptive compositions, structures, and physicochemical properties to that of living tissues. "Smart" or "intelligent" biomaterials are able to respond to either tissue-generated endogenous stimuli (e.g., temperature, pH, shear stress) or exogenic stimuli (e.g., magnetic fields, high-energy radiations, electric fields, etc.) in an instant or a programmed manner (Knipe and Peppas 2014) (Fig. 1.5).

1.7.1 Outstanding Characteristic Features of a Biomaterial Scaffold

1. The scaffolds are designed to provide a temporary three-dimensional structural support that allows the cells to adhere, leading to tissue growth and formation. The scaffolds should mimic the biological environment essential for complete growth, migration, differentiation, and proliferation of cells and enable the formation of tissue.
2. Scaffolds should be biocompatible, i.e., they should neither be recognized as a foreign material by the body, nor should it cause any inflammation or unwanted immune response by antibody-carrying cells in response to antigens.

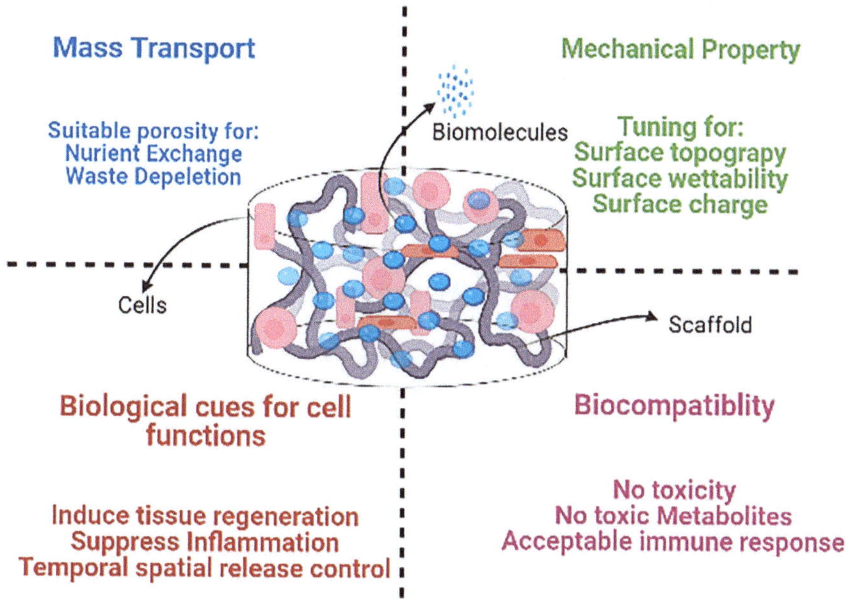

Fig. 1.5 Representation of polymeric biocompatible scaffold and various properties for tissue engineering use and preventing immune rejection. (Image created using Biorender software)

3. Ideally, scaffolds should provide a larger surface area with a high degree of porosity that can assure maximum cell stacking, cell-surface adherence, interaction, and eventually, feasible ingrowth of tissue. They should provide access for efficient transportation of oxygen, nutrients, excess by-products, or cellular waste.
4. They should be biodegradable; additionally, the rate of degradation should be proportional to the speed of regeneration of new tissue to keep their functional potential intact (Knipe and Peppas 2014).
5. Scaffolds must possess certain mechanical traits like flexibility or elasticity, tensile strength, etc., since biological processes (e.g., protein expression and differentiation) rely on mechanical stimuli and provide adequate mechanical support to resist endogenous biological forces.
6. They should induce the cells to synthesize native ECM and growth factors essential for specific tissue growth.
7. They must be sterile to counter any contamination by retaining their intact morphology and mechanical characteristics.
8. Another essential feature of the scaffolds is their specificity which allows selective adhesion of cells (Knipe and Peppas 2014).

Tissue engineering ensures immediate treatment of a particular damaged organ or area of the body. It involves target-specific delivery of cells encapsulated in scaffolds or directly implanting a scaffold to a specific location (Boehler et al. 2011). For instance, cartilage and bone tissue engineering have gained much attention to mitigate the hard tissue damage associated with rheumatoid arthritis. Similarly, pancreatic tissue engineering is one of the most favorable alternatives for autoimmune diabetes (type 1) treatment. Two main types of scaffolds have been exploited in tissue engineering: naturally derived biomaterial-based and synthetic-derived polymer-based. Naturally, derived polymers are predominantly advantageous due to their excellent biocompatibility.

1.7.2 Hyaluronic Acid (HA) or Hyaluronan

It is a naturally occurring non-sulfated glycosaminoglycan found ubiquitously in the extracellular matrix of the skin, connective tissues, umbilical cord, vitreous humor, and synovia/synovial fluids (Yoo et al. 2005). Hyaluronan can interfere with matrix architecture, even cellular interactions, by either binding to proteins or interacting with growth factors, proteoglycans, etc. (Tognana et al. 2007). It has become one of the most versatile scaffold materials in the genre of tissue engineering due to its abundant distribution throughout the body, excellent biocompatibility, and nontoxic degradation products (Burdick and Prestwich 2011). The traditional method of treating RA includes mosaicplasty or ACI (autologous chondrocyte implantation) that results in the production of fibroblasts (non-collagenous) upon chondrocyte differentiation (Behrens et al. 2006). Surgical interventions such as subchondral bone plate surgery based on microfracture technique or cartilage replacement by allogeneic grafting/transplantation may often be risky due to immune rejection, donor scarcity, and transplant infection. Thus, it is of utmost importance for scaffold-supported chondrocyte implantation for advanced treatment. In this regard, HA serves as a suitable scaffold for the growth of chondrocytes. Commercially available Hyalograft® C (made in Italy by Abano Terme) is an example of an HA-based tissue engineering scaffold that is cultured with autologous chondrocyte cells to treat cartilage defects in RA patients (Tognana et al. 2007).

1.7.3 Chitosan

Chitosan is a natural copolymer of carbohydrates and a deacetylated derivative of chitin that can be degraded by chitosanase (Lodhi et al. 2014). Chitosan has attained a high degree of interest as a promising scaffold candidate in the area of tissue engineering for being biocompatible, nontoxic, and highly biodegradable (undergoes both thermal and enzymatic degradation). Furthermore, chitosan possesses a high degree of water absorption capacity that enables quick absorption of body fluid and adequate distribution of nutrients, growth factors, and metabolites through extracellular media (Sánchez Brenes et al. 2007). The degree of deacetylation plays a crucial role for chitosan scaffolds to be considered as ideal structural support for tissue regeneration; cell proliferation and adhesion are much more accelerated, with chitosan possessing higher degrees of deacetylation.

Moreover, large-scale processing is easier and can be molded into diverse shapes, like films, hydrogels, fibers, and sponges. Chitosan composites in the form of hydrogels have been extensively used in cartilage tissue replacement to treat patients with lupus and RA, both of which lead to severe impairment of bone and cartilage in joints. Chitosan composites can be made up of a combination of biomaterials such as silk fibroin, polycaprolactone, chondroitin sulfate, polyester, and genipin (Sánchez Brenes et al. 2007). This serves as a suitable platform resembling the properties of cartilage and allows a favorable environment for extracellular matrix deposition and cellular growth, functions, and differentiation.

1.7.4 Collagen

Collagen is considered the most copious structural protein (that accounts for approximately 30% of the total body protein) of soft and hard tissues (Burgeson and Nimni 1992) in mammals. Collagen also contributes to preserving the integrity of the structural and biological environment of the extracellular matrix. Favorable criteria, for example, porous construct, low immunogenicity, high permeability, biodegradability, pertinent biocompatibility in conjugation with the ability to regulate cell adhesion, migration, and proliferation, make collagen a scaffold of choice for tissue engineering. However, the only drawback limiting its application in tissue regeneration is its substandard durability and structural instability (often degrades upon hydration). These shortcomings can be checked by either intermolecular cross-linking of collagen by physical or chemical methods or blending with numerous natural polymers (e.g., chitosan, silk fibroin), synthetic polymers (poly(ethylene glycol), polylactic acid, polycaprolactone, polyglycolide, polyvinyl alcohol, poly (lactide-*co*-glycolide)), and inorganic materials (e.g., silicate and β-tricalcium phosphate, hydroxyapatite) that improve mechanical property as well as survivability of the transplanted cells (Hunckler and García 2020; Parenteau-Bareil et al. 2010).

1.7.5 Starch

Starch is another most abundant natural polymer, chemically a complex chain of polysaccharides, that is extensively employed to fabricate scaffolds for regenerative tissue approach. Starch has been a suitable scaffold in cartilage and bone tissue engineering because of its variation in structure (Sionkowska 2011). This can result in desired biocompatibility, low cost, great mechanical properties, hydrophilicity, salient porosity for cell permeation, and easy movement along with least/no toxicity. Moreover, a scaffold with a polymeric blend of gelatin and starch elicits excellent mechanical characteristics and appropriate porous nature (almost 82.51%); this helps in cellular growth by inducing cell interaction, differentiation, proliferation, and vascularization (Sundaram et al. 2008).

1.7.6 Silk Fibroin

It is another natural macromolecular polymer with salient mechanical traits like flexibility, low cytotoxicity, porosity enabling feasible nutrient supply, and excellent biocompatibility (Sun et al. 2015). The property of biodegradability with a slow rate of degradation makes silk fibroin a potential choice for scaffold fabrication. Silk fibroin is capable of enhancing cell affinity to scaffold materials, thereby improving cell adherence (Di Felice et al. 2015). Salivary gland hypofunction is associated with Sjogren's syndrome and few other systemic ADs. A culture system was established and characterized by Zhang et al. using a 3D silk fibroin scaffold (free from sericin which is identified as an allergen in humans) containing primary salivary gland epithelial cells (pSGECs) extracted from murine submandibular and parotid glands. The culture of pSGECs on silk fibroin scaffold was reported to form aggregates or clusters and retained the structural and functional features of native salivary glands. Microscopic images (scanning electron microscopy and transmission electron microscopy) of cytosol and the cell surface confirmed the presence of secretory granule-like structures. In addition, immunofluorescence and phase-contrast microscopy confirmed that the cells on the scaffold, mainly composed of collagen type IV, constituted an extracellular matrix that resembles the indigenous environment of the salivary gland (Zhang et al. 2015b).

1.7.7 Synthetic Polymers

Synthetic scaffolds are advantageous over natural biomaterials due to their predictable and reproducible mechanical properties such as tensile strength and elasticity. However, they lack the criteria of biocompatibility, nontoxicity, and biodegradability. To overcome this, synthetic polymers are often used in combination with natural scaffolds to improve the scaffold properties to make them more appropriate candidates for tissue engineering applications. For instance, the degradation rate of polycaprolactone (PCL) can be altered by blending with natural polymer starch with better degradation properties. These copolymer scaffolds have been reported to provide structural and functional support to cells in nanofiber meshes or microporous architectures.

The most commonly used synthetic polymers in tissue engineering are polylactic acid (PLA), polyglycolide (PGA), and poly(lactide-*co*-glycolide) (PLGA). PLGA is one of the approved biomaterials by the US-FDA for clinical use. It has attracted considerable interest as a scaffold as it possesses suitable criteria such as biocompatibility, tailored biodegradation rate (depending on crystallinity, molecular weight, copolymer ratio), and a high surface area that promotes host cell infiltration, adhesion, and vascularization of transplanted cells (Blomeier et al. 2006; Graham et al. 2013). An essential application of PLGA for the treatment of ADs involves the use of non-hepatic islet cell transplantation. Diabetic patients need islet cells from more than one donor pancreas to achieve euglycemia. Moreover, insulin independence is not observed for a period of over 5 years. Furthermore, if these cells are transplanted via the portal system, the islet cells will be damaged by circulating toxins or the circulating resident macrophages in hepatic sinusoids. PLGA scaffold-based cell transplantation can thus be achieved in an extrahepatic site, such as the peritoneal cavity, intraperitoneal fat, the renal subcapsular space, or the omentum (Graham et al. 2013). It has been reported that islet cells transplanted in intraperitoneal fat of mice with streptozotocin-induced type 1 diabetes exhibited morphological similarities to that of native pancreatic islet cells, and initial transplantation of 125 cells had successfully elicited euglycemia.

Polyethylene glycol (PEG) is often used in conjunction with smart biomaterials for scaffold fabrication. It is an attractive choice for many tissue engineering applications because it is biocompatible and biologically inert. Ideally, an inert scaffold prevents protein and cell adhesion; however, it becomes useful when incorporating other materials that contain relevant functional groups or bioactivity. PEG-PLGA system is a pertinent example of smart materials included in inert scaffolds, potentially exploited in cartilage and bone tissue engineering to treat RA, SLE, or other ADs to prevent inflammation (Manzo et al. 2010).

1.8 Conclusion and Future Prospectus

Autoimmunity is a global concern. There has been a drastic improvement in the prognosis of such diseases by developing new, improved technologies and advances in science over the last few decades. Biomaterials have evolved as a new paradigm in diagnosing and treating autoimmune diseases, leading to better overall global health and well-being. However, the successful translation of biomaterials from benchtop to bedside is facing many challenges. In the context of disease detection and monitoring, several biomarkers have been projected, although it has been reported that multiple nonspecific factors with no significant link to the actual diseased condition often control their efficiency. Therefore, further research needs to be focused on identifying more specific biomarkers that can pave the path to the design and fabrication of more sensitive and accurate biosensors.

Implementation of "smart" biomaterials as a next-generation treatment strategy for various ADs is a remarkable success. Preclinical models of ulcerative colitis and rheumatoid arthritis had exerted target-specific release of anti-inflammatory drug using biomaterial scaffolds. Likewise, delivery of siRNA supported with biomaterials and scaffold assisted delivery of biological macromolecules such as IFN-β, insulin, as well as antibodies have shown promising success. On the other hand, biomaterial scaffold-based tissue engineering offers tremendous potential for the regeneration of damaged tissue or organ resulting from various ADs like RA, type 1 diabetes, multiple sclerosis, etc. However, some risk factors associated with scaffold-based cell transplantation are microbial contamination, processing error, unknown cell-substrate interaction, and loss of viability upon prolonged use, along with immunosuppression, which certainly require more attention in future research. Continued efforts are being made to investigate and understand a specific etiology of these often idiopathic disorders, especially on a molecular level, to develop more sophisticated and practical treatment approaches.

Acknowledgments Authors are thankful for funding provided by the Science and Engineering Research Board (SERB) research grants (CRG/2020/000113).

Declaration of Interest The authors have no conflict of interest to produce this work.

References

Aarvak T, Chabaud M, Thoen J, Miossec P, Natvig J (2000) Changes in the Th1 or Th2 cytokine dominance in the synovium of rheumatoid arthritis (RA): a kinetic study of the Th subsets in one unusual RA patient. Rheumatology 39(5):513–522

Ahearn JM, Liu C-C, Kao AH, Manzi S (2012) Biomarkers for systemic lupus erythematosus. Transl Res 159(4):326–342

Aletaha D, Smolen J (2002) The rheumatoid arthritis patient in the clinic: comparing more than 1300 consecutive DMARD courses. Rheumatology 41(12):1367–1374

Angum F, Khan T, Kaler J, Siddiqui L, Hussain A (2020) The prevalence of autoimmune disorders in women: a narrative review. Cureus 12(5):e8094

Ashraf R, Sofi HS, Beigh MA, Majeed S, Arjamand S, Sheikh FA (2018) Prospects of natural polymeric scaffolds in peripheral nerve tissue-regeneration. Adv Exp Med Biol 1077:501–525

Behrens P, Bitter T, Kurz B, Russlies M (2006) Matrix-associated autologous chondrocyte transplantation/implantation (MACT/MACI)—5-year follow-up. Knee 13(3):194–202

Blomeier H, Zhang X, Rives C, Brissova M, Hughes E, Baker M, Powers AC, Kaufman DB, Shea LD, Lowe WL Jr (2006) Polymer scaffolds as synthetic microenvironments for extrahepatic islet transplantation. Transplantation 82(4):452

Boehler RM, Graham JG, Shea LD (2011) Tissue engineering tools for modulation of the immune response. BioTechniques 51(4):239–254

Burdick JA, Prestwich GD (2011) Hyaluronic acid hydrogels for biomedical applications. Adv Mater 23(12):H41–H56

Burgeson RE, Nimni ME (1992) Collagen types. Molecular structure and tissue distribution. Clin Orthop Relat Res (282):250–272

Caffery B, Joyce E, Boone A, Slomovic A, Simpson T, Jones L, Senchyna M (2008) Tear lipocalin and lysozyme in Sjögren and non-Sjogren dry eye. Optom Vis Sci 85(8):661–667

Campuzano S, Pedrero M, González-Cortés A, Yáñez-Sedeño P, Pingarrón JM (2019) Electrochemical biosensors for autoantibodies in autoimmune and cancer diseases. Anal Methods 11(7):871–887

Chan AW, Neufeld RJ (2010) Tuneable semi-synthetic network alginate for absorptive encapsulation and controlled release of protein therapeutics. Biomaterials 31(34):9040–9047

Chandrashekara S (2012) The treatment strategies of autoimmune disease may need a different approach from conventional protocol: a review. Indian J Pharmacol 44(6):665

Charlton B, Lafferty KJ (1995) The Th1/Th2 balance in autoimmunity. Curr Opin Immunol 7(6): 793–798

Chen C-R, Pichurin P, Nagayama Y, Latrofa F, Rapoport B, McLachlan SM (2003) The thyrotropin receptor autoantigen in Graves disease is the culprit as well as the victim. J Clin Invest 111(12): 1897–1904

Chen L, Xia N, Li T, Bai Y, Chen X (2016) Aptasensor for visual and fluorometric determination of lysozyme based on the inner filter effect of gold nanoparticles on CdTe quantum dots. Microchim Acta 183(11):2917–2923

Chen P, Zhou G, Lin J, Li L, Zeng Z, Chen M, Zhang S (2020) Serum biomarkers for inflammatory bowel disease. Front Med 7:123

Chertok B, David AE, Yang VC (2010) Polyethyleneimine-modified iron oxide nanoparticles for brain tumor drug delivery using magnetic targeting and intra-carotid administration. Biomaterials 31(24):6317–6324

Christiane S, Hampe B (2012) B cells in autoimmune diseases. Scientifica (Cairo) 2012:215308

Clegg JR, Wechsler ME, Peppas NA (2017) Vision for functionally decorated and molecularly imprinted polymers in regenerative engineering. Regen Eng Transl Med 3(3):166–175

Danjo Y, Lee M, Horimoto K, Hamano T (1994) Ocular surface damage and tear lactoferrin in dry eye syndrome. Acta Ophthalmol 72(4):433–437

Davidson A, Diamond B (2001) Autoimmune diseases. N Engl J Med 345(5):340–350

Derkus B, Bozkurt PA, Tulu M, Emregul KC, Yucesan C, Emregul E (2017) Simultaneous quantification of Myelin Basic Protein and Tau proteins in cerebrospinal fluid and serum of multiple sclerosis patients using nanoimmunosensor. Biosens Bioelectron 89:781–788

Di Felice V, Serradifalco C, Rizzuto L, De Luca A, Rappa F, Barone R, Di Marco P, Cassata G, Puleio R, Verin L (2015) Silk fibroin scaffolds enhance cell commitment of adult rat cardiac progenitor cells. J Tissue Eng Regen Med 9(11):E51–E64

Dong H, Hao K, Tian Y, Jin S, Lu H, Zhou S-F, Zhang X (2014) Label-free and ultrasensitive microRNA detection based on novel molecular beacon binding readout and target recycling amplification. Biosens Bioelectron 53:377–383

Duddy ME, Alter A, Bar-Or A (2004) Distinct profiles of human B cell effector cytokines: a role in immune regulation? J Immunol 172(6):3422–3427

Dykman L, Khlebtsov N (2012) Gold nanoparticles in biomedical applications: recent advances and perspectives. Chem Soc Rev 41(6):2256–2282

Ekimov AI, Onushchenko AA (1981) Quantum size effect in three-dimensional microscopic semiconductor crystals. JETP Lett 34(6):345–349

Fagúndez P, Brañas G, Cairoli E, Laíz J, Tosar JP (2018) An electrochemical biosensor for rapid detection of anti-dsDNA antibodies in absolute scale. Analyst 143(16):3874–3882

Ghorbani F, Abbaszadeh H, Mehdizadeh A, Ebrahimi-Warkiani M, Rashidi M-R, Yousefi M (2019) Biosensors and nanobiosensors for rapid detection of autoimmune diseases: a review. Microchim Acta 186(12):1–11

Graham JG, Zhang X, Goodman A, Pothoven K, Houlihan J, Wang S, Gower RM, Luo X, Shea LD (2013) PLG scaffold delivered antigen-specific regulatory T cells induce systemic tolerance in autoimmune diabetes. Tissue Eng A 19(11–12):1465–1475

Harris VK, Donelan N, Yan QJ, Clark K, Touray A, Rammal M, Sadiq SA (2013) Cerebrospinal fluid fetuin-A is a biomarker of active multiple sclerosis. Mult Scler J 19(11):1462–1472

Hayter SM, Cook MC (2012) Updated assessment of the prevalence, spectrum and case definition of autoimmune disease. Autoimmun Rev 11(10):754–765

Hegen H, Auer M, Deisenhammer F (2015) Pharmacokinetic considerations in the treatment of multiple sclerosis with interferon-β. Expert Opin Drug Metab Toxicol 11(12):1803–1819

Heidari B (2011) Rheumatoid arthritis: early diagnosis and treatment outcomes. Caspian J Intern Med 2(1):161

Hu M, Chen J, Li Z-Y, Au L, Hartland GV, Li X, Marquez M, Xia Y (2006) Gold nanostructures: engineering their plasmonic properties for biomedical applications. Chem Soc Rev 35(11): 1084–1094

Hu S, Wang J, Meijer J, Ieong S, Xie Y, Yu T, Zhou H, Henry S, Vissink A, Pijpe J (2007) Salivary proteomic and genomic biomarkers for primary Sjögren's syndrome. Arthrit Rheumat 56(11): 3588–3600

Hunckler MD, García AJ (2020) Engineered biomaterials for enhanced function of insulin-secreting β-cell organoids. Adv Funct Mater 30(48):2000134

Irvine DJ, Stachowiak AN, Hori Y (2008) Lymphoid tissue engineering: invoking lymphoid tissue neogenesis in immunotherapy and models of immunity. Semin Immunol 20:137–146

Javierre BM, Hernando H, Ballestar E (2011) Environmental triggers and epigenetic deregulation in autoimmune disease. Discov Med 12(67):535–545

King-Heiden TC, Wiecinski PN, Mangham AN, Metz KM, Nesbit D, Pedersen JA, Hamers RJ, Heideman W, Peterson RE (2009) Quantum dot nanotoxicity assessment using the zebrafish embryo. Environ Sci Technol 43(5):1605–1611

Klippstein R, Pozo D (2010) Nanotechnology-based manipulation of dendritic cells for enhanced immunotherapy strategies. Nanomedicine 6(4):523–529

Knipe JM, Peppas NA (2014) Multi-responsive hydrogels for drug delivery and tissue engineering applications. Regen Biomater 1(1):57–65

Knipe JM, Peters JT, Peppas NA (2013) Theranostic agents for intracellular gene delivery with spatiotemporal imaging. Nano Today 8(1):21–38

La Belle JT, Bhavsar K, Fairchild A, Das A, Sweeney J, Alford T, Wang J, Bhavanandan VP, Joshi L (2007) A cytokine immunosensor for multiple sclerosis detection based upon label-free electrochemical impedance spectroscopy. Biosens Bioelectron 23(3):428–431

Lewis JS, Allen RP (2016) An introduction to biomaterial-based strategies for curbing autoimmunity. Exp Biol Med 241(10):1107–1115

Li H, Wei Q, He J, Li T, Zhao Y, Cai Y, Du B, Qian Z, Yang M (2011) Electrochemical immunosensors for cancer biomarker with signal amplification based on ferrocene functionalized iron oxide nanoparticles. Biosens Bioelectron 26(8):3590–3595

Liao H, Wu J, Kuhn E, Chin W, Chang B, Jones MD, O'Neil S, Clauser KR, Karl J, Hasler F (2004) Use of mass spectrometry to identify protein biomarkers of disease severity in the synovial fluid and serum of patients with rheumatoid arthritis. Arthritis Rheum 50(12):3792–3803

Liechty WB, Scheuerle RL, Peppas NA (2013) Tunable, responsive nanogels containing t-butyl methacrylate and 2-(t-butylamino) ethyl methacrylate. Polymer 54(15):3784–3795

Lodhi G, Kim Y-S, Hwang J-W, Kim S-K, Jeon Y-J, Je J-Y, Ahn C-B, Moon S-H, Jeon B-T, Park P-J (2014) Chitooligosaccharide and its derivatives: preparation and biological applications. Biomed Res Int 2014:654913

Longo CM, Higgins PJ (2019) Molecular biomarkers of Graves' ophthalmopathy. Exp Mol Pathol 106:1–6

Lovrić J, Bazzi HS, Cuie Y, Fortin GR, Winnik FM, Maysinger D (2005) Differences in subcellular distribution and toxicity of green and red emitting CdTe quantum dots. J Mol Med 83(5): 377–385

Lund FE, Garvy BA, Randall TD, Harris DP (2005) Regulatory roles for cytokine-producing B cells in infection and autoimmune disease. Curr Dir Autoimmun 8:25–54

Lv S, Tang Z, Li M, Lin J, Song W, Liu H, Huang Y, Zhang Y, Chen X (2014) Co-delivery of doxorubicin and paclitaxel by PEG-polypeptide nanovehicle for the treatment of non-small cell lung cancer. Biomaterials 35(23):6118–6129

Maddur MS, Miossec P, Kaveri SV, Bayry J (2012) Th17 cells: biology, pathogenesis of autoimmune and inflammatory diseases, and therapeutic strategies. Am J Pathol 181(1):8–18

Manzo A, Bombardieri M, Humby F, Pitzalis C (2010) Secondary and ectopic lymphoid tissue responses in rheumatoid arthritis: from inflammation to autoimmunity and tissue damage/ remodeling. Immunol Rev 233(1):267–285

Mathieu C, Lahesmaa R, Bonifacio E, Achenbach P, Tree T (2018) Immunological biomarkers for the development and progression of type 1 diabetes. Diabetologia 61(11):2252–2258

Matsumura Y, Byrne SN, Nghiem DX, Miyahara Y, Ullrich SE (2006) A role for inflammatory mediators in the induction of immunoregulatory B cells. J Immunol 177(7):4810–4817

Mavragani CP, Moutsopoulos HM (2014) Sjögren syndrome. CMAJ 186(15):E579–E586

Medzhitov R (2008) Origin and physiological roles of inflammation. Nature 454(7203):428–435

Mock J, Barbic M, Smith D, Schultz D, Schultz S (2002) Shape effects in plasmon resonance of individual colloidal silver nanoparticles. J Chem Phys 116(15):6755–6759

Murphy C, Gole A, Stone J, Sisco P, Akilany AM, Goldsmith EC, Baxter SC (2008) Gold nanoparticles in biology: beyond toxicity to cellular imaging. Acc Chem Res 41:1721

Nakae S, Nambu A, Sudo K, Iwakura Y (2003) Suppression of immune induction of collagen-induced arthritis in IL-17-deficient mice. J Immunol 171(11):6173–6177

Nathan C (2006) Neutrophils and immunity: challenges and opportunities. Nat Rev Immunol 6(3): 173–182
Nicodemus G, Bryant S (2008) Cell encapsulation in biodegradable hydrogels for tissue engineering applications. Tissue Eng B Rev 14:149
Otterness IG (1994) The value of C-reactive protein measurement in rheumatoid arthritis. Semin Arthrit Rheumat 24:91–104
Parenteau-Bareil R, Gauvin R, Berthod F (2010) Collagen-based biomaterials for tissue engineering applications. Materials 3(3):1863–1887
Park H, Li Z, Yang XO, Chang SH, Nurieva R, Wang Y-H, Wang Y, Hood L, Zhu Z, Tian Q (2005) A distinct lineage of CD4 T cells regulates tissue inflammation by producing interleukin 17. Nat Immunol 6(11):1133–1141
Perumal V, Hashim U (2014) Advances in biosensors: principle, architecture and applications. J Appl Biomed 12(1):1–15
Pickup JC (2012) Insulin-pump therapy for type 1 diabetes mellitus. N Engl J Med 366(17): 1616–1624
Piliarik M, Kvasnička P, Galler N, Krenn JR, Homola J (2011) Local refractive index sensitivity of plasmonic nanoparticles. Opt Express 19(10):9213–9220
Pinaud F, Michalet X, Bentolila LA, Tsay JM, Doose S, Li JJ, Iyer G, Weiss S (2006) Advances in fluorescence imaging with quantum dot bio-probes. Biomaterials 27(9):1679–1687
Pons T, Pic E, Lequeux N, Cassette E, Bezdetnaya L, Guillemin F, Marchal F, Dubertret B (2010) Cadmium-free CuInS2/ZnS quantum dots for sentinel lymph node imaging with reduced toxicity. ACS Nano 4(5):2531–2538
Quan Q, Xie J, Gao H, Yang M, Zhang F, Liu G, Lin X, Wang A, Eden HS, Lee S (2011) HSA coated iron oxide nanoparticles as drug delivery vehicles for cancer therapy. Mol Pharm 8(5): 1669–1676
Rook GA (2012) Hygiene hypothesis and autoimmune diseases. Clin Rev Allergy Immunol 42(1): 5–15
Rose NR, Mackay IR (2006) The autoimmune diseases. Elsevier
Ryu O, Atkinson J, Hoehn G, Illei G, Hart T (2006) Identification of parotid salivary biomarkers in Sjögren's syndrome by surface-enhanced laser desorption/ionization time-of-flight mass spectrometry and two-dimensional difference gel electrophoresis. Rheumatology 45(9):1077–1086
Sánchez Brenes A, Sibaja Ballestero M, Vega Baudrit J, Rojas Chaves M (2007) Utilización de soportes de hidrogel de quitosano obtenidos a partir de desechos del camarón langostino (pleuroncodes planipes) para el crecimiento "in vitro" de fibroblastos humanos
Sandborn WJ, Fazio VW, Feagan BG, Hanauer SB (2003) AGA technical review on perianal Crohn's disease. Gastroenterology 125(5):1508–1530
Sands BE, Kaplan GG (2007) The role of TNFα in ulcerative colitis. J Clin Pharmacol 47(8): 930–941
Santhoshkumar S, Ramasamy U, Mansuour RF, Ramaraj E (2021) A review on statistical importance and biomarkers identification in Hashimoto thyroiditis disease. In: 2021 11th international conference on cloud computing, data science & engineering (confluence). IEEE, pp 1040–1045
Schweingruber N, Haine A, Tiede K, Karabinskaya A, van den Brandt J, Wüst S, Metselaar JM, Gold R, Tuckermann JP, Reichardt HM (2011) Liposomal encapsulation of glucocorticoids alters their mode of action in the treatment of experimental autoimmune encephalomyelitis. J Immunol 187(8):4310–4318
Sheikh FA, editor. Engineering materials for stem cell regeneration. Springer; 2021 Oct 23.
Shi C, Pamer EG (2011) Monocyte recruitment during infection and inflammation. Nat Rev Immunol 11(11):762–774
Shodeinde AB, Murphy AC, Oldenkamp HF, Potdar AS, Ludolph CM, Peppas NA (2020) Recent advances in smart biomaterials for the detection and treatment of autoimmune diseases. Adv Funct Mater 30(37):1909556

Singer PA, Cooper DS, Levy EG, Ladenson PW, Braverman LE, Daniels G, Greenspan FS, McDougall IR, Nikolai TF (1995) Treatment guidelines for patients with hyperthyroidism and hypothyroidism. JAMA 273(10):808–812

Sionkowska A (2011) Current research on the blends of natural and synthetic polymers as new biomaterials. Prog Polym Sci 36(9):1254–1276

Smolen JS, Aletaha D, Bijlsma JW, Breedveld FC, Boumpas D, Burmester G, Combe B, Cutolo M, De Wit M, Dougados M (2010) Treating rheumatoid arthritis to target: recommendations of an international task force. Ann Rheum Dis 69(4):631–637

Sofi HS, Ashraf R, Khan AH, Beigh MA, Majeed S, Sheikh FA (2019) Reconstructing nanofibers from natural polymers using surface functionalization approaches for applications in tissue engineering, drug delivery and biosensing devices. Mater Sci Eng C 94:1102–1124

Soler M, Estevez M-C, de Lourdes Moreno M, Cebolla A, Lechuga LM (2016) Label-free SPR detection of gluten peptides in urine for non-invasive celiac disease follow-up. Biosens Bioelectron 79:158–164

Sun K, Li H, Li R, Nian Z, Li D, Xu C (2015) Silk fibroin/collagen and silk fibroin/chitosan blended three-dimensional scaffolds for tissue engineering. Eur J Orthop Surg Traumatol 25(2):243–249

Sundaram J, Durance TD, Wang R (2008) Porous scaffold of gelatin–starch with nanohydroxyapatite composite processed via novel microwave vacuum drying. Acta Biomater 4(4):932–942

Tadi KK, Alshanski I, Mervinetsky E, Marx G, Petrou P, Dimitrios KM, Gilon C, Hurevich M, Yitzchaik S (2017) Oxytocin-monolayer-based impedimetric biosensor for zinc and copper ions. ACS Omega 2(12):8770–8778

Tang L-Y, Wang Y-C, Li Y, Du J-Z, Wang J (2009) Shell-detachable micelles based on disulfide-linked block copolymer as potential carrier for intracellular drug delivery. Bioconjug Chem 20(6):1095–1099

Targan SR (2006) Current limitations of IBD treatment: where do we go from here? Ann N Y Acad Sci 1072(1):1–8

Tognana E, Borrione A, De Luca C, Pavesio A (2007) Hyalograft® C: hyaluronan-based scaffolds in tissue-engineered cartilage. Cells Tissues Organs 186(2):97–103

Truong VX, Ablett MP, Richardson SM, Hoyland JA, Dove AP (2015) Simultaneous orthogonal dual-click approach to tough, in-situ-forming hydrogels for cell encapsulation. J Am Chem Soc 137(4):1618–1622

Vega B, Calle A, Sánchez A, Lechuga LM, Ortiz AM, Armelles G, Rodríguez-Frade JM, Mellado M (2013) Real-time detection of the chemokine CXCL12 in urine samples by surface plasmon resonance. Talanta 109:209–215

Vives-Pi M, Takasawa S, Pujol-Autonell I, Planas R, Cabre E, Ojanguren I, Montraveta M, Santos AL, Ruiz-Ortiz E (2013) Biomarkers for diagnosis and monitoring of celiac disease. J Clin Gastroenterol 47(4):308–313

Vrolix K, Fraussen J, Molenaar PC, Losen M, Somers V, Stinissen P, De Baets MH, Martínez-Martínez P (2010) The auto-antigen repertoire in myasthenia gravis. Autoimmunity 43(5–6): 380–400

Waite JC, Skokos D (2012) Th17 response and inflammatory autoimmune diseases. Int J Inflamm 2012:819467

Wilson L, Vam Wyk J, Rassie C, Ross N, Sunday C, Makelane HR, Bilibana M, Waryo T, Mapolie S, Baker PG (2015) Electrochemical immunosensor based on the interactions between polypyrrole and cobalt (II) salicylaldiimine dendrimer. Int J Electrochem Sci 10:3207–3222

Xie J, Chen K, Huang J, Lee S, Wang J, Gao J, Li X, Chen X (2010) PET/NIRF/MRI triple functional iron oxide nanoparticles. Biomaterials 31(11):3016–3022

Yeh Y-C, Creran B, Rotello VM (2012) Gold nanoparticles: preparation, properties, and applications in bionanotechnology. Nanoscale 4(6):1871–1880

Yoo HS, Lee EA, Yoon JJ, Park TG (2005) Hyaluronic acid modified biodegradable scaffolds for cartilage tissue engineering. Biomaterials 26(14):1925–1933

Zasońska BA, Hlídková H, Petrovský E, Myronovskij S, Nehrych T, Negrych N, Shorobura M, Antonyuk V, Stoika R, Kit Y (2018) Monodisperse magnetic poly (glycidyl methacrylate) microspheres for isolation of autoantibodies with affinity for the 46 kDa form of unconventional Myo1C present in autoimmune patients. Microchim Acta 185(5):1–7

Zhang T, Stilwell JL, Gerion D, Ding L, Elboudwarej O, Cooke PA, Gray JW, Alivisatos AP, Chen FF (2006) Cellular effect of high doses of silica-coated quantum dot profiled with high throughput gene expression analysis and high content cellomics measurements. Nano Lett 6(4):800–808

Zhang H, Marmin T, Cuierrier É, Soldera A, Dory Y, Zhao Y (2015a) A new comonomer design for enhancing the pH-triggered LCST shift of thermosensitive polymers. Polym Chem 6(37): 6644–6650

Zhang B-X, Zhang Z-L, Lin AL, Wang H, Pilia M, Ong JL, Dean DD, Chen X-D, Yeh C-K (2015b) Silk fibroin scaffolds promote formation of the ex vivo niche for salivary gland epithelial cell growth, matrix formation, and retention of differentiated function. Tissue Eng A 21(9–10): 1611–1620

Zhang X, Zambrano A, Lin Z-T, Xing Y, Rippy J, Wu T (2017) Immunosensors for biomarker detection in autoimmune diseases. Arch Immunol Ther Exp 65(2):111–121

Zhao Y, Liu Y, Li X, Wang H, Zhang Y, Ma H, Wei Q (2018) Label-free ECL immunosensor for the early diagnosis of rheumatoid arthritis based on asymmetric heterogeneous polyaniline-gold nanomaterial. Sensors Actuators B Chem 257:354–361

Sampita Pal was born in India in 1994. She obtained an MS (Pharm) degree from the National Institute of Pharmaceutical Education and Research (NIPER), Kolkata, in 2019. She worked under the supervision of Dr. Umesh Prasad Singh as a master's intern (2018–2019) at CSIR-Indian Institute of Chemical Biology, Kolkata, on phytochemical screening of *Albizia lebbeck* bark to evaluate its anti-infective property. She was also associated with Dr. Amit Misra as a project associate (2020–2021) at CSIR-Central Drug Research Institute, Lucknow. Currently, she is pursuing Ph.D. at Bar-Ilan University, Tel Aviv, Israel, in the Faculty of Life Science. She is a coauthor of two scientific articles and one book chapter published in international journals. She has achieved the "**College Blue-2017**" award for being the most versatile student of the batch (2013–2017) and "**JIS SAMMAN**" in 2015 for academic excellence. She also bagged the Certificate of Award (**first**) for **Poster Presentation** in 2016 at an **International Seminar** on *Global Advancement in Pharmaceutical Research and Startup*. Her current research area is focused on biopolymer chemistry, biomaterials for drug delivery, food science, and nanotechnology.

Hasham S. Sofi, M.Pharm., is a Senior Research Fellow of the Indian Council of Medical Research and pursuing a Ph.D. at the Department of Pharmaceutical Sciences, University of Kashmir, Srinagar, India. He has pursued a Bachelor's in Pharmaceutical Sciences and completed his Master in Pharmaceutics from the University of Kashmir. He has qualified to the National Graduate Pharmacy Aptitude Test (GPAT) with an All-India Rank of 75. He has received fellowships for pursuing research from various funding agencies, including the All-India Council of Technical Education, the Department of Science and Technology, and the Indian Council of Medical Research. He has held junior/senior research fellow positions at the Department of Nanotechnology, University of Kashmir, and Senior Research fellow at the Division of Pharmaceutics and Pharmacokinetics at Central Drug Research Institute. His primary research interest is developing fast-dissolving delivery systems for rapid action, studying inhaled therapies for managing tuberculosis, and using nanofibers in regenerative medicine. He is currently studying the mechanisms involved in forming mycobacterial biofilms, their role in resistance to antitubercular drugs, and the development/repurposing of drugs that can aid the dispersal of biofilms. He has developed formulations that can preferentially target antimalarial drugs to plasmodium-infected RBCs and nanofiber-based dressings for early skin tissue regeneration. He has published research and review articles in well-reputed journals and book chapters with international publishers and received more than 500 citations.

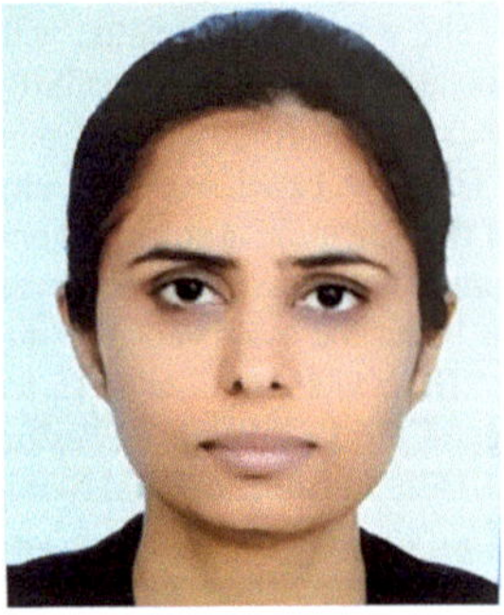

Lubna Azmi, Ph.D., is currently working as an Assistant Professor in pharmaceutical chemistry at Lucknow University. Before working in this position, she worked as a postdoc fellow at the Division of Pharmaceutics, Central Drug Research Institute, Lucknow. She has taught pharmaceutical and analytical at Lucknow University for the past year. Born in India, Dr. Lubna Azmi received a Bachelor's in Pharmacy and a Master in Pharmaceutical Chemistry from Abdul Kalam Technical University. She has expertise in analytical methods like HPLC and LC-MS/MS, synthetic chemistry, inorganic/organic chemistry, and formulation development. Her current research interests are studying enteric-coated microspheres for gastrointestinal tuberculosis, developing animal models of intestinal tuberculosis, and formulating dry powders for pulmonary tuberculosis. She has published more than 30 research papers in national and international journals. She is also authoring many book chapters and review articles.

Nisar Ahmad Khan, Ph.D., is an Associate Professor at the Department of Pharmaceutics, University of Kashmir, Srinagar, India. Dr. Khan was awarded a Bachelor of Science and Bachelor of Pharmacy from the University of Kashmir. He pursued a Master in Pharmacy in Pharmaceutics from Rajiv Gandhi Proudyogiki Vishwavidyalaya. He was awarded Ph.D. in Pharmaceutics from the University of Kashmir while working on gastro retentive drug delivery systems. He is an active member of the International Association for Pharmaceutical Technologist and the International Association of Physical Chemists. He has received funding from the University Grants Commission (UGC) and the Indian Council of Medical Research (ICMR) for various research projects. He has guided many research scholars pursuing master's and doctorate degrees in pharmacy. His primary research interests are formulation design of oral drug-controlled release systems, gastro retentive drug delivery systems, and enhancement of bioavailability of poorly soluble drugs. Dr. Khan has published more than 40 research papers and many book chapters. He has received more than 600 citations.

Shafquat Majeed, Ph.D., is currently working as an Assistant Professor in the Department of Nanotechnology, University of Kashmir, India. Dr. Majeed did his Master in Chemistry from the Department of Chemistry, University of Kashmir. He did his Ph.D. at the prestigious Indian Institute of Science Bangalore (2008–2015). He has a postdoctoral research experience in rare-earth-based nanomaterials for biomedical applications and fabrication of Perovskite-based solar cells from the Centre for Nanoscience and Engineering, IISc Bangalore (2015–2016). He, at present, heads a SCAN (Synthesis, Characterization and Application of Nanomaterials) laboratory at the University of Kashmir, where they work on synthesis, characterization, and optical and magnetic properties of nanomaterials prepared through various solution-based routes. More specifically, they are developing new and enhanced multimodal imaging modalities for biomedical and healthcare applications.

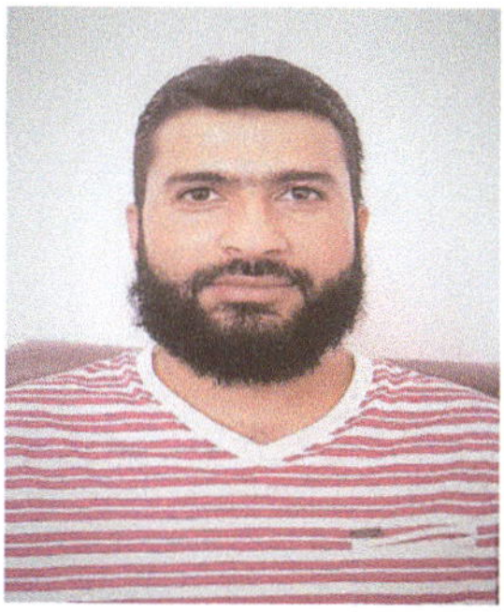

Mushtaq A. Beigh, Ph.D., is working as an Assistant Professor at the Department of Nanotechnology, University of Kashmir, India. Dr. Beigh has postdoctoral research experience in cellular signaling and disease modeling (2013–2016). He had completed his Ph.D. from the University of Kashmir, India (2008–2012). He heads a laboratory at the University of Kashmir, where they work on a bio-nano interface to understand the molecular basis of nanoconjugate internalization. They also try to understand the growth factor signaling inputs directed at multiple cellular receptors like NRP1 and integrins in order to work out the mechanistic details of receptor-based internalization.

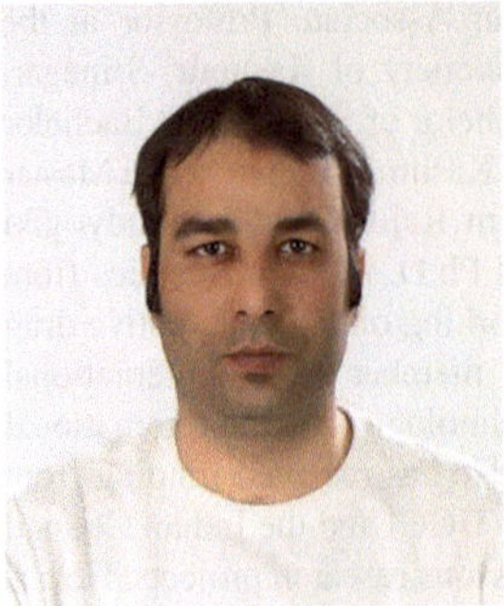

Faheem A. Sheikh, Ph.D., is an Assistant Professor at the Department of Nanotechnology, University of Kashmir, India (2015). He served as an Assistant Professor at the Department of Biotechnology at the Central University of Kashmir, India (2015–2016); Research Professor, Myongji University, South Korea (2014–2015); Assistant Professor, Hallym University, South Korea (2012–2014); Postdoc/Research Fellow, the University of Texas Rio Grande Valley, Texas, USA (2010–2012); and Research Professor at Myongji University, South Korea (2010). His research mainly focuses on fabricating nanomaterials used in tissue engineering.

Nanotechnology-Based Biosensors in Medicine 2

Anita Patel, Nisarg C. Patel, Jayvadan K. Patel, and Saurin Amin

Abstract

Over the past 20 years, the development of biosensors has significantly benefited from nanotechnology. The creation of nanomaterials (NMs), their use in novel biosensing and diagnostic applications, and their thorough characterization for in vitro and in vivo applications are a few of these. The introduction of numerous new signal transduction technologies has been made possible by using nanomaterials to create biosensors, which has increased their sensitivity and performance. Mechanical devices based on nanobiosensor architecture, optical resonators, functionalized nanoparticles, nanowires, nanotubes, and nanofibers have all been utilized. Remarkably, the potential medical applications of nanomaterials, including gold nanoparticles, carbon nanotubes, magnetic nanoparticles, and quantum dots, have been extensively researched. Therefore, we cover many advanced nanomaterials that have been widely applied in recent years and discuss how they were fabricated. The main objective of this chapter is to offer an overview of recently created advanced nanomaterial-based biosensors used to diagnose a range of diseases as well as therapeutic purposes.

A. Patel (✉)
Sankalchand Patel University, Visnagar, Gujarat, India

N. C. Patel
APMC College of Pharmaceutical Education and Research, Himatnagar, Gujarat, India

J. K. Patel
Aavis Pharmaceuticals, Hoschton, GA, USA

Faculty of Pharmacy, Sankalchand Patel University, Visnagar, Gujarat, India

S. Amin
Aavis Pharmaceuticals, Hoschton, GA, USA

F. A. Sheikh et al. (eds.), *Interaction of Nanomaterials With Living Cells*,
https://doi.org/10.1007/978-981-99-2119-5_2

Keywords

Nanotechnology · Biosensors · Diagnosis

2.1 Introduction

As per IUPAC, the biosensor is defined as "a self-contained integrated device, which is capable of providing specific quantitative or semi-quantitative analytical information using a biological recognition element that is in direct spatial contact with a transducer element." The sensor comprises a transducer, a readout system, and a detection system known as a receptor (Fig. 2.1). In biosensors, the acceptor is a biological component that is variously fixed to the converter. The method by which the bioreceptor layer (biological component) is mounted on the transducer significantly impacts biosensors' success. Selectivity is a crucial sensor property that shows how well the system can separate the analyte from other components in the sample. Sensitivity is the next critical feature. High sensitivity indicates that a considerable change in the output signal from the sensor can be seen even with

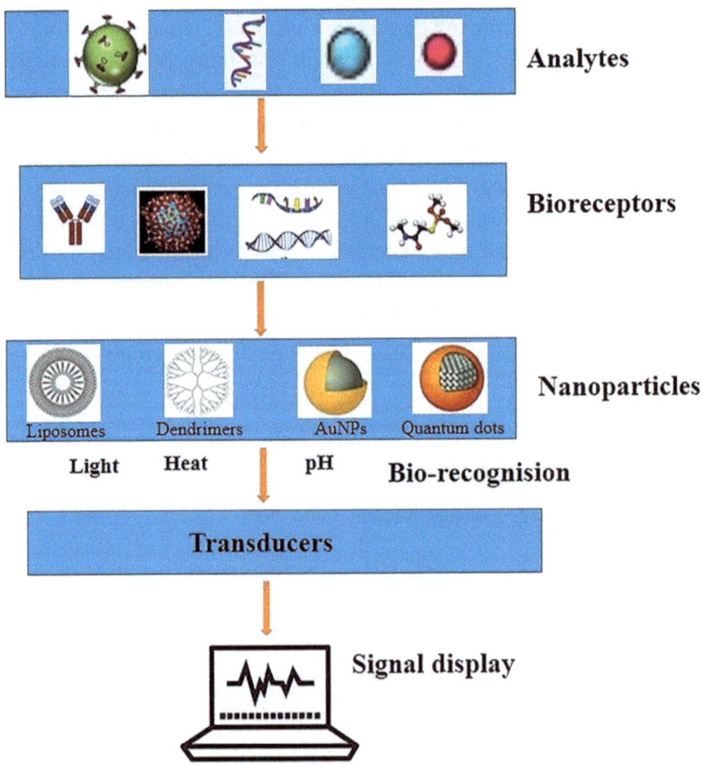

Fig. 2.1 Major components of a typical biosensor

slight differences in analyte concentration. Such a sensor has a high bit resolution. Another crucial aspect that is included in the definition of accuracy is the repeatability (measurability) of the measurement results (Ge et al. 2014; Sangadevan and Periasamy 2014; Vashist et al. 2012; Huang et al. 2021; Pandit et al. 2016). The development of biosensors is increasingly dependent on nanotechnology. Biosensors are becoming more sensitive and effective, thanks to nanomaterials' use in their production. By combining sensors, fluidics, and signal-processing circuits, nanotechnology will make it possible to further miniaturize bioanalytical systems. This will allow for the large-scale integration of various biological reactions on a more compact scale (Vo-Dinh et al. 2001).

Due to their substantial specific surface area and high free surface energy, nanoparticles play a significant role in the surface adsorption of biomolecules. Materials with 1 and 100 nm diameters are commonly referred to as nanomaterials. Due to their small size, these particles have interesting physicochemical characteristics (Wang et al. 2014; Bai and Jiang 2013; Gholipour et al. 2020). Going nano means a matter can be altered at the molecular and atomic levels in addition to its reduced size. The developing ability to manipulate the patterns of matter on the nanoscale scale will fundamentally alter the ability to sense and detect the condition of biological systems and live organisms in the case of biosensing (Alivisatos 2004). It is anticipated that such a drastic transformation will make it possible for living cells to detect several signals simultaneously and at the single molecular level. Going nano will help to enhance sensitivity and lower the detection limit, reduce the volume of analyte reagent, and shorten the detection time at the systems level. Biosensors can be reduced in size so that they can be inserted at any desired position within the body.

2.2 Various Detection Methods

These methods can be generally categorized into mechanical, optical, electromagnetic, electrical, thermal, magnetic, and electrochemical processes.

2.2.1 Mechanical Detection

Mechanical-based transducers typically use mechanical deformations or mechanical waves (or acoustic waves) as their sensing mechanisms. In order to use this technique, the underlying transducer is frequently a mechanical structure in the form of a cantilever beam, a double-clamped beam, or a disc. The surface of the transducer is functionalized by immobilizing a layer of a sensitive element (e.g., antibodies or enzymes) on it for target binding. Ex vivo applications are more suited for this kind of detection (Stenger et al. 2001).

A conventional method of detection in the case of a cantilever beam involves measuring the cantilever deflection brought on by the surface stresses produced as a result of molecule binding. Intermolecular forces like electrostatic contact or van der

Waals force of attraction are the leading causes of surface stress. The cantilever will deflect due to the generation of differential stress (Berger et al. 1997). Two methods are frequently used to measure cantilever deflection. In the first, the cantilever deflection is measured using a four-segment photodetector, while a laser beam is focused on the free end of the cantilever. The second method uses electrical equipment to measure the cantilever deflection using resistive or capacitive circuitry (Porter et al. 2003).

Advantages of mechanical detection included the ability to perform protein recognition with extremely high sensitivity and the ability to identify mismatches in oligonucleotide hybridization without labeling. Additionally, this technique works with various analyte species in gaseous or aqueous forms. The method faces some limitations as well, such as the fact that if the molecule binding events are exothermic, the heat created may make it difficult to detect them since differential thermal stress might cause the cantilever to deflect (Mertens et al. 2003). Another problem is that the molecular structures are nonlinear and viscoelastic, which might make them impossible to utilize.

Changes in acoustic properties, such as the resonant frequency, attenuation, and phase of wave propagation, are the typical approach of detection in the circumstances of a double-clamped beam or a disc structure. A molecule binding event serves as mass loading in this mode of detection, where mechanical structures behave like oscillators. This mass loading frequently causes a shift in the resonant frequency, an increase in amplitude attenuation, or a delay in the phase of wave propagation. However, the complexity of detecting the type and uniformity of the bound species makes this type of mechanical detection challenging, making it less precise in biological sensing (Rao and Zhang 2006).

2.2.2 Optical and Electromagnetic Detection

Optical detection is one of the most commonly utilized mechanisms for biosensing. This technique may be integrated into a wide range of spectroscopic techniques, which include luminescence, absorption, polarization, and fluorescence. This detection technology allows for the measurement of a target analyte's amplitude, energy, polarization, decay time, and phase, among other spectrochemical parameters. The most prominent optical technology is undoubtedly fluorescence-based detection. In this method, fluorescent markers that produce light at particular wavelengths are utilized as detection labels for the target analytes. The presence of the targets or the binding of the targets to the probes is determined by measurements of fluorescence intensity (Vo-Dinh and Cullum 2000).

Another common approach of optical detection is the evanescent wave-based sensing approach. By total internal reflection, an optical waveguide is utilized in this technique to restrict the light passing through it. Evanescent wave-based sensors are ideal for measuring molecular interactions in situ and in real time since they are susceptible to the detection of biological and chemical species at lower levels (Liu and Tan 1999). The theory of surface-enhanced Raman spectroscopy (SERS) serves as the foundation for a widely used electromagnetic detection technique. Because of surface plasmon resonance, molecules positioned near to a rough metal surface with silver or gold nanostructures (such as nanoparticles or nanowires) can exhibit surface-enhanced Raman scattering.

2.2.3 Electrical Detection

Electrical detection offers several desired characteristics as an underlying transducer because of its simplicity of use, label-free detecting capability, portability, and downsizing, even if it has not been as commonly employed as the mechanical or optical detection methods. When a biological binding event takes place at the electrode surfaces, conductometric and potentiometric techniques, two popular types of electrical detection, primarily rely on the measurement of changes in conductance (or impedance) and potential. Changes in the electrical resistance or impedance between two electrodes are detected by conductometric sensors (Chen et al. 2004). In this instance, the variations in resistance or impedance are brought on by metabolite excretion in the vicinity of the electrode surfaces or in the surrounding medium or by molecular interactions between nucleotides, proteins, antigens, and antibodies. This method of detection is appealing because, unlike electrochemical detection, it does not call for a particular reference electrode. Potentiometric sensors quantify the potential changes between electrodes. Ion-sensitive field-effect transistors or chemical field-effect transistors are used in the most popular potentiometric sensor design. Due to the inherent molecular charges on the DNA, potentiometric sensors have been employed to perform label-free detection of DNA hybridization by detecting the field effect in silicon (Fritz et al. 2002).

2.2.4 Electrochemical Detection

The electrical responses produced by the electrochemical reactions of the target redox species catalyzed by the enzymatic-sensitive elements are frequently measured by biosensors that use an electrochemical technique as the underlying transducer. Typically, these biosensors are set up in a three-electrode arrangement: a working electrode, a counter electrode, and a reference electrode. Amperometric, voltammetric, and impedimetric modes of operation are the most frequently utilized for biological detections. The electrical current produced by the exchange of electrons between the electrodes and ionic species in response to electrode polarization at a constant potential is measured by amperometric biosensors. Voltammetric biosensors evaluate the current-potential relationships (sometimes called voltammograms) brought on by a redox process. Upon cyclic excitations of the working electrode at a predetermined frequency range, impedimetric biosensors monitor changes in the complex impedance of an electrochemical process. The detection of glucose, lactose, urea, lactate, and DNA hybridization has all been done using electrochemical-based biosensors (Zhu and Snyder 2003; Zhang 2009).

2.3 Development in Nanotechnology-Based Biosensors

However, biosensors based on nanotechnology are a more comprehensive word that can be seen from three perspectives. **Type A** nanotechnology-based biosensors are those that contain nanomaterials in their structural design, according to the first point of view. These might include various nanomaterials that have been coated with bioreceptors and vary in size, shape, and construction materials. These nanoparticles increase the surface area of these biosensors and increase their sensitivity. **Type B** is concerned with biosensors that include nanostructured structures and are made using self-assembly or nanofabrication techniques. In this type, electrodes are typically nanoscale in size and coupled to a processor that is larger, like a computer. The last type (**type C**) indicates the biosensors in nanoscale, which means a device for sensing usage with at least one dimension below 100 nm. This type will be a highly packed and complex device than integrate sensing and processing in one device (Liu et al. 2009).

In the last 10 years, the fields of in vitro diagnostics, imaging, and therapies have seen the extensive application of NMs. They have made it possible to diagnose diseases at a very early stage and to simultaneously multiplex detect a large number of disease biomarkers (Cheng et al. 2011). They have also made it possible to investigate the detection of extremely low concentrations of the target analytes and have produced ultrasensitive, quick, and economical assays that only need a small sample.

The capacity to detect single molecules can also be achieved by using nanostructure (nanopores, nanowires, nanopillars, and nanogaps)-based devices, which utilize gold nanoparticles (GNPs) that have been labeled with brief DNA segments. In the past 10 years, a number of intriguing NMs have been applied to diagnostics and biosensors, including carbon nanotubes (CNTs), graphene, quantum dots (QDs), nanoparticles (NPs), and nanocomposites. The NMs with improved performance and functionality, like QDs and NPs, make effective imaging agents (Kairemo et al. 2008). Nanotechnology advancements will be very helpful in moving from late-stage diagnosis (including expensive and socially demanding therapy) to early-stage diagnosis (relatively less costly and less invasive). Due in large part to the multibillion-dollar glucose monitoring market, the significant application has nearly invariably been glucose sensing.

2.4 Nanotechnology-Based Biosensors

2.4.1 Thin Films

Due to inherent features related to their nanoscale dimensions, nanostructured thin films have made it possible to create electrochemical sensors and biosensors with high detection powers. Additionally, a wide variety of organic and inorganic components can be used for the growth of films. Additionally, the use of suitable materials such as natural polymers can help to explain the potential for improving the detection limit in biosensing devices. The purpose of using these materials is to maintain the structural integrity of biomolecules while combining a high level of detection with their ability to function as biocatalysts (Pandit et al. 2016).

2.4.2 Nanomaterial-Based Biosensors

In order to exploit certain nanomaterials in better biological signaling and transduction pathways, their electrical and mechanical properties have been investigated. Nanotubes, nanowires, nanorods, nanoparticles, and thin films comprised of crystalline solids are some of these often used materials. Colloidal nanoparticles can be employed for immunosensing and immunolabeling applications by conjugating with antibodies. Additionally, metal-based nanoparticles are excellent materials for electronic and optical applications, and by taking advantage of their optoelectronic characteristics, they can be effectively used for the detection of nucleic acid sequences. Nanomaterials like metal nanoparticles, oxide nanoparticles, magnetic nanomaterials, carbon materials, quantum dots, and metallophthalocyanines are now used due to significant advancements in the field of nanotechnology to enhance the electrochemical signals of biocatalytic events that take place at the electrode/electrolyte interface. For use in biosensors to detect biological molecules, functional nanoparticles that bind to biological molecules (such as peptides, proteins, and nucleic acids) have been produced (Pandit et al. 2016).

2.4.2.1 Carbon Nanotubes (CNTs)

CNTs have been one of the most widely employed NMs in the last 10 years for biosensors, diagnostics, tissue engineering, cell tracking and labeling, and medication and biomolecule delivery. Carbon nanotubes are a perfect material for use in chemical and biological sensing due to their nano-dimensions, graphitic surface chemistry, and electrical characteristics. Biosensors have utilized both single-walled nanotubes (SWNT) and multiwalled carbon nanotubes (MWNT). They are single-, double-, or multiwalled CNTs, which are hollow cylindrical tubes made of one, two, or more concentric graphitic layers and capped by fullerene spheres. For the susceptible detection of analytes in healthcare, industry, environmental monitoring, and food quality analysis, CNT-based biosensors and diagnostics have been used (Li et al. 2008; Muguruma et al. 2008). They have primarily been employed in electrochemical sensing, principally for the detection of glucose, but also for the detection of fructose, galactose, neurotransmitters, neurochemicals, amino acids, immunoglobulin, albumin, streptavidin, insulin, human chorionic gonadotropin, C-reactive protein, cancer biomarkers, cells, microorganisms, DNA, and other biomolecules (Davis et al. 2003). One instance involved coating glucose oxidase onto the surface of single-walled nanotubes (SWNT), which prevented a significant loss of enzyme function (Azamian et al. 2002). Similar to this, it was shown that *horseradish peroxidase* (HRP) adsorbed on a carbon nanotube microelectrode could transmit electrons straight to the electrode and still function as a catalytic enzyme for H_2O_2 (Zhao et al. 2002).

2.4.2.2 Graphene

Graphene, an atomically thin layer of sp^2-hybridized carbon, is another NM that has recently been used extensively for diagnostics and biosensors due to its captivating and exciting properties, such as high mechanical strength, high thermal conductivity, high elasticity, tunable optical properties, tunable bandgap, very high room temperature electron mobility, and demonstration of the room temperature quantum Hall effect. It is a translucent material with extremely little environmental impact and minimal production costs. For the detection of a variety of analytes, including glucose, cytochrome *c*, NADH, hemoglobin, cholesterol, ascorbic acid, dopamine, uric acid, hydrogen peroxide, HRP, catechol, DNA, heavy metal ions, and gases, it has been widely used in electrochemical, impedance, fluorescence, and electrochemiluminescence biosensors (Bai et al. 2020).

2.4.2.3 Quantum Dots

Nanoscale semiconductor crystals called quantum dots (QDs) shine or fluoresce when activated by a light source like a laser. The fluorescence of QDs is incredibly persistent, and they have a high level of resistance to deterioration. QDs are effective tagging agents because they can be selectively attached to biological components like cells, proteins, and nucleic acids. Any wavelength of light, from infrared to ultraviolet to visible, can be built into QDs. This makes it possible to employ a lot of different colors, which makes it possible to run multiplexed assays. The potential for a wide range of applications, including diagnostic tools, has been created by the development of a novel coating for inorganic particles at the nanoscale that may be able to disguise QDs as proteins. This technique enables particles to function as probes that can enter into cells and illuminate specific proteins therein (Jaiswal et al. 2004). Due to their distinct photoluminescent characteristics and possible uses, quantum dots have been the focus of extensive research (Lu et al. 2003). In cellular imaging (Kaul et al. 2003), immunoassays (Medintz et al. 2005), DNA hybridization (Hoshino et al. 2005), biosensors, and optical barcoding (Han et al. 2001), for instance, quantum dots, have been employed successfully.

Fluorescence resonance energy transfer (FRET) and the dynamic course of signal transduction in living cells have both been utilized to analyze the interaction between protein molecules using quantum dots (Jares-Erijman and Jovin 2003). In comparison with conventional fluorescent dyes, these synthesized quantum dots have a number of advantages, including improved stability, increased fluorescence intensity, and a variety of colors that may be achieved by varying the size of the quantum dots (Medintz et al. 2005). They have also been used to identify cancer target areas in vivo.

2.4.2.4 Chitosan

Due to its outstanding biocompatibility, complete biodegradability, and nontoxic makeup, chitosan is one of the most favorable NMs for the assimilation of biological components in medical devices. Natural metabolites that result from the breakdown of chitosan are completely safe. Because it is transparent, optical sensors can use it. Due to the porous nature of the chitosan films and their high ion permeability, it is also suitable for electrochemical sensors. The amine groups of chitosan make it easier for biomolecules to attach to it covalently and for polymers or NPs to form nanocomposites, and its p^H-dependent solubility makes it easier for stable films to form at neutral and basic p^H levels. But in order to make it more soluble in water and other common solvents, it needs to undergo chemical alteration, such as carboxymethylation. It has seen considerable use in lab-on-a-chip devices, biosensors, diagnostics, and other biomedical or bioanalytical applications (Sashiwa and Aiba 2004; Koev et al. 2010).

2.4.2.5 Dendrimers

The three-dimensional macromolecules known as dendrimers are hyperbranched, monodispersed, star-shaped, nanometer-scale and have a very high density of surface functional groups. They consist of three distinct components: the core, the inside dendron, and the outside surface containing terminal functional groups. They are widely used in a variety of biosensors and diagnostics, including those based on electrochemistry, fluorescence, surface-enhanced Raman scattering, impedimetry, and surface plasmon resonance. This is mainly because they improve analytical sensitivity, stability, and reproducibility while lowering nonspecific interactions. They have also been applied to various bioanalytical processes, including gene transfection, medication delivery, and catalysis (Satija et al. 2011; Shen and Shi 2010).

2.4.2.6 Nanoparticles

Due to their distinct optical and other features, NPs have also been widely exploited in various bioanalytical applications, particularly for the creation of biosensors, diagnostics, imaging, drug administration, and treatment. The attachment of molecules to their surface causes them to change color. For a variety of bioanalytical applications, the ability of nanoparticles to change their properties by altering their size or shape has been used (Parveen et al. 2012).

Gold Nanoparticles

By electrochemical stabilization, gold nanoparticles can also be deposited on the electrode. To finish preparing the enzymatic electrode, the electrode is submerged in a bath containing par benzoquinone, chitosan, glucose oxidase, and ionic liquid. By varying the solution concentration, the working potential, and the precipitation duration in this process, nanoparticle size and morphology can be adjusted.

The effective early-stage detection and photothermal treatment of cancer and other disorders are made possible by the remarkable plasmon absorption and scattering capabilities of GNPs. They have been employed as nanocarriers for the delivery of medicines, DNA, and genes for the treatment of cancer and other disorders due to their preferred accumulation at the tumor sites (Boisselier and Astruc 2009). Because they can be found using a number of analytical methods, including optical absorption, fluorescence, Raman scattering, atomic and magnetic force, and electrical conductivity, gold nanoparticles make excellent labels for sensors. The most popular nanoparticles for transferring electrons between the active core of redox proteins and the electrode surface are gold nanoparticles. The electron transfer rate constant between the enzyme glucose oxidase and the substrate in glucose detection is increased seven times by using gold nanoparticles on the electrode surface. This asymmetric biosensor for glucose responds in less than 5 s, and when compared to the same correction performed on a smooth gold electrode (without the inclusion of gold nanoparticles), the sensitivity is increased 2.8 times and the detection limit is increased by 20 times (Yang and Cui 2008).

A layer-by-layer (LBL) method may be used to assemble GNPs and methylene blue (MB) into films over a glassy carbon electrode (GCE) that had been adapted for the detection of human chorionic gonadotrophin (HCG) (He et al. 2008). Due to the enormous specific surface area and excellent biocompatibility of GNPs, HRP can be adsorbed onto a GNP layer for the detection of H_2O_2 without losing any biological activity (Gao et al. 2007). The ZrO_2/Au film electrode surface may firmly adsorb organophosphate pesticides (OPs), offering a quantitative method for OP detection that is effective. Wang and Li (2008) detailed the synthesis of ZrO_2/Au nanocomposite films using a sol-gel process and electroless plating.

Silver Nanoparticles

Similar to how gold nanoparticles are utilized, silver nanoparticles are also used to alter the surface of electrode. The high electrocatalytic reactivity and bioactivity of biomolecules are preserved by biocompatibility of silver nanoparticles. Hepatitis B antibodies have been fixed on platinum electrodes using silver nanoparticles as a substrate. Sulfur atoms stabilizing them on silver nanoparticles cause oligonucleotide receptors to assemble on the surface of their monolayer electrode. High levels of sensitivity, selectivity, and stability are displayed by the sensors made using this technique (Qian et al. 2020). Some electrochemical reactions can also be catalyzed by silver nanoparticles. These nanoparticles have been utilized to create enzymatic sensors for glucose and hydrogen peroxide. Cysteine has been detected and measured using a glass carbon electrode enhanced with silver nanoparticles and doped with mercury (Kianfar Farshad et al. 2015).

Platinum and Copper Nanomaterials

Due to their appropriate catalytic characteristics, platinum nanoparticles are used in electrochemical studies. High-sensitivity and low-detection sensors are created as a result of the catalytic role that platinum nanoparticles play in redox processes and the hybridization of DNA strands. To create sensors for the detection of NO, nitrite, and uric acid, Pt-Fe(III) nanoparticles have also been placed on the electrode surface in addition to platinum nanoparticles. Due to the potential for asymmetric technique detection at the constant potential for analyzing carbohydrates and amino acids, the copper electrode is of great importance. In addition to the previously indicated benefit, using electrodes enhanced with copper nanoparticles in electrochemical sensors speeds up the reaction because of the catalytic capabilities of the nanoparticles. With great sensitivity and constant potential in phosphate buffer solution with $p^H = 8$, all 20 amino acids may be identified on the surface of the electrode coated with copper nanoparticles. Nanoparticles of other metals, such as palladium, nickel, and iridium, have also been utilized as catalysts in electroanalytical systems in addition to gold, silver, platinum, and copper nanoparticles (Özbek et al. 2021; Sivasankar et al. 2018).

Magnetic Nanoparticles

Due to their unique magnetic properties, magnetic nanoparticles (MNP) have been extensively researched for usage in applications like hyperthermia (Kim et al. 2005), MRI contrast agent (Lee et al. 2006), tissue repair (Ito et al. 2005), immunoassay (Sincai et al. 2005), drug/gene delivery (Morishita et al. 2005), cell separation (Guedes et al. 2005), GMR sensor (Rife et al. 2003), etc. They can typically be manufactured as single domain or superparamagnetic ferrites ($MeO{\cdot}Fe_2O_3$, where Me = Ni, Co, Mg, Zn, Mn, etc.), greigite (Fe_3S_4), maghemite (g-Fe_2O_3), and other forms of ferrites (Fe_3O_4). When exposed to a magnetic field, antibodies that have been marked with magnetic nanoparticles emit magnetic signals. As loose antibodies spread in all directions and do not provide a net magnetic signal, antibodies bound to targets can be distinguished from those that are not. A novel type of magnetic dextran microsphere (MDMS) was synthesized by Zhang et al. (2008a) by suspending cross-linking iron nanoparticles and dextran. Then, a GCE modified with MDMS was used to immobilize HRP. An amperometric H_2O_2 biosensor was created using the immobilized HRP-modified electrode and hydroquinone (HQ) as the mediator. A magnetic chitosan microsphere (MCMS) was created by Lai et al. (2008) utilizing carbon-coated MNPs and chitosan. With the help of glutaraldehyde cross-linking, hemoglobin (Hb) was successfully immobilized on the surface of MCMS-modified GCE.

In fact, numerous techniques have been created for detecting and counting specific micron-scale magnetic particles. The Maxwell bridge, frequency-dependent magnetometer, superconducting quantum interference device (SQUID), and magnetoresistance techniques are all used for direct detection of magnetic particle labels. The force-amplified biological sensor (FABS), a microcantilever-based device, and magnetic relaxation switches (MRS) are examples of indirect detection. A highly accurate, linear, and hysteresis-low giant magnetoresistance-spin valve (GMR-SV) biosensing device was recently created using photolithography (Park et al. 2008). For its detection and analysis, even a single drop of human blood or nanoparticles in distilled water provided enough signal.

2.4.2.7 Protein Chips

At the Fraunhofer Institute for Interfacial Engineering and Biotechnology IGB, protein chips based on protein-binding silica nanoparticles are being developed (Stuttgart, Germany). Many different capture proteins can be arranged on the surface of a silica nanoparticle with a diameter of under 100 nm. The chips can be examined using cutting-edge mass spectrometry (MS)-matrix-assisted laser desorption/ionization (MALDI)-time-of-flight (TOF) MS after coming into contact with a sample. It is possible to identify bound proteins directly by knowing their masses (Hsu and Huang 2004).

2.4.2.8 Silicon Nanowires

Cui et al. announced the development of sensitive, real-time electrically based sensors for chemical and biological species using boron-doped silicon nanowires (SiNWs). The p^H-dependent conductance of the SiNWs that had been functionalized with amine and oxide may be explained in terms of the change in surface charge during protonation and deprotonation. Streptavidin was detected using biotin-modified SiNWs down to at least a picomolar concentration range. Additionally, antigen-functionalized SiNWs demonstrated real-time concentration-dependent detection and reversible antibody binding. These semiconductor nanowires' small size and capacity to detect a wide spectrum of chemical and biological species in real time without labels can be used in array-based screening and in vivo diagnostics (Cui et al. 2001).

In order to identify the presence of hazardous substances inside individual cells, Cullum et al. employed optical fibers with a distal-end diameter of less than 1 μm that were coated with antibodies. They were able to gauge the amount of benzo (a)pyrene tetrol (BPT) in both rat liver epithelial cells and human mammary cancer cells (Cullum et al. 2000).

2.4.2.9 Nanobarcodes

Nanobarcodes are metallic barcodes that are submicron in size and have stripes created by successive electrochemical deposition of metal ions. By using traditional light microscopy, it is possible to identify the striped patterns due to the varying reflectivities of neighboring stripes (Walton et al. 2002). As shown by DNA and protein bioassays, this readout process does not obstruct the use of fluorescence for the detection of analytes bound to particles by affinity capture. SNP mapping and multiplexed assays for protein diagnostics can both be done using nanobarcodes. These particles are valuable for bioanalytical analyses because their surfaces can be chemically altered. Larger numbers of striping patterns and more accurate identification will result from improvements in production and identification procedures.

2.4.2.10 Biomimic-Based Biosensors or Molecular Self Assembly

Molecular self-assembly is a crucial link between physics, chemistry, and biology that mirrors natural systems. Utilizing molecular self-assembly, new devices, materials, and architectures can be produced for biosensors (Boozer et al. 2003). Thin lipid films and liposomes are the self-assembled structures gaining the most attention in terms of biosensors. Lipid films and liposomes are made of phospholipids or other amphiphiles, just like a cell membrane is. They can spontaneously assemble structures due to their hydrophilic/hydrophobic properties. The supported bilayer lipid membrane (BLM) offers a suitable setting for embedding proteins, receptors, membrane/tissue fragments, and whole cells in a well-defined orientation and under nondenaturing circumstances. Because of this, BLMs are particularly appealing for use in biosensors. The ion channel switch biosensor described by Cornell et al. was a successful biomimetically built device based on BLMs (Cornell et al. 1997). Two gramicidin molecules were used to create ion channels in the membrane: one in the lower layer linked to a gold electrode and the other in the upper layer tethered to biological receptors like antibodies or nucleotides (Cornell et al. 2001). The supporting substrate for immobilizing the biorecognition molecules is commonly liposomes. Additionally, visual, sound wave, and electrochemical signals are amplified using liposomes (Baeumner et al. 2003).

2.4.2.11 Nanofabrication

Utilizing techniques developed exclusively for micromachining and integrated circuit manufacture, nanofabrication produces things with a nanoscale resolution. The four fundamental operations of photolithography, thin film growth/deposition, etching, and bonding are frequently used in combination during nanofabrication processes. By using impedimetric measurements, nanofabricated electrodes enable the detection of affinity binding of biomolecular structures (such as DNA or antigens). For instance, measuring the double-layer impedance might be used to keep track of the immobilization of glucose oxidase. Based on an ultrathin platinum film on an oxidized silicon base, a sensitive conductometric immunosensor has been shown (Pak et al. 2001). Researchers from IBM have also described using a microarray of cantilevers to simultaneously detect several unlabeled biomolecules at nanomolar concentrations in a matter of minutes (McKendry et al. 2002). This array allowed for numerous binding experiments to be run simultaneously and was capable of detecting femtomoles of DNA on the cantilever at a concentration of 75 nM in solution.

A force-amplified biological sensor (FABS) was created by the Naval Research Laboratory in the USA and is capable of identifying biological species like cells, proteins, poisons, and DNA at concentrations as low as 10^{-18} M (Baselt et al. 1996).

2.4.2.12 Ion Channel Switch (ICS) Biosensor Technology

The ion channel switch (ICSk), an advanced biosensor developed by Ambri Ltd. (Chatswood, Australia), is based on a synthetic membrane that self-assembles and functions like a biological switch to detect the presence of particular chemicals by igniting an electrical current (Wright and Harding 2000). By delivering precise and quantitative test results in an immediate time frame, the system reduces the time of emergency diagnoses from hours down to minutes.

2.4.2.13 Electronic-Based Nanobiosensors

The Biodetect™ technology detects the binding of a target DNA molecule to sensors on a microchip electronically. The target molecules connect two wires that are electrically isolated by forming a bridge. The chemical development of the attached target molecules into conductive DNA wires, which are metalized and visible by electron microscopy, produces a strong, clear signal. By analyzing the resistance or other electrical characteristics of the sensor, the bridges, which may be seen using fluorescence imaging techniques, can be quickly found. These DNA lines function like an on/off switch to "turn on" a sensor. Multiple sensors are present on each chip, and each sensor can be individually addressed using capture probes for various target DNA molecules from the same or other organisms (Jain 2005).

2.4.2.14 Viral Nanosensor

As a nanosensor for therapeutically important viruses, magnetic nanobeads have been constructed in response to HSV and adenovirus. The nanobeads have a dextran-coated supramagnetic iron oxide core (Perez et al. 2003). As a partner in the binding of antiviral antibodies, protein G is attached. Using a bifunctional linker, anti-HSV antibodies are directly attached to nanobeads in order to prevent protein G and medium components from interacting in an unintended way. As few as five virus particles can be captured by a magnetic field, and because it is less expensive, quicker, and detects fewer artifacts in a 10 mL blood sample than PCR-based detection methods, this system is more sensitive than ELISA-based approaches.

2.4.2.15 PEBBLE-Based Nanosensors

The nanosensors known as probes encapsulated by biologically localized embedding (PEBBLE) are spherical sensors with a size range of 20–200 nm that are made of sensor molecules trapped in a chemically inert matrix (Sumner et al. 2002). These sensors are immune to protein interference and can image ions and chemicals both inside and outside of cells in real time. The early identification of cancer is another application for PEBBLE. Additionally, PEBBLE nanosensors exhibit very quick reaction times, no protein interference, and excellent reversibility and resilience to leaching and photobleaching. They exhibit strong oxygen detecting ability in human plasma, unaffected by light scattering and autofluorescence (Cao et al. 2004).

2.4.2.16 Optical Biosensors

Many commercially available biosensors use the optical characteristics of lasers to track and measure interactions between biomolecules that take place on specifically designed surfaces or biochips. The most well-known instance of this technology is surface plasmon resonance. An optical-electrical phenomena called surface plasmon resonance (SPR) involves the interaction of light with the electrons of a metal. Researchers can profile and describe biomolecular interactions in parallel with the aid of the next-generation microarray-based SPR devices. There is an excellent demand for tiny optical sensors that can selectively detect low amounts of environmental and biological chemicals. There is currently no optical sensor that can identify the aforementioned species at naturally occurring concentrations without using amplification methods (Haes and Duyne 2004).

2.4.2.17 Laser Biosensors

A laser nanosensor works by firing laser light into a fiber, creating an evanescent field at the tip of fibers, and then using that field to excite target molecules attached to antibody molecules. The optical signal (such as fluorescence) coming from the analyte molecules or from the analyte-bioreceptor reaction is detected using a photometric detection device (Vo-Dinh 2005). Laser nanosensors can be used for in vivo analysis of proteins and biomarkers in individual living cells.

2.5 Applications of Nanobiosensors

Enhanced detection sensitivity and specificity are two benefits that have considerable promise in applications, including detecting DNA, RNA, proteins, glucose, pesticides, and other small molecules from clinical samples, food industry samples, and environmental sensing. The scopes of nanobiosensor are shown in Fig. 2.2.

2.5.1 Detection of Glucose

The glucose biosensor is frequently employed as a clinical diabetes indication. The performance of glucose sensors is greatly influenced by nanoscale materials as GNPs, CNTs, magnetic nanoparticles, platinum nanoparticles, quantum dots, etc. Fibrous morphology and wrapping polydiallyl dimethylammonium chloride (PDDA) over multiwalled carbon nanotubes (MWCNTs) cause a significant loading of *glucose oxidase* into the electrospun matrix (Manesh et al. 2008).

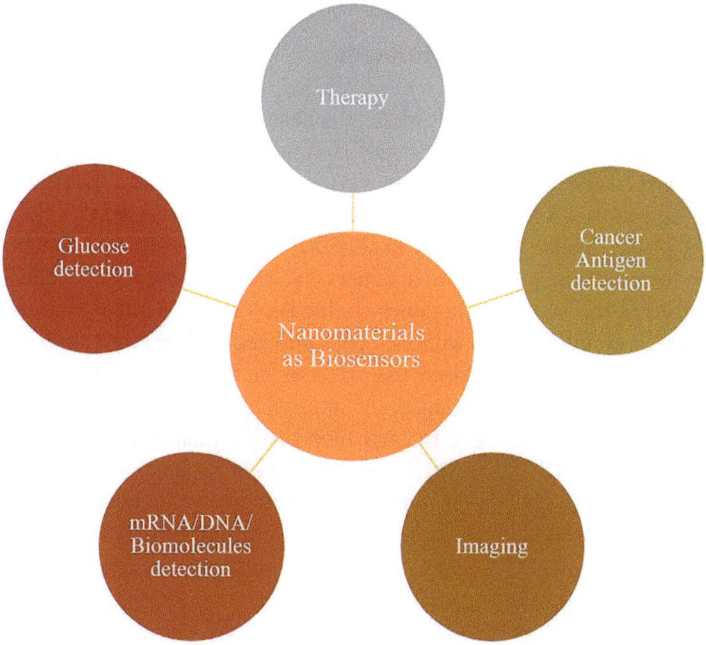

Fig. 2.2 Scopes of nanomaterials as biosensor

2.5.2 Detection of DNA and Protein

As optical biosensors, single-stranded DNA-CNT probes may be utilized to identify particular categories of DNA oligonucleotides (Cao et al. 2008). Single-stranded DNA probes can be efficiently immobilized on MWNTs/ZnO/chitosan composite film-modified glass carbon electrode (GCE) to distinguish between various DNA sequences (Zhang et al. 2008b). The bionanocomposite layer of MWNT in chitosan placed on a screen-printed carbon electrode (SPCE) is used to create a biosensor for the detection of deep DNA damage (Galandova et al. 2008). By utilizing the interactions between chemical substances and DNA, GNPs can also be utilized to identify DNA sequences. GNPs functionalized with alkanethiol-capped LNA/DNA chimaeras in a tail-to-tail hybridization mode might also perform well for single-stranded DNA (McKenzie et al. 2007), and these probes exhibit amazing discriminating between a complementary target and one that contains a single base mismatch. The first nanowire field-effect transistor-based biosensor, described by Maki et al. (2008), offers straightforward and susceptible electronic DNA methylation detection without the need for time-consuming bisulfite treatment or PCR amplification.

2.5.3 Detection of Other Molecules

Zhang et al. (2007) created a method for the detection of *Escherichia coli* (*E. coli*) utilizing modified bismuth nanofilm and flow injection analysis (FIA). A biochip sensor system made of two titanium (Ti) contact pads and a 150-nm-wide Ti nanowell device on a $LiNbO_3$ substrate was created by Seo et al. (2008). A colorimetric assay was created by Medley et al. (2008) for the quick identification of sick cells. The selectivity and affinity of aptamers and the spectroscopic benefits of GNPs are combined in this test using aptamer-conjugated GNPs. While nontarget samples did not change color, samples with sick cells showed a clear color shift. One of the initial insults in diabetes has been theorized to be mitochondrial oxidative stress (MOS). According to some evidence, the antioxidant response element (ARE) and MOS are both more responsive to high blood sugar than each other (Prow et al. 2008). According to Li et al. (2007), an electrochemical aptamer biosensor for the detection of adenosine based on impedance spectroscopy measurement provides a label-free and reusable platform for making the detection of smaller molecules easy and effective.

2.5.4 In Immunohistochemistry

Using immunostaining of cytokeratin in skin basal carcinoma as an example, nanocrystal (NC)-antibody conjugates prepared using the method described can be utilized for the targeted detection of antigens in paraffin-embedded formaldehyde-fixed cancer tissue specimens. Thus, NC-Ab conjugates could be used as extremely sensitive, photostable markers for immune histochemical detection (Sukhanova et al. 2004).

2.5.5 Detection of Disease Biomarkers

The discovery of disease biomarkers, particularly in body fluid, is of great interest and is being researched extensively. Biomarker detection uses genomics, proteomics, and related technologies. Clinically significant prostate-specific antigen (PSA) concentrations can be found using microcantilever technology, even in the presence of other proteins. Due to the lack of labeling and the ability to complete the procedure in a single reaction, the technique is possibly more efficient and less complicated than other diagnostic techniques. False positives, which are frequently brought on by the nonspecific binding of other proteins, are less likely to occur. A biobarcode assay can detect PSA at a concentration of 30 attomolars since it allows for significant amplification (Nam et al. 2004).

As a biomarker for Alzheimer's disease, the concentration of amyloid h-derived diffusible ligands (ADDLs) in the cerebrospinal fluid (CSF) was determined using the nanoparticle-based biobarcode assay (Kaul et al. 2003). While commercial enzyme-linked immunoassays (ELISA) can only identify ADDLs in brain tissue where the biomarker is most concentrated, this test can detect as few as 50 molecules of Aβ in the CSF of patients with Alzheimer's disease (Georganopoulou et al. 2005).

2.5.6 Detection of Single Nucleotide Polymorphism

A microarray-based method for multiplex single nucleotide polymorphism (SNP) genotyping in the entire human genomic DNA is made possible by nanosphere's ClearRead™ nanoparticle technology without the use of target amplification (Bao et al. 2005). The probe, a very sensitive gold nanoparticle, generates a strong signal precisely showing the presence of a particular target SNP, while one section detects any DNA changes.

2.5.7 Detection of Disease Genes

The identification of mutant genes linked to human disease can be done with DNA-based electrochemical sensors since they are sensitive, selective, and inexpensive. Electrochemical amplifications using nanoparticles are one of the innovations in this field (Drummond et al. 2003).

2.5.8 Detection of Microorganisms

2.5.8.1 Bacteria

Rapid and accurate pathogenic bacterial identification is crucial for medical diagnostics and bioterrorism defenses. Most standard diagnostic techniques have limitations, such as a lack of ultrasensitivity or a delay in results. Within 20 min, a single bacterium can be found using a bioconjugated nanoparticle-based bioassay for in situ pathogen quantification (Zhao et al. 2004). The nanoparticle is easily integrated into a biorecognition molecule, such as an antibody, and offers an extraordinarily strong fluorescent signal for bioanalysis. Through interaction and recognition between the antibody and the antigen, the antibody-conjugated nanoparticles can quickly and accurately identify a range of bacteria, including *Escherichia coli* O157:H7. By utilizing a 384-well microplate format, this technology can be applied to several bacterial samples with high throughput. The quick detection of mecA gene in methicillin-resistant *Staphylococcus aureus* genomic DNA samples has been carried out using the nanomaterial-based colorimetric assay, which enhances detection sensitivity by over four orders of magnitude when compared to a previously published absorbance-based technique (Storhoff et al. 2004). A change in the surface stress on the silicon nitride cantilever surface in situ upon bacterial adhesion allows for the detection of a small number of *Salmonella enterica* bacteria.

2.5.8.2 Viruses

Plaque tests, immunological assays, transmission electron microscopy, and PCR-based viral nucleic acid testing are examples of well-developed techniques for viral investigation. These techniques frequently call for a considerable degree of sample manipulation, which is problematic for infectious materials, and have not been successful in achieving quick detection at the single virus level. Nanowire field-effect transistors provide highly selective direct, real-time electrical detection of single virus particles (Patolsky et al. 2004). The presence of influenza A caused discrete conductance changes indicative of binding and unbinding, but not paramyxovirus or adenovirus, according to tests using nanowire arrays modified with anti-influenza A antibodies. The presence of 100 or more distinct viruses might be simultaneously screened by larger arrays of repeatable nanowire devices. Ultimately, research using nanowire devices modified with antibodies specific for either influenza or adenovirus has demonstrated the selective parallel detection of numerous viruses.

2.5.9 Cancer Diagnosis

Quantum dots (QD) bioconjugates are highly stable and luminous due to recent advancements. By enabling the imaging of cancer cells in living animals, these bioconjugates provide alternative opportunities for investigating genes, proteins, and therapeutic targets in single cells, tissue specimens, and even living organisms (Gao and Nie 2003). Magnetic microparticle probes with antibodies that mainly bind a target of interest, such as prostate-specific antigen (PSA) in the case of prostate cancer, have been used to develop a method for detecting protein analytes. For the immunofluorescent tagging of breast cancer markers, QDs covered with a polyacrylate cap and covalently coupled to antibodies or streptavidin have been employed (Wu et al. 2003). A method for investigating extravasation in vivo has been developed using QDs and emission spectrum scanning multiphoton imaging. Mice were intravenously injected with QD-labeled tumor cells, and their extravasation into lung tissue was monitored. In vivo, the behavior of tumor cells that had been QD-labeled was identical to that of unlabeled cells. Based on circulating cancer cells in the blood, magnetic nanoparticle-based assays are being developed to screen, diagnose, stage, and track cancer (Voura et al. 2004).

2.6 Challenges and Future Trends

2.6.1 Challenges

Applications of nanomaterials in biosensors have opened up new possibilities for creating a new era of biosensor technologies in recent years. Nanomaterial-based biosensors are progressing toward single-molecule biosensors and high throughput biosensor arrays, and they can enhance the mechanical, electrochemical, optical, and magnetic features of biosensors (Kerman et al. 2008). But they encounter numerous challenges, just like any newly developing field. The construction of single-molecule multifunctional nanocomposites, nanofilms, and nanoelectrodes requires the utilization of biomolecules, which have unique structures and functionalities, and nanomaterials. This presents a significant difficulty. Additionally, the mechanism governing the interaction of proteins with nanomaterials is still not fully elucidated. Another challenging task is figuring out how to innovate new, multifunctional, or homogeneous nanofilms or how to alter electrodes using these rules and principles of an optimal biosystem. Other significant challenges for the currently available techniques include processing, characterization, interface issues, the availability of high quality nanomaterials, tailoring of nanomaterials, and the mechanisms governing the behavior of these nanoscale composites on the surface of electrodes. For instance, it is very difficult to align nanomaterials like CNTs in a polymer matrix in the same direction. Another major difficulty is figuring out how to improve the signal-to-noise ratio as well as signal transduction and amplification. Due to their inability to meet the quality control requirements set by certifying organizations like the FDA, items based on nanotechnology are still only permitted in research and development environments. Since the current market price of nanomaterials is too high for any practical commercial application, there is an urgent need to create cost-effective and reproducible manufacturing procedures to lower the cost of producing nanomaterials. Nevertheless, these difficulties would soon be effectively solved because of ongoing technological advancements.

2.6.2 Future Trends

It is expected that nanotechnology-based biosensors will continue to advance and increase their application in many fields of the life sciences in the future years, particularly in the areas of biomedical diagnosis and drug delivery. These biosensors are anticipated to have some ideal characteristics for drug delivery applications, including high sensitivity, high specificity, quick response and action, low detection limit (so that a timely screening of clinically significant proteins and cancer markers is feasible), continuous and long-term monitoring capability, carrier of personalized medicine for site-specific and rate-controlled delivery, passive operation, and high specificity (possess the ability to communicate wirelessly with external monitoring devices). Some nanosensors are already developed, and they hold promise for detecting bioterrorism chemicals that are currently undetectable by molecular diagnostic methods. Nanodiagnostics will soon shorten the time spent waiting for test results. For instance, individuals with sexually transmitted diseases may provide a urine sample when they first visit an outpatient clinic or the office of the doctor, and the findings may be ready when they arrive for their appointment. The patient would have less time to worry, and the process would be less expensive if they were given the prescription right away.

The downsizing of biochip technology to the nano-range will continue to influence diagnostic trends in the future. Additionally, there is a growing desire to create diagnostic tools from the smallest building blocks. It remains to be seen if interest and use of nanomechanical detection will sustain over the long term. Another trend is to avoid fluorescent labeling since miniaturization weakens the signal. However, there have been advancements that make fluorescence with nanoparticles possible.

2.7 Conclusion

The creation of biosensors is being revolutionized by nanotechnology. More and more innovative biosensors are being created using nanomaterials and nanofabrication methods. The study of the many nanoeffects, which are peculiar to nanomaterials and are actually their most alluring feature (e.g., quantum size effect, micro size effect, surface effect, macro-quantum tunnel effect), is sadly receiving very little attention. For usage in biosensors, novel nanomaterials and nanostructures must be researched. Nanotechnology-based biosensors should ideally be built into tiny biochips with integrated electronics for handling and analyzing samples. By offering devices that are compact, portable, simple to use, inexpensive, disposable, and extremely flexible diagnostic devices, functionality will be considerably improved.

Aggregated biochips can be created by decreasing the size of the biosensor and the requirement for extremely small sample sizes that come from the usage of nanostructures. These systems are crucial for clinical diagnostic applications because they contain all the components required for sample presentation, sensing, and data processing. The use of biochip technology, which has a small sample size requirement, obviates the need for medical diagnostic laboratories, which use expensive equipment and waste time and money, and allows for the simultaneous analysis of multiple analytes in a single sample, enabling accurate and rapid diagnosis. These technology and tools will benefit patients by being less intrusive, making them more comfortable, improving the accuracy of sensing results, and improving the site, amount, and pace of controllable drug delivery.

References

Alivisatos P (2004) The use of nanocrystals in biological detection. Nat Biotechnol 22(1):47–52

Azamian BR, Davis JJ, Coleman KS, Bagshaw C, Green MLH (2002) Bioelectrochemical single walled carbon nanotubes. J Am Chem Soc 124(43):12664–12665

Baeumner AJ, Cohen RN, Miksic V, Min J (2003) RNA biosensor for the rapid detection of viable Escherichia coli in drinking water. Biosens Bioelectron 18:405–413

Bai J, Jiang X (2013) A facile one-pot synthesis of copper sulfide decorated reduced graphene oxide composites for enhanced detecting of H_2O_2 in biological environments. Anal Chem 85:8095–8101

Bai Y, Xu T, Zhang X (2020) Graphene-based biosensors for detection of biomarkers. Micromachines (Basel) 11(1):60

Bao YP, Huber M, Wei TF, Marla SS, Storhoff JJ, Muller UR (2005) SNP identification in unamplified human genomic DNA with gold nanoparticle probes. Nucleic Acids Res 33(2):e15

Baselt DR, Lee GU, Colton RJ (1996) Biosensor based on force microscope technology. J Vac Sci Technol B 14(2):789–793

Berger R, Delamarche E, Lang HP, Gerber C, Gimzewski JK, Meyer E et al (1997) Surface stress in the self-assembly of alkanethiols on gold. Science 276:2021–2023

Boisselier E, Astruc D (2009) Gold nanoparticles in nanomedicine: preparations, imaging, diagnostics, therapies and toxicity. Chem Soc Rev 38:1759–1782

Boozer C, Yu Q, Chen S, Lee C, Homola J, Yee SS et al (2003) Surface functionalization for self-referencing surface plasmon resonance (SPR) biosensors by multi-step self-assembly. Sensors Actuators B Chem 90:22–30

Cao Y, Lee Koo YE, Kopelman R (2004) Poly(decyl methacrylate)-based fluorescent PEBBLE swarm nanosensors for measuring dissolved oxygen in biosamples. Analyst 129:745–750

Cao C, Kim JH, Yoon D, Hwang ES, Kim YJ, Baik S (2008) Optical detection of DNA hybridization using absorption spectra of single-walled carbon nanotubes. Mater Chem Phys 112:738–741

Chen RJ, Choi HC, Dai H (2004) An investigation of the mechanism of electronic sensing of protein adsorption on carbon nanotube devices. J Am Chem Soc 126:1563–1568

Cheng Y, Zhao L, Li Y, Xu T (2011) Design of biocompatible dendrimers for cancer diagnosis and therapy: current status and future perspectives. Chem Soc Rev 40:2673–2703

Cornell BA, Braach-Maksvytis VLB, King LG, Osman PDJ, Raguse B, Wieczorek L et al (1997) A biosensor that uses ion-channel switches. Nature 387:580–583

Cornell BA, Krishna G, Osman PD, Pace RD, Wieczorek L (2001) Tethered bilayer lipid membranes as a support for membrane-active peptides. Biochem Soc Trans 29:613

Cui Y, Wei Q, Park H, Lieber CM (2001) Nanowire nanosensors for highly sensitive and selective detection of biological and chemical species. Science 293:1289–1292

Cullum B, Griffin G, Miller G, Vo-Dinh T (2000) Intracellular measurements in mammary carcinoma cells using fiberoptic nanosensors. Anal Biochem 277:25–32

Davis JJ, Coleman KS, Azamian BR, Bagshaw CB, Green MLH (2003) Chemical and biochemical sensing with modified single walled carbon nanotubes. Chem Eur J 9:3732–3739

Drummond TG, Hill MG, Barton JK (2003) Electrochemical DNA sensors. Nat Biotechnol 21: 1192–1199

Fritz J, Cooper EB, Gaudet S, Sorger PK, Manalis SR (2002) Electronic detection of DNA by its intrinsic molecular charge. Proc Natl Acad Sci U S A 99:14142–14146

Galandova J, Ziyatdinova G, Labuda J (2008) Disposable electrochemical biosensor with multiwalled carbon nanotubes - chitosan composite layer for the detection of deep DNA damage. Anal Sci 24:711–716

Gao X, Nie S (2003) Molecular profiling of single cells and tissue specimens with quantum dots. Trends Biotechnol 21:371–373

Gao F, Yuan R, Chai Y, Chen S, Cao S, Tang M (2007) Amperometric hydrogen peroxide biosensor based on the immobilization of HRP on nano-Au/Thi/poly (p-aminobenzene sulfonic acid)-modified glassy carbon electrode. J Biochem Biophys Methods 70:407–413

Ge X, Asiri AM, Du D, Wen W, Wang S, Lin Y (2014) Nanomaterial-enhanced paper-based biosensors. TrAC Trends Anal Chem 58:31–39

Georganopoulou DG, Chang L, Nam JM, Thaxton CS, Mufson EJ, Klein WL et al (2005) Nanoparticle based detection in cerebral spinal fluid of a soluble pathogenic biomarker for Alzheimer's disease. Proc Natl Acad Sci U S A 102:2273–2276

Gholipour G, Zhang C, Mousavi AA (2020) Numerical analysis of axially loaded RC columns subjected to the combination of impact and blast loads. Eng Struct 219:110924

Guedes MHA, Sadeghiani N, Peixoto DLG, Coelho JP, Barbosa LS, Azevedo RB et al (2005) Effects of AC magnetic field and carboxymethyldextran-coated magnetite nanoparticles on mice peritoneal cells. J Magn Magn Mater 293:283–286

Haes AJ, Duyne RP (2004) Preliminary studies and potential applications of localized surface plasmon resonance spectroscopy in medical diagnostics. Expert Rev Mol Diagn 4:527–537

Han M, Gao X, Su JZ, Nie S (2001) Quantum-dot-tagged microbeads for multiplexed optical coding of biomolecules. Nat Biotechnol 19:631–635

He X, Yuan R, Chai Y, Shi Y (2008) A sensitive amperometric immunosensor for carcinoembryonic antigen detection with porous nanogold film and nano-Au/chitosan composite as immobilization matrix. J Biochem Biophys Methods 70:823–829

Hoshino A, Fujioka K, Manabe N, Yamaya SI, Goto Y, Yasuhara M et al (2005) Simultaneous multicolor detection system of the single-molecular microbial antigen with total internal reflection fluorescence microscopy. Microbiol Immunol 49:461–470

Hsu HY, Huang YY (2004) RCA combined nanoparticle-based optical detection technique for protein microarray: a novel approach. Biosens Bioelectron 20:123–126

Huang X, Zhu Y, Kianfar E (2021) Nano biosensors: properties, applications and electrochemical techniques. J Mater Res Technol 12:1649–1672

Ito A, Ino K, Kobayashi T, Honda H (2005) The effect of RGD peptide-conjugated magnetite cationic liposomes on cell growth and cell sheet harvesting. Biomaterials 26:6185–6193

Jain KK (2005) Nanotechnology in clinical laboratory diagnostics. Clin Chim Acta 358(1–2):37–54

Jaiswal JK, Goldman ER, Mattoussi H, Simon SM (2004) Use of quantum dots for live cell imaging. Nat Methods 1:73–78

Jares-Erijman EA, Jovin TM (2003) FRET imaging. Nat Biotechnol 21:1387–1395

Kairemo K, Erba P, Bergström K, Pauwels EKJ (2008) Nanoparticles in cancer. Curr Radiopharm 1:30–36

Kaul Z, Yaguchi T, Kaul SC, Hirano T, Wadhwa R, Taira K (2003) Mortalin imaging in normal and cancer cells with quantum dot immuno-conjugates. Cell Res 13:503–507

Kerman K, Saito M, Tamiya E, Yamamura S, Takamura Y (2008) Nanomaterial-based electrochemical biosensors for medical applications. TrAC Trend Anal Chem 27:585–592

Kianfar Farshad K, Reza MMS, Ehsan K (2015) Synthesis of spiro pyran by using silica-bonded N-propyldiethylenetriamine as recyclable basic catalyst. Indian J Sci Technol 8(11):68669

Kim DH, Lee SH, Kim KN, Kim KM, Shim IB, Lee YK (2005) Cytotoxicity of ferrite particles by MTT and agar diffusion methods for hyperthermic application. J Magn Magn Mater 293:287–292

Koev ST, Dykstra PH, Luo X, Rubloff GW, Bentley WE, Payne GF et al (2010) Chitosan: an integrative biomaterial for lab-on-a-chip devices. Lab Chip 10:3026–3042

Lai GS, Zhang HL, Han DY (2008) A novel hydrogen peroxide biosensor based on haemoglobin immobilized on magnetic chitosan microspheres modified electrode. Sens Actuators B Chem 129:497–503

Lee H, Lee E, Kim DK, Jang NK, Jeong YY, Jon S (2006) Antibiofouling polymer-coated superparamagnetic iron oxide nanoparticles as potential magnetic resonance contrast agents for in vivo cancer imaging. J Am Chem Soc 128:7383–7389

Li B, Du Y, Wei H, Dong S (2007) Reusable, label-free electrochemical aptasensor for sensitive detection of small molecules. Chem Commun 36:3780–3782

Li G, Xu H, Huang W, Wang Y, Wu Y, Parajuli R (2008) A pyrrole quinoline quinone glucose dehydrogenase biosensor based on screen-printed carbon paste electrodes modified by carbon nanotubes. Meas Sci Technol 19:065203

Liu X, Tan W (1999) A fiber-optic evanescent wave DNA biosensor based on novel molecular beacons. Anal Chem 71:5054–5059

Liu Y, Zhu Y, Zeng Y, Xu F (2009) An effective amperometric biosensor based on gold nanoelectrode arrays. Nanoscale Res Lett 4(3):210–215

Lu W, Ji Z, Pfeiffer L, West KW, Rimberg AJ (2003) Real-time detection of electron tunneling in a quantum dot. Nature 423:422–425

Maki WC, Mishra NN, Cameron EG, Filanoski B, Rastogi SK, Maki GK (2008) Nanowire-transistor based ultra-sensitive DNA methylation detection. Biosens Bioelectron 23:780–787

Manesh KM, Kim HT, Santhosh P, Gopalan AI, Lee KP (2008) A novel glucose biosensor based on immobilization of glucose oxidase into multiwall carbon nanotubes-polyelectrolyte-loaded electrospun nanofibrous membrane. Biosens Bioelectron 23:771–779

McKendry R, Zhang J, Arntz Y, Strunz T, Hegner M, Lang HP et al (2002) Multiple label-free biodetection and quantitative DNA-binding assays on a nanomechanical cantilever array. Proc Natl Acad Sci U S A 99:9783–9788

McKenzie F, Faulds K, Graham D (2007) Sequence-specific DNA detection using high-affinity LNA-functionalized gold nanoparticles. Small 3:1866–1868

Medintz IL, Uyeda HT, Goldman ER, Mattoussi H (2005) Quantum dot bioconjugates for imaging, labelling and sensing. Nat Mater 4:435–446

Medley CD, Smith JE, Tang Z, Wu Y, Bamrungsap S, Tan W (2008) Gold nanoparticle-based colorimetric assay for the direct detection of cancerous cells. Anal Chem 80:1067–1072

Mertens J, Finota E, Thundat T, Fabrea A, Nadal MH, Eyraud V et al (2003) Effects of temperature and pressure on microcantilever resonance response. Ultramicroscopy 97:119–126

Morishita N, Nakagami H, Morishita R, Takeda SI, Mishima F, Terazono B et al (2005) Magnetic nanoparticles with surface modification enhanced gene delivery of HVJ-E vector. Biochem Biophys Res Commun 334:1121–1126

Muguruma H, Shibayama Y, Matsui Y (2008) An amperometric biosensor based on a composite of single-walled carbon nanotubes, plasma-polymerized thin film, and an enzyme. Biosens Bioelectron 23:827–832

Nam JM, Stoeva SI, Mirkin CA (2004) Bio-bar-code-based DNA detection with PCR-like sensitivity. J Am Chem Soc 126:5932–5933

Özbek MA, Yaşar A, Çete S, Er E, Erk N (2021) A novel biosensor based on graphene/platinum nanoparticles/Nafion composites for determination of glucose. J Solid State Electrochem 25: 1601–1610

Pak SC, Penrose W, Hesketh PJ (2001) An ultrathin platinum film sensor to measure biomolecular binding. Biosens Bioelectron 16:371–379

Pandit S, Dasgupta D, Dewan N, Prince A (2016) Nanotechnology based biosensors and its application. Pharma Innov 5(6):18–25

Park SH, Soh KS, Hwang DG, Rhee JR, Lee SS (2008) Detection of magnetic nanoparticles and Fe-haemoglobin inside red blood cells by using a highly sensitive spin valve device. J Magnetics 13:30–33

Parveen S, Misra R, Sahoo SK (2012) Nanoparticles: a boon to drug delivery, therapeutics, diagnostics and imaging. Nanomedicine 8:147–166

Patolsky F, Zheng G, Hayden O et al (2004) Electrical detection of single viruses. Proc Natl Acad Sci U S A 101:14017–14022

Perez JM, Simeone FJ, Saeki Y, Josephson L, Weissleder R (2003) Viral-induced self-assembly of magnetic nanoparticles allows the detection of viral particles in biological media. J Am Chem Soc 125:10192–10193

Porter TL, Eastman MP, Macomber C, Delinger WG, Zhine R (2003) An embedded polymer piezoresistive microcantilever sensor. Ultramicroscopy 97:365–369

Prow TW, Bhutto I, Grebe R, Uno K, Merges C, McLeod DS et al (2008) Nanoparticle-delivered biosensor for reactive oxygen species in diabetes. Vis Res 48:478–485

Qian J, Feng S, Li Y, Tao T, Han J, Chen Q et al (2020) Single-shot absolute 3D shape measurement with deep-learning-based color fringe projection profilometry. Opt Lett 45(7):1842–1845

Rao LR, Zhang G (2006) Enhancing the sensitivity of SAW sensors with nanostructures. Curr Nanosci 2(4):311–318

Rife JC, Miller MM, Sheehan PE, Tamanaha CR, Tondra M, Whitman LJ (2003) Design and performance of GMR sensors for the detection of magnetic microbeads in biosensors. Sens Actuators A Phys 107:209–218

Sangadevan S, Periasamy M (2014) Recent trends in biosensors and their application. Rev Adv Mater Sci 36:62–69

Sashiwa H, Aiba SI (2004) Chemically modified chitin and chitosan as biomaterials. Prog Polym Sci 29:887–908

Satija J, Sai VVR, Mukherji S (2011) Dendrimers in biosensors: concepts and applications. J Mater Chem 21:14367–14386

Seo S, Dobozi King M, Young RF, Kish LB, Cheng M (2008) Patterning a nanowell sensor biochip for specific and rapid detection of bacteria. Microelectron Eng 85:1484–1489

Shen M, Shi X (2010) Dendrimer-based organic/inorganic hybrid nanoparticles in biomedical applications. Nanoscale 2:1596–1610

Sincai M, Ganga D, Ganga M, Argherie D, Bica D (2005) Antitumor effect of magnetite nanoparticles in cat mammary adenocarcinoma. J Magn Magn Mater 293:438–441

Sivasankar K, Rani KK, Wang SF, Devasenathipathy R, Lin CH (2018) Copper nanoparticle and nitrogen doped graphite oxide based biosensor for the sensitive determination of glucose. Nanomaterials (Basel) 8(6):429

Stenger DA, Gross GW, Keefer EW, Shaffer KM, Andreadis JD, Ma W et al (2001) Detection of physiologically active compounds using cell-based biosensors. Trends Biotechnol 19(8): 304–309

Storhoff JJ, Lucas AD, Garimella V, Bao YP, Mqller UR (2004) Homogeneous detection of unamplified genomic DNA sequences based on colorimetric scatter of gold nanoparticle probes. Nat Biotechnol 22:883–887

Sukhanova A, Devy J, Venteo L et al (2004) Biocompatible fluorescent nanocrystals for immunolabeling of membrane proteins and cells. Anal Biochem 324:60–67

Sumner JP, Aylott JW, Monson E, Kopelman R (2002) A fluorescent PEBBLE nanosensor for intracellular free zinc. Analyst 127:11–16

Vashist SK, Venkatesh AG, Mitsakakis K, Czilwik G, Roth G, Stetten FV et al (2012) Nanotechnology-based biosensor and diagnostics: technology push versus industrial/healthcare requirements. BioNano Sci 2:115–126

Vo-Dinh T (2005) Optical nanosensors for detecting proteins and biomarkers in individual living cells. Methods Mol Biol 300:383–402

Vo-Dinh T, Cullum B (2000) Biosensors and biochips: advances in biological and medical diagnostics. Fresenius J Anal Chem 366:540–551

Vo-Dinh T, Cullum BM, Stokes DL (2001) Nanosensors and biochips: frontiers in biomolecular diagnostics. Sens Actuators B Chem 74:2–11

Voura EB, Jaiswal JK, Mattoussi H, Simon SM (2004) Tracking metastatic tumor cell extravasation with quantum dot nanocrystals and fluorescence emission-scanning microscopy. Nat Med 10: 993–998

Walton ID, Norton SM, Balasingham A et al (2002) Particles for multiplexed analysis in solution: detection and identification of striped metallic particles using optical microscopy. Anal Chem 74:2240–2247

Wang M, Li Z (2008) Nano-composite ZrO_2/Au film electrode for voltammetric detection of parathion. Sens Actuators B Chem 133:607–612

Wang H, Li S, Si Y, Zhang N, Sun Z, Wu H et al (2014) Platinum nanocatalysts loaded on graphene oxide dispersed carbon nanotubes with greatly enhanced peroxidase-like catalysis and electrocatalysis activities. Nanoscale 6:8107–8116

Wright LS, Harding H (2000) Detection of DNA via an ion channel switch biosensor. Anal Biochem 282:70

Wu X, Liu H, Liu J et al (2003) Immunofluorescent labeling of cancer marker Her2 and other cellular targets with semiconductor quantum dots. Nat Biotechnol 21:41–46

Yang D, Cui D (2008) Advances and prospects of gold nanorods. Chem Asian J 3:2010–2022

Zhang G (2009) Nanotechnology-based biosensors in drug delivery. In: de Villiers MM, Aramwit P, Kwon GS (eds) Nanotechnology in drug delivery. Biotechnology: pharmaceutical aspects, vol X. Springer, New York, pp 163–189

Zhang W, Tang H, Geng P, Wang Q, Jin L, Wu Z (2007) Amperometric method for rapid detection of Escherichia coli by flow injection analysis using a bismuth nano-film modified glassy carbon electrode. Electrochem Commun 9:833–838

Zhang HL, Lai GS, Han DY, Yu AM (2008a) An amperometric hydrogen peroxide biosensor based on immobilization of horseradish peroxidase on an electrode modified with magnetic dextran microspheres. Anal Bioanal Chem 390:971–977

Zhang W, Yang T, Huang DM, Jiao K (2008b) Electrochemical sensing of DNA immobilization and hybridization based on carbon nanotubes/nano zinc oxide/chitosan composite film. Chin Chem Lett 19:589–591

Zhao Y-D, Zhang W-D, Chen H, Luo Q-M, Li SFY (2002) Direct electrochemistry of horseradish peroxidase at carbon nanotube powder microelectrode. Sens Actuators B Chem 87:168–172

Zhao X, Hilliard LR, Mechery SJ et al (2004) A rapid bioassay for single bacterial cell quantitation using bioconjugated nanoparticles. Proc Natl Acad Sci U S A 101:15027–15032

Zhu H, Snyder M (2003) Protein chip technology. Curr Opin Chem Biol 7(1):55–63

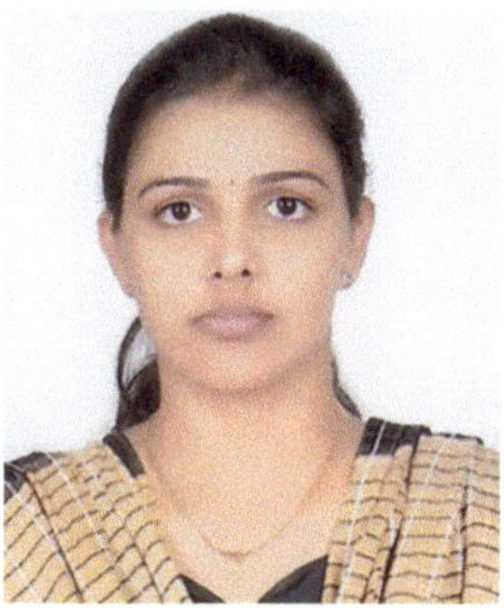

Anita Patel is an Associate Professor at Nootan Pharmacy College, Sankalchand Patel University, Visnagar, Gujarat. She has more than 13 years of research and academic experience and has published 35 research and review papers in international and national journals. She is the author of 21 book chapters and has guided more than 20 PG students and 4 Ph.D. students.

Nisarg C. Patel is working as a Professor at APMC College of Pharmaceutical Education and Research, Himatnagar, Gujarat. He has 11 years of teaching experience. He has completed his UG (Pharmacy) from Hemchandracharya North Gujarat University, Patan, and PG (Pharmacognosy) from Gujarat Technological University, Ahmedabad, and doctorate from Sankalchand Patel University, Visnagar, India. He has guided two PG students. He has published 12 national and international research papers and various presentations at national conferences and seminars.

Jayvadan K. Patel is a Formulation Scientist at Aavis Pharmaceuticals, USA. He has more than 26 years of research, academic, and industry experience and has published more than 250 research and review papers in international and national journals. He is an author/editor of 17 books and 115 book chapters. He has more than 5000 citations, 35 *h*-index, and 116 *i10*-index for his credit. He receives the very prestigious APTI Young Pharmacy Teacher Award-2014 by APTI and the AICTE-Visvesvaraya Best Teachers Award-20 by AICTE. He has completed 12 industry- and government-sponsored research projects.

Saurin Amin is an AGM, Research and Development, at Aavis Pharmaceuticals, USA. He has more than 15 years of research and industry experience and has published quality research and review papers in international and national journals. He is an editorial board member in well-reputed journals.

Materials from Natural Resources for the Application of Bone Tissue Engineering

3

Howa Begam, Sayantika Sanyal, Ananya Barui, and Samit Kumar Nandi

Abstract

Natural materials have distinctive qualities like the ability to communicate with the biological environment, especially during the regeneration process. In this perspective, natural materials exhibit significant bioactive attributes because they were synthesized from nature to carry out a particular task. Natural materials can be extracted from various sources, but their chemistry and bioactivity may differ noticeably depending on the organism under consideration and the extraction technique employed. Composite scaffolds for bone tissue engineering have been developed from natural origin. The balancing act of mechanical strength, porosity, and growth factors in bone scaffolds is still challenging. This chapter brings light to the composition of bone and the importance of bone and its tissue engineering. Further, this chapter will discuss the materials, e.g., calcium phosphorous, bioglass, collagen, chitosan hyaluronic acid, cellulose, starch, and alginate, that can be used as efficient materials for bone tissue engineering.

Keywords

Calcium phosphorous · Bioglass · Collagen · Chitosan · Hyaluronic acid · Cellulose

H. Begam · S. Sanyal · A. Barui
Center for Healthcare Science and Technology, Indian Institute of Engineering Science and Technology, Howrah, India
e-mail: 2021hcm010.sayantika@students.iiests.ac.in

S. K. Nandi (✉)
Department of Veterinary Surgery and Radiology, West Bengal University of Animal & Fishery Sciences, Kolkata, India

F. A. Sheikh et al. (eds.), *Interaction of Nanomaterials With Living Cells*,
https://doi.org/10.1007/978-981-99-2119-5_3

3.1 Introduction

Bone is vascularized tissue composed of outer hard, compact, and inner soft cancellous tissue. Collagen makes up the majority of bone's organic component (type-I), which supports the bone structure. The inorganic phases are mainly calcium phosphate. The four kinds of bone cells are osteoblast, osteoclast, osteocytes, and osteoprogenitor cells. The repair of bone defects is facilitated by osteoclast and osteoblast cell activity by balancing bone resorption and bone formation. Tissue engineering is an emerging approach to developing a new methodology for bone regeneration (Guo et al. 2021).

Mainly studied synthetic scaffold materials are ceramics, polymers, and their composites. The synthetic materials showed many disadvantages, like unsatisfactory cell growth and inadequate cell adhesion ability due to the unavailability of functional groups on their surface. These shortcomings can be compensated by the use of natural-based biomaterials. There are many benefits of the use of natural biomaterials, such as their biomimetic properties, ecological safety, and huge natural resource with minimal processing and synthesis cost. Also, natural biomaterials show good biodegradable and biocompatible properties (Barua et al. 2018). Two types of natural materials are utilized in tissue engineering applications: biopolymers and bioceramics.

3.2 Bone Tissue Engineering

The production of biological substitutes that rebuild, retain, and enhance the activity of organs or tissue is the aim of tissue engineering, which combines engineering and life science. Since the second most transplanted tissue is bone, it has garnered much interest among the many tissue types. To enable their tasks, the organic and inorganic components of bone interact and maintain physiochemical balance in a complex tissue. To resemble bone, a scaffold for bone tissue creation should have characteristics similar to those of bone extracellular matrix. Bone structure, bone mechanics, and bone tissue development are studied in bone tissue engineering, in other words, understanding the biology of bone and its ability to regenerate or repair bone.

Long bones are developed through endochondral ossification, which results in the mineralization of cartilage. Contrarily, flat bones are created through an intramembranous ossification process where mesenchymal stem cells develop osteoblasts that build bone. Osteoblasts release alkaline phosphatase, which encourages the retention of calcium and phosphate ions along with collagen fibers and unmineralized collagen (osteoid) (Zeng et al. 2017). After maturing into osteocytes, which monitor the homeostasis of bone, detect regional mechanical atmosphere, and react through balancing bone formation and resorption, osteoblasts are finally developed which are surrounded by a mineralized bone matrix (Yang et al. 2014). Bone growth and fracture healing need bone remodeling, the technique by which functional modification of the bone tissue framework takes place in

Table 3.1 Design conditions of scaffolds in bone tissue engineering (Thavornyutikarn et al. 2014)

Design criteria	Conditions
Porosity	Without compromising mechanical properties, as much as possible
Pore architecture	Interconnected pores with a diameter of 200–400 μm
Degradation behavior	The degradation time must be adaptable according to the requirements
Biocompatibility	No persistent inflammation
Sterilizability	Without affecting the material's characteristics

response to mechanical stress variations. Osteoclasts are multinucleated cells that can develop from progenitor monocytes into macrophages (immune cells) and perform bone resorption. Macrophages take part in the inflammatory response that encourages MSCs to migrate to the fracture site and consequently improves the repair of the fracture (Pajarinen et al. 2019).

Additionally, macrophages play important roles in bone formation and homeostasis, and MSCs demonstrate immunomodulatory functions. Bioceramics and polymers are potential candidates for biomaterials that resemble the organic (collagen) and inorganic (hydroxyapatite) components of the extracellular matrix (ECM) of bone. Before mineralization, successive cross-linking of collagen fibrils occurs during ECM production, gradually improving the mechanical behavior of ECM (Dhand et al. 2016). Crystalline bioactive ceramics and amorphous bioactive glasses are both examples of bioceramics. Polymers can be created artificially or naturally, like collagen. Due to their higher mechanical performance, biodegradable metals and nanomaterials made of carbon are also employed. To stimulate developmental signaling cascades biochemically, mechanically, or piezoelectrically, biomaterials can be created for bone tissue engineering. The nanotopography of an immunomodulatory substance can simulate bone nanointerfaces by steering macrophages toward bone-stimulating activity (Damaraju et al. 2017). Bone tissue engineering aims to create biomaterials that can take the place of allografts and autografts. The material should help in bone remodeling when implanted in the defect area (Koons et al. 2020). The basic building blocks of bone tissue engineering are progenitor cells (that express tissue matrix), scaffolds (provisionally support bone formation), and growth factors (stimulate osteoblast regeneration). Table 3.1 summarizes the particular standards for the optimum scaffold in bone tissue engineering.

3.3 Bone Regeneration

In the healing response after an injury or during skeletal growth, bone can naturally regenerate. Bone regeneration is a series of organized biological events, i.e., bone induction and conduction, that involve many cells and intracellular and extracellular molecular signaling pathways. Fracture repair, including intramembranous and endochondral ossification, imitates the embryonic skeletogenesis process in its normal state.

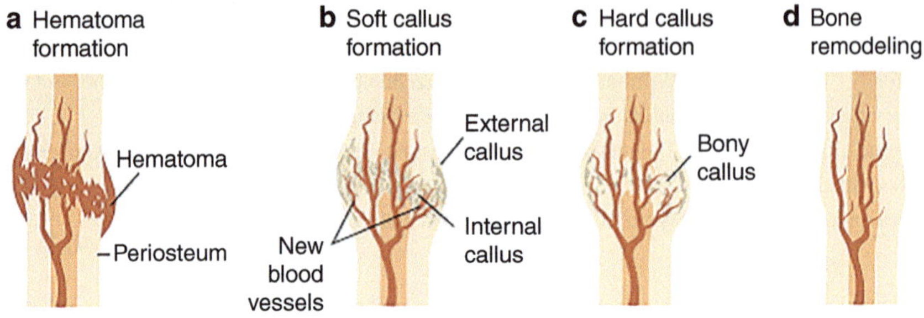

Fig. 3.1 The stages of fracture repair (Carano and Filvaroff 2003). (Reproduced from open access publication)

Several elements work together in bone formation technique: (1) a variety of cells types like osteoclasts, MSCs, etc.; (2) biomaterial (scaffold); (3) effects of growth factors, vitamins, cytokines, hormones, ions, etc.; and (4) several impulses (mechanical). Figure 3.1 depicts the four overlapping phases that the callus formation process uses to repair fractures. After musculoskeletal system damage, blood vessel disruption causes the coagulation cascade to be activated, which then results in the development of hematoma around the fracture site (Fig. 3.1a). In line with the hematoma's angiogenic activity, removal of the hematoma considerably reduces healing, whereas transplantation of the hematoma results in new bone formation. In response to the attraction of stem cells, inflammatory cells, and fibroblasts to location, new blood vessels are formed from already-existing vessels, which is known as angiogenesis. The inflammatory reaction is accompanied by discomfort, warmness, redness, and the release of many cytokines and growth factors that are crucial for repair. At the extremities of the bones, initially, granulation tissue develops before progressively giving way to fibrocartilage in a manner that appears to be tied to the vascular pattern. While this is happening, an external callus is being formed on the periosteum through direct bone production, also known as intramembranous ossification (Fig. 3.1b). After that, the rigid callus of braided bone is formed from the internal callus after it has calcified with calcium hydroxyapatite (Fig. 3.1c). The fracture callus is replaced by secondary lamellar bone during the remodeling phase, which also sees the callus' size reduced to that of the preexisting bone in the damaged area and the vascular supply return to normal (Fig. 3.1d) (Carano and Filvaroff 2003).

3.3.1 Calcium Phosphate

Large bone defects, such as those caused by complex bone fractures and incurable bone diseases, pose particular difficulties for bone regeneration. It is essential for healing that implanted materials fill the bone deficiencies. Grafting with autologous bone is the most efficient approach to repairing bone defects (Park et al. 2022).

Bioceramics are a kind of biomaterials for biomedical applications. Because of their superior biocompatibility, mechanical properties, and other attributes, these types of biomaterials are used in bone tissue engineering. Calcium phosphates are widely used in orthopedic implants, which are readily available in the human bone and tooth. Calcium phosphates are bioactive, biocompatible, and osteoconductive (Mohd Pu'ad et al. 2019).

Due to its prevalence and significance among calcium compounds, hydroxyapatite is a mineral phase that is naturally present in hard tissues. The rigidity of bone, dentin, and enamel is a result of hydroxyapatite's role as reinforcement in hard tissues. HAp crystals are present in collagen as needlelike structures that are aligned in the direction of the fibers in cortical (compact) bone. However, in reality, it has a complex chemical makeup that is nonstoichiometric as well as calcium-deficient (Ca:P molar ratio 1.67) as a result of vacancies and foreign cations and anions that are absorbed from the nearby body fluids at the time of bone metabolism, including sodium, zinc, magnesium, iron, and carbonate (Begam et al. 2017). Currently, HAp has been investigated for applications in different areas like drug delivery, collagen stimulation, skin regeneration, sun protection, water purification, wastewater treatment, and so on (Tan et al. 2020; Steffens et al. 2015; Ibrahim et al. 2020; Ideia et al. 2021). As a result, numerous techniques for synthesizing HAp with specific properties have been explored. These can be divided into three categories: dry methods (solid-state and mechanochemical), wet-chemical methods (sol-gel method, chemical-precipitation method, hydrothermal reaction, hydrolysis method, polymer-assisted route, etc.), and high-temperature methods (combustion method and pyrolysis route).

The two types of carbonate HAp that mimic biological apatite are A-type and B-type carbonate HAp with hydroxyl (OH^-) and phosphate (PO_4^{3-}) group replacements, respectively. The structural, optical, and morphological properties of HAp are greatly influenced by the presence of ions (Vinoth Kumar et al. 2021). Natural HAp is typically taken from organic materials or waste materials like plants, algae, cockles, clams, fish scales, and mammalian bones (such as those from cattle, camels, and horses). It can also be extracted from mineral materials (e.g., limestone). It has been demonstrated that this ratio works best to encourage bone repair. Because natural HAp is not stoichiometric, it either has a calcium or phosphorus deficiency (Ratner et al. 2012). The most frequent vacancies in HAp are in the calcium sites, which are filled by cations including Na^+, Mg^{2+}, and Al^{3+} (Ratner et al. 2012). The trace elements speed up the process of bone production and are crucial for bone repair.

In general, HAp crystallites can exist in a variety of shapes, including needles, rods, spheres, and others. Several variables, including extraction method, calcination temperature, and bone type, affect the characteristics, efficacy, phase purity, and particle size of HAp derived from a biological source (bone). Usually, the cortical region of the femoral bones is used to prepare scaffolds. To eradicate dust and proteins from the bone surface, bones are washed thoroughly, boiled in distilled water, and then treated with hypochlorite or NaOH. Following this initial cleaning and drying, bones are crushed or subjected to ball milling, impacting the end product's morphology and particle size. To eliminate the organic components from the bone and produce HAp, bones are treated in a furnace (600 and 1400 °C) during the calcination process. To get rid of the organic material and improve HAp's crystallinity while preventing thermal degradation of the finished product, a carefully chosen calcination regime is used. The germs that could spread diseases from the cattle to the patient are also eliminated by calcination treatment (Akram and Ahmed 2014).

The calcium carbonate content in cockle, clam, and eggshells is high, which can be transformed into HAp. Calcium can be obtained from eggshells, which have a calcium content of 94% $CaCO_3$. Typically, the shells were cleaned by first washing them in distilled water to get rid of sand and grime and then cleaning them again with bleach. CaO is obtained by calcining shells which will be used, along with surplus $CaCO_3$, for additional chemical processing to produce high-purity HAp (Mustaffa et al. 2015) (Table 3.2).

Usually, the form and sizes of the HAp produced from various sources differ. The extraction technique had an impact on the properties of the HAp that was produced. In contrast to chemical precipitation, heat methods (calcination) produce crystalline HAp. Some resources, like seashells, algae, etc., are abundant in $CaCO_3$ and are processed to prepare phase pure HAp. Therefore, a variety of techniques have been utilized to prepare phase pure HAp identical to the minerals found in human bone. Understanding HAp's characteristics, such as its phase purity and thermal stability, is crucial given the growing demand for it in biomedical applications, especially when isolated from natural sources. Furthermore, using natural resources to extract HAp ensures sustainability because they enable waste materials to be converted into value-added commodities by recovering nutrients from them.

3.3.2 Bioglass

Bioglass, also known as bioactive glass (BG), is widely used for clinical therapeutic purposes, notably in the orthopedic profession. The prerequisite for BG to heal bone fractures has increased as the human population has grown (Jones 2013). Since the landmark study by Hench et al. in 1972, BGs have attracted great interest for medical applications because of their high osteoconductive, bioactive, and biodegradable features (Rizwan et al. 2017). The implanted BGs have the ability to establish bond between implant materials and host tissues by creating an appetite layer on their surface that is comparable to bone minerals and can integrate with the host bone (Ibrahim et al. 2017). When new tissue is growing, bioglass releases physiologically active ions that encourage osteogenesis and induce vascular ingrowth (Gorustovich et al. 2009).

For the melt quench technique of 45S5 synthesis, silica, phosphorus pentoxide, calcium carbonate, and sodium carbonate were required. Possibly the most common type of waste produced worldwide is rice husk. When rice husk is heated to high

Table 3.2 Hydroxyapatite derived from natural sources

Sources	Extraction process	Phase	Crystallinity	Shape	References
Bighead carp (*Aristichthys nobilis*) scales	Deep eutectic solvents	HAp	43.13%	Irregular	Liu et al. (2020)
Rapana thomasiana shells	Hydrothermal method	HAp	79%	Irregularly shaped microcrystalline aggregates	Madalina Cursaru et al. (2022)
Eggshell	Hydrothermal method	HAp	Not available	Highly agglomerated with no regular shape	Oladele et al. (2019)
Chicken bone Black Sumatra Fighting cock	Thermal annelation (700, 900, 1100 °C)	HAp	2.348544 0.762829	Porous structure	Vinoth Kumar et al. (2021)
Meretrix meretrix clam shells	Calcination (900 °C) followed by a wet precipitation reaction	HAp	75.8%	Spherical shape Average particle size 40 nm	Sindhya et al. (2022)
Anadara granosa waste	Calcination (900 °C) followed by a wet precipitation reaction	HAp	Not available	Aggregated particles with a size range of 65–48 nm have cubic form	Dhanaraj and Suresh (2018)
Human Bovine Porcine	Calcination at 600, 900, 1200 °C	HAp	96% 95% 96%	Not available	Figueiredo et al. (2010)
Bovine	Treatment in aqueous NaOH and KOH solutions	HAp	Not available	Rectangular shape	Brzezińska-Miecznik et al. (2015)
Catfish bone Tilapia bone Seabass bone Yellowfin tuna	Calcination at 700 °C	HAp HAp HAp HAp	Highly crystalline	Nonuniform particle Uniform particles (150 nm) Fine particles (50–70 nm) Uniform particles (150 nm)	Nam et al. (2019)

(continued)

Table 3.2 (continued)

Sources	Extraction process	Phase	Crystallinity	Shape	References
Bovine bone	Subcritical water process Alkaline hydrothermal hydrolysis Thermal decomposition (850 °C)	HAp HAp HAp	Not available Poor crystallinity Highly crystalline	Nanorod shape Nano flake shape Nonuniform agglomerated big particle	Barakat et al. (2008)
Atlantic cod fish bones	Calcination (700–1200 °C)	HAp, β-TCP	Not available	Spherical-shaped (100–200 nm diameter)	Piccirillo et al. (2015)

temperatures, it can be easily removed to create rice husk ash (RHA), which contains a significant quantity of silica (Bakar et al. 2016). Additionally, millions of tons of eggshells are produced as biowaste. When eggshells are heated to a high temperature, they produce 99% or more calcium oxide and a small amount of biologically beneficial ions like Mg^{2+} and Sr^{2+} (Jayasree et al. 2017). Additionally, the composition displays antibacterial properties for calcium oxide extracted from eggshells. Therefore, using biological forms of silica and calcium oxides to synthesize biomaterials for biomedical purposes decreases total manufacturing costs and offers several beneficial biological impacts (Palakurthy et al. 2020).

Employing biowastes like RHA and eggshells, Palakurthy et al. used the melt-quenching process to create SiO_2-CaO-Na_2O bioglass (Palakurthy et al. 2020). After 7 days of SBF incubation, the SiO_2-CaO-Na_2O bioglass displayed significant bioactivity as evidenced by the deposition of a hydroxyapatite on surface. The study effectively showed that RHA and eggshells are possible beneficial, affordable alternatives to CaO and SiO_2 to prepare bioglasses.

By using the sol-gel process, Nayak et al. isolated sodium silicate from RHA for use in the fabrication of BG (soda-lime SiO_2 based) (Nayak et al. 2010). Gel powder underwent a 2-h calcination process at 700 °C to create BG powder. Crystalline Combeite-I, distributed in an amorphous glass matrix, makes up the majority of the calcined BG powder. It was discovered that the crystalline combeite phase of glass-ceramics dissolves easily in SBF and Tris buffer solution. The glass-ceramics surface began to form carbonated hydroxyapatite after only 3 days of incubation in SBF. Using a simple, low-cost extraction method, Kaya et al. extracted novel porous biosilica-based beads from the marine sponge *Geodia macandrewii* (Kaya et al. 2021). According to bioactivity investigations, hydroxyapatite developed on the beads' surface. Additionally, employing the biosilica beads as controlled cargo release carriers for anticancer drugs is encouraged by the hollow routes leading to the center cavity (Fig. 3.2).

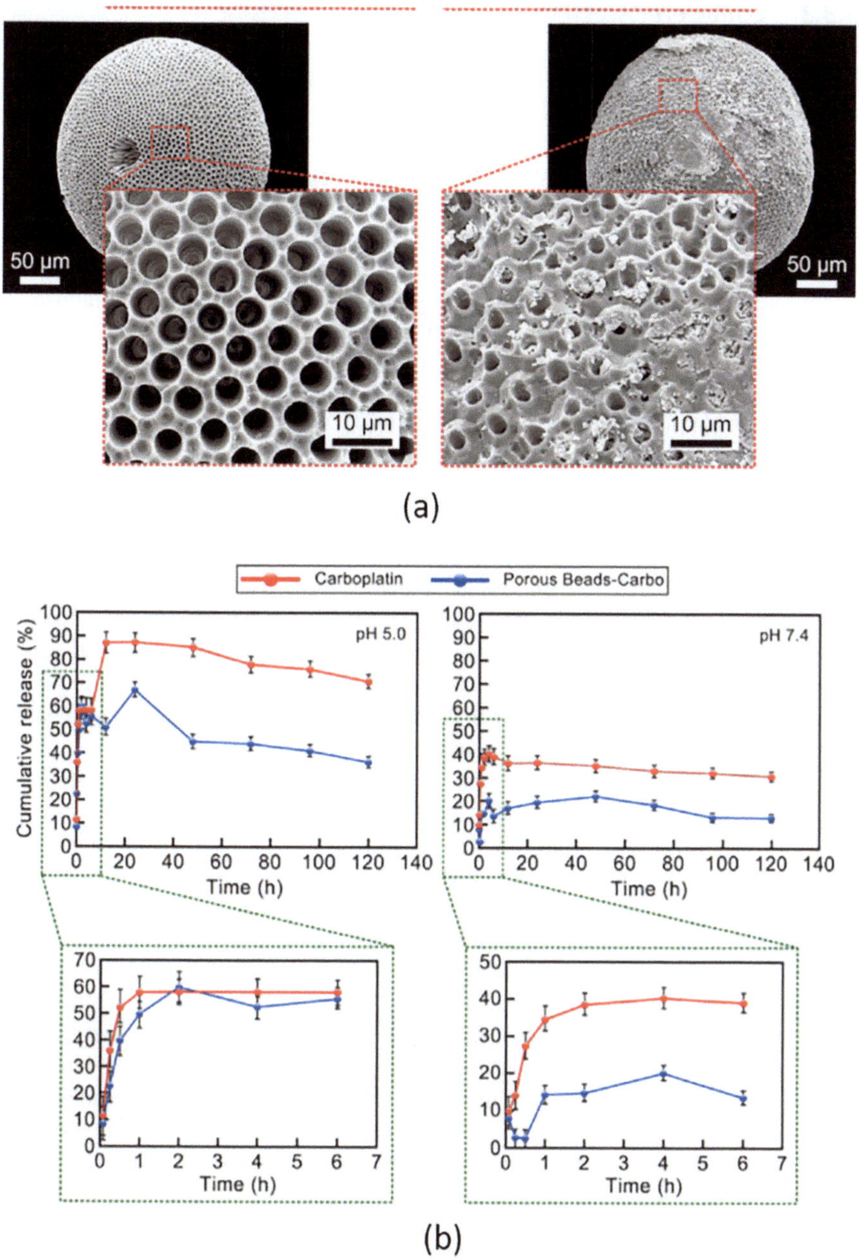

Fig. 3.2 (**a**) SEM images of porous biosilica beads with carboplatin loaded (right) and unloaded (left). (**b**) Study on medication release between pH 5.0 and 7.4 (Kaya et al. 2021). (Reproduced from open access publication)

3.3.3 Collagen

The main extracellular protein in our body is collagen, which helps to protect tissues and organs mechanically. Following hydroxyapatite as the second-largest component by weight, collagen makes up 36% of bone. Type-I collagen is available in our skin, tendons, and ligament bones. Three tropocollagen molecules (Gly-X-Y-) are found in collagen, where X is proline and Y is hydroxyproline residue. The protein can create triple helical shapes because of its structure (Shoulders and Raines 2009). Collagen extraction is expanding daily due to its growing demand and use in dental and tissue engineering applications, bone graft, food and pharmaceutical industries, and more. Currently, 29 different kinds of collagen have been recognized. Type-I is found in the ligaments, bones, and tendons; type-II is found in cartilage and vitreous body; type-III is found in reticular fibers of the lungs, liver, vessel wall, and spleen; type-IV is found in basement membranes; and type-V is distributed along with type-I (cornea). Because type-I collagen has a greater capacity for cell adhesion and is less antigenic than the other forms, it has mostly been used in biomedicine for wound dressing, tissue engineering construction, and cosmetics (Ahmed et al. 2020).

Collagen is mostly found in the skin and bone of bovine species, including cows, oxen, buffalo, and cattle. Bovine collagen that has been hydrolyzed and taken from various tissues exhibits antibacterial, antioxidant, and antihypertensive properties. Bovine lung hydrolyzed collagen also exhibited anti-inflammatory and antioxidant properties. Due to the spread of numerous illnesses like infection in the mouth and foot, transmissible spongiform encephalopathy, and bovine spongiform encephalopathy, the collagen from this origin is not extensively used.

Porcine skin and bones are also employed as sources of collagen. The majority of the uses for these sources are industrial. This type of collagen is identical to human collagen; it is regarded as safe because it cannot elicit any adverse reaction. Two essential phases are involved in collagen extraction: (1) pretreatment of starting material and (2) collagen isolation. Prior to collagen isolation, the primary goal of pretreatment is to eliminate contaminants. Skin, bone, swim bladders, and scales were among the by-products that were divided into several groups prior to pretreatment of the raw materials. This facilitates quick cleanup, contamination cleanup, and size reduction. Most of the time, fats, pigments, and other proteins are present in the raw materials utilized for extraction. Other inorganic substances, including calcium, are also found in fish bones and scales. These bones and scales are typically demineralized using inorganic materials like EDTA (ethylene diamine tetra-acetic acid). Sodium chloride, sodium hydrochloride, and *n*-butanol are used to remove the non-collagenous parts and lipids. Collagens are extracted using a variety of techniques, and according to these techniques, they can be divided into four categories: acid-soluble collagen (ASC), salt-soluble collagen (SSC), ultrasound-aided collagen (UAC), and pepsin-soluble collagen (PSC). Depending on the type of extraction procedures used, different collagens have different yields and

Table 3.3 Summary of literature about the source, extraction method, and collagen yield

Source	Method used	Yield	References
Skin of *Centrolophus niger* (black ruff)	ASC method	25%, 32%, 31%, 31%, and 45% ASC, respectively, were produced by acetic, citric, tartaric, formic, and lactic acid	Bhuimbar et al. (2019)
Tilapia skin	Acetic acid method Hot water method Sodium hydroxide method	11.7% 10.7% 8.5%	Bi et al. (2019)
Snakehead skins	PSC extraction	34.91–48.17%	Liu et al. (2019)
Sheep Lamb	ASC extraction	18.5% 12.5%	Vidal et al. (2020)
Chicken sternal cartilage	Type-II pepsin-soluble collagen subjected to ultrasonic treatment	40%	Akram and Zhang (2020)
Silver carp fish skin by-product	ASC PSC	42.85% 58.75%	Faralizadeh et al. (2021)
Asian bullfrog (*Rana tigerina*) skin	ASC PSC	22.48–31.12% 22.59–28.30%	Indriani et al. (2022)
Nile tilapia skin (*Oreochromis niloticus*)	ASC	15.3–18.3%	Menezes et al. (2020)
Common starfish	ASC (pretreatment steps employed ultrasound (US) and high shear mechanical homogenization (HSMH))	80.30 ± 0.61, (collagen extracted by classic method) 81.47 ± 0.45 (collagen extracted with the aid of HSMH) 83.18 ± 1.74% (collagen extracted with the aid of HSMH and US)	Vate et al. (2022)

physiochemical characteristics (Table 3.3). Bhuimbar et al. (col1) extracted ASC from the skin of *Centrolophus niger* (black ruff). Figure 3.3 from their research demonstrates the impact of various acids and time on yield. The pretreated skin was given a 72-h treatment with 0.5 M lactic acid to achieve the highest yield.

Liu et al. (col 3) used snakehead skin to extract collagen and studied the effect of H_2O_2 pretreatment and pepsin hydrolysis strategies. They reported that both the parameters of the pepsin hydrolysis and H_2O_2 pretreatment processes substantially influenced the production of collagen from snakehead skin. The pH of H_2O_2 bleaching significantly affects the color and structure of PSC.

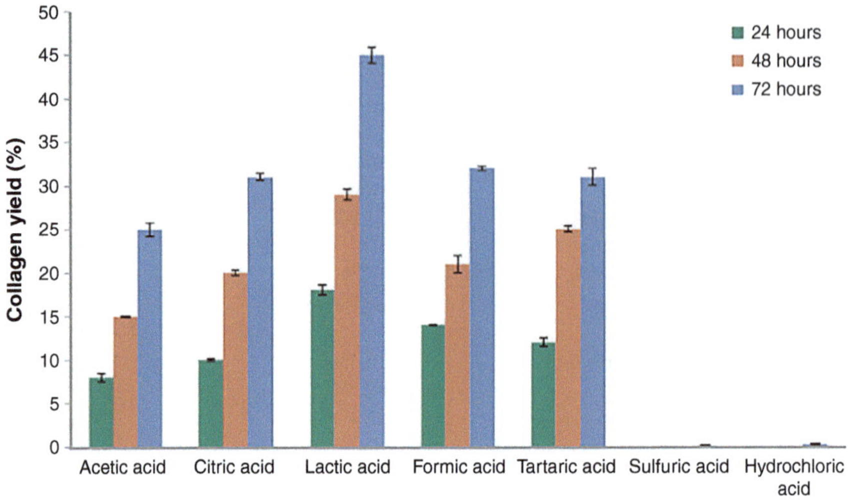

Fig. 3.3 Effects of different acids and time on collage yield. (Reproduced with permission from Bhuimbar et al. 2019)

3.3.4 Chitosan

The second most prevalent natural polymer after cellulose is chitin, which is made up of a linear chain constructed of poly (1,4) *N*-acetyl-D-glucosamine (Rinaudo 2006). Chitin makes up the majority of the exoskeletons seen on marine creatures and insect shells. Biopolymers called chitin and chitosan are derived from many terrestrial and aquatic resources, such as grasshoppers, insects, fungi, shrimp, crab, conus, lobster, and squid pens (Jantzen da Silva Lucas et al. 2021; Kumar and Shahid 2020; Sayari et al. 2016). Chitin's crystal structure was dependent on the arrangement of polymer chains, mainly where the 2-*N*,*N*′-diacetylchitobiose units were located in relation to one another. Chitin has been divided into three distinct forms based on the crystalline allomorphs: α, β, and γ. The α-chitin are made of antiparallel chains, β-chitin consist of parallel chains, and γ-chitin are made of parallel and antiparallel chains (Kaya et al. 2017). The reaction conditions, chitin source, and degree of deacetylation are the factors that influence the molecular weight of chitosan. Slight variations in these parameters can have a major impact on the characteristics of chitosan, like surface properties, solubility, cellular activity, and crystallinity (El Knidri et al. 2018; Qinna et al. 2015). The surface morphology of chitosan extracted from shrimp shells was significantly affected by the concentration of acid, as reported by Hisham et al. (2021). Demineralization, deproteinization, and deacetylation are the three fundamental steps for the isolation of chitosan. Most of the crustacean shells are made of chitin, protein, minerals, etc., but they may also have carotenoid colors. In order to eliminate these carotenoid colors, another step, i.e., the decolorization process, can be added. This process makes use of several

Table 3.4 Different natural sources for extraction of collagen

Source	Process	Degree of deacetylation	References
Fungal source (*Ganoderma lucidum* spore powders)	Dual-frequency ultrasound irradiation	81.1–81.3%	Zhu et al. (2019)
Antarctic krill (*Euphausia superba*)	Treatment in 1.7 M HCl at room temperature for 6 h, followed by 1 h of 2.5 M NaOH treatment at 75 °C	11.28 ± 0.86	Wang et al. (2013)
Fungal biomass of *Rhizopus oryzae* NRRL1526	Microwave-assisted extraction	94.6 ± 0.9%	Sebastian et al. (2019)
Shrimp shell powder	Co-fermentation involving *Acetobacter pasteurianus* and *Bacillus subtilis*	19.6%	Zhang et al. (2021)
Two common spider species (*Geolycosa vultuosa* and *Hogna radiata*)	Treatment in 4 M HCl solutions and 1 M NaOH solutions	97–99%	Kaya et al. (2014)
Shrimp shells	Enzymatic deproteinization	9%	Younes et al. (2014)
Antarctic krill shell	Using lactic acid and dispase	80.8%	Yu et al. (2020)

inorganic and organic solvents like sodium hypochlorite, hydrogen peroxide, and acetone (El Knidri et al. 2018) (Table 3.4).

3.3.5 Hyaluronic Acid

Hyaluronic acid (HA) is the only non-sulfated glycosaminoglycan found in almost all extracellular matrix (approximately 0.5 mg/mL in human body ECM) (Becker et al. 2009) of soft connective tissues. The structural basis of HA is a linear anionic polysaccharide chain of alternative disaccharide units composed of β (1,4) glucuronic acid and *N*-acetylglucosamine (GlcNAc). It has been differentiated as higher molecular weight HA (more than 100 kDa) and low molecular weight HA (below 100 kDa); the former is denser as compared to the latter forms, and it is prevalently used in bone resorption rate modification and to form tissue scaffolds with optimized mechanical stability. This compound is also found in plants. The concentration of hyaluronic acid in animal tissues varies based on their composition and biochemical pathways. HA of higher molecular weight gives the scaffold viscoelasticity and helps to stimulate the chondrocyte for regeneration. It also increases the ECM synthesis in the tissue by preventing cell proliferation. Due to its non-immunogenicity, biocompatibility, and high water-retention capacity, it has become helpful in tissue engineering (Sionkowska et al. 2020). This also has various biomedical and pharmaceutical applications such as lubrication medium, cell-signaling factor, and other biophysical properties. The natural synthesis of hyaluronic acid in the body occurs with the help of hyaluronan synthase, degraded by hyaluronidase, and bound proteins are digestible by protease.

The synthesis of hyaluronic acid is developed by the extraction from marine sources and cattle animal tissues or fermentation process or enzymatic process and biosynthesis from different bacterial strains like *Streptococci* sp., *Corynebacterium* sp., etc. (Fig. 3.3) (Boeriu et al. 2013; Srisantisaeng et al. 2013). It was first discovered and extracted from the vitreous humor of bovine eyes in 1934 by Meyer et al. The key steps of synthesis are extraction and purification; depending on the ratio of these two steps, the property of HA is determined (Karami et al. 2021). Generally, a range of molecular weight varies from 20,000 to 13×10^6 Da, which forms the nature of its activity (Karami et al. 2021). The weight modifications by different elongation systems due to alternative purification processes mainly cause variations in rheological characteristics of HA, such as viscosity and elasticity. Usually, the method for the HA extraction includes both enzymes and organic solvents like ethanol, acetone, and propanol. After extraction, processing, and purification were carried out through physical separation methods (such as ultrafiltration, electrodeposition of protein, or adsorption) to separate HA from the broth. Biotechnological processing for commercial production of HA is drawing more attention due to the high yield and high purity over some time. Animal-derived HA is mainly used for industrial production purposes where the purity is less than that of bacterial (Fallacara et al. 2018). Sources used so far in various researches include tissues from cattle, pigs, rabbits, fish, chickens, etc. Table 3.5 describes a few natural sources used in different studies made on HA production so far. One major source used for extraction is synovial fluid and cartilage, which is an important component in a matrix composed of collagen and hyaluronan. Some other principal waste parts are chicken crest or comb, eggshells, human umbilical cord, visceral tissues, bovine and fish vitreous humor, and ovine synovial fluid (Abdallah et al. 2020; Ibragimova et al. 2022). The primary goal was to obtain the optimum concentration of hyaluronic acid possible from animal waste or by-products to reduce production costs (Rossatto et al. 2022).

The chicken comb is the most commonly available source of hyaluronic acid. Various reports have shown that this production method requires cheaper setup and chemicals. Although there are chances of immunological activity and other binding proteins, the presence of viruses (Kang et al. 2010; Murado et al. 2012), through this procedure, significant yields have been obtained. An extraction process by (Kulkarni et al. 2018) has followed the isolation process through delipidation and dehydration before obtaining protein-eliminated sodium hyaluronate powder. The confirmatory tests for protein removal and SDS-PAGE have shown that the final yield of HA was relatively pure for clinical and research applications. The viscosity measured in HA was 2.55 poise. The reason for using sodium citrate was to separate extracted HA from sodium hyaluronate. Similar strategies were taken in other reports, but the purification strategy determines the product purity. To obtain 1600–2000 kDa[1] of

[1] kDa: kilo Dalton

Table 3.5 Various sources used for A synthesis

Source animal tissue/microbes		Extraction process	Purity	References
Poultry	Chicken comb	Organic solvent	Standard	Kang et al. (2010), Tovar et al. (2012)
	Eggshells			
Animal	Mice	Organic solvent	Moderate	Lambe et al. (2021)
	Porcine		Moderate	Abdallah et al. (2020)
	Bovine synovial fluid	Chemical, enzymatic	Moderate	Abdallah et al. (2020)
	Bovine nasal cartilage	Organic solvent		Boeriu et al. (2013), Lambe et al. (2021)
	Sheep synovial fluid	Enzymatic, solvent extraction	Standard	Abdallah et al. (2020), Ibragimova et al. (2022), Lambe et al. (2021)
	Sheep lung			
	Rabbit liver, cortex, vitreous, lung			
Human	Umbilical cord, placenta, visceral fluid, serum, epidermis	Enzymatic, solvent extraction	Moderate	Fallacara et al. (2018), Abdallah et al. (2020), Ibragimova et al. (2022), Lambe et al. (2021)
Marine sources	Fish eyes	Chemical, organic solvent, enzymatic separation	High	Abdallah et al. (2020), Murado et al. (2012), Lambe et al. (2021)
	Fish scales			
	Fish extracts			
Microbial sources	*Streptococcus* sp., *Bacillus*, *E. coli*, *Corynebacterium*, *L. acidophilus*, and yeasts (saccharomyces, *Pichia pastoris*)	Enzymatic, batch culture, genetic manipulation	High	Karami et al. (2021), Fallacara et al. (2018), Rossatto et al. (2022), Rodriguez-Marquez et al. (2022)

HA, Murado et al. used marine fish eyeballs (vitreous humor) with consecutive separation and purification through enzymatic digestion and chemical processes (Murado et al. 2012).

Enzymes like papain, trypsin, pepsin, and proteases are essential for the degradation of tissue. It helps to break the glycosidic bonds of the total protein to isolate the undamaged HA polymer from raw materials. Among these, papain is the most used enzyme for HA isolation. In a study, enzymatic hydrolysis is where enzymes such as papain, pepsin, and trypsin were applied on eggshell membranes to find the ideal physicochemical conditions like temperature and pH for extracting high molecular weight HA (Ibragimova et al. 2022). Microbial fermentation: The most promising source for bacterial hyaluronic acid production is the *Streptococcaceae* family, and among them, strains of *Streptococcus zooepidemicus* have given the highest yield (6 g/L) of HA as reported in some research (Ibragimova et al. 2022; Rodriguez-Marquez et al. 2022). As there is a difference in the extraction process for high

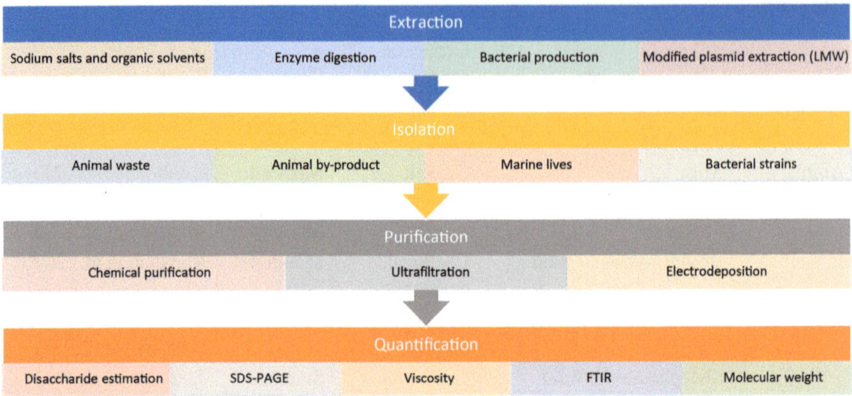

Fig. 3.4 Various methods for HA synthesis

molecular weight HA and low molecular weight HA, microbial fermentation can also be modified to obtain the desired HA. Using *Lactococcus acidophilus* HA synthesis was safe as compared to *Streptococcus zooepidemicus* due to the genetic manipulation (Rossatto et al. 2022), but it does not cut down the cost of production either. Nowadays genetically modified strains of microbes are used such as *Streptococcus thermophilus* YIT 2084, *Streptococcus zooepidemicus* ATCC 35426, ATCC 35246, *Corynebacterium glutamicum*, recombinant *Bacillus*, etc. The rate of production can be increased by using these genetically modified bacteria, but the use of enzyme inhibitors bypassing the genetic change also helps to increase the rate as stated by Samadi et al. (2022). Another study shows that *Streptococcus zooepidemicus* alters the production rate when 0.15 g/L of hyaluronidase is added in bacterial culture and thus increased dissolved oxygen concentration to enhance HA yield (6 g/L) more compared to yield without enzyme addition (5 g/L). Albeit the increase in yield, it has decreased the molecular weight from 1300 kDa to 45 kDa (Karami et al. 2021; Samadi et al. 2022) (Fig. 3.4).

3.3.6 Cellulose

One of the naturally derived fibers is cellulose. Generally, it is found as a fiber composite with hemicellulose, lignin, and other protein molecules. This most abundant biopolymer of nature is increasingly being used in tissue engineering, as a scaffold component, in substrate preparation, and for producing other composite nano-polymers. These scaffolds can be suitable material for 3D tissue formation (nerve inside the bone tissue) and cell growth for having surface modification properties and mechanical stability (Hickey and Pelling 2019). Cellulose is abundant in many forms. Most of it comes from wood and plant sources like plant fibers such as jute, hemp, kenaf, and cotton (Fig. 3.5). Other than the direct plant sources, agricultural wastes, composite biomass, and processing residues from industries also

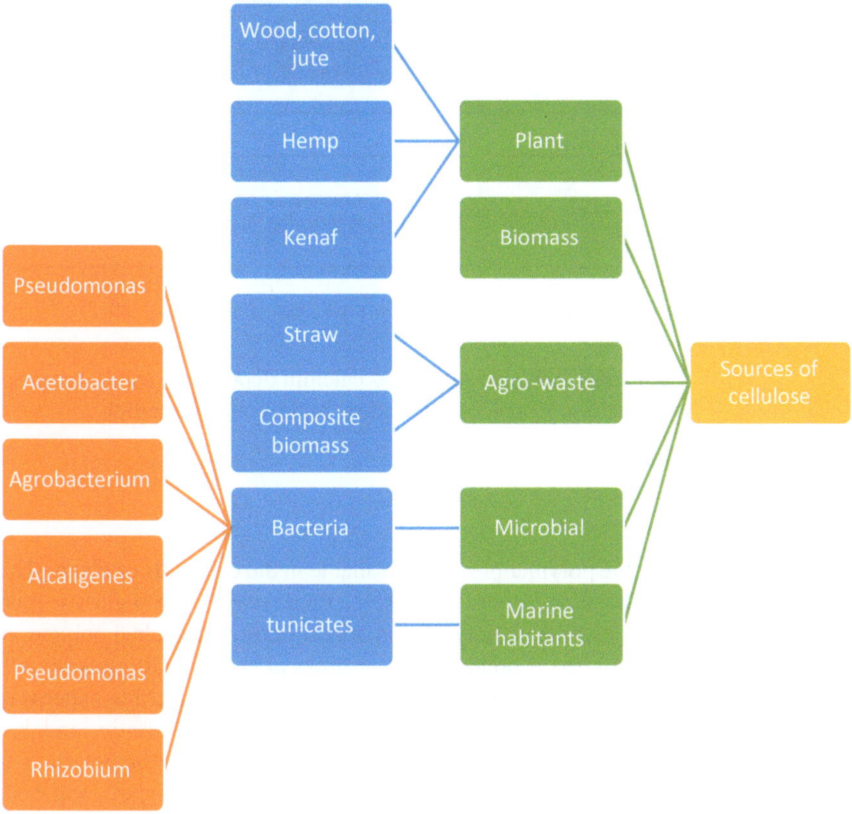

Fig. 3.5 Different common sources of cellulose

contain a large amount of cellulose. The microbial sources of cellulose are mostly bacteria. Beside this, marine biomass also has plenty of cellulose sources. The functional characteristics of cellulose depend on its chain length, polymerized units, and presence of glucose that forms molecular units of cellulose. Different forms of cellulose are shown in Fig. 3.5. The structural basis is cellulose fibril, which resulted from biosynthesis. The outline of the cellulose extraction from sources other than bacteria is described in Fig. 3.6.

The application of cellulose and its derivatives is widespread, and currently, researchers are focusing on the production and dissociation of cellulose nanoparticles (CNP) (Jonoobi et al. 2015). These are highly applicable to drug delivery systems as bacterial cellulose offers localized delivery systems in cells by means of increasing cytokine concentration. There is a similarity between cellulose fibers and collagen fibers of bone tissue; this can be utilized in bone tissue engineering. It is also found in many studies that the biocompatible scaffolds composed of cellulosic composites enhance bone regeneration (Hickey and Pelling 2019; Seddiqi

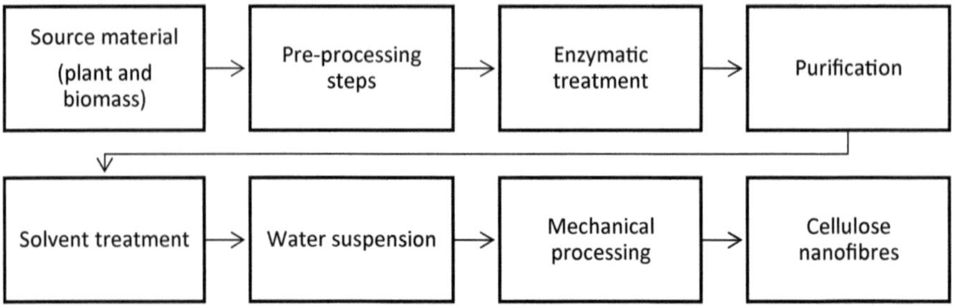

Fig. 3.6 Stepwise process of nanoscale cellulose fiber production (Jonoobi et al. 2015)

et al. 2021; Teeri et al. 2007). It also stimulates calcium deposition, and hydroxyapatite composites with cellulose nanocrystals allow rapid bone integration.

3.3.7 Starch

Starch is a biodegradable polymer primarily available in the cell walls of plants. This polysaccharide is composed mostly of amyloses, a linear polymer of D-glucose units attached through α-1,4 glycosidic linkage, and amylopectin, a branched polymer of D-glucose units joined by α-1,4 bonds and cross-links at each 25–30 glucose units joined by α-1,6 bonds. The use of starch is determined by the branching patterns of its units, and the physicochemical properties like higher water affinity and low tensile strength restrict its uses. Starch is a good candidate among biopolymers to produce biopolymer products for short-term stability like molded parts and bone scaffolds that need to be biologically active and degradable. For preparing biocomposites, starch can be mixed with other polymers as well as raw materials (such as cassava tree bark and gelatinized starch) to enhance the physicochemical properties of the desired polymer or scaffold (Engel et al. 2021). Also, it has high requirements in the food and food processing industries, where amylose and amylopectin are extracted as an additive in food processing.

Different plant sources are widely used for starch production, among which corn starch is mostly known and produced (about 90–95% of total starch production) (Shevkani et al. 2016). Though fruits and food grains (except corn) are less common in the production route for their food value, rather a novel approach using rotten and waste fruits has been recently tested where they have utilized a variety of wasted fruits like kiwifruit, avocado, cassava, mango, litchi, tamarind, annatto, jackfruit, apple, pineapple, banana, etc. (Kringel et al. 2020); a common production method is shown in Fig. 3.7. Currently, alternative extraction methods are developed to replace conventional maceration, as the new approach has better acceptance in terms of eco-friendly procedure and faster production using the ultrasound-assisted extraction system (Setyaningsih et al. 2021). The UAE[2] technique uses ultrasonic waves to

[2]UAE: ultrasonic assisted extraction

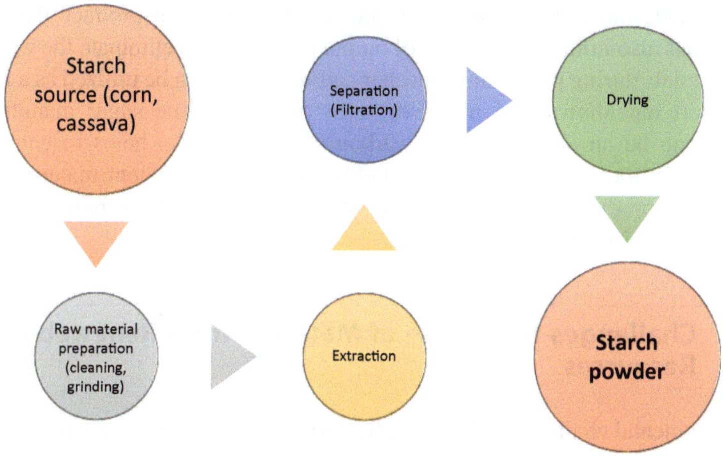

Fig. 3.7 Starch production method. (Adapted from Akram and Ahmed 2014; Madalina Cursaru et al. 2022)

produce a cavitation effect through a medium. As a result, the cell walls are broken down and allowed to release in the solvent medium. Because the starch granules in cassava are trapped by cellulose fibers compacted with pectin inside the cell wall, another study has shown that sago (*Metroxylon sagu*) and taro (*Colocasia esculenta*) are very promising alternatives for starch production (Ishak et al. 2020).

3.3.8 Alginate

Alginate polymer is abundant in nature, a polysaccharide of two isomer residues, β-D-mannuronic acid (M) and α-L-guluronic acid (G) with 1,4 glycosidic bonds in three different conformations (homogeneously or heterogeneously) of poly mannuronic (M) blocks, poly guluronic (G) blocks, and blocks of MG isomers. It is predominantly present in various algae (mostly brown and red algae), bacteria (*Azotobacter*, *Pseudomonas*, etc.), and plants. Depending on the sources and the growth conditions, the structural and functional properties of alginate vary in different residual parts.

Numerous industries have made substantial use of alginate-like food and bioprocess industries, as well as in biomaterial applications, due to their high stability and biocompatibility. The production of alginate also has a commercial background due to its popular application. Besides its physicochemical properties, it has potentially been used in pharmaceutical areas. Currently, more acceptable derivatives from alginate are gaining attention due to the environmental acceptance and easier extraction methods. The degradation requires enzymes to perform rapid production of alginate oligosaccharide and desired degree of polymerization for desired mechanical and chemical properties (Cheng et al. 2020; Kivilcimdan Moral and Yildiz 2016). The effect of sources used for production through bacterial synthesis determines the purity, cost, and yield of the final product. The culture medium has also influenced the rate of output per batch. To eliminate the wastage of pure materials during processing, cheaper carbo sources can be utilized as a nutrient. One report has shown that molasses, a by-product of the sugar manufacturing process, can be an alternative as a carbon source and also helps to enhance the rate, but it is not used in commercial fields. Starch is an effluent material in other biopolymer production processes, and it can be utilized to reduce cost also (Kivilcimdan Moral and Yildiz 2016).

3.4 Challenges to the Use of Materials from Renewable Resources

The fundamental challenge in biologically derived compounds compared to synthetically or chemically produced materials is the impurity in composition and concentration. The downstream processing of materials causes high prices for lower yields of products. To eliminate this problem, comparatively new synthesis methods are being experimented with where the post-extraction procedure should be less (Rossatto et al. 2022). Also, there is significantly less amount of information about the compositions and advantages of using animal and agricultural wastes and by-products. As a result, a huge percentage of raw materials are almost out of the processing circle. Proper channeling of these materials can save resources and produce more such polymers in an eco-friendly way (Motaung and Linganiso 2018). The aspects of production challenges are shown in Fig. 3.8. For their optimal use, it is imperative to explore these abundant marine biomaterials in the study. Researchers have recently given a great deal of attention to its widespread use in

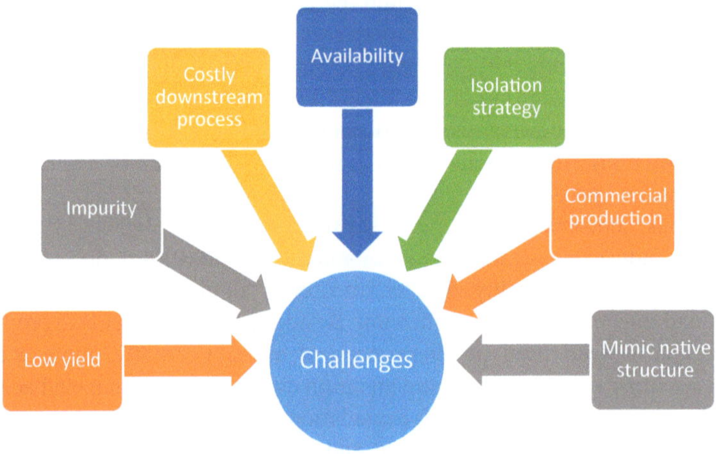

Fig. 3.8 Challenges in material production

numerous applications. These biomaterials are essential for bone tissue engineering because they are affordable, secure, and biocompatible (Lalzawmliana et al. 2019).

In the case of hydroxyapatite, excellent performance has been demonstrated despite the limited number of comparison studies that are currently available. The optimal tissue engineering scaffolds using hydroxyapatite as the nano-building block still present considerable challenges, despite substantial advancements in the development of materials with sufficient mechanical strength. The development of various hydroxyapatite-based scaffolds will also be suggested using fresh biomimetic or bioinspired methods. In comparison with marine sources, extracting chitin from insect biomass is clearly more difficult. Green technologies or process optimization may result in items that are produced in large quantities, but this may only be done after substantial research (Mohan et al. 2020). In the case of hyaluronic acid, its production faces multifaceted issues like a low amount of final product from bulk animal waste as a report shows, though the purity was 99.4% (Murado et al. 2012). This indicates that the downstream processing cost cannot be avoided for bulk raw materials. Another challenge is the isolation and enzyme digestion. For microbial production procedures, contamination is a primary issue, and maintaining culture for a long time also has drawbacks, such as unexpected enzymatic degradation or molecular weight degradation (Ibragimova et al. 2022; Samadi et al. 2022). In the case of cellulose, the commercial production from these plant residues has not been encouraged much as there is no established detailed isolation and processing method in the current research scenario, and it is less known to a large part of consumers. Agricultural wastage management is not adequately formulated for cellulose production, as found in a few papers (Motaung and Linganiso 2018).

Starch production from conventional sources creates environmental problems due to the formation of methane, and low yield from cassava is also a problem. However, it is still a commercially used process (Ishak et al. 2020). The raw plant sources, without proper treatment, may give side products like methane and other starch composites, which are not desirable for production (Kivilcimdan Moral and Yildiz 2016; Rojas-bringas et al. 2021). For alginate production, by-products of other biopolymers like starch wastes are also used, where impurities are more significant. In some methods, direct plant or animal sources have been used where the chance of side products increases due to minimized downstream processing (Kringel et al. 2020; Kivilcimdan Moral and Yildiz 2016). Preclinical tests and extensive studies on bone tissue engineering have been conducted and are still being done. The goal of all of these studies is to create the perfect bone tissue engineering scaffold that will meet the basic needs of a bone substitute, including high mechanical strength with adequate elasticity, perfect biocompatibility to support cell adhesion, migration and proliferation, sufficient porosity for facilitate nutrient and oxygen transfer, the ideal degradation rate to support the formation of new bone and cartilage, and controlled release of growth factors.

3.5 Conclusion

Bone tissue engineering is an appropriate substitute for autograft and allograft to treat bone defects. The design of scaffolds using biomaterials and their synthesis are an important part of bone tissue engineering. The characteristics of the biomaterials utilized determine whether an implant will be successful. Naturally, origin biomaterials are more advantageous compared to synthetic materials because of biocompatibility, biodegradability, and nontoxicity. Natural biomaterials are still being explored extensively for their possible applications, despite the fact that the existing results do not entirely meet clinical needs. As a result, research in this area is currently being conducted all over the world.

Acknowledgments SKN acknowledges the support of the Honorable Vice-Chancellor of the West Bengal University of Animal and Fishery Sciences, Kolkata, India.

References

Abdallah MM, Fernández N, Matias AA, Bronze MR (2020) Hyaluronic acid and chondroitin sulfate from marine and terrestrial sources: extraction and purification methods. Carbohydr Polym 243:116441. https://doi.org/10.1016/j.carbpol.2020.116441

Ahmed M, Verma AK, Patel R (2020) Collagen extraction and recent biological activities of collagen peptides derived from sea-food waste: a review. Sustain Chem Pharm 18:100315. https://doi.org/10.1016/j.scp.2020.100315

Akram M, Ahmed R (2014) Extracting hydroxyapatite and its precursors from natural resources. J Mater Sci. https://doi.org/10.1007/s10853-013-7864-x

Akram AN, Zhang C (2020) Extraction of collagen-II with pepsin and ultrasound treatment from chicken sternal cartilage; physicochemical and functional properties. Ultrason Sonochem 64: 105053. https://doi.org/10.1016/j.ultsonch.2020.105053

Bakar RA, Yahya R, Gan SN (2016) Production of high purity amorphous silica from rice husk. Proc Chem 19:189–195. https://doi.org/10.1016/J.PROCHE.2016.03.092

Barakat NAM et al (2008) Physiochemical characterizations of hydroxyapatite extracted from bovine bones by three different methods: extraction of biologically desirable HAp. Mater Sci Eng C 28(8):1381–1387. https://doi.org/10.1016/j.msec.2008.03.003

Barua E, Deoghare AB, Deb P, Lala SD (2018) Naturally derived biomaterials for development of composite bone scaffold: a review. IOP Conf Ser Mater Sci Eng 377(1):012013–012013. https://doi.org/10.1088/1757-899X/377/1/012013

Becker LC et al (2009) Final report of the safety assessment of hyaluronic acid, potassium hyaluronate, and sodium hyaluronate. Int J Toxicol 28(4_Suppl):5–67. https://doi.org/10.1177/1091581809337738

Begam H, Kundu B, Chanda A, Nandi SK (2017) MG63 osteoblast cell response on Zn doped hydroxyapatite (HAp) with various surface features. Ceram Int 43(4):3752. https://doi.org/10.1016/j.ceramint.2016.12.010

Bhuimbar MV, Bhagwat PK, Dandge PB (2019) Extraction and characterization of acid-soluble collagen from fish waste: development of collagen-chitosan blend as food packaging film. J Environ Chem Eng 7(2):102983. https://doi.org/10.1016/j.jece.2019.102983

Bi C et al (2019) Effect of extraction methods on the preparation of electrospun/electrosprayed microstructures of tilapia skin collagen. J Biosci Bioeng 128(2):234–240. https://doi.org/10.1016/j.jbiosc.2019.02.004

Boeriu CG, Springer J, Kooy FK, van den Broek LAM, Eggink G (2013) Production methods for hyaluronan. Int J Carbohydr Chem 2013:1–14. https://doi.org/10.1155/2013/624967
Brzezińska-Miecznik J, Haberko K, Sitarz M, Bućko MM, Macherzyńska B (2015) Hydroxyapatite from animal bones – extraction and properties. Ceram Int 41(3 Pt B):4841–4846. https://doi.org/10.1016/j.ceramint.2014.12.041
Carano RAD, Filvaroff EH (2003) Angiogenesis and bone repair. Drug Discov Today 8(21): 980–989. https://doi.org/10.1016/S1359-6446(03)02866-6
Cheng D, Jiang C, Xu J, Liu Z, Mao X (2020) Characteristics and applications of alginate lyases: a review. Int J Biol Macromol 164:1304–1320. https://doi.org/10.1016/j.ijbiomac.2020.07.199
Damaraju SM et al (2017) Three-dimensional piezoelectric fibrous scaffolds selectively promote mesenchymal stem cell differentiation. Biomaterials 149:51–62. https://doi.org/10.1016/J.BIOMATERIALS.2017.09.024
Dhanaraj K, Suresh G (2018) Conversion of waste sea shell (Anadara granosa) into valuable nanohydroxyapatite (nHAp) for biomedical applications. Vaccum 152:222–230. https://doi.org/10.1016/j.vacuum.2018.03.021
Dhand C et al (2016) Bio-inspired in situ crosslinking and mineralization of electrospun collagen scaffolds for bone tissue engineering. Biomaterials 104:323–338. https://doi.org/10.1016/J.BIOMATERIALS.2016.07.007
El Knidri H, Belaabed R, Addaou A, Laajeb A, Lahsini A (2018) Extraction, chemical modification and characterization of chitin and chitosan. Int J Biol Macromol 120:1181–1189. https://doi.org/10.1016/j.ijbiomac.2018.08.139
Engel JB, Luchese CL, Tessaro IC (2021) Food hydrocolloids how are the properties of biocomposite foams influenced by the substitution of cassava starch for its residual sources ? Food Hydrocoll 118:106790. https://doi.org/10.1016/j.foodhyd.2021.106790
Fallacara A, Baldini E, Manfredini S, Vertuani S (2018) Hyaluronic acid in the third millennium. Polymers 10(7):701. https://doi.org/10.3390/polym10070701
Faralizadeh S, Rahimabadi EZ, Bahrami SH, Hasannia S (2021) Extraction, characterization and biocompatibility evaluation of silver carp (Hypophthalmichthys molitrix) skin collagen. Sustain Chem Pharm 22:100454. https://doi.org/10.1016/j.scp.2021.100454
Figueiredo M, Fernando A, Martins G, Freitas J, Judas F, Figueiredo H (2010) Effect of the calcination temperature on the composition and microstructure of hydroxyapatite derived from human and animal bone. Ceram Int 36(8):2383–2393. https://doi.org/10.1016/j.ceramint.2010.07.016
Gorustovich AA, Roether JA, Boccaccini AR (2009) Effect of bioactive glasses on angiogenesis: a review of in vitro and in vivo evidences. Tissue Eng Part B Rev 16(2):199–207. https://doi.org/10.1089/TEN.TEB.2009.0416
Guo L et al (2021) The role of natural polymers in bone tissue engineering. J Control Release 338: 571–582. https://doi.org/10.1016/j.jconrel.2021.08.055
Hickey RJ, Pelling AE (2019) Cellulose biomaterials for tissue engineering. Front Bioeng Biotechnol 7:1–15. https://doi.org/10.3389/fbioe.2019.00045
Hisham F, Akmal M, Ahmad F, Ahmad K (2021) Facile extraction of chitin and chitosan from shrimp shell. Mater Today Proc 42:2369. https://doi.org/10.1016/j.matpr.2020.12.329
Ibragimova DSH, Davletova XSH, Khayitbayev AKH, Qizi SMA (2022) Extracting of hyaluronic acid from different sources. Eurasian J Phys Chem Math 3:50–54
Ibrahim NF, Mohamad H, Mohd Noor SNF, Ahmad N (2017) Apatite formation on melt-derived bioactive glass powder based on SiO_2-CaO-Na_2O-P_2O_5 system. Ceram Int 43(15): 11676–11685. https://doi.org/10.1016/J.CERAMINT.2017.05.356
Ibrahim M, Labaki M, Giraudon JM, Lamonier JF (2020) Hydroxyapatite, a multifunctional material for air, water and soil pollution control: a review. J Hazard Mater 383:121139–121139. https://doi.org/10.1016/J.JHAZMAT.2019.121139
Ideia P, Degli Esposti L, Miguel CC, Adamiano A, Iafisco M, Castilho PC (2021) Extraction and characterization of hydroxyapatite-based materials from grey triggerfish skin and black

scabbardfish bones. Int J Appl Ceram Technol 18(1):235–243. https://doi.org/10.1111/IJAC.13625

Indriani S, Benjakul S, Kishimura H, Karnjanapratum S, Nalinanon S (2022) Impact of extraction condition on the yield and molecular characteristics of collagen from Asian bullfrog (Rana tigerina) skin. LWT 162:113439. https://doi.org/10.1016/j.lwt.2022.113439

Ishak SD, Yusof YA, Kamarudin MS (2020) Different starch sources affect physical characteristics of extruded feeds prepared for the Malaysian mahseer, Tor tambroides. J Appl Aquac 34:1–18. https://doi.org/10.1080/10454438.2020.1822979

Jantzen da Silva Lucas A, Quadro Oreste E, Leão Gouveia Costa H, Martín López H, Dias Medeiros Saad C, Prentice C (2021) Extraction, physicochemical characterization, and morphological properties of chitin and chitosan from cuticles of edible insects. Food Chem 343:128550. https://doi.org/10.1016/j.foodchem.2020.128550

Jayasree R, Velkumar J, Sampath Kumar TS (2017) Egg shell derived apatite cement for the treatment of angular periodontal defects: a preliminary clinical and radiographic assessment. Dent Oral Craniofacial Res 4(2):1–2. https://doi.org/10.15761/DOCR.1000238

Jones JR (2013) Review of bioactive glass: from Hench to hybrids. Acta Biomater 9(1):4457–4486. https://doi.org/10.1016/J.ACTBIO.2012.08.023

Jonoobi M et al (2015) Different preparation methods and properties of nanostructured cellulose from various natural resources and residues: a review. Cellulose 22(2):935–969. https://doi.org/10.1007/s10570-015-0551-0

Kang DY, Kim W-S, Heo IS, Park YH, Lee S (2010) Extraction of hyaluronic acid (HA) from rooster comb and characterization using flow field-flow fractionation (FlFFF) coupled with multiangle light scattering (MALS). J Sep Sci 33(22):3530–3536. https://doi.org/10.1002/jssc.201000478

Karami M, Shahraky MK, Ranjbar M, Tabandeh F, Morshedi D, Aminzade S (2021) Preparation, purification, and characterization of low-molecular-weight hyaluronic acid. Biotechnol Lett 43(1):133–142. https://doi.org/10.1007/s10529-020-03035-4

Kaya M, Seyyar O, Baran T, Erdoğan S, Kar M (2014) A physicochemical characterization of fully acetylated chitin structure isolated from two spider species: with new surface morphology. Int J Biol Macromol 65:553–558. https://doi.org/10.1016/j.ijbiomac.2014.02.010

Kaya M et al (2017) On chemistry of γ-chitin. Carbohydr Polym 176:177–186. https://doi.org/10.1016/j.carbpol.2017.08.076

Kaya M et al (2021) Sponge-derived natural bioactive glass microspheres with self-assembled surface channel arrays opening into a hollow core for bone tissue and controlled drug release applications. Chem Eng J 407:126667–126667. https://doi.org/10.1016/J.CEJ.2020.126667

Kivilcimdan Moral Ç, Yildiz M (2016) Alginate production from alternative carbon sources and use of polymer based adsorbent in heavy metal removal. Int J Polym Sci. https://doi.org/10.1155/2016/7109825

Koons GL, Diba M, Mikos AG (2020) Materials design for bone-tissue engineering. Nat Rev Mater 5(8):584–603. https://doi.org/10.1038/s41578-020-0204-2

Kringel DH, Dias ARG, Zavareze ER, Gandra EA (2020) Fruit wastes as promising sources of starch: extraction, properties, and applications. Starch/Staerke 72(3–4):403–417. https://doi.org/10.1002/star.201900200

Kulkarni SS, Patil SD, Chavan DG (2018) Extraction, purification and characterization of hyaluronic acid from Rooster comb. J Appl Nat Sci 10(1):313–315. https://doi.org/10.31018/jans.v10i1.1623

Kumar D, Shahid M (2020) Natural materials and products from insects: chemistry and applications. Springer. https://link.springer.com/book/10.1007/978-3-030-36610-0. Accessed 19 Sept 2022

Lalzawmliana V et al (2019) Marine organisms as a source of natural matrix for bone tissue engineering. Ceram Int 45(2 Part A):1469–1481. https://doi.org/10.1016/j.ceramint.2018.10.108

Lambe S, Ghogare P, Sonawane S, Shinde L, Prashant D (2021) Isolation, purification and characterization of hyaluronic acid: a concise review. J Pharmacogn Phytochem 10(3):500–506

Liu W, Zhang Y, Cui N, Wang T (2019) Extraction and characterization of pepsin-solubilized collagen from snakehead (Channa argus) skin: effects of hydrogen peroxide pretreatments and pepsin hydrolysis strategies. Process Biochem 76:194–202. https://doi.org/10.1016/j.procbio.2018.10.017

Liu Y et al (2020) Enhanced extraction of hydroxyapatite from bighead carp (Aristichthys nobilis) scales using deep eutectic solvent. J Food Sci 85(1):150–156. https://doi.org/10.1111/1750-3841.14971

Madalina Cursaru L et al (2022) Hydroxyapatite from natural sources for medical applications. Materials. https://doi.org/10.3390/ma15155091

Menezes MdLLR, Ribeiro HL, Abreu FdOMdS, Feitosa JPdA, Filho MdSMdS (2020) Optimization of the collagen extraction from Nile tilapia skin (Oreochromis niloticus) and its hydrogel with hyaluronic acid. Colloids Surf B Biointerfaces, 189:110852. https://doi.org/10.1016/j.colsurfb.2020.110852

Mohan K et al (2020) Recent insights into the extraction, characterization, and bioactivities of chitin and chitosan from insects. Trends Food Sci Technol 105:17–42. https://doi.org/10.1016/j.tifs.2020.08.016

Mohd Pu'ad NAS, Koshy P, Abdullah HZ, Idris MI, Lee TC (2019) Syntheses of hydroxyapatite from natural sources. Heliyon 5(5):e01588. https://doi.org/10.1016/J.HELIYON.2019.E01588

Motaung TE, Linganiso LZ (2018) Critical review on agrowaste cellulose applications for biopolymers, vol 22(2). Springer India. https://doi.org/10.1007/s12588-018-9219-6

Murado MA, Montemayor MI, Cabo ML, Vázquez JA, González MP (2012) Optimization of extraction and purification process of hyaluronic acid from fish eyeball. Food Bioprod Process 90(3):491–498. https://doi.org/10.1016/j.fbp.2011.11.002

Mustaffa R, Mohd Yusof MR, Abdullah Y (2015) A novelty of synthetic hydroxyapatite from cockle shell and characterization. Adv Mater Res 1087:429–433. https://doi.org/10.4028/WWW.SCIENTIFIC.NET/AMR.1087.429

Nam PV, Hoa NV, Trung TS (2019) Properties of hydroxyapatites prepared from different fish bones: a comparative study. Ceram Int 45(16):20141–20147. https://doi.org/10.1016/j.ceramint.2019.06.280

Nayak JP, Kumar S, Bera J (2010) Sol-gel synthesis of bioglass-ceramics using rice husk ash as a source for silica and its characterization. J Non-Cryst Solids 356(28–30):1447–1451. https://doi.org/10.1016/J.JNONCRYSOL.2010.04.041

Oladele IO, Agbabiaka OG, Adediran AA, Akinwekomi AD, Balogun AO (2019) Structural performance of poultry eggshell derived hydroxyapatite based high density polyethylene bio-composites. Heliyon 5(10):e02552. https://doi.org/10.1016/J.HELIYON.2019.E02552

Pajarinen J et al (2019) Mesenchymal stem cell-macrophage crosstalk and bone healing. Biomaterials 196:80–89. https://doi.org/10.1016/J.BIOMATERIALS.2017.12.025

Palakurthy S, Venugopal Reddy K, Patel S, Abdul Azeem P (2020) A cost effective SiO_2-CaO-Na_2O bio-glass derived from bio-waste resources for biomedical applications. Prog Biomater 9:239–248. https://doi.org/10.1007/s40204-020-00145-0

Park SS, Park M, Lee BT (2022) Autologous stromal vascular fraction-loaded hyaluronic acid/gelatin-biphasic calcium phosphate scaffold for bone tissue regeneration. Mater Sci Eng C 132:112533–112533. https://doi.org/10.1016/J.MSEC.2021.112533

Piccirillo C, Pullar RC, Costa E, Santos-silva A, Pintado MME, Castro PML (2015) Hydroxyapatite-based materials of marine origin: a bioactivity and sintering study. Mater Sci Eng C 51:309–315. https://doi.org/10.1016/j.msec.2015.03.020

Qinna NA et al (2015) Influence of molecular weight and degree of deacetylation of low molecular weight chitosan on the bioactivity of oral insulin preparations. Mar Drugs 13(4):1710. https://doi.org/10.3390/md13041710

Ratner B, Hoffman A, Schoen F, Lemons J (2012) Biomaterials science, 3rd edn. Elsevier. https://www.elsevier.com/books/biomaterials-science/ratner/978-0-08-087780-8

Rinaudo M (2006) Chitin and chitosan: properties and applications. Prog Polym Sci 31(7):603–632. https://doi.org/10.1016/j.progpolymsci.2006.06.001
Rizwan M, Hamdi M, Basirun WJ (2017) Bioglass® 45S5-based composites for bone tissue engineering and functional applications. J Biomed Mater Res A 105(11):3197–3223. https://doi.org/10.1002/JBM.A.36156
Rodriguez-Marquez CD, Arteaga-Marin S, Rivas-Sánchez A, Autrique-Hernández R, Castro-Muñoz R (2022) A review on current strategies for extraction and purification of hyaluronic acid. Int J Mol Sci 23(11):6038. https://doi.org/10.3390/ijms23116038
Rojas-bringas PM, De-la-torre GE, Torres FG (2021) Science of the total environment influence of the source of starch and plasticizers on the environmental burden of starch-Brazil nut fiber biocomposite production: a life cycle assessment approach. Sci Total Environ 769:144869. https://doi.org/10.1016/j.scitotenv.2020.144869
Rossatto A, Trocado dos Santos J, Zimmer Ferreira Arlindo M, Saraiva de Morais M, Denardi de Souza T, Saraiva Ogrodowski C (2022) Hyaluronic acid production and purification techniques: a review. Prep Biochem Biotechnol 53:1–11. https://doi.org/10.1080/10826068.2022.2042822
Samadi M, Khodabandeh Shahraky M, Tabandeh F, Aminzadeh S, Dina M (2022) Enhanced hyaluronic acid production in Streptococcus zooepidemicus by an optimized culture medium containing hyaluronidase inhibitor. Prep Biochem Biotechnol 52(4):413–423. https://doi.org/10.1080/10826068.2021.1955710
Sayari N, Sila A, Abdelmalek BE, Abdallah RB, Ellouz-Chaabouni S, Bougatef A, Balti R (2016) Chitin and chitosan from the Norway lobster by-products: antimicrobial and anti-proliferative activities. Int J Biol Macromol 87:163–171. https://www.sciencedirect.com/science/article/abs/pii/S014181301630188X. Accessed 19 Sept 2022
Sebastian J, Rouissi T, Brar SK, Hegde K, Verma M (2019) Microwave-assisted extraction of chitosan from Rhizopus oryzae NRRL 1526 biomass. Carbohydr Polym 219:431–440. https://doi.org/10.1016/j.carbpol.2019.05.047
Seddiqi H et al (2021) Cellulose and its derivatives: towards biomedical applications, vol 28(4). Springer, Netherlands. https://doi.org/10.1007/s10570-020-03674-w
Setyaningsih W, Karmila, Fathimah RN, Cahyanto MN (2021) Process optimization for ultrasound-assisted starch production from cassava (Manihot esculenta Crantz) using response surface methodology. Agronomy 11(117):13. https://doi.org/10.3390/agronomy11010117
Shevkani K, Singh N, Bajaj R, Kaur A (2016) Wheat starch production, structure, functionality and applications—a review. Int J Food Sci Technol 52:38–58. https://doi.org/10.1111/ijfs.13266
Shoulders MD, Raines RT (2009) Collagen structure and stability. Annu Rev Biochem 78:929–958. https://doi.org/10.1146/annurev.biochem.77.032207.120833
Sindhya A et al (2022) Synthesis and characterization of nanohydroxyapatite (nHAp) from Meretrix Meretrix Clam shells and its in-vitro studies for biomedical applications. Vacuum 204:111341–111341. https://doi.org/10.1016/J.VACUUM.2022.111341
Sionkowska A, Gadomska M, Musiał K, Piatek J (2020) Hyaluronic acid as a component of natural polymer blends for biomedical applications: a review. Molecules 25(18). https://doi.org/10.3390/molecules25184035
Srisantisaeng P, Garnjanagoonchorn W, Thanachasai S, Choothesa A (2013) Proteolytic activity from chicken intestine and pancreas: extraction, partial characterization and application for hyaluronic acid separation from chicken comb. J Sci Food Agric 93(13):3390–3394. https://doi.org/10.1002/jsfa.6217
Steffens D, Mathor MB, Santi BTS, Luco DP, Pranke P (2015) Development of a biomaterial associated with mesenchymal stem cells and keratinocytes for use as a skin substitute. Regen Med 10(8):975–987. https://doi.org/10.2217/RME.15.58
Tan F, Zhu Y, Ma Z, Al-Rubeai M (2020) Recent advances in the implant-based drug delivery in otorhinolaryngology. Acta Biomater 108:46–55. https://doi.org/10.1016/J.ACTBIO.2020.04.012
Teeri TT, Brumer H, Daniel G, Gatenholm P (2007) Biomimetic engineering of cellulose-based materials. Trends Biotechnol 25(7):299–306. https://doi.org/10.1016/j.tibtech.2007.05.002

Thavornyutikarn B, Chantarapanich N, Sitthiseripratip K, Thouas GA, Chen Q (2014) Bone tissue engineering scaffolding: computer-aided scaffolding techniques. Prog Biomater 3:61–102. https://doi.org/10.1007/s40204-014-0026-7

Tovar AF, Pereira R, Barreto P (2012) Purification and characterization of hyaluronic acid from chicken combs. Ciência Rural 42(9):1682–1687

Vate NK, Undeland I, Abdollahi M (2022) Resource efficient collagen extraction from common starfish with the aid of high shear mechanical homogenization and ultrasound. Food Chem 393: 133426. https://doi.org/10.1016/j.foodchem.2022.133426

Vidal AR et al (2020) Extraction and characterization of collagen from sheep slaughter by-products. Waste Manag 102:838–846. https://doi.org/10.1016/j.wasman.2019.12.004

Vinoth Kumar KC et al (2021) Spectral characterization of hydroxyapatite extracted from Black Sumatra and Fighting cock bone samples: a comparative analysis. Saudi J Biol Sci 28(1): 840–846. https://doi.org/10.1016/J.SJBS.2020.11.020

Wang Y et al (2013) Crystalline structure and thermal property characterization of chitin from Antarctic krill (Euphausia superba). Carbohydr Polym 92(1):90–97. https://doi.org/10.1016/j.carbpol.2012.09.084

Yang L, Tsang KY, Tang HC, Chan D, Cheah KSE (2014) Hypertrophic chondrocytes can become osteoblasts and osteocytes in endochondral bone formation. Proc Natl Acad Sci U S A 111(33): 12097–12102. https://doi.org/10.1073/PNAS.1302703111/SUPPL_FILE/PNAS.201302703SI.PDF

Younes I, Hajji S, Frachet V, Rinaudo M, Jellouli K, Nasri M (2014) Chitin extraction from shrimp shell using enzymatic treatment. Antitumor, antioxidant and antimicrobial activities of chitosan. Int J Biol Macromol 69:489–498. https://doi.org/10.1016/j.ijbiomac.2014.06.013

Yu Y, Liu X, Miao J, Leng K (2020) Chitin from Antarctic krill shell: eco-preparation, detection, and characterization. Int J Biol Macromol 164:4125–4137. https://doi.org/10.1016/j.ijbiomac.2020.08.244

Zeng HC et al (2017) MicroRNA miR-23a cluster promotes osteocyte differentiation by regulating TGF-β signalling in osteoblasts. Nat Commun 8:15000. https://doi.org/10.1038/NCOMMS15000

Zhang Q, Wang L, Liu S, Li Y (2021) Establishment of successive co-fermentation by Bacillus subtilis and Acetobacter pasteurianus for extracting chitin from shrimp shells. Carbohydr Polym 258:117720. https://doi.org/10.1016/j.carbpol.2021.117720

Zhu L-F, Li J-S, Mai J, Chang M-W (2019) Ultrasound-assisted synthesis of chitosan from fungal precursors for biomedical applications. Chem Eng J 357:498–507. https://doi.org/10.1016/j.cej.2018.09.183

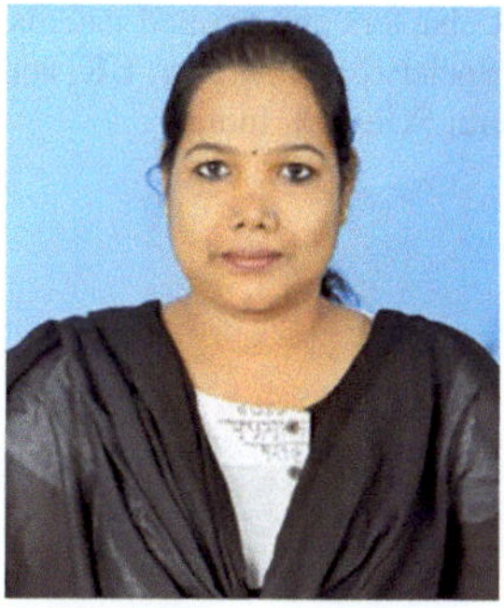

Howa Begam is a Research Associate (ICMR) of the Centre for Healthcare Science and Technology, Indian Institute of Engineering Science and Technology, Shibpur, Howrah, India. She completed her Ph.D. (Engineering) in 2016 and M.E. in Biomedical Engineering in 2009 from Jadavpur University. She received CSIR Senior Research Fellowship for the conduction of her Ph.D. thesis work. Her research interests are calcium phosphate-based scaffolds for bone tissue engineering, polymeric scaffold for drug delivery systems, etc. She published 6 SCI/Scopus indexed journals; 2 book chapters in *Lecture Notes in Bioengineering*, Springer, and 1 in *Springer Series in Biomaterials Science and Engineering*; and 15 conference proceedings.

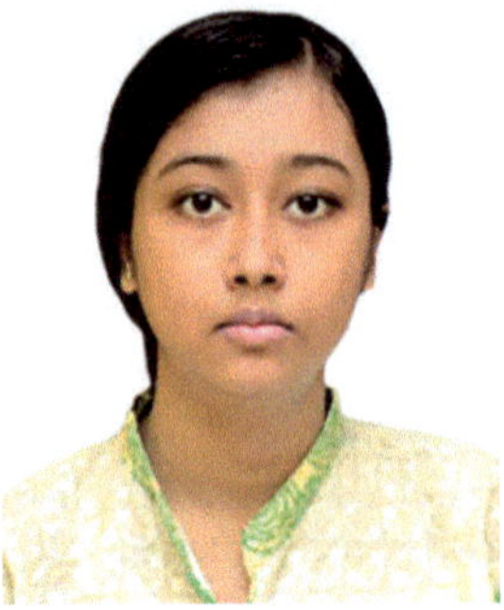

Sayantika Sanyal is pursuing a master's degree in biomedical engineering at the Indian Institute of Engineering Science and Technology, Shibpur. She graduated from biotechnology as her prime interest was researching in this field. Previously, she engaged in projects and workshops to build up basic knowledge about research works. Later on, she worked as part of a project associated with environmental biotechnology during the final year of her B.Tech degree. She has received an award from MAKAUT for that project for its easy applicability. One conference paper is also published in a reputed international journal. Now she is working on a project which is typically related to tissue engineering research; it deals with the regeneration methods for particular tissues as well as resolves some severely painful common diseases. Besides she is looking into the applications of novel strategies in research on tissue degeneration-related problems.

Ananya Barui is currently an Assistant Professor with the Centre for Healthcare Science and Technology, Indian Institute of Engineering Science and Technology Shibpur, Howrah, India. Her field of research interest is an early cancer diagnosis, stem cell mechanotransduction, biomedical imaging, and analysis. She authored research articles in reputed journals and book chapters. She is a regular reviewer of a good number of peer-reviewed international journals.

Samit Kumar Nandi is a Full Professor and former Head of the Veterinary Surgery and Radiology Department at West Bengal University of Animal and Fishery Sciences (WBUAFS), Kolkata, India. He worked as Adjunct Faculty at the School of Mechanical and Materials Engineering, Washington State University, Pullman, USA. He received several national awards from his country. He has contributed more than 172 scientific research papers, 3 books, and 25 international book chapters and granted patents. Dr. Nandi is a Fellow of the Royal Society of Chemistry, UK, and the National Academy of Agricultural Sciences, India.

Hydroxyapatite is a Next-Generation Theranostic Probe for Tissue Engineering and Biomedical Application

4

Sudip Mondal, Sumin Park, Jaeyeop Choi, and Junghwan Oh

Abstract

Hydroxyapatite (HAp), a synthetic analog of biogenic apatite, has several physicochemical characteristics that make it a desirable choice for disease diagnostics, therapy, and enhancement of biological tissues. In this article, we discuss remarkable recent research on HAp that could serve as the foundation for several novel medical applications. The review's content is organized into various HAp synthesis routes, HAp structure, and HAp-based medical application modes, such as bioimaging, controlled medication administration, gene treatments, and tissue engineering. This discussion highlights several benefits of HAp over the existing biomaterials, such as facile synthesis with tailored morphologies, biocompatibility, bioactivity, functionalization and adaptive surface modification, drug conjugation and delivery applications, etc. The novelty of this chapter is the investigation of particulate HAp's safety as a component of parenterally administered drugs, and their potential biomedical applications are covered.

S. Mondal · J. Choi
Smart Gym-Based Translational Research Center for Active Senior's Healthcare, Pukyong National University, Busan, Republic of Korea

S. Park
Industry 4.0 Convergence Bionics Engineering, Department of Biomedical Engineering, Pukyong National University, Busan, Republic of Korea

J. Oh (✉)
Smart Gym-Based Translational Research Center for Active Senior's Healthcare, Pukyong National University, Busan, Republic of Korea

Industry 4.0 Convergence Bionics Engineering, Department of Biomedical Engineering, Pukyong National University, Busan, Republic of Korea

Ohlabs Corp., Busan, Republic of Korea
e-mail: jungoh@pknu.ac.kr

F. A. Sheikh et al. (eds.), *Interaction of Nanomaterials With Living Cells*,
https://doi.org/10.1007/978-981-99-2119-5_4

Keywords

Hydroxyapatite · Tissue engineering · Biomaterials · Drug delivery · Theranostic · Bioimaging · Contrast agent

4.1 Introduction

The main component of mammalian hard tissues is hydroxyapatite (HAp), $Ca_{10}(PO_4)_6(OH)_2$, a naturally occurring biological nanomaterial (Mondal et al. 2018a). HAp could be found in enamel with a diameter of 20–40 nm or in dentin and bone as 5–20-nm-thick flakes. Additionally, it permits the integration of various cations and anions in its arrangement, which facilitates the functionalization of the system (Kim et al. 2018a; Park et al. 2022a). HAp is a wonderful and distinctive option for multiple applications in the field of nanomedicine since it differs from other ceramic nanoparticles while sharing several similar characteristics with them (Uskoković 2015). The calcium phosphate (CaP) family of biomaterials includes HAp, which is the most stable material. Dentists, orthopedic surgeons, and bone tissue engineers were the first to notice the exceptional biocompatibility and bioactivity of HAp nanoparticles. Since then, HAp's range of possible uses has widened beyond these particular industries. Today, it is understood that HAp possesses a variety of physicochemical traits that make it a desirable candidate for medical applications such as diagnostics, therapy, and tissue augmentation. According to the numerous publications on HAp's uses that have been published, the main use for HAp is in the field of tissue engineering, namely, for the regeneration and repair of bone deformities. However, the primary mechanical issues associated with the use of HAp in hard tissue engineering include inadequate fracture toughness, failure owing to fatigue, and brittleness. Like a natural bone, an ideal bone substitute should progressively resorb, preferably at a pace that corresponds to the rate of new bone development. However, the resorption rate of compact HAp blocks might be unreasonably slow, lasting for years or even decades (Proussaefs et al. 2002). The most common strategy for overcoming these obstacles is combining HAp with other substances that have comparable qualities and produce synergistic effects (Ghiasi et al. 2019). The superior osteoconductivity and lack of immunological rejection make HAp appropriate for implants in addition to applications in tissue engineering. Additionally, HAp is used as a drug delivery vehicle to carry medications to bones, where they can benefit from their natural osteoconductivity, as well as to other organs and tissues, typically by attaching the appropriate targeting tags to the HAp. Due to the pH-sensitive dissolving characteristics of HA, medications can be released intelligently in acidic environments, such as those near tumors and infection sites (Lelli et al. 2016). The ability to regulate particle size and the binding of functional groups by physisorption or surface phosphate substitution to allow more accurate cell/tissue targeting is another benefit of HAp. The very flexible crystal structure of HAp also makes it possible to include ions with a variety of atomic radii with ease (Uskoković 2020). Doped HAp for bioimaging applications has therefore been developed as a result of the coprecipitation of HAp with luminous or other photoactive ions (Zeng et al. 2017; Ignjatović et al. 2019). For instance, the presence of lanthanide ions can impart the matrix with a photoluminescence property, turning it into a fluorescent probe. This led to the development of the HAp (Eu^{3+}, Gd^{3+}, Er^{3+}, etc.) system and the suggestion that it might be suitable for use as a stable biological probe in imaging research (Mondal et al. 2020a, b). The ability of HAp to adsorb both hydrophilic and hydrophobic molecules is crucial for its use as a drug carrier, given the variety of chemical and pharmacological compositions that might be found (Kim et al. 2018b; Mondal et al. 2017a, 2019a, b). Due to its resemblance to the inorganic phase of bones, it also stands out as a biomaterial that is frequently employed in bone treatments (Mondal et al. 2016, 2018b, 2020c; Mondal and Pal 2019). Tumor excision in conditions like bone cancer leaves flaws that need to be filled up with grafts or prosthetics (Mondal et al. 2017b, c, 2019c; Park et al. 2022b) (Fig. 4.1).

4.2 Structure of Hydroxyapatite

In the hexagonal and monoclinic systems with the space groups P63/m and P21/b, respectively, HAp can crystallize. While the lattice parameters of HAp's monoclinic structure are $a = b/2 = 9.421$, $c = 6.881$, and $= 120°$, the lattice parameters of HAp in hexagonal systems are $a = b = 9.432$, $c = 6.881$, and $= 120°$. In the hexagonal structure of HAp, neighboring hydroxyl (OH) groups face one another in the opposite direction. In contrast, in the monoclinic structure, all of the OH groups in a given column face the same way (Ikoma et al. 1999). Posner et al. first refined the crystal system of HAp with the hexagonal system, suggesting a distribution of 10 Ca^{2+}, 6 PO_4^{3-}, and 2 OH^- in each unit cell of HAp (Posner et al. 1958).

4.3 Synthesis Routes

HAp NPs with a specific microstructure have so far been prepared using a variety of chemical techniques. All of the synthesis methods can be divided into five categories: (1) dry methods (solid-state and mechanochemical routes), (2) wet methods (chemical precipitation, hydrolysis, sol-gel, hydrothermal, emulsion, and sonochemical routes), (3) high-temperature processes (combustion and pyrolysis methods), (4) methods based on biogenic sources (biogenic wastes), and (5) combination techniques. The aforementioned techniques can be used to create HAp nanostructures of various sizes and shapes, including spheres, rods, needles, flakes, flowers, mesoporous spheres, bowknots, dumbbells, etc. (Mondal et al. 2018a). It should be noted that HAp NPs' shape, crystallinity, and size are some of the most important variables impacting their anticancer activities. The aforementioned characteristics had a significant impact on the anionic and cationic sites of HAp that contained hydroxyl, amino, and carboxyl groups, as well as calcium and phosphorus ions. The interaction of HAp with a biomolecule is greatly influenced

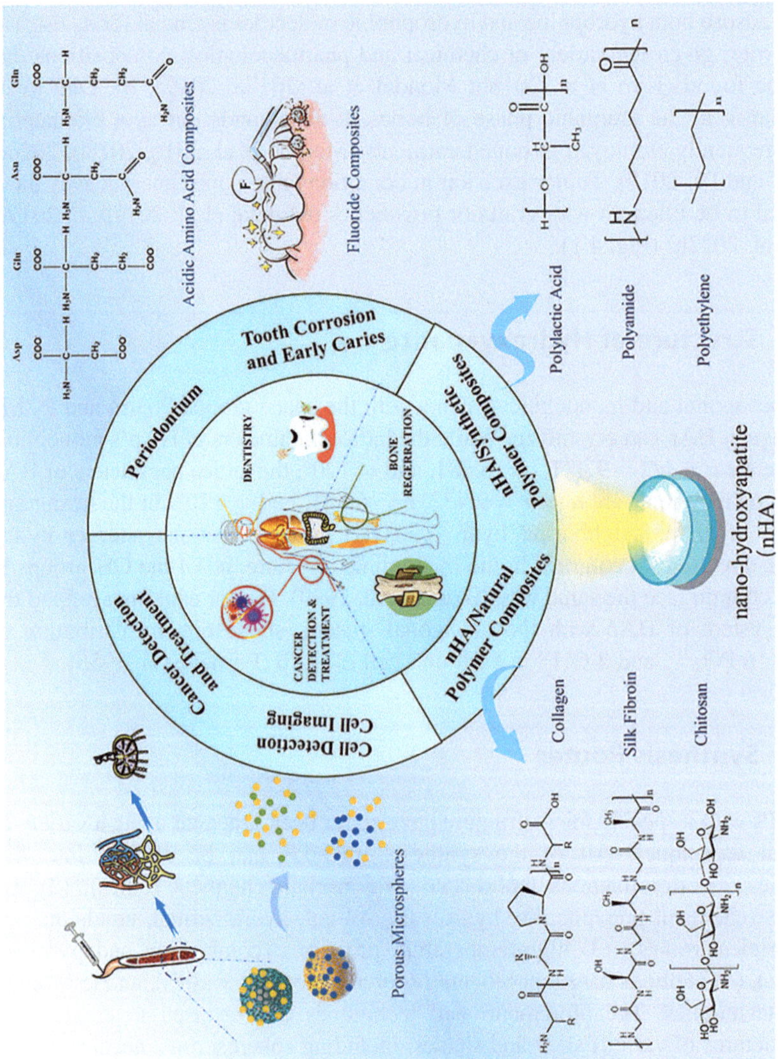

Fig. 4.1 Novel applications of hydroxyapatite in biomedical engineering, such as bone healing, cancer therapy, and dentistry. (Represented with permission from Du et al. 2021)

by the positive and negative surface charges of HAp, which are dependent on its physicochemical characteristics. HAp NPs can be used for cancer cell imaging in addition to structurally modified HAp nanostructures through a variety of well-established methods, such as (1) combining them with organic fluorophores, (2) doping them with different lanthanide ions, and (3) creating composites with other inorganic nanomaterials (such as carbon-quantum and chalcogenide-quantum dots) (Machado et al. 2019).

4.4 HAp in Tissue Engineering and as an Implant

As new technologies are created to develop efficient treatments for degenerative diseases that affect many types of tissues, tissue engineering is expanding. Recently, there has been a discernible increase in the need for bioactive, biodegradable, biocompatible, and multifunctional materials. To consolidate ceramic particles to join together and lessen interparticle gaps, a high-temperature stage (sintering) is often necessary for making HAp products, such as porous or nonporous blocks, coatings, etc. The appealing characteristics of nano-sized HAp, such as topography, geometry, high specific surface area, etc., would be irreparably lost after this heating phase, though. The integration of ceramic NPs within a matrix is a good choice for producing composite biomaterials. Consequently, if employing nano-sized HAp is the goal, scaffolds should be prepared, avoiding the final sintering stage. Mondal et al. reported a composite scaffold of HAp, bioglass, and alumina with enhanced mechanical stability. The study showed enhanced biological activity when cells were seeded on a scaffold surface (Mondal et al. 2018b) (Fig. 4.2).

The inhibition of cancer cell growth and migration by HAp NPs have been shown in several in vitro and in vivo investigations (Han et al. 2014). The shape, size, and crystallinity of Hap NPs are among the physicochemical characteristics that are acknowledged as being important determinants of their anticancer effects (Yuan et al. 2010). The simplest method for incorporating extra anticancer capabilities into a polymer that is otherwise biologically inert is to add HAp NPs to the biocompatible polymer. The first study in this field was published by Pathi et al. (2011), who introduced hydrothermally produced HAp NPs to CO_2-foamed poly(lactide-co-glycolide) (PLGA) scaffolds that were later implanted with metastatic breast cancer cells. The results showed that smaller, less crystalline HAp particles promoted greater adsorption of adhesive serum proteins and enhanced breast tumor cell adhesion and growth, in contrast to larger, more crystalline NPs, which, on the other hand, stimulated a higher expression of the osteolytic factor interleukin-8 (IL-8). Therefore, it was proposed that bone development, which supports improved cell colonization and proliferation, depends critically on modulating the nanoscale characteristics of the bone mineral found in the microenvironment.

Mondal et al. reported 3D printed PLA scaffold modified with HAp nanomaterials with enhanced cell attachment properties (Figs. 4.3 and 4.4) (Mondal et al. 2020c). More recently, a research group from Poland used 3D printed poly(L-lactic acid) (PLLA) scaffolds with HAp NPs doped with europium (III) ions (Eu^{3+})

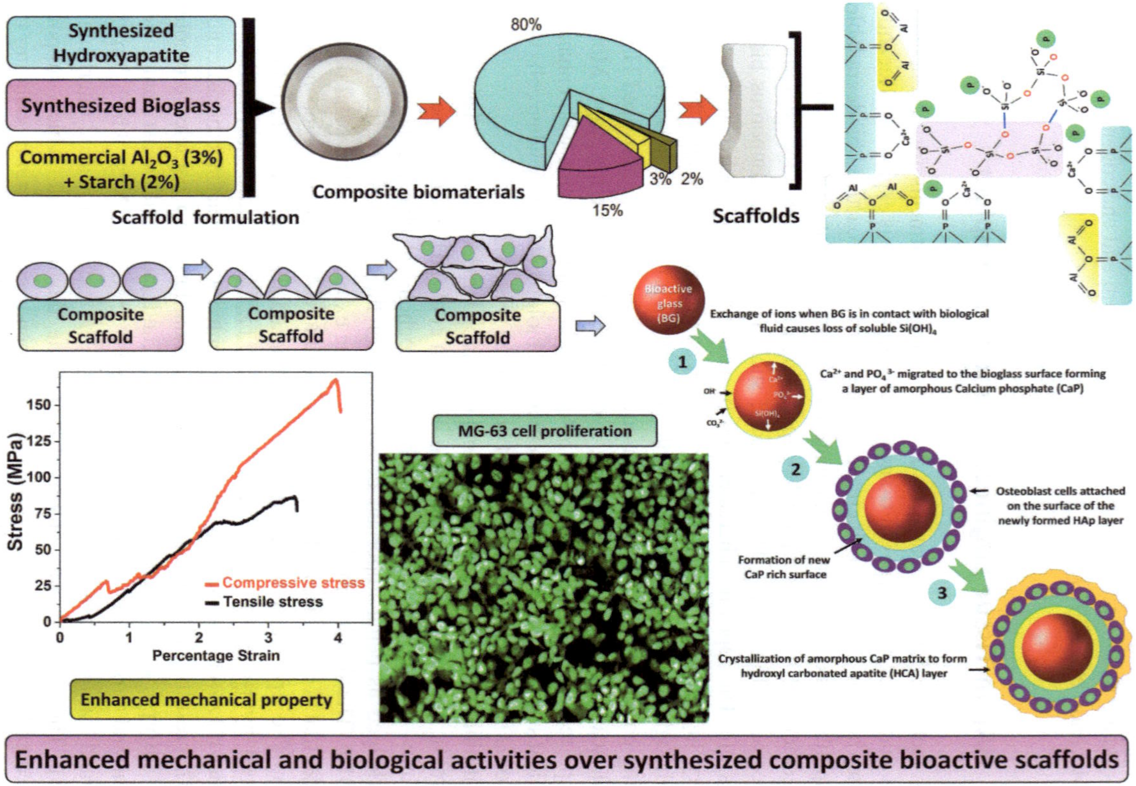

Fig. 4.2 For tissue engineering applications, composite scaffold fabrication with improved mechanical and biological performance. (Represented with permission from Mondal et al. 2018b)

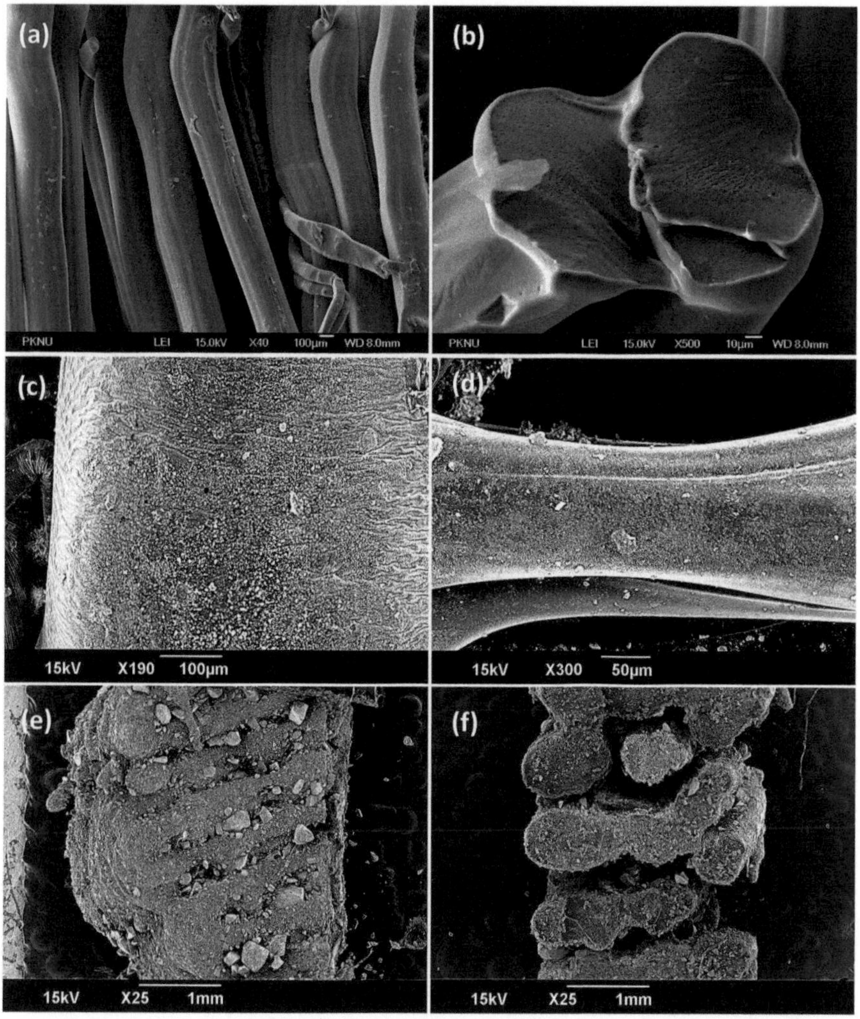

Fig. 4.3 SEM images of PLA scaffold modified with HAp nanoparticles. (Represented with permission from Mondal et al. 2020c)

to conduct several in vitro investigations. In this study, the hydrothermal technique-based microwave stimulation was used to prepare the Eu-doped HAp NPs. According to a study (Sikora et al. 2019), these composite scaffolds led an osteosarcoma cell line to go into apoptosis while not affecting the viability of unaltered adipose-derived human mesenchymal stromal cells. PLLA/HAp NP composite scaffolds boosted the viability as well as the osteogenic and chondrogenic differentiation capacity of adipose-derived human mesenchymal stromal cells, according to a second study from the same group (Marycz et al. 2020). This was correlated to increased protein and mRNA expression of the osteogenic and chondrogenic

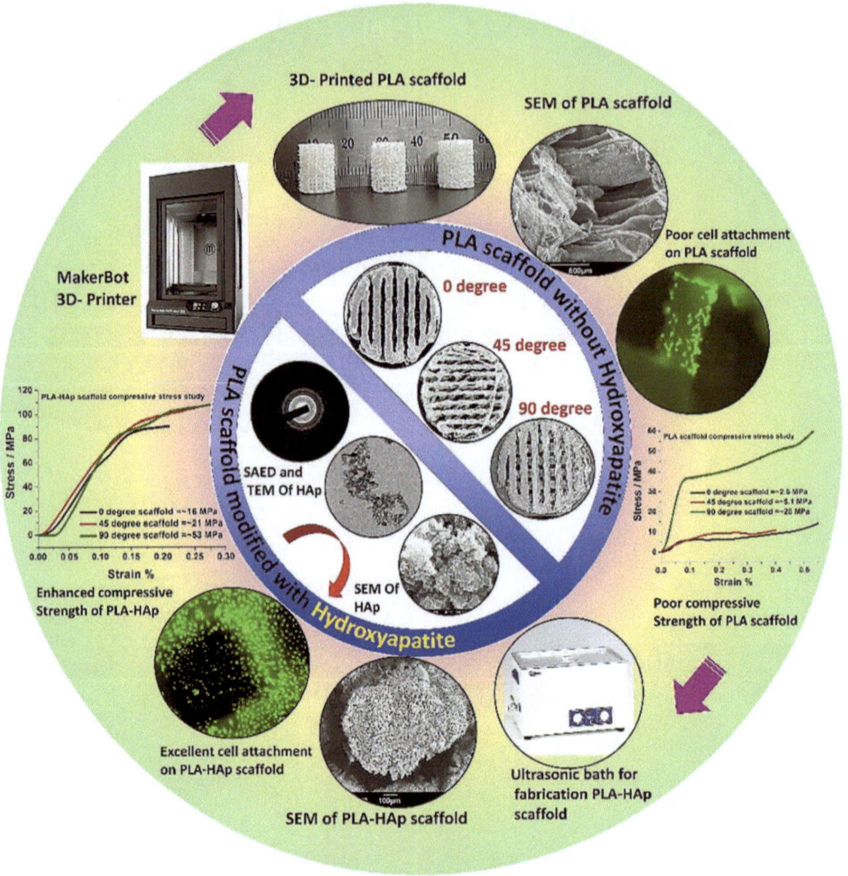

Fig. 4.4 Hydroxyapatite nanoparticle surface modification of 3D printed PLA scaffolds for improved bone tissue engineering applications. (Represented with permission from Mondal et al. 2020c)

markers. The positive impact of these biomaterials on the osteogenic and chondrogenic processes was further confirmed by the increased expression of bone morphogenic protein 2 (BMP-2) and BMP-7, as well as their receptor (i.e., BMP receptor type 1B (BMPR-1B)), suggesting a potential use in those applications where osseous regeneration is required following the removal of bone cancer tumor. Additionally, scaffolds carrying HAp NP were employed to simulate non-osseous tumors like neuroblastoma. To do this, Gallagher et al. employed neuroblastoma cell lines and collagen-based scaffolds supplemented with either HAp NPs or glycosaminoglycans, which are naturally found in bone and bone marrow, the most common metastatic sites for neuroblastoma. In contrast to 2D cultures, these composite scaffolds allowed neuroblastoma cells to attach, proliferate, migrate, and form cell clusters. This resulted in a cell response that was more like that of the in vivo environment. Additionally, compared to traditional 2D cell culture, this scaffold-based culture technique maintained higher cell densities (Gallagher et al. 2021). It is also important to note that studies using tumor model systems have used polymeric scaffolds that contain HAp NPs. Tornin et al. showed the effects of plasma-based therapies on MG-63 cells in a 3D tissue-engineered osteosarcoma model based on a highly porous HAp NPs/collagen scaffold that allowed cancer cell proliferation and the acquisition of a similar gene expression profile as compared to the original tumors. Cancer cells were less sensitive to treatment with cold plasma-activated Ringer's solution when grown in a 3D scaffold-based model as opposed to 2D cultures because the scaffold's three dimensions induced the expression of several genes that protect against reactive oxygen and nitrogen species in MG-63 cells, promoting cell proliferation and adaptation to oxidative stress (Tornín et al. 2021). According to Nakayama et al., the self-organization of HAp with poly(acrylic acid) (PAA) results in the production of liquid-crystalline hybrid nanorods with desired properties as a drug release platform for photodynamic therapy of cancer (methylene blue was used as a model drug). The spin-coated 2D-oriented scaffolds were also shown to elicit cellular alignment and elongation in the direction of the hybrid nanorods (Nakayama et al. 2019).

In addition to polymer/nano-sized HAp and graphene/nano-sized HAp scaffolds, Zhang et al. also reported the potential of merging metal with nanostructured HAp. They applied a layer of wet-fabricated HAp nanorods on titanium cylindrical scaffolds manufactured by selective laser sintering (porosity 65%, pore size 500 m) (length 46.6 nm, width 13.3 nm). The titanium scaffolds were acid-alkali treated before coating to generate a microporous TiO_2 network on the surface, improving bonding with nano-HAp. The porous metal was next covered three times with a slurry composed of HAp nanorods, methylcellulose, and H_2O_2 before being allowed to dry. The scaffolds were then heated for 1 h at 300 °C. The authors described in vitro and in vivo experiments that demonstrated the nHAp particles' dual action as a bone-regenerating substance and an anticancer agent. In tumor-bearing rabbits treated with HAp nanorods, tumor growth was inhibited, metastases were prevented, and the survival rate was improved. In particular, the nanomaterial induced an anticancer immune response and promoted mitochondrial-dependent tumor death in vivo. Histological studies conducted over a month revealed that the use of HAp nanorods prevented metastasis to the lung effectively but not when micro-sized HAp or nano-sized TiO_2 coatings were used as controls. Abnormalities were not observed in the liver, heart, kidney, and spleen, further demonstrating the biosafety of the nanorods in use. These results confirm and extend the findings of the research by Pathi et al. on the effect of HAp NPs on the growth cycles of cancer cells (Pathi et al. 2011).

4.5 HAp Nanostructures for the Delivery of Anticancer Drugs

In the past, NPs like HAp nanostructures were suggested to concurrently target some of cancer characteristics and transport and release several cytotoxic chemicals safely and effectively (VanDyke et al. 2016). The effectiveness of drug-loaded multimodal HAp NPs as effective nanomaterials for tumor metastasis-resisting therapy (TMRT) and tumor metastasis targeting therapy (TMTT) has been demonstrated in earlier experiments (Xiong et al. 2018). However, there are very few studies in the literature that concentrate on employing nano-sized HAp particles, either in the pristine or in the functionalized and drug-loaded kinds, to target and image the various indicators of cancer. For instance, HAp NPs can reduce the FAK/PI3K/Akt cell signaling pathway both in vitro and in vivo, which suppresses cancer cell proliferation, migration, and invasion (Wang et al. 2020). Through using lysosomal- and mitochondrial-dependent pathways, such as caspase-3, HAp NPs can also induce oxidative stress-induced apoptosis in cancer cells (Jin et al. 2017).

For use as a drug carrier in cancer theranostic applications, nanostructured HAp provides several advantages (Saber-Samandari et al. 2016). These benefits can be summed up as follows:

1. HAp could be formed with substantially similar characteristics to the main components of the targeted tissues, particularly bone, including chemistry, crystalline structure, and size.
2. HAp alone does not expand or change porosity, preventing the burst release of medicines, and is reasonably stable with fluctuations in the solution/environment (e.g., pH, temperature).
3. Both positively and negatively charged compounds can surface-functionalize Hap.
4. By doping HAp with various elements, NPs with desirable electrical, mechanical, magnetic, and optical properties can be prepared (Fig. 4.5).

To treat tumors connected to both hard and soft tissues, a variety of natural and synthetic anticancer drugs have been loaded onto and delivered using HAp NPs (Maia et al. 2018). Mesoporous HAp NPs are a promising class of drug delivery systems that can be used to transport a variety of anticancer cargos, including chemicals, small compounds, and generic medications. Due to their outstanding drug adsorption, storage, and release capabilities, these systems hold promise for cancer therapy methods (Meshkini and Oveisi 2017).

In this study, ribose is used as a highly biocompatible material to cross-link hybrid mineralized composites for the first time. The purpose of this study was to explore the viability of novel, bone-like scaffolds made from type I collagen matrix mineralized with magnesium-doped hydroxyapatite nanophase (MgHA/Coll) through ribose glycation in a pH-driven manner, inspired by biological processes. According to the literature, a significant portion of the experimental research use pH-sensitive nanostructures (Fig. 4.6). A suitable material for improved anticancer medication delivery to cancer cells was also mentioned as being nanostructured

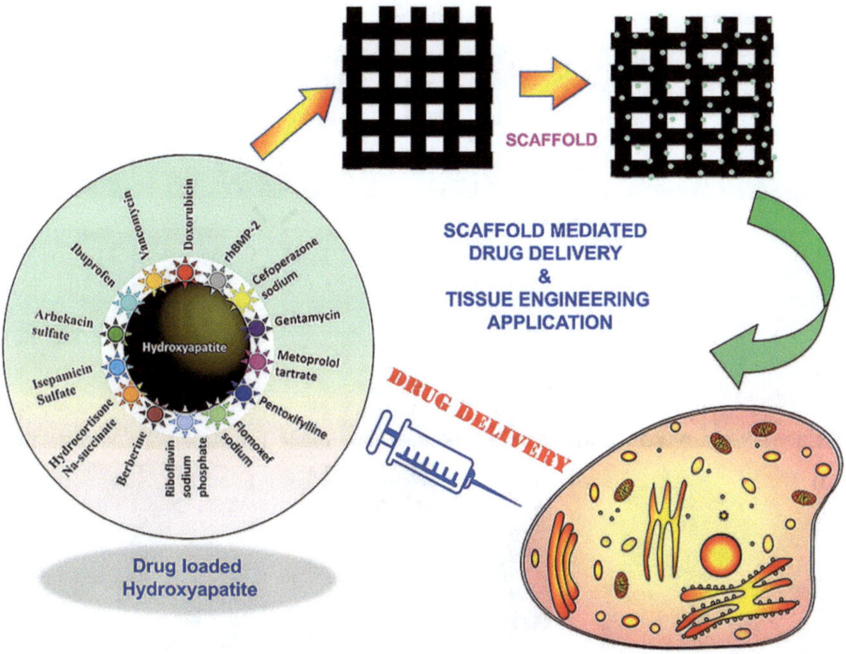

Fig. 4.5 Schematic representation of hydroxyapatite-based scaffold and drug delivery agents. (Represented with permission from Mondal et al. 2017b)

hollow HAp (Mondal et al. 2019b). A lot more focus is now being paid to larger-scale hollow micro-sized HAp structures (like microspheres), even if using nano-sized hollow HAp structures for the delivery of anticancer drugs appears promising in the early stages (Qiao et al. 2017). The outcomes showed that DOX (9.1%) and siRNA (2.0%), as well as the efficient transport of anticancer drugs into drug-resistant breast cancer cells (MCF-7/ADR cells), were loaded appropriately. In addition, after 48 h in cancer cells, the drug-loaded samples started to degrade. A biopolymer coating on HAp can enhance the bioactivity and regulate the release of anticancer medications (Padmanabhan et al. 2020). According to Li et al., the hydrothermal method followed by freeze-drying produced nano-sized HAp (20 wt. %)/GO composite scaffolds that shared the same dual functionality (photothermal anticancer effect and osteogenesis) (Li et al. 2018). Twenty minutes of in vitro exposure to 808 nm NIR radiation on osteosarcoma cells (MG-63) resulted in the death of all but 8% of the cells. Tumor xenografts implanted with the scaffolds in mice reached 60 °C after 4 min of radiation exposure and ceased growing or even shrunk in size following photothermal therapy. In addition, micro-tomographic and histological evaluations supported the notion that nano-sized HAp/GO scaffolds encouraged bone regeneration in rat cranial defects: At 8 weeks, new tissue had grown by over 65% in the cranial defect area for the scaffold-implanted group, whereas only 20% regeneration had been seen for the control (empty defect). The

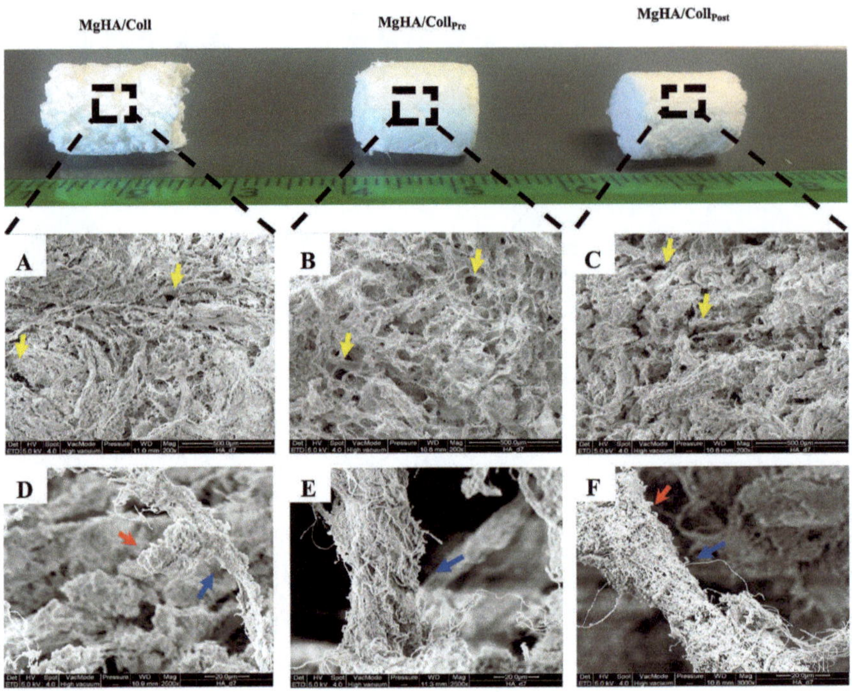

Fig. 4.6 Using the matching magnifications, SEM images of the scaffold are shown in (**a**, **d**) MgHA/Coll, (**b**, **e**) MgHA/CollPre, and (**c**, **f**) MgHA/CollPost. Red arrows denote the binding mineral phase on fibrils, yellow arrows point to open surface pores on the fibrous matrix, and blue arrows denote a small or large bundle of self-assembled collagen fibers with visible mineral deposition. Scale bars: 500 mm for (**a**–**c**) and 20 mm for (**d**–**f**). (Reprinted with permission from Krishnakumar et al. 2017)

scaffold-implanted group's bone mineral density also reached about 285 mg/cm^3, indicating fresh bone mineral deposition, compared to the control at 96 mg/cm^3. The properties of HAp-NPs, which specifically affect the drug/gene/adsorption protein's capacity and solubility in physiological fluids, subsequently affect how HAp interacts with cancer cells. Additionally, it was claimed that HAp NPs encourage p53 production and its downstream genes, which causes apoptosis (programmed cell death) in tumor cells (Sun and Ding 2009). Tang et al. found that HAp reduces cell proliferation and promotes apoptosis in several cancer cells via activating caspase-3 and caspase-9 (but not caspase-8) (gastric cancer cells [MGC80-3], cervical adenocarcinoma epithelial cells [HeLa], and hepatoma cells [HepG2]) (Tang et al. 2014). Other experimental results indicate that HAp NPs are formed by and contribute to the intracellular accumulation of reactive oxygen species (ROS), which damage the DNA of cancer cells (Xu et al. 2012). Due to their high ribosome adsorption ability, HAp NPs can be internalized in large quantities by endocytosis in cancer cells in the endoplasmic reticulum, inhibiting protein synthesis by reducing the binding of mRNA to the ribosomes in cells and stopping the cell cycle in the G0/G1 phase (Han et al. 2014).

4.6 HAp for Advanced (PTT/PDT/Hyperthermia) Cancer Therapies

Despite numerous treatments, cancer remains the second most significant cause of mortality on Earth. The next generation of anticancer agents will focus on supporting healthy cell proliferation while causing malignant cell death. As a result, HAp NPs doped with a variety of therapeutic ions have been created and used for theranostic applications for the treatment and diagnostics of cancer (Šupová 2015). Targeted magnetic hyperthermia for cancer treatment has attracted significant interest in HAp NPs as well (Abdel-Hamid et al. 2017) (Fig. 4.7). A well-known method for triggering cancer cell death without harming healthy cells is hyperthermia, which relies on the modest rise of temperature to 40–43 °C upon application of an external magnetic field. Hyperthermia also improves the effects of radiation therapy and chemotherapy (Sedighi et al. 2022). Furthermore, by providing an alternating magnetic field to the tumor, cancer-targeting magnetic HAp NPs can gather there. Additionally, the effectiveness of magnetic HAp NPs for synergistic chemo-hyperthermia therapy has been demonstrated in the past. At this point, chemotherapeutic drug-loaded magnetic nanocomposites of cobalt ferrite/HAp were created using the microwave-assisted wet precipitation method (5-fluorouracil, FU) (Sangeetha et al. 2019). With a magnetic saturation value of roughly 2.5–8.2 emu/g, the produced nanocomposites displayed ferromagnetic properties. After being exposed to an alternating magnetic field, they were able to release the encapsulated FU and raise the temperature (producing a hyperthermic effect) quickly (43 °C in 4.5 min). The samples also showed a suitable ability to inhibit the growth of osteosarcoma cells (MG-63), demonstrating their suitability for synergistic chemo-hyperthermia therapy.

One of the most recent techniques for treating tumors using external radiation sources is photodynamic therapy (PDT), which is also known as photothermal therapy (PTT) (Santha Moorthy et al. 2018). The NPs are illuminated with a suitable near-infrared (NIR) wavelength in PTT or methods incorporating NP-mediated hyperthermia (the electromagnetic radiations are NIR with a wavelength within 780–2526 nm) (Mondal et al. 2022). Under the influence of NIR, the treated NPs' conduction band electrons experience synchronized oscillations. It kills cancer cells by destroying cell membranes, tumoral DNA denaturation, and angiogenesis inhibiting processes and converts light into heat (45–50 °C) (Mondal et al. 2022; Doan et al. 2021; Vo et al. 2021; Phan et al. 2019). When a photosensitizer (PS) is activated by specific wavelength irradiation in photodynamic treatment (PDT), reactive oxygen species (ROS) and free radicals are created (620–690 nm). When exposed PS interacts with oxygen ($^3O_2$) and a specific light wavelength to produce lethal singlet oxygen ($^1O_2$ and O_2), the known pathways for cancer cell death in PDT are direct (necrosis and apoptosis) and indirect (microvascular damage and antitumor immune responses) (Pinto and Pocard 2018).

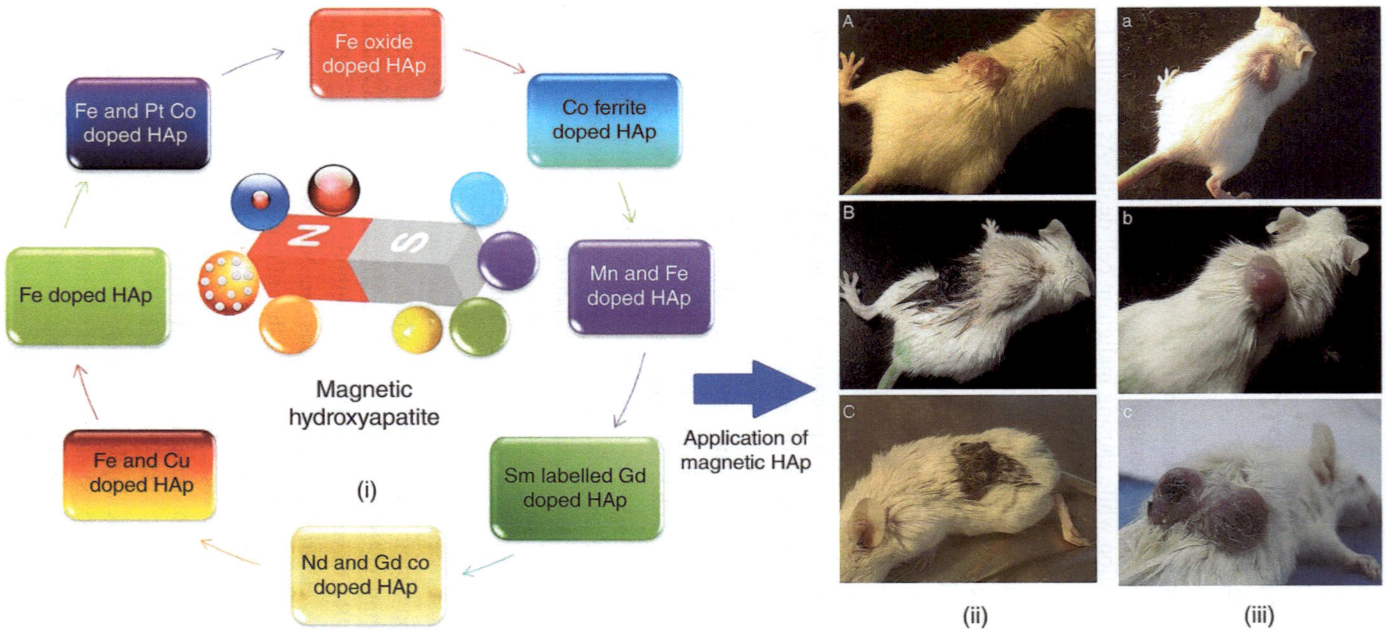

Fig. 4.7 Magnetic hydroxyapatite (mHAp) nanoparticles are based on several dopants. (ii) The tumor's diagnostic photos (mHAp with magnetic field). Tumors on days 1 (A), 5 (B), and 14 (C). The tumor's clinical pictures (iii) (mHAp without magnetic field). The tumor was on days 1 (a), 5, and 14 (b and c). (Reprinted with permission from Mondal et al. 2018a; Hou et al. 2009)

According to Zhang et al., a biocompatible system incorporating HAp and graphene could efficiently convert NIR into heat to hasten bone regeneration following photothermal therapy of malignant bone lesions. Their findings demonstrated that following three cycles of treatment at low photothermal temperatures (40–43 ° C), the growth of MC3T3-E1 cells was dramatically boosted. The findings showed that for pure HAp and HAp scaffolds, including 1 wt% graphene, the observed temperature after exposing Hap NPs to NIR was 31.6 and 45.2 °C, respectively. The technique being used appears promising for NP-mediated hyperthermia (PTT) (Zhang and Ma 2021). There are several obstacles that prevent anticancer chemicals from being loaded and desirably delivered to tumor cells, including ineffective targeting and a lack of drug solubility. To overcome these limitations, several novel techniques have been used to create HAp nanostructures as smart theranostic platforms that can target cancer cells without harming healthy cells. One of the most alluring methods for supporting sustained anticancer medication release is the surface functionalization of HAp nanostructures (Verma et al. 2016). To successfully load and distribute curcumin to MCF-7 cells, for instance, HAp NPs functionalized with several carboxylic acids (lactic acid, tartaric acid, and citric acid) were used (Lee et al. 2019). Based on the electrostatic interactions of opposite charges between the medication and the bioceramic particles, carboxylic modifiers may increase the binding affinity between curcumin molecules and the HAp surface. The results showed that, in comparison with unmodified samples, surface-functionalized HAp NPs exhibited stronger antiproliferative and apoptotic actions against cancer cells. For instance, several types of cancer cells overexpress folate receptors (such as FR), making it viable to target tumors with drug delivery systems conjugated with folic acid, a ligand with a high affinity for folate receptors (Cheung et al. 2016).

4.7 Nano-Sized HAp in Imaging

Currently, a variety of biomedical imaging approaches are being used to find cancer in all stages of development. Imaging techniques can be used to identify a variety of features of cancer, such as morphological, structural, metabolic, and functional data. These methods are divided into those that use (1) nonionizing electromagnetic radiation, including magnetic resonance imaging (MRI), electrical impedance spectroscopy, and near-infrared spectroscopy, and (2) ionizing radiation (such as gamma rays, X-rays or UV light, CT scans, and PET scans) (Fass 2008). To visualize tumors using particular imaging modalities, such as MRI, the choice of appropriate contrast agents is essential. Over the past few decades, contrast agents based on nanotechnology have been created and used for cancer imaging. Different kinds of nano-sized HAp have been applied in this way for the molecular imaging of solid tumors. For optical, magnetic resonance (MR), and multimodal imaging of malignancies, luminescent, magnetic, and luminomagnetic HAp NPs exhibit the necessary properties (Kataoka et al. 2019). The effect of the calcination temperature and the amount of Eu^{3+} doping were previously evaluated about the luminescence characteristics and

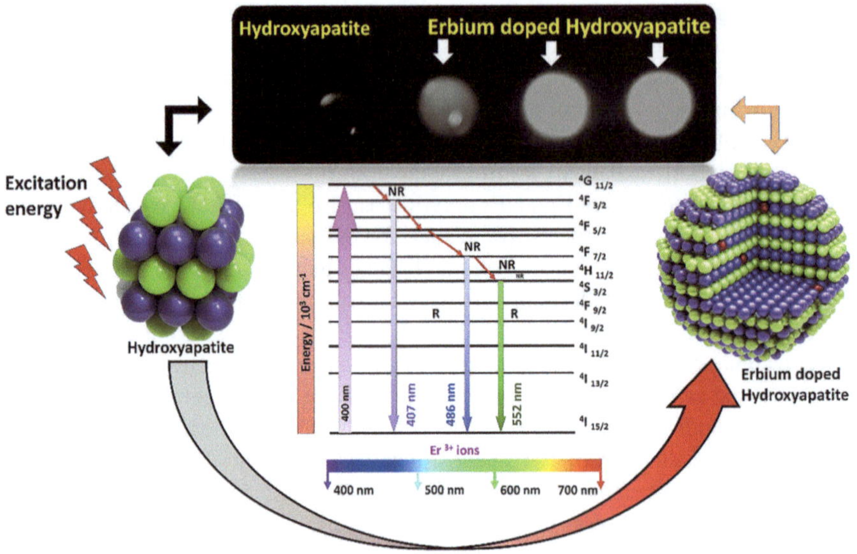

Fig. 4.8 Schematic representation of erbium-doped hydroxyapatite as a bioactive luminescent agent. (Represented with permission from Mondal et al. 2020d)

phase composition, crystal size, and crystallinity of Eu^{3+}-doped HAp NPs (Han et al. 2013). To manufacture nanocrystalline Eu-doped HAp (20–40 nm in diameter) with high luminescence, the thermal treatment (600 °C) and 2% Eu^{3+} doping content were optimized. In a different study, HAp nanocrystals (Er-HAp) were combined with luminescent erbium (Er^{3+}) ions at various concentrations (0.1, 0.25, 0.5, and 1.0 mol %) to create a nontoxic luminescent agent for biomedical imaging applications (Mondal et al. 2020d). The outcomes demonstrated that after being stimulated at 400 nm, 1.0 mol% Er-doped HAp nanocrystals (50 nm size distribution) possessed high-efficiency light emission (Fig. 4.8).

It is usual practice to categorize the contrast materials utilized in MRI scans into (1) positive (T1, Gd-based agents, and brightness contrast) and (2) negative (T2, Fe-based agents, and darkness contrast) categories. The most often utilized substances for MRI to increase diagnostic precision are hydrophilic Gd(III)-based chelates. The long-term safety of these materials for the human body, however, is still a matter of concern (Wahsner et al. 2018). Due to their distinctive characteristics, such as their wide surface area and effective contrasting impact, inorganic NPs have been identified as suitable contrast agents for MRI over the years (Na et al. 2009). Superparamagnetic iron oxide nanoparticles (SPIONs), of them, have garnered a lot of interest as MRI contrast agents. However, the buildup of Fe in soft tissues harms SPIONs as contrast agents. The solution to all of the aforementioned problems was suggested to be the addition of contrast agents to HAp NPs. For instance, Fe-doped HAp NPs have been described as appropriate contrast agents that can produce more contrast for MRI than SPIONs (Adamiano et al. 2018). In addition to MRI, the efficacy of 99mTc-MDP-labeled Fe-doped HAp NPs as a scintigraphy

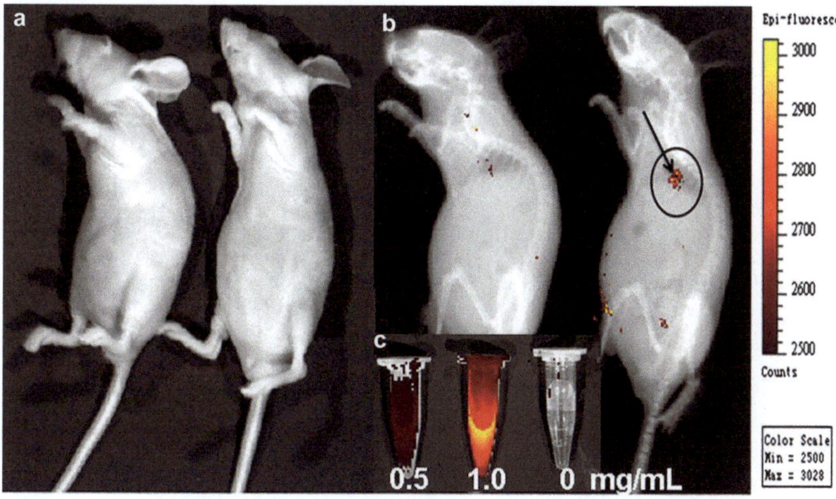

Fig. 4.9 Mice following without (**a**) and with (**b**) Eu^{3+}/Gd^{3+}-HAp ($Eu^{3+}:Gd^{3+}$ = 1:2) nanorods, subcutaneous injection. Imaging using PL in vivo. (**c**) Photographs of the PL emission from Eu^{3+}/Gd^{3+}-HAp nanorods at various concentrations. The excitation wavelength was 430 nm. (Reprinted with permission from Chen et al. 2011)

imaging agent for PET and single-photon emission computed tomography was proven (SPECT). The preparation of paramagnetic HAp NPs for use as PL contrast agents for cancer imaging by substituting other elements (such as Eu^{3+}/Gd^{3+}) was also shown to be beneficial (Fig. 4.9).

For MRI imaging, the luminomagnetic HAp may offer a very promising multimodal imaging probe. In this manner, microwave-assisted synthesis of multifunctional Eu^{3+}/Gd^{3+} dual-doped HAp nanorods was previously achieved (Chen et al. 2011). This system (Eu^{3+} to Gd^{3+} ratio 1:2, Ms = 0.15) demonstrated useful features as a T1 and T2 contrast agent of MRI after being subcutaneously injected into nude mice, as the attenuation value increased from 26 to 96 HU (Fig. 4.9). The magnetization of the samples increased along with more significant concentrations of the Gd^{3+} dopant, and the photoluminescence intensity was dependent on the Eu^{3+}/Gd^{3+} ratio in the HAp nanorods. In particular, it should be noted that bio-nanoplatforms made of doped HAp NPs and other substances, such as carbon dots, have been suggested as suitable tools for cancer cell imaging because they can produce bright blue fluorescence under UV illumination with excellent photostability and colloidal stability (Zhao et al. 2015).

4.8 Limitations of HAp and Future Directions

The biological activity, degradability, and osteoconductivity of HAp are all significant. In recent years, it has seen widespread usage in orthopedic repair, antitumor medication carriers, and dentistry (Fig. 4.10). However, hydroxyapatite has

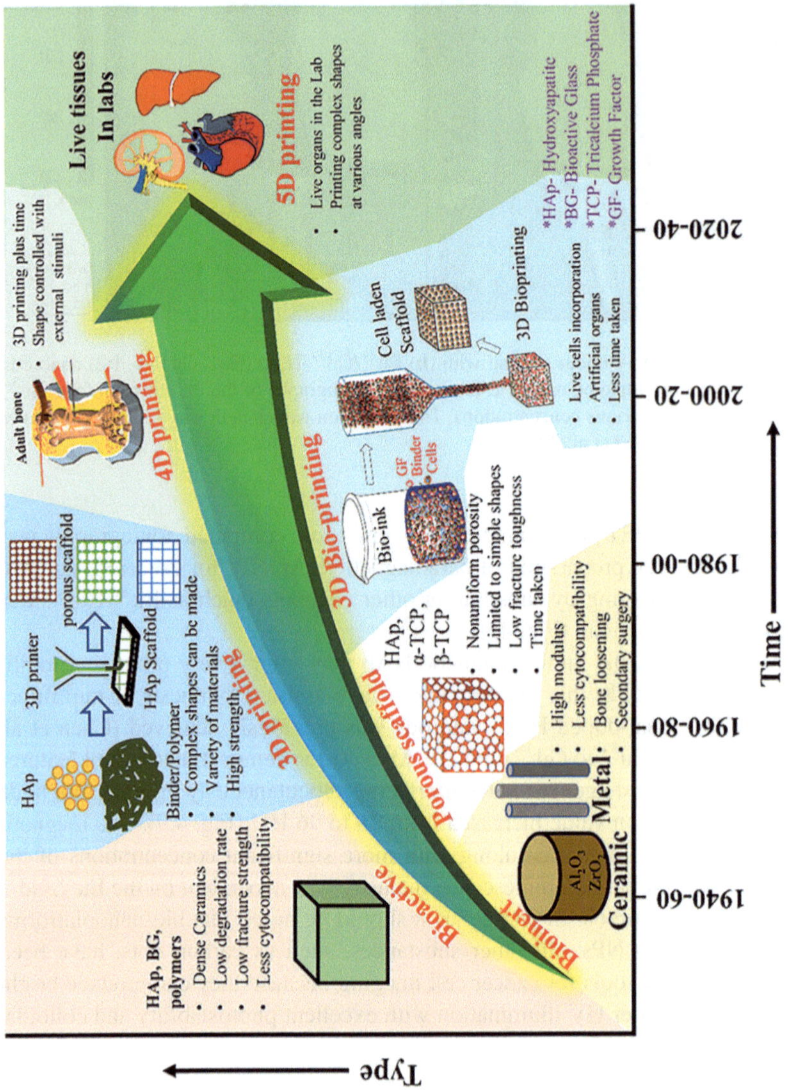

Fig. 4.10 The evolution of HAp-based composites from the early 1940s to the near future is depicted diagrammatically. (Represented with permission from George et al. 2022)

limitations, including brittleness for bone transplantation and weak mechanical characteristics for directly triggering dental enamel remineralization. The difficulties that have been experienced in the preparation, use, and modification of hydroxyapatite in the aforementioned sectors are discussed in this article along with the state of the study. Because of this, the most important contribution of this chapter is to thoroughly integrate and defend the forms and effects of HAp-based biomaterials, which offers a strong foundation for their use and gives readers clear insights into future research. We must overlook the fact that nano theranostic is still a young scientific discipline that has, for the most part, not yet attained the minimal requirements for clinical translation. The intricacy and synergistic mechanisms of the materials employed for such multifunctional applications are the cause of this reality. The more complicated a system is (e.g., when nanostructured HAp is used as a vehicle for the simultaneous release of ions and drugs with anticancer properties), the more variables must be considered, which makes it more challenging to disentangle the therapeutic effects or determine the precise magnitude of synergistic contributions.

4.9 Concluding Remarks

Over the past few years, HAp has significantly broadened the scope of biomedical applications beyond bone restoration, demonstrating considerable promise in cancer theranostics. Due to its polar surface, HAp with nanoscale dimensions is ideally adapted to interact with biomolecules in cells and tissues as well as with medicines. This characteristic offers the possibility for targeted cell- and tissue-specific applications, such as cancer treatment and imaging, in addition to the biocompatibility and biodegradability of HAp upon interaction with bodily fluids (diagnosis). In comparison with metal or composite nanoparticles, multifunctional HAp nanostructures are more adaptable and secure since they may integrate hyperthermia, drug transport, photothermal/photodynamic treatment, imaging, and, generally, superior targeting capabilities.

Further investigation should be done on the combination of HAp with natural anticancer pharmaceuticals and the potential for metallic ion doping and co-doping to reduce the need for conventional chemotherapeutics, which frequently have harmful side effects on the body. By adding functionalization treatments or stimuli-responsive polymeric coatings (serving as molecular gates) on the surface of HAp, it is also possible to carefully control the release of ions and medications. By using additive manufacturing techniques to process HAp, it will be possible to fabricate exact goods and 3D porous scaffolds with excellent control over pore size, shape, and other features. If 3D printing and medical imaging are coupled to recreate the anatomy of patients' flaws left after cancer removal, it will also be feasible to customize and individually design implants.

The experimental work addressed to this purpose is quite complex, requiring the collaboration of biomaterials scientists, biomolecular chemists, biologists, and oncologists in addition to the assessment of full in vitro/in vivo toxicity profiles along with the relationships to the stated hallmarks. A deeper understanding of the biomolecular impact of HAp-based systems on various types of cancer is more important than ever for the research to advance. This study primarily focuses on the groundbreaking discoveries that foreshadowed the interesting modern drug delivery technologies based on HAp and HAp-based composite nanostructures. A particular focus has been placed on describing the use and efficacy of modified HAp as a drug carrier agent for various ailments, including carriers for antibiotics, anti-inflammatory, and cancer-causing medications, medical imaging, and protein delivery agents. This article examines a wide range of nHAp and HAp-modified inorganic drug carriers, highlighting some of their unique characteristics that should be considered for upcoming drug delivery applications.

Acknowledgments This work was supported by the National Research Foundation of Korea (NRF) grant funded by the Korea Government (MSIT) (No. 2022R1A5A8023404).

References

Abdel-Hamid Z, Rashad M, Mahmoud SM, Kandil A (2017) Electrochemical hydroxyapatite-cobalt ferrite nanocomposite coatings as well hyperthermia treatment of cancer. Mater Sci Eng C 76:827–838

Adamiano A, Iafisco M, Sandri M, Basini M, Arosio P, Canu T, Sitia G, Esposito A, Iannotti V, Ausanio G (2018) On the use of superparamagnetic hydroxyapatite nanoparticles as an agent for magnetic and nuclear in vivo imaging. Acta Biomater 73:458–469

Chen F, Huang P, Zhu Y-J, Wu J, Zhang C-L, Cui D-X (2011) The photoluminescence, drug delivery and imaging properties of multifunctional Eu^{3+}/Gd^{3+} dual-doped hydroxyapatite nanorods. Biomaterials 32:9031–9039

Cheung A, Bax HJ, Josephs DH, Ilieva KM, Pellizzari G, Opzoomer J, Bloomfield J, Fittall M, Grigoriadis A, Figini M (2016) Targeting folate receptor alpha for cancer treatment. Oncotarget 7:52553

Doan VHM, Nguyen VT, Mondal S, Vo TMT, Ly CD, Vu DD, Ataklti GY, Park S, Choi J, Oh J (2021) Fluorescence/photoacoustic imaging-guided nanomaterials for highly efficient cancer theragnostic agent. Sci Rep 11:15943. https://doi.org/10.1038/s41598-021-95660-w

Du M, Chen J, Liu K, Xing H, Song C (2021) Recent advances in biomedical engineering of nano-hydroxyapatite including dentistry, cancer treatment and bone repair. Compos Part B 215: 108790. https://doi.org/10.1016/j.compositesb.2021.108790

Fass L (2008) Imaging and cancer: a review. Mol Oncol 2:115–152

Gallagher C, Murphy C, O'Brien FJ, Piskareva O (2021) Three-dimensional in vitro biomimetic model of neuroblastoma using collagen-based scaffolds. J Vis Exp 2021:e62627

George SM, Nayak C, Singh I, Balani K (2022) Multifunctional hydroxyapatite composites for orthopedic applications: a review. ACS Biomater Sci Eng 8:3162–3186. https://doi.org/10.1021/acsbiomaterials.2c00140

Ghiasi B, Sefidbakht Y, Rezaei M (2019) Hydroxyapatite for biomedicine and drug delivery. In: Nanomaterials for advanced biological applications. Springer, pp 85–120

Han Y, Wang X, Dai H, Li S (2013) Synthesis and luminescence of Eu^{3+} doped hydroxyapatite nanocrystallines: effects of calcinations and Eu^{3+} content. J Lumin 135:281–287

Han Y, Li S, Cao X, Yuan L, Wang Y, Yin Y, Qiu T, Dai H, Wang X (2014) Different inhibitory effect and mechanism of hydroxyapatite nanoparticles on normal cells and cancer cells in vitro and in vivo. Sci Rep 4:1–8

Hou C-H, Hou S-M, Hsueh Y-S, Lin J, Wu H-C, Lin F-H (2009) The in vivo performance of biomagnetic hydroxyapatite nanoparticles in cancer hyperthermia therapy. Biomaterials 30: 3956–3960. https://doi.org/10.1016/j.biomaterials.2009.04.020

Ignjatović N, Mančić L, Vuković M, Stojanović Z, Nikolić M, Škapin S, Jovanović S, Veselinović L, Uskoković V, Lazić S (2019) Rareearth (Gd^{3+}, Yb^{3+}/Tm^{3+}, Eu^{3+}) co-doped hydroxyapatite as magnetic, up-conversion and down-conversion materials for multimodal imaging. Sci Rep 9:16305

Ikoma T, Yamazaki A, Nakamura S, Akao M (1999) Preparation and structure refinement of monoclinic hydroxyapatite. J Solid State Chem 144:272–276

Jin Y, Liu X, Liu H, Chen S, Gao C, Ge K, Zhang C, Zhang J (2017) Oxidative stress-induced apoptosis of osteoblastic MC3T3-E1 cells by hydroxyapatite nanoparticles through lysosomal and mitochondrial pathways. RSC Adv 7:13010–13018

Kataoka T, Samitsu S, Okuda M, Kawagoe D, Tagaya M (2019) Highly luminescent hydroxyapatite nanoparticles hybridized with citric acid for their bifunctional cell-labeling and cytostatic suppression properties. ACS Appl Nano Mater 3:241–256

Kim H, Mondal S, Jang B, Manivasagan P, Moorthy MS, Oh J (2018a) Biomimetic synthesis of metal–hydroxyapatite (Au-HAp, Ag-HAp, Au-Ag-HAp): structural analysis, spectroscopic characterization and biomedical application. Ceram Int 44:20490–20500. https://doi.org/10.1016/j.ceramint.2018.08.045

Kim H, Mondal S, Bharathiraja S, Manivasagan P, Moorthy MS, Oh J (2018b) Optimized Zn-doped hydroxyapatite/doxorubicin bioceramics system for efficient drug delivery and tissue engineering application. Ceram Int 44:6062–6071. https://doi.org/10.1016/j.ceramint.2017.12.235

Krishnakumar GS, Gostynska N, Campodoni E, Dapporto M, Montesi M, Panseri S, Tampieri A, Kon E, Marcacci M, Sprio S et al (2017) Ribose mediated crosslinking of collagen-hydroxyapatite hybrid scaffolds for bone tissue regeneration using biomimetic strategies. Mater Sci Eng C 77:594–605. https://doi.org/10.1016/j.msec.2017.03.255

Lee W-H, Loo C-Y, Rohanizadeh R (2019) Functionalizing the surface of hydroxyapatite drug carrier with carboxylic acid groups to modulate the loading and release of curcumin nanoparticles. Mater Sci Eng C 99:929–939

Lelli M, Roveri N, Marzano C, Hoeschele JD, Curci A, Margiotta N, Gandin V, Natile G (2016) Hydroxyapatite nanocrystals as a smart, pH sensitive, delivery system for kiteplatin. Dalton Trans 45:13187–13195

Li D, Nie W, Chen L, McCoul D, Liu D, Zhang X, Ji Y, Yu B, He C (2018) Self-assembled hydroxyapatite-graphene scaffold for photothermal cancer therapy and bone regeneration. J Biomed Nanotechnol 14:2003–2017

Machado TR, Leite I, Inada N, Li MS, Da Silva J, Andres J, Beltrán-Mir H, Cordoncillo E, Longo E (2019) Designing biocompatible and multicolor fluorescent hydroxyapatite nanoparticles for cell-imaging applications. Mater Today Chem 14:100211

Maia ALC, Ferreira CA, Barros ALB, e Silva ATM, Ramaldes GA, Silva Cunha Júnior Ad, Oliveira DCP, Fernandes C, Ferreira Soares DC (2018) Vincristine-loaded hydroxyapatite nanoparticles as a potential delivery system for bone cancer therapy. J Drug Target 26:592–603

Marycz K, Smieszek A, Targonska S, Walsh SA, Szustakiewicz K, Wiglusz RJ (2020) Three dimensional (3D) printed polylactic acid with nano-hydroxyapatite doped with europium (III) ions (nHAp/PLLA@ Eu^{3+}) composite for osteochondral defect regeneration and theranostics. Mater Sci Eng C 110:110634

Meshkini A, Oveisi H (2017) Methotrexate-F127 conjugated mesoporous zinc hydroxyapatite as an efficient drug delivery system for overcoming chemotherapy resistance in osteosarcoma cells. Colloids Surf B Biointerfaces 158:319–330

Mondal S, Pal U (2019) 3D hydroxyapatite scaffold for bone regeneration and local drug delivery applications. J Drug Deliv Sci Technol 53:101131. https://doi.org/10.1016/j.jddst.2019.101131

Mondal S, Pal U, Dey A (2016) Natural origin hydroxyapatite scaffold as potential bone tissue engineering substitute. Ceram Int 42:18338–18346. https://doi.org/10.1016/j.ceramint.2016.08.165

Mondal S, De Anda Reyes ME, Pal U (2017a) Plasmon induced enhanced photocatalytic activity of gold loaded hydroxyapatite nanoparticles for methylene blue degradation under visible light. RSC Adv 7:8633–8645. https://doi.org/10.1039/C6RA28640B

Mondal S, Manivasagan P, Bharathiraja S, Santha Moorthy M, Nguyen VT, Kim HH, Nam SY, Lee KD, Oh J (2017b) Hydroxyapatite coated iron oxide nanoparticles: a promising nanomaterial for magnetic hyperthermia cancer treatment. Nanomaterials 7:426

Mondal S, Manivasagan P, Bharathiraja S, Santha Moorthy M, Kim HH, Seo H, Lee KD, Oh J (2017c) Magnetic hydroxyapatite: a promising multifunctional platform for nanomedicine application. Int J Nanomedicine 12:8389–8410. https://doi.org/10.2147/ijn.s147355

Mondal S, Dorozhkin SV, Pal U (2018a) Recent progress on fabrication and drug delivery applications of nanostructured hydroxyapatite. WIREs Nanomed Nanobiotechnol 10:e1504. https://doi.org/10.1002/wnan.1504

Mondal S, Hoang G, Manivasagan P, Moorthy MS, Nguyen TP, Vy Phan TT, Kim HH, Kim MH, Nam SY, Oh J (2018b) Nano-hydroxyapatite bioactive glass composite scaffold with enhanced mechanical and biological performance for tissue engineering application. Ceram Int 44:15735–15746. https://doi.org/10.1016/j.ceramint.2018.05.248

Mondal S, Hoang G, Manivasagan P, Moorthy MS, Vy Phan TT, Kim HH, Nguyen TP, Oh J (2019a) Rapid microwave-assisted synthesis of gold loaded hydroxyapatite collagen nano-bio materials for drug delivery and tissue engineering application. Ceram Int 45:2977–2988. https://doi.org/10.1016/j.ceramint.2018.10.016

Mondal S, Hoang G, Manivasagan P, Kim H, Oh J (2019b) Nanostructured hollow hydroxyapatite fabrication by carbon templating for enhanced drug delivery and biomedical applications. Ceram Int 45:17081–17093. https://doi.org/10.1016/j.ceramint.2019.05.260

Mondal S, Hoang G, Manivasagan P, Moorthy MS, Kim HH, Vy Phan TT, Oh J (2019c) Comparative characterization of biogenic and chemical synthesized hydroxyapatite biomaterials for potential biomedical application. Mater Chem Phys 228:344–356. https://doi.org/10.1016/j.matchemphys.2019.02.021

Mondal S, Nguyen VT, Park S, Choi J, Thien Vo TM, Shin JH, Kang Y-H, Oh J (2020a) Rare earth element doped hydroxyapatite luminescent bioceramics contrast agent for enhanced biomedical imaging and therapeutic applications. Ceram Int 46:29249–29260. https://doi.org/10.1016/j.ceramint.2020.08.099

Mondal S, Nguyen VT, Park S, Choi J, Tran LH, Yi M, Shin JH, Lee C-Y, Oh J (2020b) Bioactive, luminescent erbium-doped hydroxyapatite nanocrystals for biomedical applications. Ceram Int 46:16020–16031. https://doi.org/10.1016/j.ceramint.2020.03.152

Mondal S, Nguyen TP, Pham VH, Hoang G, Manivasagan P, Kim MH, Nam SY, Oh J (2020c) Hydroxyapatite nano bioceramics optimized 3D printed poly lactic acid scaffold for bone tissue engineering application. Ceram Int 46:3443–3455. https://doi.org/10.1016/j.ceramint.2019.10.057

Mondal S, Park S, Choi J, Tran LH, Yi M, Shin JH, Lee C-Y, Oh J (2020d) Bioactive, luminescent erbium-doped hydroxyapatite nanocrystals for biomedical applications. Ceram Int 46:16020–16031

Mondal S, Montaño-Priede JL, Nguyen VT, Park S, Choi J, Doan VHM, Vo TMT, Vo TH, Large N, Kim C-S et al (2022) Computational analysis of drug free silver triangular nanoprism theranostic probe plasmonic behavior for in-situ tumor imaging and photothermal therapy. J Adv Res 41:23–38. https://doi.org/10.1016/j.jare.2022.02.006

Na HB, Song IC, Hyeon T (2009) Inorganic nanoparticles for MRI contrast agents. Adv Mater 21: 2133–2148

Nakayama M, Lim WQ, Kajiyama S, Kumamoto A, Ikuhara Y, Kato T, Zhao Y (2019) Liquid-crystalline hydroxyapatite/polymer nanorod hybrids: potential bioplatform for photodynamic therapy and cellular scaffolds. ACS Appl Mater Interfaces 11:17759–17765

Padmanabhan VP, Balakrishnan S, Kulandaivelu R, Sankara Narayanan SN, Lakshmipathy M, Sagadevan S, Mohammad F, Al-Lohedan HA, Paiman S, Oh WC (2020) Nanoformulations of core–shell type hydroxyapatite-coated gum acacia with enhanced bioactivity and controlled drug delivery for biomedical applications. New J Chem 44:7175–7185

Park S, Choi J, Mondal S, Vo TMT, Pham VH, Lee H, Nam SY, Kim C-S, Oh J (2022a) The impact of Cu(II) ions doping in nanostructured hydroxyapatite powder: a finite element modelling study for physico-mechanical and biological property evaluation. Adv Powder Technol 33:103405. https://doi.org/10.1016/j.apt.2021.103405

Park S, Choi J, Vo TMT, Mondal S, Vo TH, Ko N, Kim C-S et al (2022b) In vivo mimicking injectable self-setting composite bio-cement: scanning acoustic diagnosis and biological property evaluation for tissue engineering applications. Colloids Surf B Biointerfaces 218:112722. https://doi.org/10.1016/j.colsurfb.2022.112722

Pathi SP, Lin DD, Dorvee JR, Estroff LA, Fischbach C (2011) Hydroxyapatite nanoparticle-containing scaffolds for the study of breast cancer bone metastasis. Biomaterials 32:5112–5122

Phan TTV, Hoang G, Nguyen VT, Nguyen TP, Kim HH, Mondal S, Manivasagan P, Moorthy MS, Lee KD, Junghwan O (2019) Chitosan as a stabilizer and size-control agent for synthesis of porous flower-shaped palladium nanoparticles and their applications on photo-based therapies. Carbohydr Polym 205:340–352. https://doi.org/10.1016/j.carbpol.2018.10.062

Pinto A, Pocard M (2018) Photodynamic therapy and photothermal therapy for the treatment of peritoneal metastasis: a systematic review. Pleura Peritoneum 3:20180124

Posner AS, Perloff A, Diorio AF (1958) Refinement of the hydroxyapatite structure. Acta Crystallogr 11:308–309

Proussaefs P, Lozada J, Valencia G, Rohrer MD (2002) Histologic evaluation of a hydroxyapatite onlay bone graft retrieved after 9 years: a clinical report. J Prosthet Dent 87:481–484

Qiao W, Lan X, Tsoi JK, Chen Z, Su RY, Yeung KW, Matinlinna JP (2017) Biomimetic hollow mesoporous hydroxyapatite microsphere with controlled morphology, entrapment efficiency and degradability for cancer therapy. RSC Adv 7:44788–44798

Saber-Samandari S, Nezafati N, Saber-Samandari S (2016) The effective role of hydroxyapatite based composites in anticancer drug delivery systems. Crit Rev Ther Drug Carrier Syst 33(1): 41–75

Sangeetha K, Ashok M, Girija E (2019) Development of multifunctional cobalt ferrite/hydroxyapatite nanocomposites by microwave assisted wet precipitation method: a promising platform for synergistic chemo-hyperthermia therapy. Ceram Int 45:12860–12869

Santha Moorthy M, Hoang G, Subramanian B, Bui NQ, Panchanathan M, Mondal S, Thi Tuong VP, Kim H, Oh J (2018) Prussian blue decorated mesoporous silica hybrid nanocarriers for photoacoustic imaging-guided synergistic chemo-photothermal combination therapy. J Mater Chem B 6:5220–5233. https://doi.org/10.1039/C8TB01214H

Sedighi O, Alaghmandfard A, Montazerian M, Baino F (2022) A critical review of bioceramics for magnetic hyperthermia. J Am Ceram Soc 105:1723–1747

Sikora M, Marcinkowska K, Marycz K, Wiglusz RJ, Śmieszek A (2019) The potential selective cytotoxicity of poly (L-lactic acid)-based scaffolds functionalized with Nanohydroxyapatite and europium (III) ions toward osteosarcoma cells. Materials 12:3779

Sun J, Ding T (2009) p53 reaction to apoptosis induced by hydroxyapatite nanoparticles in rat macrophages. J Biomed Mater Res A 88:673–679

Šupová M (2015) Substituted hydroxyapatites for biomedical applications: a review. Ceram Int 41: 9203–9231

Tang W, Yuan Y, Liu C, Wu Y, Lu X, Qian J (2014) Differential cytotoxicity and particle action of hydroxyapatite nanoparticles in human cancer cells. Nanomedicine 9:397–412

Tornín J, Villasante A, Solé-Martí X, Ginebra M-P, Canal C (2021) Osteosarcoma tissue-engineered model challenges oxidative stress therapy revealing promoted cancer stem cell properties. Free Radic Biol Med 164:107–118

Uskoković V (2015) The role of hydroxyl channel in defining selected physicochemical peculiarities exhibited by hydroxyapatite. RSC Adv 5:36614–36633

Uskoković V (2020) Ion-doped hydroxyapatite: an impasse or the road to follow? Ceram Int 46: 11443–11465

VanDyke D, Kyriacopulos P, Yassini B, Wright A, Burkhart E, Jacek S, Pratt M, Peterson C, Rai P (2016) Nanoparticle based combination treatments for targeting multiple hallmarks of cancer. Int J Nano Stud Technol Suppl 4:1

Verma G, Barick K, Shetake NG, Pandey B, Hassan P (2016) Citrate-functionalized hydroxyapatite nanoparticles for pH-responsive drug delivery. RSC Adv 6:77968–77976

Vo TMT, Mondal S, Nguyen VT, Park S, Choi J, Bui NT, Oh J (2021) Rice starch coated iron oxide nanoparticles: a theranostic probe for photoacoustic imaging-guided photothermal cancer therapy. Int J Biol Macromol 183:55–67. https://doi.org/10.1016/j.ijbiomac.2021.04.053

Wahsner J, Gale EM, Rodríguez-Rodríguez A, Caravan P (2018) Chemistry of MRI contrast agents: current challenges and new frontiers. Chem Rev 119:957–1057

Wang R, Liu W, Wang Q, Li G, Wan B, Sun Y, Niu X, Chen D, Tian W (2020) Anti-osteosarcoma effect of hydroxyapatite nanoparticles both in vitro and in vivo by downregulating the FAK/PI3K/Akt signaling pathway. Biomater Sci 8:4426–4437

Xiong H, Du S, Zhang P, Jiang Z, Zhou J, Yao J (2018) Primary tumor and pre-metastatic niches co-targeting "peptides-lego" hybrid hydroxyapatite nanoparticles for metastatic breast cancer treatment. Biomater Sci 6:2591–2604

Xu J, Xu P, Li Z, Huang J, Yang Z (2012) Oxidative stress and apoptosis induced by hydroxyapatite nanoparticles in C6 cells. J Biomed Mater Res A 100:738–745

Yuan Y, Liu C, Qian J, Wang J, Zhang Y (2010) Size-mediated cytotoxicity and apoptosis of hydroxyapatite nanoparticles in human hepatoma HepG2 cells. Biomaterials 31:730–740

Zeng S, Zhou R, Zheng X, Wu L, Hou X (2017) Mono-dispersed Ba^{2+}-doped Nano-hydroxyapatite conjugated with near-infrared Cu-doped CdS quantum dots for CT/fluorescence bimodal targeting cell imaging. Microchem J 134:41–48

Zhang X, Ma J (2021) Photothermal effect of 3D printed hydroxyapatite composite scaffolds incorporated with graphene nanoplatelets. Ceram Int 47:6336–6340

Zhao Y, Shi L, Fang J, Feng X (2015) Bio-nanoplatforms based on carbon dots conjugating with F-substituted nano-hydroxyapatite for cellular imaging. Nanoscale 7:20033–20041

Sudip Mondal received his Ph.D. in 2015 from the CSIR-Central Mechanical Engineering Research Institute and National Institute of Technology Durgapur, India. He joined as a Postdoctoral Fellow at the Autonomous University of Puebla, Mexico (2015–2017). Currently, he works as a Research Professor at the Department of Nano-Biomedicine and Biomedical Engineering, Pukyong National University, South Korea. His research interests include nanostructured materials synthesis, bioimaging, and biomedical applications such as cancer therapy and tissue engineering.

Sumin Park received her B.S. (2020) and M.S. (2022) degrees in Biomedical Engineering from Pukyong National University, South Korea. Her research interests include synthesizing tailored nanoparticles and their applications for diagnosis and treatment and bioimaging techniques such as photoacoustic microscopy and scanning acoustic microscopy.

Jaeyeop Choi received her B.S. (2018), M.S. (2020), and Ph.D. (2022) degrees in Biomedical Engineering from Pukyong National University, Busan, Republic of Korea. He is currently doing his research as Research Professor in Smart Gym-based Translational Research Center for Active Senior's Healthcare, Pukyong National University. His research interests include high-frequency ultrasonic transducers and scanning acoustic microscopy.

Junghwan Oh received a B.S. Degree in Mechanical Engineering from Pukyong National University in 1992 and an M.S. and Ph.D. degrees in Biomedical Engineering from the University of Texas at Austin, USA, in 2003 and 2007, respectively. In 2010, he joined the Department of Biomedical Engineering at Pukyong National University, where he is a Full Professor. He also serves as CEO of OhLabs Corporation. His current research interests include ultrasonic-based diagnostic imaging modalities for biomedical engineering applications, biomedical signal processing, and healthcare systems.

Green Synthesis of Metallic Nanoparticles and Various Biomedical Applications

5

Fahima Dilnawaz and Amarendra Narayan Misra

Abstract

Nanobiotechnology has emerged as an advanced arena in conjunction with biology and nanotechnology. Biological entities involve living organisms of both prokaryotic and eukaryotic origins. Recently attention has been focused on the green synthesis of nanoparticles because of their natural availability and environmentally friendly synthesis approach. Every biological organism varies in its functional capabilities. However, only selective biological organisms can produce nanoparticles because of their enzymatic and intrinsic metabolic processes. These eligible biological entities or their extracts have been used for the green synthesis of metallic nanoparticles. These biosynthesized metallic nanoparticles have a range of unlimited biomedical applications, such as the delivery of drugs or genes, the detection of pathogens or proteins, and tissue engineering. This chapter will discuss various green synthesis routes of nanoparticles and their current use. The application of these metallic nanoparticles briefly discussed here includes antiviral, antibacterial, antioxidant, anticancer, anti-inflammatory, and antiparasitic wound-healing activities.

Sustainable Development Goals: Good health and well-being (SDG3), Responsible production and consumption (SDG12), Climate action (SDG13), Life on land (SDG15)

F. Dilnawaz (✉)
Department of Biotechnology, School of Engineering and Technology, Centurion University of Technology and Management, Bhubaneswar, India
e-mail: fahima.dilnawaz@cutm.ac.in

A. N. Misra
School of Applied Sciences, Centurion University of Technology and Management, Bhubaneswar, India
e-mail: amarendra.misra@cutm.ac.in

F. A. Sheikh et al. (eds.), *Interaction of Nanomaterials With Living Cells*,
https://doi.org/10.1007/978-981-99-2119-5_5

Keywords

Green synthesis · Metallic nanoparticles · Nanobiotechnology · Phytosynthesis · Reactive oxygen species · Antioxidant

5.1 Introduction

Nanotechnology-mediated formation of nanoparticles is currently being utilized in the various fields of human welfare (Dilnawaz et al. 2018). Nanoscale particles display unique physical and chemical properties, such as high surface-to-volume ratios, higher reactivity and strength, and colloidal stability. Mainly the metal (copper, iron, gold, silver, etc.) nanoparticles display better physical and optical characteristics (surface plasmon reverberation or surface plasmon resonance), which attracted significant attention for specific applications in medicine and science (Mauro et al. 2021). Moreover, the synthesis of nanoparticles (NPs) is being carried out through three different strategies (physical, chemical, and biological). Conventionally, in the chemical synthesis process, chemical reduction and microemulsion/colloidal, electrochemical, and thermal decomposition are used with the intervention of metal salt precursors and the addition of particular reducing agents. Although these chemical methods are economical for large-scale production, the use of toxic substances, in turn, produces harmful by-products that cause environmental damage, thereby limiting its clinical and biomedical applications (Hua et al. 2018). Recently, many research groups have utilized green synthetic routes to synthesize different nanomaterials to reduce the usage of harmful chemicals. Green synthetic methods are carried out under mild reaction condition, which consumes low energy, devoid of conditional requirement of high temperature and pressure, hazardous chemicals, and addition of external stabilizing or capping agent (Khan et al. 2020). Among various green synthetic routes, plant leaf extract-mediated approaches for the preparation of metal oxide nanomaterials have received enormous attention (Nguyen et al. 2021). Using the producer source of the biological kingdom, the green synthesis of metal oxide nanoparticles is being carried out. Scientists have given priority to plants that can execute this process due to their biomass abundance, response to stress factors (pathogens, herbivores, and climate changes), and survival modes (seasonal changes and reproductive stage variation) that contribute toward the development of plants' primary and secondary metabolites. Hence these naturally imbibed strategies of the plants serve as the main bioreactors and molecule suppliers for green synthesis (Iravani 2011). The plant leaves contain a lot of phytochemicals; the phenolic compounds (alkaloids, tannins, and flavonoids), along with the important constituents like proteins and carbohydrates, are taken into account for the synthesis of metal oxide nanoparticles (Biswal et al. 2020; Richardson et al. 2006). Further, the functional amino groups and proteins available in the plant extracts also contribute toward metal ion reduction (Li et al. 2007). In another study, Huang et al. discussed that the functional groups of alkaloids, flavones, and anthracenes, such as –C–O–C–, –C–O–, –C=C–, and –C=O–, assist the metal NP synthesis (Huang et al. 2007).

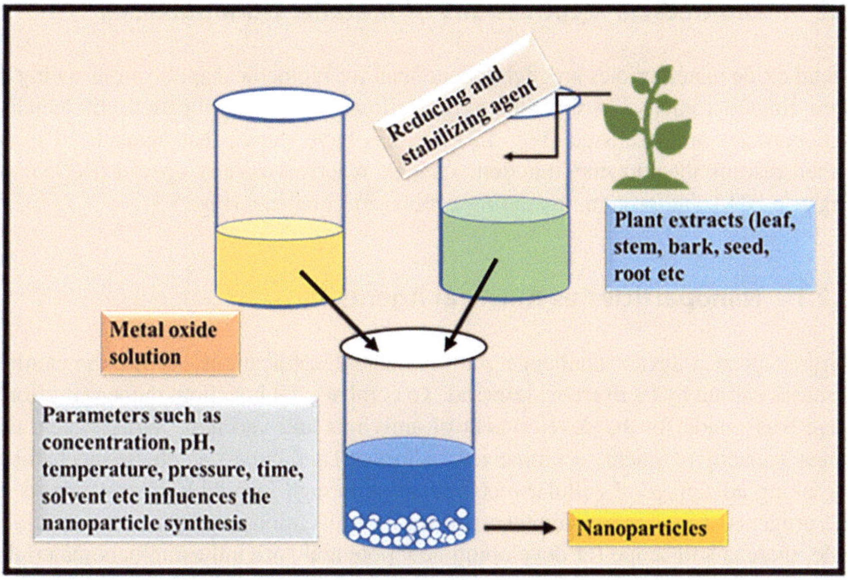

Fig. 5.1 Overview of green synthesis of nanoparticles

The metal ion reduction may also be carried out with the help of quinones and plastohydroquinone molecules present in the plant leaf extract (Kesharwani et al. 2009). All these studies indicated that extracellular metal oxide nanoparticle synthesis could be carried out by biomolecules and heterocyclic compounds present in plants. However, the vision and understanding of the metal oxide NP synthesis utilizing plant extracts are not well understood. It is comprehended that maybe the phytochemicals of the plants mediated the production of NPs through metal reduction. Thereafter, oxygen produced from either atmosphere or degrading phytochemicals links the reduced metal ions, and corresponding electrostatic attraction leads to NPs (Makarov et al. 2015). Plants, either the whole or tissues (Kumar and Yadav 2009; Marchiol 2012) or extracts of different parts (e.g., roots, barks, leaves, fruits, and seeds) (Kharissova et al. 2012; Rai et al. 2008), are taken into account for green synthesis (Fig. 5.1).

Particularly noble metals (gold and silver) are widely known for their use in medicine owing to their excellent intrinsic property, unique functional attributes, the scope for surface functionalization, etc. (Ahmad et al. 2019; Behzad et al. 2021). Therefore, the green synthesis method leading to the formation of biocompatible and environmentally friendly metal nanoparticles upsurge its applications in healthcare. The present book chapter will chronicle some of the critical biomedical applications of NPs toward better healthcare.

5.2 Biomedical Applications of Metallic Nanoparticles

Metal oxide nanoparticles are highly beneficial for biomedical applications owing to their limited toxicity. In the following sections, various applications of metallic nanoparticles are discussed (Fig. 5.2). Studies have shown that uptaken NPs can either disrupt the enzyme function, quench reactive oxygen species (ROS), or degrade DNA and protein, disrupting cellular mechanisms (Fig. 5.3).

5.2.1 Nanoparticles as Antiviral Agents

Viruses pose a severe challenge for biomedical applications, as life-threatening diseases caused by them are widespread. To combat viral infection, numerous efforts have been made for the development of antivirals and vaccines. Viruses have the innate capacity to quickly acclimatize to a host cell and bypass a defense mechanism by taking advantage of cellular metabolism. Although incredible improvements in the progress of antiviral chemotherapeutics for prevention and treatments are available, there is still scope for developing new potential antiviral using nanomaterials. Silver (Ag) nanoparticles (NPs) have been proven as potent antiviral agents against several virus families: *Retroviridae*, *Paramyxoviridae*, *Hepadnaviridae*, *Poxviridae*, *Herpesviridae*, *Arenaviridae*, and *Orthomyxoviridae* (Franci et al. 2015). With the implementation of Ag NPs, viruses are less likely to become resistant compared to conventional antiviral agents. Ag NPs have multivalent connections with surface components of virus and cell membrane receptors, by which they block their access

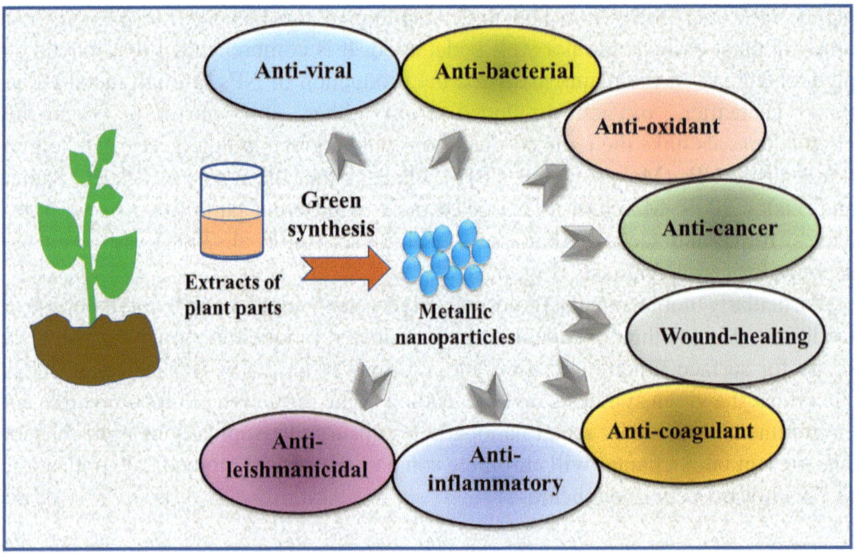

Fig. 5.2 Various modes of applications of metallic nanoparticles obtained through the green synthesis process

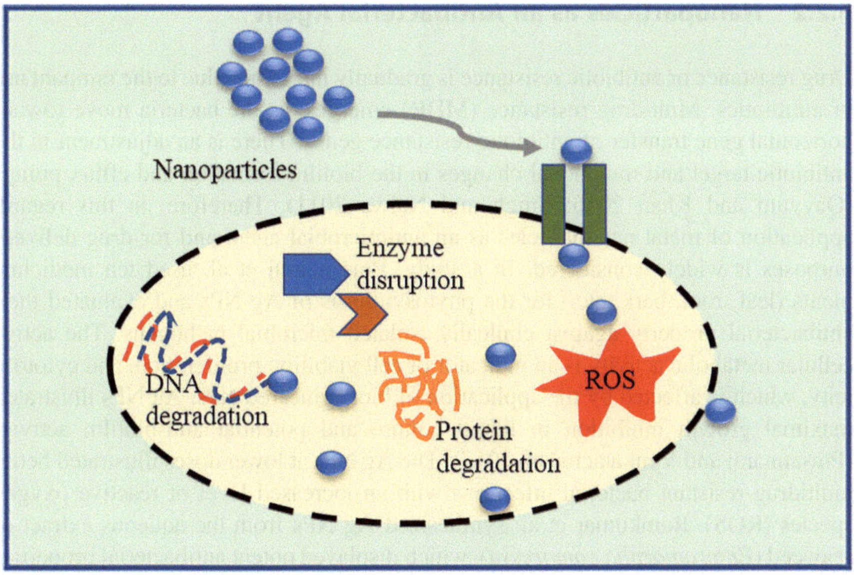

Fig. 5.3 Possible mechanism of interaction of nanoparticles with the cellular components and their effects

into the cells. The antiviral agents act directly on viral particles by binding to their viral coat proteins by disrupting their function or structure (Sharma et al. 2019). Ag NPs were synthesized by engaging the extracts of *Phyllanthus niruri*, *Andrographis paniculata*, and *Tinospora cordifolia*, and its antiviral efficacy was evaluated against the chikungunya virus. The in vitro antiviral assay of Ag NPs was evaluated based on the degree of inhibition of cytopathic effect (CPE), which showed *A. paniculata* Ag NPs to be most effective, followed by *T. cordifolia* Ag NPs and *P. niruri* Ag NPs. The cytotoxicity assay illustrated that *A. paniculata* Ag NPs inhibited the virus to a great extent. This study could indicate alternative treatment options against viral disease (Sharma et al. 2019). Orłowski et al. used tannic acid (TA) for the synthesis of Ag NPs, which served as an effective microbicide in the mucosal tissues. TA Ag NPs, when treated intravaginally in a virally infected mouse model, demonstrated better anti-herpes simplex virus-2 immune response by showing improved clinical scores with lower viral titer in the vaginal tissues (Orłowski et al. 2018). Ag NPs were synthesized utilizing the leaf extracts of *Carica papaya*, which hold antiviral activity and good binding affinity against nonstructural protein 1 of the dengue type 2 virus (Renganathan et al. 2018). In another study, Ag NPs were synthesized using the leaf extract of *Eucalyptus procera*, and its adjuvant effect on veterinary rabies vaccine was evaluated. Ag NP-loaded vaccine concentrations at 15 and 20 mg/kg showed the highest percentage of viability after injecting into the mice. At the same time, the alum-containing vaccine at the concentration of 10 mg/mL was toxic, whereas Ag NPs were nontoxic, elucidating the safety effect of the Ag NP-based vaccine (Asgary et al. 2016).

5.2.2 Nanoparticles as an Antibacterial Agent

Drug resistance or antibiotic resistance is gradually increasing due to the rampant use of antibiotics. Multidrug resistance (MDR) emerges, as the bacteria move toward horizontal gene transfer of antibiotic resistance genes. There is an adjustment in the antibiotic target and mutational changes in the biofilm formation and efflux pumps (Qayyum and Khan 2016; Singh and Nalwa 2011). Therefore, in this regard, application of metal nanoparticles as an antimicrobial agent and for drug delivery purposes is widely considered. In a study, Prasannaraj et al. used ten medicinal plants (leaf, root, bark, etc.) for the phytosynthesis of Ag NPs and evaluated their antibacterial property against clinically isolated microbial pathogens. The active cellular metabolic activity is an indicator of cell viability, proliferation, and cytotoxicity, which is affected by the application of bioengineered NPs. Ag NPs illustrated maximal growth inhibition in liquid culture and potential anti-biofilm activity (Prasannaraj and Venkatachalam 2017). The Ag NPs at lower doses illustrated better multidrug-resistant bacterial infections with an increased level of reactive oxygen species (ROS). Ramkumar et al. synthesized Ag NPs from the aqueous extract of seaweed (*Enteromorpha compressa*), which displayed potent antibacterial properties toward *E. coli*, *K. pneumoniae*, *P. aeruginosa*, *S. aureus*, and *S. paratyphi*. *Panax ginseng*, the herbal medicinal plant having an active ingredient of ginsenosides, has been explored for the synthesis of Ag NPs using its root extract. The produced Ag NPs at 3 μg/mL exhibited potent antibacterial action against *S. enterica*, *B. anthracis*, *V. parahaemolyticus*, *S. aureus*, *E. coli*, and *B. cereus* and completely inhibited the biofilm of *S. aureus* and *P. aeruginosa* at 4 μg/mL (Ramkumar et al. 2017). Phytosynthesis of Ag NPs using extracts of ten plants (*Pongamia glabra*, *Hamelia patens*, *Tectona grandis*, *Thevetia peruviana*, *Ficus petiolaris*, *Ficus busking*, *Caryota urens*, *Juniper communis*, *Calendula officinalis*, and *Bauhinia purpurea*) illustrated minimum inhibitory concentration (MIC) of 16–26 μg/mL for *E. coli*, *K. pneumoniae*, *E. cloacae*, *S. mutans*, and *S. aureus* (Qayyum and Khan 2016). A lot of studies have been carried out, and corresponding pieces of literature are available exclusively for the antibacterial effects of Ag NPs that were compiled in a review article (Lateef et al. 2019). Application of Ag NPs induces better permeability of cell membranes and produces ROS by interrupting replication of deoxyribonucleic acid (DNA) by releasing silver ions to the bacterial cell, thereby prompting cell death (Dwyer et al. 2012).

5.2.3 Nanoparticles as Antioxidant Agents

In biological systems, antioxidants play a dynamic role in scavenging toxic free radicals. The oxidative stress injuries to the cellular components such as DNA, proteins, and lipids generated by free radicals are quenched by antioxidants. Nitric oxide (NO) plays a major role in a diversity of biological functions, including neurotransmission, blood pressure regulation, smooth muscle relaxation, and antimicrobial and antitumor activities. The NO can react with the generated superoxide to form the peroxynitrite anion, which causes DNA fragmentation and initiates lipid peroxidation (Lone et al. 2013). The active oxygen species, present in the body, can cause several disorders such as carcinogenesis, aging, atherosclerosis, cataracts, inflammation, mutation, and cell death (Ragu et al. 2007). The most reactive and poisonous free radicals are hydroxyl radicals which demonstrate great oxidative power, as they can combine rapidly with almost all molecules in their immediate vicinity (Sousa et al. 2009). Gold (Au) NPs were synthesized using an aqueous extract of *Elettaria cardamomum* seeds which displayed good antioxidant activities. 2,2-Diphenylpicrylhydrazyl (DPPH) free radical scavenging is used for screening the antioxidant activity. The antioxidant activity scavenged by DPPH, NO, and OH radical methods has increased with the application of AuNPs. Au and Ag nanoparticles synthesized from seed extracts of *Embelia ribes* elucidated excellent antioxidant activity as displayed by DPPH free radical scavenging and the phosphomolybdenum assay. In another study, Ag NPs synthesized from leaf extracts of *Aristolochia indica* displayed better antioxidant activity at 100 μg/mL (Shanmugam et al. 2016). Ag NPs synthesized from pod extracts of *C. nitida* exhibited strong antioxidant activity by scavenging DPPH and ferric ions at a concentration between 20 and 100 μg/mL, similarly Bhakya et al. reported the better DPPH scavenging activity of Ag NPs formed through the root extracts of *Helicteres isora* (Bhakya et al. 2016). The fruit extract of *Couroupita guianensis* was used for the synthesis of AuNPs, which illustrated antioxidant potentials at IC50 of 36 g/mL and DPPH effects through hydroxyl radical scavenging at 37 g/mL. However, with increasing concentration, the scavenging activity upsurges for inhibition to around 90% (Sathishkumar et al. 2016a). Abbai et al. synthesized AuNPs from Siberian ginseng (*Eleutherococcus senticosus*) using its stem extract, which displayed good antioxidant potentiality (Abbai et al. 2016). Similarly in another study, *Cassia tora* leaf powder was used for the production of AuNPs. Its antioxidant potentiality was evaluated via catalase activity, and an increase in catalase activity correspondingly increased the antioxidant activity and also suppressed the release of hydrogen peroxide (Abel et al. 2016).

5.2.4 Nanoparticles as an Anticancer Agent

Cancer is a devastating disease and the leading cause of death globally. The metallic nanoparticles induce autophagy and promote cell death. Castro-Aceituno et al. synthesized Ag NPs from *Panax ginseng* fresh leaves, and its anticancer effect was evaluated on different human cancer cell lines. In human cancer cell lines (A549, MCF7, and HepG2), treatment of Ag NPs inhibited cell viability and induced oxidative stress. In A549 cells, the Ag NPs inhibited the epidermal growth factor (EGF)-enhanced migration, decreased the mRNA levels and phosphorylation of EGF receptors, and increased the apoptotic effect that was linked to the stimulation of the p38 MAPK/p53 pathway. In MCF-7 and HepG2 cells, Ag NPs induced cytotoxicity and ROS generation (Castro-Aceituno et al. 2016). In another study, Castro-Aceituno et al. synthesized Ag NPs from *Dendropanax morbifera* Léveille, which exhibited antimicrobial activity and reduced the viability of cancer cells without affecting the viability of RAW 264.7 macrophage-like cells, where it demonstrated the cytotoxic effect by generating ROS against A549 and HepG2 cell lines at different concentrations (Castro Aceituno et al. 2016). In another study, longan peel, powder-mediated Ag NPs were synthesized, and their anticancer ability was evaluated in H1299 cells as well as in the mouse model. Dose-dependent cytotoxicity and stimulation of apoptosis were observed in H1299 cells, by inhibiting the NF-κB activity, a decrease in Bcl-2, and an increase in caspase-3 and survivin expression. In a xenograft combined immunodeficient (SCID) mouse model, the tumor growth was significantly suppressed demonstrating the potential anticancer activity (He et al. 2016). Sankar et al. synthesized copper oxide NPs from *Ficus religiosa* leaf extract, which imposed an apoptotic effect by the generating ROS, in which there was disruption of mitochondrial membrane potential activity in A549 cells (Sankar et al. 2014). Wang et al., synthesized AuNPs and evaluated their anticancer activity against pancreatic cancer cell line (PANC-1), where it illustrated dose-dependent cytotoxic activity in a time- and dose-dependent manner. Further, the expression of Bcl-2 protein was decreased with an increased dose of Ag NPs, promoting the cancer cell apoptosis by increasing the apoptosis-related protein expressions in time- and dose-dependent mode (Wang et al. 2019). Cheng et al. synthesized zinc oxide (ZnO) NPs from *Rehmanniae Radix*, a Chinese herb. It exerts anticancer activity against osteosarcoma cell line MG-63, by generating more ROS, and decreased mitochondrial membrane potential (MMP) activity. The decreased MMPs tempt toward increased levels of apoptotic proteins Bax, caspase-3, and caspase-9 leading to the induction of apoptosis (Cheng et al. 2020). Dsouza et al. synthesized bimetallic (silver/zinc oxide) nanostructures from the fruit extracts of *Vateria indica*. The in vitro anticancer study performed on triple-negative breast cancer cells MDA-MB468 revealed the enhancement of antiproliferative activity compared to only Ag NPs (D'Souza et al. 2022). Satpathy et al. synthesized Ag NPs using *Pueraria tuberosa*, which demonstrated good anticancer activity in different cancer cell lines (SKOV-3, NCI/ADR, U-87, MCF-7, MDA-MB-231) ~ 29.3, 25.4, 6.05, 3.8, and 1.1 μg/mL respectively. The comparative anticancer activity showed in terms of IC_{50} lowest for MDA-MB-231 and highest for SKOV-3 (Satpathy et al. 2018). Satishkumar et al. synthesized Ag NPs using *Coriandrum sativum* leaf extract, which has shown remarkable anticancer activity in MCF-7 (Sathishkumar et al. 2016b).

5.2.5 Nanoparticles as a Wound-Healing Agent

The wound takes place due to the sharp injuries that occur to the skin where the dermal layers are cut, punctured, or torn due to trauma (Nethi et al. 2019). The wounds can be acute or chronic depending upon the healing period as well as other complications (Eming et al. 2014). Healing of the wound takes place with the interaction of various types of cells, coagulation factors, connective tissue, growth factor, and cytokines (Katas et al. 2018). Healing of the wound is dependent on four phases (hemostasis phase, inflammatory phase, proliferative phase, and maturation phase). These phases are quite complex and function in a coordinated manner. Failure or lack of functionality in any stage leads to chronic wounds (Katas et al. 2018; Martin and Nunan 2015). Nano-based approaches for wound-healing involving herbal extracts can effectively address the specificity associated with the wound. Ahn et al. generated Ag NPs from seven plant extracts (*Cratoxylum formosum, Ceratostigma minus, Phoebe lanceolate, Scurrula parasitica, Mucuna birdwoodiana, Myrsine africana*, and *Lindera strychnifolia*). Out of these studied plant extracts, Ag NPs generated from *Lindera strychnifolia* illustrated wound-healing properties. The cell scratch method utilized for wound-healing activity on mouse fibroblast cells (NIH3T3) illustrated better healing activity (Ahn et al. 2019). Ag NP hydrogel using root extract of *Arnebia nobilis* exhibits splendid antibacterial and wound-healing activity in excision albino Wistar animals. Naraginti et al. synthesized Ag NPs and AuNPs from the root extracts of *Coleus forskohlii*. These nanoparticles displayed noticeable wound-healing activity in excision wounds in albino Wistar male rats. However, topical application of formulated AuNPs is more effective in suppressing inflammation and stimulating reepithelialization compared to Ag NPs during the healing process (Naraginti and Sivakumar 2014). Titanium oxide NPs synthesized from *Origanum vulgare* efficacy were examined in the excision wound model illustrating significant wound-healing activity in albino rats (Sankar et al. 2014). Magnetic nanoparticles were biosynthesized by using *Aloe vera* extract in newly isolated bacterial nanocellulose (BNC) RM1, which was evaluated for wound-healing activity in human dermal fibroblast cells. The genes that are responsible for wound healing were TGF-B1, MMP2, MMP9, Wnt4, CTNNB1, hsa-miR-29b, and hsa-miR-29c. These genes responded in a time-dependent manner for the therapy of cutaneous wound healing (Moniri et al. 2018). ZnONPs were synthesized using an aqueous leaf extract plant of *Barleria gibsoni*. These developed ZnO-NP gels worked effectively for the healing of burn infections in rats. Copper oxide NPs were synthesized from *Ficus religious* leaf extract, and its wound-healing efficacy was evaluated. The copper oxide NPs demonstrated superior wound-healing activity by upregulation of major wound-healing proteins in the different phases of wound repair, wound contraction, and reepithelialization process (Sankar et al. 2014).

5.2.6 Nanoparticles as an Anti-inflammatory Agent

During inflammation, our body is protected from invaders through chemicals produced by the body's white blood cells that enter the blood or tissues. Inflammation is a phenomenon that occurs as a result of injury, infection, and stress, where chemicals are released through multiple mechanisms and, in turn, recruit macrophages and killer cells such as cytokines like IL-1, IL-1β, and TNF-α to the desired site (Ong et al. 2007). Metallic nanoparticles synthesized through the green route have demonstrated anti-inflammatory properties. Utilizing the aqueous extract of *Selaginella myosurus*, Ag NPs were developed, and under in vivo and in vitro conditions, it demonstrated anti-inflammatory potential. Application of Ag-NPs inhibits protein denaturation and prevents the release of acute inflammatory mediators [histamines, serotonin, kinins, prostaglandins, and cyclooxygenase products] in the Carrageenan-induced rat hind paw edema model (Kedi et al. 2018). In another study from *Prunus serrulata*, AuNPs were synthesized and evaluated against lipopolysaccharide (LPS)-induced RAW264.7 macrophage. These NPs in LPS-induced RAW264.7 cells suppressed the production of inflammatory mediators and pro-inflammatory cytokines by inhibiting NF-κB activation (Singh et al. 2018). Zinc oxide NPs developed from *Polygala tenuifolia* root extract exhibited promising anti-inflammatory activity by inhibiting the expressions of proteins iNOS, COX-2, IL-1b, IL-6, and TNF-α (Nagajyothi et al. 2017). Kup et al. used aqueous extracts of *Aesculus hippocastanum* (horse chestnut) as a reducing agent for the synthesized Ag NPs (Küp et al. 2020). With the application of Ag NPs, the superoxide radical scavenging activity is increased with increasing concentrations, and inhibition was about 62.9% as compared to the activity of plant extract (Küp et al. 2020). Green synthesized NPs act by blocking pro-inflammatory cytokines and ROS scavenging mechanisms and inhibiting the NF-κB and COX-2 pathways to minimize the inflammation with greater efficiency.

5.2.7 Nanoparticles as Antileishmanial Agent

Leishmaniasis is caused by infection with Leishmania parasites, which are spread by the bite of phlebotomine sand flies. It is another essential life-threatening disease. It is classified based on severity and intensity as visceral, cutaneous, and post kala-azar dermal leishmaniasis and mucocutaneous leishmaniasis (Arenas et al. 2017). The disease is treated with antileishmanial drugs, but in the long run, it develops resistance due to the increased efflux mechanism, inhibition of drug activation, and inactivation of active drug, which hampers the drug activity as well as decreased drug concentration inside the parasite (Mohapatra 2014). Ullah et al. synthesized from the aqueous extract of *Teucrium stocksianum* and displayed a strong antagonistic assay against *Leishmania infantum* promastigotes compared to chemically synthesized Ag NPs (Ullah et al. 2018). The cytotoxic assay based on MTT reagent (3-(4,5-dimethylthiazol-2-yl)-2,5-diphenyl-2*H*-tetrazolium bromide) revealed toxic effects of chemical Ag NPs, compared to green synthesized Ag NPs. Synthesis of AuNPs from an extract of *Maytenus royleanus* demonstrated significant antileishmaniasis against *Leishmania tropica* promastigotes (Ahmad et al. 2016). Sumbal et al. developed monometallic ZnO NPs and Ag NPs and bimetallic (ZnO/Ag) NPs using *Mirabilis jalapa* leaf extract. Bimetallic NPs displayed greater potentiality for antileishmanial activity compared to monometallic NPs (Sumbal et al. 2019). Awad et al. synthesized Ag NPs using *Commiphora molmol* (myrrh). Different concentrations (10, 50, 80, 100, and 150 μL/100 μL) of Ag NPs were applied and studied for antileishmanial activity. At higher concentrations such as 100 and 150 μL/100 μL, significantly greater inhibitory effect was observed compared to the chemical nanoparticles and pentostam at the same concentrations. After

21 days of post-administration of green synthesized Ag NPs, the lesion was healed completely, whereas the commercial NPs illustrated a moderate healing effect in vivo (Awad et al. 2021).

5.2.8 Nanoparticles as an Anticoagulant Agent

Loss of blood in the human body is protected through the clotting mechanism. The clotting mechanism is contributed by the numerous components such as platelets, coagulation factors, prostaglandins, enzymes, and proteins which act together to form clots and stop the bleeding. Clots also include stroke, pulmonary embolism, deep venous thrombosis (DVT), acute coronary syndrome (ACS), and acute myocardial infarction (AMI). A blood clot formed from infection or contamination damages the tissues leading to organ failure linked to cardiovascular disorders, autoimmune reactions, allergic responses, injuries, and the emergence of cancer (Levi et al. 2010; Davalos and Akassoglou 2012; Prandoni et al. 2007). Anticoagulants are used for the prevention of blood clots to avert the disorders associated with thrombosis. Commonly used anticoagulants are warfarin, rivaroxaban, dabigatran, and heparin products, which are closely linked with adverse drug events (Alquwaizani et al. 2013). Researchers have demonstrated the anticoagulant efficacy of green synthesized metallic NPs. Abbasi et al. synthesized Ag NPs using an aqueous extract of dried Juglans regia green husk, which has blood clot prevention efficacy within 72 h in a dose-dependent manner (Abbasi et al. 2017). Ag NPs were synthesized from *Bridelia retusa* fruit extract, which prevented blood clots in human blood samples and is used in nanomedicine (Vinayagam et al. 2017). *Peltophorum pterocarpum*-synthesized Ag NPs also showed anticoagulant activity (Raja et al. 2015). Ag NPs synthesized from cocoa beans demonstrated antiplatelet activity, and it prevented blood coagulation without disturbing the morphological features of red blood cells (Azeez et al. 2017). Lateef et al. used the seed and leaf of *Synsepalum dulcificum* for the synthesis of Ag NPs (Lateef et al. 2016), and the pods, seeds, and shell of *Cola nitida* were used for the synthesis of Ag NPs (Lateef et al. 2017), which showed anticoagulant activity and prevent clot formation for a longer period. Leaf extract of *Petiveria alliacea* was used for the synthesis of Ag NPs, which inhibited aggregation of platelets, thus preventing clot formation for a longer period, likewise the activity similar to EDTA (Lateef et al. 2018).

5.3 Conclusion

The biological synthesis of nanoparticles has received unrivaled focus to upsurge their applications in biomedicine. The green synthesis process provides a clean, nontoxic, and eco-friendly approach to the synthesis of metal NPs compared to other conventional techniques. Studies have illustrated better biomedical applications compared to their chemically synthesized counterparts. This will open up the ways for exploring the development of improved healthcare for the clinical application of mankind.

References

Abbai R, Mathiyalagan R, Markus J, Kim YJ, Wang C, Singh P, Ahn S, Farh ME, Yang DC (2016) Green synthesis of multifunctional silver and gold nanoparticles from the oriental herbal adaptogen: Siberian ginseng. Int J Nanomedicine 11:3131–3143

Abbasi Z, Feizi S, Taghipour E, Ghadam P (2017) Green synthesis of silver nanoparticles using aqueous extract of dried Juglans regia green husk and examination of its biological properties. Green Process Synth 6:477–485

Abel EE, Poonga PRJ, Panicker SG (2016) Characterization and in vitro studies on anticancer, antioxidant activity against colon cancer cell line of gold nanoparticles capped with Cassia tora SM leaf extract. Appl Nanosci 6:121–129

Ahmad A, Syed F, Imran M, Khan AU, Tahir K, Khan ZUH, Yuan Q (2016) Phytosynthesis and antileishmanial activity of gold nanoparticles. J Food Biochem 40:420–427

Ahmad N, Sharma AK, Sharma S, Khan I, Sharma DK, Shamsi A, Santhosh Kumar TR, Seervi M (2019) Biosynthesized composites of Au-Ag nanoparticles using Trapa peel extract induced ROS-mediated p53 independent apoptosis in cancer cells. Drug Chem Toxicol 42(1):43–53

Ahn EY, Jin H, Park Y (2019) Assessing the antioxidant, cytotoxic, apoptotic and wound healing properties of silver nanoparticles green-synthesized by plant extracts. Mater Sci Eng C Mater Biol Appl 101:204–216

Alquwaizani M, Buckley L, Adams C, Fanikos J (2013) Anticoagulants: a review of the pharmacology, dosing, and complications. Curr Emerg Hosp Med Rep 1(2):83–97

Arenas R, Torres-Guerrero E, Quintanilla-Cedillo MR, Ruiz-Esmenjaud J (2017) Leishmaniasis: a review. F1000Research 6:1–15

Asgary V, Shoari A, Baghbani-Arani F, Sadat Shandiz SA, Khosravy MS, Janani A, Bigdeli R, Bashar R, Cohan RA (2016) Green synthesis and evaluation of silver nanoparticles as adjuvant in rabies veterinary vaccine. Int J Nanomedicine 11:3597–3605

Awad MA, Al Olayan EM, Siddiqui MI, Merghani NM, Alsaif SSA, Aloufi AS (2021) Antileishmanial effect of silver nanoparticles: green synthesis, characterization, in vivo and in vitro assessment. Biomed Pharmacother 137:111294

Azeez MA, Lateef A, Asafa TB, Yekeen TA, Akinboro A, Oladipo IC, Gueguim-Kana EB, Beukes LS (2017) Biomedical applications of cocoa bean extract-mediated silver nanoparticles as antimicrobial, larvicidal and anticoagulant agents. J Clust Sci 28:149–164

Behzad F, Naghib SM, Kouhbanani MAJ, Tabatabaei SN, Zare Y, Rhee KY (2021) An overview of the plant-mediated green synthesis of noble metal nanoparticles for antibacterial applications. J Ind Eng Chem 94:92–104

Bhakya S, Muthukrishnan S, Sukumaran M et al (2016) Biogenic synthesis of silver nanoparticles and their antioxidant and antibacterial activity. Appl Nanosci 6:755–766

Biswal SK, Panigrahi GK, Sahoo SK (2020) Green synthesis of Fe_2O_3-Ag nanocomposite using Psidium guajava leaf extract: an eco-friendly and recyclable adsorbent for remediation of Cr (VI) from aqueous media. Biophys Chem 263:106392

Castro Aceituno V, Ahn S, Simu SY, Wang C, Mathiyalagan R, Yang DC (2016) Silver nanoparticles from Dendropanax morbifera Léveille inhibit cell migration, induce apoptosis, and increase generation of reactive oxygen species in A549 lung cancer cells. In Vitro Cell Dev Biol Anim 52(10):1012–1019

Castro-Aceituno V, Ahn S, Simu SY, Singh P, Mathiyalagan R, Lee HA, Yang DC (2016) Anticancer activity of silver nanoparticles from Panax ginseng fresh leaves in human cancer cells. Biomed Pharmacother 84:158–165

Cheng J, Wang X, Qiu L et al (2020) Green synthesized zinc oxide nanoparticles regulates the apoptotic expression in bone cancer cells MG-63 cells. J Photochem Photobiol B Biol 202: 111644

D'Souza JN, Nagaraja GK, Prabhu AK, Navada KM, Kouser S, Manasa DJ (2022) AgVI and Ag/ZnOVI nanostructures from Vateria indica (L.) exert antioxidant, antidiabetic, antiinflammatory, and cytotoxic efficacy on triple-negative breast cancer cells in vitro. Int J Pharm 615:121450

Davalos D, Akassoglou K (2012) Fibrinogen as a key regulator of inflammation in disease. Semin Immunopathol 34:43–62

Dilnawaz F, Acharya S, Sahoo SK (2018) Recent trends of nanomedicinal approaches in clinics. Int J Pharm 538(1–2):263–278

Dwyer DJ, Camacho DM, Kohanski MA, Callura JM, Collins JJ (2012) Antibiotic-induced bacterial cell death exhibits physiological and biochemical hallmarks of apoptosis. Mol Cell 46:561–572

Eming SA, Martin P, Tomic-Canic M (2014) Wound repair and regeneration mechanisms. Sci Transl Med 322:265–266

Franci G, Falanga A, Galdiero S, Palomba L, Rai M, Morelli G, Galdiero M (2015) Silver nanoparticles as potential antibacterial agents. Molecules 20:8856–8874

He Y, Du Z, Ma S, Liu Y, Li D, Huang H, Jiang S, Cheng S, Wu W, Zhang K, Zheng X (2016) Effects of green-synthesized silver nanoparticles on lung cancer cells in vitro and grown as xenograft tumors in vivo. Int J Nanomedicine 11:1879–1887

Hua S, De Matos MB, Metselaar JM, Storm G (2018) Current trends and challenges in the clinical translation of nanoparticulate nanomedicines: pathways for translational development and commercialization. Front Pharmacol 9:790

Huang J, Li Q, Sun D, Lu Y, Su Y, Yang X et al (2007) Biosynthesis of silver and gold nanoparticles by novel sundried Cinnamomum camphor leaf. Nanotechnology 18(10): 105104–105115

Iravani S (2011) Green synthesis of metal nanoparticles using plants. Green Chem 10:2638–2650

Katas H, Moden NZ, Lim CS, Celesistinus T, Chan JY, Ganasan P, Abdalla SSI (2018) Biosynthesis and potential applications of silver and gold nanoparticles and their chitosan based nanocomposites in nanomedicine. J Nanotechnol. https://doi.org/10.1155/2018/4290705

Kedi PBE, Meva FE, Kotsedi L, Nguemfo EL, Zangueu CB, Ntoumba AA, Mohamed HEA, Dongmo AB, Maaza M (2018) Eco-friendly synthesis, characterization, in vitro and in vivo antiinflammatory activity of silver nanoparticle-mediated Selaginella myosurus aqueous extract. Int J Nanomedicine 13:8537–8548

Kesharwani J, Yoon KY, Hwang J, Rai M (2009) Phytofabrication of silver nanoparticles by leaf extract of datura metal: hypothetical mechanism involved in synthesis. J Bionanosci 3(1):39–44

Khan MI, Akhtar MN, Ashraf N, Najeeb J, Munir JH, Awan TI, Tahir MB, Kabli MR (2020) Green synthesis of magnesium oxide nanoparticles using Dalbergia sissoo extract for photocatalytic activity and antibacterial efficacy. Appl Nanosci 10:2351–2364

Kharissova OV, Dias HVR, Kharisov BI, Jiménez-Pérez VM, Pérez BO, Flores BM (2012) Iron-containing nanomaterials: synthesis, properties, and environmental applications. IRSC Adv 2(25):9325–9358

Kumar V, Yadav SK (2009) Plant-mediated synthesis of silver and gold nanoparticles and their applications. J Chem Technol Biotechnol 84(2):151–157

Küp FÖ, Çoşkunçay S, Duman F (2020) Biosynthesis of silver nanoparticles using leaf extract of Aesculus hippocastanum (horse chestnut): evaluation of their antibacterial, antioxidant and drug release system activities. Mater Sci Eng C Mater Biol Appl 107:110207

Lateef AA, Akande MA, Azeez MA, Ojo SA, Folarin BI, Gueguim-Kana EB, Beukes LS (2016) Phytosynthesis of silver nanoparticles (AgNPs) using miracle fruit plant (Synsepalum dulcificum) for antimicrobial, catalytic, anticoagulant, and thrombolytic applications. Nanotechnol Rev 5:507–520

Lateef A, Ojo SA, Elegbede JA, Azeez MA, Yekeen TA, Akinboro A (2017) Evaluation of some biosynthesized silver nanoparticles for biomedical applications: hydrogen peroxide scavenging, anticoagulant, and thrombolytic activities. J Clust Sci 28:1379–1392

Lateef A, Folarin BI, Oladejo SM, Akinola PO, Beukes LS, Gueguim-Kana EB (2018) Characterization, antimicrobial, antioxidant and anticoagulant activities of silver nanoparticles synthesized from Petiveria alliacea L. leaf extract. Prep Biochem Biotechnol 48:646–652

Lateef A, Elegbede JA, Akinola OP, Ajai V (2019) Biomedical applications of green synthesized-metallic nanoparticles: a review. Pan Afr J Life Sci 3:157–182

Levi M, Schultz M, van der Poll T (2010) Disseminated intravascular coagulation in infectious disease. Semin Thromb Hemost 36:367–377

Li S, Shen Y, Xie A, Yu X, Qiu L, Zhang L et al (2007) Green synthesis of silver nanoparticles using Capsicum annuum L. extract. Green Chem 9(8):852–858

Lone AA, Ganai SA, Ahanger RA, Bhat HA, Bhat TA, Wani IA (2013) Free radicals and antioxidants: myths, facts, and mysteries. Afr J Pure Appl Chem 7:91–113

Makarov VV, Love AJ, Sinitsyna OV, Shaw J, Yaminsky IV, Kalinina NO et al (2015) A genetically modified tobacco mosaic virus that can produce gold nanoparticles from a metal salt precursor. Front Plant Sci 6:1–10

Marchiol L (2012) Synthesis of metal nanoparticles in living plants. Italian J Agron 7(37):274–282

Martin P, Nunan R (2015) Cellular and molecular mechanisms of repair in acute and chronic wound healing. Br J Dermatol 173(2):370–378

Mauro N, Utzeri MA, Varvarà P, Cavallaro G (2021) Functionalization of metal and carbon nanoparticles with potential in cancer theranostics. Molecules 26(11):3085

Mohapatra S (2014) Drug resistance in leishmaniasis: newer developments. Trop Parasitol 4(1):4–9

Moniri MM, Azizi S, Rahim RA, Saad WZ, Navaderi M, Arulselvan P, Mohamad R (2018) Molecular study of wound healing after using biosynthesized BNC/Fe_3O_4 nanocomposites assisted with a bioinformatics approach. Int J Nanomedicine 13:2955–2971

Nagajyothi PC, Pandurangan M, Kim DH et al (2017) Green synthesis of iron oxide nanoparticles and their catalytic and in vitro anticancer activities. J Clust Sci 28(1):245–257

Naraginti S, Sivakumar A (2014) Eco-friendly synthesis of silver and gold nanoparticles with enhanced bactericidal activity and study of silver catalyzed reduction of 4-nitrophenol. Spectrochim Acta A Mol Biomol Spectrosc 128:357–362

Nethi SK, Das S, Patra CR, Mukherjee S (2019) Recent advances in inorganic nanomaterials for wound-healing applications. Biomater Sci 7:2652–2674

Nguyen D, Dang HH, Vo DVN, Bach G, Nguyen TD, Tran TV (2021) Biogenic synthesis of MgO nanoparticles from different extracts (flower, bark, leaf) of Tecoma stans (L.) and their utilization in selected organic dyes treatment. J Hazard Mater 404:124146

Ong CKS, Lirk P, Tan CH, Seymour RA (2007) An evidence-based update on nonsteroidal antiinflammatory drugs. Clin Med Res 5:19–34

Orłowski P, Kowalczyk A, Tomaszewska E, RanoszekSoliwoda K, Wegrzyn A, Grzesiak J, Celichowski G, Grobelny J, Eriksson K, Krzyzowska M (2018) Antiviral activity of tannic acid modified silver nanoparticles: potential to activate immune response in herpes genitalis. Viruses 10:524

Prandoni P, Falanga A, Piccioli A (2007) Cancer, thrombosis and heparin-induced thrombocytopenia. Thromb Res 120:S137–SS40

Prasannaraj G, Venkatachalam P (2017) Enhanced antibacterial, anti-biofilm and antioxidant (ROS) activities of biomolecules engineered silver nanoparticles against clinically isolated Gram positive and Gram negative microbial pathogens. J Clust Sci 28:645–646

Qayyum S, Khan AU (2016) Biofabrication of broad range antibacterial and antibiofilm silver nanoparticles. IET Nanobiotechnol 10:349–335

Ragu S, Faye G, Iraqui I, Masurel-Heneman A, Kolodner RD, Huang ME (2007) Oxygen metabolism and reactive oxygen species cause chromosomal rearrangements and cell death. Proc Natl Acad Sci U S A 104:9747–9752

Rai M, Yadav A, Gade A (2008) Current [corrected] trends in phytosynthesis of metal nanoparticles. Crit Rev Biotechnol 28(4):277–284

Raja S, Ramesh V, Thivaharan V (2015) Antibacterial and anticoagulant activity of silver nanoparticles synthesised from a novel source-pods of Peltophorum pterocarpum. J Ind Eng Chem 29:257–264

Ramkumar VS, Pugazhendhi A, Gopalakrishnan K, Sivagurunathan P, Saratale GD, Dung TNB, Kannapiran E (2017) Biofabrication and characterization of silver nanoparticles using aqueous extract of seaweed Enteromorpha compressa and its biomedical properties. Biotechnol Rep 14: 1–7

Renganathan S, Aroulmoji V, Shanmugam G, Devarajan G, Rao KV, Rajendar V, Park S (2018) Silver nanoparticles synthesis from Carica papaya and virtual screening for anti-dengue activity using molecular docking. Mater Res Expr 6:035028

Richardson A, Chan BC, Crouch RD, Janiec A (2006) Synthesis of silver nanoparticles: an undergraduate laboratory using green approach. Chem Educ 11:331–333

Sankar R, Dhivya R, Shivashangari KS, Ravikumar V (2014) Wound healing activity of Origanum vulgare engineered titanium dioxide nanoparticles in Wistar albino rats. J Mater Sci Mater Med 25:1701–1708

Sathishkumar G, Jha PK, Vignesh V, Rajkuberan C, Jeyaraj M, Selvakumar M, Jha R, Sivaramakrishnan S (2016a) Cannonball fruit (Couroupita guianensis, Aubl.) extract mediated synthesis of gold nanoparticles and evaluation of its antioxidant activity. J Mol Liq 215:229–236

Sathishkumar P, Preethi J, Vijayan R, Yusoff ARM, Ameen F, Suresh S, Balagurunathan R, Palvannan T (2016b) Anti-acne, anti-dandruff and anti-breast cancer efficacy of green synthesised silver nanoparticles using Coriandrum sativum leaf extract. J Photochem Photobiol B Biol 163:69–76

Satpathy S, Patra A, Ahirwar B, Hussain MD (2018) Antioxidant and anticancer activities of green synthesized silver nanoparticles using aqueous extract of tubers of Pueraria tuberosa. Artif Cells 46(Suppl 3):1–15

Shanmugam C, Sivasubramanian G, Parthasarathi B, Baskaran K, Balachander R, Parameswaran VR (2016) Antimicrobial, free radical scavenging activities and catalytic oxidation of benzyl alcohol by nano-silver synthesized from the leaf extract of Aristolochia indica. Appl Nanosci 6: 711–723

Sharma V, Kaushik S, Pandit P, Dhull D, Yadav JP, Kaushik S (2019) Green synthesis of silver nanoparticles from medicinal plants and evaluation of their antiviral potential against chikungunya virus. Appl Microbiol Biotechnol 103:881–891

Singh R, Nalwa HS (2011) Medical applications of nanoparticles in biological imaging, cell labeling, antimicrobial agents, and anticancer nanodrugs. J Biomed Nanotechnol 7:489–503

Singh P, Ahn S, Kang JP, Veronika S, Huo Y, Singh H, Chokkaligam M, ElAgamy Farh M, Aceituno VC, Kim YJ, Yang DC (2018) In vitro antiinflammatory activity of spherical silver nanoparticles and monodisperse hexagonal gold nanoparticles by fruit extract of Prunus serrulata: a green synthetic approach. Artif Cells Nanomed Biotechnol 46:2022–2032

Sousa C, Nadeem S, Pereira DM, Valentao P, Ferreres F, Pereira JA, Seabra RM, Andrade PB (2009) Pieris brassicae inhibits xanthine. J Agric Food Chem 57:2288–2294

Sumbal A, Nadeem S, Naz JS, Ali A, Mannan M, Zia M (2019) Synthesis, characterization and biological activities of monometallic and bimetallic nanoparticles using Mirabilis Jalapa leaf extract. Biotechnol Rep 22:e00338

Ullah I, Coszar G, Abamor ES, Bagirova M (2018) Comparative study on the antileishmanial activities of chemically and biologically synthesized silver nanoparticles (AgNPs). 3 Biotech 8(2). https://doi.org/10.1007/s13205-018-1121-6

Vinayagam R, Varadavenkatesan T, Selvaraj R (2017) Evaluation of the anticoagulant and catalytic activities of the Bridelia retusa fruit extract-functionalized silver nanoparticles. J Clust Sci 28: 2919–2912

Wang L, Xu J, Yan Y, Liu H, Karunakaran T, Li F (2019) Green synthesis of gold nanoparticles from Scutellaria barbata and its anticancer activity in pancreatic cancer cell (PANC-1). Artif Cells Nanomed Biotechnol 47:1617–1627

Fahima Dilnawaz is an Associate Professor at Centurion University of Technology and Management, Bhubaneswar. She has earned her doctorate, postgraduate, and M.Phil. in Plant Biology. She won her ITC Fellowship from the Hungarian Academy of Sciences and National Postdoctoral Fellowship from the Department of Biotechnology. She worked in the convergence of science, particularly in the nanomedicinal approach in biomedical science as Women Scientist (WOS-A) at the Institute of Life Sciences (an autonomous institute of the Department of Biotechnology, GOI) Bhubaneswar, Odisha, India, for more than 10 years. To her credit, she has published more than 30 scientific papers and review articles and more than 20 book chapters in reputed journals and publishing houses. Apart from that, she has coauthored one book (*Remedial Biology*) for pharmaceutical students and coauthored two patents, one of which has been acclaimed in the USA, Europe, and Australia and the other has an Indian patent. Her research interest is basically focused on polymeric nanoparticles, magnetic nanoparticles, and mesoporous silica nanoparticle-based drug delivery systems for cancer therapy. She has been invited to various scientific organizations to deliver talks in India and abroad. She is serving as a reviewer for multiple journals as well as an associated editorial board member.

Amarendra Narayan Misra is a distinguished professor at the Centurion University of Technology and Management, Bhubaneswar. He has teaching experience of more than 40 years and research experience of more than 45 years. He served as a Full Professor (20 years), out of which 16 years serving as Head, Dean (Academic), Dean of School, Chairman-PG Council, Director-IQAC, Member of the Academic Council, Senate, Syndicate, Executive Council, and Vice-Chancellor. His research collaborations with US, European, and Japanese universities availed funds from the Commission of European Communities (CEC), Japanese Societies for Promotion of Science (JSPS), Tokyo University, Geneva University, DBT, INSA, FICCI, DST, etc. He worked at the best international universities like Imperial College, London, University of Wales, Tokyo University, Osaka University, Geneva University etc. As a Fulbright Scholar, he interacted with the Presidents, Vice-Chancellors, and Senior Administrative Personnel of more than 25 American universities and colleges under Fulbright Academic Administrator's Seminar 2013. Many scholars from Europe, America, Canada, Africa, and Asia have visited and worked in his laboratory as Guest Professors, Exchange Scientists, and Researchers since 1998. As a Vice-Chancellor of Khallikote University (state), he signed MOUs with ten foreign universities for faculty and student exchange.

He has earned his doctorate in the field of plant biology and biochemistry. His research field is bioenergetics, bio-signals and biosensors, molecular biology, and bioinstrumentation. He used luminescence techniques like fast fluorescence kinetics, thermoluminescence, positron-annihilation spectroscopy, flash photolysis of water and oxy-graphs, proteomics, and radioisotopes in plant biology, collaborating with electronics, physics, chemistry, and instrumentation professionals.

Functionalized Carbon Nanotube for Various Disease Treatment

6

Komal Iranna Savadatti, Asha Puthuvilayil Johnson, and Hosahalli Veerabhadrappa Gangadharappa

Abstract

The delivery of therapeutic medications employing nanoparticles has been used in pharmacy and the medical field because of their vast electrical, superior chemical, and enormous surface area. It has been shown that carbon nanotubes (CNTs) are an excellent drug delivery system because they efficiently enter cells as well as because of their polyaromatic nature, and they keep the medication intact during transit in the body without metabolizing it. Drug-CNT conjugates for drug delivery are safer and more effective than the drug used alone in conventional manufacturing as functionalized CNTs can safely carry significant molecules across nuclear and cytoplasmic membranes as well as reduce the toxicity associated with CNT as such. They are chosen as a basis for attaching antibiotics and anticancer medications to carbon nanotubes for the treatment of infections and cancer. Subsequently, other biomolecules were joined to CNTs and investigated for a variety of uses, including gene therapy, immunotherapy, tissue repair, and disease diagnostics and treatment.

Keywords

Functionalized carbon nanotubes · Toxicity · Cancer therapy · Drug delivery

K. I. Savadatti · A. P. Johnson · H. V. Gangadharappa (✉)
Department of Pharmaceutics, JSS College of Pharmacy, JSS Academy of Higher Education and Research, Mysuru, Karnataka, India
e-mail: hvgangadharappa@jssuni.edu.in

F. A. Sheikh et al. (eds.), *Interaction of Nanomaterials With Living Cells*,
https://doi.org/10.1007/978-981-99-2119-5_6

6.1 Introduction

Nanomaterials, which exist as a form of carbon allotrope graphite and exhibit a nanometer-scale diameter and a few millimeters in length (Hirlekar et al. 2009; Singh et al. 2012), have been built in cylindrical tubes. Their outstanding structural, mechanical, and electronic properties are the result of their small size and mass, incredible mechanical strength, and exceptional electrical and thermal conductivity (Usui et al. 2012; Zhang et al. 2010). Ever since the beginning of the twenty-first century, nanoparticles have been utilized in pharmacy and medicine as a therapeutic medication delivery system owing to their large surface area, superior chemical stability, and extensive electrical. It has been demonstrated that carbon nanotubes (CNTs) are an ideal vehicle for drug administration since they immediately penetrate cells and maintain the medication intact without metabolism during transit in the body (Hirlekar et al. 2009; Singh et al. 2012; Usui et al. 2012; Zhang et al. 2010) due to their polyaromatic nature. Numerous studies have shown that these compounds may be transported into cells more efficiently and securely when attached to CNTs (Singh et al. 2012; Usui et al. 2012; Zhang et al. 2010). To cure cancer and infections, it was first used to bind anticancer medicines and antibiotics to carbon nanotubes. Then additional biomolecules have been linked to CNTs and tested for various applications, such as gene therapy, immunotherapy, tissue repair, and disease diagnosis (Kateb et al. 2010; Liu et al. 2007a; Zhang et al. 2011; Rosen and Elman 2009; Bekyarova et al. 2005).

6.1.1 Carbon Nanotubes: Configurational Structures, Types, and Preparation

All of the carbon atoms that make up carbon nanotubes (CNTs) are organized in a sequence of condensed benzene rings and wrapped up into a tube-like form. This novel artificial nanomaterial is found in the native sp2 (planar) and sp3 (cubic) forms in the class of fullerenes, the third allotropic carbon structure after diamond and graphite (Hirlekar et al. 2009; Singh et al. 2012; Liu et al. 2007a). The architectures of CNTs are divided into two groups depending on the number of layers: single-walled carbon nanotubes (SWCNTs) and multiwalled carbon nanotubes (MWCNTs). SWCNTs often appear as tightly packed hexagonal bundles and are made up of an individual graphene cylinder with a diameter that ranges from 0.4 to 2 nm. Each of the coaxial cylinders present in the MWCNTs are constructed of a single layer of graphene encircling a narrow center. MWCNTs' outer diameter is between 2 and 100 nm, while the inner diameter is between 1 and 3 nm, and their length is from 0.2 to several millimeters (Singh et al. 2012; Bekyarova et al. 2005). The tips and the side walls of CNTs are two distinct zones. Rolling the graphene sheet into a tube results in a variety of tubule shapes, which play a significant role in regulating these peculiar features. Depending on its direction, the molecule can roll in one of three different ways: chiral, zigzag, and armchair (Singh et al. 2012; Usui et al. 2012; Zhang et al. 2010; Kateb et al. 2010; Liu et al. 2007a).

6.1.2 Applications of Carbon Nanotubes in the Pharmaceutical and Medical Fields

The functionalized CNTs are coated on the surface or loaded inside with the drug. Functionalized CNTs can transport desired molecules beyond the cytoplasmic and nuclear membranes without creating a harmful outcome. The conjugates of carbon nanotubes reveal to be both safer and much more potent than the medications used alone via conventional preparation (Zhang et al. 2010; Kateb et al. 2010; Liu et al. 2007a).

6.2 Properties of Carbon Nanotubes

6.2.1 Physical Property

6.2.1.1 Young's Modulus and Tensile Strength

Tensile strength and Young's modulus are measured using a stress-strain puller to drag long ropes holding plenty of aligned nanotubes. The rope's Young's modulus, its tensile strength, and the associated rope elongation can all be determined at once by exerting an axial force on the rope to confirm the number of broken tubes a load was applied along with continuous monitoring of electric conductivity. The sample's cross-section and the tubes' filling factor were determined from scanning (SEM) and transmission electron microscopy (TEM) observations to produce the stress-strain curve. While the tensile strength of the tube cannot be determined by merely averaging the values of all the tubes mostly from the data, the Young's modulus can. According to the parallel structure model of the samples, the first tube to break in a rope made up of numerous parallel and isolated tubes of varying strengths is always the weakest one. As a result, as the load is redistributed, the remaining, undamaged tubes are put under more stress, which eventually leads to the breakage of the second weak tube. This technique will reduce the apparent value of the rope's tensile strength and will undoubtedly reduce the tube's derived value (Daniels 1945).

6.2.2 Mechanical Property

6.2.2.1 CNT Deformation Under Stress

The deformation of carbon nanotubes using various methods is being studied, and in recent attempts, carbon nanotubes distributed in a polymeric film were subjected to significant compressive strains (Lourie et al. 1998). The results for the buckling of thick tubes show that the axial compression deformation of CNT when compared to single-layer CNT was less due to the hollow shape and high aspect ratio. Buongiorno Nardelli et al. (1998); however, there are differences in the plastic collapse or breakage of thin tubes. For thin tubes, the critical stress for inward collapse or fracture is anticipated to be between 100 and 150 GPa, with the compressive strain assessed to be greater than 5%. The fundamental finding of those investigations is that the compression strength of both thin-walled and thick-walled carbon nanotubes is orders of magnitude greater than those of any other known fiber. The toughest tubes are zigzag and armchair forms, with the 0 K stress being very sensitive to helicity, as shown by the molecular dynamics calculations of SWNTs in a TB-large-scale model under significant applied strains (both elongation and compression) (Ozaki et al. 2000).

In all methods, carbon nanotubes transform into new shapes with a sudden release of stress energy when subjected to substantial deformations (Buongiorno Nardelli et al. 1998). The bending is reversible up to very large bending degrees, despite imperfections and highly strained tubule regions. This flexibility property comes from the ability of the C-C bonds in the sp2 network to reversibly change the hybridization when deformed out of the plane. With increased curvature, the sp3 nature of C-C bonds in the deformed region becomes stronger. The mechanisms of strain produced in carbon nanotubes under uniaxial tension have also been investigated in order to address the question of the ultimate tensile strength of these nanoparticles (Buongiorno Nardelli et al. 1998; Lourie et al. 1998; Ozaki et al. 2000; Yakobson 1998; Srivastava et al. 1999).

6.2.3 Thermal Property

6.2.3.1 Specific Heat

In a carbon nanotube, a single graphene sheet has been wound into a cylinder as a carbon nanotube. Covering the sheet twice has a considerable impact on the phonon band structure. The two-dimensional (2D) photon band structure of the sheet first “folds” into the 1D band structure of the tube. Additionally, the distribution of the lowest-lying modes changes because the tube is firmer than the sheet due to its cylindrical shape (Saito et al. 1998; Benedict et al. 1996). As a result, changing the dimensionality of a system can significantly affect the low-energy projected density of states (PDOS) and, consequently, the low-temperature-specific heat. The magnitude of this effect in graphite is related to the up to 10 meV Debye energy of the interlayer modes. It is expected that forming 3D crystalline arrays of tubes will reduce low-energy PDOS in nanotubes (Mizel et al. 1999).

6.2.3.2 Thermal Conductivity

The materials with the highest-documented thermal conductivity at ambient temperature are diamond and graphite; consequently, nanotubes do need to function appropriately here as well. A recent theoretical study (Berber et al. 2000; Hone et al. 2000a) indicated that the thermal conductivity of nanotubes at ambient temperature might reach as high as 6600 W/m K. The consequences of 1D quantization should also be seen in the thermal conductivity at low temperatures, just like they are in the specific heat. The thermal conductivity of a highly anisotropic material is particularly susceptible to phonons with high velocities and extended scattering wavelengths. It follows that thermal conductivity should directly probe on tube phonons and be robust to intertube coupling even in nanotube bundles. Thermal conductivity in all samples has a completely linear temperature dependence below 40 K. Given that only the tube's acoustic modes transport any heat flow during 1D quantization, this temperature dependence is most likely caused by this phenomenon. The impact of intertube connections on *K*(*T*) (temperature)'s dependency is unknown, though. To more precisely assess whether the linear *K*(*T*) is due to 1D quantization, we evaluated *K*(*T*) in samples with different nanotube diameters. Since the phonon subband separation rises with a decreasing tube diameter (Llaguno et al. 2002; Hone et al. 1999), we predict that the linear *K*(*T*) could reach a wider temperature range in samples with a smaller average tube diameter. Composite materials with high thermal conductivity have a variety of possible applications, particularly in heat lowering for electronics and motors.

Carbon nanotubes' distinctive structure as well as small size are closely related to their thermal characteristics. These characteristics make nanotubes potentially the best material for the study of low-dimensional phonon physics and controlling the temperature on both a macro- and microscale the specific heat and thermal conductivity of bulk SWNT samples were examined to investigate the thermal characteristics of nanotubes. Single-walled nanotubes have a different specific heat than both 2D and 3D graphenes, particularly at low temperatures, where the 1D quantization of the phonon band structure is seen. Nanotubes have high thermal conductivity even in bulk materials; aligned bundles of SWNTs display thermal conductivity of >200 W/m K at room temperature. According to the measurements of *K*(*T*) of samples with various average nanotube diameters, a linear *K*(*T*) up to about 40 K may be the result of 1D quantization (Maarouf et al. 2000; Biercuk et al. 2002; Hone et al. 2000b; Kahn and Lu 1999; Teizer et al. 1999).

6.2.4 Optical Property

CNT has shown scope in several applications. Aligned carbon nanotube arrays in particular have displayed a variety of intriguing optical characteristics, including photonic crystal phenomena (Zhao et al. 2006; Kempa et al. 2003; Lidorikis and Ferrari 2009), directed wavelength-selective emission, polarization-dependent (Wang et al. 2004a; Shoji et al. 2008) reflection, emission (Slepyan et al. 2006; Kempa et al. 2007; Wang et al. 2009), and increased absorptivity (Yang et al. 2008; Mizuno et al. 2008).

The atomic structure of each CNT determines the optical characteristics of CNT arrays before their collective configuration. The optical characteristics of MWCNTs and SWCNTs should be taken into account separately for each CNT. The detailed atomic structure (chirality) of SWCNTs, particularly those with diameters smaller than 1 nm, exhibits a very high dependence on optical properties (Lin 1994; Bachilo et al. 2002; Guo et al. 2004a). In experiments, it is still challenging to achieve chirality-controlled CNT growth or chirality separation. As a result, SWCNT arrays often have a wide range of chirality with random distribution, which makes it more difficult to create optical characteristics and functionalities (Liu et al. 2002).

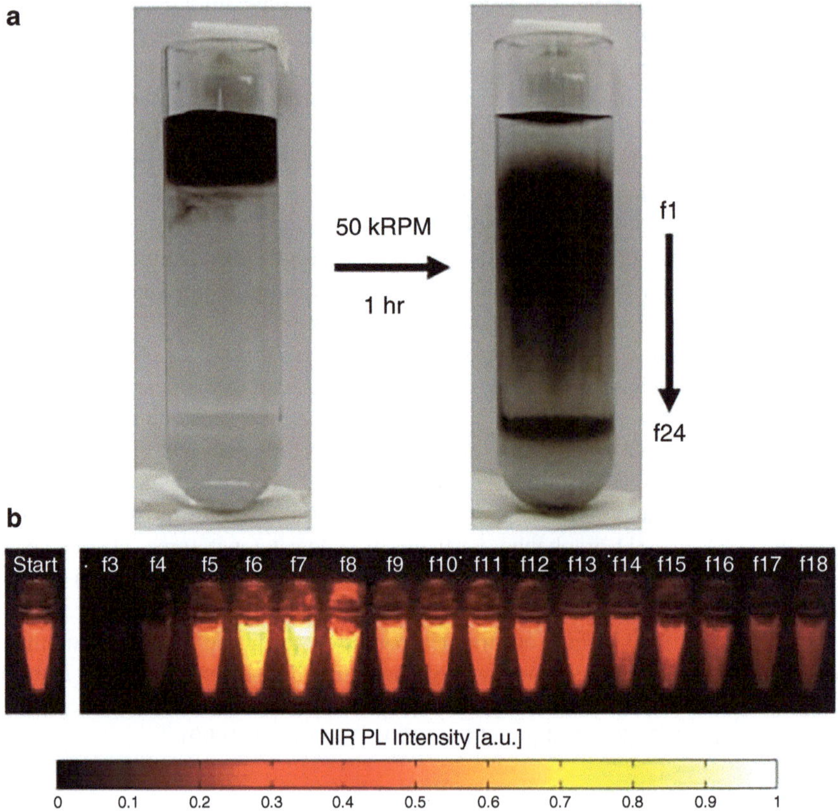

Fig. 6.1 The centrifugation of sodium-cholate-suspended single-walled carbon nanotubes (SWNTs) through a density gradient containing 1% sodium cholate, with discontinuous steps of 5%/10%/15%/20%/60% iodixanol at 50,000 rpm for 1 h, yielded a continuous distribution of SWNTs as well as a band formed at the 60% iodixanol boundary, as is clear from (**a**) photographs taken before and after DGC. Following the aliquoting of 100 μL fractions (f #), as shown in (**a**), and normalization to the same optical density, photoluminescence under 808 nm excitation (**b**) showed varying quantum yields relative to the starting material ("start"), increasing from f3 to f6–7 and decreasing monotonically thereafter (Lin 2000)

MWCNTs, in contrast, exhibit more consistent and uniform optical characteristics because of their bigger size (Lin 2000) (Fig. 6.1).

6.2.5 Electrical

Improved electrochemical properties can be obtained from CNTs loaded with metals like Si (Ganesh 2013), Sn, and Pd, along with transition metal oxides (Li et al. 2017) and sulfides (Poudel and Li 2018). It was demonstrated that irregularities, such as structural flaws produced during synthesis procedures, or physical flaws, such as

Table 6.1 Difference between SWCNT and MWCNT (Singh et al. 2012; Saifuddin et al. 2013; He et al. 2013)

SWCNT	MWCNT
• Single layer of graphene. • Not as much accumulation in the body. • Catalyst is essential for the synthesis. • Characterization and evaluation are easy. • It is more flexible and can be twisted more effortlessly. • Form bundled aggregates that are not completely distributed. • Chance of defect is higher during functionalization. • Resistivity usually in the range of 10^{-4} to 10^{-3} Ω m. • Bulk synthesis is challenging because it needs the careful management of growth and ambient conditions. • Purity is low; samples created using the chemical vapor deposition (CVD) process typically contain between 30 and 50 wt.% of SWCNTs. However, employing the arc discharge synthesis process has been shown to produce high purity of up to 80%.	• Multiple layers of graphene. • Progressively accumulates in the body. • Synthesis without a catalyst is possible. • Difficult due to its complex structure. • It cannot be easily twisted. • Homogeneously dispersed with no apparent bundled formation. • A chance of defect is less, especially when synthesized by the arc-discharged method. • Resistivity usually in the range of 1.8×10^{-5} to 6.1×10^{-5} Ω m. • Bulk synthesis is easy. • Purity is high. Typical MWCNT content in as-prepared samples by the CVD method is about 35–90 wt%.

those brought on by extremely high mechanical pressures, could change the way electrons move through CNTs. On the other hand, it has been found that semiconductor nanotubes are the most sensitive. The electrical properties of CNTs are significantly impacted by the presence of doping substances. Dopants can interact physically with the CNT electrical structure or chemically with the CNT framework (Janas et al. 2017). Many inexpensive synthetic approaches, such as laser ablation, arc discharge, solvothermal processing, etc., for CNT lately have made bulk production easy and cost-effective. CNTs have exceptional electrical properties, with a carrying capacity 1000 times greater than that of copper cables, as shown by theoretical and experimental data. CNTs are therefore probably going to have a big impact as additives in enhancing the electrical properties of composite materials (Nakano et al. 2003; Liu and Gao 2005). The difference between SWCNT and MWCNT is given in Table 6.1.

6.3 Characterization

6.3.1 Carbon Nanotubes: Structures, Types, and Preparation

Carbon nanotubes (CNTs) are tubular structures made of all the carbon atoms organized in a series of condensed benzene rings. This novel synthetic nanomaterial belongs to the family of fullerenes, which also includes graphite and diamond, which are respectively sp2 (planar) and sp3 (cubic) forms of naturally occurring carbon (Hirlekar et al. 2009; Liu et al. 2007a; Pang et al. 1993). The third allotropic type of carbon is fullerene. Based on the number of layers, the models of CNTs are separated into two categories: single-walled carbon nanotubes (SWCNTs) and multiwalled carbon nanotubes (MWCNTs).

SWCNTs typically appear as hexagonal, close-packed bundles and are made up of a single graphene cylinder with a diameter anywhere from 0.4 to 2 nm (Fig. 6.2). MWCNTs are built up of two to three coaxial cylinders that each have a hollow core and are covered with a single graphene sheet. MWCNTs have an outside diameter that ranges from 2 to 100 nm, an inner diameter that is between 1 and 3 nm, and a length that spans from 0.2 to several meters (Singh et al. 2012; Bekyarova et al. 2005). The tips and the sidewalls of CNTs are two distinct zones. Rolling the graphene sheet into a tube results in a variety of tubule shapes, which play a significant role in regulating these peculiar features. Depending on its direction, the molecule can roll in one of three different ways: chiral, zigzag, and armchair.

The arc-discharge approach, which involves the arc vaporization of two carbon rods; the laser ablation method, which uses graphite; and chemical vapor deposition are indeed the three primary methods that are frequently used to produce SWCNTs and MWCNTs (using hydrocarbon sources: carbon monoxide (CO), methane,

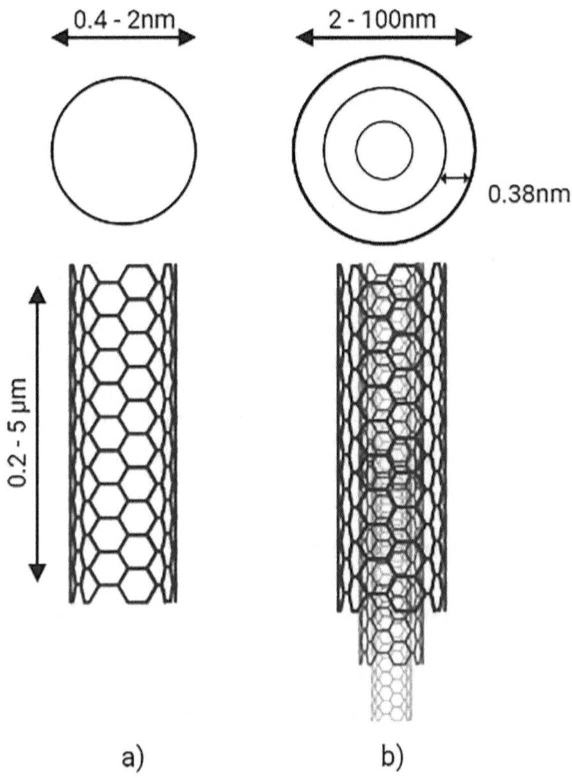

Fig. 6.2 (**a**) Conceptual diagram of single-walled carbon nanotubes (SWCNTs) and (**b**) multiwalled carbon nanotubes (MWCNTs) (He et al. 2013)

ethylene, acetylene). Amorphous carbon, fullerenes, and transition metals supplied as catalysts during the synthesis are examples of defects or impurities that are removed from CNTs after manufacture using acid refluxing, surfactant-aided sonication or the air oxidation technique (Singh et al. 2012; Usui et al. 2012; Digge et al. 2012).

6.3.2 Carbon Nanotube Structure

Graphite sheets are curled into tubes to form carbon nanotubes (Thostenson et al. 2001). Due to their large length-to-diameter ratio, nanotubes are regarded as having almost one-dimensional structures. Single-walled nanotubes (SWNTs) and multiwalled nanotubes are the most significant structures (MWNTs). An SWNT is seen as a cylinder with a single graphene sheet wrapped around it. Concentric SWNT clusters resemble multiwalled nanotubes (MWNTs). These structures are considerably different from SWNTs in terms of their length and diameter, and they also have quite different characteristics. A single vector C describes all properties of single-walled nanotubes except for their length (called a chiral vector). One of the two selected atoms in a planar graphene sheet serves as the origin. Different models that describe multiwalled carbon nanotubes are in good accord with experiments, particularly with the images from electron microscopy. Coaxially curved, coaxially polygonal, or scrolled graphene sheets can be used to make CNTs (Amelinckx et al. 1999). Although the coaxial cylindrical model for CNTs is largely accepted, polygonized tubes have also been found. These are often limited to large tube sizes, allowing three-dimensionally correlated regions; this allows for the observation of a low-angle tilt and well-aligned borders (Ajayan and Ebbesen 1997).

6.3.3 Morphological and Structural Characterizations

Similarly to CNTs, fullerenes are closed spheres made of pentagons and hexagons that are carbon-based compounds. The arrangement of these pentagons and hexagons determines their curvatures. A reversible diatomic exchange results in Stone-Wales transformation in carbon nanotubes. Two pentagons and two heptagons in pairs make up the final construction. The heptagon, a novel CNT defect that enables concave regions within the nanotube, is brought about by this change. Thus, different equilibrium shapes can be created and not just straight tubes with hemispherical crowns (Chen et al. 1999). CNTs have special mechanical and electrical features. Nevertheless, CNTs need to undergo chemical processing to be purified and to be given the required functionalization to acquire these qualities. Common treatments include oxidative techniques with nitric acids. These purification methods involve removing the caps from the ends of the CNTs, which reveal flaws such as carboxylic acid groups on the surface (Hu et al. 2001). The locations of these flaws on the walls and at the ends affect the nanotubes' characteristics (Mawhinney et al. 2000). For the study of the characteristics of nanotubes, a determination of the concentration of these flaws would be beneficial. There are several ways to measure the concentration of carboxylic acid groups produced by purifying processes.

6.3.4 Photoluminescence Spectroscopy

Both metallic and semiconducting types of SWNTs are possible. The semiconducting tubes' gap energy is inversely proportional to the tube's diameter and is correlated with chirality (Ouyang et al. 2001; Lauret et al. 2004). It is to be assumed that the recombination of electron-hole pairs near the bandgap will result in photoluminescence. The SWNTs are typically bundled together, as was observed. Nanotubes in a bundle engage in Van der Waals force interactions with one another. These bundles of nanotubes contain a few metallic nanotubes that resemble nonradiative channels. Inside these channels, the semiconducting tubes in these bundles relax their luminosity. Frequently, no photoluminescence signal is recorded because of this interaction between semiconducting and metallic nanotubes. The bundles must be divided into individual tubes to view the photoluminescence phenomenon. This separation might be accomplished using some procedures. The ultrasonication of nanotubes with surfactants in aqueous suspensions, such as SDS (sodium dodecyl sulfate), seems to be one of the most well-liked methods (Lauret et al. 2004; Lefebvre et al. 2004; Weisman et al. 2004; O'Connell et al. 2003; Lebedkin et al. 2003). Different CNT samples could be used for this procedure. Utilizing individual nanotubes grown in zeolite channels, photoluminescence has also been demonstrated (Guo et al. 2004b). The photoluminescence method could provide access to nature (whether semiconducting or not), geometries, and diameters. Additionally, it appears that the luminescence spectra are particularly sensitive to both the purity of the materials and the existence of chemical flaws (Lauret et al. 2004).

6.3.5 X-Ray Photoelectron Spectroscopy (XPS)

The XPS method can provide details about the chemical composition of carbon nanotubes. However, the most frequently cited information focuses on how the structure of CNT walls has changed as a result of chemical interactions with organic molecules or gas adsorption. XPS research on nitrogen incorporated into carbon nanotubes was carried out to understand the chemical changes formed due to the incorporation of nitrogen. There has been significant research on N1s and C1s peaks. C1s peak shows both a shift and an asymmetric widening at higher binding energies when compared to nonnitrogenated samples. Because of the polar nature of the carbon-nitrogen bond (Hammer et al. 2000), this peak's tip shift serves as proof that nitrogen was incorporated into the nanotube structure. On carbon nanofibers, XPS research was also carried out (Pham-Huu et al. 2002), and the C1s peak was compared to one discovered on a spotless highly ordered pyrolytic graphite (HOPG) (0 0 0 1) surface. The relative concordance between the highest binding energy and the surface plasmon peak at 291 eV indicates that the carbon nanofibers are graphitic. However, flaws and C H terminating surfaces are blamed for a broadening of the peak. There is a little widening at about 286.5 eV that can be attributed to surface oxygen groups with single bonds. The description of surface oxygen groups with many carbon-oxygen bonds was provided by a contribution at about 289.5 eV, though. As a result, it can be said that the nanofiber material is more similar to carbon oxide than a specific type of graphite (Pham-Huu et al. 2002).

6.3.6 X-Ray Diffraction

Using this nondestructive characterization method, some information on impurities, structural strain, and interlayer spacing is obtained. However, carbon nanotubes can be orientated in a wide variety of ways in contrast to the X-ray incident beam. Along with varying diameters, MWNTs also exhibit a range in layer counts and chirality distribution. This leads to the statistical characterization of carbon nanotubes. Due to CNTs inherent properties, the principal X-ray diffraction patterns are almost similar with those of graphite. (Figure 6.3): As CNT and graphite show similar results i.e., (1) The Bragg rule can be used to calculate the interlayer spacing from the location of a peak that resembles graphite (0 0 2 l).; (2) The individual graphene sheet's honeycomb lattice results in a family of (h k 0) peaks. As a result, the X-ray diffraction profile (Zhu et al. 2003) can be utilized to assess sample purity but is unsuitable for differentiating microstructural characteristics between CNTs and the graphite structure (catalyst, functional groups). As calculated in turbostratic graphite using the (0 0 2 l) peak location, the interlayer spacing is typically seen to be larger than in HOPG and near that value (Lambin et al. 2002; Saito et al. 1993). When compared to graphite, the peak position for SWNTs moved from 26.5° to 26° in 2θ. Additionally, the (0 0 2 l) peaks' line form is weaker and slightly widened in the low diffraction angle area as compared to graphite. The asymmetry is brought on by the presence of distinct crystalline species. At least two of the common and difficult-to-distinguish forms are the coiled graphitic plane, which is made up of a carbon nanotube, and pure graphite particles, comprised of a stack of graphene sheets. On the contrary side, the shape of the (0 0 2 l) peaks is influenced by both inner diameter distribution (Lambin et al. 2002) and the reduction in interlayer spacing with increasing shell diameter (Kiang et al. 1998). The amplitude and width of the (0 0 2 l) peaks are related to the number of layers, variations in interlayer spacing, lattice distortions, and the alignment of the carbon nanotube with regard to the incoming X-ray beam (Reznik et al. 1995; Burian et al. 1999). This is primarily because of the nanotube's curvature: the (h k 0) peaks are asymmetric (Lambin et al. 2002), and the (h k l) reflections only appear in X-ray diffraction (Liu and Cowley 1994; Bernaerts et al. 1998a). These examples include leftover carbon particles and flat graphitic layers in polygonal tubes. It has been shown that no (0 0 2) peak may be recorded by X-ray diffraction with well-aligned straight nanotubes on the substrate surface when X-ray diffraction is utilized in particular investigations on the alignment of CNTs

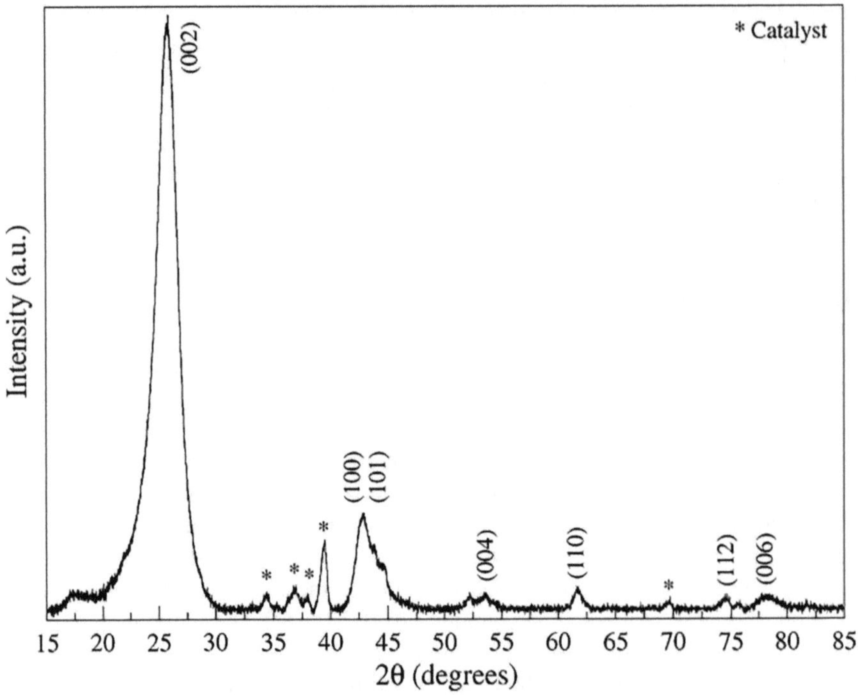

Fig. 6.3 XRD pattern of MWNTs synthesized by CVD at the CSIC laboratory (diameter of about 60 nm). The incident X-ray wavelength is $\lambda = 0.154056$ nm. The most significant Bragg peaks are noticed with Miller indices. The presence of catalysts (Co and Mo) in the CNT sample is shown by stars (Zhu et al. 2003)

(Cao et al. 2001). In the case of carbon nanotubes with a tube axis perpendicular to the substrate surface, the arriving X-ray beam is scattered within the sample and is not captured. As a result, the intensity of the (0 0 2) peak decreases monotonically as CNT alignment improves. The mean diameter of CNTs can be calculated using the Debye-Scherrer relation on the (0 0 2) peak. As this (0 0 2) peak incorporates contributions from residual graphite and nanotubes, the values are calculated through peak decomposition employing pseudo-Voigt profiles.

The consecutive SWNT diffraction amplitudes must be added in order to determine the MWNTs' diffraction patterns. The powder diffraction spectra of SWCNT bundles were computed using general X-ray diffraction formulas (Rols et al. 1999; Kuzmany et al. 2001). Numerous characteristics have been explored, including the impact of the mean tube diameter, the finite size of the bundles, and the diameter dispersivity of the tubes. Each of these factors has a big impact on the locations and sizes of the (1 0) peak. It was concluded, as a result, that these traits lead to a consistent underestimation of the tube diameter.

6.3.7 Transmission Electronic Microscopy

Transmission electronic microscopy (TEM) is an image analysis technique created to analyze multiwalled nanotubes (Gommes et al. 2003). Following numerical treatments, a model based on Lambert's law is used to fit the intensity throughout a nanotube section. Therefore, it is acceptable to calculate the outer and inner radii, as well as the linear electron absorption coefficient. When MWNTs were previously studied both before and after annealing, the electron absorption coefficient increased noticeably. It was concluded that this rise might be explained by a better-organized arrangement of the wall material. High-resolution TEM images were used in the study (Kiang et al. 1998) of the intershell spacing of MWNTs. The intershell gap is found to fluctuate with nanotube diameter and ranges from 0.34 to 0.39 nm (Fig. 6.4a–c). The graphite interplanar distance, which is 0.336 nm, is a little bit bigger than these values, however (Charlier et al. 1999). The curvature of the graphene sheets, which is altered by the tube radius, is likely to blame for this rise in intershell space. The size effect is stronger in the tiny diameter due to this curvature's increase in repelling force (below 10 nm). These authors also noted that altering the intershell gap will alter the physical and chemical properties. Increasing reactivities or creating practical features for storage media can both benefit from this. Interlayer interactions in double-wall nanotubes have been studied theoretically (Tanaka et al. 1997). The results show that a larger tube's diameter results in a weaker interlayer connection. It is anticipated that the energy gap between the two adjacent occupied and/or vacant orbitals would narrow. The nature of interlayer interaction does not change, though. A modest Van der Waals force holds the graphene layer planes together in the graphite. The arrangement of the tubes in MWNTs was Similar to the graphene layers seen in turbostratic graphite. Furthermore, there is no correlation between the many honeycomb sheets of varying widths that make up the MWNTs. As has already been seen with intercalation chemicals in graphites, dopant atoms or molecules could be inserted between neighboring nanotubes in MWNTs (Dresselhaus and Endo 2001). The intershell gap also increases as a result of this. Whenever the nanotube is parallel to this one, two or more graphitic layers are bridged by the incident electron beam (Qin 1998; Qin et al. 1997a, b) (Fig. 6.5b). When the tube is helical, these layers are out of alignment. The angle of this misalignment is therefore twice as large as a helical angle. The two sets of the resulting electron diffraction pattern are then separated. It has been shown that the true helicity of CNTs can be inferred from electron diffraction patterns by neglecting the nanotube's curvature and using half of the misalignment angle as the actual helical angle as a first approximation.

As is commonly seen, nanotubes occasionally appear independent but are typically bundled together by van der Waals forces. A bundle of nanotubes is arranged in a close-packed hexagonal pattern (Kuhlmann et al. 1998). It can be said that this structure is a two-dimensional hexagonal lattice (Fig. 6.5a). TEM in conjunction with electron diffraction can be used to determine the architectures of CNTs within the bundles. To measure the helicity of SWNTs within the bundles, many works

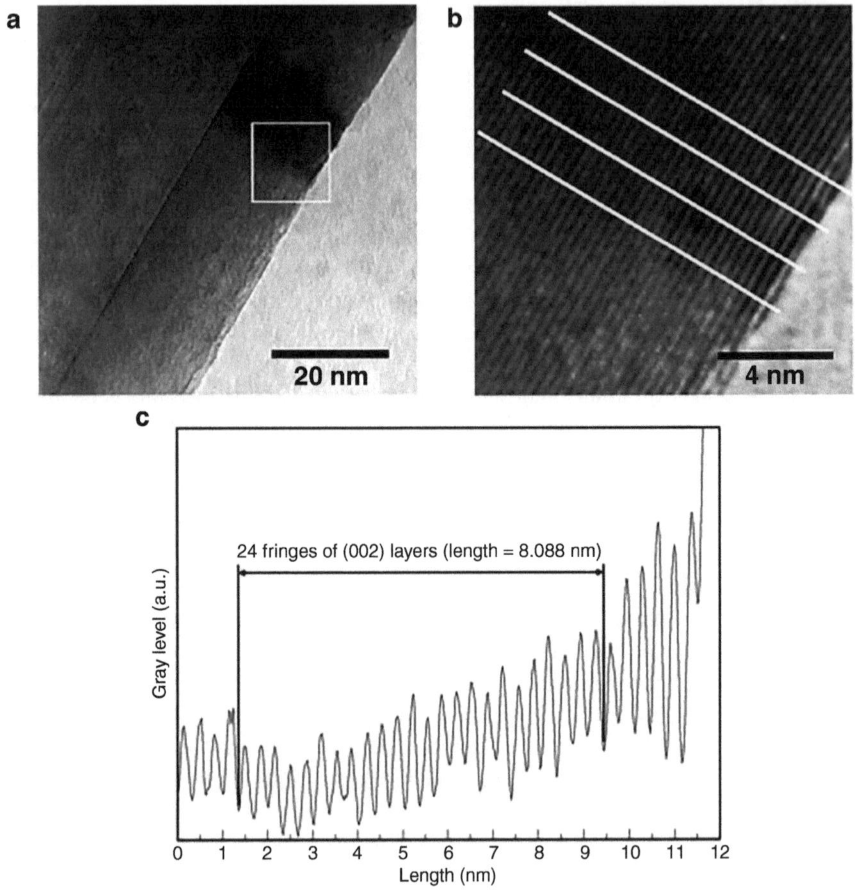

Fig. 6.4 (**a**) TEM image of a multiwalled nanotube synthesized by the CSIC laboratory. These MWNTs were produced by CVD, followed by several oxidation processes. The contrast of the walls is visible. (**b**) Enlargement of the walls of the nanotube. White lines are used in the determination of the intershell spacing. (**c**) Mean profile of the intensity levels of the walls showing the fringes of the (0 0 2) layers used in the determination of the intershell spacing. Here, the value of the intershell spacing is 0.337 ± 0.023 nm and is really close to the graphite one (Kiang et al. 1998)

have been done (Qin et al. 1997b; He et al. 1998; Cowley et al. 1997; Bernaerts et al. 1998b; Cowley and Sundell 1997).

6.3.8 Infrared Spectroscopy (FTIR)

FTIR is widely used to evaluate functionalization, which could also be categorized as a fault. The level of functionalization will modify the wettability of the nanotubes in various surfactants, which might also modify their toxicity. A trustworthy model for the dielectric function of nanotubes was developed by using the intrinsic

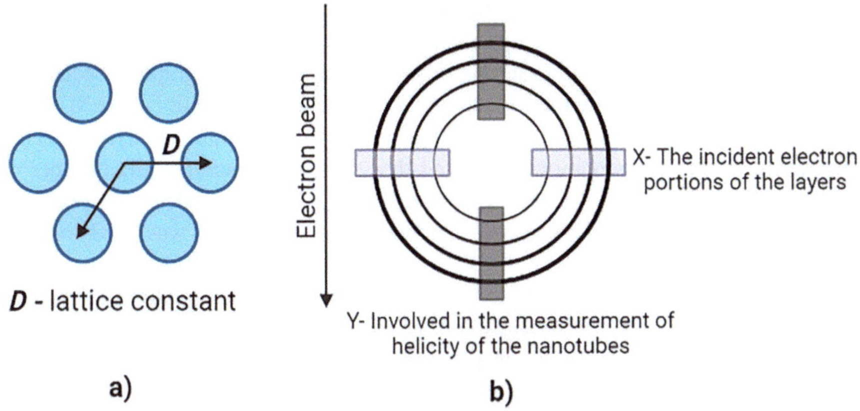

Fig. 6.5 (**a**) Two-dimensional hexagonal lattice of SWNTs within a bundle. *D* is given by the diameter of the nanotube and the intertube spacing. (**b**) Cross-section of a multiwalled nanotube. Scheme of the conditions of observation: the nanotube axis is perpendicular with respect to *X*. The portions of the layers in *Y* are involved in the measurement of the helicity of the nanotubes. The misalignment of these layers leads to the existence of a helical angle (Kuhlmann et al. 1998)

polarization-dependent properties of graphite and considering the tube in two separate situations: a cylinder and a hollow cylinder (Garcia-Vidal et al. 1997). The optical properties of graphite have been well documented and are classified as birefringent. The macroscopic optical properties of MWCNTs will therefore be constrained by tube orientation with regard to the direction of beam propagation, namely, tubes aligned with the tube axis perpendicular to or parallel to the electric field of incoming radiation. SWNTs that depend on symmetry are chiral, zigzag, and armchair and have seven to nine infrared active modes (Kuhlmann et al. 1998). The A2u and E1u modes are the major active modes for carbon nanotubes in infrared spectroscopy (Kuzmany et al. 1998). These phonon modes are seen in MWNTs at around 868 and 1575 cm^{-1}, respectively (Kastner et al. 1994; Eklund et al. 1995); nevertheless, it was discovered that these modes (at approximately 850 and 1590 cm^{-1}) appear in all CNTs symmetrically regardless of the diameters. These findings explain two structures with dimensions between 874 ± 2 and 1598 ± 3 in samples that primarily included SWNTs. The frequencies, however, depart by 5 and 8 cm^{-1} regions, respectively, from the graphite frequencies to higher values. To identify contaminants left over from production or compounds capped on the nanotube surface, infrared spectroscopy is frequently utilized in the characterization of CNTs. Numerous studies have been conducted on CNTs and organic molecules. Infrared spectroscopy shows all the structural changes made to CNTs as well as the types of compounds that have been added. Chemical modifications utilizing amino compounds are used to describe the reaction products of MWNTs (Saito et al. 2002). The characterization of molecules connected to CNTs (He et al. 2004; Aizawa and Shaffer 2003) and the catalytic characteristics of CNTs (Singh et al. 2012; Mbuyise et al. 2017) can be divided into two groups. Multiwalled carbon nanotubes were found to have catalytic properties in the oxidative dehydrogenation of ethylbenzene (ODE), according to a study published by Pereira et al. (2004). They observed that CNTs performed better than samples of activated carbon and graphite because of their inherent structure, which makes them more resistant to oxidation. The MWNTs that have been oxidized before the catalytic studies have maximum catalytic activity. The results of the investigations indicated that the crucial factor is the number of oxygenated surface groups. Alternatively, (Wang et al. 2004b) the potential applications of CNTs in the catalytic oxidation of NOx. FT-IR was used to track the changes in the nitric oxide (NO), nitrite (NO_2), nitrate (NO_3), and carbon-containing CNT signals. These tests were run using CNTs that contained 1 weight percent Pd for each CNT. In this system, the hydrogenation of CNTs is accelerated by the palladium particles. The hydrogenated CNTs act as a reducing agent to supply carbon and hydrogen again for the reduction of NO.

6.3.9 Raman Spectroscopy

Raman spectroscopy is among the best techniques for identifying carbon nanotubes. Without actually prepping the sample, a rapid, nondestructive analysis is still possible. Raman spectroscopy is active in all allotropic forms of carbon, including fullerenes, carbon nanotubes, amorphous carbon, polycrystalline carbon, etc. (Arepalli et al. 2004). The locations, widths, and relative intensities of the bands vary with the carbon forms (Ferrari and Robertson 2000) (Fig. 6.6). The characteristics that stand out the most are (1) a low-frequency peak of the SWNT with a wavelength of 200 cm^{-1} (or many peaks for polydisperse samples when resonant conditions are satisfied), attributed to the A1g “breathing” mode of the tubes, whose frequency is mostly dependent on tube diameter (RBM: radial breathing mode); (2) a significant structure (1340 cm^{-1}) attributed to remaining disordered graphite, the so-called D line; (3) a high-frequency bunch (between 1500 and 1600 cm^{-1}) known as the G band, which is likewise distinctive of nanotubes and corresponds to a splitting of the E2g stretching mode of graphite (Mamedov et al. 2002). This technique produces bands by changing the polarizability of the molecules in the presence of light. Therefore, when these molecules get into touch with light, they could be screened by specific vibrations that they emit. Due to a distinctive pattern, single-walled and multiwalled CNT quality and distribution may be determined by Raman spectroscopy (Saito et al. 2002).

6.3.10 Thermogravimetric Analysis (TGA) and Purity

A variety of techniques, including Raman spectroscopy, imaging, thermogravimetric analysis (TGA), and X-ray microanalysis, can be used to assess the sample’s general quality. Impurities include metal impurities, additional chemical species connected to the nanotube, and other forms of carbon (such as amorphous or graphitic carbons or other structured carbons like SWCNTs or fullerenes). TGA has been used for bulk

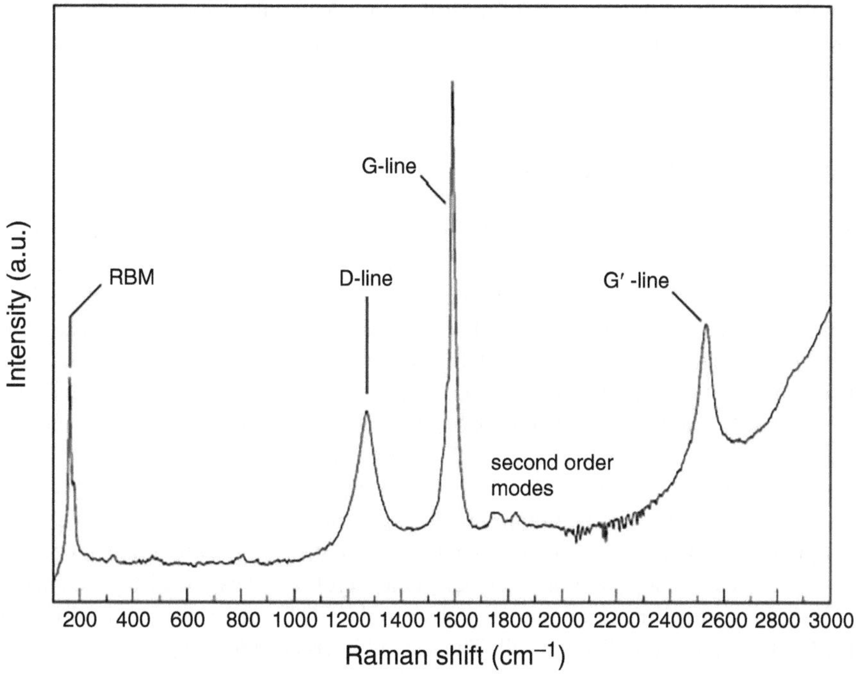

Fig. 6.6 Raman spectrum showing the most characteristic features of CNTs: radial breathing mode (RBM), D band, G band, and G′ band. Second-order modes are also observed. Spectrum obtained from an SWCNT sample (diameter of about 1.07 nm) mixed with KBr using $E_{max} = 1.16$ eV ($\lambda = 1064.5$ nm) excitation. This sample is produced by an electric arc-discharge method, followed by air oxidation at the CSIC laboratory (Ferrari and Robertson 2000)

samples to confirm batch-to-batch uniformity and the quality control of carbon nanotube populations if we are convinced that MWCNTs are presently based on sampling. TGA can be used to examine the thermal stability and purity of a material. The weight-loss curve measures important variables, such as start temperature, oxidation temperature, and residual mass. Commencement temperature is the temperature at which a substance begins to disintegrate. Oxidation temperature, which is referred to as the peak in the derivative of the weight loss as a function of temperature, is where the greatest weight loss takes place. Oxidation temperature is widely used to characterize a material's thermal stability. The mass that has remained after heating is the residual mass. The metal catalyst employed to create carbon nanotubes and its oxidation by-products is typically blamed for the residual bulk of carbon nanotubes in TGA. Based on the consistency and quality of the material, residual masses can range from almost 0% to 50%. According to studies, highly crystalline MWCNTs are more oxidation resistant than other types of carbon, such as diamond, soot, graphite, and C60 (Pang et al. 1993). The aromatic bonding of the MWCNT structure is directly responsible for its thermal stability, but other factors, such as the number of walls of the sample, catalyst presence and composition, tube flaws, and the presence of other materials, can also have an impact (i.e., amorphous carbon, graphitic particles).

6.4 CNT Functionalization

The medicine is immobilized on the surface or within the functionalized carbon nanotubes in a functionalized carbon nanotube delivery system. Functionalization is done for a variety of reasons, including targeted delivery, increased stability, diagnosis of diseases, or other reasons. The resulting conjugate is then administered to the animal via traditional routes (oral, injectable), directly to the target region using a magnetic conjugate, to the target organ, and so on. After the cell absorbs the drug's CNT capsule, the drug is delivered when the contents of the nanotube leak into the cell (Singh et al. 2012; Usui et al. 2012; Zhang et al. 2010, 2011; Kateb et al. 2010; Liu et al. 2007a, b). Functionalized CNTs can transport important molecules across nuclear and cytoplasmic membranes without causing toxicity; as a result, drug CNT conjugates are both safer and more potent than the drug employed alone in conventional synthesis. Once the drug has reached the target cell, it can be distributed in one of two ways: either the drug internally absorbs the CNT carrier, or both the drug and the CNT carrier internalize the cell. The second internalization step is more efficient than the first because the intracellular environment breaks down the drug carrier conjugate after it enters the cells, releasing drug molecules in situ, or inside the cells. There is a potential that the drug may break down during this penetration on its own even though, in the noninternalization procedure, the extracellular environment aids in the degradation of drug carrier conjugates before the drug crosses the lipid membrane to enter the cells (Fig. 6.7).

6.4.1 Functionalized Carbon Nanotubes Used for Cancer Therapy

The functionalized carbon nanomaterials have proven to be very effectively taken up by cancer cells and tissues to enable cancer treatments (Yang et al. 2010a; Liu et al. 2007b).

Methotrexate and a fluorescein probe are both present in f-CNT (Pastorin et al. 2006). A popular and effective anticancer drug, methotrexate, is also used to treat autoimmune illnesses (Wong and Esdaile 2005). However, methotrexate has severe side effects and limited bioavailability (Pignatello et al. 2004). Therefore, a tailored administration and enhanced bioavailability are both highly preferred. If there is a targeting unit present, f-CNT may be able to increase bioavailability and target only cancer cells. According to the results of a cell culture study, methotrexate conjugated with nanotubes is just as effective as methotrexate alone. The absence of increased efficacy between the f-CNT and nonconjugated medicine may be attributed to the stable amide bond between methotrexate and the nanotubes. In fact, it is possible that the drug is released from the tubes into the cytoplasm too slowly for it to effectively

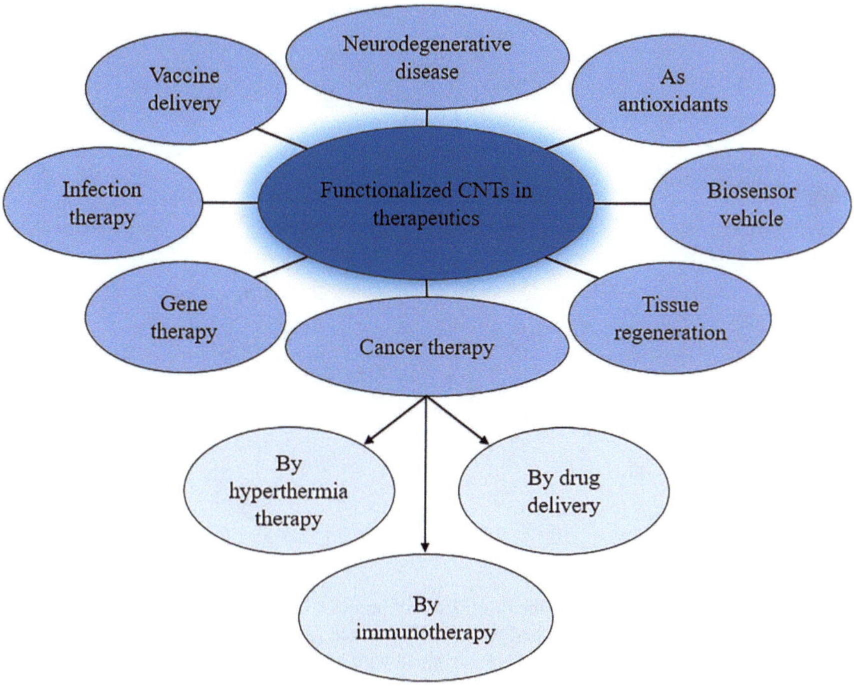

Fig. 6.7 Schematic representation of carbon nanotube application in therapeutics and biomedical diagnosis and analysis

reach its receptor. Single-walled CNT was functionalized with a substituted carborane cage for boron neutron capture treatment. The biodistribution study on different tissues revealed that intravenously injected water-soluble carborane nanotubes were more abundant in tumor cells than in other organs.

6.4.1.1 By Drug Delivery

To treat tumors, CNTs can act as medication transporters (Zhang et al. 2011; Liao et al. 2011; Liu et al. 2009a; Digge et al. 2012; Yang et al. 2007; Al-Jamal et al. 2011; Madani et al. 2011; Lay et al. 2011; Elhissi et al. 2012). Drug resistance and poor cellular penetration are additional factors that restrict the effectiveness of anticancer medications when taken alone, in addition to their systemic toxicity and small therapeutic window. Since CNTs may readily pass through the cytoplasmic and nuclear membranes, the anticancer medication delivered by this vehicle will be released in situ with intact concentration. As a result, it will have a greater impact on the tumor cell than standard therapy alone. To improve the cellular uptake of already strong medications, effective delivery mechanisms have to be developed. Because CNTs have a large surface area and numerous attachment sites for medicines, they have a high aspect ratio compared to other delivery vectors (Chen et al. 2011). A combination made of CNT and antibodies against antigens overexpressed on the

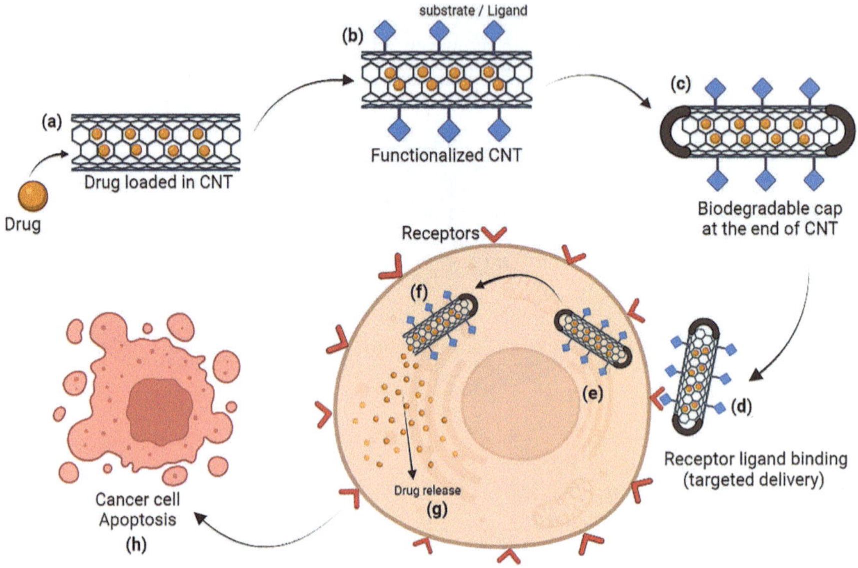

Fig. 6.8 Schematic illustration of the drug delivery process. (**a**) CNT loaded with the drug to be delivered. (**b**) substrate/ligand attached to the CNT surface as functionalization along with drug-loaded inside (**c**) the open end of the CNT is capped with a biodegradable polymer, (**d**) drug-CNT carrier is introduced in the body and reaches the target cells due to receptor on cells and ligands on CNT surface, (**e**) cell internalizes CNT by cell receptors (V) via endocytosis pathway for example, (**f**) cap is removed or biodegrades inside the cell, then drugs are released, (**g**) drug release from CNT in the cell. (**h**) Cancer cell death

surface of malignant cells can bind chemotherapy drugs (Abu Lila et al. 2021). Targeting delivery is made possible by the attraction of antigen-antibody, which allows the tumor cell to receive the CNTs only before the anticancer medicine is released from the CNTs (Madani et al. 2011; Lay et al. 2011; Elhissi et al. 2012). The multidrug resistance brought on by the increased efflux of anticancer medications by the overexpressed p-glycoprotein, which results in poor anticancer efficacy, is a significant barrier to successful anticancer therapy (Elhissi et al. 2012). SWCNT paclitaxel conjugate has been administered in vivo in a mouse breast cancer model, and it is more effective at reducing tumor development and is less harmful to healthy organs (Zhang et al. 2011; Madani et al. 2011). Longer blood circulation, greater tumor absorption, and a slower rate of drug release from SWCNTs could be the causes of increased therapeutic efficacy and fewer negative effects (Digge et al. 2012) (Fig. 6.8).

6.4.1.2 By Antitumor Immunotherapy

CNTs can be successfully used as carriers in antitumor immunotherapy (Singh et al. 2012; Digge et al. 2012; Yang et al. 2007; Al-Jamal et al. 2011; Madani et al. 2011; Lay et al. 2011; Elhissi et al. 2012; Chen et al. 2011; Li et al. 2010; Pantarotto et al.

2004a). In order to combat the cancerous tumor cells, this treatment entails activating the patient's immune system. This reaction can be triggered by the administration of a therapeutic antibody or a cancer vaccine as medicine. The use of CNTs in vaccine delivery devices has been authorized (Pantarotto et al. 2004a). The combination of CNTs and tumor immunogens can act in vitro as naturally occurring antigen-presenting cells (such as mature dendritic cells) by delivering tumor antigens to immune effector T cells because of the high avidity of the antigen on the surface and the negative charge. CNTs' effects on the complement system and their adjuvant properties may help advance anticancer immunotherapy; however, the exact mechanism is uncertain (Yang et al. 2007; Pantarotto et al. 2004a).

6.4.1.3 By Local Antitumor Hyperthermia Therapy

Exceptional near-infrared absorption is exhibited by SWCNTs (NIR; 700–1100 nm). These nanomaterials are thought to be excellent candidates for hyperthermia therapy because they generate a significant amount of heat when activated by NIR light (Madani et al. 2011; Lay et al. 2011; Elhissi et al. 2012). The photothermal effect can lead to the local thermal destruction of tumor cells by scorching SWCNTs trapped in tumor cells, such as in pancreatic cancer.

6.4.2 Carbon Nanotubes for Infection Therapy

CNTs have been tested to find solutions to issues like infectious agent resistance to various antiviral and antibacterial medications or due to a particular vaccine's ineffectiveness in the body. It has been shown that functionalized CNTs can serve as carriers for antibacterial substances such as antifungal amphotericin B (Rosen and Elman 2009; Rosen et al. 2011). Amphotericin B can bind covalently to CNTs, which can then carry it into mammalian cells. Compared to free medication, this combination displays a 40% reduction in antifungal toxicity (Rosen et al. 2011). Functionalized CNTs can also be used as delivery systems for vaccines (Liao et al. 2011; Rosen et al. 2011). A bacterial or viral antigen can be linked to CNTs to maintain antigen conformation, which leads to the proper kind of particular antibody response (Digge et al. 2012). Fixing B- and T-cell peptide epitopes to functionalized CNTs can create a multivalent system that can elicit a potent immunological response, making it a promising option for vaccine administration (Usui et al. 2012; Yang et al. 2007). Since microorganisms, like *E. coli*, may be adsorbed onto the surfaces of CNTs, CNTs themselves may have antibacterial properties. The antibacterial action was thought to be caused by the intracellular antioxidant glutathione being oxidized by carbon nanotubes, which raised oxidative stress on the bacterial cells and ultimately led to cell death (Digge et al. 2012). Antibiotics may be used to functionalize carbon nanotubes. An antimycotic drug called amphotericin B is used to treat exceptionally hardy fungi strains (Zotchev 2003). Due to its severe toxicity to mammalian cells and low solubility in water, as well as its propensity to clump and to create gaps in cell membranes, it is only marginally useful. It is believed that the toxicity and antimycotic effectiveness of amphotericin B may be modified by conjugating it into carbon nanotubes (Wu et al. 2005). Based on the cytotoxicity of f-CNT against mammalian cells, it is discovered that CNT-conjugated amphotericin B, employed at varying concentrations (up to 40 g/mL, equal to a concentration of amphotericin B attached to the tubes of 10 g/mL), was not hazardous, although amphotericin B is very toxic at 10 g/mL concentration, reaching 40% cell mortality. The ability of f-CNT containing both amphotericin B and fluorescein to penetrate cells was then examined. The identification of f-CNT in the cells is made possible by the latter component. It was easy to see the fluorescence inside the cell compartments. The toxicity mostly against yeasts as well as fungi was also increased at the same time.

6.4.3 Carbon Nanotubes for Gene Therapy by DNA Delivery

A new sector is developing around the use of carbon nanotubes for biomedical purposes (Erol et al. 2016). The primary factors attracting interest in the creation of photonic and electronic devices are CNTs' excellent electrical conductivity and mechanical stability (Mbuyise et al. 2017; Oseni et al. 2018). CNTs have indeed been designed through numerous groups as effective gene-delivery vehicles (Kiran and Gangadharappa 2019). The deoxyribonucleic acid (DNA)-SWCNT complex demonstrates improved biostability and enhances the self-delivery capability of DNA when compared to DNA employed alone because DNA probes are shielded from enzymatic cleavage and interference from nucleic acid binding proteins when linked to SWCNTs (Usui et al. 2012; Bekyarova et al. 2005; Li et al. 2008). Stable interactions between plasmid DNA and cationic CNTs have shown improved gene therapy potential in comparison to bare DNA. DNA-conjugated CNTs were discovered to release DNA before it was destroyed by the cell's defense mechanism, dramatically increasing transfection (Liu et al. 2007a; Liao et al. 2011; Li et al. 2013). Because the CNT-gene complex has preserved the ability to express proteins, it has been demonstrated that these designed structures can successfully transport the genes inside mammalian cells and retain them intact (Li et al. 2008). New SWCNT-DNA complexes that have been functionalized have been created (Pantarotto et al. 2004a) and have shown to have higher DNA expression than bare DNA. In the initial investigations examining the capacity of f-CNT to deliver genes, it was discovered that they also effectively complex (Lacerda et al. 2006) and translocate DNA inside cells (Pantarotto et al. 2004b; Singh et al. 2005) because they are cationic under physiological conditions. We have created supramolecular complexes, including the beta-galactosidase marker gene on plasmid DNA and f-CNT. This unique method of gene delivery can be viewed as promising because the expression of the marker gene utilizing f-CNT was five to ten times higher than that of the plasmid DNA given alone. For applications involving gene silencing, the potential of gene therapies based on carbon nanotubes has been further investigated (Kam et al. 2005; Zhang et al. 2006). To specifically target and eliminate cancer cells, complexes of single-walled carbon nanotubes with small interfering ribonucleic acid (siRNA) strands modified with a hydrocarbon tail were employed.

6.4.4 Carbon Nanotubes for Tissue Regeneration and Artificial Implants

Considering they are biocompatible, are resistant to biodegradation, and could be functionalized with proteins to enhance organ regeneration, carbon nanotubes may be the best tissue engineering candidate among a variety of substitute materials, including natural and synthetic polymers for tissue scaffolds. CNTs can be used as additions in this field to increase the mechanical strength of tissue scaffolding and conductivity by integrating them with the host's body (Usui et al. 2012; Zhang et al. 2010; Kateb et al. 2010; Bekyarova et al. 2005; Liao et al. 2011; MacDonald et al. 2005). A composite nanomaterial that acts as a scaffold for tissue regeneration has been successfully created by combining polymer or collagen (poly-L-lactide or poly-D,L-lactide-co-glycolide) with carboxylated SWCNTs (MacDonald et al. 2005). Other CNT tissue engineering applications, such as cell tracking and labeling, sensing cellular behavior, and enhancing tissue matrices, have also been the subject of a recent study (Singh et al. 2012; Usui et al. 2012; Zhang et al. 2010; Kateb et al. 2010; MacDonald et al. 2005). For instance, it has been observed that CNTs effectively enhance in vitro mouse bone tissue regeneration and neurogenic cell differentiation caused by embryonic stem cells (Singh et al. 2012; Zhang et al. 2010).

6.4.5 Carbon Nanotubes for Neurodegenerative Diseases and Alzheimer's Syndrome

CNTs have been applied in neurosciences as a promising biological material (Singh et al. 2012; Zhang et al. 2010; Liao et al. 2011; Yang et al. 2010b). Due to their small size and openness to external alterations, CNTs can penetrate the blood-brain barrier via a variety of targeting methods and serve as efficient delivery vehicles for the target brain. Many more functionalized SWCNTs or MWSCNTs have been employed as effective delivery vehicles for the treatment of brain tumors or neuro-degenerative illnesses (Bekyarova et al. 2005; Liao et al. 2011; Digge et al. 2012). Overall, these research findings showed that CNT-therapeutic molecule conjugates had superior impacts on neuronal development to medicines taken on their own as an individual drug compared to treating with a combination of medicine.

6.4.6 Carbon Nanotubes as Antioxidants

CNTs, specifically carboxylated SWCNTs, are antioxidants by nature and may be used medicinally to prevent chronic diseases, slow the aging process, and preserve food (Daniels 1945; Galano 2008). SWCNTs' ability to scavenge free radicals was proven to be increased by the presence of carboxylic acid (-COOH) groups, and it was found that carboxylated SWCNTs are at least as good as, if not superior to, their nonfunctionalized counterparts (Francisco-Marquez et al. 2010). To protect the skin from free radicals produced by the body or by ultraviolet (UV) light, their antioxidant activities have been used in sunscreen lotions and antiaging cosmetics (Singh et al. 2012; Digge et al. 2012). More research will be needed in the future to improve the useful effect of different CNT forms as free radical scavengers because free radicals are well known to be a very damaging species for biomedical and environmental applications (Galano 2008; Pham-Huy et al. 2008).

6.4.7 Carbon Nanotubes as Biosensor Vehicles for Diagnostic and Detection

A very intriguing application area for therapeutic monitoring and in vitro and in vivo diagnostics is the use of CNTs in biosensing nanotechnology. For better accuracy and easier manipulation than using biosensors alone, CNTs and glucose-oxidase biosensors were coupled for blood sugar control in diabetic patients (Usui et al. 2012; Digge et al. 2012; Wang 2005). For various treatment monitoring and diagnostics, further CNT-based dehydrogenase biosensors, peroxidase biosensors, and catalase biosensors have also been produced (Wang 2005; Zhu et al. 2011). Alkaline phosphatase (ALP) enzyme linked to CNTs had a greater test sensitivity for electrical DNA detection than ALP alone. In comparison to conventional fluorescence and hybridization experiments, the sensitivity of the SWCNT-DNA sensor assay was significantly higher, thanks to the integration of SWCNTs with single-strand DNAs (ssDNA). By utilizing particular antibody-antigen recognition, this CNT-biosensor-linked assay can be customized for antigen detection. As a result, it might offer a quick and easy method for molecular diagnostics in illnesses that have molecular markers, like DNA or protein (Liao et al. 2011; Wang 2005). It is strongly advised to use CNTs as biosensor vehicles to build sensitive approaches for diagnostics and analysis from the laboratory to the clinic due to their length scale and distinctive structure.

6.4.8 Carbon Nanotubes for Therapeutic and Diagnostic Applications

To produce CNT conjugates with pharmacological action, therapeutic agents can further modify any new functional group, including amines and carboxylates. Nanotubes with the capacity to carry one or more therapeutic moieties with optical or other (e.g., magnetic) probes for imaging and/or specific recognition signals for targeting may offer multimodal options in the treatment of cancer and other multifaceted diseases in which activity is only required at specific sites in the body. Normally, even after these synthetic goals are accomplished, there will still be a number of technical issues to be resolved, mostly in the area of pharmaceutical development. The stability of the complexes under physiological conditions, the degree of in vivo aggregation, the right timing, and the place of drug release are a few of these concerns. Despite this, CNT is a valuable technological foundation for the development of candidates for simultaneous diagnostics, transportation, and medication delivery due to the large range of conceivable combinations (Bianco et al. 2005a, b). The possibility of developing CNT for biomedical applications emerged as a reality after several effective ways for their functionalization were made public (Tasis et al. 2006). The nanotubes have been rendered soluble and suitable for physiological conditions utilizing a variety of techniques. This is a major issue for its integration into environments with biological systems. The level of toxicity of all CNT materials must be taken into consideration while establishing their biocompatibility. The research so far suggests that the cytotoxic effects of CNTs are greatly reduced by functionalization, while their biocompatibility is increased (Sayes et al. 2006; Singh et al. 2006). The possibility of utilizing nanotubes for medication administration is increased by evidence to date that CNTs are safer the more functionalized they are, especially when compared to pristine, purified CNT.

6.4.9 Functionalized Carbon Nanotubes for Vaccine Delivery

Their immunostimulatory peptide-based constructions fall under a different category of carbon-nanotube-based therapeutic prospects. By combining functionalized nanotubes with B- and T-cell peptide epitopes, it is possible to create a multivalent system that can vigorously elicit an immune response (Pantarotto et al. 2003a, b). Peptides can be attached to tubes using chemo-selective techniques (Goodman et al. 2002; Muller et al. 1999). Using peptides having a cysteine residue at one end, functionalized carbon nanotubes with a maleimide group that easily combines with it are created. The thiol group of the cysteine with the maleimide forms a long-lasting covalent bond. This approach has the advantage of having the peptide, produced through solid phase synthesis, completely deprotected and defined before being joined to the nanotubes. By adding a lysine branch to the f-CNT, The antigenic and immunogenic qualities of these conjugates were evaluated. In contrast to the nonconjugated peptide, the antibody responses to f-CNT were particularly high. The ability of the produced antibodies to neutralize the virus was also demonstrated, highlighting the potential of carbon nanotubes as ingredients in the production of synthetic vaccines.

6.5 Toxicity

See Fig. 6.9.

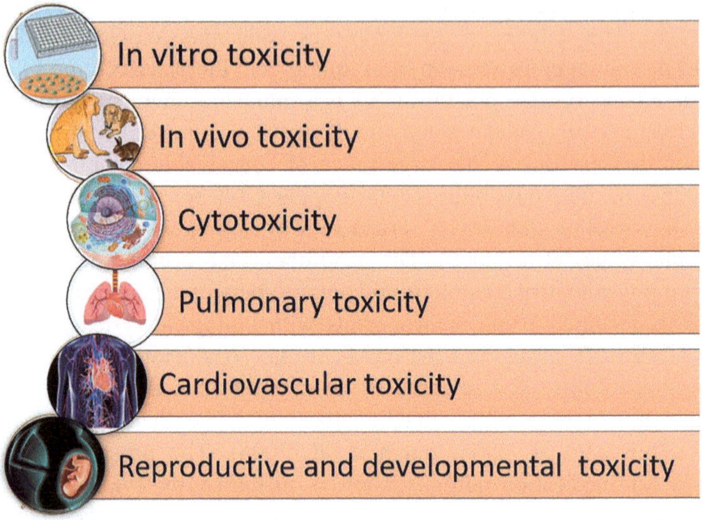

Fig. 6.9 Toxicity associated with carbon nanotube as a drug delivery system

6.5.1 In Vivo Toxicity of CNTs

The level of toxicity of a substance determines how much harm it can do to something with a metabolic function. The term "toxicity" can refer to the impact on an entire organism, such as an animal, bacterium, or plant, as well as the impact on an organism's cells or organs, such as the liver (hepatotoxicity). As a result, not all changes in organisms may be categorized as harmful reactions. When triggered by alien items, organisms will naturally react (Francis and Devasena 2018). When ingested, even flour powder can cause pulmonary alterations (Ren et al. 2010). Additionally, a variety of parameters, such as dosage, contaminants, pretreatment procedures, physical shape, surface chemistry, degree of aggregation, etc., appear to affect how hazardous CNTs are. Recent years have seen extensive investigation into CNT in vivo toxicology. In humans, cutaneous toxicity may result in an inflammatory response through skin contact. Mice exposed to unpurified CNTs typically develop localized alopecia, oxidative stress, glutathione depletion, an increase in dermal cell number, and thickened skin (Koyama et al. 2009; Murray et al. 2009). The number of metals, particularly iron, may have an impact on CNT toxicity. By interacting mostly with the skin, producing oxidative stress, and activating redox-sensitive transcription factors, metals can affect or trigger inflammation. However, the absence of skin hair loss in mice implanted with incredibly pure and uncontaminated tubes suggests that purification is an effective strategy for improving the biocompatibility of CNTs. The most likely site of CNT exposure is the lung. Inhalation is the most appropriate method for determining the toxicity of CNTs for a wide range of reasons. First, airborne CNTs are subjected to the physics of impaction, sedimentation, and diffusion, whereas implanted CNTs are not. Second, inhaled CNTs pass the lung's distal regions. Third, compared to an instilled bolus dose in an aqueous medium, inhaled CNTs are much more diffused and less aggregated. Because CNTs and asbestos fibers both have large aspect ratios (length to width), there is fear that inhaling CNTs could result in similar lung diseases (Crouzier et al. 2010; Elgrabli et al. 2008; Tantra and Cumpson 2007). The pleural pathology brought on by asbestos fibers is distinct from the pathology caused by MWCNTs. MWCNTs generated mononuclear cell agglomeration and focal subpleural fibrosis (Ryman-Rasmussen et al. 2009), while asbestos induces diffuse pleural fibrosis and pleural inflammation (granulomas) (Choe et al. 1997; Kane 2006).

Occupational situations, which require special consideration from an exposure standpoint, may include high concentrations of CNTs. For handling CNTs in professional environments and research labs, there are not many resources available. Therefore, the study (Pauluhn 2010) is very valuable for establishing standards for occupational exposure in the workplace environment and determining a fair occupational exposure limit (OEL), which will aid the government in developing relevant regulations.

6.5.2 In Vitro Toxicity of CNTs

The impact of the potentially toxic effects of CNTs on cells is a crucial factor in determining and understanding CNT compatibility vs. toxicity (Cheng et al. 2009; Davoren et al. 2007; Gellein et al. 2009; Thurnherr et al. 2009; Tutak et al. 2009). Cellular uptake and the processing of CNTs via various pathways; effects on cell signaling; lipid bilayer disruptions; production of cytokines, chemokines, and reactive oxygen species (ROS); overt toxic reactivity; cell apoptosis; and no obvious toxicity are among the interactions with both cells and CNTs that are being studied (Hillegass et al. 2010). The most common approaches typically involve the in vitro culturing of primary cells or cell lines (tissue harvested) on plastic plates, with or without serum, with bolus dosing of CNTs and, subsequently, monitoring of cell activity. Various cellular types, including cancer cell lines and neuronal, phagocytic, and other cell types, have also been chosen for study. It was found that the toxicity of human lung cancer cell lines increased from evenly distributed CNTs to asbestos and then to agglomerated CNTs (Wick et al. 2007a). CNTs are more hazardous than metal oxide nanoparticles and do not demonstrate length-dependent cytotoxicity, even in the presence of metal catalyst impurities (Simon-Deckers et al. 2008a). The proportion of CNTs that are too long for cells to absorb in long and short samples is likely approximately the same, and CNTs that cannot be taken by cells have effects that are equivalent regardless of length, so the length made little to no difference.

Studies with functionalized CNTs demonstrated that purification could improve their biocompatibility and reduce their cellular toxicity (Liu et al. 2007c). Even though it is generally agreed that well-dispersed CNTs are less dangerous than agglomerated ones (especially when using a bio-dispersant), various in vitro studies have found different levels of apparent toxicity for CNTs. Due to the special characteristics of CNTs, evaluations of their toxicity may also depend on the circumstances surrounding their modification and surface chemistry.

6.5.3 Cytotoxicity

An investigation of the cytotoxic potential of MWCNTs and titanium oxide nanoparticles was conducted (Simon-Deckers et al. 2008b). Investigations were also conducted into intracellular nanomaterial accumulation and cell survival. The findings of this study demonstrated that nanoparticles and nanotubes may both enter cells and disperse within the cytoplasm. Metal oxide nanoparticles exhibit reduced toxicity as compared to CNTs. The cytotoxicity of CNTs is unaffected by CNT length or the presence of metallic contaminants (Simon-Deckers et al. 2008b). The cellular membrane of rat macrophages (NR8383) was discovered to be penetrated by CNTs in a recent study, which altered the macrophages' physiology and cellular function (Pulskamp et al. 2007). Commercial CNTs were found to lower the potential of the mitochondrial membrane and increase intracellular reactive oxygen species in human A549 lung cells and rat macrophages (NR8383). In pure CNTs, which have a reduced metal content, these effects were either negligible or nonexistent (Pulskamp et al. 2007). Researchers hypothesized that the biological impacts could be caused by the metals linked with commercial CNTs based on the outcomes of a recent experiment. It was discovered that CNTs dispersed in surfactant were less hazardous than CNTs aggregated. It is interesting to note that string-like agglomerated CNTs were more solid, rigid, and voluminous than asbestos (Wick et al. 2007b). Even at small concentrations (0.001–0.1 mg/mL), CNT toxicity has been shown to trigger immune-mediated cytotoxicity against a variety of human cells. Lower concentrations of CNTs have been hypothesized to boost the release of cytokines that signal lymphocyte activation and increase the expression of NF-kB in immune cells, which results in indirect cytotoxicity (Sun et al. 2011).

6.5.4 Pulmonary Toxicity

Investigating the potentially harmful effects of these materials is strongly encouraged by the possible applications of CNTs. Due to their use in innovative products, nanotubes can now enter the body through a variety of exposure methods, including cutaneous and gastrointestinal contact. During manufacturing, CNTs may get into the workers' respiratory systems and build up in their lungs. Nanotubes could enter customers' stomachs and intestines if they are utilized as fillers in food-packaging items. SWCNTs and MWCNTs were harmful when ingested by rodents, according to several reports. These investigations implied that CNTs might pose a risk to people. Animals developed lung granulomas as a result of exposure to pure and metal-doped SWCNT preparations in a dose-dependent manner. Inducing an inflammatory response, oxidative stress, collagen deposition, and fibrosis in mice were found to be more successful with inhaled SWCNTs than pharyngeal aspiration.

In a study on the respiratory toxicity of MWCNTs, MWCNTs or ground MWCNTs suspended in sterile saline (0.9% NaCl) (Muller et al. 2005) were used. After 2 months of exposure, granulomas rich in collagen began to form in the lung lesions as a result of inhalation. MWCNTs and ground MWCNTs both encouraged tumor necrosis factor (TNF) production. Moreover, by evaluating the amount of hydroxyproline in the lung tissue, dose-related pulmonary fibrosis was identified. The outcomes supported MWCNTs' toxicity. When compared to asbestos and carbon black, the inflammatory effect of MWCNTs had a middle level of severity (Muller et al. 2005). After 14 days of whole-body inhalation exposure to MWCNTs at high dosages, immune suppression was observed in mice, but MWCNTs did not exhibit inflammation or granuloma formation, as previously reported (Mitchell et al. 2009). Depending on the dosage and delivery method, highly distributed MWCNTs may result in pulmonary lesions (Morimoto et al. 2012). Pneumonia in guinea pigs has reportedly been caused by CNTs of several sorts (Grubek-Jaworska et al. 2006).

6.5.5 Cardiovascular Effects

Initial animal investigations (Muller et al. 2005; Lam et al. 2004; Shvedova et al. 2005; Warheit et al. 2004) provided evidence that exposure to CNTs causes both immediate and long-lasting pulmonary inflammation. The results of the research suggested that CNT exposure should be assessed as a potential cardiovascular risk factor based on the oxidative and inflammatory hypothesis of atherosclerosis and air pollution (Simeonova and Erdely 2009) used acid-purified SWCNTs to examine potential cardiovascular harm. According to the study, CNT-induced lung inflammation generated inflammatory mediators and activated blood cells that had harmful consequences on the cardiovascular system. Stress and dosage both affected the damage to the mitochondrial DNA of the aorta that was seen after CNT mice were exposed through pharyngeal aspiration. In comparison to controls, mice treated with SWCNTs exhibited considerably more atherosclerotic plaque on the surface of the aorta. Additionally, brachiocephalic arteries were shown to have an increase in atherosclerotic lesions. The outcomes demonstrated that mouse atherosclerosis progresses more quickly after exposure to SWCNTs (Li et al. 2007).

6.5.6 Reproductive and Developmental Toxicity

Studies examining the impact of functionalized CNTs revealed that after exposure to very low concentrations (10 ng/mouse), resorptions and fetal abnormalities started and increased. Also noted was an increase in ROS production in the placentas of exposed dams (Pietroiusti et al. 2011). The effects of functionalized CNTs on reproduction and development were investigated in *Drosophila melanogaster* as well as in CD-1 mice. The results showed that severe morphological deformities, skeletal abnormalities, and resorption rates were all significantly higher in fetuses exposed to 10 mg/kg fCNTs. The external faults included open eyelids, abnormal leg structures, and abnormal tail structures. The percentage of fetuses with aberrant cervical vertebrae reduced phalange ossification of the pectoral and pelvic limbs, and variable sternal ossification was higher in these cases than in controls.

6.5.7 Toward the Reduction of Its Toxicity Issues

Toxicity issues with SWCNTs and MWCNTs are mostly caused by their physico-chemical characteristics, which would include their aspect ratio, size, chemical composition, form, crystal structure, stability, surface area, surface roughness, surface energy, and surface charge (Gatoo et al. 2014; Kostal et al. 2015). Modulating and/or changing these properties would therefore be the most effective strategy for lowering the allied toxicities of the CNT. There is a critical necessity guideline for building safer CNTs, given the broad range of applications for CNT-based systems in the biomedical and healthcare sectors (Rahamathulla et al. 2021). Surface modification is the widely used method; due to their hydrophobic surfaces, CNTs are insoluble in water, and direct contact of pure nanotubes with biological systems may result in interactions with different biomolecules, leading to toxicity (Fig. 6.10).

6.5.7.1 Covalent Modification

CNT sidewalls are frequently chemically altered in order to solubilize nanotubes in aqueous solutions. Typically, it is achieved by grafting functional moieties onto oxidized CNT sidewalls (Karousis et al. 2010; Banerjee et al. 2005). H_2SO_4 and/or HNO_3 are used to oxidize nanotubes in order to produce carboxylic groups. Furthermore, it was found that cells were exposed to several harmful effects when oxidized

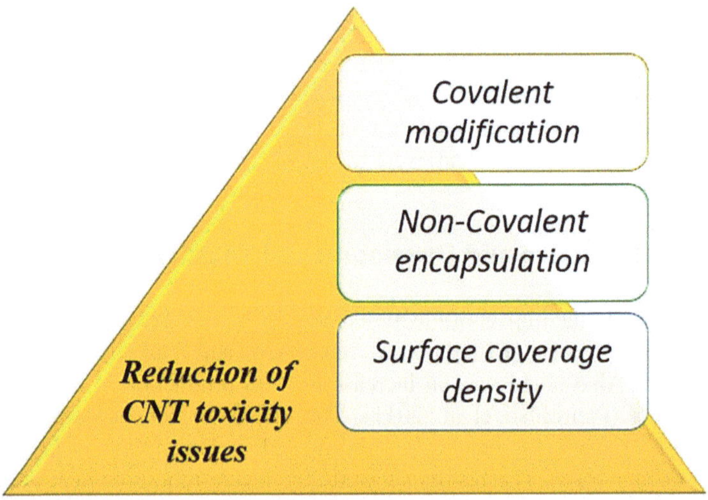

Fig. 6.10 Various methods involved in the reduction of toxicity issues associated with CNT

CNTs were employed without additional coating. The direct, extended exposure of oxidized CNTs to lung epithelial cells would stimulate the development of cancer stem cells with malignant characteristics (Luanpitpong et al. 2014), and these cells would become aggressive and form tumors. Due to surface flaws and the chemical groups that oxidation introduces, these nanotubes may be more hazardous than pure, unprocessed nanotubes (Kumarathasan et al. 2014). Therefore, oxidized CNTs without additional surface modifications should be employed with caution in both in vitro and in vivo applications. It was shown in this context that grafting antifouling polymers onto the nanotube backbone network is a successful technique for decreasing direct interactions of CNTs with biological components. Consider the often-used polyethylene glycol (PEG)-modified CNTs, which are typically created by amidating the -COOH groups of oxidized nanotubes with the -NH_2 groups of PEG. This procedure produces PEG-modified CNTs that maintain nanotubes as single and/or tiny bundles in colloidal suspensions and are stable in conditions with high salt and serum concentrations. The cellular toxicity of oxidized SWCNTs is dramatically decreased by PEG covalent grafting techniques. The branching PEG-modified CNTs appear to be more biocompatible than their linear PEG-modified counterparts due to tight binding and wide nanotube coverage. This is because fewer cellular components are in contact with the nanotube framework.

6.5.7.2 Noncovalent Encapsulation

The covalent modification of nanotube surfaces described above results in sp3 hybridization bonds by the addition of chemical groups. These changes may dramatically degrade the mechanical, physical, or chemical properties of nanotubes. While these characteristics are the basis for many bio-applications of CNTs, extensive covalent functionalization often reduces the intrinsic near-infrared (NIR) photoluminescence capabilities of nanotubes (Cognet et al. 2007; Hong et al. 2015). Using biologically acceptable amphiphilic materials, noncovalent-based nanotube solubilization is a typical approach for these purposes. In general, CNTs can be enclosed, and surface coatings with cationic, anionic, or nonionic charges can be used to control the outer charge of coated nanotubes. Because the plasma membrane is negatively charged, interactions between nanotubes and cell membranes, nanotube internalization pathways, and intracellular fate are greatly influenced by the surface charge of encapsulated CNTs. Compared to neutral and anionic nanoparticles, cationic nanoparticles frequently connect more strongly with the cell membrane and have higher absorption efficiency (Tonga et al. 2014). It is also known that negatively charged nanoparticles can be efficiently ingested through membrane diffusion or pinocytosis (Frohlich 2012; He et al. 2010; Xiao et al. 2011). Thus, maintaining neutrally charged nanotube surfaces seems to be the key to decreasing nanotubes' nonspecific binding to serum proteins and cell membranes. Additionally, nanoparticle hydrophobicity needs to be managed because it affects cellular absorption and subcellular destiny (Gupta et al. 2011; Moyano et al. 2012). In the area of possible biological applications, DNA molecules constitute a different class of biomolecules that are frequently utilized to solubilize CNTs (Zheng et al. 2003). DNA is a malleable biopolymer that can change the shape of its molecules to wrap around the exterior walls of CNTs by creating a helical configuration (Zheng et al. 2003; Battigelli et al. 2013). The aromatic nucleotides along the DNA perpendicular axis engage in noncovalent interactions, notably stacking interactions, with the CNT backbone network. This small DNA wrapping may be CNT chirality selective based on the DNA sequences (Shankar et al. 2014). Furthermore, DNA-coated CNTs are anticipated to show reduced cytotoxicity since DNA is intrinsically biocompatible. Numerous studies have suggested that the DNA sequence, chirality, and length of the nanotubes may all have an impact on how biological serum macromolecules attach to DNA-coated CNTs and how successfully they are absorbed by cells (Salem et al. 2016; Becker et al. 2007).

6.5.7.3 Surface Coverage Density

The density of chemical functionalities on the sidewall of CNTs has a big impact on how the nanotubes behave on their surface and how they interact with biological systems. Noncovalently suspended carbon nanotubes demonstrated increased solubility (Liu et al. 2009b) and decreased toxicity with increasing coverage of coating molecules, corresponding to covalently modified CNTs. In addition to improving CNT solubility in biological fluids for both covalent and noncovalent surface coverage, increasing PEG density and branching degree also made nanotubes less hazardous by lowering unintended interactions with biological elements, particularly proteins (Heister et al. 2012).

6.6 Future Prospective

Carbon nanotubes have the chance to be a more affordable alternative to metal wires because of their great electrical conductivity. They are candidates to replace current computer chips due to their semiconducting qualities. In the future, CNTs are likely to compete with carbon fiber for high-end applications, especially in applications where weight is an issue, like Kevlar. CNTs have also been discovered to be a more eco-friendly, flame-retardant component to plastics. A safer alternative to dangerous, biocide-containing paints, MWNT-containing paints have also been found to lessen the biofouling of ship hulls by preventing the attachment of algae and barnacles.

6.6.1 In 3DPC Efficiency Enhancement

CNT is anticipated to be a viable material for 3DPC. The configuration time of 3DPC can be significantly decreased and production efficiency increased by using CNT. It can increase the printing quality, the early strength of 3DPC, and the drying shrinkage of 3DPC. Last but not the least, it can fill the pores in 3DPC, increase its hardness, and then increase the shape stability following extrusion. On CNT-reinforced 3DPC, however, no related research has been done as of yet. The primary problem is that domestic 3DP technology is still in its early stages and that the 3DP method is not precisely specified. Carbon materials have a high cost of manufacturing and a difficult production procedure. These are the constraints of using CNT to improve concrete in 3DPC. But when constructing small models, CNT strengthening of the 3DPC becomes practical since 3DP can provide some intricate designs in accordance with the model.

6.6.2 In Electrochemical Sensing

Carbon nanotubes should be able to facilitate electron transfer processes with electroactive species in solutions, when utilized as electrode materials, according to delicate electronic characteristics. Electrodes made of carbon nanotubes could therefore be employed for electrochemical sensing. Carbon nanotubes demonstrated better behavior as electrode materials than conventional carbon electrodes, which would include good conducting ability and great chemical stability.

References

Abu Lila AS, Soliman MS, Kiran HC, Gangadharappa HV, Younes KM, Khafagy E-S, Shehata TM, Ibrahim MM, Abdallah MH (2021) Tamoxifen-loaded functionalized graphene nanoribbons for breast cancer therapy. J Drug Deliv Sci Technol 63:102499., ISSN 1773-2247. https://doi.org/10.1016/j.jddst.2021.102499

Aizawa M, Shaffer M (2003) Chem Phys Lett 368:121

Ajayan P, Ebbesen T (1997) Rep Prog Phys 60:1025

Al-Jamal KT, Nerl H, Muller KH et al (2011) Cellular uptake mechanisms of functionalised multi-walled carbon nanotubes by 3D electron tomography imaging. Nanoscale 3(6):2627–2635

Amelinckx S, Lucas A, Lambin P (1999) Rep Prog Phys 62:1471

Arepalli S, Nikolaev P, Gorelik O, Hadjiev V, Holmes W, Files B, Yowell L (2004) Carbon 42: 1783

Bachilo SM, Strano MS, Kittrell C, Hauge RH, Smalley RE, Weisman RB (2002) Structure-assigned optical spectra of single-walled carbon nanotubes. Science 298:2361–2366

Banerjee S, Hemraj-Benny T, Wong SS (2005) Adv Mater 17:17–29

Battigelli A, Menard-Moyon C, Da Ros T, Prato M, Bianco A (2013) Adv Drug Deliv Rev 65: 1899–1920

Becker ML, Fagan JA, Gallant ND, Bauer BJ, Bajpai V, Hobbie EK, Lacerda SH, Migler KB, Jakupciak JP (2007) Adv Mater 19:939–945

Bekyarova E, Ni Y, Malarkey EB et al (2005) Applications of carbon nanotubes in biotechnology and biomedicine. J Biomed Nanotechnol 1(1):3–17

Benedict LX, Louie SG, Cohen ML (1996) Heat capacity of carbon nanotubes. Solid State Commun 100:177

Berber S, Kwon YK, Tomanek D (2000) Phys Rev Lett 84:4613

Bernaerts D, Amelinckx S, Lambin P, Lucas A (1998a) Appl Phys A Mater Sci Process 67:53

Bernaerts D, Zettl A, Chopra N, Thess A, Smalley R (1998b) Solid State Commun 105:145

Bianco A, Kostarelos K, Partidos CD, Prato M (2005a) Biomedical applications of functionalised carbon nanotubes. Chem Commun:571–577

Bianco A, Kostarelos K, Prato M (2005b) Applications of carbon nanotubes in drug delivery. Curr Opin Chem Biol 9:674–679

Biercuk MJ, Llaguno MC, Radosavljevic M, Hyun JK, Johnson AT, Fischer JE (2002) Carbon nanotubes for thermal management. Appl Phys Lett 80(15):2767–2769

Buongiorno Nardelli M, Yakobson BI, Bernholc J (1998) Mechanism of strain release in carbon nanotubes. Phys Rev B 57(8):R4277–R4280

Burian A, Dore J, Fischer H, Sloan J (1999) Phys Rev B 59:1665

Cao A, Xu C, Liang J, Wu D, Wei B (2001) Chem Phys Lett 344:13

Charlier A, McRae E, Heyd R, Charlier M, Moretti D (1999) Carbon 37:177–179

Chen P, Wu X, Sun X, Lin J, Ji W, Tan K (1999) Phys Rev Lett 82:2548

Chen Z, Pierre D, He H et al (2011) Adsorption behavior of epirubicin hydrochloride on carboxylated carbon nanotubes. Int J Pharm 405(1–2):153–161

Cheng C, Muller KH, Koziol KKK, Skepper JN, Midgley PA, Welland ME et al (2009) Toxicity and imaging of multi-walled carbon nanotubes in human macrophage cells. Biomaterials 30: 4152–4160

Choe N, Tanaka S, Xia W, Hemenway DR, Roggli VL, Kagan E (1997) Pleural macrophage recruitment and activation in asbestos-induced pleural injury. Environ Health Perspect 105: 1257–1260

Cognet L, Tsyboulski DA, Rocha J-DR, Doyle CD, Tour JM, Weisman RB (2007) Science 316: 1465–1468

Cowley J, Sundell F (1997) Ultramicroscopy 68:1

Cowley J, Nikolaev P, Thess A, Smalley R (1997) Chem Phys Lett 265:379

Crouzier D, Follot S, Gentilhomme E, Flahaut E, Arnaud R, Dabouis V et al (2010) Carbon nanotubes induce inflammation but decrease the production of reactive oxygen species in lung. Toxicology 272:39–45

Daniels HE (1945) Proc R Soc Lond A 183:405

Davoren M, Herzog E, Casey A, Cottineau B, Chambers G, Byrne HJ et al (2007) In vitro toxicity evaluation of single walled carbon nanotubes on human A549 lung cells. Toxicol In Vitro 21: 438–448

Digge MS, Moon RS, Gattani SG (2012) Applications of carbon nanotubes in drug delivery: a review. Int J PharmTech Res 4(2):839–847

Dresselhaus MS, Endo M (2001) Top Appl Phys 80:11

Eklund P, Holden J, Jishi R (1995) Carbon 33:959

Elgrabli D, Floriani M, Abella-Gallart S, Meunier L, Gamez C, Delalain P et al (2008) Biodistribution and clearance of instilled carbon nanotubes in rat lung. Part Fibre Toxicol 5:20

Elhissi AMA, Ahmed W, Hassan IU, Dhanak VR, D'Emanuele A (2012) Carbon nanotubes in cancer therapy and drug delivery. J Drug Deliv 2012:837327., 10 pages

Erol O, Uyan I, Hatip M, Yilmaz C, Tekinay AB, Guler MO (2016) Recent advances in bioactive 1D and 2D carbon nanomaterials for biomedical applications. Nanomed Nanotechnol Biol Med 14:2433–2454. https://doi.org/10.1016/j.nano.2017.03.021

Ferrari A, Robertson J (2000) Phys Rev B 61:14095

Francis AP, Devasena T (2018) Toxicity of carbon nanotubes: a review. Toxicol Ind Health 34(3): 200–210. https://doi.org/10.1177/0748233717747472

Francisco-Marquez M, Galano A, Martínez A (2010) On the free radical scavenging capability of carboxylated single-walled carbon nanotubes. J Phys Chem C 114(14):6363–6370

Frohlich E (2012) Int J Nanomedicine 7:5577–5591

Galano A (2008) Carbon nanotubes as free-radical scavengers. J Phys Chem C 112(24):8922–8927

Ganesh EN (2013) Single walled and multi-walled carbon nanotube structure, synthesis and applications. Int J Innov Technol Explor Eng 2(4):311

Garcia-Vidal FJ, Pitarke JM, Pendry JB (1997) Effective medium theory of the optical properties of aligned carbon nanotubes. Phys Rev Lett 78(22):4289–4292

Gatoo MA, Naseem S, Arfat MY, Dar AM, Qasim K, Zubair S (2014) Physicochemical properties of nanomaterials: Implication in associated toxic manifestations. Biomed Res Int 2014:498420

Gellein K, Hoel S, Evje L, Syversen T (2009) The colony formation assay as an indicator of carbon nanotube toxicity examined in three cell lines. Nanotoxicology 3:215–221

Gommes C, Blacher S, Masenelli-Varlot K, Bossuot C, McRae E, Fonseca A, Nagy JB, Pirard JP (2003) Carbon 41:2561

Goodman M, Felix A, Moroder L, Toniolo C (2002) Methods of organic chemistry, vol E22b. Houben-Weyl, Thieme, Stuttgart

Grubek-Jaworska H, Nejman P, Czumińska K et al (2006) Preliminary results on the pathogenic effects of intratracheal exposure to one-dimensional nanocarbons. Carbon 44(6):1057–1063

Guo GY, Chu KC, Wang DS, Duan CG (2004a) Linear and nonlinear optical properties of carbon nanotubes from first-principles calculations. Phys Rev B 69:205416

Guo J, Yang C, Li ZM, Bai M, Liu HJ, Li GD, Wang EG, Chan CT, Tang ZK, Ge WK, Xiao X (2004b) Phys Rev Lett 93:017402–017401

Gupta A, Mandal D, Ahmadibeni Y, Parang K, Bothun G (2011) Eur Biophys J 40:727–736

Hammer P, Victoria N, Alvarez F (2000) J Vac Sci Technol 18:2277

He R, Jin H, Zhu J, Yan Y, Chen X (1998) Chem Phys Lett 298:170

He B, Sun W, Wang M, Liu S, Shen Z (2004) Mater Chem Phys 84:140

He C, Hu Y, Yin L, Tang C, Yin C (2010) Biomaterials 31:3657–3666

He H, Pham-Huy LA, Dramou P, Xiao D, Zuo P, Pham-Huy C (2013) Carbon nanotubes: applications in pharmacy and medicine. BioMed Res Int 2013:578290., 12 pages

Heister E, Neves V, Lamprecht C, Silva SRP, Coley HM, McFadden J (2012) Carbon 50:622–632

Hillegass JM, Shukla A, Lathrop SA, MacPherson MB, Fukagawa NK, Mossman BT (2010) Assessing nanotoxicity in cells in vitro. Wiley Interdiscip Rev Nanomed Nanobiotechnol 2: 219–231

Hirlekar R, Yamagar M, Garse H, Vij M, Kadam V (2009) Carbon nanotubes and its applications: a review. Asian J Pharm Clin Res 2(4):17–27

Hone J, Whitney M, Piskoti C, Zettl A (1999) Phys Rev B 59:R2514

Hone J, Llaguno MC, Nemes NM, Johnson AT, Fischer JE, Walters DA, Casavant MJ, Schmidt J, Smalley RE (2000a) Appl Phys Lett 77:666

Hone J, Batlogg B, Benes Z, Johnson AT, Fischer JE (2000b) Science 289:1730

Hong G, Diao S, Antaris AL, Dai H (2015) Chem Rev 115:10816–10906

Hu H, Bhowmik P, Zhao B, Hamon MA, Itkis ME, Haddon RC (2001) Chem Phys Lett 345:25

Janas D, Milowska KZ, Bristowe PD, Koziol KKK (2017) Improving the electrical properties of carbon nanotubes with interhalogen compounds. Nanoscale 9(9):3212–3221

Kahn D, Lu JP (1999) Phys Rev B 60:6535

Kam NWS, Liu Z, Dai H (2005) Functionalization of carbon nanotubes via cleavable disulfide bonds for efficient intracellular delivery of siRNA and potent gene silencing. J Am Chem Soc 127:12492–12493

Kane AB (2006) Animal models of malignant mesothelioma. Inhal Toxicol 18:1001–1004

Karousis N, Tagmatarchis N, Tasis D (2010) Chem Rev 110:5366–5397

Kastner J, Pichler T, Kuzmany H, Curran S, Blau W, Weldon DN, Delamesiere M, Draper S, Zandbergen H (1994) Chem Phys Lett 221:53

Kateb B, Yamamoto V, Alizadeh D et al (2010) Multi-walled carbon nanotube (MWCNT) synthesis, preparation, labeling, and functionalization. Methods Mol Biol 651:307–317

Kempa K, Kimball B, Ryhczynski J, Huang ZP, Wu PF, Steeves D, Sennett M, Giersig M, Rao DVGLN, Carnahan D, Wang DZ, Lao JY, Li WZ, Ren ZF (2003) Photonic crystals based on periodic arrays of aligned carbon nanotubes. Nano Lett 3:13–18

Kempa K, Ryhczynski J, Huang ZP, Gregorczyk K, Vidan A, Kimball B, Carlson J, Benham G, Wang Y, Herczynski A, Ren ZF (2007) Carbon nanotubes as optical antennae. Adv Mater 19: 421–426

Kiang C, Endo M, Ajayan P, Dresselhaus G, Dresselhaus M (1998) Phys Rev Lett 81:1869

Kiran HC, Gangadharappa HV (2019) Reinforcing nanomedicine using graphene nanoribbons. J Drug Deliv Sci Technol 49:334–344.,ISSN 1773-2247. https://doi.org/10.1016/j.jddst.2018.12.004

Kostal J, Voutchkova-Kostal A, Anastas PT, Zimmerman JB (2015) Identifying and designing chemicals with minimal acute aquatic toxicity. Proc Natl Acad Sci U S A 112:6289–6294

Koyama S, Kim YA, Hayashi T, Takeuchi K, Fujii C, Kuroiwa N et al (2009) In vivo immunological toxicity in mice of carbon nanotubes with impurities. Carbon 47:1365–1372

Kuhlmann U, Jantoljak H, Pfander M, Bernier P, Journet C, Thomsen C (1998) Chem Phys Lett 294:237

Kumarathasan P, Breznan D, Das D, Salam MA, Siddiqui Y, Mackinnon-Roy C, Guan J, de Silva N, Simard B, Vincent R (2014) Nanotoxicology 0:1–14

Kuzmany H, Burger B, Thess A, Smalley R (1998) Carbon 36:709

Kuzmany H, Plank W, Hulman M, Kramberger C, Gruneis A, Pichler T, Peterlik H, Kataura H, Achiba Y (2001) Eur Phys J B 22:307

Lacerda L, Pastorin G, Wu W, Prato M, Bianco A, Kostarelos K (2006) Functionalised carbon nanotube autofluorescence as a tool to monitor bundle formation and dissociation in aqueous phases by fluorescence spectrophotometry: the effect of plasmid DNA complexation. Adv Funct Mater 16:1839–1846

Lam CW, James JT, McCluskey R et al (2004) Pulmonary toxicity of single-wall carbon nanotubes in mice 7 and 90 days after intratracheal instillation. Toxicol Sci 77(1):126–134

Lambin P, Loiseau A, Culot C, Biro L (2002) Carbon 40:1635

Lauret JS, Voisin C, Cassabois G, Roussignol P, Delalande C, Filoramo A, Capes L, Valentin E, Jost O (2004) Phys E 21:1057

Lay CL, Liu J, Liu Y (2011) Functionalized carbon nanotubes for anticancer drug delivery. Expert Rev Med Devices 8(5):561–566

Lebedkin S, Hennrich F, Skipa T, Kappes MM (2003) J Phys Chem B 107:1949

Lefebvre J, Fraser JM, Homma Y, Finnie P (2004) Appl Phys A Mater Sci Process 78:1107

Li Z, Hulderman T, Salmen R et al (2007) Cardiovascular effects of pulmonary exposure to single-wall carbon nanotubes. Environ Health Perspect 115(3):377

Li S, He H, Jiao Q, Pham-Huy C (2008) Applications of carbon nanotubes in drug and gene delivery. Prog Chem 20(11):1798–1803

Li R, Wu R, Zhao L, Wu M, Yang L, Zou H (2010) Pglycoprotein antibody functionalized carbon nanotube overcomes the multidrug resistance of human leukemia cells. ACS Nano 4(3): 1399–1408

Li L, Lin R, He H, Jiang L, Gao MM (2013) Interaction of carboxylated single-walled carbon nanotubes with bovine serum albumin. Spectrochim Acta A 105:45–51

Li T, Tang Z, Huang Z, Yu J (2017) A comparison between the mechanical and thermal properties of single-walled carbon nanotubes and boron nitride nanotubes. Physica E 85:137–142

Liao H, Paratala B, Sitharaman B, Wang Y (2011) Applications of carbon nanotubes in biomedical studies. Methods Mol Biol 726:223–241

Lidorikis E, Ferrari AC (2009) Photonics with multiwall carbon nanotube arrays. ACS Nano 3: 1238–1248

Lin MF (1994) Plasmons and optical properties of carbon nanotubes. Phys Rev B 50:17744

Lin MF (2000) Optical spectra of single-wall carbon nanotube bundles. Phys Rev B 62:13153

Liu M, Cowley J (1994) Carbon 32:393

Liu Y, Gao L (2005) A study of the electrical properties of carbon nanotube-$NiFe_2O_4$ composites: effect of the surface treatment of the carbon nanotubes. Carbon N Y 43(1):47–52

Liu X, Pichler T, Knupfer M, Golden MS, Fink J, Kataura H, Achiba Y (2002) Phys Rev B 66: 045411

Liu Z, Sun X, Nakayama-Ratchford N, Dai H (2007a) Supramolecular chemistry on water-soluble carbon nanotubes for drug loading and delivery. ACS Nano 1(1):50–56

Liu Z, Cai W, He L, Nakayama N, Chen K, Sun X, Chen X, Dai H (2007b) In vivo biodistribution and highly efficient tumour targeting of carbon nanotubes in mice. Nat Nanotechnol 2:47–52

Liu XY, Gurel V, Morris D, Murray DW, Zhitkovich A, Kane AB et al (2007c) Bioavailability of nickel in single-wall carbon nanotubes. Adv Mater 19:2790

Liu Z, Tabakman S, Welsher K, Dai H (2009a) Carbon nanotubes in biology and medicine: in vitro and in vivo detection, imaging and drug delivery. Nano Res 2(2):85–120

Liu Z, Tabakman SM, Chen Z, Dai H (2009b) Nat Protoc 4:1372–1381

Llaguno MC, Hone J, Fischer JE, Johnson AT (2002) Thermal properties of carbon nanotubes and nanotube-based materials. Appl Phys A 74(3):339–343

Lourie E, Cox DM, Wagner HD (1998) Buckling and collapse of embedded carbon nanotubes. Phys Rev Lett 81(8):1638–1641

Luanpitpong S, Wang L, Castranova V, Rojanasakul Y (2014) Part Fibre Toxicol 11:22–22

Maarouf AA, Kane CL, Mele EJ (2000) Phys Rev B 61:11156

MacDonald RA, Laurenzi BF, Viswanathan G, Ajayan PM, Stegemann JP (2005) Collagen-carbon nanotube composite materials as scaffolds in tissue engineering. J Biomed Mater Res A 74(3): 489–496

Madani SY, Naderi N, Dissanayake O, Tan A, Seifalian AM (2011) A new era of cancer treatment: carbon nanotubes as drug delivery tools. Int J Nanomedicine 6:2963–2979

Mamedov AA, Kotov NA, Prato M, Guldi DM, Wicksted JP, Hirsch A (2002) Nat Mater 1:190

Mawhinney DB, Naumenko V, Kuznetsova A, Yates JT Jr, Liu J, Smalley RE (2000) Chem Phys Lett 324:213

Mbuyise XG, Arbab EAA, Kaviyarasu K, Pellicane G, Maaza M, Mola GT (2017) Zinc oxide doped single wall carbon nanotubes in hole transport buffer layer. J Alloy Comp 706:344–350. https://doi.org/10.1016/j.jallcom.2017.02.249

Mitchell LA, Lauer FT, Burchiel SW et al (2009) Mechanisms for how inhaled multiwalled carbon nanotubes suppress systemic immune function in mice. Nat Nanotechnol 4(7):451–456

Mizel A, Benedict LX, Cohen ML, Louie SG, Zettl A, Budraa NK, Beyermann WP (1999) Analysis of the low-temperature specific heat of multiwalled carbon nanotubes and carbon nanotube ropes. Phys Rev B 60:3264

Mizuno K, Ishii J, Kishida H, Hayamizu Y, Yasuda S, Futaba DN, Yumura M, Hata K (2008) A black body absorber from vertically aligned single-walled carbon nanotubes. Proc Natl Acad Sci U S A 106:6044–6047

Morimoto Y, Hirohashi M, Ogami A et al (2012) Pulmonary toxicity of well-dispersed multi-wall carbon nanotubes following inhalation and intratracheal instillation. Nanotoxicology 6(6): 587–599

Moyano DF, Goldsmith M, Solfiell DJ, Landesman Milo D, Miranda OR, Peer D, Rotello VM (2012) J Am Chem Soc 134:3965–3967

Muller S (1999) Synthetic peptides as antigens. In: Pillai S, van der Vliet PC (eds) Laboratory techniques in biochemistry and molecular biology, vol 28. Elsevier, Amsterdam, pp 79–131

Muller J, Huaux F, Moreau N et al (2005) Respiratory toxicity of multi-wall carbon nanotubes. Toxicol Appl Pharmacol 207(3):221–231

Murray AR, Kisin E, Leonard SS, Young SH, Kommineni C, Kagan VE et al (2009) Oxidative stress and inflammatory response in dermal toxicity of single-walled carbon nanotubes. Toxicology 257:161–171

Nakano S-i, Uotani Y, Nakashima S, Anno Y, Fujii M, Sugimoto N (2003) Large stabilization of a DNA duplex by the deoxyadenosine derivatives tethering an aromatic hydrocarbon group. J Am Chem Soc 125(27):8086–8087

O'Connell MJ, Bachilo SM, Huffman CB, Moore VC, Strano MS, Haroz EH, Rialon KL, Boul PJ, Noon WH, Kittrell C, Ma J, Hauge RH, Weisman RB, Smalley RE (2003) Science 297:503

Oseni SO, Kaviyarasu K, Maaza M, Sharma G, Pellicane G, Mola GT (2018) ZnO:CNT assisted charge transport in PTB7:PCBM blend organic solar cell. J Alloy Comp 748:216–222. https://doi.org/10.1016/j.jallcom.2018.03.141

Ouyang M, Huang JL, Cheung CL, Lieber CM (2001) Science 292:702

Ozaki T, Iwasa Y, Mitani T (2000) Stiffness of single-walled carbon nanotubes under large strain. Phys Rev Lett 84(8):1712–1715

Pang LSK, Saxby JD, Chatfield SP (1993) Thermogravimetric analysis of carbon nanotubes and nanoparticles. J Phys Chem 97(27):6941–6942

Pantarotto D, Hoebeke J, Graff R, Partidos CD, Briand J-P, Prato M, Bianco A (2003a) Synthesis, structural characterization and immunological properties of carbon nanotubes functionalized with peptides. J Am Chem Soc 125:6160–6164

Pantarotto D, Partidos CD, Hoebeke J, Brown F, Kramer E, Briand J-P, Muller S, Prato M, Bianco A (2003b) Immunization with peptide-functionalized carbon nanotubes enhances virus-specific neutralizing antibody responses. Chem Biol 10:961–966

Pantarotto D, Singh R, McCarthy D et al (2004a) Functionalized carbon nanotubes for plasmid DNA gene delivery. Angew Chem 43(39):5242–5246

Pantarotto D, Singh R, McCarthy D, Erhardt M, Briand J-P, Prato M, Kostarelos K, Bianco A (2004b) Functionalised carbon nanotubes for plasmid DNA gene delivery. Angew Chem Int Ed 43:5242–5246

Pastorin G, Wu W, Wieckowski S, Briand J-P, Kostarelos K, Prato M, Bianco A (2006) Double functionalisation of carbon nanotubes for multimodal drug delivery. Chem Commun:1182–1184

Pauluhn J (2010) Multi-walled carbon nanotubes (baytubes (R)): approach for derivation of occupational exposure limit. Regul Toxicol Pharmacol 57:78–89

Pereira M, Figueiredo J, Orfao J, Serp P, Kalck P, Kihn Y (2004) Carbon 42:2807

Pham-Huu C, Keller N, Roddatis V, Mestl G, Schlogl R, Ledoux MJ (2002) Phys Chem Chem Phys 4:514

Pham-Huy LA, He H, Pham-Huy C (2008) Free radicals, antioxidants in disease and health. Int J Biomed Sci 4(2):89–96

Pietroiusti A, Massimiani M, Fenoglio I et al (2011) Low doses of pristine and oxidized single-wall carbon nanotubes affect mammalian embryonic development. ACS Nano 5(6):4624–4633

Pignatello R, Guccione S, Forte S, Di Giacomo C, Sorrenti V, Vicari L, Uccello Barretta G, Balzano F, Puglisi G (2004) Lipophilic conjugates of methotrexate with short-chain alkylamino acids as DHFR inhibitors. Synthesis, biological evaluation, and molecular modeling. Bioorg Med Chem 12:2951–2964

Poudel YR, Li W (2018) Synthesis, properties, and applications of carbon nanotubes filled with foreign materials: a review. Mater Today Phys 7:7–34

Pulskamp K, Diabaté S, Krug HF (2007) Carbon nanotubes show no sign of acute toxicity but induce intracellular reactive oxygen species in dependence on contaminants. Toxicol Lett 168(1):58–74

Qin L (1998) Chem Phys Lett 297:23

Qin L, Ichihashi T, Iijima S (1997a) Ultramicroscopy 67:181

Qin L, Iijima S, Kataura H, Maniwa Y, Suzuki S, Achiba Y (1997b) Chem Phys Lett 268:10

Rahamathulla M, Bhosale RR, Osmani RAM, Mahima KC, Johnson AP, Hani U, Ghazwani M, Begum MY, Alshehri S, Ghoneim MM et al (2021) Carbon nanotubes: current perspectives on diverse applications in targeted drug delivery and therapies. Materials 14:6707. https://doi.org/10.3390/ma14216707

Ren HX, Chen X, Liu JH, Gu N, Huang XJ (2010) Toxicity of single-walled carbon nanotube: how we were wrong? Mater Today 13:6–8

Reznik D, Olk C, Neumann A, Copley J (1995) Phys Rev B 52:116

Rols S, Almairac R, Henrard L, Anglaret E, Sauvajol J (1999) Eur Phys J B 10:263

Rosen Y, Elman NM (2009) Carbon nanotubes in drug delivery: focus on infectious diseases. Expert Opin Drug Deliv 6(5):517–530

Rosen Y, Mattix B, Rao A, Alexis F (2011) Carbon nanotubes and infectious diseases. In: Hunter RJ (ed) Nanomedicine in health and disease. Science Publishers, London, UK, pp 249–267

Ryman-Rasmussen JP, Cesta MF, Brody AR, Shipley-Phillips JK, Everitt JI, Tewksbury EW et al (2009) Inhaled carbon nanotubes reach the subpleural tissue in mice. Nat Nanotechnol 4:747–745

Saifuddin N, Raziah AZ, Junizah AR (2013) Carbon nanotubes: a review on structure and their interaction with proteins. J Chem 2013:676815, 18 pages

Saito Y, Yoshikawa T, Bandow S, Tomita M, Hayashi T (1993) Phys Rev B 48:1907

Saito R, Dresselhaus G, Dresselhaus MS (1998) Physical properties of carbon nanotubes. Imperial College Press, London

Saito R, Matsushige K, Tanaka K (2002) Phys B Condens Matter 323:280

Salem DP, Landry MP, Bisker G, Ahn J, Kruss S, Strano MS (2016) Carbon 97:147–153

Sayes CM, Liang F, Hudson JL, Mendez J, Guo W, Beach JM, Moore VC, Doyle CD, West JL, Billups WE, Ausman KD, Colvin VL (2006) Functionalization density dependence of single-walled carbon nanotubes cytotoxicity in vitro. Toxicol Lett 16:135–142

Shankar A, Mittal J, Jagota A (2014) Langmuir 30:3176–3183

Shoji S, Suzuki H, Zaccaria RP, Sekkat Z, Kawata S (2008) Optical polarizer made of uniaxially aligned short single-wall carbon nanotubes embedded in a polymer film. Phys Rev B 77:153407

Shvedova AA, Kisin ER, Mercer R et al (2005) Unusual inflammatory and fibrogenic pulmonary responses to single-walled carbon nanotubes in mice. Am J Phys Lung Cell Mol Phys 289(5): L698–L708

Simeonova PP, Erdely A (2009) Engineered nanoparticle respiratory exposure and potential risks for cardiovascular toxicity: predictive tests and biomarkers. Inhal Toxicol 21(Suppl 1):68–73

Simon-Deckers A, Gouget B, Mayne-L'Hermite M, Herlin-Boime N, Reynaud C, Carriere M (2008a) In vitro investigation of oxide nanoparticle and carbon nanotube toxicity and intracellular accumulation in A549 human pneumocytes. Toxicology 253:137–146

Simon-Deckers A, Gouget B, Mayne-L'Hermite M et al (2008b) In vitro investigation of oxide nanoparticle and carbon nanotube toxicity and intracellular accumulation in A549 human pneumocytes. Toxicology 253(1):137–146

Singh R, Pantarotto D, McCarthy D, Chaloin O, Hoebeke J, Partidos CD, Briand J-P, Prato M, Bianco A, Kostarelos K (2005) Binding and condensation of plasmid DNA onto functionalized carbon nanotubes: towards the construction of nanotube-based gene delivery vectors. J Am Chem Soc 127:4388–4396

Singh R, Pantarotto D, Lacerda L, Pastorin G, Klumpp C, Prato M, Bianco A, Kostarelos K (2006) Tissue biodistribution and blood clearance rates of intravenously administered carbon nanotube radiotracers. Proc Natl Acad Sci U S A 103:3357–3362

Singh BGP, Baburao C, Pispati V et al (2012) Carbon nanotubes. A novel drug delivery system. Int J Res Pharm Chem 2(2):523–532

Slepyan GY, Shuba MV, Maksimenko SA (2006) Theory of optical scattering by achiral carbon nanotubes and their potential as optical nanoantennas. Phys Rev B 73:195416

Srivastava D, Menon M, Kyeongjae C (1999) Nanoplasticity of single-wall carbon nanotubes under uniaxial compression. Phys Rev Lett 83(15):2973–2976

Sun Z, Liu Z, Meng J et al (2011) Carbon nanotubes enhance cytotoxicity mediated by human lymphocytes in vitro. PLoS One 6(6):e21073

Tanaka K, Aoki H, Ago H, Yamabe T, Okahara K (1997) Carbon 35:125

Tantra R, Cumpson P (2007) The detection of airborne carbon nanotubes in relation to toxicology and workplace safety. Nanotoxicology 1:251–265

Tasis D, Tagmatarchis N, Bianco A, Prato M (2006) Chemistry of carbon nanotubes. Chem Rev 106:1105–1136

Teizer W, Hallock RB, Dujardin E, Ebbesen TW (1999) Phys Rev Lett 82:5305

Thostenson E, Ren Z, Chou T (2001) Comp Sci Technol 61:1899

Thurnherr T, Su DS, Diener L, Weinberg G, Manser P, Pfander N et al (2009) Comprehensive evaluation of in vitro toxicity of three large-scale produced carbon nanotubes on human Jurkat T cells and a comparison to crocidolite asbestos. Nanotoxicology 3:319–338

Tonga GY, Saha K, Rotello VM (2014) Adv Mater 26:359–370

Tutak W, Park KH, Vasilov A, Starovoytov V, Fanchini G, Cai SQ et al (2009) Toxicity induced enhanced extracellular matrix production in osteoblastic cells cultured on single-walled carbon nanotube networks. Nanotechnology 20:255101

Usui Y, Haniu H, Tsuruoka S, Saito N (2012) Carbon nanotubes innovate on medical technology. Med Chem 2(1):1–6

Wang J (2005) Carbon-nanotube based electrochemical biosensors: a review. Electroanalysis 17(1): 7–14

Wang Y, Kempa K, Kimball B, Carlson JB, Benham G, Li WZ, Kempa T, Rybczynski J, Herczynski A, Ren ZF (2004a) Receiving and transmitting light-like radio waves: antenna effect in arrays of aligned carbon nanotubes. Appl Phys Lett 85:2607

Wang S, Zhu W, Liao D, Ng C, Au C (2004b) Catal Today 93–95:711

Wang XJ, Flicker JD, Lee BJ, Ready WJ, Zhang ZM (2009) Visible and near-infrared radiative properties of vertically aligned multi-walled carbon nanotubes. Nanotechnology 20:215704

Warheit DB, Laurence BR, Reed KL et al (2004) Comparative pulmonary toxicity assessment of single-wall carbon nanotubes in rats. Toxicol Sci 77(1):117–125

Weisman RB, Bachilo SM, Tsyboulski D (2004) Appl Phys A Mater Sci Process 78:1111

Wick P, Manser P, Limbach LK, Dettlaff-Weglikowska U, Krumeich F, Roth S et al (2007a) The degree and kind of agglomeration affect carbon nanotube cytotoxicity. Toxicol Lett 168:121–131

Wick P, Manser P, Limbach LK et al (2007b) The degree and kind of agglomeration affect carbon nanotube cytotoxicity. Toxicol Lett 168(2):121–131

Wong JM, Esdaile JM (2005) Methotrexate in systemic lupus erythematosus. Lupus 14:101–105

Wu W, Wieckowski S, Pastorin G, Benincasa M, Klumpp C, Briand J, Gennaro R, Prato M, Bianco A (2005) Targeted delivery of Amphotericin B to cells using functionalized carbon nanotubes. Angew Chem Int Ed 44:6358–6362

Xiao K, Li Y, Luo J, Lee JS, Xiao W, Gonik AM, Agarwal RG, Lam KS (2011) Biomaterials 32: 3435–3446

Yakobson BI (1998) Mechanical relaxation and 'intramolecular plasticity' in carbon nanotubes. Appl Phys Lett 72(8):918–920

Yang W, Thordarson P, Gooding JJ, Ringer SP, Braet F (2007) Carbon nanotubes for biological and biomedical applications. Nanotechnology 18:412001., 12 pages

Yang Z-P, Ci L, Bur JA, Lin S-Y, Ajayan PM (2008) Experimental observation of extremely dark material made by a low-density nanotube array. Nano Lett 8:446

Yang K, Zhang S, Zhang G, Sun X, Lee ST, Liu Z (2010a) Graphene in mice: ultrahigh in vivo tumor uptake and efficient photothermal therapy. Nano Lett 10:3318–3323. https://doi.org/10.1021/nl100996u

Yang Z, Zhang Y, Yang Y et al (2010b) Pharmacological and toxicological target organelles and safe use of single-walled carbon nanotubes as drug carriers in treating Alzheimer disease. Nanomedicine 6(3):427–441

Zhang Z, Yang X, Zhang Y, Zeng B, Wang S, Zhu T, Roden RB, Chen Y, Yang R (2006) Delivery of telomerase reverse transcriptase small interfering RNA in complex with positively charged single-walled carbon nanotubes suppresses tumor growth. Clin Cancer Res 12:4933–4939

Zhang Y, Bai Y, Yan B (2010) Functionalized carbon nanotubes for potential medicinal applications. Drug Discov Today 15(11–12):428–435

Zhang W, Zhang Z, Zhang Y (2011) The application of carbon nanotubes in target drug delivery systems for cancer therapies. Nanoscale Res Lett 6:555–577

Zhao GL, Bagayoko D, Yang L (2006) Optical properties of aligned carbon nanotube mats for photonic applications. J Appl Phys 99:114311

Zheng M, Jagota A, Semke ED, Diner BA, McLean RS, Lustig SR, Richardson RE, Tassi NG (2003) Nat Mater 2:338–342

Zhu W, Miser D, Chan W, Hajaligol M (2003) Mater Chem Phys 82:638

Zhu Y, Wang L, Xu C (2011) Carbon nanotubes in biomedicine and biosensing. In: Naraghi M (ed) Carbon nanotubes—growth and applications. Shanghai, InTech, pp 135–162

Zotchev SB (2003) Polyene macrolide antibiotics and their applications in human therapy. Curr Med Chem 10:211–223

Komal Iranna Savadatti is a postgraduate student pursuing her M.Pharm. in the Department of Pharmaceutics at JSS College of Pharmacy, Mysuru. She secured a bachelor's degree in pharmacy from the same college. Her research interests are nanotechnology and cancer biology.

Asha Puthuvilayil Johnson obtained her bachelor's and master's degrees in pharmacy from the College of Pharmaceutical Sciences, Govt. Medical College Thiruvananthapuram. She then joined as an assistant professor at Dale View College of Pharmacy and Research Centre, Thiruvananthapuram. She is now a Ph.D. student at JSS College of Pharmacy, Mysuru. Her research interests are nanotechnology, targeted drug delivery, and cancer biology.

Hosahalli Veerabhadrappa Gangadharappa is currently working as an associate professor in the Department of Pharmaceutics, JSS College of Pharmacy, JSS Academy of Higher Education and Research, Mysuru. He has published more than 130 articles in reputed national and international journals and two Indian patents. His core research area is graphene nanoribbons and silk-fibroin-based drug delivery systems.

Nanotechnology: Changing the World of Animal Health and Veterinary Medicine

7

Zulfqar ul Haq, Henna Hamadani, Azmat Alam Khan, Abdul Majeed Ganai, Yasir Afzal Beigh, Gowher Gull Sheikh, Javaid Farooq, Imran Ahmad Ganai, and Syed Mudasir Ahmad

Abstract

Nanoparticles have long been used as diagnostic and therapeutic agents in human medicine, but their use in veterinary medicine and animal production is recent. Several laws and regulations have been implemented in response to growing concerns about antibiotic resistance, including a ban on feed-grade antibiotics in Europe. As a result of banning these antibiotics, the industry is searching for safe alternatives to fill the void left behind. The increasing global population, decreasing fertile land, increasing animal protein demand, and increasing income have further accelerated the demand for animal production. These conditions have created a dire need for technologies and substitutes that would increase the productivity and well-being of animals, improve drug delivery, and reduce the toxic effect of drugs. It appears that nanotechnology has great potential and future applications in the animal husbandry industry. According to numerous reports, nanoparticles make excellent anti-microbials, promoting animal growth and well-being. Nanotechnology research has the potential to offer revolutionary tools for improving animal health and production in the future. The purpose of this chapter is to describe the types of nano-materials used in animal sciences, as well as the unique technologies developed by nanoscience for improving animal health and production. Further, the toxicity of nano-materials and considerations for animal health and safety are also discussed.

Z. ul Haq · H. Hamadani · A. A. Khan · A. M. Ganai · Y. A. Beigh · G. Gull Sheikh · J. Farooq · S. M. Ahmad (✉)
Faculty of Veterinary Sciences, Sher-e-Kashmir University of Agricultural Sciences and Technology, Kashmir, Srinagar, Jammu and Kashmir, India
e-mail: mudasirbio@skuastkashmir.ac.in

I. Ahmad Ganai
Mountain Research Centre for Sheep and Goat, Sher-e-Kashmir University of Agricultural Sciences and Technology, Kashmir, Srinagar, Jammu and Kashmir, India

F. A. Sheikh et al. (eds.), *Interaction of Nanomaterials With Living Cells*, https://doi.org/10.1007/978-981-99-2119-5_7

Keywords

Nanotechnology · Veterinary sciences · Nanoparticles · Animal science · Diagnosis · Treatment · Vaccines · Nano-carriers · Veterinary medicine

7.1 Introduction

Nanotechnology or nanoscience refers to the study of the understanding and manipulation of the properties of nanoscale materials ranging between 1 and 100 nm in dimensions (Sim and Wong 2021). Using nanotechnology, we can design entirely new molecular structures and properties. Apart from being small, nanoparticles (NPs) have the advantage of having a large surface area-to-volume ratio, making them unique in drug delivery and anti-microbial action as well as more reproducible and reliable (Dawadi et al. 2021). Nanoparticles are finding increasing attention in improving catalytic activity, chemical stability, thermal conductivity, and non-linear optical performance (Laouini et al. 2021). By modifying their shape, surface properties, and sizes, various nanoparticles can be transformed into nano-systems that can be used in the diagnosis, imaging, and treatment of diseases, including cancer, cardiovascular, ocular, and central nervous system diseases. These functionalized nano-materials can deliver controlled-release therapy by providing drugs to specific sites or tissues (Khan et al. 2020). Most biological systems are nanoscale, making nano-materials ideal for biomedical devices. Research in alternative therapies has been driven by advances in nano-medicine and nano-devices as current methods cannot detect and treat early-stage diseases (Anjum et al. 2021).

Nanotechnology has found prospects in one health approach that encompasses animal and human health and the environment. However, several unknown probable health hazards overshadow these potential benefits, creating doubts about the successful application of technology by governments and the public worldwide. It is a delicate situation where uncertainty of behaviour of these designed nanoparticles or how their products function which restricts universal usage. As a result, it necessitates regulatory control to assure safe production processes and safe application to humans, requiring appropriate and balanced oversight (Jain et al. 2021). As the human population grows exponentially, so does the need for agricultural and animal food products, prioritizing new industry technologies to address the looming food shortage issues in the future. In attempts to increase animal production, antibiotics were used recklessly, which became a crucial factor in the evolution of cross-resistant microorganisms. By 2030, agriculture intensification will have contributed to a 67% rise in anti-microbial usage (Myers et al. 2022). Antibiotic usage climbed by over 36% in 71 countries, with the BRICS (Brazil, Russia, India, China, and South Africa) accounting for over 75% of this rise (Tiseo et al. 2020). However, measurement of anti-microbial use in animal production in low- and middle-income countries (LMICs) is difficult because of anti-microbials sold "over the counter" and of poorly enforced regulatory frameworks (Jibril et al. 2021). Nanotechnology can improve animal production qualitatively and quantitively

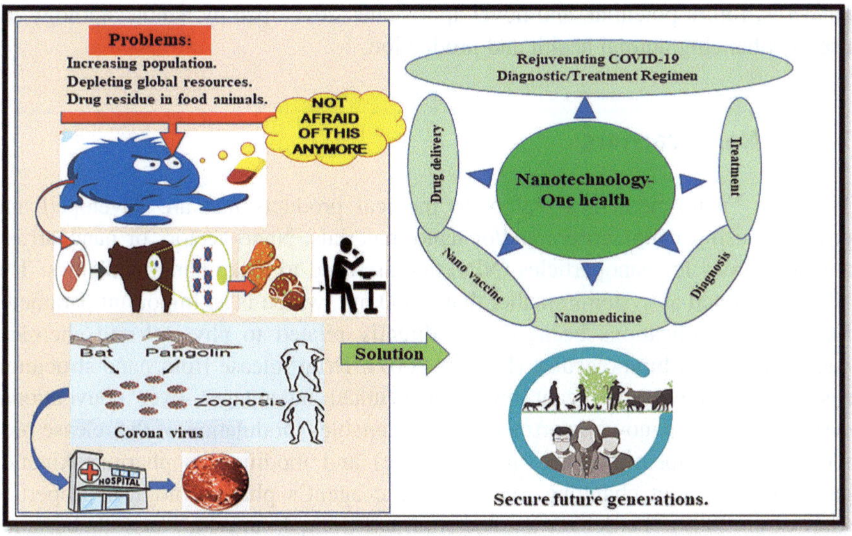

Fig. 7.1 Problems and solutions by nanotechnology

while providing many environmental, health, and economic benefits. Nanotechnology research and development are likely to facilitate and frame the next stage of development of genetically modified animal products, vaccines, precision animal feeds, and reproductive techniques (Fig. 7.1). Biosensors using nanotechnology can monitor animal health, safety, and fertility (Fesseha et al. 2020).

As the twentieth century rolled around, a pandemic seemed a long way off, despite experiencing plague and Spanish flu in the eighteenth and nineteenth centuries. However, in 2019, the entire world came to a standstill due to the COVID-19 pandemic, shattering our illusions of being invincible. The current coronavirus-associated acute respiratory disease is the third reported spillover of an animal coronavirus to humans in under two decades, resulting in an epidemic (Coronaviridae Study Group of the International Committee on Taxonomy of Viruses 2020). It has necessitated taking immediate steps to check disease screening and address new cutting-edge technologies in human therapeutics and animals as one health. It has become clear that animal and human health must be treated as one, with equal priorities, to avoid future pandemics and that new promising technologies, such as nanotechnology, which can transform diagnosis, treatment, and therapeutics in humans and animals, must be reviewed regularly. Nanomedicine and nanotechnology are emerging fields and provide excellent tools for overcoming the problem of drug resistance and dose reduction. Though nanotechnology is not a new subject but a technology with such promise for the future, it warrants periodic reviews and updates to inspire thinking and novel approaches. The primary types of nano-carriers to be used in the preparation of nano-medicines for the animal population are covered in this chapter. It will also go over toxicity concerns, future potential, and novel approaches developed by nanotechnology for use in enhancing animal health and production.

7.2 Nano-carriers

The term "nano-medicines" refers to medical products that are developed and produced using nano-carriers and/or nano-materials. Nano-vectors or nanocarriers are two names for nanoparticles (NPs) that are used to make nano-medicines. The NPs are defined as systems smaller than 1000 nm, which is an important parameter for clinical applications because it is directly related to physical and chemical interactions with biomolecules (Kreuter 2007). Drug release from nano-structured systems can have significant biopharmaceutical advantages over conventional systems because nano-compartmentalization enables modulation of the release profile of the therapeutic agent (time and site) and modifies its pharmacokinetics parameters without changing the therapeutic agent's pharmacological properties (Suri et al. 2007). To deliver antibacterial, anti-fungal, antiviral, anaesthetic, anti-parasitic, vaccine, and anti-neoplastic drugs to the animal population as well as cosmetic items to enhance the appearance of pets, nano-carriers are ideal (Carvalho et al. 2018). The therapeutic agents in nano-carriers can also be attached to their surface, dispersed in a matrix, or included in a reservoir system. Nanoparticle surfaces can be functionalized to increase drug efficacy and safety as well as to target particular tissues, organs, or cells.

7.3 Using Nano-carriers in Veterinary Sciences

Liposomes, nano-emulsions, micelles, lipidic, polymeric, mesoporous silica, metallic nanoparticles, and dendrimers are the main nano-carriers used to prepare nano-medicines for veterinary applications and are discussed in this study (Fig. 7.2) (Carvalho et al. 2020). In order to increase the bioavailability and decrease the toxicity of conventional dosage forms based on therapeutic agents with low solubility, nano-carriers are a good strategy. The primary goal of using nano-carriers in veterinary medicine is to be able to vectorize the release of the therapeutic agent with dose reduction, thereby minimizing serious adverse reactions and discomfort in long-term treatments. In general, the advantages of nano-carrier systems are extensive and include the following: (1) targeting the therapeutic agent through accumulation at the site of action; (2) lowering the required dose of the therapeutic agent, decreasing the dose-associated toxicity; (3) protecting the therapeutic agent by increasing its stability; (4) modulating the release profile of the therapeutic agent; (5) modifying the pharmacokinetic parameters of the therapeutic agent, increasing bioavailability; and (6) reduction of toxicity.

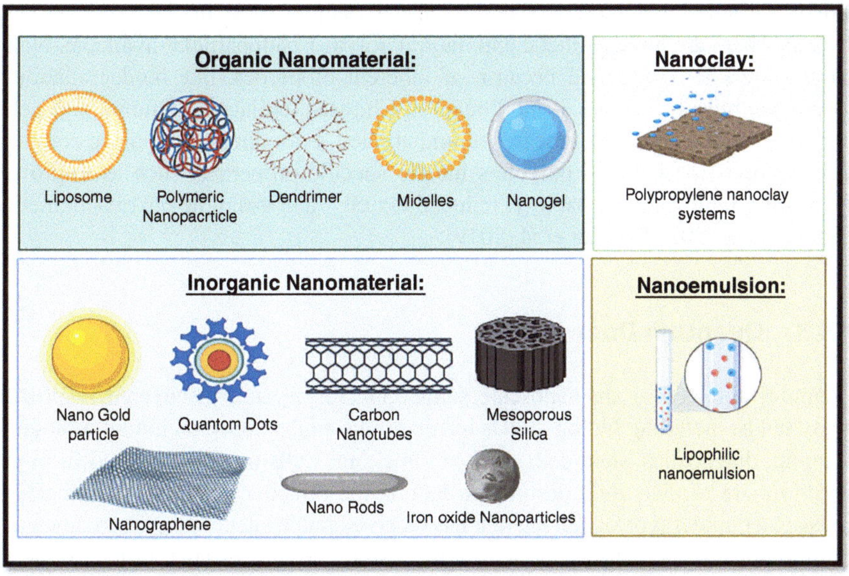

Fig. 7.2 Nano-carriers in veterinary sciences

7.4 Classification of Noncarriers

7.4.1 Liposomes

Liposomes are phospholipid and cholesterol bilayer spherical vesicles that form at least one lipid bilayer around an aqueous core in water, allowing Van der Waals forces to encapsulate both hydrophilic and hydrophobic compounds immersed in the lamellae (Gonda et al. 2019). Liposomes have unique properties due to their nano sizes, such as biocompatibility and biodegradability. They can be used in nanomedicine, cosmetics, and the food industry (Panahi et al. 2017). Liposomal formulations are less toxic and have better pharmacological parameters than drugs alone. Although they appear to be the preferred drug delivery system for many diseases, more research into dose and time regimens is needed (Beltrán-Gracia et al. 2019).

7.4.2 Polymeric Nanoparticles

With the advances in polymer science and nanotechnology, a diverse range of polymer nanoparticles has been described. The potential applications of polymeric nanoparticles are determined by their unique properties or desirable characteristics (Chenthamara et al. 2019). Important properties of polymeric nanoparticles, like biocompatibility and biodegradability, are frequently used in the drug delivery system. There are both synthetic and natural polymer nanoparticles available. Natural polymers have excelled because of inherent properties, like biodegradability, biocompatibility, and ease of modification. Chitosan, collagen, albumin, and gelatine have been extensively researched and applied. Formalized paraphrase complex modifications to synthetic polymers handle specific properties, such as improved specificity, biological availability, reduced toxicity, and desirable pharmacokinetics (Kakkar et al. 2017; Rezaei et al. 2019).

7.4.3 Quantum Dots

Quantum dots (QDs) are nanoscale semiconductor crystals discovered helpful in many fields, including biology. QDs have various applications, including biological imaging, diagnostics, stem cells, and cell tracking. Cells must be targeted from the injection site to their final destination to provide effective diagnosis and therapy. QDs are a popular type of cell tag that has received much attention. QDs are not toxic or immunogenic. QD labelling solved the tissue autofluorescence problem due to their emission properties (Kargozar et al. 2020). The usage of QDs with other methods, such as emission scanning microscopy, might aid in better detection and understanding of cellular interactions within the target tissue. Cryo-imaging can detect single-cell sensitivity cells, making it vital for stem and stromal cell therapies and regenerative medicine (Wuttisarnwattana et al. 2020).

7.4.4 Nano-shells

Nano-shells are spherical nanoparticles with dielectric silica cores covered by thin metals, mainly gold. These shells may be conjugated with other biomolecules, such as antibodies, fluorophores, oligonucleotides, targeting ligands, polymers, therapeutic drugs, and radioisotopes for more effective and better diagnostic and therapeutic goals. The synthesis of gold nano-shells can be explained in three steps: the first step includes the surface amination of silica cores, the second step involves the deposition of gold nanoparticles (2–3 nm gold seed), and the last step concerns the re-growing of these seeds (Lermusiaux et al. 2021). Nano-shells find their application in gene delivery, target therapy, biomedical imaging, tissue welding, drug delivery systems, cancer imaging, and other therapeutic applications. They have a place in medicine for such applications because of their safety, stability, biocompatibility, bioavailability, as well as high ability to bind with different therapeutic agents (Ahmadi and Arami 2014).

7.4.5 Dendrimer

The term "dendrimer" is derived from two Greek words that mean "tree" and "parts", respectively, and describes their branching form. Dendrimers are nano-metric molecules that are radially symmetrical, globular, monodispersed, and homogeneous (Sohail et al. 2020). Dendrimers have therapeutic potential because of their anti-fungal, antibacterial, and cytotoxic properties. They are helpful as drug and gene delivery vehicles, but studies have shown that some dendrimers have therapeutic potential due to their antifungal, antibacterial, and cytotoxic properties. Since their discovery, the biological utility of dendrimers has rapidly evolved, with over 100 families and thousands of chemical modifications documented. Dendrimers like Frechet-type polyether and Tomalia-type (poly)amidoamine are poly (amidoamine) dendrimers, which are widely used in biomedicine as protein, enzyme, and virus analogues (Tetteh-Quarshie et al. 2021).

7.4.6 Solid Lipid Nanoparticles

Solid lipid nanoparticles have a lipophilic centre, making them suitable for cancer treatment (Youssef et al. 2019). The outer hydrophilic shell of various hydrophilic medications or antibodies can be conjugated with them. The outer shell also improves the bio profitability of the drug. Cationic solid lipid nanoparticles can electrostatically link nucleic acid components, allowing quality control. Topical, oral, and subcutaneous injections are all options for administering this type of nanoparticle. They can pass through the blood-brain barrier, allowing them to deliver medications. They can cross the blood-brain border, allowing medicines to reach the central nervous system more efficiently (Elgqvist 2017).

7.4.7 Metallic Nanoparticles

Metallic nanoparticles made of gold, silver, manganese, and platinum have a metallic core surrounded by a protective coating. They contain antibodies and radionuclides, which can be chelated. Metallic nanoparticles, particularly bimetallic nanoparticles, have been used for disease therapy, including silver/gold nanoparticles, silver-selenium nanoparticles, and gold-platinum nanoparticles (Youssef et al. 2019).

7.4.8 Polymeric Micelles

Polymeric micelles have a hydrophobic core that facilitates the transport of hydrophobic drugs and are highly water soluble. According to their description, they were also made with amphiphilic polymers, like caprolactone or poly(lactic-co-glycolic acid) (PLGA). Less water-soluble drugs, such as paclitaxel and amphotericin B, are commonly delivered using this method (El-Sayed and Kamel 2020).

7.4.9 Polymeric Nano-spheres

Biodegradable or non-biodegradable polymers with less than a micron create uniform circular frameworks. Polymer-based drug-encapsulated nanospheres have been described for effective cancer or antigen-presenting cell (APC) targeting, and this is an exciting area of research because it allows for the controlled release of drugs in a cell- or tissue-specific manner (DeFrates et al. 2018).

7.5 Mechanism of Action of Nanoparticles Against Microorganisms

Small-sized NPs between 5 and 17 nm exhibit high anti-microbial activities (Slavin et al. 2017). There has been considerable research into various mechanisms for promoting contact between NP and bacteria, such as Van der Waals forces, electrostatic forces, and ligand-receptor and hydrophobic interactions. When NPs contact microbes, t hey can disrupt metabolic pathways and affect membrane shape and function. Inside a bacterial cell, NPs interact with cellular mechanisms to inhibit enzymes, inactivate proteins, alter gene expression, generate free radicals, and cause electrolyte imbalance (Hamida et al. 2021). The antibacterial activity of NPs is most commonly attributed to reactive oxygen species (ROS). While oxygen enters in its unwanted reduced states, it has more tendency to transform into free radicals, peroxides, and superoxides rather than to get converted into water. As the concentration of nanoparticles increases, free radicals are produced, resulting in ROS increase and cell death (Fig. 7.3) (Canaparo et al. 2021). Superoxide radical, singlet oxygen, hydroxyl radical, and hydrogen peroxide are the four types of ROS produced (Sharma et al. 2019).

Even so, as NPs concentrations rise, the cell prefers oxidation via excessive ROS production (oxidative stress), resulting in cell component damage. As a result, the cell membrane's permeability changes, resulting in cell membrane damage. Some nanoparticles have been biofunctionalized to replace chemicals like mannose, which attract bacteria via mannose receptor sites (Bondarenko et al. 2018). Carbon nanoparticles enter bacterial cells and prevent multiplying and dying, causing cell lysis. The antibacterial activity of carbon nanoparticles may be due to the electrostatic repulsion of Gram-negative microorganisms and carbon surfaces, where microbes bind to carbon particles via solid Van der Waals forces, and Gram-positive bacteria change the positive charge on the carbon surface (Quek et al. 2022). Silver nanoparticles destroy biomacromolecules by deactivating enzymes, changing protein expression, and disrupting the respiratory chain. Nanoparticles can damage or burst bacterial cell membranes by interacting with their hydrophobic chains of various lengths. Because their cationic surfaces interact strongly with bacterial

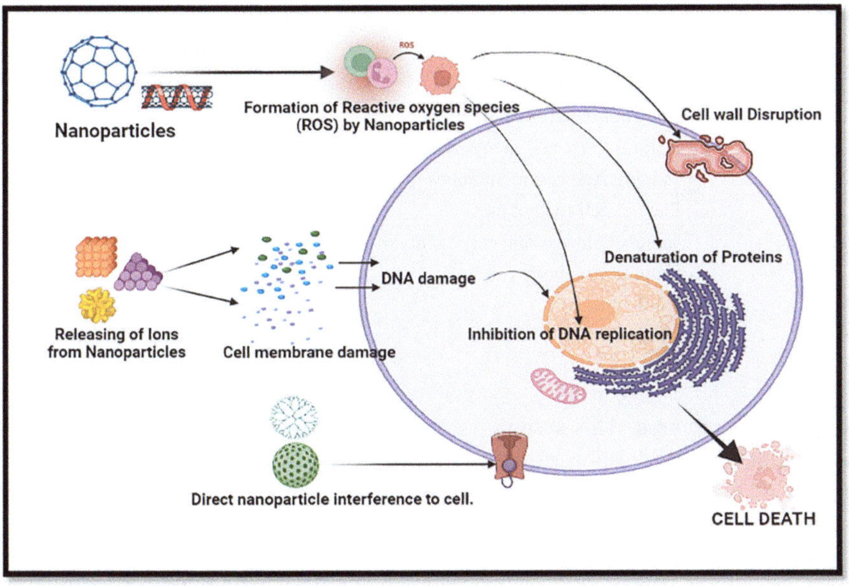

Fig. 7.3 General anti-microbial action of nanoparticles

cells, polymeric nanoparticles kill germs when they make contact them (Jiang et al. 2022).

7.6 Nanotechnology in Animal Science

7.6.1 Nano-medicine

Nano-delivery systems and nano-medicine are emerging nanoscience techniques essential in veterinary medicine. Even though such systems are relatively new, they are swiftly developing where a wide range of nanoscale biological and non-biological materials are used to serve as diagnostic tools or to deliver therapeutic agents to target sites in a regulated manner (Tewabe et al. 2021). Nano-medicine is a branch of medicine that employs nanotechnology to prevent and cure diseases using different nanoscale materials (Josyula et al. 2021). The nanoscale materials used in nano-medicine include biocompatible nanoparticles and nano-robots. Nano-medicine has various applications, like disease diagnosis, drug delivery, and sensory and actuation purposes in organisms (McNamara and Tofail 2015).

Innovative drug delivery systems can deliver therapeutic agents as small sealed packages with high efficacy to the specified target in an animal's body. Microscale and nano-mechanical systems are smart drug deliveries that make use of smaller quantities of antibiotic. These advanced drug delivery systems also possess chemical detection and decision-making capability for self-regulated drug delivery or nutrient

Table 7.1 Various types of biomaterials used to coat nano-materials

S. No.	Biomaterials	Reference
1.	Erythrocyte cell membranes	Wang et al. (2019)
2.	Neutrophil cell membranes	Zhang et al. (2018)
3.	Natural killer cell membranes	Pitchaimani et al. (2019)
4.	Macrophage cell membranes	Gong et al. (2020)
5.	Platelet cell membranes	Wang et al. (2019)
6.	Extracellular vesicle cell membranes	Kao and Papoutsakis (2019)
7.	Natural proteins	Iqbal et al. (2021)
8.	Viral capsids	Kines et al. (2021)
9.	Aptamers	Huang et al. (2021)

treatment as required. This system has the advantage of minimizing antibiotic use and, ultimately, medication expenses as well. They can also monitor the delivery effects of nutraceuticals, pharmaceuticals, nutrients, bioactive compounds, food supplements, chemicals, probiotics, and vaccines (Tewabe et al. 2021).

Although nanotechnology has provided alternative therapeutic opportunities, there are still challenges that emerge with nanoparticles. While travelling from the injection site to the target site, nanoparticles encounter various physiological and cellular barriers that recognize and eliminate foreign bodies (Chen et al. 2021). Nanoparticles are usually modified with synthetic polymers like polyethylene glycol or poly(lactic-co-glycolic acid) to extend circulation in blood. The surface of targeting ligands is grafted to bind selectively only to the specified target while preventing non-specific interactions with healthy tissues (Zhao et al. 2020). Optimizing synthetic nanoparticles has been implemented for effective targeted delivery with minimum side effects (Ryu et al. 2020). Many efforts have been made towards bioinspired nanotechnology, in which design cues are taken from the organism for effective delivery.

Biomimetic nanoparticles are an emerging class of nanoparticles whose surface is fabricated or integrated with biomaterials or other synthetic materials, able to mimic the biological functions and characteristics of native cells. Because of this property, biomimetic nanoparticles possess enhanced biocompatibility, have a long retention time and high target specificity, and induce minimal immune responses (Chen et al. 2019). Different types of biomaterials are used to embed nanoparticles to enhance their stability and efficacy, which are enlisted in Table 7.1.

7.6.2 Diagnosis and Treatment of Disease

Nanotechnology exhibits activity at the same size as that of the pathogen and therefore has an excellent capability for early diagnosis and treatment. Nanotechnology has been described as an effective technique in critical clinical diagnoses. The application of nanotechnology in diagnostic tools is remarkable, owing to enhanced specificity, sensitivity, and functionality (Zhang et al. 2019). Biomarker detection,

Table 7.2 Various nano-structured materials with application in the health sector

S. No.	Nano-structured material	Application in healthcare	Potential risks	Reference
1.	Carbon nano-tubes	Glucose monitoring	Pulmonary toxicity	Yan et al. (2015)
2.	Graphene	Biosensing, imaging	Graphene-induced toxicity, inflammatory response	Nagraik et al. (2021)
3.	Quantum dots	Optoelectronic application, detection of bio-macromolecules like proteins, enzymes, and neuro-transmitters; substitute to fluorophores; cancer diagnosis	Skin penetration	Henna and Pramod (2020)
4.	Chitosan	Used in bioanalytical device biosensors and lab-on-chip devices	Chitosan toxicity	Nagraik et al. (2021)
5.	Dendrimers	Used in various diagnostic sensors, such as those based on fluorescence, electrochemistry, impedimetric, Raman scattering or SPR transfection, drug delivery, and biocatalysis	Haemolytic toxicity, cytotoxicity, and haematological toxicity	Mahmoudpour et al. (2021)

which involves significant waiting times and high costs and requires sophisticated analyzers, needs to be replaced with more economical, faster, and more robust diagnostic tools. Nano-structured materials are promising in the advancement of diagnostics. Nanotechnology combined with proteomics, genomics, and molecular machine systems can help develop reliable, effective, and fast medical diagnostics.

Nanotechnology primarily focuses on improving conventional diagnostic methods by increasing process efficiencies and enhancing the reusability of nano-materials, therefore reducing the overall cost of diagnosis. Biomedical applications, like tissue scaffolds, drug delivery, implantable materials, or nano-devices such as biosensors, have widely used various nano-material structures. Due to its performance and sensitivity, nanotechnology plays a significant role in the development of biosensors. Biosensors that combine mechanical, biological, and physico-chemical properties of transducers have tremendous applications in healthcare and biodefence systems. Nano-structured materials in biosensors play a vital role in disease detection and diagnosis. Some commonly used nano-structured materials (Table 7.2) in biosensors are nanoparticles, carbon nano-tubes, graphene dots, quantum dots, nano-wires, and metallic sheets (Nagraik et al. 2021).

Nanotechnology offers solutions to a number of serious diseases, like tuberculosis, foot and mouth disease, methicillin-resistant *Staphylococcus aureus* (MRSA), brucellosis, and other infections due to intracellular or blood pathogens (El-Sayed and Kamel 2020), respiratory infectious diseases, melanoma, and other types of cancers. Fluorescent nanoparticles with photophysical characteristics have demonstrated excellent performance in *in vivo* imaging. In addition, nanoparticles have unique optical contrast agents in image-guided tumour surgeries. Nano-theranostics is another remarkable progress in nanotechnology, a medical technique that combines medicine and diagnostics to enhance the effectiveness of currently used medicines (Fawzy et al. 2021).

7.6.3 COVID-19 and Nanotechnology: Rejuvenating Diagnostic Regimen

Coronaviruses (CoVs) are a group of enveloped single-stranded ribonucleic acid (RNA) viruses that infect mammals and birds, causing respiratory and gastrointestinal diseases (Li et al. 2020). Alpha coronavirus (prevalent in mammals and pigs as animal hosts), beta coronavirus (prevalent in mammals and bats, mice, and cows as animal hosts), delta coronavirus (dominant in pigs and birds, with pigs as animal hosts), and gamma coronavirus (prevalent in whales and birds, with chicken and turkey as animal hosts) are the four genera of *Coronavirinae* (Krishnan et al. 2021). Before December 2019, which marked the outbreak of COVID-19, four beta coronavirus strains caused severe human diseases: HKU1, MERS-CoV, OC43, and SARS-CoV. Once again, the spillover came from the beta coronavirus strain COVID-19 in Wuhan, Hubei, Central China (Zhou et al. 2020).

At present, Omicron has accumulated over 30 mutations on its spike protein, resulting in its immune evasiveness. Omicron cannot utilize TMPRSS2-expressing cells, primarily found in the lower respiratory, explaining its alleged mildness. The matter of concern is, what if, under selection pressure, it overcomes its inability to infect TMPRSS2-expressing cells, as it has done for human angiotensin-converting enzyme-2 (ACE2)? A catastrophe like this will be beyond the capabilities of any nation. Therefore, we must review and analyze all the advanced technologies that can make a breakthrough in this area of research, and nanotechnology stands out as a promising option.

Nanotechnology has aided in developing a new generation of vaccines that are critical for the global control of the COVID-19 epidemic and the return to normalcy. Nanotechnology investments have benefited vaccine development, research, and innovation. Vaccine development has primarily taken place in countries in the northern hemisphere. South Africa, as an important emerging country, has made significant investments in nanotechnology and bioinformatics and the skills and resources needed to develop and manufacture vaccines (Dube et al. 2021). Bioinformatics tools have also aided in developing quick and cost-effective vaccines and provided recommendations for future research and development. Using an opto-electro-magnetic nano, scientists used a biosensing technique to identify the SARS-CoV-2 virus during the current COVID-19 pandemic. Viruses can be detected by using electrical, optical, or magnetic biosensors based on gene sensing and immune sensing. Paliwal et al. (2020), in their study, selectively detected the SARS-CoV-2 virus at a shallow level by using biosensors.

These efficient biosensors can be controlled with a smartphone and marketed for clinical use in COVID-19 infection diagnosis at an early stage. Furthermore, the successful integration of these SARS-CoV-2 viral sensors with artificial intelligence (AI) and the Internet of Medical Things (IoMT), effectively helped in virus detection at the point of care while also sharing bioinformatics with the medical centre for quick therapeutic decisions. This method can also manage COVID-19 infection based on patient infection profiles and keep track of chores. In addition, researchers developed stimuli-responsive nanotechnology to trap and eliminate viruses when stimulated externally to prevent human-to-human transmission of the SARS-CoV-2 virus (Kaushik 2021).

Nanotechnology, for example, could enable photo-sensitive viral destruction. Several types of clothing containing nanoparticles have been successfully used to demonstrate SARS-CoV-2 virus entrapment and eradication. The increased production and distribution of these masks for general use, on the other hand, will necessitate a significant amount of effort. One of the documented advancements in this field is microneedle-based vaccine administration to treat COVID-19 infection. Early results are encouraging, but there is still much work to be done regarding animal-model-based experiments and Food and Drug Administration (FDA) approval, both of which are required before clinical implications can be suggested (Kaushik et al. 2020).

7.6.4 Nanotechnology-Based Therapeutic Agents Against Animal Coronaviruses

Nanotechnology is crucial in combating emerging diseases, including coronaviruses like SARS-CoV-2. The similarity in the molecular sizes of nanoparticles and coronavirus enables better interactions and the potential for future applications (Medhi et al. 2020). Further biomedical properties of nanoparticles favour the development of therapeutic formulations that can have higher efficacy against viruses. Nanoparticles also have antiviral properties. Improved drug pharmacokinetic properties by applying nanoparticles with antiviral drugs have further made nanotherapeutics beneficial and effective (Chakravarty and Vora 2021). Both metallic nanoparticles and lipid-based nanoparticles have been explored against antiviral applications (Sharmin et al. 2021). Antiviral drugs such as zidovudine, maraviroc, ritonavir, and a few other pharmaceuticals are developed using multiple types of lipids for optimal cellular absorption and targeted drug delivery. Nano-formulations, which use multifunctional nanoparticles to optimize gene transport, protect cargo from nuclease degradation, and improve the targeting mechanism, are also used in non-viral gene therapy (Lin et al. 2018).

Nano-formulations outperform traditional antiviral drugs by preventing virus replication in mammals in multiple ways. First, nanoparticles can directly inactivate viruses by interacting with virus surface proteins or the viral genome, preventing replication. Gold, silver, and copper nanoparticles act as antiviral agents by preventing viral entry into host cells. These nanoparticles prevent viral replication by adhering to viruses and rendering them inactive, as well as interfering with virus-host cell interactions. Finally, virucidal zinc (Zn) nanoparticles interfere with viral DNA polymerase and block virus growth inside host cells (Jampílek and Kráľová 2019).

Nano-structured materials are used to make adjuvant-based nano-vaccines for animals to protect them from infections caused by the animal coronavirus. In recent years, there has been much interest in incorporating virus entities into the production of nano-vaccines with adjuvants. Due to improved antigen-processing pathways, nano-vaccines provide benefits such as a controlled immune response and activated B-cell immunity (Singh 2021). However, conventional vaccinations have drawbacks, such as low immunogenicity and insufficient protection. Furthermore, following the recent coronavirus disease-19 (COVID-19) pandemic, which may have started in bats (primary reservoir) and was caused by the lethal coronavirus SARS-CoV-2, it is critical to use nanotechnology to prevent viral infections from spreading to other species.

7.6.5 Anti-microbial Resistance

The emergence of anti-microbial resistance is a serious concern that has drawn attention worldwide. The issue can be aggravated by the irrational use of anti-microbial drugs, insufficient surveillance, and the inadequately controlled regulation of antibiotic use in the medication and livestock industry. Furthermore, because of the emergence of multidrug-resistant microorganisms and the scarcity of new antibiotics, there is a dire need for new anti-infectious strategies. Some of the new strategies and their action are summarized in Table 7.3.

7.6.6 Nano-vaccines

Vaccines comprise adjuvants and antigens, which trigger the defense mechanism or long-lasting protective antibody response. Live and dead organisms were used in the production of vaccines earlier. Later, the production of safe synthetic and recombinant vaccines came into vogue. These new vaccines are highly sensitive to degradation and require optimized adjuvant with good stability to augmented immunogenicity. The regulation of conventional adjuvants is complex, but advanced antigen-delivery strategies are being developed with the evolution of nanotechnology.

The immunity of vaccines is increased with nanoparticle adjuvants by imitating molecular models relating to pathogens, regulating co-stimulatory molecules on cells with antigens, maintenance of the immune system, and long-time delivery of antigens. Nanoparticles can produce virus-like particles with morphological similarity to virus capsids and can stimulate immune responses that do not cause infection

Table 7.3 Nanotechnology strategies against anti-microbial resistance

No.	Strategy	Action	Reference
1.	Use of silver nanoparticles	Antibacterial effect against pathogenic species of Gram-positive bacteria (*Bacillus licheniformis*, *Bacillus cereus*) and Gram-negative bacteria (*Escherichia coli*, *Pseudomonas aeruginosa*)	Basera et al. (2019)
		Inhibitory action against *Acidovorax oryzae* strain RS-2	Masum et al. (2019)
		Post-surgical treatment for *Corynebacterium pseudotuberculosis* infection	Santos et al. (2019)
2.	Phyto-fabrication of selenium nanoparticles	Anti-microbial, antioxidant, and biocompatibility	Gunti et al. (2019)
3.	Bio-fabrication of zinc oxide nanoparticles	Limiting the growth and mycotoxins of *Fusarium graminearum*	Lakshmeesha et al. (2019)
4.	Bio-mimetization of hydroxyapatite with *Azadirachta indica*	Therapeutic anti-inflammatory and antibacterial action	Nagaraj and Samiappan (2019)
5.	Nano-composites of nano-crystals doped with silver and silver oxide nano-crystals	Inhibition of biofilms formation	Fonseca et al. (2019)
6.	Magnetic nano-conjugated teicoplanin	Targeting of the bacterial infection site	Armenia et al. (2018)
7.	Vancomycin-loaded nanoparticles	Enhance antibacterial efficacy and sporicidal for *Clostridium difficile* infection	Chen et al. (2019)
8.	Pullulan nanoparticles	Enhance the antibacterial properties of *Lactobacillus plantarum*	Hong et al. (2019)
9.	Anti-microbial peptides and bacteriophages nano-encapsulated in liposomes	Prolonged delivery	Zambom et al. (2019)

(Ravi et al. 2021). In recent years, nanotechnology research has received massive attention for the potential to replace conventional viral vector vaccines (Fawzy et al. 2021). Various nanoparticles are used as adjuvants in nano-vaccines against livestock viral diseases (Table 7.4).

7.6.7 Feed Technology

Engineered nanoparticles could help facilitate dietary supplementation to weaned animals and boost growth rates by increasing the bioavailability and efficacy of nutrient cargoes (Hill and Li 2017). However, while developing a carrier nanoparticle for the oral administration of nutrients, certain things need to be considered. For

Table 7.4 Nanoparticle adjuvants effective against various viruses

S. No.	Virus	Potential adjuvant	Action	Reference
1.	Foot and mouth disease virus	Gold nanoparticles	Improved macrophage activation	Teng et al. (2021)
2.	Rift Valley fever virus	Silver nanoparticles	Reduction in virus infectivity	Borrego et al. (2016)
3.	Infectious bovine rhinotracheitis virus	Silver nanoparticles	Inhibited virus infection	El-Mohamady et al. (2018)
4.	Peste des petits ruminants virus	Silver nanoparticles	Inhibited virus entry	Khandelwal et al. (2014)
5.	Avian influenza virus	Silver nanoparticles	Increased protective level of antibodies	Yu et al. (2013)
		Polyanhydride nanoparticles	Prolonged release of encapsulated antigens	Kingstad-Bakke et al. (2019)
6.	Newcastle disease virus	Polyrhodanine nanoparticles	In ovo anti-Newcastle disease activity	Nazaktabar et al. (2017)
		Microalgae-mediated silver nanoparticles	In vivo anti-Newcastle disease activity	Khalid et al. (2017)
7.	Infectious bursal disease virus	PLGA nanoparticle	Increased cellular and humoral immune responses	Al-Rubaee et al. (2019)
		Silver nanoparticle-anchored graphene-oxide sheets	Antiviral activity against non-enveloped IBD virus	Chen et al. (2016)

example, it includes the challenges of resisting the action of enzymes and the pH of the gastrointestinal tract of animals.

Nanoparticles must be capable of surmounting these obstacles to supply their nutritional elements at the specified location (Ban et al. 2015). Nanoparticles developed from canola protein cruciferin are able to encapsulate both hydrophilic and hydrophobic bioactive elements, protect them from degradation in the stomach, and release them in the intestinal lumen to get absorbed and function as if they originated from the feed. Being small sized, nanoparticles gain an increased level of bioavailability in comparison to microparticles, especially in the digestive tract, as nanoparticles can pass readily through the intestinal mucosa (Akbari and Wu 2016).

Copper nanoparticles pass through intestinal mucosa more quickly than microforms, helping better absorption (Gonzales-Eguia et al. 2009). The quality and stability of livestock feed and water can be improved by the addition of nanomaterials. In a study conducted by Zha et al. (2009) on broilers, it was demonstrated that the addition of chromium nanoparticles to poultry feed increases the average daily gain, feed efficiency, thigh, and breast muscle protein content but lowers cholesterol levels. However, a comprehensive study of nanoparticle activity must be carried out to test the potency and any undesirable biological consequences.

7.6.8 Animal Breeding and Reproduction

Timely oestrus detection in dairy animals is one of the biggest challenges in dairy farming. Nano-tubes provide an effective solution for tracking oestrus in animals through real-time measurement of estradiol level changes in the blood. In addition, nano-tubes can be used to detect the oestrus period by binding to estradiol antibodies during oestrus through near-infrared fluorescence (Mekonnen 2021). Artificial insemination techniques can be advanced by incorporating nano-techniques such as nano-purification, non-invasive bioimaging of gametes, and protectants in cryo-preservation (Hill and Li 2017).

Quantum dots can help provide a better understanding of oocyte movement, mammalian spermatozoon, and their physiological interaction. These nanoparticles are of great interest to theriogenologists because of being biocompatible and photostable and due to their ability to produce more signal intensity compared to other organic fluorescent molecules, which have been used previously for imaging gametes and other types of cells in vivo. These engineered nanoparticles help visualize cellular and molecular events during fertilization, like fluorescent proteins, but at greater tissue depths (Feugang et al. 2015). The nano-purification of semen is a technique that can separate damaged sperms from healthy ones. One such technique is coating magnetic nanoparticles with antibodies against ubiquitin (a surface marker of defective spermatozoa) for a protein-based removal strategy. Another lectin-based strategy coats magnetic nanoparticles with lectins, which bind glycans exposed at the sperm surface. In a study (Odhiambo et al. 2014) on the nano-purification of bull spermatozoa, nano-purified spermatozoa achieved equal conception rates as un-purified ones. Therefore, it was concluded that more females could get inseminated from one nano-purified diluted ejaculate sample. The cryo-preservation of sperm can be enhanced by the addition of nano-protectant additives in extenders. Anti-microbial nanoparticles may also replace extender antibiotics since some antibiotics inhibit sperm motility and viability. Nanoparticles may also be used to incorporate other natural products in semen extenders to modify the properties of semen, such as to increase sperm motility (Bryła and Trzcińska 2015).

The integration of nanoparticles in molecular biology procedures may lead to further advancement in reproductive biotechnology. One such technique is the mediation of gene transfer through sperm, in which mesoporous silica nanoparticles are loaded with deoxyribonucleic acid (DNA) molecules and proteins of interest and delivered to oocytes at fertilization (Barkalina et al. 2014). Transfections with polymeric nanoparticles, like chitosan, poly(2-(dimethylamino)ethyl methacrylate (PDMAEMA), and polyethyleneimine, are reported to have more advantages over conventional viral approaches if polymers are used in low concentrations. The molecular weight of nano-polymers influences transfection efficacy and toxicity. Nanoparticles can also play a vital role in animal reproduction. However, some nanoparticles are spermatotoxic, which can have serious consequences. For example, in a study, it was reported that zinc oxide (ZnO) and titanium oxide nanoparticles reduce in vitro sperm viability (Poddar and Kishore 2022).

7.6.9 Food and Feed Safety

The meat and food industries are currently utilizing nano-material applications, such as using them as carriers for food ingredients or additives, which are added to food directly, or as a component of food packaging. Additionally, this can enhance flavour; decrease the use of fat, salt, sugar, and preservatives; and improve the ability of fat-soluble additives to disperse in food products. This can prevent hypertension and cardiovascular disease in both humans and animals. Zn NPs were once only used as a feed additive, but today, there are many uses for metal nano-materials in veterinary medicine (Rahman et al. 2022). Nevertheless, they are poisonous to microorganisms and are employed as anti-microbial agents, and their addition to cow feeds led to a marked increase in milk production. Additionally, they can be utilized in catalysis, sensors, environmental remediation, and human and animal personal care products (Reda et al. 2021).

Nutritive substances can resist protease enzymes and other desaturating agents, thanks to nano-carriers. Additionally, it might be improving its capacity to cross the intestinal membrane into blood. Additionally, nano-carriers improved the dispersion of nutrients in aqueous systems and controlled the release of nutrients in water-insoluble food ingredients (Xia et al. 2022). Nano-materials can create new tools for the rapid detection of food-borne pathogens, such as gold and iron oxide nanoparticles and nano-materials loaded with anti-*S. aureus* antibodies (Sheikhzadeh et al. 2021). The science of biomedicine, food systems, food system security, and disease treatment delivery could all benefit from the use of nanotechnology. Today, a wide range of nano-materials and nano-sensors are used as food safety additives and have significant value in all fields of animal science. Nanotechnology will soon make it possible to produce "interactive" poultry meat that alters its colour, flavour, or nutrients (Akter et al. 2022).

7.7 Toxicity Risk of Nanoparticles

The primary justification for restricting the use of biosensors in human and animal science is their toxicity risk to animals and the environment, despite the advancement and value of these tools in animal science (Hassan et al. 2020). Several aspects of biomedicine use nanotechnology, but there are also potential toxicity risks for the environment and users. Affected tissue fibres or secondary mutations cause oxidative stress and inflammation. Nanoparticle size, the health of humans and animals, administered doses, and exposure time all generally affect the toxicity of nano-materials. Additionally, the particle size and shape, crystalline form, functionalization, and purity of nanoparticles must be taken into account for an effective assessment of their risk of toxicity (Abedin et al. 2021). When sperm is exposed to certain nanoparticles in excess or for an extended period of time, such as zinc oxide and titanium oxide, it loses viability in vitro. Sperm cells that were incubated with concentrations of 100–500 g/mL of Zn NPs died within a short period of time. When buffalo sperm was incubated with 100 g/mL of titanium oxide nanoparticles, Pawar and Kaul (2014) found that the viability of the sperm was decreased. According to a number of studies, metal nanoparticles can easily enter the skin, lungs, and brain, where they can have a negative impact on biological processes.

Before using Zn NPs as feed additives, it is important to understand their short-term environmental and animal toxicity. Therefore, it is necessary to measure the effective nontoxic doses of metal nanoparticles in laboratory animal models to determine whether their field application is appropriate. There have not been any toxicological effects of nanoparticles on health to date (El-Fatah et al. 2017). To fully understand the toxicological aspects of nano-materials, more ongoing research is therefore necessary.

Ingestion is the primary method of exposure for both humans and animals when it comes to nano-materials. The liver and spleen quickly eliminate the nano-materials after ingestion, after which they enter the gastrointestinal tract and reach the intestines. The ingestion is primarily caused by the presence of nanoparticles in food as a result of direct contact between food and nano-packaging (Baltić et al. 2013). The circulatory system distributes the nanoparticles to the liver and spleen (Baltić et al. 2013). Other routes of nanoparticle toxicity include inhalation and skin contact, which are mostly found in laboratories and industrial workers in factories that produce nano-materials. According to a different study, magnesium oxide nanoparticles can enter the central nervous system and nerve cells through skin contact and inhalation.

In other studies, the Zn NPs detected the presence of skin tissue that had been penetrated, and they have cancer-causing effects, like asbestos-caused lung fibrosis. The possibility of lung cancer development also exists when high doses of nano-TiO_2 are inhaled (Aschberger and Christensen 2011). Additionally, just as with the inhalation of ambient ultrafine particles, the entry of nanoparticles via the bloodstream may have negative effects on the cardiovascular system, blood vessel function, and blood clot formation. Similar microvascular dysfunction was seen in rats exposed to low concentrations of nano-sized titanium dioxide, as well as platelet aggregation and vascular thrombosis following exposure to single-walled carbon nanotubes (SWCNTs) and multi-walled carbon nanotubes MWCNTs (Kan et al. 2018). According to some studies, nano-materials can enter the bloodstream, where they are distributed throughout body organ cell tissues and occasionally pass from the mother to the foetus through the blood supply (Baltić et al. 2013). More advanced nano-biomedical research and accessibility are required in the near future to enhance animal health. Many issues with animal disease diagnosis, animal production, reproduction, and good hygienic practices when raising and maintaining food animals could be resolved. Regarding livestock, the potential uses of the technology are almost unbelievable. However, there is still a lot of research that must be done before nanotechnology applications are used in veterinary and animal sciences, especially regarding toxicity risks.

7.8 Challenges and Strategies

Nano-materials with potential health benefits are used as feed additives in animal nutrition, but research on its validity is limited to livestock species. Extensive investigation is needed to arrive at a safe dosage for nano-feed additives because of the delicate balance between safe and toxic levels. The interaction of nanoparticles with various biological components due to large surface free energy should be well reviewed by conducting a meta-analysis to arrive at final dosages. Swain et al. (2016) reported that the actual bioavailability of a few mineral-based nanoparticles, such as ZnO nanoparticles, is still to be specified accurately. Reduced particle size with increased surface area in nano-forms makes them more toxic than their macro counterparts. Several approaches have been tried to minimize toxicity concerns of nano-materials without a change in intrinsic properties, like encapsulating inorganic nanoparticles with organic molecules like lipids, polymers, and biomolecules and coating them with silica (Yan et al. 2019). Using these surface-decorated approaches, the biocompatibility and utility of a wide range of nano-materials have been improved for several biomedical applications. Among different available mineral nanoparticles, only limited nanoparticles, like ZnO NPs, selenium (Se) NPs, and copper oxide (CuO) NPs, are studied in the context of nutritional aspects in livestock. Reports on other nano-materials as feed supplements are yet to be explored (Bhagat and Singh 2022). So there is a need for the product to provide multiple nano-minerals mainly chelated with organic elements for improved absorption and bioavailability of encapsulated minerals.

The present quantification techniques of nanoparticles in animal systems, like inductively coupled plasma mass spectroscopy (ICP-MS), atomic absorbance spectroscopy (AAS), inductively coupled plasma optical emission spectroscopy (ICP-OES), electron microscopy, histology, fluorescence microscopy, liquid scintillation counting, in vivo optical imaging, computed tomography (CT) scan, magnetic resonance imaging (MRI), positron emission tomography (PET), and single-photon emission computed tomography (SPECT) (Sarparanta and Airaksinen 2021), are limited only to small animals, such as lab animals, pigs, small ruminants, chicken, etc.

Many regulatory bodies, such as the European Union's Registration, Evaluation, Authorisation, and Restriction of Chemicals (REACH); the United States Environmental Protection Agency; FDA's Environmental Protection Agency (EPA); the International Organization for Standardization; the Occupational Health and Safety Administration (OSHA); the Organisation for Economic Cooperation and Development; the European Food Safety Authority (EFSA); India's Food Safety and Standards Authority (FSSAI); Food Standards Australia New Zealand (FSANZ); Japan's Ministry of Education, Culture, Sports, Science, and Technology (MEXT); Ministry of Economic, Trade, and Industries (METI); Ministry of Health, Labor, and Welfare (MHLW); and Ministry of the Environment (MOE); China's National Center for Nanoscience and Technology (NCNST), etc., have issued certain guidelines on the potential risks posed by nano-feed additives in food animals. Even at minimal dosages, metal-based nanoparticles are known to accumulate in body tissues. The chronic ingestion of silver compounds may accumulate in the eyes, skin, and other vital organs, such as the liver (Burduşel et al. 2018). Hence, the length of the feeding period is also an essential factor to be considered before promoting any nanotechnology-based feed additive. The anti-microbial effects of nanoparticles on beneficial bacteria are uncertain and have to be tested before employing them as feed additives. Researchers have been developing nano-packaging technologies in the food industry. However, Priyadarshi et al. (2021) raised a negative concern over the usage of nano-composites for food-packaging applications. They reported that nanoparticles would migrate into the food from the outer packaging, consequently resulting in unintentional consumer exposure to lipid peroxidation and DNA damage. These concerns led to a low public acceptance of nanotechnology-based products among consumers. Efforts need to be put in by manufacturers and food safety authorities to improve the acceptance of food nanotechnology by the public. Other bigger challenges related to nanotechnologies are environmental risks due to the release of nanoparticles into the environment.

7.9 Future Prospectus

The range of applications for nanotechnology in the animal production sector will grow as it advances and attracts more attention. It is likely possible in the near future to regularly add nano-supplements to livestock feed to improve it; however, it will take more time before nanoparticles completely replace antibiotics in feed because many biocidal candidates still need to be tested in vivo before going through clinical trials and food safety tests in accordance with government regulations. Nanoparticles have already been used outside of animal products, such as in antiseptic wound dressings, and more will come. It is crucial to look into nanoparticle cytotoxicity in both cancer cell lines and normal, healthy cell lines for studies interested in nanoparticles with anti-cancer properties.

7.10 Conclusion

This chapter highlights the numerous applications of nanoparticles in animal science and production as well as their future potential in other application areas. Nanoparticles are already on the market, and their properties will be fine-tuned for a wider range of applications with additional research and development. The use of nanotechnology in veterinary science and animal husbandry is still in its infancy, but encouraging results from nutrition, medicine, diagnostic tool, and reproductive studies are driving further investigation.

References

Abedin F, Asmatulu E, Andalib MN (2021) Nanomaterials and human health: an overview. Nanotoxicol Nanoecotoxicol 2:165–180

Ahmadi A, Arami S (2014) Potential applications of nanoshells in biomedical sciences. J Drug Target 22(3):175–190

Akbari A, Wu J (2016) Cruciferin nanoparticles: preparation, characterization and their potential application in delivery of bioactive compounds. Food Hydrocoll 54:107–118

Akter R, Bhattacharya T, Habibur Rahman M (2022) Food science nanotechnologies: implementations, recent developments, and prospects. Nanotechnol Intell Food Packag:343–360

Al-Rubaee SH, Al-Azawi TS, Taha AA (2019) Immunological effects of a single dose of PLGA nanoparticles encapsulated peptide in broilers in comparison to traditional vaccines against infectious bursal disease. Agric Sci Digest 39(4):347–352

Anjum S, Ishaque S, Fatima H, Farooq W, Hano C, Abbasi BH, Anjum I (2021) Emerging applications of nanotechnology in healthcare systems: grand challenges and perspectives. Pharmaceuticals 14(8):707

Armenia I, Marcone GL, Berini F, Orlandi VT, Pirrone C, Martegani E, Gornati R, Bernardini G, Marinelli F (2018) Magnetic nanoconjugated teicoplanin: a novel tool for bacterial infection site targeting. Front Microbiol 9:2270

Aschberger K, Christensen FM (2011) Approaches for establishing human health no effect levels for engineered nanomaterials. J Phys Conf Ser 304(1):012078

Baltić MŽ, Bošković M, Dokmanović M, Janjić J, Lončina J, Baltić T (2013) Nanotechnology and its potential applications in meat industry. Meat Technol 54(2):168–175

Ban C, Park SJ, Lim S, Choi SJ, Choi YJ (2015) Improving flavonoid bioaccessibility using an edible oil-based lipid nanoparticle for oral delivery. J Agric Food Chem 63(21):5266–5272

Barkalina N, Jones C, Kashir J, Coote S, Huang X, Morrison R, Townley H, Coward K (2014) Effects of mesoporous silica nanoparticles upon the function of mammalian sperm in vitro. Nanomedicine 10(4):859–870

Basera P, Lavania M, Agnihotri A, Lal B (2019) Analytical investigation of Cymbopogon citratus and exploiting the potential of developed silver nanoparticle against the dominating species of pathogenic bacteria. Front Microbiol 10:282

Beltrán-Gracia E, López-Camacho A, Higuera-Ciapara I, Velázquez-Fernández JB, Vallejo-Cardona AA (2019) Nanomedicine review: clinical developments in liposomal applications. Cancer Nanotechnol 10(1):1–40

Bhagat S, Singh S (2022) Nanominerals in nutrition: recent developments, present burning issues and future perspectives. Food Res Int 160:111703

Bondarenko OM, Sihtmäe M, Kuzmičiova J, Ragelienė L, Kahru A, Daugelavičius R (2018) Plasma membrane is the target of rapid antibacterial action of silver nanoparticles in Escherichia coli and Pseudomonas aeruginosa. Int J Nanomedicine 13:6779

Borrego B, Lorenzo G, Mota-Morales JD, Almanza-Reyes H, Mateos F, López-Gil E, de la Losa N, Burmistrov VA, Pestryakov AN, Brun A (2016) Potential application of silver nanoparticles to control the infectivity of Rift Valley fever virus in vitro and in vivo. Nanomedicine 12(5): 1185–1192

Bryła M, Trzcińska M (2015) Quality and fertilizing capacity of boar spermatozoa during liquid storage in extender supplemented with different antibiotics. Anim Reprod Sci 163:157–163

Burdușel AC, Gherasim O, Grumezescu AM, Mogoantă L, Ficai A, Andronescu E (2018) Biomedical applications of silver nanoparticles: an up-to-date overview. Nanomaterials 8(9):681

Canaparo R, Foglietta F, Limongi T, Serpe L (2021) Biomedical applications of reactive oxygen species generation by metal nanoparticles. Materials 14(1):53

Carvalho AM, Teixeira R, Novoa-Carballal, Pires RA, Reis RL, Pashkuleva I (2018) Redox-responsive micellar nanoparticles from Glycosaminoglycans for CD44 targeted drug delivery. Biomacromolecules 19(7):2991–2999. https://doi.org/10.1021/acs.biomac.8b00561

Carvalho SG, Araujo VHS, Dos Santos AM, Duarte JL, Silvestre ALP, Fonseca-Santos B et al (2020) Advances and challenges in nanocarriers and nanomedicines for veterinary application. Int J Pharm 580:119214

Chakravarty M, Vora A (2021) Nanotechnology-based antiviral therapeutics. Drug Deliv Transl Res 11(3):748–787

Chen Y-N, Hsueh Y-H, Hsieh C-T, Tzou D-Y, Chang P-L (2016) Antiviral activity of graphene–silver nanocomposites against non-enveloped and enveloped viruses. Int J Environ Res Public Health 13(4):430

Chen Y-H, Li T-J, Tsai B-Y, Chen L-K, Lai Y-H, Li M-J, Tsai C-Y, Tsai P-J, Shieh D-B (2019) Vancomycin-loaded nanoparticles enhance sporicidal and antibacterial efficacy for Clostridium difficile infection. Front Microbiol 10:1141

Chen L, Hong W, Ren W, Ting X, Qian Z, He Z (2021) Recent progress in targeted delivery vectors based on biomimetic nanoparticles. Signal Transduct Target Ther 6(1):1–25

Chenthamara D, Subramaniam S, Ramakrishnan SG, Krishnaswamy S, Essa MM, Lin F-H, Walid Qoronfleh M (2019) Therapeutic efficacy of nanoparticles and routes of administration. Biomater Res 23(1):1–29

Coronaviridae Study Group of the International Committee on Taxonomy of Viruses (2020) The species severe acute respiratory syndrome-related coronavirus: classifying 2019-NCoV and naming it SARS-CoV-2. Nat Microbiol 5(4):536

Dawadi S, Katuwal S, Gupta A, Lamichhane U, Thapa R, Jaisi S, Lamichhane G, Bhattarai DP, Parajuli N (2021) Current research on silver nanoparticles: synthesis, characterization, and applications. J Nanomater 2021:6687290

DeFrates K, Markiewicz T, Gallo P, Rack A, Weyhmiller A, Jarmusik B, Hu X (2018) Protein polymer-based nanoparticles: fabrication and medical applications. Int J Mol Sci 19(6):1717

Dube A, Egieyeh S, Balogun M (2021) A perspective on nanotechnology and COVID-19 vaccine research and production in South Africa. Viruses 13(10):2095

El-Fatah A, Reyad S, Bakry HH, Abo Salem MS, Hassan AAE (2017) Comparative study between the use of bulk and nanoparticles of zinc oxide in amelioration the toxic effects of aflatoxins in rats. Benha Vet Med J 33(2):329–342

Elgqvist J (2017) Nanoparticles as theranostic vehicles in experimental and clinical applications—focus on prostate and breast cancer. Int J Mol Sci 18(5):1102

El-Mohamady RS, Ghattas TA, Zawrah MF, Abd El-Hafeiz YGM (2018) Inhibitory effect of silver nanoparticles on bovine Herpesvirus-1. Int J Vet Sci Med 6(2):296–300

El-Sayed A, Kamel M (2020) Advanced applications of nanotechnology in veterinary medicine. Environ Sci Pollut Res 27(16):19073–19086. https://doi.org/10.1007/s11356-018-3913-y

Fawzy M, Khairy GM, Hesham A, Rabaan AA, El-Shamy AG, Nagy A (2021) Nanoparticles as a novel and promising antiviral platform in veterinary medicine. Arch Virol 166(10):2673–2682

Fesseha H, Degu T, Getachew Y (2020) Nanotechnology and its application in animal production: a review. Vet Med 5(2):43–50. https://doi.org/10.17140/VMOJ-5-148

Feugang JM, Youngblood RC, Greene JM, Willard ST, Ryan PL (2015) Self-illuminating quantum dots for non-invasive bioluminescence imaging of mammalian gametes. J Nanobiotechnol 13(1):1–16

Fonseca BB, Silva PLAPA, Silva ACA, Dantas NO, De Paula AT, Olivieri OCL, Beletti ME, Rossi DA, Goulart LR (2019) Nanocomposite of Ag-doped ZnO and AgO nanocrystals as a preventive measure to control biofilm formation in eggshell and Salmonella spp. entry into eggs. Front Microbiol 10:217

Gonda A, Zhao N, Shah JV, Calvelli HR, Kantamneni H, Francis NL, Ganapathy V (2019) Engineering tumor-targeting nanoparticles as vehicles for precision nanomedicine. Med One 4:e190021

Gong C, Xiaoyan Y, You B, Yan W, Rong Wang L, Han YW, Gao S, Yuan Y (2020) Macrophage-cancer hybrid membrane-coated nanoparticles for targeting lung metastasis in breast cancer therapy. J Nanobiotechnol 18(1):1–17

Gonzales-Eguia A, Chao-Ming F, Fu-Yin L, Lien T-F (2009) Effects of nanocopper on copper availability and nutrients digestibility, growth performance and serum traits of piglets. Livest Sci 126(1–3):122–129

Gunti L, Dass RS, Kalagatur NK (2019) Phytofabrication of selenium nanoparticles from Emblica officinalis fruit extract and exploring its biopotential applications: antioxidant, antimicrobial, and biocompatibility. Front Microbiol 10:931

Hamida RS, Ali MA, Goda DA, Redhwan A (2021) Anticandidal potential of two cyanobacteria-synthesized silver nanoparticles: effects on growth, cell morphology, and key virulence attributes of Candida albicans. Pharmaceutics 13(10):1688

Hassan AA, Mansour MK, El Hamaky AM, El Ahl RMS, Oraby NH (2020) Nanomaterials and nanocomposite applications in veterinary medicine. In: Multifunctional hybrid nanomaterials for sustainable agri-food and ecosystems, pp 583–638

Henna TK, Pramod K (2020) Graphene quantum dots redefine nanobiomedicine. Mater Sci Eng C 110:110651

Hill EK, Li J (2017) Current and future prospects for nanotechnology in animal production. J Anim Sci Biotechnol 8(1):1–13. https://doi.org/10.1186/s40104-017-0157-5

Hong L, Kim W-S, Lee S-M, Kang S-K, Choi Y-J, Cho C-S (2019) Pullulan nanoparticles as prebiotics enhance the antibacterial properties of Lactobacillus plantarum through the induction of mild stress in probiotics. Front Microbiol 10:142

Huang S-S, Lee K-J, Chen H-C, Prajnamitra RP, Hsu C-H, Jian C-B, Yu X-E, Chueh D-Y, Kuo CW, Chiang T-C (2021) Immune cell shuttle for precise delivery of nanotherapeutics for heart disease and cancer. Sci Adv 7(17):eabf2400

Iqbal H, Yang T, Li T, Zhang M, Ke H, Ding D, Deng Y, Chen H (2021) Serum protein-based nanoparticles for cancer diagnosis and treatment. J Control Release 329:997–1022

Jain N, Jain P, Rajput D, Patil UK (2021) Green synthesized plant-based silver nanoparticles: therapeutic prospective for anticancer and antiviral activity. Micro Nano Syst Lett 9(1):1–24

Jampílek J, Kráľová K (2019) Nanoformulations: a valuable tool in the therapy of viral diseases attacking humans and animals. In: Nanotheranostics. Springer, pp 137–178

Jiang Y, Zheng W, Tran K, Kamilar E, Bariwal J, Ma H, Liang H (2022) Hydrophilic nanoparticles that kill bacteria while sparing mammalian cells reveal the antibiotic role of nanostructures. Nat Commun 13(1):1–17

Jibril AH, Okeke IN, Dalsgaard A, Olsen JE (2021) Association between antimicrobial usage and resistance in Salmonella from poultry farms in Nigeria. BMC Vet Res 17(1):1–10

Josyula A, Parikh KS, Pitha I, Ensign LM (2021) Engineering biomaterials to prevent post-operative infection and fibrosis. Drug Deliv Transl Res 11(4):1675–1688

Kakkar A, Traverso G, Farokhzad OC, Weissleder R, Langer R (2017) Evolution of macromolecular complexity in drug delivery systems. Nat Rev Chem 1(8):1–17

Kan H, Pan D, Castranova V (2018) Engineered nanoparticle exposure and cardiovascular effects: the role of a neuronal-regulated pathway. Inhal Toxicol 30(9–10):335–342

Kao C-Y, Papoutsakis ET (2019) Extracellular vesicles: exosomes, microparticles, their parts, and their targets to enable their biomanufacturing and clinical applications. Curr Opin Biotechnol 60:89–98

Kargozar S, Hoseini SJ, Milan PB, Hooshmand S, Kim H-W, Mozafari M (2020) Quantum dots: a review from concept to clinic. Biotechnol J 15(12):2000117

Kaushik A (2021) Manipulative magnetic nanomedicine: the future of COVID-19 pandemic/endemic therapy. Expert Opin Drug Deliv 18(5):531–534

Kaushik AK, Dhau JS, Gohel H, Mishra YK, Kateb B, Kim N-Y, Goswami DY (2020) Electrochemical SARS-CoV-2 sensing at point-of-care and artificial intelligence for intelligent COVID-19 management. ACS Appl Bio Mater 3(11):7306–7325

Khalid M, Khalid N, Ahmed I, Hanif R, Ismail M, Janjua HA (2017) Comparative studies of three novel freshwater microalgae strains for synthesis of silver nanoparticles: insights of characterization, antibacterial, cytotoxicity and antiviral activities. J Appl Phycol 29(4):1851–1863

Khan AU, Khan M, Cho MH, Khan MM (2020) Selected nanotechnologies and nanostructures for drug delivery, nanomedicine and cure. Bioprocess Biosyst Eng 43(8):1339–1357

Khandelwal N, Kaur G, Chaubey KK, Singh P, Sharma S, Tiwari A, Singh SV, Kumar N (2014) Silver nanoparticles impair Peste des petits ruminants virus replication. Virus Res 190:1–7

Kines RC, Thompson CD, Spring S, Li Z, de Los E, Pinos SM, Schiller JT (2021) Virus-like particle–drug conjugates induce protective, long-lasting adaptive antitumor immunity in the absence of specifically targeted tumor antigens. Cancer Immunol Res 9(6):693–706

Kingstad-Bakke BA, Chandrasekar SS, Phanse Y, Ross KA, Masato Hatta M, Suresh YK, Osorio JE, Narasimhan B, Talaat AM (2019) Effective mosaic-based nanovaccines against avian influenza in poultry. Vaccine 37(35):5051–5058

Kreuter J (2007) Nanoparticles—a historical perspective. Int J Pharm 331(1):1–10. https://doi.org/10.1016/j.ijpharm.2006.10.021

Krishnan S, Thirunavukarasu A, Jha NK, Gahtori R, Roy AS, Dholpuria S, Kesari KK, Singh SK, Dua K, Gupta PK (2021) Nanotechnology-based therapeutic formulations in the battle against animal coronaviruses: an update. J Nanopart Res 23(10):1–16

Lakshmeesha TR, Kalagatur NK, Mudili V, Mohan CD, Rangappa S, Prasad BD, Ashwini BS, Hashem A, Alqarawi AA, Malik JA (2019) Biofabrication of zinc oxide nanoparticles with Syzygium aromaticum flower buds extract and finding its novel application in controlling the growth and mycotoxins of Fusarium graminearum. Front Microbiol 10:1244

Laouini SE, Bouafia A, Soldatov AV, Algarni H, Tedjani ML, Ali GAM, Barhoum A (2021) Green synthesized of Ag/Ag_2O nanoparticles using aqueous leaves extracts of Phoenix dactylifera L. and their azo dye photodegradation. Membranes 11(7):468

Lermusiaux L, Plissonneau M, Bertry L, Drisko GL, Buissette V, Le Mercier T, Duguet E, Tréguer-Delapierre M (2021) Seeded growth of ultrathin gold nanoshells using polymer additives and microwave radiation. Sci Rep 11(1):1–10

Li X, Song Y, Wong G, Cui J (2020) Erratum to: Bat origin of a new human coronavirus: there and back again (Science China Life Sciences, (2020), 63, 3(461–462), 10.1007/S11427-020-1645-7). Sci China Life Sci 63(7)

Lin G, Li L, Panwar N, Wang J, Tjin SC, Wang X, Yong K-T (2018) Non-viral gene therapy using multifunctional nanoparticles: status, challenges, and opportunities. Coord Chem Rev 374:133–152

Mahmoudpour M, Ding S, Lyu Z, Ebrahimi G, Du D, Dolatabadi JEN, Torbati M, Lin Y (2021) Aptamer functionalized nanomaterials for biomedical applications: recent advances and new horizons. Nano Today 39:101177

Masum M, Islam M, Siddiqa M, Ali KA, Zhang Y, Abdallah Y, Ibrahim E, Qiu W, Yan C, Li B (2019) Biogenic synthesis of silver nanoparticles using Phyllanthus emblica fruit extract and its inhibitory action against the pathogen Acidovorax oryzae strain RS-2 of rice bacterial brown stripe. Front Microbiol 10:820

McNamara K, Tofail SAM (2015) Nanosystems: the use of nanoalloys, metallic, bimetallic, and magnetic nanoparticles in biomedical applications. Phys Chem Chem Phys 17(42): 27981–27995

Medhi R, Srinoi P, Ngo N, Tran H-V, Randall Lee T (2020) Nanoparticle-based strategies to combat COVID-19. ACS Appl Nano Mater 3(9):8557–8580

Mekonnen G (2021) Review on application of nanotechnology in animal health and production. J Nanomed Nanotechnol 12:559

Myers J, Hennessey M, Arnold JC, McCubbin KD, Lembo T, Mateus A et al (2022) Crossover-use of human antibiotics in livestock in agricultural communities: a qualitative cross-country comparison between Uganda, Tanzania and India. Antibiotics 11(10):1342

Nagaraj A, Samiappan S (2019) Presentation of antibacterial and therapeutic anti-inflammatory potentials to hydroxyapatite via biomimetic with Azadirachta indica: an in vitro anti-inflammatory assessment in contradiction of LPS-induced stress in RAW 264.7 cells. Front Microbiol 10:1757

Nagraik R, Sharma A, Kumar D, Mukherjee S, Sen F, Kumar AP (2021) Amalgamation of biosensors and nanotechnology in disease diagnosis: mini-review. Sens Int 2:100089
Nazaktabar A, Lashkenari MS, Araghi A, Ghorbani M, Golshahi H (2017) In vivo evaluation of toxicity and antiviral activity of polyrhodanine nanoparticles by using the chicken embryo model. Int J Biol Macromol 103:379–384
Odhiambo JF, DeJarnette JM, Geary TW, Kennedy CE, Suarez SS, Sutovsky M, Sutovsky P (2014) Increased conception rates in beef cattle inseminated with nanopurified bull semen. Biol Reprod 91(4):91–97
Paliwal P, Sargolzaei S, Bhardwaj SK, Bhardwaj V, Dixit C, Kaushik A (2020) Grand challenges in bio-nanotechnology to manage the COVID-19 pandemic. Front Nanotechnol 2:5
Panahi Y, Farshbaf M, Mohammadhosseini M, Mirahadi M, Khalilov R, Saghfi S, Akbarzadeh A (2017) Recent advances on liposomal nanoparticles: synthesis, characterization and biomedical applications. Artif Cells Nanomed Biotechnol 45(4):788–799
Pawar K, Kaul G (2014) Toxicity of titanium oxide nanoparticles causes functionality and DNA damage in buffalo (Bubalus bubalis) sperm in vitro. Toxicol Ind Health 30(6):520–533
Pitchaimani A, Nguyen TDT, Marasini R, Eliyapura A, Azizi T, Jaberi-Douraki M, Aryal S (2019) Biomimetic natural killer membrane camouflaged polymeric nanoparticle for targeted bioimaging. Adv Funct Mater 29(4):1806817
Poddar K, Kishore AVVNGSV (2022) Nanotechnology in animal production. In: Emerging issues in climate smart livestock production. Elsevier, pp 149–170
Priyadarshi R, Roy S, Ghosh T, Biswas D, Rhim JW (2021) Antimicrobial nanofillers reinforced biopolymer composite films for active food packaging applications-a review. Sustain Mater Technol 32:e00353
Quek JY, Uroro E, Goswami N, Vasilev K (2022) Design principles for bacteria-responsive antimicrobial nanomaterials. Mater Today Chem 23:100606
Rahman HS, Othman HH, Abdullah R, Edin HYAS, AL-Haj NA (2022) Beneficial and toxicological aspects of zinc oxide nanoparticles in animals, vol 8. Vet Med Sci, p 1769
Ravi SP, Shamiya Y, Chakraborty A, Elias C, Paul A (2021) Biomaterials, biological molecules, and polymers in developing vaccines. Trends Pharmacol Sci 42(10):813–828
Reda FM, El-Saadony MT, El-Rayes TK, Attia AI, El-Sayed SA, Ahmed SY et al (2021) Use of biological nano zinc as a feed additive in quail nutrition: biosynthesis, antimicrobial activity and its effect on growth, feed utilisation, blood metabolites and intestinal microbiota. Ital J Anim Sci 20(1):324–335
Rezaei L, Safavi MS, Shojaosadati SA (2019) Protein nanocarriers for targeted drug delivery. In: Characterization and biology of nanomaterials for drug delivery. Elsevier, pp 199–218
Ryu JH, Yoon HY, Sun I-C, Kwon IC, Kim K (2020) Tumor-targeting glycol chitosan nanoparticles for cancer heterogeneity. Adv Mater 32(51):2002197
Santos LM, Stanisic D, Menezes UJ, Mendonça MA, Barral TD, Seyffert N, Azevedo V, Durán N, Meyer R, Tasic L (2019) Biogenic silver nanoparticles as a post-surgical treatment for Corynebacterium pseudotuberculosis infection in small ruminants. Front Microbiol 10:824
Sarparanta M, Airaksinen AJ (2021) Radiolabeled porous silicon for nuclear imaging and theranostic applications. In: Porous silicon for biomedical applications. Woodhead, pp 223–253
Sharma VK, Sayes CM, Guo B, Pillai S, Parsons JG, Wang C, Yan B, Ma X (2019) Interactions between silver nanoparticles and other metal nanoparticles under environmentally relevant conditions: a review. Sci Total Environ 653:1042–1051
Sharmin S, Rahaman MM, Sarkar C, Atolani O, Islam MT, Adeyemi OS (2021) Nanoparticles as antimicrobial and antiviral agents: a literature-based perspective study. Heliyon 7(3):e06456
Sheikhzadeh E, Beni V, Zourob M (2021) Nanomaterial application in bio/sensors for the detection of infectious diseases. Talanta 230:122026
Sim S, Wong NK (2021) Nanotechnology and its use in imaging and drug delivery. Biomed Rep 14(5):1–9
Singh A (2021) Eliciting B cell immunity against infectious diseases using nanovaccines. Nat Nanotechnol 16(1):16–24

Slavin YN, Asnis J, Häfeli UO, Bach H (2017) Metal nanoparticles: understanding the mechanisms behind antibacterial activity. J Nanobiotechnol 15(1):1–20

Sohail M, Naz R, Abdelsalam SI (2020) On the onset of entropy generation for a nanofluid with thermal radiation and gyrotactic microorganisms through 3D flows. Phys Scr 95(4):45206

Suri SS, Fenniri H, Singh B (2007) Nanotechnology-based drug delivery systems. J Occup Med Toxicol 2(1). https://doi.org/10.1186/1745-6673-2-16

Swain PS, Rao SB, Rajendran D, Dominic G, Selvaraju S (2016) Nano zinc, an alternative to conventional zinc as animal feed supplement: a review. Anim Nutr 2(3):134–141

Teng Z, Sun S, Luo X, Zhang Z, Seo H, Xu X, Jie Huang H, Dong SM, Du P (2021) Bi-functional gold nanocages enhance specific immunological responses of foot-and-mouth disease virus-like particles vaccine as a carrier and adjuvant. Nanomedicine 33:102358

Tetteh-Quarshie S, Blough ER, Jones CB (2021) Exploring dendrimer nanoparticles for chronic wound healing. Front Med Technol 3:661421

Tewabe A, Abate A, Tamrie M, Seyfu A, Siraj EA (2021) Targeted drug delivery—from magic bullet to nanomedicine: principles, challenges, and future perspectives. J Multidiscip Healthc 14:1711

Tiseo K, Huber L, Gilbert M, Robinson TP, Van Boeckel TP (2020) Global trends in antimicrobial use in food animals from 2017 to 2030. Antibiotics 9(12):918

Wang Y, Zhang K, Qin X, Li T, Qiu J, Yin T, Huang J, McGinty S, Pontrelli G, Ren J (2019) Biomimetic nanotherapies: red blood cell based core–shell structured nanocomplexes for atherosclerosis management. Adv Sci 6(12):1900172

Wuttisarnwattana P, Eid S, Gargesha M, Cooke KR, Wilson DL (2020) Cryo-imaging of stem cell biodistribution in mouse model of graft-versus-host-disease. Ann Biomed Eng 48(6): 1702–1711

Xia S, Tan C, Wang X, Fan C (2022) Microencapsulation liposomal technologies in bioactive functional foods and nutraceuticals. In: Liposomes for functional foods and nutraceuticals. Apple Academic Press, pp 85–143

Yan Y, Miao J, Yang Z, Xiao F-X, Yang HB, Liu B, Yang Y (2015) Carbon nanotube catalysts: recent advances in synthesis, characterization and applications. Chem Soc Rev 44(10): 3295–3346

Yan L, Zhao F, Wang J, Zu Y, Gu Z, Zhao Y (2019) A safe-by-design strategy towards safer nanomaterials in nanomedicines. Adv Mater 31(45):1805391

Youssef FS, El-Banna HA, Elzorba HY, Galal AM (2019) Application of some nanoparticles in the field of veterinary medicine. Int J Vet Sci Med 7(1):78–93

Yu S-j, Yin Y-g, Liu J-f (2013) Silver nanoparticles in the environment. Environ Sci Processes Impacts 15(1):78–92

Zambom CR, da Fonseca FH, Crusca E Jr, da Silva PB, Pavan FR, Chorilli M, Garrido SS (2019) A novel antifungal system with potential for prolonged delivery of histatin 5 to limit growth of Candida albicans. Front Microbiol 10:1667

Zha LY, Zeng JW, Chu X-W, Mao L-M, Luo H-J (2009) Efficacy of trivalent chromium on growth performance, carcass characteristics and tissue chromium in heat stressed broiler chicks. J Sci Food Agric 89(10):1782–1786

Zhang RX, Li J, Zhang T, Amini MA, He C, Lu B, Ahmed T, Lip HY, Rauth AM, Wu XY (2018) Importance of integrating nanotechnology with pharmacology and physiology for innovative drug delivery and therapy–an illustration with firsthand examples. Acta Pharmacol Sin 39(5): 825–844

Zhang Y, Li M, Gao X, Chen Y, Liu T (2019) Nanotechnology in cancer diagnosis: progress, challenges and opportunities. J Hematol Oncol 12(1):1–13

Zhao Z, Ukidve A, Kim J, Mitragotri S (2020) Targeting strategies for tissue-specific drug delivery. Cell 181(1):151–167

Zhou P, Yang X-L, Wang X-G, Hu B, Zhang L, Zhang W, Si H-R, Zhu Y, Li B, Huang C-L (2020) Addendum: A pneumonia outbreak associated with a new coronavirus of probable bat origin. Nature 588(7836):E6

Zulfkar ul Haq has been a research associate for the last year in the ICMR Project, which is in collaboration with the Division of Livestock Production and Management, Faculty of Veterinary Sciences and Animal Husbandry, Shuhama, SKUAST-Kashmir. Before that, he was SRF (Senior Research Fellow) for 5 years on a project on poultry sciences in the Division of Livestock Production and Management, Faculty of Veterinary Sciences and Animal Husbandry, Shuhama, SKUAST-Kashmir. He has a research experience of more than 10 years and has published around 30 research articles, two book chapters, and four review articles

.

Henna Hamadani has been an assistant professor in the Division of Livestock Production and Management, Faculty of Veterinary Sciences and Animal Husbandry, Shuhama, SKUAST-Kashmir for the last 5 years, having a teaching experience of 5 years and a research experience of more than 10 years. She has published around 30 research articles, two book chapters, and four review articles.

Azmat Alam Khan is a professor in the Division of Livestock Production and Management, Faculty of Veterinary Sciences and Animal Husbandry, Shuhama, SKUAST-Kashmir, currently holding charge of Officer on Special Duty to Hon'ble Vice Chancellor of SKUAST-Kashmir. His specialization is in poultry sciences, with a vast teaching experience of 22 years, and his research experience spans more than three decades, having published more than 100 research articles, seven book chapters, one book, and 12 review articles.

Abdul Majeed Ganai is a professor and head of the Division of Animal Nutrition, Faculty of Veterinary Sciences and Animal Husbandry, Shuhama, SKUAST-Kashmir, and has held this position for 10 years. He has a teaching experience of 22 years and a field experience of 15 years. His research experience spans more than three decades, and he has published 110 research articles, seven book chapters, one book, and 12 review articles.

Yasir Afzal Beigh has been an assistant professor (selection grade) in the Division of Animal Nutrition, Faculty of Veterinary Sciences and Animal Husbandry, Shuhama, SKUAST-Kashmir, for the last 13 years, having a teaching experience of 13 years and a research experience of more than 15 years. He has published around 50 research articles, three book chapters, and six review articles.

Gowhar Gull Sheikh has been an assistant professor (selection grade) in the Division of Animal Nutrition, Faculty of Veterinary Sciences and Animal Husbandry, Shuhama, SKUAST-Kashmir, for the last 7 years, having a teaching experience of 7 years. He has worked at Mountain Livestock Research Institute, Manasbal, SKUAST-Kashmir, for 5 years and at MARES, SKUAST-Kashmir, for another 2 years. He has a research experience of more than 15 years. He has published 45 research articles, three book chapters, and six review articles.

Javid Farooq has been an assistant professor (senior scale) in the Division of Animal Nutrition, Faculty of Veterinary Sciences and Animal Husbandry, Shuhama, SKUAST-Kashmir, for the last 2 years, and before that, he served at Mountain Research Centre for Sheep and Goat, SKUAST-Kashmir, for 5 years as an assistant professor cum junior scientist. He has a research experience of more than 15 years. He has published 25 research articles, one book, and four review articles.

Imran Ahmad Ganai has been an assistant professor cum junior scientist at Mountain Research Centre for Sheep and Goat, SKUAST-Kashmir, for 2 years. He has a research experience of more than 8 years. He has published 20 research articles.

Syed Mudasir Ahmad is a professor and head of the Division of Animal Biotechnology, Faculty of Veterinary Sciences and Animal Husbandry, Shuhama, SKUAST-Kashmir. He has a teaching experience of 15 years. His research experience spans more than two decades, and he has published around 100 research articles in reputed journals.

Bioinspired Materials Inherited with Antimicrobial Properties for Tissue Engineering

8

Touseef Amna, Ke Shang, Faheem A. Sheikh, Rasha Alnefaie, Jun-Feng Zhang, Abdullah A. A. Alghamdi, Somia Abd Alla Mohammed El Hassan, and M. Shamshi Hassan

Abstract

The advancement in the preparation of biomaterials that possess tissue engineering applications has predominantly concerted on developing biomimetic materials inherited with the properties of designing new tissue and very definite in cellular responses. The tissue generation is owed by identifying specific biomolecules which can be influenced by changing the microenvironment. Tissue engineering scaffolds and drug delivery systems are gaining huge interest these days. However, one of the common threats associated with the insertion of an implant is the colonization of pathogenic microbes and the development of bacterial/fungal or mixed biofilms in the implant. Theoretically, biomimetic materials mimic the functions of extracellular matrix (ECM) in tissues; therefore, biomimetic scaffolds can offer biological signals for cell-matrix interactions to promote tissue growth. Various biodegradable polymers are used as a base for

T. Amna (✉) · R. Alnefaie · A. A. A. Alghamdi · S. A. A. Mohammed El Hassan
Department of Biology, Faculty of Science, Al-Baha University, Albaha, Saudi Arabia

K. Shang
College of Animal Science and Technology/College of Veterinary Medicine, Henan University of Science and Technology, Luoyang, China

F. A. Sheikh
Nanostructured and Biomimetic Lab, Department of Nanotechnology, University of Kashmir Hazratbal, Srinagar, Jammu and Kashmir, India

J.-F. Zhang
Department of Veterinary Infectious Diseases and Avian Diseases, College of Veterinary Medicine and Center for Poultry Diseases Control, Jeonbuk National University, Iksan, South Korea

M. S. Hassan
Department of Chemistry, Faculty of Science, Al-Baha University, Albaha, Saudi Arabia
e-mail: mshasan@bu.edu.sa

F. A. Sheikh et al. (eds.), *Interaction of Nanomaterials With Living Cells*,
https://doi.org/10.1007/978-981-99-2119-5_8

local drug delivery or temporary sustenance for tissue regeneration. These polymers are disintegrated nonenzymatically through hydrolysis or by particular enzymes. Excellent biocompatibility marks them as competent material for various medical purposes. Nevertheless, the renewal of living tissue and the capability to preclude microbial colonization should be considered while fabricating materials for implant construction. This chapter gives insight into the background and applications of biomaterials with antimicrobial properties and the prospects of bioinspired materials.

Keywords

Antimicrobial activity · Cells · Biomaterials · Tissue engineering · Implants

8.1 Introduction

Nanomaterials possess outstanding properties, such as distinctive optical, magnetic, and electrical characteristics, and these materials have dimensions in the nanoscale (1–100 nm). These nanotextured materials are utilized as healing medicines and for designing devices (Ali et al. 2021). As their dimensions are reduced to the nanoscale, they exhibit plentiful exceptional properties, making these materials distinct from customary macromolecules. Representative characteristics commonly inherited by nanomaterials are high surface-to-volume ratio, enhanced electrical conductivity, superparamagnetic behavior, the spectral shift of optical absorption, and distinctive fluorescence characteristics.

Considering the versatility of nanotextured materials, they have been extensively utilized in the medical field, such as the use of nanomaterials in drug transference, well-ordered discharge systems, etc. Their amplified penetrability permits crossing biological barriers and enhances biocompatibility (Cheng et al. 2021). The interesting topological features and nanorange dimensions of nanomaterials have boosted permeability and retention effects, thus facilitating the achievement of additional entry to the cubicles of cells and tissues. Various parameters such as dimension, morphology, and surface chemistry of nanotextured materials can be automated using physiochemical tools, therefore, can provide exciting features such as the potency of nanomaterials to design many arrangements, including particles, fibers, and porous sponges. These aforementioned nanotextured materials are being utilized as scaffolds for application in medicines and the healthcare system (Zhang and Wang 2019). Furthermore, nanomaterials have been proven to demonstrate auspicious potential in crucial industries, specifically in nanomedicine, pharmacology, and the biomedical field (Albalawi et al. 2021). Profitable funding has been invested in nanotechnology research, and by the year 2024, the global nanotechnology market is expected to surpass USD 124 billion, with ~50% of the market falling in Asia Pacific states (Pinto et al. 2020).

8.2 Outline and Background of Nanomaterials and Nanofibers

Electrospun fibers are frequently described as nanofibers with dimensions roughly as thin as 500 nm. These e-spin nanofibers were primarily employed in the Soviet Union in 1938 to construct air filters to trap aerosol particles. In 1939, this effort managed to project a factory for engineering smoke filters, with nanofiber-based mats as gas masks. Throughout this period, a mechanistic understanding of electrospinning was gradually established. During 1964–1969, Geoffrey Taylor handed out several revolutionary articles, presenting mathematical models that indicate the spherical to conical silhouette alteration of polymer solutions under the impact of a robust electric current (Xue et al. 2019). Precisely, as the strength of the electric field was amplified up to a specific dire limit, the spherical droplet would progressively change into a cone (denoted as Taylor cone) and spring as a liquid jet. Later on, it would take 20 years to develop the electrospinning technique, and during this period, it did not get considerable attention from academia or industry. This inactive period was due to a lack of characterization tools; however, researchers were gifted with tools capable of measuring dimensions down to the submicrometer assortment. However, an assortment of useful applications was projected for e-spin fibers throughout this era, counting their prospective usage as wound dressing materials demonstrated by a patent (filed in 1977). During the initial 1980s period, Donaldson Co. Inc., headquartered in the United States, started to construct and sell filters consisting of e-spin fibers for air filtration. However, the company kept the manufacturing strategy of goods confidential.

Many research teams, particularly those led by Darrell Reneker and Gregory Rutledge (early 1990s), rediscovered this technique with the aid of electron microscopes possessing resolving capabilities down to the nanometer range. These scientists established that diverse organic polymers could be electrospun into nanofibers, and the term electrospinning was propagated for unfolding this practice. These investigations conveyed innovative connotations about electrospinning, and this technique ultimately converted into the method of choice for manufacturing extended and incessant fibers having dimensions down to the nanometer range.

Nevertheless, by now the electrospinning has begun to achieve accumulative consideration by designing novel materials and formulations, such as the construction of composite as well as ceramic nanofibers (Larsen et al. 2003). These unique fibers have innovative applications in various fields, such as catalysis, energy harvesting, conversion, and storage, which were previously dominated by nanoparticles. On the other hand, innovative approaches have been established to manage the assembly and arrangement of electrospun nanofibers. Nonetheless, these fascinating nanofibers flagged huge opportunities in energy-related and biomedical applications. A lot of approaches have been made for aligning the nanofibers, such as the possibility of diverse syndicate characteristics originating from size, structure, composition, morphology, porosity, and assembly of nanofibers. At the same period, coaxial electrospinning was settled to harvest continuous core-shell and hollow nanofibers and yarns of electrospun nanofibers. Currently, industrial production of electrospun nanofibers has been implemented by numerous companies throughout

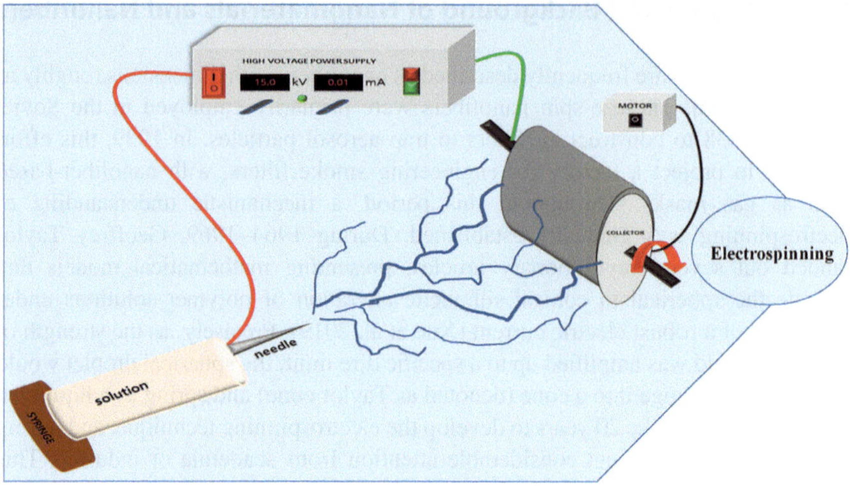

Fig. 8.1 Schematic representation of an electrospinning arrangement for nanofiber fabrication

the globe to produce nanofibers in huge volumes, approving downstream commercial products. Nowadays, these electrospun nanofibers are extensively utilized for water purification, air filtration, wound dressings, implants, etc. (Fig. 8.1). Figure 8.1 demonstrates the basic design of electrospinnig set up.

8.3 Bioinspired Nanomaterials

Nanotextured materials possess essential features, such as reduced size and shape, which are useful in various fields and trades. Nanoparticles and nanomaterials are manufactured on a considerable gauge and are necessary for several industries. This fact has supported research in various disciplines of science such as biochemistry, biophysics, and biochemical engineering and of course, the applications. The combination of nanotechnology and other disciplines has resulted in the fabrication of novel and interesting nanomaterials utilized for diagnostic tools, drug delivery systems, energy storage materials, and conservational and food processing (Barhoum et al. 2022). These nanotextured materials are diverse in size, silhouette, magnitude, configuration, permeability, stage, and consistency. Therefore, various classifications have been utilized to categorize them. Moreover, nanomaterials can be classified as natural, incidental, bioinspired, and engineered, depending on their origin and functionality. Nevertheless, it has been documented that the naturally occurring nanomaterials form all through natural physiological procedures. On the other hand, the incidental nanomaterials, also known as anthropogenic or waste particles, transpire due to simulated industrial procedures. Similarly, engineered nanomaterials are prepared in the laboratory/or in industries to attain materials with definite topography. However, bioinspired materials are nanomaterials that resemble/or imitators of those natural nanomaterials or living stuff. Nowadays engineered and bioinspired nanomaterials are attaining huge consideration compared to natural materials due to variability.

Thus expanding the progressive nanofabrication tools, various bioinspired nanomaterials with exact functions can be fabricated by curbing their configurations (Lee et al. 2017). For instance, mechanochromic elastomers have been developed that mimic the photonic structure of chameleon iridophore cells (Rasouli et al. 2018). In these sensors, rigid silica nanocrystals are rooted in an elastomer's background to form non-close-packed crystals. These sensors exhibit a color shift from red to blue during stretching and red to green during compression. Conclusively, biomimetic synthesis is aimed either by mimicking functions of natural materials, structures or the biological processes and are classified as functional biomimetic synthesis and biomimetic process synthesis (Zan and Wu 2016). Up till now, numerous bionanomaterials have been styled (Table 8.1), and many more will be established.

However, despite their particular possessions, only insufficient engineered and bioinspired nanomaterials have been permitted and utilized in the industry due to the limitations of risk assessment practices. In this chapter, our primary focus will be on novel bioinspired materials possessing antimicrobial properties, which are used in tissue engineering and other medical and biomedical fields.

8.4 Overview of Tissue Engineering

Precisely tissue engineering is an up-and-coming biotechnological field that syndicates different features of medicine, cell, and molecular biology, materials science and engineering with the aim of regeneration, restoring and substituting ailing tissues. The term tissue engineering was formally devised by Fung in October 1987 at National Science Foundation Workshop in Washington, D.C. and since the preceding epoch, this field has progressed from scientific narration to science statement possessing research-based acceptance. Tissue engineering necessitates inclusive determinations to conglomerate engineering and physical sciences with life sciences, with the objectives of repairing, interchanging, and refining the functions of impaired tissues and organs (Shafiee and Atala 2017). Tissue and organ failure due to injury or disease has been reflected as the main healthcare trials. Until the second half of the twentieth century, there were no appropriate therapies for patients with dysfunctional organs. However, in 1954, Joseph Murray, Nobel Laureate in Medicine during the 1990s, completed the first successful organ transplant, transferring a healthy kidney donated by Ronald Herrick to his identical twin brother, Richard (Renal homotransplantation in identical twins 2001). The process was guaranteed as risk-free as the donor and recipient were both genetically alike. Later on, almost after 5 years, Murray executed the world's first successful organ transplant among genetically non-identical individuals. Through his pioneering operations, many lives have been saved with organ transplantation. Meanwhile, the high number of people waiting for transplants, the scarcity of organ donors, and the massive aging population demand the progression of innovative approaches

Table 8.1 Summary of some of the recently developed bioinspired-material-based scaffolds and their respective applications published by various researchers

S. No.	Bioinspired scaffolds	Applications	Reference
1.	Pd-Ag-HAP on protruded TiO_2	Dental implant	Jang et al. (2018)
2.	Anodized titanium	Dental implant	Mühl et al. (2022)
3.	Triphasic calcium-based implant (AGNI)	Bone implant	Shaul et al. (2022)
4.	Zirconia-containing biphasic calcium phosphate	Bone implant	Youness et al. (2022)
5.	Poly(L-lactic acid)/mesoporous bioactive glass composite	Bone implant	Pant et al. (2022)
6.	Poly(xylitol-dodecanedioic acid)–co-polylactic acid	Tissue engineering	Sotoudeh et al. (2021)
7.	Hydroxyapatite nanowhisker-reinforced poly (lactic acid) composites	Bone implant	Xu et al. (2022)
8.	Chitosan/gelatin/polycaprolactone	Bone scaffold	Wulin et al. (2022)
9.	Polyurethane (PU)/chitosan (Cs)/carbon nanotubes (CNT)	Cardiac tissue engineering	Ahmadi et al. (2021)
10.	composite alginate-gelatin hydrogels incorporating PRGF	Human dental pulp and cell proliferation	Anitua et al. (2022)
11.	Curcumin-loaded mesoporous silica nanoparticles/nanofiber composites	Stem cell proliferation	Mashayekhi et al. (2020)
12.	Folic acid.MgO:ZnO/chitosan hybrid particles	Fibroblast cell proliferation	Rafie and Meshkini (2021)
13.	Chitosan/aloe vera hydrogel	Wound dressings	Movaffagh et al. (2022)
14.	*Spinacia oleracea* extract incorporated alginate/carboxymethyl cellulose microporous scaffold	Bone tissue engineering	Sharmila et al. (2020)
15.	Cellulose nanofiber scaffold	Bone tissue engineering	Chakraborty et al. (2019)

to repair the functions of impaired organs (Atala 2012). According to the US Department of Health and Human Services (https://optn.transplant.hrsa.gov), statistics indicate that in the United States, a new individual is added every 10 min to the National Transplant Waiting List and that 22 individuals pass away every day waiting for a transplant. Indeed, tissue engineering aims to alleviate the life-threatening unavailability of donor organs using in vitro production of biologically functional assemblies. To sum up, in tissue engineering, rudimentary design tactics comprise simply cell assemblies, cells and scaffolds, and solitary scaffolds. Unambiguously, autografts, allografts, and xenografts are biological paradigms constructed from a patient's cells, from other genetically non-identical organisms, as well as even non-human animal species, respectively. Alternatively, the scaffolds are developed from natural (e.g., collagen, decellularized matrices) or synthetic materials and are intended to reproduce a natural three-dimensional (3D) environment, i.e., the extracellular matrix (ECM) in order to make the cells and tissues to flourish and establish into organs that can conserve their specific conformations and topographies. Additionally, these scaffolds should be compatible with the tissue-specific cells and the preferred indigenous atmosphere within the human frame (Langer 2009). Consequently, diverse engineered tissues or/organs require distinctive preparations and constituents. Furthermore, synthetic scaffolds must be synthesized, considering important features such as pore dimension, geometry, penetrability, and spatial dissemination (Khademhosseini et al. 2006). It has been well documented that the bulk and exterior physiognomies of scaffold materials can also influence cellular performance (Khademhosseini et al. 2006). Finally, the degradation of the scaffold must be attuned to the construction of ECM by cells (Fig. 8.2).

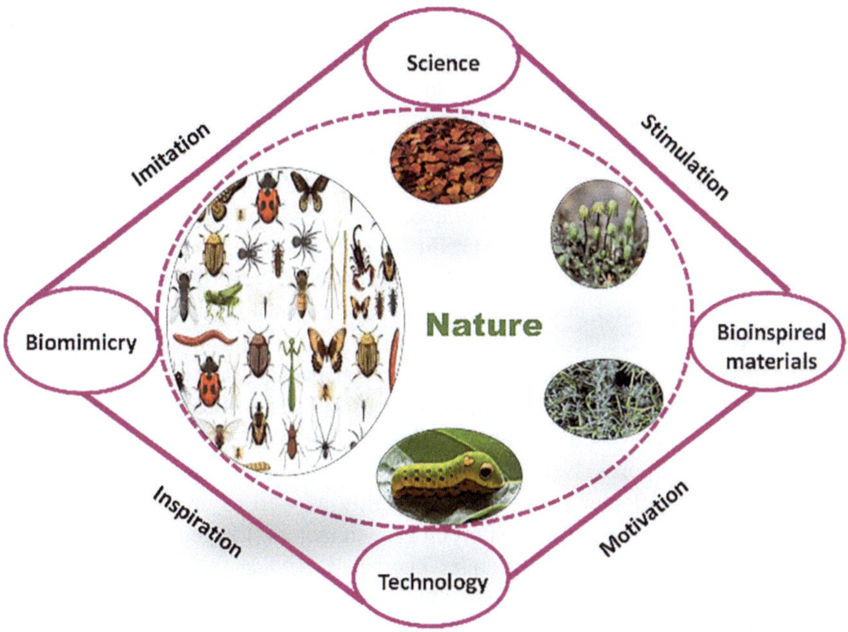

Fig. 8.2 The stepladders and biomimetic/or bioinspired materials in the tissue engineering process. Different levels of processes that can be used for biomimetic synthesis are compiled. The current progress of bioinspired/biomimetic synthesis is systematically summarized according to the following perspectives: motivation, inspiration, stimulation, imitation, and the fusion of science and technology

8.5 Tissue Engineering and Regenerative Medicine

Current progress in scaffold fabrication has boosted the arena of tissue engineering toward sophisticated objectives. A reasonably novel approach in scaffold-free tissue engineering has been presented that allows cells to yield their own ECM and self-assemble to build 3D biological configurations. Tissue engineering is one of the main subjects in regenerative medicine. For instance, the stem cell discipline, gene remedy, soluble molecules, and reprogramming of cell and tissue types are crossing points between tissue engineering and regenerative medicine. During the past 20 years, numerous accomplishments in the assembly of functional tissues and organs have aided in upgrading the quality of life of ailing individuals. The preparation of tissues and organs can be grouped into four main stages. The first and the least complex stage involves flat tissues and organs, such as the skin; the second stage includes the construction of tubular organs, for example, blood vessels and tracheas. Hollow nontubular organ assemblies such as the bladder are the second utmost organs to construct and the most multifarious structures are solid organs, such as the heart, kidney, liver, etc. Several research scientists all over the globe with different disciplines and specialties are occupied to fulfill the projected demands and encounter the challenges in this area. The initial challenge in engineering a tissue/organ is the discovery of suitable cell sources and a sufficient number of cells. The next important step is to provide refined biomaterials and scaffolds to permit all cell types in an organ to make an effort to organize in coordination in order to shape their own ECM (Atala 2009).

Significant progress in stem cell therapies, biomaterials science, and the design of delivery systems able to mimic the production of growth factors will possibly aid treatment breakthroughs for various deadly ailments. Nevertheless, encouraging outcomes of in vitro and in vivo investigations cannot always be of practical use in clinical situations. Consequently, these situations remind us to sidestep the naive interpretations as well as exaggerated optimism concerning innovative technologies (Hassanzadeh et al. 2018). In addition to scaffold and cell-centered approaches, tissue engineering also includes other methodologies to enable the repair of organs and restore their purposes (Naughton 2002). Furthermore, it has also been accredited that the usage of polymer matrix offers stability as well as an organized release profile for proteins as well as for important growth factors (Langer and Moses 1991). Tissue engineering implants are also cherished devices comprising bioactive ingredients (Blanquer et al. 2012). The fabrication of practicable constructs requires the supply of easy-to-harvest cells proficient in differentiation into particular cells which lack immunogenic properties. For instance, the mesenchymal stem cells captured in penetrable biomaterial capsules have revealed anti-osteoarthritis chattels allocated to their regenerative effects (Stock and Vacanti 2001). On the other hand, polymers (e.g., polylactic acid (PLA), polylactic-co-glycolic acid (PLGA), hydrogel, etc.) have already been utilized for the design of 3D tumor models. Meanwhile, the progress of in vitro models proficient to reconstruct the process of tumor progression is a challenging issue in cancer research (Loessner et al. 2010). However, the use of 3D cultures offers the opportunity for improvement in capturing the construction of tumors, cellular imaging as well as high quantity screening.

8.6 Tissue Engineering in Healthcare Systems

As aforementioned, tissue engineering is quickly growing scientific area that utilizes the principles and methods of physical sciences, life sciences, and engineering to apprehend physiological and pathological methods to improve current therapeutic systems. Currently, tissue/organ transplantation, surgical operation, and dialysis are common cures in most countries. But still, limitations associated with these procedures have resulted in amplified awareness in the field of tissue engineering (Lee et al. 2010). In this regard, categories such as autologous or allogeneic cells can be cast off. Alternatively, some other reports have suggested the use of exogenous cell-oriented and endogenous cell approaches (Li et al. 2017), such as the administration of chemokines as signals potentiate cell homing in an anti-inflammatory microenvironment (Aibibu et al. 2016; Andreas et al. 2014). More to the point, cells may be genetically or epigenetically reformed (Sheyn et al. 2010) to augment the efficacy of tissue regeneration. The shortage of in vitro engineered tissues is partly owing to the inability to create engineered blood vessel systems (Lovett et al. 2009). Interestingly, in certain cases, such as in tissues like the skin, cartilage, or cornea, cells can be delivered through diffusion from distant blood vessels. The engagement of endothelial cells for neoangiogenesis and revascularization by biomolecules parenthetically is a common tactic.

The inference from the aforementioned reports is that tissue engineering has the prospective future to transform approaches of health care systems to mend for the excellence of life. Additionally, it will deliver an economical and long-lasting solution to numerous age-related illnesses. Moreover, the engineered tissues may lessen the requirement for an organ transplant (Fig. 8.3).

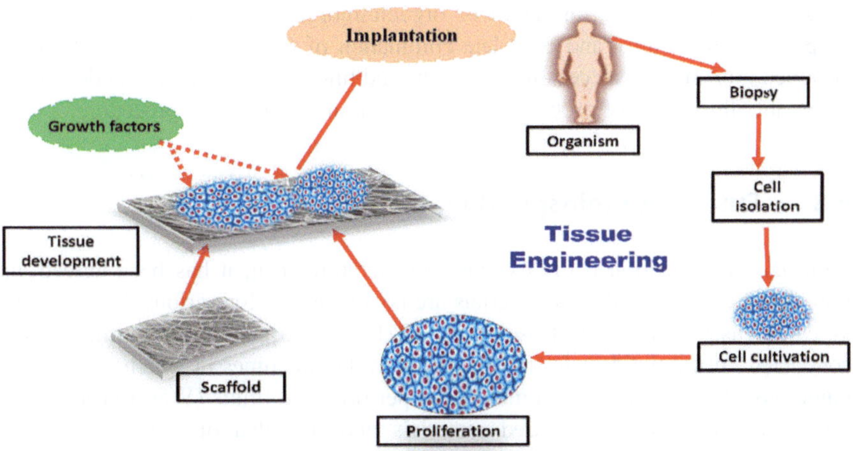

Fig. 8.3 The schematic illustration of probable steps of the tissue engineering procedure

8.7 Antimicrobial Bioinspired Materials

Undeniably a significant number of novel biomaterials and scaffolding structures are designed to exploit them in medical and biomedical areas possessing augmented healthiness potency (Spałek et al. 2022). These biomaterials are being utilized in implantable procedures or drug delivery systems that have noteworthy influences on the quality of life of diseased individuals. Conversely, their continuing exploitation and usage can cause invasion and multiplication of microbes, which may result in biofilms and consequential elicit cytotoxic reactions. These pathogenic microbial infections ultimately result in the failure of implants. Furthermore, they can hamper the distribution of drugs, thus rendering them ineffective. Numerous substitutes have been suggested over the years to avoid such complications, such as the practice of disinfectants, and antibiotics as well as the amendment of posturing lines of biomaterial, amalgamation of biomolecules. In this regard, various stimulating functionalization and alteration procedures have been engaged to encounter opportunist as well as pathogenic microbes (Spałek et al. 2022; Teixeira et al. 2021, 2022; Felgueiras 2021, 2022; Felgueiras et al. 2021; Miranda et al. 2020). Similarly, a report has been published in which collagen-constructed biomaterials fixed with nano clay depicted excellent antimicrobial potential and has been utilized for skin rejuvenation applications (Marin et al. 2021). Likewise, well-characterized skin derivative human defensins antimicrobial peptides have been investigated and their impact on the oozing of angiogenin, (a persuasive angiogenic factor) has been depicted. They discovered that numerous human defensins could arouse the delivery of angiogenin despite sustaining their antimicrobial and other immunomodulatory features (Umehara et al. 2022). Conclusively, it has been comprehended those various strategies, such as understanding of the antimicrobial activity of specific biomaterials and introduction of novel surface amendments, incorporation of bioactive molecules, formation of organic-inorganic composites, etc., can be adapted for the control and complete elimination of microbial infections from the designed implants for successful operations and functionality. Figure 8.4 depicts the schematic illustration of probable steps of the tissue engineering procedure.

8.8 Future of Bioinspired Materials

From the aforementioned discussions and literature scan, it has been agreed and implied that bioinspired nanomaterials are novel choices for various diseases, such as cardiovascular ailment (Bose et al. 2021). Subsequently, Bountiful research conducted during the decades have resulted in the outcomes in which it looks like numerous materials are overcoming their performance limits (Wegst et al. 2015). The fusion of 2D nanotextured materials and the idea of bioinspiration has stimulated the establishment of innovative materials and techniques (Zhang et al. 2020). Contrariwise, there are up now loads of trials and challenges to be resolved that bound the progress of bioinspired nanomaterials/technologies beyond 2D

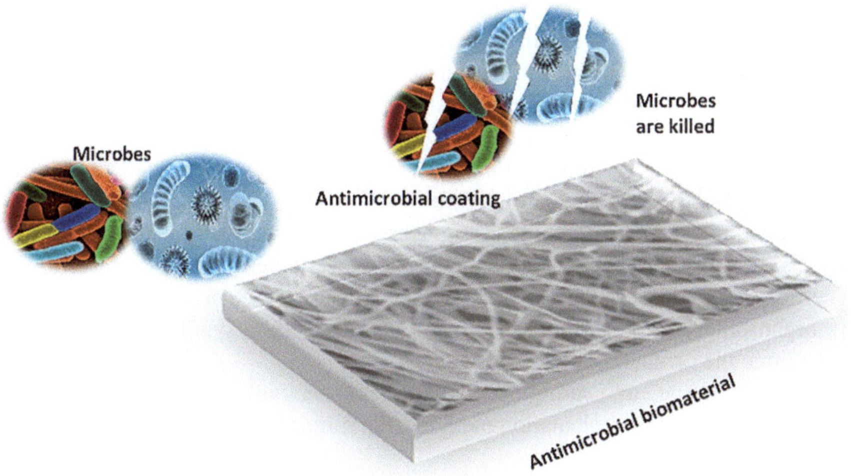

Fig. 8.4 Diagrammatic representation of antimicrobial scaffolds and bactericidal activity

bioinspired materials. Yearly, lots of investigations are being conducted to achieve success in clinical applications, and development is continuous, with the nice cooperation between researchers, clinicians, scientists, as well as engineers (Langer and Vacanti 2016). Nevertheless, bio-artificial tissue engineering has emerged as an indispensable trial. A variety of cells, growth factors, scaffolds, and stratagems are accessible and the amalgamation of these factors under in vitro and in vivo conditions is producing fruitful results. Additionally, the assortment of suitable approaches for cell stimulation, scaffold synthesis, and tissue transplantation performs a conclusive role in efficacious tissue engineering (Bakhshandeh et al. 2017). Interestingly, stem cell co-culture arrangements are remarkable and prevailing tools owing to their exceptional properties. On top, the feedback opinions and outcomes have revealed upshots success in engineering tissues (Paschos et al. 2015). However, scientists and researchers with different specializations ought to coordinate together to encourage the development of tissue engineering and regenerative medicine for practical uses.

8.9 Conclusion

The objective of tissue engineering is to generate functional and patient-specific tissues for transplantation. Diverse scientific disciplines, such as molecular biology, microbiology, cell biology, engineering, pharmacology, and medicine, as well as cellular and developmental biology, etc., have come together to work on engineering tissues with the aim of implant preparation. Virtually, almost all tissues of the human body have been investigated for the likelihood of replacement with living tissues and engineered structures. However, suitable resources of cells for tissue engineering are prerequisites and must be inevitability recognized. Additionally, several stimulating functionalization and alteration procedures are being adopted to control the growth and expansion of opportunist as well as pathogenic microbes. Screening of appropriate scaffolds for ECMs is indeed a huge and far-reaching task. Nevertheless, the optimization of delivery time and the total price of laboratory-cultivated organs are other big tasks to be considered in recruiting a marketplace for engineered organs. Currently, under in vitro conditions, engineered tissues such as skin and cartilage are being utilized in clinics in various countries. Nevertheless, tissue engineering has emerged as a swiftly growing interdisciplinary area, in which practitioners attempt to fix organ failure by implanting natural, synthetic, or semi-synthetic tissues/organs that mimic the original one.

Although tissue engineering has delivered prospective ways to overcome inadequate conventional transplantation methods, the future goal is to fabricate individual organs.

Accordingly, the progress in this auspicious field crests an innovative arrangement for the improvement of the healthcare system of the present society.

References

Ahmadi P et al (2021) Preparation and characterization of polyurethane/chitosan/CNT nanofibrous scaffold for cardiac tissue engineering. Int J Biol Macromol 180:590–598

Aibibu D et al (2016) Textile cell-free scaffolds for in situ tissue engineering applications. J Mater Sci Mater Med 27(3):63

Albalawi F et al (2021) Engineered nanomaterials: the challenges and opportunities for nanomedicines. Int J Nanomedicine 16:161–184

Ali ES et al (2021) Targeting cancer cells with nanotherapeutics and nanodiagnostics: current status and future perspectives. Semin Cancer Biol 69:52–68

Andreas K, Sittinger M, Ringe J (2014) Toward in situ tissue engineering: chemokine-guided stem cell recruitment. Trends Biotechnol 32(9):483–492

Anitua E et al (2022) Composite alginate-gelatin hydrogels incorporating PRGF enhance human dental pulp cell adhesion, chemotaxis and proliferation. Int J Pharm 617:121631

Atala A (2009) Engineering organs. Curr Opin Biotechnol 20(5):575–592

Atala A (2012) Regenerative medicine strategies. J Pediatr Surg 47(1):17–28

Bakhshandeh B et al (2017) Tissue engineering; strategies, tissues, and biomaterials. Biotechnol Genet Eng Rev 33(2):144–172

Barhoum A et al (2022) Review on natural, incidental, bioinspired, and engineered nanomaterials: history, definitions, classifications, synthesis, properties, market, toxicities, risks, and regulations. Nanomaterials (Basel) 12(2):177

Blanquer SB, Sharifi S, Grijpma DW (2012) Development of poly(trimethylene carbonate) network implants for annulus fibrosus tissue engineering. J Appl Biomater Funct Mater 10(3):177–184

Bose RJ, Ha K, McCarthy JR (2021) Bio-inspired nanomaterials as novel options for the treatment of cardiovascular disease. Drug Discov Today 26(5):1200–1211

Chakraborty PK, Adhikari J, Saha P (2019) Facile fabrication of electrospun regenerated cellulose nanofiber scaffold for potential bone-tissue engineering application. Int J Biol Macromol 122: 644–652

Cheng Z et al (2021) Nanomaterials for cancer therapy: current progress and perspectives. J Hematol Oncol 14(1):85

Felgueiras HP (2021) An insight into biomolecules for the treatment of skin infectious diseases. Pharmaceutics 13(7):1012
Felgueiras HP (2022) Frontiers in antimicrobial biomaterials. Int J Mol Sci 23:9377
Felgueiras HP et al (2021) Biodegradable wet-spun fibers modified with antimicrobial agents for potential applications in biomedical engineering. J Phys Conf Ser 1765:012007
Hassanzadeh P, Atyabi F, Dinarvand R (2018) Tissue engineering: still facing a long way ahead. J Control Release 279:181–197
Renal homotransplantation in identical twins. 1955 (2001) J Am Soc Nephrol 12(1):201–204
Jang J-M et al (2018) Ultra-fine structures of Pd-Ag-HAp nanoparticle deposition on protruded TiO_2 barrier layer for dental implant. Appl Surf Sci 432:285–293
Khademhosseini A et al (2006) Microscale technologies for tissue engineering and biology. Proc Natl Acad Sci U S A 103(8):2480–2487
Langer R (2009) Perspectives and challenges in tissue engineering and regenerative medicine. Adv Mater 21(32–33):3235–3236
Langer R, Moses M (1991) Biocompatible controlled release polymers for delivery of polypeptides and growth factors. J Cell Biochem 45(4):340–345
Langer R, Vacanti J (2016) Advances in tissue engineering. J Pediatr Surg 51(1):8–12
Larsen G et al (2003) A method for making inorganic and hybrid (organic/inorganic) fibers and vesicles with diameters in the submicrometer and micrometer range via sol-gel chemistry and electrically forced liquid jets. J Am Chem Soc 125(5):1154–1155
Lee JH et al (2010) Control of osteogenic differentiation and mineralization of human mesenchymal stem cells on composite nanofibers containing poly[lactic-co-(glycolic acid)] and hydroxyapatite. Macromol Biosci 10(2):173–182
Lee GH et al (2017) Chameleon-inspired mechanochromic photonic films composed of non-close-packed colloidal arrays. ACS Nano 11(11):11350–11357
Li X et al (2017) Administration of signalling molecules dictates stem cell homing for in situ regeneration. J Cell Mol Med 21(12):3162–3177
Loessner D et al (2010) Bioengineered 3D platform to explore cell-ECM interactions and drug resistance of epithelial ovarian cancer cells. Biomaterials 31(32):8494–8506
Lovett M et al (2009) Vascularization strategies for tissue engineering. Tissue Eng Part B Rev 15(3):353–370
Marin MM et al (2021) Development of new collagen/clay composite biomaterials. Int J Mol Sci 23(1):401
Mashayekhi S et al (2020) Curcumin-loaded mesoporous silica nanoparticles/nanofiber composites for supporting long-term proliferation and stemness preservation of adipose-derived stem cells. Int J Pharm 587:119656
Miranda CS et al (2020) Spun biotextiles in tissue engineering and biomolecules delivery systems. Antibiotics 9(4):174
Movaffagh J et al (2022) Evaluation of wound-healing efficiency of a functional Chitosan/Aloe vera hydrogel on the improvement of re-epithelialization in full thickness wound model of rat. J Tissue Viability 31:649
Mühl A et al (2022) Comparison of surface aspects of turned and anodized titanium dental implant, or abutment material for an optimal soft tissue integration. Heliyon 8(8):e10263
Naughton GK (2002) From lab bench to market: critical issues in tissue engineering. Ann N Y Acad Sci 961:372–385
Pant S et al (2022) 3D bioprinted poly (lactic acid)/mesoporous bioactive glass based biomimetic scaffold with rapid apatite crystallization and in-vitro cytocompatability for bone tissue engineering. Int J Biol Macromol 217:979–997
Paschos NK et al (2015) Advances in tissue engineering through stem cell-based co-culture. J Tissue Eng Regen Med 9(5):488–503
Pinto MP et al (2020) A molecular stratification of Chilean gastric cancer patients with potential clinical applicability. Cancers (Basel) 12(7):1863

Rafie M, Meshkini A (2021) Tailoring the proliferation of fibroblast cells by multiresponsive and thermosensitive stem cells composite F127 hydrogel containing folic acid. MgO:ZnO/chitosan hybrid microparticles for skin regeneration. Eur J Pharm Sci 167:106031
Rasouli R, Barhoum A, Uludag H (2018) A review of nanostructured surfaces and materials for dental implants: surface coating, patterning and functionalization for improved performance. Biomater Sci 6(6):1312–1338
Shafiee A, Atala A (2017) Tissue engineering: toward a new era of medicine. Annu Rev Med 68: 29–40
Sharmila G et al (2020) Fabrication and characterization of Spinacia oleracea extract incorporated alginate/carboxymethyl cellulose microporous scaffold for bone tissue engineering. Int J Biol Macromol 156:430–437
Shaul JL et al (2022) AGN1 implant material to treat bone loss: resorbable implant forms normal bone with and without alendronate in a canine critical size humeral defect model. Bone 154: 116246
Sheyn D et al (2010) Genetically modified cells in regenerative medicine and tissue engineering. Adv Drug Deliv Rev 62(7–8):683–698
Sotoudeh A et al (2021) Tissue engineering needs new biomaterials: poly(xylitol-dodecanedioic acid)–co-polylactic acid (PXDDA-co-PLA) and its nanocomposites. Eur Polym J 152:110469
Spałek J et al (2022) Biocompatible materials in otorhinolaryngology and their antibacterial properties. Int J Mol Sci 23(5):2575
Stock UA, Vacanti JP (2001) Tissue engineering: current state and prospects. Annu Rev Med 52: 443–451
Teixeira MO, Antunes JC, Felgueiras HP (2021) Recent advances in fiber–hydrogel composites for wound healing and drug delivery systems. Antibiotics 10(3):248
Teixeira MA et al (2022) Tiger 17 and pexiganan as antimicrobial and hemostatic boosters of cellulose acetate-containing poly(vinyl alcohol) electrospun mats for potential wound care purposes. Int J Biol Macromol 209:1526–1541
Umehara Y et al (2022) The antimicrobial peptides human β-defensins induce the secretion of angiogenin in human dermal fibroblasts. Int J Mol Sci 23(15):8800
Wegst UG et al (2015) Bioinspired structural materials. Nat Mater 14(1):23–36
Wulin S et al (2022) Evaluation of mechanical properties of porous chitosan/gelatin/ polycaprolactone bone scaffold prepared by microwave foaming method. Polymers 14(21): 4668
Xu H et al (2022) Pea pod-mimicking hydroxyapatite nanowhisker-reinforced poly(lactic acid) composites with bone-like strength. Int J Biol Macromol 216:114–123
Xue J et al (2019) Electrospinning and electrospun nanofibers: methods, materials, and applications. Chem Rev 119(8):5298–5415
Youness RA, Amer MS, Taha MA (2022) Tribo-mechanical measurements and in vivo performance of zirconia-containing biphasic calcium phosphate material implanted in a rat model for bone replacement applications. Mater Chem Phys 285:126085
Zan G, Wu Q (2016) Biomimetic and bioinspired synthesis of nanomaterials/nanostructures. Adv Mater 28(11):2099–2147
Zhang D, Wang Y (2019) Functional protein-based bioinspired nanomaterials: from coupled proteins, synthetic approaches, nanostructures to applications. Int J Mol Sci 20(12):3054
Zhang Y et al (2020) Bioinspired 2D nanomaterials for sustainable applications. Adv Mater 32(18): e1902806

Touseef Amna received her Ph.D. in Microbiology in 2006 from Aligarh Muslim University (AMU, Central University), India. Presently, she is working as an associate professor at Albaha University, Saudi Arabia. During her research career, she got the opportunity to spend postdoctoral stages at the University of Dortmund, Germany; Universidad de Talca, Chile; and Chonbuk National University, South Korea, and received various prestigious awards and grants. She authored two patents (one US patent and one Korean) and more than 100 research articles in journals of international repute. She is an editor, guest editor, and editorial board member of various national and international journals. Her research interest consists of the design and synthesis of biocompatible scaffolds for antimicrobial applications, cell culture, and tissue engineering.

Ke Shang received a B.S. degree in Animal and Plant Quarantine from Henan University of Science and Technology, Luoyang, China, in 2012; an M.S. degree in Preventive Veterinary Medicine from Henan University of Science and Technology, Luoyang, China, in 2015; and a Ph.D. degree in Veterinary Medicine from Jeonbuk National University, Iksan, South Korea. She is currently working as an instructor at the College of Animal Science and Technology, Henan University of Science and Technology, Luoyang, China. Her research interests include bacterial pathogenesis, antibiotic resistance, and vaccine development.

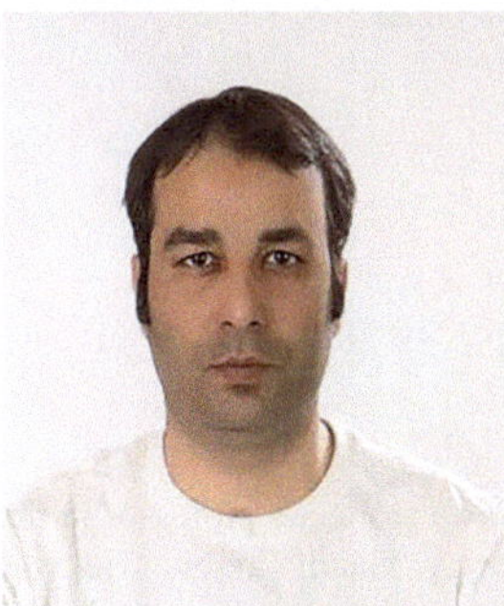

Faheem A. Sheikh is an assistant professor at the Department of Nanotechnology, University of Kashmir, India (2015). He served as an assistant professor at the Department of Biotechnology at the Central University of Kashmir, India (2015–2016); research professor at Myongji University, South Korea (2014–2015); assistant professor at Hallym University, South Korea (2012–2014); postdoc/research fellow at the University of Texas Rio Grande Valley, Texas, United States of America (2010–2012); and research professor at Myongji University, South Korea (2010). His research mainly focuses on fabricating nanomaterials used in tissue engineering.

Rasha Alnefaie received her Ph.D. degree from the Department of Biology, Colorado State University, Fort Collins, Colorado. She is currently working as an assistant professor at the Department of Biology, Albaha University, Saudi Arabia. Her research interests are mainly in the area of cell biology and genetic etiology.

Jun-Feng Zhang received a B.S. degree in Animal and Plant Quarantine from Henan University of Science and Technology, Luoyang, China, in 2012, and an M.S. degree in Preventive Veterinary Medicine from Henan University of Science and Technology, Luoyang, China, in 2015. He is working toward a Ph.D. in Veterinary Medicine at Jeonbuk National University, Iksan, South Korea. His research interests include vaccines carried by bacteria and viruses.

Abdullah A. A. Alghamdi is working as an assistant professor at Albaha University, Saudi Arabia. He did a master's study in biotechnology at New York University, USA. His Ph.D. degree in Cell Biology was received from the University of East Anglia, UK. His research interests are mainly in the field of cell biology, more specifically in cell signaling, gene expression, angiogenesis, cancer biology, oxidative stress, and the cellular biological activities of natural products.

Somia Abd Alla Mohammed El Hassan received her Masters and PhD degree in Zoology with specialization in Parasitology from Khartoum University/Sudan. She has worked in Biology Department, Faculty of Education, University of Khartoum from 1999 till now. She is currently working as an Assistant professor at Department of Biology, Albaha University, Saudi Arabia.

M. Shamshi Hassan is working as an associate professor in the Faculty of Science Albaha University, Saudi Arabia. He was awarded an international fellowship to pursue his doctoral degree and received his Ph.D. in Chemical Engineering from the Department of Chemical Engineering, Chonbuk National University, South Korea, in 2010. After obtaining his Ph.D., he was awarded Postdoctoral Fellowship (PDF) by the Textile and Fiber Engineering Department of CBNU, South Korea. His research interest includes the synthesis of pure and hybrid metal oxides (quantum dots or nanocrystals, 1D, 2D, 3D, and flower-shaped morphology), nanostructured materials, or nanofibers and their application in supercapacitors and as biological/chemical disinfectants.

3D and 4D Bioprinting Technology for Tissue Engineering Applications

9

Fatma Nur Parın

Abstract

Regenerative medicine aims to restore, regenerate, or replace tissues or organs affected by disease, trauma, or congenital disabilities. Tissue engineering is one approach used to accomplish these objectives. Tissue engineering is commonly associated with the use of cells placed on tissue scaffolds in developing new living tissue for medicinal purposes, but it is not limited to cell and tissue scaffold applications. It is the procedure of designing tissues in the body in a laboratory and implanting them in patients. Tissue engineering currently plays a minor role in clinical outcomes. People with the disease have extra bladders, tendons, skin grafts, and tiny aorta, as well as an entire bronchial tube transplant, but the therapies are experimental and costly. Regarding this, three-dimensional (3D) and 4D bioprinting methods are beneficial for the production of scaffolds with different shapes, high accuracy, high speed, and control over the size and also porosity. In this chapter, the principal and most popularly used types of 3D bioprinting methods are explained, as well as a summary of bioink compositions in 3D and 4D bioprinting. Eventually, current problems and changing demands are highlighted. Furthermore, the most recent applications in organ and tissue bioprinting are discussed. Lastly, current issues, future requirements, and the potential of bioprinting are reviewed.

Keywords

3D and 4D bioprinting · Tissue engineering · Tissue scaffold · Bio-ink · Regenerative medicine

F. N. Parın (✉)
Polymer Materials Engineering Department, Faculty of Engineering and Natural Sciences, Bursa Technical University, Bursa, Turkey
e-mail: nur.parin@btu.edu.tr

F. A. Sheikh et al. (eds.), *Interaction of Nanomaterials With Living Cells*,
https://doi.org/10.1007/978-981-99-2119-5_9

9.1 Introduction

Initially, 3D printing technology was primarily utilized in manufacturing metals and ceramics. With the rapid progress of 3D fabrication, it has wide application in the electronic and optoelectronic military and automotive industry and also, most recently, in biofabrication. The process of organizing cellular and noncellular ingredients in space to imitate the formation, form, and function of human tissue is known as biofabrication. Three-dimensional (3D)-printed materials typically have suitable mechanical performance and resolution. Appropriate biomaterials offer optimization microenvironments for implanted cells, resulting in enhanced viable cells and functionality. The high resolution of the process permits cells/matrix/biomolecules and biomaterials to be accurately distributed to simulate native tissue architecture (Huang et al. 2017; Jana and Lerman 2015). Biomaterials to be used in tissue engineering applications can be fabricated through many traditional methods, such as chemical/gas foaming, solvent casting, emulsion templating, particle/porogen leaching, electrospinning, freeze-drying, phase separation, and also 3D bioprinting. The bulk properties can be managed, while the pore size, shape, topology, and architecture cannot be determined with other techniques except 3D bioprinting. Moreover, 3D bioprinting enables concise and specific placement of biological materials, biochemicals, and living cells by layer-by-layer deposition (Huang et al. 2017; Zhu et al. 2016).

The opportunity for additive manufacturing (AM), also defined as three-dimensional (3D) printing for the production of biomaterials in 3D tissue engineering applications, has increased with rapid innovations in computer-aided design/computer-aided manufacturing (CAD/CAM) systems. In 1986, Charles W. Hull presented 3D printing for the first time. Hull's stereolithography (SLA) method benefits from ultraviolet (UV) light to cure thin layers of material on top of existing layers, resulting in a 3D pattern. Three-dimensional bioprinting is a multidisciplinary concept with roots in engineering, biology, and also material sciences, which is to manufacture 3D organ structures that protect, reestablish, and/or enhance tissue activity (Seyedmahmoud et al. 2020, 2015; Kruth et al. 1998; Langer and Vacanti 1993; Hull 1986).

The growth of such 3D in vitro applications has gained the interest of medicine. It is mainly based on two requirements: a scarcity of organs (Huang et al. 2008) and a desire for much less costly drug testing concepts (Shafiee and Atala 2016). The popularity of regenerative medicine has risen dramatically in the past years. The number of people on the organ transplant list of the next ones in the United States nearly doubled from 95,000 between 2006 and 2016. The significant increase in the list shows the need for current living donor remedies (Seyedmahmoud et al. 2020; Abouna 2008). Three-dimensional bioprinting methodologies for orthopedic tissue design are illustrated in Fig. 9.1. Moreover, the central concept of 3D bioprinting is depicted in Fig. 9.2.

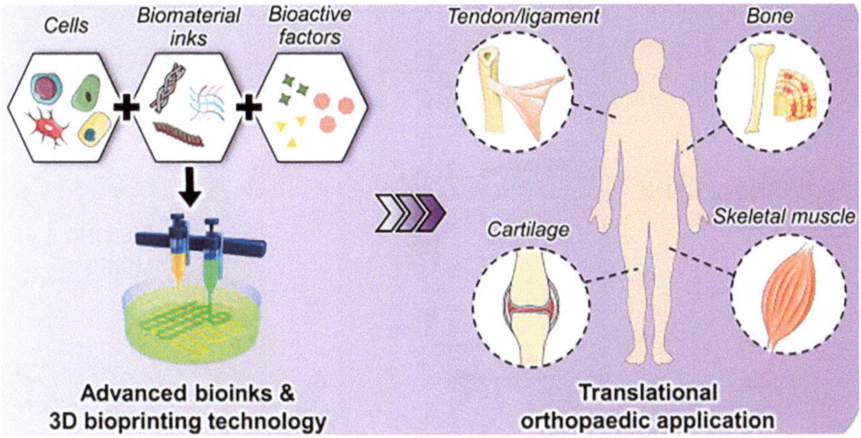

Fig. 9.1 Three-dimensional bioprinting of biomimetic functional tissue constructs and translational applications in orthopedic tissue engineering and regenerative medicine (Chae and Cho 2022)

9.2 Prebioprinting

The encompassing objective of this process is to produce a 3D vital point tissue scaffold model, which could be formed using clinical imaging techniques or computer-aided design (CAD). The most widely known imaging technology tools used to obtain data on the anatomical structure of a tissue or organ are magnetic resonance imaging (MRI), X-ray, and computed tomography (CT) (Seyedmahmoud et al. 2020; Murphy and Atala 2014; Mironov et al. 2008).

9.3 3D Bioprinting Technology

Three-dimensional bioprinting is a new form of technology used in tissue engineering and regenerative medicine to create complex tissue structures that mimic organs and tissues. This 3D bioprinting technology involves a layer-by-layer deposition of cell-loaded biomaterials in a predetermined structural architecture to create functional tissues or organs. To create complex structures, this technique integrates biomaterials, living cells, and controlled motor systems. It can produce more advanced structures than some other existing methods, such as electrospinning, emulsion templating, and freeze-drying. Computer-aided design can be used to create complex 3D tissue scaffolds (CAD) (Izbudak 2021; Chen and Shi 2013; Liu et al. 2005). The use of 3D printers enables the fabrication of scaffolds, devices, and also highly complex tissue models. Three-dimensional printers can also be utilized in medical imaging developed through computerized design. Some important features, such as patient-specific design, production on demand, obtaining high

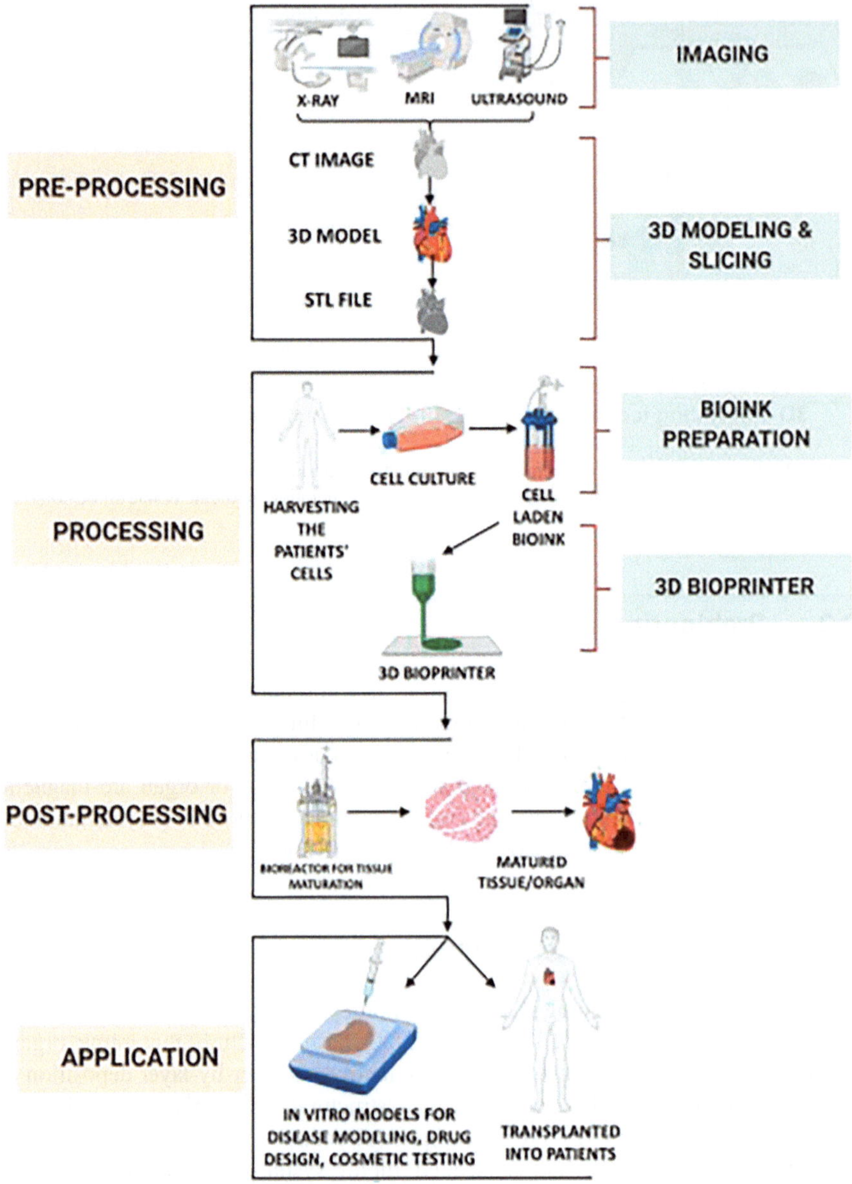

Fig. 9.2 Schematic diagram showing the three main steps in 3D bioprinting: preprocessing, processing, and postprocessing (Bejoy et al. 2021)

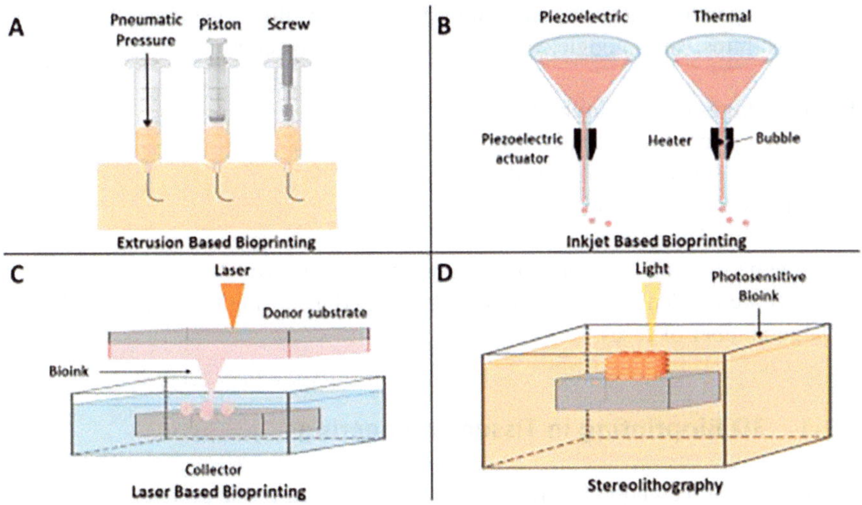

Fig. 9.3 The different 3D bioprinting techniques: (**a**) extrusion-based bioprinting, (**b**) inkjet-based bioprinting, (**c**) laser-based bioprinting, and (**d**) stereolithography (Bejoy et al. 2021)

structural complexes, low cost, and high efficiency affect the use of 3D printers in the medical field (Yüce Erarslan 2021; Guvendiren et al. 2016; Guillemot et al. 2010).

Three-dimensional bioprinting systems have opened up new valuable applications and are currently progressing as a game-changing technology in the sense of the future of human health, with studies such as artificial organ printing in the healthcare system, the repair of lacking bone parts, and the manufacturing of private implants (Ozbolat 2016; Topuz et al. 2018). Furthermore, 3D bioprinters enable the utilization of direct copies of patient architectural results attained through various scanning systems, such as X-ray CT (Topuz et al. 2018; Hockaday et al. 2012; Inzana et al. 2014) and MR, and they generate biomimetic 3D biological tissue with good precision (Topuz et al. 2018; Yanagawa et al. 2016). Bioprinting is composed of many types (see Fig. 9.3).

Advantages of 3D Bioprinting

- It is faster and mostly more concise than conventional hand-built organ strategies.
- Organs are rarely to be dismissed concerning transplantation.
- Organ trafficking has been lowered.
- Human error is decreased, and scientists will find it less complicated.
- The final products are not dependent on biomaterials or scaffolding that are not present in native tissues.
- Without human subjects, the impact of illnesses or drugs may be noted more precisely.
- Tissue reproducibility is guaranteed through strict control of both composition and shape; minimized changes.

- Tissue-specific functions can be improved with well-organized and various cell forms (https://sites.google.com/site/gsse2014b2/pros-cons).

Disadvantages of 3D Bioprinting

- When a printed object fails, liability problems occur.
- There are several moral considerations.
- It involves high pricing.
- A lot of energy is consumed.
- Some harmful particles are emitted into the atmosphere.
- There are challenges in regulating cellular media causing the death of cells in a large number of cases.

9.3.1 3D Bioprinting in Tissue Engineering

The traditional technique for developing an engineered-tissue product is associated with the first manufacturing of a unique native tissue design, usually accompanied by the supply of cells and biomolecules. Nevertheless, this strategy can play a part in two major disadvantages: restriction in cell transmission and a decrease in cell growth because of low density at the core zone (Chowdhury et al. 2020; Derakhshanfar et al. 2018). The tissue engineering innovation has influenced the present framework via computer-aided layered manufacturing, also known as 3D bioprinting. In summary, 3D bioprinting is combining the main components identified as "bioink" that serves as a biological template and numerous types of cells with the presence of chemical factors and biologically active compounds to construct a solid and functional in situ 3D living structure (Chowdhury et al. 2020; Guvendiren et al. 2016).

When printing tissue or organs, this bioink generally contains living cells; however, when printing scaffolds, this bioink does not involve living cells. Aside from cells, the bioink is composed of a variety of polymer compositions that the cells are suspended. Polymers serve as the organs' basic framework, helping to promote cell adhesion, proliferation, and growth (Munaz et al. 2016). They can exist as specific hydrogels or polymers or as a mixture of them. Moreover, bioinks containing bioceramic compounds (tricalcium phosphate and hydroxyapatite) are also used (Mukherjee et al. 2019; Pekkanen et al. 2017; Dávila et al. 2016; Jose et al. 2016). Among the different biomaterials, hydrogels are the most superior materials used as bioinks in 3D bioprinting. The main reason for this is their ability to retain living cells, their changeable chemical structure, their adjustable mechanical and biological degradation properties, and the fact that they can provide good resolution during printing (Izbudak 2021; Topuz et al. 2018). Over the last several decades, three-dimensional (3D) printing (rapid manufacturing or additive manufacturing) technologies have become increasingly popular in a variety of industries (Gu et al. 2018). However, some critical properties must be regarded in these applications (Chan and Leong 2008). First of all, the tissue scaffold to be produced must be biocompatible. The 3D scaffolds should be biodegradable or bioabsorbable, and the tissue must be able to completely substitute the scaffold. The mechanical properties that are suitable for the tissue to be implanted should be present. Moreover, the 3D scaffold should be easily fabricated in a wide range of forms and sizes (Gu et al. 2018).

Bioprinting has mainly three methodologies: imaging, computer-aided design of the tissue to be printed, and printing itself, the formation of bioink through the selection of appropriate materials, the choice of a proper bioprinter based on the product desired, for the manufacturing of scaffolds and/or tissues (Hacıoglu et al. 2018). Numerous techniques for producing 3D scaffolds from synthetic and natural polymers have been developed, such as emulsion templating, freeze-drying, solvent casting, gas foaming, phase separation, electrospinning, and melt molding (Gu et al. 2013; Kim et al. 2016a, b; Ma and Xue 2015; Oh et al. 2003). Fabrication techniques do not allow for effective control of porosity, pore size, scaffold shape, and interconnected pore morphology within the scaffold. Furthermore, the capability to make scaffolds from cells has drawbacks (Gu et al. 2018). Inkjet printing, extrusion-based methods, light-induced (photopolymerization) methods, and particle-fusion-based methods are the four techniques used in 3D bioprinting.

In the past few years, the 3D printing system has progressed quickly in the tissue engineering and regenerative medicine field (Topuz et al. 2018; Cornelissen et al. 2017). At the end of 2022, the global 3D bioprinting market is expected to reach nearly $1.82 billion (Topuz et al. 2018). Considering the future potential of bioprinting and reconstructive surgery and its objectives, it is indisputable that 3D bioprinting technology will emerge in plastic surgery (Jessop et al. 2017). Therefore, current studies will focus on improving novel bio-based materials in the 3D bioprinting sector for tissue engineering, emphasizing printable biomaterials (e.g., bioink) (Topuz et al. 2018).

9.3.2 Bioinks for 3D Bioprinting

The choice of biomaterial suitable for a bioink is a critical step in achieving a promising bioprinting product, as well as other fabrication processes. When polymers are chosen as bioinks, it is essential to know how polymer properties affect printing efficiency and cytocompatibility. These bioinks are anticipated to provide mechanical stability for the printed product due to their use during the process. In this scope, many natural polymers (gelatin, chitosan, cellulose, collagen, etc.) and some synthetic polymers, like polycaprolactone (PCL), polylactic acid (PLA), and poly(lactic acid-co-glycolic acid) (PLGA), are commonly utilized in tissue engineering applications (Hacıoglu et al. 2018).

9.3.2.1 Natural Polymers

Sodium Alginate

Sodium alginate is a polysaccharide sourced from brown algae. It is extensively used in 3D bioprinting due to its biocompatibility, low cost, and rapid gel formation

(Du 2018). According to recent studies, sodium alginate is often blended with other polymers that are easy to mold and have a preferable biological nature, and it is a dominant material in drug and cell transport and cell encapsulation (Ozbolat et al. 2014). Despite all these advantages, alginate has low cell adhesion compared to other natural biomaterials (Yüce Erarslan 2021; Ahn et al. 2012). Furthermore, low-concentration sodium alginate has poor mechanical capacity and yet is helpful in increasing cell viability and proliferation (Du 2018).

Gelatin

Gelatin is a natural protein that is formed by hydrolyzing collagen. It also has amphoteric properties due to alkaline and acidic amino acid functional groups. It can be obtained through the extraction of other sources (such as bones, skin, or tissues of various animals). (Yüce Erarslan 2021; Chiou et al. 2008). Gelatin has good cell adherence, high biocompatibility, and complete biodegradability properties in vivo conditions without immunogenic, which has increased interest (Du 2018; Kuijpers et al. 2015). Gelatin chains have helical structures at low temperatures, which combine on their own to form a gel-like structure. On the other hand, this physical formation reverts to a random helical structure at high temperatures. Therefore, gelatin is a kind of thermoreversible gel. Some studies indicated that it dissolves when incubated at 37 °C for a long time. To overcome this issue, chemical cross-linking is carried out under UV light in the presence of a photoinitiator by adding unsaturated groups to the main chain of gelatin. The unsaturated groups are usually provided by methacrylic anhydride, and such gelatin derivatives are called methacrylated gelatin “GelMA” (Yüce Erarslan 2021; Sakai et al. 2009). Aldana et al. (2021) designed GelMA-alginate-based biomaterials via 3D bioprinting. In this study, the biomaterials contained sheep-adipose-derived stem cells, and the biomaterials in different blend concentrations were cultivated in vitro for 24 h before being tested for bioactivity in 3D-bioprinted structures (Fig. 9.4).

Silk Fibroin

Fibrinogen is a protein-based polymer formed in a living body, used as a sealant and adhesive in surgery, for cartilage treatment in tissue engineering, and in wound healing applications (Fontes and Marcomini 2020; Skardal et al. 2012; Ahmed et al. 2008). Fibrinogen is required for effective blood coagulation and is utilized as a type of 3D bioink. Fibrin, like other natural polymers, has high biocompatibility, allowing for numerous high affinitie for cell adherence and growth (Tao et al. 2012; Schacht et al. 2015). However, with the fast degradation of fibrin, it is not proper for prolonged culture media to be formed in vivo tests, and low viscosity complicates processability as well (Das et al. 2013). Consequently, several studies have pointed out that fibrin can mix with other natural polymers, such as gelatin, alginate, and collagen (Yu et al. 2020; Schacht et al. 2015).

Agarose

Agarose is mainly a seaweed-derived marine polysaccharide. Because of its gelation, biocompatibility, and rheological properties, it is widely used in biological

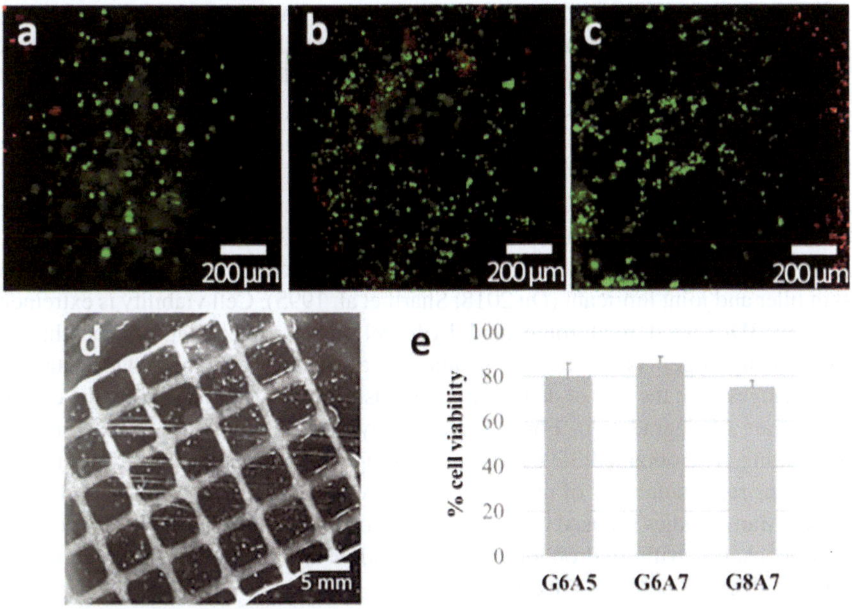

Fig. 9.4 Fluorescence microscopy images of live (green)/dead (Zhang et al. 2018) cells in (**a**) G6A5, (**b**) G6A7, and (**c**) G8A7 hydrogels at 24 h. (**d**) Image of printed G8A7 hydrogel construct. (**e**) Cell viability in gelMA-alginate constructs. Printing of stripe-patterned hydrogels: biaxial mechanical testing (Aldana et al. 2021)

applications (Jakus et al. 2016a, b; Fontes and Marcomini 2020). Agarose gelling is caused by the formation of intermolecular hydrogen bonds on cooling, which results in the agglomeration of the helical structure. Although it has excellent biological and mechanical features, its ability to enable cellular proliferation is restricted (Fontes and Marcomini 2020; Hospodiuk et al. 2017). Therefore, the use of agarose alone means that it is inappropriate for the production of cell-loaded biomaterials. Furthermore, it shows a thermo-reversible feature with a sol-gel transition between 32 and 47 °C (Du 2018; Yu et al. 2020; Medina-Esquivel et al. 2008).

Pullulan

Pullulan is a type of polysaccharide that is typically made from the yeast *Aureobasidium pullulans* (Saroia et al. 2018). It is generated on the surface of bacteria-infected cells. For the first time, studies on the physicochemical structure of pullulan began to be conducted by Bernier in 1958. It is insoluble in either organic or inorganic solvents, except water. Therefore, pullulan is preferred in food packaging and biomedical applications (Singh and Saini 2008). Additionally, pullulan can be obtained via fermentation, thanks to various types of waste (Saroia et al. 2018; Singh et al. 2009; Thirumavalavan et al. 2009). Studies have been conducted on extracting pullulan and dextran, which help the rapid development of endothelial cells. Pullulan and derived materials with osteo-communication properties were utilized in tissue engineering. The mechanical strength of pullulan was enhanced via the mixing and cross-linking of pullulan and dextran (Saroia et al. 2018; Aschenbrenner et al. 2013).

Hyaluronic Acid

Hyaluronic acid is a polysaccharide with a high molecular weight, which is one of the main components of the extracellular matrix (ECM) (Yüce Erarslan 2021; Mobaraki et al. 2020; Falcone et al. 2006). It is generally used in surgeries as a skin filler and joint lubricant (Du 2018; Sharif et al. 1995). Cell viability is extremely high in 3D-printed hyaluronic acid hydrogel sealed in cartilage tissue than in collagen hydrogel. Nevertheless, hyaluronic acid has weak mechanical stability that changes with the rate of degradation; it must be modified to allow for the control of the rate of degradation. For this reason, hyaluronic acid is not proper for 3D bioprinting. The obtained 3D hyaluronic acid hydrogel structures have low stability due to the rapid solubility of unmodified hyaluronic acid in water (Pescosolido et al. 2011). Many studies focused on the functional treatment of hyaluronic acid polysaccharide chains with hydrophobic groups and/or by photo-cross-linking methyl acrylate (MA) (Du 2018; Yüce Erarslan 2021).

Chitosan

Chitosan is a linear amino polysaccharide derived from chitin and its derivatives, consisting of (1–4)-linked D-glucosamine structures and *N*-acetyl-glucosamine groups that are randomly located (Vega-Cázarez et al. 2018). It cannot be dissolved in aqueous media with a pH greater than neutral due to its semi-crystalline structure. Thanks to its biodegradability, biocompatibility, and antimicrobial structure, it is prevalently preferred in many fields, such as bone, skin, and cartilage regeneration; the formation of sponge scaffolds; and wound dressings. Because chitosan has a slow gelation rate (10 min after injection) and low mechanical strength, just stents with a high viscosity can maintain their form for hours (Du 2018). Chitosan is applied to 3D bioprinting to develop different stents and microflow channels.

Collagen

Collagen is the most frequently known and utilized protein in tissue engineering due to its triple helix structures with self-aggregating properties via covalent and hydrogen bonds (Saroia et al. 2018). It is the principal protein constituent of ECM in real tissues/organs (Peppas et al. 2006), and collagen sources include rat and pig tendon materials (Yu et al. 2020; Osidak et al. 2019; Diamantides et al. 2019). Collagen offers excellent growth conditions for cell growth, adhesion, and function due to its abundance of integrin-binding areas (Yu et al. 2020). It can make a variety of artificial tissues, including skin, cartilage, heart valves, reconstructed breast, vocal cord, and spinal cord (Saroia et al. 2018; Tangsadthakun et al. 2017; DeLustro et al. 1986; Taylor et al. 2006; Hahn et al. 2006; Slaughter et al. 2009; Cavallo et al. 2015). Collagen can produce various types of gels, sponges, and other materials because of changes in collagen strands and the induction of cross-linking (Saroia et al. 2018). At

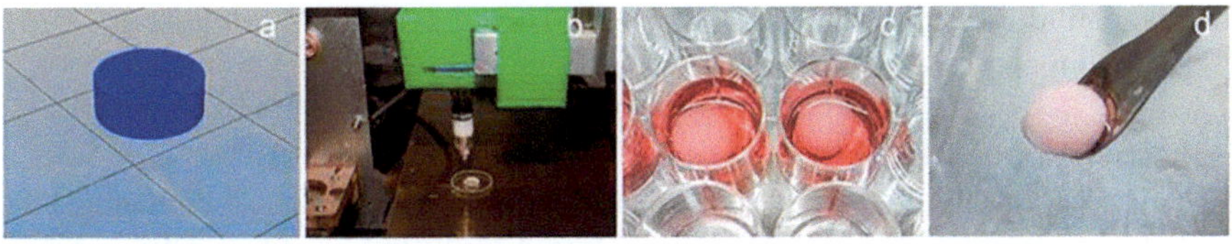

Fig. 9.5 A cylindrical CAD model (**a**) was 3D bioprinted using a chondrocyte-laden type II collagen bioink (**b**). The scaffolds were cultured for 3 weeks before harvesting to analyze chondrogenic ECM deposition and gene expression (**c**, **d**). (Reprinted from Ren et al. (2016) under Creative Commons CC BY license (Chartrain et al. 2022))

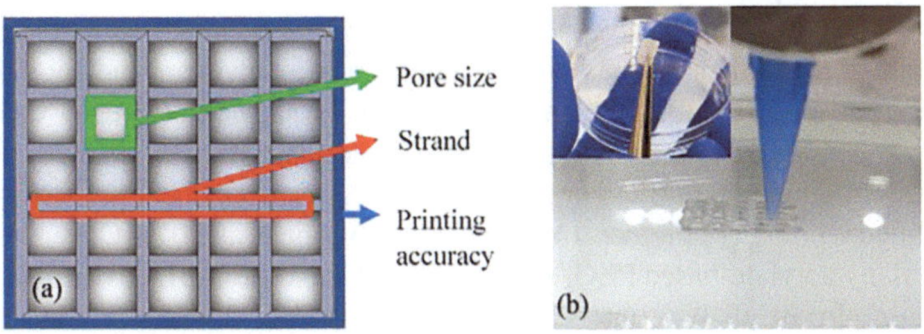

Fig. 9.6 (**a**) 3D scaffold designed in SolidWorks and (**b**) scaffold printing via a 3D bioprinter with the insight of scaffold postprinting (Kakarla et al. 2022)

low temperatures, collagen is in the shape of a pre-gel, and it may be cross-linked thermally at 37 °C. Moreover, it is cross-linked with UV, glutaraldehyde, carbodiimide, and genipin and is vulnerable to collagenase degradation. Despite its advantages, the low viscosity of collagen requires its mixing with other polymers for bioprinting (Yu et al. 2020; Weadock et al. 1995; Harriger et al. 1997; Powell and Boyce 2006; Kim et al. 2016a, b). The suitability of collagen for 3D bioprinting is determined by the concentration of collagen in a solution. Only increased collagen concentrations (higher than 20 mg/mL) in single-component collagen bioinks enable higher printing reliability (Osidak et al. 2020). Figure 9.5 shows collagen-based 3D biomaterials produced for chondral tissue engineering in a study.

Kakarla et al. (2022) designed a hydrogel based on gelatin, alginate, and boron nitride nanotubes via an extrusion 3D bioprinting method. The scaffold model was created in SolidWorks and has dimensions of 10 × 10 × 1 mm, a layer thickness of 0.17 mm (three layers), and a pore size of 0.5 mm (Fig. 9.6).

9.3.2.2 Synthetic Polymers

Synthetic polymers have superior mechanical and chemical properties compared to natural polymers. Nevertheless, synthetic polymers offer few or no cell niches performance. Generally, polycaprolactone (PCL), polylactic acid (PLA), poly(ethylene) glycol (PEG), polyvinyl alcohol (PVA), and polyglycolic acid (PGA) are used in 3D bioprinting. PLA, PGA, and PCL have outstanding biodegradability, biocompatibility, and mechanical properties (Huang et al. 2017; Mota et al. 2015). These polymers were given in detail in this section.

Polyvinyl Alcohol (PVA)

PVA is a water-soluble synthetic, one that is both biocompatible and biodegradable, and also it is confirmed by the Food and Drug Administration (FDA, USA) (Aslam et al. 2018; Marin et al. 2014). This polymer is usually blended with natural ones (gelatin, alginate, and/or chitosan) due to low cell affinity and has undergone physical modification (Yu et al. 2020).

Polyethylene Glycol (PEG)

PEG is a convenient polymer for designing 3D scaffolds, owing to its high hydrophilic nature, higher tensile strength than natural-based polymers, and good biocompatibility (Yu et al. 2020; Alcantar et al. 2000). It can be tailored by combining alginate and collagen to meet specific needs (Fontes and Marcomini 2020). Further, to increase the diacrylate (DA) or methacrylate (MA) mechanical properties, it can be modified with other polymers (Yu et al. 2020; Aduba et al. 2019; Cheng and Chen 2017). Many acrylate-based PEG, such as PEG-DA and PEG-MA hydrogels, can be printed on almost all bioprinters, e.g., extrusion-based, laser-based, and droplet-based bioprinting.

Polycaprolactone (PCL)

PCL is classified into polyesters that are biodegradable, are hydrophobic, and also has semi-crystalline features approved by the FDA (Saroia et al. 2018; Bhavsar and Amiji 2008). The incorporation of bioactive glasses and $Ca_3(PO_4)_2$-based particles into PCL in bone tissue engineering increases its many properties, especially its mechanical strength (Hajiali et al. 2018). A study showed that PCL has excellent biocompatibility with periosteal cell culture systems and human fibroblasts (Salgado et al. 2012). The mixture of PCL with other polymers or bioactive agents to develop a more appropriate media for protein progression. In many studies, PCL mixed with heparin and curdlan sulfate enhances the response of cultivated tissues and increases the protein ability of the mixture (Saroia et al. 2018). Similarly, various composites, including ceramic particles, such as calcium phosphate (CaP) and hydroxyapatite, were produced for bone regeneration (Huang et al. 2017).

Polylactic Acid (PLA) and Polyglycolic Acid (PGA)

PLA and PGA are also included in the polyester class, as is PCL. However, PLA and PGA, by themselves or in combination with other bio-based polymers, offer an excellent physiological media for cell growth because their hydrophobic nature can confine cell attachment (Saroia et al. 2018; Bee et al. 2018; Gentile et al. 2014). PLA and PGA are preferred for fabricating artificial vascular grafts, owing to their excellent mechanical strength (Parın and Terzioğlu 2022). Therefore, some modification techniques, such as plasma treatment and/or surface coating, can be applied to printed PLA and PLGA to form 3D cell culture media for the growth of different tissues (Deng et al. 2020).

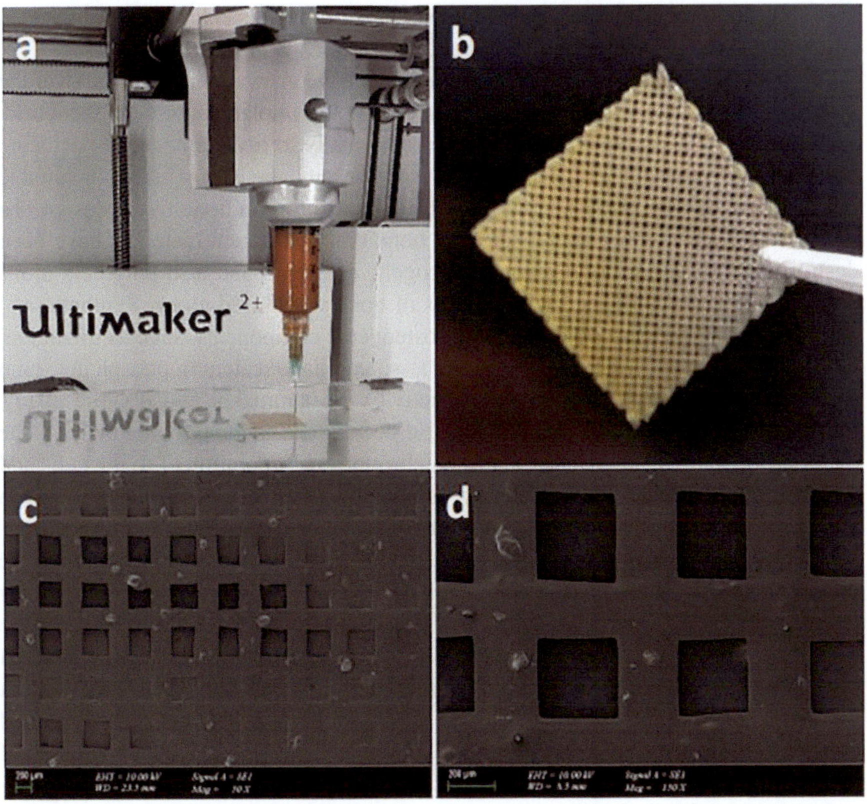

Fig. 9.7 Production of BFO-doped PLA scaffold by a modified 3D printer (**a**), BFO-doped PLA scaffold (infill 70%) (**b**), SEM images of BFO-doped PLA scaffold (infill 70%) (**c**, **d**) (Bedir et al. 2020)

Bismuth ferrite (BFO)-nanoparticle-loaded PLA scaffolds were fabricated by Bedir et al. (2020). It is clearly seen in Fig. 9.7 that the BFO is agglomerated in the neural tissue scaffolds.

9.3.3 3D Bioprinting in Tissue Engineering Applications

9.3.3.1 Bone Tissue Engineering

Bone tissue supports the mechanical strength of tissues and organs and movement in the human body. It is essential for homeostasis and blood pH regulation because it acts as a mineral store with a complicated hierarchical structure (Yilmaz et al. 2019; Sowjanya et al. 2013). Bone is comprised mainly of collagen, other proteins, water, and mineral phase. As a result, 3D-printed forms should preferably have similar properties. In this situation, composite materials, such as collagen mixed with various types of bioceramics, are being assertively utilized to enhance 3D-bioprinted structures (Yu et al. 2020; Osidak et al. 2020; Yilmaz et al. 2019).

Biomaterials and tissue-engineered structures have grown in importance over the past few decades. Bone is the second-most commonly implanted tissue in the world, with over four million operations conducted annually to heal damaged tissue using bone grafts. The development of printing processes and the improvement of suitable ink materials are the primary concerns of 3D bioprinting in bone tissue design. To produce new alternatives to conventional bone grafts, different materials have been designed to be used in bioprinting. Hydrogels, ceramics, and also polymers alone cannot completely simulate the properties of bone when used alone. Synthetic and natural polymers are usually mixed with osteogenic components, such as tricalcium phosphate (TCP), hydroxyapatite (HAP), silica, nano calcium phosphates, and bioactive glass particles to enhance bioactivity (Ashammakhi and Kaarela 2017). Furthermore, some growth factors have also been added to the polymer matrix. The major problem in 3D bioprinting is the synchronous incorporation of living cells and biostructural materials. The capability of creating cell-free scaffolds utilizing a variety of 3D-printed materials makes this method desirable for bone tissue engineering applications (Yilmaz et al. 2019). 3D bioprinted tissue structures were produced of by Matrigel TM and alginate-based hydrogels were produced by Fedorovich et al. (2008). In another study, Phillippi et al. (2008) showed the myogenic differentiation of patterned BMP-2 on fibrin-coated lamellar by inkjet bioprinting. The osteogenic and chondrogenic activities of PEGDMA-GelMA bioinks were examined in a study (Gao et al. 2014). The obtained 3D-bioprinted constructs indicated both good cell proliferation ($>80\%$) and an increased degree of differentiation in comparison with pure PGDMA (Ozbolat et al. 2016). The mechanical strength and cell viability of alginate and bone-derivatived methacrylated ECM with hASC cell-loaded mesh structures were investigated by Lee et al. (2020). Furthermore, bioprinted bone-like tissue fabricated from gelatin and alginate hybrid matrix (Zhang et al. 2020). To enhance the mechanical features, photocrosslinkable glycosaminoglycan hyaluronic acid scaffolds were fabricated by (Poldervaart et al. 2014). Skardal et al. (2012) designed bioprinted materials containing fibrous collagen gels with bone marrow mesenchymal stem cells for wound healing, and they analyzed wound closure and epithelial restoration. It demonstrated that the the in vivo bone regeneration impact was studied in a critical-size calvarial defect model of rats (Fig. 9.8).

9.3.3.2 Neural Tissue Engineering

Neural cords have been suggested as a successful neural recovery matrix that supports and helps neuron cells utilize tissue engineering strategies in the therapies of peripheral nerve and spine injuries. Bioprinting is a newer technique for fabricating manageable 3D scaffolds for neural tissues with various cellular categories and complex micro/nanoscopic functions. The main considerations for the layout of nerve guide conduits are ensuring mechanical strength while also optimizing the proximal and distal nerve ends and restricting nerve compression. Extrusion printing is one of the most widely used printing processes in nerve conduit

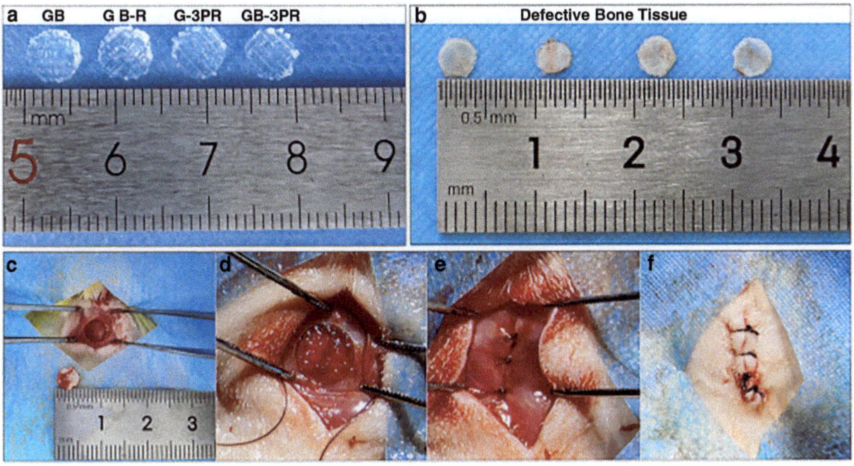

Fig. 9.8 Construction of a bone defect model and demonstration of the scaffold implant operation process. (**a**) 5 mm diameter implanted bioprinted scaffolds. (**b**) 5 mm diameter defective cranial tissue. (**c**) Isolating surrounding soft tissue to expose the skull and creating a 5 mm diameter defect in rat critical-size calvarial model. (**d**) Inserting scaffolds into a calvarial defect. (**e**) and (**f**) Suturing subcutaneous tissue and skin wounds (Shen et al. 2022)

designing because gels can be deposited quickly and easily. Nevertheless, laser-based bioprinting, like stereolithography and inkjet printing, enables higher print resolution, which is important for imitating the anisotropic design of nerve tissue. Because neural cells are comparatively more susceptible to their extracellular environment, the choice of bioink has a substantial influence on neural recovery. Generally, collagen is preferred in nerve regeneration. In collagen hydrogels produced with moderately concentrated collagen solutions, neurite outgrowth is much more noticeable. Furthermore, Hsieh et al. (2015) have bioprinted thermoresponsive polyurethane (PU) (37 °C) hydrogel with adjustable gelling ability and hardness without any cross-linker. In the study, neural stem-cell-loaded bioink has been found to benefit neural injuries. Interestingly, the bioink recovered the function of the damaged nervous system in less than 1 week. Some researchers found that alginate bioinks with the use of cell adhesion factors enhanced cell compatibility. In a study on this, the efficacy of RGD or YIGSR peptide in alginate increased cytocompatibility (Sarker et al. 2019). The scaffolds designed via an extrusion-based 3D bioink method had more cells than pure alginate scaffolds after 9 days in culture media (Sarker et al. 2019; Yilmaz et al. 2019). To assess biocompatibility for in vitro/in vivo tests, gelatin-alginate-based 3D biostructures were created by Wu et al. (2020). More than 90% of Schwann cells lived after 24 h and remained viable for 7 days (Yu et al. 2020). Meanwhile, rat PC-12 cells were investigated by Ngo et al. (2020), who found that by adjusting important characteristics of the bioink design, 3D-printed hydrogels based on hyaluronic acid methacrylate (HAMA) might increase cell survival and aid in peripheral nerve repair (Fig. 9.9).

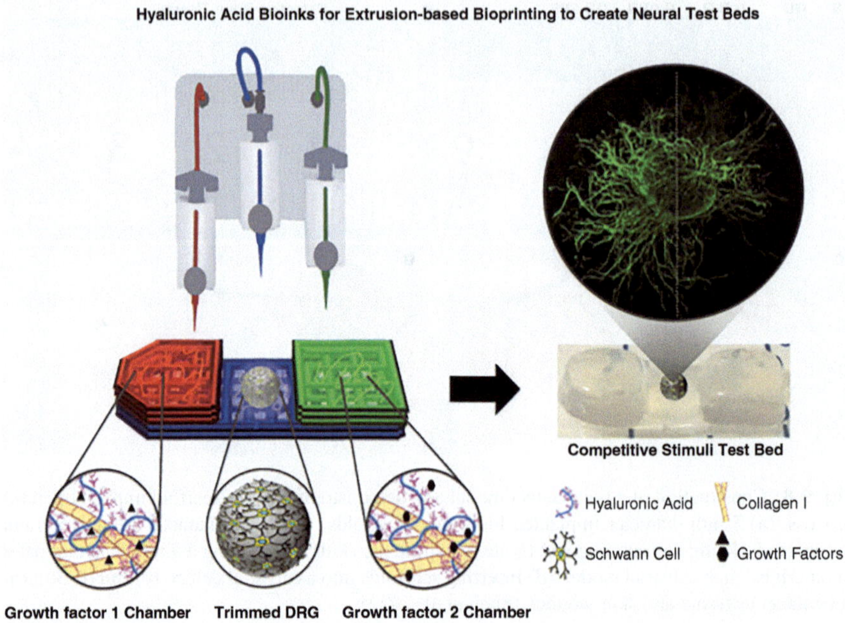

Fig. 9.9 A schematic of the 3D-bioprinted in vitro test bed design, consisting of a base printing with the main bioink and two chambers that can be loaded with different growth factors (Ngo et al. 2020). (Copyright 2020, American Chemical Society (Ding et al. 2022))

9.3.3.3 Vascular Tissue Engineering

Artificial blood vessels play a crucial role in linking metabolically demanding organs, allowing nutrients to be delivered while separating waste (Yu et al. 2020). Nutrients and oxygen are easygoing in a 2D cell culture with a cell population thickness of around 20–30 μm (Gu et al. 2018; Colton 1995). Various bioprinting methodologies, including extrusion, droplet, and laser-based bioprinting, have been used to design vascular tissue (Ozbolat et al. 2016). Accordingly, 3D bioprinting methodologies which design objects of the preferred shape utilizing various bioinks and kinds of cells have occurred as intelligent techniques ~~to~~ for designing vessels with small diameters (Gu et al. 2018). Cui and Boland (2009) designed fibrin microchannels via an inkjet-based bioprinting technique. They revealed that when they bioprinted human microvascular endothelial cells (HMVEC) laden with fibrin hydrogel, the cells tailored themselves inside these fibrin networks and reproduced. Consequently, bioprinting both cells and the scaffold at the same time promotes HMVEC proliferation and microvessel generation. Consequently, bioprinting both cells and the scaffold at the same time promotes HMVEC proliferation and microvessel generation was supported by the cells. Human neonatal dermal fibroblasts and human umbilical vein endothelial cells (HUVEC) grew significantly with time, according to the researchers (Gu et al. 2018).

9.3.3.4 Skin Tissue Engineering

Skin is the largest, most complex and outermost organ that acts as a barrier to infections and irritants, antioxidants, environmental factors, and any other externally harmful agents (Askari et al. 2021). In terms of wound size, extent, and depth, researchers have developed a variety of wound dressings or natural-based skin substitutes (Yilmaz et al. 2019; Sheridan 2009). Chronic, nonhealing wounds caused by burns, trauma, or diseases are clinically significant since they place a great deal of pressure on both the patient and the medical system. Biocompatible wound dressings, designed skin grafts, and split-thickness skin grafting from autologous skin are currently used to treat chronic wounds. Though quite efficient, these skin substitutes are generally expensive, have poor adhesiveness, are susceptible to infection, or rely on skin health donation presence. Furthermore, the majority of present clinical practice-utilized skin substitutes are made entirely of dermal fibroblasts and keratinocytes and do not enhance vasculature (Gu et al. 2018; Church et al. 2006).

Consequently, skin damage is a significant issue having in-depth impacts on other tissues (Jean et al. 2011; Metcalfe and Ferguson 2007). During skin damage, autologous grafts sourced from a patient are frequently utilized to prevent immune reactions and regulate skin activity and wound repair. However, autologous grafts do not properly mend skin damage injuries encompassing a big region or having a serious depth (Andreassi et al. 2005). Therefore, there is an urgent need to develop artificial skin replacements employing methodologies for tissue repair. The studies have resulted in complex tissue regeneration that connects with body tissue afterward in vitro regeneration and transplantation (Gu et al. 2018). Bioprinting allows the simultaneous accumulation of multiple kinds of skin cells. Also, 3D bioprinting enables the accurate placement of various types of cells and extremities within a structure (Yilmaz et al. 2019; Ng et al. 2016).

Chronic ulcers can be caused by skin damage, infections, or other genetic or physical conditions. Electrospinning, solvent casting, and freeze-drying are membrane preparation techniques that have long been used for skin graft development. Skin bioprinting is gaining popularity, owing to its intelligent and managed manufacturing qualities, which are difficult to attain with the traditional skin graft manufacturing process (Yilmaz et al. 2019). Droplet-based and laser-based bioprinting techniques are used for skin tissue substitution biofabrication. Bioprinting of skin tissue with a bioprinter with an eight-channel valve, in which a 13-layer tissue is bioprinted using collagen, was performed by Lee et al. (2013). Kim et al. (2019) explored a perfusable and preferable vascularized full-thickness skin equal made up of epidermis, dermis, and hypodermis for enhanced vascularization and effective epidermal progress. The printed HUVECs were found to be covered on the exterior of the vascular stream, forming endothelium that resembled tissue. In another study, Koch et al. (2012) used laser-assisted bioprinting to deposit 20 layers of fibroblasts (mouse NIH-3T3) and 20 layers of keratinocytes (human HaCaT) integrated into collagen gel onto a layer of decellularized dermal matrix to design dermis and epidermis layers, respectively (Yilmaz et al. 2019). Gholami et al. (2017) developed patches for wound healing, which include a 16% alginate solution (w/v)

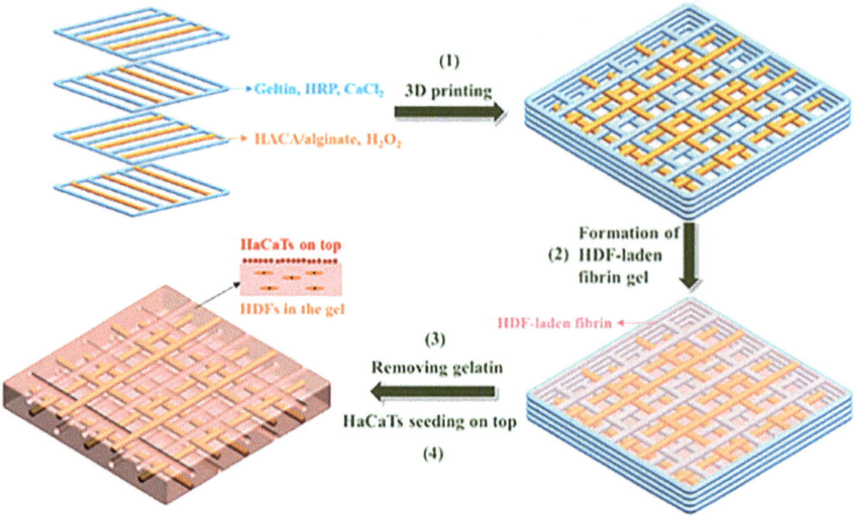

Fig. 9.10 Fabrication of a bilayer cell-laden skin-like structure model (Zhou et al. 2021). (Copyright 2022, IOP Publishing (Ding et al. 2022))

and 4% (w/v) gelatin (Yilmaz et al. 2019). In addition, Michael et al. (2013) used 3D-laser-based bioprinting to generate artificial skin. In this study, fibroblasts and keratinocytes were used to construct the skin substitutes. According to the in vivo test, the bioprinted keratinocytes obtained a multilayered epidermis with initial differentiation and stratum corneum at the end of 11 days in culture media.

Currently, nanoparticles are increasing in popularity in the field of transdermal delivery systems. Surface chemistry and form determine how well they penetrate skin tissue. In this regard, Hou et al. created an easy artificial skin model for the rapid screening of nanoparticles in terms of their transdermal penetration ability with 3D bioprinting methods. Fibroblasts were printed on the structure with collagen hydrogel and silica nanoparticles. The obtained 3D scaffolds were investigated via penetrating ability. As a result, positively charged nanoparticles penetrated deeper (Gu et al. 2018; Cubo et al. 2017). 3D bioink hydrogels consisting of two different layers of catechol-hyaluronic acid (HACA)/alginate and gelatin/horseradish peroxidase (HRP) were produced to mimic the structure of the skin in full thickness (Fig. 9.10).

9.3.3.5 Cartilage Tissue Engineering

Cartilaginous tissue is an avascular and aneural structure with low chondrogenic density and high water content (70%) (Askari et al. 2021; Sophia Fox et al. 2009). It has a functional and heterogeneous texture designed to provide a low-friction, wear-resistant, load-bearing surface for effective joint movement (Askari et al. 2021; You et al. 2017). Even if it is just a few millimeters thick, it prevents friction between the joints and provides excessive load stresses during body activities (Gu et al. 2018;

Arkel and Amis 2013). Many researches have been performed try to convey bone marrow stromal cells in alginate hydrogels for extrusion bioprinting bone and cartilage regeneration (Tan et al. 2021; Fedorovich et al. 2008).

Gruene et al. (2010) designed laser-based bioprinting of stem-cell-differentiated chondrocytes, employing a computer-aided fabrication method with the aid of LIFT. The designed material skillfully printed porcine-bone-marrow-derived mesenchymal stem cells (MSCs) with high availability. The cells retained their features and ability to differentiate into osteogenic and chondrogenic bloodlines (Ozbolat et al. 2016). To restore cartilage deficiencies, cartilage tissue has been created through inkjet-based bioprinting. The human chondrocytes integrated into PEGDMA hydrogel was bioprinted an adjusted HP desktop printer (Cui et al. 2012a, b). The mechanical and biochemical composition of the created structures were almost similar to native cartilage (Ozbolat et al. 2016). Sodium alginate has been broadly utilized to create bioprinted cartilage tissue. The hybrid bioprinting of chondrocytes loaded on printed alginate filaments in combination with the bioprinting of chondrocyte spheroids to enhance cell density are demonstrated by Ozbolat et al. (2016) and Ozbolat et al. (2014).

9.3.4 Other Requirements for Effective Scaffold Design

The achievement of the manufactured scaffold is strongly affected by the bioink used and the 3D bioprinting technique (Salah et al. 2020). It also is dependent on other variables to ensure the scaffold's achievement, which is as described in the following.

9.3.4.1 Pore Size

Porosity, pore size, and interconnected pore morphology represent critical parameters for scaffold production. All three properties allow cellular penetration, vascularization, adequate diffusion of nutrients and oxygen into the cells within the structure, and the new-formed extracellular matrix that ensures cell viability. In particular, pore size is an essential parameter for scaffold efficiency (Bružauskaitė et al. 2016). Actually, the pores must be large enough to allow cells to penetrate and migrate within the scaffold structure but small enough to allow a critical number of cells to connect (Izbudak 2021; Hutmacher 2000). A pore size that ranges from 100 m to more than 300 m is required for bone tissue engineering because it offers a hypoxic situation that improves both osteogenesis and angiogenesis for proper bone growth (Salah et al. 2020; Karageorgiou and Kaplan 2005). The effect of scaffold pore sizes on MSC differentiation by creating a gradient of oxygen required and more in the periphery and hypoxic in the center was studied by Di Luca et al. (2016). All scaffolds used for tissue engineering as a function of the host tissue type must have a specific pore size. In particular, a pore size of 20 μm is required for hepatocyte and fibroblast growth, while a pore size of around 20–150 μm is needed for soft-tissue healing. The researchers also propose a pore size between 200 and 400 μm for bone tissue engineering (Yu et al. 2015). Materials with pore sizes of 250 μm or higher are appropriate for forming blood vessels, as compared to those of smaller size (Yu et al. 2015; Babensee et al. 1998; An et al. 2015; Mukherjee et al. 2018).

A porous structure is required due to the tissue formation, vascularization, and tissue joining after implantation in an ideal tissue scaffold. In this case, the obtained scaffold must have an optimal pore structure for the intercellular nutrient and metabolic transmission to occur without adversely affecting its mechanical properties and stability. The scaffold should have pores that connect to permit cell and nutrient infiltration and waste removal (Bružauskaitė et al. 2016). The size of the pores influences cellular proliferation and attachment along the scaffold to substitute it. Small pore sizes inhibit waste removal and nutrient diffusion, whereas larger pore sizes inhibit intercellular ligand formation, which is required for cellular proliferation (Salah et al. 2020).

9.3.4.2 Surface Area

The surface area is important for cell reinforcement and proliferation. It is clearly regarded that a larger and more available surface area promotes cell and tissue connection in scaffolds. It is critical in the context of tissue or organ function restoration or replacement because a high surface-to-volume ratio allows more cells to be compensated (Mukherjee et al. 2018; Dhandayuthapani et al. 2011; Boyan et al. 1996).

9.3.4.3 Mechanical Properties

Mechanical property is also an important element in polymers as it affects the regeneration potential of hard tissues (Mukherjee et al. 2018; Dhandayuthapani et al. 2011). The structural stability of the resulting 3D shapes is influenced by the mechanical properties of the bioink. Because the scaffold must endure load and stress for new tissues to grow successfully, the rheological properties of polymers, such as maximum strain, elastic modulus, and tensile strength, are critical. Pore connectedness, orientation, form, size, and density are related elements that affect the scaffold's mechanical behavior and structural stability (Mukherjee et al. 2018).

The shape stability of the resulting 3D structures is influenced by the mechanical properties of the bioink. Natural-based bioinks have poor mechanical properties. To solve this drawback, nanoparticle reinforcement, cross-linking, and hybrid biomultiplies produced by combining synthetic bioinks are commonly used approaches (Yüce Erarslan 2021). The biostability of many scaffolds depends on factors such as strength, elasticity, absorption, and chemical degradation at the material interface (Nair and Laurencin 2007). Besides, the mechanical properties of bioinks are related to their chemical structure and molecular design (Yüce Erarslan 2021).

9.3.4.4 Biodegradability

The biodegradability of a polymer is critical because it enables the degradation of the scaffold after a particular period and substituting itself with new tissues. Separating the polymer's sensitive hydrolytic or enzymatic bonds promotes polymer biodegradation (Katti et al. 2002; Bružauskaitė et al. 2016; Mukherjee et al. 2018). The biodegradability of polymers is an intrinsic property that is affected by chemical composition, molecular weight, crystalline structure, glass transition temperature (Tg), and the spectrum of wettability (Mukherjee et al. 2018; Ye et al. 1997).

The degradation of 3D structures obtained by printing bioinks during cell culture is necessary for ideal cell differentiation. Although the shape stability of 3D structures is required for printing, controlled degradation of the structure is also vital for tissue regeneration. Microvasculature networks develop alongside tissue to form patient-specific organs through cell differentiation. However, sufficient oxygen and nutrients must be transported between cells, and metabolic waste must be removed from the structure. To successfully manage this process/issue, the cell growth rate and the degradation rate of the 3D structure must be compatible (Yüce Erarslan 2021).

9.3.4.5 Biocompatibility

Biocompatibility refers to the compatibility of a biomaterial with the body (Zhang and Zhang 2015). Biomaterials are substances that do not interfere with the expected changes in the tissues surrounding them, do not cause undesirable reactions in the tissue (inflammation, coagulation, etc.), and do not form. Biocompatibility is categorized into structural and surface compatibility. Surface compatibility refers to a biomaterial's physical, chemical, and biological suitability to body tissues. On the other hand, structural suitability refers to the material's optimal compatibility with the mechanical behavior of body tissues (Chen et al. 2016). Synthetic polymers and some natural polymers, which lack cell-binding structures, can inhibit cell adhesion and proliferation. This may cause cell death. To prevent this issue, cell-binding structures can be introduced to the matrix by blending synthetic polymers with natural polymers (Yüce Erarslan 2021).

9.3.4.6 Viscosity

Bioink viscosity has been studied extensively in 3D bioprinting and is one of the essential factors to consider when developing bioprinting methodologies (Tirella et al. 2009). Ceramics, beta-tricalcium phosphate, poly(caprolactone) (PCL), and polylactic acid (PLA) have high viscosity and high concentration, which are used in general in bone tissue engineering (bioprinting) (Theus et al. 2020). Nevertheless, because most thermoplastic materials necessitate an extreme melting temperature and do not encourage viable cell printing, hydrogel-based bioinks have become a favorable option in recent studies. For bone bioinks produced from hydrogels, gelatin methacrylate (GelMA), alginate methacrylate (AlgMA), and hyaluronic acid are utilized and supported by some osteogenic minerals and growth factors (Theus et al. 2020; Huang et al. 2019; Jakus et al. 2016a, b). In extrusion-based bioprinting systems, hydrogel viscosity varies between 30 and 60×10^7 mPa/s, with the concentration of the hydrogel affecting the viscosity level (Iordache 2019). Natural polymer alginate has low viscosity, and therefore, it is not a suitable biomaterial for extrusion-based bioprinters. Its viscosity can be enhanced by blending with materials such as cellulose, gelatin, and PVA (Iordache 2019).

Hydrogels used in inkjet-based bioprinting should be of low viscosity because, then, they can accurately flow through the piping system and nozzle without blockage. Bioink hydrogels must have rheological features that enable viscosity to rise once shear is applied. Viscosity ranges between 3.5 and 12 mPa/s based on the bioink concentration in the inkjet-based bioprinting system (Iordache 2019; Mandrycky et al. 2016). Hydrogel viscosity varies from 1 to 300 mPa/s in laser-based bioprinting. To ensure cell stability and mechanical strength, the hydrogel must be gelation capable (Iordache 2019).

In laser-assisted 3D bioprinting, the viscosity should be 1–300 mPa s. (Hospodiuk et al. 2017). It is critical that the 3D structure be sufficiently cross-linked after printing to strengthen mechanical integrity, owing to low viscosity values. Photosensitive bioinks are used in stereolithographic processes. The main advantage of this printing technique is that it reduces mechanical stresses on more viscous hydrogels containing encapsulated cells while maintaining high cell viability and functionality (Theus et al. 2020; Li et al. 2016a, b).

9.3.5 Types of 3D Bioprinting

3D bioprinting techniques are examined in detail in the following sections.

9.3.5.1 Inkjet 3D Bioprinting

Researchers have started 3D bioprinting by modifying standard 2D inkjet printers to print bioink in successive layers. Inkjet printers operate by collecting ink droplets at specific regions on a substrate. Thermal, piezoelectric, or electromagnetic forces can be used to expel droplets from the reservoir nozzle (Xu et al. 2005). Despite the fact that these forces produce severe regional conditions, the transient nature of the pressure permits the cells to remain viable with low stress. Clogging can occur even when using the best bioink for an inkjet printer (Seyedmahmoud et al. 2020; Ong et al. 2018). Inkjet bioprinting is one of the oldest printing methods. This method, which is based on the noncontact accumulation of biofunctional ink droplets, is divided into thermal, piezoelectric, and mechanical. This technique is generally preferred due to its compatibility with cells and materials, high printing speed, high cell viability, and low cost. High-viscosity materials restrict its application (Akkuş et al. 2020; Hacıoglu et al. 2018; Noh et al. 2017).

Thermal-inkjet-based bioprinters spray the bioink drop by drop from the nozzle by electrically heating the bioink cartridge. In many studies, it has been reported that this local heating has no significant negative effect on the biological molecules in the bioink. On the other hand, inkjet 3D bioprinters with piezoelectric systems have a piezoelectric crystal in the bioprinter card slot to periodically separate the bioink into droplets. When a voltage is applied to this crystal, acoustic waves form, and pressure is applied to the cartridge. The biomarker in the cartridge is sprayed from the nozzle with this pressure (Vurat 2021).

9.3.5.2 Microextrusion 3D Bioprinting

Fused deposition modeling (FDM), also defined as microextrusion-based 3D bioprinting, is an additive manufacturing process that rotates from around the accumulation of a substance in successive layers to form the ideal three-dimensional structure. In comparison to inkjet and laser-assisted bioprinters, microextrusion bioprinters can work with a broader range of viscous bioinks (Seyedmahmoud et al. 2020; Cui et al. 2012a, b; Chang et al. 2011).

The most common and low-cost printing techniques, especially nonbiological ones, are based on the principle of microextrusion. Microextrusion-based systems usually consist of one or more pulley systems that can move along the *x*, *y*, and *z* axes; a temperature-controlled material processing and dispensing system; and a light (UV) source to illuminate the deposition area or activate the photoinitiator (Vurat 2021). Microextrusion bioprinting technology is based on printing ink by mechanical force or pneumatically (with gas or pressure). In addition to being compatible with high-viscosity materials, this method provides the advantage of printing with high cell density. It is low cost and allows easy printing. As the viscosity decreases, the increased pressure negatively affects cell viability (Akkuş et al. 2020; Hacıoglu et al. 2018). Material cross-linking strategies used in extrusion printing are generally classified into three categories: (1) chemical cross-linking, such as sodium alga acid and chitosan; (2) photo-cross-linking, such as GelMA; and (3) physical cross-linking, such as agarose (Du 2018).

9.3.5.3 Laser-Assisted 3D Bioprinting

Laser-assisted bioprinting, simply identified as laser-induced forward transfer, is a droplet-based system (LIFT). Laser-based 3D bioprinters (LTB), based on the principle of laser excitation, were developed primarily for transferring metals. However, peptides have been successfully applied to biological materials, such as deoxyribonucleic acid (DNA) and cells in later times. Although LTB is less common than inkjet or microextrusion-based bioprinting, it is used increasingly for tissue and organ engineering applications. A typical LTB contains a laser-energy-absorbing layer, a pulsed laser beam, and a focusing system (Murphy and Atala 2014). Further, many factors influence the resolution of the structures produced in LTB, including laser fluence (transmitted energy per unit area), the surface tension of the bioink, the air gap between the slide and the substrate, and the viscosity of the biological substance (Yilmaz et al. 2019; Vurat 2021).

In LTB-based 3D bioprinters, the principle of operation entails applying a high-energy pulsed laser to a donor slide coated with the bioink to be printed to ensure the local spraying of small droplets. The laser light is mainly focused on a transparent laser substrate (such as glass or quartz), which is coated with a thin metal layer, such as gold or titanium, that absorbs light energy and promotes bioink transfer. Pressure is created at this step, and a small droplet is pushed toward the lower platform. Cells are printed using a laser beam, which vibrates at controlled speeds in laser-assisted bioprinting (Noh et al. 2017). Unlike inkjet bioprinting, this method can print materials with various viscosities. Since there is no nozzle in this method, there is no possibility of nozzle blockage, which is one of the problems encountered in other techniques (Akkuş et al. 2020). As a result, the viscosity on a broad scale, such as 1 to 300 MPa/s is suitable for printing biomaterials with their values (Vurat 2021; Murphy and Atala 2014). Moreover, the systems can also print bioinks with a cell density of 10^8 cells/mL at high resolutions and speeds of up to 1.6 mm/s using a laser pulse repetition rate of 5 kHz (Vurat 2021). The disadvantages of this method are that it is quite expensive and the gelling speed must be high for high resolution (Akkuş et al. 2020).

9.3.5.4 Stereolithography (SLA)

To develop 3D shapes from computer-aided design (CAD) data, a stereolithography machine (SLA-250; 3D Systems, Valencia, CA) was developed in the early 1980s. Nowadays, many 3D printers and current bioprinters accept STL files as input. Using the 3D Systems software, this file is first examined and, if necessary, modified. The device utilizes a moving helium-cadmium (HeCd) laser to produce a 250 μm UV light spot on top of the polymer vat. An optical scanning system is applied to manage the movement of this spot. The laser is permitted to finish 8–10 tracks per layer with an intensity of 14–16 mW/cm^2 at a wavelength of 365 nm, which allows the polymerization of its solution (Dhariwala et al. 2004; Yilmaz et al. 2019). The system's main components are a reservoir filled with a photosensitive polymer solution or resin, an *x*-*y*-axis controlled laser, and a fabrication stage with *z*-axis control. In summary, stereolithography is a laser-assisted production that photopolymerizes the surface of a photosensitive polymer bath using an ultraviolet (UV) laser (Yilmaz et al. 2019).

9.4 Transition from 3D Printing to 4D Printing

Four-dimensional bioprinting is a cutting-edge additive processing technique that has the inherent potential of manufacturing de novo living tissue structures that can be designed to alter different mechanical properties (Esworthy et al. 2019). To design a range of biological constructs, including bone, blood vessels, liver, and also heart tissue, many 3D bioprinting techniques are used. But, 3D bioprinting has a key disadvantage in that it just assesses the starting status a printed object and regards it to be artificial and immobile. Natural restoration entails complex 3D structures, microarchitectures, extracellular matrix compositions, and the formation of tissue with specific features obtained due to variations in tissue orientation. The majority of orientational formation is attributed to built-in strategies that reply to inherent stimuli so that 3D bioprinting cannot simulate (Wan et al. 2020; Arslan-Yildiz et al. 2016; Yu et al. 2019; Cui et al. 2016).

The first four-dimensional (4D) printing, capability of multi-material prints with over time, was developed at the Massachusetts Institute of Technology (MIT) in 2014 (Tibbits 2014). The technique has rapidly been practiced in the tissue engineering field, and the theory of time can be incorporated as the fourth dimension within 3D bioprinting technology, which led to the invention of 4D bioprinting. Four-dimensional bioprinting can be utilized to produce numerous 3D designed biologically active structures having the capability of robust orientational changes to adapt to new favored stimulation over time by using stimuli-responsive materials, describing the drawbacks of 3D bioprinting (Wan et al. 2020; Gao et al. 2016; Li et al. 2016a, b). The 4D-printed structures can alter over time in response to different stimuli and adjust to the native niches of fault fields, opening up new ways for tissue engineering, especially bone. To identify existing problems in bone tissue engineering, a sequence of accelerated 4D approaches has been suggested. In bone tissue engineering, shape recovery polymers that respond to different stimuli have been extensively researched as potential injectable hydrogels and appropriate scaffolds (Saravanan et al. 2019; Senatov et al. 2016; Graham et al. 2019; Wan et al. 2020).

The shape-conversion capability of 4D-printed bone tissue structures could satisfy the requirement for individualized bone healing, especially in the context of random bone fractures. The tensile performance of 4D-printed structures can also be regulated via a conditioned cross-linking of stimuli-responsive components (Wan et al. 2020; Suo et al. 2018). In bone tissue engineering, different cells, growth factors, or inorganic nanoparticles (hydroxyapatite and bioactive glass cement, such as calcium phosphate) can serve as support, and a heat-sensitive polysaccharide-based hydrogel that can be injected has been improved. Hydroxypropyl guar-graft-poly (*N*-vinylcaprolactam), hydroxybutyl chitosan, hydroxypropyl methylcellulose, such that the modified bio-based polymers, have a lower critical solution temperature among optimal room temperature and room temperature and can return into a gel state at skin temperature. Poly(N-isopropylacrylamide) (pNIPAM) (a conventional thermoresponsive material) has been combined with hyaluronic acid and chitosan to construct an injectable hydrogel for bone healing (Wan et al. 2020; Yang et al. 2014; Chen et al. 2013). The mechanical properties of injectable hydrogels were enhanced by the addition of mineral components like nano-hydroxyapatite, bioactive glass, and tricalcium phosphate. Inorganic-organic injectable hydrogels have appropriate rheological and in vivo properties and enhanced alkaline enzymatic activity and calcium formation in osteoblast cells for support (Wan et al. 2020; Azevedo et al. 2014; Dessi et al. 2013). PLA/hydroxyapatite porous scaffolds with increased rates of shape recovery potential could be employed as self-fitting implantable devices to restore minor bone flaws.

Renewable bio-smart scaffolds have positive shape memory effects and shape recovery at body temperature and they have has been prepared by Miao et al. (2014). In the study, PCL and cross-linkers containing predetermined amounts of castor oil were used. Meanwhile, the researchers utilized 3D laser printing to produce a biomaterial temperature-responsive shape-memory scaffold made of epoxidized acrylate materials based on renewable soybean oil. The produced biomaterials showed good mechanical strength, shape-memory effect, and biodegradable properties (Wan et al. 2020). Magneto-responsive polymer structures are polymer networks that have been physically or chemically modified with magnetic nanoparticles (MNP) made of nickel (Ni), cobalt (Co), and iron (Fe), and/or their oxides, as well. Magneto-responsive materials' possibilities in the biomedical field have been indicated in a variety of specific target pharmaceutical applications, where they provide minimally intrusive, regionally efficient, and controlled treatment

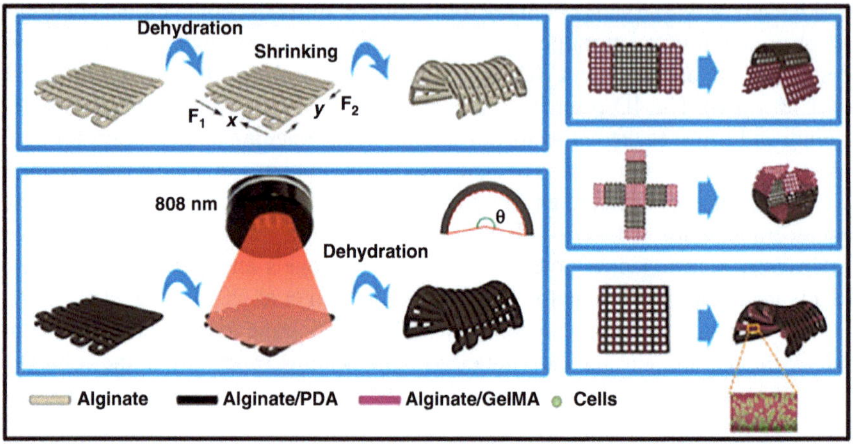

Fig. 9.11 Four-dimensional PED-cell-laden bilayered scaffolds, which transformed into saddle-like architecture upon NIR stimulation. The scaffolds were fabricated through cell-laden GelMA/alginate-based and PDA/alginate-based layers, and PDA/alginate instigated a shape morphing effect upon NIR stimulus (Luo et al. 2019). (Adapted with permission) (Arif et al. 2022)

activity. In this regard, iron(III)oxide (Fe_3O_4) nanoparticles involving mesoporous bioactive glass/poly(ε-caprolactone (Fe_3O_4/MBG/PCL) are examples of 3D-printed polymeric magneto-responsive structures used in tissue engineering applications (Zhang et al. 2014). In addition, PCL/iron-doped hydroxyapatite (PCL/FeHA) nanocomposite scaffolds and iron(III)oxide/poly(ethylene glycol diacrylate) (PEGDA) magneto scaffolds can also be given as examples (Tamay et al. 2019; D'Amora et al. 2017; De Santis et al. 2015).

Polymers of alkaline monomers act as cationic polymers in pH-responsive systems at acidic conditions, while polymers of acidic monomers act as anionic polymers at alkaline conditions. Synthetic pH-responsive polymers that are biocompatible and biodegradable involve poly(histidine) (PHIS), poly(acrylic acid) (PAA), poly(L-glutamic acid) (PGA), and poly(aspartic acid) (PASA), whereas natural pH-responsive polymers involve dextran, hyaluronic acid, alginic acid, chitosan, and gelatin (Tamay et al. 2019; Dutta and Cohn 2017; Kocak et al. 2017). Humidity responsiveness is a natural event with countless examples. Systems made of these substances can convert moisture sorption or desorption into driving forces for mobility. Some humidity-responsive materials that have been investigated include poly(ethylene glycol) diacrylate (PEGDA) cellulose (Mulakkal et al. 2018) and polyurethane copolymers (Tamay et al. 2019). A study with GelMA/alginate-based and poly (dopamine) (PDA)/alginate-based bioinks to create cell-laden scaffolds that change shape in response to near-infrared (NIR) stimuli was performed (Fig. 9.11). The overview of publications related to 3D and 4D between 2008–2019 is shown by years in Fig. 9.12.

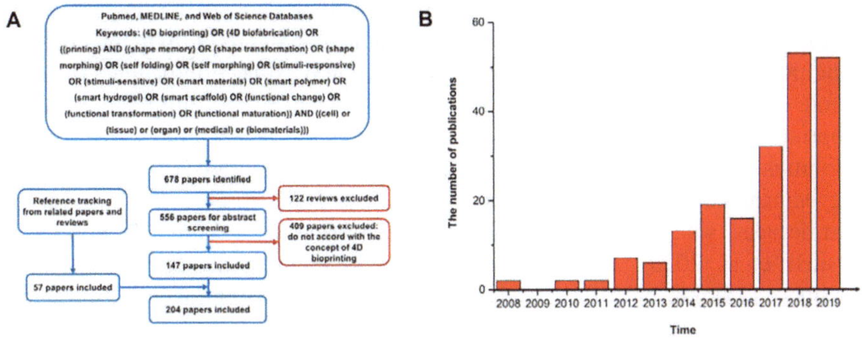

Fig. 9.12 Overview of current publications on 4D bioprinting. (**a**) The search strategy for viewing current publications on 4D bioprinting on PubMed, MEDLINE, and Web of Science databases (until 30 September 2019). (**b**) Statistics on the number of publications in recent years (Wan et al. 2020)

9.5 Challenges, Future Directions, and Conclusions

Bioprinting is still in its early stages, but there have been some notable successes in the formation of transplanted conceptual structures for a wide range of tissues. One of the most challenging aspects of 3D bioprinting is developing bioinks that are appropriate for every type of tissue and have adequate physiological, biological, and also mechanical features. The advancement and engineering of novel bioinks or bio-derived material compositions continue to be significant areas of concern and research. Further studies should be performed in developing new matrices and designs to assess and monitor the properties and processes of various bioink materials for this purpose. Bioprinting endeavors toward improved resolution, speed, and biocompatibility. Bioprinting has become popular the variety of suitable materials and methods for material accumulation with higher precision (Seyedmahmoud et al. 2020; Leberfinger et al. 2017). Vascularization is another important limitation in tissue engineering and bioprinting 3D tissues with proper functionalities. The presence of proper vascularization in bioprinted structures is essential for the long-term functionality of 3D-bioprinted tissues. Cells could die of hypoxia and exhibit stagnant growth attributable to waste and contaminant formation if sufficient cellular perfusion is not provided. The ability to effectively build a multiscale perfused vascular network and then promote its vascularization via mechanical or chemical activation is the foundation for biofabricating quite bulky tissues (Gu et al. 2019; Seyedmahmoud et al. 2020). Classical 3D bioprinting systems have mainly been used to develop, engineer 3D bioprinted structures in vitro pre—and implantation of the structure into the body. However, in vitro bioprinting methodologies could confront some logistical problems in terms of clinical applicability, such as the following: (1) 3D bioprinted structures are frequently sensitive, and inner micro-features could be damaged throughout transport from the fabrication condition to the operating condition; (2) extreme sanitized condition is needed; and (3) the bioprinted structure must be modified and trimmed before being implanted. This last problem exists when the configuration of the bioprinted structure differs from the real size of the defect due to the limited resolution ability of the CT and MRI scans used to develop the structure (Seyedmahmoud et al. 2020; Campbell and Weiss 2007; Li et al. 2015). Over the past decades, substantial development has been made in adjusting 3D bioprinting (Osidak et al. 2020). A comprehensive understanding of the effects of system parameters, bioink properties, and cellular structures on printing results offers a significant aid for the creation of new therapeutic bioinks (Ng et al. 2017; Chua et al. 2021). Despite considerable advances, bioprinting has many major challenges, such as the flexibility and function facilitates of printed biomaterials, and the ideal cell sources for innervation, vascularization, and development of printable structures. Materials with cell compatibility and appropriate mechanical properties, such as suitable modifications, the utilization of reinforced biomaterials, and protein-based and other functional group conjugation, must be specifically chosen to enable cell seeding and growth. Decellularized ECM could also be a favorable bioprinting scaffold. Various perspectives on 3D bioprinting have benefits and drawbacks. To resolve the troubles of bioprinting tissues with different properties, bioprinting methods with various mechanisms must be integrated. Moreover, appropriate stem cells, basic or mature cells, must be seriously evaluated in bioprinting to deliver proper functioning of the printable tissue (Huang et al. 2017).

Even though 3D printing has been a hot topic for many years, bioprinting is much more recent, and this field is growing rapidly. The system shows the ability and flexibility to create a variety of living systems with limited or no negative effects. This process is adaptable for the fabrication of both vascular and vascular tissues with 2D and 3D structures for tissue engineering. Nonetheless, in vivo evaluations of the process are now in their early stages, with an incredible deficiency in biological and mechanical characteristics to incorporate between printable structures and native tissue. But nevertheless, owing to its fast accuracy and continuous repeatability, bioprinting has displayed numerous opportunities in pharmaceutical research and clinical trials (Huang et al. 2017). Despite all the difficulties, however, bioprinting has a significant impact on the industry. To overcome all of the obstacles in this strongly interdisciplinary discipline, it is required to hire specialists from diverse areas such as material and biological sciences, computer engineering, and pharmaceutics. The advancement of bioink-based material applications is changed stem cell transplantation, polymer science, and fast manufacturing capabilities (Hacıoglu et al. 2018). As a result, 3D bioprinting provides the potential of manufacturing organs and, eventually, resolves the emergency of organ donor shortfall for transplantation (Huang et al. 2017).

References

Abouna GM (2008) Organ shortage crisis: problems and possible solutions. In: Transplantation proceedings, vol 40, No. 1. Elsevier, pp 34–38

Aduba DC, Margaretta ED, Marnot AEC, Heifferon KV, Surbey WR, Chartrain NA, Whittington AR, Long TE, Williams CB (2019) Vat photopolymerization 3D printing of acid-cleavable PEG-methacrylate networks for biomaterial applications. Mater Today Commun 19:204–211

Ahmed TAE, Dare EV, Hincke M et al (2008) Fibrin: a versatile scaffold for tissue engineering applications. Tissue Eng B Rev 14(2):199–215. https://doi.org/10.1089/ten.teb.2007.0435

Ahn S, Lee H, Bonassar LJ, Kim G (2012) Cells (MC3T3-E1)-laden alginate scaffolds fabricated by a modified solid-freeform fabrication process supplemented with an aerosol spraying. Biomacromolecules 13:2997–3003

Akkuş HG, Günaydin BE, Ustundag CB (2020) DOKU MÜHENDİSLİĞİNDE 3 BOYUTLU BİYO-BASKI İÇİN BİYOFONKSİYONEL MÜREKKEPLER. Int J 3D Print Technol Dig Ind 4(3):285–294

Alcantar NA, Aydil ES, Israelachvili JN (2000) Polyethylene glycol–coated biocompatible surfaces. J Biomed Mater Res 51:343–351

Aldana AA, Valente F, Dilley R, Doyle B (2021) Development of 3D bioprinted GelMA-alginate hydrogels with tunable mechanical properties. Bioprinting 21:e00105

An J, Teoh JEM, Suntornnond R, Chua CK (2015) Design and 3D printing of scaffolds and tissues. Engineering 1(2):261–268

Andreassi A, Bilenchi R, Biagioli M, D'Aniello C (2005) Classification and pathophysiology of skin grafts. Clin Dermatol 23:332–337

Arif ZU, Khalid MY, Ahmed W, Arshad H (2022) A review on four-dimensional bioprinting in pursuit of advanced tissue engineering applications. Bioprinting 2022:e00203

Arkel RV, Amis A (2013) Basics of orthopaedic biomechanics. Orthop Trauma 27(2):67–75

Arslan-Yildiz A, El Assal R, Chen P, Guven S, Inci F, Demirci U (2016) Towards artificial tissue models: past, present, and future of 3D bioprinting. Biofabrication 8(1):014103

Aschenbrenner E et al (2013) Using the polymeric ouzo effect for the preparation of polysaccharide-based nanoparticles. Langmuir 29(28):8845–8855

Ashammakhi N, Kaarela O (2017) Three-dimensional bioprinting can help bone. J Craniofac Surg 00:1

Askari M, Naniz MA, Kouhi M, Saberi A, Zolfagharian A, Bodaghi M (2021) Recent progress in extrusion 3D bioprinting of hydrogel biomaterials for tissue regeneration: a comprehensive review with focus on advanced fabrication techniques. Biomater Sci 9(3):535–573

Aslam M, Kalyar MA, Raza ZA (2018) Polyvinyl alcohol: a review of research status and use of polyvinylalcohol based nanocomposites. Polym Eng Sci 58:2119–2132

Azevedo AS, Sa MJ, Fook MV, Neto PI, Sousa OB, Azevedo SS, Teixeira MW, Costa FS, Araujo AL (2014) Use of chitosan and beta-tricalcium phosphate, alone and in combination, for bone healing in rabbits. J Mater Sci Mater Med 25(2):481–486

Babensee JE, Anderson JM, McIntire LV, Mikos AG (1998) Host response to tissue engineered devices. Adv Drug Deliv Rev 33(1–2):111–139

Bedir T, Ulag S, Ustundag CB, Gunduz O (2020) 3D bioprinting applications in neural tissue engineering for spinal cord injury repair. Mater Sci Eng C 110:110741

Bee S-L, Hamid ZAA, Mariatti M, Yahaya BH, Lim K, Bee S-T, Sin LT (2018) Approaches to improve therapeutic efficacy of biodegradable PLA/PLGA microspheres: a review. Polym Rev 58:495–536

Bejoy AM, Makkithaya KN, Hunakunti BB, Hegde A, Krishnamurthy K, Sarkar A et al (2021) An insight on advances and applications of 3d bioprinting: a review. Bioprinting 24:e00176

Bhavsar MD, Amiji MM (2008) Development of novel biodegradablepolymeric nanoparticles-in-microsphere formulation for local plasmid DNA delivery in the gastrointestinal tract. AAPS PharmSciTech 9(1):288–294

Boyan BD, Hummert TW, Dean DD, Schwartz Z (1996) Role of material surfaces in regulating bone and cartilage cell response. Biomaterials 17(2):137–146

Bružauskaitė I, Bironaitė D, Bagdonas E, Bernotienė E (2016) Scaffolds and cells for tissue regeneration: different scaffold pore sizes-different cell effects. Cytotechnology 68(3):355–369

Campbell PG, Weiss LE (2007) Tissue engineering with the aid of inkjet printers. Expert Opin Biol Ther 7:1123–1127

Cavallo JA et al (2015) Remodeling characteristics and collagen distributions of biologic scaffold materials biopsied from postmastectomy breast reconstruction sites. Ann Plast Surg 75(1):74

Chae S, Cho DW (2022) Biomaterial-based 3D bioprinting strategy for orthopedic tissue engineering. Acta Biomater 156:4–20

Chan BP, Leong KW (2008) Scaffolding in tissue engineering: general approaches and tissue-specific considerations. Eur Spine J 17:S467–S479

Chang CC, Boland ED, Williams SK, Hoying JB (2011) Direct-write bioprinting three-dimensional biohybrid systems for future regenerative therapies. J Biomed Mater Res B Appl Biomater 98: 160–170

Chartrain NA, Gilchrist KH, Ho VB, Klarmann GJ (2022) 3D bioprinting for the repair of articular cartilage and osteochondral tissue. Bioprinting 2022:e00239

Chen XL, Shi QN (2013) Research on sol-gel transition process of gelatin. In: Advanced materials research, vol 683. Trans Tech Publications Ltd., pp 474–478

Chen JP, Tsai MJ, Liao HT (2013) Incorporation of biphasic calcium phosphate microparticles in injectable thermoresponsive hydrogel modulates bone cell proliferation and differentiation. Colloids Surf B Biointerfaces 110:120–129

Chen F-M, Liu X, Polym P, ve Author, S. (2016) Advancing biomaterials of human origin for tissue engineering HHS Public Access Author manuscript. Prog Polym Sci 53:86–168. https://doi.org/10.1016/j.progpolymsci.2015.02.004

Cheng YL, Chen F (2017) Preparation and characterization of photocured poly(epsilon-caprolactone)diacrylate/poly(ethylene glycol) diacrylate/chitosan for photopolymerization-type 3D printing tissue engineering scaffold application. Mater Sci Eng C Mater Biol Appl 81:66–73

Chiou BS, AvenaBustillos RJ, Bechtel PJ et al (2008) Cold water fish gelatin films: effects of cross-linking on thermal, mechanical, barrier, and biodegradation properties. Eur Polym J 44:3748–3753

Chowdhury SR, Lokanathan Y, Xian LJ, Busra FM, Yazid MD, Sulaiman N et al (2020) 3D printed bioscaffolds for developing tissue-engineered constructs. In: Design and manufacturing. IntechOpen

Chua CK, Yeong WY, Low HY, Tran T, Tan HW (2021) 3D printing and additive manufacturing of electronics: principles and applications

Church D, Elsayed S, Reid O, Winston B, Lindsay R (2006) Burn wound infections. Clin Microbiol Rev 19(2):403–434

Colton CK (1995) Implantable biohybrid artificial organs. Cell Transplant 4(4):415–436

Cornelissen DJ, Faulkner-Jones A, Shu W (2017) Current developments in 3D bioprinting for tissue engineering. Curr Opin Biomed Eng 2:76–82

Cubo N, Garcia M, Cañizo JF, Velasco D, Jorcano JL (2017) 3D bioprinting of functional human skin: production and in vivo analysis. Biofabrication 9:015006

Cui X, Boland T (2009) Human microvasculature fabrication using thermal inkjet printing technology. Biomaterials 30:6221–6227

Cui X et al (2012a) Direct human cartilage repair using three-dimensional bioprinting technology. Tissue Eng A 18:1304–1312

Cui X, Boland T, D'Lima DD, Lotz MK (2012b) Thermal inkjet printing in tissue engineering and regenerative medicine. Recent Pat Drug Deliv Formul 6:149–155

Cui H, Nowicki M, Fisher JP, Zhang LG (2016) 3D Bioprinting for organ regeneration. Adv Healthc Mater 6(1):1601118

D'Amora U, Russo T, Gloria A, Rivieccio V, D'Antò V, Negri G et al (2017) 3D additive-manufactured nanocomposite magnetic scaffolds: effect of the application mode of a time-dependent magnetic field on hMSCs behavior. Bioact Mater 2:138–145

Das S, Pati F, Chameettachal S, Pahwa S, Ray AR, Dhara S, Ghosh S (2013) Enhanced redifferentiation of chondrocytes on microperiodic silk/gelatin scaffolds: toward tailor-made tissue engineering. Biomacromolecules 14:311–321

Dávila JL, Freitas MS, InforcattiNeto P, Silveira ZC, Silva JVL, d'Ávila MA (2016) Fabrication of PCL/β-TCP scaffolds by 3D mini-screw extrusion printing. J Appl Polym Sci 133(15):43031. 9 pp

De Santis R, D'Amora U, Russo T, Ronca A, Gloria A, Ambrosio L (2015) 3D fibre deposition and stereolithography techniques for the design of multifunctional nanocomposite magnetic scaffolds. J Mater Sci Mater Med 26:250

DeLustro F, Condell RA, Nguyen MA, McPherson JM (1986) A comparative study of the biologic and immunologic response to medical devices derived from dermal collagen. J Biomed Mater Res 20(1):109–120

Deng M, Tan J, Hu C, Hou T, Peng W, Liu J, Yu B, Dai Q, Zhou J, Yang Y et al (2020) Modification of PLGA scaffold by MSC-derived extracellular matrix combats macrophage inflammation to initiate bone regeneration via TGF-beta-induced protein. Adv Healthc Mater 9:e2000353

Derakhshanfar S, Mbeleck R, Xu K, Zhang X, Zhong W, Xing M (2018) 3D bioprinting for biomedical devices and tissue engineering: a review of recent trends and advances. Bioactive Mater 3(2):144–156

Dessi M, Borzacchiello A, Mohamed TH, Abdel-Fattah WI, Ambrosio L (2013) Novel biomimetic thermosensitive beta-tricalcium phosphate/chitosan-based hydrogels for bone tissue engineering. J Biomed Mater Res A 101(10):2984–2993

Dhandayuthapani B, Yoshida Y, Maekawa T, Kumar DS (2011) Polymeric scaffolds in tissue engineering application: a review. Int J Polym Sci 19:290602

Dhariwala B, Hunt E, Boland T (2004) Rapid prototyping of tissue-engineering constructs, using photopolymerizable hydrogels and stereolithography. Tissue Eng 10:1316–1322

Di Luca A, Ostrowska B, Lorenzo-Moldero I, Lepedda A, Swieszkowski W, Van Blitterswijk C, Moroni L (2016) Gradients in pore size enhance the osteogenic differentiation of human mesenchymal stromal cells in three dimensional scaffolds. Sci Rep 6(1):22898

Diamantides N, Dugopolski C, Blahut E, Kennedy S, Bonassar LJ (2019) High density cell seeding affects the rheology and printability of collagen bioinks. Biofabrication 11:045016

Ding YW, Zhang XW, Mi CH, Qi XY, Zhou J, Wei DX (2022) Recent advances in hyaluronic acid-based hydrogels for 3D bioprinting in tissue engineering applications. Smart Mater Med 14(4): 839

Du X (2018) 3D bio-printing review. In: IOP conference series: materials science and engineering, vol 301, no 1, p 012023. IOP Publishing

Dutta S, Cohn D (2017) Temperature and pH responsive 3D printed scaffolds. J Mater Chem B 5: 9514–9521

Esworthy TJ, Miao S, Lee SJ, Zhou X, Cui H, Zuo YY, Zhang LG (2019) Advanced 4D-bioprinting technologies for brain tissue modeling and study. Int J Smart Nano Mater 10(3):177–204

Falcone SJ, Palmeri D, Berg RA (2006) Biomedical applications of hyaluronic acid. ACS Publications, pp 2006–2014

Fedorovich NE, De Wijn JR, Verbout AJ, Alblas J, Dhert WJ (2008) Three-dimensional fiber deposition of cell-laden, viable, patterned constructs for bone tissue printing. Tissue Eng A 14: 127–133

Fontes AB, Marcomini RF (2020) 3D bioprinting: a review of materials, processes and bioink properties. J Eng Exact Sci 6(5):0617–0639

Gao G et al (2014) Bioactive nanoparticles stimulate bone tissue formation in bioprinted three-dimensional scaffold and human mesenchymal stem cells. Biotechnol J 9:1304–1311

Gao B, Yang Q, Zhao X, Jin G, Ma Y, Xu F (2016) 4D bioprinting for biomedical applications. Trends Biotechnol 158(9):746–756

Gentile P, Chiono V, Carmagnola I, Hatton PV (2014) An overview of poly(lactic-co-glycolic) acid (PLGA)-based biomaterials for bone tissue engineering. Int J Mol Sci 15(3):3640–3659

Gholami P, Ahmadi-Pajouh MA, Abolftahi N, Hamarneh G, Kayvanrad M (2017) Segmentation and measurement of chronic wounds for bioprinting. IEEE J Biomed Health Inform 22 (4):1269–1277

Graham S, Marina PF, Blencowe A (2019) Thermoresponsive polysaccharides and their thermoreversible physical hydrogel networks. Carbohydr Polym 207:143–159

Gruene M et al (2010) Laser printing of stem cells for biofabrication of scaffold-free autologous grafts. Tissue Eng C Methods 17:79–87

Gu BK, Park SJ, Kim MS, Kang CM, Kim JI, Kim CH (2013) Fabrication of sonicated chitosan nanofiber mat with enlarged porosity for use as hemostatic materials. Carbohy Polym 97:65–73

Gu BK, Choi DJ, Park SJ, Kim YJ, Kim CH (2018) 3D bioprinting technologies for tissue engineering applications. Adv Exp Med Biol 1078:15–28

Gu Z, Fu J, Lin H, He Y (2019) Development of 3D bioprinting: from printing methods to biomedical applications. Asian J Pharm Sci 15(5):529–557

Guillemot F, Souquet S, Lopez M et al (2010) High-throughput laser printing of cells and biomaterials for tissue engineering. Acta Biomater 6(7):2494–2500

Guvendiren M, Molde J, Soares RMD, Kohn J (2016) Designing bio-materials for 3D printing. ACS Biomater Sci Eng 2:1679–1693

Hacıoglu A, Yılmazer H, Ustundag CB (2018) 3D printing for tissue engineering applications. Politeknik Dergisi 21(1):221–227. https://doi.org/10.2339/politeknik.389596

Hahn MS, Teply BA, Stevens MM, Zeitels SM, Langer R (2006) Collagen composite hydrogels for vocal fold lamina propria restoration. Biomaterials 27(7):1104–1109

Hajiali F, Tajbakhsh S, Shojaei A (2018) Fabrication and properties of polycaprolactone composites containing calcium phosphate-based ceramics and bioactive glasses in bone tissue engineering: a review. Polym Rev 58(1):164–207

Harriger MD, Supp AP, Warden GD, Boyce ST (1997) Glutaraldehyde crosslinking of collagen substrates inhibits degradation in skin substitutes grafted to athymic mice. J Biomed Mater Res 35:137–145

Hockaday LA, Kang KH, Colangelo NW, Cheung PYC, Duan B, Malone E, Wu J, Girardi LN, Bonassar LJ, Lipson H, Chu CC, Butcher JT (2012) Rapid 3D printing of anatomically accurate and mechanically heterogeneous aortic valve hydrogel scaffolds. Biofabrication 4(3):005–035

Hospodiuk M, Dey M, Sosnoski D, Ozbolat IT (2017) The bioink: a comprehensive review on bioprintable materials. Biotechnol Adv 35:217–239

Hsieh F-Y et al (2015) 3D bioprinting of neural stem cell-laden thermoresponsive biodegradable polyurethane hydrogel and potential in central nervous system repair. Biomaterials 71:48–57

Huang J, Mao Y, Millis JM (2008) Government policy and organ transplantation in China. Lancet 372(9654):1937–1938

Huang Y, Zhang XF, Gao G, Yonezawa T, Cui X (2017) 3D bioprinting and the current applications in tissue engineering. Biotechnol J 12(8):1600734

Huang YH, Jakus AE, Jordan SW, Dumanian Z, Parker K, Zhao L, Patel PK, Shah RN (2019) Three-dimensionally printed hyperelastic bone scaffolds accelerate bone regeneration in critical-size calvarial bone defects. Plast Reconstr Surg 143:1397–1407

Hull CW (1986) U.S. Patent No. 4,575,330. U.S. Patent and Trademark Office, Washington, DC

Hutmacher DW (2000) Scaffolds in tissue engineering bone and cartilage. In: The biomaterials: Silver Jubilee Compendium, vol 21. https://doi.org/10.1016/B978-008045154-1.50021-6

Inzana J, Olvera D, Fuller SM, Kelly JP, Graeve O, Schwarz EM, Kates SL, Awad H (2014) 3D printing of composite calcium phosphate and collagen scaffolds for bone regeneration. Biomaterials 35(13):4026–4034

Iordache F (2019) Bioprinted scaffolds. In: Materials for biomedical engineering. Elsevier, pp 35–60

Izbudak B (2021) Osteokondral hasarların onarımına yönelik 3B-biyoyazıcı ile çift-tabakalı nanohibrit doku iskelelerinin geliştirilmesi. Master's Thesis, İstinye Üniversitesi/Sağlık Bilimleri Enstitüsü

Jakus AE, Rutz AL, Shah RN et al (2016a) Advancing the field of 3D biomaterial printing. Biomed Mater (Bristol) 11(1):1–11
Jakus AE, Rutz AL, Jordan SW, Kannan A, Mitchell SM, Yun C, Koube KD, Yoo SC, Whiteley HE, Richter CP et al (2016b) Hyperelastic "bone": a highly versatile, growth factor-free, osteoregenerative, scalable, and surgically friendly biomaterial. Sci Transl Med 8:358ra127
Jana S, Lerman A (2015) Bioprinting a cardiac valve. Biotechnol Adv 33(8):1503–1521
Jean J, Garcia-Perez ME, Pouliot R (2011) Bioengineered skin: the self-assembly approach. J Tissue Sci Eng S5:001
Jessop ZM, Al-Sabah A, Gardiner MD, Combellack E, Hawkins K, Whitaker IS (2017) 3D bioprinting for reconstructive surgery: principles, applications and challenges. J Plast Reconstr Aesthet Surg 70(9):1155–1170
Jose RR, Rodriguez MJ, Dixon TA, Omenetto F, Kaplan DL (2016) Evolution of bioinks and additive manufacturing technologies for 3D bioprinting. ACS Biomater Sci Eng 2(10): 1662–1678
Kakarla AB, Turek I, Kong C, Irving H (2022) Printable gelatin, alginate and boron nitride nanotubes hydrogel-based ink for 3D bioprinting and tissue engineering applications. Mater Des 213:110362
Karageorgiou V, Kaplan D (2005) Porosity of 3D biomaterial scaffolds andosteogenesis. Biomaterials 26(27):5474–5491
Katti DS, Lakshmi S, Langer R, Laurencin CT (2002) Toxicity, biodegradation and elimination of polyanhydrides. Adv Drug Deliv Rev 54(7):933–961
Kim JE, Kim SH, Jung Y (2016a) Current status of three-dimensional printing inks for soft tissue regeneration. Tissue Eng Regen Med 13(6):636–646
Kim YB, Lee H, Kim GH (2016b) Strategy to achieve highly porous/biocompatible macroscale cell blocks, using a collagen/genipin-bioink and an optimal 3D printing process. ACS Appl Mater Interfaces 8:32230–32240
Kim BS, Gao G, Kim JY, Cho D (2019) 3D cell printing of perfusable vascularized human skin equivalent composed of epidermis, dermis, and hypodermis for better structural recapitulation of native skin. Adv Healthc Mater 8:1801019
Kocak G, Tuncer C, Butun V (2017) pH-responsive polymers. Polym Chem 8:144–176
Koch L, Deiwick A, Schlie S, Michael S, Gruene M, Coger V, Zychlinski D, Schambach A, Reimers K, Vogt PM, Chichkov B (2012) Skin tissue generation by laser cell printing. Biotechnol Bioeng 109:1855–1863
Kruth JP, Leu MC, Nakagawa T (1998) Progress in additive manufacturing and rapid prototyping. CIRP Ann 47(2):525–540
Kuijpers AJ, van Wachem PB, van Luyn MJ et al (2015) In vivo compatibility and degradation of crosslinked gelatin gels incorporated in knitted Dacron. J Biomed Mater Res 51(1):136–145
Langer R, Vacanti JP (1993) Tissue engineering. Science 260:920–926
Leberfinger AN, Ravnic DJ, Dhawan A, Ozbolat IT (2017) Concise review: Bioprinting of stem cells for transplantable tissue fabrication. Stem Cells Transl Med 6:1940–1948
Lee V et al (2013) Design and fabrication of human skin by three-dimensional bioprinting. Tissue Eng C Methods 20:473–484
Lee J, Hong J, Kim W, Kim GH (2020) Bone-derived dECM/alginate bioink for fabricating a 3D cell-laden mesh structure for bone tissue engineering. Carbohydr Polym 250:116914
Li Q, Ma L, Gao C (2015) Biomaterials for in situ tissue regeneration: development and perspectives. J Mater Chem B 3:8921–8938
Li H, Liu S, Li L (2016a) Rheological study on 3D printability of alginate hydrogel and effect of graphene oxide. Int J Bioprint 2(2):54–66
Li YC, Zhang YS, Akpek A, Shin SR, Khademhosseini A (2016b) 4D bioprinting: the next-generation technology for biofabrication enabled by stimuli-responsive materials. Biofabrication 9(1):012001
Liu X, Won Y, Ma PX (2005) Surface modification of interconnected porous scaffolds. J Biomed Mater Res A 74(1):84–91

Luo Y, Lin X, Chen B, Wei X (2019) Cell-laden four-dimensional bioprinting using near-infrared-triggered shape-morphing alginate/polydopamine bioinks. Biofabrication 11(4):045019

Ma H, Xue L (2015) Carbon nanotubes reinforced poly(L-lactide) scaffolds fabricated by thermally induced phase separation. Nanotechnology 26:025701

Mandrycky C, Wang Z, Kim K, Kim DH (2016) 3D bioprinting for engineering complex tissues. Biotechnol Adv 34:422–434

Marin E, Rojas J, Ciro Y (2014) A review of polyvinyl alcohol derivatives: promising materials for pharmaceutical and biomedical applications. Afr J Pharm Pharmacol 8:674–684

Medina-Esquivel R, Freile-Pelegrin Y, Quintana-Owen P, Yáñez-Limón JM, Alvarado-Gil JJ (2008) Measurement of the sol–gel transition temperature in agar. Int J Thermophys 29: 2036–2045

Metcalfe AD, Ferguson MW (2007) Tissue engineering of replacement skin: the crossroads of biomaterials, wound healing, embryonic development, stem cells and regeneration. J R Soc Interface 4(14):413–437

Miao S, Wang P, Su Z, Zhang S (2014) Vegetable-oil-based polymers as future polymeric biomaterials. Acta Biomater 10(4):1692–1704

Miao S, Zhu W, Castro N, Leng JS, Zhang LG (2016) 4D printing hierarchy scaffolds with highly biocompatible smart polymers for tissue engineering applications. Tissue Eng C 22(10): 952–963

Michael S, Sorg H, Peck CT, Koch L, Deiwick A, Chichkov B, Vogt PM, Reimers K (2013) Tissue engineered skin substitutes created by laser-assisted bioprinting form skin-like structures in the dorsal skinfold chamber in mice. PLoS One 8(3):e57741

Mironov V, Kasyanov V, Drake C, Markwald RR (2008) Organ printing: promises and challenges. Regen Med 3:93

Mobaraki M, Ghaffari M, Yazdanpanah A, Luo Y, Mills DK (2020) Bioinks and bioprinting: a focused review. Bioprinting 18:00080

Mota C, Puppi D, Chiellini F, Chiellini E (2015) Additive manufacturing techniques for the production of tissue engineering constructs. J Tissue Eng Regen Med 9:174–190

Mukherjee P, Rani A, Saravanan P (2019) Polymeric materials for 3D bioprinting. In: Ahmad N, Gopinath R, Dutta R (eds) 3D printing technology in nanomedicine, pp 63–81

Mulakkal MC, Trask RS, Ting VP, Seddon AM (2018) Responsive cellulose-hydrogel composite ink for 4D printing. Mater Des 160:108–118

Munaz A, Vadivelu RK, John JS, Barton M, Kamble H, Nguyen NT (2016) Three-dimensional printing of biological matters. J Sci Adv Mater Device 1(1):1–17

Murphy SV, Atala A (2014) 3D bioprinting of tissues and organs. Nat Biotechnol 32(8):773–785

Nair LS, Laurencin CT (2007) Biodegradable polymers as biomaterials. Prog Polym Sci (Oxford) 32(8–9):762–798

Ng WL, Wang S, Yeong WY, Naing MW (2016) Skin bioprinting: impending reality or fantasy. Trends Biotechnol 34:689–699

Ng WL, Lee JM, Yeong WY, Naing MW (2017) Microvalve-based bioprinting–process, bio-inks and applications. Biomater Sci 5(4):632–647

Ngo TB, Spearman BS, Hlavac N, Schmidt CE (2020) Three-dimensional bioprinted hyaluronic acid hydrogel test beds for assessing neural cell responses to competitive growth stimuli. ACS Biomater Sci Eng 6(12):6819–6830

Noh S, Myung N, Park M, Kim S, Zhang SU, Kang HW (2017) 3D bioprinting for tissue engineering. In: Kim BW (ed) Clinical regenerative medicine in urology. Springer, Singapore, pp 105–123

Oh SH, Kang SG, Kim ES, Cho SH, Lee JH (2003) Fabrication and characterization of hydrophilic poly(lactic-co-glycolic acid)/poly(vinyl alcohol) blend cell scaffolds by melt-molding particulate-leaching method. Biomaterials 24:4011–4021

Ong CS, Yesantharao P, Huang CY, Mattson G, Boktor J et al (2018) 3D bioprinting using stem cells. Pediatr Res 83:223–231

Osidak EO, Karalkin PA, Osidak MS, Parfenov VA, Sivogrivov DE, Pereira F, Gryadunova AA, Koudan EV, Khesuani YD, Capital Ka CVA et al (2019) Viscoll collagen solution as a novel bioink for direct 3D bioprinting. J Mater Sci Mater Med 30:31

Osidak EO, Kozhukhov VI, Osidak MS, Domogatsky SP (2020) Collagen as bioink for bioprinting: a comprehensive review. Int J Bioprinting 6(3)

Ozbolat IT (2016) 3D bioprinting: fundamentals, principles and applications. Elsevier, p s.356

Ozbolat IT, Chen H, Yu Y (2014) Development of "Multi-arm Bioprinter" for hybrid biofabrication of tissue engineering constructs. Robot Comput Integr Manuf 30(3):295–304

Ozbolat IT, Peng W, Ozbolat V (2016) Application areas of 3D bioprinting. Drug Discov Today 21 (8):1257–1271

Parin FN, Terzioğlu P (2022) Electrospun porous biobased polymer mats for biomedical applications. In: Advanced functional porous materials: from macro to nano scale lengths, pp 539–586

Pekkanen AM, Mondschein RJ, Williams CB, Long TE (2017) 3D printing polymers with supramolecular functionality for biological, applications. Biomacromolecules 18(9):2669–2687

Peppas NA, Hilt JZ, Khademhosseini A, Langer R (2006) Hydrogels in biology and medicine: from molecular principles to bionanotechnology. Adv Mater 18:1345–1360

Pescosolido L, Schuurman W, Malda J, Matricardi P, Alhaique F, Coviello T et al (2011) Hyaluronic acid and dextran-based semi-IPN hydrogels as biomaterials for bioprinting. Biomacromolecules 12:1831–1838

Phillippi JA et al (2008) Microenvironments engineered by inkjet bioprinting spatially direct adult stem cells toward muscle- and bone-like subpopulations. Stem Cells 26:127–134

Poldervaart MT, Gremmels H, van Deventer K, Fledderus JO, Oner FC, Verhaar MC, Dhert WJ, Alblas J (2014) Prolonged presence of VEGF promotes vascularization in 3D bioprinted scaffolds with defined architecture. J Control Release 184:58–66

Powell HM, Boyce ST (2006) EDC cross-linking improves skin substitute strength and stability. Biomaterials 27:5821–5827

Ren X, Wang F, Chen C, Gong X, Yin L, Yang L (2016) Engineering zonal cartilage through bioprinting collagen type II hydrogel constructs with biomimetic chondrocyte density gradient. BMC Musculoskelet Disord 17:1–10

Sakai S, Hirose K, Taguchi K, Ogushi Y (2009) Injectable, in situ enzymatically gellable, gelatin derivative for drug delivery and tissue engineering. Biomaterials 30:3371–3377

Salah M, Tayebi L, Moharamzadeh K, Naini FB (2020) Three-dimensional bio-printing and bone tissue engineering: technical innovations and potential applications in maxillofacial reconstructive surgery. Maxillofac Plastic Reconstr Surg 42(1):1–9

Salgado CL, Sanchez EM, Zavaglia CA, Granja PL (2012) Biocompatibility and biodegradation of polycaprolactone-sebacic acid blended gels. J Biomed Mater Res A 100(1):243–251

Saravanan S, Vimalraj S, Thanikaivelan P, Banudevi S, Manivasagam G (2019) A review on injectable chitosan/beta glycerophosphate hydrogels for bone tissue regeneration. Int J Biol Macromol 121:38–54

Sarker MD, Naghieh S, McInnes AD, Ning L, Schreyer DJ, Chen X (2019) Bio-fabrication of peptide-modified alginate scaffolds: printability, mechanical stability and neurite outgrowth assessments. Bioprinting 14:e00045

Saroia J, Yanen W, Wei Q, Zhang K, Lu T, Zhang B (2018) A review on biocompatibility nature of hydrogels with 3D printing techniques, tissue engineering application and its future prospective. Bio-Design Manufacturing 1:265–279

Schacht K, Jungst T, Schweinlin M, Ewald A, Groll J, Scheibel T (2015) Biofabrication of cell-loaded 3D spidersilk constructs. Angew Chem Int Ed Engl 54:2816–2820

Senatov FS, Niaza KV, Zadorozhnyy MY, Maksimkin AV, Kaloshkin SD, Estrin YZ (2016) Mechanical properties and shape memory effect of 3D-printed PLA-based porous scaffolds. J Mech Behav Biomed Mater 57:139–148

Seyedmahmoud R, Mozetic P, Rainer A, Giannitelli SM, Basoli F, Trombetta M et al (2015) A primer of statistical methods for correlating parameters and properties of electrospun poly

(l-lactide) scaffolds for tissue engineering—Part 2: Regression. J Biomed Mater Res A 103(1): 103–114
Seyedmahmoud R, Messler MJ, Loboa EG (2020) 3D bioprinting technologies for tissue engineering: a mini review. HSOA J Stem Cells Res Dev Ther 6:046
Shafiee A, Atala A (2016) Printing technologies for medical applications. Trends Mol Med 22(3): 254–265
Sharif M, George E, Shepstone L et al (1995) Serum hyaluronic acid level as a predictor of disease progression in osteoarthritis of the knee. Arthrit Rheumatol 38(6):760
Shen M, Wang L, Gao Y, Feng L, Xu C, Li S et al (2022) 3D bioprinting of in situ vascularized tissue engineered bone for repairing large segmental bone defects. Mater Today Bio 16:100382
Sheridan R (2009) Closure of the excised burn wound: autografts, semipermanent skin substitutes, and permanent skin substitutes. Clin Plast Surg 36:643
Singh RS, Saini GK (2008) Pullulan-hyperproducing color variant strain of Aureobasidium pullulans FB-1 newly isolated from phylloplane of Ficus sp. Bioresour Technol 99(9):3896–3899
Singh RS, Singh H, Saini GK (2009) Response surface optimization of the critical medium components for pullulan production by Aureobasidium pullulans FB-1. Appl Biochem Biotechnol 152:42–53
Skardal A, Mack D, Kapetanovic E, Atala A, Jackson JD, Yoo J, Soker S et al (2012) Bioprinted amniotic fluid-derived stem cells accelerate healing of large skin wounds. Stem Cells Transl Med 1:792–802
Slaughter BV, Khurshid SS, Fisher OZ, Khademhosseini A, Peppas NA (2009) Hydrogels in regenerative medicine. Adv Mater 21(32–33):3307–3329
Sophia Fox AJ, Bedi A, Rodeo SA (2009) The basic science of articular cartilage: structure, composition, and function. Sports Health 1:461–468
Sowjanya JA, Singh J, Mohita T, Sarvanan S, Moorthi A, Srinivasan N, Selvamurugan N (2013) Biocomposite scaffolds containing chitosan/alginate/nano-silica for bone tissue engineering. Colloids Surf B Biointerfaces 109:294–300
Suo H, Zhang D, Yin J, Qian J, Wu ZL, Fu J (2018) Interpenetrating polymer network hydrogels composed of chitosan and photocrosslinkable gelatin with enhanced mechanical properties for tissue engineering. Mater Sci Eng C Mater Biol Appl 92:612–620
Tamay DG, Dursun Usal T, Alagoz AS, Yucel D, Hasirci N, Hasirci V (2019) 3D and 4D printing of polymers for tissue engineering applications. Front Bioeng Biotechnol 7:164
Tan B, Gan S, Wang X, Liu W, Li X (2021) Applications of 3D bioprinting in tissue engineering: advantages, deficiencies, improvements, and future perspectives. J Mater Chem B 9 (27):5385–5413
Tangsadthakun C, Kanokpanont S, Sanchavanakit N, Banaprasert T, Damrongsakkul S (2017) Properties of collagen/chitosan scaffolds for skin tissue engineering. J Metals Mater Miner 16(1):37–44
Tao H, Kaplan DL, Omenetto FG (2012) Silk materials–a road to sustainable high technology. Adv Mater 24:2824–2837
Taylor PM, Cass AE, Yacoub MH (2006) Extracellular matrix scaffolds for tissue engineering heart valves. Progr Pediatr Cardiol 21(2):219–225
Theus AS, Ning L, Hwang B, Gil C, Chen S, Wombwell A et al (2020) Bioprintability: physiomechanical and biological requirements of materials for 3d bioprinting processes. Polymers 12(10):2262
Thirumavalavan K, Manikkadan T, Dhanasekar R (2009) Pullulan production from coconut by-products by *Aureobasidium pullulans*. Afr J Biotechnol 8(2):254–258
Tibbits S (2014) 4D printing: multi-material shape change. Archit Des 84(1):116–121
Tirella A, Orsini A, Vozzi G, Ahluwalia A (2009) A phase diagram for microfabrication of geometrically controlled hydrogel scaffolds. Biofabrication 1:045002
Topuz M, Dikici B, Gavgali M, Yilmazer H (2018) A review on the hydrogels used in 3D bio-printing. Int J 3D Print Technol Dig Ind 2(2):68–75

Vega-Cázarez CA, Sánchez-Machado DI, López-Cervantes J (2018) Overview of electrospinned chitosan nanofiber composites for wound dressings. In: Chitin-chitosan-myriad functionalities in science and technology, pp 157–181

Vurat MT (2021) Üç-boyutlu Biyobasım için Yeni Kompozit Biyomürekkep Geliştirilmesi ve Karakterizasyonu. PhD Thesis, Ankara Üniversitesi/Fen Bilimleri Enstitüsü

Wan Z, Zhang P, Liu Y, Lv L, Zhou Y (2020) Four-dimensional bioprinting: current developments and applications in bone tissue engineering. Acta Biomater 101:26–42

Weadock KS, Miller EJ, Bellincampi LD, Zawadsky JP, Dunn MG (1995) Physical crosslinking of collagen fibers: comparison of ultraviolet irradiation and dehydrothermal treatment. J Biomed Mater Res 29:1373–1379

Wu Z, Li Q, Xie S, Shan X, Cai Z (2020) In vitro and in vivo biocompatibility evaluation of a 3D bioprinted gelatin-sodium alginate/rat Schwann-cell scaffold. Mater Sci Eng C Mater Biol Appl 109:110530

Xu T, Jin J, Gregory C, Hickman JJ, Boland T (2005) Inkjet printing of viable mammalian cells. Biomaterials 26:93–99

Yanagawa F, Sugiura S, Kanamori T (2016) Hydrogel microfabrication technology toward three dimensional tissue engineering. Regen Ther 3:45–57

Yang J, van Lith R, Baler K, Hoshi RA, Ameer GA (2014) A thermoresponsive biodegradable polymer with intrinsic antioxidant properties. Biomacromolecules 15(11):3942–3952

Ye WP, Du FS, Jin WH, Yang JY, Xu Y (1997) In vitro degradation of poly(caprolactone), poly (lactide) and their block copolymers: influence of composition, temperature and morphology. React Funct Polym 32(2):161–168

Yilmaz B, Tahmasebifar A, Baran ET (2019) Bioprinting technologies in tissue engineering. Curr Appl Pharm Biotechnol 171:279–319

You F, Eames BF, Chen X (2017) Application of extrusion-based hydrogel bioprinting for cartilage tissue engineering. Int J Mol Sci 18:8–14

Yu X, Tang X, Gohil SV, Laurencin CT (2015) Biomaterials for bone regenerative engineering. Adv Healthc Mater 4(9):1268–1285

Yu C, Ma X, Zhu W, Wang P, Miller KL, Stupin J, Koroleva-Maharajh A, Hairabedian A, Chen S (2019) Scanningless and continuous 3D bioprinting of human tissues with decellularized extracellular matrix. Biomaterials 194:1–13

Yu J, Park SA, Kim WD, Ha T, Xin YZ, Lee J, Lee D (2020) Current advances in 3D bioprinting technology and its applications for tissue engineering. Polymers 12(12):2958

Yüce Erarslan E (2021) Biyolojik Uygulamalarda Kullanılabilecek Polimerik Akıllı Malzemelerin Sentezi ve Karakterizasyonu PhD Thesis, İstanbul Üniversitesi/Lisansüstü Eğitim Enstitüsü

Zhang X, Zhang Y (2015) Tissue engineering applications of three-dimensional bioprinting. Cell Biochem Biophys 72:777–782

Zhang J, Zhao S, Zhu M, Zhu Y, Zhang Y, Liu Z et al (2014) 3D-printedmagnetic Fe3O4/MBG/PCL composite scaffolds with multifunctionality of bone regeneration, local anticancer drug delivery and hyperthermia. J Mater Chem B 2:7583–7595

Zhang J, Allardyce BJ, Rajkhowa R, Zhao Y, Dilley RJ, Redmond SL et al (2018) 3D printing of silk particle-reinforced chitosan hydrogel structures and their properties. ACS Biomater Sci Eng 4(8):3036–3046

Zhang J, Wehrle E, Adamek P, Paul GR, Qin XH, Rubert M, Muller R (2020) Optimization of mechanical stiffness and cell density of 3D bioprinted cell-laden scaffolds improves extracellular matrix mineralization and cellular organization for bone tissue engineering. Acta Biomater 114:307–322

Zhou Y, Fan Y, Chen Z, Yue Z, Wallace G (2021) Catechol functionalized ink system and thrombin-free fibrin gel for fabricating cellular constructs with mechanical support and inner micro channels. Biofabrication 14(1):015004

Zhu W, Ma X, Gou M, Mei D, Zhang K, Chen S (2016) 3D printing of functional biomaterials for tissue engineering. Curr Opin Biotechnol 40:103–112

Fatma Nur Parın has completed her bachelor's, master's, and Ph.D. degrees from Yalova University Polymer Engineering Department and Bursa Technical University Polymer Materials Engineering Department. She has worked at Bursa Technical University as an assistant professor. She has many papers in SCI-E and TR-indexed journals. Furthermore, she worked on TUBITAK projects and scientific research projects as a scholar, researcher, and consultant, respectively. Her research interests are porous material synthesis, drug release, emulsion polymerization, functional nanofiber fabrication, and characterization.

Hemocompatibility of Differently Modified Polymeric Nanofibers: Current Progress in the Biomedical Industry

10

Rumysa Saleem Khan, Anjum Hamid Rather, Taha Umair Wani, Muheeb Rafiq, Touseef Amna, M. Shamshi Hassan, Syed Mudasir Ahmad, Shafquat Majeed, Mushtaq A. Beigh, and Faheem A. Sheikh

Abstract

Any implant intended for tissue engineering application should be blood compatible. Since nanofibers are used as suitable candidates for applications where they are directly exposed to blood, the fabrication of nanofibers with excellent blood compatibility has been an unmet challenge for all tissue engineering applications. In this regard, various nanofibers have been fabricated through electrospinning

R. S. Khan · A. H. Rather · T. U. Wani · M. Rafiq · F. A. Sheikh (✉)
Department of Nanotechnology, University of Kashmir Hazratbal, Srinagar, Jammu and Kashmir, India
e-mail: rumysakhan.scholar@kashmiruniversity.net; anjumhamid.scholar@kashmiruniversity.net; wanitaha.scholar@kashmiruniversity.net; muheeb.ntjrf@uok.edu.in; faheemnt@uok.edu.in

T. Amna
Department of Biology, Faculty of Science, Albaha University, Albaha, Kingdom of Saudi Arabia

M. S. Hassan
Department of Chemistry, Faculty of Science, Albaha University, Albaha, Kingdom of Saudi Arabia

S. M. Ahmad
Faculty of Veterinary Sciences, Sher-e-Kashmir University of Agricultural Sciences and Technology, Srinagar, Jammu and Kashmir, India
e-mail: mudasirbio@skuastkashmir.ac.in

S. Majeed
Laboratory for Multifunctional Nanomaterials, Department of Nanotechnology, University of Kashmir Hazratbal, Srinagar, Jammu and Kashmir, India
e-mail: smshah@uok.edu.in

M. A. Beigh
Cellular Signalling and Nanotherapeutics Laboratory, Department of Nanotechnology, University of Kashmir Hazratbal, Srinagar, Jammu and Kashmir, India
e-mail: beighm@uok.edu.in

F. A. Sheikh et al. (eds.), *Interaction of Nanomaterials With Living Cells*,
https://doi.org/10.1007/978-981-99-2119-5_10

while incorporating different factors for enhancing blood compatibility. Recent studies have demonstrated the use of various contrasting agents for modifying the properties of tissue scaffolds. In this chapter, natural and synthetic polymers that are fabricated into nanofibers are discussed for their role in blood compatibility. The choice of polymer used, that is, natural or synthetic; the use of drugs, such as aspirin; the growth factors incorporated in it, like vascular endothelial growth factor; and the topography of scaffolds, for example, smooth and rough, have a significant impact on blood compatibility. This chapter focuses on the characteristics of scaffolds, the methods of preparation, and the factors incorporated in these nanofibers that influence hemocompatibility. A detailed account of different assays employed in analyzing blood compatibility has also been discussed, such as the determination of the hemolysis rate, platelet adhesion, plasma recalcification time, free hemoglobin estimation, the attachment and release of red blood cells, and the adsorption of plasma proteins.

Keywords

Blood · Hemocompatibility · Drug delivery · Nanofibers · Scaffolds · Clotting · Electrospinning · Polymers

10.1 Introduction

Nanofibrous scaffolds are highly porous with a large surface area and have excellent permeability. Henceforth, nanofibers are considered potential candidates for adsorption applications (Sahay et al. 2012). Electrospinning is one of the most robust methods of nanofiber fabrication compared to other techniques. It is easy, economical, and highly versatile in forming nanofibers using various polymers. These nanofibers can act as scaffolds, which are supposed to closely mimic the components of the extracellular matrix (ECM) (Pham et al. 2006; Sell et al. 2007). The setup for the process of electrospinning is shown in Fig. 10.1. Synthetic polymers, such as

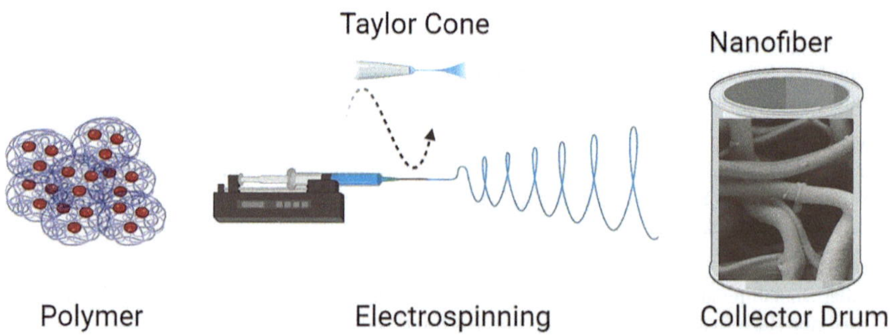

Fig. 10.1 The pictorial presentation of polymer electrospinning. The polymers can be synthetic or natural, which should be completely soluble in solvents. This polymer solution can be filled in a syringe, which, under the influence of a high voltage, is ejected into nanofibers that are collected on the metal collector

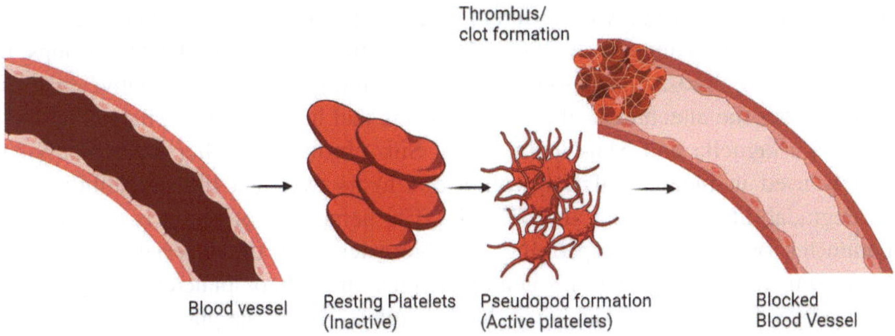

Fig. 10.2 Schematic showing the processes of platelet activation and thrombus formation

polyurethane, which has excellent biocompatibility (Lluch et al. 2013); polyvinyl alcohol, which has high mechanical strength and elasticity (Xu et al. 2008); polyether sulfone, which has good mechanical properties and high thermal stability (Hou et al. 2008); poly(vinyl pyrrolidone), having the property of amphiphilicity, inertness, and chemical stability (Kurakula and Koteswara Rao 2020); and poly (lactic acid), being thermoplastic with high biocompatibility and biodegradability, are widely used in the preparation of nanofibers through electrospinning, are sought after in biomedical applications. On the other hand, natural polymers, such as chitosan (Dash et al. 2011), silk fibroin (Nguyen et al. 2019), cellulose (Seddiqi et al. 2021), etc., are also widely used in bioengineering applications due to their biocompatible and biodegradable character. Chitosan is the only primary polysaccharide with good biocompatibility and biodegradability; additionally, it possesses antifungal properties (Li et al. 1992).

As aforementioned, these synthetic or natural polymers can be fabricated into nanofibers using the electrospinning technique for various applications. However, the fabrication of nanofibers with robust blood compatibility remains an unmet goal in the biomedical field (Soundararajan et al. 2018). On the contact of blood with foreign material (which can be undesired nanofibers), a cascade of lethal effects can occur, for instance, adsorption of plasma proteins, platelet accumulation, and, finally, thrombus or clot formation (Fig. 10.2) (Gorbet and Sefton 2006). Contrary to this, it has been observed that the plasma proteins and platelets present in the blood are readily adsorbed on these polymeric nanofiber surfaces (Gorbet and Sefton 2006; Ren et al. 2013). For example, the positively charged chitosan readily adsorbed negatively charged proteins, which caused platelet activation and thrombus formation. However, at the same time, it preferably bound human serum albumin, reducing platelet attachment and activation (Lv et al. 2017; Balan and Verestiuc 2014). Various efforts have been put forward to modify the blood compatibility of polymeric nanofibers, such as polyurethane, through the introduction of surface additives (Hsiao et al. 2015; Chen et al. 2011). To improve compatibility with blood, the electrospun mats have been functionalized by adsorbing certain moieties, like those of antithrombogenic factors (Freitas et al. 2010), moieties of bioinert and inorganic materials, (Hauert 2005), and organic coatings (Kim and Kim 2002) or by preparing cell-friendly surfaces (Qi et al. 2013). Similarly, chitosan, its composites, and derivatives modified with different functional groups are fabricated (Balan and Verestiuc 2014; Yang et al. 2007). The negative -SO_3^- or -COO^- groups in functionalized chitosan have good blood compatibility because of negative charges, which lower the attachment of plasma proteins that have negative charges on their surfaces (Sagnella and Mai-Ngam 2005). Silk fibroin is another natural polymer widely used in biomaterials fabrication due to its excellent biocompatibility and biodegradability (Naskar et al. 2017a, b). The hemolysis of blood is a significant parameter for checking compatibility with biomaterials. If the material has excellent hemocompatibility, there is no hemolytic phenomenon or platelet adsorption on the surface. Upon hemolysis, hemoglobin is released into the plasma following the damage of membranes of red blood cells, which is directly associated with the compatibility of material (Song et al. 2016). There are several methods for determining the blood compatibility of samples in a laboratory, including measuring the hemolysis rate. The hemolysis rate measures the extent of lysed red blood cells (RBCs) upon the contact of the sample with blood. The lesser the hemolytic rate is, the greater is the blood compatibility of samples. An ideal biomaterial should have a hemolysis rate of less than 5% (Song et al. 2016). Another method includes the proliferation of platelets on the nanofiber surface. The spreading of platelets and their aggregation are major confirmatory indications of platelet activation and are taken as the chief pathway of thrombosis. The attachment of blood platelets is a spontaneous approach to measuring the compatibility of materials (He et al. 2011). The thrombogenicity indicator of a material in a blood compatibility experiment is given by the number of platelets that attach on the surface and the altered shape of those platelets on coming in contact of the material with the blood. Upon contact of foreign material with blood, the first response is the adsorption of serum proteins and the subsequent platelet adhesion and activation of coagulation pathways, which finally result in thrombus formation (Huang et al. 2011). The Vroman effect is the consequence of the hydrophilicity and topography of material surfaces (De Mel et al. 2012). Moreover, the adsorption of plasma proteins is an indispensable parameter in evaluating the thrombogenicity of materials. The blood plasma consists of several proteins, such as albumin (45 mg/mL), immunoglobulin G (10 mg/mL), fibrinogen (3 mg/mL), transferrin (3 mg/mL), and immunoglobulin A (1 mg/mL) (Vogler 2012). Upon the contact of blood with the biomaterial, initially, the proteins adsorb onto the surface, which leads to the recruitment of platelet, white blood cells (WBCs), and red blood cells (RBCs). Finally, adhesion on the plasma protein layer occurs, leading to the formation of thrombin (Vogler 2012). The adsorption of protein in blood plasma increases with the hydrophilicity and causes unwanted thrombus formation events. Furthermore, the stress on the cells is dependent on the roughness of the substrate, which later increases hemolysis (Leszczak et al. 2013). Different strategies have been used to prevent protein and platelet attachment on the biomaterials to attain hemocompatibility and anticoagulation (Weber et al. 2002). Another method involves the assessment of the anticoagulant characteristics of the substrate by recording free hemoglobin in the blood (Soundararajan et al. 2018). In this method, coagulation is initiated, which leads to the formation of a stabilized fibrin clot formed by RBCs, WBCs, and platelets. Then the release of hemoglobin from the clot is measured, which determines the anticoagulant property.

Furthermore, how much the given biomaterial sample causes a delay in the clotting of platelet-poor plasma (PPP) is measured by plasma recalcification time (PRT), which causes the activation of prothrombin (factor II) on the addition of Ca^{2+}. PRT gives the point of delay in the coagulation process (Soundararajan et al. 2018). In this method, the time taken by the sample in showing the first signs of fibrin thread formation is observed. When the sample contacts with PPP, it can signal the intrinsic coagulation pathway, given that coagulation factors are present. Blood compatibility is also determined by the adsorption of blood proteins on biomaterials (Wei et al. 2009; Denis et al. 2002; Scopelliti et al. 2010). In this chapter, we discuss the impact of different electrospun nanofibers with different modifications on checking blood compatibility. This chapter will provide readers an insight into the different strategies for modifying nanofibers and their evaluation by different hemocompatibility assays for use in biological applications.

10.2 Effect of Drug-Loaded Nanofibers on Blood Compatibility

Poly(lactic acid) and silk fibroin nanofiber mats of different mass compositions loaded with aspirin for their anticoagulant property have been prepared by electrospinning (Qin et al. 2013). In designing the solutions for electrospinning, the solvent used was trifluoroacetic acid and dichloromethane in the proportion of 70 parts:30 parts. Poly(lactic acid) was used due to its potential for the controlled release of drugs in the form of fibroin scaffolds, electrospun mats, microspheres, and films (Dev et al. 2010; Kim et al. 2010; Nagarwal et al. 2011). However, due to its low biological activity and hydrophobicity, cell attachment is cumbersome. Therefore, they incorporated poly(lactic acid) in fibrous protein (i.e., silk fibroin), which is hydrophilic, inducing low to no immunogenic response. In addition, it exhibits flexible degradation rates, fluctuating from days to months in vivo due to enhanced crystallinity (Fei et al. 2011). However, regenerated silk fibroin has low mechanical properties because of the modifications in conformations and the distortion of molecules (Li et al. 2011). This led the researchers to blend poly(lactic acid) and silk fibroin into a composite nanofiber mat for the fabrication of biomaterial with the required properties. After degumming and dialysis, the silk fibroin solution was filtered and freeze-dried to get silk fibroin sponges. Then different solutions containing different compositions of poly(lactic acid)/silk fibroin (8/1, 8/3, and 8/5) and aspirin (0.7%, 0.9%, and 1.1%) were prepared. It was observed that when silk fibroin concentration was increased, the degree of compatibility in poly(lactic acid) and silk fibroin was reduced and resulted in increased diameters in the nanofibers. With the increase in aspirin concentration, the fiber diameter decreased to 80 nm from 210 nm because the charge on the surface of the jet flow increased. The aspirin that was released from fibers was observed spectrophotometrically at 270 nm. In vitro drug release studies illustrated that the highest concentration of poly (lactic acid)/silk fibroin gave the highest amount of drug release due to the bulging of fibers. When the effect on the release of aspirin content was observed, it was seen that with the increase in aspirin concentration, the amount of released drug also increased. This was due to the effect of aspirin on the diameter of the fiber mats. The thinner are the nanofibers, the larger is the surface area and the more release of the drug from the nanofibers. The morphology of 0.7% aspirin-incorporated mats before and after its release was observed by scanning electron microscopy (SEM). Prior to the release of the drug, the morphology of nanofibers was even and drug particles were imperceptible because of their incorporation inside the nanofibers. Following drug release after 72 h, the mats swelled as a result of insolubilization (Muthumanickkam et al. 2013). The appearance of the mats became rough and ruptured upon the release of aspirin, and small particles uniformly distributed on the fibers could be seen. It was concluded that after swelling, degradation of the polymer may occur and the drug may be released by diffusion and matrix erosion' (Song et al. 2012).

To check the in vitro blood compatibility of these composite aspirin-loaded nanofibers, a platelet adhesion experiment was carried out. The 2 × 2 cm samples were fitted to the bottom of a 96-well plate and equilibrated with physiological saline for 1 h. Also, fresh human blood was added to trisodium citrate dehydrate to prevent coagulation in the ratio anticoagulant: blood of 1:9 v/v was centrifuged at 800 rpm for 10 min to get platelet-rich plasma (PRP). Following this, the nanofiber samples were incubated with 0.2 mL of PRP for 1 h at 37 °C. The samples were rinsed with phosphate-buffered saline (PBS) to eliminate loosely bound or unbound platelets. Then to prepare samples for SEM, they were fixed with 2.5% glutaraldehyde solution. After dehydration with ethanol, they were permuted with tertiary butanol and incubated in a refrigerator at 4 °C for 24 h. Lastly, the attached platelets were examined by SEM. Many adherent platelets were found attached to the pristine poly (lactic acid) nanofiber membrane. The platelets were seen to aggregate, adhere, and cover the nanofibers. Moreover, they deformed and exhibited pseudopodia. This showed the noncompatibility of pristine poly(lactic acid) nanofiber mats. On the other hand, the surface of poly(lactic acid)/silk fibroin (8/3) showed a smaller number of adhered platelets, and also fewer platelets were deformed and activated. The rationale behind the platelet adhesion on these mats was the equal proportion of hydrophilic/hydrophobic regions and the micro-phase separation in the material. Given the strong hydrophobicity of poly(lactic acid) nanofibers, poly(lactic acid)/ silk fibroin nanofibers had excellent hydrophilicity/hydrophobicity balance and a good microphase separation structure. Therefore, poly(lactic acid)/silk fibroin composites had better blood compatibility. Also, some platelets attached and did not change their discoid shape on a poly(lactic acid)/silk fibroin composite having 0.9 wt.% aspirin, indicating an inactivated state. This concludes that aspirin prevented platelet stimulation and attachment and clot forming by poly(lactic acid)/silk fibroin composites.

10.3 Effect of Surface Topography of Nanofibers on Blood Compatibility

Smooth, porous, and rough nanofibers of poly(lactic acid) were fabricated in a study by electrospinning to see the effect of the topography of these nanofibers on hemocompatibility (Soundararajan et al. 2018). To prepare nanofibers with porous topography, 7% (w/v) poly(lactic acid) was dissolved in dichloromethane/acetone in a 7:3 ratio to prepare the spinning solution. The spinning solution for smooth nanofibers was prepared by using dimethyl formamide in place of acetone in the binary solvent. Other constraints, like the concentration of polymer solutions, the magnitude of voltage, the diameter of the needle, the tip-to-collector distance, and the flow rate, were constant in both cases. To prepare nanofibers having rough topography, hydrophobic superparamagnetic iron oxide nanoparticles (SPION) were produced by coprecipitation. This SPION was complexed with curcumin to form the Cur-SPION complex. This complex (5 mg/mL) was dispersed in poly(lactic acid) polymeric solution dissolved in dichloromethane/dimethyl formamide to produce curcumin-incorporated magnetic fibers with rough topography. The water contact angle was found to be lowered in porous and rough mats as compared to the mats with smooth surfaces. The electrospun nanofibers were then analyzed for hemolysis properties. The integrity of the red blood cell membrane was analyzed by exposing the cells to different topographies of nanofibers. The human blood was incubated in 0.5 mL of 0.1 M ethylenediaminetetraacetic acid (EDTA). After centrifugation at 1500 rpm for 15 min, the pellet containing RBCs was obtained. It was followed by washing with PBS three times, and the volume was made up to 3.5 mL. Next, 0.2 mL of this suspension was pipetted out and added to 0.8 mL of distilled water set as a positive control or PBS set as a negative control. Mats were incubated in the positive control at 37 °C for 2 h, followed by centrifugation at 1500 rpm for 15 min. Then the optical density of the supernatant was measured at 540 nm, and the hemoglobin released was calculated. Following this, the hemolysis percentage was calculated. To categorize the biomaterials as hemolytic or nonhemolytic, hemolysis percentage values over 5% are taken as hemolytic. The results of the hemolysis experiment showed haptoglobin (HP) of less than 2% in smooth nanofibers (1.17%), which established the nonhemolytic behavior of smooth fibers. On the other hand, porous and rough mats showed an increased hemolysis percentage of 3.80% and 4.94%, respectively. However, this was smaller than the threshold value (hemolysis percentage >5%) (Soundararajan et al. 2018). Slight hemolysis on smooth mats can be attributed to the delicateness of RBCs on attaching to nonuniform surfaces compared to smoother surfaces. The increase in hemolysis percentage in rough nanofibers was due to the impact of surface chemistry and the rough structure on hemolysis. The efficiency of scaffolds to bind and release RBCs with no deformation was evaluated through the capture and release of RBCs. For this, 80 μL diluted RBC suspension was taken and dropped on nanofiber mats, followed by incubation for 30 min at 37 °C. To capture the cells, 2.5% glutaraldehyde was used. To release RBCs, fibers cultured with cells were first fixed with 2.5% glutaraldehyde for 1 h at 37 °C and then dehydrated with ethanol, and the fibers were imaged under FE-SEM. It was seen that porous nanofibers retained more cells than smooth nanofibers. The biconcave morphology of RBCs was also retained in both scaffolds, and adhered cells were released with 100% efficiency. In contrast, rough nanofibers (with Cur-SPION) did not have many cells attached as compared to the other two combinations. Furthermore, the anticoagulant assay was carried out to confirm the clot-inhibition properties of samples via the kinetic clotting time method. Next, 20 μL of anticoagulated blood was dispensed onto the samples and glass slabs, with the latter being taken as a positive control. Then 10 μL of $CaCl_2$ solution

(0.2 mol/L) was mixed homogeneously to induce coagulation. The samples were incubated for different time intervals of up to 60 min. Lastly, 5 mL water was added to all solutions and incubated for 5 min. Following this, the free hemoglobin in water was recorded at 540 nm. It was observed that mats with smooth, porous, and rough topographies had higher absorbance than coverslips, indicating a higher concentration of free hemoglobin and a slower rate of clotting. Furthermore, stimulated platelets were confirmed by their morphological change after adhesion to the sample surfaces from a discoid to a spread structure. Smooth fibers showed a higher number of platelets attached and had an elongated appearance with no pseudopodia formation. In contrast, porous nanofibers showed lower adhesion of platelets and retained discoidal morphology due to the hydrophilicity of fibers. The cells on Cur-SPION-modified fibers had an original discoidal shape, referring to inactivated platelets. PRT was done to assess the sample-induced time of delay in the coagulation process. Centrifugation of whole blood at 3000 rpm was performed to obtain PPP. Thereafter, tubes were incubated with nanofibers and glass coverslips (positive control) and added with 0.1–0.5 mL of PPP. From these combinations, 0.1 mL of PPP was added to 0.1 mL of $CaCl_2$. The time of coagulation was measured by a steel hook, which was immersed in the solution to sense fibrin threads. It was observed that the positive control activated the coagulation cascade by adsorbing the clotting proteins. On the other hand, fibers took more time for thread development than the positive control. The porous and rough nanofibers showed improved clot-inhibiting activity (234 s) compared to smooth nanofibers as they exhibited longer clotting times of 277 s and 279 s, respectively. The researchers also investigated the adsorbed serum proteins using sodium dodecyl sulfate-polyacrylamide gel electrophoresis. Samples were incubated in PRP and were allowed to adsorb for 2 h at 37 °C. The proteins that did not attach or that attached loosely were eliminated by rinsing with PBS. Then the proteins that attached to the sample surfaces were eluted using SDS. The samples were mixed with loading buffer and boiled at 95 °C for 10 min, then loaded into wells of acrylamide gel following cooling at room temperature. Electrophoresis was carried out at a voltage of 80 V in a Tris-glycine-SDS buffer. Coomassie brilliant blue dye was used to stain the adsorbed proteins after electrophoresis. It was observed that the smooth nanofibers showed high adsorption of fibronectin (200–220 kDa), prothrombin (72 kDa), and albumin (66 kDa).

10.4 Effect of Synthetic Nanofibers on Blood Compatibility

Nanofibers of polyurethane were fabricated with polyhedral oligomeric silsesquioxanes (POSS) to impart viscoelastic properties (Song et al. 2016). POSS was used due to the hollow cage assembly of silicon (Wang et al. 2009). Furthermore, it has excellent biocompatibility (Le et al. 2013; Raghunath et al. 2009). Because the hydrophobic surface with low surface energy prevent the adsorption of plasma proteins and platelets (Hasebe et al. 2013; Bhatt et al. 2011). It was assumed that incorporating POSS into polyurethane would cause a reduction in platelets and protein attachment on polyurethane mats because of lowered surface tension (Kanna et al. 2006; Kidane et al. 2009). POSS-polyurethane spinning solutions (12%) having different concentrations of POSS (0%, 1%, and 2%) were prepared in dimethyl formamide/tetrahydrofuran (1:2). The hydrophilicity results showed that nanofibers with 2% POSS concentration had the highest water contact angle among all concentrations. The lowest contact angle was found in the case of pristine polyurethane without POSS. The reason for the increased hydrophobicity can be attributed to weak polarity due to the presence of hydrophobic POSS molecules in the polymer (Kidane et al. 2009; Teng et al. 2014). The hemolysis analysis showed that the hemolysis rate in POSS-incorporated polyurethane nanofibers reduced significantly with the increase in POSS concentration. The plasma protein adsorption assay showed a decrease in the amount of bovine serum albumin protein adsorption upon increasing POSS in the nanofibers. The adsorption amount of pristine polyurethane was 3336.36 μg/cm^2 and 159.50 μg/cm^2, and 115.70 μg/cm^2 for the lowest and highest concentrations of POSS, respectively. Platelet adhesion was studied to analyze the hemocompatibility of nanofibers. The SEM images showed that the platelets attached more to pristine polyurethane nanofibers than to POSS-incorporated nanofibers. Furthermore, with the increase in the concentration of POSS, a smaller number of platelets got attached to the nanofiber samples. This showed the improved compatibility of samples on the incorporation of POSS. The above results of decreasing protein and platelet adsorption on POSS-incorporated polyurethane nanofibers may be ascribed to the lessening of the amount of urethanes in polyurethane on the incorporation of POSS (Groth et al. 1995). POSS also lowers the surface tension of polyurethane nanofibers, which is an advantage in inhibiting protein and platelet attachment, and lowers the binding of platelets to the nanofibers (Silver et al. 1999). After microbial culturing for 12 h, the number of *E. coli* on pristine polyurethane surfaces was the highest compared to POSS-incorporated fibers, and absolutely no growth was seen on the surface with the highest content of POSS. This indicated that the presence of POSS macromolecules could suppress bacterial growth. The reason behind the antibacterial activity of these fibers may be attributed to the lower surface tension caused by incorporating POSS in polyurethane nanofibers (Knorr et al. 2005). Also, the lotus effect of hydrophobic substrates can prevent bacterial growth due to the formation of self-cleaning surfaces (Knorr et al. 2005; Page et al. 2009; Meng et al. 2015). Furthermore, hydrophobic POSS molecule decreases the free energy of membrane and because bacteria tend to adhere to a high-energy surface (Teng et al. 2014), thereby, contributing to resist the adherence of bacteria.

A wearable artificial kidney is a new intervention for renal failure patients, replacing dialysis (Moon et al. 2017; Crews et al. 2019; Voinova et al. 2019). Studies are aiming to fabricate dialyzers by modifying membranes for blood purification through convection, diffusion, or adsorption (Castro et al. 2018; Mohammadi et al. 2018; Sultan et al. 2019). Studies have proven that zeolite is a potential material for the selective adsorption of toxins in an artificial kidney (Narasimhan et al. 2013; Arstad et al. 2008). The powder form of zeolite is unfit for use in dialysis due to its loose powder-like appearance. Therefore, some hydrophilic and biocompatible materials need to be used as substrates. Poly(vinyl pyrrolidone) due to its excellent biocompatibility, and solubility in organic and inorganic solvents. This polymer has good spinnability when the electric field is applied and has been used in engineering materials (Huang et al. 2016). It is potentially apt to use along with zeolites and acts as an adhesive for binding zeolite nanoparticles. However, due to the poor mechanical strength of poly(vinyl pyrrolidone), polyether sulfone polymers are the better choice for dialysis membranes because of their desirable characteristics (Qu et al. 2010). So, in this regard, a wearable blood purification system was proposed (Haghdoost et al. 2021). In this study, polyether sulfone/poly(vinyl pyrrolidone)-zeolite was fabricated by single-step electrospinning. Core spinning was employed with polyether sulfone solution and shell spinning with poly(vinyl pyrrolidone)-incorporated zeolites. Poly(vinyl pyrrolidone) prevented zeolites from discharging into circulation. Two types of zeolites were used in their study, viz., CP8C11 (beta) and ZSM-5 zeolites, due to their ability to attach creatinine from the solution. Polyether sulfone spinning solution (25% v/v) was prepared in dimethyl formamide. Poly(vinyl pyrrolidone)/zeolite solution was prepared by dissolving (10% v/v) in binary solvent dimethyl formamide: acetone. Following this, ZSM-5 and other beta zeolites (10% and 20%) were added to the poly(vinyl pyrrolidone) solution. It was observed that the beta zeolites adsorbed the creatinine more than the ZSM-5 zeolites. Also, the silicon/aluminum ratio is an aspect of creatinine adherence (Lu et al. 2017; Namekawa et al. 2014). It is shown in Fig. 10.3 that the adsorption capacity of beta zeolite decreases with an increase in its concentration due to zeolite aggregation at high concentrations, which in turn lowers the surface area for adsorption.

The cytotoxicity of polyether sulfone/poly(vinyl pyrrolidone)-zeolite nanofibers was investigated. Figure 10.4 shows the cell viability results of commercial polyether sulfone, the core-shell nanofiber of polyether sulfone/poly(vinyl pyrrolidone), and zeolite-incorporated polyether sulfone/poly(vinyl pyrrolidone) nanofiber. The growth of fibroblasts on the core-shell nanofibers was as desired till 24 h. This study suggested the efficient use of core-shell nanofibers (viability of over 90%) for blood purification in dialysis approaches. Moreover, the optical microscope images are shown in Fig. 10.5 to observe the viability of fibroblasts after 24 and 48 h of cell culture with samples.

Figure 10.6 shows the platelet adhesion results of the samples. No pseudopod morphology was acquired by cells on the samples, which indicates no platelet stimulation and good compatibility of zeolites. These images (Fig. 10.6) illustrate none of the adsorbed platelets on zeolite-incorporated nanofibers. However, there was a significant number of platelets attached to pure nanofibers.

10.5 Effect of Natural Nanofibers on Blood Compatibility

Carboxymethyl chitosan is a widely used derivative of chitosan used in biomaterial fabrication. Its compatibility with the blood system has been reported in the literature (Fu et al. 2011). Moreover, it is used in modifying other polymers also to improve blood compatibility (Aiping and Tian 2006). Electrospun carboxymethyl chitosan and poly(lactic acid) nanofibers were prepared to test their compatibility with blood

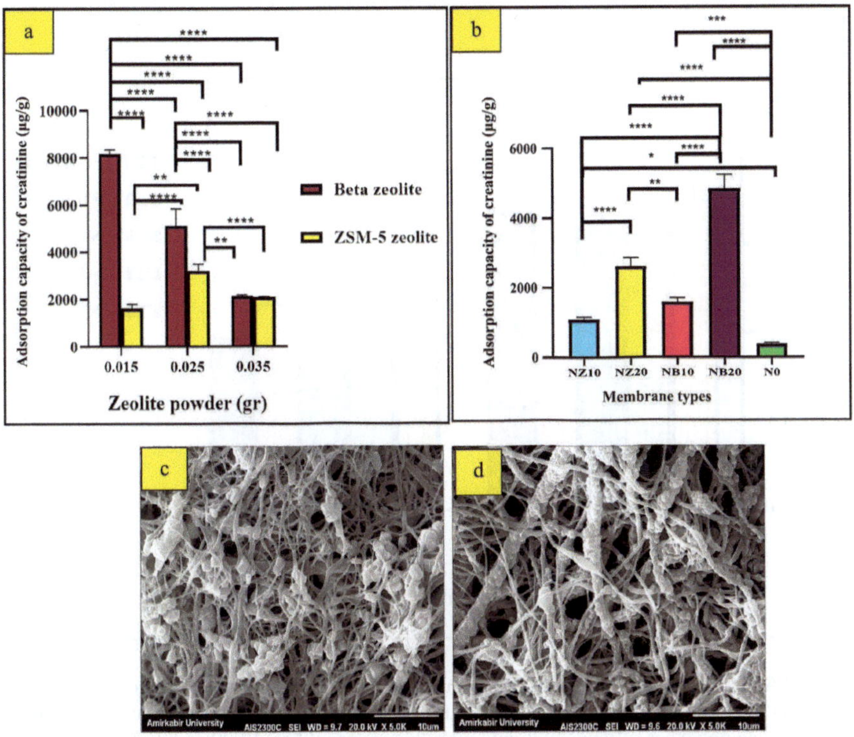

Fig. 10.3 (**a**) Solution adsorption capacity of creatinine using zeolite (powder type), (**b**) creatinine adsorption by core-shell nanofiber (membrane type), (**c**) micrographs of creatinine solution adsorption of ZSM-5 zeolite 20% incorporated polyether sulfone/poly(vinyl pyrrolidone) nanofibers, and (**d**) micrographs of creatinine solution adsorption of beta zeolite 20% incorporated polyether sulfone/poly(vinyl pyrrolidone) nanofibers. (Reproduced with permission from Haghdoost et al. 2021)

(Lv et al. 2017). Carboxymethyl chitosan nanopowder was prepared by ball milling from 400 mg carboxymethyl chitosan powder. Carboxymethyl chitosan was downsized by two steel balls with a 5 mm radius for 4 h at 20 Hz, and carboxymethyl chitosan nanopowder was obtained. To prepare the spinning solution, 8% poly(lactic acid) prepared in binary solvent dimethyl formamide and dichloromethane (v/v = 7/3) was mixed with 400 mg carboxymethyl chitosan nanopowder under constant magnetic stirring. The carboxymethyl chitosan/poly(lactic acid) membranes were linked in 50% glutaraldehyde solution at 80 °C for 12 h, followed by 0.1 mol/L glycine treatment to block unreacted aldehyde groups.

Multilayered silk fibroin was obtained from tussah cocoons incorporated in poly (lactic acid) nanofibers by electrospinning (Shao et al. 2017). The silk fibroin sponge was prepared following the treatment of cocoons, viz., degumming, dissolving, dialysis, and lyophilization. Next, the silk fibroin sponge and poly(lactic acid) were dissolved in 1,1,1,3,3,3-hexafluoro-2-propanol in different proportions to

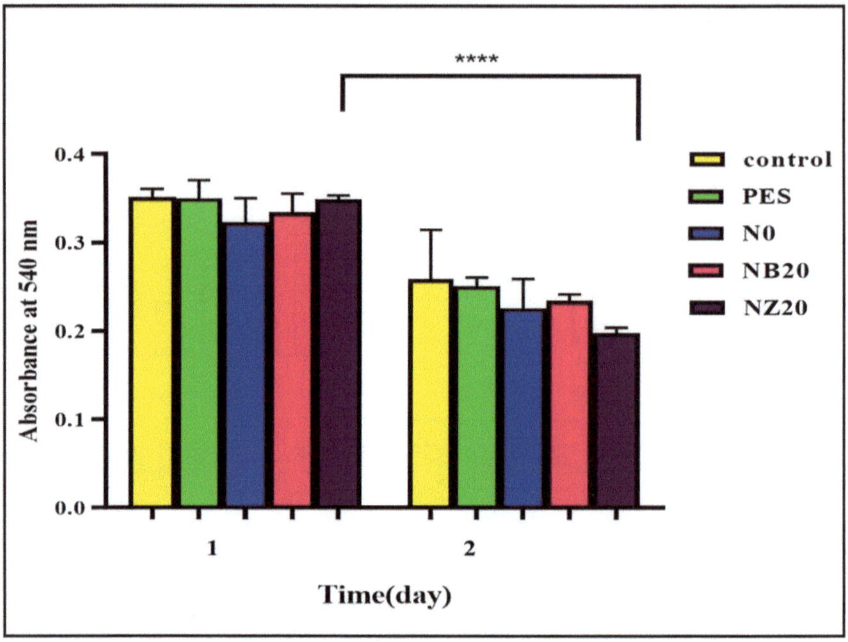

Fig. 10.4 Cell viability results at 540 nm absorbance of nanofibers with different zeolite concentrations, where control is the tissue culture plate, PES is pristine polyether sulfone nanofiber, N0 is pristine polyether sulfone/poly(vinyl pyrrolidone) nanofiber, NB20 is 20% beta zeolite, and NZ20 is 20% ZSM-5 zeolite. (Reproduced from Haghdoost et al. 2021)

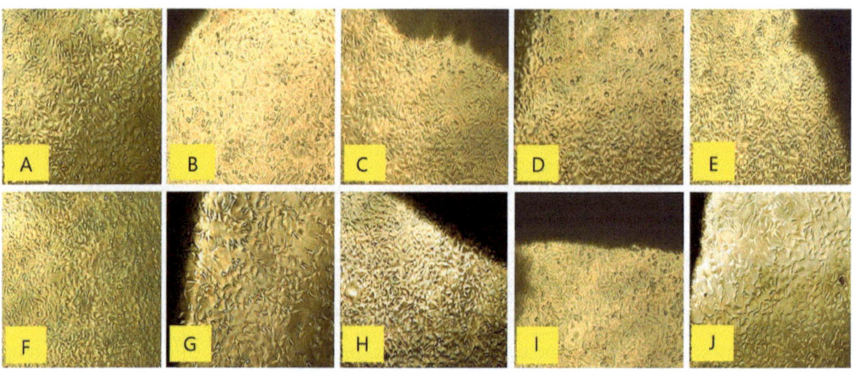

Fig. 10.5 Optical microscope images showing the proliferation of fibroblasts (**a**) and (**f**) tissue culture plate; (**b**) and (**g**) polyether sulfone; (**c**) and (**h**) pure; (**d**) and (**i**) NZ20; (**e**) and (**j**) NB20. Top (after 24 h) and bottom (after 48 h). (Reproduced with permission from Haghdoost et al. 2021)

make the final concentration of the spinning solution 8%. To prepare the nanofiber mats, the nanofiber yarns were electrospun, and a fiber web was formed by rotating them in a funnel. To fabricate multilayer fibers, parallel twisted and weft strands of these yarns were intertwined longitudinally to produce a fabric sheet. Three layers

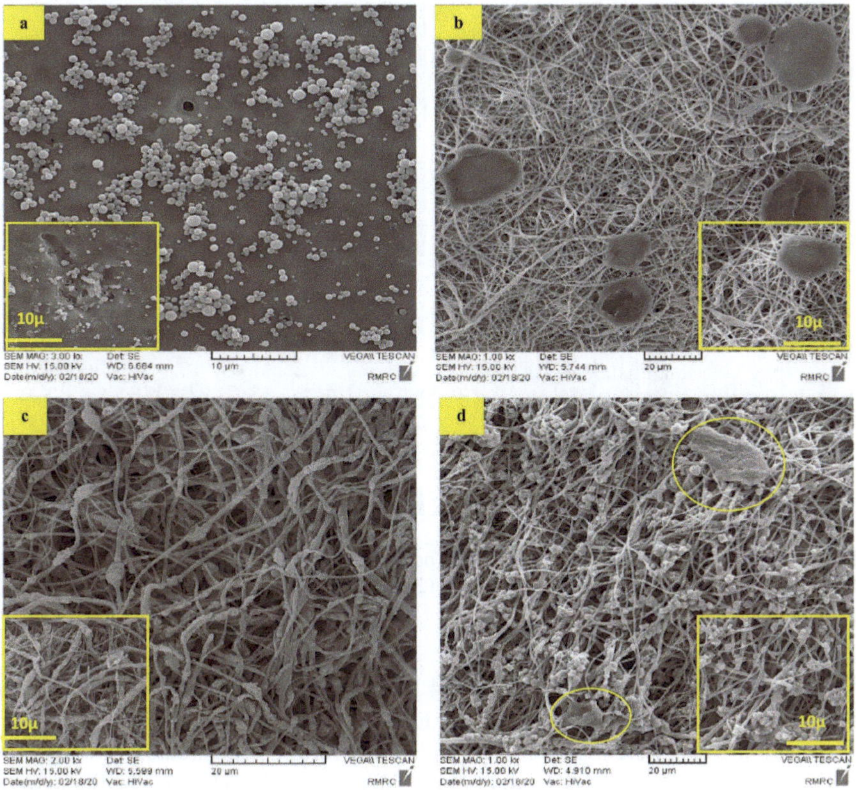

Fig. 10.6 SEM micrographs of platelet attachment on (**a**) commercial polyether sulfone, (**b**) pure polyether sulfone/poly(vinyl pyrrolidone), (**c**) beta-zeolite-incorporated nanofibers, (**d**) Z20-incorporated nanofibers. (Reproduced with permission from Haghdoost et al. 2021)

were later combined with seriatim, and a multilayered nanofiber fabric membrane was formed. To test for biomineralization, simulated body fluid was prepared, and 1% of aspartic acid (a polymer that dissolves in aqueous solvents) was mixed to bring about mineralization. The electrospun nanofiber mats were soaked in simulated body fluid and kept at 37 °C for 72 h, and the medium was replaced every day. Following this, samples were washed with distilled water and deposited on coverslips after freeze-drying. Following mineralization, the samples were observed under SEM after 6, 12, 24, and 72 h (Fig. 10.7). At 6 h, the surface of the samples showed little mineralization in addition to discontinuous mineral clusters; at some points, these particles were observed until mineralization time of 24 h. The reason may be the hydrophobic nature of the poly(lactic acid) surface, which cannot easily constrain the calcium present in simulated body fluid in the form of ions and nucleation points. The surface of multilayer poly(lactic acid)/silk fibroin nanofibers showed a higher number of spherical particles at 6 h as compared to pure poly(lactic acid). After 12 h, the mineral particles covered about all of the nanofibers,

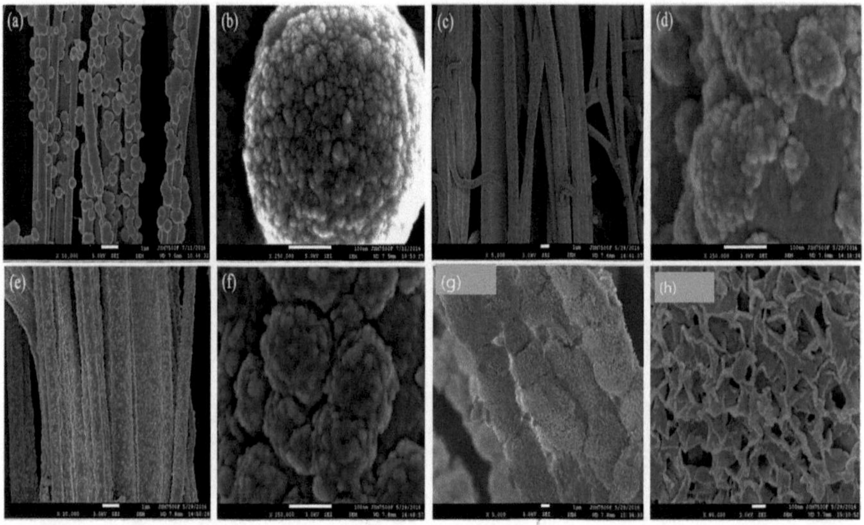

Fig. 10.7 SEM micrographs of silk fibroin and poly(lactic acid) nanofibers after biomineralization for different time intervals: (**a**, **b**) 6 h; (**c**, **d**) 12 h; (**e**, **f**) 24 h; (**g**, **h**) 72 h. (Reproduced with permission from Shao et al. 2017)

with a decrease in their diameter. At 24 h of mineralization, the particles further reduced in diameter and got compressed together to cover all nanofibers.

Nanofiber mats were placed in a 24-well tissue culture plate, treated with PBS for 2 h, and then added with 75% ethanol for sterilization. Following this, the fibers were incubated for 24 h at 37 °C in the presence of 1 mL of 10% fetal bovine serum. The absorbance was measured at 280 nm to check the concentration of fetal bovine serum proteins prior to and after adsorption. The blood compatibility was quantified by hemolysis assay, and the hemolysis rate was measured in all cases. The poly (lactic acid)/silk fibroin mats had a hemolysis rate of 0.9%, indicating their good blood compatibility. The proliferation of cells cultured on poly(lactic acid)/silk fibroin/hyaluronic acid multilayered mats was found to be enormously higher than on poly(lactic acid)/silk fibroin. It was also found that the proliferation on reference coverslips was low, pointing out that the mineral crystals on the nanofibers promote proliferation. Both nanofiber mats and nanofiber fabrics showed the same value on day 4 of cell culture. On the other hand, the proliferation on fabrics was a bit increased compared to the fiber mats on day 7. The morphology of mesenchymal stem cells (MSCs) was determined by confocal laser scanning microscopy (Fig. 10.8). A directional growth of cells was seen on the nanofiber fabric along the axial direction of the yarn; however, the cells on nanofiber mats showed a random arrangement. This phenomenon might arise due to the contact guidance of the underlying substrate, such as nanopatterned material and nanofiber (Laco et al. 2013). Also, the fluorescence intensity on the fabric group was higher compared to mats with multilayered nanofiber fabrics, which offer a much optimal environment for the proliferation of cells.

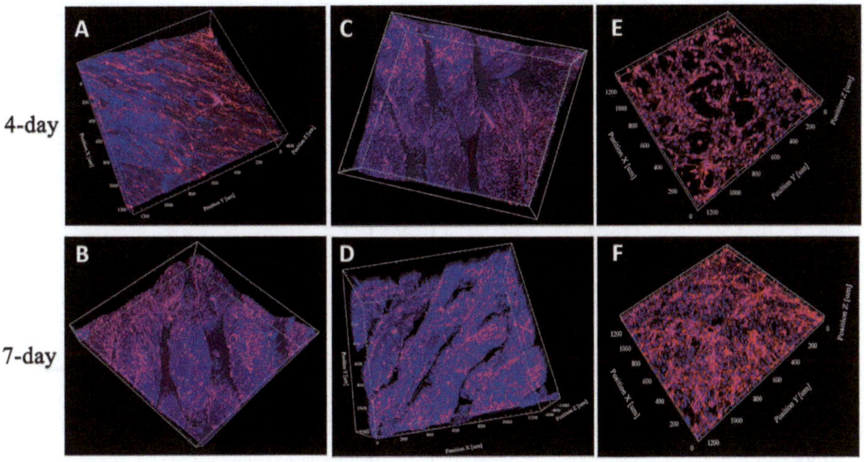

Fig. 10.8 The confocal microscope images of MSCs grown on (**a** and **b**) poly(lactic acid)/silk fibroin fabric, (**c** and **d**) poly(lactic acid)/silk fibroin/hyaluronic acid fabric, and (**e** and **f**) poly(lactic acid)/silk fibroin/hyaluronic acid mats at 4 and 7 days of culture. (Reproduced with permission from Shao et al. 2017)

The expression of marker genes on cells attached to the mats for different times (7, 14, and 21 days) was analyzed. After culture, the cells were washed with PBS and suspended in a cold TRIzol reagent. cDNA was produced using the first-strand synthesis system, which was later on followed by quantitative PCR using SYBR Green. The marker genes for osteocytes that were used were those for alkaline phosphatase (ALP), osteocalcin (OCN), osteopontin (OPN), and collagen type I (Col I). After 14 days, ALP and type I collagen expression levels were much higher in all nanofiber groups (Fig. 10.9a, b). The mineralization of nanofibers increased the osteogenic differentiation of MSCs. After 21 days of culture, the OCN gene (a marker for the late expression of osteogenesis) expression in cells increased by threefolds on poly(lactic acid)/silk fibroin/hyaluronic acid nanofiber mats than on the poly(lactic acid)/silk fibroin group, which indicated that hyaluronic acid accelerated the differentiation of osteoblasts (Fig. 10.9c). In addition to this, OPN gene expression in cells cultured on fabric scaffolds had higher levels of expression than on nanofiber scaffolds after 7 days of culture (Fig. 10.9d), indicating that the arrangement of nanofibers can promote the differentiation of MSCs into osteoblasts.

10.6 Effect of Growth-Factor-Incorporated Nanofibers on Blood Compatibility

Vascular grafts made from polymeric nanofibers have been used in treating vascular diseases (Desai et al. 2011). However, these grafts should prevent platelet aggregation and promote endothelialization on the surface of the graft. To achieve this goal, many bioactive molecules have been merged in the grafts to modify their characters.

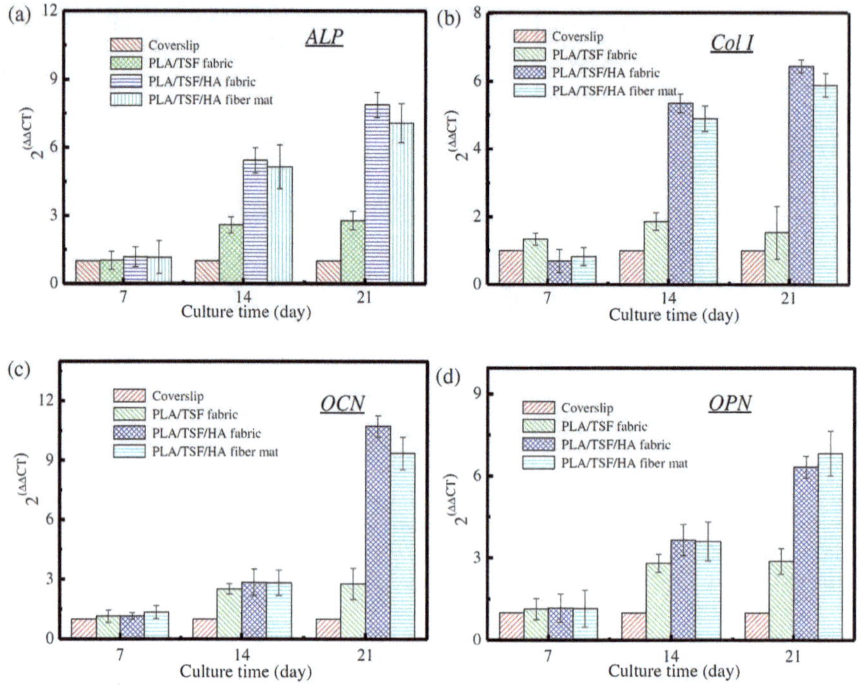

Fig. 10.9 The expression of genes obtained from RT-PCR: (**a**) alkaline phosphatase, (**b**) collagen type I, (**c**) osteocalcin, (**d**) osteopontin. (Reproduced with permission from Shao et al. 2017)

For instance, anticoagulant heparin has been routinely introduced into tissue materials to inhibit thrombogenesis (Hoshi et al. 2013; Yao et al. 2014; Seib et al. 2014). Furthermore, vascular endothelial growth factor (VEGF) is also an effective molecule for endothelialization (Coultas et al. 2005; Asahara et al. 1999). Despite the benefits of VEGF in endothelial progenitor cell (EPC) proliferation (Asahara et al. 1999), it has a rapid half-life in the biological system (Takeshita et al. 1994). Therefore, any delivery agent used for VEGF should preserve its activity and release it at a precise rate. Given this, diverse quantities of heparin were encapsulated into a poly(L-lactic acid-co-ε-caprolactone) nanofiber through emulsion electrospinning, followed by hemocompatibility analysis (Chen et al. 2015). Furthermore, the optimal concentration of heparin was selected and encapsulated, along with VEGF, to form nanofibers. For spinning solutions, aqueous solutions of heparin in distilled water with varying concentrations (5%, 10%, and 15%) were prepared. The internal structure of the nanofibers was examined with transmission electron microscopy (TEM). For this, the samples were prepared by directly depositing fibers on copper grids. The TEM images showed that the composite poly(L-lactic acid-co-ε-caprolactone) scaffolds presumed a uniform core-shell morphology (Fig. 10.10a, b). However, the fibers with heparin and those with or without VEGF had a no-uniform multicore shell conformation, and the incorporated factors

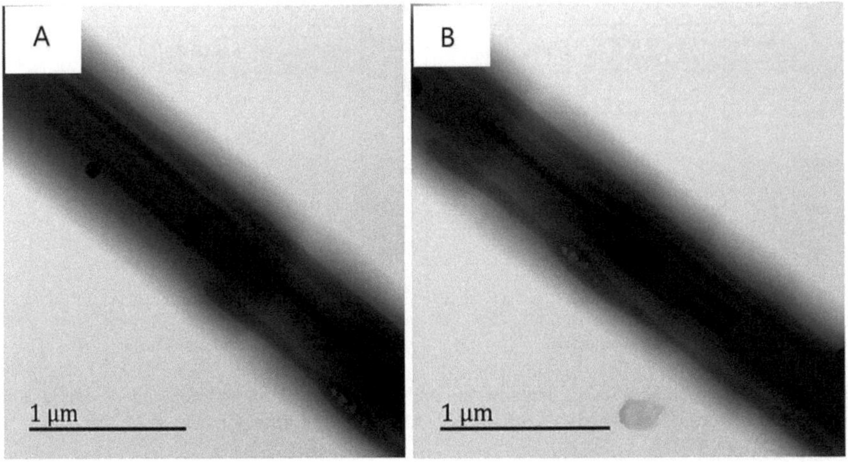

Fig. 10.10 TEM images of different electrospun poly(L-lactic acid-co-ε-caprolactone) nanofibers, (**a**) Poly(L-lactic acid-co-ε-caprolactone) incorporated heparin; (**b**) Poly(L-lactic acid-co-ε-caprolactone) incorporated heparin and VEGF. (Reproduced with permission from Chen et al. 2015)

showed a zonal distribution inside the fibers. The reason for this is that heparin, because of its strong negative charge, increased interdroplet repulsion, leading to the discontinuous core-shell morphology at the time of stretching and evaporation.

The optimum release of heparin and VEGF from the nanofibers was recorded. The poly(L-lactic acid-co-ε-caprolactone) scaffolds containing varied concentrations of heparin and VEGF were immersed in 5 mL of PBS solution and incubated. At different time intervals, 1 mL of the supernatant was pipetted out, and 1 mL of fresh medium was added. To measure the release of heparin from nanofibers, a toluidine blue test was conducted, and to measure VEGF release, an enzyme-linked immunosorbent assay kit was used. The release profiles of heparin and VEGF showed two stages. There was a constant release of heparin for 6 days; after this, a slow release was observed for 29 days. However, there was no perceptible change between the scheme for the release of heparin in the two different concentrations of composite nanofibers, which indicated that VEGF has no role in the release of heparin. On the other hand, the release profile of VEGF was different compared to that of heparin. Due to the higher amount of heparin than of VEGF, the release of VEGF was significantly affected. VEGF release was higher from both types of scaffolds until day 4, which was followed by a slow and constant release. In the scaffold with a higher concentration of VEGF (20%), more VEGF was released compared to the scaffold with 15% VEGF. The results indicate that the amount of VEGF release significantly depends on the loading amount of heparin and that both the growth factors loaded in fibers can be released in a tuned style (Fig. 10.11).

The hemolysis assay revealed hemolysis rates of less than 5%, indicating good blood compatibility of all mats. The hemolysis rate for poly(L-lactic acid-co-ε-caprolactone) with 10% heparin and 15% heparin were less than 0, indicating

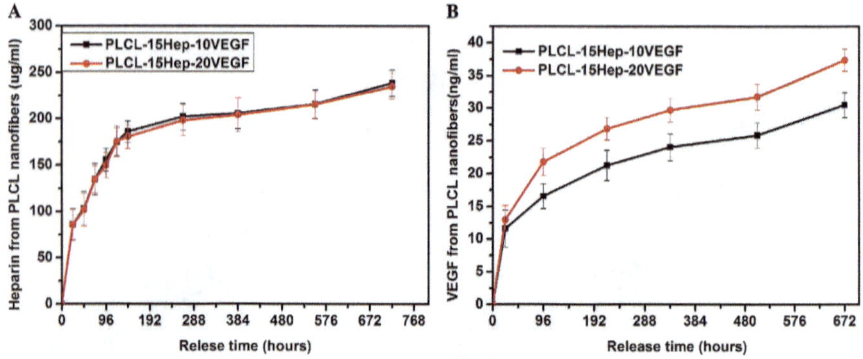

Fig. 10.11 Release profiles of heparin (**a**) and VEGF (**b**) from different poly(L-lactic acid-co-ε-caprolactone) (PLCL) mats. (Reproduced with permission from Chen et al. 2015)

very few ruptured RBCs than the negative control. The reason for this may be attributed to the anticoagulant properties of heparin and also because it acts as a regulator of the complement system (Ninomiya et al. 2000). It has also been studied that heparin has an important role in the regulation of complement-system-induced hemolysis (Mannari et al. 2008). The platelet adhesion studies showed that the number of platelets that adhered to the nanofibers with loaded heparin was lesser as compared to the pure poly(L-lactic acid-co-ε-caprolactone). For the biocompatibility assay, bone marrow cells were produced from the bones of Sprague-Dawley rats (weight = 200–250 g). The mononuclear cells having low density were separated by density gradient centrifugation in a medium exclusive for lymphocytes. The detached cells were resuspended in a culture medium (EGM-2), having the optimal composition for the lymphocytes to grow. Lastly, mononuclear cells were seeded on fibronectin-coated plates and incubated with ample CO_2 supply at a temperature of 37 °C. Fluorescence microscopy confirmed the presence of endothelial progenitor cells (EPCs) by imaging double-positive cells. Furthermore, fluorescence-activated cell sorting (FACS) confirmed the EPCs characteristics of cells by confirming the presence of CD34 and VEGFR-2(KDR) in cells. Cell growth was recorded by a cell counting kit-8 (CCK-8) assay. The samples were placed at the bottom of 24-well plates, and cells were seeded and grown for different time intervals for 7 days. CCk-8 solution was added to the wells already containing the culture medium at a concentration of 10 μL/100 μL, and the samples were raised for 2 h in this solution. Following this, absorbance was measured at 490 nm. Cell viability was determined by correlating the amount of absorbance with the number of viable cells. In addition to this, two more plates were cultured for 7 days. One plate was set for the fluorescent staining to determine the morphology and quantity of cells and the other for SEM to see the cell attachment on the surface of scaffolds. Absorbance was noticeably higher in VEGF-incorporated fibers than in pure fibers (Fig. 10.12). The results suggest that EPCs multiply faster on VEGF-incorporated fibers than those without VEGF, and this effect was concentration and time-dependent.

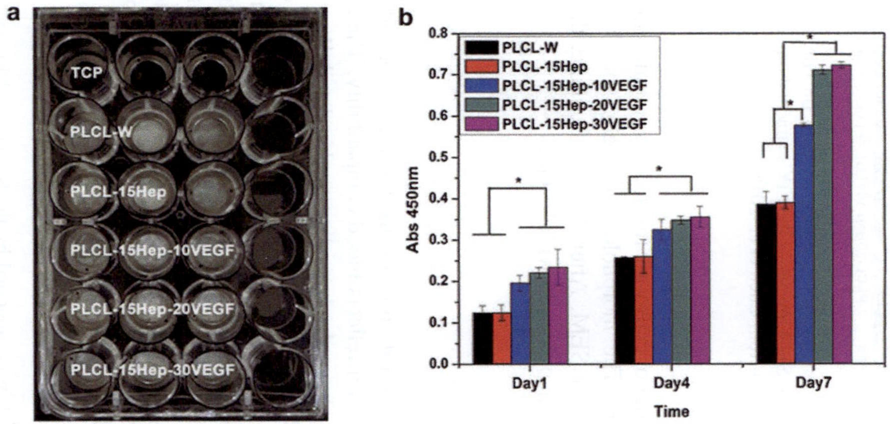

Fig. 10.12 (**a**) Image of the 24-well plate showing the distribution of five types of scaffold samples. (**b**) The CCK-8 assay for the viability of EPCs on five types of poly(L-lactic acid-co-ε-caprolactone) nanofiber mats. (Reproduced with permission from Chen et al. 2015)

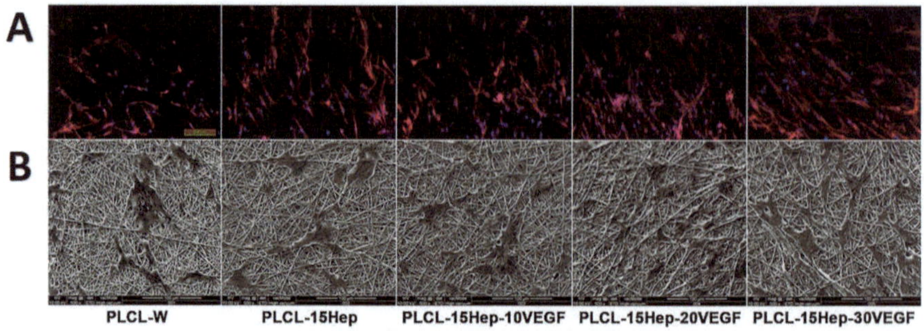

Fig. 10.13 (**a**) Immunofluorescent images of EPCs cultured on composite poly(L-lactic acid-co-ε-caprolactone) nanofibers. (**b**) SEM images of the EPCs attached on nanofibers show proliferation. (Reproduced with permission from Chen et al. 2015)

The structure and number of cells were revealed by immunofluorescence staining. shown in Fig. 10.13a, b. The morphology of cells and their interaction with nanofibers were analyzed by SEM. After 7 days of culture, the cells spread well on the nanofiber surfaces incorporated with VEGF.

10.7 Conclusion

In this chapter, the impact of different agents when incorporated into nanofiber has been discussed, which in turn reflects blood compatibility. The different assays for testing hemocompatibility have also been discussed in detail. We also discussed how different topographies of the nanofibers could impact the blood compatibility of a nanofiber sample. It was also seen that the introduction of growth factors and drugs causes variation in blood compatibility. For example, aspirin loading showed an enhanced effect of the anticoagulation of the poly(lactic acid)/silk fibroin nanofibers. Furthermore, the method of fabrication of nanofibers and the choice of polymer can significantly influence hemocompatibility. In conclusion, the chapter gives information on the biocompatible properties of different nanofibers and their use in other biomedical applications.

Acknowledgments This work was supported by the Science and Engineering Research Board (SERB) research grants (CRG/2020/000113) and Council of Scientific & Industrial Research (CSIR) (File No: 22(0846)/20/EMR-II) research grants.

References

Aiping Z, Tian C (2006) Blood compatibility of surface-engineered poly (ethylene terephthalate) via o-carboxymethylchitosan. Colloids Surf B Biointerfaces 50(2):120–125. https://www.sciencedirect.com/science/article/pii/S0927776506001536. Accessed 5 Sept 2022

Arstad B, Fjellvåg H, Kongshaug KO, Swang O, Blom R (2008) Amine functionalised metal organic frameworks (MOFs) as adsorbents for carbon dioxide. Adsorption 14(6):755–762. https://doi.org/10.1007/s10450-008-9137-6

Asahara T et al (1999) VEGF contributes to postnatal neovascularization by mobilizing bone marrow-derived endothelial progenitor cells. EMBO J 18(14):3964–3972. https://doi.org/10.1093/emboj/18.14.3964

Balan V, Verestiuc L (2014) Strategies to improve chitosan hemocompatibility: a review. Eur Polym J 53(1):171–188. https://doi.org/10.1016/j.eurpolymj.2014.01.033

Bhatt S et al (2011) Nanostructure protein repellant amphiphilic copolymer coatings with optimized surface energy by inductively excited low pressure plasma. Langmuir 27(23):14570–14580. https://doi.org/10.1021/la203256w

Castro AC, Neri M, Nayak Karopadi A, Lorenzin A, Marchionna N, Ronco C (2018) Wearable artificial kidney and wearable ultrafiltration device vascular access - future directions. Clin Kidney J 12(2):300–307. https://doi.org/10.1093/ckj/sfy086

Chen L, Han D, Jiang L (2011) On improving blood compatibility: from bioinspired to synthetic design and fabrication of biointerfacial topography at micro/nano scales. Colloids Surf B Biointerfaces 85(1):2–7. https://doi.org/10.1016/j.colsurfb.2010.10.034

Chen X et al (2015) Electrospun poly(l-lactic acid-co-ε-caprolactone) fibers loaded with heparin and vascular endothelial growth factor to improve blood compatibility and endothelial progenitor cell proliferation. Colloids Surf B Biointerfaces 128:106–114. https://doi.org/10.1016/j.colsurfb.2015.02.023

Coultas L, Chawengsaksophak K, Rossant J (2005) Endothelial cells and VEGF in vascular development. Nature 438(7070):937–945. https://doi.org/10.1038/nature04479

Crews DC, Bello AK, Saadi G (2019) 2019 World Kidney Day Editorial - Burden, access, and disparities in kidney disease. Jornal brasileiro de nefrologia 41(1):1–9. https://doi.org/10.1590/2175-8239-JBN-2018-0224

Dash M, Chiellini F, Ottenbrite RM, Chiellini E (2011) Chitosan - a versatile semi-synthetic polymer in biomedical applications. Prog Polym Sci (Oxford) 36(8):981–1014. https://doi.org/10.1016/j.progpolymsci.2011.02.001

De Mel A, Cousins BG, Seifalian AM (2012) Surface modification of biomaterials: a quest for blood compatibility. Int J Biomater. https://doi.org/10.1155/2012/707863

Denis FA et al (2002) Protein adsorption on model surfaces with controlled nanotopography and chemistry. Langmuir 18(3):819–828. https://doi.org/10.1021/la011011o

Desai M, Seifalian AM, Hamilton G (2011) Role of prosthetic conduits in coronary artery bypass grafting. Eur J Cardiothorac Surg 40(2):394–398. https://doi.org/10.1016/j.ejcts.2010.11.050

Dev A et al (2010) Preparation of poly(lactic acid)/chitosan nanoparticles for anti-HIV drug delivery applications. Carbohydr Polym 80(3):833–838. https://doi.org/10.1016/j.carbpol.2009.12.040

Fei Y, Chen Y, Wang H, Gao W, Yang R, Wan Y (2011) Preparation, characterization of antibacterial PLA/TP nanofibers. Fibers Polym 12(3):340–344. https://doi.org/10.1007/s12221-011-0340-9

Freitas SC, Barbosa MA, Martins MCL (2010) The effect of immobilization of thrombin inhibitors onto self-assembled monolayers on the adsorption and activity of thrombin. Biomaterials 31(14):3772–3780. https://doi.org/10.1016/j.biomaterials.2010.01.097

Fu D, Han B, Dong W, Yang Z, Lv Y, Liu W (2011) Effects of carboxymethyl chitosan on the blood system of rats. Biochem Biophys Res Commun 408(1):110–114. https://doi.org/10.1016/j.bbrc.2011.03.130

Gorbet MB, Sefton MV (2006) Biomaterial-associated thrombosis: roles of coagulation factors, complement, platelets and leukocytes. In: The biomaterials: Silver Jubilee Compendium, pp 219–241

Groth T, Campbell EJ, Herrmann K, Seifert B (1995) Application of enzyme immunoassays for testing haemocompatibility of biomedical polymers. Biomaterials 16(13):1009–1015. https://doi.org/10.1016/0142-9612(95)94909-5

Haghdoost F, Bahrami SH, Barzin J, Ghaee A (2021) Preparation and characterization of electrospun polyethersulfone/polyvinylpyrrolidone-zeolite core–shell composite nanofibers for creatinine adsorption. Sep Purif Technol 257:117881. https://doi.org/10.1016/j.seppur.2020.117881

Hasebe T et al (2013) Hydrophobicity and non-thrombogenicity of nanoscale dual rough surface coated with fluorine-incorporated diamond-like carbon films: biomimetic surface for blood-contacting medical devices. Diam Relat Mater 38:14–18. https://doi.org/10.1016/j.diamond.2013.06.001

Hauert R (2005) A review of DLC coatings for biological applications. Proc World Tribol Congress III 2005:681–682. https://doi.org/10.1115/wtc2005-63879

He C et al (2011) Chemically induced graft copolymerization of 2-hydroxyethyl methacrylate onto polyurethane surface for improving blood compatibility. Appl Surf Sci 258(2):755–760. https://doi.org/10.1016/j.apsusc.2011.08.074

Hoshi RA, Van Lith R, Jen MC, Allen JB, Lapidos KA, Ameer G (2013) The blood and vascular cell compatibility of heparin-modified ePTFE vascular grafts. Biomaterials 34(1):30–41. https://doi.org/10.1016/j.biomaterials.2012.09.046

Hou C, Yuan Q, Huo D, Zheng S, Zhan D (2008) Investigation on clotting and hemolysis characteristics of heparin-immobilized polyether sulfones biomembrane. J Biomed Mater Res A 85(3):847–852. https://doi.org/10.1002/jbm.a.31502

Hsiao CR, Lin CW, Chou CM, Chung CJ, He JL (2015) Surface modification of blood-contacting biomaterials by plasma-polymerized superhydrophobic films using hexamethyldisiloxane and tetrafluoromethane as precursors. Appl Surf Sci 346:50–56. https://doi.org/10.1016/j.apsusc.2015.03.208

Huang XJ, Guduru D, Xu ZK, Vienken J, Groth T (2011) Blood compatibility and permeability of heparin-modified polysulfone as potential membrane for simultaneous hemodialysis and LDL removal. Macromol Biosci 11(1):131–140. https://doi.org/10.1002/mabi.201000278

Huang S, Zhou L, Li MC, Wu Q, Kojima Y, Zhou D (2016) Preparation and properties of electrospun poly(vinyl pyrrolidone)/cellulose nanocrystal/silver nanoparticle composite fibers. Materials (Basel) 9:7. https://doi.org/10.3390/ma9070523

Kanna RY et al (2006) The antithrombogenic potential of a polyhedral oligomeric silsesquioxane (POSS) nanocomposite. Biomacromolecules 7(1):215–223. https://doi.org/10.1021/bm050590z

Kidane AG, Burriesci G, Edirisinghe M, Ghanbari H, Bonhoeffer P, Seifalian AM (2009) A novel nanocomposite polymer for development of synthetic heart valve leaflets. Acta Biomater 5(7):2409–2417. https://doi.org/10.1016/j.actbio.2009.02.025

Kim JH, Kim SC (2002) PEO-grafting on PU/PS IPNs for enhanced blood compatibility - effect of pendant length and grafting density. Biomaterials 23(9):2015–2025. https://doi.org/10.1016/S0142-9612(01)00330-1

Kim ES, Kim SH, Lee CH (2010) Electrospinning of polylactide fibers containing silver nanoparticles. Macromol Res 18(3):215–221. https://doi.org/10.1007/s13233-010-0316-4

Knorr SD, Combe EC, Wolff LF, Hodges JS (2005) The surface free energy of dental gold-based materials. Dent Mater 21(3):272–277. https://doi.org/10.1016/j.dental.2004.06.002

Kurakula M, Koteswara Rao GSN (2020) Moving polyvinyl pyrrolidone electrospun nanofibers and bioprinted scaffolds toward multidisciplinary biomedical applications. Eur Polym J 136:109919. https://doi.org/10.1016/j.eurpolymj.2020.109919

Laco F, Grant MH, Black RA (2013) Collagen-nanofiber hydrogel composites promote contact guidance of human lymphatic microvascular endothelial cells and directed capillary tube formation. J Biomed Mater Res A 101A(6):1787–1799. https://doi.org/10.1002/jbm.a.34468

Le X, Poinern G, Ali N et al (2013) Engineering a biocompatible scaffold with either micrometre or nanometre scale surface topography for promoting protein adsorption and cellular response. hindawi.com. https://www.hindawi.com/journals/ijbm/2013/782549/. Accessed 5 Sept 2022

Leszczak V, Smith BS, Popat KC (2013) Hemocompatibility of polymeric nanostructured surfaces. J Biomater Sci Polym Ed 24(13):1529–1548. https://doi.org/10.1080/09205063.2013.777228

Li Q, Dunn ET, Grandmaison EW, Goosen MFA (1992) Applications and properties of chitosan. J Bioact Compat Polym 7(4):370–397. https://doi.org/10.1177/088391159200700406

Li L et al (2011) Electrospun poly(ε-caprolactone)/silk fibroin core-sheath nanofibers and their potential applications in tissue engineering and drug release. Int J Biol Macromol 49(2): 223–232. https://doi.org/10.1016/j.ijbiomac.2011.04.018

Lluch C, Lligadas G, Ronda JC, Galià M, Cádiz V (2013) Thermoplastic polyurethanes from undecylenic acid-based soft segments: structural features and release properties. Macromol Biosci 13(5):614–622. https://doi.org/10.1002/mabi.201200433

Lu L, Chen C, Samarasekera C, Yeow JTW (2017) Influence of zeolite shape and particle size on their capacity to adsorb uremic toxin as powders and as fillers in membranes. J Biomed Mater ResB Appl Biomater 105(6):1594–1601. https://doi.org/10.1002/jbm.b.33698

Lv J, Yin X, Zeng Q, Dong W, Liu H, Zhu L (2017) Preparation of carboxymethyl chitosan nanofibers through electrospinning the ball-milled nanopowders with poly(lactic acid) and the blood compatibility of the electrospun NCMC/PLA mats. J Polym Res 24(4):60. https://doi.org/10.1007/s10965-017-1224-5

Mannari D, Liu C, Hughes D, Mehta A (2008) The role of heparin in alleviating complement-mediated acute intravascular haemolysis. Acta Haematol 119(3):166–168. https://doi.org/10.1159/000134221

Meng J et al (2015) Antibacterial cellulose membrane via one-step covalent immobilization of ammonium/amine groups. Desalination 359:156–166. https://doi.org/10.1016/j.desal.2014.12.032

Mohammadi F, Valipouri A, Semnani D, Alsahebfosoul F (2018) Nanofibrous tubular membrane for blood hemodialysis. Appl Biochem Biotechnol 186(2):443–458. https://doi.org/10.1007/s12010-018-2744-0

Moon BM et al (2017) Novel fabrication method of the peritoneal dialysis filter using silk fibroin with urease fixation system. J Biomed Mater Res B Appl Biomater 105(7):2136–2144. https://doi.org/10.1002/jbm.b.33751

Muthumanickkam A, Subramanian S, Goweri M, Sofi Beaula W, Ganesh V (2013) Comparative study on eri silk and mulberry silk fibroin scaffolds for biomedical applications. Iran Polym J (English Ed) 22(3):143–154. https://doi.org/10.1007/s13726-012-0113-3

Nagarwal RC, Kumar R, Dhanawat M, Pandit JK (2011) Modified PLA nano in situ gel: a potential ophthalmic drug delivery system. Colloids Surf B Biointerfaces 86(1):28–34. https://doi.org/10.1016/j.colsurfb.2011.03.023

Namekawa K, Tokoro Schreiber M, Aoyagi T, Ebara M (2014) Fabrication of zeolite-polymer composite nanofibers for removal of uremic toxins from kidney failure patients. Biomater Sci 2(5):674–679. https://doi.org/10.1039/c3bm60263j

Narasimhan L, Kuchta B, Schaef O, Brunet P, Boulet P (2013) Mechanism of adsorption of p-cresol uremic toxin into faujasite zeolites in presence of water and sodium cations-a Monte Carlo study. Microporous Mesoporous Mater 173:70–77. https://doi.org/10.1016/j.micromeso.2013.02.003

Naskar D, Ghosh AK, Mandal M, Das P, Nandi SK, Kundu SC (2017a) Dual growth factor loaded nonmulberry silk fibroin/carbon nanofiber composite 3D scaffolds for in vitro and in vivo bone regeneration. Biomaterials 136:67–85. https://doi.org/10.1016/j.biomaterials.2017.05.014

Naskar D, Bhattacharjee P, Ghosh AK, Mandal M, Kundu SC (2017b) Carbon nanofiber reinforced nonmulberry silk protein fibroin nanobiocomposite for tissue engineering applications. ACS Appl Mater Interfaces 9(23):19356–19370. https://doi.org/10.1021/acsami.6b04777

Nguyen TP et al (2019) Silk fibroin-based biomaterials for biomedical applications: a review. Polymers 11(12):1933. https://doi.org/10.3390/polym11121933

Ninomiya H, Kawashima Y, Nagasawa T (2000) Inhibition of complement-mediated haemolysis in paroxysmal nocturnal haemoglobinuria by heparin or low-molecular weight heparin. Br J Haematol 109(4):875–881. https://doi.org/10.1046/j.1365-2141.2000.02125.x

Page K, Wilson M, Parkin IP (2009) Antimicrobial surfaces and their potential in reducing the role of the inanimate environment in the incidence of hospital-acquired infections. J Mater Chem 19(23):3818–3831. https://doi.org/10.1039/b818698g

Pham QP, Sharma U, Mikos AG (2006) Electrospinning of polymeric nanofibers for tissue engineering applications: a review. Tissue Eng 12(5):1197–1211. https://doi.org/10.1089/ten.2006.12.1197

Qi P, Maitz MF, Huang N (2013) Surface modification of cardiovascular materials and implants. Surf Coat Technol 233:80–90. https://doi.org/10.1016/j.surfcoat.2013.02.008

Qin J et al (2013) Evaluation of drug release property and blood compatibility of aspirin-loaded electrospun PLA/RSF composite nanofibers. Iran Polym J (English Ed) 22(10):729–737. https://doi.org/10.1007/s13726-013-0171-1

Qu P, Tang H, Gao Y, Zhang LP, Wang S (2010) Polyethersulfone composite membrane blended with cellulose fibrils. Bioresources 5(4):2323–2336. http://ojs.cnr.ncsu.edu/index.php/BioRes/article/view/1151. Accessed 5 Sept 2022

Ren Z, Chen G, Wei Z, Sang L, Qi M (2013) Hemocompatibility evaluation of polyurethane film with surface-grafted poly(ethylene glycol) and carboxymethyl-chitosan. J Appl Polym Sci 127(1):308–315. https://doi.org/10.1002/app.37885

Sagnella S, Mai-Ngam K (2005) Chitosan based surfactant polymers designed to improve blood compatibility on biomaterials. Colloids Surf B Biointerfaces 42(2):147–155. https://doi.org/10.1016/j.colsurfb.2004.07.001

Sahay R et al (2012) Electrospun composite nanofibers and their multifaceted applications. J Mater Chem 22(26):12953–12971. https://doi.org/10.1039/c2jm30966a

Scopelliti PE et al (2010) The effect of surface nanometre-scale morphology on protein adsorption. PLoS One 5(7):e11862. https://doi.org/10.1371/journal.pone.0011862

Seddiqi H et al (2021) Cellulose and its derivatives: towards biomedical applications. Cellulose 28(4):1893–1931. https://doi.org/10.1007/s10570-020-03674-w

Seib FP, Herklotz M, Burke KA, Maitz MF, Werner C, Kaplan DL (2014) Multifunctional silk-heparin biomaterials for vascular tissue engineering applications. Biomaterials 35(1):83–91. https://doi.org/10.1016/j.biomaterials.2013.09.053

Sell S et al (2007) Extracellular matrix regenerated: tissue engineering via electrospun biomimetic nanofibers. Polym Int 56(11):1349–1360. https://doi.org/10.1002/pi.2344

Shao W, He J, Wang Q, Cui S, Ding B (2017) Biomineralized poly(l-lactic-co-glycolic acid)/graphene oxide/tussah silk fibroin nanofiber scaffolds with multiple orthogonal layers enhance osteoblastic differentiation of mesenchymal stem cells. ACS Biomater Sci Eng 3(7):1370–1380. https://doi.org/10.1021/acsbiomaterials.6b00533

Sheikh FA, Barakat NA, Kim BS, Aryal S, Khil MS, Kim HY (2009) Self-assembled amphiphilic polyhedral oligosilsesquioxane (POSS) grafted poly (vinyl alcohol)(PVA) nanoparticles. Mater Sci Eng: C 29(3):869–876. https://doi.org/10.1016/j.msec.2008.07.029

Silver JH, Lin JC, Lim F, Tegoulia VA, Chaudhury MK, Cooper SL (1999) Surface properties and hemocompatibility of alkyl-siloxane monolayers supported on silicone rubber: effect of alkyl chain length and ionic functionality. Biomaterials 20(17):1533–1543. https://doi.org/10.1016/S0142-9612(98)00173-2

Song X, Gao Z, Ling F, Chen X (2012) Controlled release of drug via tuning electrospun polymer carrier. J Polym Sci B Polym Phys 50(3):221–227. https://doi.org/10.1002/polb.23005

Song X, Li T, Cheng B, Xing J (2016) POSS-PU electrospinning nanofibers membrane with enhanced blood compatibility. RSC Adv 6(70):65756–65762. https://doi.org/10.1039/c6ra07681e

Soundararajan A et al (2018) Surface topography of polylactic acid nanofibrous mats: influence on blood compatibility. J Mater Sci Mater Med 29(9):145. https://doi.org/10.1007/s10856-018-6153-2

Sultan MT et al (2019) Recirculating peritoneal dialysis system using urease-fixed silk fibroin membrane filter with spherical carbonaceous adsorbent. Mater Sci Eng C 97:55–66. https://doi.org/10.1016/j.msec.2018.12.021

Takeshita S et al (1994) Therapeutic angiogenesis. A single intraarterial bolus of vascular endothelial growth factor augments revascularization in a rabbit ischemic hind limb model. J Clin Invest 93(2):662–670. https://doi.org/10.1172/JCI117018

Teng CP et al (2014) Star-shaped polyhedral oligomeric silsesquioxane-polycaprolactone-polyurethane as biomaterials for tissue engineering application. NPG Asia Mater 6(11):e142. https://doi.org/10.1038/am.2014.102

Vogler EA (2012) Protein adsorption in three dimensions. Biomaterials 33(5):1201–1237. https://doi.org/10.1016/j.biomaterials.2011.10.059

Voinova M, Repin N, Sokol E, Tkachuk B, Gorelik L (2019) Physical processes in polymeric filters used for dialysis. Polymers 11(3):389. https://doi.org/10.3390/polym11030389

Wang W, Lin Guo Y, Otaigbe JU (2009) The synthesis, characterization and biocompatibility of poly(ester urethane)/polyhedral oligomeric silesquioxane nanocomposites. Polymer (Guildf) 50(24):5749–5757. https://doi.org/10.1016/j.polymer.2009.05.037

Weber N, Wendel HP, Ziemer G (2002) Hemocompatibility of heparin-coated surfaces and the role of selective plasma protein adsorption. Biomaterials 23(2):429–439. https://doi.org/10.1016/S0142-9612(01)00122-3

Wei J et al (2009) Influence of surface wettability on competitive protein adsorption and initial attachment of osteoblasts. Biomed Mater 4(4):045002. https://doi.org/10.1088/1748-6041/4/4/045002

Xu F, Li Y, Deng Y, Xiong J (2008) Porous nano-hydroxyapatite/poly(vinyl alcohol) composite hydrogel as artificial cornea fringe: characterization and evaluation in vitro. J Biomater Sci Polym Ed 19(4):431–439. https://doi.org/10.1163/156856208783719473

Yang Y, Zhou Y, Chuo H, Wang S, Yu J (2007) Blood compatibility and mechanical properties of oxidized-chitosan films. J Appl Polym Sci 106(1):372–377. https://doi.org/10.1002/app.25399

Yao Y et al (2014) Effect of sustained heparin release from PCL/chitosan hybrid small-diameter vascular grafts on anti-thrombogenic property and endothelialization. Acta Biomater 10(6):2739–2749. https://doi.org/10.1016/j.actbio.2014.02.042

Rumysa Saleem Khan was born in Srinagar, J&K, India, in 1995. She received a B.S. degree in Biochemistry in 2016 and an M.S. degree in Biotechnology from the University of Kashmir, Srinagar, in 2018 and was receiving DBT Fellowship. From 2019 to 2020, she was a JRF with a biochemical engineering lab at NIT, Srinagar, where she received MHRD Junior Research Fellowship.

She is currently pursuing a Ph.D. degree under the supervision of Dr. Faheem A. Sheikh at the University of Kashmir, Srinagar, J&K, India, and is receiving the Science and Engineering Research Board (SERB) Fellowship. She is currently the author of 16 scientific articles published in international journals. Ms. Khan's research interests include the biomimicking of biological molecules in the lab, the fabrication of biocompatible scaffolds for tissue engineering, nanofiber formation for wound healing and bone regeneration, and exploiting the roles of different nanoparticles in tissue engineering. She also has expertise in mammalian cell culture and molecular biology techniques.

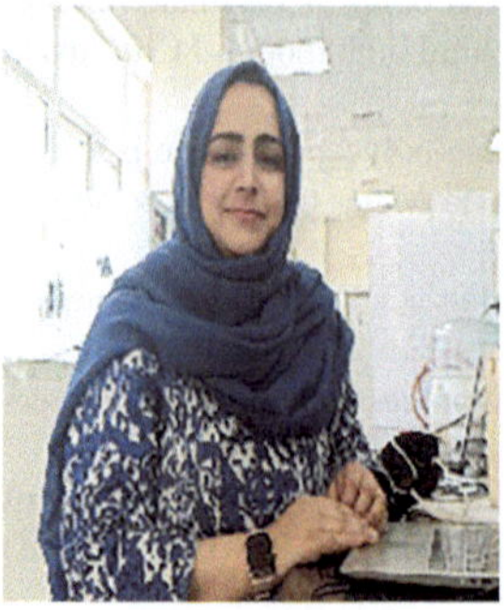

Anjum Hamid Rather was born in J&K, India, and studied for her M.Sc. in the Department of Bioresources, University of Kashmir, Srinagar, JK. She qualified for the CSIR-UGC NET in 2019 and was awarded a National Fellowship by the University Grants Commission (UGC) in 2020 to pursue her Ph.D. research for 5 years.

She is currently pursuing her Ph.D. degree in the Department of Nanotechnology, University of Kashmir, Srinagar, J&K, under the supervision of Dr. Faheem A. Sheikh. Her interests include developing polymeric nanofibers for advanced tissue engineering, wound healing applications, and exploiting the role of different nanoparticles and essential oils in tissue engineering. She has contributed to 13 peer-reviewed articles and 4 book chapters.

Taha Umair Wani received his bachelor's degree in Pharmacy from the University of Kashmir, India, in 2012 and his master's in Pharmaceutics from Punjab Technical University, Punjab, India, in 2014. He completed his Ph.D. in Pharmaceutics from the University of Kashmir, India, in 2020.

He is currently working as a research associate in the Department of Nanotechnology, University of Kashmir. He is the author of more than 40 publications. His research interests include drug formulation development, preclinical research, nanomedicine, and biomaterial fabrication. He has expertise in cell culture studies, antimicrobial assays, drug delivery, in vivo experiments, and pharmacokinetic studies.

Muheeb Rafiq was born and raised in Srinagar, J&K, India. He received his B.Sc. degree from the University of Kashmir in 2017. Later in 2012, he completed his M.Sc. in Nanotechnology from the University of Kashmir. During his master's, he gained a hands-on experience in the research area as he worked as an intern for 6 months under Dr. Mushtaq (Asst. Prof., Dept of Nanotechnology, University of Kashmir). After completing his master's, he started working on a project funded by the Council of Scientific and Industrial Research (CSIR) under the supervision of Dr. Faheem A. Sheikh (Asst. Prof., Dept of Nanotechnology, University of Kashmir).

He is currently enrolled in a Ph.D. program and is working under the mentorship of Dr. Faheem A. Sheikh. He has authored a 5 publications, including research papers and review articles. His research interests include the fabrication of biocompatible nanofibers and nanoporous scaffolds for hard and soft tissue engineering, the biological synthesis of nanoparticles, and post-modification of nanofibers.

Touseef Amna received her Ph.D. in Microbiology in 2006 from Aligarh Muslim University (AMU, Central University), India. Presently, she is working as an associate professor at Albaha University, Saudi Arabia. During her research career, she got the opportunity to spend postdoctoral stages at the University of Dortmund, Germany; Universidad de Talca, Chile; and Chonbuk National University, South Korea, and received various prestigious awards and grants. She authored two patents (one US patent and one Korean) and more than 100 research articles in journals of international repute. She is an editor, guest editor, and editorial board member of various national and international journals. Her research interests consist of the design and synthesis of biocompatible scaffolds for antimicrobial applications, cell culture, and tissue engineering.

M. Shamshi Hassan is working as an associate professor in the Faculty of Science Albaha University, Saudi Arabia. He was awarded an international fellowship to pursue his doctoral degree and received his Ph.D. in Chemical Engineering from the Department of Chemical Engineering, Chonbuk National University, South Korea, in 2010. After his Ph.D. he has been awarded Postdoctoral Fellowship (PDF) by the Textile and Fiber Engineering Department of CBNU, South Korea. His research interests include the synthesis of pure and hybrid metal oxides (quantum dots or nanocrystals, 1D, 2D, 3D, and flower-shaped morphology), nanostructured materials, or nanofibers and their application in supercapacitors and as biological/chemical disinfectants.

Syed Mudasir Ahmad is a professor and head of the Division of Animal Biotechnology, Faculty of Veterinary Sciences and Animal Husbandry, Shuhama, SKUAST-Kashmir. He has a teaching experience of 15 years. His research experience spans more than two decades, and he has published around 100 research articles in reputed journals.

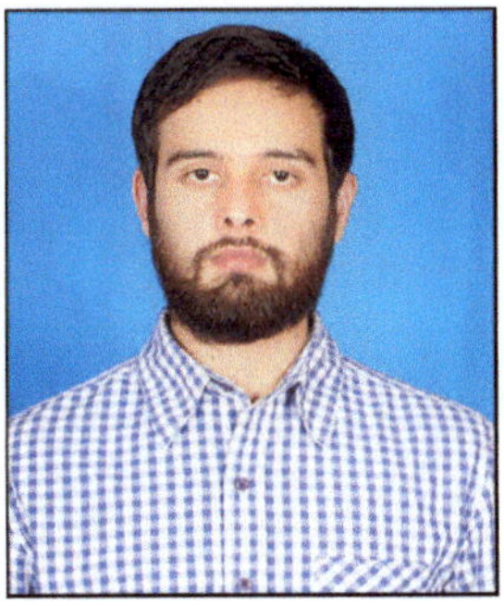

Shafquat Majeed is currently working as an assistant professor in the Department of Nanotechnology, University of Kashmir, India. Dr. Majeed did his master's in Chemistry from the Department of Chemistry, University of Kashmir. He did his Ph.D. at the prestigious Indian Institute of Science Bangalore (2008–2015). He has a postdoctoral research experience in rare-earth-based nanomaterials for biomedical applications and the fabrication of perovskite-based solar cells from the Centre for Nanoscience and Engineering, IISc Bangalore (2015–2016). He heads the laboratory, where they work on the synthesis, characterization, and optical and magnetic properties of nanomaterials prepared through various solution-based routes. More specifically, they are working on developing new and enhanced multimodal imaging modalities for biomedical and healthcare applications.

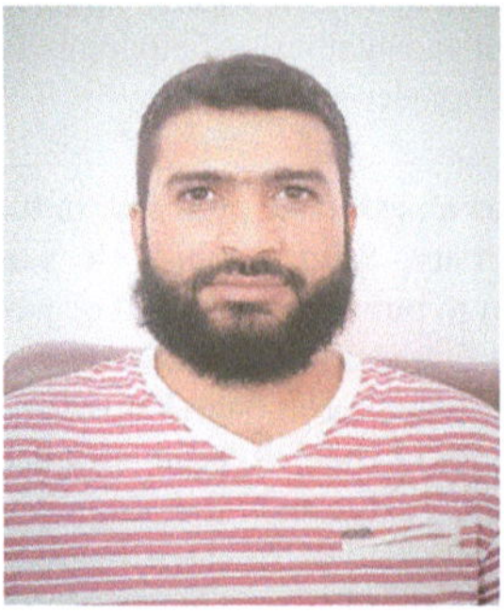

Mushtaq A. Beigh is working as an assistant professor at the Department of Nanotechnology, University of Kashmir, India. Dr. Beigh has postdoctoral research experience in cellular signaling and disease modeling (2013–2016). He completed his Ph.D. at the University of Kashmir, India (2008–2012). He heads a laboratory at the University of Kashmir, where they work on a bio-nano interface to understand the molecular basis of nanoconjugate internalization. They also try to understand the growth factor signaling inputs directed at multiple cellular receptors, like NRP1 and integrins, in order to work out the mechanistic details of receptor-based internalization.

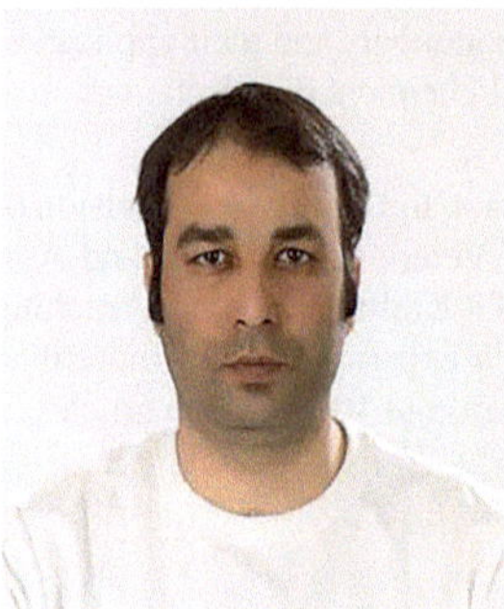

Faheem A. Sheikh is an assistant professor at the Department of Nanotechnology, University of Kashmir, India (2015). He served as an assistant professor at the Department of Biotechnology at the Central University of Kashmir, India (2015–2016); a research professor at Myongji University, South Korea (2014–2015); an assistant professor at Hallym University, South Korea (2012–2014); a postdoc/research fellow at the University of Texas Rio Grande Valley, Texas, United States of America (2010–2012); and a research professor at Myongji University, South Korea (2010). His research mainly focuses on fabricating nanomaterials used in tissue engineering.

Polyurethane Nanofibers Fabricated by Electrospinning as Drug Carrier Systems for the Treatment of Cancer

11

Anjum Hamid Rather, Rumysa Saleem Khan, Taha Umair Wani, Muheeb Rafiq, Touseef Amna, M. Shamshi Hassan, Shafquat Majeed, Mushtaq A. Beigh, and Faheem A. Sheikh

Abstract

Nanofibers fabricated by electrospinning as drug-delivery systems provide high therapeutic efficiency at a lower dose. The literature discussed in this chapter sheds light on the research innovations for developing multifunctional polyurethane nanofibers for the triggered delivery of drugs to treat various cancers. Being a hydrophobic polymer, it is easy to electrospun, and these nanofibers exhibit high tensile strength and elastic properties. Moreover, incorporating anticancer drugs into polyurethane and its derivatives may boost the anticancer agents' accessibility to tumor sites. This may further reduce cancer cells that the first dose of the medicines does not manage to kill and lessen the adverse side effects.

A. H. Rather · R. S. Khan · T. U. Wani · M. Rafiq · F. A. Sheikh (✉)
Nanostructured and Biomimetic Lab, Department of Nanotechnology, University of Kashmir, Srinagar, Jammu and Kashmir, India
e-mail: anjumhamid.scholar@kashmiruniversity.net; rumysakhan.scholar@kashmiruniversity.net; wanitaha.scholar@kashmiruniversity.net; muheeb.ntjrf@uok.edu.in; faheemnt@uok.edu.in

T. Amna
Department of Biology, Faculty of Science, Albaha University, Albaha, Kingdom of Saudi Arabia

M. S. Hassan
Department of Chemistry, Faculty of Science, Albaha University, Albaha, Kingdom of Saudi Arabia

S. Majeed
Laboratory for Multifunctional Nanomaterials, Department of Nanotechnology, University of Kashmir Hazratbal, Srinagar, Jammu and Kashmir, India
e-mail: smshah@uok.edu.in

M. A. Beigh
Cellular Signalling and Nanotherapeutics Laboratory, Department of Nanotechnology, University of Kashmir Hazratbal, Srinagar, Jammu and Kashmir, India
e-mail: beighm@uok.edu.in

F. A. Sheikh et al. (eds.), *Interaction of Nanomaterials With Living Cells*,
https://doi.org/10.1007/978-981-99-2119-5_11

This chapter will highlight general nanotechnological tools that can help to diagnose cancers, e.g., quantum dots, nanoshells, and gold nanoparticles. Particular emphasis is placed on using polyurethane nanofibers for targeted drug delivery and controlled release. Finally, we will learn how polyurethane nanofibers can be incorporated with various compounds to induce hyperthermia effectively for cancer treatment.

Keywords

Electrospinning · Nanofibers · Drug delivery · Polyurethane · Cancer

11.1 Introduction

Cancer is defined by uncontrolled multiplication and the absence of cell death, resulting in an abnormal mass or tumor (Bertolaso 2016). The main difference between cancer and a tumor is that the latter is localized; these cells grow in a specific location and can be removed surgically. In contrast, cancerous cells spread from one place to another and kill the host. Cancer is caused by a variety of factors, including genetic factors (mutations, translocations, and hereditary) and environmental factors (UV, chemicals, viral infections) (Parsa 2012). All cells have protooncogenes; if these genes are transformed into oncogenes, the normal cell will be changed into a cancer cell (Bishop 1987). Furthermore, mutations in the tumor suppressor gene increase the risk of evolving a tumor or cancer. As a result, the growth-promoting protooncogene and the growth-restricting tumor suppressor genes must be appropriately balanced for cell growth control.

Various treatments are available for curing cancer, such as surgery, radiation, chemotherapy, hormone therapy, immune therapy, etc. (Abbas et al. 2018). All the treatments aim to remove cancerous tissues altogether without harming healthy tissues, which is difficult to achieve. Among existing cancer treatments, chemotherapy is the most employed form of intervention. However, this is associated with detrimental side effects caused by chemotherapeutic agents, such as drug resistance and high cost (Mitra et al. 2015). These treatments may cause kidney failure, nerve injury, nausea, cell toxicity, and death if high doses are administrated (Baudino 2015). To address the limitations of current therapy, researchers are looking for new promising anticancer agents with higher efficacy and fewer side effects.

Nanotechnology has improved drug-delivery routes in recent years by reducing the adverse effects of delivering a minute concentration of drugs (Patra et al. 2018). In general, a nano-drug-delivery system allows for more controlled delivery (release rate and time) and the maintenance of drug concentration at a certain level within the valid therapeutic window. It is regarded that nanotechnology-based targeted drug-delivery systems such as liposomes, hydrogels, nanoparticles (NPs), and nanofibers could provide anticancer drugs with sustained release (Kumar et al. 2012). However, in recent years, there has been a surge in interest in using electrospun nanofibers for drug delivery (Contreras-Cáceres et al. 2019). These nanofibers' popularity can be attributed to their high porosity, large surface area, tiny size, and interconnecting channels, which offer a valuable strategy for high loading and the release of drugs. These ID nanostructures, e.g., the nanofibers in current times, are mainly produced by the electrospinning technique. Electrospinning offers greater flexibility in selecting biodegradable or nonbiodegradable materials to form nanofibers resulting in remarkable properties such as more advanced control over drug release kinetics (Elsadek et al. 2022). Moreover, the site-specific delivery of drugs using a patch system in a particular area is the main benefit of the fibrous carriers. Additionally, many drugs may be directly contained within the fibers and then released sustainably. Because of their large surface area, these nanofibers can cause a water-insoluble drug to decompose slowly and then release and protect it from corrosion by stomach acid and enzymes, improving drug stability (Castillo-Henríquez et al. 2020).

Several anticancer drugs have been loaded into various electrospun nanofibers for cancer treatment, including paclitaxel (PTX) (Ma et al. 2011; Faraji Dizaji et al. 2020), doxorubicin (DOX) (Abasian et al. 2019), 5-fluorouracil (5-FU) (Grant et al. 2021), and camptothecin (Amna et al. 2020). The nanofibers' high surface-to-volume ratio is a unique structural feature; however, it causes undesirable initial burst release of drugs. The drug loading may require a different approach depending on the type of drug incorporated, such as small molecule drugs, proteins, and genes, to achieve a successful release profile. Medications can be simply mixed with the polymer solution (Hu et al. 2015), physically or chemically surface immobilized (Yoo et al. 2009), or indirectly loaded onto the nanofiber (Thakkar and Misra 2017). The simplest way to incorporate drugs into nanofibers is to dissolve the drugs and polymer in a single solvent and then co-electrospin them. Nanofibers have high drug-loading efficiency due to a high surface-to-volume ratio, which can further increase when combined with self-assembled NPs or microspheres that are evenly distributed within the nanofiber mats (Forouharshad and Ajalloueian 2022; Fan et al. 2016).

Drug-loaded nanofibers can be fabricated from both natural and synthetic polymers; particularly, biodegradable polymers are preferred. Biodegradable polymers are particularly interested in electrospinning because they eliminate the need for a second surgery to remove the implanted carrier (Tian et al. 2012). Various synthetic polymers such as polylactic acid (PLA) (Abasian et al. 2019; Dai et al. 2017), polycaprolactone (PCL) (Yan et al. 2020), and polyethylene oxide (PEO) (Darbasizadeh et al. 2021; Rengifo et al. 2019) have been extensively studied to fabricate nanofibers with desired properties for drug-delivery applications. However, hydrophobic polymers are accessible to electrospin because of the highly volatile solvents available, and the nanofibers exhibit high tensile strength and elastic properties (Son et al. 2014). One of the most well-liked polymers is biodegradable polyurethane, in which each molecule comprises a hard segment, a soft segment, and a chain extender. It is inexpensive, highly biocompatible, and thus proper for drug administration systems (Bahadur et al. 2017). For instance, incorporating anticancer drugs into polyurethane and its derivatives may boost the anticancer agents' accessibility to tumor areas, reduce cancer cells that the first dose of the drugs did not manage to kill, and lessen the adverse side effects (Krukiewicz and Zak 2016). However, polyurethane nanofibers for drug-delivery applications have received less focus. Therefore, this chapter aims to examine the uses of electrospun polyurethane nanofibers for cancer drug delivery.

11.2 Nanotechnology in Cancer Diagnosis

Currently, we can diagnose cancer using imaging tools and the morphological examination of tissues or cells. Moreover, imaging procedures like endoscopy, ultrasonography, computed tomography, and magnetic resonance imaging are often employed (Frangioni 2008). However, these traditional approaches to cancer diagnosis have numerous drawbacks because they cannot identify cancer at a very early stage, lack specificity, and have severe side effects. By that time, innumerable tumors may have increased and even metastasized. The current monsterlike situation of cancer has forced researchers to create several methods for accurate diagnosis. As a result, researchers turn to a different technique: a nanotechnology-based diagnostic technique that is being developed as a potentially helpful tool for speedy, practical, and affordable cancer detection (Zhang et al. 2019).

Nanotechnology involves creating and assembling NPs, which are submicroscopic objects with diameters between 1 and 100 nm. A subfield of nanotechnology known as nanomedicine has evolved as a cutting-edge technique for using NPs to detect and treat human disease (Lee et al. 2012). Nanomedicine enables early diagnosis, treatment with few side effects, and noninvasive evaluation of treatment effectiveness. Cancer nanotechnology aims for cellular and molecular elements to interact with nanoscale devices, specifically concerning cancer diagnosis and treatment (Cuenca et al. 2006). The potential of cancer nanotechnology resides in the ability to create therapeutic vehicles with distinct features that, due to their small size, can profoundly and specifically enter tumors. One of the most critical steps in any diagnosing process is screening. Effective screening at the beginning of the disease can help to decrease its adverse effects. Nanotechnology makes it possible to swiftly and precisely detect cancer-related substances, allowing for observing molecular changes even when they are only present in a small percentage of cells. NPs can bind cancer biomarkers, such as proteins, carbohydrates, and nucleic acids, which are expressly released by the cancer cells and thus help diagnose and target drug delivery to the tumor site (Wu and Qu 2015). This would make it possible to diagnose cancer early, which is essential for enhancing cancer treatment. Nanotechnology will also reduce screening tools, meaning many tests can be run on a single device. As a result, cancer screening is quicker and more economical. Quantum dots, nanoshells, and gold nanoparticles (Au NPs) are among the significant nanomaterials involved in cancer diagnosis that are covered here.

11.2.1 Quantum Dots

Quantum dots are highly fluorescent nanometer-size crystals of semiconductor materials that exhibit strong fluorescence emitting different colors concerning their size (Zhang et al. 2008). QDs are resistant to photobleaching, have a high detection intensity, and are thus frequently used for diagnostic applications (Smith et al. 2014). Depending on their size and composition, they have an intrinsic fluorescence emission spectrum with a wavelength ranging from 400 to 2000 nm and are made up of an inorganic elemental core (such as cadmium or mercury) and an outer layer of metal (Nikalaje and Shende 2018). The optical characteristics of NPs are strongly influenced by their structure, particularly the color (wavelength) and diameter of quantum dots. The quantum dots can be injected into a subject and then detected by exciting them to emit light. For instance, cadmium selenide NPs (also known as quantum dots) illuminate when exposed to UV light. They seep into cancerous tumors when injected. The glowing tumor is visible to the surgeon, who can utilize it as a reference point for more precise tumor excision. Nida et al. investigated the use of quantum dots conjugated to anti-epidermal growth factor receptor antibodies to detect precancerous biomarkers in vitro. They found illumination of SiHa cervical cancer cells due to epidermal growth factor receptor overexpression (Nida et al. 2005).

11.2.2 Nanoshells

Nanoshells may be spherical in shape with a diameter of 10–300 nm, have a dielectric core often made of silica, and are encased in a thin metal shell made of gold (Zhao et al. 2014). Nanoshells have the ability to either absorb or scatter light. While scattering nanoshells are utilized as contrast agents, absorbing nanoshells are primarily used to induce hyperthermia (increase in body temperature) (Lal et al. 2008). Hyperthermia, commonly referred to as thermal therapy, is one of the main techniques to eradicate tumors where tumor cells are killed by subjecting them to high temperatures using nanoshells. Hyperthermia involves heating tumor cells to a temperature of 40–45 °C to inhibit tumor cells by initiating a sequence of thermally induced metabolic processes, such as apoptosis (Beik et al. 2016). Compared to normal cells, malignant cells are more sensitive to high temperatures and, therefore, can be killed explicitly by hyperthermia using nanoshells.

11.2.3 Gold (Au) NPs

With their distinctive unusual optical and electrical properties and minimal toxicity, Au NPs are one of the most alluring families used in cancer diagnostics (Fan et al. 2020). This is because Au can be easily synthesized and has been approved for use in the treatment of human disease due to its high biocompatibility. These NPs have been investigated as a delivery system for therapeutic agents, in photodynamic therapy for cancer treatment, and as diagnostic tools to find disease-related biomarkers (Singh et al. 2018). These Au NPs act as contrast agents by scattering visible light in vitro. Au NPs can also be used in conjugation with antibodies for biopsies and identification of cervical and pancreatic cancers (Jain et al. 2007). Smilowitz et al. enhanced radiation and X-ray imaging in mice by using Au NPs (Hainfeld et al. 2004). They discovered that because the tumor's vascular permeability was higher, Au NPs aggregated there, and a higher contrast was visible. These findings demonstrate the potential of Au NPs to increase tumor diagnostic rates. Thus, Au NPs are expected to play a more critical role in X-ray scattering imaging. A different work by Lu et al. showed multifunctional oval-shaped Au NPs for the targeted detection of breast cancer using a straightforward colorimetric and highly sensitive two-photon scattering assay (Lu et al. 2010). Oval-shaped Au NPs caused a noticeable color change and a 13-fold increase in two-photon scattering intensity compared to the breast cancer SK-BR-3 cell line. The technique successfully identified cancerous and noncancerous cells and set them apart from other breast cancer cell lines that exhibit low levels of human epidermal growth factor receptor 2. These discoveries have significant implications for early detection because the technology can detect malignancies inside the body that are only a few millimeters in diameter. As a result, Au NPs are a priority-based diagnostic tool for many malignancies.

11.3 Electrospinning

A number of advanced techniques have been developed for the generation of 1D nanostructures, including electron beam or focused ion beam (Tseng 2005), hydrothermal synthesis (Hu et al. 2007), chemical vapor deposition (Zang et al. 2009), electrospinning (Shafi et al. 2023), etc. Among these methods, electrospinning is the simplest method to produce nanofibers with unique properties such as small diameter, high surface area, adjustable porosity, and small fiber-to-fiber distance (Jeckson et al. 2021). This technique was demonstrated more than 90 years ago and was first patented in the 1930s (Subbiah et al. 2005). However, it did not receive much attention till the early 1990s. It has a low start-up cost and is compatible with other pre-/posttreatment methods. To date, the electrospinning technique has evolved as the ideal technique used to make nanofibers. Therefore, in this chapter's coming text, we will elaborate the electrospinning technique that has been used to deliver anticancer drugs. Unlike other methods for forming nanostructures, electrospinning is based on the electric field force acting on the polymer solution (this solution can inherit anticancer drugs). In electrospinning, solidified fiber can be achieved by stretching the electrified jet for the electrostatic repulsions between the surface charges and the eventual evaporation of solvent (Li and Wang 2013). A basic setup of electrospinning is shown in Fig. 11.1. This mainly consists of syringe pump, a spinneret, a high-voltage power supply, and a collector (Long et al. 2019). During the process, the polymer solution is placed into a syringe, and then it is pushed to the tip of the spinneret by external pumping (desired flow rate) to provide a continuous

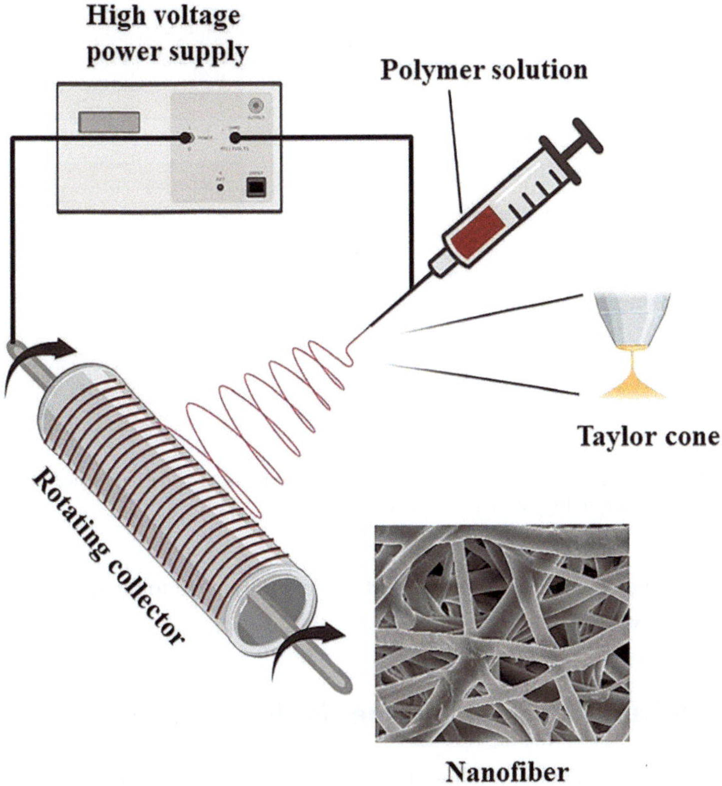

Fig. 11.1 A straightforward illustration of the electrospinning setup used to create nanofibers using a high-voltage power source and the Taylor cone's creation

flow of solution. When a droplet is formed at the spinneret, and electrical voltage (typically 1–30 kV) is applied between the tip of the spinneret and the collector placed in front of it, the solvent quickly evaporates from the jets, and solid nanofibers are finely deposited on the collector. Typically, the diameters of the electrospun fibers can be controlled in the range of tens of nanometers to micrometers, and the fibers can be deposited as nonwoven mats or aligned into uniaxial arrays and further stacked into multilayered architectures. Here collector plate should be of good electric conductivity to neutralize the charge carried by the polymer nanofiber. This technique could be applied to synthetic and natural polymers, polymer alloys, and polymer-incorporated functional nanomaterials. These unique advantages of electrospinning impart multifunctional properties to nanofibers for diverse applications. To achieve the desired fiber diameter, bead-free fibers, as well as a porous and uniform fibrous mat, optimization of the electrospinning parameters is necessary (Ibrahim and Klingner 2020). These electrospinning parameters include three main components—solution, process, and ambient—as shown in Fig. 11.2. All

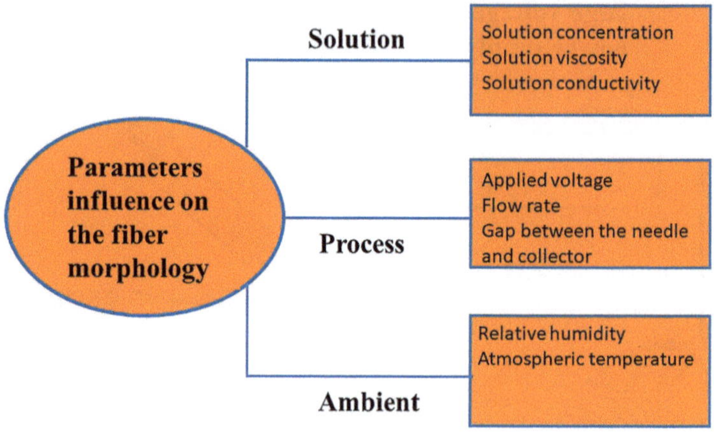

Fig. 11.2 Mentioned in this figure are the various parameters that influence the resulting nanofibers' final morphology during electrospinning

of these parameters are extremely important, and they have to be adjusted carefully for successful electrospinning.

11.4 Polyurethane as a Polymer for Electrospinning

Polyurethane is a synthetic polymer formed by the polycondensation of isocyanates and polyols (alcohol that has two or more hydroxyl groups within a molecule), with a third component acting as an extender in some cross-linked products, as shown in Fig. 11.3 (Akindoyo et al. 2016). Dr. Otto Bayer invented and studied polyurethane for the first time in 1937 (Joseph et al. 2018). It is one of the most adaptable macromolecular compounds, with soft and hard segments. The former is created through the isocyanate-polyol reaction, while the latter is created through the isocyanate-chain extender reaction (Das and Mahanwar 2020). Due to the versatility of this polymer, it has practically been used in every field of material engineering (e.g., coatings for aircraft to all types of shoes) (Sikdar et al. 2022). The main benefits of polyurethane are its durability, toughness, and chemical resistance. The urethane group is repeating unit in polyurethane and is formed by the reaction of alcohol and isocyanate; however, this polymer also contains ethers, esters, urea, and some aromatic compounds. The properties of a polyurethane are typically determined by the polyols and isocyanates from which they are derived. Generally, stretchy polymers can be obtained from long segments of polyols with low cross-linking, whereas rigid polymers can be obtained from shorter chains with high cross-linking. Moreover, another group of compounds that often play essential roles are chain extenders and cross-linkers (Bin Ying et al. 2020). These are highly beneficial in improving the morphology of polyurethane adhesives, elastomers, fibers, and a variety of other significant microcellular and skin foams. The hard segments,

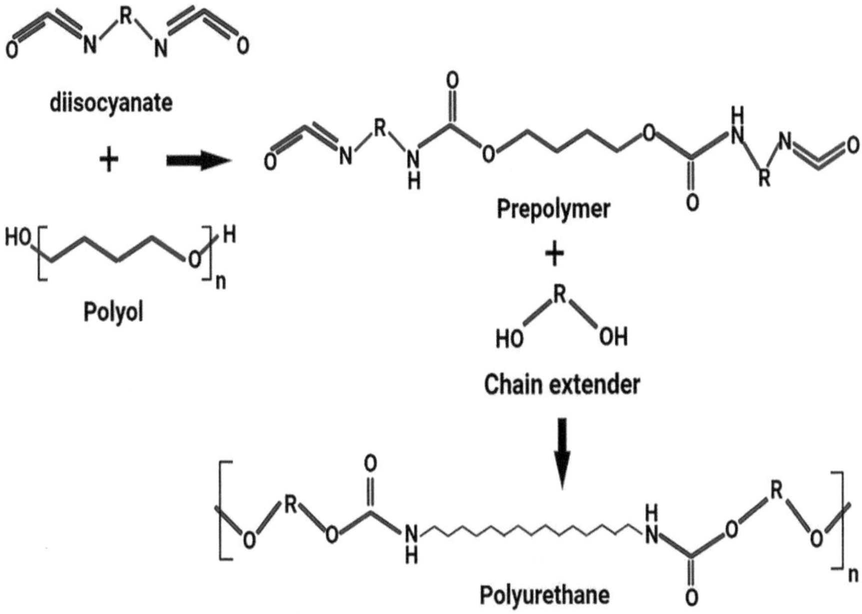

Fig. 11.3 Traditional polyurethane synthesis method uses polyol, diisocyanate, and a low molecular weight chain extender created using Bio Render

which are made of isocyanate and chain extenders, are immobile and stiff, whereas the soft segments, which are made of polyols, can move freely and frequently appear in foil forms (Mohamad Sadeghi and Sayaf 2012). Covalent bonding between the hard and soft segments inhibits plastic flow within the polymer chains, resulting in elastomeric resilience (Prisacariu 2011).

11.5 Electrospun Polyurethane Nanofibers for Drug Delivery Against Cancer

Compared to conventional drug-delivery strategies, such as using films and gels, the nanofibers have a large surface area to volume ratio and enough porosity, making it feasible to deliver homogeneous, significant, and regulated dosages of pharmacological preparation at the target site. Nanofibrous drug-delivery systems also provide high therapeutic efficiency at the lower dose, reduced toxicity (since the drug is delivered to the precise action site), accessibility to a broad application area, and a low incidence of side effects (Joshi et al. 2014). Polyurethane has been chosen as the preferred polymer for creating nanofibers due to several benefits, including good strength, chemical resistance, biocompatibility, and a high elastic memory for sustaining tension (Vaithylingam et al. 2017). Pertinently, it can also support patients in the sense of comfort due to its high moisture transmission rate, which is important for breathability and softening at body temperature, both of which are required to retain their qualities if used as a patch system delivery.

To load and regulate the release of doxorubicin, Kilic et al. fabricated polyurethane nanofibers (Kiliç et al. 2018). When the kinetics of the release of 3 and 6 mg of doxorubicin was investigated at pH 4.5 and 7.5, it was found that the drug loaded to drug released was inverse. At pH 4.5, polyurethane nanofibers loaded with doxorubicin exhibited immediate release behavior and more than 70% in the initial 10 min. In contrast, at pH 7.5, doxorubicin-loaded nanofibers exhibited slower release, and 80% of the drug was released in 50 min. Most drugs are said to accumulate on the fiber surface during the electrospinning process, which is why the literature claims that the electrospun fibers typically exhibit an initial surge release (Yu et al. 2014). Drug release at pH 4.5 was seen to be in this situation, whereas pH 7.5 saw a more regulated release. However, the initial burst release of drugs from the nanofiber surfaces is the primary problem with drug-loaded nanofibrous systems. An easy, unique solution to these issues is the combination of drug-loaded NPs inside the nanofibers. The drug nanocarriers then incorporated into electrospun nanofibers can result in prolonged and sustained release.

Shahrousvand et al. used a technique to create the poly (ε-caprolactone) (PCL)-diol based (b)-polyurethane (Shahrousvand et al. 2016). In short, hexamethylene diisocyanate and 12 g PCL-diol were combined in a glass vial reactor and heated to 85 °C for 3 h. The reaction was then continued for another 20 min after 1.08 g of 1,4-butanediol was added to the reactor's contents. The reactor's contents were then dried for 24 h at 70 °C in a vacuum oven. A three-time deionized water wash was performed on the synthesized PCL-diol-b-polyurethane. Then, PCL-diol-b-polyurethane/Au composite nanofibers were made by electrospinning PCL-diol-b-polyurethane solutions, which were combined with Au NPs (Irani et al. 2017). The Au NPs were used to increase the anticancer activity of nanofibers against glioblastoma cells because of their small size, low toxicity, and high potential to pass the blood-brain barrier. Successful encapsulation of TMZ is one of the most effective treatments for glioblastoma because of its capacity to pass through the blood-brain barrier. A sustained delivery method of TMZ from nanofibers was observed, allowing for a continuous release for the treatment of glioblastoma (Fig. 11.4). Because TMZ has a short plasma half-life of 1.8 h, polymeric micro/NPs have been used to encapsulate it for controlled release. The cytotoxicity of the produced nanofibers against U-87 human glioblastoma cells showed that, in contrast to free TMZ, which essentially does not affect the percentage of cell multiplication, the Au coating on the nanofibers' surface improved the cytotoxicity of PCL-diol-b-polyurethane/Au@TMZ nanofibers (Fig. 11.5).

In another study, Seyyedi et al. used acrylated polyurethane/nanohydroxyapatite (n-Hap) nanocomposites to incorporate the anticancer drug PTX (Seyyedi and Molajou 2021). The results showed that the synthetic fibers are excellent candidates for drug loading as the loading efficiency was more than 90% for all manufactured nanofibers. However, the average fiber diameter increased from 290 to 360 nm by loading PTX into the nanofibers. Although the hydrophilic nature of n-Hap NPs could lead to a faster release of drug molecules, the filling of some nanofiber pores

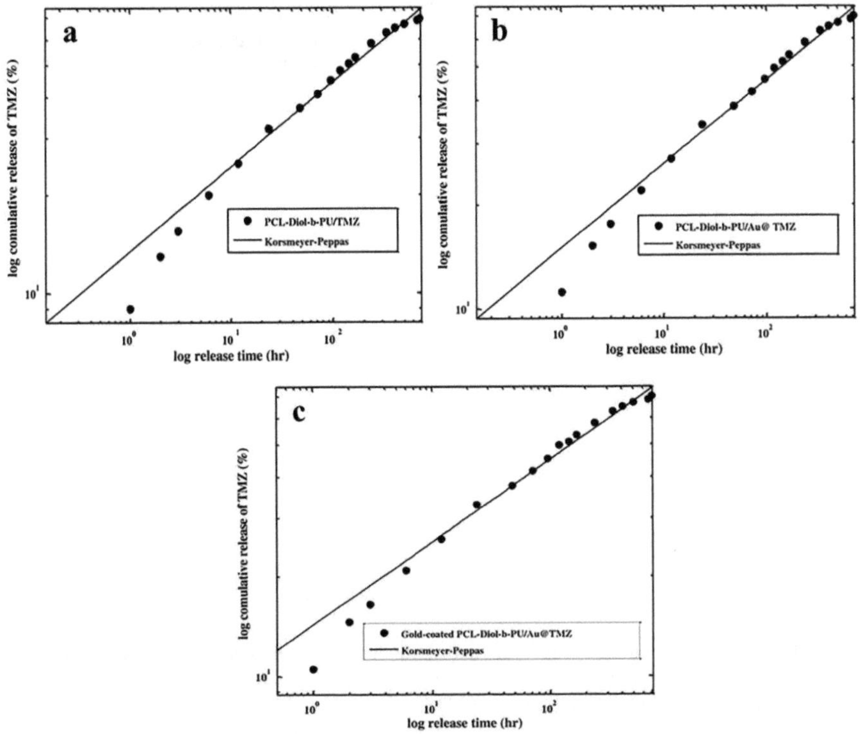

Fig. 11.4 TMZ release from (**a**) PCL-diol-b-polyurethane/TMZ, (**b**) PCL-diol-b-polyurethane/Au@TMZ, (**c**) Au-coated PCL-diol-b-polyurethane/Au@TMZ nanofibers, and fitting data with Korsmeyer-Peppas kinetic model. During the first 24 h, Au-coated PCL-diol-b-polyurethane/Au@TMZ nanofibrous formulations showed lower burst release (24%) in comparison with PCL-diol-b-polyurethane/Au@TMZ (32%) and PCL-diol-b-polyurethane/TMZ (30%). *PU* polyurethane. (Reproduced from Irani et al. 2017 with copyright permission from Elsevier's)

with n-HAp NPs prevented PTX molecules from being released from the nanofibers, which neutralized the effect of hydrophilicity on the release rate of PTX and led to a gradually increasing drug release percentage.

For the prolonged release of DOX and folic acid (FA) against breast cancer, UiO-66 metal-organic framework (MOF) NPs loaded with carboxymethyl chitosan (CMC)/poly(ethylene oxide) (PEO)/polyurethane core-shell nanofibers were fabricated by Farboudi et al. (2020a). However, due to their hydrophilicity, CMC/PEO nanofibers could not release anticancer medications over an extended period. The hydrophobic polymer polyurethane was applied to the chitosan nanofibrous layer to increase the hydrophobicity of the nanofibers. FA and DOX had better than 95% drug-loading efficiency for the fabricated MOF-loaded nanofibers. Neither the DOX nor the FA drugs were released in a burst from the FA-DOX/UiO-66 loaded CMC/PEO nanofibers throughout 72 h. The co-delivery of DOX and FA from DOX-FA/UiO-66 nanofiber led to the cell death rise from

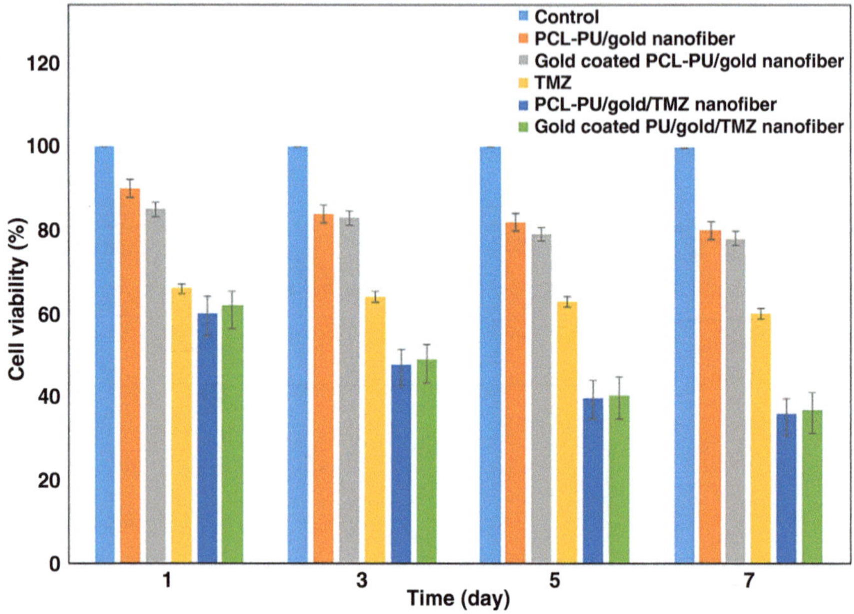

Fig. 11.5 In vitro cytotoxicity of TMZ and TMZ-loaded synthesized nanofibers. PCL-diol-b polyurethane/Au@TMZ and gold-coated PCL-diol-b-polyurethane/Au@TMZ nanofibers exhibited more cytotoxicity against U-87 cells than pure TMZ after 3, 5, and 7 days, which could be attributed to the slower release of TMZ from nanofibers. *PU* polyurethane. (Reproduced from Irani et al. 2017 with copyrights permission of Elsevier's)

41 ± 1% to 54 ± 2%, as shown in Fig. 11.6. The DAPI staining also revealed nuclear fragmentation in the chromatin and apoptosis in MCF-7 cells as indicated by the reduction in the intensity of the dye. As a result, simultaneous administration of DOX and FA into the core-shell fibers may considerably improve the apoptotic nuclei (Fig. 11.7).

Moreover, the antitumor effectiveness of anticancer drug-delivery systems can be increased by including mixtures of cytotoxic drugs within the drug carriers. For instance, Farboudi et al. synthesized magnetic Au-coated PCL-b-polyurethane/poly (*N*-isopropylacrylamide)-grafted-chitosan core-shell nanofibers for controlled release of PTX and 5-FU toward 4T1 breast cancer cells (Farboudi et al. 2020b). Coating core-shell nanofibers with magnetic Au NPs caused a progressive rise in surface roughness. Therefore, the drug-loaded nanofibers coated with magnetic Au NPs were a good substrate for 4T1 cancer cell proliferation and treatment of breast cancer cells (Fig. 11.8). Tumor growth was continuously inhibited by the prolonged release of anticancer drugs from drug-loaded nanofibers and magnetic gold-coated nanofibers, respectively. After 20 days, the presence of drug-loaded nanofibers coated with magnetic Au NPs resulted in the least amount of tumor growth, as shown in Fig. 11.9. The outcomes demonstrated that the simultaneous insertion of PTX and 5-FU into the core-shell fibers and coating of nanofibers with magnetic Au

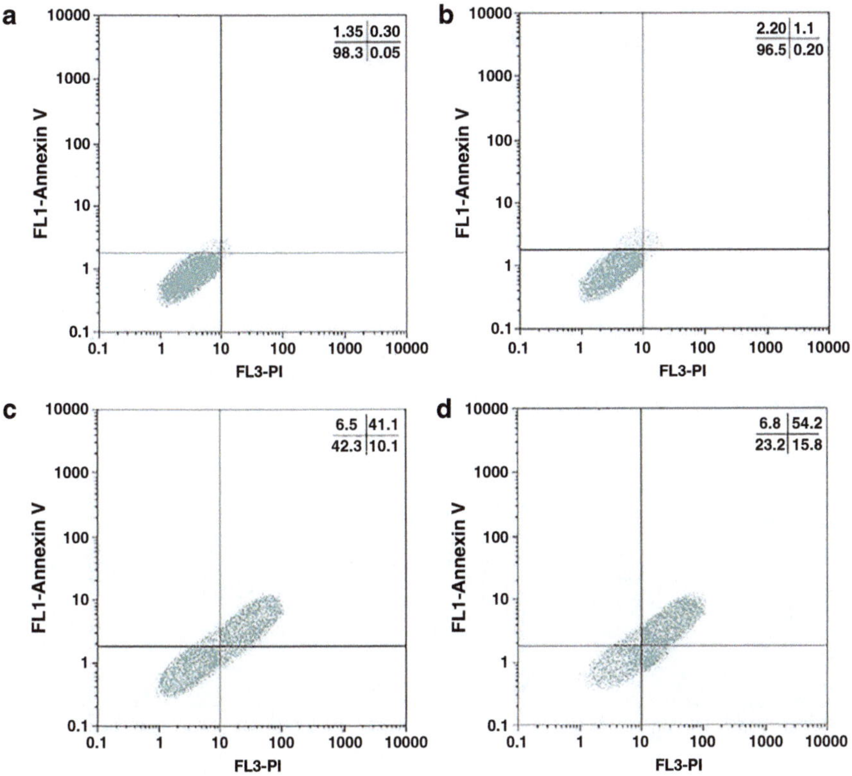

Fig. 11.6 Flow cytometry analyses of MCF-7 cells treated with the (**a**) control group, (**b**) core-shell nanofibers, (**c**) DOX/UiO-66 loaded core-shell fibers, and (**d**) DOX-FA/UiO-66 loaded core-shell fibers after 24 h. *PU* polyurethane. (Reproduced from Farboudi et al. 2020a with copyright permission of Elsevier)

NPs were beneficial for boosting the anticancer activity of produced nanofibrous carrier toward 4T1 breast cancer cells (Fig. 11.10). During the G2-M phases, PTX encourages microtubule assembly and hinders the formation of mitotic spindles. These activities reduce the growth of tumor cells. 5-FU, on the other hand, can stop the growth of tumor cells by interfering with nucleic acid metabolism during the G1-S phase of the cell cycle. In order to prevent the dispersion of drug-loaded nanofibrous carriers onto normal cells, the external magnetic field can control the residence time of drug-loaded nanofibers coated with magnetic Au NPs on the surface of cancer cells. This could result in the presence of a high concentration of drugs in the target area. Therefore, in the presence of a magnetic Au NPs coated-core-shell nanofibrous carrier, the external magnetic field causes the apoptosis of breast cancer cells. Thus, the localized delivery of anticancer drugs through the nanofibrous drug-delivery systems has more significant benefits in treating breast cancer.

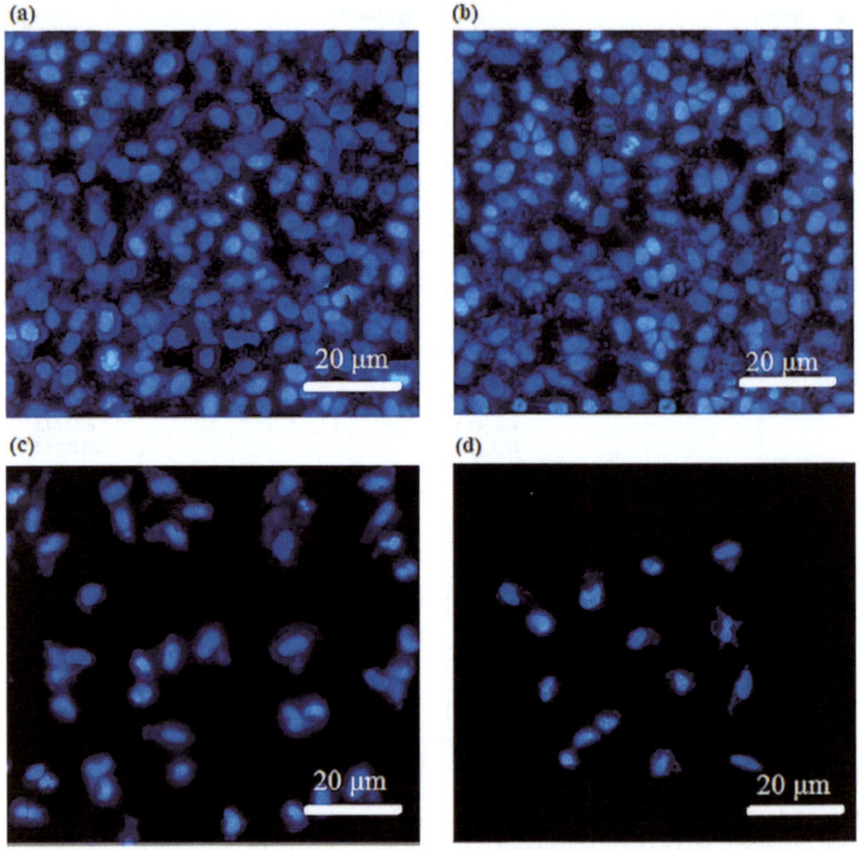

Fig. 11.7 DAPI stained images of (**a**) untreated MCF-7 cells, (**b**) MCF-7 cells treated with core-shell nanofiber, (**c**) MCF-7 cells treated with DOX/UiO-66 loaded core-shell fibers, and (**d**) MCF-7 cells treated with DOX-FA/UiO-66 loaded core-shell fibers after 24 h. (Reproduced from Farboudi et al. 2020a with copyright permission of Elsevier's)

11.6 Thermo-Responsive Nanofibers

The cancer cells are sensitive to body temperatures above normal (hyperthermia), making them susceptible to cell death. Given this, one of the main techniques for treating a cancer cell is hyperthermia. In order to provide simultaneous release of PTX and 5-FU, Erik et al. created poly-(*N*-isopropylacrylamide-co-*N*-(hydroxymethyl)acrylamide) (p-NP-HM)/polyurethane core-shell nanofibers (Aguilar et al. 2017). Figure 11.11 demonstrates that the drug release profile of the composites has a temperature-sensitive response. The drug release was observed to be zero at 25 °C (room temperature) and 36.5 °C, corresponding to normal body temperature. However, the release of 28% and 29.5% PTX and 2.3% and 23.7%

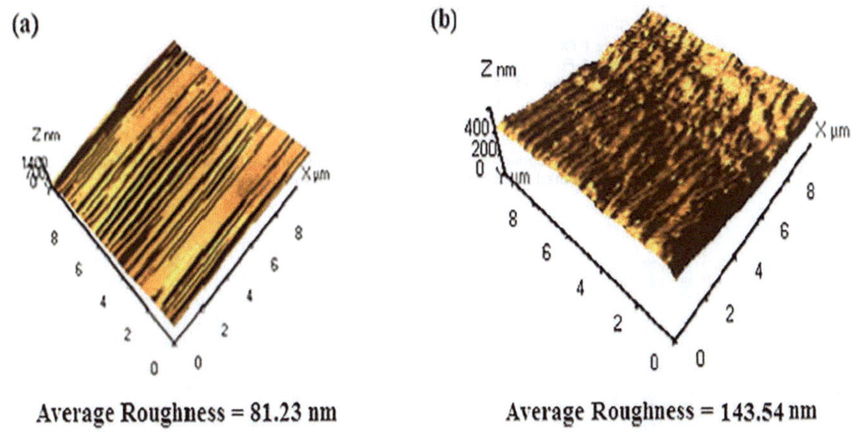

Fig. 11.8 AFM analysis of (**a**) pure nanofibers and (**b**) Au NP-coated nanofibers. The surface roughness of the core-shell nanofibers was less rough, as shown in figure; however, as the nanofibers were coated with magnetic Au NPs, the surface roughness increased. (Reproduced from Farboudi et al. 2020b with copyrights permission of Elsevier)

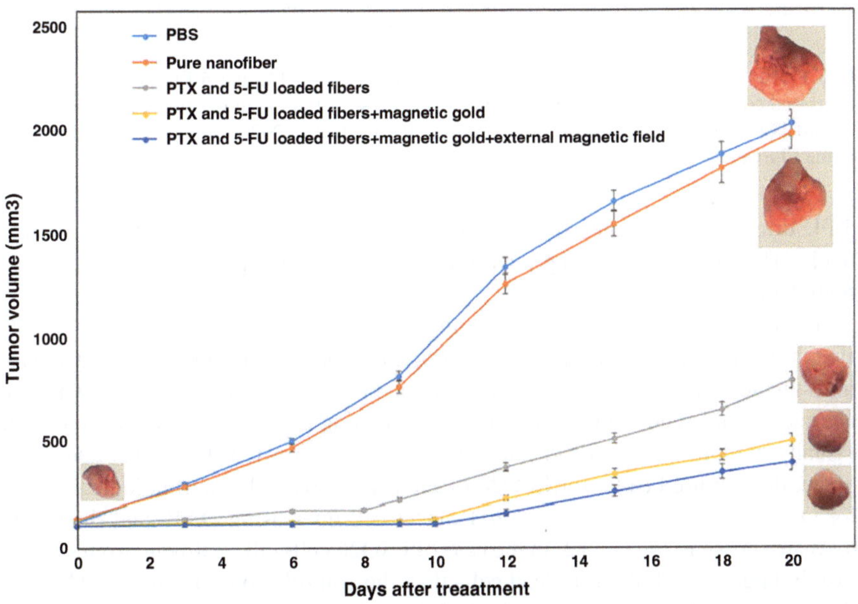

Fig. 11.9 The figure showed that in the presence of an external magnetic field, the PTX and 5-FU loaded nanofibers coated with magnetic Au NPs resulted in the most nominal tumor growth after 20 days, whereas the pure nanofibers did not have obvious cytotoxicity toward 4T1 breast cancer cells and the tumor volume increased rapidly. (Reproduced from Farboudi et al. 2020b with copyright permission of the Elsevier's)

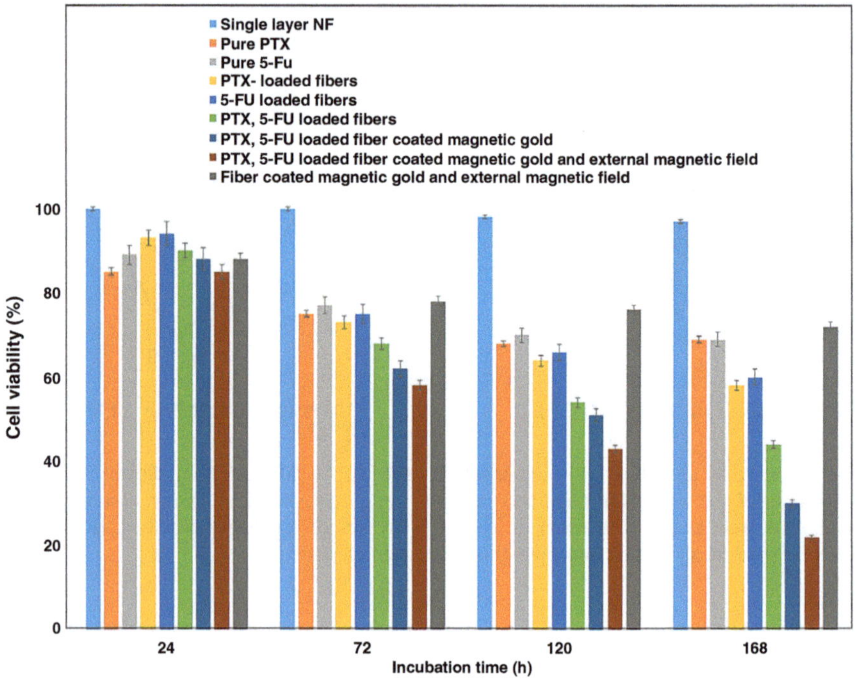

Fig. 11.10 Cell viability testing revealed that incorporating PTX and 5-FU into the core-shell fibers and coating nanofibers with magnetic Au NPs resulted in the most remarkable cytotoxicity toward 4T1 breast cancer cells. (Reproduced from Farboudi et al. 2020b with copyright permission of the Elsevier's)

5-FU drugs occurred when the medium temperature reached 40 and 50 °C, respectively. This suggests that drug release increases as the temperature exceeds normal body temperature.

The primary mechanism for releasing the drugs was the component of p-NP-HM switching from a hydrophilic to a hydrophobic phase. The in vitro drug release investigation for PTX and 5-FU showed good controlled release profiles when the alternating magnetic field was applied. Cell staining experiments have been used to visually explore the live/dead ratio utilizing (Resazurin red, live cells, and Sytox green, dead/injured cells) for ESO26 (adenocarcinoma). The morphological studies for OE21 (squamous cell carcinoma) cancer cell lines using DAPI and rhodamine dyes are shown in Fig. 11.12. In the case of ESO26 cells, all treatment groups have a visibly larger distribution of dead cells than the control group. By using DAPI and rhodamine labeling, the morphological investigation of OE21 revealed nucleus blebbing, restricted cell spreading, and stunted actin expansion. This suggests cytotoxicity across all treatment groups. Favorable results can be achieved especially with HHRx-hyperthermia (45 °C) with dual drug release and LHRx-hyperthermia (40 °C) with single drug release treatment groups. Thus, it was shown that the drug-loaded nanofibers were highly effective against the cancer cell lines ESO26 and

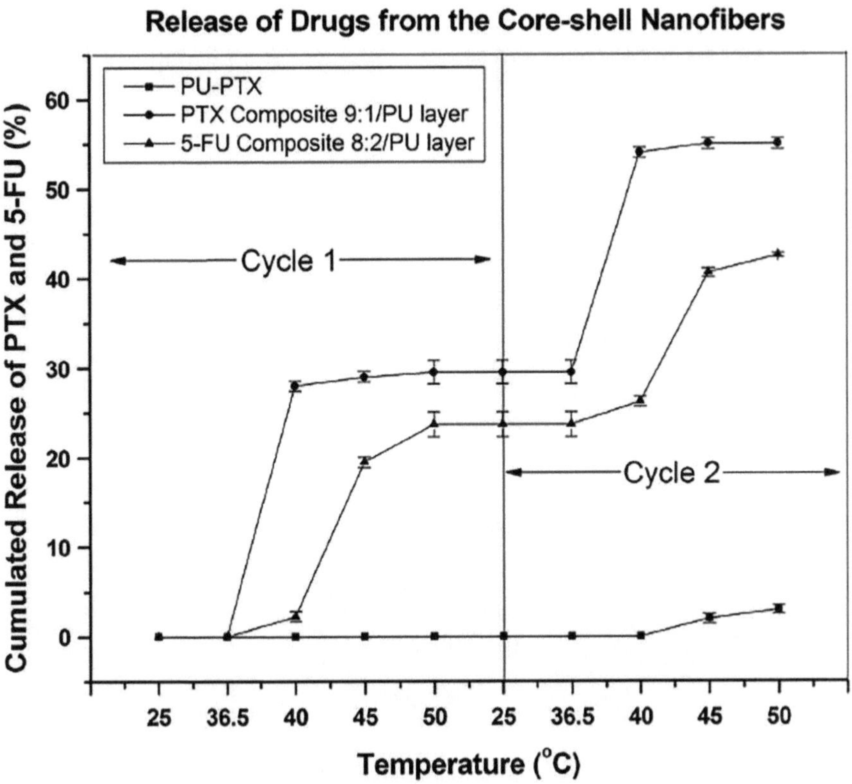

Fig. 11.11 In vitro drug release profiles of both PTX and 5-FU from the core-shell nanofiber composite mat at different temperatures, time points, and cycles. Each temperature point was maintained for 10 min. *PU* polyurethane. (Reproduced from Aguilar et al. 2017 with copyright permission of Elsevier's)

OE21, and they might be used as a promising material for cancer therapy by combining the effects of hyperthermia and chemotherapy.

In a different study, superparamagnetic iron NPs (γ-Fe_2O_3) were coupled with polyurethane nanofibers to treat cancer by heating the body to a high temperature (Song et al. 2018). The composite fibrous membrane created by in situ electrospinning demonstrated effective heating ability and well-maintained cycle heating performance in the presence of an alternating magnetic field when using the auxiliary electrode. The magnetic composite fibrous membrane created in situ by the auxiliary electrode is a remarkable option for magnetic hyperthermia in cancer treatment. This encouraging result came about because of the inclusion of a conical auxiliary electrode made of aluminum. Superparamagnetic iron oxide NPs were enclosed in polyurethane nanofiber by Amarjargal et al. (2013). The amount of magnetic Fe_3O_4 NPs in/on the membranes increased along with a steady increase in the heating rate. This electrospun nanofiber may be a possible candidate for a novel heat-generating substrate for localized hyperthermia cancer therapy.

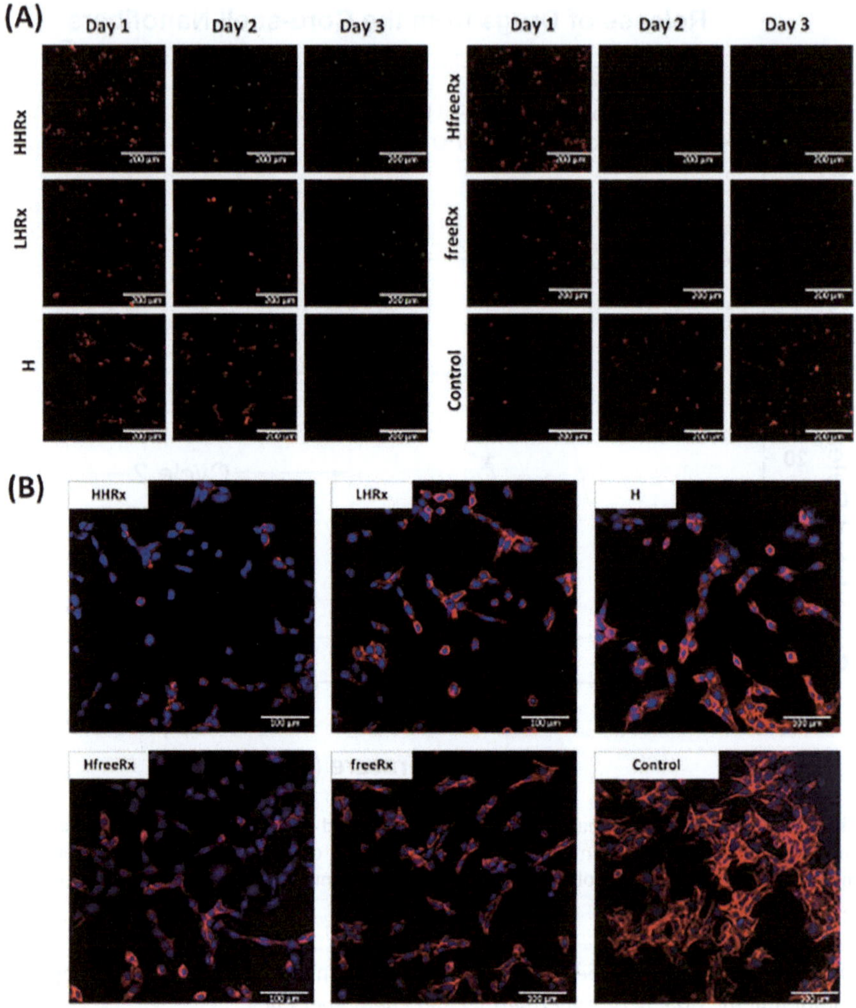

Fig. 11.12 Confocal images of (**a**) ESO26 cell line live/dead staining and (**b**) OE21 cell line DAPI and rhodamine stains. Treatment group abbreviations: HHRx = hyperthermia (45 °C) with dual drug release, LHRx = hyperthermia (40 °C) with single drug release, H = hyperthermia (45 °C), HfreeRx = hyperthermia (45 °C) with free combination drugs, freeRx = free combination drugs. (Reproduced from Aguilar et al. 2017 with copyrights permission of Elsevier's)

11.7 Conclusion

Nanotechnology enables early cancer diagnosis and prevention in the fight against the disease's pain, suffering, and mortality. We may be able to prevent cancer even before it begins with nanomedicine. With such technology, nanomedicine has the potential to increase the life span of human beings. The literature included in this chapter emphasizes the importance of electrospun polyurethane nanofibers as a drug-delivery mechanism for cancer treatment. Different delivery carriers can be created by mixing and matching various polymers. The electrospun polyurethane nanofibers have a variety of benefits as drug-delivery vehicles. These polyurethane drug products can provide patients with a sensation of comfort because of their high moisture vapor transmission rate, which is crucial for breathability and softening at body temperature, both of which are necessary for maintaining their properties. Additionally, polyurethane has drawn attention among the numerous polymeric architectures for electrospinning because of its higher strength, chemical resistance, and biocompatibility. Drugs can be incorporated into the electrospun nanofibers by simply mixing with the polymer solution, physically or chemically (surface immobilized), or indirectly loaded onto the nanofiber. Although nanofibers possess high drug incorporation efficiency, it further increases when combined with self-assembled NPs or microspheres that are evenly distributed within the nanofibers. Thus, polyurethane based drug-delivery devices in nanofibers are promising systems that can provide vast benefits and drug therapy for cancer.

Acknowledgments This work is supported by the research grants received by Dr. Faheem A. Sheikh from the Science and Engineering Research Board (SERB), Grant/Award Number: CRG/220/000113, and a Council of Scientific and Industrial Research (CSIR) sponsored project (22(0846)/20/EMR-II). Anjum Hamid is grateful for the fellowship she received from the Maulana Azad National Fellowship (MANF).

References

Abasian P et al (2019) Incorporation of magnetic NaX zeolite/DOX into the PLA/chitosan nanofibers for sustained release of doxorubicin against carcinoma cells death in vitro. Int J Biol Macromol 121:398–406. https://doi.org/10.1016/J.IJBIOMAC.2018.09.215

Abbas Z, Rehman S, Abbas Z, Rehman S (2018) An overview of cancer treatment modalities. In: Neoplasm. InTech. https://doi.org/10.5772/INTECHOPEN.76558

Aguilar LE, GhavamiNejad A, Park CH, Kim CS (2017) On-demand drug release and hyperthermia therapy applications of thermoresponsive poly-(NIPAAm-co-HMAAm)/polyurethane core-shell nanofiber mat on non-vascular nitinol stents. Nanomed Nanotechnol Biol Med 13(2): 527–538. https://doi.org/10.1016/J.NANO.2016.12.012

Akindoyo JO, Beg MDH, Ghazali S, Islam MR, Jeyaratnam N, Yuvaraj AR (2016) Polyurethane types, synthesis and applications—a review. RSC Adv 6(115):114453–114482. https://doi.org/10.1039/C6RA14525F

Amarjargal A, Tijing LD, Park CH, Im IT, Kim CS (2013) Controlled assembly of superparamagnetic iron oxide nanoparticles on electrospun PU nanofibrous membrane: a novel heat-generating substrate for magnetic hyperthermia application. Eur Polym J 49(12): 3796–3805. https://doi.org/10.1016/J.EURPOLYMJ.2013.08.026

Amna T, Hassan MS, Sheikh FA (2020) Nanocamptothecins as new generation pharmaceuticals for the treatment of diverse cancers: overview on a natural product to nanomedicine. Appl Nanotechnol Biomed Sci:39–49. https://doi.org/10.1007/978-981-15-5622-7_3

Bahadur A et al (2017) Biocompatible waterborne polyurethane-urea elastomer as intelligent anticancer drug release matrix: a sustained drug release study. React Funct Polym 119:57–63. https://doi.org/10.1016/J.REACTFUNCTPOLYM.2017.08.001

Baudino TA (2015) Targeted cancer therapy: the next generation of cancer treatment. Curr Drug Discov Technol 12:3–20

Beik J et al (2016) Nanotechnology in hyperthermia cancer therapy: from fundamental principles to advanced applications. J Control Release 235:205–221. https://doi.org/10.1016/J.JCONREL.2016.05.062

Bertolaso M (2016) Cancer biology. Hist Philos Theory Life Sci 18:1–16. https://doi.org/10.1007/978-94-024-0865-2_1/COVER

Bin Ying W et al (2020) Waterproof, highly tough, and fast self-healing polyurethane for durable electronic skin. ACS Appl Mater Interfaces 12(9):11072–11083. https://doi.org/10.1021/ACSAMI.0C00443

Bishop JM (1987) The molecular genetics of cancer. Science 235(4786):305–311. https://doi.org/10.1126/SCIENCE.3541204

Castillo-Henríquez L, Vargas-Zúñiga R, Pacheco-Molina J, Vega-Baudrit J (2020) Electrospun nanofibers: a nanotechnological approach for drug delivery and dissolution optimization in poorly water-soluble drugs. ADMET DMPK 8:325–353. https://doi.org/10.5599/admet.844

Contreras-Cáceres R et al (2019) Electrospun nanofibers: recent applications in drug delivery and cancer therapy. Nanomaterials (Basel) 9:656. https://doi.org/10.3390/nano9040656

Cuenca AG, Jiang H, Hochwald SN, Delano M, Cance WG, Grobmyer SR (2006) Emerging implications of nanotechnology on cancer diagnostics and therapeutics. Cancer 107(3):459–466. https://doi.org/10.1002/CNCR.22035

Dai J, Jin J, Yang S, Li G (2017) Doxorubicin-loaded PLA/pearl electrospun nanofibrous scaffold for drug delivery and tumor cell treatment. Mater Res Express 4(7):075403. https://doi.org/10.1088/2053-1591/AA7479

Darbasizadeh B et al (2021) Electrospun doxorubicin-loaded PEO/PCL core/sheath nanofibers for chemopreventive action against breast cancer cells. J Drug Deliv Sci Technol 64:102576. https://doi.org/10.1016/J.JDDST.2021.102576

Das A, Mahanwar P (2020) A brief discussion on advances in polyurethane applications. Adv Ind Eng Polym Res 3(3):93–101. https://doi.org/10.1016/J.AIEPR.2020.07.002

Elsadek NE et al (2022) Electrospun nanofibers revisited: an update on the emerging applications in nanomedicine. Materials (Basel) 15(5):1934. https://doi.org/10.3390/MA15051934

Fan R et al (2016) Dual drug loaded biodegradable nanofibrous microsphere for improving anti-colon cancer activity. Sci Rep 6(1):1–13. https://doi.org/10.1038/srep28373

Fan M et al (2020) Ultrasmall gold nanoparticles in cancer diagnosis and therapy. Theranostics 10(11):4944. https://doi.org/10.7150/THNO.42471

Faraji Dizaji B et al (2020) Synthesis of PLGA/chitosan/zeolites and PLGA/chitosan/metal organic frameworks nanofibers for targeted delivery of paclitaxel toward prostate cancer cells death. Int J Biol Macromol 164:1461–1474. https://doi.org/10.1016/J.IJBIOMAC.2020.07.228

Farboudi A et al (2020a) UiO-66 metal organic framework nanoparticles loaded carboxymethyl chitosan/poly ethylene oxide/polyurethane core-shell nanofibers for controlled release of doxorubicin and folic acid. Int J Biol Macromol 150:178–188. https://doi.org/10.1016/J.IJBIOMAC.2020.02.067

Farboudi A et al (2020b) Synthesis of magnetic gold coated poly (ε-caprolactonediol) based polyurethane/poly(N-isopropylacrylamide)-grafted-chitosan core-shell nanofibers for controlled release of paclitaxel and 5-FU. Int J Biol Macromol 150:1130–1140. https://doi.org/10.1016/J.IJBIOMAC.2019.10.120

Forouharshad M, Ajalloueian F (2022) Tunable self-assembled stereocomplexed-polylactic acid nanoparticles as a drug carrier. Polym Adv Technol 33(1):246–253. https://doi.org/10.1002/PAT.5510

Frangioni JV (2008) New technologies for human cancer imaging. J Clin Oncol 26(24):4012. https://doi.org/10.1200/JCO.2007.14.3065

Grant JJ et al (2021) Electrospun fibres of chitosan/PVP for the effective chemotherapeutic drug delivery of 5-fluorouracil. Chemosensors 9(4):70. https://doi.org/10.3390/CHEMOSENSORS9040070

Hainfeld JF, Slatkin DN, Smilowitz HM (2004) The use of gold nanoparticles to enhance radiotherapy in mice. Phys Med Biol 49(18):N309. https://doi.org/10.1088/0031-9155/49/18/N03

Hu H, Huang X, Deng C, Chen X, Qian Y (2007) Hydrothermal synthesis of ZnO nanowires and nanobelts on a large scale. Mater Chem Phys 106(1):58–62. https://doi.org/10.1016/J.MATCHEMPHYS.2007.05.016

Hu J, Prabhakaran MP, Tian L, Ding X, Ramakrishna S (2015) Drug-loaded emulsion electrospun nanofibers: characterization, drug release and in vitro biocompatibility. RSC Adv 5(121): 100256–100267. https://doi.org/10.1039/C5RA18535A

Ibrahim HM, Klingner A (2020) A review on electrospun polymeric nanofibers: production parameters and potential applications. Polym Test 90:106647. https://doi.org/10.1016/J.POLYMERTESTING.2020.106647

Irani M, Sadeghi GMM, Haririan I (2017) The sustained delivery of temozolomide from electrospun PCL-diol-b-PU/gold nanocompsite nanofibers to treat glioblastoma tumors. Mater Sci Eng C 75:165–174. https://doi.org/10.1016/J.MSEC.2017.02.029

Jain PK, ElSayed IH, El-Sayed MA (2007) Au nanoparticles target cancer. Nano Today 2(1):18–29. https://doi.org/10.1016/S1748-0132(07)70016-6

Jeckson TA, Neo YP, Sisinthy SP, Gorain B (2021) Delivery of therapeutics from layer-by-layer electrospun nanofiber matrix for wound healing: an update. J Pharm Sci 110(2):635–653. https://doi.org/10.1016/J.XPHS.2020.10.003

Joseph J, Patel RM, Wenham A, Smith JR (2018) Biomedical applications of polyurethane materials and coatings. Int J Surf Eng Coat 96(3):121–129. https://doi.org/10.1080/00202967.2018.1450209

Joshi M, Butola BS, Saha K (2014) Advances in topical drug delivery system: micro to nanofibrous structures. J Nanosci Nanotechnol 14(1):853–867. https://doi.org/10.1166/JNN.2014.9083

Kiliç E, Yakar A, Pekel Bayramgil N (2018) Preparation of electrospun polyurethane nanofiber mats for the release of doxorubicine. J Mater Sci Mater Med 29(1):1–9. https://doi.org/10.1007/S10856-017-6013-5/FIGURES/7

Krukiewicz K, Zak JK (2016) Biomaterial-based regional chemotherapy: local anticancer drug delivery to enhance chemotherapy and minimize its side-effects. Mater Sci Eng C 62:927–942. https://doi.org/10.1016/J.MSEC.2016.01.063

Kumar S, Kushwaha S, Awani R (2012) Novel drug delivery system for anticancer drug: a review. Int J PharmTech Res 4:542–553. Available: https://www.researchgate.net/publication/287478414. Accessed 10 Oct 2022

Lal S, Clare SE, Halas NJ (2008) Nanoshell-enabled photothermal cancer therapy: impending clinical impact. Acc Chem Res 41(12):1842–1851. https://doi.org/10.1021/AR800150G

Lee DE, Koo H, Sun IC, Ryu JH, Kim K, Kwon IC (2012) Multifunctional nanoparticles for multimodal imaging and theragnosis. Chem Soc Rev 41(7):2656–2672. https://doi.org/10.1039/C2CS15261D

Li Z, Wang C (2013) Introduction of electrospinning. In: Springer briefs in materials. Springer, pp 1–13. https://doi.org/10.1007/978-3-642-36427-3_1/COVER

Long YZ, Yan X, Wang XX, Zhang J, Yu M (2019) Electrospinning: the setup and procedure. In: Electrospinning: nanofabrication and applications, pp 21–52. https://doi.org/10.1016/B978-0-323-51270-1.00002-9

Lu W et al (2010) Multifunctional oval-shaped gold-nanoparticle-based selective detection of breast cancer cells using simple colorimetric and highly sensitive two-photon scattering assay. ACS Nano 4(3):1739–1749. https://doi.org/10.1021/NN901742Q

Ma G, Liu Y, Peng C, Fang D, He B, Nie J (2011) Paclitaxel loaded electrospun porous nanofibers as mat potential application for chemotherapy against prostate cancer. Carbohydr Polym 86(2): 505–512. https://doi.org/10.1016/J.CARBPOL.2011.04.082

Mitra AK et al (2015) Novel delivery approaches for cancer therapeutics. J Control Release 219: 248. https://doi.org/10.1016/J.JCONREL.2015.09.067

Mohamad Sadeghi GM, Sayaf M (2012) From PET waste to novel polyurethanes. In: Material recycling—trends and perspectives. InTech. https://doi.org/10.5772/31642

Nida DL, Rahman MS, Carlson KD, Richards-Kortum R, Follen M (2005) Fluorescent nanocrystals for use in early cervical cancer detection. Gynecol Oncol 99(3):S89–S94. https://doi.org/10.1016/J.YGYNO.2005.07.050

Nikalaje Y, Shende V (2018) Recent development applications of nanotechnology in cancer: a literature review of imaging and treatment. World J Pharm Pharm Sci. https://doi.org/10.20959/wjpps20182-10868

Parsa N (2012) Environmental factors inducing human cancers. Iran J Public Health 41(11):1. Available: /pmc/articles/PMC3521879/. Accessed 10 Oct 2022

Patra JK et al (2018) Nano based drug delivery systems: recent developments and future prospects. J Nanobiotechnol 16(1):1–33. https://doi.org/10.1186/S12951-018-0392-8

Prisacariu C (2011) Polyurethane elastomers. Springer, Vienna. https://doi.org/10.1007/978-3-7091-0514-6

Rengifo AFC et al (2019) PEO-chitosan nanofibers containing carboxymethyl-hexanoyl chitosan/dodecyl sulfate nanoparticles loaded with pyrazoline for skin cancer treatment. Eur Polym J 119:335–343. https://doi.org/10.1016/J.EURPOLYMJ.2019.08.001

Seyyedi M, Molajou A (2021) Nanohydroxyapatite loaded-acrylated polyurethane nanofibrous scaffolds for controlled release of paclitaxel anticancer drug. J Res Sci Eng Technol 9(1):50–61. https://doi.org/10.24200/JRSET.VOL9ISS01PP50-61

Shafi H et al (2023) Super disintegrating oromucosal nanofiber patch of zolmitriptan for rapid delivery and efficient brain targeting. Chem Eng J 463:142481. https://doi.org/10.1016/j.cej.2023.142481

Shahrousvand M, Sadeghi GMM, Salimi A (2016) Artificial extracellular matrix for biomedical applications: biocompatible and biodegradable poly (tetramethylene ether) glycol/poly (-ε-caprolactone diol)-based polyurethanes. J Biomed Sci Polym Ed 27(17):1712–1728. https://doi.org/10.1080/09205063.2016.1231436

Sikdar P, Dip TM, Dhar AK, Bhattacharjee M, Hoque MS, Bin Ali S (2022) Polyurethane (PU) based multifunctional materials: emerging paradigm for functional textiles, smart, and biomedical applications. J Appl Polym Sci 139(38):e52832. https://doi.org/10.1002/APP.52832

Singh P, Pandit S, Mokkapati VRSS, Garg A, Ravikumar V, Mijakovic I (2018) Gold nanoparticles in diagnostics and therapeutics for human cancer. Int J Mol Sci 19(7):1979. https://doi.org/10.3390/IJMS19071979

Smith AM, Dave S, Nie S, True L, Gao X (2014) Multicolor quantum dots for molecular diagnostics of cancer. Expert Rev Mol Diagn 6(2):231–244. https://doi.org/10.1586/14737159.6.2.231

Son YJ, Kim WJ, Yoo HS (2014) Therapeutic applications of electrospun nanofibers for drug delivery systems. Arch Pharm Res 37(1):69–78. https://doi.org/10.1007/S12272-013-0284-2/FIGURES/3

Song C et al (2018) Electric field-assisted in situ precise deposition of electrospun γ-Fe2O3/polyurethane nanofibers for magnetic hyperthermia. Nanoscale Res Lett 13(1):1–11. https://doi.org/10.1186/S11671-018-2707-Y/FIGURES/9

Subbiah T, Bhat GS, Tock RW, Parameswaran S, Ramkumar SS (2005) Electrospinning of nanofibers. J Appl Polym Sci 96(2):557–569. https://doi.org/10.1002/APP.21481

Thakkar S, Misra M (2017) Electrospun polymeric nanofibers: new horizons in drug delivery. Eur J Pharm Sci 107:148–167. https://doi.org/10.1016/J.EJPS.2017.07.001

Tian H, Tang Z, Zhuang X, Chen X, Jing X (2012) Biodegradable synthetic polymers: preparation, functionalization and biomedical application. Prog Polym Sci 37(2):237–280. https://doi.org/10.1016/J.PROGPOLYMSCI.2011.06.004

Tseng AA (2005) Recent developments in nanofabrication using focused ion beams. Small 1(10):924–939. https://doi.org/10.1002/SMLL.200500113

Vaithylingam R, Ansari MNM, Shanks RA (2017) Recent advances in polyurethane-based nanocomposites: a review. Polym Plast Technol Eng 56(14):1528–1541. https://doi.org/10.1080/03602559.2017.1280683

Wu L, Qu X (2015) Cancer biomarker detection: recent achievements and challenges. Chem Soc Rev 44(10):2963–2997. https://doi.org/10.1039/C4CS00370E

Yan E et al (2020) pH-sensitive core-shell electrospun nanofibers based on polyvinyl alcohol/polycaprolactone as a potential drug delivery system for the chemotherapy against cervical cancer. J Drug Deliv Sci Technol 55:101455. https://doi.org/10.1016/J.JDDST.2019.101455
Yoo HS, Kim TG, Park TG (2009) Surface-functionalized electrospun nanofibers for tissue engineering and drug delivery. Adv Drug Deliv Rev 61(12):1033–1042. https://doi.org/10.1016/J.ADDR.2009.07.007
Yu H, Jia Y, Yao C, Lu Y (2014) PCL/PEG core/sheath fibers with controlled drug release rate fabricated on the basis of a novel combined technique. Int J Pharm 469(1):17–22. https://doi.org/10.1016/J.IJPHARM.2014.04.045
Zang CH et al (2009) Optical properties of a ZnO/P nanostructure fabricated by a chemical vapor deposition method. J Phys Chem C 113(43):18527–18530. https://doi.org/10.1021/JP905648M
Zhang H, Yee D, Wang C (2008) Quantum dots for cancer diagnosis and therapy: biological and clinical perspectives. Nanomedicine (London) 3(1):83–91. https://doi.org/10.2217/17435889.3.1.83
Zhang Y, Li M, Gao X, Chen Y, Liu T (2019) Nanotechnology in cancer diagnosis: progress, challenges and opportunities. J Hematol Oncol 12(1):1–13. https://doi.org/10.1186/S13045-019-0833-3/FIGURES/2
Zhao J, Wallace M, Melancon MP (2014) Cancer theranostics with gold nanoshells. Nanomedicine 9(13):2041–2057. https://doi.org/10.2217/NNM.14.136

Anjum Hamid Rather was born in J&K, India, and studied for her M.Sc. in the Department of Bioresources, University of Kashmir, Srinagar, JK. She qualified for the CSIR-UGC NET in 2019 and was awarded a National Fellowship by the University Grants Commission (UGC) in 2020 to pursue her Ph.D. research for 5 years.

She is currently pursuing her Ph.D. degree in the Department of Nanotechnology, University of Kashmir, Srinagar, J&K, under the supervision of Dr. Faheem A. Sheikh. Her interests include developing polymeric nanofibers for advanced tissue engineering, wound healing applications, and exploiting the role of different nanoparticles and essential oils in tissue engineering. She has contributed to nine peer-reviewed articles and four book chapters.

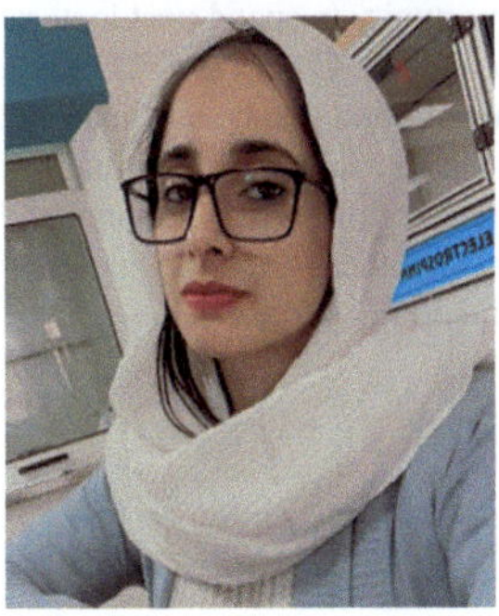

Rumysa Saleem Khan was born in Srinagar, J&K, India, in 1995. She received a B.S. Degree in Biochemistry in 2016 and an M.S. Degree in Biotechnology from the University of Kashmir, Srinagar, in 2018, and was receiving DBT Fellowship. From 2019 to 2020, she was a JRF with Biochemical Engineering Lab in NIT, Srinagar, where she received MHRD Junior Research Fellowship.

She is currently pursuing a Ph.D. degree under the supervision of Dr. Faheem A Sheikh at the University of Kashmir, Srinagar, J&K, India, and is receiving the Science and Engineering Research Board (SERB) Junior Research Fellowship. She is currently the author of 13 scientific articles published in international journals. Ms. Khan's research interests include biomimicking of biological molecules in the lab, fabrication of biocompatible scaffolds for tissue engineering, nanofiber formation for wound healing and bone regeneration, and exploiting the roles of different nanoparticles in tissue engineering. She also has an expertise in mammalian cell culture and molecular biology techniques.

Taha Umair Wani received his Bachelor's Degree in Pharmacy from the University of Kashmir, India, in 2012 and his Master's Degree in Pharmaceutics from Punjab Technical University, Punjab, India, in 2014. He completed his Ph.D. in Pharmaceutics from the University of Kashmir, India, in 2020.

He is currently working as a Research Associate in the Department of Nanotechnology, University of Kashmir. He is the author of more than 40 publications. His research interest includes drug formulation development, preclinical research, nanomedicine, and biomaterial fabrication. He has got expertise in cell culture studies, antimicrobial assays, drug delivery, in vivo experiments, and pharmacokinetic studies.

Muheeb Rafiq was born and raised in Srinagar, J&K, India. He received his B.Sc. degree from the University of Kashmir, in 2017. Later he completed his M.Sc. in Nanotechnology from the University of Kashmir, in 2021. During his master's, he gained a hands-on experience in the research area as he worked as an intern for 6 months under Dr. Mushtaq (Asst. Prof. Dept. of Nanotechnology, University of Kashmir). After completing his master's, he started working in a project funded by the Council of Scientific and Industrial Research (CSIR) under the supervision of Dr. Faheem A. Sheikh (Asst. Prof. Dept. of Nanotechnology, University of Kashmir).

He is currently enrolled in a Ph.D. program and is working under the mentorship of Dr. Faheem A. Sheikh. He has authored a couple of publications including research papers and review articles. His research interest includes the fabrication of biocompatible nanofibers and nano porous scaffolds for hard and soft tissue engineering, biological synthesis of nanoparticles, and nanocomposite formations.

Touseef Amna received her Ph.D. in 2006 from Aligarh Muslim University (AMU, Central University), India, in Microbiology. Presently she is working as an Associate Professor at Albaha University, Saudi Arabia. During her research career, she got the opportunity to spend postdoctoral stages at the University of Dortmund, Germany, Universidad de Talca, Chile, and Chonbuk National University, South Korea, and received various prestigious awards and grants. She authored 2 patents (one US patent and one Korean) and more than 100 research articles in journals of international repute. She is editor, guest editor, and editorial board member of various national and international journals. Her research interest consists of design and synthesis of biocompatible scaffolds for antimicrobial applications, cell culture, and tissue engineering.

M. Shamshi Hassan is working as an Associate Professor in the Faculty of Science Albaha University, Saudi Arabia. He was awarded an international fellowship to pursue his doctoral degree and received his Ph.D. in Chemical Engineering from the Department of Chemical Engineering, Chonbuk National University, South Korea, in 2010. After Ph.D. he has been awarded Postdoctoral Fellowship (PDF) by the Textile and Fiber Engineering Department of CBNU, South Korea. His research interest includes the synthesis of pure and hybrid metal oxides (quantum dots or nanocrystals, 1D, 2D, 3D, and flower-shaped morphology), nanostructured materials, or nanofibers and their application in supercapacitor and as biological/chemical disinfectant.

Shafquat Majeed is currently working as an Assistant Professor in the Department of Nanotechnology, University of Kashmir, India. Dr. Majeed did his Master's in Chemistry from the Department of Chemistry, University of Kashmir. He did his Ph.D. from the prestigious Indian Institute of Science Bangalore (2008–2015). He has a postdoctoral research experience in rare-earth based nanomaterials for biomedical applications and fabrication of Perovskite based solar cells from the Centre for Nanoscience and Engineering, IISc, Bangalore (2015–2016).

He at present heads a SCAN (*Synthesis, Characterization and Application of Nanomaterials*) laboratory at the University of Kashmir where they work on synthesis, characterization, and optical and magnetic properties of nanomaterials prepared through various solution-based routes. More specifically, they are working on developing new and enhanced multimodal imaging modalities for biomedical and healthcare applications.

Mushtaq A. Beigh is working as an Assistant Professor at the Department of Nanotechnology, University of Kashmir, India. Dr. Beigh has postdoctoral research experience in cellular signaling and disease modeling (2013–2016). He had completed his Ph.D. from the University of Kashmir, India (2008–2012). He heads a laboratory at the University of Kashmir, where they work on a bio-nano interface to understand the molecular basis of nanoconjugate internalization. They also try to understand the growth factor signaling inputs directed at multiple cellular receptors like NRP1 and integrins in order to work out the mechanistic details of receptor-based internalization.

Faheem A. Sheikh is an Assistant Professor at the Department of Nanotechnology, University of Kashmir, India (2015). He served as an Assistant Professor at the Department of Biotechnology at the Central University of Kashmir, India (2015–2016); Research Professor at Myongji University, South Korea (2014–2015); Assistant Professor at Hallym University, South Korea (2012–2014); Postdoc/Research Fellow at the University of Texas Rio Grande Valley, Texas, USA (2010–2012); and Research Professor at Myongji University, South Korea (2010). His research mainly focuses on fabricating nanomaterials used in tissue engineering.

12 Recent Trends in the Application of Materials for Cancer Therapy and Diagnosis

Behnaz Bakhshandeh, Zohreh Jahanafrooz, Dorrin Mohtadi Haghighi, Nasim Rahmani-Kukia, Ardeshir Abbasi, and Armaghan Pourramezanali

Abstract

The application of nanomaterials has been expanded in various medical fields. One of these important fields in the use of nanoparticles is cancer therapy and its diagnosis. Nanoparticles have unique properties due to their small size. Therefore, nanoparticles are now widely used to treat or diagnose cancer. The key factor in efficient cancer treatment is the detection of it in its early stages. In addition to the usual methods, many methods have been found for the early detection of cancer in which nanoparticles have been used to increase efficiency.

B. Bakhshandeh (✉)
Department of Biotechnology, College of Science, University of Tehran, Tehran, Iran
e-mail: b.bakhshandeh@ut.ac.ir; https://profile.ut.ac.ir/en/~b.bakhshandeh

Z. Jahanafrooz
Department of Biology, Faculty of Sciences, University of Maragheh, Maragheh, Iran
e-mail: jahanafrooz@maragheh.ac.ir; https://res.maragheh.ac.ir/~z.jahanafrooz/en/

D. M. Haghighi
Department of Pharmaceutics, Faculty of Pharmacy, Tehran University of Medical Sciences, Tehran, Iran

N. Rahmani-Kukia
Department of Biochemistry, School of Medicine, Shiraz University of Medical Sciences, Shiraz, Iran

A. Abbasi
Department of Immunology, Faculty of Medical Sciences, Tarbiat Modares University, Tehran, Iran
e-mail: Ardeshir.Abbasi@modares.ac.ir; https://www.researchgate.net/profile/Ardeshir-Abbasi

A. Pourramezanali
Department of Microbiology, School of Biology, Faculty of Science, Tehran University, Tehran, Iran
e-mail: A.pourramezanali@ut.ac.ir

F. A. Sheikh et al. (eds.), *Interaction of Nanomaterials With Living Cells*,
https://doi.org/10.1007/978-981-99-2119-5_12

It is also known that nanoparticles have recently been used to treat different types of cancers. Accordingly, in this chapter, we focus on the applications and advantages of nanotechnology in cancer diagnosis and treatment. The goal of this chapter is the introduction of nanotechnology as a new approach for this purpose. Moreover, challenges in nanotechnology and future aspects will be discussed.

Keywords

Cancer · Diagnosis · Cancer therapy · Nanotechnology · Nanomaterials

12.1 Introduction

The application of nanomaterials has been expanded in various fields such as drug and gene delivery, bio-detection of pathogens, protein detection, fluorescent biological labels, DNA structure probing, tissue engineering, tumor eradication through hyperthermia, separation and purification of cells and biological molecules, and improvement of MRI contrast (Bakhshandeh et al. 2017a, 2022). Growing attention is attracted to nanomaterials in cancer therapy and diagnosis (Salata 2004). Nanomaterials possess inimitable characteristics such as small size, large surface area, versatility, targeted action, and stability that favor their use in biomedicine. Their small size leads to their participation in physicochemical pathways owing to their comparable size with biological molecules. Nanomaterials can transfer a large amount of cargo due to the large surface area. They own high sensitivity, specificity, and affinity to target cells that cause the reduction of "off-target" effects of therapeutic agents and consequently decrease the toxicity of cargo. Their stability tailored them for systemic administration and optimized bio-distribution. In addition, nanomaterials can overcome drug resistance mechanisms (Barkalina et al. 2014).

Cancer is a disease that can influence all tissues of the body. According to the World Health Organization, cancer has caused 7.4 million deaths in 2007, and it is expected this number will rise to 12 million in 2030 (Jahanafrooz et al. 2016). The main cause of cancer is an irregularity in the control of cell proliferation cycle. Once the cell size reaches 1–2 mm, the formation of new blood vessels occurs, called angiogenesis which is mediated by releasing of substances from the microenvironment of tumor, such as the vascular endothelial growth factor. Angiogenesis is a result of pro and anti-angiogenic molecule imbalance. These new blood vessels have abnormal spaces between endothelial cells that facilitate substance entry and exit. These spaces are formed as a result of some proteolytic enzymes named matrix metalloprotease and plasminogen activator plasmid system. The higher permeability of tumor vessels compared to normal tissues and lower clearance of macromolecules from interstitial spaces is known as enhanced permeability and retention (EPR). All of the abovementioned factors lead to the accumulation of macromolecules and hydrogen ion in cancerous tissue at concentrations five to ten times higher than in normal tissues. Because of hydrogen ion reposition, the extracellular pH of tumor is

5.8–7.8. Another reason for the decrease in pH of tumor tissue is associated with the production of lactic acid and carbonic acid due to aerobic/anaerobic glycolysis and ATP synthesis. As the tumor cells grow, the inner cells deprive from nutrition and O_2. Angiogenesis promotes their access to necessary substances and accelerates tumor growth (Barreto et al. 2011). There are several biological barriers in the way of drug delivery to cancerous tissues, including cellular membranes, the structure of blood vessels, the cancer microenvironment, the phagocytic system, and multi-drug resistant system. There are said to be more lactate release within tumor cells that derives the charge of cell surface to negative. Therefore, more nanomaterials uptake happens via their positive surface charge. Phagocytic cells only uptake the cationic nanomaterials that can affect their efficacy. In particle size below 200 nm, nanomaterials uptake is assisted by clathrin-mediated endocytosis, but once the particle size reaches above 200 nm then, caveolin-mediated endocytosis strikes out. Nanomaterials also bear some overexpressed receptors such as folate receptors, integrin, glucose transporters, transferrin receptors, and growth factor receptors. Extracellular matrix impedes the nanomaterials internalization because the existence of cross-linked structures prevents the formation of internalization process. On the other hand, hypoxia that occurs in cancerous tissues hinders chemotherapy and radiation process. Overexpression of chemokine-ligand 8 that takes place as a result of hypoxia forces angiogenesis and inhibits immune cells. After reaching nanomaterials to the systemic circulation, some proteins such as albumin, fibrinogen, and globulin attach to nanomaterials in order to form protein corona. These structures play a critical role in the determination of nanomaterials destination because they can alter the opsonization process and, consequently, desired drug delivery. ATP-binding cassettes are the most important family of multi-drug resistant system. These efflux transporters excrete exogenous compounds such as chemotherapeutic agents against the gradient of concentration which figures out drug resistance (Choudhury et al. 2019).

Key factor in efficient cancer treatment is detection of it in early stages. Mammography for breast cancer (Nishikawa 2007), blood tests such as prostate-specific antigen measurement for prostate cancer (Salami et al. 2013), and stool test for colorectal cancer (Winawer 2007) are among traditional methods in cancer detection. Nowadays cancer diagnosis methods named histopathology, cytology, X-ray, computed tomography (CT), magnetic resonance imaging (MRI), ultrasound, and endoscopy are able to diagnose cancer in terminal stages when tissues underwent widespread changes. These methods cannot distinguish between benign and malignant tumors as well. Therefore, it seems essential to find a method that enable us to detect cancer in primary stages. Nanotechnology is one of successful techniques that provide detection of cancer in early stages. The main feature of nanomaterials, as mentioned before, is high surface to volume ratio that appropriates them to recognize cancer biomarkers which include proteins, carbohydrates, and nucleic acids released by cancerous tissues. Various types of biosensors are employed in this field. Their function can be improved by nanomaterials. Quantum dots, gold nanoparticles, and polymer dots can be mentioned as these nanomaterials (Zhang et al. 2019a).

History of chemotherapy goes back to the 1940s when observing of toxic effects of nitrogen mustard led to the use of this agent in lymphoma treatment. This drug acts through DNA alkalization. Other groups of cytotoxic drugs are antimetabolites such as methotrexate and 5-flurouracil. In the 1960s–1970s natural antibiotics named actinomycine D and vinca alkaloid emerged. Recent decades accompany with appearing taxanes and topoisomerase I inhibitors. Finally, to improve the efficacy liposomal form of some drugs was prepared. Chemotherapeutic agents are usually administrated via intravascular and oral route. However, for some tumors limited to specific region, their administration is carried out into that area. These drugs are used in combination with each other to overcome tumor resistance. These agents typically reduce cancer symptoms, but the complete tumor cure rate is only 20% with serious side effects. Resistance to cytotoxic agents originates from biological barriers that are present in the way of drug delivery system that was mentioned before. In order to optimize chemotherapy, there are some suggestions. Because of low selectivity of these agents, liposomal forms are prepared that are useful in some cancers. Prodrugs activated by enzymes specifically present in tumors site have led to enhance therapeutic index. Increasing of dose for drugs that possess positive dose-response relationship can improve efficiency in breast, testis, and hematological cancers. In order to increase dose with lower toxicity, employing of bone marrow stimulating growth factors has been suggested. Personalized medicine might be helpful as well. In this approach tumor cells of individuals are exposed to different cytotoxic agents, and in the end drugs that lack effect on tumor cells will be exploded. The milestone of the last approach is based on applying some agents such as cyclosporine A and calcium channel inhibitors that can confront with drug resistance (Nygren 2001).

Radiotherapy is employed in two-third of cancer treatment regimen in western countries. Recently efficacy of radiotherapy has been increased from 30% to 80% in head and neck cancers. However, radiotherapy needs to be improved due to intrinsic and extrinsic resistance that are present in tumors. Mechanisms that radiotherapy works through are reactive oxygen species (ROS) targeting, hypoxia targeting, and DNA damage induction. In the first mechanism ionization radiation induces water hydrolysis that triggers ROS production from mitochondria that damage DNA. In the second mechanism hypoxic radiosensitizers such as hyperbaric oxygen and nitroimidazole lead to higher efficacy compared to alone radiotherapy. In the third mechanism induction of DNA damage takes place through several pathways (Chen and Kuo 2017).

Excisional surgery is the first intervention in cancer treatment. Minimal residual disease (MDR) is a lesion that remains in situ after surgery. Progressed cancers are more influenced by tumor removal. One of these effects is absence of angiogenesis inhibition that causes intermediate effect when used alone. Various surgery methods such as open resection, laparotomy, and laparoscopy affect growth of residual tumors in different ways. After surgery, tumor manipulation leads to dissemination of cells, and in the next phase the patient is more susceptible to tumorigenesis and MRD expands as a result. Finally, surgery alters biological properties of neoplastic cells and leads to increase of cell proliferation and lessen the cell death. Considering the above factors it needs to use chemotherapy agents after surgery (Coffey et al. 2003).

Breast cancer in women and prostate cancer in men are the most prevalent cancers in both sexes. In each disease tumors express some hormone receptors that can be a target of endocrine therapy. However, most treatments cause decrease in bone density. One goal in breast cancer treatment is reduction of estrogen by ovaries through GnRH agonists. In addition, interaction of estrogen with receptor can be affected by selective estrogen receptor modulator. Another approach is employing of pure anti-estrogen that acts via down regulation of estrogen receptors. In prostate cancer the purpose is the reduction of androgen hormone levels. In this disease GnRH agonists can be used. Furthermore, GnRH antagonists can be applied that their effect is significantly higher than agonists. Another option is anti-androgen agents. All of these drugs are not able to reduce testosterone secretion from adrenal glands. To resolve this problem, inhibitors of cyp17 enzyme are useful as they block adrenal glands (Rachner et al. 2018).

The aim of using of viral vectors is to increase chemotherapy efficacy as these vectors increase cytotoxic drug concentration in the site of malignant cells. In this approach, oncolytic viruses that grow specifically in tumor cells are utilized to activate immune response. Conventional drugs, unlike viruses, are not able to augment themselves, and it is essential to use higher concentration of them. Generally, RNA viruses have quicker replication rate compared to DNA viruses. Studies elicited that INF-sensitive viruses replicate in tumor cells; therefore some RNA viruses such as reovirus, paramyxovirus, and vesicular stomatitis virus possess natural propensity for tumor cells because of their sensitivity to INF. Some DNA viruses such as adenovirus reach to the nucleus and undertake cell cycle through inhibition of tumor suppressor genes. The main barriers against virotherapy are tumor stroma and immune system. Viruses can be considered as anticancer drugs when used in combination with other treatments (Urruticoechea et al. 2010).

Gene therapy is based on insertion of normal gene into the cell of body in order to correct protein deficiency. In this method, normal gene can be inserted in presence of abnormal gene or can be replaced with it. One of gene therapy's applications is in cancer treatment. Tumor cells express some oncogenes or are not able to express tumor suppressor genes (Gutierrez et al. 1992). In a new approach known as gene-directed-enzyme prodrug therapy, a foreign gene that is specific to the tumor cells or has higher concentration within cancer cells is delivered to the tumor site. The next step includes prodrug administration that will be activated by produced enzyme to transform to cytotoxic drug (Greco and Dachs 2001).

Studies showed a relationship between dysbiosis (disruption in human microbiome) with several diseases such as neurological, metabolic, cardiovascular diseases and cancers. *Salmonella typhi* and *Helicobacter pylori* play a part in biliary and gastric adenocarcinoma, respectively. Bacteria act through several mechanisms including production of proinflammatory factors, ROS production, and alteration in signaling pathways as a carcinogen agent. In addition to carcinogenicity, they affect cancer therapy process. Therefore, nowadays there is a focus on treatments that target microbiome. Fecal microbiota transplantation for diarrhea, employing

Table 12.1 Small molecules and their mechanisms

Small molecule	Pathway
Imiquimod, Resiquimod	Toll-like receptor antagonist
Indoximod, Epacadostat	Indoleamine-2,3 dioxygenase 1 inhibitor
Celexocib, Melafolone	Prostaglandin pathway inhibition
Cabozantinib	Cellular-mesenchymal-epithelial transition factor inhibition
Crizotinib	Anaplastic lymphoma kinase inhibition
Gefitinib, Erlotinib	Epidermal growth factor receptor inhibition

probiotics, and targeting of tumor microbiota in colorectal and pancreas cancers can be named as these treatments (Helmink et al. 2019). Small molecules target several pathways and some of them are summarized in Table 12.1 (van der Zanden et al. 2020; Zhong et al. 2021).

Metabolism alteration as a cancer indicator is a useful tool for diagnosis and treatment of cancer. These changes are upregulation of enzymes that play a role in pentose phosphate pathway, glutaminolysis, glycolysis and lipogenesis, and down regulation of mitochondrial proteins that their function is oxidative phosphorylation. Accordingly, targeting of mitochondrial metabolism is considered as a treatment option in cancer. Uses of glycolysis inhibitors such as 3-bromopyruvate and idoacetate in colon cancer and isoflavonoid genistein in hepatocellular carcinoma are examples of metabolic therapy. Metformin which inhibits ATP synthesis can be employed in pancreatic cancer. Another ATP synthesis inhibitor is gboxin that suppresses growth of glioblastoma (Torresano et al. 2020).

The first monoclonal antibodies (mabs) that emerged are bevacizumab (anti-vascular endothelial growth factor antibody) and cetuximab (anti-epidermal growth factor antibody). In addition, anti Her2/neu antibody such as trastazumab reduces mortality in breast cancer when used in combination with chemotherapy adjuvant. First generations of mabs have murine origin and because of immunogenicity have limited clinical application. Then next generations of mabs such as chimeric antibody and humanized antibody were introduced that bypass previous problems. There are several mechanisms that mabs work through including interaction with immune system via antibody-dependent cellular cytotoxicity, alteration of signal transduction, complement dependent cytotoxicity, and elimination of cell surface antigens. Furthermore, they can target payloads such as toxins, drugs, and radioisotopes. Similar to gene therapy there are systems known as antibody-directed enzyme prodrug therapy that leads to activation of prodrug in tumor site (Adams and Weiner 2005).

Tumor-associated antigens (TATs) and tumor-specific antigens (TSAs) are proteins related to tumor that can be a target for vaccination. TSAs are molecules that specifically express in tumor cells while TAAs are molecules expressed in both normal and tumor cells but in different patterns. Cancer vaccines can be used to prevent (prophylactic) and cure cancer. When the purpose is prophylaxis, TAAs must be targeted. Vaccines used to prophylaxis such as hepatitis B virus vaccine and human papilloma virus vaccine can prevent from hepatic cancer and cervical carcinoma in women, respectively. On the other hand, when the goal is cancer treatment, TSAs must be targeted. An ideal vaccine must distinguish between normal and cancer cells. They must be able to extirpate primary and residual tumors and create long memory with acceptable adverse effects (Aly 2012). Nanotechnology has several advantages in cancer therapy including early diagnosis and prediction, prevention of cancer, personalized medicine, overcome drug resistance, and improvement of cytotoxic drugs bioavailability, residency at target site, and stability (Lee et al. 2015; Gao et al. 2012; Misra et al. 2010).

In this chapter, we focus on application and advantages of nanotechnology in cancer diagnosis and treatment. The aim of this chapter is introduction of nanotechnology as a new approach for this purpose. Finally, challenges in nanotechnology and future aspects will be discussed.

12.2 Types of Nanomaterials for Biological Application

Several types of nanoparticles employed in biology with unique properties are discussed below.

12.2.1 Polymeric Nanoparticles

Polymeric nanoparticles (PNs) referred to sub-micron sized particles (below 100 nm). Natural polymers such as heparin, chitosan, gelatin, albumin, collagen, and dextran and synthetic polymers including polyethylene glycon (PEG), polyglutamic acid, poly lactide, poly lactide-co-glycolide, polycaprolactone, and polyhydroxyalkanoate are widely used for biological applications (Ziraksaz et al. 2013). PNs are categorized into nanocapsule and nanosphere. In the former type substance is surrounded in the core by a shell while in the second type drug is dispersed in polymer matrix (Barreto et al. 2011). There are several methods for PNs fabrication named emulsification, nanoprecipitation, salting out, solvent evaporation/extraction, and supercritical antisolvent (Masood 2016). The main characteristics that should be considered in PNs design are particle size, distribution of size, morphology, loading and encapsulation efficiency, drug release, and stability. PNs possess several biological applications in the following designs:

1. Long-circulating stealth PNs: One of the limitations of PNs application is elimination from systemic blood circulation via mononuclear phagocyte system (MPS) uptake. To resolve this issue, they must be coated with PEG to become undetectable by MPS.
2. Environment-sensitive PNs: Some PNs are sensitive to several factors such as pH, redox, temperature, light, ultrasound, and ionic strength. Drug release from PNs is assisted by alterations that take place in abnormal tissue. pH drop in cancer tissue leads to drug release from pH-sensitive polymers.

3. Multifunctional PNs: These multifunctional polymers have several roles. Drug protection, tissue and cell targeting, and controlled release of drug tailor them for various biological applications like drug delivery, gene therapy, and disease diagnosis. The major point in these PNs' design is interaction of functional groups with each other that might influence their action (Lu et al. 2011).

12.2.2 Liposome

Liposomes are spherical vesicles in size of 30 nm to several micrometers that comprise cholesterol and natural phospholipids. Size, hydrophobic and hydrophilic properties, and biocompatibility of liposomes offer biological applications of them. Components, size, preparation method, and surface charge determine fluidity and rigidity of liposomes. Liposomes increase drug efficacy and stability, and owing to their flexibility, biocompatibility, biodegradability, lack of toxicity, and immunogenicity they are suitable for systemic and non-systemic administration. Liposomes lessen exposure of sensitive tissues to toxic drugs. In addition, they decrease toxicity of entrapped drugs. Binding to specific ligands, they play important role in active targeting. On the other hand, they possess several limitations as well. Low solubility and short half-life are of these limitations. Phospholipids of liposome are susceptible to oxidation and hydrolysis like reactions. There is possibility of drug leakage from their structure, and their production cost is high. Preparation methods of liposome can be divided into mechanical methods (such as sonication, french pressure cell (extrusion), freeze thawed liposomes, lipid film hydration by shaking or freeze drying, membrane extrusion, and micro-emulsification) and solvent dispersion methods including ethanol injection, ether injection, and reverse phase evaporation. Biological applications of liposome are anticancer therapy, parasitic disease and infection treatment, gene delivery, vaccine formulation, ophthalmology, and wound healing (Ahmed et al. 2019; Akbarzadeh et al. 2013).

12.2.3 Magnetic Nanoparticles

In the 1950s lymphatic node is treated by applying metallic nanoparticles heated via magnetic field. After that magnetic nanoparticles (MNPs) have been developed for drug delivery, enzyme immobilization, and other biological applications. Characteristics such as size uniformity, superparamagnetic, bioavailability, adsorption kinetic, magnetic moment, and large surface area that can be adjusted expand their applications. Among diverse MNPs types, iron oxide nanoparticles have been employed for diagnosis thanks to their lack of magnetic effect after magnetic field removal. MNPs could be designed in different shapes such as nanorods, nanowires, and nanocubes through various types of preparation methods known as wet chemistry approach. Hydrothermal, solvothermal, sol-gel, coprecipitation, electrochemical pyrolysis, laser pyrolysis, and flow injection synthesis can be mentioned as examples of wet chemistry approach. New methods named laser ablation, microbial method, and evaporation synthesis could be applied for production of MNPs. Magnetic field parameters including frequency, duration, and amplitude affect MNPs toxicity. Particle size must be between 10 and 200 nm to be eliminated by hepatic filtration and are not influenced by renal filtration and finally leads to rapid penetration avoidance. Surface modification with biocompatible and biodegradable polymers such as polysaccharides and linoleic acid leads to MNPs properties improvement. These MNPs are able to bind to various receptors and therefore used for several goals including drug and bacterial attachment and imaging. Other applications of MNPs are hyperthermia, drug release, lab-on chip, immunoassay, and tissue engineering. The major benefit of MNPs is no harmful effect for normal tissues due to their controlled localization and thermal activation by applying magnetic field. Cancer cells can be bound to their surface functional groups and eliminated from blood. In magnetic fluid hyperthermia, magnetic field causes nanoparticles movement in diverse directions and heat releasing that can eradicate tumor cells (Materón et al. 2021).

12.2.4 Gold Nanoparticles

The use of gold nanoparticles goes back to ancient time because of optical properties. To synthesize gold nanoparticles, salt is reduced in the presence of a reducing agent such as citric acid, and then nucleation leads to the conversion of gold ions into nanoparticles. Another component that is applied in gold nanoparticles synthesis is stabilizer. The mechanism of stabilizer is binding chemically or through adsorption to nanoparticle surface from one endpoint, and another point places toward solution and stabilizes the gold colloid. Gold nanoparticles can be found in spherical, rod, and hallow shape. The most applicable stabilizers are thiol-based surfactants with carboxylic acid functional groups that in addition to gold nanoparticle stabilizing offer a surface for other biological molecules binding. Although gold nanoparticles are inert, they can accumulate inside the cell. Actin fiber can be damaged in presence of gold nanoparticles. Gold cluster can fit to DNA grooves and cause cytotoxicity. Gold nanoparticles have several applications including labeling and visualization (immunostaining, single particle tracking, contrast agent for X-ray, and transmission electron microscopy), vehicle for gene delivery, source of heat, opening of chemical bond, opening of container wall of drug, and as a sensor (Sperling et al. 2008).

12.2.5 Polymeric Micelles

Polymeric micelles (PMs) are comprised of self-assembled ampiphilic co-polymers (10–200 nm) that provide specific properties such as high molecular weight, low rate of dissociation, better drug encapsulation, proper stability, high accumulation in tumor site, low critical micellar concentration (CMC), better penetration to tissues, suitable biodegradability, pharmacokinetic (PK) and pharmacodynamics (PD), easy preparation method, and reproducibility. In addition, thanks to their size they are invisible by MPM and therefore have high circulation time. PMs possess several applications such as nonomedicine, biotechnology, catalysis, and nano-patterning. Polymeric components of PMs determine toxicity, biodistribution, and clinical compatibility. Hydrophobic compartment of PMs (core) plays an important role in bioactive solubilizing and entrapment, controlled release, and stimuli responsive bioactive release. Furthermore, they determine PK and PD profiles. On the other hand, stealth properties and PK are affected by molecular weight and chemical structure of hydrophilic compartment (shell). Encapsulation efficiency relies on hydrophilic-lipophilic balance (HLB) and polymer-to-drug ratio. Corona functions include cell penetration function, antibody function, receptor-specific ligand, and stimuli sensitivity. Hydrophobic drug entrapment, SiRNA linkage, and imaging agent can be named as core functions. Traditional methods of PMs utilize organic solvents that impede clinical study because of toxic initiators and residual solvents. Radiation-induced copolymerization via gamma ray, UV, electron beam, microwave, and plasma are other preparation methods with higher efficiency and the absence of toxic initiators. For bioactive encapsulation, physical entrapment or chemical conjugation can be employed. A poorly soluble drug, such as anticancer agents, can be entrapped by oil in water emulsion, solid dispersion method, dialysis, and emulsion solvent evaporation. For the characterization of PMs CMC, morphology, size, molecular weight, and stability must be determined. Common toxicity of PMs includes inflammation response, increase of cell volume, apoptosis induction, and blood components alteration (Deshmukh et al. 2017).

12.2.6 Polymer-Drug Conjugates

Polymer-drug conjugates were exerted in the 1950s and 1960s for the first time. This system demonstrates a long half-life, high water solubility and stability, low immunogenicity and antigenicity, and specific targeting properties. This technology is employed to improve proteins, small molecules, peptide and oligonucleotide properties, and delivery of drugs, proteins, imaging agents, and targeting moieties. The most applicable polymers are *N*-(2-hydroxypropyl) methacrylamide, polyglutamic acid, poly lactide-co-glycolide, poly ethylene glycol. These polymers might have degradable or non-degradable bonds. The design must avoid steric hindrance caused by polymers. In addition, the structure-activity relationship must be illuminated for conjugated drugs because the mechanism and efficacy of them may vary (Khandare and Minko 2006; Pasut and Veronese 2007).

12.2.7 Lipid-Drug Conjugates

Lipid drug conjugates (LDC) are the system in which a bioactive compound is bound to a lipid moiety such as free acids, glycerides, and phospholipids in covalent or non-covalent manner. The lipid part of this system improves solubility, penetration, and therefore bioavailability due to their similarity to biological lipids. This system is utilized for targeted drug delivery by exploiting of metabolic pathways. Selection of suitable lipid depends upon desired function and drug structure. As mentioned before, one of the lipid moieties employed to fabricate LDC is free acids. If the free acid length is sufficient, then it bypasses first-pass metabolism, and absorption happens via the lymphatic route. Therefore, it can be applied for lymphatic cancers. Drugs with carboxylate functional groups bind to diglyceride via an ester bond. For drugs without carboxylate, coupling with glyceride is carried out by spacers such as succinic acid. Nucleoside agents are suitable candidates for coupling to phospholipids. Majority of nucleoside agents are antiviral or antineoplastic agents. However, studies depicted that drugs formulated in this way have poor absorption and PK profile. The first step in LDC preparation is the fabrication of insoluble lipid conjugates through the salt formation with desired lipid or covalent binding with ester or ether. Then drug attachment to lipid conjugate takes place with one of these approaches: LDC with a covalent bond without a spacer, LDC with a spacer, or LDC with noncovalent bond.

The last step includes the processing of LDC to nanoparticles via various methods such as microemulsion technique, high-pressure homogenization, solvent-emulsification evaporation, solvent injection, wet milling, and solvent-emulsification diffusion. Mechanism of drug release from LDC accomplishes in two steps. The first step is releasing of LDC from nanoparticles through diffusion or erosion, and then physiologic enzymes lead to the cleavage of LDC and release of the drug to body fluid. Released lipids enter metabolic pathways or are consumed as a nutrient.

For LDC characterization, attention must be drawn to identification, partition coefficient, successful conjugation, melting point and crystallinity, size and morphology, particle size, entrapment efficacy, distribution, and in vitro drug release (Adhikari et al. 2017).

12.3 Nanomaterials Used for Cancer Diagnosis

Cancers are a group of diseases that include the abnormal growth of cells with the potential to invade or spread to other body parts. In summary, the steps of diagnosing cancer in an individual are such that if a person has a symptom or the result of a screening test that indicates cancer, the doctor must find out whether it is caused by cancer or another cause. These steps include:

1. **Medical examination**: It is done by asking for personal and family medical history and a physical exam.
2. **Laboratory tests**: Laboratory tests of blood, urine, or other body fluids measure high or low levels of some substances in the body as signs of cancers.
3. **Imaging tests** visualize the areas inside the body that help the doctor see if there is a tumor or not.
4. **Biopsy** is often the only way to diagnose cancer definitively.

In most cancers, more than 60% have metastasized or remained hidden until cancer diagnosis. In the case of these patients, the diagnosis is so late, and generally, the treatment does not achieve a good response in them (Biller and Schrag 2021; Dass et al. 2021; Mottaghitalab et al. 2019). Therefore, the current formidable state of cancer has urged researchers to develop different techniques for accurate cancer diagnosis and treatment, so nano-oncology is one of the emerging treatment approaches that have the potential to contest cancer. Nanotechnology-assisted cellular and molecular detection has improved cancer biomarker detection; for example, nanobiosensors detect multiple protein biomarkers in seconds (Tran et al. 2017). Various approaches to nanomedicine, including dendrimers, polymeric/non-polymeric nanoparticles, carbon nanotubes, quantum dots, lipid-based nanoparticles, and micelles, have been extensively investigated (Wang et al. 2010).

On the other hand, recent imaging strategies are not very diverse to provide complete clinical information about different tumor types and their stages, which is why successful treatment and patient outcomes have become burdensome. Early detection of metastatic cancers is essential for treatment planning and treatment. It can be challenging due to the distance of metastases from the primary tumor site, undetectable primary size, and low vascularity (Kim et al. 2010a). Therefore, the most critical thing in cancer treatment is its early and accurate diagnosis, which is generally divided into four parts here, and each region has been examined.

12.3.1 Nanotechnology-Assisted Cancer Imaging

Recently the potential of nanoparticles in cancer detection and monitoring has led to developing different types of nanoparticles for cellular and molecular imaging. Among the valuable potentials that highlight nanoparticles in imaging for cancer detection, we can mention their biocompatibility, small size, and high atomic number (Nie et al. 2007). Quantum dots, semiconductors, and iron oxide nanocrystals have some properties that are less common in other molecules. The conjugation of nanoparticles with various biomolecules, such as antibodies, peptides, or other chemicals, as well as anti-cancer drugs, is utilized to label and detect cancer cells (Popescu et al. 2015; Singh 2019). Continued advances in combining MRI/PET/CT (standard-of-care imaging modalities for the diagnosis of cancers), along with NP imaging, have made significant contributions in early-stage cancer diagnosis.

12.3.1.1 Magnetic Resonance Imaging (MRI)

MRI uses strong magnetic fields and radio waves to visualize specific organs and tissues and produce 3D images, high accuracy, excellent spatial resolution, good contrast with soft tissue, excellent signal-to-noise ratio, and lack of harmful radiation (Cruz et al. 2016). However, it has disadvantages such as a large capacity of generated data and long processing time (Zhao et al. 2016). NPs can be used to identify and visualize metastatic cells that are very difficult to detect with conventional imaging technology. Superparamagnetic iron oxide nanoparticles (SPIONs) are targets in MRI imaging with cancer cell lines (Wan et al. 2016). The high specificity of SPIONs with no known side effects makes them suitable building blocks for aerosols in MRI imaging (Sherry and Woods 2008; Kim et al. 2010b; Jafari et al. 2015). To increase MRI contrast, multifunctional SPIO NP containing Gd or antibodies or other target ligands have been used to detect metastases in different organs (Kievit et al. 2012).

On the other hand, it has been found that SPIONs have been used to predict the presence of brain metastases in melanoma, specifically regarding cancer vessels and their heterogeneity (Sundstrøm et al. 2013). In diagnosing brain cancer metastasis, using a multifunctional nanotherapeutic system based on poly (methacrylic acid)-polysorbate 80 grafted starch as surfactant coatings has enabled the delivery of MRI imaging and therapeutic agents (Li et al. 2014a; Patil et al. 2015). Additionally, in multimodal imaging, where MRI is integrated with another imaging modality towards vascular targeting to track brain, lung, and liver metastases of breast cancer, the absorption properties of iron oxide nanochain particles have been used for molecular fluorescence tomography (Peiris et al. 2012). In this regard, these advances in nanotechnology have created revolutionary development in cancer diagnosis and treatment.

12.3.1.2 Positron Emission Tomography (PET)

Positron emission tomography (PET) is a compassionate non-invasive technology suitable for imaging many cancers. Radiolabeled detectors are injected to obtain 3D images. The molecules to be evaluated are marked with an isotope capable of producing two gamma rays by emitting a positron from its nucleus (Gambhir 2002). The currently used PET tracer [F^{18}] fluorodeoxyglucose (FDG) provides a general approach to cancer imaging with 85% of sensitivity and specificity in different applications (Gambhir 2002). In a study, radionuclide imaging with [F^{18}] tetrafluoroborate ([F^{18}]BF_4) (PET/CT) was performed on the model in a breast cancer model. The results showed that the PET radiotracer was sensitive and specific for detecting metastasis with excellent contrast (Diocou et al. 2017). The human sodium iodide marker was expressed as a reporter, and [F^{18}] BF_4 was found to be superior compared to the conventional [I^{123}] iodide tracer. Also, PET/MRI agents with intrinsically multimodal and all-organic NPs have also been fabricated as “porphysomes” and directed to evaluate bone metastases of prostate cancer in vivo (Liu et al. 2013).

12.3.1.3 Computed Tomography (CT) and Single-Photon Emission Computed Tomography (SPECT)

CT is an X-ray-based method for creating detailed images of soft tissue, internal organs, blood vessels, and bones (Anderla et al. 2013). Computed tomography (CT) can be used to obtain three-dimensional imaging. In general, cross-sectional images using X-ray radiation contain larger doses of radiation than conventional X-ray imaging methods, which have the potential to be hazardous to public health. X-ray is one of the significant disadvantages of the CT method that is obtained (Power et al. 2016; Brenner and Hall 2007). Nanotechnology is effective in CT quality, as it has been found that using Au NPs coated with 30 nm PEG for CT contrast material in vivo has increased the image contrast, which allows for the reduction of the required radiation dose. In this case, it overcomes the limitations of conventional contrast agents (such as iodine-based compounds), such as rapid renal clearance, nephrotoxicity, and vascular permeability (Kim et al. 2007). For weakness compensation, composite nanoparticles such as a superparamagnetic iron oxide/ silica core and a gold nanoshell are used in the body as dual contrast agents for CT and magnetic resonance imaging (MRI) (Kim et al. 2011).

SPECT is a nuclear medicine tomography imaging method that works with gamma rays which is similar to conventional CT imaging, except for using a gamma camera with nuclear energy detection (Spanoudaki and Ziegler 2008). In this regard, a gamma-emitting radioisotope material (usually gallium(III) isotope) in combination with a targeting ligand is injected into the bloodstream; the concentration of the ligand is detected by a gamma camera (Spanoudaki and Ziegler 2008; Bateman 2012). The use of imaging agents such as MNPs overcomes their limitations and can act as a contrast agent for the imaging process. Au nanoparticles doped with Au^{199} atoms (due to the stability of the radiolabel) are helpful for targeted cancer imaging by SPECT in the murine triple-negative breast cancer (TNBC) model (Zhao et al. 2016). Another study suggested that multifunctional polyethyleneimine-encapsulated gold nanoparticles (Au PENPs) were designed and valuable for targeted radionuclide SPECT/CT imaging of glioma cancer (Zhao et al. 2019). As a result, it has been determined that the formed nanosystem has a significant potential to be used in the targeted diagnosis of cancers.

12.3.1.4 Optical Fluorescence Imaging

This optical method, defined as fluorescence imaging, differs from other imaging techniques using strong magnetic fields along with radio frequencies, such as gamma rays and ionized X-rays or MRI (Joshi and Wang 2018). These targeted probes consist of a carrier molecule (e.g., a peptide, antibody, or small molecule) with an attached fluorescent probe directed to a disease biomarker (Zhang et al. 2017). Optical fluorescence imaging techniques have been applied in various applications, including microscopy, endoscopy, and image-guided surgery (Schouw et al. 2021). Using multifunctional graphene oxide/iron oxide nanoparticles (Fe_3O_4 NPs) for magnetically targeted drug delivery has led to dual fluorescence/magnetic resonance imaging in cancer diagnosis (Gonzalez-Rodriguez et al. 2019). In another study, it has been shown that the presented AuNC/BSA-NPs formulation can serve as a potential platform for the diagnosis, light visualization, and treatment of colon cancer (Park et al. 2019). Therefore, optical imaging based on targeted fluorescence is rapidly evolving in diagnosing various cancers. However, this imaging method still needs further development and evolution to overcome some significant challenges to be considered part of the standard of care in the clinic.

12.3.1.5 Ultrasound Imaging

Ultrasound imaging uses high-frequency sound waves to view the inside of the body. Because ultrasound images are taken in real-time, blood flow in blood vessels and the movement of internal organs could be recognized. Unlike some imaging that uses X-rays, ultrasound imaging does not expose patients to ionizing radiation. In the ultrasound method, a transducer is placed directly on the skin with a thin layer of gel so that the ultrasound waves are transferred from the transducer to the body through the gel (Zlitni and Gambhir 2018). The ultrasound image is produced based on the reflection of waves from body structures. This technique is used to detect cancers such as skin, breast, prostate, etc. (Dencks et al. 2018; Catalano et al. 2019). The limitations of ultrasound contrast agents include high background signal, short half-life inside the body, and limited distribution in the blood circulation due to their micrometer dimensions (Borden et al. 2021). To overcome these problems, nanotechnology has helped a lot. It has been shown that the use of perfluorooctylbromide nanoparticles (PFOB) as a carrier has the potential advantage of selective delivery of cargo to the target site while improving site visualization using ultrasound imaging (Li et al. 2018a). Echogenic glycol chitosan nanoparticles containing iodine have been developed for cancer diagnosis in ultrasound imaging and X-ray computed tomography (CT). Using GC-DTA-PFP nanoparticles showed that dual CT/US X-ray imaging using echogenic iodinated nanoparticles provides more accurate diagnostic information for tumor diagnosis (Choi et al. 2018a). With the advancement of nanotechnology, ultrasound imaging is also developing in the diagnosis of various types of cancer.

12.3.1.6 Photoacoustic Imaging

Photoacoustic imaging or optoacoustic imaging is based on the use of laser-generated ultrasound such that non-ionizing laser pulses are delivered to biological tissues, part of the energy being absorbed and is converted to heat, which leads to transient thermoelastic expansion, resulting in the emission of broadband (i.e., MHz) ultrasound waves. Finally, the ultrasonic waves generated by the detected ultrasonic transducers are analyzed in the images (Beard 2011). It has been suggested that photoacoustic imaging is well-suited for evaluating cancer and other specified pathologies. Among the potential applications of photoacoustic imaging, one can examine the imaging of breast cancer and skin abnormalities in the morphology and function of vessels, as well as the evaluation of superficial soft tissue injuries (such as microcirculation abnormalities and burns) in diabetic patients (Nyayapathi and Xia 2019; Stridh et al. 2022; Mennes et al. 2018). Like the methods mentioned, nanotechnology has aided this method in better-diagnosing cancers and other diseases. One of the nanoparticles used in this method is gold nanoparticles (Li and Chen 2015). For example, in a study to increase the sensitivity and contrast, gold nanoparticles were coated with Prussian blue dye, which increased the high optical stability and high molar extinction coefficient in the near-infrared region, which detected cancer in nude mice in the method photoacoustic/CT combination which has made bimodal imaging much more accurate (Jing et al. 2014). On the other hand, using Au-PLGA hybrid nanoparticles has increased the near-infrared photothermal activities and catalase-mimicking in guiding photoacoustic imaging for diagnosing and treating colon cancer (Xi et al. 2018). As a result, these nanoparticles can help diagnose and treat cancers more effectively and may create opportunities for cancer treatment in the future.

12.3.1.7 Nanotechnology for In Vivo Imaging

Fluorescent imaging provides a visual picture of dynamic processes in living animals and is widely used to detect and study molecular and biological processes (Guo and Cui 2021). It has been determined that many organic fluorescent materials show no emission or weak emission when accumulated or at high concentrations. The accumulation-induced quenching effect is an essential obstacle in fluorescence imaging, which is solved by the aggregation-induced emission (AIE) process. AIEs have butterfly-shaped rotor structures that give rise to more intense fluorescent emissions observed when they are in the aggregation state (Mei et al. 2014). Based on the specific fluorescence property of aggregation-induced emission (AIE) and nanoengineering optimization techniques, many AIE nanoparticles have been used for tumor-targeted bioimaging in vivo. In addition, in a study, BSA-coated AIE nanoparticles showed better efficacy than blank AIE nanoparticles in tumor targeting in mice due to the EPR effect, which indicated low cytotoxicity and very high fluorescence intensity (Qin et al. 2012).

12.3.2 Nanotechnology Tools Used in Cancer Diagnosis

Recent research has confirmed that nanotechnology can detect cancer imaging at the tissue in cellular and molecular levels (Garrigue et al. 2018). This approach is achieved by understanding the capacity of nanotechnology to explore the cancer environment. For example, a pH-responsive fluorescence nanoprobe can detect fibroblast activator protein-a on the membrane of cancer-associated fibroblasts (Kievit et al. 2012). In addition to the previously discussed NPs properties, there are three aspects of NPs-based cancer diagnostic systems that specifically affect cancer diagnosis, including signal enhancement, miniaturization, and multiple detections (Minelli et al. 2010). In the following, we will discuss some temporal and spatial techniques based on nanotechnology that can significantly help accurately evaluate and track living cells.

12.3.2.1 Near Infrared (NIR) Quantum Dots

A new class of fluorescent labels with excellent bioimaging properties is near-infrared (NIR) quantum dots, which have unique features such as sufficient electron density, good fluorescence stability, high fluorescence intensity, and strong tissue penetration ability (Brunetti et al. 2018). One of the limitations of visible spectrum imaging is the inability to penetrate objects. To overcome this challenge, quantum dots with fluorescence emission in the near-infrared spectrum (700–1000 nm range) have been developed, making them more suitable for imaging liver cancer, pancreatic cancer, colorectal cancer, and lymphomas (Parungo et al. 2004; Gao et al. 2004). On the other hand, the second NIR window (NIR-II, 900–1700 nm) with newer features, such as higher spatial and temporal resolution and higher tissue penetration depth, has been developed and introduced to help cancer imaging. They have great potential for in vivo cancer imaging, cancer diagnosis, and high-resolution electron microscopy studies of cancer cells (Brunetti et al. 2018). A study has shown that using NIR-II silver sulfide (Ag_2S) quantum dots synthesized directly in colon cancer cells led to the accurate diagnosis of colorectal cancer in vivo (Deng et al. 2019). So, with the advancement of nanotechnology in the future, we can see a change in the next-generation NIR quantum dots system, which can detect all types of cancers and malignancies with greater accuracy and speed.

12.3.2.2 Nanoshells

Nanoshells or nanoshell plasmon are spherical nanoparticles with a dielectric core of 10–300 nm and are made of silica covered by a thin gold shell (Loo et al. 2004). These nanoshells, which contain a quasi-particle called a plasmon, convert electrical energy into light. Its ability to regulate light with an absorption/emission array from UV to infrared light has made them unique (Grodzinski et al. 2019). Another characteristic of nanoshells is their lack of toxicity (heavy metals) in imaging (Grodzinski et al. 2019). Scientists widely use gold nanoshells as contrast agents in optical coherence tomography to evaluate and detect cancer cells. The widespread use of these nanoshells is that the optical resonance of gold nanoshells can be precisely tuned in a wide range, including the near-infrared, where tissue transmission is higher (Abbasi et al. 2017). Gold nanoshells are cost-effective, non-invasive, and safe, enabling high-resolution imaging (Helm 2010).

12.3.2.3 Colloidal Gold Nanoparticles

Gold nanoparticles are one of the most exciting classes facilitating cancer diagnosis. Colloidal nanoparticles have unique physicochemical properties of great interest for cancer diagnosis and treatment purposes (Mottram 2003). In the nanotechnology industry, they have been used as antennas for external light, X-rays, magnetic fields, or ultrasound to generate reactive oxygen species or heat, sensitizing target tissue to radiation (Feliu and Parak 2020). These nanoparticles act as a good contrast agent because they scatter visible light in the body. In addition, it has been found that gold nanoparticles, in combination with antibodies, have been used for the biopsy and detection of pancreatic and cervical cancers and photoacoustic tomography (Purohit and Singh 2018). In this way, gold nanoparticles are a beneficial factor in the diagnosis and monitoring of the treatment of various cancers.

12.3.2.4 Detection of Circulating Tumor Cells

One of the problems of solid tumors, which is the cause of death of about 90% of cancer patients, is the metastasis of cancer cells (Gupta and Massagué 2006). A cancer cell from the primary tumor attacks the surroundings and enters the lymphatic system and blood vessels. Finally, it is transferred through the bloodstream to small vessels in distant tissues (known as circulating tumor cells (CTCs)) (Priestley et al. 2019). Liquid biopsy (CTCs as a part of liquid biopsy) enables the non-invasive and dynamic analysis of molecular or cellular biomarkers. It has a high potential for prognosis, diagnosis, understanding of the molecular organization of cancer, monitoring cancer progression and treatment effectiveness, and ultimately identifying it as an anticancer drug (Li et al. 2019). The detection of CTC, due to the unique structural and functional characteristics of nanomaterials (high surface-to-volume ratio), can overcome the challenges of insufficient absorption and low purity of CTCs and offers strong hope for the detection of CTCs (Bhana et al. 2015). Various types of nanomaterials, such as gold nanoparticles (Au NPs), quantum dots, magnetic nanoparticles (MNPs), nanopillars, nanowires, dendrimers, silicon nanopillars, and polymers, are developed for CTC detection (Huang et al. 2018). Among nanotechnology applications in the detection of CTCs, for example, MNPs attached to antibodies (immunomagnetic nanoparticles) for CTC detection; usually, CTCs that express EpCAM are identified explicitly by anti-EpCAM functionalized MNPs (Burinaru et al. 2019). Therefore, applications of liquid biopsy based on nanotechnology bring a new perspective on cancer diagnosis, monitoring, and treatment.

12.3.2.5 Detection Through Cell Surface Protein Recognition

In this method, nanoparticle probes conjugated with antibodies, oligonucleotide aptamers, short peptides, or proteins connect to surface markers and thus detect cancer cells (Harris et al. 2019). The cell surface marker epithelial cell adhesion molecule (EpCAM) is highly expressed in CTCs of many cancers (colorectal, breast, head, and neck). Hence, nanoparticles attached to anti-EpCAM antibodies or aptamers are often used to screen CTCs (Schneck et al. 2015). Many cell surface markers, such as vimentin (gastrointestinal) (Li et al. 2018b), androgen receptor (metastatic breast cancer) (Zhang et al. 2019a), glycan (breast cancer) (Wang et al. 2017), major cavity protein (MVP) (hepatocellular carcinomas) (Lee et al. 2017), and protein fibroblast activator α (FAPα) (non-small cell lung cancer) (Chen et al. 2018), have been studied to detect CTCs. In this way, nanotechnology has helped to detect cancer cells.

12.3.2.6 Detection Based on mRNA

The only coding RNA in organisms is messenger RNA (mRNA). So, we can predict the expression of genes by detecting the mRNA level (Choi et al. 2018b). Due to their high proliferation, the number of mRNA changes in cancer cells can be used as a marker for cancer diagnosis. But the identification of sensitive mRNA is challenging due to its inherent characteristics, such as sequence similarities and low expression levels (Zheng et al. 2021). Therefore, nanoparticles have been developed to detect extracellular and intracellular nucleic acids. For example, gold nanoparticle probes modified by oligonucleotides hybridized to fluorophore-labeled complements can detect mRNA in living cells (Seferos et al. 2007). Au NPs were functionalized with 2–3 DNA detection strands and later hybridized with short complementary reporter strands to detect the desired gene (Choi et al. 2013). Another nanotechnology-based approach is a plasmonic nanoparticle network structure that produces a plasmon-coupled dimer capable of detecting mRNA types. This method was used to detect and quantify BRCA1 mRNA binding species in vitro and in vivo. As a result, nanotechnology can make great strides in achieving many successes in the evaluation, identification, and treatment of cancers.

12.3.3 Nanotechnology for Detection of Extracellular Cancer Biomarkers

Before the treatment of any cancer, it should be detected, and early detection improves the efficiency of therapy. Cancer detection relies on the specific features or component of cancer cell that cannot be found in normal cell (specificity), are measurable, and named biomarker. Cancer biomarkers provide the possibility of not only early detection but also monitoring of recurrence or efficiency of treatment which improve the achievable targeted therapy. Some cancer biomarkers can be distinguished in extracellular spaces such as cancer tissue and body fluids like blood, urine, and saliva. Although low concentration, timing, and heterogeneity in the abundance of cancer biomarkers limit the admissible detection, nanotechnology can propose high sensitivity, selectivity, and facility in simultaneous measurement of them. Even specific targeting can be carried out with NP/nanomaterial-based biosensors, because nanomaterials enhance the biosensors' surface-to-volume ratio (Doria et al. 2012). The most common nanoparticle probes applied in cancer diagnosis are gold NPs, quantum (QD), and polymer dots (PDs).

12.3.3.1 Protein Detection

Proteins can be regarded as a biomarker. Extracellular protein biomarkers are secreted with cancer cells or body when cancer exists. CEA, CA-125, PSA, and AFP are a number of FDA-approved cancer-detection protein biomarkers. Their specific interplays with antibodies, antibody fragments, and aptamers can be calculated after being converted to quantifiable signals. Generally, a staining or fluorescence secondary antibody involved in a sandwich-type test as a general strategy produces measurable signals for the detection of protein biomarkers. Recently, QD-based biosensors in a sandwich assay have been used for this strategy in several studies. In a way that QDs are conjugated to antibodies to detect biomarkers like neuron-specific enolase (NSE) and carcinoembryonic antigen (CEA) biomarkers. These markers are secreted from cancer cells, and upon secretion, their concentration could reach over 15 ng/mL. CEA as a most general biomarker in cancer is extremely studied to improve the efficiency of monitoring and prediction of cancer recurrence after treatment or resection with surgery (Gold and Freedman 1965; Shitrit et al. 2005). The elevation of NSE, an enzyme that converts 2-phosphoglycerate to phosphoenolpyruvate (Tapia et al. 1981), correlates with cancers such as small-cell lung carcinoma (Harding et al. 1990) and islet cell cancer (Portela-Gomes et al. 2004). In a homogeneous in-solution QD-based assay limit of detection (LOD) for both CEA and NSE reached 1.0 ng/mL (Li et al. 2011). Theses sandwich-type techniques are composed of some components such as capture antibody, biomarker (as an analyte), a second capture antibody, and a secondary-antibody which is frequently fluorescent probe tagged. These types of assays have great specificity because of the great selectivity and affinity that a primary antibody provides for its analyte (Engvall 1980). Whenever a QD is applied to the fluorophore-label secondary-antibody, high sensitivity is generated as a consequence of QD-coupled antibodies' intense signal (Chinen et al. 2015). Until now, many studies according to this assay have been designed in which some will be brought in the following.

One study, instead of streptavidin beads, immobilized the capture antibodies on polystyrene microspheres to develop this fluoroimmunoassay to decrease the LOD of each CEA and NSE biomarkers to 0.625 ng/mL (Cao et al. 2011). It is significant to improve the assay to a lower LOD and truthfully screen the patients' biomarkers due to quite a low level of cancer biomarkers.

Next, QD-based anti-PSA micro and nano-arrays were applied to determine prostate-specific antigen (PSA). It was fabricated with dip pen nanolithography and presented to PEGylated QDs coated with an antibody against PSA. The fluorescence signals were assessed with a microarray scanner. With micro- and nanoarrays many compounds can be rapidly monitored which is regarded as a desirable feature for clinical use (Gokarna et al. 2008). PSA as a glycoprotein is a member of the serine proteases family secreted from the prostate gland. It is found in various forms in the body (such as free and bound to serum proteins) (Kuriyama et al. 1981). PSA concentration needs long-time monitoring due to biodiversity in the population; its high level doesn't permanently indicate prostate cancer (Thompson et al. 2004; Andriole et al. 2012), a point that was considered in the abovementioned assay. Thereafter, CdSe/ZnS core/shell QD-conjugated antibodies were used in a waveguide-based biosensor to detect and develop the sensitivity of CEA to less than 0.1 ng/mL. In the presence of CEA, immobilized anti-CEA monoclonal antibodies bind to CEA. Then, this complex sandwiches with CdSe/ZnS core/shell QD-conjugated monoclonal antibodies to generate a fluorescent signal (Mukundan et al. 2009a, 2009b).

Another type of QD-probed anti-CEA or AFP (alpha-fetoprotein as a plasma protein produced by liver can be an indicator of liver cancer) antibodies assay was conducted to sandwich the immobilized capture CEA and AFP antibodies, respectively, to generate a fluorescent signal which can be imaged through fluorescence microscopy. This assay was done on micro or nano biochips and named the lab-on-a-chip technique that enable the detection of biomarkers on the small amount of patient sample and in turn the chance of performing laboratory experiments on minor scales. In these biochips the LOD for each analyte reached 0.25 nM (Yan et al. 2008; Hu et al. 2010a). Another example of QD-based sandwich assay is for CEA and cancer antigen 125, which is a cell membrane glycoprotein upregulated in ovarian cancer patients and applied as a late-stage monitoring and prognosis marker (Canney et al. 1984; Kenemans et al. 1993). In this method, on agarose bead immobilized capture antibodies are introduced to antigens which then sandwich with QD-probes antibodies and base on the analyte concentration led to the imaging of fluorescence signals by photomicrographs of the fluorophore-labeled beads. Actually, the microfluidic microporous agarose bead arrays were developed to lower the cancer analytes' LOD that reached 20 pg/mL (0.11 pM) for CEA (Jokerst et al. 2009); nevertheless with an analogous assay, it achieved 50 fM (Hu et al. 2010b).

In another study a QD-based immunochromatography test strip was set up to detect AFP rapidly, quantitatively, and sensitively (Yang et al. 2011). This device, with a capillary action, carries the sample through the strip (Li et al. 2010). Then, this test strip was developed for detecting C-reactive protein (CRP) with a LOD of 0.63 U/mL (Cheng et al. 2014). CRP, as an inflammatory protein secreted from the liver, correlates with lung and colorectal cancer (Allin and Nordestgaard 2011). To improve the detection assay, FRET-based biosensors were thus generated. FRET is a mechanism extremely sensitive to small changes, transfers the energy from a donor to an acceptor fluorophore, and thereby is ideal for QD-based assays, which named in-solution sandwich fluoroimmunoassay (Wei et al. 2006). Hence, an advanced form of this assay with a luminescent terbium chelate-QD FRET pair was utilized to detect AFP with a LOD of 0.4 ng/mL (Chen et al. 2012). Following constructing an immunoassay with a terbium–QD FRET for cancer biomarkers, PSA was perceived at a concentration of 1.6 ng/L (Wegner et al. 2013). For more precise and sensitive detection, a QD-based sandwich immunoassay was sat on a glass substrate involving a ZnO nanowire array. A fluorescence microscopy is also used in this method to detect the enhanced fluorescence produced with FRET. This technique was utilized for CEA in a range from 0.001 to 100 ng/mL. In addition, a ZnO-QD-based anti-immunosensor coated with carbohydrate antigen 19-9 antibody has been applied for the detection of pancreatic ductal adenocarcinoma. In an electrochemical assay, the dynamic range was 0.1–180 U/mL with a LOD of 0.04 U/mL, whereas in an optical spectral detection, the dynamic range was 1–180 U/mL with a LOD of 0.25 U/mL (Kim et al. 2014).

Subsequently, a paper-based immunosensor system using a QD-FRET was developed to detect CA 125, AFP, and CA 15-3 to determine the tumor recurrence in patients with breast cancer (Shitrit et al. 2005; Duffy et al. 2000). In this device, antibodies are captured on a paper sensor, and CuO-conjugated antibodies bind the analyte. In the presence of biomarkers, Cu^{2+} is released from CuO NPs to make fluorescence which can be then recovered. The LOD for CA 125, AFP, and CA 15-3 were 6.1×10^{-5} U/mL, 0.3 pg/mL, and 2.9×10^{-4} U/L, respectively (Ge et al. 2013). As QDs can be applied as an oxidizing agent, this feature can be employed in a sandwich assay for the detection of cancer biomarkers. As a result, an immunosensor employing CuS QDs to oxidize a small molecule was used to detect PSA with a LOD of 0.1 pg/mL. In the presence of PSA, the small molecule becomes oxidized to produce fluorescent signals detected with fluorescence spectroscopy (Zhu et al. 2014; Ward et al. 2001). These indicate that CuS-based fluorescent immunosensors are positively comparable with non-fluorescence-based immunosensors (Gao et al. 2013). Lately, a gold NP-coated graphene nanosheet (Au NPs/GN) from CdS QD-functionalized mesoporous CdS-QDs/TiO_2 for a sensitive photoelectrochemical PSA aptasensing over the exciton–plasmon interplay between CdS-QDs and Au NPs was developed. Accordingly, a sandwiched aptamer between CdS QDs/TiO_2 and Au NPs/GN was carried out in the presence of PSA, which in turn led to the dissociation of Au NPs/GN from the CdS QDs/TiO_2 to elevate the photocurrent and detect PSA with a LOD as low as 0.52 pg/mL (Cai et al. 2018a).

Determining several cancer markers simultaneously in body fluid can be beneficial for the detection of numerous targets in clinical applications. In this regard, an ultrasensitive multiplex electrochemical immunoassay utilizing biofunctionalized polyethylenimine-Au NPs tagged with metals as signal probes was developed to simultaneously detect CEA, PSA, AFP, and IL-8 with LOD of 1.6, 0.9, 1.7, and 1.0 fg/mL, respectively (Putnin et al. 2019). Subsequently, an electrochemical immunosensor according to the sandwich method has been designed for the simultaneous detection of three cancer biomarkers. This multiplexed immunosensor determined PSA, AFP, and CA 125 with LOD of 86 fg/mL, 14 fg/mL, and 0.0019 U/mL, respectively (Wang et al. 2021). Recently, a personal glucometer is proposed for detection of AFP at a lower concentration as 5.4 pg/mL. It is consistent with sandwich-type immunoassay antibody-invertase cross-linkage NPs as the signal producing-tags for routine use. Antibody-invertase cross-linkage NP-tags are generated through reverse micellar technique with glutaraldehyde. As it targets AFP, sucrose hydrolyses into fructose and glucose, which can be detected on a transportable personal glucometer (Li et al. 2022).

Moreover, hydrogel nanoparticles (NPs) as an innovative sample preparation technology were applied to recognize early-stage biomarkers of breast cancer in serum. This NP-based protein capture technique was precisely handled to catch and detect low existing and low-molecular-weight proteins of the proteome in body fluid. Consequently, proteins, protein fragments, and peptides were identified as serum biomarkers in breast cancer (Fredolini et al. 2020).

12.3.3.2 Circulating Tumor DNA (ctDNA) Detection

ctDNA refers to DNA fragments of primary or circulating tumor cells, found in the blood stream as a result of cancer cells apoptosis, autophagy, and necrosis (Thierry et al. 2016). ctDNA can carry cancer particularly genetic and epigenetic alterations (Aravanis et al. 2017). Determination of this alteration can be helpful in early cancer detection. By using complementary nucleic acid sequences that are probed and can specifically be hybridized with DNA, ctDNA can be identified in liquid biopsy (Das et al. 2016). Because of low concentration and short half-life of ctDNAs, their rapid detection with ultrasensitive method is urgent. NP-conjugated fluorescent probes as a biosensor are developed in this field (Goon et al. 2010). DNA-conjugated silver nanocluster fluorescent probes were developed to identify a single exon of the BRCA1 gene in breast cancer. A high fluorescence intensity is produced with DNA-Ag nanoclusters and leads to the detection of a deletion with a LOD of 6.4×10^{-11} M in optimized conditions (Borghei et al. 2017). Recently, a network of probe DNA improved gold-coated magnetic NPs (DNA-Au MNPs) in a new electrochemical sensing assay which has provided an accurate technique in short time (20 min) for cDNA detection with a LOD of 5 fM. This DNA-Au MNPs biosensor can be used for short as well as long strands DNA targets (Chen et al. 2021). In addition, an improved DNA-rN1-DNA-regulated surface-enhanced Raman scattering (SERS) frequency shift experiment in ctDNA has enabled a sensitive diagnosis of one single-base pair-mutation in mutated KARS G12D gene of lung cancer in femtomolar concentration of ctDNA. In this assay a hairpin

DNA-rN1-DNA probe and an employed RNase HII enzyme for specific hydrolyzation the DNA-rN1-DNA/ctDNA hybrid has been designed to let ctDNA recycling in the method for quantifying signal magnification (Zhang et al. 2019b).

12.3.3.3 microRNA (miRNA) Detection

miRNAs as circulating biomarkers can be applied for the detection and quantification of multiple cancers using body fluids (Hafizi et al. 2013). Plasmonic origin, several metal NPs, QDs, and their composites in magnetic particles have been applied in miRNA-sensors. Gold NPs (Coutinho and Somoza 2019), Ag/Pt nanoclusters for sensing miRNA-21 in human urine samples (Fakhri et al. 2020), a newly advanced Au/Ag nanospheres via SERS as an extremely sensitive detection system for miRNA-133a (Sun and Li 2018), and Fe3O4@Au nanomaterials using a DNA four-way junction-based electrochemical determination of target miRNA are some examples (Pang et al. 2016).

Recently, nanomaterial-based biosensors especially magnetic materials have been developed as the base for some highly efficient miRNA sensing methods in body fluids like serum and urine. In this regard, one study improved a device to increase the sensitivity of detecting miR-141 by employing the benefits of both FRET and chemiluminescence resonance energy transfer. FRET quencher was conjugated to nucleic acid functionalized CdSe/ZnS QDs. Upon miR-141 binding, the FRET was released and led to high fluorescence emotion to detect miR-141 with a LOD of 1 pM. Then, QDs were presented to G-quadruplex-involving telomerase and dNTPs. To further improve the sensitivity, hemin was incorporated to G-quadruplexes on QD to accelerate the process of luminol oxidation by H_2O_2 and decrease the LOD to 0.28 pM for miR-141 (Jou et al. 2015). miR-141 as a biomarker for ovarian, gastric, and prostate cancers is identified in the blood (Gao and Wu 2015; Zuo et al. 2015). Two-dimensional nano-structures and their composites in magnetic particles can also be used in miRNA sensors (Hizir et al. 2014; Cai et al. 2018b).

12.3.3.4 DNA Methylation Detection

DNA methylation can specifically alter in multiple cancers and play a significant role in the progression of cancers. These alterations are known as epigenetic markers; as a result, they can be employed as a biomarker for early detection, management of treatment, and prognosis monitoring (Jones 2002). Ultrasensitive biosensors have been developed to rapidly and selectively quantify DNA methylation in blood or saliva samples since it has been observed that cancer and normal genomes have different DNA solvation and DNA-gold affinity. These devices propose a simple, sensitive, rapid, and selective colorimetric or electrochemical one-step assay to diagnose DNA methylation in cancer (Sina et al. 2018). Therefore, an enzyme-free and label-free one-step quick colorimetric detection technique along with unmodified Au NPs has been suggested. In this method, the color of gold nanoparticle solution is changed to detect methylated DNA without any covalent modification of Au NPs. The LOD for DNA methylated was 8.47 nM, and it is particularly possible to distinguish methylated DNA with the naked eye at low concentration as 20 nM (Li et al. 2020). In another study, hybridization on a network of probe DNA-altered

Au-coated magnetic NPs (DNA-Au MNPs) and an enzymatic cleavage for differentiation of methylated DNA strands from unmethylated ones in ovarian cancer have been used. The dynamic range in this biosensor for 110 nucleotide DNA sequences involving single-site methylation was from 2 aM to 20 nM, and the distinguished concentration was as low as 2 aM to quickly identify ovarian cancer in 35 min (Chen et al. 2022). Moreover, an ultrasensitive DNA methylation ratio detection has recently been applied according to target-induced NP-coupled and site-specific base oxidation damage in colorectal cancer. For this nanosensor, the sensitivity for diagnosing methylated DNA was lower as 32×10^{-17} M. Besides, a high specificity with this method was attained for identifying 0.001% methylation ratio without operating the amplification process (Luo et al. 2022).

12.3.3.5 Extracellular Vesicle Detection

Extracellular vesicles (EVs) are nanoscale-sized particles released from normal and cancer cells. As EVs carry molecular information of mother cells like DNA, protein, miRNA, and mRNA, they have the potential to be used as a biomarker to detect the molecular state of cancer cells, which are problematic to achieve (Bakhshandeh et al. 2017b). Compared with other biomarkers, EVs have more stability and high abundance in body fluids (Dakubo 2016). During the past decades several biosensors with different sensitivities for EV detection have been applied. However, an extremely sensitive biosensor is significant for developing the sensitivity of cancer detection. The optical sensors such as fluorescent and spectral signals are utilized in EV detection. The fluorescent signals are sensed by conjugating the fluorescent probes with aptamers or antibodies. Multiple fluorescent signals' providing modes like fluorescence quenching, fluorescence polarization, QDs, and molecular beacons have been developed to improve the LOD in EVs. Besides, spectral signal sensing containing surface plasmon resonance, near-infrared, and SERS is beneficial for EV detection (Min et al. 2021). A number of miRNA molecular beacons in combination with antibodies have been designed to simultaneously distinguish EV-derived miRNAs and surface proteins (Cho et al. 2019). Then, an antibody-coupled multifunctional magneto-plasmonic nanorod has been used to amplify fluorescent signals with metal. In this method, EVs are captured with magnetic nanorods through antibody affinity, and molecular beacons bound to the nanorods detect the released miRNAs (Lee et al. 2019a). QDs have also been labeled to antibodies to quantify the multiplex EV surface proteins to detect pancreatic cancer with two correlated EV biomarkers (Rodrigues et al. 2019).

Several machine learning and RNA-sequencing algorithms have revealed improved performance in EVs analysis toward a single biomarker. In this regard, a deep learning-based SERS has been suggested for exosome detection to accurately diagnose lung cancer in early-stage. This deep learning model was developed with SERS signals of exosomes from normal and lung cancer cell lines to differentiate them with a precision of 95%. In fact, CNN algorithms were applied in the device to improve the machine learning tool (Shin et al. 2020). Recently, a europium-NP-based assay for the detection and characterization of various surface glycoconjugates of EVs has been used for five different colorectal cancer cells and HEK cells (Vinod et al. 2021).

12.3.4 Nanoparticles-Based Biosensors for Cancer Biomarker Screening

During cancer generation and progression, some cellular features and process alter. If these alterations become correctly and entirely identified, cancer can better be understood, and some stable, measurable, and detectable alterations can be regarded as indicators of cancer and help to verify cancer cells presence. DNA, proteins, and protein fragments, which are detected in body fluids, can be used as screening biomarkers, therefore, having the potential to be applied for screening people for cancer risk (Jin et al. 2020). Most of these biomarkers are low-molecular-weight proteins and may be removed during detection as a result of depletion of the high-molecular-weight proteins such as albumin, which acts as a carrier for small molecules (Kuk et al. 2009). Nevertheless, some detection methods have been developed with NPs to highly detect biomarkers from biological liquids. In these assays, low molecular weight proteins are captured by NPs (Luchini et al. 2010). These NPs, such as hydrogel NPs, mesoporous silica particles, and carbon nanotubes by owning electric charge and functional biomolecules act as nanocarriers and reduce the chance of proteins binding to carrier proteins (Najam-ul-Haq et al. 2007; Geho et al. 2006).

Besides, screening-nanocarrier methods have also been improved to develop the sensitivity of mass spectrometry techniques. In this regard, carbon nanotubes have been applied to increase the energy transfer efficiency of the analyte and facilitate ionization (Najam-ul-Haq et al. 2007; Wang et al. 2007). In addition, lab-on-a-chip microfluidics tools have been developed using QDs like CdSe and ZnS which are conjugated to antibodies against CEA, AC 125, and Her-2/Neu to function as multiplexed protein detection devices (Jokerst et al. 2009). A platform on the surface of various sized nanometers has also been used to differentiate between multiple cancer cells and evaluate their cell growth (Hung et al. 2010).

12.3.5 Clinical Trials of Nanotechnology-Based Applications in Cancer Diagnosis

As mentioned above, many NP-based assays and methods have been designed and developed in multiple studies to quickly, sensitively, and selectively detect cancer biomarkers in the laboratory. Nevertheless, the final destination is the application of these techniques in clinics. In clinical application, we may face challenges which should be resolved. NP probes may be aggregated, have nonspecific binding or toxicity for systemic application in in vivo imaging, produce fluctuated signals due to body fluid composition, and are not reproducible. In addition, minimizing these NP-base devices to batch-to-batch volume for large-scale applications can be challenging, which should be considered. Meanwhile, the point of care of nanomaterial-based tools should be developed to accelerate their clinical application. Nevertheless, several nanotechnology-based methods have achieved the clinical application for cancer diagnosis that are summarized in Table 12.2.

12.4 Nanomaterials Used for Cancer Therapy

Magnetic nanoparticles (MNPs) have the potential to be used in various biomedical fields. The focus on MNPs is rising due to significant physical assets, nanoscale size, and increasing surface/volume ratio (Sun et al. 2008; Shabestari Khiabani et al. 2016; Torchilin 2006). Nanomaterial-based drug delivery systems can greatly enhance the efficacy of chemotherapeutics treatments. Cancers affect the quality of life and can be linked with short survival rates (Gellhorn 1958). In case of insoluble anti-cancer drugs, nanomaterials can improve both the solubility and the activity of the drug (Dean et al. 2005). A nano-sized drug can be formulated so it can fit different dosage methods such as injection, nasal, and oral consumption (Kipp 2004). The other factor affecting chemotherapy is the method of delivering drugs into its target site (Kumar and Mohammad 2011). The combination of nanomaterials with anti-cancer drugs can enhance drug selectivity and reduce the toxicity level again in normal non-cancerous cells (Zhang et al. 2013).

Drug delivery systems can be made with absorption or conjugation of the drug on the particle, encapsulation of the drug inside the lipid or polymer, and dissolving the drug within the matrix of the particle, in which an unfavorable drug can be masked, or the drug can be protected from a dangerous surrounding.

Moreover, nanocarriers can be gathered favorably at tumor sites by the benefit of the enhanced permeability and retention (EPR) effect. The EPR effect includes characteristics that cannot be seen in normal cells and are specific to the tumor, thus enhancing better target selection (Shenoy et al. 2005; Kingsley et al. 2006). A lot of effort has been made to improve the stability of the drug in action. One such effort is a surface modification by adding PEG (Papahadjopoulos et al. 1991; Bamrungsap et al. 2012). To make nanocarriers work well for drug delivery, they need to meet two basic requirements. First, there should be the least amount of volume loss and activity loss in the blood. Second, drugs should not target healthy cells of our body. They should target just the cancerous cells. There are two ways to meet these needs: drugs can be passively or actively targeted. Under certain conditions, like low oxygen, the endothelium layer of blood vessels becomes more penetrable. To deal with not getting enough oxygen, tumor cells tend to make more blood vessels or swallow up the ones that are already there. This process is called "neovascularization." These blood vessels that were formed recently are leakier because they have large pores. Passive targeting is mostly based on how the tumor works and how the size and circulation time of carrier differs. The EPR effect depends a lot on basic tumor biology, such as (1) how much angiogenesis and lymph angiogenesis are going on, (2) how much perivascular tumor invasion is

Table 12.2 A list of some nanotechnology-based methods/materials with clinical applications for cancer diagnosis

Composition	Condition	Target	Status	NCT number
Carbon nanoparticle (CNP)	Rectal cancer	Lymph nodes	Not applicable	NCT03550001
Iron oxide nanoparticle	Head and neck cancer	Imaging	Early phase 1	NCT01895829
Sentinel™ PCC4	Prostate cancer	Diagnostic test	–	NCT04661176
CNP	Breast cancer	Tumor location and lymph nodes	Not applicable	NCT04482803
CNP	Colorectal cancer	Tumor localization and lymph nodes	Not applicable	NCT03350945
CNP	Endometrial neoplasms	Lymph nodes	Not applicable	NCT03778255
CNP	Uterine cervical neoplasms	Lymph nodes	Not applicable	NCT03778268
Indium-labeled PSMA	Prostate cancer/lymph node metastases	Imaging	Phase 1/2	NCT04300673
Magnetic nanoparticles coated with antibodies	Leukemia	Imaging	–	NCT04290923
Fluorescent cRGDY-PEG-Cy5.5-QD	Head and neck cancer/ melanoma/breast cancer/ colorectal cancer	Imaging	Phase 1/2	NCT02106598
Ferumoxtran-10	Prostate cancer	Imaging	Phase 3	NCT04261777
Microfluidic and Raman spectrum	Breast neoplasms/ circulating tumor cells	Detection of circulating tumor cells	–	NCT04239105
Immuno-tethered lipoplex nanoparticle	Lymphoma	Monitoring/ detection	Not applicable	NCT03656835
64Cu-NOTA-PSMAi-PEG-Cy5.5-QD	Prostate cancer	PET/MRI	Phase 1	NCT04167969
Nanosensors	Stomach diseases/gastric cancer	Diagnostic test	–	NCT01420588
Gold nanomaterials	Solid tumors/benign disease/normal	Diagnostic test	–	NCT03967652
Iron oxide nanoparticle	Head and neck squamous cell carcinoma	Lymph nodes	Not applicable	NCT03817307
124I-cRGDY-PEG-dots	Metastatic melanoma/ malignant brain tumors	Localization	Not applicable	NCT01266096

(continued)

Table 12.2 (continued)

Composition	Condition	Target	Status	NCT number
CNP	Rectal cancer	Lymph nodes	Not applicable	NCT03550001
Ferumoxytol	Rectal cancer	Lymph nodes	Early phase 1	NCT03280277
89Zr-DFO-cRGDY-PEG-Cy5-C′ dots	Brain cancer	PET imaging	Phase 1	NCT03465618
IRDye800CW	Head and neck squamous cell carcinoma	Imaging	Phase 1/2	NCT03134846
Ferumoxytol	Esophageal cancer	MRI	Early phase 1	NCT02857218
USPIO	Rectal neoplasms/breast neoplasms	MRI	Phase 3	NCT02751606
Electrosensing antibody probing system	Non-small cell lung cancer	Device	Not applicable	NCT01359436
Iron oxide NPs	Leukemia	Device	Not applicable	NCT01411904

going on, and (3) how much pressure is inside the tumor (Kubik et al. 2005; Pelicano et al. 2006).

Active targeting or ligand-mediated targeting needs specific ligands or molecules, like transferrin, that bind to molecules or receptors that are expressed on the target cells. The active targeting mechanism uses interactions between the targeting ligand and specific cell surface markers to increase the number of cells that take up the drug and keep it in the tumor. Through conjugation, it is possible to control how many targeting ligands are on the surface of the nanoparticles. In general, small molecule ligands like sugars and peptides are more appealing than antibodies because they are more stable, pure, easy to make through synthetic methods, and don't cause immune responses (Kingsley et al. 2006; Farokhzad and Langer 2009).

12.4.1 Multifunctional Nanosystems for Cancer Therapy

Nanotechnology (the study of particles and devices in the nanoscale, dimensions of 1–100 nm) opens up new avenues for cancer therapy and diagnostics (theragnostic). Some major classifications are described as follows:

12.4.1.1 Core-Shell Nanostructure

Because of their inclination to agglomerate and their short blood half-life, they must be functionalized with the help of certain polymers. Core-shell NPs, which generally represent composite nanomaterials, are made up of shells that cover the exterior layer and cores as the inner core material. The notion of core-shell NPs has recently been modified to include a class of nanomaterials in which the inner particles are entirely or partially covered by a layer with a defined boundary. Core-shell NPs have developed as an important class of nonstructural materials due to their extensive uses in biology, sensors and catalysis, as well as their appealing characteristics. A metal core surrounded by shells made of metal, nonmetal, and polymer could be utilized to build a variety of core-shell structures. Nonmetal or polymer shells can be used in polymer-core systems, and the two polymers are not the same (Lenin et al. 2018; Sur et al. 2019).

12.4.1.2 Polymers

Polymer nanoparticles are systems made of natural or synthetic polymers. They offer significant advantages over other nanocarriers, such as liposomes and micelles. Other remarkable aspects of polymeric nanoparticles include their outstanding durability in biological fluids, as well as the vast availability of diverse polymers, the ability to functionalize their surfaces, and adjust polymer degradation and leakage of the encapsulated compound(s) as a function of specific stimuli. To improve antitumor activity, suppress metastasis, and reduce effective dose and side effects, several cancer treatments have been encased in polymeric delivery methods. Polymers have the ability to encapsulate or adsorb active compounds within their structure (Guo et al. 2016).

12.4.1.3 Liposome/Magnetic Nanoparticles Hybrid Nanoparticles

When combined with liposomes, functional inorganic nanoparticles can be used in many different ways in biomedical research. Magnetoliposomes (MLs), which are liposomes with magnetic nanoparticles inside, are a promising platform for bioimaging, drug delivery, and controlling cell signaling. They are made of a mixture of organic and inorganic materials. MLs are often used in MRI as contrast agents and as drug carriers in chemotherapy. The size of MNPs is one of the most important things that affects how magnetic they are. Traditional ways to put inorganic or metal nanoparticles into liposomes depend on hydrophobic interactions between the aliphatic hydrocarbons of the phospholipid bilayers and the lipid surface of the nanoparticles. When the lipid film gets wet, nanoparticles with hydrophobic coatings on their surfaces naturally embed themselves in the phospholipid bilayers of liposomes.

12.4.1.4 Micelles

Micelles are formed by the self-assembly of lipids or other amphiphilic molecules, such as polyamino acids or polymers, into small nanoparticles with a hydrophobic core. This structure has been established as a hydrophobic drug delivery carrier. Hydrophobic medicines are enclosed within the hydrophobic core of the micelle. These copolymers are biodegradable and biocompatible; also, they are cell-specifically targetable (Choi et al. 2019).

12.4.1.5 Photodynamic Therapy

Photodynamic therapy (PDT) is an exciting way to treat cancer that uses a light source, a photosensitizer, and molecular oxygen. PDT is a way to treat cancer by giving a PS and shining light on the tumor to cause photooxidative reactions in the PS. After tumor tissue takes in PS, a certain wavelength of light is needed to get PS excited and cause it to make highly cytotoxic reactive oxygen species (ROS). PDT has several clear advantages over other ways of treating cancer, such as not being invasive, having low side effects, being sure to work, and not causing drug resistance. Because of this, PDT has been used in clinics to treat many types of tumors, such as bladder cancer, lung cancer, cancer of the head and neck, and skin cancer. So, targeted delivery systems are needed to get PSs to their destinations while causing as little damage as possible to normal cells and tissues.

These nanocarriers make it possible for PSs to be selectively gathered in tumors using passive targeting methods that depend on the increased permeability and retention (EPR) effect of nanocarriers in tumor tissues. Grafting targeted ligands onto the surface of nanocarriers is another common way to reduce side effects by reducing interactions between nanocarriers and healthy cells. Targeting ligands like antibodies, peptides, and folate can be grafted onto the surface of nanocarriers to make them more effective at finding and killing cancer cells.

By putting the PS in nanocarriers that have been modified with targeted ligands, the PS builds up in tumors only. This is called tumor selectivity. When it comes to targeting, third-generation PSs do better on average than second-generation PSs (Sun et al. 2019).

In the last 10 years, it has become more important for the drug delivery system to make smart materials that can respond to environmental changes. Some people think that being able to control how drug carriers act from afar could make drug delivery to target sites more effective and efficient. In many ways, liposomes are better than other drug carriers. Liposomes are made of two layers of lipids that are made in a lab. Their sizes range from a few hundred nanometers to a few hundred micrometers. Liposomes are biocompatible and biodegradable, and they can encapsulate both drugs that are water-loving and drugs that don't like water. This keeps the drugs safe from the environment. Researchers have recently found a smart way to treat people with a new type of liposome. This makes it possible for the liposome to release its contents in response to things like temperature, pH, light, ultrasound, magnetic field, and so on (Zou et al. 2011).

12.4.2 Magnetic Nanoparticles in Cancer Therapy

Magnetic nanoparticles (MNPs) have been used in nanomedicine for some time now. MNPs particularly superparamagnetic iron oxide nanoparticles, Fe_3O_4/Fe_2O_3, are prominent among nanoparticles due to their unique physicochemical properties such as high magnetic saturation, biocompatibility, and excellent heating ability (when exposed to an Alternating Magnetic Field (AMF)). There is a controlled interaction between the external magnetic fields by permanent magnet MNPs. As a result, the position of an MNP could be used to pinpoint the exact location of a medical problem. An alternating magnetic field causes MNPs to heat up, making them useful in nuclear MRI and as a drug delivery vehicle (Mulens et al. 2013).

12.4.2.1 Hyperthermia

Hyperthermia, a therapy that raises the temperature of malignant areas of the body to 40–43 °C, has indeed been found to trigger apoptosis in cancer cells by increasing the efficiency of radiotherapy and chemotherapy. Despite the ability of hyperthermia to improve radio- and chemotherapy treatments, toxicity due to the similar hyperthermia responses of malignant and healthy tissues remains a barrier to clinical application. Magnetic hyperthermia, a type of hyperthermia that is currently conducting clinical trials, is one promising method to overcome this barrier. Nanometer-sized (10–100 nm) ferrite nanoparticles, particularly magnetite (Fe_3O_4) or maghemite (–Fe_2O_3), are the most frequently used materials for magnetic hyperthermia. The properties of magnetic iron oxide nanoparticles (MIONs) are due to the presence of ions with different valencies in their crystalline structure (Chang et al. 2018; Kirschning et al. 2012).

MIONs may be delivered to the tumor via intra-peritoneal, intra-tumoral, intra-arterial, intra-cavitary, or intravenous routes. MIONs cannot be administered orally because the majority of the nanoparticles will be excreted focally due to their large size. After intra-tumoral delivery MIONs successfully localized in the tumor, which can result in efficient heating of primary tumors. Malignancies that frequently spread to the peritoneum, including ovarian and gastric cancers, benefit from intra-peritoneal administration. The intra-peritoneal route achieved a threefold greater intra-tumoral level compared to intravenous delivery. This was also capable of transporting chemotherapeutic drugs and inhibiting pancreatic tumors while causing no systemic toxicity (Chang et al. 2018; Gao et al. 2017).

12.4.2.2 MNPs for Hyperthermia-Based Therapy

MNPs used in hyperthermia therapeutics are made from a variety of materials, including Zn, Mn, Fe, Mg, Co, Ni, Gd, and their oxides. Pure metal applications in biomedical engineering have been limited due to their potential toxicity and instability in the human body. Metal oxides are more appealing for medical applications because they are more stable. Magnetite (Fe_3O_4) is a well-known magnetic material that can be stabilized by a variety of ligands like dextran (Farzin et al. 2020). The essential factors in determining MNP hyperthermic strength are the chemistry, particle size, NP dispersity, and magnetic moment.

12.4.2.3 Delivery of MNPs to the Tumor Site

It is possible to maintain MNPs in a consistent manner and to localize them in the cancerous site by exposing them to an external magnetic field close to the target tissue. The direct injection technique involves injecting cancerous tissue with magnetic fluid that has been formulated to contain a certain amount of MNPs (Ito et al. 2003; Zhai et al. 2009).

A further benefit of direct injection is that the concentration of MNPs at the tumor site can be controlled. However, this method is just effective when the tumor site is easily accessible (Huang and Hainfeld 2013).

MNPs can be actively delivered to the tumor site for improved localization. By utilizing these interactions, higher NP concentrations and cellular uptake in targeted tumor cells can be achieved. The development of ferric oxide nanoparticles with a dextran coating and a surface conjugation of a human epidermal growth factor receptor aptamer (HER2) is described. This was used as mediator of hyperthermia. The results demonstrated that NPs tagged with aptamers specifically targeted HER2-expressing cells. By manipulating the external magnetic field, tumor MNP aggregation could be enhanced. Overall, active delivery of MNPs results in improved tumor particle distribution (Farzin et al. 2020).

12.4.2.4 Mechanism of Heat Generation Using MNPs

Weiss domains are multi-magnetic domains that can generate heat during a magnetization cycle due to hysteresis loss. Applying alternating magnetic field (AMF) to ferromagnetic structures drives all magnetic domains to be in parallel with AMF direction, which locates them in a lower energy state. Two models can represent the heat generated in ferromagnetic materials subjected to an AMF: relaxation (which can be categorized into Néel and Brownian) and hysteresis loss. The Néel relaxation theory is the relaxing magnetism induced by rapidly altering magnetization orientation (internal dynamics). Brownian method refers to the friction that a particle rotational dispersion can cause in a medium. In this method, viscosity inhibits nanoparticles from freely rotating in the external medium (Shah et al. 2016; Wust et al. 2002).

12.4.3 Drug Delivery Vehicles

A sophisticated drug delivery vehicle is necessary for cancer. The drug delivery vehicles increase the therapeutic index of chemotherapeutic drugs and decrease the chemo-resistance state. Various materials, including polymers, inorganic carriers, hydrogels, lipids, peptides, and macromolecular scaffolds, have been used to design a drug delivery vehicle. Nanoparticles-derived nanoscale drug vehicles have even more demonstrated potential because of high selective accumulation in tumors. The large surface-area-to-volume ratio and small size of nanoscale vehicles allow them to bind, absorb, and carry anticancer agents with better efficiency. Both organic and inorganic nanoscale vehicles have been widely used in cancer therapy (Felice et al. 2014).

Figure 12.1 shows varieties of fabricated materials in sizes, shapes, architectures, chemical properties, and targeting strategies. To enhance the bioactivity of inorganic nanocarriers as well as water solubility, their surfaces can be coated with polar species along with mono- or multi-layer ligand shells. In addition to surface decoration, shape, net surface charge, and size of the vehicle are an important parameter in its delivery efficacy. Moreover, various strategies can be used to increase vehicle-targeted delivery, such as the application of specific antibodies, vitamins, aptamers, peptides, and proteins. In the following, an explanation of various nanoscale drug delivery vehicles has been provided (Senapati et al. 2018).

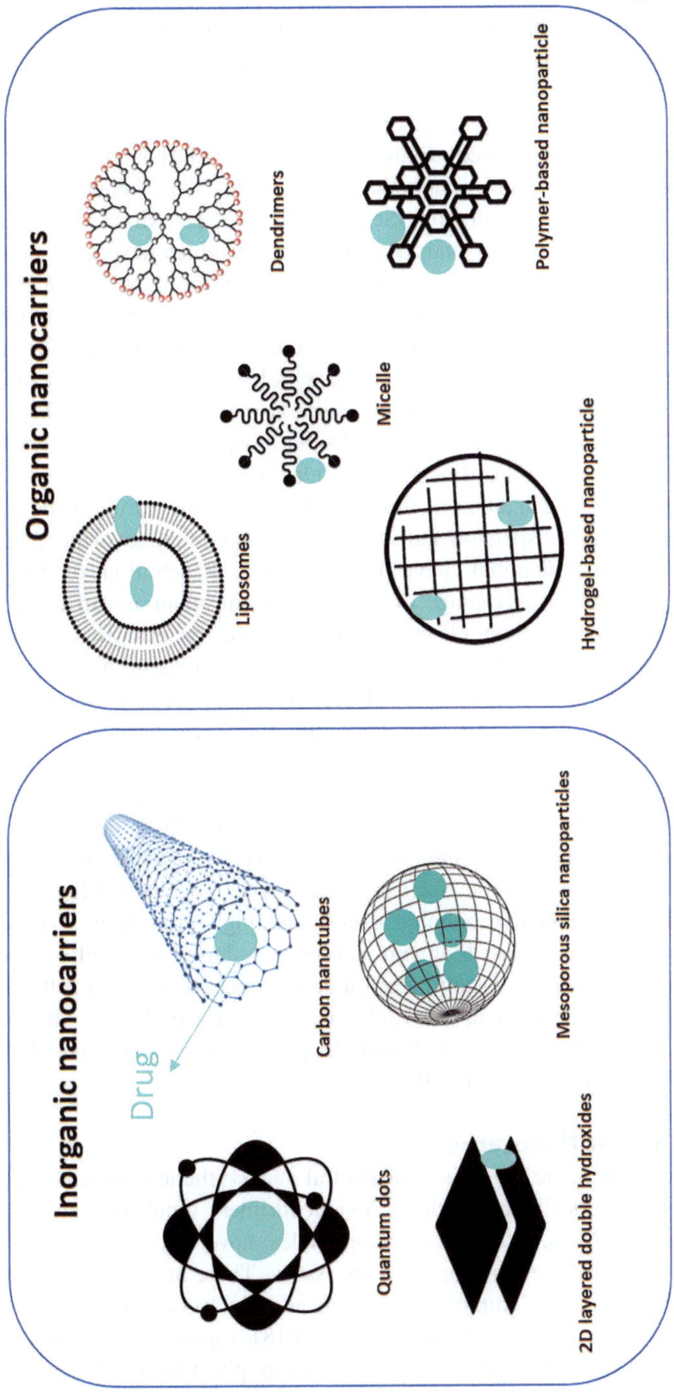

Fig. 12.1 Different types of drug delivery vehicles for cancer treatment

12.4.3.1 Inorganic Nanocarriers

Inorganic nanoscale vehicles have some advantages, including surface-area-to-volume ratio, higher bioavailability, higher capacity of drug loading, controlled drug release, lower side effects, and tolerance towards most organic solvents. Quantum dots (QDs), calcium phosphate nanoparticles (CPN), carbon nanotubes (CNTs), two-dimensional (2D) layered double hydroxides (LDHs), mesoporous silica nanoparticles (MSNs), and magnetic nanoparticles are frequently inorganic materials used in cancer treatment (Senapati et al. 2018; Mombini et al. 2019). QDs or semiconductor nanocrystals are nanoparticles which have great potential in drug delivery approaches. They increase the cellular uptake of the drug, leading to an increase in the anticancer efficacy of drugs (Pardo et al. 2018). CNTs, as a group of synthetic 1D nanomaterials, are made from carbon. In CNTs, sp^2 hybridized carbon atoms into hollow tubes and built rolled sheets of graphene rings. Because of the unique optical properties of CNTs, they enable using photo-thermal ablation therapy. In addition, the CNTs afford the real-time tracking of drug delivery efficacy in vivo (Abedi et al. 2021). Bioactive peptides, drugs, and nucleic acids can be applied for the functionalization of the CNTs to deliver their drug to specific cells. To reach the desirable dispersion and solubility of CNTs, non-covalent and covalent functionalization/modifications have been applied. Penetration of the plasma membranes is the main advantage of CNTs for using them as carriers to deliver therapeutic agents into the cytoplasm as well as into the nucleus (Zhang et al. 2011). LDHs have high biocompatibility, pH-responsive drug release, high drug loading efficacy, low cost, biodegradation in pH between 4 and 6 (the cellular cytoplasm) efficient endosomal escape, and controlled drug release rate capacity. LDHs consist of layers with a net positive charge; therefore, it has anion exchange capacity and anionic molecules (such as peptides, proteins, and genetic materials) and drugs can easily be inserted in the LDHs interlayer (Senapati et al. 2018). MSNs are another type of nanoscale vehicle with unique properties, such as large pore volume, easy functionalizing, large specific surface area, good biocompatibility, delivering membrane-impermeable hydrophobic drugs, and high protection of drugs from degeneration or denaturation. Magnetic nanoparticle-based drug delivery relies on external magnetic field guidance to reach their target cells, and their cellular uptake has not been dependent on nanoparticle size and surface charge. The enhanced efficacy of magnetic nanoparticle-based drug delivery is predominantly due to magnetic effects (Senapati et al. 2018).

12.4.3.2 Organic Nanocarriers

Organic nanocarriers are comprised of natural and synthetic polymers, liposomes, proteins, lipids, extracellular vesicles (EVs), dendrimers, emulsions, and hydrogels. Over the last decades, polymeric nanoparticles have been among the widely investigated nanoparticles in drug delivery issues. There are some FDA-approved biodegradable polymeric nanoparticles, such as poly lactic acid (PLA) and poly lactide-co-glycolide (PLGA) (Senapati et al. 2018). Liposomes are small, sphere-shaped, self-closed structures with one or more phospholipid bilayers and an encapsulated hydrophilic phase in the center. To reduce the percentage of liposomal uptake by macrophages, their surface has been decorated by polyethylene glycol (PEG). Hydrophobic drugs can be directly combined into liposomes, and hydrophilic drugs can be encapsulated in their inner aqueous core. Niosomes are modified liposomes with nonionic surfactants instead of phospholipids and can have a longer circulation time than liposomes. Transfersomes are elastic liposomes with more penetration ability. Ethosomes are also modified liposomes that enhance the permeation. Protein-based nanocarriers are biodegradable, biocompatible, and non-immunogenic. They have an efficient binding capacity for different drugs and a long half-life in circulating plasma. Albumin-based nanocarriers are a common type of protein-based nanoscale vehicles, and some of its derivations, such as paclitaxel-albumin-based nanoparticles, are FDA-approved. Polymersomes are hollow shell nanocarriers, and, similar to liposomes, consist of the internal and external hydrophilic context with a hydrophobic middle context. The thick external membrane of polymersomes provides easy reaction with surface agents (Singhvi et al. 2020). Polymeric micelles have sustained drug release, composed of an inner hydrophobic core and an outer hydrophilic shell formed by the self-assembly of block copolymers in an aqueous solution. Therefore, micelles-based nanoparticles can incorporate both hydrophobic and hydrophilic drugs. A hydrogel is a 3D polymer, flexible, and cross-linked hydrophilic polymer with a porous structure and high encapsulating efficiency. Drug delivery or controlled release is among the important therapeutic applications of the hydrogel. Both natural and synthetic hydrogels have been widely used in medicine. EVs are naturally liposome-like nanoscale nanoparticles, which play a crucial role in cell-cell communication. EVs include exosomes, microvesicles, and apoptotic bodies with tissue-specific contents. EVs transfer various macromolecules between cells, and various studies have mimicked their natural function to deliver their desired drug by them. As an example, mesenchymal stem cells-derived EVs containing miR-379 caused a significant decrease in tumor volume in mice model (O'Neill and Dwyer 2020).

12.4.3.3 Hybrid Materials for Drug Delivery

Hybridization of multiple molecules can create a multifunctional carrier and improve the biodegradability, stability, biocompatibility, and other characteristics of the vehicle. Gold, silica, iron, copper, hydroxy apatite, and layered double hydroxide-based nanohybrids are among the more applicable materials in drug delivery. For example, magnetic MSNs, as inorganic–inorganic nanohybrids, showed a pH-sensitive controlled release, which is beneficial for lower pH of the tumor microenvironment (TME) (Choi et al. 2021). Lipid-polymer hybrid nanoparticles are a type of common hybrid nanomaterials which have higher loading capacity as well as higher stability because of their lipid-polymer composition. As an example, a hybrid nano-hydrogels composed of carboxymethylcellulose–silver nanoparticles–DOX showed a controlled release of DOX against melanoma cancer cells (Choi et al. 2021; Capanema et al. 2019).

12.4.4 Types of Utilized Therapies Combined with Nanoscale Vehicles

Anticancer molecules such as paclitaxel, vincristine, doxorubicin (DOX), nucleic acids, peptides, and antibodies have been the main cargoes of the drug delivery nanoscale vehicles. In the following, examples for delivering each type of drug have been provided. Therapeutic drugs have various inhibition mechanisms; DOX and cisplatin cause chemical damage to cellular DNA; paclitaxel disrupts spindle formation and thereby causes M phase cell cycle arrest. 5-fluorouracil (5-FU), methotrexate, mercaptopurine, and cytarabine by inhibition of DNA synthesis caused S phase cell cycle arrest; vinblastine, vincristine, and etoposide cause M phase cell cycle arrest (Singhvi et al. 2020).

12.4.4.1 Chemotherapeutics

Cai et al. synthesized QDs (ultra-small ~3 nm) functionalized with poly (ethylene glycol) (PEG) and hyaluronic acid to target a high-expressed glycoprotein CD44 on cancer cells surface and DOX (Cai and Chen 2008). Chiu et al. developed a good stable, biocompatible and low toxic nanoscale vehicle based on carbon QDs (CQDs) doped with S, N, and Gd (GdNS@CQDs) with ~80% drug-loading capacity. To reach a dual mode fluorescence/magnetic resonance, they functionalized GdNS@CQDs with folic acid (FA) (as a ligand for frequently expressed receptors in cancer cells) through ECD/Sulfo-NHS reaction. This carrier showed the targeting capacity on HeLa and HepG2 cancerous cell lines (Chiu et al. 2016). Yang et al. synthesized quercetin (QE)-loaded CdSe@ZnS QDs and showed its increased anticancer capability to two- to sixfold compared to raw QE and CdSe QDs. In one study, Mn:ZnS QDs were encapsulated in chitosan and functionalized with FA and decreased tumor size and metastasis in the lung-tumor-bearing mice group compared to 5-FU treated group (Yang et al. 2017). Sui et al. showed that graphene QDs (GQDs) increased the anticancer activity of cisplatin (Sui et al. 2016).

Cisplatin has been loaded into single-walled CNTs (SWCNTs), and showed a toxic effect on DU145 and PC3 human prostate cancer cell lines. Zhang et al. have synthesized DOX-loaded polysaccharide functionalized SWCNTs, with stimuli-responsive drug release, biocompatible, water-dispersible multifunctional drug delivery system, and simultaneous targeting characteristics. They wrapped chitosan/alginate biopolymer chains around the SWCNTs by sonication and stirring of a chitosan/alginate solution containing SWCNTs. Therefore, $pH < 7$ of tumor tissues, the hydrophilicity of DOX is increased, and facilitates its release from the surface of SWCNT. To increase the nanoscale carrier's targeting activity, they incorporated NH_2 groups of chitosan with FA. It was reported that interacted raloxifene hydrochloride (RH), as an anticancer drug, with a series of magnesium aluminum LDHs with different interlayer exchangeable anions (NO_3^-, CO_3^{2-}, and PO_4^{3-}) released in a controlled manner (Zhang et al. 2016). MSNs have been used for loading various anticancer agents such as docetaxel, paclitaxel, and DOX. Large surface and tunable porous architecture is the main advantage of MSNs which can convert them to a controlled release carrier when the vehicle reaches the tumor.

Various gatekeepers or stimuli can be used to a controlled release of the drug. DOX conjugated with dextran and encapsulated in a hydrogel has fewer side effects and high efficacy in solid tumors therapy (Iturrioz-Rodríguez et al. 2019). Yang et al. showed that cell-penetrating peptide functionalized polymersomes caused efficient and controlled delivery of methotrexate disodium to A549 human lung cancer in vivo compared to non-functionalized polymersomes and free drugs (Yang et al. 2018). Tamoxifen embedded PLGA nanoparticles exhibited higher DNA cleavage potential compared to the pure drug. Paclitaxel-loaded bovine serum albumin-based nanoparticles followed by FA decoration targeted the PC3 prostate cancer cell line effectively (Pandey et al. 2016). A micelle composed of encapsulated DOX into cationic 1,2-dioleoyl-3-trimethylammonium propane/methoxy PEG-(DPP) nanoparticles showed a cytotoxic effect against bladder cancer. The therapeutic effect of paclitaxel-loaded natural polysaccharide nanoscale hydrogel (nanogels) was higher than paclitaxel alone, in tumor mice breast cancer models (Kumari et al. 2017; Jang et al. 2020).

12.4.4.2 Radiotherapeutics

Radiation therapy has a long history and uses radiation, usually X-rays, which can eradicate or control many tumors. However, its side effects need to be decreased, and its therapeutic effects need to be increased. Nanotechnology has reprogrammed radiotherapy, and it can increase radiosensitization. Nanoparticles have higher permeability and retention (EPR) effects; therefore, they increase the tumor-specific release of radiosensitizing drugs. Recently, radiopharmaceuticals have been developed that can deliver radiation directedly to tumor tissues. Nanoscale vehicles hold promising prospects, including liposomes, MSNs, bovine serum albumin- and polymeric-based nanocarriers that can incorporate radiosensitizers such as cisplatin, selenium (Se), DNA repair inhibitors, catalases, siRNA, and miRNA. For instance, cisplatin, by binding to the DNA of cancer cells, concentrates radiation in the nucleus of targeted tumor cells and contributes to DNA damage and cell death. Notably, the PLGA-PEG nanocarrier containing both the cisplatin and wortmannin (as a DNA repair inhibitor) has been approved by FDA for ovarian cancer therapy. Liposomal-based nanocarrier coated with cisplatin and catalase (as a decomposer of H_2O_2 into H_2O and O_2) enhanced the radiotherapy effect and impeded tumor growth. It is worth mentioning that O_2 is essential for radiotherapy-induced DNA breaks. A combination of radiation and drug-loaded nanocarriers can especially control tumors with high numbers of macrophages, which are usually difficult to treat (Jin and Zhao 2020).

High-atomic number of used nanoparticles causes an increase in the radio-enhancement effect. According to studies, high-atomic number metallic nanoparticles such as gold (Au), hafnium (Hf), gadolinium (Gd), and bismuth (Bi) have higher radiosensitization potential. Over the past two decades, there has been extensive research on the benefits of nanotechnology application in clinical radiation oncology (Lin 2015).

12.4.4.3 Immunotherapeutic

Cancer immunotherapy helps the host immune system to find cancer cells and kill them. There have been various types of immunotherapy, including antibodies, immune checkpoint inhibitors, chimeric antigen receptor (CAR) T-cell therapy, and cancer vaccines. In cancer immunotherapy, antibodies act as therapeutic agents and do not target ligands. Immune checkpoint inhibitors are also antibodies against the immunosuppressing receptors either on cancer cells, such as programmed cell death protein-ligand 1 (PD-L1), or on immune cells, such as programmed cell death protein 1 (PD-1). CAR T-cell therapy modifies the patient's own T cells ex vivo to act against the cancer cells. Cancer vaccines deliver the tumor antigens or their genetic information to the host, which then induces immune responses against cancer cells that express those antigens (Jahanafrooz et al. 2020).

Despite the high efficiency of various types of immunotherapies, only a small portion of patients respond ideally to them. Immunotherapies combined with nanoscale vehicles containing different drugs have shown to be promising candidates for cancer immune-combined therapy. For example, encapsulated MSN with DOX showed the ability to promote dendritic cells (DCs) maturation and higher tumor-infiltrating cytotoxic T-cell anti-tumor cytokines release. Notably, DOX chemotherapy alone stimulates immunosuppression response by INF-γ secretion. Indeed, cancer immunotherapy has been enhanced with nanomedicine; it was shown that liposomes decorated with anti-CD137 antibodies caused activation of cytotoxic T cells tenfold more than anti-CD137 antibodies in soluble form. According to one study, a combination of checkpoint inhibitors anti-PD-1 antibody with a polymer-based nanocarrier containing DOX showed improved anticancer properties. As a side note, following the interaction of PD-1 is an immune checkpoint on immune cells that interact with PD-L1 on tumor cells, and immunosuppression happens (Gao et al. 2018).

12.4.4.4 Peptides

The common anticancer peptides include cell-penetrating peptides (CPPs), tumor-targeting peptides (TTPs), and pore-forming proteins (PFPs) (Hafezi Ghahestani et al. 2017). CPPs are hydrophobic short lengths of amino acids with the ability to easily translocate through the plasma membrane. CPPs target many molecules involved in growth, survival, and proliferation signaling pathways. CPPs are an efficient approach for the intracellular delivery of anticancer drugs; for instance, trans-activator of transcription (Tat), obtained from human immunodeficiency virus (HIV), conjugated to DOX, and caused an increase in DOX uptake by the drug-resistant cancer cells (Hafezi Ghahestani et al. 2017; Gui et al. 2020). TTPs bind the especially expressed molecules on cancer cell's surface and internalize into the cells; there have been many reported TTPs-based drug deliveries. Various anticancer agents have been conjugated into TTPs to increase their efficacy and cellular uptake (Zhu et al. 2018). For example, in one study, the peptide NGR (Asn-Gly-Arg) was conjugated to a PEGylated liposome-polycation-DNA containing DOX or C-Myc siRNA and produced significant cytotoxic effects both in vitro and in a fibrosarcoma mouse model (Hafezi Ghahestani et al. 2017). PFPs are present in all living creatures. PFPs are usually short cationic peptides that have a natural tendency to negative cell surfaces of cancer cells because of electrostatic interactions (Mokhtarzadeh et al. 2016). Pore formation, as the crucial mechanism in membrane permeabilization, is a commonly used manner by PFPs, and induces cell death in the target cells (Dissanayake et al. 2017).

12.4.4.5 Oligonucleotides

Various types of oligonucleotides have been used as a drug. Small oligonucleotides such as siRNAs and miRNAs are the commonly used nucleic acids in cancer therapy. FDA approved siRNAs drugs in 2018. Both siRNAs and miRNAs prevent mRNA translation either by its degradation or translation inhibition, respectively. Xie et al. synthesized a hybrid anionic nanoparticle composed of PEGylated carboxymethyl chitosan/calcium phosphate (PEG-CMCS-CaP), which was an effective delivery vehicle for siRNA. It induced significant apoptosis in liver carcinoma HepG2 cells and in liver tumor-bearing mice model similar to commercial Lipofectamine 2000. Sometimes, to reach an effective therapy, a co-delivery of oligonucleotide along with an anticancer drug has been applied (Xie et al. 2014). In a study, Li et al. used an LDH nanocarrier encapsulated with 5-FU and conjugated with AllStars Cell Death siRNA on its surface. The co-delivery of the mentioned drugs showed more cytotoxicity on cancer cell lines (because of simultaneous effects on mitochondrial permeability) compared to separate treatment of 5-FU or CD-siRNA. In addition, PEGylated calcium phosphate hybrid micelles-based nanoparticles showed high efficiency in siRNA delivery to pancreatic-tumor-bearing transgenic mice model (Li et al. 2014b).

Despite siRNAs, miRNA can target hundreds of mRNAs, and shut down so many genes. Chemical modification and encapsulation of miRNAs can both stabilize the miRNAs. In addition, nanoparticles can improve miRNAs delivery in the body. Our previous study showed that compared to dendrimer, liposome is a better delivery vehicle for miRNA transfection in T47D breast cancer cell line (Jahanafrooz et al. 2022). Both natural and synthetic biomaterials have been used as a vehicle for miRNAs such as PLGA, gold, liposome, PEG-lipid, chitosan, hyaluronic acid, and silica. For example, hyaluronic acid-based nanoscale vehicle loaded with miR-145 inhibited tumor growth in the colon cancer xenograft model (Lee et al. 2019b).

12.4.5 Remote Controlled Pulsatile Drug Release

This type of cancer therapy is a non-invasive and non-continuous therapy with an external controlling agent. Here, the smart drug delivery vehicles are the triggered drug-releasing vehicles that prompt a controlled drug release at a specific time or site in response to external or remote stimuli, including magnetic fields, temperature changes, light, ultrasound, and/or electric fields (Fig. 12.2). These vehicles are pulsatile drug delivery systems and could have the maximum drug toxicity and the minimum side effects. This type of drug release can overcome the drug resistance characteristic of cancer. It was shown that the addition of photodynamic and

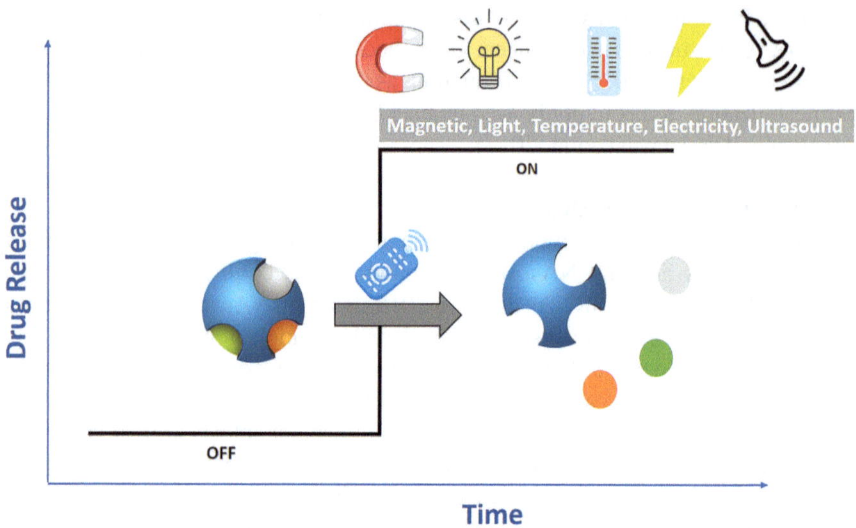

Fig. 12.2 Graph illustrating the drug-releasing profile in a remotely controlled manner in response to remote stimuli such as magnetic fields, light, temperature changes, electric fields, or ultrasound

photothermal therapies to the immobilized curcumin on silica-coated Fe_3O_4 nanocomposite increased its anticancer effects and caused a significant decrease in tumor volume in the breast cancer mice model (Ashkbar et al. 2020). As well as pulsatile delivery of mitoxantrone (a chemotherapeutic) by magnetically responsive biphasic ferrogels is more effective at eliminating melanoma cells in vitro (Emi et al. 2018). In addition to site-specific accumulation, magnetic nanocarrier could reach the hyperthermia temperature and provide the hyperthermia therapy. As an example, a high-frequency magnetic field created the hyperthermia state (approximately 42 ° C) in DOX-loaded magnetic liposomes and caused cytotoxic effects on L-929 colorectal cancer cells (Hardiansyah et al. 2014).

12.4.6 Drug Targeting Approaches for Cancer Therapy

There are two main approaches undertaken in drug delivery to specific target sites, including active and passive targeting strategies (Fig. 12.3). Active targeting relies on the binding of specific ligands to the cell surface receptor in target cells. Passive targeting is based solely on the EPR in pathological sites (Attia et al. 2019). Therefore, the preparation of active targeting nano-vehicle is relatively complex manufacturing.

12.4.6.1 Active Targeting

In active targeting cellular uptake is more than passive targeting because of specific homing as a result of the specific ligand-receptor interaction. These receptors are

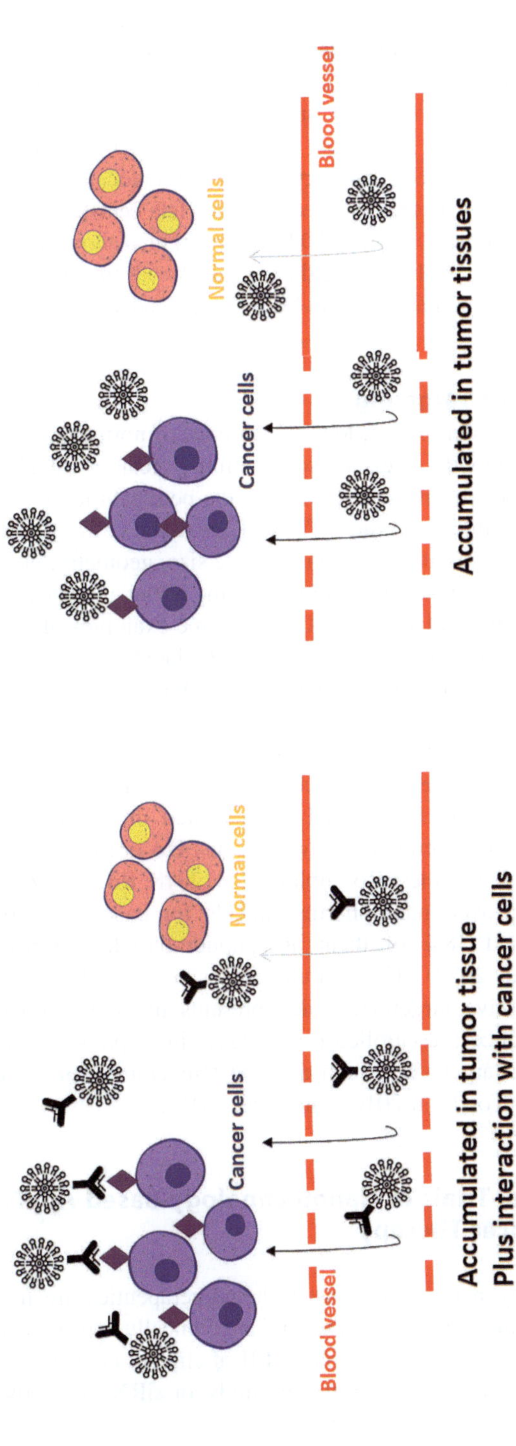

Fig. 12.3 Comparison of active and passive targeting. As shown in the figure, the leakage of the tumor vessels causes the accumulation of drug delivery vehicles in tumor tissue both in passive and active targeting; however, cancer-specific ligands even more result in conducting the vehicles toward the cancer cells only in active targeting

cancer-specific or overexpressed in cancer cells. Several nanocarriers formulations derived from active compounds have been developed. Numerous ligands have been used for nanoparticle platforms, including peptides, hyaluronic acid, folate or FA, antibodies or antibody fragments, aptamers, and carbohydrates or polysaccharides. For instance, the FA receptor is overexpressed in several cancers, and FA-conjugated PEG-PLGA nanoparticle delivering docetaxel induced more apoptotic and cytotoxic effects than the free drug on Hela cervical carcinoma cell line. Or receptor of hyaluronic acid (i.e., CD44 receptor) also showed a high expression on several cancer cells; hyaluronic acid conjugated with micelles delivering paclitaxel was shown a more cellular uptake in MCF-7 breast adenocarcinoma cells (Muhamad et al. 2018).

12.4.6.2 Passive Targeting

Hypoxic media, inflammation, low pH, absence of normal lymphatic drainage, and more permeable blood vessels are some common features of TME. The mentioned characteristics have been used to enhance the permeability and retention period of anticancer agents in tumor tissues. The leakage of the newly formed tumor vessels is even more used in passive targeting. Particle size, geometry, elasticity, and surface charge affect the retention and circulation time of the anticancer agents. PEGylated micelles delivering cisplatin and DOX are some examples of formulations derived from passive targeting (Torchilin 2010). PEGylated liposomal DOX was the first passive targeting approved nanoscale vehicle. Other approved ones include Doxil™, Abraxane™, Marqibo™, Vyxeos™, DaunoXome™, Onivyde™, Myocet™, Mepact™, Genexol-PM™, and SMANCS™ (Rosenblum et al. 2018). All the mentioned passive targeting approach demonstrated significantly more response rates compared to the free-related encapsulated drugs. Table 12.3 summarizes some of the passive targeting (approved by the FDA or European Medicines Agency (EMA)) nanoscale vehicles (Rosenblum et al. 2018; Rodríguez et al. 2022). Totally 56% of the approved nano-pharmaceuticals for cancer therapy are lipid-based nanoparticles, and 38% of them are peptide-based, and 6% are metal-based (Rodríguez et al. 2022). The stimuli-responsive drug delivery vehicles are also classified in passive targeting, which presents the most efficient drug delivery vehicles that induce a controlled drug release in response to the physical stimuli provided by the target sites, such as local temperature, pH, and the presence of oxidative stress (Torchilin 2010; Said et al. 2019).

12.4.7 Clinical Trials of Nanotechnology-Based Applications in Cancer Therapy

At present, many nanoparticle-based chemotherapeutics are in various stages of preclinical or clinical studies, and several are clinically approved. Every year several new nano-pharmaceuticals enter phase I/II of clinical investigation for cancer therapy. Their cargo is chemical drugs, plasmids, or siRNAs. Liposome-, polymeric-,

Table 12.3 Some of the approved nano-pharmaceuticals for cancer therapy (Rosenblum et al. 2018; Rodríguez et al. 2022)

Product	Company	Vehicle	Drug	Indication
FDA-approved				
Doxil	Ortho Biotech	PEGylated liposome	DOX	Kaposi's sarcoma, ovarian cancer, multiple myeloma
DaunoXome	Galen	Liposome	Daunorubicin	Kaposi's sarcoma
Marqibo	Spectrum		Vincristine	Acute lymphoid leukemia
Onivyde	Merrimack		Irinotecan	Pancreatic and colorectal cancer
Oncaspar	Les Laboratoires Servier	Protein-drug conjugates	PEGylated L-asparaginase	Acute lymphoblastic leukemia
Ontak	Les Laboratoires Servier		Denileukin Diftitox	Cutaneous T-cell lymphoma
Eligard	Recordati Industria Chimica e Farmaceutica		Leuprorelin acetate	Prostate cancer
Abraxane	American Biosciences, Inc.	Albumin-based nanocarrier	Paclitaxel	Breast cancer, non-small lung cancer, pancreatic cancer
Kadcyla	Genentech	Protein-drug conjugates	Trastuzumab emtansine	Metastatic breast cancer
Genexol	Samyang Holdings	Polymeric micelle	Paclitaxel	Metastatic breast cancer, pancreatic cancer
EMA-approved				
Caelyx	Schering-Plough	PEGylated liposome	DOX	Metastatic breast cancer, ovarian cancer, Kaposi's sarcoma, multiple myeloma
Myocet	Teva UK	Liposome	DOX	Metastatic breast
Mepact	Millenium	Liposome	Mifamurtide	Osteosarcoma
Ameluz	Biofrontera Bioscience GmbH	Lipid-based nanocarrier	5-Aminolevulinic acid	Superficial and/or nodular basal cell carcinoma
Vyxeos	Jazz Pharmaceuticals	Liposome	Daunorubicin Cytarabine	Acute myeloid leukemia
Pazenir	Ratiopharm GmbH	Albumin-based nanocarrier	Paclitaxel	Metastatic breast cancer, metastatic adenocarcinoma of the pancreas, non-small cell lung cancer

(continued)

Table 12.3 (continued)

Product	Company	Vehicle	Drug	Indication
NanoTherm	Magforce	Metallic nanoparticles	Iron nanoparticles (Fe_2O_3)	Glioblastoma, prostate, pancreatic cancer
FDA-EMA-approved				
Kadcyla	Roche Genentech	Protein-drug conjugates	DM1 (or Emtansine)	$HER2^+$ breast cancer

and micelle-based vehicles are the commonly used nanoscale vehicle in clinical trials (Table 12.4) (Riggio et al. 2011).

12.5 Consideration of Nanomaterial: Advantages and Challenges

Over the past few decades (2000–2020), nanotechnology progress has provided a novel hope for next-generation cancer diagnosis and treatment. Using nanomaterial in cancer issues can overcome the drawbacks posed by conventional chemotherapy and have some advantages, including noninvasive, high bioavailability, low degradation in the body fluids, the high solubility of the drug, high stability of drug, specific targeting, and requirement of lower drug dose. Therefore, they can yield a high therapeutic index with no/fewer side effects. In addition to therapeutic issues, nanomaterial-based strategies for cancer diagnosis have shown a rapid increase over the last few decades (Iturrioz-Rodríguez et al. 2019). Despite all of these promises, there are important concerns that reside in some drawbacks and challenges, which are mentioned in the following. Potential toxicity of nanoparticles (such as blood clots, nephrotoxicity, and hepatotoxicity) raises concerns about their safety, especially for long-term administration. Pre-mature release of drugs and barrier to delivery are the important remaining challenges for nano-pharmaceuticals. Enzymatic degradation, accumulation/aggregation of nanocarrier, and physical barrier of skin are some barriers to the delivery route (Singhvi et al. 2020). Moreover, the coating formed by serum proteins and opsonins or protein corona could affect the stability, size, and other physiochemical properties of nanocarrier (Rosenblum et al. 2018; Sengupta and Balla 2018). Therefore, protein corona could compromise targeting delivery, cellular uptake, internalization of nanocarrier, and its pharmacokinetics and toxicity (Rosenblum et al. 2018). It was shown that the first injection of PEGylated liposomes can cause the emergence of anti-PEG immunity in some patients which causes quick elimination and increased hepatic uptake of the next dose of PEGylated liposomes. It is worth mentioning that immunogenicity against the nanocarrier itself is another possible undesirable effect of nano-pharmaceuticals for cancer therapy (Kozma et al. 2020).

A tumor is composed of a heterogeneous population of cancerous and noncancerous cells. Therefore, a nanoscale vehicle with a specific ligand can destroy only a

Table 12.4 Examples of anticancer nanoscale vehicles in clinical trials (Riggio et al. 2011)

Cancer type	Vehicle	Therapeutic agent	Status	Conjugated ligand
Lung, head, and neck cancer	Liposome	Cisplatin	Phase II	–
Non-small cell lung cancer			Phase III	–
Pancreatic cancer				–
Non-Hodgkin lymphoma		Interleukin 2	Phase I	–
Acute lymphocytic leukemia		Annamycin	Phase I	–
Advanced colorectal cancer		Oxaliplatin	Phase II	–
Ovarian cancer		Lurtotecan	Phase II	–
Various cancers		siRNA	Phase I/II	–
Various cancers		DNA plasmid (with p53 gene)	Phase II	Transferrin receptor antibody fragment
Various cancers	PEG immunoliposome	DOX	Phase I	F(ab′)2
Various cancers	*N*-glutaryl phosphatidylethanolamine-liposome	Oxaliplatin	Phase II	Transferrin
Various solid tumors	Cyclodextrin polymer-based nanoparticles	siRNA	Phase I	Transferrin
Neuroendocrine tumors	Lipid nanoparticle	PLK1 siRNA	Phase I/II	–
Various solid tumors	Liposome	PKN3 siRNA	Phase I/II	–
Liver cancer	Liposomal nanoparticle	CEBPA siRNA	Phase I/II	–
Various cancers	Polymeric nanoparticles	Paclitaxel	Phase I	–
Advanced stomach cancer	Polymeric micelles	Paclitaxel	Phase II	–
Adenocarcinoma of esophagus	Polymeric micelles	DOX	Phase III	–
Solid tumors	Polymeric micelles	Cisplatin	Phase I/II	–
Breast cancer			Phase I	
Pancreatic cancer			Phase V	
Non-small cell lung cancer			Phase II	

few subpopulations of tumors. In addition, the permeability/penetration of the hypoxic core and periphery of tumors are different. The low efficiency of endosomal escape of nanocarriers also limits their therapeutic benefit (Jahanafrooz et al. 2020). To reach realistically translation of nano-pharmaceuticals in clinical applications, several aspects still need to be addressed, such as their overall cytotoxicity and immunogenicity, EPR, body clearance, synthesis protocol scalability, biodegradability, environmental effects, manufacturing costs, and other issues.

Conflict of Interest The authors declare no conflicts of interest.

Funding The authors received no specific funding for this work.

References

Abbasi A, Park K, Bose A, Bothun GD (2017) Near-infrared responsive gold–layersome nanoshells. Langmuir 33(21):5321–5327

Abedi A, Bakhshandeh B, Babaie A, Mohammadnejad J, Vahdat S, Mombeiny R et al (2021) Concurrent application of conductive biopolymeric chitosan/polyvinyl alcohol/MWCNTs nanofibers, intracellular signaling manipulating molecules and electrical stimulation for more effective cardiac tissue engineering. Mater Chem Phys 258:123842

Adams GP, Weiner LM (2005) Monoclonal antibody therapy of cancer. Nat Biotechnol 23(9): 1147–1157

Adhikari P, Pal P, Das AK, Ray S, Bhattacharjee A, Mazumder B (2017) Nano lipid-drug conjugate: an integrated review. Int J Pharm 529(1–2):629–641

Ahmed KS, Hussein SA, Ali AH, Korma SA, Lipeng Q, Jinghua C (2019) Liposome: composition, characterisation, preparation, and recent innovation in clinical applications. J Drug Target 27(7): 742–761

Akbarzadeh A, Rezaei-Sadabady R, Davaran S, Joo SW, Zarghami N, Hanifehpour Y et al (2013) Liposome: classification, preparation, and applications. Nanoscale Res Lett 8(1):1–9

Allin KH, Nordestgaard BG (2011) Elevated C-reactive protein in the diagnosis, prognosis, and cause of cancer. Crit Rev Clin Lab Sci 48(4):155–170

Aly HA (2012) Cancer therapy and vaccination. J Immunol Methods 382(1–2):1–23

Anderla A, Culibrk D, Delso G, Mirkovic M (2013) MR image based approach for metal artifact reduction in X-ray CT. Sci World J 2013:524243

Andriole GL, Crawford ED, Grubb RL III, Buys SS, Chia D, Church TR et al (2012) Prostate cancer screening in the randomized prostate, lung, colorectal, and ovarian cancer screening trial: mortality results after 13 years of follow-up. J Natl Cancer Inst 104(2):125–132

Aravanis AM, Lee M, Klausner RD (2017) Next-generation sequencing of circulating tumor DNA for early cancer detection. Cell 168(4):571–574

Ashkbar A, Rezaei F, Attari F, Ashkevarian S (2020) Treatment of breast cancer in vivo by dual photodynamic and photothermal approaches with the aid of curcumin photosensitizer and magnetic nanoparticles. Sci Rep 10(1):1–12

Attia MF, Anton N, Wallyn J, Omran Z, Vandamme TF (2019) An overview of active and passive targeting strategies to improve the nanocarriers efficiency to tumour sites. J Pharm Pharmacol 71(8):1185–1198

Bakhshandeh B, Zarrintaj P, Oftadeh MO, Keramati F, Fouladiha H, Sohrabi-Jahromi S et al (2017a) Tissue engineering; strategies, tissues, and biomaterials. Biotechnol Genet Eng Rev 33(2):144–172

Bakhshandeh B, Kamaleddin MA, Aalishah K (2017b) A comprehensive review on exosomes and microvesicles as epigenetic factors. Curr Stem Cell Res Ther 12(1):31–36

Bakhshandeh B, Ranjbar N, Abbasi A, Amiri E, Abedi A, Mehrabi MR et al (2022) Recent progress in the manipulation of biochemical and biophysical cues for engineering functional tissues. Bioeng Transl Med 8:e10383

Bamrungsap S, Zhao Z, Chen T, Wang L, Li C, Fu T et al (2012) Nanotechnology in therapeutics: a focus on nanoparticles as a drug delivery system. Nanomedicine 7(8):1253–1271

Barkalina N, Charalambous C, Jones C, Coward K (2014) Nanotechnology in reproductive medicine: emerging applications of nanomaterials. Nanomedicine 10(5):e921–e938
Barreto JA, O'Malley W, Kubeil M, Graham B, Stephan H, Spiccia L (2011) Nanomaterials: applications in cancer imaging and therapy. Adv Mater 23(12):H18–H40
Bateman TM (2012) Advantages and disadvantages of PET and SPECT in a busy clinical practice. J Nucl Cardiol 19(1):3–11
Beard P (2011) Biomedical photoacoustic imaging. Interface Focus 1(4):602–631
Bhana S, Wang Y, Huang X (2015) Nanotechnology for enrichment and detection of circulating tumor cells. Nanomedicine 10(12):1973–1990
Biller LH, Schrag D (2021) Diagnosis and treatment of metastatic colorectal cancer: a review. JAMA 325(7):669–685
Borden MA, Dayton PA, Slagle C, Walmer RW (2021) Ultrasound contrast agents. In: Molecular imaging, pp 639–653
Borghei Y-S, Hosseini M, Ganjali MR (2017) Detection of large deletion in human BRCA1 gene in human breast carcinoma MCF-7 cells by using DNA-Silver Nanoclusters. Methods Appl Fluoresc 6(1):015001
Brenner DJ, Hall EJ (2007) Computed tomography—an increasing source of radiation exposure. N Engl J Med 357(22):2277–2284
Brunetti J, Riolo G, Gentile M, Bernini A, Paccagnini E, Falciani C et al (2018) Near-infrared quantum dots labelled with a tumor selective tetrabranched peptide for in vivo imaging. J Nanobiotechnol 16(1):1–10
Burinaru T, Volmer M, Avram M, Ţucureanu V, Avram A, Ţîncu B et al (2019) Antibody functionalized magnetic nanoparticles for circulating tumor cells detection and capture using magnetophoresis, IOP conference series: materials science and engineering. IOP Publishing
Cai W, Chen X (2008) Preparation of peptide-conjugated quantum dots for tumor vasculature-targeted imaging. Nat Protoc 3(1):89–96
Cai G, Yu Z, Ren R, Tang D (2018a) Exciton–plasmon interaction between AuNPs/graphene nanohybrids and CdS quantum dots/TiO2 for photoelectrochemical aptasensing of prostate-specific antigen. ACS Sensors 3(3):632–639
Cai B, Guo S, Li Y (2018b) MoS 2-based sensor for the detection of miRNA in serum samples related to breast cancer. Anal Methods 10(2):230–236
Canney PA, Moore M, Wilkinson PM, James RD (1984) Ovarian cancer antigen CA125: a prospective clinical assessment of its role as a tumour marker. Br J Cancer 50(6):765–769
Cao Z, Li H, Lau C, Zhang Y (2011) Cross-talk-free simultaneous fluoroimmunoassay of two biomarkers based on dual-color quantum dots. Anal Chim Acta 698(1–2):44–50
Capanema NS, Carvalho IC, Mansur AA, Carvalho SM, Lage AP, Mansur HS (2019) Hybrid hydrogel composed of carboxymethylcellulose–silver nanoparticles–doxorubicin for anticancer and antibacterial therapies against melanoma skin cancer cells. ACS Appl Nano Mater 1(11): 7393–7408
Catalano O, Roldán FA, Varelli C, Bard R, Corvino A, Wortsman X (2019) Skin cancer: findings and role of high-resolution ultrasound. J Ultrasound 22(4):423–431
Chang D, Lim M, Goos J, Qiao R, Ng YY, Mansfeld FM et al (2018) Biologically targeted magnetic hyperthermia: potential and limitations. Front Pharmacol 9:831
Chen HH, Kuo MT (2017) Improving radiotherapy in cancer treatment: promises and challenges. Oncotarget 8(37):62742
Chen M-J, Wu Y-S, Lin G-F, Hou J-Y, Li M, Liu T-C (2012) Quantum-dot-based homogeneous time-resolved fluoroimmunoassay of alpha-fetoprotein. Anal Chim Acta 741:100–105
Chen L, Peng M, Li N, Song Q, Yao Y, Xu B et al (2018) Combined use of EpCAM and FRα enables the high-efficiency capture of circulating tumor cells in non-small cell lung cancer. Sci Rep 8(1):1–10
Chen D, Wu Y, Hoque S, Tilley RD, Gooding JJ (2021) Rapid and ultrasensitive electrochemical detection of circulating tumor DNA by hybridization on the network of gold-coated magnetic nanoparticles. Chem Sci 12(14):5196–5201

Chen D, Wu Y, Tilley RD, Gooding JJ (2022) Rapid and ultrasensitive electrochemical detection of DNA methylation for ovarian cancer diagnosis. Biosens Bioelectron 206:114126

Cheng X, Pu X, Jun P, Zhu X, Zhu D, Chen M (2014) Rapid and quantitative detection of C-reactive protein using quantum dots and immunochromatographic test strips. Int J Nanomedicine 9:5619

Chinen AB, Guan CM, Ferrer JR, Barnaby SN, Merkel TJ, Mirkin CA (2015) Nanoparticle probes for the detection of cancer biomarkers, cells, and tissues by fluorescence. Chem Rev 115(19): 10530–10574

Chiu SH, Gedda G, Girma WM, Chen JK, Ling YC, Ghule AV et al (2016) Rapid fabrication of carbon quantum dots as multifunctional nanovehicles for dual-modal targeted imaging and chemotherapy. Acta Biomater 46:151–164

Cho S, Yang HC, Rhee WJ (2019) Simultaneous multiplexed detection of exosomal microRNAs and surface proteins for prostate cancer diagnosis. Biosens Bioelectron 146:111749

Choi CHJ, Hao L, Narayan SP, Auyeung E, Mirkin CA (2013) Mechanism for the endocytosis of spherical nucleic acid nanoparticle conjugates. Proc Natl Acad Sci 110(19):7625–7630

Choi D, Jeon S, You DG, Um W, Kim J-Y, Yoon HY et al (2018a) Iodinated echogenic glycol chitosan nanoparticles for X-ray CT/US dual imaging of tumor. Nanotheranostics 2(2):117

Choi J, Grosely R, Prabhakar A, Lapointe CP, Wang J, Puglisi JD (2018b) How mRNA and nascent chain sequences regulate translation elongation. Annu Rev Biochem 87:421

Choi WI, Sahu A, Wurm FR, Jo S-M (2019) Magnetoliposomes with size controllable insertion of magnetic nanoparticles for efficient targeting of cancer cells. RSC Adv 9(26):15053–15060

Choi G, Rejinold NS, Piao H, Choy JH (2021) Inorganic–inorganic nanohybrids for drug delivery, imaging and photo-therapy: recent developments and future scope. Chem Sci 12(14): 5044–5063

Choudhury H, Gorain B, Pandey M, Khurana RK, Kesharwani P (2019) Strategizing biodegradable polymeric nanoparticles to cross the biological barriers for cancer targeting. Int J Pharm 565: 509–522

Coffey JC, Wang J, Smith M, Bouchier-Hayes D, Cotter T, Redmond H (2003) Excisional surgery for cancer cure: therapy at a cost. Lancet Oncol 4(12):760–768

Coutinho C, Somoza Á (2019) MicroRNA sensors based on gold nanoparticles. Anal Bioanal Chem 411(9):1807–1824

Cruz LJ, Que I, Aswendt M, Chan A, Hoehn M, Löwik C (2016) Targeted nanoparticles for the non-invasive detection of traumatic brain injury by optical imaging and fluorine magnetic resonance imaging. Nano Res 9(5):1276–1289

Dakubo GD (2016) Extracellular vesicles as cancer biomarkers in circulation. In: Cancer biomarkers in body fluids. Springer, pp 233–260

Das J, Ivanov I, Sargent EH, Kelley SO (2016) DNA clutch probes for circulating tumor DNA analysis. J Am Chem Soc 138(34):11009–11016

Dass SA, Tan KL, Selva Rajan R, Mokhtar NF, Mohd Adzmi ER, Wan Abdul Rahman WF et al (2021) Triple negative breast cancer: a review of present and future diagnostic modalities. Medicina 57(1):62

Dean M, Fojo T, Bates S (2005) Tumour stem cells and drug resistance. Nat Rev Cancer 5(4): 275–284

Dencks S, Piepenbrock M, Opacic T, Krauspe B, Stickeler E, Kiessling F et al (2018) Clinical pilot application of super-resolution US imaging in breast cancer. IEEE Trans Ultrason Ferroelectr Freq Control 66(3):517–526

Deng Z, Jiang M, Li Y, Liu H, Zeng S, Hao J (2019) Endogenous H2S-triggered in situ synthesis of NIR-II-emitting nanoprobe for in vivo intelligently lighting up colorectal cancer. iScience 17: 217–224

Deshmukh AS, Chauhan PN, Noolvi MN, Chaturvedi K, Ganguly K, Shukla SS et al (2017) Polymeric micelles: basic research to clinical practice. Int J Pharm 532(1):249–268

Diocou S, Volpe A, Jauregui-Osoro M, Boudjemeline M, Chuamsaamarkkee K, Man F et al (2017) [18F] tetrafluoroborate-PET/CT enables sensitive tumor and metastasis in vivo imaging in a sodium iodide symporter-expressing tumor model. Sci Rep 7(1):1–13

Dissanayake S, Denny WA, Gamage SA, Sarojini V (2017) Recent developments in anticancer drug delivery using cell penetrating and tumor targeting peptides. J Control Release 250:62–76
Doria G, Conde J, Veigas B, Giestas L, Almeida C, Assunção M et al (2012) Noble metal nanoparticles for biosensing applications. Sensors 12(2):1657–1687
Duffy M, Shering S, Sherry F, McDermott E, O'Higgins N. (2000) CA 15-3: a prognostic marker in breast cancer. Int J Biol Markers 15(4):330–333
Emi TT, Barnes T, Orton E, Reisch A, Tolouei AE, Madani SZM et al (2018) Pulsatile chemotherapeutic delivery profiles using magnetically responsive hydrogels. ACS Biomater Sci Eng 4(7): 2412–2423
Engvall E (1980) [28] Enzyme immunoassay ELISA and EMIT. In: Methods in enzymology, vol 70. Elsevier, pp 419–439
Fakhri N, Abarghoei S, Dadmehr M, Hosseini M, Sabahi H, Ganjali MR (2020) Paper based colorimetric detection of miRNA-21 using Ag/Pt nanoclusters. Spectrochim Acta A Mol Biomol Spectrosc 227:117529
Farokhzad OC, Langer R (2009) Impact of nanotechnology on drug delivery. ACS Nano 3(1): 16–20
Farzin A, Etesami SA, Quint J, Memic A, Tamayol A (2020) Magnetic nanoparticles in cancer therapy and diagnosis. Adv Healthc Mater 9(9):e1901058
Felice B, Prabhakaran MP, Rodriguez AP, Ramakrishna S (2014) Drug delivery vehicles on a nano-engineering perspective. Mater Sci Eng C 41:178–195
Feliu N, Parak WJ (2020) Colloids for nanobiotechnology: an introduction. Front Nanosci 16:1–7
Fredolini C, Pathak KV, Paris L, Chapple KM, Tsantilas KA, Rosenow M et al (2020) Shotgun proteomics coupled to nanoparticle-based biomarker enrichment reveals a novel panel of extracellular matrix proteins as candidate serum protein biomarkers for early-stage breast cancer detection. Breast Cancer Res 22(1):1–16
Gambhir SS (2002) Molecular imaging of cancer with positron emission tomography. Nat Rev Cancer 2(9):683–693
Gao Y-c, Wu J (2015) MicroRNA-200c and microRNA-141 as potential diagnostic and prognostic biomarkers for ovarian cancer. Tumor Biol 36(6):4843–4850
Gao X, Cui Y, Levenson RM, Chung LW, Nie S (2004) In vivo cancer targeting and imaging with semiconductor quantum dots. Nat Biotechnol 22(8):969–976
Gao Z, Zhang L, Sun Y (2012) Nanotechnology applied to overcome tumor drug resistance. J Control Release 162(1):45–55
Gao Z, Xu M, Hou L, Chen G, Tang D (2013) Magnetic bead-based reverse colorimetric immunoassay strategy for sensing biomolecules. Anal Chem 85(14):6945–6952
Gao N, Bozeman EN, Qian W, Wang L, Chen H, Lipowska M et al (2017) Tumor penetrating theranostic nanoparticles for enhancement of targeted and image-guided drug delivery into peritoneal tumors following intraperitoneal delivery. Theranostics 7(6):1689–1704
Gao F, Zhang C, Qiu WX, Dong X, Zheng DW, Wu W et al (2018) PD-1 blockade for improving the antitumor efficiency of polymer–doxorubicin nanoprodrug. Small 14(37):1802403
Garrigue P, Tang J, Ding L, Bouhlel A, Tintaru A, Laurini E et al (2018) Self-assembling supramolecular dendrimer nanosystem for PET imaging of tumors. Proc Natl Acad Sci 115(45):11454–11459
Ge S, Ge L, Yan M, Song X, Yu J, Liu S (2013) A disposable immunosensor device for point-of-care test of tumor marker based on copper-mediated amplification. Biosens Bioelectron 43:425–431
Geho DH, Jones CD, Petricoin EF, Liotta LA (2006) Nanoparticles: potential biomarker harvesters. Curr Opin Chem Biol 10(1):56–61
Gellhorn A (1958) Recent studies on pathophysiologic mechanisms in human neoplastic disease. J Chronic Dis 8(1):158–170
Gokarna A, Jin LH, Hwang JS, Cho YH, Lim YT, Chung BH et al (2008) Quantum dot-based protein micro-and nanoarrays for detection of prostate cancer biomarkers. Proteomics 8(9): 1809–1818

Gold P, Freedman SO (1965) Specific carcinoembryonic antigens of the human digestive system. J Exp Med 122(3):467–481

Gonzalez-Rodriguez R, Campbell E, Naumov A (2019) Multifunctional graphene oxide/iron oxide nanoparticles for magnetic targeted drug delivery dual magnetic resonance/fluorescence imaging and cancer sensing. PLoS One 14(6):e0217072

Goon IY, Lai LM, Lim M, Amal R, Gooding JJ (2010) 'Dispersible electrodes': a solution to slow response times of sensitive sensors. Chem Commun 46(46):8821–8823

Greco O, Dachs GU (2001) Gene directed enzyme/prodrug therapy of cancer: historical appraisal and future prospectives. J Cell Physiol 187(1):22–36

Grodzinski P, Liu CH, Hartshorn CM, Morris SA, Russell LM (2019) NCI alliance for nanotechnology in cancer—from academic research to clinical interventions. Biomed Microdevices 21(2):1–8

Gui L, Zhang XH, Qiao ZY, Wang H (2020) Cell-penetrating peptides and polymers for improved drug delivery. ChemNanoMat 6(8):1138–1148

Guo Z, Cui Z (2021) Fluorescent nanotechnology for in vivo imaging. Wiley Interdiscip Rev Nanomed Nanobiotechnol 13(5):e1705

Guo X, Wang L, Wei X, Zhou S (2016) Polymer-based drug delivery systems for cancer treatment. J Polym Sci A Polym Chem 54(22):3525–3550

Gupta GP, Massagué J (2006) Cancer metastasis: building a framework. Cell 127(4):679–695

Gutierrez AA, Lemoine NR, Sikora K (1992) Gene therapy for cancer. Lancet 339(8795):715–721

Hafezi Ghahestani Z, Alebooye Langroodi F, Mokhtarzadeh A, Ramezani M, Hashemi M (2017) Evaluation of anti-cancer activity of PLGA nanoparticles containing crocetin. Artif Cells Nanomed Biotechnol 45(5):955–960

Hafizi M, Atashi A, Bakhshandeh B, Kabiri M, Nadri S, Hosseini RH et al (2013) MicroRNAs as markers for neurally committed CD133+/CD34+ stem cells derived from human umbilical cord blood. Biochem Genet 51(3–4):175–188

Hardiansyah A, Huang LY, Yang MC, Liu TY, Tsai SC, Yang CY et al (2014) Magnetic liposomes for colorectal cancer cells therapy by high-frequency magnetic field treatment. Nanoscale Res Lett 9(1):1–13

Harding M, McAllister J, Hulks G, Vernon D, Monie R, Paul J et al (1990) Neurone specific enolase (NSE) in small cell lung cancer: a tumour marker of prognostic significance? Br J Cancer 61(4):605–607

Harris JC, Scully MA, Day ES (2019) Cancer cell membrane-coated nanoparticles for cancer management. Cancers 11(12):1836

Helm L (2010) Optimization of gadolinium-based MRI contrast agents for high magnetic-field applications. Future Med Chem 2(3):385–396

Helmink BA, Khan M, Hermann A, Gopalakrishnan V, Wargo JA (2019) The microbiome, cancer, and cancer therapy. Nat Med 25(3):377–388

Hizir MS, Balcioglu M, Rana M, Robertson NM, Yigit MV (2014) Simultaneous detection of circulating oncomiRs from body fluids for prostate cancer staging using nanographene oxide. ACS Appl Mater Interfaces 6(17):14772–14778

Hu M, Yan J, He Y, Lu H, Weng L, Song S et al (2010a) Ultrasensitive, multiplexed detection of cancer biomarkers directly in serum by using a quantum dot-based microfluidic protein chip. ACS Nano 4(1):488–494

Hu M, He Y, Song S, Yan J, Lu H-T, Weng L-X et al (2010b) DNA-bridged bioconjugation of fluorescent quantum dots for highly sensitive microfluidic protein chips. Chem Commun 46(33):6126–6128

Huang HS, Hainfeld JF (2013) Intravenous magnetic nanoparticle cancer hyperthermia. Int J Nanomedicine 8:2521–2532

Huang Q, Wang F-B, Yuan C-H, He Z, Rao L, Cai B et al (2018) Gelatin nanoparticle-coated silicon beads for density-selective capture and release of heterogeneous circulating tumor cells with high purity. Theranostics 8(6):1624

Hung Y-C, Pan H-A, Tai S-M, Huang GS (2010) A nanodevice for rapid modulation of proliferation, apoptosis, invasive ability, and cytoskeletal reorganization in cultured cells. Lab Chip 10(9):1189–1198

Ito A, Tanaka K, Honda H, Abe S, Yamaguchi H, Kobayashi T (2003) Complete regression of mouse mammary carcinoma with a size greater than 15 mm by frequent repeated hyperthermia using magnetite nanoparticles. J Biosci Bioeng 96(4):364–369

Iturrioz-Rodríguez N, Correa-Duarte MA, Fanarraga ML (2019) Controlled drug delivery systems for cancer based on mesoporous silica nanoparticles. Int J Nanomedicine 14:3389

Jafari A, Salouti M, Shayesteh SF, Heidari Z, Rajabi AB, Boustani K et al (2015) Synthesis and characterization of Bombesin-superparamagnetic iron oxide nanoparticles as a targeted contrast agent for imaging of breast cancer using MRI. Nanotechnology 26(7):075101

Jahanafrooz Z, Motameh N, Bakhshandeh B (2016) Comparative evaluation of silibinin effects on cell cycling and apoptosis in human breast cancer MCF-7 and T47D cell lines. Asian Pac J Cancer Prev 17(5):2661–2665

Jahanafrooz Z, Baradaran B, Mosafer J, Hashemzaei M, Rezaei T, Mokhtarzadeh A et al (2020) Comparison of DNA and mRNA vaccines against cancer. Drug Discov Today 25(3):552–560

Jahanafrooz Z, Bakhshandeh B, Shirzadi E, Jahanafrooz Z (2022) Comparative evaluation of lipofectamine and dendrimer for transfection of short RNA into human T47D and MCF-10A cell lines. Adv Pharm Bull. https://doi.org/10.34172/apb.2023.022

Jang H, Zhi K, Wang J, Zhao H, Li B, Yang X (2020) Enhanced therapeutic effect of paclitaxel with a natural polysaccharide carrier for local injection in breast cancer. Int J Biol Macromol 148: 163–172

Jin J, Zhao Q (2020) Engineering nanoparticles to reprogram radiotherapy and immunotherapy: recent advances and future challenges. J Nanobiotechnol 18(1):1–17

Jin C, Wang K, Oppong-Gyebi A, Hu J (2020) Application of nanotechnology in cancer diagnosis and therapy-a mini-review. Int J Med Sci 17(18):2964

Jing L, Liang X, Deng Z, Feng S, Li X, Huang M et al (2014) Prussian blue coated gold nanoparticles for simultaneous photoacoustic/CT bimodal imaging and photothermal ablation of cancer. Biomaterials 35(22):5814–5821

Jokerst JV, Raamanathan A, Christodoulides N, Floriano PN, Pollard AA, Simmons GW et al (2009) Nano-bio-chips for high performance multiplexed protein detection: determinations of cancer biomarkers in serum and saliva using quantum dot bioconjugate labels. Biosens Bioelectron 24(12):3622–3629

Jones PA (2002) DNA methylation and cancer. Oncogene 21(35):5358–5360

Joshi BP, Wang TD (2018) Targeted optical imaging agents in cancer: focus on clinical applications. Contrast Media Mol Imaging 2018:2015237

Jou AF-J, Lu C-H, Ou Y-C, Wang S-S, Hsu S-L, Willner I et al (2015) Diagnosing the miR-141 prostate cancer biomarker using nucleic acid-functionalized CdSe/ZnS QDs and telomerase. Chem Sci 6(1):659–665

Kenemans P, Yedema C, Bon G, von Mensdorff-Pouilly S (1993) CA 125 in gynecological pathology—a review. Eur J Obstet Gynecol Reprod Biol 49(1–2):115–124

Khandare J, Minko T (2006) Polymer–drug conjugates: progress in polymeric prodrugs. Prog Polym Sci 31(4):359–397

Kievit FM, Stephen ZR, Veiseh O, Arami H, Wang T, Lai VP et al (2012) Targeting of primary breast cancers and metastases in a transgenic mouse model using rationally designed multifunctional SPIONs. ACS Nano 6(3):2591–2601

Kim D, Park S, Lee JH, Jeong YY, Jon S (2007) Antibiofouling polymer-coated gold nanoparticles as a contrast agent for in vivo X-ray computed tomography imaging. J Am Chem Soc 129(24): 7661–7665

Kim D, Jeong YY, Jon S (2010a) A drug-loaded aptamer–gold nanoparticle bioconjugate for combined CT imaging and therapy of prostate cancer. ACS Nano 4(7):3689–3696

Kim HS, Oh SY, Joo HJ, Son KR, Song IC, Moon WK (2010b) The effects of clinically used MRI contrast agents on the biological properties of human mesenchymal stem cells. NMR Biomed 23(5):514–522

Kim D, Yu MK, Lee TS, Park JJ, Jeong YY, Jon S (2011) Amphiphilic polymer-coated hybrid nanoparticles as CT/MRI dual contrast agents. Nanotechnology 22(15):155101

Kim J, Kwon S, Park J-K, Park I (2014) Quantum dot-based immunoassay enhanced by high-density vertical ZnO nanowire array. Biosens Bioelectron 55:209–215

Kingsley JD, Dou H, Morehead J, Rabinow B, Gendelman HE, Destache CJ (2006) Nanotechnology: a focus on nanoparticles as a drug delivery system. J Neuroimmune Pharmacol 1(3): 340–350

Kipp JE (2004) The role of solid nanoparticle technology in the parenteral delivery of poorly water-soluble drugs. Int J Pharm 284(1–2):109–122

Kirschning A, Kupracz L, Hartwig J (2012) New synthetic opportunities in miniaturized flow reactors with inductive heating. Chem Lett 41(6):562–570

Kozma GT, Shimizu T, Ishida T, Szebeni J (2020) Anti-PEG antibodies: Properties, formation, testing and role in adverse immune reactions to PEGylated nano-biopharmaceuticals. Adv Drug Deliv Rev 154:163–175

Kubik T, Bogunia-Kubik K, Sugisaka M (2005) Nanotechnology on duty in medical applications. Curr Pharm Biotechnol 6(1):17–33

Kuk C, Kulasingam V, Gunawardana CG, Smith CR, Batruch I, Diamandis EP (2009) Mining the ovarian cancer ascites proteome for potential ovarian cancer biomarkers. Mol Cell Proteomics 8(4):661–669

Kumar CSSR, Mohammad F (2011) Magnetic nanomaterials for hyperthermia-based therapy and controlled drug delivery. Adv Drug Deliv Rev 63(9):789–808

Kumari P, Muddineti OS, Rompicharla SVK, Ghanta P, Adithya Karthik BBN, Ghosh B et al (2017) Cholesterol-conjugated poly (D, L-lactide)-based micelles as a nanocarrier system for effective delivery of curcumin in cancer therapy. Drug Deliv 24(1):209–223

Kuriyama M, Wang MC, Lee C-l, Papsidero LD, Killian CS, Inaji H et al (1981) Use of human prostate-specific antigen in monitoring prostate cancer. Cancer Res 41(10):3874–3876

Lee W-H, Loo C-Y, Traini D, Young PM (2015) Inhalation of nanoparticle-based drug for lung cancer treatment: advantages and challenges. Asian J Pharm Sci 10(6):481–489

Lee HM, Joh JW, Seo S-R, Kim W-T, Kim MK, Choi HS et al (2017) Cell-surface major vault protein promotes cancer progression through harboring mesenchymal and intermediate circulating tumor cells in hepatocellular carcinomas. Sci Rep 7(1):1–15

Lee J-H, Choi J-H, Chueng S-TD, Pongkulapa T, Yang L, Cho H-Y et al (2019a) Nondestructive characterization of stem cell neurogenesis by a magneto-plasmonic nanomaterial-based exosomal mirna detection. ACS Nano 13(8):8793–8803

Lee SWL, Paoletti C, Campisi M, Osaki T, Adriani G, Kamm RD et al (2019b) MicroRNA delivery through nanoparticles. J Control Release 3123:80–95

Lenin N, Sakthipandi K, Rajesh Kanna R, Rajkumar G (2018) Electrical, magnetic and structural properties of polymer-blended lanthanum-added nickel nano-ferrites. Ceram Int 44(17): 21866–21873

Li W, Chen X (2015) Gold nanoparticles for photoacoustic imaging. Nanomedicine 10(2):299–320

Li Z, Wang Y, Wang J, Tang Z, Pounds JG, Lin Y (2010) Rapid and sensitive detection of protein biomarker using a portable fluorescence biosensor based on quantum dots and a lateral flow test strip. Anal Chem 82(16):7008–7014

Li H, Cao Z, Zhang Y, Lau C, Lu J (2011) Simultaneous detection of two lung cancer biomarkers using dual-color fluorescence quantum dots. Analyst 136(7):1399–1405

Li J, Cai P, Shalviri A, Henderson JT, He C, Foltz WD et al (2014a) A multifunctional polymeric nanotheranostic system delivers doxorubicin and imaging agents across the blood–brain barrier targeting brain metastases of breast cancer. ACS Nano 8(10):9925–9940

Li L, Gu W, Chen J, Chen W, Xu ZP (2014b) Co-delivery of siRNAs and anti-cancer drugs using layered double hydroxide nanoparticles. Biomaterials 35(10):3331–3339

Li X, Sui Z, Li X, Xu W, Guo Q, Sun J et al (2018a) Perfluorooctylbromide nanoparticles for ultrasound imaging and drug delivery. Int J Nanomedicine 13:3053

Li H, Meng QH, Noh H, Somaiah N, Torres KE, Xia X et al (2018b) Cell-surface vimentin–positive macrophage-like circulating tumor cells as a novel biomarker of metastatic gastrointestinal stromal tumors. Onco Targets Ther 7(5):e1420450

Li W, Wang H, Zhao Z, Gao H, Liu C, Zhu L et al (2019) Emerging nanotechnologies for liquid biopsy: the detection of circulating tumor cells and extracellular vesicles. Adv Mater 31(45): 1805344

Li Z-M, Pi T, Yan X-L, Tang X-M, Deng R-H, Zheng X-J (2020) Label-free and enzyme-free one-step rapid colorimetric detection of DNA methylation based on unmodified gold nanoparticles. Spectrochim Acta A Mol Biomol Spectrosc 238:118375

Li L, Liang D, Guo W, Tang D, Zeng Y (2022) Antibody-invertase cross-linkage nanoparticles: a new signal tag for point-of-care immunoassay of alpha-fetoprotein for hepatocellular carcinoma with personal glucometer. Electroanalysis 34(2):246–251

Lin Y (2015) Internal radiation therapy: a neglected aspect of nuclear medicine in the molecular era. J Biomed Res 29(5):345

Liu TW, MacDonald TD, Jin CS, Gold JM, Bristow RG, Wilson BC et al (2013) Inherently multimodal nanoparticle-driven tracking and real-time delineation of orthotopic prostate tumors and micrometastases. ACS Nano 7(5):4221–4232

Loo C, Lin A, Hirsch L, Lee M-H, Barton J, Halas N et al (2004) Nanoshell-enabled photonics-based imaging and therapy of cancer. Technol Cancer Res Treat 3(1):33–40

Lu X-Y, Wu D-C, Li Z-J, Chen G-Q (2011) Polymer nanoparticles. Prog Mol Biol Transl Sci 104: 299–323

Luchini A, Fredolini C, Espina B, Meani F, Reeder A, Rucker S et al (2010) Nanoparticle technology: addressing the fundamental roadblocks to protein biomarker discovery. Curr Mol Med 10(2):133–141

Luo B, Zhou J, Li Z, Song J, An P, Zhang H et al (2022) Ultrasensitive DNA methylation ratio detection based on the target-induced nanoparticle-coupling and site-specific base oxidation damage for colorectal cancer. Anal Chem 94(16):6261–6270

Masood F (2016) Polymeric nanoparticles for targeted drug delivery system for cancer therapy. Mater Sci Eng C 60:569–578

Materón EM, Miyazaki CM, Carr O, Joshi N, Picciani PH, Dalmaschio CJ et al (2021) Magnetic nanoparticles in biomedical applications: a review. Appl Surf Sci Adv 6:100163

Mei J, Hong Y, Lam JW, Qin A, Tang Y, Tang BZ (2014) Aggregation-induced emission: the whole is more brilliant than the parts. Adv Mater 26(31):5429–5479

Mennes OA, van Netten JJ, Slart RH, Steenbergen W (2018) Novel optical techniques for imaging microcirculation in the diabetic foot. Curr Pharm Des 24(12):1304–1316

Min L, Wang B, Bao H, Li X, Zhao L, Meng J et al (2021) Advanced nanotechnologies for extracellular vesicle-based liquid biopsy. Adv Sci 8(20):2102789

Minelli C, Lowe SB, Stevens MM (2010) Engineering nanocomposite materials for cancer therapy. Small 6(21):2336–2357

Misra R, Acharya S, Sahoo SK (2010) Cancer nanotechnology: application of nanotechnology in cancer therapy. Drug Discov Today 15(19–20):842–850

Mokhtarzadeh A, Parhiz H, Hashemi M, Abnous K, Ramezani M (2016) P53-Derived peptides conjugation to PEI: an approach to producing versatile and highly efficient targeted gene delivery carriers into cancer cells. Expert Opin Drug Deliv 13(4):477–491

Mombini S, Mohammadnejad J, Bakhshandeh B, Narmani A, Nourmohammadi J, Vahdat S et al (2019) Chitosan-PVA-CNT nanofibers as electrically conductive scaffolds for cardiovascular tissue engineering. Int J Biol Macromol 140:278–287

Mottaghitalab F, Farokhi M, Fatahi Y, Atyabi F, Dinarvand R (2019) New insights into designing hybrid nanoparticles for lung cancer: diagnosis and treatment. J Control Release 295:250–267

Mottram PL (2003) Past, present and future drug treatment for rheumatoid arthritis and systemic lupus erythematosus. Immunol Cell Biol 81(5):350–353

Muhamad N, Plengsuriyakarn T, Na-Bangchang K (2018) Application of active targeting nanoparticle delivery system for chemotherapeutic drugs and traditional/herbal medicines in cancer therapy: a systematic review. Int J Nanomedicine 13:3921

Mukundan H, Xie H, Anderson AS, Grace WK, Shively JE, Swanson BI (2009a) Optimizing a waveguide-based sandwich immunoassay for tumor biomarkers: evaluating fluorescent labels and functional surfaces. Bioconjug Chem 20(2):222–230

Mukundan H, Kubicek JZ, Holt A, Shively JE, Martinez JS, Grace K et al (2009b) Planar optical waveguide-based biosensor for the quantitative detection of tumor markers. Sensors Actuators B Chem 138(2):453–460

Mulens V, Morales MP, Barber DF (2013) Development of magnetic nanoparticles for cancer gene therapy: a comprehensive review. ISRN Nanomaterials 2013:1–14

Najam-ul-Haq M, Rainer M, Szabó Z, Vallant R, Huck C, Bonn G (2007) Role of carbon nanomaterials in the analysis of biological materials by laser desorption/ionization-mass spectrometry. J Biochem Biophys Methods 70(2):319–328

Nie S, Xing Y, Kim GJ, Simons JW (2007) Nanotechnology applications in cancer. Annu Rev Biomed Eng 9:257–288

Nishikawa RM (2007) Current status and future directions of computer-aided diagnosis in mammography. Comput Med Imaging Graph 31(4–5):224–235

Nyayapathi N, Xia J (2019) Photoacoustic imaging of breast cancer: a mini review of system design and image features. J Biomed Opt 24(12):121911

Nygren P (2001) What is cancer chemotherapy? Acta Oncol 40(2–3):166–174

O'Neill CP, Dwyer RM (2020) Nanoparticle-based delivery of tumor suppressor microRNA for cancer therapy. Cell 9(2):521

Pandey SK, Patel DK, Maurya AK, Thakur R, Mishra DP, Vinayak M et al (2016) Controlled release of drug and better bioavailability using poly(lactic acid-co-glycolic acid) nanoparticles. Int J Biol Macromol 89:99–110

Pang Y, Wang C, Wang J, Sun Z, Xiao R, Wang S (2016) Fe3O4@Ag magnetic nanoparticles for microRNA capture and duplex-specific nuclease signal amplification based SERS detection in cancer cells. Biosens Bioelectron 79:574–580

Papahadjopoulos D, Allen TM, Gabizon A, Mayhew E, Matthay K, Huang SK et al (1991) Sterically stabilized liposomes: improvements in pharmacokinetics and antitumor therapeutic efficacy. Proc Natl Acad Sci 88(24):11460–11464

Pardo J, Peng Z, Leblanc RM (2018) Cancer targeting and drug delivery using carbon-based quantum dots and nanotubes. Molecules 23(2):378

Park S, Kim H, Lim SC, Lim K, Lee ES, Oh KT et al (2019) Gold nanocluster-loaded hybrid albumin nanoparticles with fluorescence-based optical visualization and photothermal conversion for tumor detection/ablation. J Control Release 304:7–18

Parungo CP, Ohnishi S, De Grand AM, Laurence RG, Soltesz EG, Colson YL et al (2004) In vivo optical imaging of pleural space drainage to lymph nodes of prognostic significance. Ann Surg Oncol 11(12):1085–1092

Pasut G, Veronese F (2007) Polymer–drug conjugation, recent achievements and general strategies. Prog Polym Sci 32(8–9):933–961

Patil R, Ljubimov AV, Gangalum PR, Ding H, Portilla-Arias J, Wagner S et al (2015) MRI virtual biopsy and treatment of brain metastatic tumors with targeted nanobioconjugates: nanoclinic in the brain. ACS Nano 9(5):5594–5608

Peiris PM, Toy R, Doolittle E, Pansky J, Abramowski A, Tam M et al (2012) Imaging metastasis using an integrin-targeting chain-shaped nanoparticle. ACS Nano 6(10):8783–8795

Pelicano H, Martin DS, Xu RH, Huang P (2006) Glycolysis inhibition for anticancer treatment. Oncogene 25(34):4633–4646

Popescu RC, Fufă M, Grumezescu AM (2015) Metal-based nanosystems for diagnosis. Romanian J Morphol Embryol 56(2 Suppl):635–649

Portela-Gomes GM, Hacker GW, Weitgasser R (2004) Neuroendocrine cell markers for pancreatic islets and tumors. Appl Immunohistochem Mol Morphol 12(3):183–192

Power SP, Moloney F, Twomey M, James K, O'Connor OJ, Maher MM (2016) Computed tomography and patient risk: facts, perceptions and uncertainties. World J Radiol 8(12):902

Priestley P, Baber J, Lolkema MP, Steeghs N, de Bruijn E, Shale C et al (2019) Pan-cancer whole-genome analyses of metastatic solid tumours. Nature 575(7781):210–216

Purohit R, Singh S (2018) Fluorescent gold nanoclusters for efficient cancer cell targeting. Int J Nanomedicine 13:15

Putnin T, Ngamaroonchote A, Wiriyakun N, Ounnunkad K, Laocharoensuk R (2019) Dually functional polyethylenimine-coated gold nanoparticles: a versatile material for electrode modification and highly sensitive simultaneous determination of four tumor markers. Microchim Acta 186(5):1–12

Qin W, Ding D, Liu J, Yuan WZ, Hu Y, Liu B et al (2012) Biocompatible nanoparticles with aggregation-induced emission characteristics as far-red/near-infrared fluorescent bioprobes for in vitro and in vivo imaging applications. Adv Funct Mater 22(4):771–779

Rachner TD, Coleman R, Hadji P, Hofbauer LC (2018) Bone health during endocrine therapy for cancer. Lancet Diab Endocrinol 6(11):901–910

Riggio C, Pagni E, Raffa V, Cuschieri A (2011) Nano-oncology: clinical application for cancer therapy and future perspectives. J Nanomater 2011:164506

Rodrigues M, Richards N, Ning B, Lyon CJ, Hu TY (2019) Rapid lipid-based approach for normalization of quantum-dot-detected biomarker expression on extracellular vesicles in complex biological samples. Nano Lett 19(11):7623–7631

Rodríguez F, Caruana P, De la Fuente N, Español P, Gámez M, Balart J et al (2022) Nano-based approved pharmaceuticals for cancer treatment: present and future challenges. Biomolecules 12(6):784

Rosenblum D, Joshi N, Tao W, Karp JM, Peer D (2018) Progress and challenges towards targeted delivery of cancer therapeutics. Nat Commun 9(1):1–12

Said SS, Campbell S, Hoare T (2019) Externally addressable smart drug delivery vehicles: current technologies and future directions. Chem Mater 31(14):4971–4989

Salami SS, Schmidt F, Laxman B, Regan MM, Rickman DS, Scherr D et al (2013) Combining urinary detection of TMPRSS2: ERG and PCA3 with serum PSA to predict diagnosis of prostate cancer. Urol Oncol 31:566

Salata OV (2004) Applications of nanoparticles in biology and medicine. J Nanobiotechnol 2(1): 1–6

Schneck H, Gierke B, Uppenkamp F, Behrens B, Niederacher D, Stoecklein NH et al (2015) EpCAM-independent enrichment of circulating tumor cells in metastatic breast cancer. PLoS One 10(12):e0144535

Schouw H, Huisman L, Janssen Y, Slart R, Borra R, Willemsen A et al (2021) Targeted optical fluorescence imaging: a meta-narrative review and future perspectives. Eur J Nucl Med Mol Imaging 48(13):4272–4292

Seferos DS, Giljohann DA, Hill HD, Prigodich AE, Mirkin CA (2007) Nano-flares: probes for transfection and mRNA detection in living cells. J Am Chem Soc 129(50):15477–15479

Senapati S, Mahanta AK, Kumar S, Maiti P (2018) Controlled drug delivery vehicles for cancer treatment and their performance. Signal Transduct Target Ther 3(1):1–19

Sengupta S, Balla VK (2018) A review on the use of magnetic fields and ultrasound for non-invasive cancer treatment. J Adv Res 14:97–111

Shabestari Khiabani S, Farshbaf M, Akbarzadeh A, Davaran S (2016) Magnetic nanoparticles: preparation methods, applications in cancer diagnosis and cancer therapy. Artif Cells Nanomed Biotechnol 45(1):6–17

Shah RR, Dombrowsky AR, Paulson AL, Johnson MP, Nikles DE, Brazel CS (2016) Determining iron oxide nanoparticle heating efficiency and elucidating local nanoparticle temperature for application in agarose gel-based tumor model. Mater Sci Eng C Mater Biol Appl 68:18–29

Shenoy D, Little S, Langer R, Amiji M (2005) Poly(ethylene oxide)-modified poly(beta-amino ester) nanoparticles as a pH-sensitive system for tumor-targeted delivery of hydrophobic drugs: part 2. In vivo distribution and tumor localization studies. Pharm Res 22(12):2107–2114

Sherry AD, Woods M (2008) Chemical exchange saturation transfer contrast agents for magnetic resonance imaging. Annu Rev Biomed Eng 10:391

Shin H, Oh S, Hong S, Kang M, Kang D, Ji Y-g et al (2020) Early-stage lung cancer diagnosis by deep learning-based spectroscopic analysis of circulating exosomes. ACS Nano 14(5): 5435–5444

Shitrit D, Zingerman B, Shitrit AB-G, Shlomi D, Kramer MR (2005) Diagnostic value of CYFRA 21-1, CEA, CA 19-9, CA 15-3, and CA 125 assays in pleural effusions: analysis of 116 cases and review of the literature. Oncologist 10(7):501–507

Sina AAI, Carrascosa LG, Liang Z, Grewal YS, Wardiana A, Shiddiky MJ et al (2018) Epigenetically reprogrammed methylation landscape drives the DNA self-assembly and serves as a universal cancer biomarker. Nat Commun 9(1):1–13

Singh R (2019) Nanotechnology based therapeutic application in cancer diagnosis and therapy. 3 Biotech 9(11):1–29

Singhvi G, Rapalli VK, Nagpal S, Dubey SK, Saha RN (2020) Nanocarriers as potential targeted drug delivery for cancer therapy. Nanosci Med 1:51–88

Spanoudaki VC, Ziegler SI (2008) PET & SPECT instrumentation. In: Molecular imaging I. Springer, pp 53–74

Sperling RA, Gil PR, Zhang F, Zanella M, Parak WJ (2008) Biological applications of gold nanoparticles. Chem Soc Rev 37(9):1896–1908

Stridh MT, Hult J, Merdasa A, Albinsson J, Pekar-Lukacs A, Gesslein B et al (2022) Photoacoustic imaging of periorbital skin cancer ex vivo: unique spectral signatures of malignant melanoma, basal, and squamous cell carcinoma. Biomed Opt Express 13(1):410–425

Sui X, Luo C, Wang C, Zhang F, Zhang J, Guo S (2016) Graphene quantum dots enhance anticancer activity of cisplatin via increasing its cellular and nuclear uptake. Nanomedicine 12(7):1997–2006

Sun Y, Li T (2018) Composition-tunable hollow Au/Ag SERS nanoprobes coupled with target-catalyzed hairpin assembly for triple-amplification detection of miRNA. Anal Chem 90(19): 11614–11621

Sun C, Lee JSH, Zhang M (2008) Magnetic nanoparticles in MR imaging and drug delivery. Adv Drug Deliv Rev 60(11):1252–1265

Sun W, Zhao X, Fan J, Du J, Peng X (2019) Boron dipyrromethene nano-photosensitizers for anticancer phototherapies. Small 15(32):1804927

Sundstrøm T, Daphu I, Wendelbo I, Hodneland E, Lundervold A, Immervoll H et al (2013) Automated tracking of nanoparticle-labeled melanoma cells improves the predictive power of a brain metastasis model nanoparticle tracking of melanoma brain metastasis. Cancer Res 73(8): 2445–2456

Sur S, Rathore A, Dave V, Reddy KR, Chouhan RS, Sadhu V (2019) Recent developments in functionalized polymer nanoparticles for efficient drug delivery system. Nano-Struct Nano-Objects 20:100397

Tapia FJ, Barbosa A, Marangos P, Polak J, Bloom S, Dermody C et al (1981) Neuron-specific enolase is produced by neuroendocrine tumours. Lancet 317(8224):808–811

Thierry A, El Messaoudi S, Gahan P, Anker P, Stroun M (2016) Origins, structures, and functions of circulating DNA in oncology. Cancer Metastasis Rev 35(3):347–376

Thompson IM, Pauler DK, Goodman PJ, Tangen CM, Lucia MS, Parnes HL et al (2004) Prevalence of prostate cancer among men with a prostate-specific antigen level$\leq$ 4.0 ng per milliliter. N Engl J Med 350(22):2239–2246

Torchilin VP (2006) Multifunctional nanocarriers. Adv Drug Deliv Rev 58(14):1532–1555

Torchilin VP (2010) Passive and active drug targeting: drug delivery to tumors as an example. Drug Deliv 197:3–53

Torresano L, Nuevo-Tapioles C, Santacatterina F, Cuezva JM (2020) Metabolic reprogramming and disease progression in cancer patients. Biochim Biophys Acta 1866(5):165721

Tran S, DeGiovanni P-J, Piel B, Rai P (2017) Cancer nanomedicine: a review of recent success in drug delivery. Clin Transl Med 6(1):1–21

Urruticoechea A, Alemany R, Balart J, Villanueva A, Vinals F, Capella G (2010) Recent advances in cancer therapy: an overview. Curr Pharm Des 16(1):3–10 van der Zanden SY, Luimstra JJ, Neefjes J, Borst J, Ovaa H (2020) Opportunities for small molecules in cancer immunotherapy. Trends Immunol 41(6):493–511

Vinod R, Mahran R, Routila E, Leivo J, Pettersson K, Gidwani K (2021) Nanoparticle-aided detection of colorectal cancer-associated glycoconjugates of extracellular vesicles in human serum. Int J Mol Sci 22(19):10329

Wan X, Song Y, Song N, Li J, Yang L, Li Y et al (2016) The preliminary study of immune superparamagnetic iron oxide nanoparticles for the detection of lung cancer in magnetic resonance imaging. Carbohydr Res 419:33–40

Wang C-h, Li J, Yao S-j, Guo Y-l, Xia X-h (2007) High-sensitivity matrix-assisted laser desorption/ionization Fourier transform mass spectrometry analyses of small carbohydrates and amino acids using oxidized carbon nanotubes prepared by chemical vapor deposition as matrix. Anal Chim Acta 604(2):158–164

Wang J, Sui M, Fan W (2010) Nanoparticles for tumor targeted therapies and their pharmacokinetics. Curr Drug Metab 11(2):129–141

Wang D, Wu L, Liu X (2017) Glycan markers as potential immunological targets in circulating tumor cells. In: Isolation and molecular characterization of circulating tumor cells. Springer, Cham, pp 275–284

Wang N, Wang J, Zhao X, Chen H, Xu H, Bai L et al (2021) Highly sensitive electrochemical immunosensor for the simultaneous detection of multiple tumor markers for signal amplification. Talanta 226:122133

Ward AM, Catto J, Hamdy F (2001) Prostate specific antigen: biology, biochemistry and available commercial assays. Ann Clin Biochem 38(6):633–651

Wegner KD, Jin Z, Linden S, Jennings TL, Hildebrandt N (2013) Quantum-dot-based Forster resonance energy transfer immunoassay for sensitive clinical diagnostics of low-volume serum samples. ACS Nano 7(8):7411–7419

Wei Q, Lee M, Yu X, Lee EK, Seong GH, Choo J et al (2006) Development of an open sandwich fluoroimmunoassay based on fluorescence resonance energy transfer. Anal Biochem 358(1):31–37

Winawer SJ (2007) Colorectal cancer screening. Best Pract Res Clin Gastroenterol 21(6):1031–1048

Wust P, Hildebrandt B, Sreenivasa G, Rau B, Gellermann J, Riess H et al (2002) Hyperthermia in combined treatment of cancer. Lancet Oncol 3(8):487–497

Xi J, Wang W, Da L, Zhang J, Fan L, Gao L (2018) Au-PLGA hybrid nanoparticles with catalase-mimicking and near-infrared photothermal activities for photoacoustic imaging-guided cancer therapy. ACS Biomater Sci Eng 4(3):1083–1091

Xie Y, Qiao H, Su Z, Chen M, Ping Q, Sun M (2014) PEGylated carboxymethyl chitosan/calcium phosphate hybrid anionic nanoparticles mediated hTERT siRNA delivery for anticancer therapy. Biomaterials 32(27):7978–7991

Yan J, Hu M, Li D, He Y, Zhao R, Jiang X et al (2008) A nano-and micro-integrated protein chip based on quantum dot probes and a microfluidic network. Nano Res 1(6):490–496

Yang Q, Gong X, Song T, Yang J, Zhu S, Li Y et al (2011) Quantum dot-based immunochromatography test strip for rapid, quantitative and sensitive detection of alpha fetoprotein. Biosens Bioelectron 30(1):145–150

Yang X, Zhang W, Zhao Z, Li N, Mou Z, Sun D et al (2017) Quercetin loading CdSe/ZnS nanoparticles as efficient antibacterial and anticancer materials. J Inorg Biochem 167:36–48

Yang W, Xia Y, Fang Y, Meng F, Zhang J, Cheng R et al (2018) Selective cell penetrating peptide-functionalized polymersomes mediate efficient and targeted delivery of methotrexate disodium to human lung cancer in vivo. Adv Healthc Mater 7(7):1701135

Zhai Y, Xie H, Gu H (2009) Effects of hyperthermia with dextran magnetic fluid on the growth of grafted H22 tumor in mice. Int J Hyperthermia 25:65–71

Zhang W, Zhang Z, Zhang Y (2011) The application of carbon nanotubes in target drug delivery systems for cancer therapies. Nanoscale Res Lett 6(1):1–22

Zhang G, Zeng X, Li P (2013) Nanomaterials in cancer-therapy drug delivery system. J Biomed Nanotechnol 9(5):741–750

Zhang R, Xing R, Jiao T, Ma K, Chen C, Ma G et al (2016) Carrier-free, chemophotodynamic dual nanodrugs via self-assembly for synergistic antitumor therapy. ACS Appl Mater Interfaces 8(21):13262–13269

Zhang RR, Schroeder AB, Grudzinski JJ, Rosenthal EL, Warram JM, Pinchuk AN et al (2017) Beyond the margins: real-time detection of cancer using targeted fluorophores. Nat Rev Clin Oncol 14(6):347–364

Zhang Y, Li M, Gao X, Chen Y, Liu T (2019a) Nanotechnology in cancer diagnosis: progress, challenges and opportunities. J Hematol Oncol 12(1):1–13

Zhang J, Dong Y, Zhu W, Xie D, Zhao Y, Yang D et al (2019b) Ultrasensitive detection of circulating tumor DNA of lung cancer via an enzymatically amplified SERS-based frequency shift assay. ACS Appl Mater Interfaces 11(20):18145–18152

Zhao Y, Pang B, Luehmann H, Detering L, Yang X, Sultan D et al (2016) Gold nanoparticles doped with 199Au atoms and their use for targeted cancer imaging by SPECT. Adv Healthc Mater 5(8):928–935

Zhao L, Li Y, Zhu J, Sun N, Song N, Xing Y et al (2019) Chlorotoxin peptide-functionalized polyethylenimine-entrapped gold nanoparticles for glioma SPECT/CT imaging and radionuclide therapy. J Nanobiotechnol 17(1):1–13

Zheng Y, Luo Y, Chen X, Li H, Huang B, Zhou B et al (2021) The role of mRNA in the development, diagnosis, treatment and prognosis of neural tumors. Mol Cancer 20(1):1–16

Zhong L, Li Y, Xiong L, Wang W, Wu M, Yuan T et al (2021) Small molecules in targeted cancer therapy: Advances, challenges, and future perspectives. Signal Transduct Target Ther 6(1):1–48

Zhu Y-D, Peng J, Jiang L-P, Zhu J-J (2014) Fluorescent immunosensor based on CuS nanoparticles for sensitive detection of cancer biomarker. Analyst 139(3):649–655

Zhu Y, Jiang Y, Meng F, Deng C, Cheng R, Zhang J et al (2018) Highly efficacious and specific anti-glioma chemotherapy by tandem nanomicelles co-functionalized with brain tumor-targeting and cell-penetrating peptides. J Control Release 278:1–8

Ziraksaz Z, Nomani A, Soleimani M, Bakhshandeh B, Arefian E, Haririan I et al (2013) Evaluation of cationic dendrimer and lipid as transfection reagents of short RNAs for stem cell modification. Int J Pharm 448(1):231–238

Zlitni A, Gambhir SS (2018) Molecular imaging agents for ultrasound. Curr Opin Chem Biol 45: 113–120

Zou J, Jafr G, Themistou E, Yap Y, Wintrob ZAP, Alexandridis P et al (2011) pH-Sensitive brush polymer-drug conjugates by ring-opening metathesis copolymerization. Chem Commun 47(15):4493–4495

Zuo Q, Zhang R, Li B, Zhao Y, Zhuang Y, Yu T et al (2015) MicroRNA-141 inhibits tumor growth and metastasis in gastric cancer by directly targeting transcriptional co-activator with PDZ-binding motif, TAZ. Cell Death Dis 6(1):e1623

Behnaz Bakhshandeh is an associate professor of medical biotechnology at the University of Tehran, since my PhD thesis at 2007, I have focused on biomaterials and epigenetic modifications of stem cells to find more effective cell differentiation protocols for tissue regeneration purposes. In this regard, I have investigated different induction factors and mechanisms such as microvesicles, small molecules, miRNAs and electrical stimulation, conductive biopolymers, hydrogels, and quantum dots to achieve better functional differentiated cells for more effective clinical outcomes. Meanwhile, studying cancer-related pathways deepened my mastery of the molecular pathways of the cell.

Zohreh Jahanafrooz is an assistant professor of cell and molecular biology at the University of Maragheh. In my PhD project, silibinin was investigated as an anticancer agent. During my sabbatical in Austria in 2016, I worked on DMAS and silibinin effects on chordoma cells. Now I resumed my research on biocompatible compounds and natural biomaterials application in cancer therapy. Studying about cancer genetics and cell and molecular biology of stem cells is the other focuses of my research. The applications of biomaterials in a wide range of drug and cell delivery, macromolecule labeling, tissue engineering, regenerative medicine, diagnosis, and therapy are my other interests.

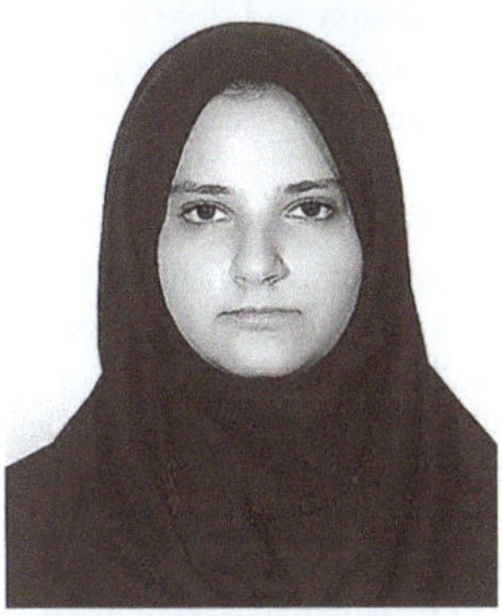

Dorrin Mohtadi Haghighi is a pharmacist, after my graduation at Tehran University in 2017, I studied pharmaceutics and formulation of dosage forms during my PhD period. I worked on the synthesis method of acetylcholine esterase inhibitors to treat Alzheimer disease, and the formulation of an inhalation form of cefixime to treat respiratory infections. Meanwhile, studying cancer-related pathways and applying nanoparticles in cancer therapy deepened my mastery of targeted drug delivery and effective drug formulation.

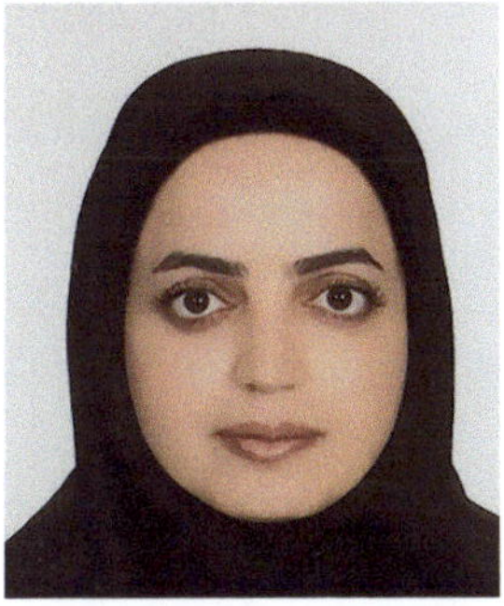

Nasim Rahmani-Kukia is a student of clinical biochemistry, I selected my master thesis, in 2013, on cancer research focusing on cancer immunotherapy with the title of "Mesenchymal stromal stem cell-derived microvesicles enhance tumor lysate-pulsed dendritic cell stimulated autologous T lymphocyte cytotoxicity." In 2019, I started working on my PhD thesis with a title "The effect of reducing and increasing the expression of ERMP1 as an endoplasmic reticulum metallopeptidase 1 on the cell proliferating pathways in colorectal cancer cell line." In summary, I have frequently focused on cancer treatment protocols using nanoparticles, investigating the effect of the employed agent on underlying mechanisms in cancer cells and cancer microenvironment. I hope to manage new research projects and develop novel assays in these fields to limit cancer formation, progression, and metastasis.

Ardeshir Abbasi is a senior Scientist at Tarbiat Modares University and ACTOVERCO pharmaceutical company, Tehran, Iran. I am a Cancer Immunologist with a long-lasting interest in cancer immunotherapies, cancer vaccines, and T-cell immunology. As an independent investigator, my research focuses on immunotherapies for breast cancer. I have successfully collaborated across institutions on multiple projects. I hold a PhD in Medical Immunology from Tarbiat Modares University. In my PhD thesis, I investigated the effect of immunotherapy using exosomes containing HSP-70 and HSP-90 following treatment with hydrogen peroxide in a mice model of breast cancer. In this project, I considered two important aspects of cancer, such as the tumor microenvironment and its immunotherapy. Overall, I will contribute to the cancer therapy mission and, together with basic and clinical investigators, foster the translation of novel laboratory findings into novel immunotherapy for cancers.

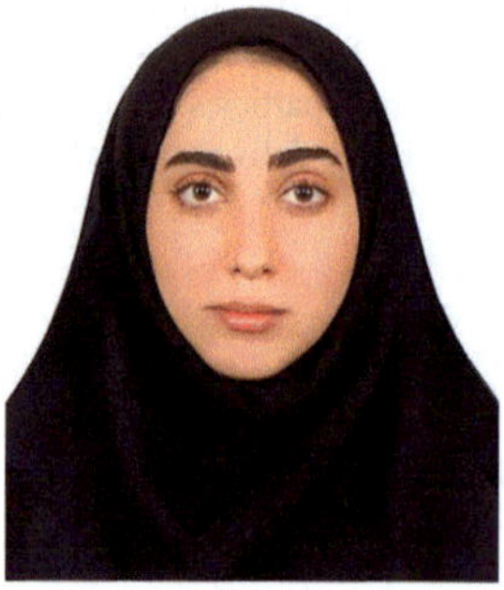

Armaghan Pourramezanali study MScs in Microbial Biotechnology at the University of Tehran, which is one of my favorite fields regarding Biology. My master's thesis is about the role of secondary microbial metabolites in the differentiation of stem cells. I am also interested in gene and cell therapy in cancer issues. In the mentioned issues, biomaterial/nanomaterial is frequently applied. Notably, I prefer nature-based biomaterials compared to synthetic ones for medical applications including cancer diagnosis and treatment. In these fields, I would like to collaborate with other researchers all around the world.

Application of Bioactive Compounds and Biomaterials in Promoting Cell Differentiation, Proliferation, and Tissue Regeneration

13

Azadeh Izadyari Aghmiuni, Arezoo Ghadi, Mahdieh Asadi, and Azim Akbarzadeh Khiyavi

Abstract

Cells, mainly stem cells, possess immense potential in regenerative medicine. Stem cells (SCs) can also play a crucial role in treating diseases due to their remarkable ability of self-renewal, proliferation, and differentiation into germ layers or gastrula (i.e., internal layer (ectoderm), middle layer (mesoderm), and outer layer (endoderm)). In this regard, pluripotent stem cells like induced pluripotent stem cells (iPSCs) and embryonic stem cells (ESCs) can differentiate into all three germ layers. However, targeted cellular differentiation is not possible without a suitable induction environment, microenvironment, or substrate. Based on the reports, the group of polymers or biopolymers, bioactive compounds, and biomaterials can provide such a microenvironment or substrate and lead to an increase in the interactions of cell-to-cell and cell-to-substrate, better differentiation, and secretion of growth factors to accelerate tissue regeneration. These materials can synthesize by living organisms such as animals, bacteria, fungi or algae, and plants. In this field, polysaccharides (such as hyaluronic acid, galactans, poly-galactosamine, chitosan, etc.) and proteins (such as silk fibroin, silk sericin, collagen, etc.) are known as natural biomaterials or biopolymers. Moreover, polyphenols, flavonoids, mucilage, pectin, apigenin, quercetin, galangin, curcumin, etc. obtained from medicinal herbs are

A. I. Aghmiuni (✉) · A. A. Khiyavi
Department of Nanobiotechnology, Pasteur Institute of Iran, Tehran, Iran
e-mail: azimakbarzadeh@pasteur.ac.ir

A. Ghadi
Department of Chemical Engineering, Islamic Azad University, Ayatollah Amoli Branch, Amol, Iran

M. Asadi
Department of Sport Physiology, Islamic Azad University, Sari Branch, Sari, Iran

F. A. Sheikh et al. (eds.), *Interaction of Nanomaterials With Living Cells*,
https://doi.org/10.1007/978-981-99-2119-5_13

phytochemicals and bioactive compounds which can be led to an increase in cellular proliferation and differentiation, induction of cellular signaling, promotion of regeneration rate, and immunomodulation. Hence, the present chapter is aimed to study the applications and consequences of bioactive compounds and biomaterials on stem cell proliferation and differentiation, as well as to highlight their role in tissue regeneration.

Keywords

Stem cells · Regenerative medicine · Biopolymers · Bioactive compounds · Biomaterials

13.1 Introduction

Stem cells are known as precursor biological cells that can be classified into totipotent,[1] pluripotent,[2] multipotent,[3] oligopotent,[4] and unipotent[5] stem cells, depending upon the differentiation capacities. These cells possess a high potential for self-renewal and differentiation into other cells, and as a promising resource for medicine, applications can play an important role in regenerative medicine or tissue engineering and cellular therapies (Chagastelles and Nardi 2011). However, applying stem cells to treat diseases and regenerate damaged tissues will be successful when suitable conditions are available to improve cellular interaction control and differentiate into the target cells (Izadyari Aghmiuni et al. 2021). Indeed, the creation of environments for therapeutic cloning that can well control cellular behavior can be the main factor in this regard (Fig. 13.1). Based on the reports, the use of biomaterials or biopolymers and bioactive compounds obtained from medicinal plants is considered a fundamental strategy for this purpose, due to their excellent bioactivity and role in the improvement of cellular biological response as well as in mimicking functional and structural properties of the target tissues (Yu et al. 2019). Such that, recent efforts in the design of biomaterial-based scaffolds, hydrogels, substrates, etc. (Chaudhari et al. 2016; Shafei et al. 2017; Ahmed 2013; Talebian et al. 2019; Zhao et al. 2020; Wang et al. 2018a) have provided significant opportunities to use these materials in regenerative medicine and generate novel

[1]Totipotent is embryonic stem cells of 1–3 days (e.g., zygotes)—differentiation into any cell types.

[2]Pluripotent is embryonic stem cells of 5–14 days (e.g., blastula) and induced pluripotent stem cells (iPSCs)—differentiation into cells from any of the three germ layers.

[3]Multipotent is adult stem cells—differentiation into a limited range of cell types such as neural stem cells, epithelial stem cells, hematopoietic stem cells, etc.

[4]Oligopotent is adult stem cells—differentiation into a limited number of cell types such as neural progenitor cells, myeloid stem cells, lymphoid stem cells, etc.

[5]Unipotent is adult stem cells—differentiation into single cell type such as gut cell, erythrocytes, neurons, etc.

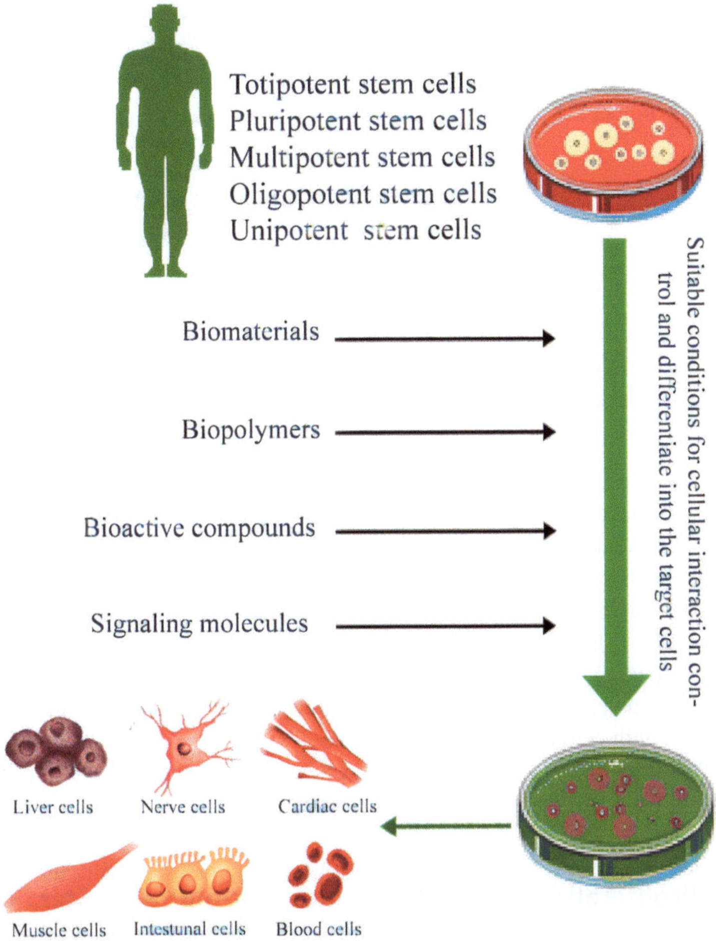

Fig. 13.1 The effective materials for the proliferation and differentiation of stem cells

substitutes for tissue engineering applications (Ramalingam et al. 2019; Aghmiuni et al. 2020).

There are biomaterials or bioactive materials with different natures which seem to be effective in targeting cell differentiation and tissue regeneration. Medicinal herbs are one of these bioactive materials gaining significant attention among researchers and scientific communities due to the promotion of cell proliferation and controlled differentiation (Izadyari Aghmiuni et al. 2020; Dey et al. 2010). The biopolymers similar to signaling molecules such as glycoproteins, proteoglycans, and glycosaminoglycans which exist in the network of the tissue extracellular matrix (ECM) are other samples from these biomaterials that can promote cell-cell signaling on the engineered substrates to modulate cellular functions (i.e., adhesion, proliferation, differentiation, and morphogenesis). Silk fibers (SF) are one of the biomaterials in this matter that owing to the amino acids of alanine and glycine can mimic the

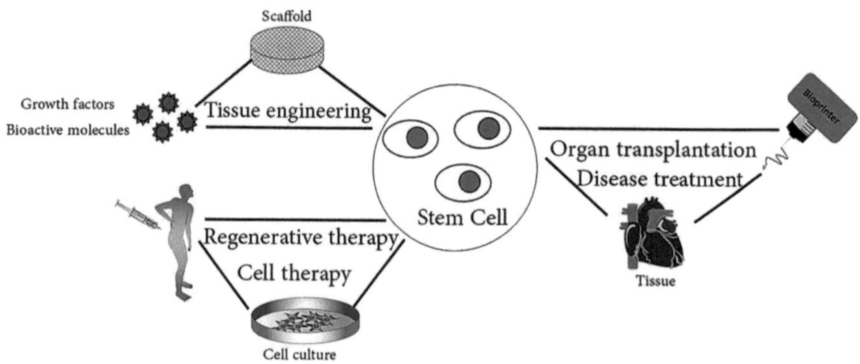

Fig. 13.2 Potential applications of stem cells

proteoglycans' behavior in the tissue ECMs (Omenetto and Kaplan 2010). Moreover, hyaluronic acid (HA), as a linear polysaccharide and signaling molecule, can act like glycosaminoglycans of the ECM (Kogan et al. 2006; Volpi et al. 2009; Petrey and de la Motte 2014) and play a crucial role in cell-cell and cell-substrate interactions via communication between cell-surface receptors (Volpi et al. 2009; Lam et al. 2014; Dicker et al. 2014). According to the importance of subject and attention to personalized therapies in tissue engineering or regenerative medicine, the present chapter is aimed to assess the association of bioactive materials and biomaterials with cellular functions (such as cell signaling, proliferation, differentiation, migration, etc.) and their role in increasing effectiveness and safety of therapeutic methods and regenerating damaged tissues.

13.2 Proliferation and Differentiation Potential of Stem Cells

Stem cells are known as cells with extensive biological functions to differentiate into cell types, as well as to enable the growth, healing, or replacement of cells (Fig. 13.2) (Zakrzewski et al. 2019). These cells are found both in embryos (uncontrolled range of differentiation) and adult or somatic cells. Although adult stem cells possess a restricted range of differentiation, however, it is possible to reprogram these cells back to their pluripotent states or improve their capacity of self-renewal (Wu et al. 2018). Therefore, to be useful in therapy and be converted into target cells, as well as avoid teratoma formation, the creation of similar body conditions or imitation of the extracellular microenvironments plays an important role in controlling cell behavior (cellular interaction, proliferation, and differentiation). In this regard, understanding signaling pathways and identifying effective substances to improve these pathways are also the main factors in successful regenerative medicine. In the following, the different materials that can bring us closer to this aim, as well as their role in inducting cell signaling and improving proliferation and differentiation, have been mentioned.

13.3 Herbal Bioactive Compounds to Induce Differentiation of Stem Cells

The tremendous differentiation of stem cells is one of the desirable properties of these cells. However, this property can act as a double-edged sword, which means an increase in the risk of tumorigenicity. Thus, therapeutic strategies which led to the complete and irreversible differentiation of stem cells play an important role in controlling differentiation.

One of these strategies includes the use of biomaterials or bioactive materials to design scaffolds, substrates, or tissue substitutes (Izadyari Aghmiuni and Heidari Keshel 2022; Aghmiuni et al. 2022). This approach can provide suitable physicomechanical properties to differentiate stem cells and lead to a sustained delivery and local of bioactive molecules for tissue regeneration. Indeed, bioactive compounds or biomaterials play the main role in providing the cell differentiation microenvironment, so that they can control cellular functions and interactions via creating an ideal substrate. In this matter, Izadyari Aghmiuni's research has shown that the phytochemical materials in herbs possess high therapeutic and regenerative effects (Izadyari Aghmiuni et al. 2020). However, their physical forms like gel, mucilage, and the extract or essential oil from different parts of the plants cannot be applied on damaged tissue (such as the bone, skin, cartilage, blood vessels, nerve, cornea, tendon, etc.) directly. This research team showed that quince seed mucilage is one of the bioactive components that can turn an engineered scaffold into a smart biological substrate. They stated that this herbal bioactive material is classified into the group of polysaccharide and composed of xylose(glucuronoxylan) and glucuronic acid; however, swell in water due to the existence of hydrophilic groups (i.e., amino, hydroxyl, amide, and carboxylic acid), unlike some of the polysaccharides such as chitosan which suspended in the acidic solvent. The results of this study indicated that quince seed mucilage-based hybrid scaffolds can create a better porous network compared to chitosan-based scaffolds. Moreover, these scaffolds support the proliferation of dermal fibroblasts and provide high water absorption capacity for the scaffold. Based on this research, the combination of quince seed mucilage and PEG leads to an increase in the transduction of mechanical force-induced signals and improves the biological signals to induce stem cell differentiation (with the same differentiation patterns) into targeted cells such as keratinocytes.

Another study in this regard is *Aloe vera*-based scaffolds that possess wide applications in biomedical and pharmaceutical sciences. The popularity of these herbal polysaccharides is inducted by their bioactive components (i.e., anthraquinones, anthrones, carbohydrates, proteins, vitamins, etc.) that lead to antimicrobial, anti-inflammatory, and antiviral properties, antioxidant effects, as well as immunomodulatory, neo-angiogenesis, and tissue repair effects (Darzi et al. 2021; Rahman et al. 2017). In this regard, Kallyanashis Paul et al. reported that *Aloe vera*-based hydrogel could alleviate maternal birth injuries (Paul et al. 2021). Based on this study, *Aloe vera*-alginate hydrogels can be an immediate treatment by delivering stem cells to regenerate damaged tissue. The results of this research team showed that local injection of hydrogel with/without stem cells can be significantly effective in repairing birth injuries, so that it leads to the improvement of elastin content and smooth muscle. They stated that such hydrogels can be a suitable therapeutic strategy for preventing pelvic organ prolapse or healing birth injuries.

The electrospun mesh of *Aloe vera* is another sample that can act as a transdermal therapeutic factor. Suganya et al. showed that fibers of poly(caprolactone) containing *Aloe vera* powder (10 wt%) possess better hydrophilic properties along with higher tensile strength (6.28 MPa) and elastic modulus similar to skin tissue, as well as more desirable cell proliferation, secretion of collagen, and expression of F-actin compared to polycaprolactone-collagen fibers (Suganya et al. 2014a). Based on the reports of Tahmasebi et al., nanofibrous scaffolds of poly(3-hydroxybutyrate-co-3-hydroxy valerate) blended with *Aloe vera* gel can be also useful for applications of bone tissue engineering (Tahmasebi et al. 2020). The results of this study indicated the higher biocompatibility of the mentioned nanofibrous scaffold when blended with *Aloe vera*. Moreover, alkaline phosphatase activity, amount of mineralization, and expression of bone-related genes or proteins increase compared to *Aloe vera*-free nanofibrous scaffold. They stated that *Aloe vera* gel possesses the osteoinductive potential and can be used for acceleration of bone regeneration as a bio-implant.

According to the reports, the use of *Aloe vera*, along with other herbal bioactive compounds, can accelerate wound healing processes. In this matter can mention to hybrid nanofibrous scaffolds based on *Aloe vera* and curcumin were designed by Ezhilarasu et al. to increase their synergistic effects on the proliferation of fibroblasts and antimicrobial activities (Ezhilarasu et al. 2019).

The study of Oryan et al. demonstrates that adipose-derived stem cell-loaded *Aloe vera* hydrogels can be also effective in the burn wound models (Oryan et al. 2019). The results of this study showed that hydrogels containing *Aloe vera* and stem cells can significantly increase the rate of burn wound healing, lead to improvement of angiogenesis and reepithelialization, as well as decrease TGF-β_1[6] and interleukin-1β levels. *Urtica dioica* L. (nettle) is one of the other herbs in this matter that can be led to osteogenic differentiation when blended with biomaterials such as silk fibroin. Zadegan et al. indicated that silk fibroin-nettle nano-fiber possesses better water uptake and cellular attachment as well as higher cellular proliferation than that of silk fibroin nano-fiber. Moreover, nettle-based nano-fibers can express both early and late markers of osteoblast differentiation (Zadegan et al. 2019).

Urtica dioica, along with ZnO nanoparticles, can also provide the synergy effects for the increase of antibacterial activities in the electrospun scaffolds. Ghiyasi et al. stated that incorporation of *Urtica dioica* and ZnO nanoparticles to poly (caprolactone) scaffolds can be led to an increase in the tensile strength up to 2.54 MPa and improvement of water uptake ability and promotion of fibroblast L929 cell proliferation (Ghiyasi et al. 2021). Another study relates to the induction of

[6]Transforming growth factor-β_1.

periosteal cell differentiation and proliferation by *Urtica dioica* extract. According to the report of Bing et al., the extract of this herb led to an increase in alkaline phosphatase and calcium nodule levels, as well as induction of osteoblast differentiation and proliferation on the polyclonal lactone scaffolds (Xu and Liu 2018).

Moreover, the study of Hajiali et al. showed that nanofibrous dressings based on sodium alginate-lavender essential oil possess high efficacy for the treatment of UVB-induced skin burns with antibacterial activities (Hajiali et al. 2016). These dressings are also able to decrease and control the inflammatory responses induced in the skin fibroblasts due to UVB exposure. Some reports showed animal fats could also increase the differentiation and proliferation of stem cells. In this regard, the design of emu oil-based electrospun nanofibrous by Pilehvar-Soltanahmadi et al. illustrates that emu oil not only promotes differentiation and proliferation of adipose tissue-derived stem cells into keratinocytes but also can lead to increase in cell adherence and cytoprotection (Pilehvar-Soltanahmadi et al. 2017). Such that, it can be a suitable candidate to fabricate wound dressings and/or bio-engineered substitutes containing stem cells for skin tissue repair.

There are many studies on this matter and some examples are listed in Table 13.1.

Specifically, studies on herbal bioactive compounds have illustrated that herb-derived materials not only can increase the proliferation and differentiation of adult stem cells but also inhibit the proliferation of cancer cells (Saud et al. 2019; Olatunbosun et al. 2012; Kornicka et al. 2017; Potu et al. 2009; Gao et al. 2013). However, studies have shown that the proliferation and/or differentiation ability of stem cells is influenced by the doses of the stimulant compounds. It means that a specific dose of herbal extracts can promote stem cell proliferation and induce its differentiation into the targeted cell. In this regard, we can refer to Zhang's study on the effects of naringin on human bone mesenchymal stem cell proliferation and osteogenic differentiation (Zhang et al. 2009a). Zhang et al. showed that 1–100 μg/mL concentrations of extract from this citrus lead to an increase in the proliferation of human BM-MSCs and their osteogenic differentiation, while, 200 μg/mL concentration can decrease the growth of these cells. Similar research in this field is related to the study of Yu et al. They showed that naringin in 50 μg/mL concentration can activate the Notch signaling pathway and lead to stimulation of osteogenic differentiation of BMSCs.[7] However, higher concentrations than 100 μg/mL can suppress the proliferation rate of these cells (Yu et al. 2016).

Zhang et al. also reported that naringin can induce osteogenic activity and differentiation of canine bone marrow stromal cells (Zhang et al. 2021). To this end, they added different concentrations of this bioactive to the mentioned cells. Their results showed that the 10^{-6} mol/L concentration can promote cell proliferation and be led to an increase in cellular proliferation and calcium nodules, as well as induction of bone marrow stromal cell differentiation into osteoblasts.

[7] Bone marrow stromal cells.

Table 13.1 Herbs-based scaffolds, substrates, or gels for improvement of cellular interactions and functions

Herb	Scaffolds, substrates, or gels	Function	Reference
Curcumin	Collagen-alginate scaffold containing curcumin-loaded chitosan nanoparticles	Acceleration in wound closure; the complete epithelialization along with the formation of thick granulation tissue; the decrease of inflammation; regeneration of diabetic wounds	Venkata et al. (2016)
	Poly(ε-caprolactone) nano-fibers loaded by curcumin	More than 70% viability on human foreskin fibroblast cells; anti-inflammatory property; high antioxidant activity; the decrease of oxidative stress and IL-6 release; wound healing capability by increasing rate of wound closure in a diabetic mice model induced by streptozotocin	Merrell et al. (2009)
	Electrospun poly (ε-caprolactone)-poly (ethylene glycol)-poly (ε-caprolactone) fibrous mat containing curcumin	Improvement of antioxidant properties and low cytotoxicity; the increase of wound closure; suitable candidate for wound dressings	Fu et al. (2014)
	The thermosensitive hydrogel containing encapsulated curcumin	Biodegradable gel with suitable tissue adhesiveness for release of curcumin along with antioxidant and anti-inflammatory properties; accelerating agent in wound closure; improvement of collagen content, granulation, and wound maturity; reduction of superoxide dismutase; the higher thicker epidermis and tensile strength in regenerated skin; suitable for	Gong et al. (2013)

(continued)

Table 13.1 (continued)

Herb	Scaffolds, substrates, or gels	Function	Reference
		wound healing as a wound dressing	
	Nano-graphene oxide reinforced fish scale collagen-based 3D scaffold containing curcumin	No toxicity against NIH 3T3 fibroblast cells; suitable antimicrobial properties against the growth of gram-positive and gram-negative bacteria; acceleration of wound healing; suitable for skin tissue engineering applications	Mitra et al. (2015)
	Chitosan-gelatin composite sponge containing curcumin	Sponge possessed a high capacity for water absorption, antibacterial activity, and drug release; the increase of wound closure on rabbit wound model; suitable for wound healing applications	Nguyen et al. (2013)
	Collagen matrix incorporated by curcumin	Increase of cellular proliferation and wound healing; efficient free radical inhibiting and decrease of oxidative stress; suitable for supporting dermal wound healing	Gopinath et al. (2004)
	Curcumin-treated pluronic F-127 gel (25%) and curcumin (0.3%) in pluronic gel	Increase of the wound contraction; reduction in the expressions of inflammatory cytokines or enzymes such as TNF-α[a], IL-1β[b], and MMP-9[c]; increase in the level of anti-inflammatory cytokines such as IL-10 and antioxidant enzymes (like dismutase, catalase superoxide, and glutathione peroxidase); promotion of fibroblast proliferation and	Kant et al. (2014)

(continued)

Table 13.1 (continued)

Herb	Scaffolds, substrates, or gels	Function	Reference
		improvement of collagen deposition; covering wound via creating a layer of thick epithelial and diabetic wound healing	
Aloe vera	Electrospun polycaprolactone mat incorporated with *Aloe vera*	The control of scaffold degradation rate; improvement of wettability behavior and increase of hydrophilicity of the substrate; promotion of fibroblast proliferation	Agnes Mary and Giri Dev (2015)
	Electrospun silk fibroin-hydroxyapatite scaffold containing *Aloe vera*	The creation of biomimicry similar to the natural bone constitution in scaffold; enhancement of osteocalcin expression and osteogenesis; increase in proliferation of human mesenchymal stem cells, mineral deposition, and osteogenic differentiation	Suganya et al. (2014b)
	Electrospun polycaprolactone-silk fibroin nanofibrous scaffolds containing *Aloe vera*	The increase of adipose-derived stem cell proliferation on scaffolds; promotion in the expression of osteogenic markers such as osteocalcin and alkaline phosphatase; enhancement of osteogenic differentiation and mineralization; suitable for bone tissue regeneration applications	Shanmugavel et al. (2014)
	Bacterial nano-cellulose-*Aloe vera* composites	Improvement of mechanical properties, suitable for biomedical application, design of engineered scaffolds or substitutes, and cell	Godinho et al. (2016)

(continued)

Table 13.1 (continued)

Herb	Scaffolds, substrates, or gels	Function	Reference
		culture substrate applications	
	Eye drops derived from *Aloe vera* gel	The decrease of the corneal epithelial defect area; no developed hypersensitivity reaction, corneal perforation or descemetocele, and limbal ischemia; control of inflammatory responses	Rezaei Moghadam et al. (2020)
	Nanofibrous poly (L-lactic acid)-collagen scaffold coated with *Aloe vera* gel and chitosan	The improvement of mouse fibroblast (L929) behaviors (adhesion, viability, and proliferation) on the scaffold and physicomechanical properties of the substrate; suitable for skin tissue engineering	Salehi et al. (2016)
	Core-shell electrospun mat containing *Aloe vera* extract	The improvement of mechanical and physicochemical properties in the substrate; the increase of cellular adhesion and growth on the mat; high potential for wound healing	Zahedi et al. (2019)
Azadirachta indica, *Indigofera aspalathoides*, *Myristica andamanica*, and *Memecylon edule*	Electrospun polycaprolactone nanofibrous scaffolds containing herbal extracts	The promotion of human dermal fibroblast proliferation; induction of epidermal differentiation of adipose-derived stem cells (both early and intermediate stages of differentiation), suitable for skin tissue engineering	Jin et al. (2013)
Cissus quadrangularis	Polycaprolactone nanofibrous scaffold containing hydroxyapatite and *Cissus quadrangularis*	The increase in osteogenic activity and bone tissue regeneration, promotion of cell	Suganya et al. (2014c)

(continued)

Table 13.1 (continued)

Herb	Scaffolds, substrates, or gels	Function	Reference
		proliferation, and mineralization	
Xylan (natural polysaccharide in plants)	Polyvinyl alcohol nano-fibers containing xylan	Promotion of fibroblast proliferation on the nanofibrous scaffold; the improvement of mechanical properties and increase in the natural biodegradable rate of the scaffold; enhancement of fibroblast adhesion; better interactions of cell-to-matrix to regenerate skin tissue	Krishnan et al. (2012)
Althea officinalis	Electrospun poly (ε-caprolactone)-gelatin scaffold containing *Althea officinalis*	Expedition of the therapy duration and acceleration of wound healing; the increase of anti-inflammatory and antimicrobial properties of the scaffold; improvement of mechanical and physicochemical properties of the scaffold; promotion of cellular proliferation	Ghaseminezhad et al. (2020)
Arnebia euchroma	The nanofibrous scaffold of polycaprolactone-chitosan-polyethylene oxide containing *Arnebia euchroma*	Scaffolds possess high potential in burn wound healing; control of biodegradation rate of the scaffold; improvement of swelling and mechanical properties of the scaffold; increase of antibacterial activity and human dermal fibroblast cell proliferation; suitable for skin tissue engineering applications	Asghari et al. (2022)
Spinacia oleracea	Alginate-carboxymethyl cellulose scaffold incorporated with	The mechanical stability of the scaffold; improvement of biocompatibility of scaffold; control of	Sharmila et al. (2020)

(continued)

Table 13.1 (continued)

Herb	Scaffolds, substrates, or gels	Function	Reference
	Spinacia oleracea extract	biodegradation rate; promotion of MG-63 human osteosarcoma cell proliferation; suitable for bone tissue engineering applications	
Angiogenin and curcumin	Polyethyleneimine-carboxymethyl chitosan-poly (D, L-lactic-co-glycolic acid)-cellulose nanocrystal electrospun nano-fiber containing angiogenin and curcumin	Nano-fibers possessed excellent biocompatibility, anti-infection, and angiogenesis properties; suitable for skin regeneration	Mo et al. (2017)
Gum tragacanth	Gum tragacanth-poly (ε-caprolactone) electrospun nano-fiber containing curcumin	High antibacterial properties against gram-positive and gram-negative bacteria; reduction of the epithelial gap; fast wound closure along with the formation of granulation tissue, hair follicles, and sweat glands; proliferation of fibroblasts; creation of collagen deposition and the layer of early regenerated epithelial completely; increase of angiogenesis number	Ranjbar-Mohammadi et al. (2016)

[a] Tumor necrosis factor-alpha
[b] Interleukin-1beta
[c] Matrix metalloproteinase-9

Beom Su Kim et al. indicated that the extract of brown algae *Laminaria japonica* (fucoidan) in 0.1–10 μg/mL concentrations can lead to JNK- and ERK[8]-dependent BMP2[9]–Smad 1/5/8 signaling and induces osteoblast differentiation of human mesenchymal stem cells (Kim et al. 2015). They also found that this osteogenesis bioactive can increase ALP activities and accumulation of calcium and leads to the expression of the osteoblast-specific gene.

[8] Extracellular signal-regulated protein.

[9] Bone morphogenetic protein.

Moreover, MaríaSatué et al. reported that some flavonoids lead to the stimulation of MC3T3-E1 cell differentiation into osteoblast and inhibition of osteoclastogenesis in RAW 264.7 cells (Satué et al. 2013). The results of this study illustrated that doses greater than 100 μM of diosmetin, galangin, and chrysin, as well as 500 μM doses of taxifolin, possess a toxic effect on cells. However, quercitrin with safe doses of 200 and 500 μM and taxifolin in safe doses of 100 and 200 μM can induce osteocalcin mRNA and bone sialoprotein expression and lead to higher osteocalcin levels. Fei Li et al. also showed that the echinacoside as phenylethanoid glycosides can promote bioactivities of cell line MC3T3-E1 (i.e., proliferation, differentiation, and mineralization of osteoblastic) (Li et al. 2012). To this end, this bioactive component isolated from Cistanches Herba stems and the amount of the secretion of osteoprotegerin, osteocalcin, and collagen I were evaluated. Based on this study's results, concentrations between 0.01 and 10 nmol/L can significantly promote cell proliferation, osteocalcin levels, and collagen I content can lead to an increase in the mineralization of osteoblastic. They stated that echinacoside possesses a stimulatory effect on the formation of osteoblastic bone and can potentially be effective against osteoporosis.

Likewise, Hui-HuiXiao et al. observed that vanillic acid isolated from *Sambucus williamsii* Hance possesses estrogen-like activity in the rat UMR 106 cells that can be due to MAP[10] kinase (MEK/ERK)-mediated specific estrogen receptor (ER) antagonist signaling pathways (Xiao et al. 2014). The results of this study showed that this phenolic acid stimulates the mentioned cell proliferation and alkaline phosphatase (ALP) activity. Vanillic acid can also increase Runx2,[11] osteocalcin, and the ratio of osteoprotegerin-receptor activator of nuclear factor kB ligand [i.e., OPG-RANKL] mRNA expression.

Kim's study is one of the other samples hereof (Kim et al. 2014). Kim et al. illustrated that kirenol extracted from Herba Siegesbeckia can promote osteoblast differentiation via activation of the BMPs and Wnt/β-catenin signaling pathways in MC3T3-E1 cells. Moreover, this natural diterpenoid compound can lead to the promotion of mineralization and ALP activity, as well as the increase in osteopontin, collagen I, and expression of OPG/RANKL. Puerarin and phytoestrogens isolated from *Pueraria mirifica* are the other bioactive compounds in this field that can lead to an increase in cellular proliferation and the expression of osteoblastic differentiation markers in osteoblast-like UMR106 cells (Tiyasatkulkovit et al. 2012). Such that, Tiyasatkulkovit et al. indicated that puerarin can increase the mRNA expression of ALP, as well as lead to a decrease in the mRNA expression of RANKL and induction of bone gain via increasing osteoblast differentiation in rat osteoblast-like UMR106 cells and suppressing osteoclast functions. They stated that puerarin can induce the differentiation of osteoblast rather than the proliferation of osteoblast in the estrogen receptor-dependent manner; hence, it can be the appropriate option to prevent and treat postmenopausal osteoporosis. Choi et al. stated that honokiol

[10]Mitogen-activated protein.

[11]Runt-related transcription factor 2.

isolated from the bark of *Magnolia officinalis* not only can stimulate osteoblast MC3T3-E1 cell function (i.e. increase in cell growth, improvement of ATP activity, promotion of collagen synthesis and glutathione content, and release of osteoprotegerin in the cells) but also decrease or inhibit the production of bone-resorbing mediators. They suggested that this phenolic compound can be effective in natural therapies for osteoporosis (Choi 2011).

There are many studies on stem cell differentiation into the progenitor cells of endothelial or vascular, cardiomyocyte, neuronal, osteogenic, neurogenic, etc. that their stimulation factor is bioactive components extracted from herbs. Indeed, these bioactive components can not only promote cellular proliferation and differentiation but also decrease the time required for tissue regeneration. Table 13.2 illustrates some effective herbs in this regard.

13.4 Biomaterials and Their Characteristic to Design Ideal Substrates and Induce Cell Differentiation

The ideal substrates possess different characteristics which originate from their materials and lead to the targeted differentiation of stem cells (Discher et al. 2005).

Regardless of the tissue type, some of the key characteristics of bio- or active material-based substrates when designing and determining the suitability of the scaffolds for regenerative medicine applications include biocompatibility, mechanical properties, biodegradability, the architecture of the scaffold, and manufacturing technology. One of the requirements of a scaffold to regenerate tissue is biocompatible properties; according to Williams, "The biocompatibility of a scaffold or matrix for a tissue engineering product refers to the ability to perform as a substrate that will support the appropriate cellular activity" (Williams 2008). Indeed, the scaffold structure must possess suitable stimuli or recognizable stimuli by cells to colonize scaffolds and regenerate (Parisi et al. 2018). In other words, the cells must first adhere to the scaffold, proliferate, differentiate, and then migrate onto the surface of the scaffold. Finally, differentiated cells should proliferate on the scaffold to create a new matrix (Fig. 13.3).

In this regard, natural polymers are considered one of the main candidates that not only can increase cell biological activities on the scaffold (i.e., adhesion or attachment, spreading, proliferation, differentiation, and migration) but also imitate the structure and function of the native extracellular matrix (ECM) and lead to controlled induction of cell functions (Izadyari Aghmiuni et al. 2020; Izadyari Aghmiuni and Heidari Keshel 2022). Indeed, biomaterials are agents that can provide an ideal microenvironment with suitable physicomechanical and physicochemical properties to mimic the structure of the ECM network and functions of targeted cells and tissues (Izadyari Aghmiuni et al. 2021; Izadyari Aghmiuni and Heidari Keshel 2022). Collagen is one of these biopolymers that is found in most soft and hard tissues such as the blood vessels, nerves, cornea, skin, tendon, cartilage, bone, etc. Given that, this biomaterial leads to maintaining the structural and biological integrity of the native ECM network and can help in the physiological functions of the

Table 13.2 The effect of herbal bioactive compounds on the cellular functions

Bioactive compound name/herb name	Classification	Source of cells/cell lines	Function	Effective dose	Reference
Ugonin K/*Helminthostachys zeylanica* L.	Flavonoid	MC3T3-E1 mouse osteoblast-like cells	Induction of differentiation and mineralization of cells, the increases in alkaline phosphatase (ALP) activity, expressions of osteocalcin and bone sialoprotein, as well as formation of bone nodule via estrogen receptor antagonist ICI 182,780 that led to upregulation in the expressions of osterix and Runx2, increase of the phosphorylated level of c-Src, promotion of estrogen receptor-α protein level, stimulation of osteoblastic differentiation of rat primary osteoblasts and promotion of osteoblastic differentiation of human MG-63 osteosarcoma cells via estrogen receptor-dependent activation in the pathways of nonclassical signaling mediated by c-Src phosphorylation	10 μM	Lee et al. (2012a)

Icariin, baohuoside-I, epimedin-B, and sagittatoside-A/Herba epimedii	Flavonoid	Rat osteoblastic-like UMR-106 cells	The stimulation of the proliferation rate of the cells; the increase of ALP activity and osteoprotegerin (OPG)/RANKL mRNA expression in cells Sagittatoside A also possesses selectively activated ERE[a]-luciferase activity by ERα[b] Moreover, all four flavonoids can stimulate estrogen receptor-dependent osteoblastic function	10^{-8} to 10^{-10} M	Xiao et al. (2014)
Neobavaisoflavone/*Psoralea corylifolia* L.	Isoflavone	Murine calvaria-derived osteoblastic cell line MC3T3-E1	Increase of ALP activity, enhancement in the expression of bone-specific matrix proteins like collagen I, bone sialoprotein, and osteocalcin; promotion of osteogenesis in cells; formation of bone nodules; regulation of Runx2[c] and osterix expression	10–15 μM	Don et al. (2012)
Poncirin/*Poncirus trifoliata*	Flavanone glycoside	C3H10T1/2 mesenchymal stem cells and primary bone marrow mesenchymal stem cells	The prevention of adipocyte differentiation in C3H10T1/2 cells via inhibiting accumulation of cytoplasm lipid droplet and	1 and 10 μM	Yoon et al. (2011)

(continued)

Table 13.2 (continued)

Bioactive compound name/herb name	Classification	Source of cells/cell lines	Function	Effective dose	Reference
			PPAR-γ[d] and protein-β (C/EBP-β) mRNA expression via binding CCAAT enhancer; promotion of osteoblast differentiation in mesenchymal stem cells; the increase in expression of the key osteogenic transcription factors, ALP, Runx2, and osteocalcin; the increase in the formation of the mineral nodule in primary bone marrow mesenchymal stem cells		
Genistein/lupin, fava beans, and soya beans	Phytoestrogen	MC3T3-E1 cells and primary rat calvarial osteoblasts	Increase of the phosphorylation of kinase 1/2 regulated with the extracellular signal; the increase of MAPK[e] activated with p38 mitogen and c-Jun N-terminal kinase 1/2; the increase in translocation of c-Jun and NF-κB; induction of ERα mRNA expression and increase of its level and content of nuclear ERα protein in rat calvarial	10 μM	Liao et al. (2014)

			osteoblast; induction of expressions of the genes related to osteoblast differentiation; increase in osteoblast mineralization		
Quercetin and rutin/fruit and vegetables	Flavonoid	Mouse bone marrow-derived mesenchymal stem cells	The increase of ATP activity and mineralization; promotion of the prominent marker expression (such as RunX2, osteopontin, osteoprotegerin, osterix, and osteocalcin) to differentiate osteoblast; effective in controlling osteodegenerative disorders and effective in bone development	50 μM for quercetin and 25 μM for rutin	Srivastava et al. (2013)
Puerarin/*Pueraria lobata* (Willd.)	Isoflavone glycoside	Human osteoblastic MG-63 cell	Stimulation of osteoblast proliferation and differentiation via both ERα and ERβ, as well as activation of estrogen receptor-dependent MEK/ERK[f] and PI3K[g]/Akt signaling (partially); opposition with apoptosis induced from cisplatin in osteoblastic MG-63 cells	0.1 μM	Wang et al. (2013)
Sweroside/Fructus corni	Iridoid glycoside	Human MG-63 cells and rat osteoblasts	Increase in osteogenic effect on the proliferation and differentiation of rat	10^{-5} and 10^{-6} g/mL	Sun et al. (2013)

(continued)

Table 13.2 (continued)

Bioactive compound name/herb name	Classification	Source of cells/cell lines	Function	Effective dose	Reference
			osteoblasts and MG-63 cells; the increase of ALP activity and osteocalcin level; attenuation and inhibition of apoptosis		
Salidroside/*Rhodiola rosea* L.	Phenylpropanoid glycosides	Murine pluripotent mesenchymal cell line C3H10T1/2 and osteoblastic cell line MC3T3-E1	The increase of the ALP activity and the mRNA expressions of osteoblast marker genes; enhancement of C3H10T1/2 cell mineralization; the increase of the mRNA level in genes involved in regulating BMP[h] signaling pathways (such as BMP2, BMP6, and BMP7) and regulation of bone metabolism; enhancement of the Smad1/5/8 and ERK1/2 phosphorylation; promotion of the osteoblast proliferation and differentiation	0.5–10 μM	Chen et al. (2013a)
Ginsenoside-Rb_2/ginseng	20(S)-protopanaxadiol glycoside	Osteoblastic MC3T3-E1 cells	Promotion of MC3T3-E1 cell proliferation; improvement of ALP expression; the increase of calcium mineralization and osteogenic gene expression (like	0.1–10 μM	Huang et al. (2014)

			osteopontin, Col1a1, and osteocalcin) against H_2O_2-induced oxidative damage; reduction of the RANKL and IL-6[i] expression level; inhibition of the ROS[j] production induced by H_2O_2; improvement of a micro-architecture of trabecular bone; the increase of BMD[k] of the L4[l] and distal femur		
Gastrodin/*Gastrodia elata*	Phenolic glycoside	Human bone marrow mesenchymal stem cells (hBMMSCs) and macrophage cell line (RAW264.7 cells)	Promotion of the hBMMSC proliferation; improvement of osteogenic markers; reduction of lipid generation and inhibition of mRNA expression of adipogenic genes; reduction in the number of osteoclasts, TRAP, and CTX-1 activity (serum bone degradation markers) and osteoclast-specific gene expression; suppression of the production of ROS in both studied cells; prevention and treatment of osteoporosis	0.1, 1, 10 μM	Huang et al. (2015a)

(continued)

Table 13.2 (continued)

Bioactive compound name/herb name	Classification	Source of cells/cell lines	Function	Effective dose	Reference
Ophiopogonin D/*Ophiopogon japonicus* Radix	Steroidal glycoside	Mouse pre-osteoblast cell line MC3T3-E1 subclone 4 cells and a macrophage cell line RAW264.7 cells	Promotion of MC3T3-E1 cell proliferation; improvement of the osteogenic markers; reduction of TRAP and CTX-1 activities (as the serum bone degradation markers) and mRNA expression in the osteoclastic gene of RAW264.7 cell; suppression of ROS produced in both studied cells; anti-osteoporosis effect through ROS reduction by the FoxO3a-β-catenin signaling pathways	1, 10, and 100 μM	Huang et al. (2015b)
2,3,5,4′-Tetrahydroxystilbene-2-*O*-β-D-glucoside/*Polygonum multiflorum* Thunb	Stilbene glycoside oligomers	Osteoblastic MC3T3-E1 cells	Increase of cellular survival, calcium deposition; promotion of ALP activities and the expression of ALP, collagen I, and osteocalcin induced by H_2O_2; reduction of the RANKL production, IL-6, intracellular reactive oxygen species, and MDA[m] of studied cells induced by H_2O_2;	1–10 μM	Zhang et al. (2012)

			protective effect on the cells through inhibiting the release of bone-resorbing mediator and cell oxidative damage		
Imperatorin and bergapten (coumarin derivatives)/ *Cnidiummonnieri* and *Angelica pubescens*	Furanocoumarin	Primary osteoblastic cell	Increase of ALP activity and collagen I synthesis; promotion of bone nodule formation; the increase of BMP-2 expression; enhancement of the phosphorylation of SMAD[n] 1/5/8, p38, and ERK[o] and consequently increase of bone formation	0.3–10 μM	Tang et al. (2008)
Albiflorin/Paeoniae radix [root of *Paeonia lactiflora* Pallas (Paeoniaceae)]	Monoterpene	Osteoblast-like MC3T3-E1 cells	Reduction of ROS/RNS level; the decrease of apoptosis and lactate dehydrogenase release; enhancement of mineralization; reduction of cardiolipin peroxidation and mitochondrial cytochrome c loss induced by antimycin A; improvement of mitochondrial function by preserving cytochrome c and cardiolipin	0.01–1 μM	Suh et al. (2013)
Salvianolic acid B/*Salvia miltiorrhiza*	Phenolic acid	Human mesenchymal stem cells	Increase of ALP activity; stimulation of Runx2, osteopontin, and osterix expressions in studied cells; promotion of	5 μM	Xu et al. (2014)

(continued)

Table 13.2 (continued)

Bioactive compound name/herb name	Classification	Source of cells/cell lines	Function	Effective dose	Reference
			mineralization and osteogenesis of cells via activating the ERK[p] signaling pathways		
Psoralen (coumarin-like derivative)/*Psoralea corylifolia* fruit (Buguzhi)	Derivative of umbelliferone (structurally)/linear furanocoumarins	Primary mouse calvarial osteoblasts	Stimulation of osteoblast differentiation via activation of BMP signaling; increase in expressions of osteocalcin, collagen I, and bone sialoprotein; improvement of ALP activity; upregulation and increase of the Bmp2 and Bmp4 expression, phospho-Smad1/5/8 level, and BMP (12xSBE-OC-Luc) activity	1–100 μM	Tang et al. (2011)
Resveratrol (3,4′,5-trihydroxystilbene)/mulberries, peanuts, and the berry skins of most grape cultivars	Polyphenolic phytoestrogen	Human bone marrow-derived mesenchymal stem cell	The increase of cell growth/proliferation; stimulation of osteoblastic differentiation and osteoblastic maturation inducted by ALP activity; promotion of calcium deposition; the expression of osterix, osteocalcin, and RUNX2/CBFA1; rapid activation of MAPK[q] and ERK1/2 signaling at 10–6 M concentration	10^{-8} to 10^{-5} M	Dai et al. (2007)

Costunolide/*Saussurea lappa* (Compositae)	Sesquiterpene lactone	Osteoblastic MC3T3-E1 cells	Increase of MC3T3-E1 cell proliferation induced by PI3K, ER, ERK, PKC[r], and mitochondrial ATP-sensitive K^+ channels; promotion of collagen content, ALP activity, and mineralization in cells	10 μg/mL	Lee and Choi (2011a)
Apocynin (4-hydroxy-3-methoxy-acetophenone)/isolated from a variety of plant sources	Methoxy-substituted catechol	Osteoblastic MC3T3-E1 cells	Increase of collagen content, ALP activity, and mineralization; promotion of cell survival/proliferation; the increase of calcium deposition and release of osteoprotegerin; the decrease of ROS production and TNF-α, IL-6, and RANKL expressions	0.01–1 μM	Lee and Choi (2011b)
Osthole (7-methoxy-8-isopentenoxycoumarin)/*Cnidium monnieri* and *Angelica pubescens*	Coumarin-like derivative	Primary mouse calvarial osteoblasts	Increase in the formation of new bone via local injection of the bioactive compound; improvement of bone micro-architecture parameter, and biomechanical and histomorphometric properties; increase in upregulation of collagen I, osteocalcin, and BSP[s]; activating Wnt/β-catenin	10–100 μM	Tang et al. (2010)

(continued)

Table 13.2 (continued)

Bioactive compound name/herb name	Classification	Source of cells/cell lines	Function	Effective dose	Reference
			signaling; increase in the expression of Bmp2; stimulation of osteoblast differentiation via β-catenin-BMP signaling		
Kobophenol A/*Caragana sinica* (Buc'hoz) Rhed	Tetrameric stilbene	Human osteoblast-like MG-63 cells	Inhibition of downregulation of Bcl-XL and Bcl-2; inhibition of JNK and c-Jun phosphorylation induced by SNP; the decrease of NF-κB and AP-1 activities induced by SNP; reduction of death inducted by nitric oxide; protective effects against osteoblast apoptosis via regulating NF-κB, JNK, and AP-1 signaling pathway	25 and 50 μg/mL	Lee et al. (2011)
Emodin/roots/bark of herbs such as the genus *Rhamnus*	Anthraquinone	Mouse osteoblastic MC3T3-E1 subclone 4 cells	The induction of ALP expression; induction of the PI3K, Akt, and MAP kinase activation; regulation of BMP-2 expression; the increase of mineralization; acceleration of osteoblast differentiation; effective on bone health	Up to 5 μM	Lee et al. (2008)

Acerogenin A/*Acer nikoense* Maxim	Diarylheptanoid	MC3T3-E1 osteoblastic cells and RD-C6 osteoblastic cells	Increase of ALP activity and mRNA expression related to osteoblast differentiation (such as primary osteoblasts, osteocalcin, Runx2, and osterix); increase of mRNA expression level of Bmp-2, Bmp-4, and Bmp-7; stimulation of the studied cell proliferation and differentiation via BMP activity	30 μM	Kihara et al. (2011)
Icariine/*Epimedium pubescens*	Flavonoid	Human marrow mesenchymal stem cells	Promotion of ALP activity and the calcified nodule level of osteoblast; enhancement of BMP-2 mRNA synthesis; increase in the human osteoblast proliferation and differentiation	20 μg/mL	Leishi (2007)
Soy peptide/*Glycine max* var.	Isoflavones	Human adipose tissue-derived mesenchymal stem cells and cord blood-derived mesenchymal stem cells as adult stem cells	Increase of TGF-β_1[t], VEGF[u], and IL-6 in the studied cells; activation of the mTOR signaling pathways and ERK; stimulation of cellular proliferation rate	Preparation of 1% medium in DMEM (serum-free)	Lee et al. (2012b)
Extracts of the leaves containing linalool, eugenol, methyl cinnamate, methyl chavicol, steroidal glycoside, ferulate, methyl eugenol, and	Natural terpene alcohol, allylbenzene, methyl ester, phenylpropene, steroidal glycoside, hydroxycinnamic acid,	Human dental pulp-derived mesenchymal stem cells and bone marrow-derived mesenchymal stem cells	Reduction of level of osteonectin; the decrease of doubling time in both cells; induction of cell proliferation; acceleration	10 μg/mL concentration	Mendi et al. (2017)

(continued)

Table 13.2 (continued)

Bioactive compound name/herb name	Classification	Source of cells/cell lines	Function	Effective dose	Reference
triterpenoids/*Ocimum basilicum*	isopentenyl pyrophosphate oligomers		of the osteogenic differentiation; *O. basilicum* acts as a smart osteoinductive agent		
Hydro-alcoholic guaraná extract/*Paullinia cupana*	Rich in catechin and caffeine	Human lipoaspirate-derived senescent adipocyte stem cells	The increase of cellular proliferation; reduction in lipoperoxidation, ROS level, DNA damage, and oxidative stress marker; the increase of catalase activity/gene expression; reversion of the initial senescence processes in adipocyte-mesenchymal cells and consequently tissue regeneration	5 mg/mL	Machado et al. (2015)
Ethyl acetate extract of root containing glabridin and glabrene/*Glycyrrhiza glabra*	Rich in phytoestrogen	Human bone marrow mesenchymal stem cells	The increase in cellular proliferation; the effect on the osteogenesis via ALP activity, stimulation of estrogen receptor-mediated mechanism, and the bone-specific gene expression (like BMP-2, Runx2, and osteocalcin); promotion of calcium deposition	10–50 μg/mL	Azizsoltani et al. (2018)
Seed extract containing L-DOPA[v]/*Mucuna gigantea*	Amino acid	Mesenchymal stem cells	Stimulation in mRNA expression of nestin,	50 μg/mL	Kongros (2012)

			MAP-2[w], and β-III tubulin messenger ribonucleic acid (as neural markers and neural protein)		
β-Mercaptoethanol/*Salvia miltiorrhiza*	Biological antioxidant	Human umbilical cord Wharton's jelly-derived mesenchymal stem cells	Induction of expression of neural protein markers; induction of nestin and pleiotrophin genes, β-tubulin III, NF[x], and GFAP[y]; stimulation of the differentiation into nerve-like cells (neuronal and glial cells); increase in proliferation of the neural cells	2–4 mmol/L	Leishi (2005)
Herb-derived small molecules such as resveratrol and stilbene (as WNT/β-catenin stimulators)/*Trifolium pratense* (leaves), *Salix aegyptiaca* (leaves and bark), and *Morus alba* (root)	Phytoestrogen and phenolic compounds, isoflavonoid derivate	Adipose tissue-derived mesenchymal stem cells	Stimulation of signaling pathways during stem cell differentiation of myocardial cells; improvement of stem cell differentiation into cardiomyocytes	1–3 μM	Azadian et al. (2019)
Quercetin and its metabolites (i.e., quercetin 3-glucuronide and 3′-*O*-methyl quercetin)/fruit and vegetables	Flavonoid	Mitotic rat embryonic cardiomyoblast-derived H9c2 cells	The differentiation of cardiomyocyte-like phenotype; increase in the activation of PKB, JNK, ERK1/2, and p38 MAPK induced by H_2O_2; protective of	30–100 μM	Daubney et al. (2015)

(continued)

Table 13.2 (continued)

Bioactive compound name/ herb name	Classification	Source of cells/cell lines	Function	Effective dose	Reference
			cardioprotection against cell death and cardiotoxicity induced by oxidative stress		

[a] Estrogen response element
[b] Estrogen receptor alpha
[c] Runt-related transcription factor 2
[d] Peroxisome proliferator-activating receptor-γ
[e] Mitogen-activated protein kinase
[f] Extracellular signal-regulated protein
[g] Phosphatidylinositol 3-kinase
[h] Bone morphogenetic proteins
[i] Interleukin-6
[j] Reactive oxygen species
[k] Bone mineral density
[l] Fourth lumbar vertebrae
[m] Malondialdehyde
[n] Transcription factor activated by TGF-β
[o] Extracellular signal-regulated protein
[p] Extracellular signal-regulated kinase signaling
[q] p38 mitogen-activated protein kinase
[r] Protein kinase C
[s] Bone sialoprotein
[t] Transforming growth factor-beta1
[u] Vascular endothelial growth factor
[v] L-3,4-Dihydroxyphenylalanine
[w] Microtubule associated protein-2
[x] Neurofilament
[y] Glial fibrillary acidic protein

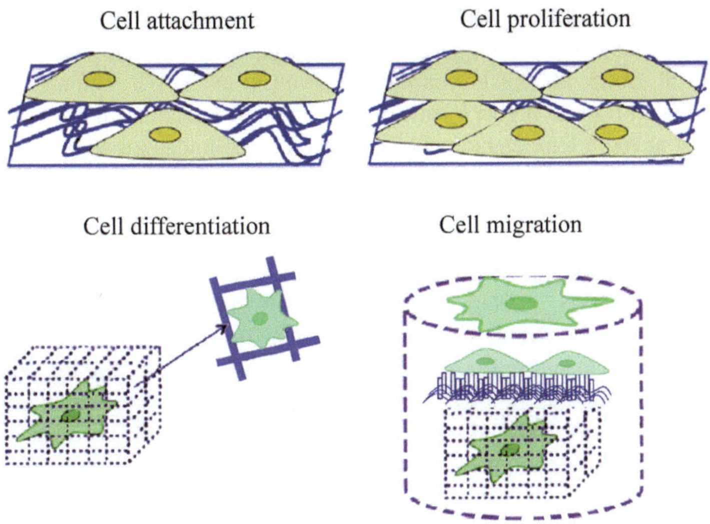

Fig. 13.3 The cellular interaction on the scaffold

engineered scaffolds as dynamic material (Barnes et al. 2007; Sell et al. 2010). Moreover, low antigenicity, low cytotoxic and inflammatory response, and biodegradability are other properties of this biomaterial (Farach-Carson et al. 2007; Sell et al. 2007, 2009). The study of Shih et al. indicates that electrospun collagen fibrous mat can promote growth, proliferation, adhesion, motility, and osteogenic differentiation of human MSCs[12] (Mano et al. 2007).

Many reports have shown that the blends of collagen with other biomaterials can increase the proliferation rate of cells and be used for regeneration templates in different tissues (Aghmiuni et al. 2020; Izadyari Aghmiuni et al. 2020). In this matter, Zhong et al. showed that collagen glycosaminoglycan (GAG)-based blended scaffolds can be led to an increase in the proliferation of dermal fibroblasts (FB) (Zhong et al. 2005). Moreover, the substrates based on chondroitin sulfate-collagen, collagen-nanohydroxyapatite, as well as aggrecan (chondroitin sulfate, dermatan sulfate, keratin sulfate)-collagen can be applied in the regenerating skin, bone, and cartilage, respectively (Choi et al. 2004; Thomas et al. 2007). Based on Chen's report, scaffolds of collagen/chitosan-PEO (as wound dressing) possessed good in vitro biocompatibility and led to the promotion of 3T3 FBs (Chen et al. 2008). In another study, collagen-chitosan scaffolds also illustrated an increase in the proliferation of smooth muscle cells (SMCs) and endothelial cells (ECs) (Chen et al. 2010). Indeed, mixing collagen with other biopolymers can lead to a decrease in the dimensions of the fibers or an increase in the porosity and tensile strength of the scaffold and consequently promotion of cellular attachment or adhesion on the scaffold surface (Barnes et al. 2007).

[12]Bone marrow-derived mesenchymal stem cells.

Gelatin also is known as an attractive biopolymer for tissue engineering applications due to its biological and biomechanical similarities to collagen (Zhang et al. 2005, 2009b; Heydarkhan-Hagvall et al. 2008). Although fibers of this biopolymer possess higher tensile moduli compared to collagen fibers, however, its gelation at room temperature and dissolution as colloidal-sol at 37 °C or > are major drawbacks of this biomaterial which is often solved via combining with other natural or synthetic polymers (Sell et al. 2010). Given the gelatin's similarity to collagen, electrospun gelatin-based blended scaffolds have been developed to apply in regenerative medicine (Zhang et al. 2005, 2009b; Heydarkhan-Hagvall et al. 2008; Li et al. 2005, 2006a; Gauthaman et al. 2009; Gupta et al. 2009a, b; Songchotikunpan et al. 2008; Song et al. 2008; Gui-Bo et al. 2010). The electrospun gelatin-polyurethane and poly (ε-caprolactone)-gelatin scaffolds are the samples of these substrates that can be used in wound healing and skin regeneration (Chong et al. 2007; Kim et al. 2009). Based on the reports, such scaffolds can lead to better migration of fibroblasts and decrease therapeutic costs. Poly(ε-caprolactone)-gelatin scaffolds can also act as a positive factor for supporting neurite outgrowth and nerve differentiation, proliferation, and nerve regeneration (Ghasemi-Mobarakeh et al. 2008). Likewise, conductive nanofibrous scaffold of polyaniline-poly(ε--caprolactone)/gelatin can lead to nerve stem cell attachment and proliferation, as well as neurite outgrowth via electrical stimulation of scaffold (Ghasemi-Mobarakeh et al. 2009). Moreover, an electrospun composite based on the synthetic polypeptide-gelatin can lead to the induction of calcium phosphate mineralization and be used as dental biomaterials or a biocompatible substitute for the regeneration of the hard tissue ECM (Ohkawa et al. 2009). Nanohydroxyapatite-gelatin nanofibrous scaffolds are one of the other substrates that are capable of imitation of natural ECM of bone tissue along with effective mineralization, the proliferation of osteoblasts, and successful bone regeneration (Francis et al. 2010).

Indeed, electrospun gelatin or gelatin-based scaffolds can be also used in the regeneration of cardiac tissue. According to the study of Li et al., the combination of gelatin with conductive polymers such as polyaniline leads to the improvement of cardiac myoblast attachment or adhesion, spreading, and migration, as well as an increase in cellular proliferation on the scaffold, modulus, and tensile strength (Li et al. 2006b). Studies have shown that blended gelatin with natural and synthetic polymers can significantly improve mechanical properties and facilitate the excellent proliferation of cardiac myoblasts. In this field, Li et al. indicated that gelatin composites composed of poly(lactic-co-glycolic acid) and α-elastin can be effective in soft tissue engineering applications, such as heart, blood vessels, and lung (Li et al. 2006c). Such composites possess low average diameters; however, upon hydration of the scaffold, the average diameter increases due to the swelling of fibers, without disintegrating the scaffold.

Elastin is one of the main biomaterials in this regard that naturally is found in the wall of many tissues such as vessel walls; hence, the use of fibrous structures of this biopolymer in tissue engineering applications can be favorable (Daamen et al. 2007; Bailey et al. 2003; Deborde et al. 2016). Based on the studies, elastin or elastin-like polypeptides can be effective in the regeneration of cartilage, heart valves, and skin tissues (Neuenschwander and Hoerstrup 2004; Nettles et al. 2010; McHale et al. 2005; Betre et al. 2002). Wang et al. showed that collagen and elastin as the main components of ECM in tissues can improve the proliferation of valve interstitial cells, when used as the engineered 2D or 3D substrates (Wang et al. 2018b). Moreover, such substrates constitute effective tools to engineer 3D tissues and increase endothelial-mesenchymal transition. Chen et al. also illustrated that bilayer collagen-elastin scaffolds play an important role in mimicking the mechanical and biological activities of heart valves (Chen et al. 2013b). Chitosan/γ-poly(glutamic acid) scaffolds modified by albumin, elastin, and poly-L-lysine are another sample of elastin-based scaffold application in cartilage tissue engineering (Kuo et al. 2017). The study of Kuo et al. shows that such a scaffold can provide an effective approach to inhibition of chondrocyte apoptosis and lead to promotion in the growth of chondrocytes, secretion of ECM, and improvement of cartilaginous tissue regeneration. Moreover, the existence of single bond $-NH_3^+$ and single bond $-COO^-$ in the structure of this scaffold plays an important role in its porous morphology and interconnected network, as well as the mechanical properties of the scaffold.

Chitosan is one of the important polysaccharides in the design of tissue-engineered scaffolds due to the structural similarity to the glycosaminoglycans (GAGs) of ECM of tissues (Huang et al. 2015c). The functional groups of hydroxyl (–OH) and amine (–NH) in this biopolymer can link to other materials to make a new scaffold (He et al. 2011; Du et al. 2016). In this regard, Izadyari Aghmiuni et al. stated that polyethylene glycol-chitosan-poly(ε-caprolactone) copolymers containing collagen can be considered as base bio-composites for the design of dermal substrates. Such substrates can imitate dermis properties and lead to the induction of keratinocyte differentiation from stem cells (Izadyari Aghmiuni et al. 2021; Aghmiuni et al. 2020). The design of injectable chitosan-based hydrogels is another example in this field that can be used in cartilage tissue regeneration (Jin et al. 2009). Jin et al. showed that the hydrogels of chitosan-glycolic acid/phloretic acid that are designed via enzymatic crosslinking can increase chondrocyte proliferation and then be degraded by a hydrolytic enzyme such as lysozyme. Likewise, chitosan hydrogels modified by cartilaginous ECM are another type of hydrogel that can be used in cartilage tissue engineering applications (Choi et al. 2014). Mori et al. also stated that spongelike dressings containing chitosan and sericin can treat chronic skin ulcers. Such a substrate possesses suitable mechanical resistance and high hydration, along with cellular proliferation and antioxidant activities on human fibroblast cells (Mori et al. 2016).

Hyaluronic acid (HA) also is a linear polysaccharide and is classified in the group of glycosaminoglycans of native ECMs and can maintain the structural integrity of tissue ECMs when used in the structure of engineered scaffolds (Izadyari Aghmiuni et al. 2021; Kogan et al. 2006; Volpi et al. 2009; Petrey and de la Motte 2014). This biomaterial possesses various biological activities (like modulation of immune cell function, synthesis of proteoglycan, reduction of pro-inflammatory cytokine activities, etc.) and can act as the signaling factor for improvement of cell-to-cell and cell-to-scaffold interactions and acceleration of tissue regeneration (Izadyari Aghmiuni et al. 2021; Lam et al. 2014; Li et al. 2018). Based on the study of Shirzaei

Sani et al., HA/elastin-like polypeptide-based hydrogels can act as antimicrobial substrates for tissue engineering applications (Shirzaei Sani et al. 2018). This study illustrated that these hydrogels possess high adhesive strength to attach to the tissue and can support cellular spreading, growth, and proliferation. Izadyari et al. also demonstrated that control of HA content can provide a microporous environment along with higher tensile strength and modulus (Izadyari Aghmiuni et al. 2021). According to the reports, given that hyaluronic acid plays an important role in osmotic balance and regulation of tissue hydration, HA-based substrates can significantly mimic the physical and mechanical properties of ECM (Chan and Tayama 2002).

Silk fibroin (SF) is another biopolymer that can play an important role in cell attachment and spreading when mixed with collagen (Yeo et al. 2008). This biomaterial also possesses high elasticity, resistance to failure (under compression), minimal inflammatory response, slow degradation rate, as well as high biocompatibility, toughness, and strength (Pérez-Rigueiro et al. 2001; Liu et al. 2007; Zhang et al. 2009c). Notably, the existence of hydrophobic domains in the protein's random coil network of this biomaterial has led to the formation of a β-sheet structure, which is responsible for the elastic properties and tensile strength of SF (Aghmiuni et al. 2020; Zhang et al. 2009c). This property plays a crucial role in the scaffold structure and improves the module and stress strain of scaffolds (Zhang et al. 2010). Moreover, the β-sheet structures of SF can affect the biodegradation rate of the scaffold, so that the biodegradation rate of the scaffold matches with the rate of tissue repair (Izadyari Aghmiuni et al. 2021; Sell et al. 2010; Alessandrino et al. 2008; Bayraktar et al. 2005; Liu et al. 2008; Silva et al. 2008). Hence, silk-based scaffolds can be designed for the ligament, bone, and vascular skin applications (Izadyari Aghmiuni et al. 2020, 2021).

Generally, in our opinion, the features of these biomaterials and other bio- or active materials can provide an exciting opportunity to design engineered hybrid/composite substrates with a high ability to differentiate stem cells and regenerate tissue.

13.5 Future Prospective and Conclusion

Nowadays, the advancement in the field of phyto-sciences, technologies, and the development of studies on herbal extracts have revealed their excellent repairing properties and role in regenerative medicine. Based on the studies, herbal bioactive compounds or active ingredients can play a crucial role in the process of tissue regeneration. Such that, the cellular and molecular researches on herbal extracts indicate that herbs possess positive effects on the promotion, proliferation, and differentiation of types of stem cells. It can be due to the existence of flavonoids; coumarins; glycosides; terpenoids such as mono-terpenoids, sesquiterpenoids, and diterpenoids; as well as anthraquinones, phenolic acids, diarylheptanoids, phenols, and tetrameric stilbene. Hence, if protocols or methods can be created via the use of herbal extracts to proliferate and differentiate stem cells into targeted cells in damaged tissues, it will offer hope to cure many incurable diseases. In this field, knowledge of herbs and the effects of their extracts and therapeutic doses can play a crucial role in the determination of suitable therapeutic methods, such as the design of substrates, encapsulation, gel, etc. Moreover, the studies of tissue engineering demonstrate that a combination of modern and traditional medicine (i.e., use of biomaterials along with herbal bioactive compounds) can provide new developments in the fields of regenerative tissue techniques and stem cell differentiation into targeted tissue cells. This not only can decrease the therapeutic economic burden and healthcare problems but also develop new drugs or methods with easier availability and least/no side effect.

References

Aghmiuni AI, Baei MS, Keshel SH, Khiyavi AA (2020) Design of novel 3D-scaffold as a potential material to induct epidermal-dermal keratinocytes of human-adipose-derived stem cells and promote fibroblast cells proliferation for skin regeneration. Fibers Polym 21:33–44. https://doi.org/10.1007/s12221-020-9402-1

Aghmiuni AI, Ghadi A, Azmoun E, Kalantari N, Mohammadi I, Kordmahaleh HH (2022) Electrospun polymeric substrates for tissue engineering: viewpoints on fabrication, application, and challenges. In: Electrospinning—material technology of the future. InTech, pp 1–26

Agnes Mary S, Giri Dev VR (2015) Electrospun herbal nanofibrous wound dressings for skin tissue engineering. J Textile Inst 106:886–895. https://doi.org/10.1080/00405000.2014.951247

Ahmed LA (2013) Stem cells and cardiac repair: alternative and multifactorial approaches. J Regen Med Tissue Eng 2:10–7243

Alessandrino A, Marelli B, Arosio C, Fare S, Tanzi MC, Freddi G (2008) Electrospun silk fibroin mats for tissue engineering. Eng Life Sci 8:219–225

Asghari F, Rabiei Faradonbeh D, Malekshahi ZV, Nekounam H, Ghaemi B, Yousefpoor Y, Ghanbari H, Faridi-Majidi R (2022) Hybrid PCL/chitosan-PEO nanofibrous scaffolds incorporated with A. euchroma extract for skin tissue engineering application. Carbohydr Polym 278:118926. https://doi.org/10.1016/j.carbpol.2021.118926

Azadian Z, Shafiei M, Hosseini S, Kazemi J, Alipour A, Shahsavarani H (2019) Efficient in vitro differentiation of adipose tissue-derived mesenchymal stem cells into the cardiomyocyte using plant-derived natural compounds. Proc Singap Natl Acad Sci 13:47–63. https://doi.org/10.1142/s2591722619400064

Azizsoltani A, Piri K, Behzad S, Soleimani M, Nekouei M, Mahmoudi Z, Kazemi A (2018) Ethyl acetate extract of licorice root (Glycyrrhiza glabra) enhances proliferation and osteogenic differentiation of human bone marrow mesenchymal stem cells. Iran J Pharm Res 17:1057–1067

Bailey MT, Pillarisetti S, Xiao H, Vyavahare NR (2003) Role of elastin in pathologic calcification of xenograft heart valves. J Biomed Mater Res A 66:93–102. https://doi.org/10.1002/jbm.a.10543

Barnes CP, Sell SA, Boland ED, Simpson DG, Bowlin GL (2007) Nanofiber technology: designing the next generation of tissue engineering scaffolds. Adv Drug Deliv Rev 59:1413–1433

Bayraktar O, Malay Ö, Özgarip Y, Batıgün A (2005) Silk fibroin as a novel coating material for controlled release of theophylline. Eur J Pharm Biopharm 60:373–381

Betre H, Setton LA, Meyer DE, Chilkoti A (2002) Characterization of a genetically engineered elastin-like polypeptide for cartilaginous tissue repair. Biomacromolecules 3:910–916. https://doi.org/10.1021/bm0255037

Chagastelles PC, Nardi NB (2011) Biology of stem cells: an overview. Kidney Int Suppl 1:63–67. https://doi.org/10.1038/kisup.2011.15

Chan RW, Tayama N (2002) Biomechanical effects of hydration in vocal fold tissues. Otolaryngol Head Neck Surg 126:528–537

Chaudhari AA, Vig K, Baganizi DR, Sahu R, Dixit S, Dennis V, Singh SR, Pillai SR (2016) Future prospects for scaffolding methods and biomaterials in skin tissue engineering: a review. Int J Mol Sci 17:1974. https://doi.org/10.3390/ijms17121974

Chen J-P, Chang G-Y, Chen J-K (2008) Electrospun collagen/chitosan nanofibrous membrane as wound dressing. Colloids Surf A Physicochem Eng Asp 313:183–188

Chen ZG, Wang PW, Wei B, Mo XM, Cui FZ (2010) Electrospun collagen–chitosan nanofiber: a biomimetic extracellular matrix for endothelial cell and smooth muscle cell. Acta Biomater 6: 372–382

Chen JJ, Zhang NF, Mao GX, He XB, Zhan YC, Deng HB, Song DQ, Li DD, Li ZR, Si SY, Qiu Q, Wang Z (2013a) Salidroside stimulates osteoblast differentiation through BMP signaling pathway. Food Chem Toxicol 62:499–505. https://doi.org/10.1016/j.fct.2013.09.019

Chen Q, Bruyneel A, Carr C, Czernuszka J (2013b) Bio-mechanical properties of novel bi-layer collagen-elastin scaffolds for heart valve tissue engineering. Proc Eng 59:247–254. https://doi.org/10.1016/j.proeng.2013.05.118

Choi EM (2011) Honokiol isolated from Magnolia officinalis stimulates osteoblast function and inhibits the release of bone-resorbing mediators. Int Immunopharmacol 11:1541–1545. https://doi.org/10.1016/j.intimp.2011.05.011

Choi JS, Lee SW, Jeong L, Bae S-H, Min BC, Youk JH, Park WH (2004) Effect of organosoluble salts on the nanofibrous structure of electrospun poly (3-hydroxybutyrate-Co-3-hydroxyvalerate). Int J Biol Macromol 34:249–256

Choi B, Kim S, Lin B, Wu BM, Lee M (2014) Cartilaginous extracellular matrix-modified chitosan hydrogels for cartilage tissue engineering. ACS Appl Mater Interfaces 6:20110–20121. https://doi.org/10.1021/am505723k

Chong EJ, Phan TT, Lim IJ, Zhang YZ, Bay BH, Ramakrishna S, Lim CT (2007) Evaluation of electrospun PCL/gelatin nanofibrous scaffold for wound healing and layered dermal reconstitution. Acta Biomater 3:321–330. https://doi.org/10.1016/j.actbio.2007.01.002

Daamen WF, Veerkamp JH, Van Hest JCM, Van Kuppevelt TH (2007) Elastin as a biomaterial for tissue engineering. Biomaterials 28:4378–4398

Dai Z, Li Y, Quarles LD, Song T, Pan W, Zhou H, Xiao Z (2007) Resveratrol enhances proliferation and osteoblastic differentiation in human mesenchymal stem cells via ER-dependent ERK1/2 activation. Phytomedicine 14:806–814. https://doi.org/10.1016/j.phymed.2007.04.003

Darzi S, Paul K, Leitan S, Werkmeister JA, Mukherjee S (2021) Immunobiology and application of aloe vera-based scaffolds in tissue engineering. Int J Mol Sci 22:1–19. https://doi.org/10.3390/ijms22041708

Daubney J, Bonner PL, Hargreaves AJ, Dickenson JM (2015) Cardioprotective and cardiotoxic effects of quercetin and two of its in vivo metabolites on differentiated H9c2 cardiomyocytes. Basic Clin Pharmacol Toxicol 116:96–109. https://doi.org/10.1111/bcpt.12319

Deborde C, Simionescu DT, Wright C, Liao J, Sierad LN, Simionescu A (2016) Stabilized collagen and elastin-based scaffolds for mitral valve tissue engineering. Tissue Eng A 22:1241–1251. https://doi.org/10.1089/ten.tea.2016.0032

Dey SK, Banerjee D, Chattapadhyay S, Karmakar KB (2010) Antimicrobial activities of some medicinal plants of West Bengal. Int J Pharm Bio Sci 1:1–10

Dicker KT, Gurski LA, Pradhan-Bhatt S, Witt RL, Farach-Carson MC, Jia X (2014) Hyaluronan: a simple polysaccharide with diverse biological functions. Acta Biomater 10:1558–1570. https://doi.org/10.1016/j.actbio.2013.12.019

Discher DE, Janmey P, Wang Y (2005) Tissue cells feel and respond to the stiffness of their substrate. Science 310:1139–1143

Don MJ, Lin LC, Chiou WF (2012) Neobavaisoflavone stimulates osteogenesis via p38-mediated up-regulation of transcription factors and osteoid genes expression in MC3T3-E1 cells. Phytomedicine 19:551–561. https://doi.org/10.1016/j.phymed.2012.01.006

Du T, Chen Z, Li H, Tang X, Li Z, Guan J, Liu C, Du Z, Wu J (2016) Modification of collagen-chitosan matrix by the natural crosslinker alginate dialdehyde. Int J Biol Macromol 82:580–588. https://doi.org/10.1016/j.ijbiomac.2015.10.039

Ezhilarasu H, Ramalingam R, Dhand C, Lakshminarayanan R, Sadiq A, Gandhimathi C, Ramakrishna S, Bay BH, Venugopal JR, Srinivasan DK (2019) Biocompatible aloe vera and tetracycline hydrochloride loaded hybrid nanofibrous scaffolds for skin tissue engineering. Int J Mol Sci 20:5174. https://doi.org/10.3390/ijms20205174

Farach-Carson MC, Wagner RC, Kiick KL (2007) Extracellular matrix: structure, function, and applications to tissue engineering. In: Tissue engineering. CRC Press, pp 35–56

Francis L, Venugopal J, Prabhakaran MP, Thavasi V, Marsano E, Ramakrishna S (2010) Simultaneous electrospin-electrosprayed biocomposite nanofibrous scaffolds for bone tissue regeneration. Acta Biomater 6:4100–4109. https://doi.org/10.1016/j.actbio.2010.05.001

Fu SZ, Meng XH, Fan J, Yang LL, Wen QL, Ye SJ, Lin S, Wang BQ, Chen LL, Wu JB, Chen Y, Fan JM, Li Z (2014) Acceleration of dermal wound healing by using electrospun curcumin-loaded poly(ε-caprolactone)-poly(ethylene glycol)-poly(ε-caprolactone) fibrous mats. J Biomed Mater Res B Appl Biomater 102:533–542. https://doi.org/10.1002/jbm.b.33032

Gao L-N, An Y, Lei M, Li B, Yang H, Lu H, Chen F-M, Jin Y (2013) The effect of the coumarin-like derivative osthole on the osteogenic properties of human periodontal ligament and jaw bone marrow mesenchymal stem cell sheets. Biomaterials 34:9937–9951

Gauthaman K, Venugopal JR, Yee FC, Peh GSL, Ramakrishna S, Bongso A (2009) Nanofibrous substrates support colony formation and maintain stemness of human embryonic stem cells. J Cell Mol Med 13:3475–3484

Ghasemi-Mobarakeh L, Prabhakaran MP, Morshed M, Nasr-Esfahani M-H, Ramakrishna S (2008) Electrospun poly(ε-caprolactone)/gelatin nanofibrous scaffolds for nerve tissue engineering. Biomaterials 29:4532–4539. https://doi.org/10.1016/j.biomaterials.2008.08.007

Ghasemi-Mobarakeh L, Prabhakaran MP, Morshed M, Nasr-Esfahani MH, Ramakrishna S (2009) Electrical stimulation of nerve cells using conductive nanofibrous scaffolds for nerve tissue engineering. Tissue Eng A 15:3605–3619. https://doi.org/10.1089/ten.tea.2008.0689

Ghaseminezhad K, Zare M, Lashkarara S, Yousefzadeh M, Aghazadeh Mohandesi J (2020) Fabrication of Althea officinalis loaded electrospun nanofibrous scaffold for potential application of skin tissue engineering. J Appl Polym Sci 137:48587. https://doi.org/10.1002/app.48587

Ghiyasi Y, Salahi E, Esfahani H (2021) Synergy effect of Urtica dioica and ZnO NPs on microstructure, antibacterial activity and cytotoxicity of electrospun PCL scaffold for wound dressing application. Mater Today Commun 26:102163. https://doi.org/10.1016/j.mtcomm.2021.102163

Godinho JF, Berti FV, Müller D, Rambo CR, Porto LM (2016) Incorporation of aloe vera extracts into nanocellulose during biosynthesis. Cellulose 23:545–555. https://doi.org/10.1007/s10570-015-0844-3

Gong CY, Wu QJ, Wang YJ, Zhang DD, Luo F, Zhao X, Wei YQ, Qian ZY (2013) A biodegradable hydrogel system containing curcumin encapsulated in micelles for cutaneous wound healing. Biomaterials 34:6377–6387. https://doi.org/10.1016/j.biomaterials.2013.05.005

Gopinath D, Ahmed MR, Gomathi K, Chitra K, Sehgal PK, Jayakumar R (2004) Dermal wound healing processes with curcumin incorporated collagen films. Biomaterials 25:1911–1917. https://doi.org/10.1016/S0142-9612(03)00625-2

Gui-Bo Y, You-Zhu Z, Shu-Dong W, De-Bing S, Zhi-Hui D, Wei-Guo F (2010) Study of the electrospun PLA/silk fibroin-gelatin composite nanofibrous scaffold for tissue engineering. J Biomed Mater Res A 93:158–163

Gupta D, Venugopal J, Prabhakaran MP, Dev VRG, Low S, Choon AT, Ramakrishna S (2009a) Aligned and random nanofibrous substrate for the in vitro culture of schwann cells for neural tissue engineering. Acta Biomater 5:2560–2569

Gupta D, Venugopal J, Mitra S, Dev VRG, Ramakrishna S (2009b) Nanostructured biocomposite substrates by electrospinning and electrospraying for the mineralization of osteoblasts. Biomaterials 30:2085–2094

Hajiali H, Summa M, Russo D, Armirotti A, Brunetti V, Bertorelli R, Athanassiou A, Mele E (2016) Alginate–lavender nanofibers with antibacterial and anti-inflammatory activity to effectively promote burn healing. J Mater Chem B 4:1686–1695

He L, Mu C, Shi J, Zhang Q, Shi B, Lin W (2011) Modification of collagen with a natural cross-linker, procyanidin. Int J Biol Macromol 48:354–359. https://doi.org/10.1016/j.ijbiomac.2010.12.012

Heydarkhan-Hagvall S, Schenke-Layland K, Dhanasopon AP, Rofail F, Smith H, Wu BM, Shemin R, Beygui RE, MacLellan WR (2008) Three-dimensional electrospun ECM-based hybrid scaffolds for cardiovascular tissue engineering. Biomaterials 29:2907–2914

Huang Q, Gao B, Jie Q, Wei BY, Fan J, Zhang HY, Zhang JK, Li XJ, Shi J, Luo ZJ, Yang L, Liu J (2014) Ginsenoside-Rb2 displays anti-osteoporosis effects through reducing oxidative damage and bone-resorbing cytokines during osteogenesis. Bone 66:306–314. https://doi.org/10.1016/j.bone.2014.06.010

Huang Q, Shi J, Gao B, Zhang HY, Fan J, Li XJ, Fan JZ, Han YH, Zhang JK, Yang L, Luo ZJ, Liu J (2015a) Gastrodin: an ancient Chinese herbal medicine as a source for anti-osteoporosis agents via reducing reactive oxygen species. Bone 73:132–144. https://doi.org/10.1016/j.bone.2014.12.059

Huang Q, Gao B, Wang L, Zhang HY, Li XJ, Shi J, Wang Z, Zhang JK, Yang L, Luo ZJ, Liu J (2015b) Ophiopogonin D: a new herbal agent against osteoporosis. Bone 74:18–28. https://doi.org/10.1016/j.bone.2015.01.002

Huang R, Li W, Lv X, Lei Z, Bian Y, Deng H, Wang H, Li J, Li X (2015c) Biomimetic LBL structured nanofibrous matrices assembled by chitosan/collagen for promoting wound healing. Biomaterials 53:58–75. https://doi.org/10.1016/j.biomaterials.2015.02.076

Izadyari Aghmiuni A, Heidari Keshel S (2022) The role of the extracellular matrix (ECM) and ECM-like polymeric substrates in health and disease. In: Berhardt LV (ed) Advances in medicine and biology. NOVA Medicine and Health, pp 145–175

Izadyari Aghmiuni A, Heidari Keshel S, Sefat F, Akbarzadeh Khiyavi A (2020) Quince seed mucilage-based scaffold as a smart biological substrate to mimic mechanobiological behavior of skin and promote fibroblasts proliferation and H-ASCs differentiation into keratinocytes. Int J Biol Macromol 142:668–679. https://doi.org/10.1016/j.ijbiomac.2019.10.008

Izadyari Aghmiuni A, Heidari Keshel S, Sefat F, AkbarzadehKhiyavi A (2021) Fabrication of 3D hybrid scaffold by combination technique of electrospinning-like and freeze-drying to create mechanotransduction signals and mimic extracellular matrix function of skin. Mater Sci Eng C 120:111752. https://doi.org/10.1016/j.msec.2020.111752

Jin R, Moreira Teixeira LS, Dijkstra PJ, Karperien M, van Blitterswijk CA, Zhong ZY, Feijen J (2009) Injectable chitosan-based hydrogels for cartilage tissue engineering. Biomaterials 30:2544–2551. https://doi.org/10.1016/j.biomaterials.2009.01.020

Jin G, Prabhakaran MP, Kai D, Annamalai SK, Arunachalam KD, Ramakrishna S (2013) Tissue engineered plant extracts as nanofibrous wound dressing. Biomaterials 34:724–734. https://doi.org/10.1016/j.biomaterials.2012.10.026

Kant V, Gopal A, Pathak NN, Kumar P, Tandan SK, Kumar D (2014) Antioxidant and anti-inflammatory potential of curcumin accelerated the cutaneous wound healing in streptozotocin-induced diabetic rats. Int Immunopharmacol 20:322–330

Kihara T, Ichikawa S, Yonezawa T, Lee JW, Akihisa T, Woo JT, Michi Y, Amagasa T, Yamaguchi A (2011) Acerogenin A, a natural compound isolated from Acer nikoense Maxim, stimulates osteoblast differentiation through bone morphogenetic protein action. Biochem Biophys Res Commun 406:211–217. https://doi.org/10.1016/j.bbrc.2011.02.017

Kim SE, Heo DN, Lee JB, Kim JR, Park SH, Jeon SH, Kwon IK (2009) Electrospun gelatin/polyurethane blended nanofibers for wound healing. Biomed Mater 4. https://doi.org/10.1088/1748-6041/4/4/044106

Kim MB, Song Y, Hwang JK (2014) Kirenol stimulates osteoblast differentiation through activation of the BMP and Wnt/β-catenin signaling pathways in MC3T3-E1 cells. Fitoterapia 98:59–65. https://doi.org/10.1016/j.fitote.2014.07.013

Kim BS, Kang HJ, Park JY, Lee J (2015) Fucoidan promotes osteoblast differentiation via JNK- and ERK-dependent BMP2-Smad 1/5/8 signaling in human mesenchymal stem cells. Exp Mol Med 47:1–9. https://doi.org/10.1038/emm.2014.95

Kogan G, Šoltés L, Stern R, Gemeiner P (2006) Hyaluronic acid: a natural biopolymer with a broad range of biomedical and industrial applications. Biotechnol Lett 29:17–25. https://doi.org/10.1007/s10529-006-9219-z

Kongros K (2012) The effects of seed extract of Mucuna gigantea on the expression of neural markers in mesenchymal stem cells. J Med Plants Res 6:1297–1303. https://doi.org/10.5897/jmpr11.1406

Kornicka K, Kocherova I, Marycz K (2017) The effects of chosen plant extracts and compounds on mesenchymal stem cells—a bridge between molecular nutrition and regenerative medicine-concise review. Phytother Res 31:947–958

Krishnan R, Rajeswari R, Venugopal J, Sundarrajan S, Sridhar R, Shayanti M, Ramakrishna S (2012) Polysaccharide nanofibrous scaffolds as a model for in vitro skin tissue regeneration. J Mater Sci Mater Med 23:1511–1519. https://doi.org/10.1007/s10856-012-4630-6

Kuo YC, Ku HF, Rajesh R (2017) Chitosan/γ-poly(glutamic acid) scaffolds with surface-modified albumin, elastin and poly-L-lysine for cartilage tissue engineering. Mater Sci Eng C 78:265–277. https://doi.org/10.1016/j.msec.2017.04.067

Lam J, Truong NF, Segura T (2014) Design of cell-matrix interactions in hyaluronic acid hydrogel scaffolds. Acta Biomater 10:1571–1580. https://doi.org/10.1016/j.actbio.2013.07.025

Lee YS, Choi EM (2011a) Costunolide stimulates the function of osteoblastic MC3T3-E1 cells. Int Immunopharmacol 11:712–718. https://doi.org/10.1016/j.intimp.2011.01.018

Lee YS, Choi EM (2011b) Apocynin stimulates osteoblast differentiation and inhibits bone-resorbing mediators in MC3T3-E1 cells. Cell Immunol 270:224–229. https://doi.org/10.1016/j.cellimm.2011.05.011

Lee SU, Shin HK, Min YK, Kim SH (2008) Emodin accelerates osteoblast differentiation through phosphatidylinositol 3-kinase activation and bone morphogenetic protein-2 gene expression. Int Immunopharmacol 8:741–747. https://doi.org/10.1016/j.intimp.2008.01.027

Lee SR, Kwak JH, Park DS, Pyo S (2011) Protective effect of kobophenol A on nitric oxide-induced cell apoptosis in human osteoblast-like MG-63 cells: involvement of JNK, NF-κB and AP-1 pathways. Int Immunopharmacol 11:1251–1259. https://doi.org/10.1016/j.intimp.2011.04.004

Lee CH, Huang YL, Liao JF, Chiou WF (2012a) Ugonin K-stimulated osteogenesis involves estrogen receptor-dependent activation of non-classical Src signaling pathway and classical pathway. Eur J Pharmacol 676:26–33. https://doi.org/10.1016/j.ejphar.2011.12.001

Lee J, Roh KB, Kim SC, Lee J, Park D (2012b) Soy peptide-induced stem cell proliferation: involvement of ERK and TGF-β1. J Nutr Biochem 23:1341–1351. https://doi.org/10.1016/j.jnutbio.2011.08.003

Leishi L (2005) Human umbilical cord Wharton's jelly-derived mesenchymal stem cells differentiation into nerve-like cells. Chin Med J 118:1987–1993

Leishi L (2007) Icariine stimulates proliferation and differentiation of human osteoblasts by increasing production of bone morphogenetic protein 2. Chin Med J 120:204–210

Li M, Mondrinos MJ, Gandhi MR, Ko FK, Weiss AS, Lelkes PI (2005) Electrospun protein fibers as matrices for tissue engineering. Biomaterials 26:5999–6008

Li J, He A, Zheng J, Han CC (2006a) Gelatin and gelatin-hyaluronic acid nanofibrous membranes produced by electrospinning of their aqueous solutions. Biomacromolecules 7:2243–2247

Li M, Guo Y, Wei Y, MacDiarmid AG, Lelkes PI (2006b) Electrospinning polyaniline-contained gelatin nanofibers for tissue engineering applications. Biomaterials 27:2705–2715. https://doi.org/10.1016/j.biomaterials.2005.11.037

Li M, Mondrinos MJ, Chen X, Gandhi MR, Ko FK, Lelkes PI (2006c) Elastin blends for tissue engineering scaffolds. J Biomed Mater Res A 79:963–973. https://doi.org/10.1002/jbm.a

Li F, Yang Y, Zhu P, Chen W, Qi D, Shi X, Zhang C, Yang Z, Li P (2012) Echinacoside promotes bone regeneration by increasing OPG/RANKL ratio in MC3T3-E1 cells. Fitoterapia 83:1443–1450. https://doi.org/10.1016/j.fitote.2012.08.008

Li Q, Niu Y, Xing P, Wang C (2018) Bioactive polysaccharides from natural resources including Chinese medicinal herbs on tissue repair. Chin Med (United Kingdom) 13:1–11. https://doi.org/10.1186/s13020-018-0166-0

Liao MH, Tai YT, Cherng YG, Liu SH, Chang YA, Lin PI, Chen RM (2014) Genistein induces oestrogen receptor-α gene expression in osteoblasts through the activation of mitogen-activated protein kinases/NF-κB/activator protein-1 and promotes cell mineralisation. Br J Nutr 111:55–63. https://doi.org/10.1017/S0007114513002043

Liu H, Ge Z, Wang Y, Toh SL, Sutthikhum V, Goh JCH (2007) Modification of sericin-free silk fibers for ligament tissue engineering application. J Biomed Mater Res B Appl Biomater 82: 129–138

Liu L, Liu J, Wang M, Min S, Cai Y, Zhu L, Yao J (2008) Preparation and characterization of nano-hydroxyapatite/silk fibroin porous scaffolds. J Biomater Sci Polym Ed 19:325–338

Machado AK, Cadoná FC, Azzolin VF, Dornelles EB, Barbisan F, Ribeiro EE, Mânica-Cattani MF, Duarte MMMF, Saldanha JRP, da Cruz IBM (2015) Guaraná (Paullinia cupana) improves the proliferation and oxidative metabolism of senescent adipocyte stem cells derived from human lipoaspirates. Food Res Int 67:426–433. https://doi.org/10.1016/j.foodres.2014.11.056

Mano JF, Silva GA, Azevedo HS, Malafaya PB, Sousa RA, Silva SS, Boesel LF, Oliveira JM, Santos TC, Marques AP (2007) Natural origin biodegradable systems in tissue engineering and regenerative medicine: present status and some moving trends. J R Soc Interface 4:999–1030

McHale MK, Setton LA, Chilkoti A (2005) Cartilaginous tissue repair. Tissue Eng 11:1768–1779

Mendi AH, Gökçınar Yağcı B, Sarac N, Kızıloğlu M, Yılmaz D, Uçkan D (2017) Effect of Ocimum basilicum on mesenchymal stem cell proliferation and differentiation: does the effect change according to niches? Int J Second Metab 4:1–10. https://doi.org/10.21448/ijsm.356244

Merrell JG, Mclaughlin SW, Tie L, Laurencin CT, Chen AF, Nair LS, Phil M, Exp C, Author PP (2009) Curcumin loaded poly(ε-caprolactone) nanofibers: diabetic wound dressing with antioxidant and anti-inflammatory properties NIH public access author manuscript. Clin Exp Pharmacol Physiol 36:1149–1156. https://doi.org/10.1111/j.1440-1681.2009.05216.x. Curcumin

Mitra T, Manna PJ, Raja STK, Gnanamani A, Kundu PP (2015) Curcumin loaded nano graphene oxide reinforced fish scale collagen-a 3D scaffold biomaterial for wound healing applications. RSC Adv 5:98653–98665. https://doi.org/10.1039/c5ra15726a

Mo Y, Guo R, Zhang Y, Xue W, Cheng B, Zhang Y (2017) Controlled dual delivery of angiogenin and curcumin by electrospun nanofibers for skin regeneration. Tissue Eng A 23:597–608. https://doi.org/10.1089/ten.tea.2016.0268

Mori M, Rossi S, Ferrari F, Bonferoni MC, Sandri G, Chlapanidas T, Torre ML, Caramella C (2016) Sponge-like dressings based on the association of chitosan and sericin for the treatment of chronic skin ulcers. I. Design of experiments–assisted development. J Pharm Sci 105:1180–1187. https://doi.org/10.1016/j.xphs.2015.11.047

Nettles DL, Chilkoti A, Setton LA (2010) Applications of elastin-like polypeptides in tissue engineering. Adv Drug Deliv Rev 62:1479–1485. https://doi.org/10.1016/j.addr.2010.04.002

Neuenschwander S, Hoerstrup SP (2004) Heart valve tissue engineering. Transpl Immunol 12:359–365. https://doi.org/10.1016/j.trim.2003.12.010

Nguyen VC, Nguyen VB, Hsieh MF (2013) Curcumin-loaded chitosan/gelatin composite sponge for wound healing application. Int J Polym Sci 2013. https://doi.org/10.1155/2013/106570

Ohkawa K, Hayashi S, Kameyama N, Yamamoto H, Yamaguchi M, Kimoto S, Kurata S, Shinji H (2009) Synthesis of collagen-like sequential polypeptides containing O-phospho-L-hydroxyproline and preparation of electrospun composite fibers for possible dental application. Macromol Biosci 9:79–92. https://doi.org/10.1002/mabi.200800122

Olatunbosun LO, Caxton-Martin A, Jimoh OR, Biliaminu SA, Ghazal OK (2012) Proliferative effect of aqueous extracts of Parquetina nigrescens on haemopoietic multipotent stem cells in irradiated guinea pigs. J Appl Pharm Sci 2:108–111

Omenetto FG, Kaplan DL (2010) New opportunities for an ancient material. Science (New York, N. Y.) 329:528–531. https://doi.org/10.1126/science.1188936

Oryan A, Alemzadeh E, Mohammadi AA, Moshiri A (2019) Healing potential of injectable aloe vera hydrogel loaded by adipose-derived stem cell in skin tissue-engineering in a rat burn wound model. Cell Tissue Res 377:215–227. https://doi.org/10.1007/s00441-019-03015-9

Parisi L, Toffoli A, Ghiacci G, Macaluso GM (2018) Tailoring the interface of biomaterials to design effective scaffolds. J Funct Biomater 9:1–31. https://doi.org/10.3390/jfb9030050

Paul K, Darzi S, Del Borgo MP, Cousins FL, Werkmeister JA, Gargett CE, Mukherjee S (2021) Vaginal delivery of tissue engineered endometrial mesenchymal stem/stromal cells in an aloe vera-alginate hydrogel alleviates maternal simulated birth injury. Appl Mater Today 22:100890. https://doi.org/10.1016/j.apmt.2020.100890

Pérez-Rigueiro J, Elices M, Llorca J, Viney C (2001) Tensile properties of silkworm silk obtained by forced silking. J Appl Polym Sci 82:1928–1935

Petrey AC, de la Motte CA (2014) Hyaluronan, a crucial regulator of inflammation. Front Immunol 5:1–14. https://doi.org/10.3389/fimmu.2014.00101

Pilehvar-Soltanahmadi Y, Nouri M, Martino MM, Fattahi A, Alizadeh E, Darabi M, Rahmati-Yamchi M, Zarghami N (2017) Cytoprotection, proliferation and epidermal differentiation of adipose tissue-derived stem cells on emu oil based electrospun nanofibrous mat. Exp Cell Res 357:192–201. https://doi.org/10.1016/j.yexcr.2017.05.015

Potu BK, Bhat KMR, Rao MS, Nampurath GK, Chamallamudi MR, Nayak SR, Muttigi MS (2009) Petroleum ether extract of Cissus quadrangularis (Linn.) enhances bone marrow mesenchymal stem cell proliferation and facilitates osteoblastogenesis. Clinics 64:993–998

Rahman S, Carter P, Bhattarai N (2017) Aloe vera for tissue engineering applications. J Funct Biomater 8:6

Ramalingam V, Raja S, Sundaramahalingam S, Rajaram R (2019) Chemical fabrication of graphene oxide nanosheets attenuates biofilm formation of human clinical pathogens. Bioorg Chem 83:326–335

Ranjbar-Mohammadi M, Rabbani S, Bahrami SH, Joghataei MT, Moayer F (2016) Antibacterial performance and in vivo diabetic wound healing of curcumin loaded gum tragacanth/poly (ε-caprolactone) electrospun nanofibers. Mater Sci Eng C 69:1183–1191. https://doi.org/10.1016/j.msec.2016.08.032

Rezaei Moghadam M, Jafarinasab M-R, Yousefi Z, Sanjari Moghaddam A, Memarzadeh H, Rezaei Kanavi M (2020) Aloe vera gel-derived eye drops for alkaline corneal injury in a rabbit model. J Ophthal Vis Res. https://doi.org/10.18502/jovr.v15i1.5932

Salehi M, Farzamfar S, Bastami F, Tajerian R (2016) Fabrication and characterization of electrospun PLLA/collagen nanofibrous scaffold coated with chitosan to sustain release of aloe vera gel for skin tissue engineering. Biomed Eng Appl Basis Commun 28:1650035. https://doi.org/10.4015/S1016237216500356

Satué M, Arriero MDM, Monjo M, Ramis JM (2013) Quercitrin and taxifolin stimulate osteoblast differentiation in MC3T3-E1 cells and inhibit osteoclastogenesis in RAW 264.7 cells. Biochem Pharmacol 86:1476–1486. https://doi.org/10.1016/j.bcp.2013.09.009

Saud B, Malla R, Shrestha K (2019) A review on the effect of plant extract on mesenchymal stem cell proliferation and differentiation. Stem Cells Int 2019. https://doi.org/10.1155/2019/7513404

Sell S, Barnes C, Smith M, McClure M, Madurantakam P, Grant J, Mcmanus M, Bowlin G (2007) Extracellular matrix regenerated: tissue engineering via electrospun biomimetic nanofibers. Polym Int 56:1349–1360

Sell SA, McClure MJ, Garg K, Wolfe PS, Bowlin GL (2009) Electrospinning of collagen/biopolymers for regenerative medicine and cardiovascular tissue engineering. Adv Drug Deliv Rev 61:1007–1019

Sell SA, Wolfe PS, Garg K, McCool JM, Rodriguez IA, Bowlin GL (2010) The use of natural polymers in tissue engineering: a focus on electrospun extracellular matrix analogues. Polymers 2:522–553. https://doi.org/10.3390/polym2040522

Shafei S, Foroughi J, Stevens L, Wong CS, Zabihi O, Naebe M (2017) Electroactive nanostructured scaffold produced by controlled deposition of PPy on electrospun PCL fibres. Res Chem Intermed 43:1235–1251

Shanmugavel S, Reddy VJ, Ramakrishna S, Lakshmi B, Dev VG (2014) Precipitation of hydroxyapatite on electrospun polycaprolactone/aloe vera/silk fibroin nanofibrous scaffolds for bone tissue engineering. J Biomater Appl 29:46–58. https://doi.org/10.1177/0885328213513934

Sharmila G, Muthukumaran C, Kirthika S, Keerthana S, Kumar NM, Jeyanthi J (2020) Fabrication and characterization of Spinacia oleracea extract incorporated alginate/carboxymethyl cellulose microporous scaffold for bone tissue engineering. Int J Biol Macromol 156:430–437. https://doi.org/10.1016/j.ijbiomac.2020.04.059

Shirzaei Sani E, Portillo-Lara R, Spencer A, Yu W, Geilich BM, Noshadi I, Webster TJ, Annabi N (2018) Engineering adhesive and antimicrobial hyaluronic acid/elastin-like polypeptide hybrid hydrogels for tissue engineering applications. ACS Biomater Sci Eng 4:2528–2540. https://doi.org/10.1021/acsbiomaterials.8b00408

Silva SS, Maniglio D, Motta A, Mano JF, Reis RL, Migliaresi C (2008) Genipin-modified silk-fibroin nanometric nets. Macromol Biosci 8:766–774

Song J-H, Kim H-E, Kim H-W (2008) Production of electrospun gelatin nanofiber by water-based co-solvent approach. J Mater Sci Mater Med 19:95–102

Songchotikunpan P, Tattiyakul J, Supaphol P (2008) Extraction and electrospinning of gelatin from fish skin. Int J Biol Macromol 42:247–255

Srivastava S, Bankar R, Roy P (2013) Assessment of the role of flavonoids for inducing osteoblast differentiation in isolated mouse bone marrow derived mesenchymal stem cells. Phytomedicine 20:683–690. https://doi.org/10.1016/j.phymed.2013.03.001

Suganya S, Venugopal J, Agnes Mary S, Ramakrishna S, Lakshmi BS, Giri Dev VR (2014a) Aloe vera incorporated biomimetic nanofibrous scaffold: a regenerative approach for skin tissue engineering. Iran Polym J 23:237–248. https://doi.org/10.1007/s13726-013-0219-2

Suganya S, Venugopal J, Ramakrishna S, Lakshmi BS, Dev VRG (2014b) Aloe vera/silk fibroin/hydroxyapatite incorporated electrospun nanofibrous scaffold for enhanced osteogenesis. J Biomater Tissue Eng 4:9–19. https://doi.org/10.1166/jbt.2014.1139

Suganya S, Venugopal J, Ramakrishna S, Lakshmi BS, Giri Dev VR (2014c) Herbally derived polymeric nanofibrous scaffolds for bone tissue regeneration. J Appl Polym Sci 131. https://doi.org/10.1002/app.39835

Suh KS, Choi EM, Lee YS, Kim YS (2013) Protective effect of albiflorin against oxidative-stress-mediated toxicity in osteoblast-like MC3T3-E1 cells. Fitoterapia 89:33–41. https://doi.org/10.1016/j.fitote.2013.05.016

Sun H, Li L, Zhang A, Zhang N, Lv H, Sun W, Wang X (2013) Protective effects of sweroside on human MG-63 cells and rat osteoblasts. Fitoterapia 84:174–179. https://doi.org/10.1016/j.fitote.2012.11.010

Tahmasebi A, Shapouri Moghadam A, Enderami SE, Islami M, Kaabi M, Saburi E, Daei Farshchi A, Soleimanifar F, Mansouri V (2020) Aloe vera-derived gel-blended PHBV nanofibrous scaffold for bone tissue engineering. ASAIO J 66:966–973. https://doi.org/10.1097/MAT.0000000000001094

Talebian S, Mehrali M, Taebnia N, Pennisi CP, Kadumudi FB, Foroughi J, Hasany M, Nikkhah M, Akbari M, Orive G (2019) Self-healing hydrogels: the next paradigm shift in tissue engineering? Adv Sci 6:1801664

Tang CH, Yang RS, Chien MY, Chen CC, Fu WM (2008) Enhancement of bone morphogenetic protein-2 expression and bone formation by coumarin derivatives via p38 and ERK-dependent pathway in osteoblasts. Eur J Pharmacol 579:40–49. https://doi.org/10.1016/j.ejphar.2007.10.013

Tang DZ, Hou W, Zhou Q, Zhang M, Holz J, Sheu TJ, Li TF, Cheng SD, Shi Q, Harris SE, Chen D, Wang YJ (2010) Osthole stimulates osteoblast differentiation and bone formation by activation of β-catenin-BMP signaling. J Bone Miner Res 25:1234–1245. https://doi.org/10.1002/jbmr.21

Tang DZ, Yang F, Yang Z, Huang J, Shi Q, Chen D, Wang YJ (2011) Psoralen stimulates osteoblast differentiation through activation of BMP signaling. Biochem Biophys Res Commun 405:256–261. https://doi.org/10.1016/j.bbrc.2011.01.021

Thomas V, Dean DR, Jose MV, Mathew B, Chowdhury S, Vohra YK (2007) Nanostructured biocomposite scaffolds based on collagen coelectrospun with nanohydroxyapatite. Biomacromolecules 8:631–637

Tiyasatkulkovit W, Charoenphandhu N, Wongdee K, Thongbunchoo J, Krishnamra N, Malaivijitnond S (2012) Upregulation of osteoblastic differentiation marker mRNA expression in osteoblast-like UMR106 cells by puerarin and phytoestrogens from Pueraria mirifica. Phytomedicine 19:1147–1155. https://doi.org/10.1016/j.phymed.2012.07.010

Venkata V, Reddy S, Kuppusamy G, Talluri SV, Mannemala SS, Kollipara R, Wadhwani AD, Mulukutla S, Rama K, Raju S, Malayandi R (2016) Curcumin loaded chitosan nanoparticles impregnated into collagen-alginate scaffolds for diabetic wound healing. Int J Biol Macromol. https://doi.org/10.1016/j.ijbiomac.2016.05.038

Volpi N, Schiller J, Stern R, Soltes L (2009) Role, metabolism, chemical modifications and applications of hyaluronan. Curr Med Chem 16:1718–1745. https://doi.org/10.2174/092986709788186138

Wang Y, Wang WL, Xie WL, Li LZ, Sun J, Sun WJ, Gong HY (2013) Puerarin stimulates proliferation and differentiation and protects against cell death in human osteoblastic MG-63 cells via ER-dependent MEK/ERK and PI3K/Akt activation. Phytomedicine 20:787–796. https://doi.org/10.1016/j.phymed.2013.03.005

Wang Y, Adokoh CK, Narain R (2018a) Recent development and biomedical applications of self-healing hydrogels. Expert Opin Drug Deliv 15:77–91

Wang X, Ali MS, Lacerda CMR (2018b) A three-dimensional collagen-elastin scaffold for heart valve tissue engineering. Bioengineering 5. https://doi.org/10.3390/bioengineering5030069

Williams DF (2008) On the mechanisms of biocompatibility. Biomaterials 29:2941–2953

Wu C, Chen L, Huang Y, Huang Y, Parolini O, Zhong Q, Tian X, Deng L (2018) Comparison of the proliferation and differentiation potential of human urine-, placenta decidua basalis-, and bone marrow-derived stem cells. Stem Cells Int 2018:1–11. https://doi.org/10.1155/2018/7131532

Xiao HH, Gao QG, Zhang Y, Wong KC, Dai Y, Yao XS, Wong MS (2014) Vanillic acid exerts oestrogen-like activities in osteoblast-like UMR 106 cells through MAP kinase (MEK/ERK)-mediated ER signaling pathway. J Steroid Biochem Mol Biol 144:382–391. https://doi.org/10.1016/j.jsbmb.2014.08.002

Xu B, Liu Y (2018) Urtica extracts induce periosteal cell proliferation and differentiation: tissue-engineered bone construction and ultrastructural changes. Chin J Tissue Eng Res 22:1805–1810. https://doi.org/10.3969/j.issn.2095-4344.0196

Xu D, Xu L, Zhou C, Lee WYW, Wu T, Cui L, Li G (2014) Salvianolic acid B promotes osteogenesis of human mesenchymal stem cells through activating ERK signaling pathway. Int J Biochem Cell Biol 51:1–9. https://doi.org/10.1016/j.biocel.2014.03.005

Yeo I-S, Oh J-E, Jeong L, Lee TS, Lee SJ, Park WH, Min B-M (2008) Collagen-based biomimetic nanofibrous scaffolds: preparation and characterization of collagen/silk fibroin bicomponent nanofibrous structures. Biomacromolecules 9:1106–1116

Yoon HY, Yun SI, Kim BY, Jin Q, Woo ER, Jeong SY, Chung YS (2011) Poncirin promotes osteoblast differentiation but inhibits adipocyte differentiation in mesenchymal stem cells. Eur J Pharmacol 664:54–59. https://doi.org/10.1016/j.ejphar.2011.04.047

Yu GY, Zheng GZ, Chang B, Hu QX, Lin FX, Liu DZ, Wu CC, Du SX, Li XD (2016) Naringin stimulates osteogenic differentiation of rat bone marrow stromal cells via activation of the notch signaling pathway. Stem Cells Int. https://doi.org/10.1155/2016/7130653

Yu Q, Chang J, Wu C (2019) Silicate bioceramics: from soft tissue regeneration to tumor therapy. J Mater Chem B 7:5449–5460

Zadegan S, Nourmohammadi J, Vahidi B, Haghighipour N (2019) An investigation into osteogenic differentiation effects of silk fibroin-nettle (Urtica dioica L.) nanofibers. Int J Biol Macromol 133:795–803. https://doi.org/10.1016/j.ijbiomac.2019.04.165

Zahedi E, Esmaeili A, Eslahi N, Shokrgozar MA, Simchi A (2019) Fabrication and characterization of core-shell electrospun fibrous mats containing medicinal herbs for wound healing and skin tissue engineering. Mar Drugs 17:27. https://doi.org/10.3390/md17010027

Zakrzewski W, Dobrzyński M, Szymonowicz M, Rybak Z (2019) Stem cells: past, present, and future. Stem Cell Res Ther 10:68. https://doi.org/10.1186/s13287-019-1165-5

Zhang Y, Ouyang H, Lim CT, Ramakrishna S, Huang Z-M (2005) Electrospinning of gelatin fibers and gelatin/PCL composite fibrous scaffolds. J Biomed Mater Res 72B:156–165. https://doi.org/10.1002/jbm.b.30128

Zhang P, Dai KR, Yan SG, Yan WQ, Zhang C, Chen DQ, Xu B, Xu ZW (2009a) Effects of naringin on the proliferation and osteogenic differentiation of human bone mesenchymal stem cell. Eur J Pharmacol 607:1–5. https://doi.org/10.1016/j.ejphar.2009.01.035

Zhang S, Huang Y, Yang X, Mei F, Ma Q, Chen G, Ryu S, Deng X (2009b) Gelatin nanofibrous membrane fabricated by electrospinning of aqueous gelatin solution for guided tissue regeneration. J Biomed Mater Res A 90A:671–679. https://doi.org/10.1002/jbm.a.32136

Zhang Q, Yan S, Li M (2009c) Silk fibroin based porous materials. Materials 2:2276–2295

Zhang Y, Yang H, Shao H, Hu X (2010) Antheraea pernyi silk fiber: a potential resource for artificially biospinning spider dragline silk. J Biomed Biotechnol 2010:1–8. https://doi.org/10.1155/2010/683962

Zhang JK, Yang L, Meng GL, Fan J, Chen JZ, He QZ, Chen S, Fan JZ, Luo ZJ, Liu J (2012) Protective effect of tetrahydroxystilbene glucoside against hydrogen peroxide-induced dysfunction and oxidative stress in osteoblastic MC3T3-E1 cells. Eur J Pharmacol 689:31–37. https://doi.org/10.1016/j.ejphar.2012.05.045

Zhang J, Xu DP, Shang J, Liang XJ, Zhang XB (2021) Effects of naringin on the proliferation and osteogenic differentiation of canine bone marrow stromal cells in vitro. J Hard Tissue Biol 30: 73–78. https://doi.org/10.2485/jhtb.30.73

Zhao Z, Vizetto-Duarte C, Moay ZK, Setyawati MI, Rakshit M, Kathawala MH, Ng KW (2020) Composite hydrogels in three-dimensional in vitro models. Front Bioeng Biotechnol 8:611

Zhong S, Teo WE, Zhu X, Beuerman R, Ramakrishna S, Yung LYL (2005) Formation of collagen-glycosaminoglycan blended nanofibrous scaffolds and their biological properties. Biomacromolecules 6:2998–3004

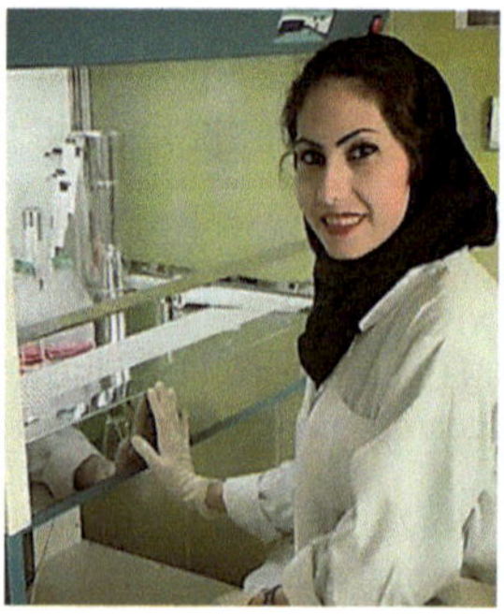

Azadeh Izadyari Aghmiuni has a Ph.D. in Chemical Engineering-Tissue Engineering. Now, she is working on stem cells, biological macromolecules, and natural/synthetic polymers for the construction of engineered scaffolds and smart biological substrates to regenerate hard and soft tissues. In this field, she has one patent entitled “Hybrid nano-substrate containing collagen to regenerate skin wounds, IR101120.” Moreover, she investigates on the application of the herbal active ingredient and targeted drug delivery (TDD) method in tissue engineering and regenerative medicine to treat diseases.

Arezoo Ghadi has a Ph.D. in Chemical Engineering. She is working on the design of polymeric substrates and the differentiation of stem cells into soft tissues. Now, she is Assist. Professor at the Islamic Azad University-Ayatollah Amoli Branch, Department of Chemical Engineering, Amol, Iran.

Mahdieh Asadi has a Ph.D. in Sports Metabolism and Biochemistry. She worked on the effects of swimming, cell, and laser therapy on the expression of autophagocytosis-involved genes. Now, she is working in the specialized center of Seir Salamat.

Azim Akbarzadeh Khiyavi did his Ph.D. in Chemistry from Ankara University, Ankara, Turkey, and Postdoctorate Degree in Clinical Biochemistry, Uksak Ehtesas Hospital, Ankara, Turkey. He worked as Head of Pilot Biotechnology Department of Pasteur Institute of Iran. Now, he is a Professor of the Institute Pasteur of Iran in the Department of Nanobiotechnology, Tehran, Iran; and he is working in the fields of chemistry, cell culture, and targeted drug delivery.

Materials for Gene Delivery Systems

14

S. K. Manohar, M. P. Gowrav, and H. V. Gangadharappa

Abstract

Gene therapy has garnered a lot of interest in the recent past for the treatment of various life-threatening genetic diseases. Gene therapy involves using genes that should be delivered to the infected cells to treat diseases. The materials used for gene delivery play a vital role in successful gene therapy. The common viruses used in gene delivery are adenovirus, adeno-associated virus, vaccinia virus, retrovirus, etc. This chapter will discuss the different viral and nonviral vectors that are used in gene delivery. Furthermore, lipid-, polymer-, and peptide-based gene delivery methods are discussed in this chapter. The physical method of gene delivery uses various techniques such as electroporation, sonoporation, needle injection, hydroboration, and magneto-fiction. In the future, standard DNA and RNA molecular techniques can be used as the principal mode of treatment in biomedical applications.

Keywords

Vectors · Gene therapy · Nucleic acid delivery · Virus

14.1 Introduction

Targeted gene delivery is the way of delivering genes to cells, tissues, and organs through local or systemic blood circulation. This enables the interaction of genes directly on the sites of the intended diseases and produces therapeutic benefits. By

S. K. Manohar · M. P. Gowrav · H. V. Gangadharappa (✉)
Department of Pharmaceutics, JSS College of Pharmacy, JSS Academy of Higher Education and Research, Mysuru, Karnataka, India
e-mail: hvgangadharappa@jssuni.edu.in

F. A. Sheikh et al. (eds.), *Interaction of Nanomaterials With Living Cells*,
https://doi.org/10.1007/978-981-99-2119-5_14

enhancing therapeutic molecular activity at certain regions while lowering toxic side effects at normal sites, this selective delivery keeps the systemic effect to a lower level. This type of delivery is used in the therapy known as gene therapy (Zhou et al. 2017; Han et al. 2016; Kuang et al. 2017; Liu et al. 2016; Wang et al. 2016; Chen et al. 2017; Saha et al. 2017; Kemp et al. 2016). Gene therapy has earned considerable interest over the last 20 years as a favorable future treatment of choice for major diseases like cancer, AIDS, cardiovascular or neuronal disorders, as well as hereditary single gene abnormalities. For gene therapy to be effective, the therapeutic gene must be delivered to the infected cells of the patient (Mulligan 1993; Verma et al. 2000; Wirth et al. 2013). By injecting a gene into a patient's cell, gene therapy may allow medical practitioners to treat disorders without the use of medications or surgery. Numerous techniques of therapy are being investigated by some scientists and medical professionals, including (1) substituting a disease-causing mutated gene with a healthy gene, (2) "knocking out" or inactivating a mutated gene, and (3) inserting new genes into the cells to help defend against the ailments. The interaction between the gene delivery system and the target cell must be thoroughly understood to develop an efficient gene delivery system. A plasmid-based gene expression system that controls a gene's activity within the target cell, a gene that encodes a particular therapeutic protein, and a gene delivery system that regulates the transfer of the gene expression plasmid to a particular site inside the body include the three elements that form the gene delivery systems (Han et al. 2000; Mahato et al. 1999) (Fig. 14.1).

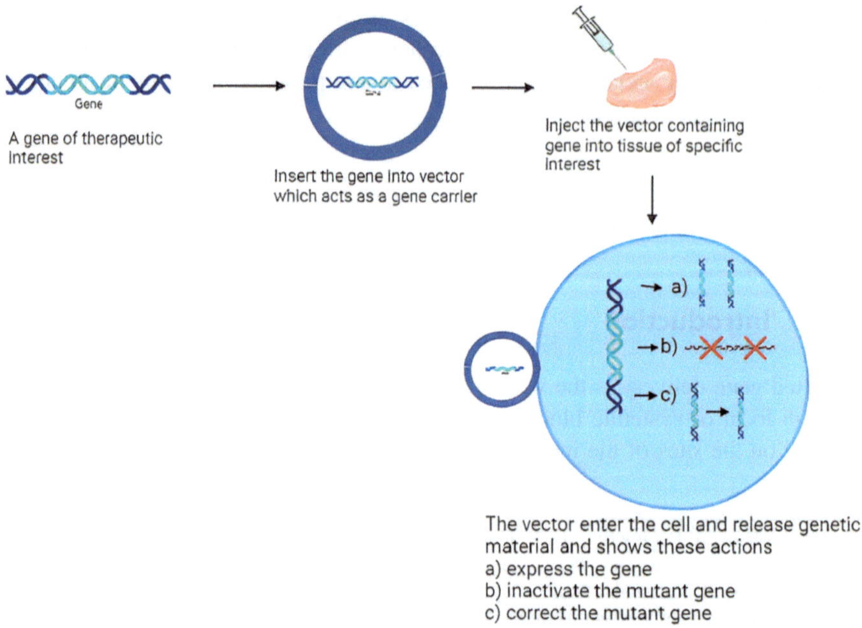

Fig. 14.1 Schematic representation of gene therapy

Since the first human gene therapy study was launched in 1990 (Blaese et al. 1995; Basarkar and Singh 2007), there has been massive interest in delivery materials for better gene delivery. The availability of reliable and secure carrier for the delivery of genes plays a vital role in gene delivery systems.

14.2 Gene Delivery Materials

The materials used for gene delivery are categorized into three types:

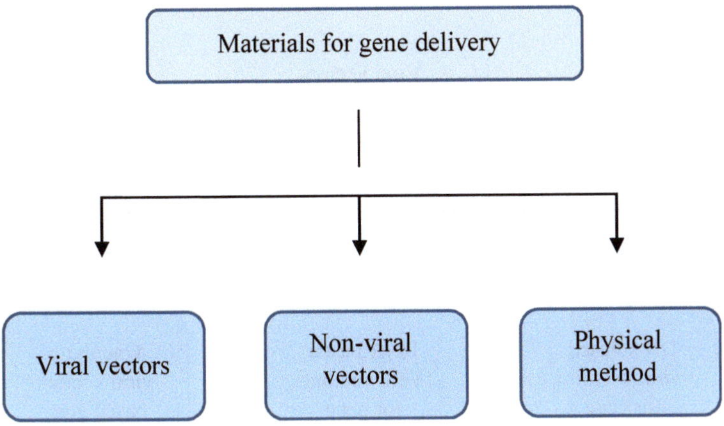

14.2.1 Viral Vectors

These are the vectors in which viruses act as delivery systems that deliver genes to the target cells. Depending on their capacity to inject their DNA into host cells, these vectors deliver genes. These capitalize on a virus's capacity to replicate its own genetic material. Because the shape of the virus prevents DNA liposome breakdown, hence viruses are an excellent method for delivering genes (Kamimura et al. 2011; Van Nies et al. 2018; Singh et al. 2017). Viral-based vectors were first developed as a transgene expression technique in the 1980s. In order to protect chimpanzees against hepatitis B, the vaccinia virus was utilized as a vaccine vector in 1984 (Moss et al. 1984). Seventy percent of all gene therapy clinical trials conducted globally as of 2015 involved viral vectors for gene delivery (Ako-Adounvo et al. 2017). The replication gene of viral vectors used for the delivery of genes is removed during genetic engineering and replaced with the therapeutic gene (gene of interest) (Thomas et al. 2003). This approach retained the virus's capacity to infect host cells (Basarkar and Singh 2007).

The main characteristics to be considered for choosing a viral vector are as follows:

1. Vector should be safe to be handled.
2. Vector should show less toxic effects in infected cells.
3. Vector should be stable enough to ensure the repeatability of the work.
4. Vector should be capable of being altered for cell-type specificity.
5. Vector should be able to integrate marker genes for facilitating identification.
6. Vector should be capable of carrying significant foreign genes.
7. Vector should be effective in the transducing and transfecting process (Ako-Adounvo et al. 2017).

Viral vectors can be classified into three categories.

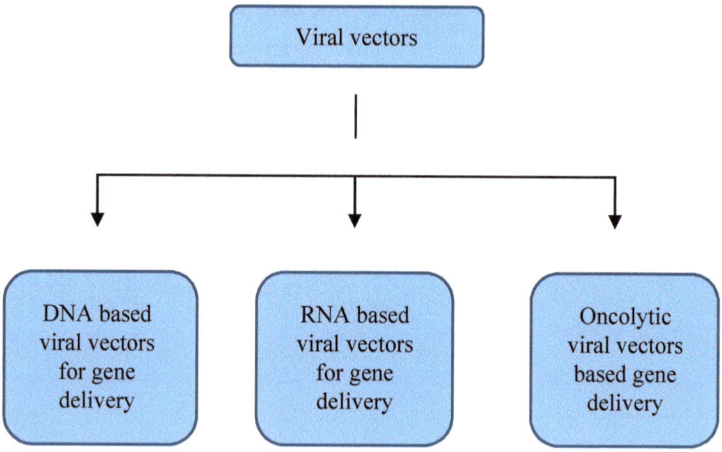

14.2.1.1 DNA-Based Viral Vectors for Gene Delivery

DNA-based viral vectors that deliver genes generally last longer and are incorporated into the genes. Viral vectors are used in DNA-based gene delivery systems to transfer genetic material to the host cells. The genetic components are effectively delivered to the host cell by the viral vectors (Wivel and Wilson 1998). Plasmids delivering transgenes for gene therapy are among the viral vectors based on the DNA for the delivery of genes (Crooke 1998). The group of materials known as DNA-based viral vectors has evolved as a promising option for gene carriers for gene therapy in a variety of diseases, including cardiovascular disorders, neurological diseases like Parkinson's and Alzheimer's disease, AIDS, and cancer (Stull and Szoka 1995; Patil et al. 2005a).

Some of the main viral vectors based on the DNA used for delivery of genes are as follows:

Adenovirus

In 1953, human adenoid tissue cultures yielded adenoviruses, which are linear double-stranded and non-enveloped DNA viruses (Rowe et al. 1953; Campos and

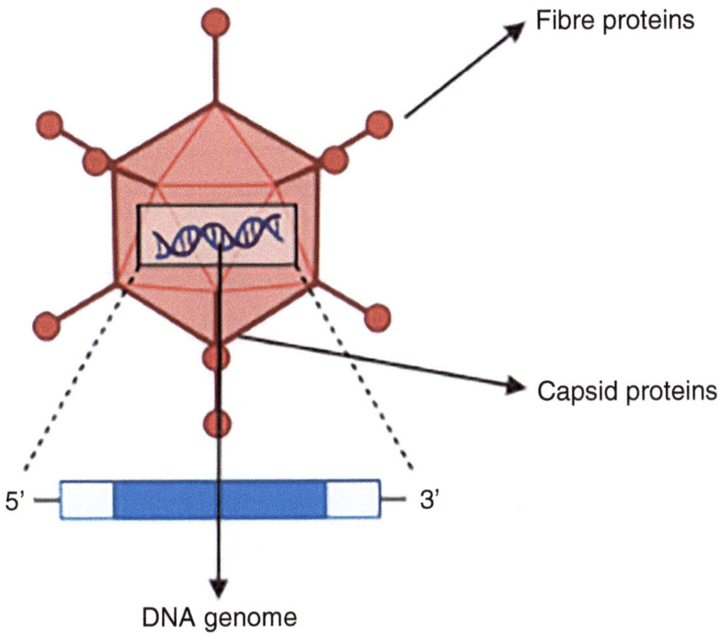

Fig. 14.2 Structure of adenovirus

Barry 2007; Majhen and Ambriović-Ristov 2006) (Fig. 14.2). Adenoviruses can be genetically modified to have a very large capacity for the introduction of transgenes due to their large viral genomes (36–38 kb). The high-capacity "gutless" Helper-dependent adenovirus, for instance, may deliver 37 kb of the transgene (Kamimura et al. 2011; Volpers and Kochanek 2004). Although adenoviral vector systems have a high transgenic capacity, the host immunological response and subsequent transitory expression of gene severely restrict their usage as delivery systems (Kamimura et al. 2011).

Poxvirus

Poxvirus is a desirable option for immune-based cancer treatment because of the absence of viral integration into the host cellular genome, the maximum size (25 kb) of the gene insert, and it elicits high immune activation. One of the main examples of poxvirus is the vaccinia virus.

Vaccinia Virus

All cells are infected by the vaccinia virus, but even after multiple injections, the tumor immune response is not completely suppressed by the host immunological response to the vector (Fig. 14.3). The administration of vaccinia in immunocompromised cancer patients is made possible by the availability of attenuated viruses, and the data suggests that this carrier improves tumor immune rejection. The major adverse effects were moderate flulike symptoms that lasted for 1–2 days and

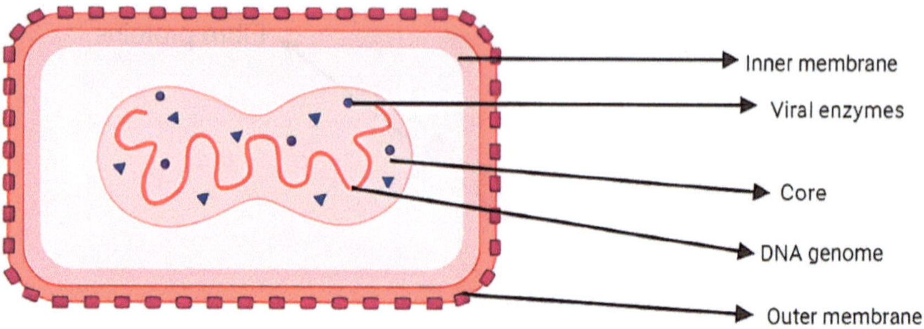

Fig. 14.3 Structure of vaccinia virus

Fig. 14.4 Structure of adeno-associated virus

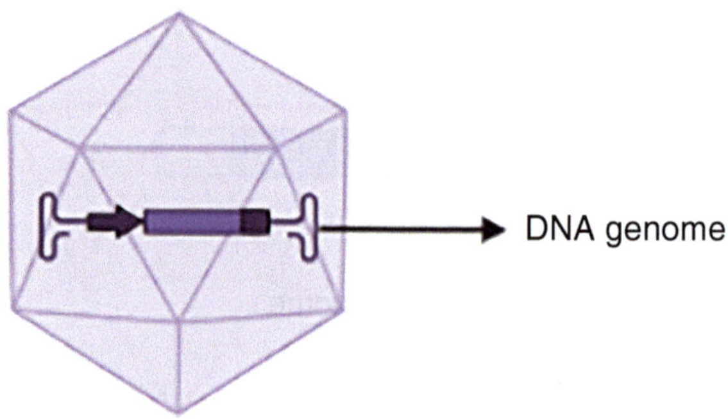

localized skin rashes/irritation at the injection site that typically lasted 4–5 days. The immune response at the cellular level and the clinical response were not correlated (Gardlík et al. 2005; Ayllón Barbellido et al. 2008).

Adeno-Associated Virus

These viruses are having properties like targeting non-proliferating cells, having discrete genome insertion sites, exhibiting little immunogenicity, and having no known toxicity (Fig. 14.4). Using an AAV vector, "suicide" gene therapy was demonstrated to be effective in oral cancer cells. Additionally, antisense or ribozyme genes have been effectively transferred using AAV vectors in preclinical models of cancer. It has been shown that AAV vectors can successfully transduce CD34+ blood cells, the brain, and the liver.

There are numerous limitations in using the adeno-associated virus as a viral vector.

1. Some cells need to be infected at very high multiplicities.
2. The AAV genome is compact, with only space for an additional 4.8 kb of DNA.
3. Since effective packing cells have not yet been devised, the manufacturing of viral particles is still extremely labor-intensive (Gardlík et al. 2005; Kay et al. 1997; Biçeroğlu and Memiş 2005; Mulherkar 2001; Xi and Grandis 2003; Zhou et al. 2004).

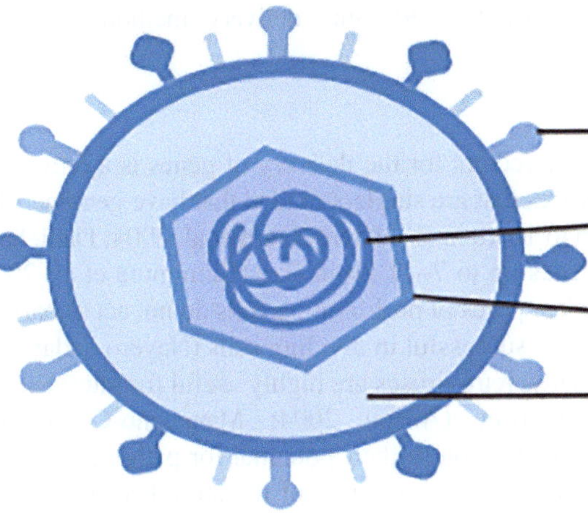

Fig. 14.5 Structure of herpes simplex virus

Herpes Simplex Virus
It is a virus that is large in size with a broad range of actions that show continued gene expression from protracted infection (Fig. 14.5). The HSV poses slight chances of insertional mutagenesis since it stays outside the nucleus (episomal). HSV type 1 (HSV-1) strains form the basis for the majority of herpes virus vectors. This double-stranded DNA virus has various unique characteristics, such as the capacity to stay dormant in tissues and to revive at an infection site. HSV-1 replicates once it has infected a cell, leading to cell and infection of neighboring cells. Furthermore, HSV-1 is a widespread human infection that infrequently results in serious illnesses. HSV vectors can quickly and effectively transmit genes while accommodating significant amounts of foreign DNA.

Herpes simplex virus has some limitations, such as the following:

1. Poor transfection efficiency.
2. Large genome size.
3. The function of HSV is confined due to its high affinity toward neuronal cells. However, some researchers are taking advantage of this limitation to target the neurons (Ayllón Barbellido et al. 2008; Gardlik et al. 2011; Roman et al. 2011).

14.2.1.2 RNA-Based Viral Vectors for Gene Delivery
Infectious RNA transcripts can now be directly translated by RNA-based viral vectors for the delivery of genes. Gene delivery with RNA is typically temporary and not permanent. Oncoretro-viral vectors, human foamy virus, and lentiviral vectors are RNA-based viral vectors for delivery of genes used in the treatment of genetic diseases. The advanced system provides negative-strand RNA templates with RNA-dependent polymerase complexes (Mogler and Kamrud 2015). Patients who received transplantation for AIDS-related lymphoma have used lentiviral

vector-altered CD34(+) cells as RNA-based gene delivery methods for HIV (DiGiusto et al. 2010).

Retrovirus

One of the main RNA-based viral vectors for the delivery of genes is a retrovirus. Retroviruses are RNA-based viruses that are single-stranded that have genomes that can accommodate transgenes up to 10 kb in size (Barquinero et al. 2004; Daniel and Smith 2008) and can contain genes up to 7–11 kb in size (Kamimura et al. 2011; Barquinero et al. 2004). The nuclear pores of proliferating cells do not act as barriers to retroviral vectors, which are very successful in dividing cells (Nayerossadat et al. 2012; Bushman 2007). Additionally, retroviruses are highly useful for interventions that favor permanent gene transfer (Anson 2004; Mancheño-Corvo and Martín-Duque 2006). However, in addition to their potential for pathogenicity and immunogenicity, retroviral vectors also carry the risk of mutation because of how efficiently they incorporate into the DNA of the host cell (Basarkar and Singh 2007; Anson 2004). Additionally, target cells may become randomly infected when retroviral vector systems are used (Yi et al. 2011).

The limitations of retrovirus are as follows:

1. Less vector titer
2. Lesser transfection efficiency exhibited in in vitro studies
3. Particle instability
4. Difficult to transduce nondividing postmitotic cells

It is important to remember that each viral vector system has specific advantages and disadvantages, requiring a distinct study to determine the applications for which each is most appropriate (Mancheño-Corvo and Martín-Duque 2006; Siemens et al. 2003).

14.2.1.3 Oncolytic Viral Vectors for Gene Delivery

These viruses are used as a novel type of treatment for diseases related to cancer. Lately, they have emphasized oncology in an attempt to improve the effectiveness of their treatment interventions (Howells et al. 2017). The benefits and drawbacks of the many modifications made to them to improve infectivity and therapeutic safety for the interaction of tumor cells and oncolytic viruses (OVs) have been discussed. Through the development of T cells expressing IL-18R or IL-12R2, oncolytic adenoviruses co-expressing IL-18 and IL-12 enhance tumor-specific immunity (Choi et al. 2011). The ultimate objective is to develop a virus that can successfully multiply inside the host, find a specific target, and kill cancerous cells. Adenovirus-induced decorin expression triggers p53 activation and mitochondrial apoptosis, which kill cancer cells (Yoon et al. 2017). With the support of gene therapy, oncolytic adenovirus vectors provide a promising treatment option for cancer (Choi et al. 2015). IFN- and TNF-producing T cells that express IL-23 and p35 are activated to provide antitumor immunity. According to studies, cytokine immune-gene therapy is one of the most effective treatments for cancer (Choi et al. 2013; Hernandez-Gea et al. 2013; El-Aneed 2004; Baban et al. 2010).

14.2.2 Nonviral Vectors

In substitution for typical viral-based vectors, the nonviral vector strategy was developed. Lipids, polymers, and peptides are nonviral vector approaches explored for gene delivery (Godbey and Mikos 2001; Zhi et al. 2013; Eliyahu et al. 2005).

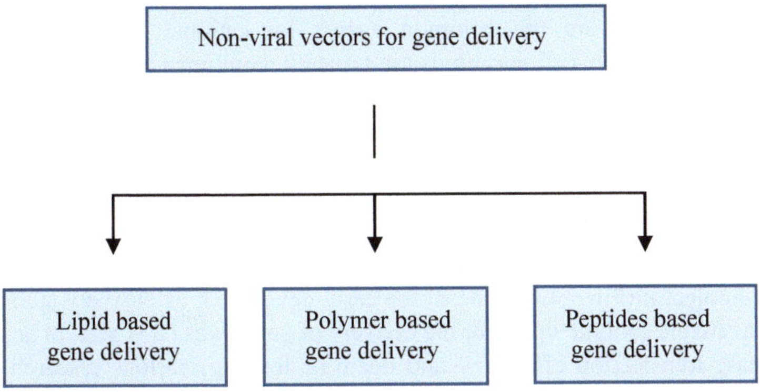

14.2.2.1 Lipid-Based Gene Delivery

Currently, the nonviral vector that has received the most emphasis is lipid-based gene delivery, which has shown promise for controlling cellular gene expression in both research and therapeutic contexts. They self-build with DNA having a negative charge to produce a cationic-charged complex, which leads to the production of a complex of plasmid DNA and cationic lipids (Wasungu and Hoekstra 2006). They are cationic at normal pH and are made up of a neutral lipid or cholesterol and cationic lipid (de Ilarduya et al. 2010). DNA is delivered into the cytoplasm when lipoplex comes in contact with cellular plasma membrane and internalize into the cell through endocytosis. This causes the lipoplex to become unstable. However, the delivery of other medicinal macromolecules has been the predominant use for anionic liposomes (Mayhew and Papajadjopoulos 1983).

They are classified into two types:

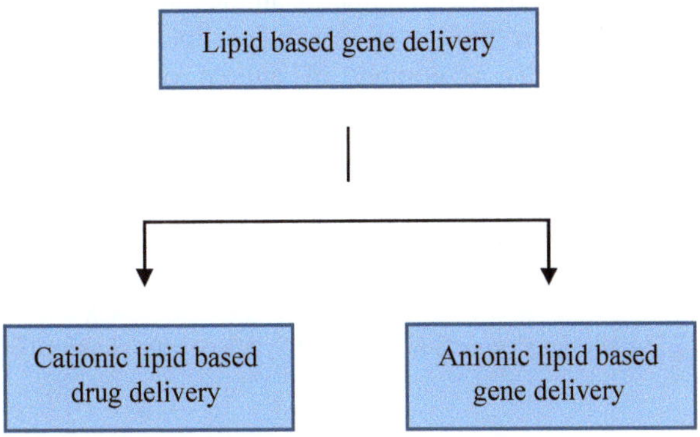

Cationic Lipid-Based Gene Delivery

Researchers have been working on designing and examining cationic lipid formulations for effective gene transfer since 1983; [1,2-bis(oleoyloxy)-3-(trimethylammonio)propane] (DOTAP) (Leventis and Silvius 1990), *N*-[1-(2,3-dioleyloxy)propyl]-*N*,*N*,*N*-trimethylammonium chloride (DOTMA) (Felgner et al. 1987), 3β[*N*-(*N*′,*N*′-dimethylaminoethane)-carbamoyl] cholesterol (DC-Chol) (Gao and Huang 1991), and dioctadecylamido glycylspermine (DOGS) (Behr et al. 1989) are the common agents for cationic lipids. To facilitate endolysosomal escape, cationic lipids are typically combined with the neutral lipid dioleoylphosphatidylethanolamine (DOPE) (Farhood et al. 1995).

N-[1-(2, 3-Dioleyloxy)Propyl]-*N*,*N*,*N*-Trimethylammonium Chloride (DOTMA)

N-[1-(2, 3-dioleyloxy)propyl]-*N*,*N*,*N*-trimethylammonium chloride, or DOTMA, was used as a lipofectin. DOTMA was combined in a 1:1 ratio with a neutral lipid called dioleoylphosphatidylethanolamine (DOPE) to enhance the transfection efficacy of lipofectin (Fig. 14.6). One of the early developed, extensively studied, and widely available cationic lipids for the delivery of genes was DOTMA. In an attempt to increase transfection efficiency and decrease toxicity, various research groups prepared modified DOTMA by altering its key functional components, such as linker, its head group, hydrocarbon chains, and linkage bonds (Vaheri and Pagano 1965; Ren et al. 2000). It was discovered that the cytotoxicity linked to the synthesized monovalent lipids depended on the density of the plated cells and the lipids' structural properties (Vaheri and Pagano 1965).

2,3-Dioleyloxy-*N*[2(Sperminecarbaxamido)Ethyl]-*N*,*N*-Dimethyl-1-Propaminium Trifluoroacetate (DOSPA)

Cationic lipid developed as a derivative of DOTMA is 2,3-dioleyloxy-*N*[2 (sperminecarboxamido)ethyl]-*N*,*N*-dimethyl-1-propanaminium trifluoroacetate, or DOSPA. DOSPA and DOTMA are almost similar in structure (Fig. 14.7). The main distinction between them is that DOSPA contains a spermine functional group that is linked to the hydrophobic chains by a peptide bond, enabling more effective DNA packing (Jain et al. 1989).

N-[1-(2,3-Dioleyloxy)-Propyl]-*N*,*N*,*N*-Trimethylammonium Chloride (DOTAP)

N-[1-(2,3-dioleyloxy)-propyl]-*N*,*N*,*N*-trimethylammonium chloride (DOTAP) (Felgner et al. 1987) is the most well-researched lipid which has been used to in vivo genetically modify a number of animal organs (Zhu et al. 1993)

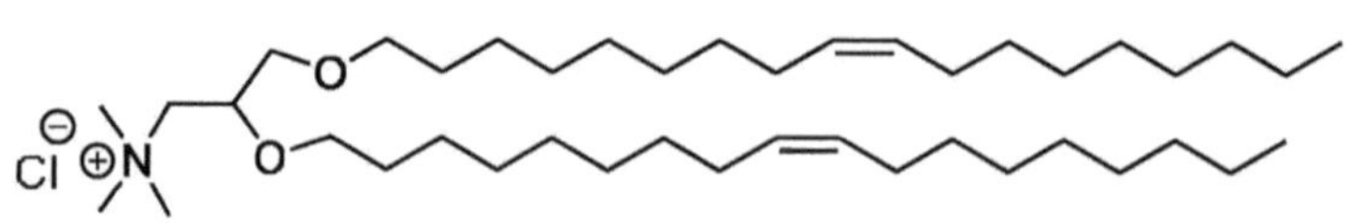

Fig. 14.6 *N*-[1-(2, 3-dioleyloxy)propyl]-*N*,*N*,*N*-trimethylammonium chloride (DOTMA) (Vaidya et al. 2022)

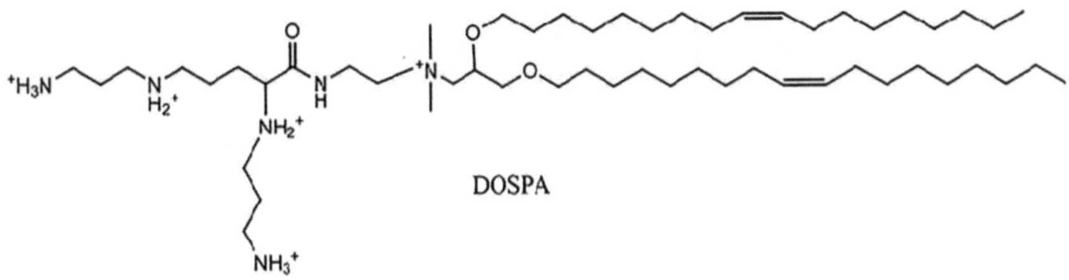

Fig. 14.7 2,3-Dioleyloxy-*N*[2(sperminecarbaxamido)ethyl]-*N*,*N*-dimethyl-1-propaminium trifluoroacetate (Li et al. 2015)

Fig. 14.8 *N*-[1-(2,3-dioleyloxy)-propyl]-*N*,*N*,*N*-trimethylammonium chloride (DOTAP)

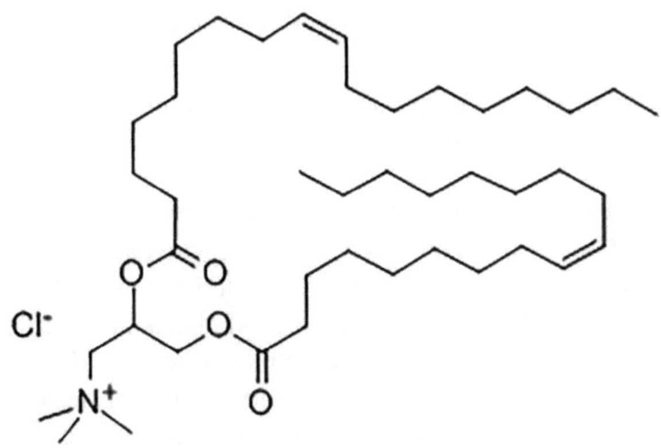

Fig. 14.9 3[*N*-(*N′*, *N′*-dimethylaminoethane)-carbamoyl]cholesterol (DC-Chol) (Vaidya et al. 2022)

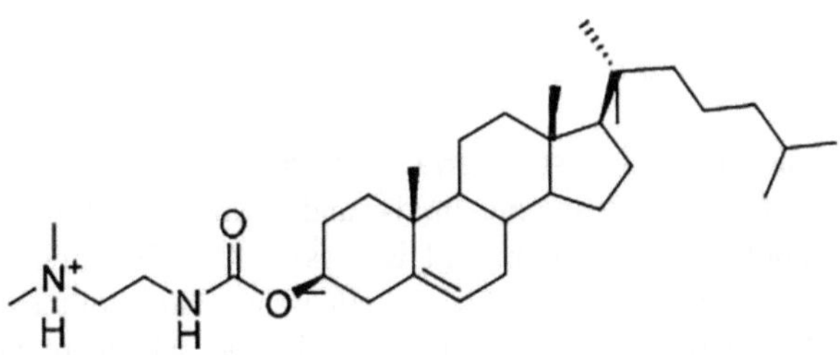

(Fig. 14.8). The key distinction among DOTMA and DOTAP is that DOTAP has ester bonds instead of ether bonds connecting the backbones, which can be hydrolyzed to help break down lipids and lessen toxicity. Using a cationic lipid complex containing cholesterol and DOTAP, high tumor selectivity can be achieved as per the experimental studies. When compared to liver tissue, this approach yields a minimum of ten times more expression for every milligram of tumor tissue (Wang et al. 2013).

3[*N*-(*N′*,*N′*-Dimethylaminoethane)-Carbamoyl]Cholesterol (DC-Chol)

3[*N*-(*N′*,*N′*-dimethylaminoethane)-carbamoyl]cholesterol was discovered in 1991 (Gao and Huang 1991) (Fig. 14.9). Unlike DOTAP and DOTMA, DC-Chol has a

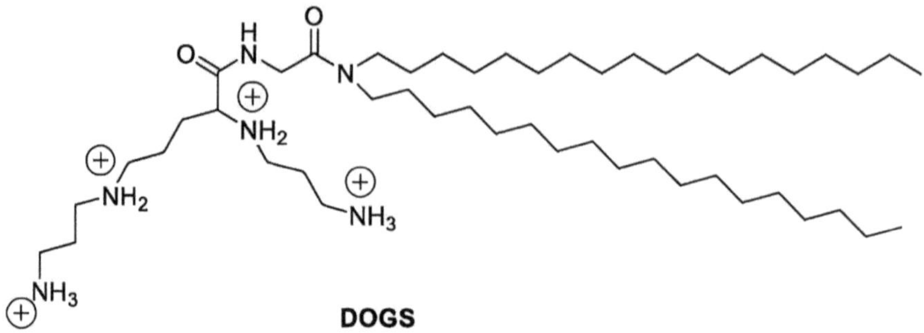

Fig. 14.10 Di-octadecyl-amido-glycyl-spermine (DOGS) (Zhi et al. 2018)

tertiary amine group that may assist to prevent lipoplex aggregation and promote increased expression of transgene (Ajmani and Hughes 1999).

Di-Octadecyl-Amido-Glycyl-Spermine (DOGS)

DOGS, also known as di-octadecyl-amido-glycyl-spermine, is marketed as transfectam and has similar characteristics to DOSPA (Fig. 14.10). Both DOGS and DOSPA have 2 18-carbon alkyl chains in addition to a spermine group. These two are different from one another since DOSPA contains a quaternary amine. Contrarily, DOGS lacks quaternary amines; instead, it has saturated chains and is connected to the head group by a peptide linkage. Numerous cell lines have been transfected using DOGS. The delivery of the chloramphenicol acetyltransferase (CAT) reporter plasmid without any cytotoxic effects was demonstrated using DOGS (Behr et al. 1989).

Anionic Lipid-Based Gene Delivery

Plasmid DNA, anionic lipids, and cations constitute the components of anionic lipoplexes (Srinivasan and Burgess 2009). Gene delivery using anionic lipids is often not very desirable or effective. Phospholipids like phosphatidylglycerol, phosphatidylserine, and phosphatidic acid, which are naturally present in biological membranes, are often employed as anionic lipids in gene delivery. As a nonviral means of delivery, DNA encapsulation into anionic and neutral liposomes has also been investigated. Owing to electrostatic repulsive interactions that exist among an anionic head group of the lipids and the phosphate backbone of DNA, an anionic liposome cannot effectively attach to anionic DNA. To use anionic lipids for cell-specific targeting, DNA must be enclosed. However, the DNA matrix's size and shape restrict the uses of its encapsulated form (Ledley 1995).

A few number of cell types, including hippocampal neurons and CHO cells, have been recorded for gene delivery using a variety of anionic liposomes (Ledley 1995; Patil et al. 2004, 2005b). Nevertheless, despite extremely positive and hopeful outcomes, our total understanding of anionic lipofection is still restricted due to numerous challenges, few of which will definitely test our scientific acumen. The absence of reproducibility is one of the primary causes. Furthermore, because of their systematic administration, it is linked to undesirable side effects from nonspecific immune system cytotoxicity that results in a variety of unfavorable consequences.

14.2.2.2 Polymer-Based Gene Delivery

In an attempt to overcome the immunogenic and carcinogenic issues related with viral vectors, carriers based on the polymer for the delivery of genes have been designed as their alternative. Due to its tremendous scope for the design of safe and reliable vectors, it has garnered a lot of interest over the past two decades (Kang et al. 2012).

Researchers studying gene delivery methods have investigated the possible advantages of polymeric vehicles for genes in cationic biopolymers including chitosan derivatives and liposomes (Ginn et al. 2018; Hulin-Curtis et al. 2016). Polymeric materials can conceal the DNA of negative charges and compress the big genes into smaller tiny molecules when used with cationic polymers for gene delivery. The polyplex is the complex of cationic polymer-based nucleic acid. One important targeting delivery system for gene therapy is the gene complex. The majority of studies concentrate on the impact of ligands that are attached to the DNA complex by covalent bonding. The targeted ligands can combine with the cationic polymers. Poly(L-lactide) is a polymer that can be a potential alternative that is frequently used to link the targeted ligands (Stone 2010; Manno et al. 2006; Kabanov and Kabanov 1995; Buwalda et al. 2012). Under a wide range of circumstances, compaction occurs during cationic polymer condensation of plasmid DNA. Condensing agents mainly utilized are multivalent cationic polymers (Bloomfield 1991, 1996).

One of the main classes of cationic polymers used as condensing agent is polyamidoamine (PAMAM) dendrimer.

Polyamidoamine Dendrimers (PAMAM Dendrimers)

A group of highly branched cationic polymers known as polyamidoamine dendrimers may condense DNA and deliver it to a number of cell types with minimal cytotoxic effects (Kukowska-Latallo et al. 1996). These are circular polymers that are highly branched and are frequently used in transfer of genes using nonviral vectors.

The most widely utilized dendrimers for the delivery of genes using nonviral vectors are the 6-generation Starburst™ PAMAM dendrimers, either in fragmented or intact form. The first polyamidoamine (PAMAM) dendrimers were developed by Tomalia et al. (1985) (Fig. 14.11). The process used to form this type of starburst dendrimer involves adding methyl acrylate by Michael addition to an original core (such as ethylenediamine or ammonia) that has a number of branching points at its center. Next, the resulting esters are amidated with ethylenediamine. Later investigations made use of PAMAM-OH dendrimers with hydroxyl ends (Fig. 14.11b). PAMAMOH that had been neutralized on the surface showed advantages, including less toxicity and lower transfection efficiency (Lee et al.

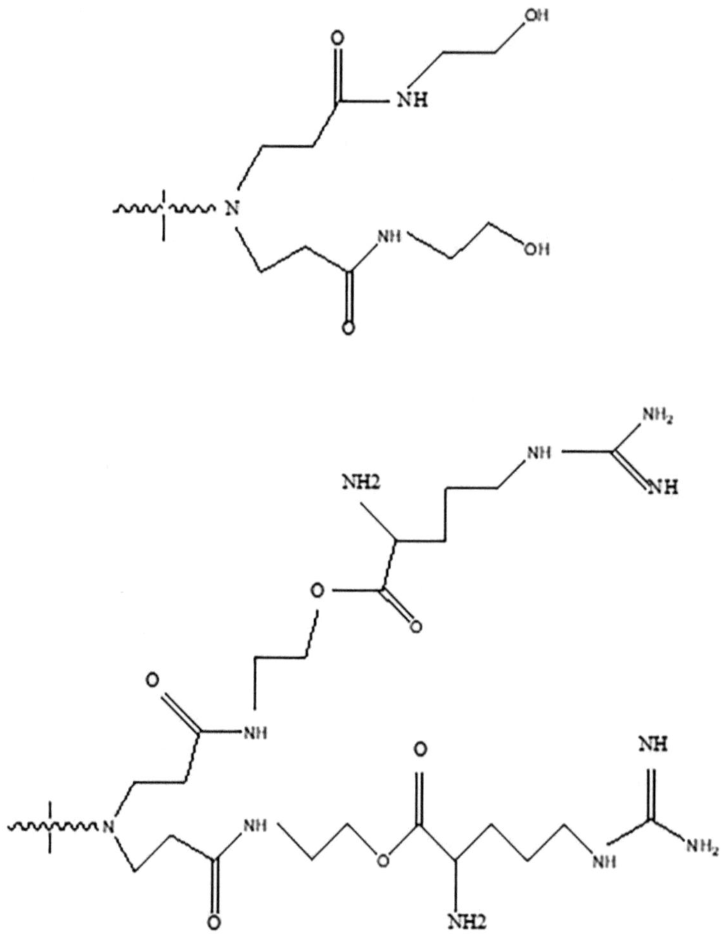

Fig. 14.11 Polyamidoamine dendrimers, (**a**) G3 PAMAM dendrimer (Tomalia et al. 1985). (**b**) Hydroxyl-terminated PAMAM-OH dendrimer (Lee et al. 2003). (**c**) L-Arginine ester-modified PAMAM-OH dendrimer (Nam et al. 2009)

2003). L-Arginine was used to modify these dendrimers by forming degradable ester linkages (Fig. 14.11c). In fact, the L-arginine ester-grafted PAMAM-OH G4 had a higher transfection and reduced toxic effects than the earlier L-arginine amide-grafted PAMAM G4 (Nam et al. 2009).

The primary benefit of the fractured dendrimer's structural design is the presence of highly dense amine at the molecule's periphery, which facilitates effective condensation of nucleic acids while releasing up the amino group present inside for a proton sponge throughout endolysosomal acidification, facilitating most effective endosomal escape. In comparison with the intact polymer, these dendrimers exhibit much increased (>50-fold) levels of expression of reporter gene. This finding's cause is unknown, although it is possible that one of the key causes is that the polymer's enhanced flexibility and improved capacity for complexing with DNA serve to promote gene expression (Tang et al. 1996). Researchers have looked at the tertiary structures that resemble noncondensing plasmid DNA in complexes with less hydrophobized stearyl-poly(L-lactide) (Kim et al. 1998a). The process of adding hydrophilic components, such as or hydrophobic stearyl chains, dextran, polyethylene glycol, hydrophobic stearyl chains, or hyaluronic acid, has however been facilitated by DNA condensation (Toncheva et al. 1998; Maruyama et al. 1997a; Katayose and Kataoka 1997; Kim et al. 1997).

Polymer-based gene delivery has been divided into three categories:

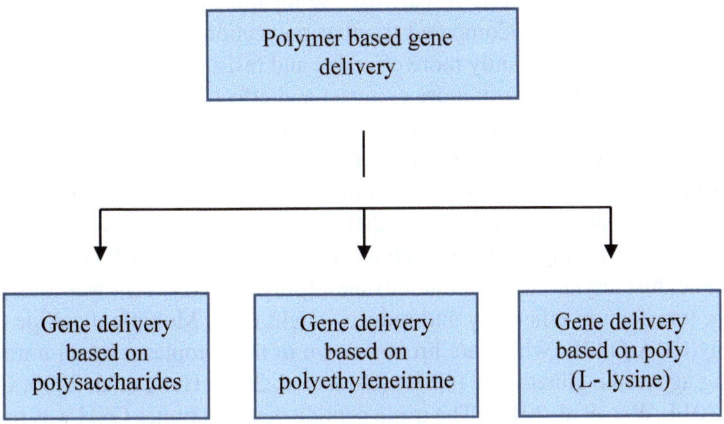

Polysaccharide-Based Gene Delivery Systems

On the basis of grafted oligoamine residues for natural polysaccharides, a range of biodegradable cationic polymers were identified for the delivery of genes (Azzam et al. 2002). The grafting concept enables 3D contact with anionic surfaces of double-stranded DNA chains by attaching side chain oligomers to hydrophilic polysaccharides, which are branched or linear in nature. The polysaccharide-based gene delivery technology suggests that the cationic polysaccharide's structure has a key impact in the transfection activity for delivery of genes. The delivery of compounds such as plasmids and oligonucleotides through mucosa had been explored using colloidal polysaccharide particles (Janes et al. 2001). Nanoparticles made of natural polysaccharides have also been investigated as methods of delivering drugs and genes (Liu et al. 2008). The mechanism for manufacturing polysaccharide-based nanoparticles included electrolyte complexing, ionic and covalent cross-linking, and hydrophobic polysaccharide self-assembly. Natural polysaccharides like chitosan have been evaluated as vehicles for drugs or genes (Morris et al. 2010).

Chitosan

It is a biodegradable polysaccharide that is linear in nature comprising of *N*-acetyl-D-glucosamine and b-1,4-linked D-glucosamine residues. Owing to its non-allergenicity, biodegradability, biocompatibility, mucoadhesive property, and great binding with DNA, it is one of the highly reported naturally produced cationic gene polymers used in nonviral gene transfer. In a series of experiments using both experimental animals and humans, chitosan displayed little cytotoxicity and increased transfection effectiveness (Rao and Sharma 1997; Aspden et al. 1997).

Gene Delivery Systems Based on Polyethyleneimine

In most cases, polyamines that turn cationic under physiologic conditions are the cationic polymers used for gene complexation. Due to their great complex stability, polyethyleneimines (PEIs), which were initially proposed by Boussif et al. (1995), are one of the most extensively researched and regarded as the reference of nonviral vectors for gene delivery. Compared to other polycations like PLL, PEIs provide a transfection that is significantly more effective and resistant to nuclease degradation. It might be because PEIs form more compact and effective complexations and have higher charge densities. Amino groups present in the PEIS have the ability to increase their ability to act as buffers. These result in the rupture of lysosome and then permeate into the cytoplasm through the effect of the proton sponge (Benjaminsen et al. 2013; Luu et al. 2012). The other functional molecules can receive some buffering capacity enhancement from PEI. PEI conjugated poly (cystamine bis(acrylamide)-diaminohexane) [poly(CBA-DAH)] was designed to improve transfection efficiency and reduce weight ratio. Multiple disulfides constitute poly(CBA-DAH), which are broken down in the cytoplasm by an intracellular reducing agent like glutathione (GSH) (Doss et al. 2013; Hong et al. 2006; Oupický and Li 2014; Wen et al. 2011). The triple peptide that constitutes GSH was formed in the cytosol from precursor amino acids (Chakravarthi et al. 2006).

The improved buffering ability of other compounds including imidazole, histidine, and PEI was also demonstrated in other investigations (Bello Roufaï and Midoux 2001; Pack et al. 2000; Pires et al. 2011; Yang et al. 2008; Zhang et al. 2015). Additionally, water-soluble, branching poly(ethylenimine)-cholesterol lipopolymers for gene transport were produced (Wang et al. 2002). Gene delivery uses altered linear polyethylenimine-cholesterol conjugates for DNA complexation (Furgeson et al. 2003). A gene delivery approach using polyethylenimine-cholesterol/DNA complexes which are linear in nature had been investigated for tumor effectiveness and biodistribution (Furgeson et al. 2004). A biodegradable gene carrier made of polyethyleneimine with acid-labile links had been developed and evaluated (Kim et al. 2005). Gene delivery technologies consisting of reducible poly (amidoethylenimine)s have been developed and evaluated (Christensen et al. 2006, 2007; Jeong et al. 2007, 2010).

Poly(L-Lysine)-Based Gene Delivery Systems

The main cationic polymer used for the transfer of genes is poly(L-lysine) (PLL). PLL falls under the category of cationic-charged polymers at neutral pH. A hydrophilic cationic amino group is mainly found in PLL. PLL has the capacity to join with DNA to build a polyelectrolyte complex. In physiologic conditions, PLL (pK_a 10.5) is protonated, which ionically interacts with DNA's negatively charged phosphate groups and builds a polyelectrolyte complex of nanoparticulate range (Laemmli 1975). To facilitate the co-adsorption of plasmid DNA, PLL-based replica particles are cross-linked using a homo-bifunctional linker have been developed for the delivery of genes (Zhang et al. 2010). For the in vivo delivery of genes to the liver, poly-L-lysine/DNA polyplexes have been stabilized (Kwoh et al. 1999). There have been reports on the development and evaluation of poly-L-lysine-based carriers for the delivery of genes (Choi et al. 2000). Modified poly(L-lysine) such as *N*-terminal modified poly(L-lysine) antibody conjugate is utilized as a vehicle for delivering specific genes in endothelial cells present in the lungs of the mouse (Trubetskoy et al. 1992). Terplex DNA delivery systems and DNA nanoparticle carriers with grafted poly(L-lysine) polysaccharide copolymers were initially developed and recognized as a carrier of genes (Kim et al. 1998a; Maruyama et al. 1997b). For gene delivery systems, terplex delivery systems and polyethylene glycol grafted poly(L-lysine) were also designed (Kim et al. 1998b; Choi et al. 1998a,b, 1999). In the area of biomedical applications, clinical evaluations are now being conducted (Park et al. 2006; Meel et al. 2016; Bodles-Brakhop et al. 2009; Pulkkanen and Yla-Herttuala 2005; Rainov 2000; Young et al. 2006; Yoshida et al. 2004; Weichselbaum et al. 2002; Amer 2014; Jayant et al. 2016; Nam et al. 2015).

14.2.2.3 Peptide-Based Gene Delivery

Genomic carriers based on peptides provide a number of advantageous properties. With the aid of the naturally occurring amino acids ornithine, arginine, and lysine, which provide positive charges, electrostatic interaction can condense the pDNA (Plank et al. 1999) and perform a variety of other functions, including endosomal escape and receptor-targeted delivery. Oligolysine peptides are a substitute for heterogeneous polylysine peptides that include lysines of a fixed length (L-lysine). It is also possible to carry out site-specific adjustments because of the specific structure. According to studies, pDNA can be compacted by oligolysine molecules with 13 or more lysine monomers (Wadhwa et al. 1997), and a peptide with 18 lysines can form stable polyplexes with pDNA that are guarded against breakdown (Adami et al. 1998).

Cross-linking has been shown to significantly enhance the stability of DNA complexes. In order to establish bioreversible disulfide linkages through oxidation, cysteine has been added to the peptide sequence. They studied modifications of the Trp-Lys20 peptide by replacing one to four of the lysines with cysteines. The best transfection efficiency was demonstrated by a peptide containing two terminal cysteines (McKenzie et al. 2000a). Similar transfection effectiveness was attained using reduced lysine chains that contained only two terminal cysteines and four lysines (McKenzie et al. 2000b).

Fusogenic peptides, such as amphipathic peptides with large amounts of basic amino acids, like Tat (Fawell et al. 1994), melittin (Boeckle et al. 2006; Chen et al. 2006), KALA (Wyman et al. 1997), and others (Lehto et al. 2010; El Andaloussi et al. 2011), were used to mediate endosomal escape as an alternate choice to the mechanism of proton sponge (Zorko and Langel 2005). These peptides can damage endosomes by interacting with their lipid membrane. Numerous polymers, cationic peptides, or other gene delivery vehicles were combined with fusogenic peptides.

14.2.3 Physical Method of Gene Delivery

The physical method of transferring genes is not vector-mediated and does not require carriers.

It involves various methods such as electroporation, sonoporation, photoporation, magneto-fection, needle injection, hydroporation, and gene guns.

Electroporation: This method involves the use of electromagnetic pulse, which helps in the formation of pores in the membrane of the cell to enable the genetic materials into the cell.

Sonoporation: This technique involves the usage of sound waves for the creation of pores in the membrane of the cell for the movement of genetic materials into the cell.

Photoporation: The method of photoporation helps in the formation of pores in the cell membrane by using a laser pulse for entering the genetic materials into the cell.

Magneto-fection: It involves utilizing magnetic particles that have complexed with DNA and an external magnetic field to concentrate nucleic acid particles inside target cells.

Needle injection: This method involves the usage of needles to directly inject genetic materials into the cell.

Hydroporation: This is the technique that involves the use of the hydrodynamic capillary effect to alter the permeability of the cell membrane for the entry of genetic materials into the cell (Sung and Kim 2018).

14.3 Conclusion

In recent years, gene therapy has garnered significant interest and can be used as a promising treatment of choice for chronic conditions caused by genetic abnormalities. Gene therapy is a type of treatment in which genes are mainly used for the treatment of diseases. In gene therapy, components used for gene delivery are the essential components for the treatment of genetic diseases. The major aim of the research on gene delivery systems is to design proper vectors for the treatment of serious life-threatening diseases like cancer, Alzheimer's, and AIDS. Several types of materials used in gene delivery systems are summarized and discussed here.

As all we know, viral vectors are playing a vital role in clinical trials of gene delivery such as adenovirus, herpes simplex virus, adeno-associated virus, retrovirus, etc. These vectors are efficient to carry the genome into target cells. On the other hand, nonviral vectors are also gaining attention in gene therapy. The materials utilized for nonviral vectors are lipids, peptides, and polymers. There are different categories of lipids such as anionic lipids and cationic lipids which are used for the delivery of genes. This lipid-based gene delivery acts as a promising approach to altering gene expression at a cellular level in therapeutic and research applications. Polymeric gene delivery systems have gained interest in basic research and clinical applications. These gene delivery systems have several merits such as cost-effectiveness, easy scale-up production, and easy modulation of DNA loading capacity. However, materials that are stable and monodisperse in size must be used for the effective delivery of genes. Hence the main advantages of gene delivery systems should be assessed carefully to reach the desired target by building an effective system with proper biodistribution to first-pass organs and rapid clearance of complexes for efficient gene delivery. In the future, our main research has to focus on DNA and RNA molecular techniques to become the main treatment in the biomedical field.

References

Adami RC, Collard WT, Gupta SA, Kwok KY, Bonadio J, Rice KG (1998) Stability of peptide-condensed plasmid DNA formulations. J Pharm Sci 87(6):678–683

Ajmani PS, Hughes JA (1999) 3β [N-(N′, N′-Dimethylaminoethane)-carbamoyl] cholesterol (DC-Chol)-mediated gene delivery to primary rat neurons: characterization and mechanism. Neurochem Res 24(5):699–703

Ako-Adounvo AM, Marabesi B, Lemos RC, Patricia A, Karla PK (2017) Drug and gene delivery materials and devices. In: Emerging nanotechnologies for diagnostics, drug delivery and medical devices, pp 375–392. https://doi.org/10.1016/B978-0-323-42978-8.00015-2

Amer MH (2014) Gene therapy for cancer: present status and future perspective. Mol Cell Therap 2(1):1–9. https://doi.org/10.1186/2052-8426-2-27

Anson DS (2004) The use of retroviral vectors for gene therapy-what are the risks? A review of retroviral pathogenesis and its relevance to retroviral vector-mediated gene delivery. Genet Vaccines Ther 2(1):1–3. https://doi.org/10.1186/1479-0556-2-9

Aspden TJ, Mason JD, Jones NS, Lowe J, Skaugrud Ø, Illum L (1997) Chitosan as a nasal delivery system: the effect of chitosan solutions on in vitro and in vivo mucociliary transport rates in human turbinates and volunteers. J Pharm Sci 86(4):509–513

Ayllón Barbellido S, Campo Trapero J, Cano Sánchez J, Perea García MA, Escudero-Castaño N, Bascones MA (2008) Gene therapy in the management of oral cancer: review of the literature. Med Oral Patol Oral Cir Bucal 13(1):E15–E21

Azzam T, Eliyahu H, Shapira L, Linial M, Barenholz Y, Domb AJ (2002) Polysaccharide-oligoamine based conjugates for gene delivery. J Med Chem 45(9):1817–1824

Baban CK, Cronin M, O'Hanlon D, O'Sullivan GC, Tangney M (2010) Bacteria as vectors for gene therapy of cancer. Bioeng Bugs 1(6):385–394

Barquinero J, Eixarch H, Perez-Melgosa M (2004) Retroviral vectors: new applications for an old tool. Gene Ther 11(1):S3–S9

Basarkar A, Singh J (2007) Nanoparticulate systems for polynucleotide delivery. Int J Nanomedicine 2(3):353

Behr JP, Demeneix B, Loeffler JP, Perez-Mutul J (1989) Efficient gene transfer into mammalian primary endocrine cells with lipopolyamine-coated DNA. Proc Natl Acad Sci 86(18): 6982–6986

Bello Roufaï M, Midoux P (2001) Histidylated polylysine as DNA vector: elevation of the imidazole protonation and reduced cellular uptake without change in the polyfection efficiency of serum stabilized negative polyplexes. Bioconjug Chem 12(1):92–99

Benjaminsen RV, Mattebjerg MA, Henriksen JR, Moghimi SM, Andresen TL (2013) The possible "proton sponge" effect of polyethylenimine (PEI) does not include change in lysosomal pH. Mol Ther 21(1):149–157

Biçeroğlu S, Memiş A (2005) Gene therapy: applications in interventional radiology. Diagn Interv Radiol 11(113):8

Blaese RM, Culver KW, Miller AD, Carter CS, Fleisher T, Clerici M, Shearer G, Chang L, Chiang Y, Tolstoshev P, Greenblatt JJ (1995) T lymphocyte-directed gene therapy for ADA–SCID: initial trial results after 4 years. Science 270(5235):475–480

Bloomfield VA (1991) Condensation of DNA by multivalent cations: considerations on mechanism. Biopolymers 31(13):1471–1481

Bloomfield VA (1996) DNA condensation. Curr Opin Struct Biol 6(3):334–341

Bodles-Brakhop AM, Heller R, Draghia-Akli R (2009) Electroporation for the delivery of DNA-based vaccines and immunotherapeutics: current clinical developments. Mol Ther 17(4):585–592

Boeckle S, Fahrmeir J, Roedl W, Ogris M, Wagner E (2006) Melittin analogs with high lytic activity at endosomal pH enhance transfection with purified targeted PEI polyplexes. J Control Release 112(2):240–248

Boussif O, Lezoualc'h F, Zanta MA, Mergny MD, Scherman D, Demeneix B, Behr JP (1995) A versatile vector for gene and oligonucleotide transfer into cells in culture and in vivo: polyethylenimine. Proc Natl Acad Sci 92(16):7297–7301

Bushman FD (2007) Retroviral integration and human gene therapy. J Clin Invest 117(8): 2083–2086

Buwalda SJ, Dijkstra PJ, Feijen J (2012) Poly (ethylene glycol)–poly (l-lactide) star block copolymer hydrogels crosslinked by metal–ligand coordination. J Polym Sci A Polym Chem 50(9): 1783–1791

Campos SK, Barry MA (2007) Current advances and future challenges in adenoviral vector biology and targeting. Curr Gene Ther 7(3):189–204

Chakravarthi S, Jessop CE, Bulleid NJ (2006) The role of glutathione in disulphide bond formation and endoplasmic-reticulum-generated oxidative stress. EMBO Rep 7(3):271–275

Chen CP, Kim JS, Steenblock E, Liu D, Rice KG (2006) Gene transfer with poly-melittin peptides. Bioconjug Chem 17(4):1057–1062

Chen D, Yang D, Dougherty CA, Lu W, Wu H, He X, Cai T, Van Dort ME, Ross BD, Hong H (2017) In vivo targeting and positron emission tomography imaging of tumor with intrinsically radioactive metal–organic frameworks nanomaterials. ACS Nano 11(4):4315–4327

Choi YH, Liu F, Kim JS, Choi YK, Park JS, Kim SW (1998a) Polyethylene glycol-grafted poly-L-lysine as polymeric gene carrier. J Control Release 54(1):39–48

Choi YH, Liu F, Park JS, Kim SW (1998b) Lactose-poly (ethylene glycol)-grafted poly-L-lysine as hepatoma cell-targeted gene carrier. Bioconjug Chem 9(6):708–718

Choi YH, Liu F, Choi JS, Kim SW, Park JS (1999) Characterization of a targeted gene carrier, lactose-polyethylene glycol-grafted poly-L-lysine, and its complex with plasmid DNA. Hum Gene Ther 10(16):2657–2665

Choi JS, Joo DK, Kim CH, Kim K, Park JS (2000) Synthesis of a barbell-like triblock copolymer, poly (L-lysine) dendrimer-block-poly (ethylene glycol)-block-poly (L-lysine) dendrimer, and its self-assembly with plasmid DNA. J Am Chem Soc 122(3):474–480

Choi IK, Lee JS, Zhang SN, Park J, Lee KM, Sonn CH, Yun CO (2011) Oncolytic adenovirus co-expressing IL-12 and IL-18 improves tumor-specific immunity via differentiation of T cells expressing IL-12Rβ2 or IL-18Rα. Gene Ther 18(9):898–909

Choi IK, Li Y, Oh E, Kim J, Yun CO (2013) Oncolytic adenovirus expressing IL-23 and p35 elicits IFN-γ-and TNF-α-co-producing T cell-mediated antitumor immunity. PLoS One 8(7):e67512. https://doi.org/10.1371/journal.pone.0067512

Choi JW, Lee YS, Yun CO, Kim SW (2015) Polymeric oncolytic adenovirus for cancer gene therapy. J Control Release 219:181–191. https://doi.org/10.1016/jconrel.2015.10.009
Christensen LV, Chang CW, Kim WJ, Kim SW, Zhong Z, Lin C, Engbersen JF, Feijen J (2006) Reducible poly (amido ethylenimine) s designed for triggered intracellular gene delivery. Bioconjug Chem 17(5):1233–1240
Christensen LV, Chang CW, Yockman JW, Conners R, Jackson H, Zhong Z, Feijen J, Bull DA, Kim SW (2007) Reducible poly (amido ethylenediamine) for hypoxia-inducible VEGF delivery. J Control Release 118(2):254–261
Crooke ST (1998) An overview of progress in antisense therapeutics. Antisense Nucleic Acid Drug Dev 8(2):115–122
Daniel R, Smith JA (2008) Integration site selection by retroviral vectors: molecular mechanism and clinical consequences. Hum Gene Ther 19(6):557–568
de Ilarduya CT, Sun Y, Düzgüneş N (2010) Gene delivery by lipoplexes and polyplexes. Eur J Pharm Sci 40(3):159–170
DiGiusto DL, Krishnan A, Li L, Li H, Li S, Rao A, Mi S, Yam P, Stinson S, Kalos M, Alvarnas J (2010) RNA-based gene therapy for HIV with lentiviral vector–modified CD34+ cells in patients undergoing transplantation for AIDS-related lymphoma. Sci Transl Med 2(36):36ra43
Doss C, Debottam S, Debajyoti C (2013) Glutathione-responsive nano-transporter-mediated siRNA delivery: silencing the mRNA expression of Ras. Protoplasma 250(3):787–792
El Andaloussi S, Lehto T, Mäger I, Rosenthal-Aizman K, Oprea II, Simonson OE, Sork H, Ezzat K, Copolovici DM, Kurrikoff K, Viola JR (2011) Design of a peptide-based vector, PepFect6, for efficient delivery of siRNA in cell culture and systemically in vivo. Nucleic Acids Res 39(9): 3972–3987
El-Aneed A (2004) Current strategies in cancer gene therapy. Eur J Pharmacol 498(1–3):1–8
Eliyahu H, Barenholz Y, Domb AJ (2005) Polymers for DNA delivery. Molecules 10(1):34–64
Farhood H, Serbina N, Huang L (1995) The role of dioleoyl phosphatidylethanolamine in cationic liposome mediated gene transfer. Biochim Biophys Acta 1235(2):289–295
Fawell S, Seery J, Daikh Y, Moore C, Chen LL, Pepinsky B, Barsoum JA (1994) Tat-mediated delivery of heterologous proteins into cells. Proc Natl Acad Sci 91(2):664–668
Felgner PL, Gadek TR, Holm M, Roman R, Chan HW, Wenz M, Northrop JP, Ringold GM, Danielsen M (1987) Lipofection: a highly efficient, lipid-mediated DNA-transfection procedure. Proc Natl Acad Sci 84(21):7413–7417
Furgeson DY, Chan WS, Yockman JW, Kim SW (2003) Modified linear polyethylenimine-cholesterol conjugates for DNA complexation. Bioconjug Chem 14(4):840–847
Furgeson DY, Yockman JW, Janat MM, Kim SW (2004) Tumor efficacy and biodistribution of linear polyethylenimine-cholesterol/DNA complexes. Mol Ther 9(6):837–845
Gao X, Huang L (1991) A novel cationic liposome reagent for efficient transfection of mammalian cells. Biochem Biophys Res Commun 179(1):280–285
Gardlík R, Pálffy R, Hodosy J, Lukács J, Turna J, Celec P (2005) Vectors and delivery systems in gene therapy. Med Sci Monit 11(4):110–121
Gardlik R, Celec P, Bernadic M (2011) Targeting angiogenesis for cancer (gene) therapy. Bratisl Lek Listy 112(8):428–434
Ginn SL, Amaya AK, Alexander IE, Edelstein M, Abedi MR (2018) Gene therapy clinical trials worldwide to 2017: an update. J Gene Med 20(5):e3015
Godbey WT, Mikos AG (2001) Recent progress in gene delivery using non-viral transfer complexes. J Control Release 72(1–3):115–125
Han SO, Mahato RI, Sung YK, Kim SW (2000) Development of biomaterials for gene therapy. Mol Ther 2(4):302–317
Han L, Zhang Y, Lu X, Wang K, Wang Z, Zhang H (2016) Polydopamine nanoparticles modulating stimuli-responsive PNIPAM hydrogels with cell/tissue adhesiveness. ACS Appl Mater Interfaces 8(42):29088–29100
Hernandez-Gea V, Toffanin S, Friedman SL, Llovet JM (2013) Role of the microenvironment in the pathogenesis and treatment of hepatocellular carcinoma. Gastroenterology 144(3):512–527

Hong R, Han G, Fernández JM, Kim BJ, Forbes NS, Rotello VM (2006) Glutathione-mediated delivery and release using monolayer protected nanoparticle carriers. J Am Chem Soc 128(4): 1078–1079
Howells A, Marelli G, Lemoine NR, Wang Y (2017) Oncolytic viruses—interaction of virus and tumor cells in the battle to eliminate cancer. Front Oncol 7:195. https://doi.org/10.3389/fonc.2017.00195
Hulin-Curtis SL, Uusi-Kerttula H, Jones R, Hanna L, Chester JD, Parker AL (2016) Evaluation of CD46 re-targeted adenoviral vectors for clinical ovarian cancer intraperitoneal therapy. Cancer Gene Ther 23(7):229–234
Jain S, Zon G, Sundaralingam M (1989) Base only binding of spermine in the deep groove of the A-DNA octamer d (GTGTACAC). Biochemistry 28(6):2360–2364
Janes KA, Calvo P, Alonso MJ (2001) Polysaccharide colloidal particles as delivery systems for macromolecules. Adv Drug Deliv Rev 47(1):83–97
Jayant RD, Sosa D, Kaushik A, Atluri V, Vashist A, Tomitaka A, Nair M (2016) Current status of non-viral gene therapy for CNS disorders. Expert Opin Drug Deliv 13(10):1433–1445. https://doi.org/10.1080/17425247.2016.1188802
Jeong JH, Christensen LV, Yockman JW, Zhong Z, Engbersen JF, Kim WJ, Feijen J, Kim SW (2007) Reducible poly (amido ethylenimine) directed to enhance RNA interference. Biomaterials 28(10):1912–1917
Jeong JH, Kim SH, Christensen LV, Feijen J, Kim SW (2010) Reducible poly (amido ethylenimine)-based gene delivery system for improved nucleus trafficking of plasmid DNA. Bioconjug Chem 21(2):296–301
Kabanov AV, Kabanov VA (1995) DNA complexes with polycations for the delivery of genetic material into cells. Bioconjug Chem 6(1):7–20
Kamimura K, Suda T, Zhang G, Liu D (2011) Advances in gene delivery systems. Pharmaceut Med 25(5):293–306
Kang HC, Huh KM, Bae YH (2012) Polymeric nucleic acid carriers: current issues and novel design approaches. J Control Release 164(3):256–264
Katayose S, Kataoka K (1997) Water-soluble polyion complex associates of DNA and poly (ethylene glycol)-poly (l-lysine) block copolymer. Bioconjug Chem 8(5):702–707
Kay MA, Liu D, Hoogerbrugge PM (1997) Gene therapy. Proc Natl Acad Sci U S A 94:12744–12746
Kemp JA, Shim MS, Heo CY, Kwon YJ (2016) “Combo” nanomedicine: co-delivery of multi-modal therapeutics for efficient, targeted, and safe cancer therapy. Adv Drug Deliv Rev 98:3–18
Kim JS, Maruyama A, Akaike T, Kim SW (1997) In vitro gene expression on smooth muscle cells using a terplex delivery system. J Control Release 47(1):51–59
Kim JS, Maruyama A, Akaike T, Kim SW (1998a) Terplex DNA delivery system as a gene carrier. Pharm Res 15(1):116–121
Kim JS, Kim BI, Maruyama A, Akaike T, Kim SW (1998b) A new non-viral DNA delivery vector: the terplex system. J Control Release 53(1–3):175–182
Kim YH, Park JH, Lee M, Kim YH, Park TG, Kim SW (2005) Polyethylenimine with acid-labile linkages as a biodegradable gene carrier. J Control Release 103(1):209–219
Kuang Y, Zhang K, Cao Y, Chen X, Wang K, Liu M, Pei R (2017) Hydrophobic IR-780 dye encapsulated in cRGD-conjugated solid lipid nanoparticles for NIR imaging-guided photothermal therapy. ACS Appl Mater Interfaces 9(14):12217–12226
Kukowska-Latallo JF, Bielinska AU, Johnson J, Spindler R, Tomalia DA, Baker JR Jr (1996) Efficient transfer of genetic material into mammalian cells using Starburst polyamidoamine dendrimers. Proc Natl Acad Sci 93(10):4897–4902
Kwoh DY, Coffin CC, Lollo CP, Jovenal J, Banaszczyk MG, Mullen P, Phillips A, Amini A, Fabrycki J, Bartholomew RM, Brostoff SW (1999) Stabilization of poly-L-lysine/DNA polyplexes for in vivo gene delivery to the liver. Biochim Biophys Acta 1444(2):171–190
Laemmli UK (1975) Characterization of DNA condensates induced by poly (ethylene oxide) and polylysine. Proc Natl Acad Sci 72(11):4288–4292

Ledley FD (1995) Nonviral gene therapy: the promise of genes as pharmaceutical products. Hum Gene Ther 6(9):1129–1144

Lee JH, Lim YB, Choi JS, Lee Y, Kim TI, Kim HJ, Yoon JK, Kim K, Park JS (2003) Polyplexes assembled with internally quaternized PAMAM-OH dendrimer and plasmid DNA have a neutral surface and gene delivery potency. Bioconjug Chem 14(6):1214–1221

Lehto T, Abes R, Oskolkov N, Suhorutšenko J, Copolovici DM, Mäger I, Viola JR, Simonson OE, Ezzat K, Guterstam P, Eriste E (2010) Delivery of nucleic acids with a stearylated (RxR) 4 peptide using a non-covalent co-incubation strategy. J Control Release 141(1):42–51

Leventis R, Silvius JR (1990) Interactions of mammalian cells with lipid dispersions containing novel metabolizable cationic amphiphiles. Biochim Biophys Acta 1023(1):124–132

Li L, He ZY, Wei XW, Gao GP, Wei YQ (2015) Challenges in CRISPR/CAS9 delivery: potential roles of nonviral vectors. Hum Gene Ther 26(7):452–462

Liu Z, Jiao Y, Wang Y, Zhou C, Zhang Z (2008) Polysaccharides-based nanoparticles as drug delivery systems. Adv Drug Deliv Rev 60(15):1650–1662

Liu LX, Li BX, Wang QY, Dong ZP, Li HM, Jin QM, Hong H, Zhang J, Wang Y (2016) An integrative folate-based metal complex nanotube as a potent antitumor nanomedicine as well as an efficient tumor-targeted drug carrier. Bioconjug Chem 27(12):2863–2873

Luu QP, Shin JY, Kim YK, Islam MA, Kang SK, Cho MH, Choi YJ, Cho CS (2012) High gene transfer by the osmotic polysorbitol-mediated transporter through the selective caveolae endocytic pathway. Mol Pharm 9(8):2206–2218

Mahato RI, Smith LC, Rolland A (1999) Pharmaceutical perspectives of nonviral gene therapy. Adv Genet 41:95–156

Majhen D, Ambriović-Ristov A (2006) Adenoviral vectors—how to use them in cancer gene therapy? Virus Res 119(2):121–133

Mancheño-Corvo P, Martín-Duque P (2006) Viral gene therapy. Clin Transl Oncol 8(12):858–867

Manno CS, Pierce GF, Arruda VR, Glader B, Ragni M, Rasko JJ, Ozelo MC, Hoots K, Blatt P, Konkle B, Dake M (2006) Successful transduction of liver in hemophilia by AAV-factor IX and limitations imposed by the host immune response. Nat Med 12(3):342–347

Maruyama A, Katoh M, Ishihara T, Akaike T (1997a) Comb-type polycations effectively stabilize DNA triplex. Bioconjug Chem 8(1):3–6

Maruyama A, Ishihara T, Kim JS, Kim SW, Akaike T (1997b) Nanoparticle DNA carrier with poly (L-lysine) grafted polysaccharide copolymer and poly (D, L-lactic acid). Bioconjug Chem 8(5): 735–742

Mayhew E, Papajadjopoulos D (1983) Therapeutic applications of liposomes. In: Ostro MJ (ed) Liposomes. Marcel Dekker, New York

McKenzie DL, Kwok KY, Rice KG (2000a) A potent new class of reductively activated peptide gene delivery agents. J Biol Chem 275(14):9970–9977

McKenzie DL, Smiley E, Kwok KY, Rice KG (2000b) Low molecular weight disulfide cross-linking peptides as nonviral gene delivery carriers. Bioconjug Chem 11(6):901–909

Meel RV, Vehmeijer LJ, Kok RJ, Storm G, van Gaal EV (2016) Ligand-targeted particulate nanomedicines undergoing clinical evaluation: current status. Intracellular Deliv III:163–200

Mogler MA, Kamrud KI (2015) RNA-based viral vectors. Expert Rev Vaccines 14(2):283–312. https://doi.org/10.1586/14760584.2015.979798

Morris GA, Kök SM, Harding SE, Adams GG (2010) Polysaccharide drug delivery systems based on pectin and chitosan. Biotechnol Genet Eng Rev 27(1):257–284

Moss B, Smith GL, Gerin JL, Purcell RH (1984) Live recombinant vaccinia virus protects chimpanzees against hepatitis B. Nature 311(5981):67–69

Mulherkar R (2001) Gene therapy for cancer. Curr Sci 81:555–560

Mulligan RC (1993) The basic science of gene therapy. Science 260(5110):926–923

Nam HY, Nam K, Hahn HJ, Kim BH, Lim HJ, Kim HJ, Choi JS, Park JS (2009) Biodegradable PAMAM ester for enhanced transfection efficiency with low cytotoxicity. Biomaterials 30(4): 665–673

Nam K, Jung S, Nam JP, Kim SW (2015) Poly (ethylenimine) conjugated bioreducible dendrimer for efficient gene delivery. J Control Release 220:447–455

Nayerossadat N, Maedeh T, Ali PA (2012) Viral and nonviral delivery systems for gene delivery. Adv Biomed Res 1. https://doi.org/10.4103/2277-9175.98152

Oupický D, Li J (2014 Jul) Bioreducible polycations in nucleic acid delivery: past, present, and future trends. Macromol Biosci 14(7):908–922

Pack DW, Putnam D, Langer R (2000) Design of imidazole-containing endosomolytic biopolymers for gene delivery. Biotechnol Bioeng 67(2):217–223

Park TG, Jeong JH, Kim SW (2006) Current status of polymeric gene delivery systems. Adv Drug Deliv Rev 58(4):467–486

Patil SD, Rhodes DG, Burgess DJ (2004) Anionic liposomal delivery system for DNA transfection. AAPS J 6(4):13–22

Patil SD, Rhodes DG, Burgess DJ (2005a) DNA-based therapeutics and DNA delivery systems: a comprehensive review. AAPS J 7(1):E61–E77

Patil SD, Rhodes DG, Burgess DJ (2005b) Biophysical characterization of anionic lipoplexes. Biochim Biophys Acta 1711(1):1

Pires LR, Oliveira H, Barrias CC, Sampaio P, Pereira AJ, Maiato H, Simões S, Pêgo AP (2011) Imidazole-grafted chitosan-mediated gene delivery: in vitro study on transfection, intracellular trafficking and degradation. Nanomedicine 6(9):1499–1512

Plank C, Tang MX, Wolfe AR, Szoka FC (1999) Branched cationic peptides for gene delivery: role of type and number of cationic residues in formation and in vitro activity of DNA polyplexes. Hum Gene Ther 10(2):319–332

Pulkkanen KJ, Yla-Herttuala S (2005) Gene therapy for malignant glioma: current clinical status. Mol Ther 12(4):585–598

Rainov NG (2000) A phase III clinical evaluation of herpes simplex virus type 1 thymidine kinase and ganciclovir gene therapy as an adjuvant to surgical resection and radiation in adults with previously untreated glioblastoma multiforme. Hum Gene Ther 11(17):2389–2401. https://doi.org/10.1089/10430340075003849 9

Rao SB, Sharma CP (1997) Use of chitosan as a biomaterial: studies on its safety and hemostatic potential. J Biomed Mater Res 34(1):21–28

Ren T, Song YK, Zhang G, Liu D (2000) Structural basis of DOTMA for its high intravenous transfection activity in mouse. Gene Ther 7(9):764–768

Roman G, Ima D, YH CJ. (2011) Approaches in gene therapy of cancer and cardiovascular diseases. In: Gene therapy applications. IntechOpen. Available from: http://www.intechopen.com/source/pdfs/17760/InTechApproaches_in_gene_therapy_of_cancer_and_cardiovascular_diseases.pdf. Last accessed on 6 Sept 2022

Rowe WP, Huebner RJ, Gilmore LK, Parrott RH, Ward TG (1953 Dec) Isolation of a cytopathogenic agent from human adenoids undergoing spontaneous degeneration in tissue culture. Proc Soc Exp Biol Med 84(3):570–573

Saha A, Mohanta SC, Deka K, Deb P, Devi PS (2017) Surface-engineered multifunctional Eu: Gd2O3 nanoplates for targeted and pH-responsive drug delivery and imaging applications. ACS Appl Mater Interfaces 9(4):4126–4141

Siemens DR, Crist S, Austin JC, Tartaglia J, Ratliff TL (2003) Comparison of viral vectors: gene transfer efficiency and tissue specificity in a bladder cancer model. J Urol 170(3):979–984

Singh BN, Gupta VK, Chen J, Atanasov AG (2017) Organic nanoparticle-based combinatory approaches for gene therapy. Trends Biotechnol 35(12):1121–1124

Srinivasan C, Burgess DJ (2009) Optimization and characterization of anionic lipoplexes for gene delivery. J Control Release 136(1):62–70

Stone D (2010) Novel viral vector systems for gene therapy. Viruses 2(4):1002–1007

Stull RA, Szoka FC Jr (1995) Antigene, ribozyme and aptamer nucleic acid drugs: progress and prospects. Pharm Res 12(4):465–483

Sung YK, Kim SW (2018) The practical application of gene vectors in cancer therapy. Integrat Cancer Sci Therap 5:1–5

Tang MX, Redemann CT, Szoka FC (1996) In vitro gene delivery by degraded polyamidoamine dendrimers. Bioconjug Chem 7(6):703–714

Thomas CE, Ehrhardt A, Kay MA (2003) Progress and problems with the use of viral vectors for gene therapy. Nat Rev Genet 4(5):346–358

Tomalia DA, Baker H, Dewald J, Hall M, Kallos G, Martin S, Roeck J, Ryder J, Smith P (1985) A new class of polymers: starburst-dendritic macromolecules. Polym J 17(1):117–132

Toncheva V, Wolfert MA, Dash PR, Oupicky D, Ulbrich K, Seymour LW, Schacht EH (1998) Novel vectors for gene delivery formed by self-assembly of DNA with poly (L-lysine) grafted with hydrophilic polymers. Biochim Biophys Acta 1380(3):354–368

Trubetskoy VS, Torchilin VP, Kennel SJ, Huang L (1992) Use of N-terminal modified poly (L-lysine)-antibody conjugate as a carrier for targeted gene delivery in mouse lung endothelial cells. Bioconjug Chem 3(4):323–327

Vaheri A, Pagano JS (1965) Infectious poliovirus RNA: a sensitive method of assay. Virology 27(3):434–436

Vaidya S, Dangi A, Panda K, Jeengar MK (2022) Therapeutic potential of cationic liposome-based nucleic acid delivery for the treatment of cancer and various human ailments. Mol Biol 11:315

Van Nies P, Westerlaken I, Blanken D, Salas M, Mencía M, Danelon C (2018) Self-replication of DNA by its encoded proteins in liposome-based synthetic cells. Nat Commun 9(1):1–2

Verma IM, Naldini L, Kafri T, Miyoshi H, Takahashi M, Blömer U, Somia N, Wang L, Gage FH (2000) Gene therapy: promises, problems and prospects. In: Genes and resistance to disease. Springer, Berlin, pp 147–157

Volpers C, Kochanek S (2004) Adenoviral vectors for gene transfer and therapy. J Gene Med 6(S1): S164–S171

Wadhwa MS, Collard WT, Adami RC, McKenzie DL, Rice KG (1997) Peptide-mediated gene delivery: influence of peptide structure on gene expression. Bioconjug Chem 8(1):81–88

Wang DA, Narang AS, Kotb M, Gaber AO, Miller DD, Kim SW, Mahato RI (2002) Novel branched poly (ethylenimine)-cholesterol water-soluble lipopolymers for gene delivery. Biomacromolecules 3(6):1197–1207

Wang Y, Su HH, Yang Y, Hu Y, Zhang L, Blancafort P, Huang L (2013) Systemic delivery of modified mRNA encoding herpes simplex virus 1 thymidine kinase for targeted cancer gene therapy. Mol Ther 21(2):358–367

Wang S, Zhao X, Wang S, Qian J, He S (2016) Biologically inspired polydopamine capped gold nanorods for drug delivery and light-mediated cancer therapy. ACS Appl Mater Interfaces 8(37):24368–24384

Wasungu L, Hoekstra D (2006) Cationic lipids, lipoplexes and intracellular delivery of genes. J Control Release 116(2):255–264

Weichselbaum RR, Kufe DW, Hellman S, Rasmussen HS, King CR, Fischer PH, Mauceri HJ (2002) Radiation-induced tumour necrosis factor-α expression: clinical application of transcriptional and physical targeting of gene therapy. Lancet Oncol 3(11):665–671

Wen HY, Dong HQ, Xie WJ, Li YY, Wang K, Pauletti GM, Shi DL (2011) Rapidly disassembling nanomicelles with disulfide-linked PEG shells for glutathione-mediated intracellular drug delivery. Chem Commun 47(12):3550–3552

Wirth T, Parker N, Ylä-Herttuala S (2013) History of gene therapy. Gene 525(2):162–169

Wivel NA, Wilson JM (1998) Methods of gene delivery. Hematol Oncol Clin North Am 12(3): 483–501

Wyman TB, Nicol F, Zelphati O, Scaria PV, Plank C, Szoka FC (1997) Design, synthesis, and characterization of a cationic peptide that binds to nucleic acids and permeabilizes bilayers. Biochemistry 36(10):3008–3017

Xi S, Grandis JR (2003) Gene therapy for the treatment of oral squamous cell carcinoma. J Dent Res 82(1):11–16

Yang Y, Xu Z, Chen S, Gao Y, Gu W, Chen L, Pei Y, Li Y (2008) Histidylated cationic polyorganophosphazene/DNA self-assembled nanoparticles for gene delivery. Int J Pharm 353(1–2):277–282

Yi Y, Jong Noh M, Hee LK (2011) Current advances in retroviral gene therapy. Curr Gene Ther 11(3):218–228

Yoon AR, Hong J, Yun CO (2017) Adenovirus-mediated decorin expression induces cancer cell death through activation of p53 and mitochondrial apoptosis. Oncotarget 8(44):76666

Yoshida J, Mizuno M, Wakabayashi T (2004) Interferon-β gene therapy for cancer: basic research to clinical application. Cancer Sci 95(11):858–865

Young LS, Searle PF, Onion D, Mautner V (2006) Viral gene therapy strategies: from basic science to clinical application. J Pathol 208(2):299–318. https://doi.org/10.1002/path.1896

Zhang X, Oulad-Abdelghani M, Zelkin AN, Wang Y, Haîkel Y, Mainard D, Voegel JC, Caruso F, Benkirane-Jessel N (2010) Poly (L-lysine) nanostructured particles for gene delivery and hormone stimulation. Biomaterials 31(7):1699–1706

Zhang X, Duan Y, Wang D, Bian F (2015) Preparation of arginine modified PEI-conjugated chitosan copolymer for DNA delivery. Carbohydr Polym 122:53–59

Zhi D, Zhang S, Cui S, Zhao Y, Wang Y, Zhao D (2013) The headgroup evolution of cationic lipids for gene delivery. Bioconjug Chem 24(4):487–519

Zhi D, Bai Y, Yang J, Cui S, Zhao Y, Chen H, Zhang S (2018) A review on cationic lipids with different linkers for gene delivery. Adv Colloid Interf Sci 253:117–140

Zhou HS, Liu DP, Liang CC (2004) Challenges and strategies: the immune responses in gene therapy. Med Res Rev 24(6):748–761

Zhou Y, Chen M, Zhuo Y, Chai Y, Xu W, Yuan R (2017) In situ electrodeposited synthesis of electrochemiluminescent Ag nanoclusters as signal probe for ultrasensitive detection of Cyclin-D1 from cancer cells. Anal Chem 89(12):6787–6793

Zhu N, Liggitt D, Liu Y, Debs R (1993) Systemic gene expression after intravenous DNA delivery into adult mice. Science 261(5118):209–211

Zorko M, Langel Ü (2005) Cell-penetrating peptides: mechanism and kinetics of cargo delivery. Adv Drug Deliv Rev 57(4):529–545

S.K. Manohar is presently pursuing Ph.D. in the Department of Pharmaceutics at JSS College of Pharmacy, Mysuru. He completed M. Pharm in Quality Assurance at Krupanidhi College of Pharmacy in 2021. His research interest is on orodispersible drug delivery system.

M.P. Gowrav is currently working as Assistant Controller of Examinations at JSS Academy of Higher Education and Research, Mysuru. He has completed M. Pharm and Ph.D. at JSS College of Pharmacy, Mysuru. He has worked as a Lecturer at Sharada Vilas College of Pharmacy for 1 year 5 months. He has 9 years of research experience and about 10 years of teaching experience. He has published 65 scientific papers in peer-reviewed national and international journals. His thrust areas include 3D printing in healthcare, multiparticulate drug delivery systems, and regulatory sciences.

H.V. Gangadharappa is currently working as an Associate Professor in the Department of Pharmaceutics, JSS College of Pharmacy, JSS Academy of Higher Education and Research, Mysuru. He has published more than 130 articles in reputed national and international journals and 2 Indian patents. His core research area is graphene nanoribbon and silk fibroin-based drug delivery systems.

Natural Hydrogels as Wound Dressing for Skin Wound-Healing Applications

15

Gabriela Fletes-Vargas, Sergio Yair Rodríguez-Preciado, Mariana Díaz-Zaragoza, and Rogelio Rodríguez-Rodríguez

Abstract

Natural polymers are widely used to produce hydrogels for skin wound-healing applications. Hydrogels possess porosity, water absorption and water retention capability, mechanical properties, and biocompatibility. Also, bioactive molecules and metal nanoparticles can be added into hydrogels to improve antimicrobial and wound-healing properties, which are necessary for dressing. This chapter reviews the main physicochemical and biological properties of the natural hydrogels used as a wound dressing. The different natural polymers such as chitosan, alginate, cellulose, and gelatin and fabrication methods to produce hydrogels are described. This chapter will contribute to a better understanding of natural hydrogels as a potential dressing for skin wound-healing applications.

Keywords

Hydrogels · Bioactive components · Regeneration · Healing

G. Fletes-Vargas
Departamento de Ciencias Clínicas, Centro Universitario de los Altos (CUALTOS), Universidad de Guadalajara, Tepatitlán de Morelos, Jalisco, Mexico

Tecnología de Alimentos, Centro de Investigación y Asistencia en Tecnología y Diseño del Estado de Jalisco A.C. (CIATEJ, A.C.), Zapopan, Jalisco, Mexico

S. Y. Rodríguez-Preciado · M. Díaz-Zaragoza
Departamento de Ciencias de la Salud, Centro Universitario de los Valles (CUVALLES), Universidad de Guadalajara, Ameca, Jalisco, Mexico

R. Rodríguez-Rodríguez (✉)
Departamento de Ciencias Naturales y Exactas, Centro Universitario de los Valles (CUVALLES), Universidad de Guadalajara, Ameca, Jalisco, Mexico
e-mail: rogelio.rodriguez4085@academicos.udg.mx

F. A. Sheikh et al. (eds.), *Interaction of Nanomaterials With Living Cells*,
https://doi.org/10.1007/978-981-99-2119-5_15

15.1 Introduction

Hydrogels are porous hydrophilic biomaterials with a tridimensional cross-linked network structure capable of absorbing and retaining considerable water into a polymer structure (Tang et al. 2022; Xiao et al. 2022). A cross-linker agent is a molecule that interconnects polymer chains via functional groups such as amine, carboxyl, or hydroxyl. The cross-linking process increases molecular weight, provides higher mechanical properties, improves stability, and impacts the hydrogels' physical properties (Zainal et al. 2021). Hydrogels can expand and absorb several folds of water into their structures without disintegrating, providing a favorable environment for the survival of various cells and mimicking the natural tissue (Saravanan et al. 2019; Xiao et al. 2022; Xu et al. 2022a). Also, hydrogels possess exciting properties such as sol-gel behavior in response to stimuli from the external environment, biocompatibility, nontoxicity, and biodegradability (Do et al. 2022; Xu et al. 2022a). The interaction between polymeric chain networks and water or biological fluids occurs through different phenomena: capillary, osmotic, and hydration forces, which are counterbalanced, causing the expansion of polymer chain networks (Varaprasad et al. 2017). The hydrogels can respond to several stimuli, such as temperature, pressure, pH, ionic charge, or antigens, with changes in specific characteristics. Then, when the stimulus finishes, hydrogels can return to their original structure. This class of hydrogels is denominated as "smart" materials (Stan et al. 2021).

Hydrogels possess significant benefits for wound dressing applications due to their mild processing conditions and ability to combine bioactive agents that help the healing process. The molecules added to hydrogels can be delivered with more accurate and progressive control than the topical or dermal application (Fan et al. 2021). For example, hydrogels can deliver specific molecules, i.e., antiseptics, antibiotics, anti-inflammatories, and antioxidants (Stan et al. 2021).

15.2 Wound Healing

15.2.1 Skin: Structure and Function

The skin is considered the largest organ and the physical barrier of the human body. It accomplishes many critical functions, such as protecting internal organs from mechanical damage and ultraviolet radiation, preventing fluid loss and controlling the body temperature, and protecting the host from microbial infections (Rodrigues et al. 2019; Nguyen and Soulika 2019). The skin is formed of three layers: epidermis, dermis, and hypodermis. The epidermis is the outermost layer formed by corneocytes and keratinocytes, providing barrier protection from environmental conditions (Vig et al. 2017). Furthermore, the epidermis is constantly renewed due to the proliferation of keratinocytes, which lose their nuclei and migrate from the basement membrane to the surface skin, creating the cornified stratum (Chu 2012). The dermis is located below the epidermis and contains nerve endings, microvascular vessels, and higher content of proteins such as proteoglycans and collagen fibers (Rippa et al. 2019). The cells that conform to the dermis are myofibroblasts, resident immune cells such as macrophages, Langerhans cells, dendritic cells, and fibroblasts (Woodley 2017). The fibroblasts are abundant cells capable of synthesizing collagen type I and supporting the remodeling of the extracellular matrix (ECM) (Sorrell and Caplan 2009). Subjacent to the dermis, we could find the hypodermis or subcutaneous fat tissue, abundant fibrocytes, and adipocytes whose principal functions are storing energy as fatty acids, thermal isolation, and endocrine, regulating glucose and lipid metabolism (Tavakoli and Klar 2020). Additionally, the subcutaneous fat tissue includes copious blood vessels and lymph vessels and produces crucial mediators such as growth factors, adipokines, and cytokines (Cildir et al. 2013).

15.2.2 Wound-Healing Process

Wound healing is a dynamic and complex process that coordinates the cells in the different skin layers to restore homeostasis. It consists of four phases that overlap in time and space (Fig. 15.1): hemostasis, inflammation, proliferation, and tissue remodeling (Tavakoli and Klar 2020).

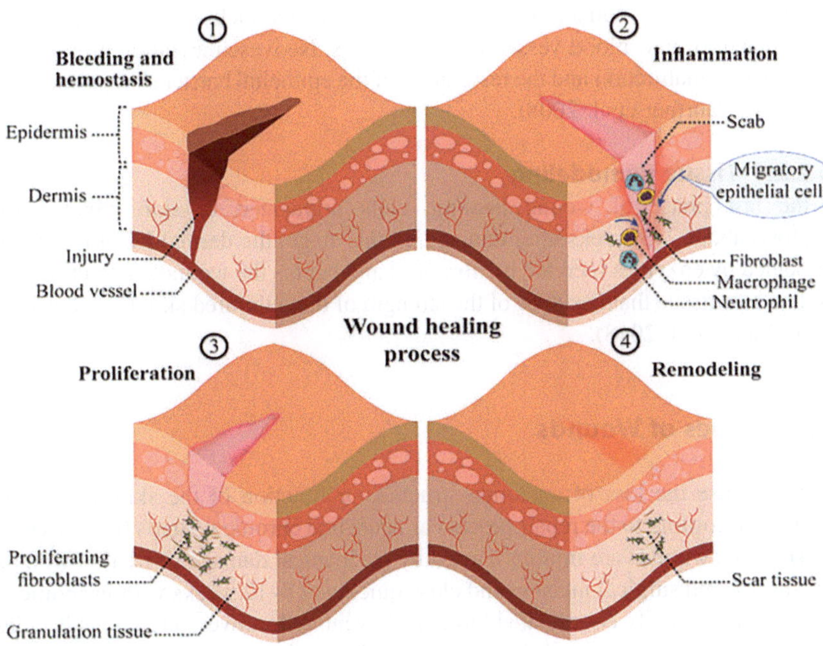

Fig. 15.1 Scheme of the wound-healing process. (Reprinted from Abazari et al. 2022, copyright 2022, with permission of Elsevier)

15.2.2.1 Hemostasis

After the vascular damage, the subendothelial matrix is disrupted and exposed. Platelets initiate the vasoconstriction process attached to vessel walls and aggregate with each other to form the blood clot and stop the bleeding (Rumbaut and Thiagarajan 2010). Also, the activated platelets release growth factors and cytokines that act as mediators, such as transforming growth factor-β (TGF-β), epidermal growth factor (EGF), and vascular endothelial growth factor (VEGF), even 7 days after injury (Qing 2017). Following this, the coagulation cascade activation converts the fibrinogen into a fibrin mesh to form the thrombus as a temporal scaffold for critical cells for wound healing.

15.2.2.2 Inflammation

During the inflammatory phase, the neutrophils are the predominant immunity cells during the first 48 h after the damage. Neutrophils phagocyte dead cells and destroy bacteria, release chemokines, and attract macrophages to the wound site. Also, the circulating monocytes migrate and maturate into tissue macrophages (Pereira et al. 2017).

15.2.2.3 Proliferation

The proliferation phase happens about 2–10 days after the insult and is characterized by fibroblasts proliferating in the wound area, forming the granulation tissue, and depositing new ECM proteins like collagen. The granulation tissue will be subsequently replaced by connective tissue. Growth factors such as VEGF induce the development of new blood vessel or angiogenesis. Neovascularization provides the keratinocytes maturation and the restoration of the epithelial barrier (Desjardins-Park et al. 2018; Gurtner et al. 2008).

15.2.2.4 Tissue Remodeling

In the last phase of wound healing, cells implicated in skin repair suffer apoptosis about 2–3 weeks after damage. In this stage, the dermal ECM is actively remodeled by enzymes secreted by fibroblast and begins the wound contraction. The result is scar tissue that has 80% of the strength of the uninjured skin (Bowden et al. 2016; Wang et al. 2018).

15.2.3 Types of Wounds

Wounds cause the loss of anatomic structure or function of the skin and can be classified according to the repair process as acute and chronic wounds (Lazarus et al. 1994). Acute wounds are injuries caused by mechanical damage or friction with the skin surface and surgical incision and close quickly at 8–12 weeks with insignificant scarring due to highly coordinated biological events. Moreover, alterations in cellular signaling and excessive inflammation in the wound-healing process (Berman et al. 2017) can induce abnormalities such as excessive scarring or wounds that do not heal, even after 12 weeks. These injuries are classified as chronic wounds (Lindholm and Searle 2016; Wilkinson and Hardman 2020).

15.2.4 Causes of Chronic Wounds

Chronic wounds heal poorly and are associated with underlying pathological conditions such as hemoglobinopathies, diabetes, vascular disease, cancer, and malnutrition (Han and Ceilley 2017). Studies have shown that inflammatory environments of chronic wounds are related to high expression of reactive oxygen species (ROS) interrupting the cellular redox balance, associated with metabolic disorder and compromising the integrity of blood vessels, and avoiding the normal transition between the inflammatory to proliferation phases (Malone-Povolny et al. 2019; Xu et al. 2020). Also, this offers a proper natural environment for bacterial infections, which prolongs the damage and hypoxic conditions (Tandara and Mustoe 2004).

15.2.4.1 Diabetic Wounds

Diabetes mellitus is a chronic metabolic disease linked to hyperglycemia and foot ulcers that do not heal. Among the physiological complications of diabetic foot ulcers include (1) infection and barrier injury, (2) excessive oxidative stress, (3) neuropathy, (4) microvascular difficulties, and chronic inflammation (Burgess et al. 2021). Generally it initiates as foot deformity, and consequently, the nerves are damaged, reducing skin sensitivity. These alterations exacerbate vascular injury causing gangrene, arterial obstruction, and ischemia (Blanco-Fernandez et al. 2020).

15.2.4.2 Pressure Ulcers

Wound pressure is an injury caused by the localized destruction of skin integrity or underlying tissue because a body area constantly interacts with an external surface driving to pressure damage or ulcers. Pressure ulcers typically happen in old/paralyzed patients and are favored by devices such as nasal cannulas and nasogastric tubes (Maaz Arif et al. 2021; Bowers and Franco 2020). Pressure ulcers not treated can harm deep soft tissue and develop complications such as osteomyelitis. The pathophysiological components accompanying pressure ulcers are ischemia-reperfusion wounds, inadequate lymphatic drainage, cellular apoptosis, and failure to heal (Niemiec et al. 2020).

15.3 The Bacterial Population on Wounds

15.3.1 Skin Microbiota

Microorganisms can be found in many environments (e.g., water, soil, and the atmosphere), including animals, plants, and humans (Ederveen et al. 2020). Within humans, microorganisms can colonize different areas such as the nose, throat, mouth, vagina, intestine, and skin, giving rise to a bacterial community that is part of the human microbiota (Da Silva and Domingues 2017). The microbiota located on the skin is made up of bacteria from four main phyla: *Actinobacteria* (51.8%), *Bacteroidetes* (6.3%), *Firmicutes* (24.4%), and *Proteobacteria* (16.5%); however, the presence of these bacteria will differ throughout the skin. For example, *Streptococcus* is one of the bacteria in higher proportion in the forehead and behind the ears, while *Corynebacterium* is present in the armpits. In the moist areas of the skin, the most abundant species are *Staphylococcus* and *Corynebacterium*, and in the sebaceous sites, some *Propionibacterium* has been reported (Gao et al. 2010; Sanford and Gallo 2013). In addition to bacteria, commensal fungi and viruses are part of the skin microbiota. For example, *Aspergillus*, *Rhodotorula*, *Cryptococcus*, and *Epicoccum* are some fungi species that are part of this microbiota and have been found mainly in the foot area (Adamczyk et al. 2020).

15.3.2 The Role of the Microbiota in the Skin

The skin microbiota plays an essential role in maintaining homeostasis, producing proteases that participate in the desquamation and renewal of the stratum corneum. In addition, they retain a slightly acidic pH, reducing triglyceride levels in sebaceous areas and favoring the production of fatty acids. The microbiota's generation of antimicrobial compounds inhibits opportunistic microorganisms' growth, thus preventing the onset of infectious processes (Boxberger et al. 2021; Gribbon et al. 1993). On the other hand, the microbiota interacts with the host's immune system, generating an innate and adaptive immune response, thus reinforcing itself due to the detection process of the various bacterial populations (Park and Lee 2018). Studies have reported that this interaction with the immune system favors wound repair (Lai et al. 2009; Linehan et al. 2018). However, other reports mention the opposite effect, where the absence of the microbiota in the skin tends to the healing process; therefore, more research is necessary to understand the role of the microbiota in the wound repair process (Canesso et al. 2014). For example, *C. striatum* generates a factor that inhibits the Agr gene regulatory system, which controls the virulence factors of *S. aureus*, thus avoiding infections by this bacterium (Ramsey et al. 2016).

15.3.3 Factors That Modify the Skin Microbiota

The microbiota can be altered by a wide variety of intrinsic and extrinsic factors to which the human being is exposed; these changes will also depend on the time of exposure to these factors (Moskovicz et al. 2020). Within the intrinsic factors, we find the area of the skin. As mentioned above, the microbiota will depend on the conditions of the skin being colonized (e.g., moist, dry, or sebaceous sites) (Grice et al. 2009). Another factor is ethnicity; Li et al. (2019) found that the microbiota of East Asians is different from that of Caucasian and Latino populations (Li et al. 2019). In another study conducted by Perez Perez et al. (2016), the skin microbiota of African Americans differed from other population groups (Latinos, Caucasians, and Asians) (Perez Perez et al. 2016).

Other factors that contribute to variations in the microbiota are gender and age. Regarding gender, the most abundant bacteria in men are *Enhydrobacter*, *Cutibacterium*, *Corynebacterium amycolatum*, and *Corynebacterium kroppenstedtii*, while in sebaceous sites are *Epicoccum* and *Cryptococcus*. In women, the most abundant microbiota are *Staphylococcus*, *Streptococcus*, *Enterobacteriales*, *Moraxellaceae*, *Lactobacillaceae*, *Corynebacterium urealyticum*, *Corynebacterium variabile*, and *Pseudomonadaceae*, while in sebaceous sites, there is a more significant presence of *Malassezia* (Callewaert et al. 2013; Fierer et al. 2008; Jo et al. 2016; Leung et al. 2015; Prohic et al. 2014; Shami et al. 2019; Zhai et al. 2018). Even though it has been seen that the presence of bacteria does not vary with age, the amount of microorganisms does decrease as age advances (Jo et al. 2016; Dimitriu et al. 2019).

The extrinsic factors influencing the first type of microbiota composition are the type of childbirth, the postpartum environment, and the health staff. However, this microbiota is temporary, as it will later be influenced by other extrinsic factors described below (Chu et al. 2017; Dominguez-Bello et al. 2010).

Lifestyle, hygiene, and cosmetics are factors that also influence the skin microbiota. Reports have indicated that makeup inhibits the growth of *S. aureus* and *C. acne*, whereas the use of emulsifiers favors the growth of *S. aureus* (Gannesen et al. 2019; Nielsen et al. 2016; Staudinger et al. 2011). Other factors that affect the microbiota on the skin are geographic location, climate, seasonality, and air pollution, the latter of which has been seen to degrade the diversity of the microbial population (Boxberger et al. 2021).

15.3.4 Skin Diseases Caused by Microorganisms

Although the microbiota confers various benefits to the host, changes in the microbiota alter host-microbiome interactions, resulting in multiple diseases (Schommer and Gallo 2013). These alterations in the microbiota generate dysbiosis, defined as the loss of balance in the composition of the microbiota or changes in the metabolic activities of the microbiota (Degruttola et al. 2016). Some diseases that are generated when the balance is lost are mentioned below. Acne vulgaris is a disease due to *Propionibacterium acnes* and *Cutibacterium acnes*; these bacteria colonize the sebaceous follicles producing enzymes, such as hyaluronidases, lipases, and proteases, causing local injuries and inflammations (Byrd et al. 2018; Flowers and Grice 2020; Schommer and Gallo 2013). Rosacea is a chronic skin condition involving the central part of the face with transient or persistent erythema, telangiectasias, inflammatory papules and pustules, or connective tissue hyperplasia (Oge' et al. 2015). The presence of different microorganisms, such as *Staphylococcus epidermidis*, *Helicobacter pylori*, *Chlamydophila pneumonia*, and *Demodex folliculorum*, has been associated with this disease (Murillo et al. 2014).

Atopic dermatitis is a chronic and highly pruritic inflammatory skin disease (Kapur et al. 2018). People with this disease are susceptible to infections by *Staphylococcus aureus* and the herpes virus; this is attributed to the decrease in antimicrobial proteins. The severity of the disease is also associated with the loss of diversity of the microbiota, so one of the most effective treatments is the increase in the presence of bacteria of *Corynebacterium*, *Streptococcus*, and *Propionibacterium* genus (Sanford and Gallo 2013; Schommer and Gallo 2013).

Psoriasis is a chronic proliferative and inflammatory dermatosis of the skin in which firmicutes have been found to be predominant. Another study also reports a high presence of *Corynebacterium* and a reduction of *Staphylococcus* and *Cutibacterium*. However, it is unknown if the changes in the microbiome are caused by the disease or vice versa (Loesche et al. 2018; Schommer and Gallo 2013). The infections due to opportunistic pathogens occur mainly in people with primary immunodeficiency. These individuals are more susceptible to fungal infections such as *Candida* spp. and *Aspergillus* spp. and bacteria such as *Serratia marcescens* (Byrd et al. 2018). Lastly, many of the bacteria in the normal microbiota can eventually cause infection in nonhealing or poorly healing wounds (diabetic foot ulcers, postsurgical wounds, or decubitus ulcers), occurring more frequently in elderly or diabetic people (Sanford and Gallo 2013). Bacteria such as *Staphylococcus aureus* and *S. epidermis* have been isolated from superficial wounds, while bacteria such as *P. aeruginosa*, *Finegoldia*, *Peptoniphilus*, and *Peptostrptococcus* have been found in deeper wounds and with longer healing time (Ederveen et al. 2020). Infections due to *Staphylococcus* spp. and *Streptococcus* spp. have been reported in the wounds of diabetics (Gardner et al. 2013) in addition to opportunistic fungal infections such as *Cladosporium* spp. and *Candida* spp. (Swaney and Kalan 2021). In wounds caused by burns, the presence of infections caused by thermophilic bacteria (*Aeribacillus*, *Caldalkalibacillus*, *Nesterenkonia*, and *Halomonas*) and a decrease in commensal bacteria of the genus *Cutibacterium* and *Corynebacterium* have been reported (Rensburg et al. 2015).

15.3.5 Conventional Antimicrobial Agents

The skin is the human body's largest organ and forms an integral part of the immune system. In this sense, the skin is the first line of defense against microbial infections. One strategy that may reduce the risk of bacterial infection is applying an antibacterial dressing (Yang et al. 2022c). Antibiotics are antimicrobial compounds used to kill bacteria and fight bacterial infections, for instance, tetracycline, ciprofloxacin, gentamicin, and sulfadiazine (Hauser et al. 2016). However, there are other compounds with an antimicrobial effect used for the treatment of bacterial infections, such as nanoparticles (Mussin et al. 2021) and natural products such as honey, essential oils, and chitosan (Yang et al. 2022c) (Simões et al. 2018). These materials have shown promising antibacterial activities following their application after a wound surgery and provided the potential ability to reduce wound infections (Yang et al. 2022c).

15.3.5.1 Silver Nanoparticles

Silver is a metal with a long history in traditional medicine because it has a high antimicrobial activity and low toxicity in animal cells (Rai et al. 2009, 2012). These nanoparticles may be toxic in humans, but effects can be attenuated when silver is used to form nanoparticles (Ferdous and Nemmar 2020). Some silver compounds (e.g., silver nitrate and silver sulfadiazine) have been used to treat burns, wounds, and several bacterial infections to reduce skin infections. In recent years, research has increased on the antimicrobial effect of silver nanoparticles in treating wounds and skin infections (Mussin et al. 2021). It has been reported that these nanoparticles have antimicrobial activity against different bacterial species such as *Escherichia coli*, *Enterococcus faecalis*, *Pseudomonas aeruginosa*, *Staphylococcus aureus*, and *Streptococcus mutans* (Brunauer et al. 2021; Bruna et al. 2021; Yin et al. 2020).

15.3.5.2 Essential Oils

Essential oils are secondary metabolites, volatile, natural, complex compounds characterized by a strong odor produced by aromatic plants. They have been widely used for antioxidant, virucidal, fungicidal, antiparasitic, insecticidal, medicinal, and bactericidal applications. The biological activity of the oils is compared with synthetic pharmaceutical compounds (Hamdy 2020). Essential oils possess antibacterial properties, e.g., terpenes and terpenoids show inhibitory activity against *Staphylococcus aureus* (Safaei-Ghomi and Ahd 2010), carvacrol has specific effects on *S. aureus* and *Staphylococcus epidermidis*, and perilla oil suppresses the expression of α-toxin of *Staphylococcus* enterotoxin A and B and toxic shock syndrome toxin. By last, geraniol shows promising activity in modulating drug resistance in several gram-negative species (Ning Chen 2021; Solórzano-Santos and Miranda-Novales 2012).

Reports indicate that essential oils have antibacterial properties against many bacterial strains, such as *Listeria monocytogenes*, *L. innocua*, *Salmonella typhimurium*, *Escherichia coli*, *Shigella dysenteria*, *Bacillus cereus*, *Staphylococcus aureus*, and *Salmonella typhimurium* (Chouhan et al. 2017; Man et al. 2019). The mechanism of action of essential oils of plants includes attacking the cell membrane, disrupting enzyme systems, damaging the bacteria's genetic material, and forming fatty acid hydroperoxides caused by the oxygenation of unsaturated fatty acids (Turgis et al. 2009). With high antimicrobial activity, these essential oils are natural phenolics used as antibacterial ingredients in hydrogel dressing (Ning Chen 2021).

15.4 Natural Hydrogels as a Wound Dressing

Natural polymers, including polysaccharides and proteins, are the most used for producing hydrogels since they are biocompatible and can be obtained easily from natural resources, e.g., polysaccharides from plants, algae, and microorganisms like fungi and bacteria (Raina et al. 2022). Moreover, polysaccharides possess abundant functional groups, such as hydroxyl, carboxyl, and amine groups, for chemical modification and induce the high-water retention property (Stan et al. 2021).

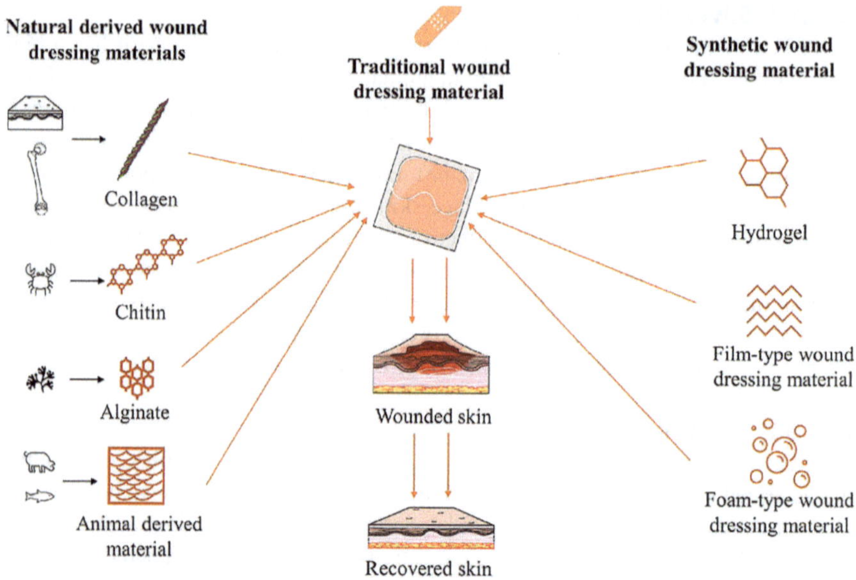

Fig. 15.2 Classification and molecular structure of different skin and wound dressing materials. (Reprinted from Peng et al. 2022, copyright 2022, with permission of Elsevier)

Wound dressings are classified into two classes: passive and interactive. Passive dressings act only to protect the wound area but they do not directly affect the wound (Prasathkumar and Sadhasivam 2021). Traditional dressings such as medical skimmed cotton gauze, cotton pads, and Vaseline gauze are the most widely used for skin wounds in clinical practice. Traditional dressings are still commonly used in skin wounds due to their low price, relatively simple manufacturing process, ease of use, and protective effect on wound healing. However, traditional dressings also have obvious shortcomings (Fig. 15.2) (Xu et al. 2022a). Interactive dressings are the most modern dressing products as they interact with the wound surface area to produce an optimum environment at the dressing interface (Prasathkumar and Sadhasivam 2021).

Wound dressings are required to provide a barrier between the wound and the external environment. The "ideal" hydrogel for wound management should (1) have antibacterial activity, (2) absorb all excess exudate and toxins on the wound surface, (3) keep good moisture between the wound and the dressing, (4) offer mechanical protection, (5) preserve the wound from external sources of infection, (6) prevent excess heat at the wound, (7) have good permeability to gases, (8) be easy to remove after healing without further trauma to the wound, (9) be sterile, (10) be biocompatible, and (11) be nonallergenic (Fig. 15.3) (Rodríguez-Rodríguez et al. 2020; Stan et al. 2021). Due to their high moisture content, these dressings also provide a cooling, soothing effect and reduce the pain associated with dressing changes. In addition, the limited adhesion of hydrogels means that they can be easily removed

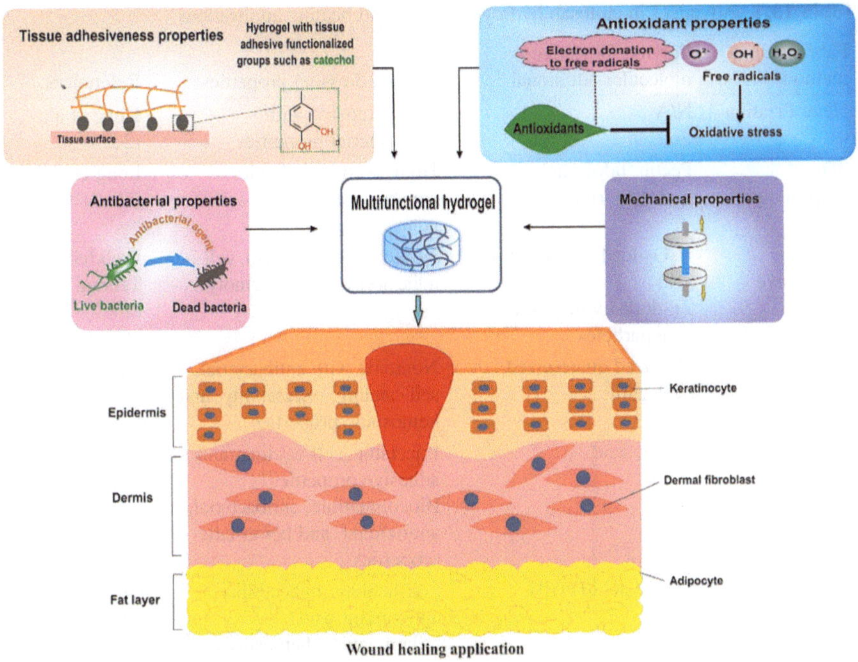

Fig. 15.3 Properties of multifunctional hydrogels for wound healing: antioxidant effects, antibacterial activities, tissue adhesiveness, and mechanical properties. (Reprinted from Asadi et al. 2021, copyright 2022, with permission of Elsevier)

from the wound without causing further trauma to the healing tissue. The transparent nature of some hydrogel dressings also allows clinical assessment of the healing process without the need to remove the dressing (Gupta et al. 2019).

Thus, the physicochemical properties of the hydrogels, such as mechanical, rheological, swelling, moisturizing, and heat absorption properties, are relevant for their applicability (Cui et al. 2022). Table 15.1 displays the advances in natural hydrogels used for wound-healing applications. In literature, hydrogels have been widely used as a polymer dressing. Bioactive molecules, metal nanoparticles, or other compounds can be added to hydrogels to improve their properties (Koehler et al. 2018; Raina et al. 2022; Zhang et al. 2020).

15.4.1 Chitosan Hydrogels as a Wound Dressing

Chitosan is a cationic linear polysaccharide composed of β-(1-4)-2-amino-2-deoxy-D-glucopyranose structure obtained from chitin, the second most prevalent natural polysaccharide in nature (Ji et al. 2022). Chitosan can be processed using various methods such as casting, fiber spinning, supercritical fluid processing, and electrospinning to produce different forms like films, microparticles, or nanofibers

Table 15.1 Advances in chitosan hydrogels for wound healing applications

Fabrication method	Polymer/bioactive molecules additional	Wound dressing properties	References
Neutralization	N/A	Biodegradable with antibacterial properties	Kong et al. (2020)
3D bioprinting	Pectin, lidocaine hydrochloride	High swelling behavior, suitable drug release, self-adhesion to skin.	Long et al. (2019)
Stimuli-responsive	H_2O_2-loaded polylactic acid Zn-doped whitlockite nanoparticles	Nontoxic, cell growth, cell adhesion, and low hemolysis	Dadkhah Tehrani et al. (2022), Yang et al. (2022a)
Gelification in situ	Oxidized quaternized guar gum	Nontoxic with antibacterial, self-healing, injectability, and hemostatic properties	Yu et al. (2022)
	Tannic acid	Injectable, self-healing, and adhesive properties Biocompatible, antibacterial, antioxidant, and hemostatic properties	Guo et al. (2022b)
	Tannic acid/Fe(III)	Self-healing, injectability, antioxidant, anti-inflammatory, hemostasis, biocompatibility, and wound healing ability	Guo et al. (2022c)
	N/A	Biocompatibility and wound-healing properties	Luo et al. (2022)
	Carboxymethyl chitosan, heparin	Cell migration and proliferation Deposition of collagen fibers and the formation of blood vessels	Chang et al. (2022)
	Adenine	Self-healing, biocompatibility, and hemostatic	Deng et al. (2022b)
Photo-cross-linking	F127/chlorhexidine NPs	Antibacterial, antioxidant, and anti-inflammatory properties	Xu et al. (2022b)
Double-cross-linking GA/$CaCl_2$	Alginate/curcumin-β-cyclodextrin inclusion	Antibacterial properties and nontoxicity	Kiti and Suwantong (2020)
Freezing/thawing	PVA and silver nanoparticles	Antibacterial properties and nontoxicity	Nešović et al. (2019)

(Kou et al. 2022). The molecular weight and acetylation degree influence several critical properties of chitosan for biomedical applications. For example, the acetylation degree affects the antimicrobial properties of chitosan by increasing its solubility and positive charge (Matica et al. 2019). Chitosan has interesting properties such as biocompatibility, biodegradability, antibacterial, hemostasis, anti-inflammatory, good absorption of exudate, and tissue regeneration and skin collagen fiber growth (Xu et al. 2022a).

The solubility, viscosity, biocompatibility, antimicrobial, analgesic, antioxidant, hemostatic, and mucoadhesive properties of chitosan increase with decreasing degree acetylation, while crystallinity and biodegradability increase with increasing degree of acetylation (Matica et al. 2019). In this sense, chitosan exerts its wound-healing effect by promoting hemostasis, antimicrobial activity, and free radical scavenging activity and regulating the inflammatory response (Loo et al. 2022). However, chitosan has disadvantages as dressings, including moisture sensitivity, poor mechanical performance, and insolubility in water and solvents (Ji et al. 2022). Chitosan hydrogels can be produced commonly by physical or chemical cross-linking. Physical hydrogels involve the formation of electrostatic, hydrophobic, and hydrogen bonding forces between polymer chains. In this sense, chitosan can form a hydrogel without adding any additive. For example, chitosan hydrogel can be produced using a neutralization process of their amino groups, which prevents repulsion between the polymer chains and hydrogen bonds; hydrophobic interactions and chitosan crystallites are formed (Rodríguez-Rodríguez et al. 2020; Pita-López et al. 2021). Chemical cross-linking leads to hydrogels with improved mechanical properties and chemical stability. To create these hydrogels, the polymer chains are covalently bonded by small cross-linker molecules, secondary polymerizations, or irradiation (Rodríguez-Rodríguez et al. 2020). Table 15.1 displays the significant advances of chitosan hydrogels for wound-healing applications. Long et al. (2019) developed 3D-printed chitosan-pectin hydrogel incorporating lidocaine. The hydrogels produced displayed suitable printability and structural integrity. Also, hydrogels swelled quickly and reached an equilibrium between 2 and 4 h. The hydrogels showed adhesive strength between 0.85 and 1.24 N, similar to commercial wound dressings fabricated of silicone, polyurethane, and acrylate. In this sense, a model wound dressing should be self-adhesive, easily detachable, and painless. The authors described that appropriate adherence and easily removable dressings could protect the wound against trauma and prevent tissue harm.

Dadkhah Tehrani et al. (2022) produced thermosensitive chitosan hydrogels covered with a decellularized human amniotic membrane and H_2O_2-loaded polylactic acid microparticles. The porous hydrogel displayed low hemolysis (5%) and was nontoxic, favoring cell growth and adhesion for fibroblasts. Similar results were reported by Yang et al. (2022a) on multifunctional methacrylate anhydride quaternized chitosan hydrogel incorporating Zn-doped whitlockite nanoparticles. The hydrogels displayed antibacterial activity against *Staphylococcus aureus* and *Escherichia coli* (Fig. 15.4).

Guo et al. (2022b) developed a porous multifunctional injectable quaternary ammonium chitosan hydrogel for wound-healing applications. The chitosan hydrogels were obtained using tannic acid as an ionic cross-linker agent. For the gelation time of 8 and 21 min, water content of chitosan hydrogels decreases with increasing tannic acid concentrations (1.25, 2.5, and 5 wt%). In contrast, mechanical properties increased with the cross-linking degree. The authors related these results with the cross-linking degree. The adhesive property of hydrogels is an important parameter that helps to adhere and seal wounds, preventing bacterial infection. The

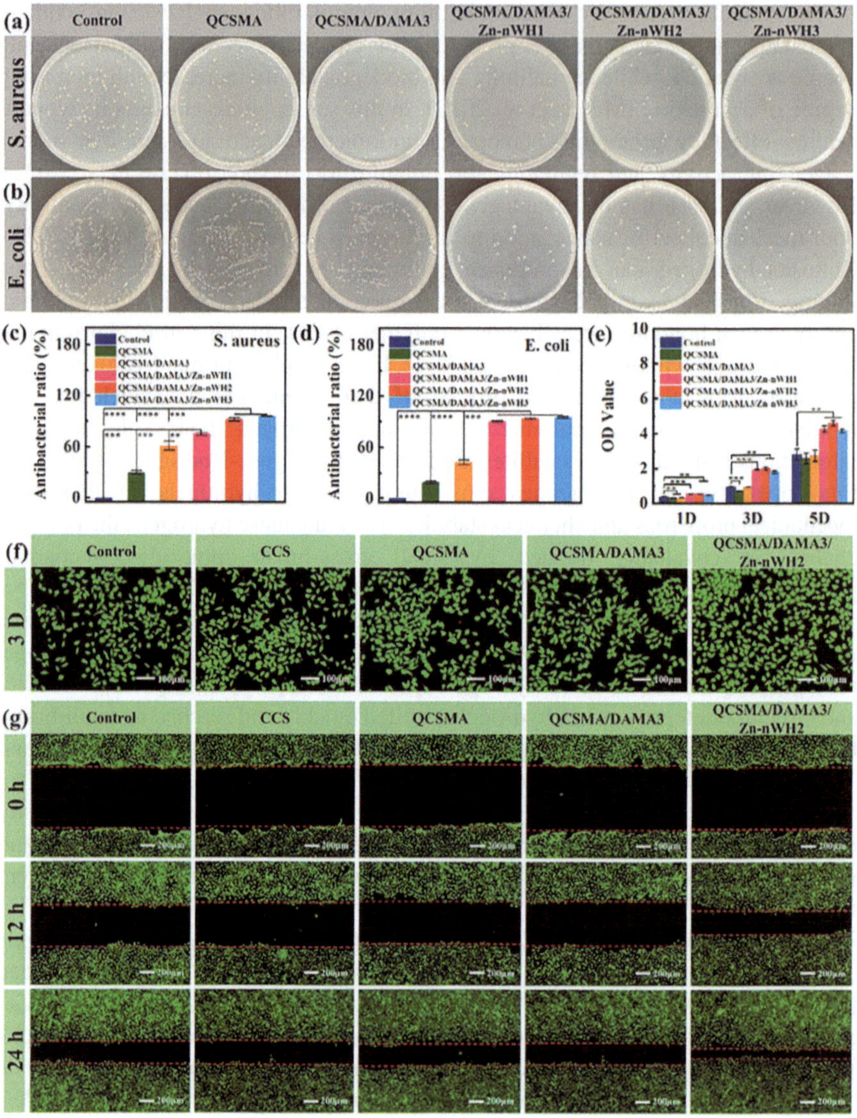

Fig. 15.4 Antibacterial property and cell compatibility of hydrogels on agar plates after contact with hydrogels using *S. aureus* (**a**) and *E. coli* (**b**). The antibacterial rate of hydrogels to *S. aureus* (**c**) and *E. coli* (**d**) by direct contact method. (**e**) The OD value of L929 cells was obtained by direct contact method for 1 day, 3 days, and 5 days. (**f**) Live/dead staining of L929 cells after coculture with hydrogels for 3 days. Scale bar: 100 μm. (**g**) HUVECs cell migration at different times (0, 12, and 24 h). Scale bar: 200 μm. (Reprinted from Yang et al. 2022a, copyright 2022, with permission of Elsevier)

chitosan hydrogels produced in this study displayed an adhesive strength that increased with tannic acid concentration. Also, chitosan hydrogels killed more than 99% of *S. aureus* and *E. coli* using the surface antibacterial activity and the zone of inhibition test after interacting with chitosan hydrogels for 2 h. On the other hand, hydrogels displayed low hemolysis ratios (<5%) and high cell viability (>70%).

Similar results were reported by Guo et al. (2022c). They developed a multifunctional chitosan hydrogel using a one-step free radical polymerization reaction. Tannic acid and different concentrations of $FeCl_3$-$6H_2O$ were added (0, 9, 18, and 36 mM). The hydrogels displayed a porous structure with high water content (92%), which would help the adsorption of tissue exudates and the exchange of nutrients and gases through the wound. Also, the increase of Fe(III) concentration decreased the elastic behavior of the chitosan hydrogels, which can be related to the decrease in cross-linking degree. Interestingly, chitosan hydrogels displayed suitable adhesiveness in different biological tissues (heart, liver, spleen, lung, kidney, skin) and non-biological materials (wood, iron, plastic, glass, and rubber). The hydrogels showed excellent antibacterial properties against *Escherichia coli* and *Staphylococcus aureus*, with mortality values of 95%.

Luo et al. (2022) produced physical chitosan hydrogels using an alkaline aqueous solution (7% NaOH and 12% urea), followed by thermal gelling and solvent change. The transparent chitosan solutions gelled at temperatures higher than ~40 °C. The chitosan hydrogels display a translucent appearance, interconnected porous structure, and elastic behavior. Compared with hydrogels produced using the acid method, the hydrogels made using the alkaline methods displayed higher mechanical properties (tensile strength, Young's modulus, and elongation). The leaching solutions from the chitosan hydrogels did not show cytotoxicity for L929 fibroblasts. The cells were seeded into chitosan hydrogels, demonstrating that chitosan hydrogels were noncytotoxic. The chitosan hydrogels induced a faster wound closure than gauze and improved reepithelialization and granulation tissue formation.

Xu et al. (2022b) produced a chitosan methacrylate-gallic acid hydrogel loaded with nanoparticles with antioxidant and antimicrobial activity. The porous hydrogels displayed a suitable water vapor transmission property similar to those obtained for normal skin, a critical property as a wound dressing. The NIH 3T3 cells exhibited excellent biocompatibility using chitosan hydrogel extracts, while hydrogels support cell adhesion. For wound dressing applications, chitosan hydrogels displayed antibacterial properties. The chitosan hydrogels containing higher NP concentrations (F127/chlorhexidine) showed the highest bactericidal efficiency against *S. aureus* and *E. coli* (99.9%).

15.4.2 Cellulose Hydrogels as a Wound Dressing

Cellulose is the most abundant bioavailable and cost-effective polymer on Earth, mainly produced from various agricultural wastes (Thivya et al. 2022). Cellulose represents about 40% of the concentration of carbon in plants, providing their mechanical and structural integrity (Liu et al. 2022a). Wood pulps (85–88%) and cotton linter represent the primary source of cellulose (Wong et al. 2021). Also, cellulose is nontoxic and biodegradable and is a biocompatible polymer with a stable structure (Liu et al. 2022a). Cellulose is a linear polymer composed of a long chain of basic monomeric units of D-glucose joined together through β-(1,4) glycosidic linkages (Wong et al. 2021). Cellulose possesses numerous hydroxyl groups, which can form polymer networks linked by hydrogen bonds. Thus, hydrogels can be produced by establishing intermolecular hydrogen bonding within the polymer chains and/or covalent bonding with functionalized cross-linkers (Wong et al. 2021). Conversely, cellulose is insoluble in common solvents and thus poses a significant threat in the preparation of hydrogels. In this sense, cellulose can be chemically modified to break hydrogen bonds and improve hydrophilicity, increasing their solubility (Liu et al. 2022a). Cellulose cannot be used naturally due to its high concentration of hydroxyl groups. Commonly, cellulose is modified using different chemical reactions to form cellulose hydrogel (Kundu et al. 2022). Also, cellulose did not possess antimicrobial properties (Table 15.2).

Deng et al. (2022a) developed cellulose composite hydrogels with chitosan by covalent self-cross-linking through Schiff base reaction. Hydrogels displayed excellent biocompatibility with higher than 90% cell viability values. The hydrogels showed a homogeneous tridimensional polymer structure with a surface roughness favorable for cell adhesion, high water absorption properties, and equilibrium swelling ratios above 1000%. In this sense, hydrogels should have good adsorption to guarantee the appropriate absorption of additional liquid on the wound surface. Similarly, hydrogels must generate a humid atmosphere to avoid hydrogel adherence to the wound (Song et al. 2021). The author describes that the antibacterial activity of hydrogels is a critical property as a suitable wound dressing, preventing wound infection and favoring the healing process (Deng et al. 2022a). Hydrogels containing cellulose and chitosan displayed suitable antibacterial properties with an efficient killing rate between 75.8% and 96%.

Silver nanoparticles have been used in biomedical applications because of their antimicrobial properties, which can be incorporated into cellulose hydrogels (Song et al. 2021; Gupta et al. 2020). Cellulose hydrogels with silver nanoparticles displayed antimicrobial activity, while that of cellulose hydrogels containing curcumin did not exhibit activity. Forero-Doria et al. (2020) produced cellulose hydrogels with multiwalled carbon nanotubes and bioactive compounds enhancing antimicrobial and wound-healing properties. Similar results were reported by Koneru et al. (2020) and Dharmalingam and Anandalakshmi (2020) using grapefruit seed extract.

Yang et al. (2022b) developed resveratrol-cellulose nanofibrils with PVA-borax. The porous hydrogel displayed suitable healing ability, where it can be strained to more than ten times its original length. This property was corroborated using a strain amplitude sweep. Similar to chitosan hydrogels previously revised, cellulose-based hydrogel showed excellent adhesion to wood, metal, plastic, and glass. Also, cellulose hydrogels showed intense tissue-adhesive activity, allowing them to be

Table 15.2 Advances in cellulose hydrogels for wound-healing applications

Fabrication method	Polymer/bioactive molecules additional	Wound dressing properties	References
Schiff base reaction	Quaternized chitosan Carboxymethyl chitosan	Water retention capacity, cell proliferation, cell spreading, self-healing, and antibacterial properties Hemostatic effects	Deng et al. (2022a), Yin et al. (2022)
Dual light-responsive	Prussian blue nanoparticles and Pluronic® F127	Hemostatic effects and antibacterial properties	Shi et al. (2022)
Coagulation	Rifampicin	Healing and antibacterial properties, cell proliferation	Zhang et al. (2022)
Freeze-thawing process	Polyvinyl alcohol, silver nanoparticles	Antibacterial, wound-healing, and biocompatibility properties	Song et al. (2021)
Chalcone cross-linking	Allantoin, dexpanthenol, resveratrol, and linezolid Multiwalled carbon nanotubes, chalcone	Wound-healing and antibacterial properties	Forero-Doria et al. (2020)
Solvent casting	Grapefruit seed extract nanoparticles Zinc oxide nanoparticles	Antimicrobial activity	Koneru et al. (2020), Dharmalingam and Anandalakshmi (2020)
	Reduced graphene oxide	Antibacterial properties and low citotoxicity	Ali et al. (2019)
	Tungsten oxide	Anti-inflammatory and antibacterial properties	El Fawal et al. (2018)
Ionic cross-linking	Collagen	Cell adhesion and proliferation. Wound healing properties	Basu et al. (2018)
Gelation in situ	Resveratrol-polyethylene glycol-cellulose nanofibril conjugate, PVA, Borax	Antibacterial, antioxidante, self-healing properties	Yang et al. (2022b)
Electron beam irradiation and neutralization	Acrylic acid	Cell adhesion and biocompatibility	Loh et al. (2018)
Derivatization process	N/A	Biocompatibility and antibacterial properties	Orlando et al. (2020)

directly attached to the human skin. Also, they displayed high water vapor permeability, the critical parameter that maintains the equilibrium between fluids on the wound site. Lastly, cellulose-based hydrogels did not show cytotoxicity using L929 cells, wound closure capabilities, and antioxidant and antibacterial properties against *S. aureus* as a bacteria model.Loh et al. (2018) developed bacterial cellulose/acrylic acid hydrogels as cell carriers for wound healing applications. Dermal cells (dermal fibroblasts and epidermal keratinocytes) attached to cellulose hydrogels increase the number of cells with time. Also, cellulose-based hydrogels induced high cell viability and low cytotoxicity at 1 and 3 days of cell culture. The use of the cellulose hydrogels reduced the animal wound area over time, while the healing rate was different. On day 13, the wound treated with cells was reepithelialized and healed wholly compared to other groups.

Orlando et al. (2020) synthesized bacterial cellulose using a derivatization process to add active functional groups through covalent attachment to the polymer structure. The modified bacterial cellulose films did not have cytotoxicity for keratinocytes. At the same time, cell morphology on monolayer culture was preserved, demonstrating that the cellulose hydrogels maintained cell growth and cell proliferation. Also, cellulose hydrogels displayed higher antibacterial ability against *Escherichia coli* and *Staphylococcus aureus* than those for unmodified bacterial cellulose. The modified cellulose films decreased at 53% the bacterial cells growing by more than half as compared to the unmodified bacterial cellulose films.

Several authors have reported cellulose hydrogels adding inorganic molecules such as graphene oxide (Ali et al. 2019) and tungsten oxide (El Fawal et al. 2018). For example, El Fawal et al. (2018) developed hydroxyethyl cellulose films with tungsten oxide for wound treatment. The films displayed a sponge-like structure with high porosity and high swelling capacity. Also, cells seeded on the films did not show morphological changes. The addition of tungsten oxide (0.04%) favored the cell migration toward the scratched area to almost closure of the wound. The results displayed that the cellulose membranes had an anti-inflammatory and antibacterial efficacy against *Salmonella* sp., *P. aeruginosa*, and *E. coli*.

15.4.3 Alginate Hydrogels as a Wound Dressing

Alginate is an anionic biopolymer composed of β-L-guluronic acid (G) and (1–4) related α-D-mannuronic acid (M), commercially isolated from the marine brown algae class of *Phaeophyceae* such as *Ascophyllum nodosum*, *Macrocystis pyrifera*, *Laminaria digitata*, and *Laminaria hyperborea*. Wound dressings developed from alginate are characterized by nontoxicity, biocompatibility, reduced wound odor and pain, oxygen permeability, and hemostatic and antimicrobial properties, which are significant roles for acute and chronic wound healing such as surgical infection wounds, pressure sores, and leg ulcers (Prasathkumar and Sadhasivam 2021). Table 15.3 displays the significant advances of alginate hydrogels for wound-healing applications. Alginate is a polymer that can be readily cross-linked using calcium ions to produce physical hydrogels. Li et al. (2022a) fabricated alginate hydrogels loaded with deferoxamine and copper nanoparticles. The hydrogels were noncytotoxic and demonstrated their effectiveness against *E. coli* and *S. aureus*. Also, adding deferoxamine and copper nanoparticles into hydrogels accelerated the wound-healing activity compared with the control, reducing the wound area after 10 days using in vivo model.

Table 15.3 Advances in alginate hydrogels for wound-healing applications

Fabrication method	Polymer/bioactive molecules additional	Wound dressing properties	References
Ionic cross-linking	Deferoxamine and copper nanoparticles	Nontoxic, cell migration and proliferation with antibacterial and wound-healing activity	Li et al. (2022a)
	Platelet-rich plasma fibrin	Nontoxic, wound-healing properties	Gao et al. (2022)
	Vitamin D3, D-glucono-δ-lactone	Hemo- and cytocompatible, wound-healing properties, reepithelialization and granular tissue formation	Ehterami et al. (2020)
Ionic cross-linking, 3D printing	ZnO	Nontoxic, antibacterial properties	Cleetus et al. (2020)
Solvent cast	Chlorogenic acid, *Eucommia ulmoides* rubber	Wound-healing and antibacterial properties	Guo et al. (2022a)
Oxidation cross-linking	Dopamine	Nontoxic, wound-healing properties	Chi et al. (2022)
Solvent casting method	Amikacin, poloxamer 407, pluronic F127, and polyvinyl alcohol	Antibacterial properties and wound-healing capacity	Abbasi et al. (2020)
Complexation and ionic gelation ($CaCl_2$)	Carboxymethyl chitosan, hyaluronic aldehyde acid, $ZnCl_2$	In vivo biodegradation, wound-healing properties	Yan et al. (2022)
	Chitosan, *Aloe vera*, honey	Cell adhesion, antibacterial properties	Saberian et al. (2021)
Gelation in situ	Chitosan oligosaccharide and zinc oxide nanoparticles	Biocompatible with antibacterial and wound-healing activity	Zhang et al. (2021)
Gelation in situ	Aldehyde alginate and polyetherimide, strontium-ion-doped	Self-adhesion wound-healing properties	Lu et al. (2020)

Zhang et al. (2021) developed sodium alginate-chitosan oligosaccharide hydrogels containing zinc oxide nanoparticles by spontaneous Schiff base reaction. The porous hydrogels with high swelling degree displayed antibacterials activity against four microorganisms: *Escherichia coli*, *Staphylococcus aureus*, *Candida albicans*, and *Bacillus subtilis*. These hydrogels display low hemolysis with hemolysis rates of 1.3–2.4%, comparable to the negative control PBS group. The authors described that the hemolysis rate is directly associated with the blood compatibility of polymer hydrogels. Also, the results demonstrated that hydrogels enhanced the wound-healing process due to the synergistic effects of zinc oxide nanoparticles and chitosan oligosaccharide and the water retention properties of alginate hydrogel.

Saberian et al. (2021) produced alginate hydrogels with chitosan (2%), *Aloe vera* (2.5%), and honey (20%) and their different blends. The porous alginate hydrogels displayed suitable water vapor transmission and high hydrophilicity. Also, composite hydrogels showed excellent antibacterial properties against *Staphylococcus aureus* and *Pseudomonas aeroginosa* with an inhibitory zone of 23 mm and 14 mm, respectively. The extracts from the hydrogels did not show cytotoxicity at 7 days of incubation, while hemolytic activity (red blood cells) was lower than 5%, which is acceptable. Ehterami et al. (2020) produced alginate hydrogels cross-linked with calcium carbonate/D-glucono-δ-lactone loaded with vitamin D. Alginate hydrogels displayed a highly interconnected and porous structure. The addition of vitamin D increased the porosity of the alginate hydrogels reaching values of 91%. The alginate-vitamin D hydrogels displayed low hemolysis values compared with the control group. Also, hydrogels loaded with vitamin D induced a high proliferation rate of L929 cells at 24 and 72 h of cell culture. The hydrogels displayed suitable properties of wound closure compared with the negative control (gauze-treated wound). The authors described that these results are related to the proliferation rate of the cells seeded on alginate hydrogels.

The addition of nanoparticles has been reported to improve the wound-healing properties of hydrogels. For example, Cleetus et al. (2020) added zinc and titanium nanoparticles into 3D-alginate hydrogels. The 3D-alginate hydrogels loaded with zinc nanoparticles (ZnO, 0.5% and 1%) displayed antibacterial properties against *S. epidermidis*, similar to the erythromycin activity. However, alginate hydrogels loaded with titanium nanoparticles did not show antibacterial activity. Lastly, alginate hydrogels were noncytotoxic using fibroblasts.

Lu et al. (2020) produced multifunctional alginate hydrogels with self-healing properties. Also, strontium ions were incorporated into the alginate hydrogel to favor tissue repair. The authors reported excellent self-healing properties using a continuous step strain test. The hydrogels immediately recovered their original values after the strain returned from 60% to 1%. Also, the hydrogels loaded with strontium ions showed proliferation cell capacity and chemotactic effect, favoring the migration of vascular endothelial cells.

15.4.4 Gelatin Hydrogels as a Wound Dressing

Gelatin is a biopolymer protein obtained from collagen thermal denaturation, the primary component of connective tissue (Prasathkumar and Sadhasivam 2021). Since gelatin is a collagen derivative, it possesses similar properties (Naomi et al. 2021). Gelatin is used in tissue engineering to produce biomaterials since it has excellent biological properties, including high biocompatibility, low antigenicity, biodegradability, and the ability to enhance cell attachment. Gelatin contains repeating amino acid sequences of Gly-X-Y, where X and Y are mainly proline and hydroxyproline (Prasathkumar and Sadhasivam 2021). Table 15.4 displays the significant advances of gelatin hydrogels for wound healing applications. Thi et al. (2020) produced an injectable hydrogel composed of gallic acid-conjugated gelatin. The porous gelatin hydrogels displayed pore sizes from 50 to 150 μm and antioxidant properties. The authors evaluated the effect of the gelatin hydrogels on the

Table 15.4 Advances in gelatin hydrogels for wound-healing applications

Fabrication method	Polymer/bioactive molecules additional	Wound dressing properties	References
Gelation in situ	Horseradish peroxidase, H_2O_2	Wound-healing activity	Thi et al. (2020)
	Poly (γ-glutamic acid)	Cell adhesion and proliferation	Dou et al. (2022)
	Tannic acid, gellan gum	Antibacterial properties and wound-healing activity	Zheng et al. (2018)
Schiff base and chelating with Fe_3^+ ions	2-(4′-aldehydephenyl)-4-(2′,3′,4′-trihydroxyphenyl)-2,3-phthalazine-1(2H)-one	Tissue adhesion and self-healing properties. biocompatibility, hemostatic, antibacterial activity, and wound healing	Li et al. (2022b)
Solvent cast	PVA, ginger	Wound-healing properties	Khan et al. (2020)
Electrospinning and photo-cross-linking	Dopamine	Cell growth and wound-healing activity	Liu et al. (2022b)
Chemical cross-linking	*E. adenophorum* emulsion (Pluronic F68®)	Antibacterial activity	Chuysinuan et al. (2019)

inhibition of the reactive oxygen species. The authors found that the antioxidant capacity of gelatin hydrogel was improved after the conjugation with antioxidant molecules. The results were related to cell survival, obtaining values of about 86% on cells cultured with 0.75 mM H_2O_2 into gallic acid-conjugated gelatin hydrogels.

Dou et al. (2022) developed a porous and transparent physical gelatin hydrogel containing a covalently cross-linked poly (γ-glutamic acid) network. The hydrogels displayed self-healing properties since both sections were re-bonded after cutting them. After 30, 60, and 120 min of healing, the tensile strength of hydrogel was 0.08, 0.13, and 0.14 MPa, respectively. These values were similar to those obtained for the original hydrogel (0.23 MPa). Lastly, the gelatin hydrogels displayed excellent biocompatibility for L929 cells. The cells were viable after 3 days of incubation, demonstrating that the cells had good activity. The in vivo evaluation for accelerating wound healing showed that the wound area of rats treated with gelatin hydrogels was less than that of those treated with gauze.

15.5 Conclusions

The hydrogels produced from natural polymers are potential candidates for skin wound-healing applications. The physicochemical, mechanical, and biological properties of hydrogels stimulate the wound-healing process. Also, the incorporation of bioactive molecules and nanoparticles in hydrogels enhances these properties.

Acknowledgments The authors thank the Universidad de Guadalajara-Centro Universitario de los Valles (UdG-CUVALLES) for their support through Programa PROSNI-Apoyo a la Mejora en las Condiciones de Producción SNI y SNCA (2023).

Funding This research was funded by "Fomento a la Investigación 2022-2023 Centro Universitario de los Valles (CUVALLES)-Universidad de Guadalajara" grant number 267758.

References

Abazari M, Akbari T, Hasani M, Sharifikolouei E, Raoufi M, Foroumadi A, Sharifzadeh M, Firoozpour L, Khoobi M (2022) Polysaccharide-based hydrogels containing herbal extracts for wound healing applications. Carbohydr Polym 294:119808

Abbasi AR, Sohail M, Minhas MU, Khaliq T, Kousar M, Khan S, Hussain Z, Munir A (2020) Bioinspired sodium alginate based thermosensitive hydrogel membranes for accelerated wound healing. Int J Biol Macromol 155:751–765

Adamczyk K, Garncarczyk A, Antończak P, Wcisło-Dziadecka D (2020) The foot microbiome. J Cosmet Dermatol 19:1039–1043

Ali NH, Amin MCIM, Ng S-F (2019) Sodium carboxymethyl cellulose hydrogels containing reduced graphene oxide (rGO) as a functional antibiofilm wound dressing. J Biomater Sci Polym Ed 30:629–645

Asadi N, Pazoki-Toroudi H, Del Bakhshayesh AR, Akbarzadeh A, Davaran S, Annabi N (2021) Multifunctional hydrogels for wound healing: special focus on biomacromolecular based hydrogels. Int J Biol Macromol 170:728–750

Basu A, Celma G, Strømme M, Ferraz N (2018) In vitro and in vivo evaluation of the wound healing properties of nanofibrillated cellulose hydrogels. ACS Appl Biomater 1:1853–1863

Berman B, Maderal A, Raphael B (2017) Keloids and hypertrophic scars: pathophysiology, classification, and treatment. Dermatol Surg 43:S3

Blanco-Fernandez B, Castaño O, Mateos-Timoneda MÁ, Engel E, Pérez-Amodio S (2020) Nanotechnology approaches in chronic wound healing. Adv Wound Care 10:234–256

Bowden LG, Byrne HM, Maini PK, Moulton DE (2016) A morphoelastic model for dermal wound closure. Biomech Model Mechanobiol 15:663–681

Bowers S, Franco E (2020) Chronic wounds: evaluation and management. Am Fam Physician 101: 159–166

Boxberger M, Cenizo V, Cassir N, La Scola B (2021) Challenges in exploring and manipulating the human skin microbiome. Microbiome 9:125

Bruna T, Maldonado-Bravo F, Jara P, Caro N (2021) Silver nanoparticles and their antibacterial applications. Int J Mol Sci 22:7202

Brunauer A, Verboket RD, Kainz DM, Von Stetten F, Früh SM (2021) Rapid detection of pathogens in wound exudate via nucleic acid lateral flow immunoassay. Biosensors (Basel) 11:74

Burgess JL, Wyant WA, Abdo Abujamra B, Kirsner RS, Jozic I (2021) Diabetic wound-healing science. Medicina 57:1072

Byrd AL, Belkaid Y, Segre JA (2018) The human skin microbiome. Nat Rev Microbiol 16:143–155

Callewaert C, Kerckhof F-M, Granitsiotis MS, Van Gele M, Van De Wiele T, Boon N (2013) Characterization of Staphylococcus and Corynebacterium clusters in the human axillary region. PLoS One 8:e70538

Canesso MCC, Vieira AT, Castro TBR, Schirmer BGA, Cisalpino D, Martins FS, Rachid MA, Nicoli JR, Teixeira MM, Barcelos LS (2014) Skin wound healing is accelerated and scarless in the absence of commensal microbiota. J Immunol 193:5171–5180

Chang G, Dang Q, Liu C, Wang X, Song H, Gao H, Sun H, Zhang B, Cha D (2022) Carboxymethyl chitosan and carboxymethyl cellulose based self-healing hydrogel for accelerating diabetic wound healing. Carbohydr Polym 292:119687

Chi J, Li A, Zou M, Wang S, Liu C, Hu R, Jiang Z, Liu W, Sun R, Han B (2022) Novel dopamine-modified oxidized sodium alginate hydrogels promote angiogenesis and accelerate healing of chronic diabetic wounds. Int J Biol Macromol 203:492–504

Chouhan S, Sharma K, Guleria S (2017) Antimicrobial activity of some essential oils-present status and future perspectives. Medicines (Basel) 4:58

Chu DH (2012) Chapter 7. Development and structure of skin. In: Goldsmith LA, Katz SI, Gilchrest BA, Paller AS, Leffell DJ, Wolff K (eds) Fitzpatrick's dermatology in general medicine, 8th edn. The McGraw-Hill Companies, New York

Chu DM, Ma J, Prince AL, Antony KM, Seferovic MD, Aagaard KM (2017) Maturation of the infant microbiome community structure and function across multiple body sites and in relation to mode of delivery. Nat Med 23:314–326

Chuysinuan P, Chimnoi N, Reuk-Ngam N, Khlaychan P, Makarasen A, Wetprasit N, Dechtrirat D, Supaphol P, Techasakul S (2019) Development of gelatin hydrogel pads incorporated with Eupatorium adenophorum essential oil as antibacterial wound dressing. Polym Bull 76:701–724

Cildir G, Akıncılar SC, Tergaonkar V (2013) Chronic adipose tissue inflammation: all immune cells on the stage. Trends Mol Med 19:487–500

Cleetus CM, Alvarez Primo F, Fregoso G, Lalitha Raveendran N, Noveron JC, Spencer CT, Ramana CV, Joddar B (2020) Alginate hydrogels with embedded ZnO nanoparticles for wound healing therapy. Int J Nanomedicine 15:5097–5111

Cui R, Zhang L, Ou R, Xu Y, Xu L, Zhan X-Y, Li D (2022) Polysaccharide-based hydrogels for wound dressing: design considerations and clinical applications. Front Bioeng Biotechnol 10: 845735

Da Silva GJ, Domingues S (2017) We are never alone: living with the human microbiota. Front Young Minds 5:35

Dadkhah Tehrani F, Shabani I, Shabani A (2022) A hybrid oxygen-generating wound dressing based on chitosan thermosensitive hydrogel and decellularized amniotic membrane. Carbohydr Polym 281:119020

Degruttola AK, Low D, Mizoguchi A, Mizoguchi E (2016) Current understanding of dysbiosis in disease in human and animal models. Inflamm Bowel Dis 22:1137–1150

Deng L, Wang B, Li W, Han Z, Chen S, Wang H (2022a) Bacterial cellulose reinforced chitosan-based hydrogel with highly efficient self-healing and enhanced antibacterial activity for wound healing. Int J Biol Macromol 217:77–87

Deng P, Yao L, Chen J, Tang Z, Zhou J (2022b) Chitosan-based hydrogels with injectable, self-healing and antibacterial properties for wound healing. Carbohydr Polym 276:118718

Desjardins-Park HE, Foster DS, Longaker MT (2018) Fibroblasts and wound healing: an update. Regen Med 13:491–495

Dharmalingam K, Anandalakshmi R (2020) Functionalization of cellulose-based nanocomposite hydrogel films with zinc oxide complex and grapefruit seed extract for potential applications in treating chronic wounds. Polymer 202:122620

Dimitriu PA, Iker B, Malik K, Leung H, Mohn WW, Hillebrand GG (2019) New insights into the intrinsic and extrinsic factors that shape the human skin microbiome. mBio 10:e00839–e00819

Do NHN, Truong QT, Le PK, Ha AC (2022) Recent developments in chitosan hydrogels carrying natural bioactive compounds. Carbohydr Polym 294:119726

Dominguez-Bello MG, Costello EK, Contreras M, Magris M, Hidalgo G, Fierer N, Knight R (2010) Delivery mode shapes the acquisition and structure of the initial microbiota across multiple body habitats in newborns. Proc Natl Acad Sci 107:11971–11975

Dou C, Li Z, Luo Y, Gong J, Li Q, Zhang J, Zhang Q, Qiao C (2022) Bio-based poly (γ-glutamic acid)-gelatin double-network hydrogel with high strength for wound healing. Int J Biol Macromol 202:438–452

Ederveen THA, Smits JPH, Boekhorst J, Schalkwijk J, Van Den Bogaard EH, Zeeuwen PLJM (2020) Skin microbiota in health and disease: from sequencing to biology. J Dermatol 47:1110–1118

Ehterami A, Salehi M, Farzamfar S, Samadian H, Vaez A, Sahrapeyma H, Ghorbani S (2020) A promising wound dressing based on alginate hydrogels containing vitamin D3 cross-linked by calcium carbonate/d-glucono-δ-lactone. Biomed Eng Lett 10:309–319

El Fawal GF, Abu-Serie MM, Hassan MA, Elnouby MS (2018) Hydroxyethyl cellulose hydrogel for wound dressing: fabrication, characterization and in vitro evaluation. Int J Biol Macromol 111:649–659

Fan F, Saha S, Hanjaya-Putra D (2021) Biomimetic hydrogels to promote wound healing. Front Bioeng Biotechnol 9:718377

Ferdous Z, Nemmar A (2020) Health impact of silver nanoparticles: a review of the biodistribution and toxicity following various routes of exposure. Int J Mol Sci 21:2375

Fierer N, Hamady M, Lauber CL, Knight R (2008) The influence of sex, handedness, and washing on the diversity of hand surface bacteria. Proc Natl Acad Sci 105:17994–17999

Flowers L, Grice EA (2020) The skin microbiota: balancing risk and reward. Cell Host Microbe 28: 190–200

Forero-Doria O, Polo E, Marican A, Guzmán L, Venegas B, Vijayakumar S, Wehinger S, Guerrero M, Gallego J, Durán-Lara EF (2020) Supramolecular hydrogels based on cellulose for sustained release of therapeutic substances with antimicrobial and wound healing properties. Carbohydr Polym 242:116383

Gannesen AV, Borrel V, Lefeuvre L, Netrusov AI, Plakunov VK, Feuilloley MGJ (2019) Effect of two cosmetic compounds on the growth, biofilm formation activity, and surface properties of acneic strains of Cutibacterium acnes and Staphylococcus aureus. MicrobiologyOpen 8:e00659

Gao Z, Perez-Perez GI, Chen Y, Blaser MJ (2010) Quantitation of major human cutaneous bacterial and fungal populations. J Clin Microbiol 48:3575–3581

Gao Y, Kang Y, Wang T, Li C, Shen S, Qu C, Gong S, Liu P, Yang L, Liu J, Han B, Li C (2022) Alginate microspheres-collagen hydrogel, as a novel 3D culture system, enhanced skin wound healing of hUCMSCs in rats model. Colloids Surf B Biointerfaces 219:112799

Gardner SE, Hillis SL, Heilmann K, Segre JA, Grice EA (2013) The neuropathic diabetic foot ulcer microbiome is associated with clinical factors. Diabetes 62:923–930

Gribbon EM, Cunliffe WJ, Holland KT (1993) Interaction of Propionibacterium acnes with skin lipids in vitro. J Gen Microbiol 139:1745–1751

Grice EA, Kong HH, Conlan S, Deming CB, Davis J, Young AC, Bouffard GG, Blakesley RW, Murray PR, Green ED, Turner ML, Segre JA (2009) Topographical and temporal diversity of the human skin microbiome. Science 324:1190–1192

Guo C, Zhou Z, Zhang S, Peng P, Liu J, Yan Y, Dong J (2022a) Sulfonated, quaternized, and chlorogenic acid composited sodium alginate hydrogels/Eucommia ulmoides rubber films as in vitro antibacterial wound dressings for accelerating wound healing. Ind Crop Prod 190: 115885

Guo S, Ren Y, Chang R, He Y, Zhang D, Guan F, Yao M (2022b) Injectable self-healing adhesive chitosan hydrogel with antioxidative, antibacterial, and hemostatic activities for rapid hemostasis and skin wound healing. ACS Appl Mater Interfaces 14:34455–34469

Guo S, Yao M, Zhang D, He Y, Chang R, Ren Y, Guan F (2022c) One-step synthesis of multifunctional chitosan hydrogel for full-thickness wound closure and healing. Adv Healthc Mater 11:2101808

Gupta A, Kowalczuk M, Heaselgrave W, Britland ST, Martin C, Radecka I (2019) The production and application of hydrogels for wound management: a review. Eur Polym J 111:134–151

Gupta A, Briffa SM, Swingler S, Gibson H, Kannappan V, Adamus G, Kowalczuk M, Martin C, Radecka I (2020) Synthesis of silver nanoparticles using curcumin-Cyclodextrins loaded into bacterial cellulose-based hydrogels for wound dressing applications. Biomacromolecules 21: 1802–1811

Gurtner GC, Werner S, Barrandon Y, Longaker MT (2008) Wound repair and regeneration. Nature 453:314–321

Hamdy AS (2020) Essential oil as antimicrobial agents: efficacy, stability, and safety issues for food application. In: de Oliveira MS, da Costa WA, Silva SG (eds) Essential oils. Intech Open, Rijeka

Han G, Ceilley R (2017) Chronic wound healing: a review of current management and treatments. Adv Ther 34:599–610

Hauser AR, Mecsas J, Moir DT (2016) Beyond antibiotics: new therapeutic approaches for bacterial infections. Clin Infect Dis 63:89–95

Ji M, Li J, Wang Y, Li F, Man J, Li J, Zhang C, Peng S, Wang S (2022) Advances in chitosan-based wound dressings: modifications, fabrications, applications and prospects. Carbohydr Polym 297:120058

Jo JH, Deming C, Kennedy EA, Conlan S, Polley EC, Ng WI, Segre JA, Kong HH (2016) Diverse human skin fungal communities in children converge in adulthood. J Invest Dermatol 136: 2356–2363

Kapur S, Watson W, Carr S (2018) Atopic dermatitis. Allergy, Asthma Clin Immunol 14:52

Khan BA, Ullah S, Khan MK, Uzair B, Menaa F, Braga VA (2020) Fabrication, physical characterizations, and in vitro, in vivo evaluation of ginger extract-loaded gelatin/poly(vinyl alcohol) hydrogel films against burn wound healing in animal model. AAPS PharmSciTech 21: 323

Kiti K, Suwantong O (2020) Bilayer wound dressing based on sodium alginate incorporated with curcumin-β-cyclodextrin inclusion complex/chitosan hydrogel. Int J Biol Macromol 164:4113–4124

Koehler J, Brandl FP, Goepferich AM (2018) Hydrogel wound dressings for bioactive treatment of acute and chronic wounds. Eur Polym J 100:1–11

Koneru A, Dharmalingam K, Anandalakshmi R (2020) Cellulose based nanocomposite hydrogel films consisting of sodium carboxymethylcellulose–grapefruit seed extract nanoparticles for potential wound healing applications. Int J Biol Macromol 148:833–842

Kong Y, Tang X, Zhao Y, Chen X, Yao K, Zhang L, Han Q, Zhang L, Ling J, Wang Y, Yang Y (2020) Degradable tough chitosan dressing for skin wound recovery. Nanotechnol Rev 9:1576–1585

Kou S, Peters L, Mucalo M (2022) Chitosan: a review of molecular structure, bioactivities and interactions with the human body and micro-organisms. Carbohydr Polym 282:119132

Kundu R, Mahada P, Chhirang B, Das B (2022) Cellulose hydrogels: green and sustainable soft biomaterials. Curr Res Green Sustain Chem 5:100252

Lai Y, Di Nardo A, Nakatsuji T, Leichtle A, Yang Y, Cogen AL, Wu Z-R, Hooper LV, Schmidt RR, Von Aulock S, Radek KA, Huang C-M, Ryan AF, Gallo RL (2009) Commensal bacteria regulate Toll-like receptor 3–dependent inflammation after skin injury. Nat Med 15:1377–1382

Lazarus GS, Cooper DM, Knighton DR, Margolis DJ, Pecoraro RE, Rodeheaver G, Robson MC (1994) Definitions and guidelines for assessment of wounds and evaluation of healing. Arch Dermatol 130:489–493

Leung MHY, Wilkins D, Lee PKH (2015) Insights into the pan-microbiome: skin microbial communities of Chinese individuals differ from other racial groups. Sci Rep 5:11845

Li M, Budding AE, Van Der Lugt-Degen M, Du-Thumm L, Vandeven M, Fan A (2019) The influence of age, gender and race/ethnicity on the composition of the human axillary microbiome. Int J Cosmet Sci 41:371–377

Li S, Wang X, Chen J, Guo J, Yuan M, Wan G, Yan C, Li W, Machens H-G, Rinkevich Y, Yang X, Song H, Chen Z (2022a) Calcium ion cross-linked sodium alginate hydrogels containing deferoxamine and copper nanoparticles for diabetic wound healing. Int J Biol Macromol 202: 657–670

Li Y, Liu C, Cheng X, Zhang A, Liu W, Zhang S, Jian X (2022b) Tunicate inspired gelatin-based tough hydrogel wound dressing containing twisted phthalazinone with adhesive, self-healing and antibacterial properties. Int J Biol Macromol 218:639–653

Lindholm C, Searle R (2016) Wound management for the 21st century: combining effectiveness and efficiency. Int Wound J 13:5–15

Linehan JL, Harrison OJ, Han SJ, Byrd AL, Vujkovic-Cvijin I, Villarino AV, Sen SK, Shaik J, Smelkinson M, Tamoutounour S, Collins N, Bouladoux N, Dzutsev A, Rosshart SP, Arbuckle JH, Wang CR, Kristie TM, Rehermann B, Trinchieri G, Brenchley JM, O'Shea JJ, Belkaid Y (2018) Non-classical immunity controls microbiota impact on skin immunity and tissue repair. Cell 172:784–796.e18

Liu Y, Li S, Wang Z, Wang L (2022a) Ultrasound in cellulose-based hydrogel for biomedical use: from extraction to preparation. Colloids Surf B Biointerfaces 212:112368

Liu Y, Wang Q, Liu X, Nakielski P, Pierini F, Li X, Yu J, Ding B (2022b) Highly adhesive, stretchable and breathable gelatin methacryloyl-based nanofibrous hydrogels for wound dressings. ACS Appl Biomater 5:1047–1056

Loesche MA, Farahi K, Capone K, Fakharzadeh S, Blauvelt A, Duffin KC, Deprimo SE, Muñoz-Elías EJ, Brodmerkel C, Dasgupta B, Chevrier M, Smith K, Horwinski J, Tyldsley A, Grice EA (2018) Longitudinal study of the psoriasis-associated skin microbiome during therapy with Ustekinumab in a randomized phase 3b clinical trial. J Investig Dermatol 138:1973–1981

Loh EYX, Mohamad N, Fauzi MB, Ng MH, Ng SF, Mohd Amin MCI (2018) Development of a bacterial cellulose-based hydrogel cell carrier containing keratinocytes and fibroblasts for full-thickness wound healing. Sci Rep 8:2875

Long J, Etxeberria AE, Nand AV, Bunt CR, Ray S, Seyfoddin A (2019) A 3D printed chitosan-pectin hydrogel wound dressing for lidocaine hydrochloride delivery. Mater Sci Eng C 104: 109873

Loo HL, Goh BH, Lee L-H, Chuah LH (2022) Application of chitosan-based nanoparticles in skin wound healing. Asian J Pharm Sci 17:299–332

Lu W, Bao D, Ta F, Liu D, Zhang D, Zhang Z, Fan Z (2020) Multifunctional alginate hydrogel protects and heals skin defects in complex clinical situations. ACS Omega 5:17152–17159

Luo Y, Cui L, Zou L, Zhao Y, Chen L, Guan Y, Zhang Y (2022) Mechanically strong and on-demand dissoluble chitosan hydrogels for wound dressing applications. Carbohydr Polym 294:119774

Maaz Arif M, Khan SM, Gull N, Tabish TA, Zia S, Ullah Khan R, Awais SM, Arif Butt M (2021) Polymer-based biomaterials for chronic wound management: promises and challenges. Int J Pharm 598:120270

Malone-Povolny MJ, Maloney SE, Schoenfisch MH (2019) Nitric oxide therapy for diabetic wound healing. Adv Healthc Mater 8:1801210

Man A, Santacroce L, Jacob R, Mare A, Man L (2019) Antimicrobial activity of six essential oils against a group of human pathogens: a comparative study. Pathogens 8:15

Matica MA, Aachmann FL, Tøndervik A, Sletta H, Ostafe V (2019) Chitosan as a wound dressing starting material: antimicrobial properties and mode of action. Int J Mol Sci 20:5889

Moskovicz V, Gross A, Mizrahi B (2020) Extrinsic factors shaping the skin microbiome. Microorganisms 8:1023

Murillo N, Aubert J, Raoult D (2014) Microbiota of Demodex mites from rosacea patients and controls. Microb Pathog 71–72:37–40

Mussin J, Robles-Botero V, Casañas-Pimentel R, Rojas F, Angiolella L, San Martín-Martínez E, Giusiano G (2021) Antimicrobial and cytotoxic activity of green synthesis silver nanoparticles targeting skin and soft tissue infectious agents. Sci Rep 11:14566

Naomi R, Bahari H, Ridzuan PM, Othman F (2021) Natural-based biomaterial for skin wound healing (gelatin vs. collagen): expert review. Polymers 13:2319

Nešović K, Janković A, Radetić T, Vukašinović-Sekulić M, Kojić V, Živković L, Perić-Grujić A, Rhee KY, Mišković-Stanković V (2019) Chitosan-based hydrogel wound dressings with electrochemically incorporated silver nanoparticles—in vitro study. Eur Polym J 121:109257

Nguyen AV, Soulika AM (2019) The dynamics of the skin's immune system. Int J Mol Sci 20:1811

Nielsen CK, Kjems J, Mygind T, Snabe T, Meyer RL (2016) Effects of tween 80 on growth and biofilm formation in laboratory media. Front Microbiol 7:01878

Niemiec SM, Louiselle AE, Liechty KW, Zgheib C (2020) Role of microRNAs in pressure ulcer immune response, pathogenesis, and treatment. Int J Mol Sci 22:64

Ning Chen W (2021) A novel biosustainable durian antimicrobial gel dressing. J Wound Care 30: 330

Oge' LK, Muncie HL, Phillips-Savoy AR (2015) Rosacea: diagnosis and treatment. Am Fam Physician 92:187–196

Orlando I, Basnett P, Nigmatullin R, Wang W, Knowles JC, Roy I (2020) Chemical modification of bacterial cellulose for the development of an antibacterial wound dressing. Front Bioeng Biotechnol 8:557885

Park YJ, Lee HK (2018) The role of skin and orogenital microbiota in protective immunity and chronic immune-mediated inflammatory disease. Front Immunol 8:1955

Peng W, Li D, Dai K, Wang Y, Song P, Li H, Tang P, Zhang Z, Li Z, Zhou Y, Zhou C (2022) Recent progress of collagen, chitosan, alginate and other hydrogels in skin repair and wound dressing applications. Int J Biol Macromol 208:400–408

Pereira RF, Sousa A, Barrias CC, Bayat A, Granja PL, Bártolo PJ (2017) Advances in bioprinted cell-laden hydrogels for skin tissue engineering. Biomanuf Rev 2:1

Perez Perez GI, Gao Z, Jourdain R, Ramirez J, Gany F, Clavaud C, Demaude J, Breton L, Blaser MJ (2016) Body site is a more determinant factor than human population diversity in the healthy skin microbiome. PLoS One 11:e0151990

Pita-López ML, Fletes-Vargas G, Espinosa-Andrews H, Rodríguez-Rodríguez R (2021) Physically cross-linked chitosan-based hydrogels for tissue engineering applications: a state-of-the-art review. Eur Polym J 145:110176

Prasathkumar M, Sadhasivam S (2021) Chitosan/hyaluronic acid/alginate and an assorted polymers loaded with honey, plant, and marine compounds for progressive wound healing—know-how. Int J Biol Macromol 186:656–685

Prohic A, Simic D, Sadikovic TJ, Krupalija-Fazlic M (2014) Distribution of Malassezia species on healthy human skin in Bosnia and Herzegovina: correlation with body part, age and gender. Iran J Microbiol 6:253–262

Qing C (2017) The molecular biology in wound healing & non-healing wound. Chin J Traumatol 20:189–193

Rai M, Yadav A, Gade A (2009) Silver nanoparticles as a new generation of antimicrobials. Biotechnol Adv 27:76–83

Rai MK, Deshmukh SD, Ingle AP, Gade AK (2012) Silver nanoparticles: the powerful nanoweapon against multidrug-resistant bacteria. J Appl Microbiol 112:841–852

Raina N, Pahwa R, Thakur VK, Gupta M (2022) Polysaccharide-based hydrogels: new insights and futuristic prospects in wound healing. Int J Biol Macromol 223:1586–1603

Ramsey MM, Freire MO, Gabrilska RA, Rumbaugh KP, Lemon KP (2016) Staphylococcus aureus shifts toward commensalism in response to Corynebacterium species. Front Microbiol 7. https://doi.org/10.3389/fmicb.2016.01230

Rensburg JJV, Lin H, Gao X, Toh E, Fortney KR, Ellinger S, Zwickl B, Janowicz DM, Katz BP, Nelson DE, Dong Q, Spinola SM (2015) The human skin microbiome associates with the outcome of and is influenced by bacterial infection. mBio 6:e01315–e01315

Rippa AL, Kalabusheva EP, Vorotelyak EA (2019) Regeneration of dermis: scarring and cells involved. Cell 8:607

Rodrigues M, Kosaric N, Bonham CA, Gurtner GC (2019) Wound healing: a cellular perspective. Physiol Rev 99:665–706

Rodríguez-Rodríguez R, Espinosa-Andrews H, Velasquillo-Martínez C, García-Carvajal ZY (2020) Composite hydrogels based on gelatin, chitosan and polyvinyl alcohol to biomedical applications: a review. Int J Polym Mater Polym Biomater 69:1–20

Rumbaut RE, Thiagarajan P (2010) Integrated systems physiology: from molecule to function to disease. In: Platelet-vessel wall interactions in hemostasis and thrombosis. Morgan & Claypool Life Sciences, San Rafael, CA

Saberian M, Seyedjafari E, Zargar SJ, Mahdavi FS, Sanaei-Rad P (2021) Fabrication and characterization of alginate/chitosan hydrogel combined with honey and aloe vera for wound dressing applications. J Appl Polym Sci 138:51398

Safaei-Ghomi J, Ahd A (2010) Antimicrobial and antifungal properties of the essential oil and methanol extracts of *Eucalyptus largiflorens* and *Eucalyptus intertexta*. Pharmacogn Mag 6: 172–175

Sanford JA, Gallo RL (2013) Functions of the skin microbiota in health and disease. Semin Immunol 25:370–377

Saravanan S, Vimalraj S, Thanikaivelan P, Banudevi S, Manivasagam G (2019) A review on injectable chitosan/beta glycerophosphate hydrogels for bone tissue regeneration. Int J Biol Macromol 121:38–54

Schommer NN, Gallo RL (2013) Structure and function of the human skin microbiome. Trends Microbiol 21:660–668

Shami A, Al-Mijalli S, Pongchaikul P, Al-Barrag A, Abdurahim S (2019) The prevalence of the culturable human skin aerobic bacteria in Riyadh, Saudi Arabia. BMC Microbiol 19:189

Shi X, Chen Z, He Y, Lu Q, Chen R, Zhao C, Dong D, Sun Y, He H (2022) Dual light-responsive cellulose nanofibril-based in situ hydrogel for drug-resistant bacteria infected wound healing. Carbohydr Polym 297:120042

Simões D, Miguel SP, Ribeiro MP, Coutinho P, Mendonça AG, Correia IJ (2018) Recent advances on antimicrobial wound dressing: a review. Eur J Pharm Biopharm 127:130–141

Solórzano-Santos F, Miranda-Novales MG (2012) Essential oils from aromatic herbs as antimicrobial agents. Curr Opin Biotechnol 23:136–141

Song S, Liu Z, Abubaker MA, Ding L, Zhang J, Yang S, Fan Z (2021) Antibacterial polyvinyl alcohol/bacterial cellulose/nano-silver hydrogels that effectively promote wound healing. Mater Sci Eng C 126:112171

Sorrell JM, Caplan AI (2009) Chapter 4: Fibroblasts—a diverse population at the center of it all. In: International review of cell and molecular biology. Academic Press

Stan D, Tanase C, Avram M, Apetrei R, Mincu N-B, Mateescu AL, Stan D (2021) Wound healing applications of creams and "smart" hydrogels. Exp Dermatol 30:1218–1232

Staudinger T, Pipal A, Redl B (2011) Molecular analysis of the prevalent microbiota of human male and female forehead skin compared to forearm skin and the influence of make-up. J Appl Microbiol 110:1381–1389

Swaney MH, Kalan LR (2021) Living in your skin: microbes, molecules, and mechanisms. Infect Immun 89:e00695–e00620

Tandara AA, Mustoe TA (2004) Oxygen in wound healing—more than a nutrient. World J Surg 28: 294–300

Tang Y, Wang H, Liu S, Pu L, Hu X, Ding J, Xu G, Xu W, Xiang S, Yuan Z (2022) A review of protein hydrogels: protein assembly mechanisms, properties, and biological applications. Colloids Surf B: Biointerfaces 220:112973

Tavakoli S, Klar AS (2020) Advanced hydrogels as wound dressings. Biomol Ther 10:1169

Thi PL, Lee Y, Tran DL, Thi TTH, Kang JI, Park KM, Park KD (2020) In situ forming and reactive oxygen species-scavenging gelatin hydrogels for enhancing wound healing efficacy. Acta Biomater 103:142–152

Thivya P, Akalya S, Sinija VR (2022) A comprehensive review on cellulose-based hydrogel and its potential application in the food industry. Appl Food Res 2:100161

Turgis M, Han J, Caillet S, Lacroix M (2009) Antimicrobial activity of mustard essential oil against Escherichia coli O157:H7 and Salmonella typhi. Food Control 20:1073–1079

Varaprasad K, Raghavendra GM, Jayaramudu T, Yallapu MM, Sadiku R (2017) A mini review on hydrogels classification and recent developments in miscellaneous applications. Mater Sci Eng C 79:958–971

Vig K, Chaudhari A, Tripathi S, Dixit S, Sahu R, Pillai S, Dennis VA, Singh SR (2017) Advances in skin regeneration using tissue engineering. Int J Mol Sci 18:789

Wang P-H, Huang B-S, Horng H-C, Yeh C-C, Chen Y-J (2018) Wound healing. J Chin Med Assoc 81:94
Wilkinson HN, Hardman MJ (2020) Wound healing: cellular mechanisms and pathological outcomes. Open Biol 10:200223
Wong LC, Leh CP, Goh CF (2021) Designing cellulose hydrogels from non-woody biomass. Carbohydr Polym 264:118036
Woodley DT (2017) Distinct fibroblasts in the papillary and reticular dermis: implications for wound healing. Dermatol Clin 35:95–100
Xiao Z, Li Q, Liu H, Zhao Q, Niu Y, Zhao D (2022) Adhesion mechanism and application progress of hydrogels. Eur Polym J 173:111277
Xu Z, Han S, Gu Z, Wu J (2020) Advances and impact of antioxidant hydrogel in chronic wound healing. Adv Healthc Mater 9:1901502
Xu J, Zhang M, Du W, Zhao J, Ling G, Zhang P (2022a) Chitosan-based high-strength supramolecular hydrogels for 3D bioprinting. Int J Biol Macromol 219:545–557
Xu Z, Liu G, Zheng L, Wu J (2022b) A polyphenol-modified chitosan hybrid hydrogel with enhanced antimicrobial and antioxidant activities for rapid healing of diabetic wounds. Nano Res 16:905
Yan Y, Ren P, Wu Q, Zhang T (2022) Precise design of alginate hydrogels crosslinked with microgels for diabetic wound healing. Biomolecules 12:1582
Yang E, Hou W, Liu K, Yang H, Wei W, Kang H, Dai H (2022a) A multifunctional chitosan hydrogel dressing for liver hemostasis and infected wound healing. Carbohydr Polym 291: 119631
Yang G, Zhang Z, Liu K, Ji X, Fatehi P, Chen J (2022b) A cellulose nanofibril-reinforced hydrogel with robust mechanical, self-healing, pH-responsive and antibacterial characteristics for wound dressing applications. J Nanobiotechnol 20:312
Yang X, Ma W, Lin H, Ao S, Liu H, Zhang H, Tang W, Xiao H, Wang F, Zhu J, Liu D, Lin S, Zhang Y, Zhou Z, Chen C, Liang H (2022c) Molecular mechanisms of the antibacterial activity of polyimide fibers in a skin-wound model with Gram-positive and Gram-negative bacterial infection in vivo. Nanoscale Adv 4:3043–3053
Yin IX, Zhang J, Zhao IS, Mei ML, Li Q, Chu CH (2020) The antibacterial mechanism of silver nanoparticles and its application in dentistry. Int J Nanomedicine 15:2555–2562
Yin H, Song P, Chen X, Huang Q, Huang H (2022) A self-healing hydrogel based on oxidized microcrystalline cellulose and carboxymethyl chitosan as wound dressing material. Int J Biol Macromol 221:1606–1617
Yu X, Cheng C, Peng X, Zhang K, Yu X (2022) A self-healing and injectable oxidized quaternized guar gum/carboxymethyl chitosan hydrogel with efficient hemostatic and antibacterial properties for wound dressing. Colloids Surf B: Biointerfaces 209:112207
Zainal SH, Mohd NH, Suhaili N, Anuar FH, Lazim AM, Othaman R (2021) Preparation of cellulose-based hydrogel: a review. J Mater Res Technol 10:935–952
Zhai W, Huang Y, Zhang X, Fei W, Chang Y, Cheng S, Zhou Y, Gao J, Tang X, Zhang X, Yang S (2018) Profile of the skin microbiota in a healthy Chinese population. J Dermatol 45:1289–1300
Zhang A, Liu Y, Qin D, Sun M, Wang T, Chen X (2020) Research status of self-healing hydrogel for wound management: a review. Int J Biol Macromol 164:2108–2123
Zhang M, Qiao X, Han W, Jiang T, Liu F, Zhao X (2021) Alginate-chitosan oligosaccharide-ZnO composite hydrogel for accelerating wound healing. Carbohydr Polym 266:118100
Zhang S, Shan S, Zhang H, Gao X, Tang X, Chen K (2022) Antimicrobial cellulose hydrogels preparation with RIF loading from bamboo parenchyma cells: a green approach towards wound healing. Int J Biol Macromol 203:1–9
Zheng Y, Liang Y, Zhang D, Sun X, Liang L, Li J, Liu Y-N (2018) Gelatin-based hydrogels blended with gellan as an injectable wound dressing. ACS Omega 3:4766–4775

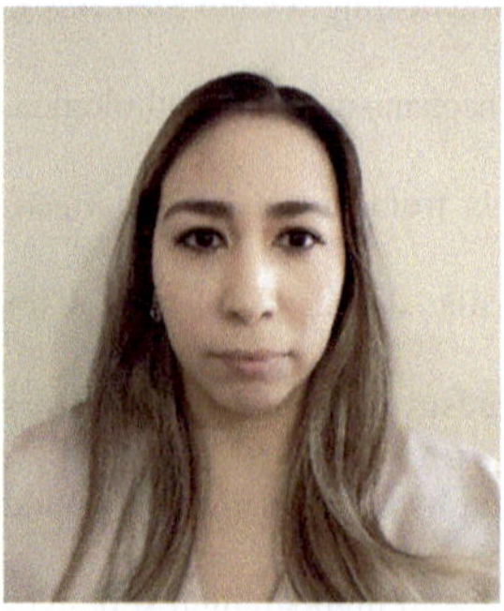

Gabriela Fletes-Vargas received his Bachelor's Degree in Biology and a Master's in Science from the University of Guadalajara (Mexico). She is pursuing a Ph.D. in Biosciences at the University of Guadalajara in cooperation with CIATEJ A.C. (Mexico). His research is focused on neurodegeneration and the evaluation of biomaterials for tissue engineering and wound healing applications.

Sergio Yair Rodríguez-Preciado is a research professor of Centro Universitario de los Valles at Universidad de Guadalajara where he is in charge of subjects of genetics and biomedicine. Dr. Rodríguez graduated from Universidad de Guanajuato, and he has a B.S. Degree in Biological and Pharmaceutical Chemistry (2010), M.S. Degree (2016), and Ph. D. (2018) in Human Genetics from the Universidad de Guadalajara. At present, he is a Member of the National Researcher System (Sistema Nacional de Investigadores—CONACyT) level I, in Mexico; and he is also a member of the research group of pediatric infectious diseases (UDG-CA-777). Also, he is a member of the Bioethics Committee of the Centro Universitario de los Valles of the Universidad de Guadalajara and Colegio mexicano de ciencias de Laboratorio clínico (CMCLab). His research focuses on infectious diseases and microbiology, mainly respiratory diseases and antibiotic resistance. He has also collaborated in projects related with the study of molecular markers in chronic diseases (chronic lymphocytic leukemia and dyslipidemias). Currently he is interested in including the study of the microbiome within his research. Dr. Rodríguez has different publications in index and propagation journals. In the same way, several posters have been presented by Dr. Rodríguez in international and national conferences.

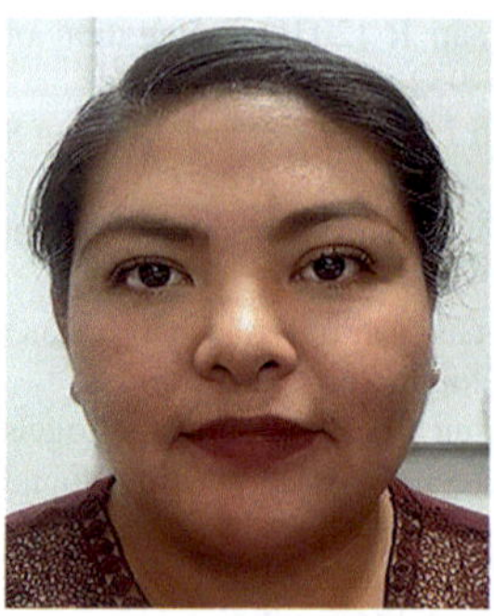

Mariana Díaz-Zaragoza is a full-time Associate Professor-Researcher B at the Centro Universitario de los Valles, of the University of Guadalajara. She has a degree in biology from the Faculty of Higher Studies Iztacala, of the National Autonomous University of Mexico. In 2010, she obtained the degree of Master of Science, Experimental Biology orientation, from the Institute of Biology, UNAM. In 2015 she received the Doctor of Science, Biomedicine orientation, from the Institute of Biomedical Research, UNAM. During 2017 and 2018, she did a postdoctoral stay in the Department of Microbiology and Parasitology at the Faculty of Medicine, UNAM. She has published various scientific articles internationally. She is a member of the National System of Researchers level I. During her academic career, she has participated in various investigations, such as the identification and quantification of mycotoxins in different foods for human consumption and in the formation and detection of aflatoxin

B1-ADN adducts as an etiological factor in the development of liver cancer in humans. With the elaboration of his doctoral project, she specialized in the immunology of breast cancer studied with proteomic techniques for the identification of proteins that could be used as disease biomarkers, with the objective of knowing the role of immunosurveillance and natural antibodies such as IgM and if they could be used to establish a tool for early immunodiagnostic of breast cancer through antigen-IgM interactions. During postdoctoral stay, she used proteomics to characterize the protein expression profiles in cysticerci during the TH1 and Th2 immune response in mice infected with *Taenia crassiceps*, both in mice susceptible or resistant to parasite infection. She is currently dedicated to investigating the resistance to antibiotics of bacteria that inhabit bodies of water influenced by wastewater from towns and hospitals.

Rogelio Rodriguez-Rodríguez is a biologist and pharmaceutical chemist from Nayarit University (Mexico). He received his Ph.D. (2020) in Biotechnology Innovation from CIATEJ, AC (Mexico). At present, Rogelio is a professor at the Centro Universitario de los Valles, University of Guadalajara (Mexico). He is a National System of Researchers (SNI) level I member. His research focuses on the synthesis and physicochemical, thermal, and mechanical characterization of chitosan-based hydrogels for biological applications such as tissue engineering, drug delivery, and wound healing. The chitosan-based hydrogels produced by Rogelio are mainly synthesized by physical cross-linking methods.

Nanomaterial Applications in Cancer Therapy and Diagnosis

16

Akanksha Deshwal, Ekta Arjundas Kukreja, Neeraj Shrivastava, Faheem A. Sheikh, Touseef Amna, and Ravi Mani Tripathi

Abstract

Cancer is one of the leading health challenges and a significant cause of death globally which is characterized by uncontrolled, random cell division and invasiveness. Conventional cancer therapeutics have been decried due to several unpleasant side effects and inadequate mortality of cancer cells. Nanomaterial-based cancer diagnosis and therapy opens up a novel procedure that offers advancement over traditional cancer theranostics. The significance of nanomaterials in cancer theranostics is indisputable. Still, the conversion into actual medical practice has been a major challenge due to a sequence of concerns related to human toxicity and a lack of comprehensive reports on trustworthy animal models. However, the potential action of the nanomaterials approach in cancer theranostics is interesting, and outstanding results are reported yearly. This chapter reviews the application of nanomaterials that have emerged in the diagnosis and therapy of cancers. The objective of this chapter is to offer a better

A. Deshwal · E. A. Kukreja · R. M. Tripathi (✉)
Amity Institute of Nanotechnology, Amity University Uttar Pradesh, Noida, Uttar Pradesh, India
e-mail: rmtripathi@amity.edu

N. Shrivastava
Amity Institute of Microbial Technology, Amity University Uttar Pradesh, Noida, Uttar Pradesh, India
e-mail: nsrivastava1@amity.edu

F. A. Sheikh
Nanostructured and Biomimetic Lab, Department of Nanotechnology, University of Kashmir Hazratbal, Srinagar, Jammu and Kashmir, India
e-mail: faheemnt@uok.edu.in

T. Amna
Department of Biology, Faculty of Science, Albaha University, Albaha, Saudi Arabia

F. A. Sheikh et al. (eds.), *Interaction of Nanomaterials With Living Cells*,
https://doi.org/10.1007/978-981-99-2119-5_16

understanding of nanomaterial-based cancer diagnosis and therapy and accelerate innovations in this specific area.

Keywords

Cancer · Diagnosis · Therapy · Nanomaterials · Drug delivery · Biosensors

16.1 Introduction

Cancer is one of the leading health problems which is described by uncontrolled, random cell division and invasiveness (Hari et al. 2022). According to a survey by GLOBOCAN in 2018, there are an estimated 18.1 million new cancer cases and 9.6 million cancer-related deaths (Bray et al. 2018). According to the current statistics on cancer, it has been estimated that the cancer-related death rate will increase and going to strike 13.1 million by 2030 (Chaturvedi et al. 2019). Cancer is characterized by a decrease or failure of normal maturation mechanisms and cellular control (Hari et al. 2022; Tripathi et al. 2015). Genetic manipulation caused by carcinogens and their asymptomatic nature make cancer a complex disease. There are various stages involved in the development of cancer, and the most dreaded stage is metastasis which makes the disease lethal (Conde et al. 2012; Misra et al. 2010). Early diagnosis is one of the crucial aspects of the effective treatment of cancer. The traditional methods of treatment include chemotherapy, radiation, and surgery which are time-consuming, costly, and complicated (Misra et al. 2010). The conventional detection procedures are inefficient in detecting tumor at the initial stage.

Nowadays, nanomaterials have been applied in various fields, such as drug delivery (Tripathi et al. 2015), biosensors (Bisht et al. 2021), colorimetric detection (Tripathi et al. 2020; Tripathi and Chung 2020), nanozymes (Mishra et al. 2022; Tripathi and Chung 2021), water treatment (Agarwal et al. 2016), antibacterial agents (Saxena et al. 2012; Yadav et al. 2017), and catalysis (Mehrotra et al. 2017). Nanoparticles are now being applied for the diagnosis because of their novel properties. These particles can easily reach the inner regions of the body and detect the presence of tumor (Hari et al. 2022; Mehrotra and Tripathi 2015; Tripathi et al. 2015). Nanomedicine is a new field that unites nanopharmaceuticals and nanobiotechnology for the enhancement of medical applications (Chaturvedi et al. 2019; Tripathi and Chung 2019; Tripathi et al. 2022). The application of nanomaterial extends to a variety of tools for the detection and treatment of cancer. Nanomaterials have been employed in cancer theranostics, such as imaging agents and targeted delivery of therapeutic molecules to the cells and tissues which show cancer growth and metastasis (Conde et al. 2012). Diverse range of nanomaterials have been used in oncology, for example, lipid nanoparticles, nanocrystals, lipid drug conjugates, nanostructured lipid carriers, nanoliposomes, nanoshells, emulsions, dendrimers, quantum dots, nanotubes, metallic nanoparticles, magnetic nanoparticles, etc. (Chaturvedi et al. 2019). In recent years, nanomedicine developments have shown that nanomaterials can be used for drug delivery owing to their ability to cross biological barriers and release the drug at the target site without affecting the neighboring healthy cells. Researchers have tried hard to develop biocompatible nanoparticles that could carry therapeutic and imaging chemicals while also targeting specific cancer indicators to aid in the diagnosis and treatment of cancer (Chaturvedi et al. 2019). The use of nanotechnology has made it possible to treat cancer at cellular levels with almost no side effects. Nanomaterials are also used in post-anticancer therapy to reduce the toxicity caused by the drugs during the treatment. Nanomaterials are being used to boost outcomes of techniques like photothermal therapy, sonochemical therapy, cell imaging, etc. The release of the drug is facilitated by the effect of temperature and pH (Chaturvedi et al. 2019). Nanomaterials have been widely used as vehicles for drug delivery by conjugating various therapeutic molecules and releasing the drug at targeted sites using active and passive mechanisms (Chaturvedi et al. 2019). The various factors have facilitated the delivery of drugs, such as permeability, binding of nanoparticles with the membrane receptors, size and surface charge on the nanoparticles, biocompatibility of the nanomaterials, cytotoxicity, etc. Whereas, various linking agents such as aptamers, antibodies, specific proteins, etc. functionalized nanomaterials have been employed development of detection tools such as biosensors, colorimetric assay, etc., for the selective and sensitive detection of biological analytes.

In this chapter, we discussed the application of nanomaterials that develop the diagnosis and therapy for cancer theragnostic. Presently, cancer is a leading disease and a big challenge for scientists as the elevated mortality rate is reportedly high. The traditional cancer therapeutics have been criticized due to several unpleasant causes and inadequate mortality of cancer cells. Therefore, there is a demand to develop new and advanced cancer diagnoses and therapy, which demands new and innovative methods. This chapter aims to offer a better understanding of nanomaterial-based cancer diagnosis and therapy and accelerate innovations in this area.

16.2 Nanotechnological Approach for Cancer Detection

Early detection of cancer increases the possibility of survival which is achievable by nanotechnological approaches. These nano-diagnostics are cost-effective, more straightforward, and highly selective and sensitive. The distinctive properties of nanoparticles, such as optical properties, surface-to-volume ratio, and surface chemistry, have been exploited to get efficient methods for cancer detection. Specific nanoparticles have extraordinary therapeutic abilities and can also convert light waves of different frequencies into heat. The most exciting feature of nanoparticles is that their shapes and sizes are tuneable and can be functionalized and synthesized as per the requirement. The use of nanoparticles with several different molecules in various techniques has exceptional results as functionalized or conjugated nanoparticles increase the sensitivity and selectivity of the method, making a point-of-care diagnosis possible. The different nanomaterial-based diagnostic tools have been discussed below in detail.

16.2.1 Colorimetric Detection

Visual diagnosis of cancer without any complex instruments can be made by colorimetric detection. The use of nanoparticles for this technique has been proven very useful for early detection and point-of-care diagnosis (Medley et al. 2008). Nanomaterial-based colorimetric detection mainly works on two principles: agglomeration/aggregation and catalytic activity (Tripathi et al. 2019; Tripathi and Chung 2020). Colorimetric detection using nanoparticles can be done in a variety of approaches depending upon the properties and reactive activities of different nanoparticles. The most widely used nanoparticles for colorimetric detection are Au and Ag because of their unique surface plasmon resonance property. When the nanoparticles come in the vicinity of each other, the absorption spectra change; hence, the change in color is observed. Au and Ag NPs are conjugated with different types of molecules such as enzymes, antibodies, nucleic acids, aptamers, etc. Aptamers are oligonucleotide strands known to detect and bind to specific proteins present in the target cells. In an experiment, aptamer conjugated Au NPs were used, and in the presence of tumor cells, Au NPs bind to tumor cells and show no color change, while in the presence of normal cells, Au NPs aggregate and the color of the solution changes to blue (Medley et al. 2008). In another experiment, a colorimetric assay was performed by the non-cross-linking aggregation mechanism using citrate-coated Au NPs. In this assay, an enzyme protein kinase is used, which causes phosphorylation of the peptides present in the target cells, and an increase in the anionic charge is observed at the site of phosphorylation. With normal cells, the Au NPs aggregate and the color changes from red to blue, while with the target cells, due to a decrease in the net cationic charge, the color remains red (Kang et al. 2010). The addition of enzymes, polymers, or other biomolecules for the functionalization of nanoparticles is done depending upon the type of protein or nucleic acid present in the cancer cells which is to be detected.

16.2.2 Biosensors

Biosensors are analytical devices that include a biological material, a bio-based material, or a biomimetic that is intimately associated without or integrated into a physicochemical transducer or transducing microsystem (Bisht et al. 2021). Biosensors enable this same measurement of such a single analyte or a cluster of analytes. The analyte in question might be a cancer biomarker that reveals a specific stage of the disease. The analyte might also be a protein or receptor that is abundantly expressed upon that cancer cell's surface. The analyte (i.e., cancer biomarkers or protein on the surface of tumor cells) in a biosensor could be detected via the receptor fixed to an edge of a biosensor transducer (Perfézou et al. 2012).

The inclusion of nanomaterials could well improve a biosensor's sensitivity along with its precision and reliability. Furthermore, even though mass transport occurs over shorter distances, nano-sized devices usually provide quicker responses (Perfézou et al. 2012). Such biosensors have been based on electrochemical techniques wherein analyte sensing is accomplished by observing the electrical response as an analyte electrochemically interacts with the surface of the sensor's working electrode. Carbon nanoparticles have excellent electrical properties and a high specific surface area, making them suitable components of biosensor platforms. For sensing applications, there are well-established methods for fabrication. Au NPs have been frequently used throughout biosensor construction, even though Au NPs are excellent electric conducting materials and chemisorptions can quickly immobilize thiolated molecules on their surface. Carbon nanomaterials were typically utilized (1) in the sensor's sensing platform, where they act as binding sites for desired biomarkers or molecules capturing target biomarkers, (2) within the transducer component, which directly converts the detected molecular interaction on the electrode surface into a measurable signal, and (3) in signal acquisition as labels for desired biomarkers (Pasinszki et al. 2017). Electrochemical biosensors were intended to convert physiochemical signals into electrical signals that can be evaluated. Because of their low cost, ease of fabrication while using, and limited range, these sensors have been widely used. CNTs, like Au NPs, can enhance electron transfer, making them ideal expansions to electrochemical biosensors (Banerjee et al. 2021). The far more widely utilized biomarker for prostate cancer is prostate-specific antigen (PSA), a single chain glycol-protein. Several CNT-based biosensors for PSA detection were developed, with anti-PSA antibodies serving as recognition elements on the sensor surface. The sensor platforms usually involve SWCNT-modified microelectrode arrays, cross-linked starch functionalized MWCNT-Au NP nanocomposite film, SWCNT forest-primary antibody platform to multi-label secondary antibody-MWCNT-horseradish peroxidase (HRP) bioconjugate labels, and Au NP functionalized polypyrrole (PPy)@MWCNT nanocomposite with HRP-conjugated (Pasinszki et al. 2017).

Carcinoembryonic antigen (CEA) is a biomarker being used to diagnose cervical carcinomas, as well as pancreatic, gastric, colorectal, and lung cancer. Several CNT-based electrochemical sensors for CEA detection were developed, with anti-CEA antibodies serving as target recognition elements (Pasinszki et al. 2017). Mucin 1 (MUC 1), an O-glycosylated protein, is highly expressed in early-stage breast cancer. Chen et al. created a sandwich-type aptasensor for detecting MUC 1. The sensor was made by electropolymerizing poly(o-phenylenediamine) (oPD) on a Au electrode and then depositing Au NP and casting thiolated primary aptamers on the electrode surface. Tracing tags were created by depositing Au NPs, thionine, and thiolated aptamers onto SiO_2@MWCNT nanocomposites (Pasinszki et al. 2017). Graphene can be integrated well with electrodes of electrochemical biosensors for enhanced biosensing owing to its favorable electrical properties. In addition to its enhanced conductivity, graphene is chemically resistant, inexpensive, and electrochemically inert. Another biosensor for *E. coli* sensing that does not rely on antibodies is focused on the adsorption of slightly negatively charged *E. coli* on the surface of p-type graphene. *Staphylococcus aureus* is yet another bacterium that has been detected utilizing graphene-based biosensors in a similar manner (using either antibodies or aptamers) (Banerjee et al. 2021). To recognize the cancer biomarker specifically, the highest recognition materials should be used as the surface receptor in the biosensor layout. This is especially useful in health diagnosis because the specificity and sensitivity of the detecting molecules will be critical to the sensor device's success. Antibodies (monoclonal and polyclonal) have been utilized in cancer treatment as therapeutics that target tumors within the body, as well as in diagnostic tests that target cancer cells and biomarkers. Polyclonal antibodies can be raised against any biomarker or organelle, and with the development of high throughput techniques, such molecules have been successfully used in sensing devices (Tothill 2009).

16.2.3 Cell Imaging

Cell imaging is done to know the chemical composition of the cells, as the activity of the cancer cells differs from that of normal cells. Imagining techniques like Raman imaging, photoacoustic imaging, dark-field microscopy, computed tomography, and magnetic resonance imaging use Au NPs for detecting signals because of their electrochemical properties (Li and Chen 2015). Photoacoustic imaging technique is based on the absorption of light and provides information about the molecular characteristics of the cells. Au NPs are used due to their biocompatibility and optical properties, and as they are responsive to acidic surroundings, they can be used for imaging. The use of Au NPs with Raman imaging is helpful in detecting specific components of the cells and distinguishing between cancer cells and normal cells depending on the surface-enhanced Raman spectroscopy (SERS) spectra (Ravanshad et al. 2018).

Au NPs are also used in dark-field microscopy which is based on image contrast of a particular sample and as non-bleaching markers. In computed tomography, cells are inspected at each layer to know the interior of the tissue using X-rays. Au NPs increase the efficiency of the technique and could detect the distribution, migration, and kinetics of T-cells when injected intravenously. Magnetic resonance imaging has also benefited from Au NPs as contrast agents, which aggregate at a particular site and facilitate targeted imaging.

16.3 Nanotechnological Approach for Cancer Therapy

16.3.1 Drug Delivery

Nanoparticles are specially designed to pass through the fenestrated capillaries, and the nanoparticles are preferentially concentrated at tumor sites. The drug exhibits enhanced permeability and retention (EPR) due to a deficient tumor lymphatic system (Haley and Frenkel 2008). Such particles can penetrate tissues by passing through the fenestration of such small blood vessel epithelial tissue. They could indeed enter the systemic bloodstream without aggregating blood platelets. Their smaller particles result in a larger surface area and, as a result, a quicker drug release strategy. Engineering carriers could be modulated and controlled so that they can be activated by changes in the environmental pH, chemical stimuli, the application of a rapidly oscillating magnetic field, or the application of an external heat source (Conti et al. 2006).

Regarding drug targeting, the NP may be biodegradable as a vehicle for delivering drugs. As a result, it is deteriorated and metabolized, and it does not accumulate or interfere with cellular processes. This is significant because increased levels of nonbiodegradable particles can have toxic effects (Haley and Frenkel 2008). The drug would either be physically entrapped in it or covalently bound to the polymer matrix, based on the preparation technique. The eventual resulted compounds might have a capsule structure (polymeric nanoparticles), an amphiphilic core/shell structure (polymeric micelles), or a polymerized macromolecule structure. Natural and synthetic polymers are the two types of polymers used as drug conjugates (Cho and Wang 2008).

Natural polymers can also be employed to develop drug delivery nanocarriers. Gelatin, dextran, and chitosan are the most commonly used polymers. Such NPs have just a drug encapsulation efficiency in general (Conti et al. 2006). The PGA has been the first biodegradable polymer to be employed in conjugate formation among synthetic polymers, including *N*-(2-hydroxypropyl)-methacrylamide copolymer (HPMA), polystyrene-maleic anhydride copolymer, polyethylene glycol (PEG), and poly-L-glutamic acid (PGA) (Cho and Wang 2008). The use of Au NPs is a novel technology in the field of tumor-targeted drug delivery. Paciotti et al. described the use of these carriers for such targeted delivery of tumor necrosis factor-alpha (TNF-α) to solid tumors (Conti et al. 2006). Micelle nanoparticles (MNPs) have been prepared to deliver various therapeutic drugs using their three principle characteristics: the capability to solubilize hydrophobic drugs, the release of drugs in a sustained manner, and easy alteration according to the targeted organ (Hari et al. 2022). Anticancer drugs can be delivered by two strategies, namely, active and passive targeting. The passive targeting method is mainly assisted by the enhanced permeability and retention effect (Fig. 16.1). The blood-brain barrier (BBB) is a crucial challenge for treating brain tumors and neurodegenerative disorders. MNPs have the ability to cross the BBB using one of the mechanisms such as transcellular lipophilic pathway, paracellular aqueous pathway, receptor-mediated transcytosis, efflux pump, carrier-mediated pathway, and adsorptive transcytosis (Kim et al. 2010). The drug delivery into the brain by crossing the BBB was represented in diagrammatic manner (Fig. 16.2).

Liposomes are spherical self-assembling colloidal structures composed of lipid bilayers with an outer lipid bilayer with a central aqueous space. A lipid bilayer encompasses an aqueous space in the center. So many types of cancer drugs are being applied to this lipid-based system through various preparation techniques. Liposomal formulations of both the anthracyclines doxorubicin (Doxil, Myocet) and daunorubicin (DaunoXome) are among them. They are confirmed for metastatic breast cancer treatment and AIDS-related Kaposi's sarcoma (Cho and Wang 2008).

Nowadays, quantum dots are trendy among scientists due to their facile and rapid synthesis method using biological routes (Mahajan et al. 2017). Quantum dots now have the potential to significantly improve clinical diagnostic tests for cancer

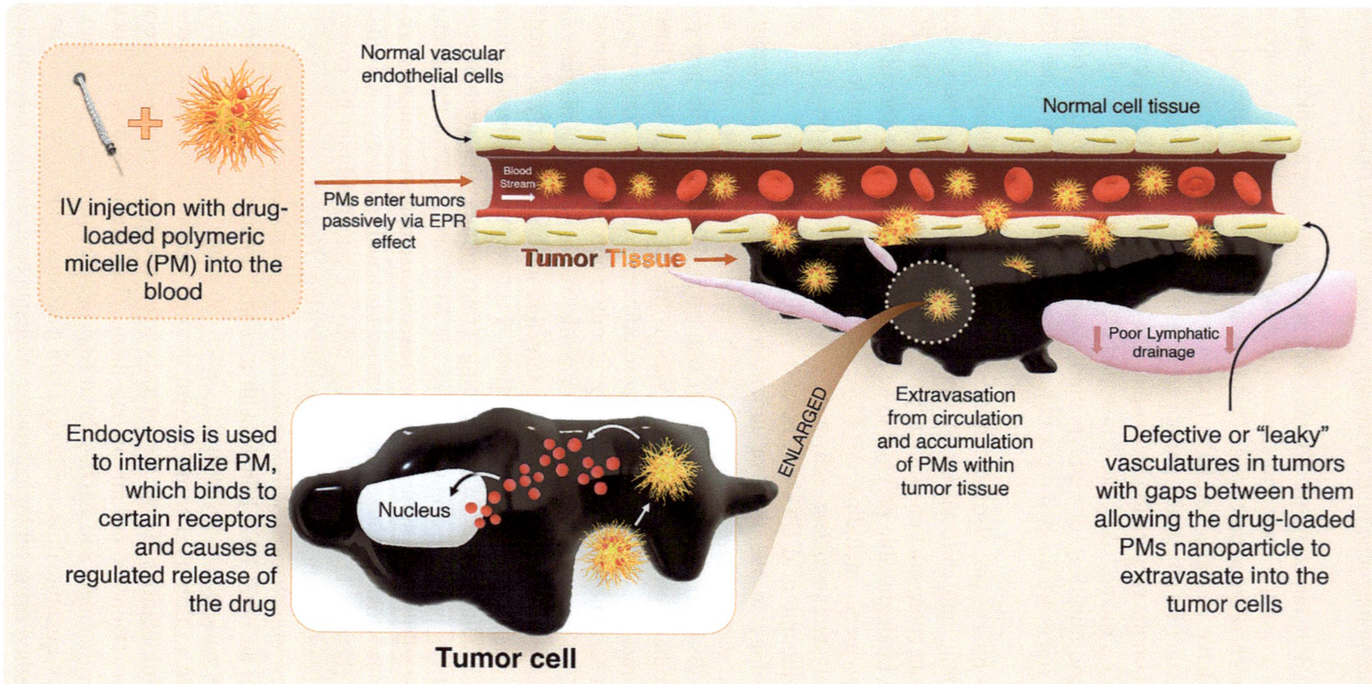

Fig. 16.1 Diagrammatic representation of drug delivery using enhanced permeability and retention effect. (The figure was reproduced from Hari et al. 2022 with the permission of Springer Nature)

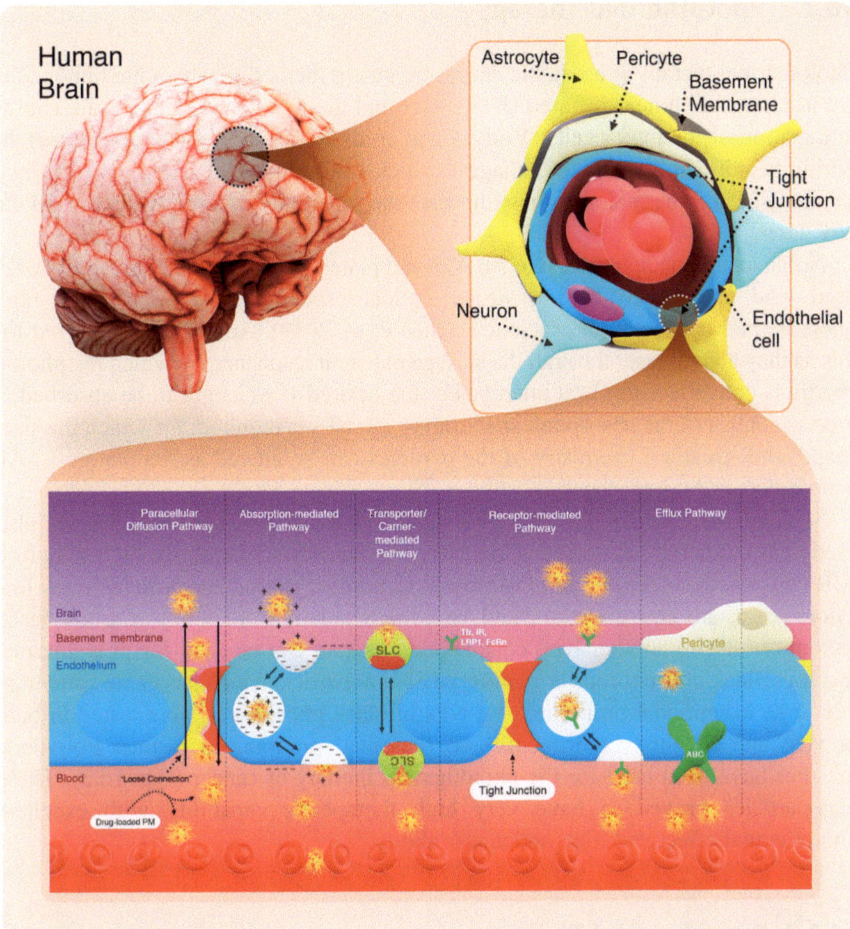

Fig. 16.2 Representation of the crossing of the blood-brain barrier for drug delivery into brain tumor using MNPs. (The figure was reproduced from Hari et al. 2022 with the permission of Springer Nature)

detection. These engineered semiconductor molecules combine cadmium as well as selenide in a densely populated atomic structure, which emits light in six colors up to four near-infrared colors as the dots get smaller. Thousands of subtle color variations might be created by fine-tuning the size of the dots. Once conjugated with antibodies, peptides, proteins, or DNA, such tiny glowing particles establish bioconjugated dots that can act as markers on molecules and genes, allowing scientists to rapidly but instead differentially highlight pathologic tissues (Conti et al. 2006).

16.3.2 Photothermal Therapy

Photothermal therapy is a reliable cancer treatment method in which the tumor cells are irradiated with near-infrared (NIR) laser and the increase in temperature causes termination of the tumor cells (Vodyashkin et al. 2021). In this process, the neighboring healthy cells are also damaged, which can be avoided using photothermal absorbers. Photothermal absorbers increase the sensitivity of target cells toward the NIR laser.

Noble metal nanoparticles like Au NPs and platinum nanoparticles are commonly used, and apart from having strong optical properties, they also have high photothermal conversation efficiency (Chen et al. 2019). Au NPs are preferred for this as they absorb laser through the surface plasmon resonance by which the photon energy is converted into heat energy. For the desired wavelength to be absorbed, a precise control of the absorption spectrum of Au NPs is required, for which the size, shape, and structural chemistry of the particles can be altered as Au nanorods, Au nanoshells, or Au nanocages.

Semiconducting nanoparticles like copper chalcogenide nanoparticles are also used as photothermal absorbers. Copper-based nanomaterial requires higher limit of NIR and penetrates deeper into the biological tissues. These nanoparticles show low cytotoxicity, are cost-effective, and have high ablation efficiency. Carbon-based nanomaterials are another type of nanoparticles that can be used for photothermal therapy. Black phosphorous is a new type of nanoparticle which has gained attention due to its exceptional properties. Au NPs in black phosphorus were tested in both in vivo and in vitro experiments and presented pronounced results. Polymeric micelles have been used for the photothermal therapy of cancer (Fig. 16.3). These nanoparticles proved to have very high therapeutic capability with excellent photothermal effect and high anticancer ability.

16.3.3 Sonochemical Therapy

Oncologists have been working to find new methodologies for the treatment of cancer and so have explored the use of ultrasound waves. Sonochemical therapy uses ultrasonic waves that are illuminated on the tumor cells (Mason 2003). The effect of these waves on the human body is in the form of heat, mechanical stress, and cavitation.

There are different methods of therapy using ultrasonic waves. In high intensity focused ultrasound (HIFU), the tumor cells are destroyed by the effect of heat, and the temperature is increased to about 60–85 °C. In low-intensity ultrasound (LIU), this is sonoporation method, and here the permeability of the membrane is increased to increase the chemotherapeutic ability of the drugs. The last is the sonodynamic therapy (SDT), which is related to the cavitation effect of the ultrasonic waves, and this therapy seems promising as the ultrasonic waves can reach intensely localized target cells. Nanoparticles especially Au NPs owing to their therapeutic effects are mainly used with SDT for recovery after anticancer therapy (Vodyashkin et al.

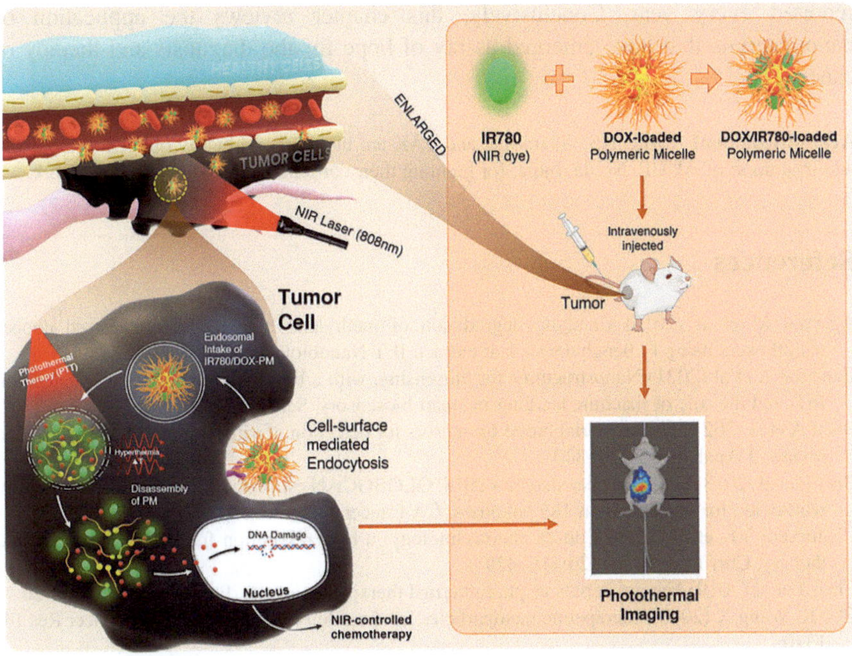

Fig. 16.3 Diagrammatic representation of photothermal treatment using micelles. (The figure was reproduced from Hari et al. 2022 with the permission of Springer Nature)

2021). Au NPs with ultrasound waves are used to control the damage of the affected cells as Au NPs reduce the activity of reactive forms of oxygen present in the damaged cells. The use of nanoparticles and ultrasonic waves for the treatment of cancer is being explored, and the properties of both can be used for encouraging outputs.

16.4 Conclusion

In this chapter, we reviewed the current innovations of nanomaterials for cancer diagnosis and therapy. Currently, cancer is a leading disease and a big challenge to the scientists as the elevated mortality rate has been reported. The traditional cancer therapeutics have been criticized due to several unpleasant causes and inadequate death of cancer cells. Therefore, there is a demand to develop new and advanced cancer diagnosis and therapy, which requires new and innovative methods. Nanomaterial-based cancer theragnostic were found to be accurate, timely tools for cancer detection and extremely effective approaches for cancer therapy due to their unique physicochemical characteristics, comprising ultrasmall size, high reactivity, and tunable functional alteration. The prospective action of nanomaterials approach in cancer theragnostic is interesting, and outstanding results are being reported every year. Conclusively, this chapter reviews the application of nanomaterials that have emerged as ray of hope for the diagnosis and therapy of cancers.

Acknowledgment RMT, NS, Akanksha, and EAK are thankful and highly obliged to the entire administration of AUUP, Noida, India, for granting their constant support for the above work.

References

Agarwal M et al (2016) Catalytic degradation of methylene blue by biosynthesised copper nanoflowers using F. benghalensis leaf extract. IET Nanobiotechnol 10:321–325

Banerjee A et al (2021) Nanostructures for biosensing, with a brief overview on cancer detection, IoT, and the role of machine learning in smart biosensors. Sensors 21:1253

Bisht A et al (2021) Nanomaterial based biosensors for detection of viruses including sars-cov-2: a review. J Anal Test 5:327–340

Bray F et al (2018) Global cancer statistics 2018: GLOBOCAN estimates of incidence and mortality worldwide for 36 cancers in 185 countries. CA Cancer J Clin 68:394–424

Chaturvedi VK et al (2019) Cancer nanotechnology: a new revolution for cancer diagnosis and therapy. Curr Drug Metab 20:416–429

Chen J et al (2019) Nanomaterials as photothermal therapeutic agents. Prog Mater Sci 99:1–26

Cho K, Wang X (2008) Therapeutic nanoparticles for drug delivery in cancer. Clin Cancer Res 14: 1310

Conde J et al (2012) Noble metal nanoparticles applications in cancer. J Drug Deliv 2012:751075

Conti M et al (2006) Anticancer drug delivery with nanoparticles. In Vivo 20:697–701

Haley B, Frenkel E (2008) Nanoparticles for drug delivery in cancer treatment. Urol Oncol 26:57

Hari SK et al (2022) Polymeric micelles and cancer therapy: an ingenious multimodal tumor-targeted drug delivery system. Drug Deliv Transl Res:1–29

Kang J-H et al (2010) Gold nanoparticle-based colorimetric assay for cancer diagnosis. Biosens Bioelectron 25:1869–1874

Kim JY et al (2010) Hydrotropic solubilization of poorly water-soluble drugs. J Pharm Sci 99: 3953–3965

Li W, Chen X (2015) Gold nanoparticles for photoacoustic imaging. Nanomedicine 10:299–320

Mahajan R et al (2017) Green synthesis of highly stable carbon nanodots and their photocatalytic performance. IET Nanobiotechnol 11:360–364

Mason TJ (2003) Sonochemistry and sonoprocessing: the link, the trends and (probably) the future. Ultrason Sonochem 10:175–179

Medley CD et al (2008) Gold nanoparticle-based colorimetric assay for the direct detection of cancerous cells. Anal Chem 80:1067–1072

Mehrotra N, Tripathi RM (2015) Short interfering RNA therapeutics: nanocarriers, prospects and limitations. IET Nanobiotechnol 9:386–395

Mehrotra N et al (2017) Catalytic degradation of dichlorvos using biosynthesized zero valent iron nanoparticles. IEEE Trans Nanobioscience 16:280–286

Mishra S et al (2022) Recent advances in silver nanozymes: concept, mechanism, and applications in detection. Adv Mater Interfaces. https://doi.org/10.1002/admi.202200928

Misra R et al (2010) Cancer nanotechnology: application of nanotechnology in cancer therapy. Drug Discov Today 15:842–850

Pasinszki T et al (2017) Carbon nanomaterial based biosensors for non-invasive detection of cancer and disease biomarkers for clinical diagnosis. Sensors 17:1919

Perfézou M et al (2012) Cancer detection using nanoparticle-based sensors. Chem Soc Rev 41: 2606–2622

Ravanshad R et al (2018) Application of nanoparticles in cancer detection by Raman scattering based techniques. Nano Rev Exp 9:1373551

Saxena A et al (2012) Green synthesis of silver nanoparticles using aqueous solution of Ficus benghalensis leaf extract and characterization of their antibacterial activity. Mater Lett 67:91–94

Tothill IE (2009) Biosensors for cancer markers diagnosis. Semin Cell Dev Biol 20:55–62

Tripathi R, Chung SJ (2019) Biogenic nanomaterials: synthesis, characterization, growth mechanism, and biomedical applications. J Microbiol Methods 157:65–80

Tripathi RM, Chung SJ (2020) Phytosynthesis of palladium nanoclusters: an efficient nanozyme for ultrasensitive and selective detection of reactive oxygen species. Molecules 25:3349

Tripathi RM, Chung SJ (2021) Eco-friendly synthesis of SnO2-Cu nanocomposites and evaluation of their peroxidase mimetic activity. Nanomaterials 11:1798

Tripathi R et al (2015) Biogenic gold nanoparticles: as a potential candidate for brain tumor directed drug delivery. Artif Cells Nanomed Biotechnol 43:311–317

Tripathi R et al (2019) Metal-induced redshift of optical spectra of gold nanoparticles: an instant, sensitive, and selective visual detection of lead ions. Int Biodeterior Biodegradation 144:104740

Tripathi RM et al (2020) Enzyme mimetic activity of ZnO-Pd nanosheets synthesized via a green route. Molecules 25:2585

Tripathi RM et al (2022) Biogenic ZnO nanoflowers: as an alternative antibacterial nanomedicine. Curr Nanomed (Formerly: Recent Patents Nanomed) 12:76–81

Vodyashkin AA et al (2021) Application of gold nanoparticle-based materials in cancer therapy and diagnostics. Chem Eng 5:69

Yadav L et al (2017) Antibacterial activity of Cu nanoparticles against E. coli, Staphylococcus aureus and Pseudomonas aeruginosa. Nano Biomed Eng 9:9–14

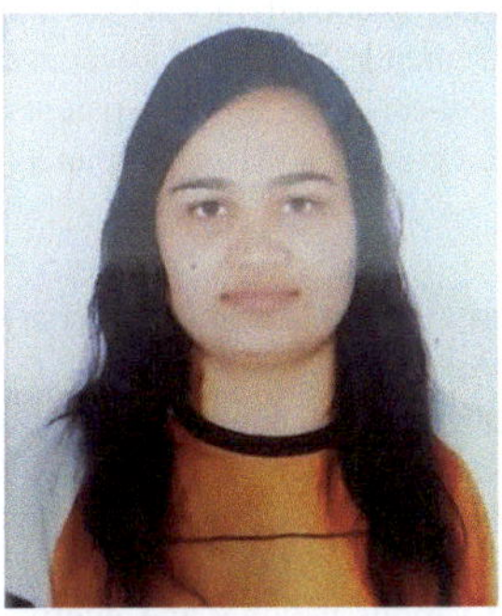

Akanksha Deshwal is a Ph.D. Scholar at Amity University, Noida. She is currently pursuing Ph.D. under the supervision of Dr. Ravi Mani Tripathi, Associate Professor, Amity University, Noida. Her Ph.D. research work is focused on the development of nanozyme-based colorimetric assay for the detection of cancer.

Ekta Arjundas Kukreja is a student at Amity Institute of Nanotechnology, Amity University, Uttar Pradesh, Noida, Uttar Pradesh, India, pursing her Integrated B. Tech-M. Tech in Nanotechnology. Her field of interest is biosensors and colorimetric assay for detection of biological analytes.

Neeraj Shrivastava is Assistant Professor III at Amity University Uttar Pradesh, Noida, India. Earlier, he was a Visiting Scientist and Postdoctoral Researcher at the Institute of Biotechnology, Zhejiang University, China. He was awarded fellowships from the Chinese Scholarship Council (CSC) and National Postdoctoral Scholarship Foundation China and Senior Research Fellowship from the CSIR, MHRD Govt. of India. He has published more than 48 research papers and book chapters in peer-reviewed journals and reputed international publishers. He has handled two research projects from CSIR-HRDG and DST SEED worth 60 lacs. He has delivered several talks at national and international conferences and also published abstracts and papers. He has successfully guided nine Ph.D. and two M.Phil. students during his tenure. He is serving as Editorial Board Member of journals and Expert Reviewer of various national and international high impact journals.

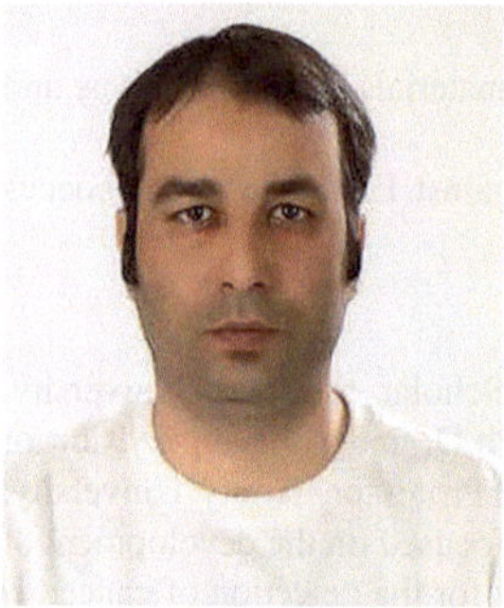

Faheem A. Sheikh is working as Assistant Professor in the Department of Nanotechnology at the University of Kashmir. He served as an Assistant Professor in the Central University of Kashmir (2015–2016), Assistant Research Professor at the Myongji University (2014–2015), Assistant Professor Research at the Hallym University (2012–2014), Postdoc at the University of Texas Rio Grande Valley (2010–2012), and Research Professor at Myongji University (2010). He received his Ph.D. in Bionanosystem Engineering from Jeonbuk National University. He has published more than 100 peer-reviewed articles, patents, and book chapters. He has a total citation of 4251 to his papers, 77 to h-index, and 35 to 10-index.

Touseef Amna received her Ph.D. in Microbiology in 2006 from Aligarh Muslim University (AMU, Central University), India. Presently she is working as an Associate Professor at Albaha University, Saudi Arabia. During her research career, she got the opportunity to spend postdoctoral stages at the University of Dortmund, Germany, Universidad de Talca, Chile, and Chonbuk National University, South Korea, and received various prestigious awards and grants. She authored 2 patents (1 US patent and 1 Korean) and more than 100 research articles in journals of international repute. She is Editor, Guest Editor, and Editorial Board Member of various national and international journals. Her research interest consists of design and synthesis of biocompatible scaffolds for antimicrobial applications, cell culture, and tissue engineering.

Ravi Mani Tripathi is working as Associate Professor at Amity Institute of Nanotechnology, Amity University. Dr. Ravi received his Ph.D. in Nanobiotechnology from Jiwaji University. He did his postdoctoral research at the School of Pharmacy, Sungkyunkwan University, South Korea. He received the Young Scientist Award in 2019 by the International Scientist Award for Engineering, Science, and Medicine. He developed an innovative method for the biological synthesis of nanomaterials like particles, nanodots, flowers, rods, fibers, etc. He has authored more than 45 articles in peer-reviewed journals of international repute. He is serving as Associate Editor of *Frontiers in Nanotechnology* and Editorial Board Member of various international journals.

Nanocellulose as a Sustainable Nanomaterial for Films and Coating Layers via Spray-Coating and Applications

17

Kirubanandan Shanmugam, Narendhar Chandrasekar, and Ramachandran Balaji

Abstract

Cellulose fibrous substrates such as paper and paper board are widely used mainly in packaging applications and also in various functional applications. The cellulose materials are also biodegradable and, therefore, perfectly safe for the environment. The hydrophilic nature of cellulose fibrils in the paper and paper board limits their water vapor-barrier properties. To mitigate these limitations, paper is often associated with other materials, such as plastics, wax, and aluminum, for their excellent barrier properties. However, these materials suffer from serious environmental issues, as they are difficult and inefficient to recycle. Recently, nanocellulose has been considered as an alternative natural nanomaterial to produce green and eco-friendly barrier materials for various applications such as food packaging materials, and substrates can be used for developing functional materials. Existing techniques to prepare free-standing nanocellulose films and barrier layers on paper substrates are commercially not feasible, and the production cost was prohibitive. Therefore, other cost-effective and readily implementable methodologies are required to achieve nanocellulose barrier layers and free-standing nanocellulose films. Recently, a novel approach has been conceptualized using spray-coating techniques to produce cellulose nanomaterials film with excellent barrier properties. In spray-coating, the significant reduction in the operation time for producing the nanocellulose films/

K. Shanmugam (✉)
Saveetha School of Engineering, SIMATS, Chennai, Tamil Nadu, India

N. Chandrasekar
Department of BioNano Technology, Gachon University, Seongnam-si, Gyeonggi-do, Republic of Korea

R. Balaji
Department of Electronics and Communication, K. L. University, Vijayawada, India

F. A. Sheikh et al. (eds.), *Interaction of Nanomaterials With Living Cells*,
https://doi.org/10.1007/978-981-99-2119-5_17

nanocellulose barrier layers on the paper substrates recommends that this technique can be utilized for the development of a scalable process for the fabrication of self-standing nanocellulose films, barrier coating, and various cellulose-based nanocomposite.

Keywords

Nanocellulose · Spray-coating · Vacuum filtration · Suspension consistency · Uniformity · Basis weight and thickness

17.1 Introduction

Past decades synthetic plastics have been the base material for developing various functional materials such as food packaging, flexible electronics, and coating on paper substrates (Muncke 2021). These materials are very odd to recycle, or nonbiodegradable, and, their waste contaminates the ecosystems (Phelan et al. 2022). Recently, these materials are also one of the important sources for micro plastics and nano plastics contaminating the environment via land, water, and soil pollution (Jadhav et al. 2021). These microplastics and nano plastics are severe pollutants and have the ability to enter into humans via various sources such as food chain (Jin et al. 2021). To mitigate these issues in the current world, biopolymers are the considerable base biomaterials for developing functional materials in the food packaging sector (Aulin and Lindström 2011). In general, these biopolymers are good in biodegrdability, ecofriendly with environment, the presence of potential functional groups in the biopolymers for chemical modification for various applications (Rouf and Kokini 2018). Cellulose is the predominant fibrous-based biopolymer and provides a possible platform for various modifications to fit numerous functional applications (Shanmugam 2019). Cellulose was used for the development of paper and paper boards for various applications (Aulin and Lindström 2011; Kirubanandan 2022).

Paper and paper board are the predominant cellulose fibrous materials and are used mainly in the sector of the packaging application and also the base fibrous substrates for printed flexible electronics and packaging wrap. These substrates are cellulose macrofibers having large pores in the fibrous structure. As a consequence, water vapor and oxygen are easily passing through the cellulose fibrous matrix and across the fibrous matrix. This phenomenon lowers the barrier performance of the paper substrates, and due to wetting with water vapor under high humidity, paper substrates are wetted and loose fibrous matrix resulting in widening the pores in the fibrous matrix and thereby their low mechanical properties. Furthermore, it was discussed that the cellulose fibrils are hydrophilic and susceptible to water vapor and water molecules. Especially paper and paper board have limitations in the packaging sector though these are biodegradable and perfectly safe for the environment (Qureshi et al 2022; Shanmugam 2020; 2021c, d; 2022a, b). To increase the barrier and mechanical properties of paper and paper board, coating techniques have been developed for the surface of the paper substrates. These coating techniques engineer the surface of the paper substrates and also enhance various properties such as barrier performance, mechanical properties, and surface roughness. Wax and synthetic plastics are used as feed stock for coating the paper substrates via extrusion and lamination. However, these are not recyclable and could cause pollution to the environment (Kirubanandan 2022). To improve the properties of paper and paper board, coating nanocellulose on the paper and paper board is a novel approach, and cellulose nanofibers can cover the surface of the paper substrates via forming the barrier film on the paper surface via filling the surface pores of the paper and alter properties drastically than that of uncoated paper substrates (Aulin and Lindström 2011; Kirubanandan 2022; Beneventi et al. 2015).

Nanocellulose is a nanomaterial that is made up of cellulose nanofibers and nanoparticles and produced from the pulps of various cellulose sources such as wood, hemp, grass, and cotton, via various methods such as mechanical, chemical, and enzymatic processes (Dufresne 2013). Nanocellulose is a carbohydrate-based fibrous nanomaterial and renewable, biodegradable, and ecofriendly nanomaterial used for various applications in the current era (Abraham et al. 2011). It is made by the breakdown of either cellulose macrofibers or microfibers via mechanical, chemical, or enzymatic process (Shanmugam 2019; Souza et al. 2020). The diameter of cellulose nanofibers varies from 5 to 100 nm with an average length of 8 μm (Dufresne 2013; Wang et al. 2018). It is a building block for the development of various functional materials such as the coating on paper substrates (Shanmugam 2020) and paper board, thin films for biological applications, flexible sheets (Shanmugam 2021a) as substrates for flexible printed electronics, tissue engineering applications (Subha et al. 2022), composites (Maliha et al. 2020), and food packaging (Azeredo et al. 2017).

Nanocellulose is a potential alternative for synthetic plastics as this nanomaterial is a renewable, recyclable, compostable, and biodegradable nanofibers (Azeredo et al. 2017; Ahankari et al. 2021). Because of these unique properties, free-standing film is also developed and acts as a base substrate for developing various novel materials such as food packaging (Souza et al. 2020), flexible electronics substrates (Shanmugam 2021a), membranes for waste water and water treatment (Norfarhana et al. 2022; Onur et al. 2019), virus removal membrane, tissue engineering substrates (Subha et al. 2022), and nanocomposites (Shanmugam 2021b). The coating of paper substrates with nanocellulose was performed by the solvent casting, which is normally the time-consuming process for forming the barrier layers on the paper substrates; vacuum filtration, a conventional process for paper producing process and can be used for coating nanocellulose on the papers; roll-to-roll coating process; rod coating; bar coating; and recently reported spray-coating process (Shanmugam 2021c). Like barrier coating nanocellulose on the paper substrates, the above techniques were implemented to fabricate the free-standing nanocellulose films (Shanmugam 2021c). Vacuum filtration is the most common and conventional technique for the preparation of nanocellulose film and has the disadvantage of significant time consumption. Vacuum filtration is the laboratory version of fourdrinier machine used for paper production at a conventional scale. In vacuum filtration, time consumed for forming the nanocellulose film varied from 10 min to 24 h. Time consumption has been increased with nanocellulose suspension consistency for film formation. Furthermore, the film has imprints of the filter mesh from the vacuum filtration apparatus and affects the uniformity of the film and elevates the surface roughness of the film. Due to time constraints in the formation of film, this process is very odd to scale up and commercialization for large-scale production of free-standing nanocellulose films. This method has high dewatering time or time consumed for drainage of the water from nanocellulose suspension (Varanasi and Batchelor 2013).

The method available for the preparation of free-standing nanocellulose film has several limitations, either drawbacks in the film's properties or the process inabilities for scale-up and commercialization. This is why the rapid and flexible process for coating nanocellulose on the paper substrates is required for coating process commercialization (Shanmugam et al. 2017). Spraying or spray-coating was reported as a rapid and flexible process for either creating nanocellulose barrier layer on the paper substrates or producing free-standing nanocellulose film. It has many advantages, such as contour coating, forming homogenous layer on the base surface and contactless coating with substrates (Shanmugam and Browne 2021).

This book chapter reveals the development of spraying process as a rapid and flexible method to prepare both free-standing nanocellulose films and its composites. This chapter deals with the literature review on the limitations of nanocellulose films preparation, concept, flexible process on spraying to make NC films, and application of spray-coating to make composites and smooth NC films for specific functionality. In addition to that, spraying nanocellulose on the paper substrates also implemented to enhance the barrier performance of the paper. This chapter has broadly dealt (1) the concept of developing free-standing nanocellulose film via spraying of NC on a polished impermeable surface and its operational limit, (2) convert a base concept into a full-fledged process of spray-coating of nanocellulose to produce free-standing NC film and composites, and (3) to apply full potential spray-coating process to make a barrier coating on the paper substrates.

17.2 Nanocellulose

Recently, the nanotechnology approach in the development of packaging materials (Paul and Robeson 2008) is booming and increased the attention on nanotechnology from the industry. Implementing nanocellulose as nanomaterials for packaging applications is a novel approach to replacing synthetic plastics, which causes a threat to the environment (Bras and Saini 2017). The nanocellulose-based packaging materials and other functional materials have biodegradability, biocompatibility, and good mechanical strength for a potential fit for a specific application (Azeredo et al. 2017). Nanocellulose is a carbohydrate fibrous-based biopolymer and this nanomaterial produced from the various pulps of wood, hemp, cotton, grass, etc. (Dufresne 2013). Nanocellulose can be produced from wood and also from other cellulosic material via various methods of high-pressure homogenization,

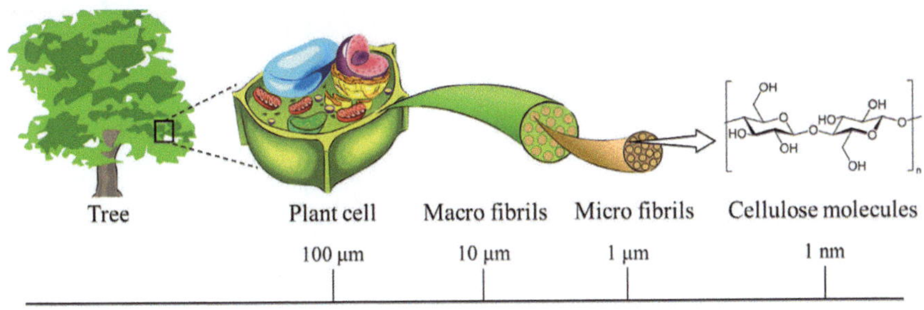

Fig. 17.1 Hierarchical structure of wood fibers, network of microfibrils. (Miyashiro, D.; Hamano, R.; Umemura, K. A Review of Applications Using Mixed Materials of Cellulose, Nanocellulose and Carbon Nanotubes. *Nanomaterials* **2020**, *10*, 186. https://doi.org/10.3390/nano10020186 under Open access)

high-intensity ultrasonication, micro-grinding, microfluidization, electrospinning, steam explosion, and chemical processes such as acid hydrolysis, enzymatic hydrolysis, and 2,2,6,6-tetramethylpiperidine 1-oxyl radical (TEMPO) oxidation. This nanomaterial has good biocompatibility, biodegradability, and chemical stability (Shanmugam and Browne 2021). Recently, the research on the use of NC in developing barrier materials for packaging applications is increasing exponentially. Due to the high availability of NC, it could minimize the cost of the starting material for packaging products from NC and also offers a platform for recyclability and reusability, leading to a sustainable pathway (Shanmugam et al. 2019; Vartiainen et al. 2018).

Nanocellulose (NC) also called as cellulose nanofibers (CNF) is one of the bio-based sustainable materials and has diameter and width of the fiber in nanometer (Dufresne 2013). The cellulose pulps from wood or cotton are subjected to delamination of cellulose fibers into microfibers or nanofibers (Nechyporchuk et al. 2016). It is a long fiber with an average length of 8 µm with an aspect ratio of 142 for the microfibrillated cellulose. NC is a kind of fibrous matrix made up of various size distributions of cellulose nanofibrils and microfibrils with the entanglement of various sizes (Abraham et al. 2011). Figure 17.1 reveals the hierarchy of the cellulose fibers from the wood to nanofibers. This figure explains how nanofibers defibrillate from the cellulose macrofibers via various mechanical and chemical means such as homogenization, ball milling, and acid hydrolysis.

17.2.1 Characteristics of Nanocellulose

Having long and flexible nanocellulose fibers, this fibrous entanglement consists of two regions, namely, amorphous and crystalline regions. These fibers' distribution and regions play a major role in the contribution of various functional properties, especially barrier performance of the nanocellulose films and coatings. Figure 17.2 reveals the surface morphology of the cellulose nanofibrils/nanocellulose after two

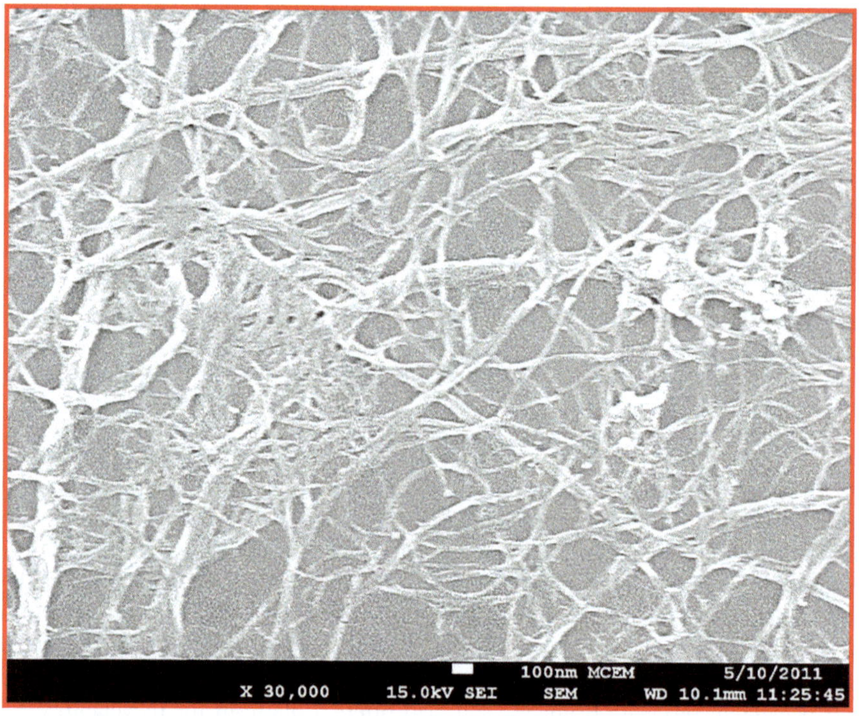

Fig. 17.2 Cellulose nanofiber or nanocellulose containing various size distributions of cellulose nanofibrils

passes high-pressure homogenization. The diameter of cellulose nanofibrils after high-pressure homogenization is evaluated to be 20 nm. The mechanical strength of NC is high elastic modulus (150 GPa) and high tensile strength (10 GPa) and is controlled by their cellulose nanofibrils' diameter and length. The source, processing method, and particle type of the feed stock for cellulose nanofibers decide the diameter and aspect ratio of the cellulose nanofibrils in the nanocellulose matrix and also decide applicability and functionality in various areas (Abitbol et al. 2016; Shanmugam 2019).

Figures 17.2 and 17.3 show the surface morphology and topography of cellulose nanofibers/nanocellulose after high-pressure homogenization. The SEM micrographs confirm the cellulose nanofibrils are entangled together to form a fibrous mesh. In addition to that, the nanofibrils are distributed in various sizes, and some of the fibrils are aggregated via forming hydrogen bonding between hydroxyl group to adjacent nanofibrils. Figure 17.4 shows the size distribution of cellulose nanofibrils and confirms that the average diameter of cellulose nanofibrils is evaluated to be 73 nm, and most of the size of nanofibrils falls below 100 nm. Some of the cellulose nanofibrils showed greater than 100 nm due to the aggregation of cellulose nanofibrils (Shanmugam 2021c; Shanmugam et al. 2019).

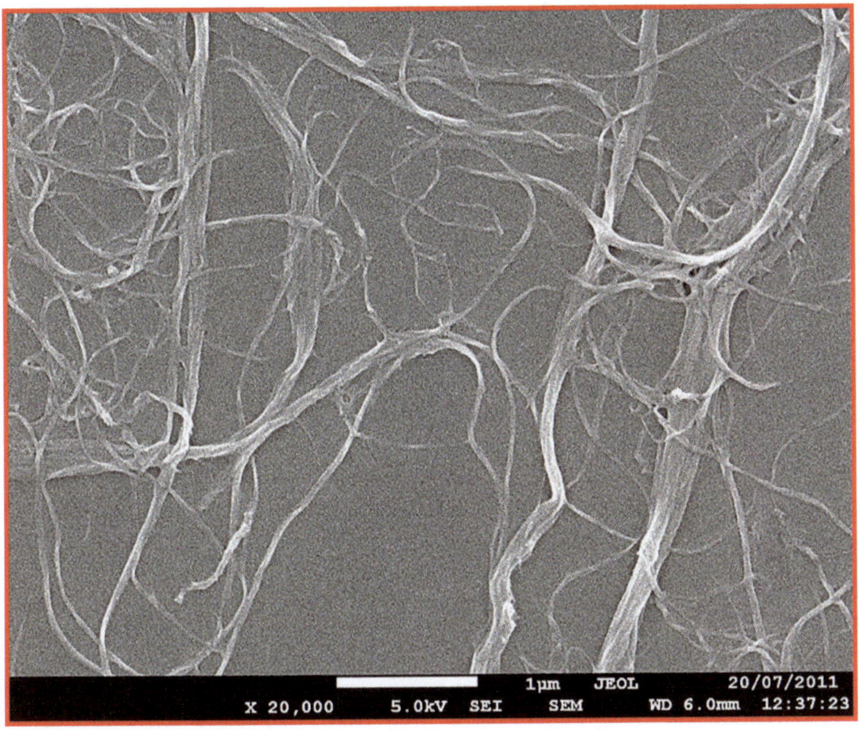

Fig. 17.3 Surface topography and morphology of cellulose nanofiber or nanocellulose

The diameter of cellulose nanofibrils ranges typically from 5 to 100 nm, and the length of each fibril is in several microns (Nair et al. 2014). Both crystalline and amorphous regions are the main regions in the fiber chain (Nair et al. 2014). The function of crystalline region contributes mostly to various functionalities of materials such as barrier and composites materials. For example, the barrier application of nanocellulose film has a tortuous pathway produced by the crystalline regions in the films. As a result, the barrier performance of the NC film boosted and increased the diffusion pathway for water vapor, oxygen, and other gaseous molecules. The cellulose nanofibril diameter distribution of NC and its aspect ratio depends on the source, pretreatment processing, and fibrillation processes, and fiber diameter and length indirectly control its rheological and interfacial properties of NC suspension (Shanmugam 2021c; Shanmugam et al. 2019).

The evaluation of diameter and aspect ratio of NC was performed with an image analysis of micrographs collected from SEM, TEM, and AFM. Sedimentation method is also used to evaluate their gel point of NC fibers. The gel point is defined as the point at which the lowest fiber concentration in the nanocellulose suspension formed and also a transition point from a highly dilute solution of NC fibers into a dense, thick, concentrated suspension. At this point, nanofibrils in the NC

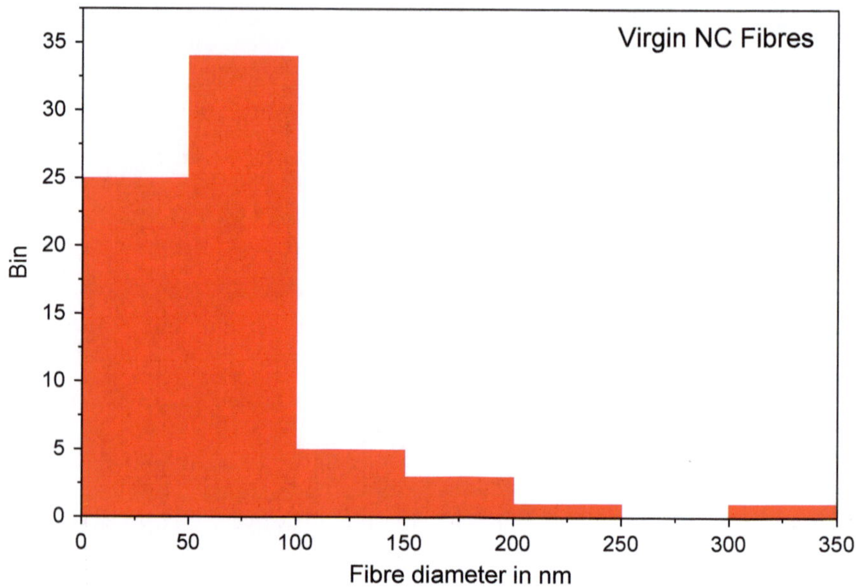

Fig. 17.4 Size distribution of cellulose nanofibers or nanocellulose

suspension form a compact network of fibers due to hydrogen bonding between fibers. This network is known as the threshold of connectivity. Beyond this concentration, the NC suspension loses its mechanical strength due to insufficient contact of fibers with each other. This property depends on the aspect ratio of nanocellulose fibers. This characteristic has an important effect on the drainage time in vacuum filtration for NC film fabrication (Shanmugam and Browne 2021).

The notable properties of nanocellulose are light in weight, optical transparency, chemical functionality and modification, dimensional stability, and good barrier properties. The compatibility of NC is excellent with other materials like bio-natural polymers, proteins, and living cells for developing scaffolds for tissue engineering and drug delivery/drug targeting vehicles. The presence of surface hydroxyl groups in the cellulose nanofibers provides a platform for chemical modification and functionalization such as hydroxy and amine terminals which can further the binding targets to the polymers. The viscosity of the NC suspension is one of the main criteria for coating nanocellulose on the paper substrates and other base surfaces. The NC suspension behaves as a non-Newtonian fluid such as pseudoplastics behavior under shear thinning fluids (Shanmugam 2021c).

17.2.2 Production of Nanocellulose

Wood, seed fibers (e.g., cotton, coir, etc.), bast fibers (flax, hemp, jute, ramie, etc.), and grasses (bagasse, bamboo, etc.) are the feed stocks for the production of nanocellulose. Recently, carrot and bagasse are the source for the production of nanocellulose. In addition to that, the recycled cellulose fibers, such as waste paper, are subjected to homogenization for the production of nanocellulose. There are so many researches going on for the development of low-cost feed stock. Marine animals (tunicate), algae, fungi, invertebrates, and bacteria are less considered for the production of cellulose nanofiber nanomaterials (Nechyporchuk et al. 2016). The pulp from the wood is one of the most important feed stocks for NC production. The wood for pulp production was classified into hard wood and soft wood. The main difference between hard wood and soft wood is the structural difference in the hierarchy of cellulose fibers in the wood. Cellulose is the main component of wood, and the other components are hemicellulose, lignin, and inorganic substances present in the wood. Normally, hard wood is more rigid than that of soft wood. This is why soft wood is the most preferred for the nanocellulose production due to the least energy consumed for defibrillation or homogenization process. Similarly, the energy consumed for non-woody cellulose feed stock such as grass, carrot, and marine sources is low for the production of nanocellulose. The common problem for wood as feed stock for nanocellulose is the contamination of lignin with nanocellulose. Therefore, a few additional steps for the purification of nanocellulose are required to remove the lignin from the cellulose nanofibrils. Recently agricultural and food waste were used as a feed stock for the manufacture of nanocellulose. It consumes less energy for nanocellulose fibrillation free from other cellulose substances such as hemicelluloses and lignin. Refining processes are the challenging step for fibrillation from cellulose microfibrils to cellulose nanofibrils known as nanocellulose. Refining processes are classified into mechanical, chemical, and enzymatic processes. Prior to refining processes, microfibrils are subjected to the purification process, such as delignification, and it reduced the energy consumption for fibrillation processes and produced high-quality nanocellulose fibers.

Disintegration is one of the most important processes in mechanical fibrillation of cellulose microfiber into cellulose nanofiber. Grinding comes under disintegration of fibers and the most efficient method for the production of nanocellulose. High-pressure homogenization is the recently reported method for fibrillation of microfibrils into nanofibrils. The disadvantage of this method is a highly energy-consuming process and requires 25,000 kWh/tonne of the production of microfibrillated cellulose. The mechanical process is a good fit for scale-up and commercialization for the large-scale production of nanocellulose. Especially, pressure, NC suspension concentration, and number of passes in the method of high-pressure homogenization strongly decide the energy consumption in the process. The various methods of mechanical disintegration of fibers are reported in Table 17.1. It summarizes the concept of cellulosic fibers reduction, its isolation methods, and its advantages and disadvantages (Osong et al. 2016; Shanmugam 2019).

As discussed earlier, mechanical fibrillation process from microfibrils into nanofiber is an energy-consuming process, and it is one of the process constraints for scaling up and commercialization. In recent decades, the demand for cellulosic

Table 17.1 Mechanical refining for production of nanocellulose (Osong et al. 2016)

Methods for isolation of cellulosic fibers	The concept of reduction of size	Advantages	Disadvantages
High-pressure homogenization	High-impact shearing forces reduce fiber size	Quick, effective, and continuous process Good reproducibility Could control the degree of defibrillation	Clogging Pretreatment of fibers required High passing time and high energy consumption Rise in temperature of suspension
Microfluidization	An intense collision with high impact led to the splitting of macrofibers into nanofibrils	Less clogging Uniformity in size Lesser cycles	Not suitable for scale-up
Micro grinding	The macrofibers are pressed in the gap between the stator and rotor disc in the grinder. High impact and frictional forces disintegrate the fibers into fibrils	Less energy and cycle No pretreatment of fibers	Maintenance cost is expensive Difficult in replacement of internal parts such as disk The crystalline nature of nanocellulose is reduced
High-intensity ultrasonication	Sound energy is utilized to disintegrate macrofibers into nanofibrils	High-power output and the high efficiency of defibrillation	Heat generation and noise pretreatment only lab scale application
Refining	High shearing forces used for disintegration		
Cryo-crushing	The refined macrofibers are treated with liquid nitrogen, and it freezes the water in the fibers and then subjected to high-impact grinding for disintegration	High disintegration performance	Ice formation
Steam explosion	The suspension is heated at high pressure and then vented into a vessel with low pressure. It is a type of explosion for reduction of fibers	–	Chemical pretreatment required

nanomaterials is exponentially increasing its application in both research on developing new products and industries. Therefore, an energy-efficient process needs to scale up for large production of nanocellulose (Nechyporchuk et al. 2016). Chemical or enzymatic processes are proven low-energy processes compared with mechanical

Table 17.2 Chemical processing for nanocellulose (Osong et al. 2016)

Serial no.	Chemical methods	Reaction mechanism	Advantages	Disadvantages
1.	TEMPO-oxidation	Oxidation of C6 hydroxyl groups into carboxyl groups and partially into aldehydes	Shorter reaction time	TEMPO is a poisonous and expensive chemical reagent
2.	Periodate chlorite oxidation	Oxidation of the vicinal hydroxyl groups in the C2 and C3 positions	Increases carboxyl group	Weakens the structure of cellulose Long reaction time
3.	Alkaline extraction	Assistance with initializations of fibrils	Lignin degraded Enhanced fibrillation	Cellulose can be degraded High alkali concentration
4.	Carboxymethylation	Incorporating carboxymethylated cellulose to the fibers. Enhancing anionic groups, reduction of fiber friction, disintegration into fibrils	Improved fibrillation Reduced energy consumption	Thinner NFC produced
5.	Acid hydrolysis	The breaking of glycoside bonds	High crystallinity or NC	Use of strong mineral acids High controlled reaction

disintegration processes and have improved fibrillation efficiency (Osong et al. 2016). Various types of chemical processing of cellulosic nanofibers are presented, with their disadvantages and advantages in Table 17.2.

Enzymatic processes were used to produce cellulose nanofibers/nanocellulose from the cellulose biomass. In this method, the cleavage of glycosidic bonds in the cellulose molecules was achieved via enzymatic hydrolysis. Cellobiohydrolases and endoglucanases are the common enzymes for the production of NC, and the former attacks the crystalline region of cellulose and later attacks the amorphous or disordered region of cellulose. The yield of NC production via the enzymatic process depends on hydrolysis time, the concentration of enzymes in the reaction mixture, and types of enzymes used for hydrolysis. Normally, chemical and enzymatic process is performed before the mechanical process so that the clogging of mechanical device can be avoided and less energy consumed. As a result, good fibrillation of cellulose pulp was achieved, and high-quality fibers were produced (Shanmugam and Browne 2021; Ribeiro et al. 2019).

17.3 Nanocellulose Films

Free-standing nanocellulose film is one of the predominant products of the nanocellulose nanomaterial (Varanasi and Batchelor 2013). This film plays a predominant role in various fields of application, because the film derived from nanocellulose has notable properties such as mechanical strength, barrier performance, and surface properties (Aulin et al. 2010). Due to the outstanding performance of nanocellulose films properties, the films were used as food packaging materials (Nair et al. 2014), base substrates for flexible printed electronics (Shanmugam 2021a), membranes for water and waste water treatment (Norfarhana et al. 2022), filtration media (Manukyan et al. 2019) for virus removal from the air, fabricating the nanocomposites (Garusinghe et al. 2018), tissue engineering scaffolds, and other biomedical applications (Werrett et al. 2018). Nanocellulose is biocompatible biopolymer with tissues, and the cellulose structure mimics the glycoaminoglycans in the extracellular matrix (ECM). This is why that this material can be used as base biomaterial for tissue engineering constructs (Luo et al. 2019), biomedical nanocomposite with antimicrobial agents for wound dressings (Maliha et al. 2020), drug delivery vehicle, and drug targeting applications (Shanmugam 2021b). As discussed earlier in this chapter, nanocellulose is an ecofriendly nanomaterial with capability of biodegradability in the environment. So that, nanocellulose film can be used for laminating paper substrates via various methods for increasing the barrier performance of the paper against water vapor and oxygen (Kirubanandan 2022; Beneventi et al. 2014).

The abovementioned is not limited in the application of nanocellulose films. There is many ongoing research in the free-standing nanocellulose films nowadays. This is why the demand of nanocellulose films increases exponentially, and nanocellulose can be a good alternative for synthetic plastics. The process constraints in the preparation of free-standing nanocellulose films are high energy consumption for nanocellulose production and time consumption in the film formation and its drying. There are many methods available for the fabrication of free-standing nanocellulose film. Solvent casting, vacuum filtration, hot pressing, layer by layer coating, and spraying are the notable methods for developing self-standing nanocellulose films (Shanmugam 2019). The detailed description of each process will be discussed in the following headings. Due to process constraints in the production's methods, the commercialization of nanocellulose films in the market is delayed. The main problem in the fabrication of film is the large processing time in the formation of film. This is why, the rapid method for forming the film is required and to speed up the production rate of the free-standing nanocellulose films. Therefore, a method should have ability to handle high suspension concentration of nanocellulose fibers. So that, the basis weight and thickness of the nanocellulose film can be tailored, and its effects could be observed in the mechanical strength, barrier performance, and various surface and bulk properties of the film. To conclude, a fast and flexible process is required to fabricate the nanocellulose films and to tailor the properties of the film through adjusting the process parameters in the film formation processes (Shanmugam 2021c).

17.3.1 Solvent Casting

The laboratory-scale technique for the fabrication of nanocellulose film is solvent casting. In this method, the nanocellulose suspension was poured into the Petri dish and allowed to dry the suspension for many hours. The drying consumed more than 24 h to days, and the dried nanocellulose film was formed from the casting and it was peeled from the Petri dish (Syverud and Stenius 2009). The casted film has poor uniformity and shrinkages which affects various properties of nanocellulose films. In addition to that, the suspension concentration of nanocellulose decides the thickness and basis weight of the film which are important criteria for the film to control their barrier performance and mechanical strength of the films. Moreover, increasing the suspension concentration of nanocellulose would increase the drying time for the solvent casting to form the film. The disadvantages of this method are only laboratory scale performance in the film formation and formation of shrinkage and wrinkles on the nanocellulose film resulting in the weak mechanical strength and poor barrier performance. Due to the slow evaporation of the solvent from nanocellulose suspension casted on the Petri dish, it is inability to scale up for pilot scale and large-scale production of the films to meet the huge demand in the current era (Shanmugam 2019).

17.3.2 Spin Coating

Spin coating is a technique for fabricating these films on the surface and can be used for various research works on the thin film. This method is used to fabricate the free-standing nanocellulose films with nanometric thickness. The films prepared via spin coating can be used for the study of the biological molecules' interaction with the film and behavior of various biomolecules with hydroxyl group of the nanocellulose in the film. In this technique, high-speed spinning is utilized to remove the excess suspension of NC, thus leaving ultra-thin NC films. This is a laboratory scale method that can meet the research requirements of NC film as per the current demand in the laboratory research on nanocellulose films. It is the perfectly complete process to fabricate thin films for investigating at biomolecule interaction with cellulose fibrils (Shanmugam 2019; Shanmugam and Browne 2021).

17.3.3 Roll-to-Roll Printing

Roll-to-roll coating/printing is a rapid and fast process for the fabrication of nanocellulose films in a semi-batch or continuous operation mode. In addition to that, this method is also applicable for the coating of nanocellulose on the cellulose substrates such as paper and other paper board in a continuous board. The coated nanocellulose on the paper substrates formed as barrier film to increase the barrier performance of the paper with the elevated mechanical strength. In a similar manner, the nanocellulose coating on the plastic substrates was performed, and the nanocellulose film on the plastic can be peeled. The film can be peelable from the plastic substrates and can be used as the base material for developing various functional materials. The surface roughness of the nanocellulose film from this method was 400 nm, and it was replicated from the plastic substrates. This surface roughness of the film can be used for developing flexible and printed electronics substrates. The thickness and basis weight of the nanocellulose film can be controlled by nanocellulose suspension concentration, types of plastic substrates, and velocity of the webs in the coating process (Shanmugam 2019; Shanmugam and Browne 2021).

17.3.4 Layer by Layer Assembly

This is the method of fabricating the nanocellulose film from nanometer to micron thickness via coating. By this technique, this method is applicable for thin functional film and also coating on the substrates for specific application. This method is normally used in the surface engineering of the substrates or film via coating. This method was also used in the surface functionalization of the substrates in the nanometer range of thickness of the functional materials (Shanmugam 2019).

17.3.5 Vacuum Filtration

Currently, this is the method for conventional practice for the production of free-standing nanocellulose films (Varanasi and Batchelor 2013). This method is a laboratory version of fourdrinier machine for large scale production of paper and its related cellulose fiber materials. In vacuum filtration process, the nanocellulose suspension was poured into the column, and there is a metallic mesh at the bottom of the column. The water in the nanocellulose suspension was drained through the mesh when applying the vacuum at the bottom of the column. As a result, the nanocellulose fibers form a sheet on the metallic mesh, and the wet film was peeled from the mesh after couching with blotting paper. The wet film was peeled for the subjection to drying via drum drier. The retention of nanocellulose fibers on the mesh depends on the type of mesh and their pore size. If the size of mesh is small, it leads to the longest drainage time for dewatering from the suspension. In filtration process, the drainage time increased exponentially with nanocellulose suspension. As a consequence, the time taken for wet film formation on the mesh varied from 10 min to 24 h. Time consumption in the filtration to form the wet film of nanocellulose can be reduced by fibrillation of nanocellulose via high-pressure homogenization, ball milling, and acid hydrolysis of cellulose nanofibers. As a result, the drainage or dewatering time was achieved to 10 min (Varanasi and Batchelor 2013). Apart from time consumption as one of the process constraints, the basis weight and thickness of wet film increased when filtration time progresses. The progressed wet film on the mesh acts a resistance for further filtration processes. The nanocellulose film after drying has shortcomings such as filter markings on the film,

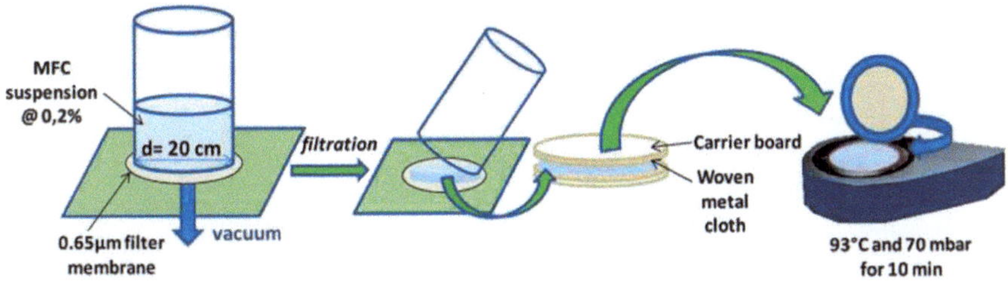

Fig. 17.5 Vacuum filtration for NC film fabrication. (Reprinted adapted with permission from Sehaqui et al. 2010)

and it affects the uniformity and surface roughness of the film (Shanmugam 2019). Various stages of vacuum filtration have been shown in Fig. 17.5.

17.3.6 Spraying Process

Spray-coating is a process of formation of mist or atomized particles on the base surface (Czerwonatis 2008). This coating has many advantages than that of other coating process, such as rod coating, bar coating, etc. (Shanmugam et al. 2017). This method has been reported recently for the development of free-standing nanocellulose films and nanocellulose coating on the paper substrates as well (Shanmugam et al. 2017). It has the concept of spraying nanocellulose fibers suspension on the solid surface and results in the formation of film after drying process (Magnusson 2016). The solid surface as a base surface is used as either permeable substrates such as paper and paper boards or impermeable substrates such as stainless steel plate, fabric media, and silicon wafers (Shanmugam 2021c). Spraying provides various advantages such as contour coating, contactless coating with a base surface, and topography of the base surface not influencing the coating process (Shanmugam 2021c).

The spraying fibers is a novel approach for the fabrication of nanocellulose films in a free-standing mode (Shanmugam et al. 2017). Figure 17.6 shows various sequences of the spraying process. In spray-coating process, the sequences consist of the formation of spray jet of cellulose nanofibers from the nanocellulose suspension and then atomization of the sprayed jet into the mist of the cellulose nanofibers suspension. The atomization consists of liquid lamella disintegration and disintegration of the jet of nanocellulose suspension. The fine droplets formed from atomization coalesce on the contact surface, and a film is formed on the base surface. Spraying of these nanofibers onto the solid surface causes coalescence of sprayed droplets due to the film-forming properties of the cellulose polymers via hydrogen bonding between hydroxyl groups between adjecent fibres. The film that is formed on the surface can be peeled from the substrates after drying (Shanmugam and Browne 2021) (Table 17.3).

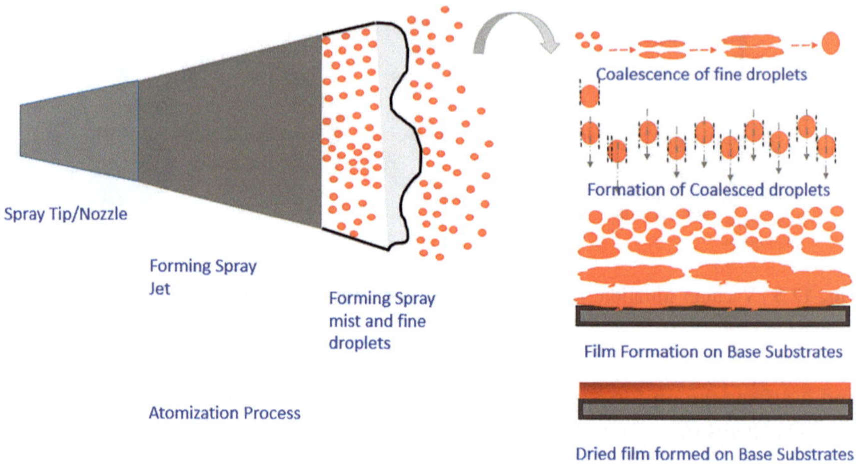

Fig. 17.6 Sequences in the spraying process on the base surface. (With permission from Shanmugam, K.; Browne, C; *Nanoscale Process.* **2021**, 247–297, Elsevier)

Table 17.3 Processing time in conventional method for nanocellulose films

Processing method	Processing time	Basis weight (g/m^2)
Spray deposition	10–27 min	13.7–124
Vacuum filtration	3 min	56.4
Membrane filtration	55 min	56
Filter paper-based filtration	~48 h	NA
Fabric filtration	1–3 h	55
Current spray-coating[a]	<1 min	52.8–193.0

[a]Shanmugam et al. (2017)

Spraying nanocellulose fiber on the base surface was classified into two process, namely, (1) spraying nanocellulose fibers on the permeable substrates such as paper substrates and paper boards. Recently, spraying cellulose nanofibers on the packaging paper, blotting paper, and news print paper was carried out, and it was increased their barrier performance of the paper substrates with increased mechanical strength. (2) Spraying cellulose nanofibers on the impermeable solid substrates such as stainless steel plate, super polished stainless steel plates, and silicon wafers for the production of free-standing nanocellulose films (Shanmugam 2021c). In the case of spraying cellulose nanofibers on the permeable substrates, this method provides high coat weight on the paper substrates than that of any other method such as rod coating, bar coating, and vacuum filtration. The spraying nanocellulose on the surface is independent on suspension concentration and dependent on the viscosity of the nanocellulose suspension. This method is applicable for production of nanocellulose barrier layers on the paper substrates (Beneventi et al. 2014). It was observed that spraying high solid content of nanocellulose suspension on the paper substrates could reduce the water content for reducing the drying load in the drying process of nanocellulose coating on the paper surface (Shanmugam 2022b).

In the process of spraying fibers on the impermeable substrates, the deposition of nanocellulose fiber on the stainless steel plate was carried out to fabricate the free-standing cellulose film (Shanmugam 2021c). The other impermeable substrates are fabric surface (Beneventi et al. 2015), stainless steel plates (Shanmugam et al. 2017), and silicon wafers (Shanmugam et al. 2020). In addition to that, spraying microfibrillated cellulose on 3D brass architecture was attempted to fabricate 3D dimensional free-standing film. In this approach, the wrinkles and scratches on the spray-coated film were observed. There are pros and cons for each method for fabrication of free-standing nanocellulose film (Magnusson 2016). Based on the pros and cons in the reported spraying process, the new method on spraying was developed and discussed this method how it would be the performance in rapidity and flexible when compared with other conventional methods (Shanmugam 2021c). To conclude, solvent casting is a slow method in the fabrication of free-standing nanocellulose films. It consumes unlimited time in the evaporation of solvent from nanocellulose suspension, and also the film has poor uniformity due to the formation of wrinkles on the films. In the case of vacuum filtration process, the drainage time was increased with the solid content in the nanocellulose suspension. The peeling of the nanocellulose wet film from the filter mesh is a crucial task and also marks on the films which were shortcomings in this method. This is why the following criteria on the preparation of the free-standing nanocellulose films and nanocellulose barrier coat on the paper substrates have been derived. Based on these criteria, the spray-coating was developed (Shanmugam 2019).

17.4 Criteria for Fabrication of Free-Standing Nanocellulose Films and Barrier Coating on the Paper Substrates

The preparation of free-standing nanocellulose films or barrier coating on the paper substrate is slow in the existing methods and inability to scale up into conventional scale to meet the large scale production. The rate of manufacturing of self-standing nanocellulose films or coating on the paper substrates is very slow when compared to the manufacture of plastic films or synthetic plastic coating on the paper substrates via extrusion. Speed and flexibility in the manufacture of nanocellulose films and barrier coating on the paper is required to meet the demand of the films and increased its application in various fields.

- Increase film properties via rapid preparation method.
- Tailor the thickness and basis weight of the NC film via processing method without increasing the operation time.
- Engineer the strength and bulk properties of NC film by varying the NC suspension concentration without affecting operation time in the method.
- Produce a film with an outstanding uniformity comparable to the conventional method such as vacuum filtration.

- Produce NC film in a single step to avoid further processing such as dewatering, vacuum drying, and couching of the wet film.
- Produce a smooth nanocellulose film and engineer their roughness for developing potential material without any additional treatment in the methods for the engineering of the surface roughness like any physical and chemical treatments (Shanmugam and Browne 2021).

17.4.1 Proof of Concept of Spray-Coating

The concept was developed on the spraying or spray deposition of nanocellulose suspension on the polished metal surface to fabricate the wet nanocellulose film (Magnusson 2016). Through the literature review studies, spraying nanocellulose on the base surface causes the replication of the surface roughness of the base surface to the nanocellulose films (Shanmugam et al. 2017). As a result, the surface roughness of the film can be tailored or lowered to produce smooth surface. As per the requirement for the new spraying process, the question is that "is it possible to scalable process for continuous production of nanocellulose films like synthetic plastic films" or "nanocellulose coating on the paper substrates in a continuous mode to produce barrier coating on the paper substrates." During the spraying process itself, the process should have a capability to tailor the mechanical and bulk properties of the film. Spraying is an answerable process for satisfying these requirements for the production of nanocellulose films (Shanmugam 2021c) (Fig. 17.7).

The spraying of nanocellulose suspension was carried out on the surface of polished circular stainless steel plate. The base surface of the polished stainless steel plate was very smooth and had very minimal scratches. Figure 17.8 shows the experimental setup for the proof of concept for spraying nanocellulose on the stainless steel plate. In this experimental setup, there are professional spray system, conveyor, and seating arrangement for keeping the stainless steel plate after sprayed nanocellulose suspension. Nanocellulose (NC) supplied from DAICEL Chemical Industries Limited (Celish KY-100S) was utilized to prepare free-standing films. NC sample was used at consistencies ranging from 0.5 to 2.0 wt.%, prepared by diluting the original concentration of 25 wt.% with distilled water and mixing for 15,000 revolutions in a disintegrator. The viscosity of the NC suspension was evaluated by the flow cup method which evaluates the process of coating fluid flow through an orifice to be used as a relative measurement of kinematic viscosity, with the results expressed in seconds of flow time in DIN-Seconds (Shanmugam et al. 2017).

In this concept, the nanocellulose was sprayed on the circular stainless steel plate which is kept on the conveyor. The speed of the conveyor and nanocellulose suspension were considered as process variables. These parameters can influence the bulk properties of free-standing nanocellulose films. In this approach, the nanocellulose suspension was varied from 0.25 to 2.00 wt.% for spraying suspension on the stainless steel plate kept on the conveyor at constant speed or velocity. The nanocellulose suspension is fixed for spraying on the plate, and the conveyor's speed

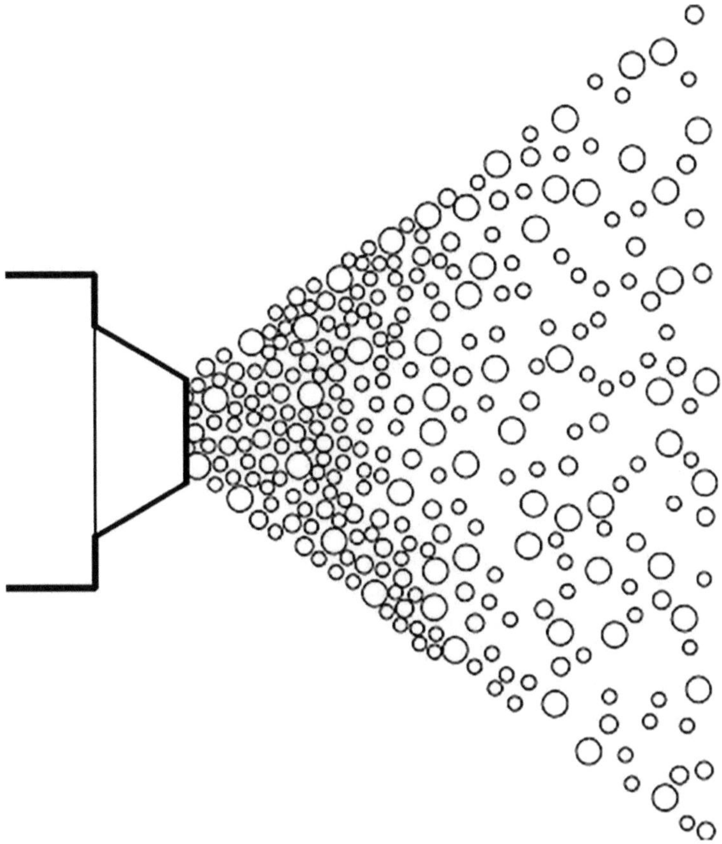

Fig. 17.7 Function of spray gun for spraying nanocellulose fibers of droplets. (Modified image from Wagner, *WAGNER ProjectPro 117 0418B Owner's Manual*, n.d.)

can be varied in the experimental setup. As a result, the quantity of spray deposited nanocellulose was high at the lowest velocity than that of higher velocity. As a consequence, very thick and high basis weight of the film can be fabricated. By this way, the properties of the nanocellulose films can be tailored (Shanmugam et al. 2017; Alsaiari et al. 2022). In order to achieve the uniform and good quality nanocellulose films, there are number of variables hidden in the spray-coating process and also in the spray-coating experimental setup. Atomization is the integral part of spraying fiber on the base surface. Atomization depends on the viscosity of the sprayable liquid and the orifice in the spray nozzle. The formation of spray jet also depends on the diameter of orifice and nozzle type in the spray gun. There are three spray patterns, namely, circular, elliptical, and rectangular patterns. These spray patterns can be created by either spray gun or orifice position and fan types in the spray system (Wagner n.d.-a). Apart from the equipment side, the viscosity of the nanocellulose suspension should be in the sprayable range for developing films. In this case, the nanocellulose suspension was varied from 0.25 to 2.00 wt.% fiber

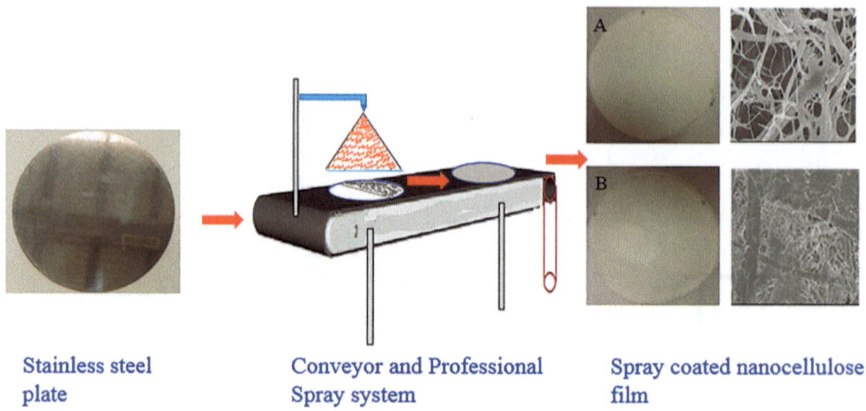

Fig. 17.8 Proof of concept for spray-coating for the production of nanocellulose films. (With permission from Shanmugam, K et al., *Cellulose* **2017**, *24*, 2669–2676)

content for optimum spraying operation. The sprayable behavior of nanocellulose suspension depends on their solid content and crucial step in the formation of nanocellulose films (Shanmugam et al. 2018).

The spraying of nanocellulose suspension on a polished circular stainless steel plate which is kept on the variable speed conveyor was carried out. The professional Wagner spray system (Model Number 117) was used to spray the nanocellulose suspension at a pressure of 200 bar. In this spray system, the type 517 spray tip was used and produced an elliptical spray jet of 22.5 cm width and spray jet angle of 50°. The vertical distance between the spray nozzle to the circular steel plate was 30.0 ± 1.0 cm known as spray distance. The velocity of the conveyor was operated at 0.32 cm/s during the spraying of NC on the plate. The professional Wagner spray system is a pressure-driven spray system and requires 30 s run before coating nanocellulose on the base surface. This was allowed to reach the equilibrium in the spraying process (Shanmugam et al. 2017).

After spray-coating of nanocellulose on the base surface, such as stainless steel, the wet film on the stainless steel plate was dried under standard laboratory conditions. It is suggested that the wet film can be dried in the fume hood, and the air flow can help the drying film easily. Normally the drying of the wet nanocellulose film will be taken more than 24 h at a temperature of 25 °C. The dried film was stored at 23 °C and 50% RH for evaluation of mechanical and barrier properties. As spraying is a new method, the performance of this method and the spray-coated film was compared with the existing method. The vacuum filtration was considered to be a conventional method for the fabrication of the nanocellulose films. The film also was made via vacuum filtration process and can be considered as an ideal film for comparison of spray-coated nanocellulose film (Shanmugam et al. 2017) (Fig. 17.9).

Vacuum filtration is used to fabricate the standard nanocellulose films for using as standard film to compare with spray-coated nanocellulose film. In short summary,

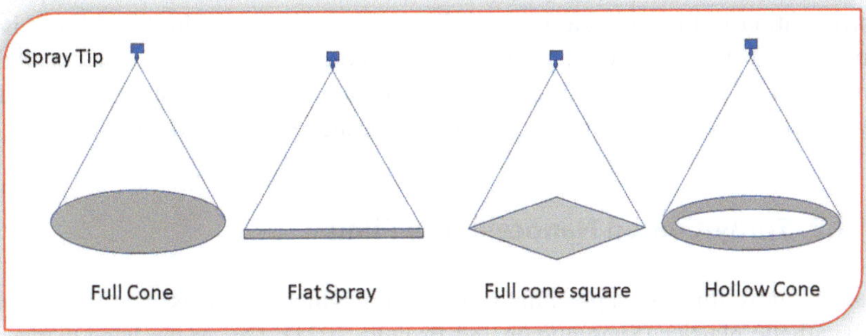

Fig. 17.9 The effect of fan and orifice in the spray patterns and spray jets for spraying nanofibers. (Modified from Wagners Manual, n.d.)

600 mL of NC suspension with 0.2 wt.% concentration was poured into a cylindrical container having a 125-mesh filter at the bottom and then filtered until it formed a wet film on the mesh. The wet film was carefully separated using blotting papers and then dried at 105 °C in drum drier for around 10 min. The film prepared by this method is used as a reference film to compare the uniformity and thickness of the spray-coated film. In addition to that, the tensile and barrier performance of the spray-coated film was compared with the film prepared via filtration. Based on the laboratory scale to achieve the proof of concept for the preparation of nanocellulose films, the operation time for forming wet film was less than 1 min. When comparing with vacuum filtration, the operation time for spraying nanocellulose film is quite very low. In vacuum filtration, the drainage time or dewatering time increased with nanocellulose suspension consistency (Varanasi and Batchelor 2013). As a result, the film formation on the mesh consumed varied from 10 min to 24 h (Shanmugam et al. 2017). Similarly, the water removal from the spray-coated nanocellulose wet films is one of the shortcomings in the spray-coating process (Shanmugam 2019). When comparing with filtration, water consumption in spray-coating process is very minimal. It was explained with various examples and studies in the spray-coating for production of free-standing nanocellulose films. In this spraying operation, the solid content in nanocellulose suspension was varied from 1 to 2.00 wt.% for the preparation of thin films to thick nanocellulose films. The developed laboratory scale spray-coating experimental setup is able to handle up to 2.5 wt.% of nanocellulose suspension. 2.5 wt.% of nanocellulose suspension contains 97.5 wt.% water and rest nanocellulose fibers, so that the concentration becomes 39 g of water per g of cellulose nanofibers, which is much lower than of 499 g of water per g of cellulose fibers in the vacuum filtration process. As a consequence, the drainage time for forming nanocellulose film was increased in the filtration process. In the case of spraying process, the operation time is independent of nanocellulose suspension concentration. To conclude, the spraying operation provides minimal operation time for the formation of nanocellulose films. The basis weight and thickness of the

nanocellulose films were easily tailored via adjusting the nanocellulose suspension concentration in the spraying process. The biggest limitation in this process is the formation of wet film, and drying consumes more time to get dried nanocellulose film. It can be resolved by using infra-red radiation drying or waste heat utilized for drying the nanocellulose film at conventional scale (Shanmugam 2019).

17.4.2 Spray-Coated Nanocellulose Films

Nanocellulose films are a potential nanomaterial which can be used as an alternative for synthetic plastics (Shanmugam and Browne 2021). Due to the huge demand of nanocellulose films to replace synthetic plastics, the fast and flexible method was required to meet large-scale manufacturing capabilities. Spraying is a fast and flexible process for the production of free-standing nanocellulose films (Shanmugam 2019). In this approach, spraying nanocellulose suspension on the polished metal surface such as stainless steel produces a unique nanocellulose film with two distinct surfaces (Shanmugam 2021c). The surface exposed to the air is called free surface or rough surface. The film via spraying was adhered to the metal surface and peeled from the metal surface easily and also replicated the surface smoothness from the metal side. As a result, the nanocellulose film from metal side was very smooth (Shanmugam et al. 2017, 2020).

Figures 17.10 and 17.11 show the spray-coated nanocellulose films in the form of circular and rectangular shapes. The film was very compact and had two unique surfaces. Through spray-coating, the stable and homogeneous film was produced. The basis weight and thickness of the film was tailored via process variables such as nanocellulose suspension consistency and engineering variables such as velocity of the conveyor. The scanning electron microscopy (SEM) micrographs of spray-coated nanocellulose film reveal the structure, morphology, and topography of cellulose nanofibers with fiber diameter distribution in the film (Shanmugam 2021c; Shanmugam et al. 2017).

17.4.3 Scanning Electron Microscopy of Spray-Coated Nanocellulose Films

SEM micrograph reveals that spray-coated nanocellulose films have two unique surfaces, namely, rough and smooth surface. The film was very compact and glossy in the smooth side. It is one of the added advantages in the spraying nanocellulose suspension on the polished metal surface, and this method has capacity to replicate the part of the surface roughness from the base surface to the film. As a result, the film was glossy and shiny, and the smoothness can be used in the construction of various functional materials such as printed and flexible electronics, OLED devices, solar cell construction, and sensors. The rough side of the film has porous and more surface roughness than that of the smooth side. The surface roughness of the spray-

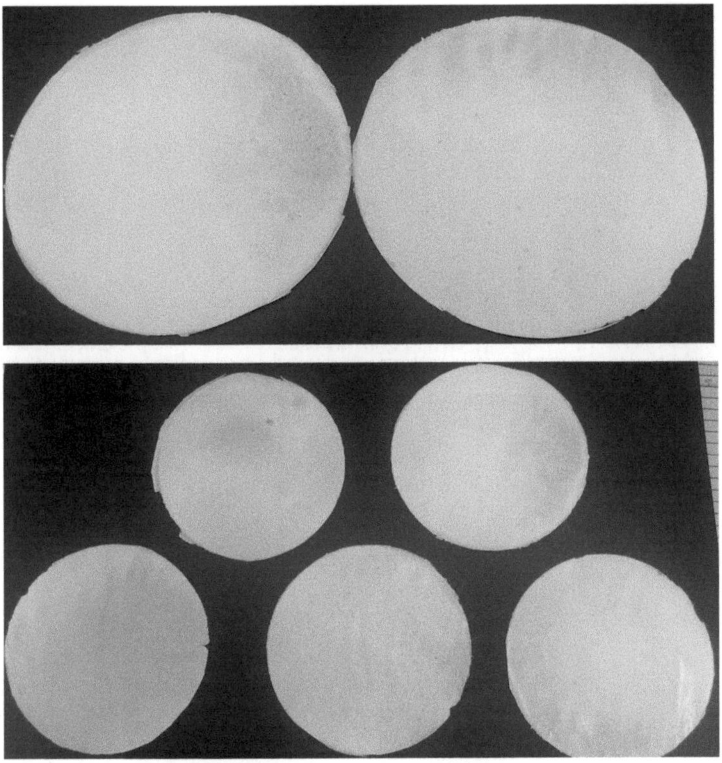

Fig. 17.10 Spray-coated nanocellulose films—circular shape

coated nanocellulose films was evaluated through various methods and discussed in the later part of this chapter (Shanmugam et al. 2020) (Fig. 17.12).

Figure 17.13 shows the cross-sectional view of spray-coated nanocellulose films. It reveals the cellulose nanofibers in the spray-coated are compressed and intertwined with neighboring fibers. The spray-coated nanocellulose films have many layers in the film and compressed well revealed in the SEM micrographs. It is also revealed that the film would have a complex tortuous pathway for diffusion of water vapor, air, and oxygen. As a consequence, the barrier performance of the film would be improved, and the film would become an impermeable sheet which can be a potential alternative packaging material for synthetic plastics (Nair et al. 2014).

The rough side of the nanocellulose film has surface roughness similar to the surface of the film via vacuum filtration. The rough surface of the film confirms that the film contains densely packed various sizes of free fibers and clumps of cellulose fibers and is further free from pinhole. Figure 17.14 shows the rough side of the nanocellulose film (free surface) which is the side directly to contact with atmospheric. The surface topography of the film is asymmetric structure exclusively the pore size on the rough surface differed from the smooth surface. The fibers are well connected between different fibers size and form fibrous matrix of various pore sizes.

Fig. 17.11 Spray-coated nanocellulose films—rectangular shape

In the rough side of the film, there are free fibers towards the atmospheric and high surface roughness.

17.4.4 Thickness Investigation and Thickness Mapping of Spray-Coated Nanocellulose Films

Thickness is the crucial parameter for spray-coated nanocellulose films. The variation in thickness not only affects the uniformity of the film but also causes their effects on the tensile properties and barrier properties. As discussed earlier, thickness of the film/sheets can be tailored by adjusting the suspension consistency to be sprayed on the metal plate. Figure 17.15 shows the thickness detail of 2.00 wt. % microfibrillated cellulose/nanocellulose film prepared via spray-coating.

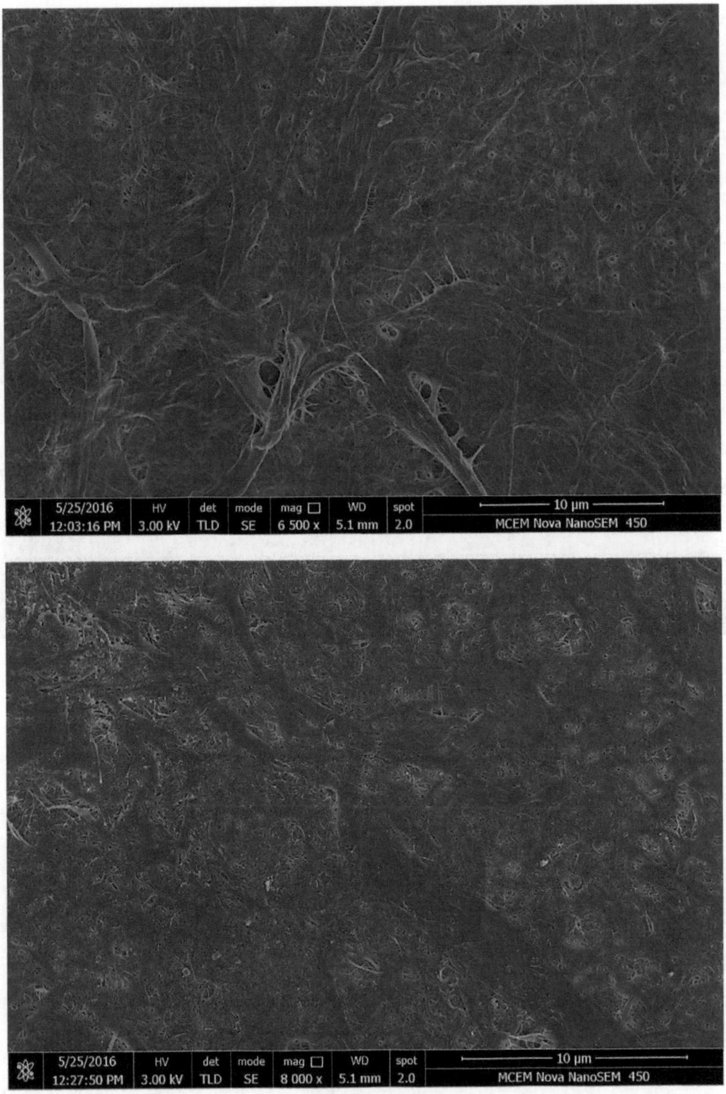

Fig. 17.12 The micrograph of spray-coated sheet at 10 μm

Figure 17.15 shows thickness details about seven films and maximum and minimum thickness of each film mentioned in the plot. The average thickness of each film has also been mentioned in the plot. The thickness of the 2.00 wt.% varies from 200 to 250 μm. * signs in the plot signifies that there is an outlier in the thickness data which can be omitted for evaluation.

Figure 17.15 reveals the consistency of the production of nanocellulose films via spray-coating and shows minimal variation in thickness. When spray the 2.00 wt.%

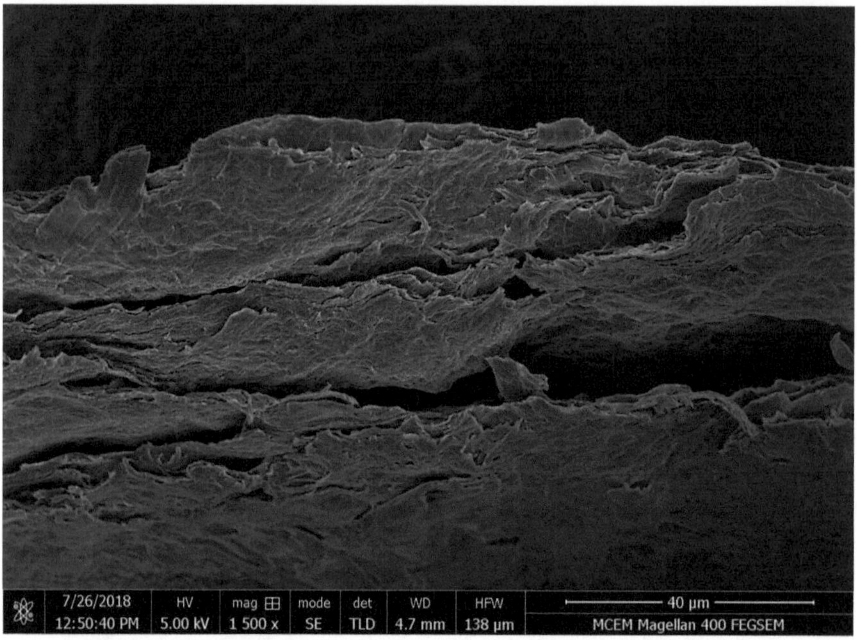

Fig. 17.13 Cross-section of spray-coated nanocellulose film

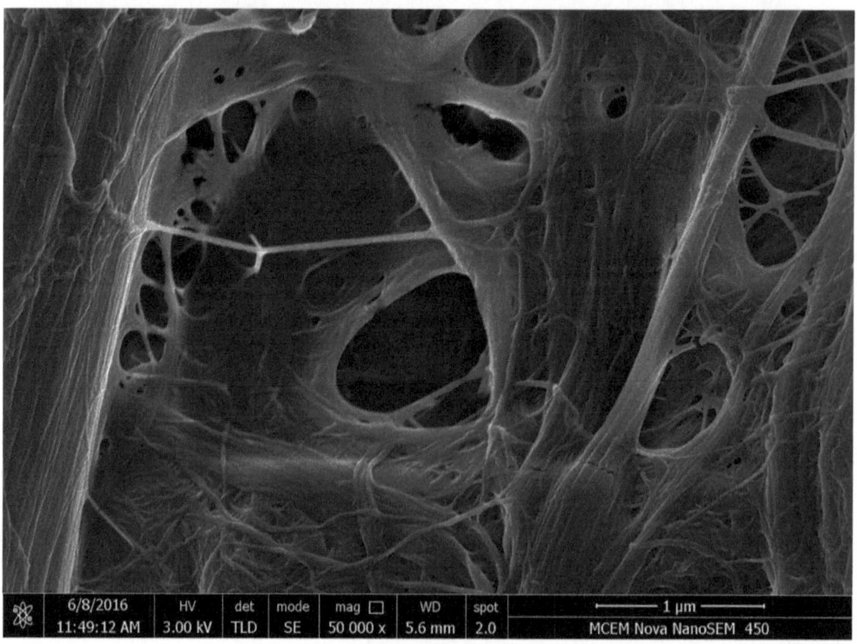

Fig. 17.14 Smooth side of spray-coated nanocellulose film

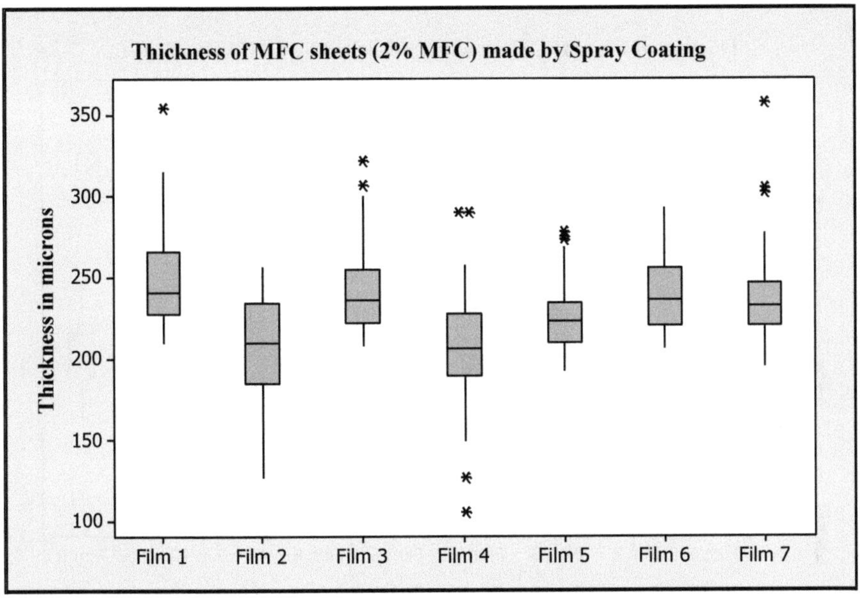

Fig. 17.15 Thickness of the 2.00 wt.% nanocellulose/MFC film via spray-coating

on the stainless steel plate, the thick and good basis weight of the film was formed. Due to the handling of wet film on the plate, the movement of NC suspension in wet film causes the variation in thickness (Fig. 17.16).

Figures 17.17 and 17.18 reveal a thickness of 1.25 and 1.00 wt.% of spray-coated nanocellulose films. In these plots, there were maximum variations in thickness of the films. 1.25 and 1.00 wt.% nanocellulose suspension behave watery suspension due to the low solid fiber content in the suspension. Due to the sprayed watery suspension on the plate, there was chance to move the suspension in the wet film resulting in the thickness variation on the film. This problem was resolved by the spraying of high solid content in the nanocellulose suspension. The spraying above 1.5 wt.% nanocellulose suspension can minimize thickness variation of the film and improves the uniformity of the films. The spray system in the current experimental setup has limitations in the spraying of nanocellulose suspension up to 2.00 wt.% NC suspension. The addition of CMC or MMT into nanocellulose suspension decreases the viscosity of the suspension (Shanmugam 2021b; Shanmugam et al. 2020). So that, the spray behavior of the nanocellulose suspension would be increased beyond 2.0 wt.%. Alternatively, high performance spray system was recommended to handle the high viscous nanocellulose suspension for producing high thickness and basis weight nanocellulose films (Shanmugam 2019) (Fig. 17.19).

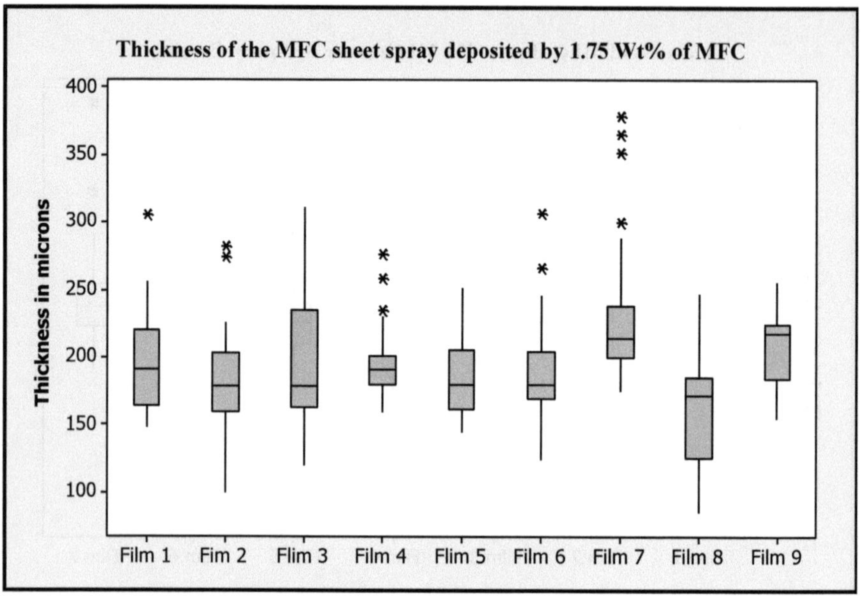

Fig. 17.16 Thickness variation of 1.75 wt.% nanocellulose film

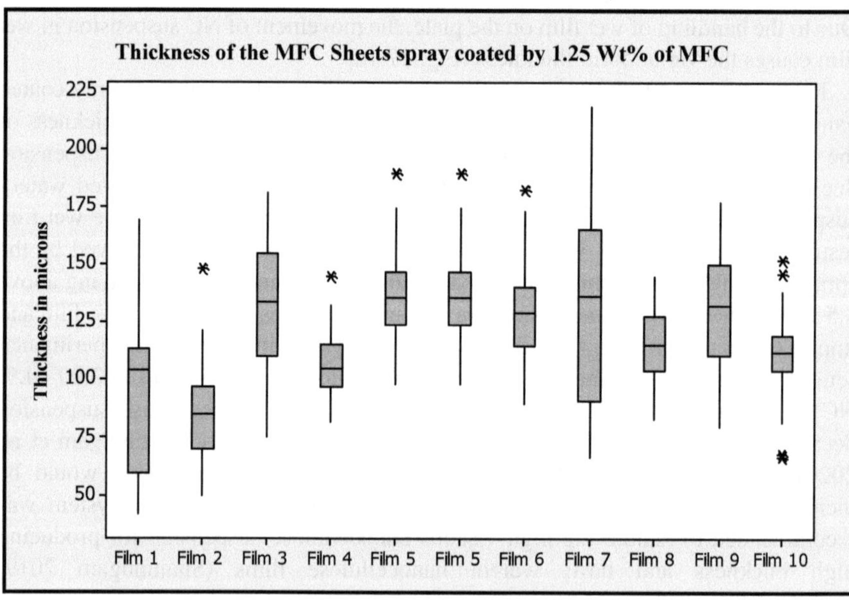

Fig. 17.17 Thickness variation of 1.25 wt.% nanocellulose film

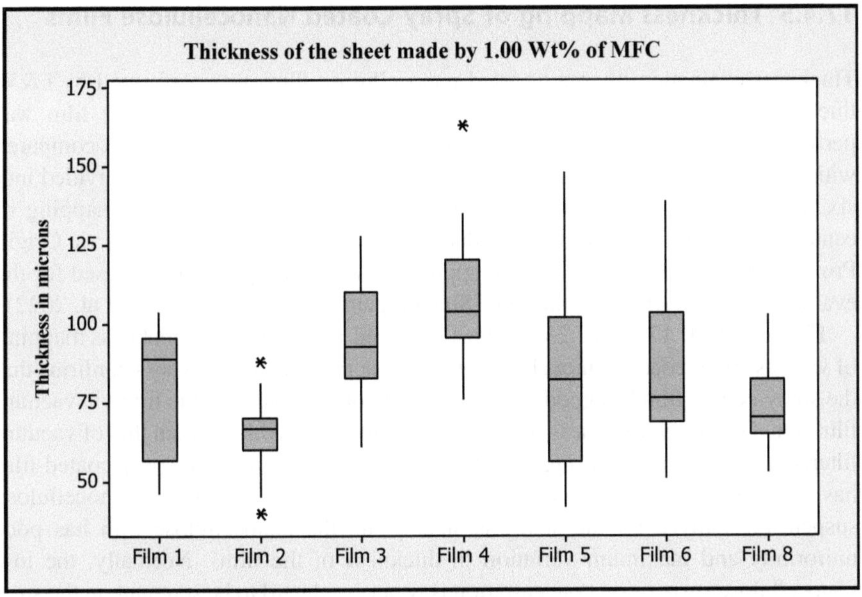

Fig. 17.18 Thickness variation of 1.00 wt.% nanocellulose film

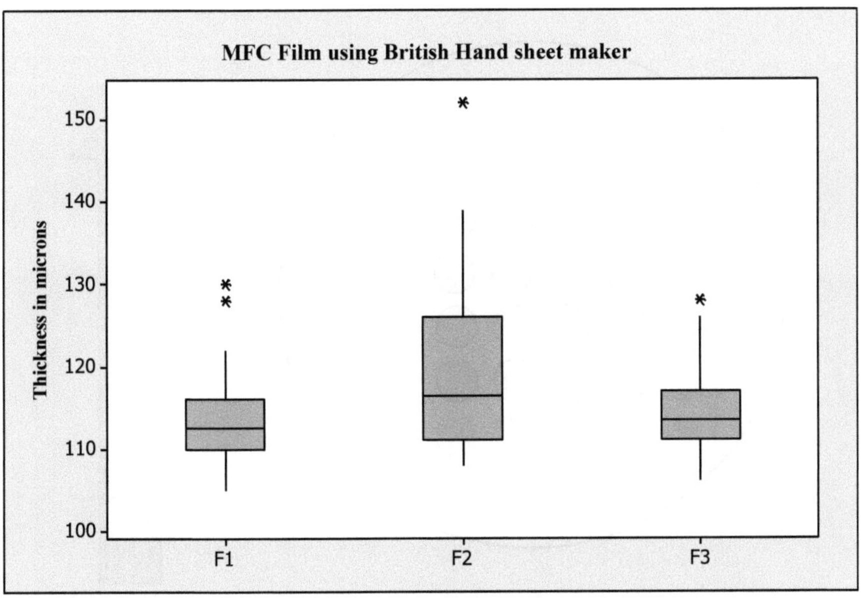

Fig. 17.19 Thickness variation of nanocellulose film prepared via vacuum filtration

17.4.5 Thickness Mapping of Spray-Coated Nanocellulose Films

Thickness evaluation of spray-coated nanocellulose films was performed by L&W thickness analyzer. Thickness mapping of spray-coated nanocellulose film was performed with thickness data to evaluate the uniformity of the film and compared with the film via vacuum filtration. The circular portion of the film was divided into six regions and thickness measured on these regions. The thickness mapping of center rectangular region of the circular film was plotted by contour plot via Origin Pro 9.1. The concept of thickness mapping is shown in Fig. 17.20 and used for the evaluation of uniformity of the film (Shanmugam et al. 2017; Alsaiari et al. 2022).

Figures 17.21, 17.22, 17.23, 17.24, 17.25, and 17.26 show the thickness mapping of various spray-coated nanocellulose films. The mapping of thickness confirms that the spray-coated film has good uniformity and comparable with the film via vacuum filtration. Generally, thickness of the spray-coated film is thicker than that of vacuum filtered film. When comparing the film via vacuum filtration, the spray-coated film has good uniformity in thickness. At the lowest concentration of nanocellulose suspension sprayed on the stainless steel plate, the nanocellulose film has poor uniformity and maximum variation in thickness of the films. Normally, the low nanocellulose suspension concentration is watery and hard to be stagnant on the steel plate. As a result, the film becomes very thin and has poor uniformity due to the huge variation in thickness. The suspension consistency increased for spraying, the wet film was improved stamina for stagnant film and gives good thickness and basis

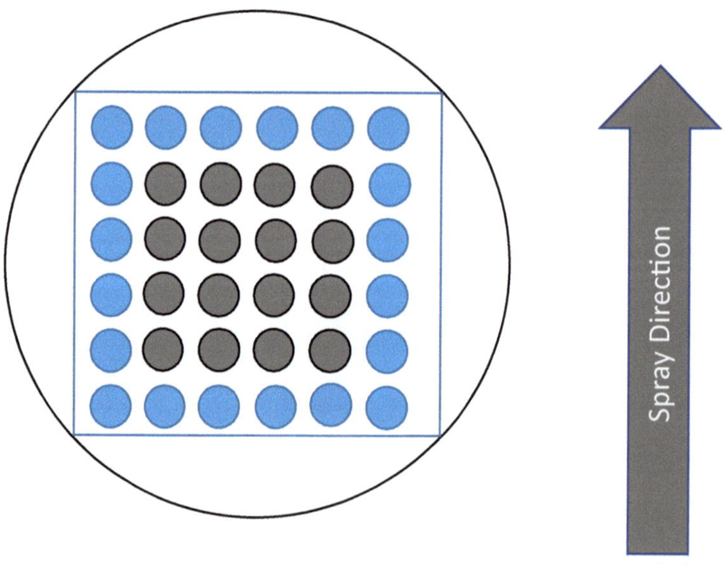

Fig. 17.20 Mapping of the thickness of the NC film. The thickness is measured in the center region of the film. The square section of the center part of the film is used for contour plotting. The gray point of thickness used for mapping to confirm the uniformity of the film. (With permission from Shanmugam, K et al., *Cellulose* **2017**, *24*, 2669–2676)

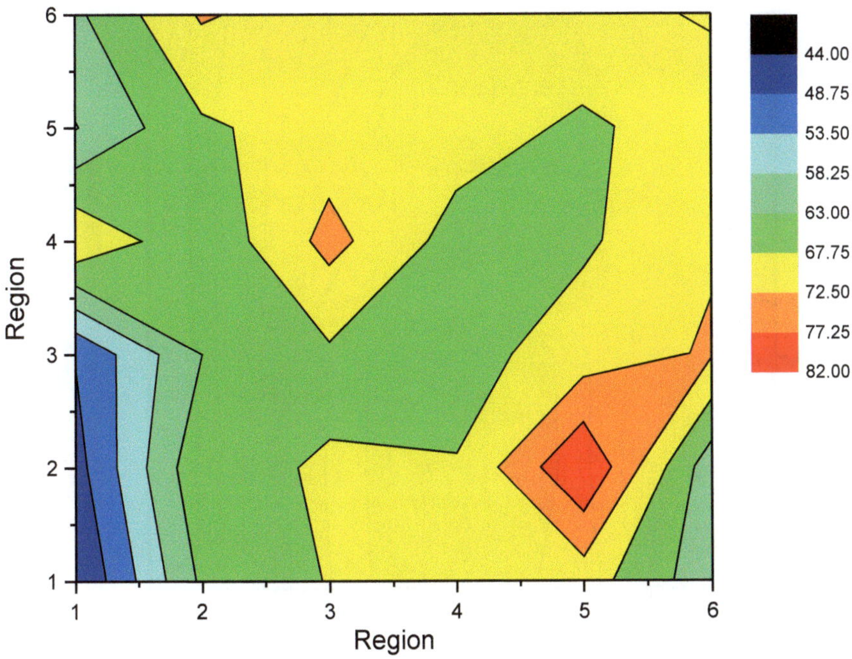

Fig. 17.21 Thickness mapping of 1.00 wt.% of spray-coated nanocellulose films

weight film after drying. This is why the contour plot on thickness mapping confirms good uniformity in the high concentration suspension sprayed than that of the film made via the lowest concentration of NC. The spray-coated nanocellulose film has comparable uniformity with the film prepared via vacuum filtration (Shanmugam 2021c) (Fig. 17.27).

17.4.6 Uniformity of Spray-Coated Nanocellulose Films

The uniformity of the nanocellulose film was evaluated by the instrument known as paper perfect formation tester. This instrument measures the optical uniformity of cellulose substrates such as papers and paper board and was implemented to evaluate the uniformity of spray-coated nanocellulose films and compared with the film via vacuum filtration. This instrument directly measures the quality of formation of the film via passing the light across the film. So the quality of the film was evaluated by the length scales from 0.5 to 60 mm. The final data was reported in Relative Formation Value (RFV) of each component in spray-coated nanocellulose films when comparing with the standard film from vacuum film. RFV less than/equal to/greater than 1 signifies poor/better/good uniformity of the film. Figure 17.28 shows the optical uniformity of spray-coated nanocellulose film performed by paper formation tester (Shanmugam et al. 2018). Figure 17.29 is the standard or reference film prepared via vacuum filtration. The image reveals that the film has

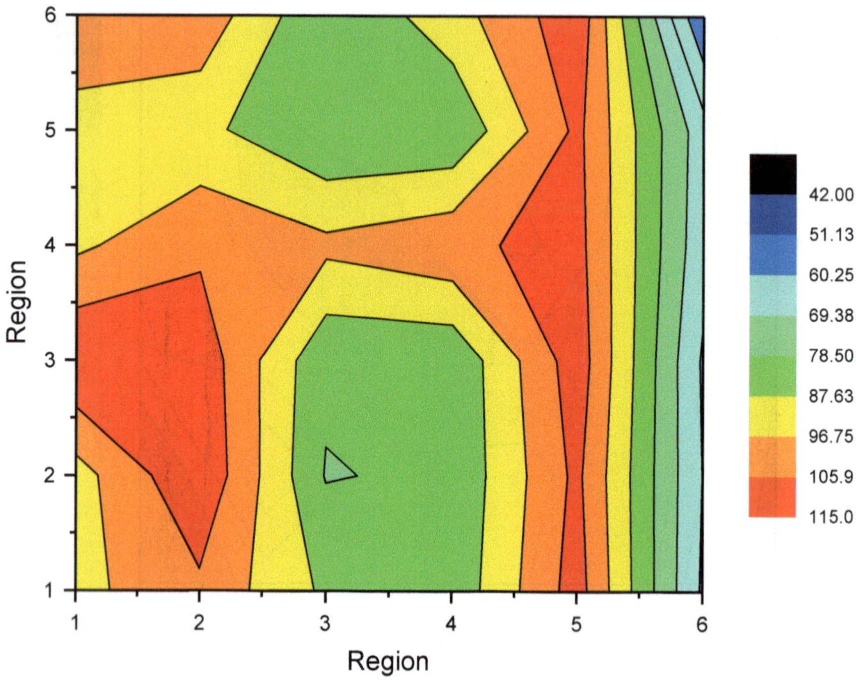

Fig. 17.22 Thickness mapping of 1.25 wt.% of spray-coated nanocellulose films

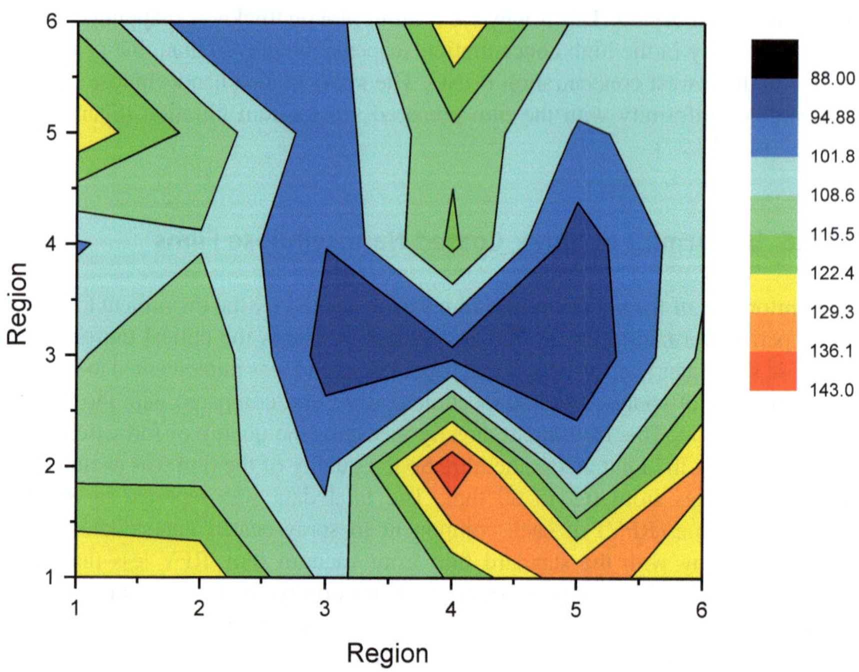

Fig. 17.23 Thickness mapping of 1.5 wt.% of spray-coated nanocellulose films

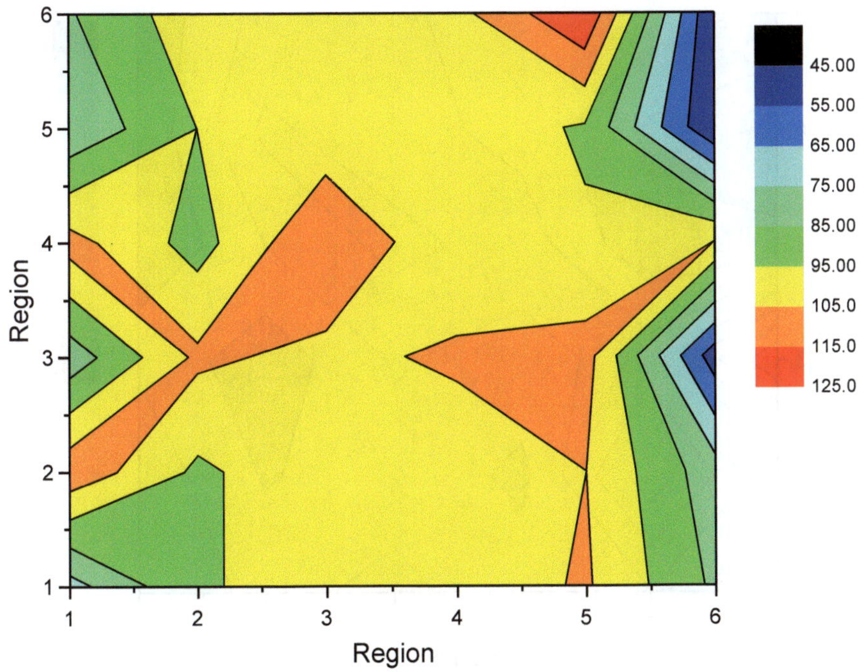

Fig. 17.24 Thickness mapping of 1.25 wt.% of spray-coated nanocellulose films

poor uniformity and RFV value less than 1 and required increased suspension consistency for spraying to make thick film with cellulose nanofibers. When spraying nanocellulose suspension on the stainless steel plate, there is a chance of entrainment of air bubbles in the film. It will affect the uniformity of the film. The uniformity of the nanocellulose is highly interlinked with tensile properties, barrier performance, and other bulk properties such as basis weight and thickness. Increasing the suspension concentration would increase the uniformity of the film. In addition to that, the reengineering of spray process and spray system drastically improves the uniformity of the spray-coated nanocellulose films (Shanmugam 2021c; Shanmugam et al. 2018).

17.4.7 Surface Roughness of Spray-Coated Nanocellulose Films

The spraying nanocellulose suspension on the polished metal surface is a novel approach for the development of smooth nanocellulose films. The film via spray-coating has two unique surfaces, namely, (1) rough surface which the film exposed to the air and (2) the smooth surface which the film adhered to the metal side. In this spraying process, the replication of surface smoothness of the base surface such as polished stainless steel plate was transferred to the nanocellulose film resulting in smoothness of the film (Shanmugam 2021c; Shanmugam et al. 2017). The rough

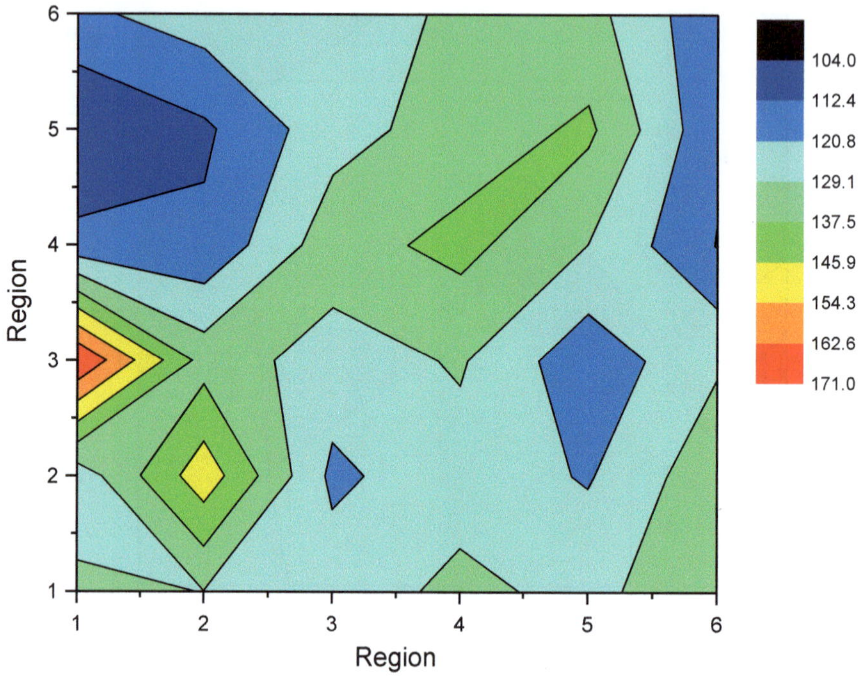

Fig. 17.25 Thickness mapping of 1.5 wt.% of spray-coated nanocellulose films

side and smooth side of the film have played a major role in the construction of various functional materials such as flexible electronics, printed electronics, solar cells, and microfluidic devices (Shanmugam 2021a). The surface roughness of the film was a crucial parameter in these applications. Therefore, the measurement of surface roughness of the film was performed via Atomic Force Microscopy (AFM), Optical Profilometry (OP), and Parker Surface Print Instrument (PPI). AFM and OP deal about the nanoscale surface roughness of the film, and PPI mentions the macroscale roughness of the films.

17.4.7.1 Atomic Force Microscopy of Nanocellulose Films

17.4.7.2 Atomic Force Microscopy of Nanocellulose Films

17.4.7.3 Atomic Force Micrographs of Film Prepared by Vacuum Filtration

17.4.7.4 Atomic Force Micrographs of Sheet Prepared by Vacuum Filtration

AFM micrographs reveal the surface roughness of the spray-coated film and vacuum-filtered film. The spray-coated nanocellulose film is very smooth replicated

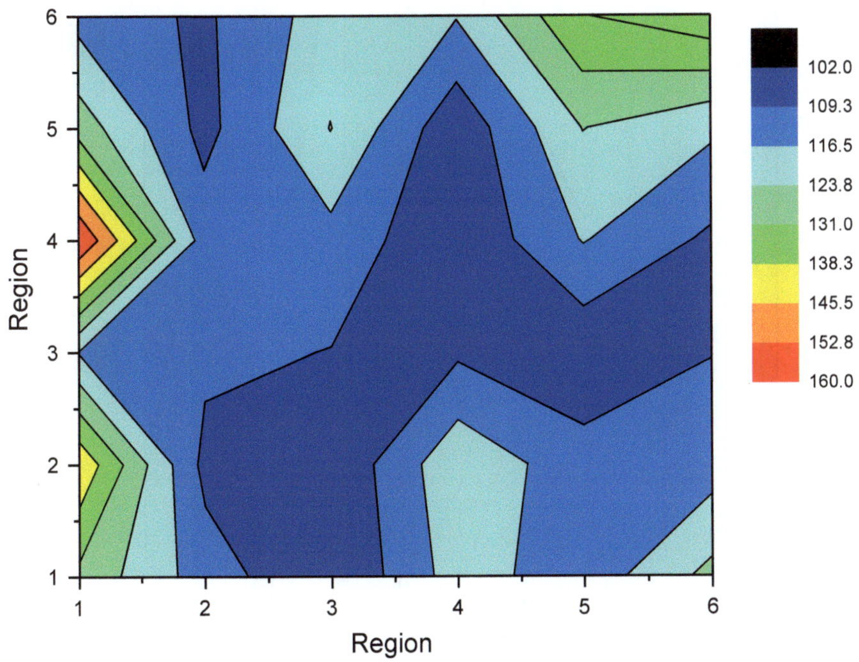

Fig. 17.26 Thickness mapping of 1.5 wt.% of spray-coated nanocellulose films

from the polished metal surface and minimum porous than that of the rough side of the film. Sometimes fiber aggregates on the smooth side or rough side of the film elevate the surface roughness of the film. The formation of fiber aggregates can be resolved by the addition of rheology modifier into the nanocellulose suspension. The most common rheology modifiers are nanoclay, such as bentonite and carboxy methyl cellulose (CMC) (Shanmugam 2019). AFM micrographs reveal that the smooth surface of the sprayed film has a glossy, shiny appearance. The RMS roughness from the rough side is 414.0 nm for 10 μm × 10 μm film area and 51.4 nm for 2 μm × 2 μm film area, whereas the RMS surface roughness of the smooth side is only 81.1 nm for 10 μm × 10 μm film area and 16.7 nm for 2 μm × 2 μm film area. AFM measurements of the film via vacuum filtration show the sides had a surface roughness of 417.7 nm (Side 1) and 330.8 nm (Side 2) at an inspection area of 10 μm × 10 μm, which is approximately the same as the rough side of the film prepared by spray-coating. Figures 17.30 and 17.31 show the AFM micrographs of spray-coated nanocellulose films, and Figs. 17.32 and 17.33 show the AFM micrographs of the vacuum-filtered nanocellulose film (Shanmugam et al. 2017).

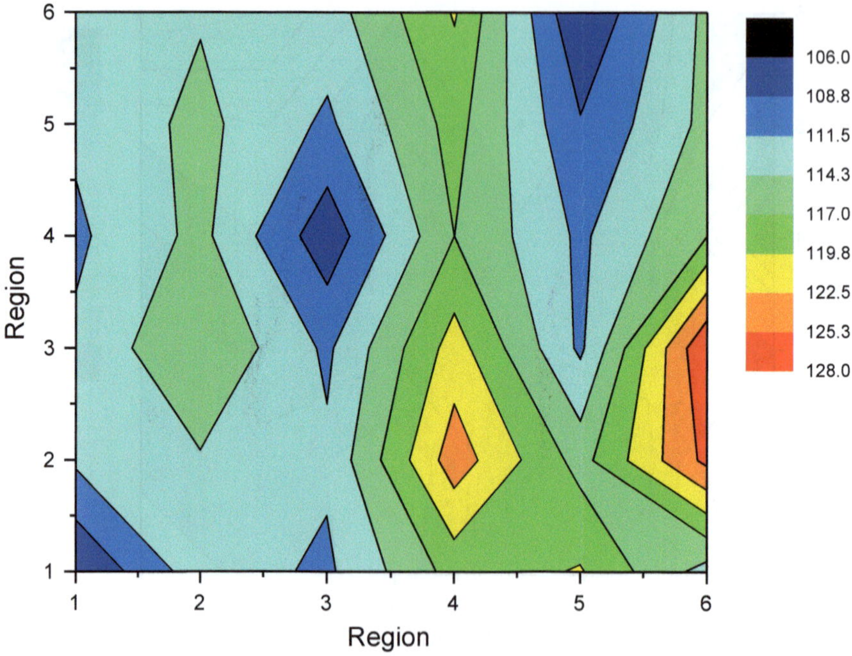

Fig. 17.27 Thickness mapping of nanocellulose films prepared via vacuum filtration

17.4.7.5 Optical Profilometry Images for Spray-Coated Nanocellulose Films

17.4.7.6 Optical Profilometry Images of Vacuum Filtered Nanocellulose Films

17.4.7.7 Optical Profilometry Images of Base Surface: Circular Polished Stainless Steel Plate

The optical profilometry is one of the non-destructive measurements of surface roughness of the nanocellulose film prepared via spray-coating and vacuum filtration. Figures 17.34 and 17.35 show optical profilometry images of nanocellulose films. Similarly, the RMS roughness of the NC film measured using the optical profilometry at an inspection area of 1 cm × 1 cm is 2087 nm on the rough side and 389 nm on the spray-coated side when compared with NC film prepared via vacuum filtration with RMS value of 2673 nm on side 1 and 3751 nm on side 2. The RMS measurement on these specimen has performed on the 50 × magnified optical profilometry images. From 5 × magnified optical profilometry images, the RMS value was found to be 5000 nm on rough side and 2500 nm on smooth side for spray-coated nanocellulose film (Figs. 17.36 and 17.37). In the case of vacuum filtered nanocellulose film, the RMS was determined 5900 nm on side 1 and 6400 nm on

Fig. 17.28 Uniformity of spray-coated nanocellulose film

side 2. These values conclude that the spraying nanocellulose suspension on the polished metal surface produces the film with a significant smoothness which was replicated from the stainless plate. The spraying provides one-step method for tailoring the surface roughness/smoothness of the film without any further/chemical treatment. Figure 17.38 shows the optical profile images of the circular polished steel plate. The RMS of the base surface was evaluated to be 195 ± 33 nm, and this part of the surface roughness has been replicated to the nanocellulose films. By the spray-coating method, the surface roughness of the nanocellulose films can be tailored via spraying nanocellulose suspension on various smooth substrates (Shanmugam et al. 2017).

17.4.7.8 Parker Surface Print Instrument for Evaluation of Macroscale Roughness

Parker surface print instrument was used to evaluate the surface roughness of the paper and paper boards at macroscale. This same instrument can be applicable for the measurement of surface roughness of nanocellulose film at macroscale level.

Fig. 17.29 Uniformity of vacuum filtered nanocellulose film

The surface roughness of the spray-coated film was evaluated to be 10.8 ± 0.17 μm on the rough side of the film and 5.5 ± 1.4 μm on the smooth side. The roughness of the vacuum filtered nanocellulose film was determined to be 10.6 ± 0.3 μm on the filter side and 9.9 ± 0.1 μm on the free side of the film. These investigations conclude that the spray-coated films have lowest surface roughness on the spray-coated side even at macroscale. It is another advantage in the spraying method which provides the finishing quality of the film on the coated side without any addition treatment. The spray-coated side of the film is very shiny, glossy, and reflective in the film and a kind of finishing quality provided by spraying on the polished base surface (Shanmugam 2021c). Table 17.4 summarizes the surface roughness of nanocellulose films predicted from various approaches.

17.4.8 Bulk Properties of Nanocellulose Films

The bulk properties of the nanocellulose film can be tailored by the process parameters and engineering parameters in the experimental setup of spray system. Thickness and basis weight of the film are the bulk properties and play a major role in the tensile properties and barrier performance of the film. The process parameters

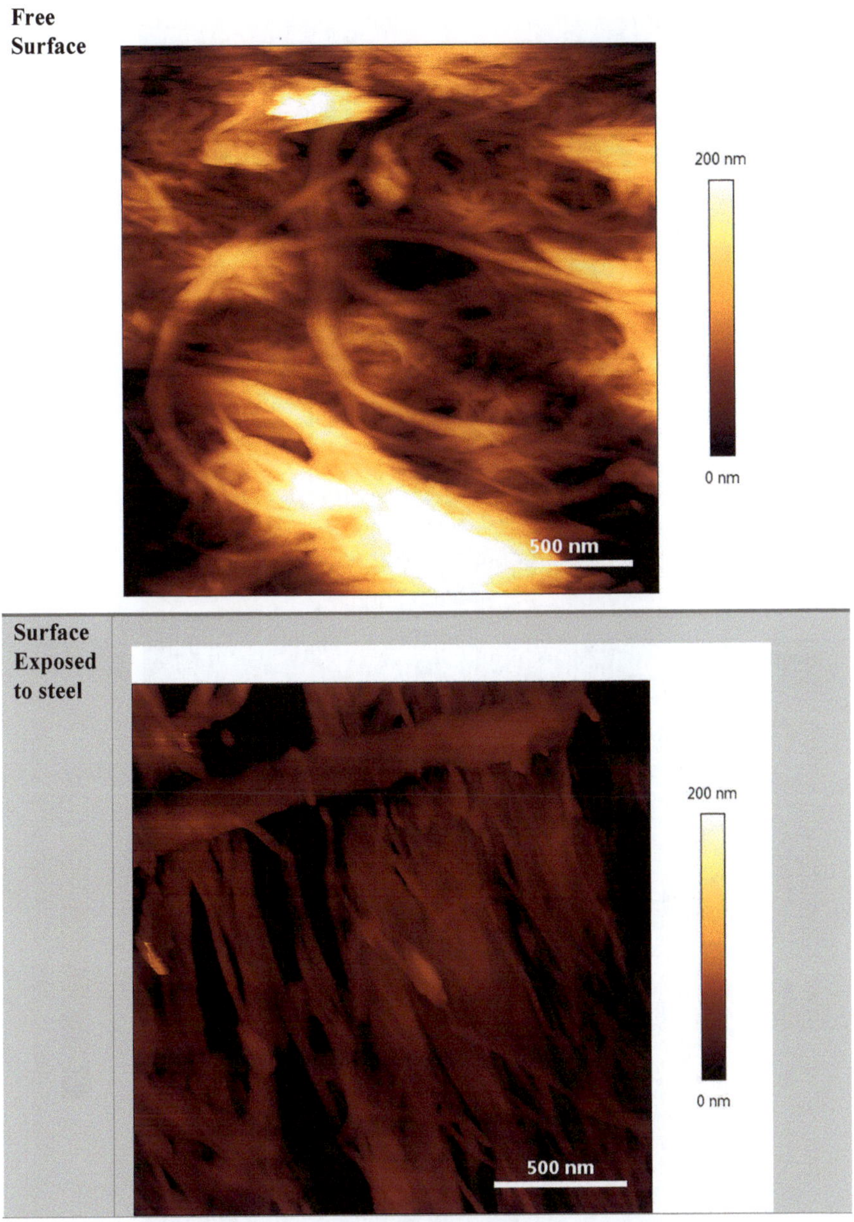

Fig. 17.30 AFM image of spray-coated sheet. The sheet prepared by 1.5 wt.% NC spraying on the base surface

such as effect of nanocellulose suspension consistency were only discussed in this section. The nanocellulose suspension consistency plays a major role in the controlling thickness and basis weight of the nanocellulose films. Figure 17.39

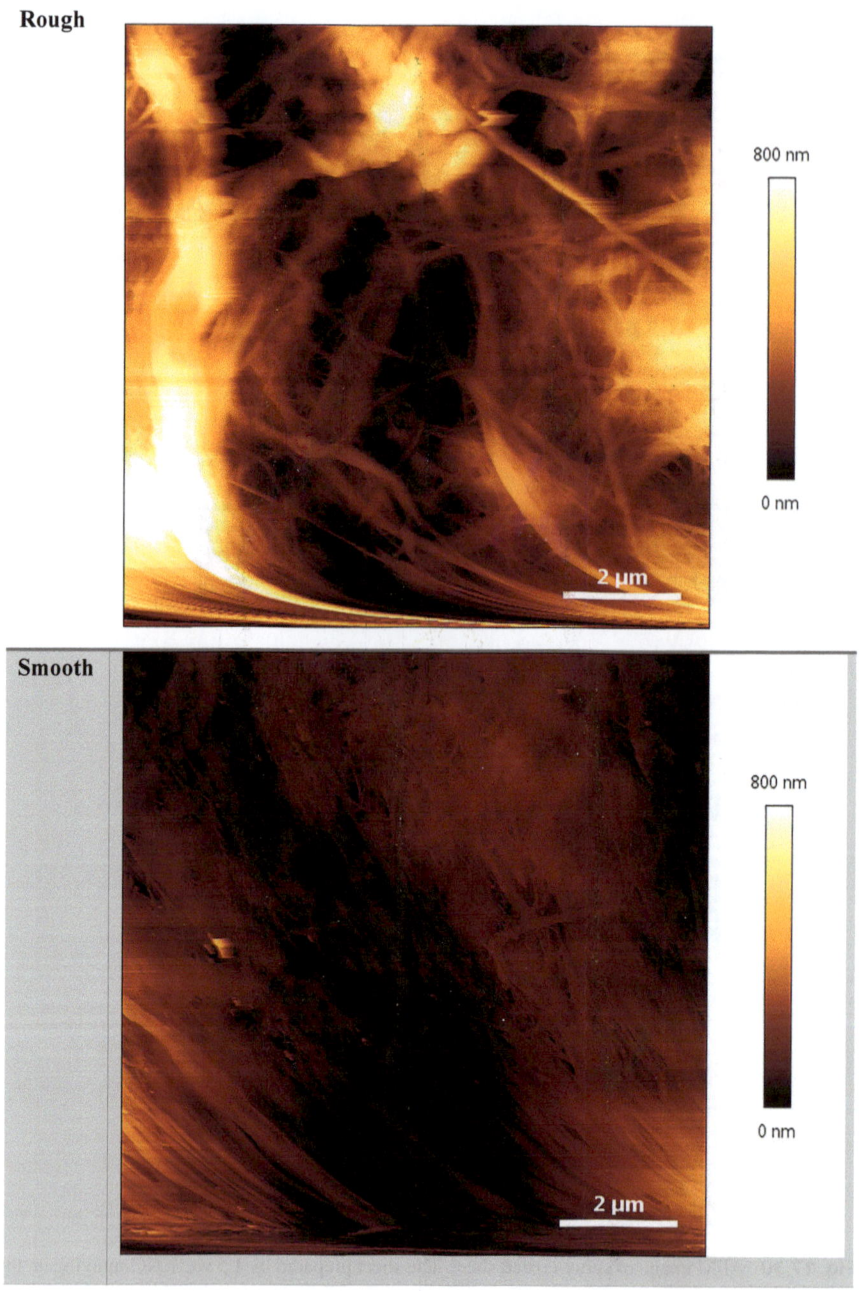

Fig. 17.31 AFM image of spray-coated film

side 1

Side 2

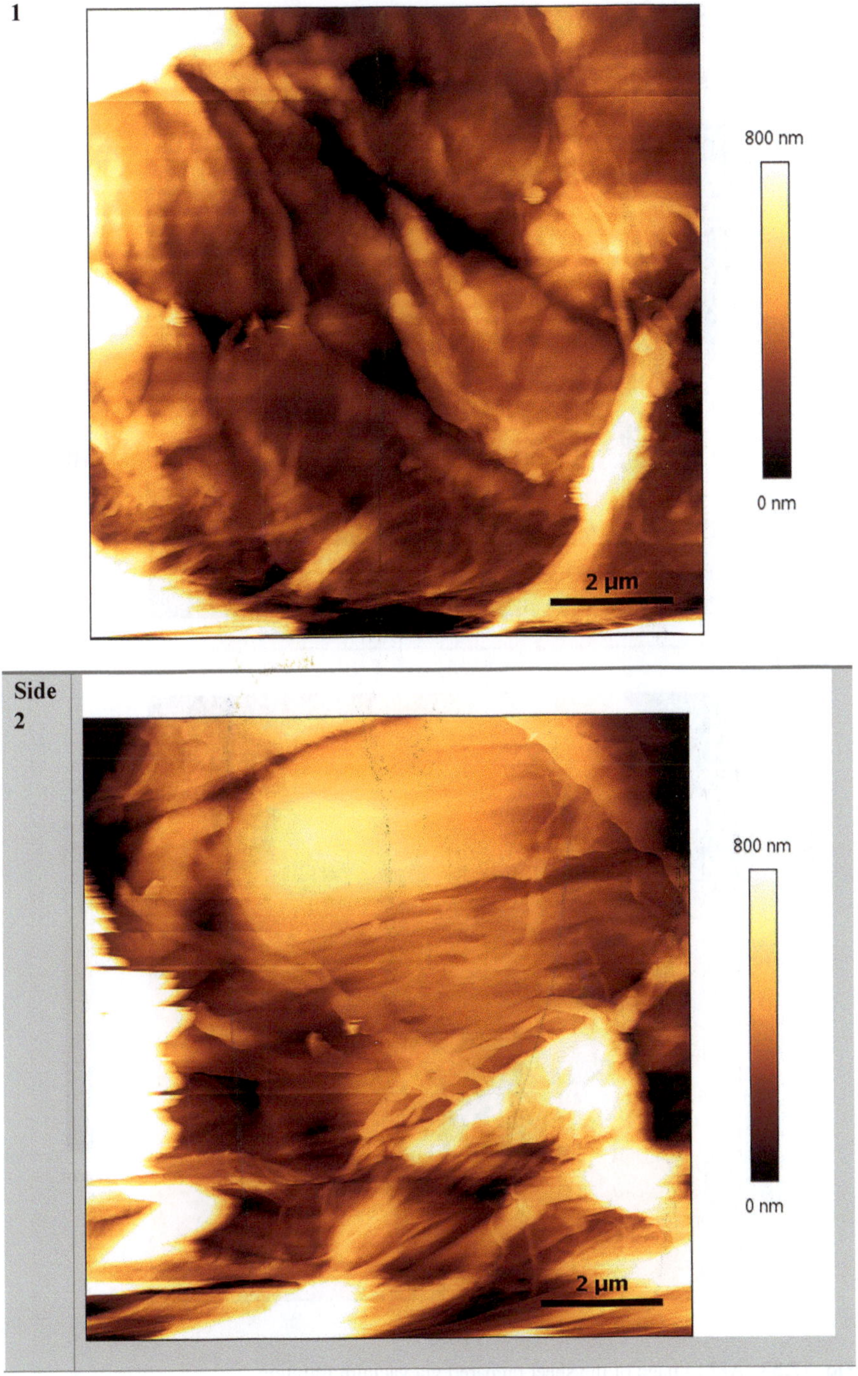

Fig. 17.32 AFM image of the sheet prepared via vacuum filtration

Side 1

Side 2

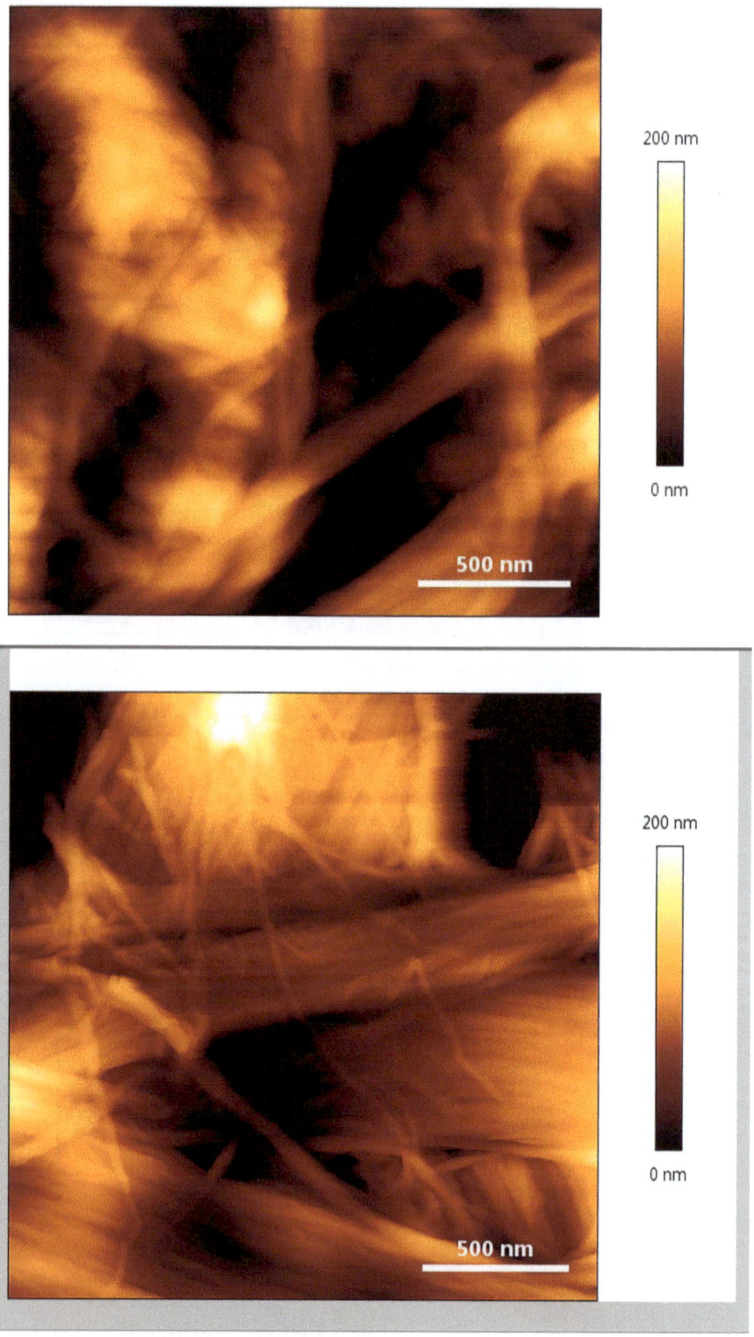

Fig. 17.33 AFM image of the sheet prepared via vacuum filtration

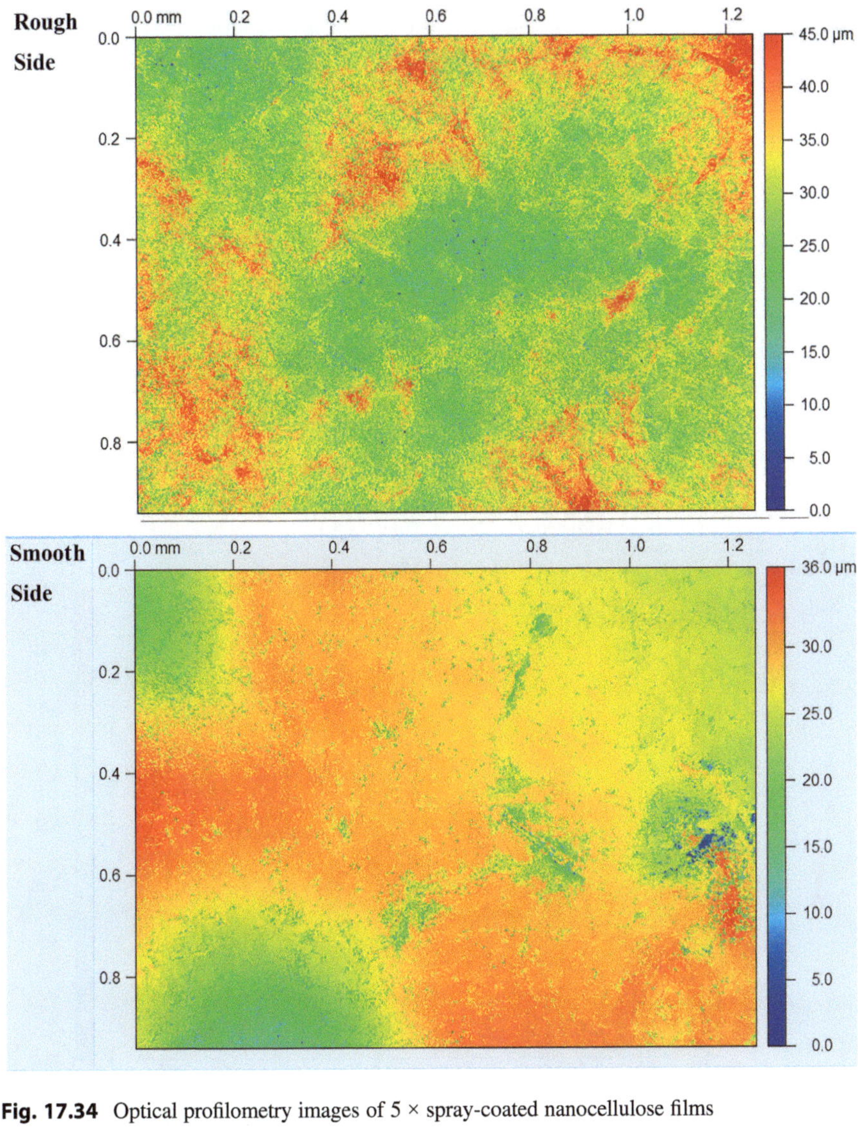

Fig. 17.34 Optical profilometry images of 5 × spray-coated nanocellulose films

shows the effect of nanocellulose suspension consistency on the basis weight and thickness of the film.

Figure 17.39 shows that the mass per unit area of the nanocellulose film exhibited a linear increase when the suspension concentration of NC increased from 1.00 to 2.00 wt.%. The mass of the film per unit area was mainly controlled by the solid content in the suspension and unlikely by the velocity of the conveyor (Shanmugam et al. 2017, 2018). Moreover, the film obtained at optimized velocity was free from the pinholes in the sheet and has high thickness and contains densely packed fibers of

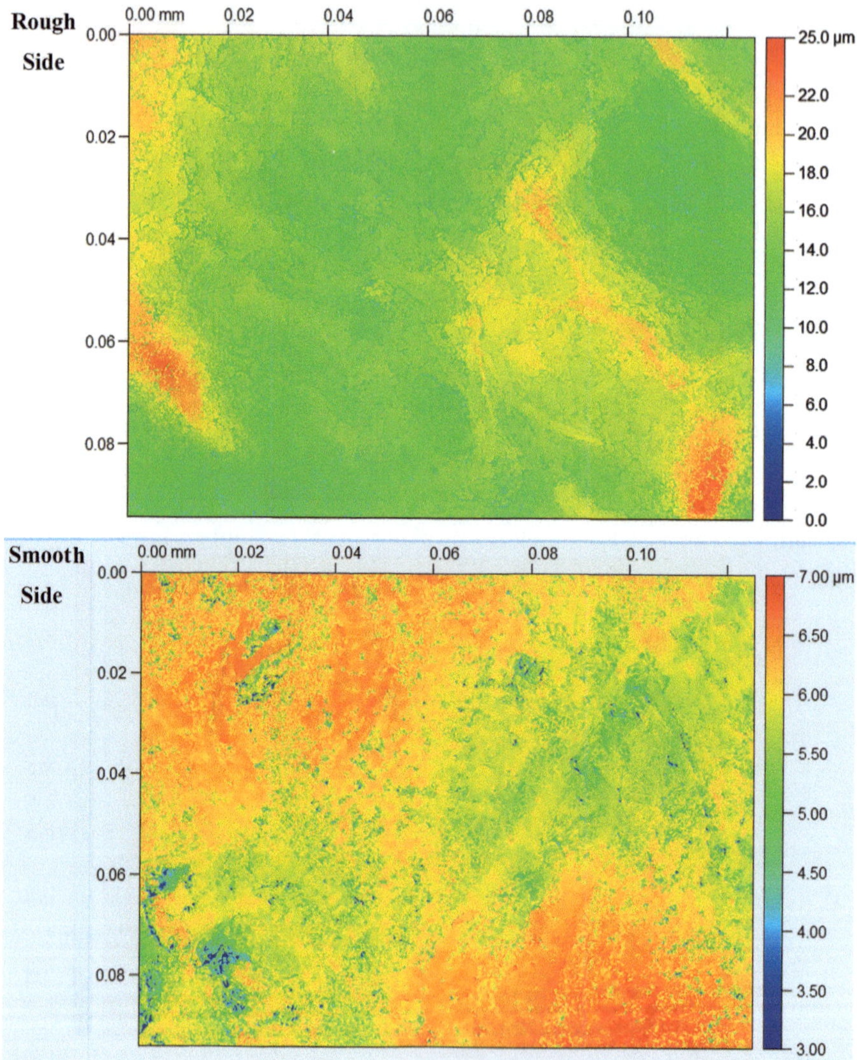

Fig. 17.35 Optical profilometry images of 50 × spray-coated nanocellulose films

various sizes and particles (Shanmugam et al. 2018). The maximum mass of spray-coated nanocellulose film is 195 g/m^2. The film has thickness of 240 μm and homogenous in nature (Shanmugam et al. 2018). The professional Wagner spray systems are capable to spray these MFC/NC suspensions beyond this concentration. The nanocellulose suspension shows non-Newtonian behavior and has time-dependent viscosity. The fibers/particles increased in the suspension increase their viscosity. As a result, the spraying of high solid content in the suspension is really challenging, and the high shear force required to spray the suspension through the

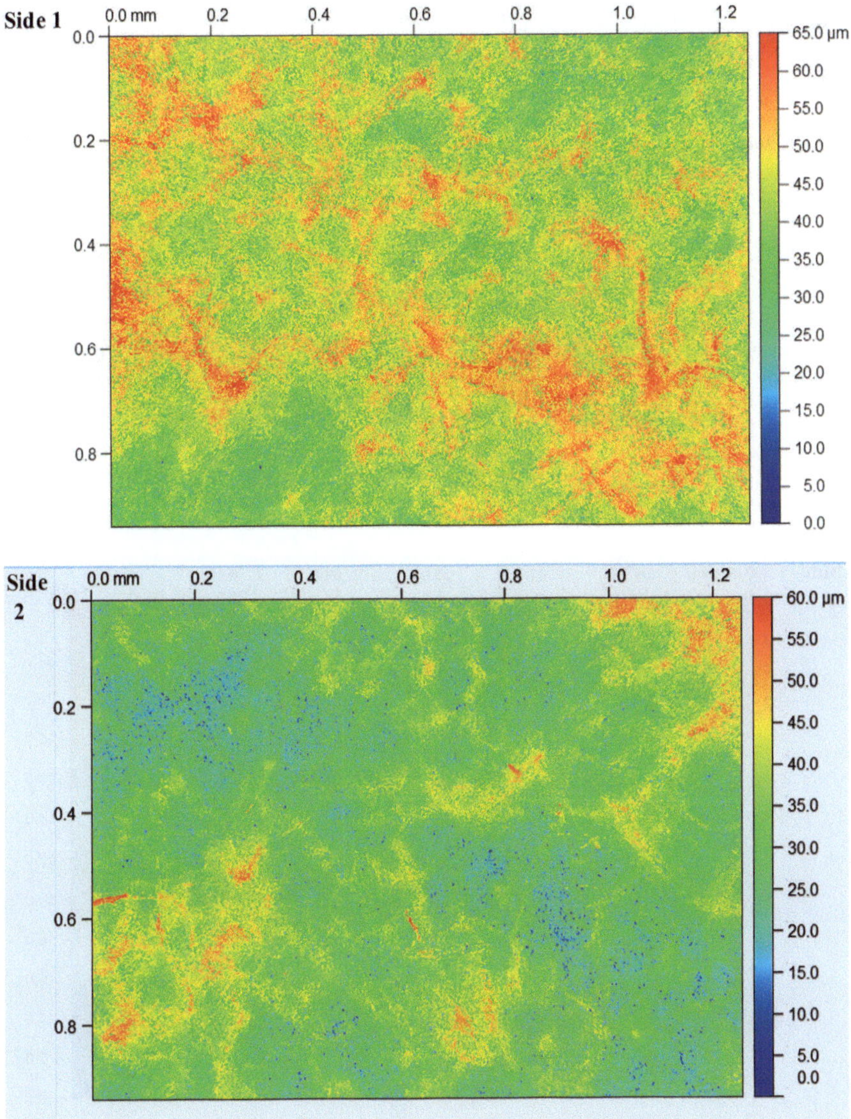

Fig. 17.36 Optical profilometry images of 5 × vacuum filtered nanocellulose films

nozzle is needed. However, while increasing the solid/fiber content in the suspension, the disintegration of the fibers in the suspension is difficult. As a consequence, the fluidity of the suspension would be lost, and handling of these suspensions through nozzle is difficult and has more chances of clogging the nozzle. Furthermore, the high shear force is required to pump and spray the high fiber content of the slurry (Shanmugam 2021c).

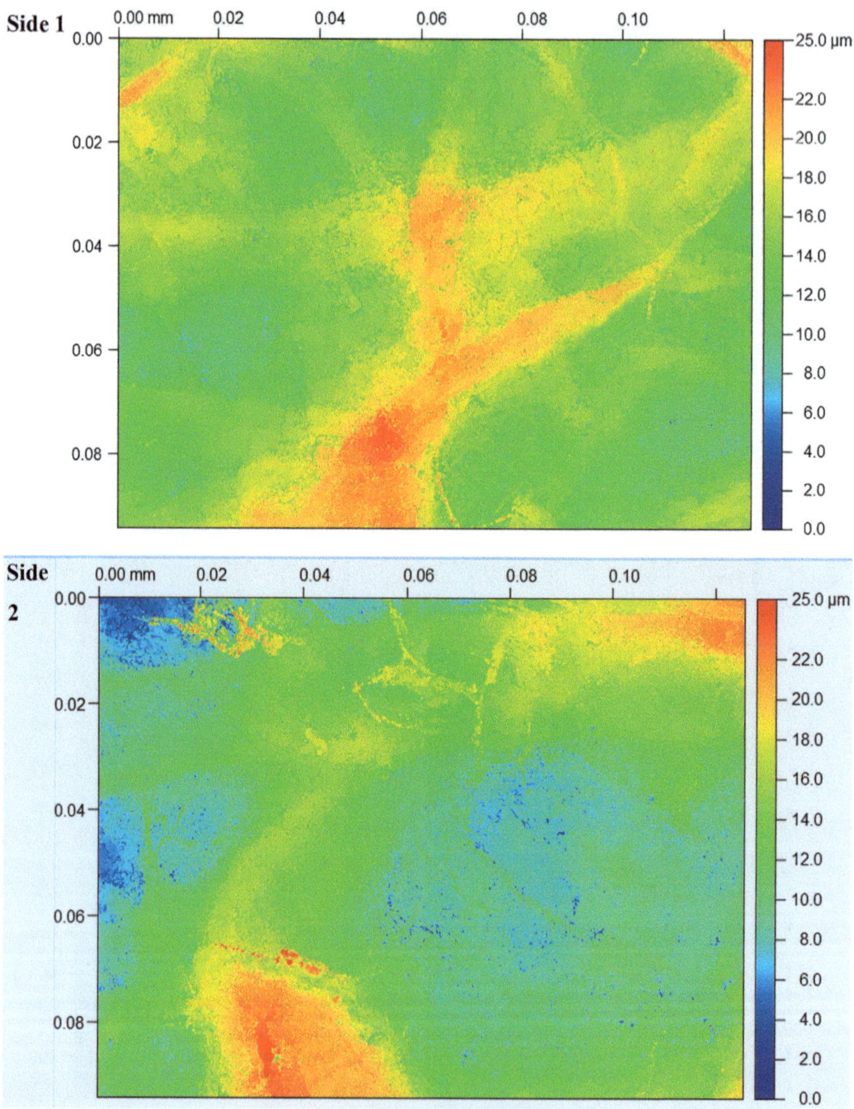

Fig. 17.37 Optical profilometry images of 50 × vacuum filtered nanocellulose films

17.4.9 Mechanical and Barrier Performance of Nanocellulose Film

The spray-coated nanocellulose film has compact structure and two unique surfaces, namely, rough surface and smooth surface. The rough surface is normally porous whereas the smooth surface is glossy and shiny. The effect of the surface roughness and smoothness of the film on the barrier performance of the film remains obscure. The rapidity of spraying process increases the potentiality of the nanocellulose films

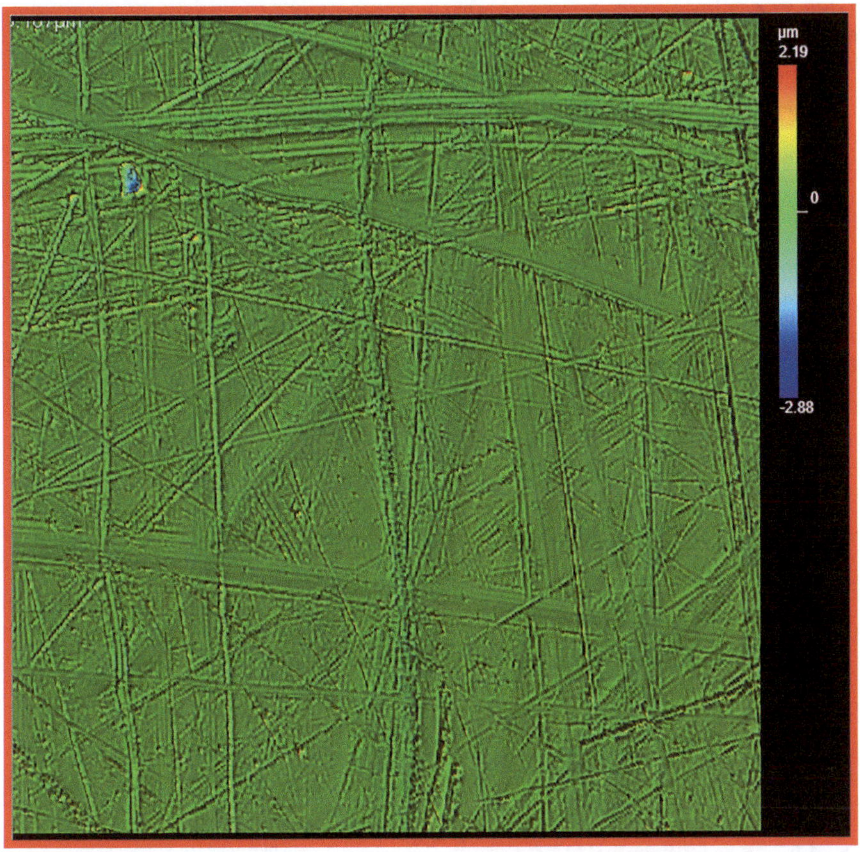

Fig. 17.38 Optical profilometry images of base surface—circular stainless steel plate (50× magnification)

to replace synthetic plastics for barrier applications. The air permeance of the spray-coated nanocellulose film was evaluated to be less than <0.003 µm/Pa s, concluding that the film was impermeable and a good barrier against air, oxygen, and other gaseous substances. When compared with the film via vacuum filtration, the spray-coated film has comparable air permeance and fit to be a potential packaging material (Shanmugam et al. 2017).

The barrier against water vapor is an important property of the nanocellulose films (Nair et al. 2014; Belbekhouche et al. 2011). Normally synthetic plastics have excellent barrier against water vapor and play as conventional packaging materials (Jadhav et al. 2021; Ilyas et al. 2018). The plastic film is nonbiodegradable, good recyclability into various materials, cause the threat to the environment via land disposal of plastics, generation of microplastics and penetration into the food chain (Rajmohan et al. 2019). To mitigate these problems, nanocellulose films can be used as a packaging material (Kirubanandan 2022). Cellulose nanofiber is a hydrophilic

Table 17.4 Surface roughness of nanocellulose films prepared via spray-coating and vacuum filtration

Films preparation method	Technique used	Surface	RMS roughness	
Spray-coating	Atomic force microscopy	Rough	414.0 nm (10 μm × 10 μm)	51.4 nm (2 μm × 2 μm)
		Smooth	81.1 nm (10 μm × 10 μm)	16.7 nm (2 μm × 2 μm)
	Optical profiler 50× image	Rough	2086.69 nm	
		Smooth	389 nm	
Vacuum filtration	Atomic force microscopy	Free side	417.7 nm (10 μm × 10 μm)	102.3 nm (2 μm × 2 μm)
		Filter side	330.8 (10 μm × 10 μm)	70.7 nm (2 μm × 2 μm)
	Optical profiler 50× image	Free side	2673 nm	
		Filter side	3751 nm	

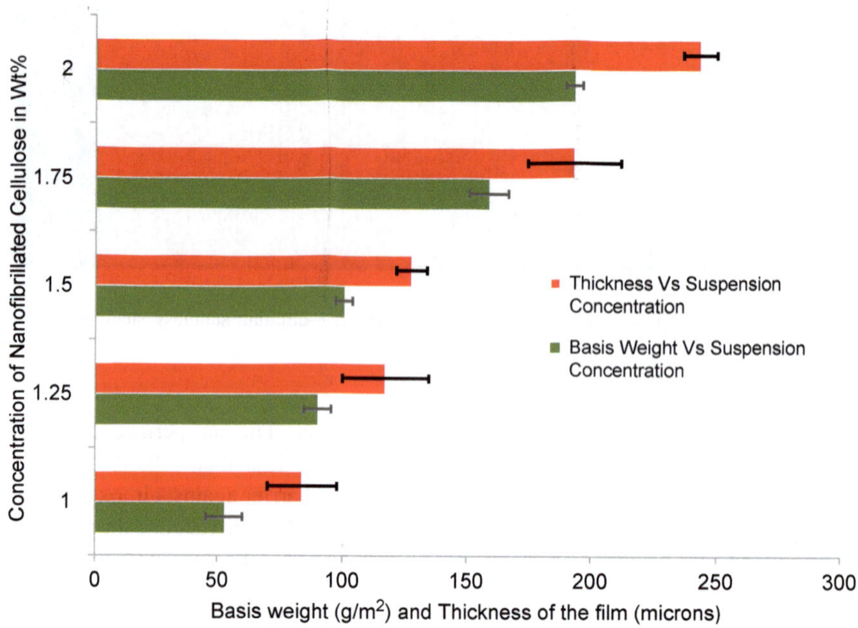

Fig. 17.39 Effect of suspension concentration on the mass per unit area of the nanocellulose film using spray-coating at a constant velocity of 0.32 cm/s. Each point is a minimum of four replicates. Replicates were prepared sequentially

nanomaterial capable for resisting water vapor transmission across the film (Ilyas et al. 2018). Under high moisture environment, cellulose nanofibrils are susceptible to water and water vapor, and fibers become well wetted and loosen cellulose nanofiber matrix resulting in widening the pores. As a consequence, the water vapor and other gaseous molecules are transferred more amount and increased their permeability (Garusinghe et al. 2018; Garusinghe 2017). To tailor the barrier performance of nanocellulose films, spray-coating can be very helpful to fulfill this criterion as the operation time in this method is independent of the nanocellulose suspension concentration (Shanmugam 2021c). Therefore, thickness and basis weight of the film can be tailorable easily (Maliha et al. 2020; Alsaiari et al. 2022). The spraying nanocellulose suspension on the polished metal surface produces a compact film that is glossy and shiny (Shanmugam et al. 2017). The water vapor transmission rate (WVTR) of the spray-coated nanocellulose film was reported to be 35.6 ± 8.5 g/m^2 day whereas the WVTR of the vacuum filtered nanocellulose film was found to be 33.9 ± 3.5 g/m^2 day (Shanmugam et al. 2017). These values confirm the spray-coated nanocellulose films have potential barrier against water vapor, and these values can be reduced further and brought nearly into the barrier properties of synthetic plastics through the incorporation of nanoclay into the nanocellulose matrix (Shanmugam et al. 2021). Nanocellulose film consists of tortuous pathway for transfer of gaseous and water vapor molecules across the film. Due to the complex cellulose nanofibrous structures, the diffusion pathway for water vapor and gaseous substances was increased and barrier performance was enhanced. Another consideration is the reduction of pore size in the nanocellulose films and increased barrier performance than that of normal cellulose substrates such as paper and paper boards (Nair et al. 2014).

Mechanical properties of nanocellulose films are very important in the design of packaging materials and barrier coating on the paper substrates (Beneventi et al. 2015; Shanmugam 2021c). The tensile properties of the film depend on the fabrication methods, types of cellulose fibers from the feed stock, diameter and length of nanofibers, and their aspect ratio (Shanmugam 2019). The tensile properties of spray-coated nanocellulose films were comparable with the film via vacuum filtration (Shanmugam et al. 2018). The tensile index of the spray-coated nanocellulose film and film via vacuum filtration was evaluated to be 60.2 ± 1.5 and 62.3 ± 3.4 Nm/g, respectively (Shanmugam et al. 2017). The tensile stiffness of these films was found to be 946 ± 86 kN/m for spray-coated film and 941 ± 78 kN/m for vacuum filtered film, respectively (Shanmugam et al. 2017). These values suggest that the spray-coated nanocellulose films have comparable strength with the film prepared via vacuum filtration. The cellulose nanofibers in the nanocellulose films form the compactness via hydrogen bonding with neighboring fibers and resulting good strength and modulus. The strength of the film also depends on their thickness and basis weight of the film. As discussed earlier, spraying can only provide the opportunity for tailoring thickness and basis weight of the film via simply adjusting the nanocellulose suspension concentration and varying engineering parameters in the spray-coating experimental system. As a result, the engineering of strength of nanocellulose films can be achieved via spraying effectively (Shanmugam et al. 2018).

17.4.10 Critical Parameters in Spray-Coating

Spraying nanocellulose suspension on the polished metal surface is a rapid process for fabrication of nanocellulose films. The operation time for forming wet film on the plates consumes less than a minute. In the practice to confirm the proof of concept, the processing time was consumed around 50.2 s to fabricate the 15.9 cm diameter of the film. The time consumed to form the same diameter of film and basis weight of 60 g/m^2 in filtration processes was around 10 min. In vacuum filtration process, the drainage time exponentialy increased with NC suspension consistency. In spraying, the operation time was independent of suspension consistency to fabricate the film. However, there are a number of critical parameters involved in the spraying process to effect on the formation of nanocellulose films. The following critical parameters in the spraying process were discussed (Shanmugam 2019).

17.4.11 Nanocellulose Suspension Consistency

NC Suspension consistency is one of the critical parameters in spraying process (Shanmugam et al. 2017). Increasing the suspension consistency for spraying would elevate the thickness and basis weight of the film linearly (Shanmugam et al. 2017; Alsaiari et al. 2022). The current spray system has inability to spray the suspension beyond 2.00 wt.% nanocellulose suspension consistency (Shanmugam et al. 2018). This problem can be solved by the addition of rheology modifiers such as CMC (carboxy methyl cellulose) (Shanmugam et al. 2020) and MMT (Montmorillonite) nanoclay (Shanmugam 2021b) into the nanocellulose suspension. As a result, the suspension can be sprayed at higher solid concentration to form the composite films. Another observation on the spraying nanocellulose suspension was the flow behavior of spray jet from the spray nozzle. The spraying of low suspension concentration (0.25–0.5 wt.%) produce wide and wavery spray jet due to the low viscosity of the suspension. In this case, the spray jet was reflective when the spray jet hits the base surface (Kirubanandan 2022). As a consequence, the poor uniformity of the film was formed (Shanmugam et al. 2018). This case can be suitable for coating nanocellulose as barrier layers on the paper substrates (Kirubanandan 2022; Qureshi et al. 2022). The paper substrates absorbed the water in the nanocellulose suspension and formed cellulose nanofibers coating on the paper surface and minimized the reflection of the spray jet. In the case of spraying high suspension concentration (0.75–2.25 wt.%), the reflection of spray jet from the base surface was minimized with increased solid concentration of cellulose nanofibers. Due to the increased viscosity of suspension at higher concentration, the spray jet velocity reduced and less wide of the spray jet. As a result, the uniformity of the film formed was increased. Normally, spraying high concentration of nanocellulose suspension will be useful for fabrication of thick and dense free-standing nanocellulose films (Shanmugam et al. 2018).

17.4.12 Adhesion Between Nanocellulose Suspension and Base Surface

The adhesion between nanocellulose suspension and base surface and wettability of suspension on the base surface play a major role in the formation of film or barrier layer on the substrates. Normally, the circular stainless steel plate was highly polished and surface was hydrophobic. Spraying low solid nanocellulose suspension concentration on the plates forms liquid droplets which are highly moveable on the plate resulting in the poor formation of the films. The contact angle between sprayed nanocellulose droplets and circular stainless steel plate was large as visually observed. To mitigate this problem, spraying high concentration nanocellulose suspension on the plate easily wet the plate, and the spray-coated suspension was very stagnant (Shanmugam et al. 2018). However, spraying low NC suspension concentration on the paper substrates easily wets the paper surface and forms the barrier coating on the paper surface effectively (Qureshi et al. 2022). The paper substrates should absorb water from the spray-coated nanocellulose suspension immediately (Shanmugam 2021c). So, the adhesion and wettability of the coating will be improved (Shanmugam 2019).

17.4.13 Spray Distance in the Experimental Setup

Spray distance between spray tip to the base surface is a very important parameter which affects the spray pattern, spray width, and spray jet from the spray nozzle. The optimum distance is required to develop uniform film formation and coating on the substrates. The optimization of spray distance depends on the nanocellulose suspension consistency. If the viscosity of the suspension is low, the distance should be sufficiently large around 50 cm, so that the reflectivity of spray jet from the base surface should be minimized and the formation of the film will be good. The distance should be small and maintained around 30 cm in the case of high suspension consistency. So that high suspension nanocellulose easily deposits on the plate and forms the film effectively without any reflection from the plate (Alsaiari et al. 2022; Shanmugam et al. 2018).

17.4.14 Base Surface

The formation of barrier layer on the substrates and film on the base surface depends on the surface morphology and topography of the base surface (Shanmugam et al. 2020). The wettability of the base surface is very important in the formation of films. There are many base surfaces classified into permeable substrates and impermeable substrates. Permeable substrates such as news print paper, packaging paper, and blotting paper are used as a base surface for coating nanocellulose on their surface for improving their barrier performance (Shanmugam 2022). Blotting paper and news print paper are easily wetted by the water and adhesion force easily controlled between paper surface and nanocellulose coating. Low concentration of nanocellulose suspension was required for barrier coating on the paper substrates. The impermeable substrates are circular stainless steel plates, ordinary stainless steel plate, super polished stainless steel plate, and silicon wafer. The smoothness of these substrates increased the same order in the above materials. Due to the smoothness of the substrate, the surface was more hydrophobic, and the spray-coated nanocellulose suspension forms droplets with high contact angle and poor wettability. This problem was resolved by spraying high suspension consistency of nanocellulose (Shanmugam et al. 2020).

17.4.15 Spray Systems

In the current investigation, the spray system used is professional Wagner system which is used in paint spray practice. The performance of spraying nanocellulose suspension can be increased by the spray system improvement. The lowest version of spray system has inability to spray high suspension concentration beyond 2.00 wt. %, because the formation of cellulose nanofibers aggregated in the spray nozzle blocks the spray pattern and avoids formation of spray jet. As a result, the poor uniformity of the film was formed with less thickness and basis weight. With increase in the version of spray system, the equipment was capable to handle the high suspension nanocellulose concentration without any block of fiber aggregates in the spray nozzle. Nanocellulose suspension shows non-Newtonian behavior and has time-dependent viscosity. The fibers/particles increased in the suspension increase its viscosity. As a result, the spraying of high solid content in the suspension is really challenging, and the high shear force required to spray the suspension through the nozzle is needed. However, while increasing the solid/fiber content in the suspension, the disintegration of the fibers in the suspension is difficult. As a consequence, the fluidity of the suspension would be lost, and handling of this suspension through nozzle is difficult and more chances of clogging the nozzle. Furthermore, the high shear force is required to pump and spray the high fiber content of the slurry (Shanmugam 2019).

17.4.16 Velocity of the Conveyor

The velocity of the conveyor is another important parameter in the experimental setup of spray system, and this parameter effects on the basis weight and thickness of the nanocellulose film (Shanmugam et al. 2017). Once the optimum NC concentration with good solid content was evaluated and fixed for spraying on the base surface, the changes in the velocity of the conveyor produce thickness and basis weight of the interest of the film. At the low velocity of the conveyor, the NC suspension was deposited more on the base surface and results in good thickness and basis weight of the film formed with excellent uniformity. At higher velocity of the conveyor, less amount of nanocellulose was deposited via spraying and resulted in thin film formation. The challenge is the operation of conveyor at the lowest velocity or speed in the experimental setup spray system. The tailoring between suspension consistency and velocity of the conveyor is a key to success in the fabrication of good nanocellulose film with an excellence in uniformity (Shanmugam et al. 2018).

17.4.17 Other Engineering Parameters for Improving the Spray System

Spraying nanocellulose suspension on the polished metal surface produces a wet film of nanocellulose. The drying of nanocellulose film requires fast process for the removal of water from wet spray-coated NC film. In paper and pulp industry, there are some waste heat that can be utilized for the drying processes. In the case of the continuous processes, the roll-to-roll coating integrated with spray system can be implemented for the fabrication of nanocellulose film and drying continuously via infrared (IR) dryer. The computer simulation of spray pattern and jet analysis will be helpful for further improvement of the spraying processes and to achieve an excellence in the uniformity of the film (Shanmugam 2019).

17.4.18 Application for Spray-Coating

The concept of spraying nanocellulose on the base surface was applicable to the coating of paper substrate for increasing their barrier properties (Aulin and Lindström 2011; Czerwonatis 2008) and mechanical strength (Shanmugam 2019). There are many coating methods available for coating nanocellulose on the paper surface. There are bar coating, rod coating, vacuum filtration, solvent casting, and spray-coating (Shanmugam 2021c). The nanocellulose coating on the paper substrates forms a barrier layer on the paper surface (Syverud and Stenius 2009). This layer resists against the transfer of water vapor and gaseous substances air, oxygen, and others. As a result, the barrier performance and mechanical strength of the coated paper substrates are elevated. The pure paper substrate has wide pores in the cellulose fiber structure resulting in significant passage of air and water vapor (Shanmugam 2020). To mitigate this issue, the paper can be coated with the synthetic plastics/wax/extrusion with some synthetic polymers to increase the barrier performance and mechanical properties of the paper substrates (Shanmugam 2021c). However, these coatings on the paper surface are not recyclable and can cause a threat to the environment via pollution to the land. This is why the coating of the paper or paper board with nanocellulose was required to increase the material compatibility of the coating with paper materials. Moreover, the nanocellulose is a biodegradable and ecofriendly nanomaterial, and the coating gives tortuous pathway on the paper surface resulting in the enhancement of the barrier properties of the paper substrates (Nair et al. 2014).

Figure 17.40 shows SEM micrograph of the uncoated paper showing the presence of pores in the surface, and it allows the transfer of air and water vapor significantly.

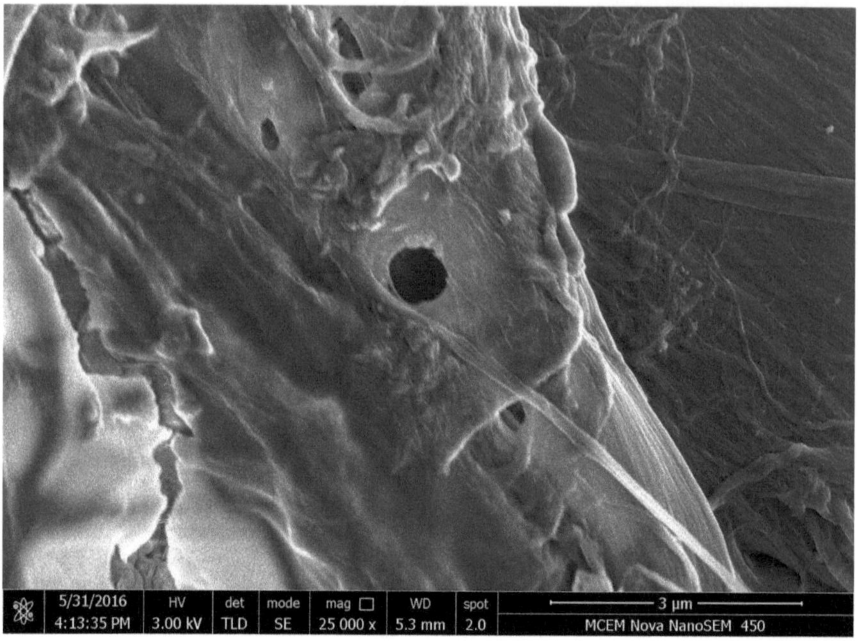

Fig. 17.40 SEM micrograph of plain packaging paper

The coating of nanocellulose on the paper can fill these pores and forms a barrier layer on the paper surface (Shanmugam 2021d). The porosity of the paper ensures good control on the transport of the gaseous substances. The pores also ensure the diffusion of air or water vapor in a controlled manner across the paper substrates. The coating NC on the paper substrate can tailor the surface, barrier, and tensile properties of the substrates (Kirubanandan 2022; Shanmugam 2021c).

Figure 17.41 shows the SEM micrograph of the paper coated with nanocellulose via spray-coating. The surface of the spray-coated paper shows the particles of nanocellulose and cellulose nanocrystal on the surface creating the complex pathway for the diffusion of water vapor, air, and oxygen. The nanocellulose consists of amorphous and crystalline region in the cellulose nanofibrous matrix (Nair et al. 2014). These regions play a major role in the creation of complex tortuous pathway for transfer of water vapor, air, and oxygen. As a result, the barrier performance of the coated substrates was increased (Hubbe et al. 2017).

Figures 17.41 and 17.42 reveal the SEM micrograph of the uncoated paper exploring that the pores on the surface of the paper surface are widely opened. The fibers in the paper substrates have good diameter and length size. The spraying nanocellulose on the paper surface forms a good cover/layer on the surface of the paper and paper substrates. The surface pores on uncoated paper substrate reveal a wide opening and allow the transfer of water vapor and oxygen easily resulting in poor barrier potential. The surface pores can be blocked by the coating paper substrate with nanocellulose (Shanmugam 2021c, 2022).

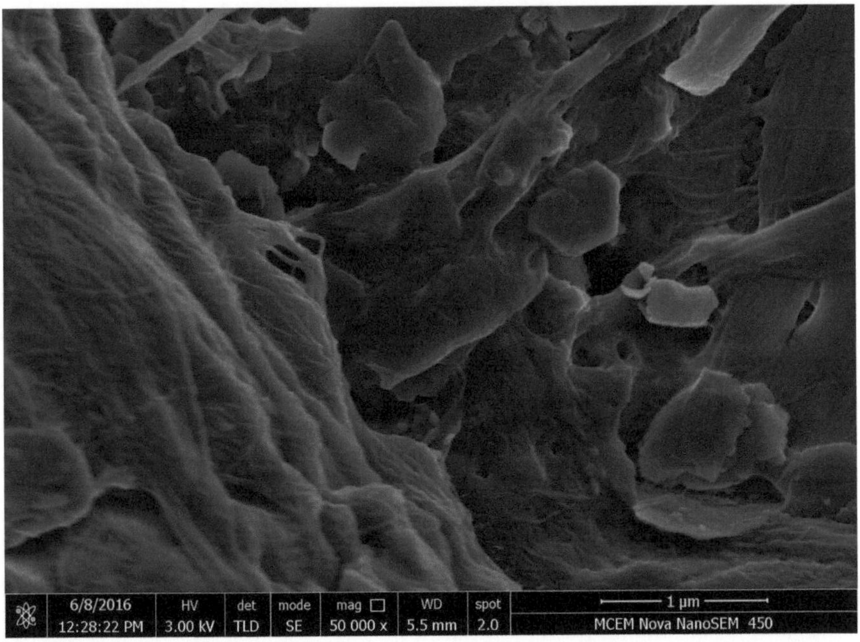

Fig. 17.41 SEM image of nanocellulose coated paper via spray-coating

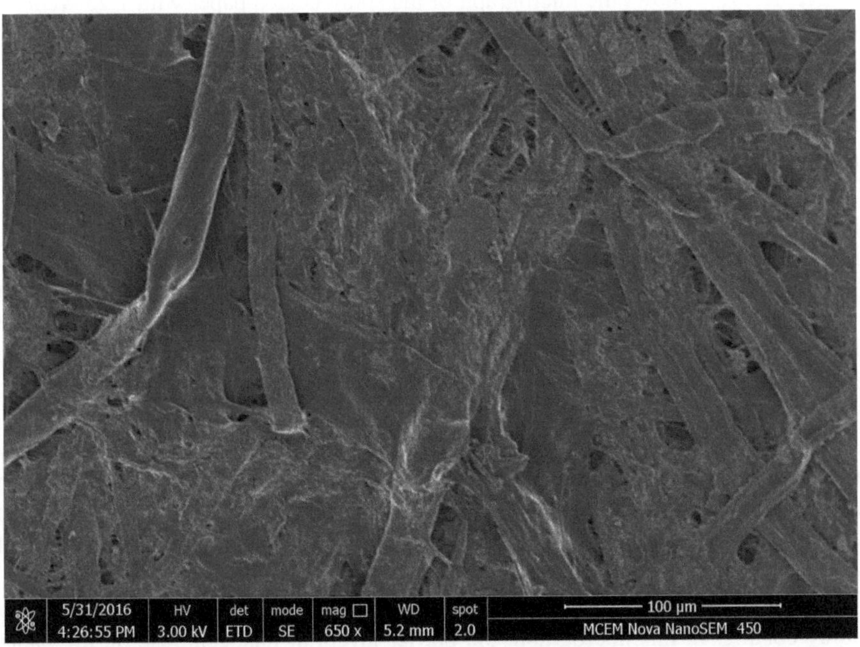

Fig. 17.42 SEM micrograph of uncoated paper

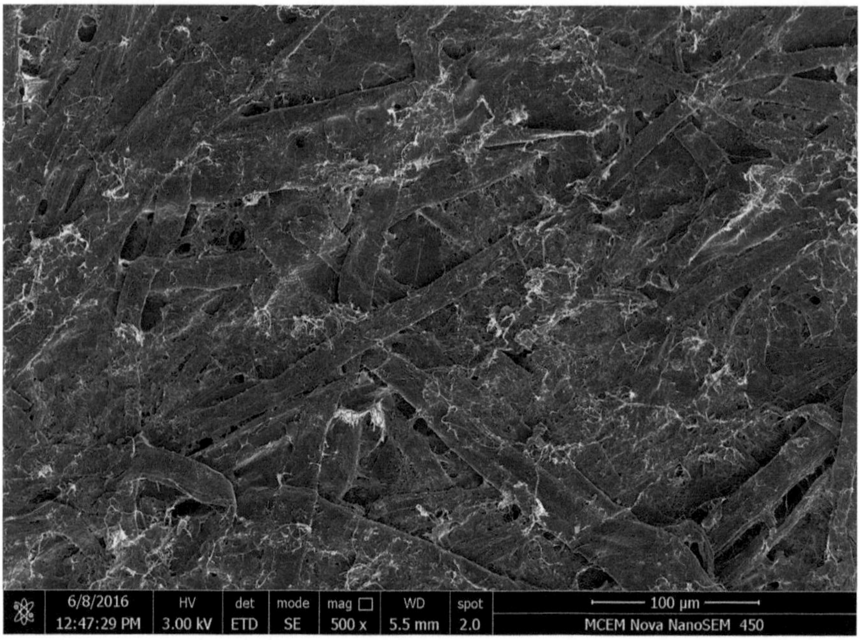

Fig. 17.43 SEM micrograph of coated paper

Figure 17.43 shows the micrograph of the spray-coated paper with 1.25 wt.% of the nanocellulose at low magnification. The micrograph (100 μm) shows the deposited cellulose fiber clumps and fibers on the surface of the base sheet. It also confirmed the different sizes of the fiber entangled with cellulose fiber clumps on the surface. Moreover, the micrograph (100 μm) confirms the complete coverage of micro-fibrillated cellulose coating formulation on the base sheet/paper substrate. When compared to the micrograph (100 μm) of the uncoated paper, the coated paper showed that the coating formulation filled many surface pores and void space between the cellulose fibers (Shanmugam 2021d).

Figure 17.44 shows the SEM micrograph of the coated paper with 0.25 wt.% of nanocellulose on the paper surface via spraying. 0.25 wt.% coating gives the fill of the pores with nanocellulose fibers. The low concentration of cellulose nanofibers in nanocellulose suspension gives pores filling than that of forming the layers on the surface of the papers (Shanmugam 2021d).

Figure 17.45 shows the SEM micrograph of the uncoated paper showing clearly the large pores on the paper surface. It is also confirmed that the fibers are big in diameter and length. The size of surface pores in paper substrates are not uniform and form a wide opening resulting in the poor performance in barrier and mechanical strength (Kirubanandan 2022).

Figure 17.46 reveals the 1.25 wt.% nanocellulose coated on the paper surface via spray-coating. In this micrograph, the barrier layers of nanocellulose on the paper surface completely covered and form a protection against water vapor and air across

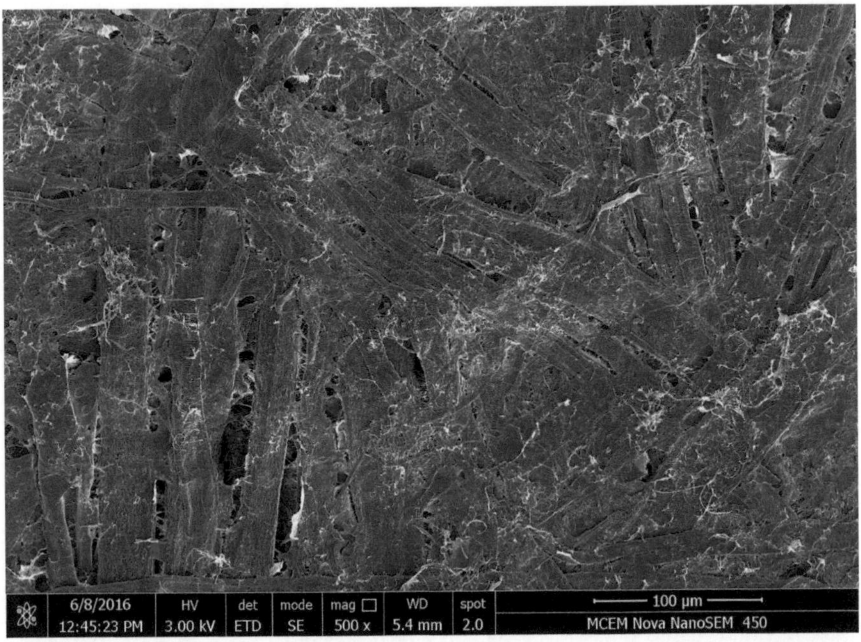

Fig. 17.44 SEM micrograph of coated paper

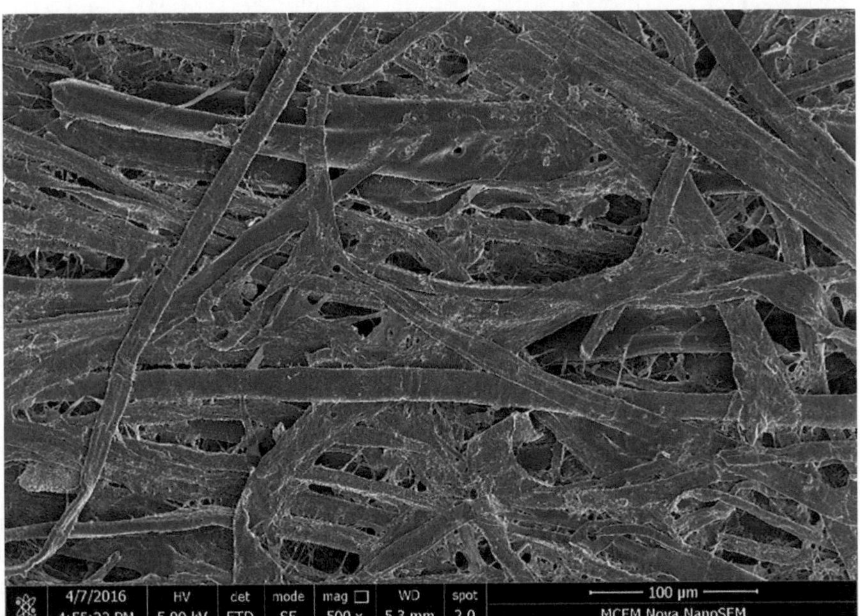

Fig. 17.45 SEM micrograph of uncoated paper

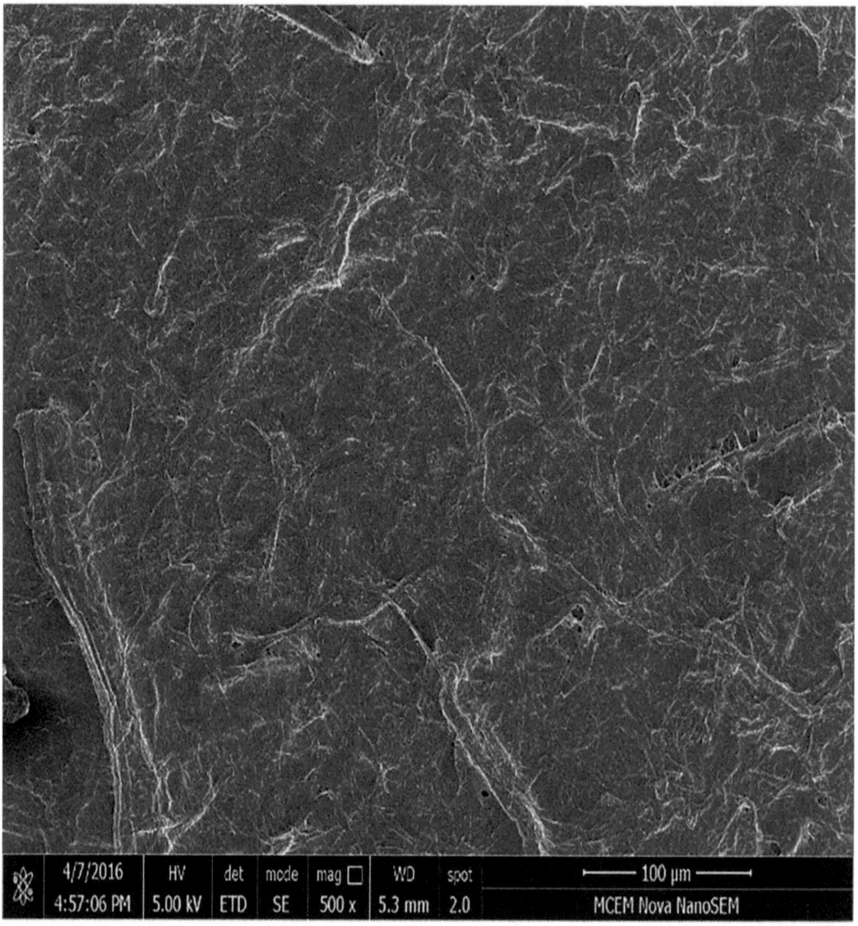

Fig. 17.46 SEM micrograph of spray-coated nanocellulose on the paper surface

the paper. The spraying provides a homogeneous coating of nanocellulose on the paper surface without any cracks and pinholes. It confirms the potential for scale up to industry for large scale production (Kirubanandan 2022).

17.4.19 Air Permeance of Spray-Coated Nanocellulose Paper

Figure 17.47 shows the air permeance of the spray-coated nanocellulose barrier layers on the paper substrates. The plot reveals that the air permeance of the coated sheet drastically reduced with the concentration of NC for spray-coating on the paper substrates. At lower percentages of NC, NC filled the surface pores of the paper substrates resulting in the air permeance of the paper substrates reduced from the

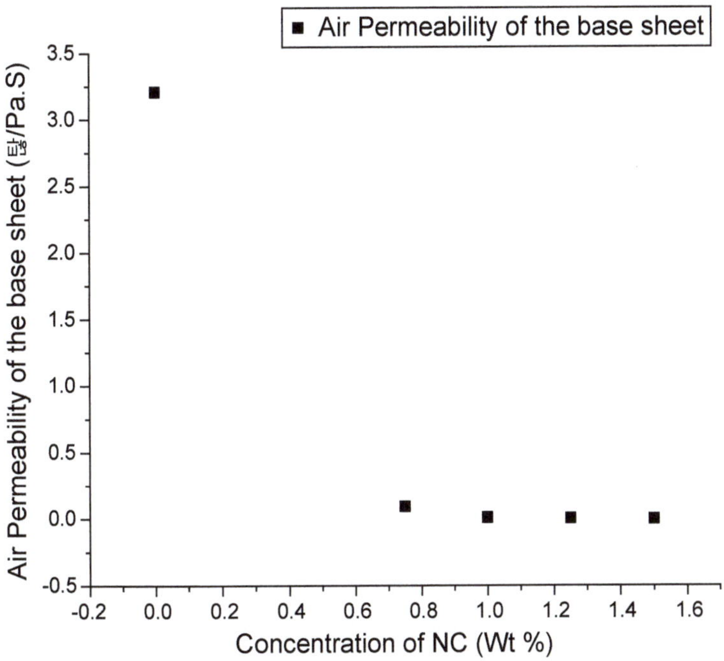

Fig. 17.47 Air permeance of the nanocellulose coated paper via spray-coating

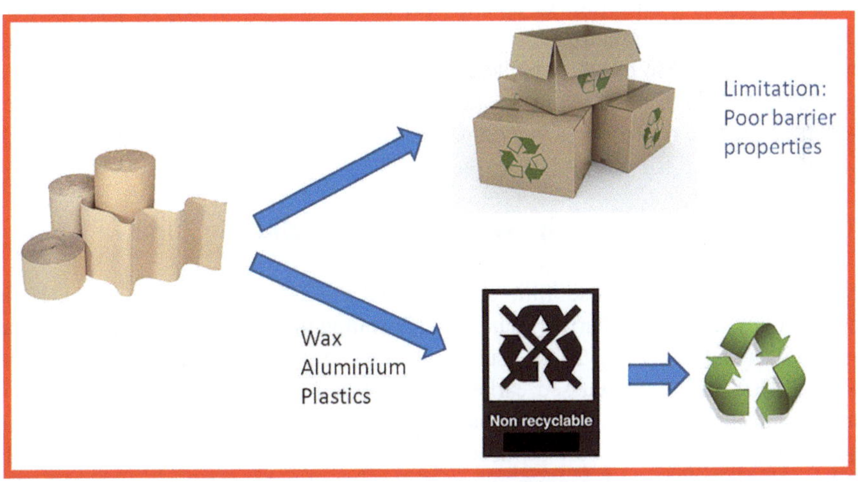

Fig. 17.48 Coating of biopolymers on the paper and paper board

initial value of 3.15 μm/Pa s. This is the value of air permeance of the uncoated paper. At higher percentages of NC coating, the spray-coated NC forms the barrier layer on the paper surface and results in the elevation of barrier properties. The coated sheet was completely impermeable against air and becomes a greener packaging material. As discussed earlier, the basis weight of the coat and thickness of the coat was linear in the case of NC coating on the paper substrates (Kirubanandan 2022).

Figure 17.48 reveals the necessary of the coating of the biopolymers on the paper and paper board. Generally, papers are cellulose macrofiber substrates which have highly porous substrate allowing air, water vapor, and oxygen in a considerable quantity. The result is that the paper has poor barrier performance and mechanical strength. To resolve this issue, the paper substrates were coated with wax or extrusion with aluminum or synthetic plastics. These coats on the paper substrates are not ecofriendly and not recyclable. This is why the biopolymer coat on the paper substrates is required to enhance the barrier performance and mechanical strength of the paper. In this attempt, nanocellulose coating on the paper substrates was performed by spray-coating and increased the barrier performance via NC barrier coat on the paper substrates (Shanmugam 2021c).

Figure 17.49 reveals the spray-coated nanocellulose barrier layers on the paper surface, and the coated surface shows nanocellulose layers on the paper substrates. Increasing the nanocellulose suspension for spray-coating would increase the dense layer on the paper substrates. As earlier discussed, low NC suspension can fill the surface pores of the paper whereas high NC suspension would form the barrier layers on the substrates to make the sheet completely impermeable against air and water vapor (Kirubanandan 2022).

Figure 17.50 shows the concept of the continuous process for spray-coating nanocellulose on the paper substrates. Roll-to-roll coating or dow web coating can be used to satisfy the proof of concept for nanocellulose coating in a continuous process. Dow web coating can be integrated with spray system for spraying nanocellulose on the paper web and then dried by infrared (IR) heater for effective removal of water from the spray-coated nanocellulose suspension. The dried paper substrates can be rolled into a web for commercial application (Shanmugam 2021c).

17.4.20 Application of Spray-Coated Nanocellulose Films

Spray-coated nanocellulose film can be used as a good substrate for various applications especially barrier materials (Shanmugam et al. 2021), antimicrobial materials (Maliha et al. 2020), substrates for flexible electronics (Shanmugam et al. 2020), etc. As discussed earlier, spray-coated nanocellulose films have two unique surfaces, namely, rough surface and smooth surface on the metal side. The smoothness of the film can be used in the construction of functional materials (Shanmugam 2021c). Moreover, nanocellulose is a biodegradable and ecofriendly material and provides a platform for recycling. So that it can be used as an alternative for synthetic plastics.

Fig. 17.49 Spray-coated nanocellulose barrier layers on the paper substrates

17.4.21 Flexible and Printed Electronics

Cellulose nanofiber is an ecofriendly nanomaterial used for fabrication of various functional materials. It is an alternative for synthetic plastic and other petroleum-derived materials (Dufresne 2013). Due to demand of CNF film, fast and rapid method for fabrication of CNF film is required. A new method on spray-coating to prepare smooth cellulose nanofiber (CNF) films was developed. In this method,

Fig. 17.50 Concept for continuous process for spray-coating nanocellulose on the paper substrates

spraying CNF suspension (Shanmugam et al. 2017) onto a smooth and polished surface was carried out and then allowed the wet film to dry in air under standard laboratory conditions. Spraying has a notable advantage such as contour coating and contactless coating with the base substrate. The basis weight and thickness of the CNF film is tailorable by adjusting CNF suspension in spraying process. CNF film prepared via spray-coating has unique two-sided surface roughness with the surface in contact with the surface much smoother than the air-contact side. The surface roughness is one of the controlling parameters in barrier performance of the CNF film, and alterations in surface roughness switch the wettability.The RMS roughness of the two surfaces investigated by optical profilometry was found to be 2087 nm on the rough side and 389 nm on the spray-coated side, respectively. The spray-coated CNF film has ultra-high smoothness on the side exposed to the polished stainless steel surface. The factors including the size of cellulose fibrils and surface smoothness of base surface control the roughness of the film which was used in the development of flexible electronics. The factors controlling surface smoothness requirements for substrate applications in flexible and printed electronics are cellulose nanofibrils diameter and types of base surface and it has been shown in Fig. 17.51.

Normally, paper is a cellulose substrate which has biodegradability and recyclability and was used as the substrate for flexible and printed electronics. It undergoes high water absorption and possesses high surface roughness. These properties limit the development of printed electronics on the paper substrates. Additionally, the surface roughness of the paper substrates varies from 2 to 10 μm and limits the spreadness of the conductive ink on the paper substrates. Spray-coated nanocellulose film has good smoothness and good barrier against the air and water vapor (Fig. 17.52).

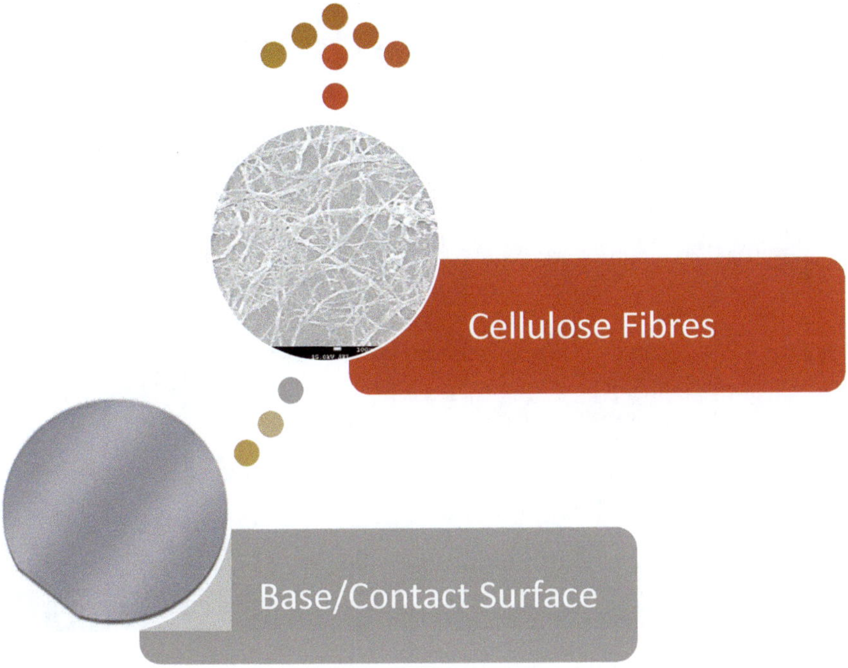

Fig. 17.51 Factors controlling the surface roughness of the films

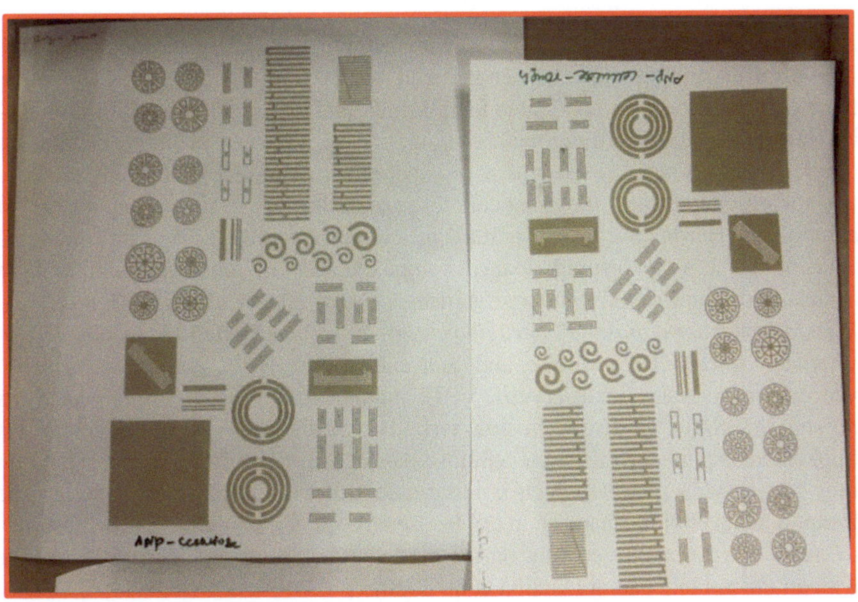

Fig. 17.52 The printed circuits on the spray-coated nanocellulose films

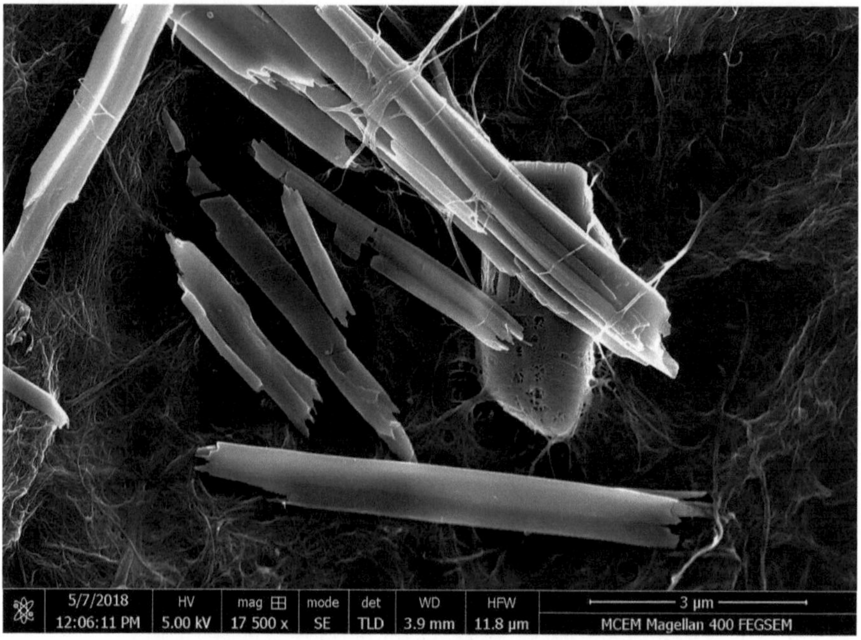

Fig. 17.53 Bismuth-based nanocellulose composite prepared via spray-coating

17.4.22 Bismuth-Based Nanocellulose Composite

Figure 17.53 shows the bismuth-based nanocellulose composite via spraying, and it is used as an antimicrobial barrier material. Apart from the barrier performance, these composites can be applicable to anti-microbial packaging materials, biomedical bandages, and anti-microbial coating on the surface. Phenyl bismuth bis (diphenylphosphinate) is used as a broad-spectrum antimicrobial agent used for the incorporation into the nanocellulose suspension for the development of composites (Herdman 2021; Shanmugam and Browne 2021). The anti-microbial zone of the 5 wt.% bismuth loaded composite was 15 mm zone of inhibition against both gram-negative microorganisms such as *Escherichia coli* (*E. coli*) and *Pseudomonas aeruginosa* (*P. aeruginosa*); gram-positive bacterium such as *Staphylococcus aureus* (*S. aureus*); and anti-microbial-resistant pathogens, namely, vancomycin-resistant *Enterococcus* (VRE) and methicillin-resistant *Staphylococcus aureus* (MRSA) via disc diffusion test. The addition of phenyl bismuth bis (diphenylphosphinate) into nanocellulose suspension did not affect the operation time for spraying process which is independent of suspension concentration. The bismuth-nanocellulose suspension can be used for spray-coating the paper substrates to create bifunctionality coating on the paper surface such as barrier performance with antimicrobial coating (Maliha et al. 2020).

17.4.23 Silver Nanowires-Nanocellulose Composites

Silver nanowire (AgNWs) is a one-dimensional nanostructure with varying diameters from 10 to 200 nm, with lengths from 5 to 100 μm. Silver nanowires were incorporated into the spray-coated nanocellulose film to fabricate the conductive composite for various applications. In this approach, the spray-coated nanocellulose films were used as base substrates for the metallization with silver wires to become the conductive surface. The adhesion between silver nanowires and spray-coated nanocellulose film was very challenging to prepare the conductive surface. This composite can be used as electrode for various applications (Czibula et al. 2017).

17.4.24 Biomedical Device

The spray-coated nanocellulose film was used as base substrate for the incorporation of antimicrobial agents such as antibiotics and nanoclay for dermal wound repair. Cellulose nanofiber is the most familiar inexhaustible, recyclable, and nanocomposite material. The spray-coated cellulose nanofibers were impregnated with silver nanoparticles and Montmorillonite (MMT) via spraying to improve the mechanical strength and liquid absorption of cellulose nanofiber film. The AgNPs show good antioxidant and antimicrobial activity for better wound closure. The cell viability assay of the composite was analyzed using 3T3 fibroblast cell by MTT assay. The antidiabetic assay for the plant extract and AgNPs was observed to be 56% and 61%. The CNF-MMT-AgNPs can be used in various wounds for soft tissue repair due to their good biocompatibility and cost-effective nature (Subha et al. 2022).

17.4.25 Nanocellulose-MMT Composite

The production of smooth Montmorillonite–Cellulose nanofiber composite functioning as a green barrier material has been developed via spray-coating process. The effect of MMT loading into the cellulose nanofiber suspension for producing nanocomposite is independent of the operation time and is important and perfect for scale up of the process. The barrier, surface, and topography of the spray-coated nanocomposites were comparable with the nanocomposites via vacuum filtration as a conventional method performed. The air permeance and water vapor permeability of the spray-coated nanocomposites are well achieved to confirm the green materials were in fact barrier material. Aggregation of MMT clays on a further increase of MMT in nanocomposite resulted in the elevation of WVP, so the effectiveness of spray-coating for producing sustainable nanocomposites as an alternative for synthetic plastics is significant. Figures 17.54 and 17.55 reveal the morphology and topography of pure nanocellulose film and nanocellulose-MMT composite via

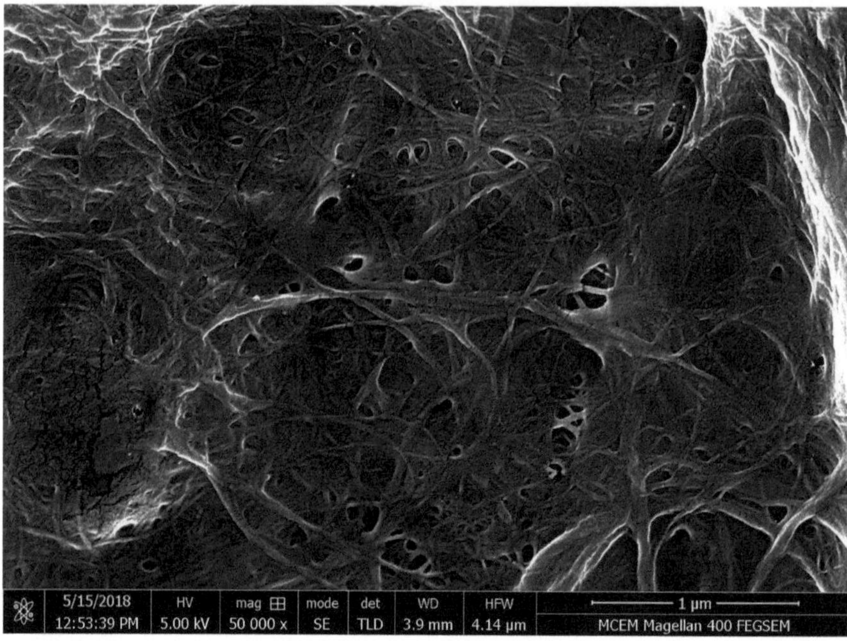

Fig. 17.54 Pure nanocellulose film via spray-coating

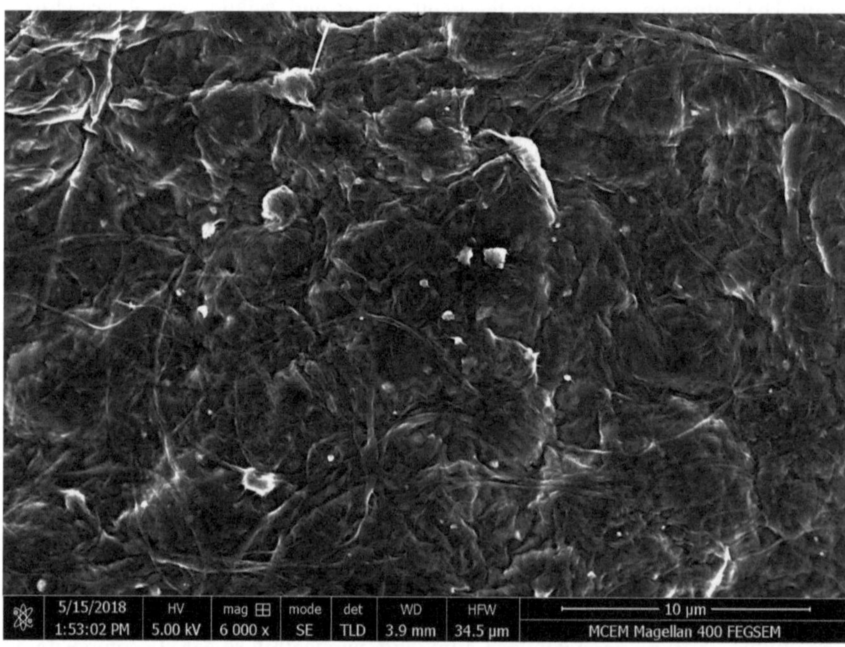

Fig. 17.55 Nanocellulose-MMT composite prepared via spray-coating

spray-coating. The SEM micrograph of the nanocellulose-MMT composite was shown that the MMT was well placed in the cellulose fibrous matrix and results in promoting a tortuous pathway for gases and water vapor. As a result, the barrier performance of the composite was best compared to that of pure nanocellulose films (Shanmugam 2021b; Shanmugam et al. 2021).

17.4.26 Spray-Coated Nanocellulose as Layers for Membrane Development

Depth type composite filters were implemented to remove various contaminants from waste water via mechanical entrapment and adsorption. Adsorption of one type of charged contaminants and microparticle filtration are treated via depth filtration. Cellulose nanofiber (CNF) layer spray-coated on the surface of the filters was a perfect fit for depth filtration. Depth filter was fabricated with cellulose, perlite, and PAE via both vacuum filtration and spray-coating to form as a base sheet. The performance of CNF top coating has molecular weight cut-off up to 80% for two different molecular weights. Furthermore, the removal of positively charged contaminants increased significantly after adding CNF layer on the membrane. The coating of the layer via spray-coating is also robust in terms of getting an adequate internal strength within the interface of base sheet and top layer. The concept of top barrier coating is a promising and easy method to alter the filter performance and used in membrane fabrication for waste water treatment (Onur et al. 2019).

17.5 Recommendations

The iterations for spray-coating will greatly vary the smoothness of the resulting films. The power applied, distance between spray nozzle and substrate, and the rate of coating can be automated by robotic arms which will greatly give an accurate detail of how to get extremely smooth surfaces. The pores size can be manipulated by the concentration of the polymer and solvent as well as the environmental conditions where the experiment is performed. The reproducibility and uniformity of the reaction conditions can lead to even more appreciable results for the films to act as base for printed electronics. Based on the proof of concept and laboratory scale experiments on spray-coating, it has the ability to scale up for conventional scale production of nanocellulose film. The version of professional spray-coating equipment increases to enhance handling of high suspension of NC for the production of good uniformity nanocellulose films.

17.6 Conclusion

In conclusion, spray-coating to produce smooth nanocellulose films was conceptualized and focuses on application of spray-coated nanocellulose film and spray-coated barrier layers on the paper substrates. The smoothness of the spray-coated nanocellulose film plays a major role in the performance of the material in the fabrication of various cellulose-based functional materials such as printed electronics and flexible electronics. Spray-coating enhances the scope of nanocellulose films as high-performance barrier materials and a potential alternative for synthetic plastic packaging. Through spray-coating, the operation time for forming nanocellulose film is less than a minute. This method has excellent potential for rapidity for manufacturing nanocellulose films when compared with vacuum filtration. The drying of spray-coated wet film takes more than 24 h under air drying in a controlled laboratory environment. Improving the drying process on the wet film is a future research work and out of the scope from the current work. The formation of film via spraying is the proof of the concept and compared the quality of the spray-coated film with the film prepared from vacuum filtration. The drying of the wet film takes much longer than the film formation. Spraying of nanocellulose on the polished impermeable surface produced the films with a shiny surface, which could be a platform for the numerous functional devices with a sustainable approach.

References

Abitbol T, Rivkin A, Cao Y, Nevo Y, Abraham E, Ben-Shalom T, Lapidot S, Shoseyov O (2016) Nanocellulose, a tiny fiber with huge applications. Curr Opin Biotechnol 39:76–88

Abraham E et al (2011) Extraction of nanocellulose fibrils from lignocellulosic fibers: a novel approach. Carbohydr Polym 86(4):1468–1475

Ahankari SS et al (2021) Nanocellulose in food packaging: a review. Carbohydr Polym 255:117479

Alsaiari NS et al (2022) Optimization of spraying process via response surface method for fabrication of cellulose nanofiber (CNF) film. J Nanomater 2022:1

Aulin C, Lindström T (2011) Biopolymer coatings for paper and paperboard

Aulin C, Gällstedt M, Lindström T (2010) Oxygen and oil barrier properties of microfibrillated cellulose films and coatings. Cellulose 17(3):559–574

Azeredo HM, Rosa MF, Mattoso LHC (2017) Nanocellulose in bio-based food packaging applications. Ind Crop Prod 97:664–671

Belbekhouche S et al (2011) Water sorption behavior and gas barrier properties of cellulose whiskers and microfibrils films. Carbohydr Polym 83(4):1740–1748

Beneventi D et al (2014) Highly porous paper loading with microfibrillated cellulose by spray-coating on wet substrates. Ind Eng Chem Res 53(27):10982–10989

Beneventi D, Zeno E, Chaussy D (2015) Rapid nanopaper production by spray deposition of concentrated microfibrillated cellulose slurries. Ind Crop Prod 72:200–205

Bras J, Saini S (2017) Nanocellulose in functional packaging. In: Cellulose-reinforced nanofibre composites. Elsevier, pp 175–213

Czerwonatis N (2008) Spray coating—a contactless coating process for paper finishing. In: Proceedings of the TECH 31 Tape Summit, pp 7–9

Czibula C et al (2017) Silver nanowires: a versatile tool for conductive paper. Adv Pulp Paper Res 2:723–737

Dufresne A (2013) Nanocellulose: a new ageless bionanomaterial. Mater Today 16(6):220–227

Garusinghe U (2017) Flexible nanocellulose-nanoparticle composites: structures and properties. Monash University

Garusinghe UM et al (2018) Nanocellulose-montmorillonite composites of low water vapour permeability. Colloids Surf A Physicochem Eng Asp 540:233–241

Herdman ME (2021) Bismuth and gallium phosphinates as antibacterial agents, and additives in Bi-cellulose composites. Doctoral dissertation, Monash University

Hubbe MA et al (2017) Nanocellulose in thin films, coatings, and plies for packaging applications: a review. Bioresources 12(1):2143–2233

Ilyas R et al (2018) Water barrier properties of biodegradable films reinforced with nanocellulose for food packaging application: a review. In: 6th Postgraduate seminar on natural fiber reinforced polymer composites 2018. Institute of Tropical Forestry and Forest Product (INTROP), Universiti Putra

Jadhav EB et al (2021) Microplastics from food packaging: an overview of human consumption, health threats, and alternative solutions. Environ Nanotechnol Monit Manag 16:100608

Jin M et al (2021) Microplastics contamination in food and beverages: direct exposure to humans. J Food Sci 86(7):2816–2837

Kirubanandan S (2022) Spray-coated cellulose nanofiber laminates on the paper to enhance its barrier and mechanical properties. J Sustain Environ Manag 1:10–17

Luo H et al (2019) Advances in tissue engineering of nanocellulose-based scaffolds: a review. Carbohydr Polym 224:115144

Magnusson J (2016) Method for spraying of free-standing 3D structures with MFC: creation and development of a method

Maliha M et al (2020) Bismuth phosphinate incorporated nanocellulose sheets with antimicrobial and barrier properties for packaging applications. J Clean Prod 246:119016

Manukyan L et al (2019) Growth media filtration using nanocellulose-based virus removal filter for upstream biopharmaceutical processing. J Membr Sci 572:464–474

Muncke J (2021) Tackling the toxics in plastics packaging. PLoS Biol 19(3):e3000961

Nair SS, Zhu JY, Deng Y, Ragauskas AJ (2014) High performance green barriers based on nanocellulose. Sustain Chem Process 2:1–7

Nechyporchuk O, Belgacem MN, Bras J (2016) Production of cellulose nanofibrils: a review of recent advances. Ind Crop Prod 93:2–25

Norfarhana A, Ilyas R, Ngadi N (2022) A review of nanocellulose adsorptive membrane as multifunctional wastewater treatment. Carbohydr Polym 291:119563

Onur A et al (2019) Cellulose fibre-perlite depth filters with cellulose nanofibre top coating for improved filtration performance. Colloids Surf A Physicochem Eng Asp 583:123997

Osong SH, Norgren S, Engstrand P (2016) Processing of wood-based microfibrillated cellulose and nanofibrillated cellulose, and applications relating to papermaking: a review. Cellulose 23:93–123

Paul DR, Robeson LM (2008) Polymer nanotechnology: nanocomposites. Polymer 49(15): 3187–3204

Phelan AA et al (2022) Plastic pollution and packaging: corporate commitments and actions from the food and beverage sector. J Clean Prod 331:129827

Qureshi WA, Radjarejesri S, Shanmugam K, Elavarasan P, Gowrishankar M, Gebeyehu KB (2022) Central composite design of spraying process to laminate the paper substrates with cellulose nanofibers (CNF) as green packaging wrap. J Nanomater 2022:1

Rajmohan KVS, Ramya C, Viswanathan MR, Varjani S (2019) Plastic pollutants: effective waste management for pollution control and abatement. Curr Opin Environ Sci Health 12:72–84

Ribeiro RS et al (2019) Production of nanocellulose by enzymatic hydrolysis: trends and challenges. Eng Life Sci 19(4):279–291

Rouf TB, Kokini JL (2018) Natural biopolymer-based nanocomposite films for packaging applications. In: Bionanocomposites for packaging applications. Springer, pp 149–177

Sehaqui H et al (2010) Fast preparation procedure for large, flat cellulose and cellulose/inorganic nanopaper structures. Biomacromolecules 11(9):2195–2198

Shanmugam K (2019) Spray-coated nanocellulose films-production, characterisation and applications. Monash University
Shanmugam K (2020) Preparation of cellulose nanofibre laminates on the paper substrate via vacuum filtration. J Res Chem 1:78
Shanmugam K (2021a) Spray-coated cellulose nanofiber (CNF) film as an eco-friendly substrate for flexible and printed electronics. J Eng Sci 1:68–81
Shanmugam K (2021b) Preparation of cellulose nanofiber (CNF)–montmorillonite (MMT) nanocomposite via spray-coating process. J Mater Sci Surf Eng 8:978–986
Shanmugam K (2021c) Development of cellulose nanofibre (CNF) coating on (1) metal surface for free-standing CNF film and (2) paper substrates for CNF barrier laminates. J Mech Eng 1:10–36
Shanmugam K (2021d) Scanning electron microscopic evaluation of spray-coated micro-fibrillated cellulose on the packaging paper. J Mater Sci Surf Eng 8:966
Shanmugam K (2022a) Cellulose nanofiber lamination of the paper substrates via spray coating–proof of concept and barrier performance. J Mater Sci 1:30–51
Shanmugam K (2022b) Spray coated cellulose nanofiber laminates on the paper to enhance its barrier and mechanical properties. J Sustain Environ Manag 1(1):10–17
Shanmugam K, Browne C (2021) Nanocellulose and its composite films: applications, properties, fabrication methods, and their limitations. In: Nanoscale processing. Elsevier, pp 247–297
Shanmugam K et al (2017) Rapid preparation of smooth nanocellulose films using spray coating. Cellulose 24(7):2669–2676
Shanmugam K et al (2018) Flexible spray-coating process for smooth nanocellulose film production. Cellulose 25(3):1725–1741
Shanmugam K et al (2019) Nanocellulose films as air and water vapour barriers: a recyclable and biodegradable alternative to polyolefin packaging. Sustain Mater Technol 22:e00115
Shanmugam K et al (2020) Engineering surface roughness of nanocellulose film via spraying to produce smooth substrates. Colloids Surf A Physicochem Eng Asp 589:124396
Shanmugam K et al (2021) High-performance homogenized and spray-coated nanofibrillated cellulose-montmorillonite barriers. Cellulose 28(1):405–416
Souza E, Gottschalk L, Freitas-Silva O (2020) Overview of nanocellulose in food packaging. Recent Pat Food Nutr Agric 11(2):154–167
Subha V et al (2022) Functionalization of spray-coated cellulose nanofiber sheet with montmorillonite (MMT) and silver nanoparticles (AgNPs) to biomedical nanocomposite as wound regeneration scaffold. Prog Org Coat 166:106782
Syverud K, Stenius P (2009) Strength and barrier properties of MFC films-table O2 permeability
Varanasi S, Batchelor WJ (2013) Rapid preparation of cellulose nanofibre sheet. Cellulose 20(1): 211–215
Vartiainen J et al (2018) Mechanical recycling of nanocellulose containing multilayer packaging films. J Appl Polym Sci 135(19):46237
Wagner (n.d.-a) Spraying manual
Wagner (n.d.-b) WAGNER ProjectPro 117 0418B Owner's Manual
Wang L, Okada K, Sodenaga M, Hikima Y, Ohshima M, Sekiguchi T, Yano H (2018) Effect of surface modification on the dispersion, rheological behavior, crystallization kinetics, and foaming ability of polypropylene/cellulose nanofiber nanocomposites. Compos Sci Technol 168:412–419
Werrett MV et al (2018) Bismuth phosphinates in bi-nanocellulose composites and their efficacy towards multi-drug resistant bacteria. Chem Eur J 24(49):12938–12949

Kirubanandan Shanmugam worked as Research and Development Associate in Vedia Solutions, a part of Lakshmi Dental Pvt Ltd, Mumbai, India where the product is biomedical devices. Alternatively, he works as a distinguished Faculty in the School of Engineering, Saveetha University, Chennai. His areas of interest are nanomaterials, food packaging, and application of machine learning in nanotechnology. He has done three post-doctoral works in various universities, both in India and abroad, and done Ph.D. in Chemical Engineering from Monash University, Australia, MASc in Chemical Engineering from Dalhousie University, Canada, and M.Tech in Biopharmaceutical Technology, Anna University, India, and B.Tech in Chemical Engineering from University of Madras, India. He has worked in various engineering colleges and universities in India and abroad. He has published more than 100 research articles in both international and national journals and filed seven patents and received three granted design patents. Dr. Shanmugam is a registered professional engineer in India and abroad.

Narendhar Chandrasekar is a Post-Doctoral Researcher in the Department of BioNanoTechnology, Gachon University, South Korea. Having completed M.Tech Nanotechnology in 2009, Dr. Narendhar received his Doctorate from Anna University, Chennai, in 2016. He has generated two Ph.D. scholars in the year 2022 at Anna University. With the experience of novel nanoparticles synthesis, characterization, and expertise in drug delivery systems, biosensors, and environmental remediation strategies, he has been working on several breakthrough methods for identifying and removing toxic contaminants from water bodies.

Dr. Narendhar had 13 years of teaching experience in India where he was working as an Assistant Professor in the Department of Nanotechnology in Sri Ramakrishna Engineering College, Coimbatore, for 10 years, as a Lecturer in Materials Science Engineering Department at CARE School of Engineering, and as a Teaching Assistant in the Department of Nanotechnology in Anna University-Tiruchirappalli. He has 31 publications to his credit, 3 books and 5 book chapters, and several international conference presentations. Dr. Narendhar had secured an Industry Project grant from Cameron Schlumberger in 2016 for developing a Nanocomposite paint. In 2022, he secured funding from the Department of Science and Technology in Category-1 of AMT Program of TDT Division of DST for developing a hemostatic nanofiber patch for rapid blood clotting. He has an International patent filed in Australia and serves as a reviewer in several scientific journals.

Ramachandran Balaji is currently working as an assistant professor at K. L. University, Vijayawada in the Department of Electronics and Communication Engineering. Dr. Balaji has a Post-Doctoral Experience in the Department of Chemical Engineering, National Taiwan University, Taiwan. He received his Ph. D. in Optoelectronics from National Taipei University of Technology. He completed his Bachelor degree (ECE) in 2014 and Master degree (Nanoscience and Technology) in 2016 from Sri Ramakrishna Engineering College, India. His research interests include nanomaterials engineering, electrochemical sensing devices, biosensors, semiconducting materials, and surface enhanced Raman spectroscopy. He has published 50+ SCI research articles in various fields of research, 5 book chapters and holds an international Australian Patent. He is a guest editor and reviewer for many nanotechnologies-related journals.

Nanoparticle-Based Drug Delivery System for Beginners

18

Timanshi Chansoriya, Barkha Khilwani, and Abdul Salam Ansari

Abstract

Nanoparticles are the simplest form of structures with sizes in the nanometer range. They possess unique physicochemical properties such as high surface area, nanoscale size, and optical characteristics. The nanoparticle synthesis and the study of their size and properties are important in medicine as well as in biotechnology applications. Nanoparticle-based delivery provides a new drug delivery method for the treatment of chronic human diseases by site-specific and target-oriented delivery of a variety of drugs. The synthesis, characteristics, and applications of different types of nanoparticles having potential in nano-drug delivery systems (NDDSs) are described. The article illustrates various properties of nanoparticles and then, based on composition, classifies these nanoparticles in multiple categories. The advantages of different types of nanoparticles are mentioned along with many applications with emphasis on drug delivery, and then we briefly describe herein the future of nanoparticles in targeted drug delivery.

Keywords

Targeted drug delivery · Liposomes · Drug carrier · Synthesis · Applications

18.1 Introduction

Nanotechnology is an emerging branch of science that created a bridge between biological and physical sciences through diverse synthetic strategies, particle structure, and size modification in different fields of science. This engineering technology deals with the preparation of nanoparticles which range from 1 to 100 nm in size and

T. Chansoriya · B. Khilwani · A. S. Ansari (✉)
Centre for Advanced Studies, Department of Zoology, University of Rajasthan, Jaipur, India

F. A. Sheikh et al. (eds.), *Interaction of Nanomaterials With Living Cells*,
https://doi.org/10.1007/978-981-99-2119-5_18

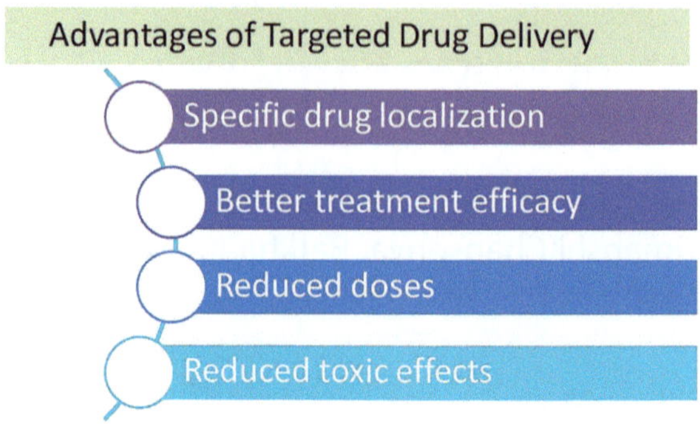

Fig. 18.1 Advantages of targeted drug delivery

primarily influences nanomedicine and nano-based drug delivery systems (Patra et al. 2018). The term nanoparticle is a combination of the words "nanos" (Greek: dwarf) and "particulum" (Latin: particle). Nanoparticles (NPs) are the particulate substances with a diameter of 100 nm (Laurent et al. 2010) with up to three-dimensional configurations depending on overall shape (Tiwari et al. 2012). By this definition, NPs have at least one nanometric dimension. For instance, 3D nanostructures and quantum dots are typically placed at the upper and lower ends of this scale, respectively. NPs have improved biological, physical, and chemical capabilities in comparison with their original equivalent materials (Dolez 2015).

Nano-drug delivery system (NDDS) is one of the promising applications of nanotechnology in medical sciences and human healthcare, which includes various classes of nanomaterials which can increase the water solubility or stability of the drug, prolong cycle time, the high uptake rate of target cells or tissues, reduction in enzyme degradation which also improves the effectiveness of many therapeutic drugs (Deng et al. 2020). In NDDS, different drug administration routes can be used, such as inhalation, oral administration, or intravenous injection. The early developed NPs were not able to cross the biological barriers to delivery, but in the present time substantial research is being directed toward the development of biodegradable polymeric nanoparticles, which incorporate complex architecture, bio-responsive moieties, and targeting agents to enhance the drug delivery system and tissue engineering. Hence, with this advancement, controlled release of drugs, stabilizing labile molecules (e.g., proteins, peptides, or DNA) from degradation, and site-specific drug targeting can be accomplished (Singh and Lillard 2009; Mitchell et al. 2021) (Fig. 18.1).

18.2 Properties of Nanoparticles

Nanoparticles are ultrafine units that are measured in nanometers (nm; 1 nm = 10^{-9} m) with submicroscopic size and unique material characteristics and thus can be classified into various types. According to the Commission of the European Union, 2011, nanoparticles can be defined as "a natural, incidental or manufactured material containing particles, in an unbound state or as an aggregate or as an agglomerate and where, for 50% or more of the particles in the number size distribution, one or more external dimensions is in size range 1–100 nm." NPs are of different shapes, sizes, and structures (www.britannica.com). The nanoparticles have three general physical properties which are highly mobile in the free states, have enormous specific surface areas, and may exhibit quantum effects.

A particle of 200 nm size or larger activates the vascular and lymphatic systems, which filter and clear out the foreign matter or chemicals and remove them from circulation quickly. Thus, the size of the ideal nanoparticles is approximately 100 nm. The particles of this size can cross the blood-brain barrier using endothelium-tight junction openings with the help of hyperosmotic mannitol, which provides target-specific controlled delivery of macro- and micromolecules of therapeutic agents which are used in the treatment of diseases like brain tumors and cancers (Prokop and Davidson 2008; Singh and Lillard 2009; Rizvi and Saleh 2018).

In drug distribution, due to hydrophobicity, surface, non-modified nanoparticles have opsonization properties, and then they also cleared by the mononuclear phagocyte system (MPS). Therefore, in the human body, deliverable drugs with their suitable nanoparticles are coated with polymers/surfactants or biodegradable copolymers which have hydrophilic characteristics, e.g., polyethylene glycol (PEG), polyethylene oxide, poloxamer, poloxamine, and polysorbate 80. This minimizes the opsonization and increases the circulation of nanoparticles (Singh and Lillard 2009; Chandrakala et al. 2022). Thus, the above description illustrates that in NDDS, drug release or delivery of any therapeutic agents into the human body depends on the size and surface area of particles. The controlled particle size and surface area properties contribute to faster polymer degradation as well as drug release.

The ideal NDDS should have a high drug-loading capacity and an efficient drug-release system. Drug loading can be accomplished by the incorporation method (incorporation of the drug at the time of nanoparticle formulation) and adsorption/absorption method (absorption of the drug after nanoparticle formation). Some studies have shown that the ionic interactions and isoelectric point (p*I*) can be very effective in increasing drug loading (Singh and Lillard 2009). The release of the drug depends on the type of nanoparticle used in the system; if nanocapsules were used, then the release is controlled by drug diffusion through the polymeric layer, and if nanospheres are used, then the drug is physically and uniformly dispersed by erosion of the matrix. The drug release from nanoparticle-based formulation is affected by many factors like pH, temperature, drug solubility, adsorbed drug, matrix swelling, and the combination of erosion and diffusion processes (Rizvi and Saleh 2018).

18.3 Classification of Nanoparticles

According to the composition of the materials, morphology, size, and physical and chemical properties, various nanomaterials can be categorized as carbon-based NPs, lipid-based NPs, metal, ceramics, and semiconductor. The following is a description of important nanomaterials and their features, which are compatible for use in NDDSs:

18.3.1 Carbon-Based NPs

Carbon nanotubes (CNTs) and fullerenes are two primary groups found in carbon-based NPs. Nanomaterials comprised of globular hollow cages, such as allotropic forms of carbon, are found in fullerenes. They have sparked significant commercial interest in nanocomposites for a variety of applications, including fillers (Saeed and Khan 2014, 2016), efficient gas adsorbents for environmental remediation (Ngoy et al. 2014), and support medium for various inorganic and organic catalysts (Mabena et al. 2011; Ngoy et al. 2014). The fullerene-based delivery system has the potential to carry multiple drug payloads with other chemotherapeutic drugs, e.g., Taxol® (Ashcroft et al. 2006) can load 40 fullerenes onto a single skin cancer antibody called ZME-108, which can be used to deliver drugs directly into melanomas. Thus, multiple drugs can be loaded into a single antibody in a spontaneous manner (Singh and Lillard 2009). Carbon-based NPs also have promising futures in different fields of nanotechnology due to their abundance in nature as various allotropes, minimal cellular toxicity, flexible engineering, and exceptional physical qualities at the atomic scale. The exceptional electrical conductivity, great anisotropic thermal conductivity, and mechanical durability of carbon-based NPs are frequently combined with polymer materials and nanocomposites (Parvej et al. 2022).

18.3.2 Liposomes and Micelles

Liposomes are made up of at least one phospholipid bilayer and an aqueous core, which can load hydrophobic drugs and carry the drug to target cell membranes. Liposome-based drug delivery system is FDA-approved and also called as "contact-facilitated drug delivery" (Nagalingam 2017; Li et al. 2019). Their morphological similarity to cellular membranes is one of the key characteristics that made them excellent as a nanocarrier for bioactive compounds or pharmaceutical agents (e.g., medicines, genes). Liposomes show many advantages as a type of drug carrier, such as they are nontoxic, non-immunogenic, sustained-release drugs, prolonging drug action time, etc. (Yingchoncharoen et al. 2016). Liposomes cannot develop for the entrapment for hydrophilic and ionic molecules (Chandrasekaran and King 2014), but by expanding their use, LNPs can form unilamellar or multilamellar vesicular structures, which allow liposomes to entrap, carry, and deliver hydrophilic,

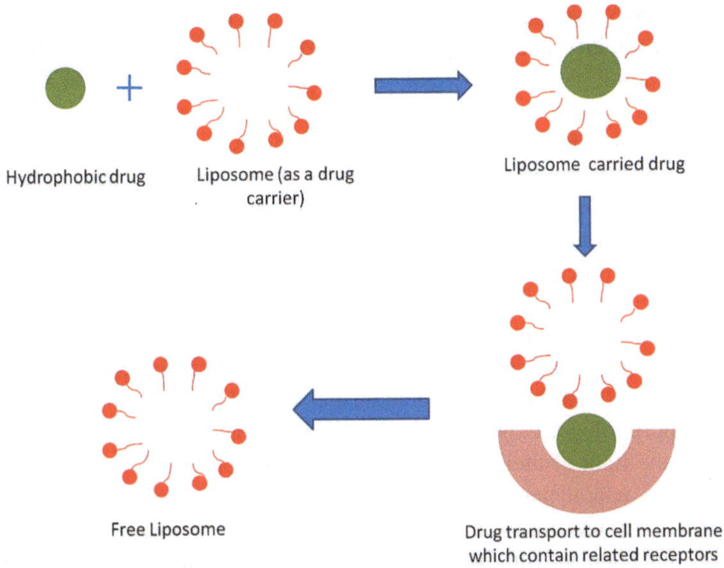

Fig. 18.2 Targeted drug delivery using liposomes as a drug carrier

hydrophobic, and lipophilic drugs (Sarfraz et al. 2018). Liposomes which deliver nucleic acids form micellar structures within the particle core. Micelles are spherical and amphiphilic copolymer assemblies that can accommodate hydrophobic drugs. Their outer shell is hydrophilic that makes the micelle water soluble. Some examples of micelle formulations are sterically stabilized micelles (SSM) which have been used as nanocarriers for CPT (CPT-SSM) for cancer treatment, SP1049C (doxorubicin-encapsulated pluronic micelles), NK911, and Genexol-PM (paclitaxel-encapsulated PEG-PLA micelle). Micelles have several advantages over other drug delivery systems but also have drawbacks (Fig. 18.2).

18.3.3 Metal NPs

Metal NPs are completely made up of metal precursors. These NPs have unique optoelectrical features due to their well-known localized surface plasmon resonance (LSPR) characteristics. In the visible zone of the electromagnetic spectrum, NPs of alkali and noble metals, such as Cu, Ag, and Au, have large absorption band. Metal NPs with regulated facet, size, and shape are important in today's cutting-edge materials (Dreaden et al. 2012). Metal NPs are used in a variety of scientific fields due to their excellent optical characteristics. Gold NP coating is commonly used for SEM sampling to improve the electronic stream, which aids in the acquisition of high-quality SEM images.

(a) **Copper NPs**: Copper NPs have received a lot of attention due to low cost, chemical stability, and simple preparation (Basher et al. 2019). They have a high melting point, excellent thermal and electrical conductivity, and high ductile strength and are strongly localized. They have been used as a coloring agent since antiquity. Even now, as pigment ingredients in inkjet printing, they are a feasible alternative to noble metal NPs (gold and silver). They are frequently employed in biological and pharmaceutical applications due to their catalytic and antibacterial properties. However, there have been reports of potential negative biological effects of copper NPs on embryogenesis (Khan et al. 2022). Copper oxide (CuO) NPs are present in spherical shape with a diameter of 1–30 nm with large specific surface area. The copper atom is linked to oxygen in a rectangular shape in the monoclinic crystal structure. Saha et al. (2018) demonstrated that the other characteristics of CuO NPs are heavily influenced by its morphology. Grigore et al. (2016) highlighted the major features of CuO NPs in the light of their synthesis process and biomedical uses. The CuO NPs with a few nanometers in diameter have been reported to have weak ferromagnetic activity (Zhang et al. 2014), but Joy et al. (1998) and Bisht et al. (2010) found that standard zero field cooled (ZFC) magnetization was not present. The antibacterial characteristics of CuO NPs are used to prevent bacterial infection in the textile industry and hospitals. Verma and Kumar (2019) demonstrated biomedical applications of CuO NPs due to its sensing and therapeutic properties.

(b) **Aluminum NPs**: Aluminum NPS are used as powder in rocket fuel to boost combustion speed and stability due to their catalytic activity and high energy release. Because they have a wide optical absorption band, the LSP resonances can be modulated from UV to NIR by manipulating their shape. They are also suited for application in photovoltaic solar cells due to their high radiative efficiency (Temple and Bagnall 2011). Aluminum NPs are known to cause cellular toxicity and DNA damage (Zhang et al. 2018).

(c) **Gold NPs**: Gold NPs are the oldest and most widely used metal NPs. Due to their better physical properties, nanoscaled gold particles have been widely used. More crucially, by manipulating their morphology (both size and shape), solvent, surface ligand, temperature, etc., these properties can be fine-tuned. Localized SPR is more pronounced in spherical gold NPs than in other plasmonic particles, resulting in significant radiative, absorption, and scattering characteristics. Gold NPs have a fluorescence quenching ability and an absorption peak at 400–550 nm, depending on particle size, making them attractive in bioimaging, probing, colorimetric sensing, and sensor construction (Yeh et al. 2012). They are used to sputter coat the material in a scanning electron microscope (SEM) in order to obtain a high-quality image by increasing the electronic stream. Furthermore, gold particles are a common vehicle for carrying therapeutic compounds, targeted medications, genes, and targeting agents on their surface due to their huge specific surface area and high electron density nanoscale.

(d) **Silver NPs**: Silver NPs have a wide range of use in biomedical devices, medication, highly conductive composites, and the textile industry due to their unique physical properties. Silver NPs have excellent SPR, strong absorption, and NP characteristic packing near 400 nm and tunable scattering capabilities at longer wavelength, making them ideal for bioimaging, molecular labeling, and improved optical spectroscopy. As reported by Zhang et al. (2016), nanoscaled silver has long been regarded as a popular biomaterial with antibacterial action. Their antimicrobial properties are commonly employed to reduce biofouling. They have also shown promise against the HIV virus and cancer cell death. Furthermore, their anti-inflammatory properties make them ideal for wound healing. The toxicity of silver NPs is mostly determined by their size. When compared to particles made from other heavy metals like gold, platinum, and zinc, silver NPs have demonstrated a high level of antibacterial activity and a low level of cytotoxicity (Crisan et al. 2021). They have the ability to adhere to cells, inhibit enzyme activity, weaken the cell membrane, and ultimately cause cell death (Tang and Zheng 2018).

(e) **Iron NPs**: Iron NPs have excellent thermal and electrical conductivities with strongest magnetic properties of any magnetic NP (Rubel and Hossain 2022). Iron NPs exhibit surface plasmon resonance (SPR) which is important in memory tape, magnetic data storage, and magnetic resonance imaging (MRI). Iron NPs with a diameter of about 2 nm have magnetic characteristics, and the magnetic anisotropy energy constant increases as the particle size increased (Bedker et al. 1994). Magnetic NPs of iron oxide of less than 10 nm in size exhibit super paramagnetic properties that play crucial role for a variety of biomedical applications. The suitability of iron oxide NPs as a contrast agent for magnetic resonance imaging (MRI) and as a nanocarrier for bio-elements such as drugs, proteins, and therapeutic genes has been investigated. However, poisoning is frequently one of the drawbacks that make large magnetic components unsuitable for medicinal applications of these important NPs.

(f) **Platinum NPs**: Platinum NPs with exceptional chemical and optical characteristics are gaining popularity in industrial and biological applications. Platinum NPs when suspended in aqueous solution create a brownish-red or black nanofluid. They have a high level of thermochemical stability, corrosion resistance, and catalytic activity. They can be used in catalytic converters, hydrogen peroxide (H_2O_2) breakdown, nitric acid synthesis, proton-exchange membrane fuel cells (Reddington et al. 1998), pollution reduction, etc. At ambient temperature, carbon-coated platinum NPs also demonstrate ferromagnetic properties. They are frequently utilized as dopants with other metallic particles to make ultraefficient alloys.

(g) **Lead NPs**: Lead NPs are black spherical powder, apparently prone to oxidation and susceptible to water and humid ammonia (Bochenkov et al. 2004). They have potential applications in electron microscopy for real-time imaging due to their optical and redox characteristics.

(h) **Cobalt NPs**: The appearance of pure cobalt NPs are in the form of gray or black granules. The high magnetic characteristic of this NP is most suitable for imaging, sensing, and targeted delivery of biological molecules and medicines. The detailed toxicity of cobalt NPs has been studied on osteoclast-like cells (Liu et al. 2015c).

18.3.4 Ceramics NPs

Ceramic NPs are inorganic nonmetallic solids and found in amorphous, polycrystalline, thick, porous, or hollow forms (Sigmund et al. 2006). These NPs are used in various applications like catalysis, photocatalysis, dye, photodegradation, and imaging (Thomas et al. 2015). Ceramic NPs possess superior mechanical strength, thermochemical stability, and environmental resistance. One significant drawback of ceramic NPs is the potential for toxicity in medicinal applications such as drug administration.

18.3.5 Semiconductor, Inorganic and Nanoshell NPs

Semiconductor materials are intermediate between metals and nonmetals (Ali et al. 2017; Khan et al. 2017). Semiconductor NPs contain broad bandgaps, and bandgap tuning causes considerable changes in their characteristics. Therefore, they are used in photocatalysis, photo optics, and electronic devices (Sun et al. 2000). Due to their optimal bandgap and band edge positions, large numbers of semiconductor NPs have been discovered in water splitting applications (Hisatomi et al. 2014). When compared to single semiconductor particles, nanoshells have a higher luminescence quantum yield. Changing the shell material and thickness, as well as the core form, improves the tunability of other optical parameters, including absorbance and scattering (Nayak et al. 2017).

(a) **Germanium NPs:** Germanium (Ge) is the second most extensively used indirect bandgap (0.66 eV at bulk scale) semiconductor in group IV. Ge is usually found in the crystal structure of diamonds, which can vary in a cluster of more than 40 atoms. Ge NPs are a grayish black powder with an average diameter of 70–120 nm. The mechanical stability of Ge clusters has been determined by the crystal structure (Pizzagalli et al. 2001). Ge has greater static dielectric constant and lower effective mass of the electron-hole pair. In addition, electrochemical etching of Ge has not been as successful. By utilizing differential surface tension and size purification, the emission spectra of Ge NPs may be fine-tuned to narrow lines. The Ge NPs have wide range of uses in microelectronics.

(b) **Magnesium oxide NPs**: Magnesium oxide (MgO) are normally white powder as NPs, but depending on the presence of foreign elements, they might be brown or black. The size of MgO NPs influences the optical characteristics. Stankic et al. (2005) used UV diffuse reflectance spectroscopy to study the optical characteristics of MgO nanocubes. The MgO NPs are effective while absorbing harmful ions from aqueous solutions (Hoque et al. 2018). Furthermore, by employing them as a chemical additive, their catalytic action can be exploited. They also have high-temperature dehydrating capabilities, reduce corrosion, and cleanse water by reducing bacterial development. At low concentrations, MgO has a strong antibacterial action at the nanoscale, making it a potential plant pathogenic antibacterial agent for disease management (Cai et al. 2018). Ceramics undergo grain development and a considerable increase in fracture toughness when treated with MgO NPs (Tan et al. 2013).

(c) **Gallium nitride NPs**: Gallium nitride (GaN) is a semiconductor material of the group III–V family with a 3.4 eV direct bandgap. It has hexagonal (wurtzite) single crystal structure and can be manufactured at the nanoscale in a variety of morphological assemblages (NPs, nanorods, nanotubes, nanowires, and so on) using various synthesis procedures. Due to quantum confinement, GaN NPs have a high mechanical toughness and outstanding thermal and optical characteristics that depend on nanocrystal size. GaN NPs are utilized in a variety of devices, including LEDs, LDs (laser diodes), biosensors, solar cells, field-effect transistors, photocatalysts for water splitting, and piezoelectric nanogenerators, due to tunable optical and dielectric properties (Lan et al. 2016).

(d) **Indium NPs**: The hexagonal (wurtzite) and cubic (zinc blende) crystal forms of indium phosphide (InP) have bulk bandgaps of 1.42 and 1.35 eV, respectively. It has better electron mobility than GaAs, making it a good contender for high-speed optoelectronic devices and digital circuits (Zafar and Iqbal 2016). Furthermore, InP quantum dots are potential to be future competitor of cadmium-based quantum dots in terms of luminescence efficiency due to their decreased toxicity (Brichkin 2015). With a direct bandgap of around 0.354 eV, indium arsenide (InAs) is one of the least extensively utilized group III–V semiconductor compounds. Gray crystals having cubic (zinc blende) structure can be found. As InAs have properties similar to GaAs, they are frequently combined with InP to get the most out of their small bandgap and strong electron mobility. InAs photodiodes are frequently used in infrared detectors and diode lasers. Microorganisms have been reported to be acutely hazardous to both InAs and GaAs (Nguyen et al. 2020).

(e) **Silicon NPs**: Silicon NPs (SiNPs) are biologically compatible, metal-free quantum dots that exhibit size and surface tailorable photoluminescence. Silica NPs are mesopores (2–50 nm pores) of silica that display unique physicochemical properties. These nanocarriers can be prepared in a variety of sizes and shapes including nanohelices, nanotubes, nanozigzags, and nanoribbons (Meier et al. 2007). The nanostructure of these materials influences their optical, chemical, and material properties and hence plays an important role in their future-generation applications in sensors, battery electrodes, optical materials, contrast agents, etc. Nanosilica or silica NPs are commercial terms for nanoscaled silicon dioxide (SiO_2) particles. There are two varieties of silica NPs based on structure: P-type and S-type NPs, both of which are white powder. The P-type particles have a higher specific surface area and porosity than S-type particles. As nanopowders, they have minimal toxicity, pozzolanic reactivity, and filling ability (Challa and Das 2019). Because P-type silica NPs have a high UV reflectivity, they can be used as a protective covering.

(f) **Titanium nitride NPs**: Mechanically, titanium nitride (TiN) NPs are extremely strong. Their hardness and wear resistance are exceptional, allowing them to be used with other ceramics in cutting equipment and bearing materials to extend their life. The transmission electron microscopic observation of TiN NPs indicates that they are virtually spherical, with sizes of 5–20 nm. Because of their high thermal conductivity and melting point (2950 °C), they can endure high temperatures. TiN has a low sintering temperature, making it ideal for embedding in nanocomposites. They have outstanding UV protection and infrared absorption. The scattering efficiency is not as great as gold NPs, and the plasmonic performance of single cubic crystal-structured TiN is nearly as good (Kaskel et al. 2003; George et al. 2009).

(g) **Alumina NPs**: Alumina NPs are white powder with a spherical shape. At a certain particle size, structures of alumina NPs are temperature dependent, and this dependence is altered by particle shrinking. At the nanoscale, the most stable-phased alumina, for instance, stabilizes at a lower temperature. This considerably improves the nanoceramic's flexural strength (Zemtsova et al. 2015). Superior qualities of alumina NPs are used in cutting tools, integrated circuits, and transparent ceramics.

(h) **Titania NPs**: Titanium dioxide (TiO_2) NPs are n-type semiconductors that occur naturally in a variety of polymorphic crystal forms. They can be made in a variety of ways, i.e., crystal, powder, nanotube or nanorods, and thin films. Titania NPs are highly effective in blocking UV rays while remaining harmless. Furthermore, their transparency makes them useful for making skin-protective cosmetics such as sunscreens, vanishing creams, and beauty creams. They can also be used to process ink and as a surface coating. They also have strong photocatalytic properties (Hossain et al. 2017, 2018a, b), making them useful for the production of disinfectants and antibacterial chemicals.

(i) **Calcium NPs**: Calcium phosphate (CaP) NPs are present in a variety of shapes and sizes. Tricalcium phosphate ($Ca_3(PO_4)_2$) and hydroxylapatite ($Ca_5(PO_4)_3$) are the most prevalent types. The ratio of Ca to P has a big impact on the properties of these NPs. CaP is a common bone substitute due to its biocompatibility and likeness to the inorganic mineral compositional constituent in the skeleton of human. As a widely utilized nonviral vector in gene therapy, nanoscale CaP can thus play an important role. Because calcium carbonate ($CaCO_3$) is abundant in nature, it is one of the cheapest inorganic materials available. Slow biodegradability is an additional benefit. Calcium carbonate exists in three polymorphic crystal forms—calcite (trigonal), aragonite (orthorhombic), and vaterite—depending on the synthesis conditions (hexagonal). Calcite is the most chemically stable of these minerals, whereas vaterite is the least stable (Biradar et al. 2011). Chemically, nanosized $CaCO_3$ particles are harmless, making them environmentally beneficial. They can also be employed for regulated and harmless drug distribution because of their biocompatibility.

According to recent pharmaceutics research, building an enteric drug delivery method using calcium carbonate in a tablet-encapsulated form is possible (Render et al. 2016). Furthermore, they have a good radiopacity characteristic.

18.4 Synthesis of NPs

The following two methods are employed for the synthesis of NPs: (1) the bottom-up or building-up method and (2) the top-down method (Daniel and Astruc 2004; Wang and Xia 2004), which are further divided into several subcategories based on the operation, reaction state, and procedures used.

18.4.1 Bottom-Up or Building-Up Synthesis

In bottom-up or building-up synthesis procedure, NPs are produced from relatively simple elements. Two methods, viz., sedimentation and reduction procedures, are included in this category. Other minor methods like spinning, green synthesis, sol-gel, and biological synthesis are also included (Iravani 2011). Using this method, TiO_2 anatase NPs containing graphene domains have been created (Mogilevsky et al. 2014). Needham et al. (2016) have used a solvent-exchange approach to create low-density lipoprotein (LDL) NPs of limit size for medical cancer medication. Nucleation is the bottom route in this procedure, followed by growth, which is the top approach. The LDL nanoparticles were made without using phospholipids having high hydrophobicity which is important for the administration of drugs (Needham et al. 2016).

18.4.2 Top-Down Synthesis

For syntheses of NPs, the destructive technique is used in top-down procedure. For the synthesis, first, the large-sized molecules are broken into smaller units, and then appropriate sized NPs are synthesized. In this direction, grinding and milling, chemical vapor deposition (CVD), physical vapor deposition (PVD), and other decomposition process technologies are generally employed (Iravani 2011). Coconut shell (CS) NPs were produced using this method. Initially the raw CS powders were finely milled for various intervals of time using ceramic balls and planetary mill as per the requirement of the size of the NPs. The reddish tint vanished with each hour increment as the NPs shrank in size (Bello et al. 2015). From the top-down methods, different approaches like mechanical milling, laser ablation, etching, sputtering, and electro-explosion are used for the production of NPs for nanostructured materials. By using a top-down laser irradiation approach, monocrystalline, well-uniform, and spherical-shaped Au nanospheres have been created (Liu et al. 2015a, b).

Both top-down and bottom-up techniques were used to prepare monodispersed spherical bismuth (Bi) NPs (Wang and Xia 2004). The colloidal characteristics of these NPs are found to be excellent. In bottom-up approach, bismuth acetate was boiled in ethylene glycol, but in the top-down approach, bismuth was converted to molten form and then emulsified within boiled diethylene glycol to synthesize NPs. The size of the NPs produced by both procedures ranged from 100 to 500 nm (Wang and Xia 2004). Because of the feasibility and less hazardous nature of processes, green and biogenic bottom-up syntheses are widely used. These techniques are both cost-effective and environmentally beneficial as they normally use plant extracts from biological systems to synthesize NPs. To prepare NPs, bacteria, yeast, fungi, *Aloe vera*, tamarind, and even human cells have been employed. The microorganisms and plant extracts were used as reducing agents to synthesize Au NPs from wheat and oat biomass (Ahmed and Ikram 2016; Parveen et al. 2016).

18.5 Applications of NPs

NPs possess a variety of potential applications, namely, in preparation of food, environmental cleanup, preclinical medicine, clinical medicine, physics, optics, and electronics. According to reports, 90% of drugs are insoluble in water and thus are not able to reach their targets (Carissimi et al. 2021). Various studies have been performed to improve drug delivery and higher efficacy of targeted drug delivery. In recent times, the development of nanoparticle-based delivery systems has instigated site-specific and target-oriented delivery of a variety of drugs or therapeutic agents or natural-based active compounds for the treatment of various chronic human diseases. In NDDSs, targeted drug delivery using nanomaterials or nanoformulations can be done actively or passively. In active targeting, therapeutic agents such as antibodies or peptides conjugated to a tissue- or cell-specific ligand, which anchors them to reach at the receptor site of the targeted cell. In passive targeting, conjugated therapeutic drug and nanocarrier reach the target organ through bloodstream, and this type of target method is influenced by pH, temperature, molecular site, and shape (Singh and Lillard 2009; Patra et al. 2018). For NDD, the main target sites in the body are lipid components or receptors of cell membrane and cell surface proteins or antigens. NDDSs show their most applications are in the delivery of chemotherapeutic agents, immunotherapeutics, anti-inflammatories, antibiotics, anesthesia, hormones, etc. The most commonly used targeting agents for drug delivery are liposomes, peptides, antibodies, designed proteins, nucleic acid aptamers, and other small organic molecules (Rizvi and Saleh 2018; Carissimi et al. 2021). For the development of immunotherapeutics, T cells are the direct targets in the pathology and in giving immunotherapies in diseases like T cell lymphocytic leukemia, T cell lymphoma, and human immunodeficiency virus (HIV) infection (Cevaal et al. 2021) (Fig. 18.3).

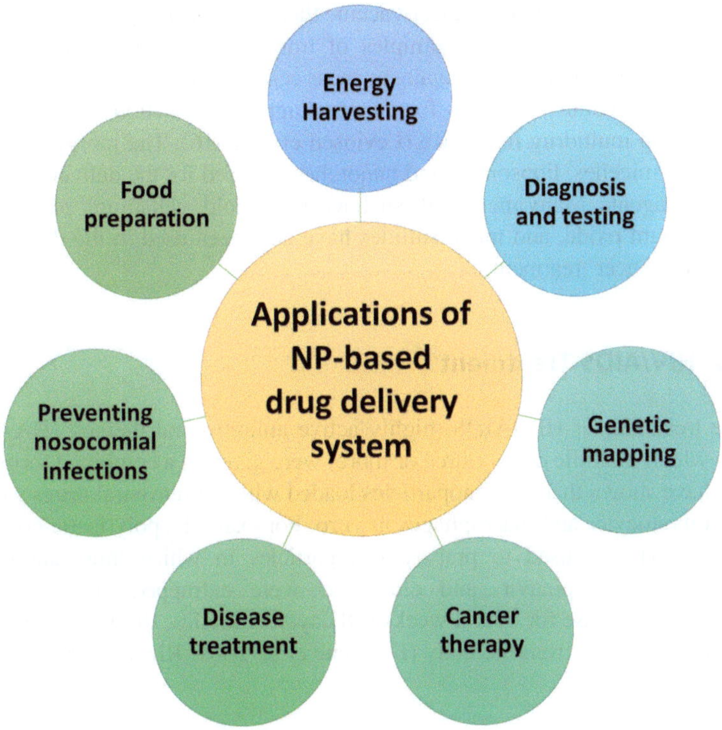

Fig. 18.3 Application of nanoparticle-based drug delivery system

18.5.1 Cancer Therapy

The use of nanotechnology in cancer detection, therapy, and management has opened a new area of research. NPs increase the intracellular concentration of medications while avoiding toxicity in healthy tissue, through either active or passive targeting. To establish and regulate drug release, targeted NPs can be developed and adjusted to be pH-sensitive or temperature-sensitive. Within the acidic TME, the pH-sensitive drug delivery system can distribute pharmaceuticals. Temperature-sensitive NPs release medications into the target site in response to temperature changes brought on by sources such as magnetic fields and ultrasonic waves. Furthermore, NPs' physicochemical properties, such as shape, size, molecular mass, and surface chemistry, play an important role in the process of immunotherapy using NPs. The immune system plays a crucial role in the establishment and growth of cancer cells. The development of immunotherapy has transformed cancer treatment. NPs have been discovered to aid in the delivery of chemotherapy to specific targets and can also be utilized in conjunction with immunotherapy. Immune checkpoint blockade therapy, cancer vaccine therapy, and chimeric antigen receptor (CAR)-T cell therapy are all examples of immunotherapy techniques aimed at stimulating the immune system against cancer cells. Moreover, the nanotechnology has given tremendous outcomes for cancer diagnosis, detection, and therapy and circumventing multidrug resistance (Levinsen et al. 2016). The inorganic NPs like dendrimers, micelles, liposomes, and nanotubes are used for the delivery of chemotherapeutic agents. Many metal NPs such as silver, gold, palladium, platinum, zinc oxide, titanium oxide, and metal sulfides have also been used in the drug delivery systems for cancer treatment.

18.5.2 HIV/AIDS Treatment

For the treatment of HIV/AIDS, highly active antiretroviral therapy (HAART) is used in which multiple drugs (three or more) were given in a combined form. Many studies have shown that the nanoparticles loaded with antiretroviral drugs were able to target monocytes and macrophages in vitro. For example, poly(lactic-co-glycolic acid) (PLGA) was used to prepare nanoparticles in which three antiretroviral drugs—ritonavir, lopinavir, and efavirenz—were entrapped, and this NDDS sustained drug release for over 4 weeks (28 days), while free drugs were eliminated within 48 h (2 days) from the body (Destache et al. 2009; Rizvi and Saleh 2018).

18.5.3 Diagnosis and Testing

The use of nanoparticles for diagnostic purposes is a highly explored area of NDDSs. Nanoparticles help in diagnosis and identification of the stage of diseases which can report the location and provide information regarding treatment responses (Patra et al. 2018; Rizvi and Saleh 2018). The development of quantum dots allows monitoring of various biological events simultaneously by tagging which can be defined by many customized specific colors. Their absorption spectrum ranges from UV to a wavelength of a visible spectrum which provides photostability, high quantum yield, and tunable emission spectrum. Size of the nanodot specifies the spectrum where individual particle falls, e.g., larger particles have longer wavelengths and narrow emission (Rizvi and Saleh 2018). According to studies, theranostic nanoparticles such as surfactant aggregates (micelles and vesicles), dendrimers, drug conjugates, core-shell particles, and carbon nanotubes are used for monitoring of pathway and localization of nanoparticles at the site of drug target and also monitor action of drug to assess therapeutic response by combining both drug and imaging agent (Bhojani et al. 2010; Janib et al. 2010).

18.6 Other Applications of NPs

18.6.1 Nosocomial Infections

Hospital acquired infections (nosocomial infections) are the greatest cause of death (Wenzel 2007). The 60–70% of nosocomial infections are linked to bacterial contamination of medical devices that have been implanted (Donlan 2001; Bryers 2008). The antimicrobial properties of a large number of synthesized NPs such as silica/iron oxide NPs, graphene, graphene oxide, bifunctional Fe_3O_4-Ag NPs, titanium, copper, zinc, silver, and gold have been investigated (Kang et al. 2008; Rodrigues and Elimelech 2010; Narayanan and Sakthivel 2011; Santos et al. 2012; Mejias Carpio et al. 2014; Musico et al. 2014;Rodrigues et al. 2015).

18.6.2 Preparation of Food

Nanotechnology-based applications are used to improve the procurement of raw materials, sorting and grading, primary processing, packing, transportation and storage, and food processing. Enhancing palatability, toxin elimination, enzyme deactivation, spoilage organisms, pathogens, and additional fortification and enrichment with micronutrients are the key deliverables of food processing in which nanotechnology based techniques are invariably used. Nanostructured food ingredients are being produced with the promise of better taste, texture, and consistency. Nanotechnology is also used to extend the shelf life of various food ingredients and reduce the amount of food waste caused by microbial infestation (Pradhan et al. 2015).

18.6.3 Solar Power

Nanotechnology-enhanced prototype solar panels convert sunlight to electricity more efficiently than normal designs, paving the way for less expensive solar power. Because they can be produced in flexible rolls rather than isolated panels, nanostructured solar cells are already cheaper to make and install. Through improved catalysis, nanotechnology is increasing the efficiency of fuel generation from regular and low-grade raw petroleum materials as well as fuel consumption (Hussein 2015).

18.6.4 Cleanup of the Environment

Nanotechnology is helpful in identification and cleaning up of environmental toxins. Through quick, low-cost detection and treatment of contaminants in water, nanotechnology could assist in addressing the demand for affordable, clean drinking water. For energy-efficient desalination, a thin film membrane incorporating nanopores has been developed. The molybdenum disulfide (MoS_2) membrane filtered two to five times the amount of water as contemporary filters. NPs are being created to remove industrial water contaminants from groundwater by chemical processes that render the pollutants harmless. This method would be less expensive than systems that require the water to be pumped out of the ground for treatment. Current cleanup technology is not significantly and economically adequate to solve all of today's cleanup needs. Nanotechnology is one of the most important trends in science and perceived as one of the key technologies of the present century (Zhang and Elliot 2006). Nanoscale iron particles are very effective for the transformation and detoxification of a wide variety of common environmental contaminants, such as chlorinated organic solvents, organochlorine pesticides, and PCBs (Rickerby and Morrison 2007).

18.6.5 Energy Harvesting

The NPs have large surface area and optical characteristics and they are catalytic in nature. NPs were frequently used to generate energy from photoelectrochemical (PEC) and electrochemical water splitting (Mueller and Nowack 2008; Avasare et al. 2015; Ning et al. 2016). There are many advanced choices for generating energy such as solar cells and piezoelectric generators (Fang et al. 2013; Lei et al. 2015; Gawande et al. 2016; Li et al. 2016). Nanotechnology can be used for affordable and safe drinking water through filtration and purification system (Mishra et al. 2012; Rabbani et al. 2016; Mobasser and Firoozi 2017).

18.6.6 Agriculture

Nanotechnology has been used to modify the genetic architecture of crop plants (Prasad et al. 2017). For targeted or controlled release of agrochemicals, nano-coated fertilizers, nano-sized nutrients, carbon-based nanomaterials, or engineered metal oxide and nano-pesticides are used, and they also have full biological effect without overdosing (Iavicoli et al. 2017).

18.6.7 Improving Life Standards with Nanoelectronics

The nanotechnology has a lot of promise for improving the capabilities of electronic components, particularly in terms of reducing their size, weight, and power consumption. Indeed, because electronic components are typically small and light, shrinking these dimensions to the nanoscale level allows for the creation of electronic devices with far greater capabilities, as it allows for the incorporation of far more components while also reducing the device's size and weight. The advances in nanotechnology may enable the development of new types of electronic components that can be employed in both traditional and modern electronic devices. Researchers are working on a memory chip that could have a memory density of one terabyte per square inch or higher (Bhatia et al. 2013).

18.7 Nanotechnology in Future

Nanotechnology has the potential to generate multifunctional materials in the construction and maintenance of safer, smarter, lighter, and more efficient vehicles like spacecraft, airplanes, and ships. Nanotechnology also provides a number of options for nanoengineered materials in automotive products such as structural parts made up of polymer nanocomposites, high-power rechargeable battery systems, thermoelectric materials which can be used for temperature control, lower rolling-resistance tires, high efficiency and low-cost sensors or electronics, smart solar panels having thin films, fuel additives, and improved catalytic converters for cleaner exhaust and extended range.

The nanoengineering of steel, aluminum, asphalt, or concrete and other cementitious materials with their recycled forms has a lot of potential as they can enhance the performance, resiliency, and life span of roadways and infrastructure of transport components while lowering their costs. Nanotechnology proves its great usage in either chemical or physical modification of individual atoms and molecules at a specific location. This new age technology also makes possible to develop devices which can scan and manipulate objects at near atomic scale (Kubik et al. 2005).

18.8 Conclusion

Nanotechnology established itself as an advanced field of science where intense research is carried out to implement the technology. The size of NPs ranges from a few nanometers to 500 nanometers. Based on material used in synthesis, NPs are classified as carbon-based, metal-based, ceramic, semiconductor, nanoshell, etc. NPs can also be categorized into organic and inorganic NPs. Morphology, structure, particle size, surface area, and optical properties are used in characterization of different NPs. Each category of NP has a significant application based on its properties. Despite the fact that NPs are valuable for a variety of applications, their unpredictable use and discharge pose significant health risks. Due to recent advances in nanotechnology, poorly soluble, poorly absorbed and labile biologically active compounds have been re-modified into viable, delivery able pharmaceuticals. In recent years, toxicity profiling of NPs has become a popular study topic all around the world. Natural NPs have been around for a long time in the ecosystem, and they contain some processes that make them less toxic to living things. In past few years, NPs are tested for many new and different applications which enhance the efficiency and performance of the object or process, and subsequently cost is reduced which makes NP-based nanotechnology accessible to everyone. Thus, nanotechnology has a great future because of its efficiency and environment-friendly properties.

References

Ahmed S, Ikram S (2016) Biosynthesis of gold nanoparticles: a green approach. J Photochem Photobiol B Biol 161:141–153

Ali S, Khan I, Khan SA, Sohail M, Ahmed R, Rehman AU, Ansari MS, Morsy MA (2017) Electrocatalytic performance of Ni@ Pt core–shell nanoparticles supported on carbon nanotubes for methanol oxidation reaction. J Electroanal Chem 795:17–25

Ashcroft JM, Tsyboulski DA, Hartman KB, Zakharian TY, Marks JW, Weisman RB, Rosenblum MG, Wilson LJ (2006) Fullerene (C60) immunoconjugates: interaction of water-soluble C60 derivatives with the murine anti-gp240 melanoma antibody. Chem Commun 26:3004–3006

Avasare V, Zhang Z, Avasare D, Khan I, Qurashi A (2015) Room-temperature synthesis of TiO_2 nanospheres and their solar driven photoelectrochemical hydrogen production. Int J Energy Res 39(12):1714–1719

Basher MK, Mishan R, Biswas S, Khalid Hossain M, Akand MAR, Matin MA (2019) Study and analysis the Cu nanoparticle assisted texturization forming low reflective silicon surface for solar cell application. AIP Adv 9(7):075118

Bedker F, Morup S, Linderoth S (1994) Surface effects in metallic iron nanoparticles. Phys Rev Lett 72(2):282–285

Bello SA, Agunsoye JO, Hassan SB (2015) Synthesis of coconut shell nanoparticles via a top down approach: assessment of milling duration on the particle sizes and morphologies of coconut shell nanoparticles. Mater Lett 159:514–549

Bhatia S, Raman A, Lal N (2013) The shift from microelectronics to nanoelectronics: a review. Int J Adv Res Comput Commun Eng 2(11):4464–4468

Bhojani MS, Van Dort M, Rehemtulla A, Ross BD (2010) Targeted imaging and therapy of brain cancer using theranostic nanoparticles. Mol Pharm 7(6):1921–1929

Biradar S, Ravichandran P, Gopikrishnan R, Goornavar V, Hall JC, Ramesh V, Baluchamy S, Jeffers RB, Ramesh GT (2011) Calcium carbonate nanoparticles: synthesis, characterization and biocompatibility. J Nanosci Nanotechnol 11(8):6868–6874

Bisht V, Rajeev KP, Banerjee S (2010) Anomalous magnetic behavior of CuO nanoparticles. Solid State Commun 150(17–18):884–887

Bochenkov VE, Zagorsky VV, Sergeev GB (2004) Chemi-resistive properties of lead nanoparticles, covered by oxide and sulfide layer. Sensors Actuators B Chem 103(1–2): 375–379

Brichkin SB (2015) Synthesis and properties of colloidal indium phosphide quantum dots. Colloid J 77(4):393–403

Bryers JD (2008) Medical biofilms. Biotechnol Bioeng 100:1–18

Cai L, Chen J, Liu Z, Wang H, Yang H, Ding W (2018) Magnesium oxide nanoparticles: effective agricultural antibacterial agent against *Ralstonia solanacearum*. Front Microbiol 9:1–19

Carissimi G, Montalban MG, Fuster MG, Víllora G (2021) Nanoparticles as drug delivery systems. In: Pham PV (ed) 21st Century nanostructured materials—physics, chemistry, classification, and emerging applications in industry, biomedicine, and agriculture. Intech Open, London, pp 227–250

Cevaal PM, Ali A, Czuba-Wojnilowicz E, Symons J, Lewin SR, Cortez-Jugo C, Caruso F (2021) *In vivo* T cell-targeting nanoparticle drug delivery systems: considerations for rational design. ACS Nano 15(3):3736–3753

Challa PR, Das BB (2019) Methods to monitor resources and logistic planning at project sites. In: Das B, Neithalath N (eds) Sustainable construction and building materials. Springer, Singapore, pp 793–802

Chandrakala V, Aruna V, Angajala G (2022) Review on metal nanoparticles as nanocarriers: current challenges and perspectives in drug delivery systems. Emerg Mater 4:1–23

Chandrasekaran S, King MR (2014) Microenvironment of tumor-draining lymph nodes: opportunities for liposome-based targeted therapy. Int J Mol Sci 15(11):20209–20239

Crisan CM, Mocan T, Manolea M, Lasca LI, Tabaran FA, Mocan L (2021) Review on silver nanoparticles as a novel class of antibacterial solutions. Appl Sci 11(3):1120

Daniel MC, Astruc D (2004) Gold nanoparticles: assembly, supramolecular chemistry, quantum-size-related properties, and applications toward biology, catalysis, and nanotechnology. Chem Rev 104(1):293–346

Deng Y, Zhang X, Shen H, He Q, Wu Z, Liao W, Yuan M (2020) Application of the nano-drug delivery system in treatment of cardiovascular diseases. Front Bioeng Biotechnol 7(489):1–18

Destache CJ, Belgum T, Christensen K, Shibata A, Sharma A, Dash A (2009) Combination antiretroviral drugs in PLGA nanoparticle for HIV-1. BMC Infect Dis 9(198):1–8

Dolez PI (2015) Nanomaterials definitions, classifications and applications. In: Dolez PI (ed) Nanoengineering: global approaches to health and safety issues. Elsevier, Amsterdam, pp 3–40

Donlan RM (2001) Biofilms and device-associated infections. Emerg Infect Dis 7(2):277–281

Dreaden EC, Alkilany AM, Huang X, Murphy CJ, El-Sayed MA (2012) The golden age: gold nanoparticles for biomedicine. Chem Soc Rev 41(7):2740–2779

Fang XQ, Liu JX, Gupta V (2013) Fundamental formulations and recent achievements in piezo-electric nano-structures: a review. Nanoscale 5(5):1716–1726

Gawande MB, Goswami A, Felpin FX, Asefa T, Huang X, Silva R, Zou X, Zboril R, Varma RS (2016) Cu and Cu-based nanoparticles: synthesis and applications in catalysis. Chem Rev 116(6):3722–3811

George PP, Gedanken A, Makhlouf SBD (2009) Synthesis and characterization of titanium nitride, niobium nitride, and tantalum nitride nanocrystals via the RAPET (reaction under autogenic pressure at elevated temperature) technique. J Nanopart Res 11:995–1003

Grigore ME, Biscu ER, Holban AM, Gestal MC, Grumezescu AM (2016) Methods of synthesis, properties and biomedical applications of Cuo nanoparticles. Pharmaceuticals 9(4):1–14

Hisatomi T, Kubota J, Domen K (2014) Recent advances in semiconductors for photocatalytic and photoelectrochemical water splitting. Chem Soc Rev 43(22):7520–7535

Hoque MA, Ahmed MR, Rahman GT, Rahman MT, Islam MA, Khan MA, Hossain MK (2018) Fabrication and comparative study of magnetic Fe and α-Fe_2O_3 nanoparticles dispersed hybrid polymer (PVA+ Chitosan) novel nanocomposite film. Results Phys 10:434–443

Hossain MK, Pervez MF, Mia MN, Mortuza AA, Rahaman MS, Karim MR, Islam JM, Ahmed F, Khan MA (2017) Effect of dye extracting solvents and sensitization time on photovoltaic performance of natural dye sensitized solar cells. Results Phys 7:1516–1523

Hossain MK, Mortuza AA, Sen SK, Basher MK, Ashraf MW, Tayyaba S, Mia MN, Uddin MJ (2018a) A comparative study on the influence of pure anatase and Degussa-P25 TiO_2 nanomaterials on the structural and optical properties of dye sensitized solar cell (DSSC) photoanode. Optik 171:507–516

Hossain MK, Pervez MF, Uddin MJ, Tayyaba S, Mia MN, Bashar MS, Jewel MK, Haque MA, Hakim MA, Khan MA (2018b) Influence of natural dye adsorption on the structural, morpho-logical and optical properties of TiO_2 based photoanode of dye-sensitized solar cell. Mater Sci 36(1):93–101

Hussein AK (2015) Applications of nanotechnology in renewable energies—a comprehensive overview and understanding. Renew Sust Energ Rev 42:460–476

Iavicoli I, Leso V, Beezhold DH, Shvedova AA (2017) Nanotechnology in agriculture: opportunities, toxicological implications and occupational risks. Toxicol Appl Pharmacol 329: 96–111

Iravani S (2011) Green synthesis of metal nanoparticles using plants. Green Chem 13(10): 2638–2650

Janib SM Moses AS MacKay JA (2010) Imaging and drug delivery using theranostic nanoparticles. Advanced drug delivery reviews 62(11):1052–1063

Joy PA, Kumar PA, Date SK (1998) The relationship between field cooled and zero field cooled susceptibilities of some ordered magnetic systems. J Phys Condens Matter 10(48):11049–11054

Kang S, Herzberg M, Rodrigues DF, Elimelech M (2008) Antibacterial effects of carbon nanotubes: size does matter. Langmuir 24(13):6409–6413

Kaskel S, Schlichte K, Chaplais G, Khanna M (2003) Synthesis and characterisation of titanium nitride-based nanoparticles. J Mater Chem 13:1496–1499

Khan I, Abdalla A, Qurashi A (2017) Synthesis of hierarchical WO_3 and Bi_2O_3/WO_3 nanocomposite for solar-driven water splitting applications. Int J Hydrog Energy 42(5): 3431–3439

Khan MI, Hossain MI, Hossain MK, Rubel MH, Hossain KM, Mahfuz AM, Anik MI (2022) Recent progress in nanostructured smart drug delivery systems for cancer therapy: a review. ACS Appl Biomater 5(3):971–1012

Kubik T, Bogunia-Kubik K, Sugisaka M (2005) Nanotechnology on duty in medical applications. Curr Pharm Biotechnol 6:17–33

Lan Y, Li J, Wong-Ng W, Derbeshi RM, Li J, Lisfi A (2016) Free-standing self-assemblies of gallium nitride nanoparticles: a review. Micromachines (Basel) 7(9):1–22

Laurent S, Forge D, Port M, Roch A, Robic C, Vander Elst L, Muller RN (2010) Magnetic iron oxide nanoparticles: synthesis, stabilization, vectorization, physicochemical characterizations, and biological applicatons. Chem Rev 108(6):2064–2110

Lei YM, Huang WX, Zhao M, Chai YQ, Yuan R, Zhuo Y (2015) Electrochemiluminescence resonance energy transfer system: mechanism and application in ratio metricaptasensor for lead ion. Anal Chem 87(15):7787–7794

Levinsen M, Harila-Saari A, Grell K, Jonsson OG, Taskinen M, Abrahamsson J, Vettenranta K, Åsberg A, Risteli J, Heldrup J, Schmiegelow K (2016) Efficacy and toxicity of intrathecal liposomal cytarabine in first-line therapy of childhood acute lymphoblastic leukemia. J Pediatr Hematol Oncol 38(8):602–609

Li D, Baydoun H, Verani CN, Brock SL (2016) Efficient water oxidation using CoMnP nanoparticles. J Am Chem Soc 138(12):4006–4009

Li M, Du C, Guo N, Teng Y, Meng X, Sun H, Li S, Yu P, Galons H (2019) Composition design and medical application of liposomes. Eur J Med Chem 164:640–653

Liu D, Li C, Zhou F, Zhang T, Zhang H, Li X, Duan G, Cai W, Li Y (2015a) Rapid synthesis of monodisperse au nanospheres through a laser irradiation-induced shape conversion, self-assembly and their electromagnetic coupling SERS enhancement. Sci Rep 5(1):1–9

Liu J, Liu Y, Liu N, Han Y, Zhang X, Huang H, Lifshitz Y, Lee ST, Zhong J, Kang Z (2015b) Metal-free efficient photocatalyst for stable visible water splitting via a two-electron pathway. Science 347(6225):970–974

Liu YK, Ye J, Han QL, Tao R, Liu F, Wang W (2015c) Toxicity and bioactivity of cobalt nanoparticles on the monocytes. Orthop Surg 7(2):168–173

Mabena LF, Sinha Ray S, Mhlanga SD, Coville NJ (2011) Nitrogen-doped carbon nanotubes as a metal catalyst support. Appl Nanosci 1(2):67–77

Meier C, Gondorf A, Lüttjohann S, Lorke A, Wiggers H (2007) Silicon nanoparticles: absorption, emission and the nature of the electronic bandgap. J Appl Phys 101(10):103–112

Mejias Carpio IE, Mangadlao J, Nguyen HN, Advincula R, Rodrigues DF (2014) Graphene oxide functionalized with ethylenediamine triacetic acid for heavy metal adsorption and anti-microbial applications. Carbon 77:289–301

Mishra Y, Chakravadhanula V, Hrkac V, Jebril S, Agarwal D, Mohapatra S, Adelung R (2012) Crystal growth behaviour in Au-ZnO nanocomposite under different annealing environments and photo switch ability. J Appl Phys 112(6):301–309

Mitchell MJ, Billingsley MM, Haley RM, Wechsler ME, Peppas NA, Langer R (2021) Engineering precision nanoparticles for drug delivery. Nat Rev Drug Discov 20:101–124

Mobasser S, Firoozi A (2017) Review of nanotechnology applications in science and engineering. J Civil Eng Urban 6(4):84–93

Mogilevsky G, Hartman O, Emmons ED, Balboa A, DeCoste JB, Schindler BJ, Iordanov I, Karwacki CJ (2014) Bottom-up synthesis of anatase nanoparticles with graphene domains. ACS Appl Mater Interfaces 6(13):10638–10648

Mueller NC, Nowack B (2008) Exposure modeling of engineered nanoparticles in the environment. Environ Sci Technol 42(12):4447–4453

Musico YLF, Santos CM, Dalida MLP, Rodrigues DF (2014) Surface modification of membrane filters using graphene and graphene oxide-based nanomaterials for bacterial inactivation and removal. ACS Sustain Chem Eng 2:1559–1565

Nagalingam A (2017) Drug delivery aspects of herbal medicines. Focus on inflammation. In: Arumugam S, Watanabe K (eds) Japanese treatment of common diseases: focus on inflammation. Elsevier, Tokyo, pp 143–203

Narayanan KB, Sakthivel N (2011) Green synthesis of biogenic metal nanoparticles by terrestrial and aquatic phototrophic and heterotrophic eukaryotes and biocompatible agents. Adv Colloid Interf Sci 169:59–79

Nayak MK, Singh J, Singh B, Soni S, Pandey VS, Tyagi S (2017) Introduction to semiconductor nanomaterial and its optical and electronics properties. In: Gupta RK, Misra M (eds) Metal semiconductor core-shell nanostructures for energy and environmental applications. Elsevier, Amsterdam, pp 1–33

Needham D, Arslanagic A, Glud K, Hervella P, Karimi L, Hoeilund-Carlsen PF, Kinoshita K, Mollenhauer J, Parra E, Utoft A, Walke P (2016) Bottom-up design of nanoparticles for anti-cancer diapeutics "put the drug in the cancer's food". J Drug Target 24(9):836–856

Ngoy JM, Wagner N, Riboldi L, Bolland O (2014) A CO_2 capture technology using multi-walled carbon nanotubes with polyaspartamide surfactant. Energy Procedia 63:2230–2248

Nguyen CH, Field JA, Sierra-Alvarez R (2020) Microbial toxicity of gallium-and indium-based oxide and arsenide nanoparticles. J Environ Sci Health A 55(2):168–178

Ning F, Shao M, Xu S, Fu Y, Zhang R, Wei M, Evans DG, Duan X (2016) TiO_2/graphene/NiFe-layered double hydroxide nanorod array photoanodes for efficient photoelectrochemical water splitting. Energy Environ Sci 9(8):2633–2643

Parveen K, Banse V, Ledwani L (2016) Green synthesis of nanoparticles: their advantages and disadvantages. AIP Conf Proc 1724(1):1–7

Parvej MS, Khan MI, Hossain MK (2022) Preparation of nanoparticles-based polymer composites. In: Rangappa SM, Parameswaranpillai J, Yashas TG, Siengchin S, Seydibeyoglu MO (eds) Metal nanoparticle-based polymer composites. Woodhead Publishing, New Delhi, pp 1–94

Patra JK, Das G, Fraceto LF, Campos EVR, Rodriguez-Torres MDP, Acosta-Torres LS, Diaz-Torres LA, Grillo R, Swamy MK, Sharma S, Habtemariam S, Shin HS (2018) Nano based drug delivery systems: recent developments and future prospects. J Nanobiotechnol 16(71):1–33

Pizzagalli L, Galli G, Klepeis JE, Gygi F (2001) Structure and stability of germanium nanoparticles. Phys Rev B 63(16):1653241–1653245

Pradhan N, Singh S, Ojha N (2015) Facets of nanotechnology as seen in food processing, packaging and preservation industry. Biomed Res Int 365672:1–17

Prasad R, Bhattacharyya A, Nguyen QD (2017) Nanotechnology in sustainable agriculture: recent developments, challenges and perspectives. Front Microbiol 8(1014):1–13

Prokop A, Davidson JM (2008) Nanovehicular intracellular delivery systems. J Pharm Sci 97(9): 3518–3590

Rabbani MM, Ahmed I, Park SJ (2016) Application of nanotechnology to remediate contaminated soils. In: Rahman AM (ed) Environmental remediation technologies for metal-contaminated soils. Springer, Tokyo, pp 219–229

Reddington E, Sapienza A, Gurau B, Viswanathan R, Sarangapani S, Smotkin ES, Mallouk TE (1998) Combinatorial electrochemistry: a highly parallel, optical screening method for discovery of better electrocatalysts. Science 280(5370):1735–1737

Render D, Samuel T, King H, Vig M, Jeelani S, Babu RJ, Rangari V (2016) Biomaterial-derived calcium carbonate nanoparticles for enteric drug delivery. J Nanomater 2016:1–8

Rickerby DG, Morrison M (2007) Nanotechnology and the environment: a European perspective. Sci Technol Adv Mater 8(1):19–24

Rizvi SAA, Saleh AM (2018) Applications of nanoparticle systems in drug delivery technology. Saudi Pharmaceutical Journal 26:64–70

Rodrigues DF, Elimelech M (2010) Toxic effects of single-walled carbon nanotubes in the development of *E. coli* biofifilm. Environ Sci Technol 44:4583–4589

Rodrigues DF, Advincula RC, Claydon F, Santos CM, Tria TRM (2015) Nanocomposite polymer-carbon based nanomaterial filters for the simultaneous removal of bacteria and heavy metals. US Patent US8:736-925

Rubel MHK, Hossain MK (2022) Crystal structures and properties of nanomagnetic materials. In: Gupta RK, Mishra SR, Nguyen TA (eds) Fundamentals of low dimensional magnets. CRC Press, Boca Raton, pp 177–200

Saeed K, Khan I (2014) Preparation and properties of single-walled carbon nanotubes/poly (butylene terephthalate) nanocomposites. Iran Polym J 23:53–58

Saeed K, Khan I (2016) Preparation and characterization of single-walled carbon nanotube/nylon 6, 6 nanocomposites. Instrum Sci Technol 44(4):435–444

Saha SMK, Ali MH, Hossen MF, Pervez MF, Mia MNH, Hossain MK (2018) Structural, morphological and optical properties of CuO thin films treated by gamma ray. In: International conference on computer, communication, chemical, material and electronic engineering (IC4ME2). Institute of Electrical and Electronics Engineers (IEEE), New York, pp 1–4

Santos CM, Mangadlao J, Ahmed F, Leon A, Advincula RC, Rodrigues DF (2012) Graphene nanocomposite for biomedical applications: fabrication, antimicrobial and cytotoxic investigations. Nanotechnology 23(39):1–10

Sarfraz M, Afzal A, Yang T, Gai Y, Raza SM, Khan MW, Cheng Y, Ma X, Xiang G (2018) Development of dual drug loaded nanosized liposomal formulation by a reengineered ethanolic injection method and its pre-clinical pharmacokinetic studies. Pharmaceutics 10:1–22

Sigmund W, Yuh J, Park H, Maneeratana V, Pyrgiotakis G, Daga A, Taylor J, Nino JC (2006) Processing and structure relationships in electrospinning of ceramic fiber systems. J Am Ceram Soc 89(2):395–407

Singh R, Lillard JW Jr (2009) Nanoparticle-based targeted drug delivery. Exp Mol Pathol 86(3): 215–223

Stankic S, Müller M, Diwald O, Sterrer M, Knözinger E, Bernardi J (2005) Size-dependent optical properties of MgO nanocubes. Angew Int Chem Int Ed 44(31):4917–4920

Sun S, Murray CB, Weller D, Folks L, Moser A (2000) Monodisperse FePt nanoparticles and ferromagnetic FePt nanocrystal superlattices. Science 287(5460):1989–1992

Tan CY, Yaghoubi A, Ramesh S, Adzila S, Purbolaksono J, Hassan MA, Kutty MG (2013) Sintering and mechanical properties of MgO-doped nanocrystalline hydroxyapatite. Ceram Int 39(8):8979–8983

Tang S, Zheng J (2018) Antibacterial activity of silver nanoparticles: structural effects. Adv Healthc Mater 7(13):1–10

Temple T, Bagnall DM (2011) Optical properties of gold and aluminium nanoparticles for silicon solar cell applications. J Appl Phys 109(8):1–13

Thomas SC, Harshita, Mishra PK, Talegaonkar S (2015) Ceramic nanoparticles: fabrication methods and applications in drug delivery. Curr Pharm Des 21(42):6165–6188

Tiwari JN, Tiwari RN, Kim KS (2012) Zero-dimensional, one dimensional, two-dimensional and three-dimensional nanostructured materials for advanced electrochemical energy devices. Prog Mater Sci 57(4):724–803

Verma N, Kumar N (2019) Synthesis and biomedical applications of copper oxide nanoparticles: an expanding horizon. ACS Biomater Sci Eng 5(3):1170–1118

Wang Y, Xia Y (2004) Bottom-up and top-down approaches to the synthesis of monodispersed spherical colloids of low melting-point metals. Nano Lett 4(10):2047–2050

Wenzel RP (2007) Health care-associated infections: major issues in the early years of the 21st century. Clin Infect Dis 45(1):S85–S88

Yeh YC, Creran B, Rotello VM (2012) Gold nanoparticles: preparation, properties and applications in bionanotechnology. Nanoscale 4(6):1871–1880

Yingchoncharoen P, Kalinowski DS, Richardson DR (2016) Lipid-based drug delivery systems in cancer therapy: what is available and what is yet to come? Pharmacol Rev 68(3):701–787

Zafar F, Iqbal A (2016) Indium phosphide nanowires and their applications in optoelectronic devices. Royal Soc. 472:1–18

Zemtsova EG, Monin AV, Smirnov VM, Semenov BN, Morozov NF (2015) Formation and mechanical properties of alumina ceramics based on Al_2O_3 micro-and nanoparticles. Phys Mesomech 8(2):134–138

Zhang WX, Elliot DW (2006) Applications of iron nanoparticles for groundwater remediation. Remed J 16(2):7–21

Zhang Q, Zhang K, Xu D, Yang G, Huang H, Nie F (2014) CuO nanostructures: synthesis, characterization, growth mechanisms, fundamental properties and applications. Prog Mater Sci 60(1):208–337

Zhang XF, Liu ZG, Shen W, Gurunathan S (2016) Silver nanoparticles: synthesis, characterization, properties, applications and therapeutic approaches. Int J Mol Sci 17(9):1–34

Zhang Y, Ding Z, Zhao G, Zhang T, Xu Q, Cui B, Liu JX (2018) Transcriptional responses and mechanisms of copper nanoparticle toxicology on zebrafish embryos. J Hazard Mater 344: 1057–1068

Timanshi Chansoriya is working as a Ph.D. scholar in the Department of Zoology, University of Rajasthan, Jaipur. She did her B.Sc. and M.Sc. (Entomology) from the University of Lucknow. She is a Life Member of ISSRF and M.P. PCOS. Research work focuses on veterinary reproductive biology and nanotechnology.

Barkha Khilwani is working as a Postdoctoral Fellow, Rashtriya Uchchatar Shiksha Abhiyan (RUSA 2.0), at the Department of Zoology, University of Rajasthan, Jaipur. She did her Ph. D. research work at the Indian Institute of Science Education and Research (IISER), Mohali. She has published more than 15 international peer-reviewed research publications, and 2 are under communication.

Abdul Salam Ansari, Ph.D., FAMS, has been working in the Department of Zoology, University of Rajasthan, Jaipur, since 1996. He has published around 80 research papers in reputed international journals and 18 textbooks/practical manuals/edited volumes. His outstanding research has significantly contributed to developing an ideal male contraceptive. He is involved in the phase III human clinical trials employing RISUG—a new perspective in nonhormonal male contraception—at Jaipur Centre. Currently, his research is focused on developing dual contraceptives, which could be applicable for both sexes with RISUG.

Osteoarthritis: Novel Insights in Treatment 19

Sabreena Aashaq, Asma Rafiq, Iqra Jan, Mehvish Ali, Bilal A. Reshi, Zaffar Amin Shah, Faheem A. Sheikh, Shafquat Majeed, and Mushtaq A. Beigh

S. Aashaq
Department of Immunology and Molecular Medicine, Sher-i-Kashmir Institute of Medical Sciences, Srinagar, Jammu and Kashmir, India

Cellular Signalling Laboratory, Department of Nanotechnology, University of Kashmir Hazratbal, Srinagar, Jammu and Kashmir, India

A. Rafiq · I. Jan · M. Ali
Cellular Signalling Laboratory, Department of Nanotechnology, University of Kashmir Hazratbal, Srinagar, Jammu and Kashmir, India

B. A. Reshi
Department of Biotechnology, University of Kashmir Hazratbal, Srinagar, Jammu and Kashmir, India
e-mail: bareshi@uok.eu.in

Z. A. Shah
Department of Immunology and Molecular Medicine, Sher-i-Kashmir Institute of Medical Sciences, Srinagar, Jammu and Kashmir, India

F. A. Sheikh
Nanostructured and Biomimetic Lab, Department of Nanotechnology, University of Kashmir Hazratbal, Srinagar, Jammu and Kashmir, India
e-mail: faheemnt@uok.edu.in

S. Majeed
Laboratory for Multifunctional Nanomaterials, Department of Nanotechnology, University of Kashmir Hazratbal, Srinagar, Jammu and Kashmir, India
e-mail: smshah@uok.edu.in

M. A. Beigh (✉)
Cellular Signalling and Nanotherapeutics Laboratory, Department of Nanotechnology, University of Kashmir Hazratbal, Srinagar, Jammu and Kashmir, India
e-mail: beighm@uok.edu.in

F. A. Sheikh et al. (eds.), *Interaction of Nanomaterials With Living Cells*,
https://doi.org/10.1007/978-981-99-2119-5_19

Abstract

Osteoarthritis is the most prevalent disease of the joints with increased prevalence and incidence upon the progression of age. It is the key source of socioeconomic cost, disability, and pain globally. It occurs as a result of the complex communion of systemic and local factors. The disease has multifactorial epidemiology comprising of biomechanical, biological, and genetic elements. In order to recognize the best therapies to prevent or inhibit the progression of the disease and to find the best remedies for symptomatic relief, epidemiologic and clinical research in the field is critical. The treatment options for the disease include non-pharmacologic, pharmacologic, surgical, and intra-articular routes or the combinatorial use of these therapies to attain optimal outcomes. Pharmacological interventions in treating the disease are mostly inclined to reduce the symptoms and improve the functional status of the affected joint. The drug regulatory agencies have not yet approved any disease-modifying drug formulation that will stop or slow-down the progression of the disease apart from relieving the symptoms.

Keywords

Osteoarthritis · Cytotoxic · Symptoms · Treatment · Diagnostic

19.1 Introduction

Osteoarthritis is a global epidemic and a devastating disease. In the United States and other developed countries, osteoarthritis is the most common disease affecting the joints and the most prevalent cause of disability and chronic pain among the elderly (Murray et al. 2013). Knees, hands, spine, and hips are osteoarthritis' most commonly affected sites. The symptoms include remarkable malfunction of the joints, pain, moving disability, and joint stiffness. Among adults aged 60 years or more, approximately 13% of women and 10% of men suffer from symptomatic knee osteoarthritis (Lawrence et al. 2008). Osteoarthritis of the knee joints affects almost 19% of the adult population in America and contributes more than 80% to the total load of the disease (Vos et al. 2012; Lawrence et al. 2008). The predominant cause of knee osteoarthritis appears to be the inflammation and disintegration of joint tissues caused by mechanical loading (Felson 2013; Robinson et al. 2016). The prevalence of knee osteoarthritis increases with increasing age (Felson et al. 2000). In the United States, the increased life expectancy since the twentieth century has contributed to knee osteoarthritis as the wear and tear of the joint tissues increases with age (Loeser 2013). Moreover, the increased body mass index (BMI) in the American population is an established risk factor contributing to knee osteoarthritis. The increase in BMI contributes to the development of knee osteoarthritis because of the adiposity inflammation and the mechanical stress of the joint tissues (Wluka et al. 2013). Hence the prevalence of osteoarthritis increases with the obesity epidemic and aging. The etiology of osteoarthritis is multifactorial and complex, with genetic, mechanical, and biological components (Loeser et al. 2016). Repetitive use of joints, intra-articular crystal deposition, knee injury, joint laxity, joint misalignment, old age, female gender, peripheral neuropathy, obesity, and muscle weakness are the key factors contributing to the development of osteoarthritis in the load-bearing joints. The changes in the structure and composition of the cartilage matrix due to aging and various genetic factors also contribute to the development of osteoarthritis (Sellam and Berenbaum 2010).

In osteoarthritis, different components of the joint, like connective tissue constituents, the lining of the synovial joint, and periarticular bone, are affected. The formation of osteophytes, degradation of cartilage, inflammation of synovial joints, and changes in the subchondral and periarticular bone are the characteristic features of osteoarthritis (Sellam and Berenbaum 2010; Zhang and Jordan 2010). During osteoarthritis, enzymes like MMP-3 (matrix metalloproteinase-3), MMP-13 (matrix metalloproteinase-13), and aggrecanases accumulate within the joint spaces and cause degradation of the cartilage matrix (Goldring and Goldring 2007; Hashimoto et al. 2013). Post-traumatic osteoarthritis (PTOA) is a type of osteoarthritis that is caused due to the mechanical injury of the joint tissue. In 78% of young adults, radiological changes emerge, and 40% develop radiological osteoarthritis in the injured knee after 14 years of tear in the anterior cruciate ligament. The disintegration of the joint is irreversible and rapid. Unfortunately, the disease cannot be reversed once the damage is radiologically evident. Post-traumatic osteoarthritis progression can be averted by immediate therapeutic intervention post-injury (Goldring and Goldring 2007).

Initially, the development of osteoarthritis involves the increased cellular proliferation and enhanced synthesis of proteinases, matrix proteins, growth factors, and proinflammatory cytokines by the chondrocytes. Therefore, the chondrocytes serve as the cellular mediators of osteoarthritis (Yang et al. 2013). The other tissues and cells within the joint, like the subchondral bone and the synovium, also contribute to the pathogenesis of the disease (Lee et al. 2005). The articular chondrocyte, which is usually involved in the maintenance of cartilage, is not able to restore the original architecture of the cartilage matrix in osteoarthritis (Bau et al. 2002). The present therapeutic interventions that address chronic pain in osteoarthritis are inadequate, and currently, no structure-reforming treatment is accessible (Brown et al. 2006).

19.2 Osteoarthritis: Pathophysiology

Age is the prime risk factor that contributes to the pathophysiology of osteoarthritis (von Porat et al. 2004). Present-day clinical research is primarily engrossed in understanding the effects of aging on the cartilage and the development of osteoarthritis (Goldring and Goldring 2007; Felson 2013). Multiple mechanisms contribute to the age-induced degradation of joints and the progression of osteoarthritis (Sacitharan 2019). During the progression of osteoarthritis, considerable apoptotic or necrotic loss of chondrocytes occurs in the middle and peripheral cartilage zones (Lotz and Carames 2011). Apoptotic loss of chondrocytes occurs due to reduced cellular metabolism, reduced mechanical defense of the joints, and the production of various proinflammatory cytokines. The loss of chondrocytes results in the reduction of cartilage volume and increased breakdown of the extracellular matrix, leading to the loss of cartilage and the progression of osteoarthritis (Heinegard and Saxne 2011; Stockwell 2011).

The principal cause of osteoarthritis is the degradation of the matrix components of the articular cartilage by the matrix-degrading enzymes like MMPs that degrade collagen and the members of ADAMTS (a disintegrin and metalloproteinase with thrombo-spondin motif) that degrade aggrecan (Buckwalter and Mankin 1998; Nagase et al. 2006). In osteoarthritis, chondrocytes produce enhanced levels of cytokines like TNF-α and IL-1β, which activate the matrix-degrading enzymes and hence lead to the degradation of matrix components (Goldring and Otero 2011). In osteoarthritis, mononuclear cell infiltration in the synovial membrane is also detected. Significant evidence further supports the activation of innate immune responses in osteoarthritis. The inflammation associated with osteoarthritis further promotes tissue breakdown and advances the painful episodes of the disease (Scanzello and Goldring 2012).

Cytokines like TNF-α and interleukin-1 also play a crucial role in the pathophysiology of rheumatoid arthritis (Orlowsky and Kraus 2015; Glyn-Jones et al. 2015). Blocking these cytokines, mainly TNF-α is an effective treatment for rheumatoid arthritis (Mclnnes and Schett 2007; Calich et al. 2010; Chevalier et al. 2013; Goldring and Berenbaum 2015). In osteoarthritis, levels of IL-1 are also enhanced in the cartilage, and both TNF-α and IL-1 are present in the synovial fluid of osteoarthritic joints. Also, the intra-articular administration of IL-1 in animal models results in the development of osteoarthritis-like disease (Feldmann and Maini 2001; Smith et al. 1997; Melchiorri et al. 1998). The introduction of the antagonist of IL-1 receptor in the animal models of osteoarthritis diminishes the loss of cartilage. Trails that investigated the efficacy of TNF-α in the treatment of osteoarthritis showed that the blockade of TNF-α modestly decreased the progression of the disease. The inhibition of IL-1β has also shown promising results in patients suffering from hand and knee osteoarthritis; however, the patients exhibited infections in the upper respiratory tract following IL-1β blockade (Smith et al. 1997). TNF-α and IL-1 also increase the levels of inducible nitric oxide synthase (iNOS) and COX-2 (Cyclooxygenase 2) in the chondrocytes and hence promote the synthesis of nitric oxide (NO) and prostaglandin E2 (PGE2) in these cells (Melchiorri et al. 1998; Pettipher et al. 1986; Pelletier et al. 1997). Nitric oxide, a cytotoxic free radical, enhances the apoptosis and ER stress associated with osteoarthritis. PEG2 counteracts IL-1-mediated effects on the synthesis of cartilage matrix by enhancing the expression of COX2, which may serve as a counteractive mechanism in chondrocyte dedifferentiation induced by IL-1 (Amin and Abramson 1998; Melchiorri et al. 1998; Heinegård and Saxne 2011).

MMPs are the main proteolytic enzymes that are involved in the etiology of osteoarthritis (Nagase et al. 2006). Collagenases, which constitute the prime class of MMPs, are involved in the progression of osteoarthritis. Collagenases catalyze the cleavage of collagen fibrils at a single point. MMP-8 catalyzes the cleavage of type I collagen, whereas MMP-13 catalyzes the cleavage of type II collagen (Neuhold et al. 2001). MMP-1 is involved in the cleavage of type III collagen. Ectopic expression of

MMP-13 in mice leads to the development of osteoarthritis-like disease (Little et al. 2009). Furthermore, the knockout of MMP-13 in mice results in protection from surgically induced osteoarthritis. In normal cartilage, MMPs are expressed as pro-proteins (pro-MMPs). However, in the osteoarthritic synovial tissue and cartilage, the expression of MMP-2, MMP-9, MMP-13, MMP-14, MMP-16, and MMP-28 is increased (Hofmann et al. 1992). Apart from MMPs, another family of metalloproteinases known as the ADAMTS (ADAMTS1–ADAMTS19) is also involved in the degradation of matrix components (Stanton et al. 2011). The substrates of this family of proteins contain a type I thrombospondin motif. The ECM proteins contain this motif and serve as the substrates for these metalloproteinases. The matrix proteins, including fibromodulin and decorin, are cleaved by these metalloproteinases (Nagase and Kashiwagi 2003). The deletion of ADAMTS5 in the mice cartilage has been shown to inhibit inflammatory arthritis and cartilage disintegration (Glasson et al. 2004). The proteolytic activity of these proteases (ADAMTS and MMPs) is regulated by TIMPs (tissue inhibitors of metalloproteinases) (Milner et al. 2006). TIMP-3 inhibits aggrecanases-induced degradation of aggrecan from retinoic acid, or IL-1-stimulated cartilage explants (Nagase and Kashiwagi 2003). It also serves as a dynamic inhibitor of ADAMTS4 and 5 under in vitro conditions (Gendron et al. 2003) (Fig. 19.1).

IGF-1 (Insulin-like growth factor 1) is also involved in the pathophysiology of osteoarthritis (Sah et al. 1994). The stimulation of the cartilage explants or bovine chondrocytes with IGF-1 has been shown to increase the synthesis of proteoglycans. In addition, the stimulation of periosteal mesenchymal cells with TGF-β and IGF-1 regulates the differentiation and proliferation of these cells during chondrogenesis (Fukumoto et al. 2003). In chondrocytes, TGF-β increases the expression of IGF1 receptors without modulating their ligand affinity. Deficiency of IGF-1 in mice results in the development of lesions in the articular cartilage (Tsukazaki et al. 1994). Moreover, adding IGF-1 to the chondrocyte grafts in equine and murine models stimulates chondrogenesis in cartilage defects (Ekenstedt et al. 2006). IGF-1 is hence essential in the maintenance of the integrity of articular cartilage. Significant evidence suggests that the potential of the chondrocytes to respond to IGF-1 diminishes in osteoarthritis and with age (Fortier et al. 2003; Goodrich et al. 2007; Loeser et al. 2000; Morales 2008).

The TGF-β family of mitogenic growth factors plays a crucial part in cell migration, apoptosis, differentiation, and cellular proliferation (Massagué 1998; Aashaq et al. 2018). Enhanced levels of TGF-β have been reported in the synovial fluid of patients suffering from osteoarthritis. The synovial tissue, bone, and cartilage express all the TGF-β isoforms (TGF-β1, TGF-β2, and TGF-β3) and the TGF-β receptors, and they are known to play diverse functions within these tissues (Blom et al. 2004; Shen et al. 2014). However, the function of TGF-β in the pathophysiology of osteoarthritis is ambiguous. The expression of the dominant negative mutant of the TGF-β type II receptor in the skeletal tissue of transgenic mice results in the progression of skeletal disintegration (Serra et al. 1997). Furthermore, the deletion of TGF-β type II receptors in the articular chondrocytes of mice results in a phenotype resembling osteoarthritis. In addition, the exogenous expression of TGF-β1 in the osteoblasts of mice results in the sclerosis of the subchondral bone and the

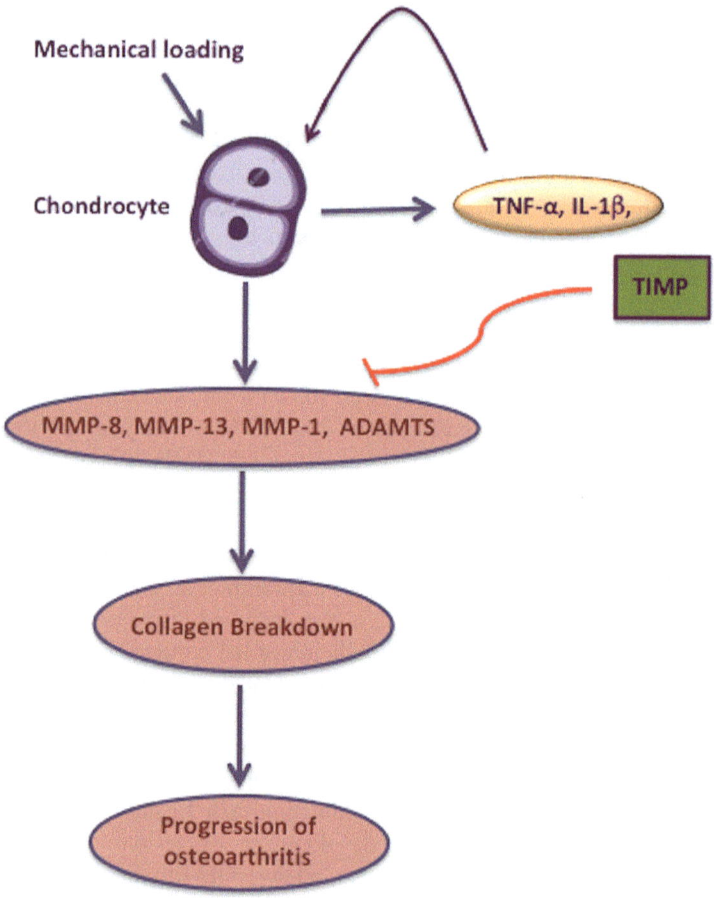

Fig. 19.1 Proteolytic degradation of collagen in osteoarthritis

development of spontaneous osteoarthritis; however, inhibiting the activity of TGF-β in the subchondral bone stabilizes its architecture and diminishes the pathology of osteoarthritis (Hiramatsu et al. 2011; Zhen et al. 2013).

19.3 Markers of Osteoarthritis

The constituents of the extracellular matrix, like the degradation products or the precursors of proteoglycan and collagen, the enzymes, and the cytokines have the prospect to serve as the biochemical markers of osteoarthritis. The concentration of these markers can be measured in the synovial fluid, urine, and blood. The concentrations are linked to tissue metabolism (Van Spil et al. 2010). Various biochemical markers have been advocated, but these have not been corroborated for use in clinical practice. The aim of the biochemical markers is to recognize the pathological changes associated with a disease. The tissue disintegration markers, including cartilage oligomeric matrix protein and C-terminal telopeptide of collagen type II (CTX-II), have been validated for clinical use and are the most widely used biomarkers for osteoarthritis (Garnero et al. 2001). The specificity and sensitivity of biochemical markers is bad compared to that of the imaging tests. The limitation associated with the use of biochemical markers in osteoarthritis is the site of origination; hence their predicted values are useless unless the disease is restricted to a particular joint. The use of synovial fluid for the assays overcomes the limitation, but its use is restricted due to the lack of patient acceptability (Hoch et al. 2011). The use of joint-specific post-translational modifications of proteins appears to be a promising area, but further research in this field may warrant their usefulness as the markers of osteoarthritis. With the progression of protein biology, the figure of the analytical biomarkers for osteoarthritis has dramatically increased over the past decade. The proteomic investigation of the cartilage has revealed several joint-specific biomarkers of osteoarthritis (Dam et al. 2009; Catterall et al. 2012). In the future, the use of biochemical markers for the diagnosis of osteoarthritis may comprise a broad range of analytical assays apart from disease phenotyping to ascertain a suitable treatment.

19.4 Risk Factors of Osteoarthritis

A number of risk factors encompassing age, mechanical injury, gender, obesity and diet, and genetics contribute to the development of osteoarthritic joints. This disease occurs as a result of complex crosstalk between local and systemic factors (Onnerford et al. 2012; Neogi and Zhang 2013). The essence of the risk factors varies for the progression and the onset of the disease, different stages of the disease, and for the different joints. For the development of osteoarthritis, aging serves as a prominent risk factor (Reynard and Loughlin 2012; Gly-Jones et al. 2015). The risk for the development of both symptomatic and radiographic osteoarthritis increases with increasing age. Increased incidence of osteoarthritis in the hand, hip, and knee joints has been observed with aging. Global estimates show that 18% of women and 9.6% of men over 60 suffer from symptomatic osteoarthritis (Felson et al. 2000; Lotz and Loeser 2012). According to the Framingham osteoarthritis study, 27% of the population in the age group of 63 to 70 have radiographic knee osteoarthritis, and the incidence increases to 44% in the population aged 80 years and above (Shane Anderson and Loeser 2010; Zhang et al. 2002). According to this study, the pervasiveness of symptomatic osteoarthritis of the knee is 9.5% in all subjects. However, it escalated in women but not in men with age (Murray and Lopez 1996). The escalated incidence and pervasiveness of osteoarthritis with age may be an outcome of enhanced predisposition to various biological changes and risk factors that accompany aging. Osteoarthritis of the hip joint is less common compared to knee osteoarthritis. Osteoarthritis in the hand joint is most common in the elderly population (Shane Anderson and Loeser 2010). Osteoarthritis of the hand does not affect the moving ability as knee and hip osteoarthritis. According to the

Framingham estimates, almost 26% of women and 13% of men over the age of 70 suffer from the symptomatic osteoarthritis of the hand affecting at least one joint (Zhang et al. 2002).

Apart from age, gender also plays a role in the development of osteoarthritis. Studies have shown that women have a greater risk of developing osteoarthritis over 50 years compared to men. However, men are more likely to develop osteoarthritis than women below 50 years of age. According to the studies conducted by Peter-Alhambra and colleagues on Spanish patients, women have a 1.52 times greater risk of developing osteoarthritis in the joints of hip, knee, and hands (Prieto Alhambra et al. 2014). The sex differences for the development of osteoarthritis in hip and knee joints peaked in the age group of 70–75 years. However, for hand osteoarthritis the difference peaked at 50–55 years of age (Srikanth et al. 2005). The prevalence of osteoarthritis in women increases around menopause, suggesting the role of hormones in the development of this disease. In contrast, several studies have shown the positive role of both exogenous and endogenous estrogen in osteoarthritis. Intake of either placebo or progestin and estrogen showed no remarkable difference in the pervasiveness of knee osteoarthritis and the physical disability associated with it (Wluka et al. 2000; Nevitt et al. 2001).

According to the Framingham study, subjects having the highest tercile of 25-hydroxyvitamin D have threefold less risk of knee osteoarthritis progression compared to those in the low (<27 ng/mL) and the middle tercile (27–33 ng/mL) (McAlindon et al. 1996a). However, this distribution does not show any effect on the development of hip osteoarthritis. The dietary insufficiency of vitamin C is also linked to the increased progression of knee osteoarthritis (McAlindon et al. 1996b). High levels of vitamin K in the serum decrease the pervasiveness of radiographic osteoarthritis of the hand. However, the supplementation of vitamin K does not show a positive effect on the radiographic osteoarthritis of the hand (Neogi et al. 2008). Metabolic syndrome and obesity also serve as the key risk factors of osteoarthritis, particularly knee osteoarthritis (Sellam and Berenbaum 2013). The odds ratio (OR) for the development of osteoarthritis is 2.96 for overweight individuals compared to normal ones (Hunter et al. 2002). Type 2 diabetes and dyslipidemia also serve as potential risk factors for osteoarthritis independent of obesity. Obesity or being overweight increases the load on the joints and may be the prime cause of hip or knee osteoarthritis. Studies have shown that a weight loss of 5 kg in women decreases the risk of symptomatic knee osteoarthritis development by 50% (King et al. 2013). Weight loss also reduces the disability and pain associated with knee osteoarthritis. The combination of exercise and weight loss has been shown to be effective in reducing the pain and functional disability associated with osteoarthritis. However, weight loss or exercise alone does not ameliorate the symptoms (Messier et al. 2004).

The history of the previous injury serves as a key risk factor for osteoarthritis. The major causes of posttraumatic arthritis include injured joint structures like meniscal tear, cartilage damage, and ligament injury (Brown et al. 2006; Thomas et al. 2017). A study conducted by Brown and colleagues revealed that almost 12% of all osteoarthritis cases are caused due to mechanical injury. Posttraumatic osteoarthritis

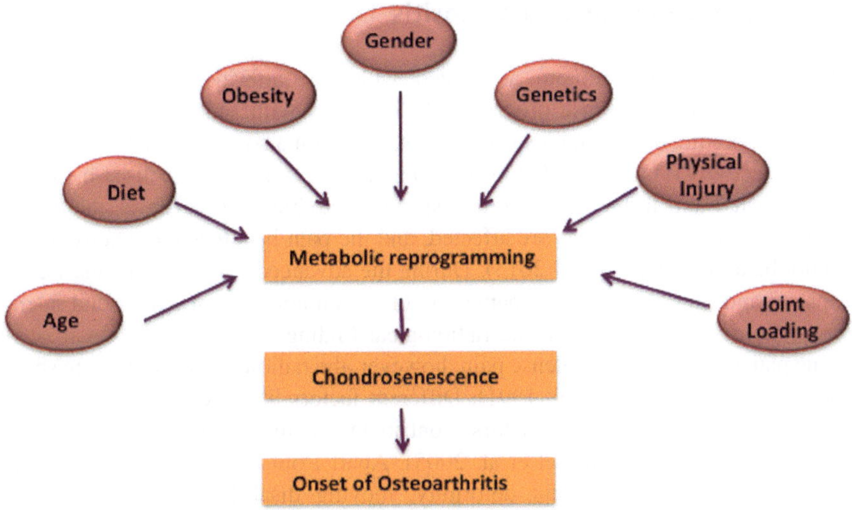

Fig. 19.2 Risk factors of osteoarthritis

constitutes 2% of hip osteoarthritis, 10% of knee osteoarthritis, and 20–78% of ankle osteoarthritis (Brown et al. 2006). Increased physical activity in occupations like lifting, jumping, bending, squatting, and kneeling also increases the risk of knee osteoarthritis. Evidence has shown that farmers, forestry and construction workers, and military personnel are at increased risk of osteoarthritis development (Coggon et al. 2000). Athletes also suffer from joint degradation due to articular cartilage damage caused by excessive loading of joints. Almost 80% of American football players with previous knee injuries suffer from osteoarthritis 10–30 years post-competing (Allen et al. 2010; Sandmark et al. 2000; Cameron et al. 2011; Kujala et al. 1995).

Apart from the risk mentioned above, a link has been established between genetic factors and osteoarthritis. Osteoarthritis susceptibility is polygenic in nature, and over 80 genes are involved in the pathogenesis of osteoarthritis (Peach et al. 2005). These genes include IGF-1 (insulin-like growth factor 1) and the receptors for vitamin D. Apart from this, osteoarthritis has been linked to the SNP in the growth and differentiation factor gene 5, which is implicated in the healthy development of cartilage and bone (Spector and MacGregor 2004). Several studies have recognized hereditary factors as the prime components for the development of primary osteoarthritis (Hunter et al. 2004; Valdes and Spector 2009). 70% of cases of hand osteoarthritis are caused due to hereditary factors. However, 50% of hip osteoarthritis and 10–30% of knee osteoarthritis are also estimated to be controlled by hereditary factors. Evidence also advocates a hereditary link to the joint that may develop osteoarthritis. Several epigenetic mechanisms like histone modification and DNA methylation are presently being examined for their role in osteoarthritis (Musemeci et al. 2015; Barter et al. 2012) (Fig. 19.2).

19.5 Symptoms of Osteoarthritis

The most eminent symptom associated with osteoarthritis is pain. The pain associated with the disease progressively increases with time. The pain may be intense, sporadic, or continuous aching pain. The pain also increases with increased physical activity (Hawker et al. 2008). During the early stages of the disease, the pain is anticipated and triggered by certain specific activities. However, as the disease progresses, routine activities are affected, and the pain becomes more constant and cannot be anticipated (Neogi 2013). During the advanced stages of the disease, the constant and dull pain is accompanied by uncertain episodes of severe and intense pain. The degree of pain and the radiological findings are not always consistent. Some patients suffer from intense pain; however, the radiological findings are scanty and vice-versa (Altman et al. 1991). Different factors like socio-cultural, environmental, and psychological factors contribute to the experience of pain in osteoarthritic patients (Fautrel et al. 2005). Apart from the pain, other symptoms of osteoarthritis include joint deformity, moving disability, locking of joints, swelling of joints, and grating and clicking of joints. Osteoarthritic patients also suffer from morning stiffness that gets better in 30 min. However, in rheumatoid arthritis, the morning stiffness lasts longer. The systemic symptoms, including abnormal serological tests, weight loss, and fever, should not accompany osteoarthritis, because these symptoms point toward other conditions like malignancy or infection. Osteoarthritis results in compromised physical activity and loss of independence to do routine activities. It significantly affects the quality of life by negatively influencing both physical and mental well-being (Fautrel et al. 2005; Salaffi et al. 2005) (Fig. 19.3).

19.6 Diagnostic Approaches of Osteoarthritis

Osteoarthritis is a vigorous and metabolically active disorder affecting joints by inducing morphological changes (van der Kraan 2012; Spector and MacGregor 2004). The morphological changes associated with the disease are assessed using radiography (Zhang et al. 2009). The main characteristics of osteoarthritis include subchondral cysts, sclerosis of subchondral bone, osteophytes, and narrowing of joint spaces (Lethbridge-Cejku et al. 1995; Peterson and Jacobson 2002). Radiographs have poor reproducibility and hence are insensitive to the early diagnosis (initial degradation of cartilage) and disease progression (Odding et al. 1998). Also, many structures of the joint, like ligaments, menisci, and cartilage, are not visualized on the radiographs (Benson and Croft 2008). The lesions of the bone marrow are also not visualized on the radiographs. The radiographs also provide scant evidence of effusion and synovitis. The MRI provides an alternative method for visualizing essential anatomic structures like soft and hard tissue (Neogi et al. 2009). It is used to observe the diffuse and focal changes in the cartilage and is resilient to changes in the position of the joint (Mobasheri 2012). The MRI also detects synovitis changes, lesions in the subchondral marrow, meniscal disruptions,

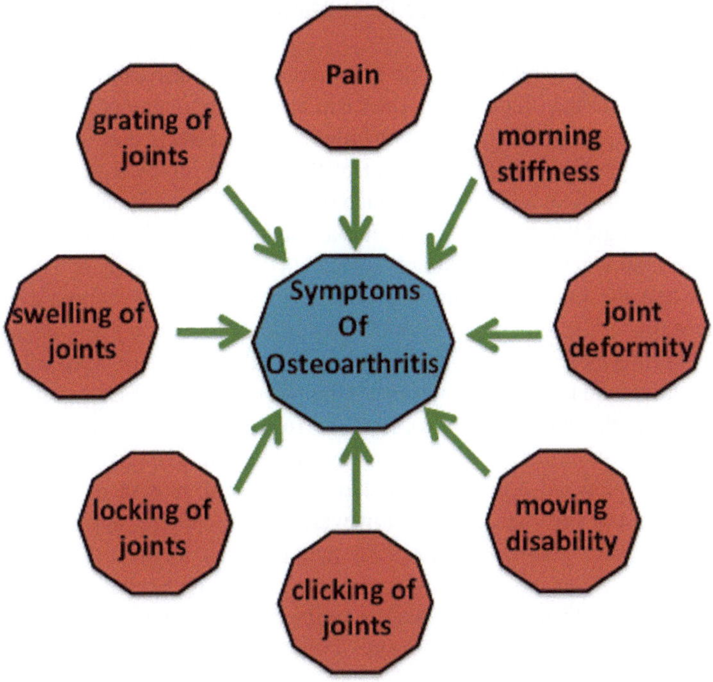

Fig. 19.3 Symptoms of osteoarthritis

and damage in the soft tissue structures within and surrounding the joint. The examination of the synovial fluid, urine, or blood is usually not required in the diagnosis of osteoarthritis. Still, it may be used to substantiate or rule out the coexistence of other inflammatory disorders like rheumatoid arthritis, gout, or pyrophosphate crystal deposition in individuals with signs and symptoms of the disease (Mobasheri 2012).

MRI (magnetic resonance imaging)	Radiography
It is a high-cost examination	It is a cost-effective examination
It allows the 3D examination of the joint structures	It is a 1D examination
It is used for imaging the ligaments, labrum, menisci, cartilage, and bone	It is used for bone imaging only
Its use is not so frequent but allows precise joint evaluation	It is the routinely used technique of imaging
The execution of the procedure and evaluation of results take a long time	The execution of the procedure and evaluation of results take a short time

19.7 Treatment Options of Osteoarthritis

The treatment options and therapeutic interventions for osteoarthritis have dramatically widened over the past decade. The main aim of the treatment is to improve the quality of life by controlling pain with minimal side effects and to improve joint function and mobility. The management of the disease involves the combinatorial use of pharmacological and non-pharmacological therapies (Hochberg et al. 2012). The pharmacological remedies for the initial management of osteoarthritis include nutritional supplements, paracetamol, topical or oral application of NSAIDs like tramadol, and the injectable intra-articular corticosteroids like chondroitin sulfate and glucosamine. Patients not responding to initial treatment therapies are given opioids and injectable intra-articular hyaluronic acid, duloxetine. Surgical intervention is recommended in patients not responding to non-invasive treatment therapies. The surgical methods include arthroplasty, correction osteotomy, and debridement/lavage. The non-pharmacological interventions include physical and educational remedies. The physical approaches include a range of motion exercises, muscle strengthening, and aerobic activity. However, the educational approaches include walking aids, joint protection strategies, and changes in lifestyle (exercise and dietary changes). The physiotherapy approaches, including manual therapy, thermal techniques, and electrotherapy, are also recommended in certain patients (Zhang and Doherthy 2006; Zhang et al. 2007, 2008, 2009).

19.7.1 Non-pharmacological Intervention

The change in lifestyle relieves many symptoms of osteoarthritis. The most important modifiable risk factor for osteoarthritis treatment is obesity (Messier et al. 2011). In obese patients, weight loss alleviates the risk of developing symptomatic osteoarthritis and reduces the symptoms once the disease has appeared (Pai et al. 1997). Studies have shown that reducing the weight by 10% results in a remarkable reduction in a load of knee joints. Also, the weight loss of 5 kg in women reduces the risk of symptomatic knee osteoarthritis by 50%. The weight loss results in a favorable change in the bone and cartilage biomarkers. It also helps in improving the structure of the cartilage (Gudbergsen et al. 2012). It has been shown that constant physical exercise improves the quality of life and physical function and also helps in pain reduction in knee osteoarthritis (Deyle et al. 2005). This improvement in symptoms is comparable to that caused by NSAIDs. Before starting the home exercise, most patients follow a therapy program from an occupational or physical therapist that gives them access to both the equipment and direction and hence helps in getting better results. Apart from exercises, bracing may also prove effective in the treatment of knee osteoarthritis (Rannou et al. 2010; Zhang et al. 2017). In the case of lateral and medial knee osteoarthritis, the relief of symptoms is observed by unloading knee braces (Anandacoomarasamy et al. 2012; Fransen et al. 2015). However, in the case of patellar osteoarthritis, taping or the use of a peripatellar device containing sleeves may prove to be useful. A medial or lateral wedge in the soles is also effective in managing osteoarthritis (Bhatia et al. 2013). The use of rigid or semi-rigid thumb base splints is used for joint immobilization in carpometacarpal osteoarthritis. Education and self-management also serve as capable approaches for the treatment of pain in osteoarthritis. In self-management, the symptom management and disease consequences are put in the patients' hands. It includes self-tailoring, problem-solving, decision-making, self-monitoring, self-efficacy, and patient-physician partnership. The self-management programs have proven to be advantageous and have been shown to alleviate disability and pain associated with arthritis (Du et al. 2011). Social isolation, anxiety, depression, and psychological dysfunction have been shown to increase the pain associated with osteoarthritis. Psychotherapy sessions and CBT (cognitive behavioral therapy) to modify negative behavior and thinking also help in the treatment of osteoarthritis (Helminen et al. 2015).

19.7.2 Pharmacological Intervention

The first line of treatment for osteoarthritis is acetaminophen and NSAIDs (Helminen et al. 2015; Derry et al. 2016). NSAIDs have proven efficient in treating osteoarthritic pain. In patients suffering from gastrointestinal disease, NSAIDs, nonselective NSAIDs, and COX-2 inhibitors should be used with care and be given with gastro-protective agents. NSAIDs are more effective than acetaminophen in the treatment of osteoarthritic pain. Acetaminophen modestly works for osteoarthritic pain. However, it is preferable and safe in case NSAIDs are contraindicated (Towheed et al. 2006). The topical application of NSAIDs also provides an attractive treatment option for managing osteoarthritic pain. The topical application of diclofenac and ketoprofen has been shown to be effective in treating the pain from osteoarthritis. Narcotic analgesics and weak opioids are also used in osteoarthritis in case the other pharmacological therapies are contraindicated or fail. However, the recommendations issued by the Centre for Disease Control and Prevention in 2016 should be followed while treating patients with opioids (Dowell et al. 2016). In addition, the inhibitors of serotonin-norepinephrine reuptake are also included in the pharmacological treatment options of osteoarthritis. The first random trial of duloxetine showed a remarkable reduction in average pain scores compared to the placebo (Chappell et al. 2011). NSAIDs along with duloxetine have been shown to be more effective than NSAIDs alone in treating osteoarthritic pain. These studies resulted in the approval of duloxetine by FDA (Food and Drug Administration) for treating chronic knee osteoarthritis (Brown and Boulay 2013; Bennett et al. 2012).

19.7.3 Surgical Intervention

Patients with severe osteoarthritis who don't respond to conservative pain management and whose quality of life is severely affected by the disease have total joint arthroplasty as an option (Skou et al. 2015). It remarkably reduces pain and improves joint function in patients with severe knee or hip osteoarthritis. Studies have shown significant relief in pain and quality of life post-1 year of surgery compared to conservative management. Patients usually stay for 2 or 3 days in the hospital although many patients leave the hospital sooner post-surgery. Post-surgical prevention of complications is critical (Chen et al. 2012). The early complications of the surgery include deep vein thrombosis, wound dehiscence, failed prosthesis, and infections. The outcomes of the surgery can be improved by early mobilization, and it may reduce the hospital stay as well. Hip precautions are to be followed in patients with total hip arthroplasty (THA) in order to prevent the dislocation of the hip. In the case of posterior approach total hip arthroplasty, the precautions include avoiding internal hip rotation, adduction, and hyperflexion (Bade and Stevens-Lapsley 2012). However, in the case of direct anterior approach total hip arthroplasty, the external hip rotation, adduction, and hyperextension should be avoided. Post-surgery the rehabilitation program needs to be followed. However, there is a lack of consensus with regard to the intensity, duration, and frequency of physical rehabilitation. The physical rehabilitation programs include strengthening the quadriceps, gait training, range of motion, and training in routine life activities. In the case of patients with morbid obesity, bilateral total joint arthroplasty, or elderly patients (age greater than 85 years), acute physical rehabilitation is recommended (Chen et al. 2012; Quinn et al. 2018). In patients with total hip arthroplasty, most of the functional gains are observed within the initial 6 months. However, with total knee replacement, the gain of function may take up to 1 year. In a total knee replacement, the strength of the quadriceps may get reduced by 60%, which may be alleviated by strengthening exercises. The exercises may enhance coordination, gait speed, and the ability to climb stairs (Skou et al. 2015).

19.7.4 Intra-articular Intervention

Delivery of drugs through an intra-articular system has several benefits compared to systemic delivery (Emami et al. 2018; Evans et al. 2014). The advantages include diminished cost, lesser adverse effects, diminished local exposure, and magnified local concentration and bioavailability. Despite the advantages, the efficacy of this mode of treatment is still controversial, and the guidelines pertaining to their use are usually contradictory. The lymphatic drainage system usually clears the intra-articular therapies from the synovial fluid, and the rate depends upon the molecular size of the drug (Evans et al. 2014). The half-life of hyaluronic acid in the joint is approximately 26 h; however, albumin takes about 1–13 h to clear the joint. Apart from that, the half-life of steroids and NSAIDs in the joint is approximately 1–4 h. Irrespective of the brief residence time of the intra-articular drugs, studies have reported outcomes lasting several months. The reason behind these long-term outcomes may be treatment specific and needs further evaluation (Gerwin et al. 2006). Despite the widespread use of intra-articular therapies for osteoarthritis treatment, there is still conflict about whether the intra-articular therapies (glucocorticoids and hyaluronic acid) are advantageous in comparison to the joint aspiration technique alone. The intra-articular injections have been shown to evoke a strong placebo effect (Nguyen et al. 2016; Larsen et al. 2008; Hameed and Ihm 2012). Parameters like stiffness and pain have been shown to respond to an intra-articular placebo (Hameed and Ihm 2012). Patients receiving the mock injections (placebo or injections lacking the therapeutic agent) have shown outcomes that are not significantly different from those receiving hydrocortisone, procaine, lactic acid, or saline. In patients not responding to non-pharmacological treatment therapies like analgesics and NSAIDs, hyaluronic acid and glucocorticoids constitute the standard intra-articular options for managing knee pain related to osteoarthritis. Significant evidence questions the efficacy of these standard intra-articular therapies, and many professional organizations question the adequacy of their use (Roseland et al. 2004; Abhishek and Doherty 2013).

Hyaluronate is a natural polysaccharide consisting of the repeating units of β-D-glucuronic acid and β-D-*N*-acetylglucosamine. It is naturally present in synovial fluid and cartilage (Ghosh and Guodilin 2002). The molecular weight of hyaluronic acid in the synovial fluid ranges from 6500 to 10,900 kDa. An adult knee contains about 2 mL of synovial fluid having a hyaluronate concentration of 2.5 to 4 mg/mL. Hyaluronate in the synovial fluid contributes to the rheological properties and hence enables it to serve as a cushion and a lubricant (Balazs and Denlinger 1993). Osteoarthritis is characterized by higher rates of hyaluronate depolymerization (molecular weight, 2700–4500 kDa) and clearing than normal. The half-life of hyaluronate in the synovial fluid is about 20 h. This half-life gets reduced to 11–12 h in the inflamed joint, resulting in the reduction of the visco-elasticity of synovial fluid (Balazs et al. 1967). Knee osteoarthritis symptoms can be relieved by the administration of exogenous intra-articular hyaluronate (Bannaru et al. 2011). It is either obtained from animal sources like rooster comb or synthesized by bacterial fermentation. The molecular weight of the injectable polymer ranges from 100 to 10,000 kDa. The exogenous administration of intra-articular hyaluronate enhances the visco-elastic properties of synovial fluid and provides lubrication to the joint. This therapy is sometimes referred to as viscosupplementation (Hunter 2015; Strauss et al. 2009). Hyaluronate has also been shown to exert chondroprotective, analgesic, and anti-inflammatory ramifications on the synovium and articular cartilage. The clinical benefits of intra-articular hyaluronate persist beyond the residence time of the polymer in the joint. It has been suggested that intra-articular hyaluronate reestablishes joint homeostasis by enhancing the endogenous hyaluronate synthesis that persists even after the injected material is cleared off the joint.

The efficacy of hyaluronate in the treatment of knee osteoarthritis is still uncertain. Intra-articular administration of a high molecular weight (6000 kDa) hyaluronate derivative, hylan G-F 20, results in notable but clinically insignificant pain reduction in knee osteoarthritis compared to the saline placebo (Bedard et al. 2018; Conrozier and Chevalier 2008; Chevalier et al. 2010). The effect of molecular weight on the efficacy of hyaluronate in the treatment of osteoarthritis is also controversial. Treatment of knee osteoarthritis using the intermediate molecular weight derivative of hyaluronate was found to be remarkably superior compared to the low molecular weight derivatives (Berenbaum et al. 2012). However, in other trials, the benefit was nearly the same for both the high and intermediate molecular weight derivatives (Ghosh and Guodilin 2012; Maheu et al. 2011; Reichenbach et al. 2007). The efficacy of intra-articular hyaluronate was found to be modest and, at worst, inappreciable from a placebo. This treatment is also characterized by mild pain and inflammation at the injection site, which goes off with analgesics and ice (Jones et al. 2019).

Corticosteroids, due to their glucocorticoid effects, serve both as immunosuppressive and anti-inflammatory agents. These drugs inhibit the activity of phospholipase A2, alter the levels of cytokines and enzymes, and modulate the traffic of WBCs (Gordon and Schumacher 1979)). These reactions result in the diminished production of proinflammatory agents like histamines, bradykinins, leukotrienes, and prostaglandins. Under physiological conditions, pain is induced by the stimulation of primary afferent nociceptive fibers by histamine and bradykinin. These nociceptors are sensitized by leukotrienes and prostaglandins. The reduction of these proinflammatory derivatives explains the corticosteroid-induced pain relief. Corticosteroids also inhibit the synthesis of immunoglobulins and neutrophil superoxide. They also interfere with the migration and adhesion associated with inflammation. They inhibit the c-fiber transmission and stabilize the neural membranes, hence serving as immediate antinociceptive agents (Uthman et al. 2003; Creamer 1999).

Corticosteroids have been shown to enhance cartilage lesions and even promote degenerative lesions within the cartilage by inhibiting the synthesis of cartilage proteoglycan in rabbits (Papacrhistou et al. 1997; Stitik et al. 2006). In contrast, administering low-dose intra-articular corticosteroids in dogs normalizes cartilage proteoglycan synthesis. It remarkably diminishes the severity and incidence of cartilage lesions and osteophyte formation (Pelletier and Martel Pelletier 1989; 1991). The use of intra-articular corticosteroids is associated with certain adverse side effects. In approximately 40% of patients, unpleasant facial flushing lasting from a few hours to several days is the most common side effect (Patrick and Doherty 1987; Wicki et al. 2000). Patients having a medical history of diabetes or hypertension should monitor their blood glucose and blood pressure for several days after the administration of intra-articular corticosteroid injections as they have been shown to increase both blood glucose levels and blood pressure (Younes et al. 2007). Apart from these side effects, intra-articular corticosteroids also suppress the hypothalamic-pituitary axis for almost 1 month; however, the effects of this suppression are not clear. Intra-articular corticosteroid injections also cause a flare in the joint within the first 24–48 h, which goes off with the use of ice and analgesics. The joint replacement surgery should not be carried out within 6 weeks of the injection as it increases the risk of post-surgery infection (McIntosh et al. 2006). Other less commonly reported side effects include avascular necrosis, tendon rupture, hypo- or hyperpigmentation around the site of injection, and fat atrophy (Fisher and Bickel 1971). Different preparations of corticosteroids are present in the markets, and the frequencies with which these preparations are used vary. The most commonly used preparations include methylprednisolone acetate, which is followed by triamcinolone acetonide and beta-methasone sodium (Pelletier et al. (1994), Pyne et al. 2004). These preparations vary in their solubility, betamethasone being the most soluble and triamcinolone acetonide being insoluble. Preparations having lesser solubility remain within the joints for a longer time compared to the more soluble preparations. Lower solubility also results in lesser absorption and hence diminished systemic side effects (Pelletier et al. (1994), Pyne et al. 2004). The dosage of intra-articular corticosteroid injections also varies depending on the size of the affected joint and the formulation used. The larger the joint, the higher is the dose of the formulation required. Comparing equal quantities of triamcinolone hexacetonide and methylprednisolone in patients suffering from knee osteoarthritis showed better results with methylprednisolone at week 8 on the visual analog scale. Two meta-analyses studies have evaluated the efficacy of intra-articular corticosteroid injections, and the results show that these medications provide short-term symptomatic relief with pain reduction lasting 2–3 weeks (Cole and Schumacher 2005; Arroll and Goodyear Smith 2004; Bellamy et al. 2005; Skedros et al. 2007).

Platelet-rich plasma is a whole blood autologous derivative that is highly enriched in platelets. It is prepared by venesection of peripheral blood, which is then centrifuged to enrich the plasma platelets. The plasma contains thrombin and various cytokines, and the platelet degranulation releases additional mitogenic factors. Various growth factors present in PRP (platelet-rich plasma) include VEGF (vascular endothelial growth factor), FGF (fibroblast growth factor), PDGF (platelet-derived growth factor), IGF (insulin-like growth factor), and TGF-β (transforming growth factor β) (Bennell et al. 2017; Zhu et al. 2013). It also contains different bioactive agents that are involved in the healing of bone, muscle, ligaments, and tendons. PRP has been shown to be efficacious in the treatment of various pathologies like osteoarthritis, degenerative spine disease, hamstring injuries, Achilles and patella tendinopathy, rotator cuff disease, and lateral epicondylitis (Kanchanatawan et al. 2016; Meheux et al. 2016; Sadoghi et al. 2013; Dragoo et al. 2014). Growth factors present in PRP positively affect mesenchymal stem cell proliferation and chondrogenesis (Akeda et al. 2006; Ishida et al. 2007; Kabiri et al. 2014; Sundman et al. 2014; Lee et al. 2012). It diminishes the production of proinflammatory cytokines and enhances the production of anti-inflammatory mediators (Wu et al. 2011). It has been shown to inhibit NF-κB (nuclear factor-kappa B) transactivation and hence inhibit inflammatory reactions (Bendinelli et al. 2010; Van Buul et al. 2011). It also diminishes the expression of inflammatory enzymes like disintegrins, metalloproteinases, cyclooxygenase 2, and cyclooxygenase 4. These potential therapeutic effects make PRP a strong injectable candidate for the treatment of osteoarthritis (Sundman et al. 2014; Pereira et al. 2013).

PRP preparation involves the use of single or double-spinning centrifugation protocol. However, the effect of these protocols on the separation process is controversial. Apart from the manual centrifugation protocols, PRP kits are also commercially available (Zhu et al. 2013). Platelet concentration in the plasma may differ depending upon the manufacturer, centrifugation time, and the gravity forces. It also depends on the physical characteristics and the sex of the patient. PRP also includes various other products containing different concentrations of fibrin and leukocytes (Dragoo et al. 2012). Plasma concentrate and autologous conditioned plasma (ACP) are the other terms used for these preparations. Apart from PRP, other blood preparations, including autologous blood injection and autologous conditioned serum, are also used as therapeutic agents in osteoarthritis. In autologous blood injection, venous blood is directly used for the treatment, and however, in autologous conditioned serum, venous blood is prepared in such a way so that the growth factors and interleukin-1 receptor antagonist are concentrated (Braun et al. 2014). Different PRP systems that vary in the concentration of growth factors and WBCs are used. Some systems are enriched in WBCs, while others are enriched in growth factors. Neutrophils being the most prominent WBCs, their excessive joint infiltration results in delayed healing and chronic inflammation (Muraglia et al. 2014; Dragoo et al. 2012). Phagocytic macrophages mediate the clearing of the particulate debris that is left behind after the activation of neutrophils and release of proteases. Leukocyte-rich PRP produces a potent inflammatory response and also enhances the death of synoviocytes compared to leukocyte-poor PRP. Both leukocyte-poor and leukocyte-rich PRP have the same safety profile and both induce short-term reactions than hyaluronate (Wu et al. 2011).

PRP is advantageous for osteoarthritis as it inhibits inflammatory and catabolic responses and promotes anabolic responses (Sundman et al. 2014; Drengk et al. 2009). Growth factors present in PRP are involved in tissue healing, collagen synthesis, vessel and bone remodeling, and inhibition of chondrocyte apoptosis. Fibrin released by platelets serves as a scaffold and also mediates stem cell migration to the affected tissue, thereby triggering a healing response. Pre-clinical studies have shown that PRP injections promote the regeneration of damaged joints in osteoarthritis (Hamid et al. 2014; Lee et al. 2012). They also enhance the synthesis of proteoglycans and type II collagen and encourage the proliferation of chondrocytes and cartilage regeneration. PRP also shows positive effects on synoviocytes and meniscal cells (Ishida et al. 2012). The role of PRP in the regulation of inflammation is complex. It initially promotes inflammation, which is followed by a decline in the production of inflammatory molecules. PRP also acts as a direct analgesic by augmenting cannabinoid receptors (Lee et al. 2012). PRP hence serves as a disease-modifying therapeutic agent in osteoarthritis by slowing tissue degeneration and promoting tissue repair (Malavolta et al. 2014; Filardo et al. 2015). Even in the absence of cartilage regeneration, PRP ensures symptomatic relief through its analgesic effects and modulation of inflammation (Kim et al. 2014).

Several advantages are associated with the use of PRP for osteoarthritis management (Meheux et al. 2016). They are easy to use and require a minimally invasive procedure for intra-articular administration. The use of intra-articular PRP injections is safe because patients own molecules are used in this treatment approach (Kwon et al. 2012). No adverse side effects are associated with the use of intra-articular PRP injections. Minor side effects, including bruising and/or swelling at the site of injection, tenderness, and bleeding, are reported with their use (Fig. 19.4).

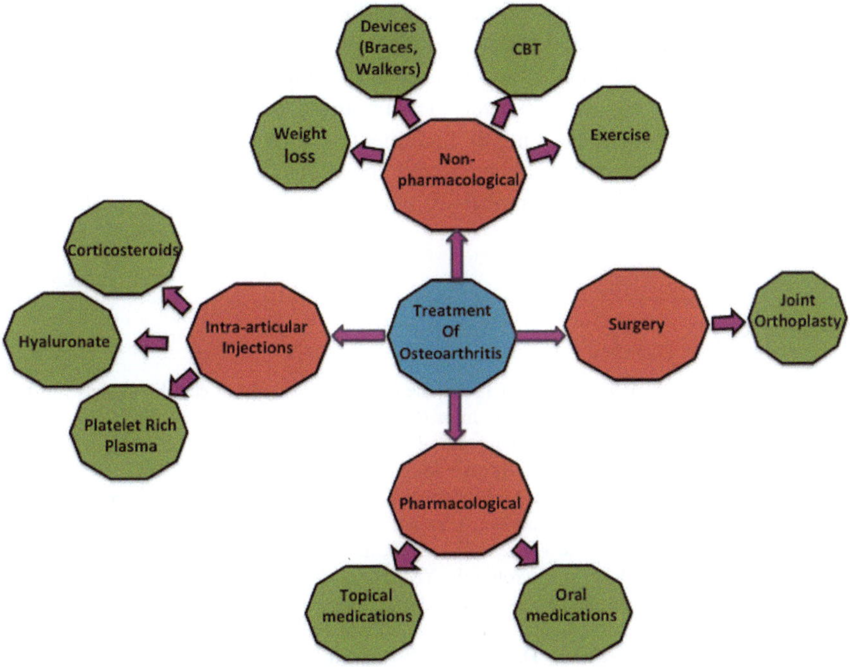

Fig. 19.4 Treatment options for osteoarthritis

Apart from the above-discussed injectable materials, several other compounds have been used to treat osteoarthritis pain (Braun et al. 2014). These include botulinum toxin type A, prolotherapy, anakinra, polynucleotides, thiotepa cytostatica, lactic acid solution, mucopolysaccharide polysulfuric acid ester, chloroquine, sodium pentosan polysulfate, somatostatin, glucosamine, morphine, saline solution, silicone, and orgotein. The use of these compounds has no definitive recommendation, as they have been examined in minor studies only. Botulinum toxin works as an antinociceptive agent by inhibiting the expression of calcitonin gene-related protein and substance P, which are the known mediators of neurogenic inflammation (Uthman et al. 2003; Singh 2010).

19.8 Conclusion

Osteoarthritis, a disease affecting the musculoskeletal system, diminishes the quality of life and imparts a huge socioeconomic cost globally. Although much progress has been made in crafting the novel therapeutic strategies for rheumatoid arthritis and in comprehending its pathphysiology, the pathogenesis of osteoarthritis holds multiple mysteries that still remain to be explored. Although first recognized as the non-inflammatory disorder affecting the joints with treatment focused mainly on disproportionate wear and tear and biomechanical features, recent advances in science helped in refining the interpretation of the disease. In order to diminish the financial burden of the disease and to prolong joint health, standard protocol must be developed to determine whether surgical reconstruction or conservative treatment is required by a patient. Also to gear the ongoing process of understanding the pathogenesis of the disease and to find the novel therapeutic treatments, it is important that the researchers be supported and encouraged, so as to improve the quality of life for the infinite number of patients surviving with the disability and pain of osteoarthritis.

References

Aashaq S, Batool A, Andrabi KI (2019) TAK1 mediates convergence of cellular signals for death and survival. Apoptosis 24(1–2):3–20. https://doi.org/10.1007/s10495-018-1490-7

Abhishek A, Doherty M (2013) Mechanisms of the placebo response in pain in osteoarthritis. Osteoarthr Cartil 21(9):1229–1235. https://doi.org/10.1016/j.joca.2013.04.018

Akeda K, An HS, Okuma M, Attawia M, Miyamoto K, Thonar EJ-MA, Lenz ME, Sah RL, Masuda K (2006) Platelet-rich plasma stimulates porcine articular chondrocyte proliferation and matrix biosynthesis. Osteoarthr Cartil 14(12):1272–1280. https://doi.org/10.1016/j.joca.2006.05.008

Allen KD, Chen J-C, Callahan LF, Golightly YM, Helmick CG, Renner JB, Jordan JM (2010) Associations of occupational tasks with knee and hip osteoarthritis: the Johnston County Osteoarthritis Project. J Rheumatol 37(4):842–850. https://doi.org/10.3899/jrheum.090302

Altman R, Alarcón G, Appelrouth D, Bloch D, Borenstein D, Brandt K, Brown C, Cooke TD, Daniel W, Feldman D (1991) The American College of Rheumatology criteria for the classification and reporting of osteoarthritis of the hip. Arthritis Rheum 34(5):505–514. https://doi.org/10.1002/art.1780340502

Amin AR, Abramson SB (1998) The role of nitric oxide in articular cartilage breakdown in osteoarthritis. Curr Opin Rheumatol 10(3):263–268. https://doi.org/10.1097/00002281-199805000-00018

Anandacoomarasamy A, Leibman S, Smith G, Caterson I, Giuffre B, Fransen M, Sambrook PN, March L (2012) Weight loss in obese people has structure-modifying effects on medial but not on lateral knee articular cartilage. Ann Rheum Dis 71(1):26–32. https://doi.org/10.1136/ard.2010.144725

Anitua E, Sánchez M, Nurden AT, Zalduendo MM, de la Fuente M, Azofra J, Andía I (2007) Platelet-released growth factors enhance the secretion of hyaluronic acid and induce hepatocyte growth factor production by synovial fibroblasts from arthritic patients. Rheumatology (Oxford, England) 46(12):1769–1772. https://doi.org/10.1093/rheumatology/kem234

Arroll B, Goodyear-Smith F (2004) Corticosteroid injections for osteoarthritis of the knee: meta-analysis. BMJ (Clin Res Ed) 328(7444):869. https://doi.org/10.1136/bmj.38039.573970.7C

Bade MJ, Stevens-Lapsley JE (2012) Restoration of physical function in patients following total knee arthroplasty: an update on rehabilitation practices. Curr Opin Rheumatol 24(2):208–214. https://doi.org/10.1097/BOR.0b013e32834ff26d

Balazs EA, Denlinger JL (1993) Viscosupplementation: a new concept in the treatment of osteoarthritis. J Rheumatol Suppl 39:3–9

Balazs EA, Watson D, Duff IF, Roseman S (1967) Hyaluronic acid in synovial fluid. I. Molecular parameters of hyaluronic acid in normal and arthritis human fluids. Arthritis Rheum 10(4):357–376. https://doi.org/10.1002/art.1780100407

Bannuru RR, Natov NS, Dasi UR, Schmid CH, McAlindon TE (2011) Therapeutic trajectory following intra-articular hyaluronic acid injection in knee osteoarthritis—meta-analysis. Osteoarthr Cartil 19(6):611–619. https://doi.org/10.1016/j.joca.2010.09.014

Barter MJ, Bui C, Young DA (2012) Epigenetic mechanisms in cartilage and osteoarthritis: DNA methylation, histone modifications and microRNAs. Osteoarthr Cartil 20(5):339–349. https://doi.org/10.1016/j.joca.2011.12.012

Bau B, Gebhard PM, Haag J, Knorr T, Bartnik E, Aigner T (2002) Relative messenger RNA expression profiling of collagenases and aggrecanases in human articular chondrocytes in vivo and in vitro. Arthritis Rheum 46(10):2648–2657. https://doi.org/10.1002/art.10531

Bedard NA, DeMik DE, Glass NA, Burnett RA, Bozic KJ, Callaghan JJ (2018) Impact of clinical practice guidelines on use of intra-articular hyaluronic acid and corticosteroid injections for knee osteoarthritis. J Bone Joint Surg Am 100(10):827–834. https://doi.org/10.2106/JBJS.17.01045

Bedson J, Croft PR (2008) The discordance between clinical and radiographic knee osteoarthritis: a systematic search and summary of the literature. BMC Musculoskelet Disord 9:116. https://doi.org/10.1186/1471-2474-9-116

Bellamy N, Campbell J, Robinson V, Gee T, Bourne R, Wells G (2005) Intraarticular corticosteroid for treatment of osteoarthritis of the knee. Cochrane Database Syst Rev 2:CD005328. https://doi.org/10.1002/14651858.CD005328

Bendinelli P, Matteucci E, Dogliotti G, Corsi MM, Banfi G, Maroni P, Desiderio MA (2010) Molecular basis of anti-inflammatory action of platelet-rich plasma on human chondrocytes: mechanisms of NF-κB inhibition via HGF. J Cell Physiol 225(3):757–766. https://doi.org/10.1002/jcp.22274

Bennell KL, Hunter DJ, Hinman RS (2012) Management of osteoarthritis of the knee. BMJ (Clin Res Ed) 345:e4934. https://doi.org/10.1136/bmj.e4934

Bennell KL, Hunter DJ, Paterson KL (2017) Platelet-rich plasma for the management of hip and knee osteoarthritis. Curr Rheumatol Rep 19(5):24. https://doi.org/10.1007/s11926-017-0652-x

Berenbaum F, Grifka J, Cazzaniga S, D'Amato M, Giacovelli G, Chevalier X, Rannou F, Rovati LC, Maheu E (2012) A randomised, double-blind, controlled trial comparing two intra-articular hyaluronic acid preparations differing by their molecular weight in symptomatic knee osteoarthritis. Ann Rheum Dis 71(9):1454–1460. https://doi.org/10.1136/annrheumdis-2011-200972

Bhatia D, Bejarano T, Novo M (2013) Current interventions in the management of knee osteoarthritis. J Pharm Bioallied Sci 5(1):30–38. https://doi.org/10.4103/0975-7406.106561

Blom AB, van Lent PLEM, Holthuysen AEM, van der Kraan PM, Roth J, van Rooijen N, van den Berg WB (2004) Synovial lining macrophages mediate osteophyte formation during experimental osteoarthritis. Osteoarthr Cartil 12(8):627–635. https://doi.org/10.1016/j.joca.2004.03.003

Braun HJ, Kim HJ, Chu CR, Dragoo JL (2014) The effect of platelet-rich plasma formulations and blood products on human synoviocytes: implications for intra-articular injury and therapy. Am J Sports Med 42(5):1204–1210. https://doi.org/10.1177/0363546514525593

Brown JP, Boulay LJ (2013) Clinical experience with duloxetine in the management of chronic musculoskeletal pain. A focus on osteoarthritis of the knee. Therap Adv Musculoskelet Dis 5(6): 291–304. https://doi.org/10.1177/1759720X13508508

Brown TD, Johnston RC, Saltzman CL, Marsh JL, Buckwalter JA (2006) Posttraumatic osteoarthritis: a first estimate of incidence, prevalence, and burden of disease. J Orthop Trauma 20(10): 739–744. https://doi.org/10.1097/01.bot.0000246468.80635.ef

Buckwalter JA, Mankin HJ (1998) Articular cartilage: tissue design and chondrocyte-matrix interactions. Instr Course Lect 47:477–486

Calich ALG, Domiciano DS, Fuller R (2010) Osteoarthritis: can anti-cytokine therapy play a role in treatment? Clin Rheumatol 29(5):451–455. https://doi.org/10.1007/s10067-009-1352-3

Cameron KL, Hsiao MS, Owens BD, Burks R, Svoboda SJ (2011) Incidence of physician-diagnosed osteoarthritis among active duty United States military service members. Arthritis Rheum 63(10):2974–2982. https://doi.org/10.1002/art.30498

Catterall JB, Hsueh MF, Stabler TV, McCudden CR, Bolognesi M, Zura R, Jordan JM, Renner JB, Feng S, Kraus VB (2012) Protein modification by deamidation indicates variations in joint extracellular matrix turnover. J Biol Chem 287(7):4640–4651. https://doi.org/10.1074/jbc.M111.249649

Chappell AS, Desaiah D, Liu-Seifert H, Zhang S, Skljarevski V, Belenkov Y, Brown JP (2011) A double-blind, randomized, placebo-controlled study of the efficacy and safety of duloxetine for the treatment of chronic pain due to osteoarthritis of the knee. Pain Pract 11(1):33–41. https://doi.org/10.1111/j.1533-2500.2010.00401.x

Chen AF, Stewart MK, Heyl AE, Klatt BA (2012) Effect of immediate postoperative physical therapy on length of stay for total joint arthroplasty patients. J Arthroplast 27(6):851–856. https://doi.org/10.1016/j.arth.2012.01.011

Chevalier X, Jerosch J, Goupille P, van Dijk N, Luyten FP, Scott DL, Bailleul F, Pavelka K (2010) Single, intra-articular treatment with 6 ml hylan G-F 20 in patients with symptomatic primary osteoarthritis of the knee: a randomised, multicentre, double-blind, placebo controlled trial. Ann Rheum Dis 69(1):113–119. https://doi.org/10.1136/ard.2008.094623

Chevalier X, Eymard F, Richette P (2013) Biologic agents in osteoarthritis: hopes and disappointments. Nat Rev Rheumatol 9(7):400–410. https://doi.org/10.1038/nrrheum.2013.44

Coggon D, Croft P, Kellingray S, Barrett D, McLaren M, Cooper C (2000) Occupational physical activities and osteoarthritis of the knee. Arthritis Rheum 43(7):1443–1449. https://doi.org/10.1002/1529-0131(200007)43:7<1443::AID-ANR5>3.0.CO;2-1

Cole BJ, Schumacher HRJ (2005) Injectable corticosteroids in modern practice. J Am Acad Orthop Surg 13(1):37–46. https://doi.org/10.5435/00124635-200501000-00006

Conrozier T, Chevalier X (2008) Long-term experience with hylan GF-20 in the treatment of knee osteoarthritis. Expert Opin Pharmacother 9(10):1797–1804. https://doi.org/10.1517/14656566.9.10.1797

Creamer P (1999) Intra-articular corticosteroid treatment in osteoarthritis. Curr Opin Rheumatol 11(5):417–421. https://doi.org/10.1097/00002281-199909000-00016

Dam EB, Loog M, Christiansen C, Byrjalsen I, Folkesson J, Nielsen M, Qazi AA, Pettersen PC, Garnero P, Karsdal MA (2009) Identification of progressors in osteoarthritis by combining biochemical and MRI-based markers. Arthritis Res Ther 11(4):R115. https://doi.org/10.1186/ar2774

Derry S, Conaghan P, Da Silva JAP, Wiffen PJ, Moore RA (2016) Topical NSAIDs for chronic musculoskeletal pain in adults. Cochrane Database Syst Rev 4(4):CD007400. https://doi.org/10.1002/14651858.CD007400.pub3

Deyle GD, Allison SC, Matekel RL, Ryder MG, Stang JM, Gohdes DD, Hutton JP, Henderson NE, Garber MB (2005) Physical therapy treatment effectiveness for osteoarthritis of the knee: a randomized comparison of supervised clinical exercise and manual therapy procedures versus a home exercise program. Phys Ther 85(12):1301–1317

Dowell D, Haegerich TM, Chou R (2016) CDC guideline for prescribing opioids for chronic pain—United States, 2016. JAMA 315(15):1624–1645. https://doi.org/10.1001/jama.2016.1464

Dragoo JL, Braun HJ, Durham JL, Ridley BA, Odegaard JI, Luong R, Arnoczky SP (2012) Comparison of the acute inflammatory response of two commercial platelet-rich plasma systems in healthy rabbit tendons. Am J Sports Med 40(6):1274–1281. https://doi.org/10.1177/0363546512442334

Dragoo JL, Wasterlain AS, Braun HJ, Nead KT (2014) Platelet-rich plasma as a treatment for patellar tendinopathy: a double-blind, randomized controlled trial. Am J Sports Med 42(3):610–618. https://doi.org/10.1177/0363546513518416

Drengk A, Zapf A, Stürmer EK, Stürmer KM, Frosch K-H (2009) Influence of platelet-rich plasma on chondrogenic differentiation and proliferation of chondrocytes and mesenchymal stem cells. Cells Tissues Organs 189(5):317–326. https://doi.org/10.1159/000151290

Du S, Yuan C, Xiao X, Chu J, Qiu Y, Qian H (2011) Self-management programs for chronic musculoskeletal pain conditions: a systematic review and meta-analysis. Patient Educ Couns 85(3):e299–e310. https://doi.org/10.1016/j.pec.2011.02.021

Ekenstedt KJ, Sonntag WE, Loeser RF, Lindgren BR, Carlson CS (2006) Effects of chronic growth hormone and insulin-like growth factor 1 deficiency on osteoarthritis severity in rat knee joints. Arthritis Rheum 54(12):3850–3858. https://doi.org/10.1002/art.22254

Emami A, Tepper J, Short B, Yaksh TL, Bendele AM, Ramani T, Cisternas AF, Chang JH, Mellon RD (2018) Toxicology evaluation of drugs administered via uncommon routes: intranasal, intraocular, intrathecal/intraspinal, and intra-articular. Int J Toxicol 37(1):4–27. https://doi.org/10.1177/1091581817741840

Evans CH, Kraus VB, Setton LA (2014) Progress in intra-articular therapy. Nat Rev Rheumatol 10(1):11–22. https://doi.org/10.1038/nrrheum.2013.159

Fautrel B, Hilliquin P, Rozenberg S, Allaert F-A, Coste P, Leclerc A, Rossignol M (2005) Impact of osteoarthritis: results of a nationwide survey of 10,000 patients consulting for OA. Joint Bone Spine 72(3):235–240. https://doi.org/10.1016/j.jbspin.2004.08.009

Feldmann M, Maini RN (2001) Anti-TNF alpha therapy of rheumatoid arthritis: what have we learned? Annu Rev Immunol 19:163–196. https://doi.org/10.1146/annurev.immunol.19.1.163

Felson DT (2006) Clinical practice. Osteoarthritis of the knee. N Engl J Med 354(8):841–848. https://doi.org/10.1056/NEJMcp051726

Felson DT (2013) Osteoarthritis as a disease of mechanics. Osteoarthr Cartil 21(1):10–15. https://doi.org/10.1016/j.joca.2012.09.012

Felson DT, Nevitt MC (2004) Epidemiologic studies for osteoarthritis: new versus conventional study design approaches. Rheum Dis Clin N Am 30(4):783–797., vii. https://doi.org/10.1016/j.rdc.2004.07.005

Felson DT, Naimark A, Anderson J, Kazis L, Castelli W, Meenan RF (1987) The prevalence of knee osteoarthritis in the elderly. The Framingham Osteoarthritis Study. Arthritis Rheum 30(8): 914–918. https://doi.org/10.1002/art.1780300811

Felson DT, Lawrence RC, Dieppe PA, Hirsch R, Helmick CG, Jordan JM, Kington RS, Lane NE, Nevitt MC, Zhang Y, Sowers M, McAlindon T, Spector TD, Poole AR, Yanovski SZ, Ateshian G, Sharma L, Buckwalter JA, Brandt KD, Fries JF (2000) Osteoarthritis: new insights. Part 1: the disease and its risk factors. Ann Intern Med 133(8):635–646. https://doi.org/10.7326/0003-4819-133-8-200010170-00016

Filardo G, Kon E, Roffi A, Di Matteo B, Merli ML, Marcacci M (2015) Platelet-rich plasma: why intra-articular? A systematic review of preclinical studies and clinical evidence on PRP for joint degeneration. Knee Surg Sports Traumatol Arthrosc 23(9):2459–2474. https://doi.org/10.1007/s00167-013-2743-1

Fisher DE, Bickel WH (1971) Corticosteroid-induced avascular necrosis. A clinical study of seventy-seven patients. J Bone Joint Surg Am 53(5):859–873

Fortier LA, Mohammed HO, Lust G, Nixon AJ (2002) Insulin-like growth factor-I enhances cell-based repair of articular cartilage. J Bone Joint Surg Br 84(2):276–288. https://doi.org/10.1302/0301-620x.84b2.11167

Fransen M, McConnell S, Harmer AR, Van der Esch M, Simic M, Bennell KL (2015) Exercise for osteoarthritis of the knee. Cochrane Database Syst Rev 1:CD004376. https://doi.org/10.1002/14651858.CD004376.pub3

Fukumoto T, Sperling JW, Sanyal A, Fitzsimmons JS, Reinholz GG, Conover CA, O'Driscoll SW (2003) Combined effects of insulin-like growth factor-1 and transforming growth factor-beta1 on periosteal mesenchymal cells during chondrogenesis in vitro. Osteoarthr Cartil 11(1):55–64. https://doi.org/10.1053/joca.2002.0869

Garnero P, Piperno M, Gineyts E, Christgau S, Delmas PD, Vignon E (2001) Cross sectional evaluation of biochemical markers of bone, cartilage, and synovial tissue metabolism in patients with knee osteoarthritis: relations with disease activity and joint damage. Ann Rheum Dis 60(6): 619–626. https://doi.org/10.1136/ard.60.6.619

Gendron C, Kashiwagi M, Hughes C, Caterson B, Nagase H (2003) TIMP-3 inhibits aggrecanase-mediated glycosaminoglycan release from cartilage explants stimulated by catabolic factors. FEBS Lett 555(3):431–436. https://doi.org/10.1016/s0014-5793(03)01295-x

Gerwin N, Hops C, Lucke A (2006) Intraarticular drug delivery in osteoarthritis. Adv Drug Deliv Rev 58(2):226–242. https://doi.org/10.1016/j.addr.2006.01.018

Ghosh P, Guidolin D (2002) Potential mechanism of action of intra-articular hyaluronan therapy in osteoarthritis: are the effects molecular weight dependent? Semin Arthritis Rheum 32(1):10–37. https://doi.org/10.1053/sarh.2002.33720

Glasson SS, Askew R, Sheppard B, Carito BA, Blanchet T, Ma H-L, Flannery CR, Kanki K, Wang E, Peluso D, Yang Z, Majumdar MK, Morris EA (2004) Characterization of and osteoarthritis susceptibility in ADAMTS-4-knockout mice. Arthritis Rheum 50(8):2547–2558. https://doi.org/10.1002/art.20558

Glyn-Jones S, Palmer AJR, Agricola R, Price AJ, Vincent TL, Weinans H, Carr AJ (2015) Osteoarthritis. Lancet (London, England) 386(9991):376–387. https://doi.org/10.1016/S0140-6736(14)60802-3

Goldring MB, Berenbaum F (2015) Emerging targets in osteoarthritis therapy. Curr Opin Pharmacol 22:51–63. https://doi.org/10.1016/j.coph.2015.03.004

Goldring MB, Goldring SR (2007) Osteoarthritis. J Cell Physiol 213(3):626–634. https://doi.org/10.1002/jcp.21258

Goldring MB, Otero M (2011) Inflammation in osteoarthritis. Curr Opin Rheumatol 23(5): 471–478. https://doi.org/10.1097/BOR.0b013e328349c2b1

Goodrich LR, Hidaka C, Robbins PD, Evans CH, Nixon AJ (2007) Genetic modification of chondrocytes with insulin-like growth factor-1 enhances cartilage healing in an equine model. J Bone Joint Surg Br 89(5):672–685. https://doi.org/10.1302/0301-620X.89B5.18343

Gordon GV, Schumacher HR (1979) Electron microscopic study of depot corticosteroid crystals with clinical studies after intra-articular injection. J Rheumatol 6(1):7–14

Gudbergsen H, Boesen M, Lohmander LS, Christensen R, Henriksen M, Bartels EM, Christensen P, Rindel L, Aaboe J, Danneskiold-Samsøe B, Riecke BF, Bliddal H (2012) Weight loss is effective for symptomatic relief in obese subjects with knee osteoarthritis independently of joint damage severity assessed by high-field MRI and radiography. Osteoarthr Cartil 20(6): 495–502. https://doi.org/10.1016/j.joca.2012.02.639

Hameed F, Ihm J (2012) Injectable medications for osteoarthritis. PM R 4(5 Suppl):S75–S81. https://doi.org/10.1016/j.pmrj.2012.02.010

Hamid MSA, Mohamed Ali MR, Yusof A, George J, Lee LPC (2014) Platelet-rich plasma injections for the treatment of hamstring injuries: a randomized controlled trial. Am J Sports Med 42(10):2410–2418. https://doi.org/10.1177/0363546514541540

Hashimoto K, Otero M, Imagawa K, de Andrés MC, Coico JM, Roach HI, Oreffo ROC, Marcu KB, Goldring MB (2013) Regulated transcription of human matrix metalloproteinase 13 (MMP13) and interleukin-1β (IL1B) genes in chondrocytes depends on methylation of specific proximal promoter CpG sites. J Biol Chem 288(14):10061–10072. https://doi.org/10.1074/jbc.M112.421156

Hawker GA, Stewart L, French MR, Cibere J, Jordan JM, March L, Suarez-Almazor M, Gooberman-Hill R (2008) Understanding the pain experience in hip and knee osteoarthritis—an OARSI/OMERACT initiative. Osteoarthr Cartil 16(4):415–422. https://doi.org/10.1016/j.joca.2007.12.017

Heinegård D, Saxne T (2011) The role of the cartilage matrix in osteoarthritis. Nat Rev Rheumatol 7(1):50–56. https://doi.org/10.1038/nrrheum.2010.198

Helminen E-E, Sinikallio SH, Valjakka AL, Väisänen-Rouvali RH, Arokoski JPA (2015) Effectiveness of a cognitive-behavioural group intervention for knee osteoarthritis pain: a randomized controlled trial. Clin Rehabil 29(9):868–881. https://doi.org/10.1177/0269215514558567

Hiramatsu K, Iwai T, Yoshikawa H, Tsumaki N (2011) Expression of dominant negative TGF-β receptors inhibits cartilage formation in conditional transgenic mice. J Bone Miner Metab 29(4): 493–500. https://doi.org/10.1007/s00774-010-0248-2

Hoch JM, Mattacola CG, Medina McKeon JM, Howard JS, Lattermann C (2011) Serum cartilage oligomeric matrix protein (sCOMP) is elevated in patients with knee osteoarthritis: a systematic review and meta-analysis. Osteoarthr Cartil 19(12):1396–1404. https://doi.org/10.1016/j.joca.2011.09.005

Hochberg MC, Altman RD, April KT, Benkhalti M, Guyatt G, McGowan J, Towheed T, Welch V, Wells G, Tugwell P (2012) American College of Rheumatology 2012 recommendations for the use of nonpharmacologic and pharmacologic therapies in osteoarthritis of the hand, hip, and knee. Arthritis Care Res 64(4):465–474. https://doi.org/10.1002/acr.21596

Hofmann C, Gropp R, von der Mark K (1992) Expression of anchorin CII, a collagen-binding protein of the annexin family, in the developing chick embryo. Dev Biol 151(2):391–400. https://doi.org/10.1016/0012-1606(92)90179-k

Hunter DJ (2015) Viscosupplementation for osteoarthritis of the knee. N Engl J Med 372(11): 1040–1047. https://doi.org/10.1056/NEJMct1215534

Hunter DJ, March L, Sambrook PN (2002) Knee osteoarthritis: the influence of environmental factors. Clin Exp Rheumatol 20(1):93–100

Hunter DJ, Demissie S, Cupples LA, Aliabadi P, Felson DT (2004) A genome scan for joint-specific hand osteoarthritis susceptibility: the Framingham Study. Arthritis Rheum 50(8): 2489–2496. https://doi.org/10.1002/art.20445

Ishida K, Kuroda R, Miwa M, Tabata Y, Hokugo A, Kawamoto T, Sasaki K, Doita M, Kurosaka M (2007) The regenerative effects of platelet-rich plasma on meniscal cells in vitro and its in vivo application with biodegradable gelatin hydrogel. Tissue Eng 13(5):1103–1112. https://doi.org/10.1089/ten.2006.0193

Jones IA, Togashi R, Wilson ML, Heckmann N, Vangsness CTJ (2019) Intra-articular treatment options for knee osteoarthritis. Nat Rev Rheumatol 15(2):77–90. https://doi.org/10.1038/s41584-018-0123-4

Kabiri A, Esfandiari E, Esmaeili A, Hashemibeni B, Pourazar A, Mardani M (2014) Platelet-rich plasma application in chondrogenesis. Adv Biomed Res 3:138. https://doi.org/10.4103/2277-9175.135156

Kanchanatawan W, Arirachakaran A, Chaijenkij K, Prasathaporn N, Boonard M, Piyapittayanun P, Kongtharvonskul J (2016) Short-term outcomes of platelet-rich plasma injection for treatment of osteoarthritis of the knee. Knee Surg Sports Traumatol Arthrosc 24(5):1665–1677. https://doi.org/10.1007/s00167-015-3784-4

Kashiwagi M, Tortorella M, Nagase H, Brew K (2001) TIMP-3 is a potent inhibitor of aggrecanase 1 (ADAM-TS4) and aggrecanase 2 (ADAM-TS5). J Biol Chem 276(16):12501–12504. https://doi.org/10.1074/jbc.C000848200

Kim H-J, Yeom JS, Koh Y-G, Yeo J-E, Kang K-T, Kang Y-M, Chang B-S, Lee C-K (2014) Anti-inflammatory effect of platelet-rich plasma on nucleus pulposus cells with response of TNF-α and IL-1. J Orthop Res 32(4):551–556. https://doi.org/10.1002/jor.22532

King LK, March L, Anandacoomarasamy A (2013) Obesity & osteoarthritis. Indian J Med Res 138(2):185–193

Kujala UM, Kettunen J, Paananen H, Aalto T, Battié MC, Impivaara O, Videman T, Sarna S (1995) Knee osteoarthritis in former runners, soccer players, weight lifters, and shooters. Arthritis Rheum 38(4):539–546. https://doi.org/10.1002/art.1780380413

Kwon DR, Park GY, Lee S-U (2012) The effects of intra-articular platelet-rich plasma injection according to the severity of collagenase-induced knee osteoarthritis in a rabbit model. Ann Rehabil Med 36(4):458–465. https://doi.org/10.5535/arm.2012.36.4.458

Larsen C, Ostergaard J, Larsen SW, Jensen H, Jacobsen S, Lindegaard C, Andersen PH (2008) Intra-articular depot formulation principles: role in the management of postoperative pain and arthritic disorders. J Pharm Sci 97(11):4622–4654. https://doi.org/10.1002/jps.21346

Lawrence RC, Felson DT, Helmick CG, Arnold LM, Choi H, Deyo RA, Gabriel S, Hirsch R, Hochberg MC, Hunder GG, Jordan JM, Katz JN, Kremers HM, Wolfe F (2008) Estimates of the prevalence of arthritis and other rheumatic conditions in the United States. Part II. Arthritis Rheum 58(1):26–35. https://doi.org/10.1002/art.23176

Lee JH, Fitzgerald JB, Dimicco MA, Grodzinsky AJ (2005) Mechanical injury of cartilage explants causes specific time-dependent changes in chondrocyte gene expression. Arthritis Rheum 52(8): 2386–2395. https://doi.org/10.1002/art.21215

Lee H-R, Park KM, Joung YK, Park KD, Do SH (2012) Platelet-rich plasma loaded hydrogel scaffold enhances chondrogenic differentiation and maturation with up-regulation of CB1 and CB2. J Control Release 159(3):332–337. https://doi.org/10.1016/j.jconrel.2012.02.008

Lethbridge-Cejku M, Scott WWJ, Reichle R, Ettinger WH, Zonderman A, Costa P, Plato CC, Tobin JD, Hochberg MC (1995) Association of radiographic features of osteoarthritis of the knee with knee pain: data from the Baltimore longitudinal study of aging. Arthritis Care Res 8(3): 182–188. https://doi.org/10.1002/art.1790080311

Loeser RF (2013) Aging processes and the development of osteoarthritis. Curr Opin Rheumatol 25(1):108–113. https://doi.org/10.1097/BOR.0b013e32835a9428

Loeser RF, Shanker G, Carlson CS, Gardin JF, Shelton BJ, Sonntag WE (2000) Reduction in the chondrocyte response to insulin-like growth factor 1 in aging and osteoarthritis: studies in a non-human primate model of naturally occurring disease. Arthritis Rheum 43(9):2110–2120. https://doi.org/10.1002/1529-0131(200009)43:9<2110::AID-ANR23>3.0.CO;2-U

Loeser RF, Collins JA, Diekman BO (2016) Ageing and the pathogenesis of osteoarthritis. Nat Rev Rheumatol 12(7):412–420. https://doi.org/10.1038/nrrheum.2016.65

Lotz MK, Caramés B (2011) Autophagy and cartilage homeostasis mechanisms in joint health, aging and OA. Nat Rev Rheumatol 7(10):579–587. https://doi.org/10.1038/nrrheum.2011.109

Lotz M, Loeser RF (2012) Effects of aging on articular cartilage homeostasis. Bone 51(2):241–248. https://doi.org/10.1016/j.bone.2012.03.023

Maheu E, Zaim M, Appelboom T, Jeka S, Trc T, Berenbaum F, Maasalu K, Berenbaum F (2011) Comparative efficacy and safety of two different molecular weight (MW) hyaluronans F60027 and Hylan G-F20 in symptomatic osteoarthritis of the knee (KOA). Results of a non inferiority, prospective, randomized, controlled trial. Clin Exp Rheumatol 29(3):527–535

Malavolta EA, Gracitelli MEC, Ferreira Neto AA, Assunção JH, Bordalo-Rodrigues M, de Camargo OP (2014) Platelet-rich plasma in rotator cuff repair: a prospective randomized study. Am J Sports Med 42(10):2446–2454. https://doi.org/10.1177/0363546514541777

Martin P, D'Souza D, Martin J, Grose R, Cooper L, Maki R, McKercher SR (2003) Wound healing in the PU.1 null mouse—tissue repair is not dependent on inflammatory cells. Curr Biol 13(13): 1122–1128. https://doi.org/10.1016/s0960-9822(03)00396-8

Massagué J (2012) TGFβ signalling in context. Nat Rev Mol Cell Biol 13(10):616–630. https://doi.org/10.1038/nrm3434

McAlindon TE, Felson DT, Zhang Y, Hannan MT, Aliabadi P, Weissman B, Rush D, Wilson PW, Jacques P (1996a) Relation of dietary intake and serum levels of vitamin D to progression of osteoarthritis of the knee among participants in the Framingham Study. Ann Intern Med 125(5): 353–359. https://doi.org/10.7326/0003-4819-125-5-199609010-00001

McAlindon TE, Jacques P, Zhang Y, Hannan MT, Aliabadi P, Weissman B, Rush D, Levy D, Felson DT (1996b) Do antioxidant micronutrients protect against the development and progression of knee osteoarthritis? Arthritis Rheum 39(4):648–656. https://doi.org/10.1002/art.1780390417

McInnes IB, Schett G (2007) Cytokines in the pathogenesis of rheumatoid arthritis. Nat Rev Immunol 7(6):429–442. https://doi.org/10.1038/nri2094

McIntosh AL, Hanssen AD, Wenger DE, Osmon DR (2006) Recent intraarticular steroid injection may increase infection rates in primary THA. Clin Orthop Relat Res 451:50–54. https://doi.org/10.1097/01.blo.0000229318.51254.79

Meheux CJ, McCulloch PC, Lintner DM, Varner KE, Harris JD (2016) Efficacy of intra-articular platelet-rich plasma injections in knee osteoarthritis: a systematic review. Arthroscopy 32(3): 495–505. https://doi.org/10.1016/j.arthro.2015.08.005

Melchiorri C, Meliconi R, Frizziero L, Silvestri T, Pulsatelli L, Mazzetti I, Borzì RM, Uguccioni M, Facchini A (1998) Enhanced and coordinated in vivo expression of inflammatory cytokines and nitric oxide synthase by chondrocytes from patients with osteoarthritis. Arthritis Rheum 41(12): 2165–2174. https://doi.org/10.1002/1529-0131(199812)41:12<2165::AID-ART11>3.0.CO;2-O

Messier SP, Loeser RF, Miller GD, Morgan TM, Rejeski WJ, Sevick MA, Ettinger WHJ, Pahor M, Williamson JD (2004) Exercise and dietary weight loss in overweight and obese older adults with knee osteoarthritis: the arthritis, diet, and activity promotion trial. Arthritis Rheum 50(5): 1501–1510. https://doi.org/10.1002/art.20256

Messier SP, Legault C, Loeser RF, Van Arsdale SJ, Davis C, Ettinger WH, DeVita P (2011) Does high weight loss in older adults with knee osteoarthritis affect bone-on-bone joint loads and muscle forces during walking? Osteoarthr Cartil 19(3):272–280. https://doi.org/10.1016/j.joca.2010.11.010

Milner JM, Rowan AD, Cawston TE, Young DA (2006) Metalloproteinase and inhibitor expression profiling of resorbing cartilage reveals pro-collagenase activation as a critical step for collagenolysis. Arthritis Res Ther 8(5):R142. https://doi.org/10.1186/ar2034

Mobasheri A (2012) Osteoarthritis year 2012 in review: biomarkers. Osteoarthr Cartil 20(12):1451–1464. https://doi.org/10.1016/j.joca.2012.07.009

Mollon B, Kandel R, Chahal J, Theodoropoulos J (2013) The clinical status of cartilage tissue regeneration in humans. Osteoarthr Cartil 21(12):1824–1833. https://doi.org/10.1016/j.joca.2013.08.024

Morales TI (2008) The quantitative and functional relation between insulin-like growth factor-I (IGF) and IGF-binding proteins during human osteoarthritis. J Orthop Res 26(4):465–474. https://doi.org/10.1002/jor.20549

Muraglia A, Ottonello C, Spanò R, Dozin B, Strada P, Grandizio M, Cancedda R, Mastrogiacomo M (2014) Biological activity of a standardized freeze-dried platelet derivative to be used as cell culture medium supplement. Platelets 25(3):211–220. https://doi.org/10.3109/09537104.2013.803529

Murray CJL, Atkinson C, Bhalla K, Birbeck G, Burstein R, Chou D, Dellavalle R, Danaei G, Ezzati M, Fahimi A, Flaxman D, Foreman, Gabriel S, Gakidou E, Kassebaum N, Khatibzadeh S, Lim S, Lipshultz SE, London S et al (2013) The state of US health, 1990-2010: burden of diseases, injuries, and risk factors. JAMA 310(6):591–608. https://doi.org/10.1001/jama.2013.13805

Musumeci G, Aiello FC, Szychlinska MA, Di Rosa M, Castrogiovanni P, Mobasheri A (2015) Osteoarthritis in the XXIst century: risk factors and behaviours that influence disease onset and progression. Int J Mol Sci 16(3):6093–6112. https://doi.org/10.3390/ijms16036093

Nagase H, Kashiwagi M (2003) Aggrecanases and cartilage matrix degradation. Arthritis Res Ther 5(2):94–103. https://doi.org/10.1186/ar630

Nagase H, Visse R, Murphy G (2006) Structure and function of matrix metalloproteinases and TIMPs. Cardiovasc Res 69(3):562–573. https://doi.org/10.1016/j.cardiores.2005.12.002

Nelson AE, Allen KD, Golightly YM, Goode AP, Jordan JM (2014) A systematic review of recommendations and guidelines for the management of osteoarthritis: the chronic osteoarthritis management initiative of the U.S. bone and joint initiative. Semin Arthritis Rheum 43(6):701–712. https://doi.org/10.1016/j.semarthrit.2013.11.012

Neogi T (2013) The epidemiology and impact of pain in osteoarthritis. Osteoarthr Cartil 21(9):1145–1153. https://doi.org/10.1016/j.joca.2013.03.018

Neogi T, Zhang Y (2013) Epidemiology of osteoarthritis. Rheum Dis Clin N Am 39(1):1–19. https://doi.org/10.1016/j.rdc.2012.10.004

Neogi T, Felson DT, Sarno R, Booth SL (2008) Vitamin K in hand osteoarthritis: results from a randomised clinical trial. Ann Rheum Dis 67(11):1570–1573. https://doi.org/10.1136/ard.2008.094771

Neogi T, Felson D, Niu J, Nevitt M, Lewis CE, Aliabadi P, Sack B, Torner J, Bradley L, Zhang Y (2009) Association between radiographic features of knee osteoarthritis and pain: results from two cohort studies. BMJ (Clin Res Ed) 339:b2844. https://doi.org/10.1136/bmj.b2844

Neuhold LA, Killar L, Zhao W, Sung ML, Warner L, Kulik J, Turner J, Wu W, Billinghurst C, Meijers T, Poole AR, Babij P, DeGennaro LJ (2001) Postnatal expression in hyaline cartilage of constitutively active human collagenase-3 (MMP-13) induces osteoarthritis in mice. J Clin Invest 107(1):35–44. https://doi.org/10.1172/JCI10564

Nevitt MC, Felson DT, Williams EN, Grady D (2001) The effect of estrogen plus progestin on knee symptoms and related disability in postmenopausal women: the Heart and Estrogen/Progestin Replacement Study, a randomized, double-blind, placebo-controlled trial. Arthritis Rheum 44(4):811–818. https://doi.org/10.1002/1529-0131(200104)44:4<811::AID-ANR137>3.0.CO;2-F

Nguyen C, Lefèvre-Colau M-M, Poiraudeau S, Rannou F (2016) Evidence and recommendations for use of intra-articular injections for knee osteoarthritis. Ann Phys Rehabil Med 59(3): 184–189. https://doi.org/10.1016/j.rehab.2016.02.008

Odding E, Valkenburg HA, Algra D, Vandenouweland FA, Grobbee DE, Hofman A (1998) Associations of radiological osteoarthritis of the hip and knee with locomotor disability in the Rotterdam Study. Ann Rheum Dis 57(4):203–208. https://doi.org/10.1136/ard.57.4.203

Önnerfjord P, Khabut A, Reinholt FP, Svensson O, Heinegård D (2012) Quantitative proteomic analysis of eight cartilaginous tissues reveals characteristic differences as well as similarities between subgroups. J Biol Chem 287(23):18913–18924. https://doi.org/10.1074/jbc.M111.298968

Orlowsky EW, Kraus VB (2015) The role of innate immunity in osteoarthritis: when our first line of defense goes on the offensive. J Rheumatol 42(3):363–371. https://doi.org/10.3899/jrheum.140382

Pai YC, Rymer WZ, Chang RW, Sharma L (1997) Effect of age and osteoarthritis on knee proprioception. Arthritis Rheum 40(12):2260–2265. https://doi.org/10.1002/art.1780401223

Papacrhistou G, Anagnostou S, Katsorhis T (1997) The effect of intraarticular hydrocortisone injection on the articular cartilage of rabbits. Acta Orthop Scand Suppl 275:132–134. https://doi.org/10.1080/17453674.1997.11744766

Pattrick M, Doherty M (1987) Facial flushing after intra-articular injection of steroid. Br Med J (Clin Res Ed) 295(6610):1380. https://doi.org/10.1136/bmj.295.6610.1380

Peach CA, Carr AJ, Loughlin J (2005) Recent advances in the genetic investigation of osteoarthritis. Trends Mol Med 11(4):186–191. https://doi.org/10.1016/j.molmed.2005.02.005

Pelletier JP, Martel-Pelletier J (1989) Protective effects of corticosteroids on cartilage lesions and osteophyte formation in the Pond-Nuki dog model of osteoarthritis. Arthritis Rheum 32(2): 181–193. https://doi.org/10.1002/anr.1780320211

Pelletier JP, Martel-Pelletier J (1991) In vivo protective effects of prophylactic treatment with tiaprofenic acid or intraarticular corticosteroids on osteoarthritic lesions in the experimental dog model. J Rheumatol Suppl 27:127–130

Pelletier JP, Mineau F, Raynauld JP, Woessner JFJ, Gunja-Smith Z, Martel-Pelletier J (1994) Intraarticular injections with methylprednisolone acetate reduce osteoarthritic lesions in parallel with chondrocyte stromelysin synthesis in experimental osteoarthritis. Arthritis Rheum 37(3): 414–423. https://doi.org/10.1002/art.1780370316

Pelletier JP, Caron JP, Evans C, Robbins PD, Georgescu HI, Jovanovic D, Fernandes JC, Martel-Pelletier J (1997) In vivo suppression of early experimental osteoarthritis by interleukin-1 receptor antagonist using gene therapy. Arthritis Rheum 40(6):1012–1019. https://doi.org/10.1002/art.1780400604

Pereira RC, Scaranari M, Benelli R, Strada P, Reis RL, Cancedda R, Gentili C (2013) Dual effect of platelet lysate on human articular cartilage: a maintenance of chondrogenic potential and a transient proinflammatory activity followed by an inflammation resolution. Tissue Eng Part A 19(11–12):1476–1488. https://doi.org/10.1089/ten.TEA.2012.0225

Petersson IF, Jacobsson LTH (2002) Osteoarthritis of the peripheral joints. Best Pract Res Clin Rheumatol 16(5):741–760. https://doi.org/10.1053/berh.2002.0266

Pettipher ER, Higgs GA, Henderson B (1986) Interleukin 1 induces leukocyte infiltration and cartilage proteoglycan degradation in the synovial joint. Proc Natl Acad Sci U S A 83(22): 8749–8753. https://doi.org/10.1073/pnas.83.22.8749

Prieto-Alhambra D, Judge A, Javaid MK, Cooper C, Diez-Perez A, Arden NK (2014) Incidence and risk factors for clinically diagnosed knee, hip and hand osteoarthritis: influences of age, gender and osteoarthritis affecting other joints. Ann Rheum Dis 73(9):1659–1664. https://doi.org/10.1136/annrheumdis-2013-203355

Pyne D, Ioannou Y, Mootoo R, Bhanji A (2004) Intra-articular steroids in knee osteoarthritis: a comparative study of triamcinolone hexacetonide and methylprednisolone acetate. Clin Rheumatol 23(2):116–120. https://doi.org/10.1007/s10067-003-0841-z

Quinn RH, Murray JN, Pezold R, Sevarino KS (2018) Surgical management of osteoarthritis of the knee. J Am Acad Orthop Surg 26(9):e191–e193. https://doi.org/10.5435/JAAOS-D-17-00424

Rannou F, Poiraudeau S, Beaudreuil J (2010) Role of bracing in the management of knee osteoarthritis. Curr Opin Rheumatol 22(2):218–222. https://doi.org/10.1097/BOR.0b013e32833619c4

Reichenbach S, Blank S, Rutjes AWS, Shang A, King EA, Dieppe PA, Jüni P, Trelle S (2007) Hylan versus hyaluronic acid for osteoarthritis of the knee: a systematic review and meta-analysis. Arthritis Rheum 57(8):1410–1418. https://doi.org/10.1002/art.23103

Reynard LN, Loughlin J (2012) Genetics and epigenetics of osteoarthritis. Maturitas 71(3): 200–204. https://doi.org/10.1016/j.maturitas.2011.12.001

Robinson WH, Lepus CM, Wang Q, Raghu H, Mao R, Lindstrom TM, Sokolove J (2016) Low-grade inflammation as a key mediator of the pathogenesis of osteoarthritis. Nat Rev Rheumatol 12(10):580–592. https://doi.org/10.1038/nrrheum.2016.136

Rosseland LA, Helgesen KG, Breivik H, Stubhaug A (2004) Moderate-to-severe pain after knee arthroscopy is relieved by intraarticular saline: a randomized controlled trial. Anesth Analg 98(6):1546–1551, table of contents. https://doi.org/10.1213/01.ane.0000112433.71197.fa

Sacitharan PK (2019) Ageing and osteoarthritis. Subcell Biochem 91:123–159. https://doi.org/10.1007/978-981-13-3681-2_6

Sadoghi P, Rosso C, Valderrabano V, Leithner A, Vavken P (2013) The role of platelets in the treatment of Achilles tendon injuries. J Orthop Res 31(1):111–118. https://doi.org/10.1002/jor.22199

Sah RL, Chen AC, Grodzinsky AJ, Trippel SB (1994) Differential effects of bFGF and IGF-I on matrix metabolism in calf and adult bovine cartilage explants. Arch Biochem Biophys 308(1): 137–147. https://doi.org/10.1006/abbi.1994.1020

Salaffi F, Carotti M, Stancati A, Grassi W (2005) Health-related quality of life in older adults with symptomatic hip and knee osteoarthritis: a comparison with matched healthy controls. Aging Clin Exp Res 17(4):255–263. https://doi.org/10.1007/BF03324607

Sandmark H, Hogstedt C, Vingård E (2000) Primary osteoarthrosis of the knee in men and women as a result of lifelong physical load from work. Scand J Work Environ Health 26(1):20–25. https://doi.org/10.5271/sjweh.505

Scanzello CR, Goldring SR (2012) The role of synovitis in osteoarthritis pathogenesis. Bone 51(2): 249–257. https://doi.org/10.1016/j.bone.2012.02.012

Sellam J, Berenbaum F (2010) The role of synovitis in pathophysiology and clinical symptoms of osteoarthritis. Nat Rev Rheumatol 6(11):625–635. https://doi.org/10.1038/nrrheum.2010.159

Sellam J, Berenbaum F (2013) Is osteoarthritis a metabolic disease? Joint Bone Spine 80(6): 568–573. https://doi.org/10.1016/j.jbspin.2013.09.007

Serra R, Johnson M, Filvaroff EH, LaBorde J, Sheehan DM, Derynck R, Moses HL (1997) Expression of a truncated, kinase-defective TGF-beta type II receptor in mouse skeletal tissue promotes terminal chondrocyte differentiation and osteoarthritis. J Cell Biol 139(2):541–552. https://doi.org/10.1083/jcb.139.2.541

Shane Anderson A, Loeser RF (2010) Why is osteoarthritis an age-related disease? Best Pract Res Clin Rheumatol 24(1):15–26. https://doi.org/10.1016/j.berh.2009.08.006

Shen J, Li S, Chen D (2014) TGF-β signaling and the development of osteoarthritis. Bone Res 2: 14002. https://doi.org/10.1038/boneres.2014.2

Singh JA (2010) Botulinum toxin therapy for osteoarticular pain: an evidence-based review. Therap Adv Musculoskelet Dis 2(2):105–118. https://doi.org/10.1177/1759720X09357113

Skedros JG, Hunt KJ, Pitts TC (2007) Variations in corticosteroid/anesthetic injections for painful shoulder conditions: comparisons among orthopaedic surgeons, rheumatologists, and physical medicine and primary-care physicians. BMC Musculoskelet Disord 8:63. https://doi.org/10.1186/1471-2474-8-63

Skou ST, Roos EM, Laursen MB, Rathleff MS, Arendt-Nielsen L, Simonsen O, Rasmussen S (2015) A randomized, controlled trial of total knee replacement. N Engl J Med 373(17): 1597–1606. https://doi.org/10.1056/NEJMoa1505467

Smith MD, Triantafillou S, Parker A, Youssef PP, Coleman M (1997) Synovial membrane inflammation and cytokine production in patients with early osteoarthritis. J Rheumatol 24(2): 365–371

Spector TD, MacGregor AJ (2004) Risk factors for osteoarthritis: genetics. Osteoarthr Cartil 12 Suppl A:S39–S44. https://doi.org/10.1016/j.joca.2003.09.005

Spector TD, Cicuttini F, Baker J, Loughlin J, Hart D (1996) Genetic influences on osteoarthritis in women: a twin study. BMJ (Clin Res Ed) 312(7036):940–943. https://doi.org/10.1136/bmj.312.7036.940

Srikanth VK, Fryer JL, Zhai G, Winzenberg TM, Hosmer D, Jones G (2005) A meta-analysis of sex differences prevalence, incidence and severity of osteoarthritis. Osteoarthr Cartil 13(9): 769–781. https://doi.org/10.1016/j.joca.2005.04.014

Stanton H, Melrose J, Little CB, Fosang AJ (2011) Proteoglycan degradation by the ADAMTS family of proteinases. Biochim Biophys Acta 1812(12):1616–1629. https://doi.org/10.1016/j.bbadis.2011.08.009

Stitik TP, Kumar A, Foye PM (2006) Corticosteroid injections for osteoarthritis. Am J Phys Med Rehabil 85(11 Suppl):S51–S65; quiz S66–8. https://doi.org/10.1097/01.phm.0000245508.35730.f3

Strauss EJ, Hart JA, Miller MD, Altman RD, Rosen JE (2009) Hyaluronic acid viscosupplementation and osteoarthritis: current uses and future directions. Am J Sports Med 37(8):1636–1644. https://doi.org/10.1177/0363546508326984

Sundman EA, Cole BJ, Karas V, Della Valle C, Tetreault MW, Mohammed HO, Fortier LA (2014) The anti-inflammatory and matrix restorative mechanisms of platelet-rich plasma in osteoarthritis. Am J Sports Med 42(1):35–41. https://doi.org/10.1177/0363546513507766

Thomas AC, Hubbard-Turner T, Wikstrom EA, Palmieri-Smith RM (2017) Epidemiology of posttraumatic osteoarthritis. J Athl Train 52(6):491–496. https://doi.org/10.4085/1062-6050-51.5.08

Towheed TE, Maxwell L, Judd MG, Catton M, Hochberg MC, Wells G (2006) Acetaminophen for osteoarthritis. Cochrane Database Syst Rev 1:CD004257. https://doi.org/10.1002/14651858.CD004257.pub2

Tsukazaki T, Matsumoto T, Enomoto H, Usa T, Ohtsuru A, Namba H, Iwasaki K, Yamashita S (1994) Growth hormone directly and indirectly stimulates articular chondrocyte cell growth. Osteoarthr Cartil 2(4):259–267. https://doi.org/10.1016/s1063-4584(05)80078-0

Uthman I, Raynauld J-P, Haraoui B (2003) Intra-articular therapy in osteoarthritis. Postgrad Med J 79(934):449–453. https://doi.org/10.1136/pmj.79.934.449

Valdes AM, Spector TD (2009) The contribution of genes to osteoarthritis. Med Clin North Am 93(1):45–66., x. https://doi.org/10.1016/j.mcna.2008.08.007

van Buul GM, Koevoet WLM, Kops N, Bos PK, Verhaar JAN, Weinans H, Bernsen MR, van Osch GJVM (2011) Platelet-rich plasma releasate inhibits inflammatory processes in osteoarthritic chondrocytes. Am J Sports Med 39(11):2362–2370. https://doi.org/10.1177/0363546511419278

van der Kraan PM (2012) Osteoarthritis year 2012 in review: biology. Osteoarthr Cartil 20(12): 1447–1450. https://doi.org/10.1016/j.joca.2012.07.010

van Spil WE, DeGroot J, Lems WF, Oostveen JCM, Lafeber FPJG (2010) Serum and urinary biochemical markers for knee and hip-osteoarthritis: a systematic review applying the consensus BIPED criteria. Osteoarthr Cartil 18(5):605–612. https://doi.org/10.1016/j.joca.2010.01.012

von Porat A, Roos EM, Roos H (2004) High prevalence of osteoarthritis 14 years after an anterior cruciate ligament tear in male soccer players: a study of radiographic and patient relevant outcomes. Ann Rheum Dis 63(3):269–273. https://doi.org/10.1136/ard.2003.008136

Vos T, Flaxman AD, Naghavi M, Lozano R, Michaud C, Ezzati M, Shibuya K, Salomon JA, Abdalla S, Aboyans V, Abraham J, Ackerman I, Aggarwal R, Ahn SY, Ali MK, Alvarado M, Anderson HR, Anderson LM, Andrews KG et al (2012) Years lived with disability (YLDs) for 1160 sequelae of 289 diseases and injuries 1990-2010: a systematic analysis for the Global Burden of Disease Study 2010. Lancet (London, England) 380(9859):2163–2196. https://doi.org/10.1016/S0140-6736(12)61729-2

Wicki J, Droz M, Cirafici L, Vallotton MB (2000) Acute adrenal crisis in a patient treated with intraarticular steroid therapy. J Rheumatol 27(2):510–511

Wluka AE, Cicuttini FM, Spector TD (2000) Menopause, oestrogens and arthritis. Maturitas 35(3): 183–199. https://doi.org/10.1016/s0378-5122(00)00118-3

Wluka AE, Lombard CB, Cicuttini FM (2013) Tackling obesity in knee osteoarthritis. Nat Rev Rheumatol 9(4):225–235. https://doi.org/10.1038/nrrheum.2012.224

Wu C-C, Chen W-H, Zao B, Lai P-L, Lin T-C, Lo H-Y, Shieh Y-H, Wu C-H, Deng W-P (2011) Regenerative potentials of platelet-rich plasma enhanced by collagen in retrieving pro-inflammatory cytokine-inhibited chondrogenesis. Biomaterials 32(25):5847–5854. https://doi.org/10.1016/j.biomaterials.2011.05.002

Yang C-C, Lin C-Y, Wang H-S, Lyu S-R (2013) Matrix metalloproteases and tissue inhibitors of metalloproteinases in medial plica and pannus-like tissue contribute to knee osteoarthritis progression. PLoS One 8(11):e79662. https://doi.org/10.1371/journal.pone.0079662

Younes M, Neffati F, Touzi M, Hassen-Zrour S, Fendri Y, Béjia I, Ben Amor A, Bergaoui N, Najjar MF (2007) Systemic effects of epidural and intra-articular glucocorticoid injections in diabetic and non-diabetic patients. Joint Bone Spine 74(5):472–476. https://doi.org/10.1016/j.jbspin.2006.10.009

Zhang W, Doherty M (2006) EULAR recommendations for knee and hip osteoarthritis: a critique of the methodology. Br J Sports Med 40(8):664–669. https://doi.org/10.1136/bjsm.2004.016840

Zhang Y, Jordan JM (2010) Epidemiology of osteoarthritis. Clin Geriatr Med 26(3):355–369. https://doi.org/10.1016/j.cger.2010.03.001

Zhang Y, Niu J, Kelly-Hayes M, Chaisson CE, Aliabadi P, Felson DT (2002) Prevalence of symptomatic hand osteoarthritis and its impact on functional status among the elderly: the Framingham Study. Am J Epidemiol 156(11):1021–1027. https://doi.org/10.1093/aje/kwf141

Zhang W, Doherty M, Arden N, Bannwarth B, Bijlsma J, Gunther K-P, Hauselmann HJ, Herrero-Beaumont G, Jordan K, Kaklamanis P, Leeb B, Lequesne M, Lohmander S, Mazieres B, Martin-Mola E, Pavelka K, Pendleton A, Punzi L, Swoboda B et al (2005) EULAR evidence based recommendations for the management of hip osteoarthritis: report of a task force of the EULAR Standing Committee for International Clinical Studies Including Therapeutics (ESCISIT). Ann Rheum Dis 64(5):669–681. https://doi.org/10.1136/ard.2004.028886

Zhang W, Doherty M, Leeb BF, Alekseeva L, Arden NK, Bijlsma JW, Dinçer F, Dziedzic K, Häuselmann HJ, Herrero-Beaumont G, Kaklamanis P, Lohmander S, Maheu E, Martín-Mola E, Pavelka K, Punzi L, Reiter S, Sautner J, Smolen J et al (2007) EULAR evidence based recommendations for the management of hand osteoarthritis: report of a task force of the EULAR Standing Committee for International Clinical Studies Including Therapeutics (ESCISIT). Ann Rheum Dis 66(3):377–388. https://doi.org/10.1136/ard.2006.062091

Zhang W, Moskowitz RW, Nuki G, Abramson S, Altman RD, Arden N, Bierma-Zeinstra S, Brandt KD, Croft P, Doherty M, Dougados M, Hochberg M, Hunter DJ, Kwoh K, Lohmander LS, Tugwell P (2008) OARSI recommendations for the management of hip and knee osteoarthritis, part II: OARSI evidence-based, expert consensus guidelines. Osteoarthr Cartil 16(2):137–162. https://doi.org/10.1016/j.joca.2007.12.013

Zhang W, Doherty M, Leeb BF, Alekseeva L, Arden NK, Bijlsma JW, Dincer F, Dziedzic K, Hauselmann HJ, Kaklamanis P, Kloppenburg M, Lohmander LS, Maheu E, Martin-Mola E, Pavelka K, Punzi L, Reiter S, Smolen J, Verbruggen G et al (2009) EULAR evidence-based recommendations for the diagnosis of hand osteoarthritis: report of a task force of ESCISIT. Ann Rheum Dis 68(1):8–17. https://doi.org/10.1136/ard.2007.084772

Zhang W, Nuki G, Moskowitz RW, Abramson S, Altman RD, Arden NK, Bierma-Zeinstra S, Brandt KD, Croft P, Doherty M, Dougados M, Hochberg M, Hunter DJ, Kwoh K, Lohmander LS, Tugwell P (2010) OARSI recommendations for the management of hip and knee osteoarthritis: part III: changes in evidence following systematic cumulative update of research published through January 2009. Osteoarthr Cartil 18(4):476–499. https://doi.org/10.1016/j.joca.2010.01.013

Zhang Q, Yue J, Golianu B, Sun Z, Lu Y (2017) Updated systematic review and meta-analysis of acupuncture for chronic knee pain. Acupunct Med 35(6):392–403. https://doi.org/10.1136/acupmed-2016-011306

Zhen G, Wen C, Jia X, Li Y, Crane JL, Mears SC, Askin FB, Frassica FJ, Chang W, Yao J, Carrino JA, Cosgarea A, Artemov D, Chen Q, Zhao Z, Zhou X, Riley L, Sponseller P, Wan M et al (2013) Inhibition of TGF-β signaling in mesenchymal stem cells of subchondral bone attenuates osteoarthritis. Nat Med 19(6):704–712. https://doi.org/10.1038/nm.3143

Zhu Y, Yuan M, Meng HY, Wang AY, Guo QY, Wang Y, Peng J (2013) Basic science and clinical application of platelet-rich plasma for cartilage defects and osteoarthritis: a review. Osteoarthr Cartil 21(11):1627–1637. https://doi.org/10.1016/j.joca.2013.07.017

Sabreena Aashaq was born in J&K, India. She completed her B.Sc. in 2011 from the University of Kashmir, India, and later joined M.Sc Biotechnology and completed the degree in 2013 from the University of Kashmir, India. She completed her doctoral program in Biotechnology from the University of Kashmir, India, in 2018 and is currently working as a Research Associate in the Department of Immunology and Molecular Medicine, Sher-i-Kashmir Institute of Medical Sciences, India. She has qualified various national-level exams, including CSIR-UGC NET & JRF, GATE, and ASRB-NET, and has been the recipient of various national-level fellowships. She has contributed extensively to world literature, with publications appearing in reputed journals. Her research interests include the regulation of various signal transduction pathways in cells, how dysregulation of these pathways results in different cancers, and the exploitation of these pathways in cancer therapeutics.

Asma Rafiq received her B.Sc in 2016 from the University of Kashmir, India, and later joined M.Sc Biochemistry. She completed her Masters program in Biochemistry at the University of Kashmir, India, in 2018. She qualified for GATE in 2020 and is currently pursuing her Ph.D. in the Department of Nanotechnology at the University of Kashmir, India. She has authored a couple of publications appearing in reputed journals. Her research interests include understanding the regulatory pathways in cancer and how these pathways help cancer cells avoid destruction by the immune attack, so these pathways can be exploited for therapeutic intervention in cancer treatment.

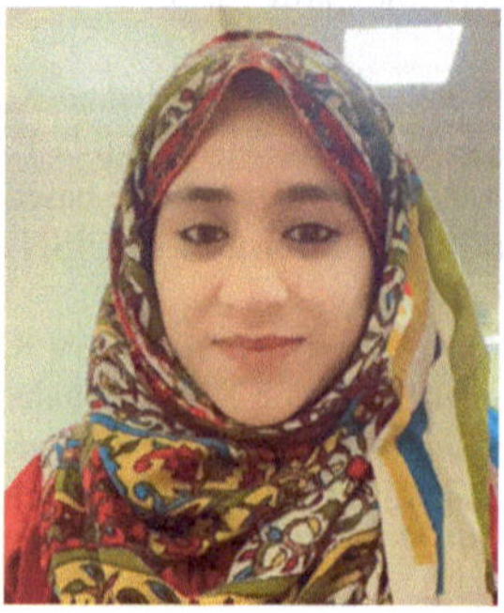

Iqra Jan was born in J&K, India. She completed her B.Sc. in 2016 from the University of Kashmir, India, and later joined her M.Sc. Biotechnology and completed the degree in 2018 from the University of Kashmir, India. She qualified for the GATE in 2020. She has been enrolled in a doctoral program in the Department of Biotechnology, University of Kashmir, India, since 2020 under Dr. Mushtaq A. Beigh. Her research interests include studying the role of post-translational modifications in cancer metastasis and exploiting these pathways in drug therapeutics.

Mehvish Ali completed her M.Sc. in Nanotechnology from the University of Kashmir in 2021, and is actively involved in understanding the nanotherapeutics for the delivery of phytochemicals from endangered plants.

Bilal A. Reshi has been working as an Assistant professor in the Department of Biotechnology, University of Kashmir, since 2005. He has done his master's (2001) and M.Phil. (2005) in Biochemistry from the University of Kashmir. He has qualified for UGC-NET (2001) and is currently pursuing Ph.D. (Part-time) in the department of Biotechnology since 2019. His area of specialization is plant science (exploring the potential of plant compounds as therapeutics). He has 17 years of teaching experience in the field.

Zaffar Amin Shah is a professor and head of Immunology and Molecular Medicine at Sher-i-Kashmir Institute of Medical Sciences, Soura, Srinagar. He holds BVSc, MVSc, and Ph.D. degrees and is actively involved in teaching and research. He has around 170 publications to his credit and has a total of 1500 citations..

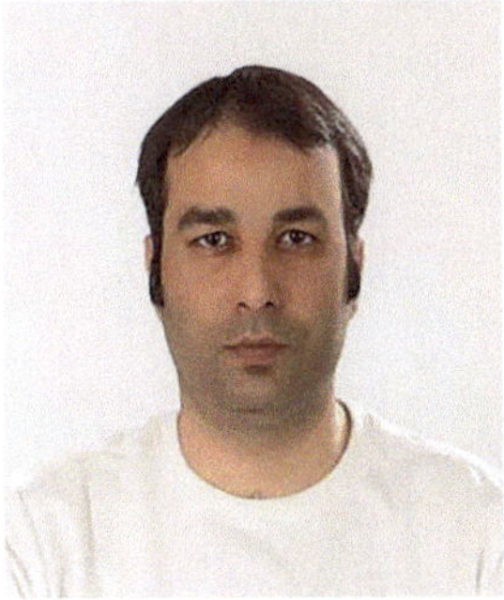

Faheem A. Sheikh is an Assistant Professor at the Department of Nanotechnology, University of Kashmir, India (2015). He served as an Assistant Professor at the Department of Biotechnology at the Central University of Kashmir, India (2015–2016); a Research Professor at Myongji University, South Korea (2014–2015); Assistant Professor at Hallym University, South Korea (2012–2014); Post-Doc/Research fellow at the University of Texas Rio Grande Valley, Texas, United States of America (2010–2012); and Research Professor at Myongji University, South Korea (2010). His research mainly focuses on fabricating nanomaterials used in tissue engineering.

Shafquat Majeed is currently working as an Assistant Professor in the Department of Nanotechnology, University of Kashmir, India. Dr. Majeed did his Masters in Chemistry from the Department of Chemistry, University of Kashmir. He did his Ph.D. at the prestigious Indian Institute of Science Bangalore (2008–2015). He has a postdoctoral research experience in rare-earth-based nanomaterials for biomedical applications and fabrication of perovskite-based solar cells from the Centre for Nanoscience and Engineering, IISc Bangalore (2015–2016). He heads the laboratory, where they work on the synthesis, characterization, and optical and magnetic properties of nanomaterials prepared through various solution-based routes. More specifically, they are working on developing new and enhanced multimodal imaging modalities for biomedical and healthcare applications.

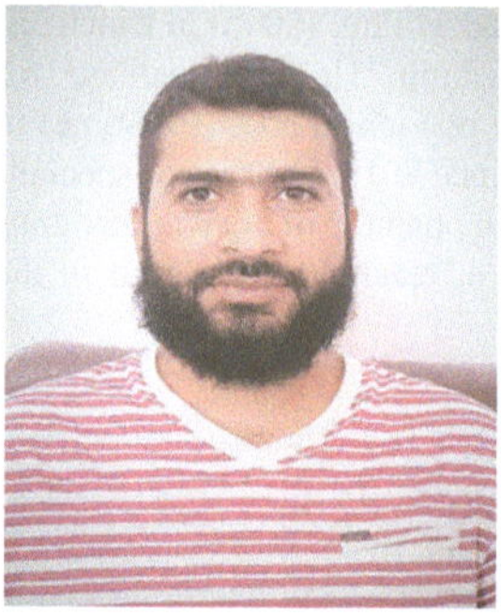

Mushtaq A. Beigh is working as an Assistant Professor at the Department of Nanotechnology, University of Kashmir, India. Dr. Beigh has postdoctoral research experience in cellular signalling and disease modelling (2013–2016). He had completed his Ph.D. at the University of Kashmir, India (2008–2012). He heads a laboratory at the University of Kashmir, where they work on a bio-nano interface to understand the molecular basis of nanoconjugate internalization. They also try to understand the growth factor signalling inputs directed at multiple cellular receptors like NRP1 and integrins in order to work out the mechanistic details of receptor-based internalization.

Promoting the Bio-potency of Bioactive Compounds Through Nanoencapsulation

20

Aasiya Majeed, Saima Masood, Sadaf Nazir, Faheem A. Sheikh, Mushtaq A. Beigh, and Shafquat Majeed

Abstract

Nanotechnology is actively used in food science and technology, and one of the evolving research areas involves the nanoencapsulation of volatile bioactive compounds. Quite often, these bioactive compounds are inappropriately absorbed, and by using nanostructured systems incorporating these compounds, several characteristics can be upgraded, such as solubility, stability, bioavailability, and protection against degradation. Essential oils (EOs) and their bioactive compounds are safe, and their formulations are used as green preservatives in the

A. Majeed
Department of Biochemistry, Sher-e-Kashmir University of Agricultural Sciences and Technology (Jammu), Chatha, Jammu and Kashmir, India

S. Masood
Department of Basic Sciences and Humanities, SKUAST-K, Srinagar, Jammu and Kashmir, India

S. Nazir
Multifunctional Nanomaterials Laboratory, Department of Nanotechnology, University of Kashmir, Srinagar, Jammu and Kashmir, India

F. A. Sheikh
Nanostructured and Biomimetic Lab, Department of Nanotechnology, University of Kashmir Hazratbal, Srinagar, Jammu and Kashmir, India
e-mail: faheemnt@uok.edu.in

M. A. Beigh
Cellular Signalling and Nanotherapeutics Laboratory, Department of Nanotechnology, University of Kashmir Hazratbal, Srinagar, Jammu and Kashmir, India
e-mail: beighm@uok.edu.in

S. Majeed (✉)
Laboratory for Multifunctional Nanomaterials, Department of Nanotechnology, University of Kashmir Hazratbal, Srinagar, Jammu and Kashmir, India
e-mail: smshah@uok.edu.in

F. A. Sheikh et al. (eds.), *Interaction of Nanomaterials With Living Cells*,
https://doi.org/10.1007/978-981-99-2119-5_20

food industry. However, their volatility and instability in fluctuating environmental conditions significantly hinder large-scale practical use. Recently, contrasting encapsulation technologies have been advocated as the promoter for improving EOs bio-potency. Encapsulation conserves and protects EOs from aggressive external conditions and, at the same time, allows their controlled release, which is helpful for applications in agronomy. This chapter focuses on some dominant and popular encapsulation techniques and their role in enhancing the bio-potency of EOs used as horticulture preservatives.

Keywords

Essential oils · Encapsulation techniques · Bio-potency · Gist material · Controlled release

20.1 Introduction

Nanotechnology is generally described as the design, creation, and utilization of the architecture, arrangement, and systems where the diameter and framework of the material is restricted to the range of 10^{-9} nm (Neethirajan and Jayas 2011). This technology has robust applications in multitude of sectors, and its speedy advancement has revolutionized numerous fields of food science, especially those engaged in refining, loading, stockpiling, transport, management, and in various aspects of food safety (Bajpai et al. 2018). Essential oils (EOs) are secondary metabolites of different savory plants biosynthesized in distinct plant parts like epidermal cells, glandular trichomes, and secretary voids or canals. EOs are a blend of diverse bioactive compounds like terpenes, phenols, ketones, aldehydes, alcohols, esters, and phenyl propanoid (Bakkali et al. 2008). These oils are volatile liquids distilled from different structures belonging to plants. Albeit plants have been extensively utilized in the pharmaceuticals, agronomy, and aroma industry for many years, the significance of their EOs has profoundly increased in recent years due to their valuable properties. For example, various studies have demonstrated their antioxidant and biocide properties, which find applications in infinite areas (El Asbahani et al. 2015; Obolskiy et al. 2011). They have a broad spectrum of biocide activities due to their numerous bioactive volatiles. Generally, 300 EOs associated with families Zingiberaceae, Asteraceae, Apiaceae, Lamiaceae, Myrtaceae, Lauraceae, Piperaceae, and Cyperaceae are economically being utilized in perfume and essence industries (Burt 2004). Since relics, diverse formulations of EOs and capricious extracts (derived from plants) have been used in conventional medicine for various health benefits. The significance of EOs is elucidated by the presence of their abundant number of bioactive ingredients (Friedman 2014). Extensive research has been carried out in order to determine their biocide, insecticidal, antioxidant, and herbicidal effects (Jing et al. 2014; Nikolić et al. 2013). For example, activity against *Escherichia coli*, *Staphylococcus aureus*, and *Bacilus subtilis* has been shown by EOs isolated from *Citrusmedica* L. var. sarcodactylis, cinnamon, thyme, and citronella (Lou et al. 2017; Timung et al. 2016). Antifungal activities have also been reported; for example, *Penicillum parasiticus* and *Aspergilus niger* growth have been retarded by *Tetraclinis articulata* (Vahl) (Bourkhiss et al. 2007). Strong antioxidant properties have also been exhibited by various plants, for example, *Juniperus phoenicea* and *Citrus medica* L. EOs showed potent antioxidant activity (Lou et al. 2017; Bouzouita et al. 2008). Some insecticidal effects have also been displayed by *Juniperus phoenicea* EO against *Tribolium confusum*, *Cymbopogon citratus* (DC.), *Ocimum canum Sims*, *Ocimum gratissimum* L. var. "gratissimum" L., and *Thymus vulgaris* L. EOs against *anopheles gambiae* (Bouzouita et al. 2008; Tchoumbougnang et al. 2009). These properties could be used to produce bio-pesticides (Sumalan et al. 2019). Other applications of EOs involve their use for improving human well-being (Chitprasert and Sutaphanit 2014; Badea et al. 2019). EOs are extensively utilized in aromatherapy, cosmetics, and massage (Lardry 2007). In the industries, EOs are also used as preservatives and essence in food (Friedman 2017). Additionally, EOs are also used as fragrances in soaps and perfumes since the nineteenth century (Baser and Buchbauer 2015; Boh and Knez 2006).

Recently, the nanoencapsulation technology has supplemented a contemporary dimension to food industries as it intensifies the bio-potency of essential oils by elevating the bio-accessibility of nano-diameter commodities that are easily capable of strengthening the bioactive volatiles in solid-liquid interface and liquid-rich state where microbes are more cordially located (Hosseini et al. 2013). EOs offer excellent antimicrobial, insecticidal, antioxidant, and pharmaceutical properties; however, their practical utilization is yet minimal as the EOs are profoundly volatile and quickly degrade upon oxidation by light, heat, oxygen, and humidity (Bilia et al. 2014). On the other hand, nanoencapsulation being a powerful, skillful, and profitable technology offers a contemporary dimension for the enrichment and stability of EOs and bioactive volatiles by safeguarding them from direct and natural climatic effects. Moreover, encapsulation also reduces the volatility and lethal effects of EOs, upgrades their hydrophilicity, and further intensifies their bio-accessibility and bio-potency due to a surge in surface-to-volume proportion, as well as deep tissue permeation capability (Gupta and Variyar 2016).

The main challenges of controlling the high volatility of essential oils has been solved by advancements in the fields of various encapsulation methods. Encapsulation can restrain EOs by a physiochemical interaction with a matrix, which retains the essential oil for an extended time (Baser and Buchbauer 2015). Relatively all applications of EOs demand an increment in retention times and release profiles (El Asbahani et al. 2015). For example, in the cosmetic industry, it is convenient to develop a technique of encapsulation in which the EOs are released by automatic effects (Kerdudo et al. 2015). On the other hand, the encapsulation of flavors for food applications calls for a limited controlled release (Madene et al. 2006). In the case of pesticides, encapsulation needs to allow for gradual and uninterrupted delivery of the active products at an optimal threshold level in the environment (Lopez et al. 2012). The delivery of the active agent after encapsulation is possible by simple diffusion, exterior factor interruption, or matrix erosion. In every case, it is necessary that neither the matrix nor its residues are lethal to the environment. As a result, many innate polymers have been utilized for pesticide encapsulation.

20.2 Encapsulation Methods of Essential Oils

Encapsulation of essential oils or their bioactive volatiles refers to the entrapment of plant products inside polymeric protective integument. Nanoencapsulation concerns the plant product's encapsulation in the nano range, i.e., 10–1000 nm (Zuidam and Shimoni 2010). Widely, these techniques are divided into three types, i.e., chemical processes (e.g., in situ polymerization, molecular insertion) and physicochemical procedures (Zuidam and Shimoni 2010). Encapsulated material is called the *core material*, intrinsic state, or active material, while the encapsulating compound or integument is known as *shell or wall material*. Some workers showed the efficacy of various wall materials, viz., starch-based (cellulose, dextrins, pectin), fats-based (phospholipids, cholesterol), protein-based (albumin, lecithin), and synthetic compounds (sodium caseinate, poly lactic acid) used in the encapsulation of bioactive agents (Zuidam and Shimoni 2010). Reports suggest that core materials like fat and water-soluble vitamins, fish oils, essential oils, and antimicrobial agents encapsulated showed promising properties like slow release, enhanced stability, aroma masking, etc. (Pandit et al. 2016).

Amidst diverse nanoencapsulation methods or techniques, ionic gelation and coacervation are the most widely used encapsulation techniques employed to promote the preservative potential of bioactive volatiles. Based on encapsulation techniques, distinct types of encapsulated bioactive volatiles have been evolved, like nanoemulsion, nanogel, nanoparticle, nanofiber, nanoliposome, and nanosponge. Among all of them, nanogel, nanoemulsion, and nanoparticles have been used as powerful distribution systems in food industries and pharmaceutical sectors (Wan et al. 2019). Encapsulation methods have been observed to elevate the antimicrobial potency of EOs. For example, encapsulated Lavender EO have shown three times more intensification in its antimicrobial efficacy (Yuan et al. 2019). Additionally, encapsulation further elevates the biocide efficiency and thermal stability of EOs. Some of the primary benefits of nanoencapsulation methods are presented in Fig. 20.1 (Tiwari et al. 2020).

Some dominant and popular encapsulation methods and their role in enhancing the bio-potency of bioactive compounds are expressed below:

20.2.1 Liposome-Based Techniques

20.2.1.1 Emulsification

This encapsulation technique is most commonly used, and it involves the fission of two immiscible solvents, and the resulting product is called an emulsion. Emulsification is also used as a preliminary step for encapsulation techniques, e.g., spray drying and freeze-drying (Laokuldilok et al. 2016; Mehyar et al. 2014). Generally, emulsions consist of a dispersed phase, a continuous phase, and a pacifier or surfactant, or stabilizer that functions as coherence between the two phases. The categorization of an emulsion is in accordance to its dispersed phase, i.e., oil dispersed in a continuous water phase is referred as an oil in water (O/W) emulsion

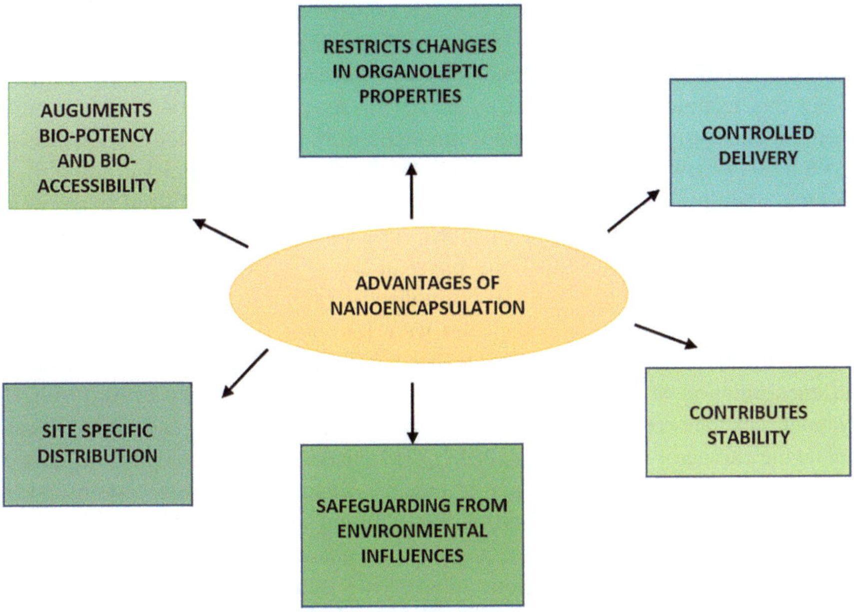

Fig. 20.1 Advantages of encapsulation of essential oils

and water dissolved in oil is called water in oil (W/O) emulsion (Saifullah et al. 2016). Aroma emulsions utilized mainly in liquid and semi-solid food products (e.g., ice cream, soft drinks) are oil-in-water emulsions because flavors are generally lipid in nature. Depending on the oil droplet size, emulsions can be categorized as macroemulsions, microemulsions, and nanoemulsions. Micro- and nanoemulsions are extremely refined emulsions displaying exceptionally improved action as compared to macroemulsions in terms of stability, optical lucidity, and controlled release. These ultra-refined emulsions can be formulated by utilizing either high-energy methods (e.g., high-pressure valve homogenizers, micro-fluidizers, and sonication methods) or low-energy methods (e.g., phase inversion and spontaneous emulsification) (Saifullah et al. 2016). The high-energy methods depend upon specifically devised automated machinery; hence these can be extravagant. On the other hand, low-energy methods are not in need of any sophisticated equipment, and they are easy and elementary, inexpensive, and extra energy-efficient (Anton and Vandamme 2011).

20.2.1.2 High-Pressure Homogenization

High-pressure homogenization (HPH) is a frequently practiced high-energy technique in the food industry. Primarily in this technique, a crude emulsion is formulated by a high-shearing mixer. After that, the resulting emulsion is allowed to pass through a confined orifice with tremendous acceleration under elevated pressure (100–2000 bar). This step unmasks the emulsion to a blend of extreme disruptive forces (i.e., instability, shear, vacuity) to promote droplet disruption converting the coarse or giant droplets into tinier ones, developed in an ultimate emulsion (Saifullah et al. 2016). Some workers prepared a nanoemulsion of a D-limonene flavored organogel of mono stearin and medium chained triglyceride oil using this technique HPH (Zahi et al. 2015). As per the study, monostearin, a medium chain triglyceride oil, and D-limonene were blended in the weight ratio of 1.5:8.5:20 (w/w/w). D-Limonene organogel was utilized as the oil phase, and Milli-Q water containing varying amounts of Tween 80 surfactant was utilized as the water phase. The study produced the smallest emulsion size of 36 nm. The resultant nanoemulsions revealed a satisfying physical stability and powerful antimicrobial action when compared with free D-limonene (Zahi et al. 2015).The carvacrol nanoemulsions(O/W) and nanocapsules were formulated by using a high-pressure homogenization method with malto-dextrin and Tween 20. The study demonstrated an encapsulation efficacy of 49.3–76.4% and a loading capacity of 48.7–69.2%. Nanoencapsulates revealed boosted efficacy than nanoemulsions in order to encapsulate the carvacrol (Hussein et al. 2017). Nevertheless, the desirable characteristics of emulsions and elevated efficacy in essence, retention can be accomplished by regulating the homogenization environment. Current studies reveal that the emulsion droplet size may diminish with expanding homogenization pressure or a number of gaps, reducing interfacial tension, intensifying emulsifier adsorption rate, and shrinks the continuous phase viscosity (Zahi et al. 2015; Yuan et al. 2008). There is generally a beeline relationship between the logarithm of homogenization pressure (P) and the logarithm of the droplet diameter (d), i.e., $\log d \alpha \log P$. In order to intensify the emulsification efficacy and stability of an emulsion, an adequate amount of emulsifier is also very crucial.

20.2.1.3 Ultrasonic Technique

High-intensity ultrasonic waves (frequency >20 kHz) can be utilized to create severe disruptive forces, culminating in fine emulsion droplet production (Jafari et al. 2007). Complementary to high-energy methods, a pre-fused crude emulsion is utilized in the ultrasound technique. An ultrasonic probe is immersed into the sample or set in the streamed flow of the sample to be homogenized. The probe produces excessive disruptive forces at its tip due to the amalgamation of vacuity, turbulence, and interfacial wave. Ultrasonic homogenization can be performed continuously or in batches (Leong et al. 2009). For a laboratory-scale generation, batch ultrasonic homogenization is appropriate. On the other hand, continuous ultrasonic homogenizers are generally utilized for the large-scale generation of fine emulsions (Laokuldilok et al. 2016). Roasted coffee bean oil mini-emulsions were formulated utilizing poly-L-lactic acid (PLLA) and poly hydroxybutyrate-co-hydroxyvalerate (PHBV) as delivery material through HPH and sonication. The end products elucidated the good encapsulation of major essence compounds of coffee bean oil in nanocapsules. Sonication produced the maximum oil restoration for PLLA systems, while HPH governed the maximum oil recovery for PHBV systems (Freiberger et al. 2015). Still, the ultrasonic method is exceptionally applicable for little viscosity fluids over elevated viscous systems. A diagrammatic representation of ultrasonic method is being depicted in Fig. 20.2 (Kumar et al. 2019).

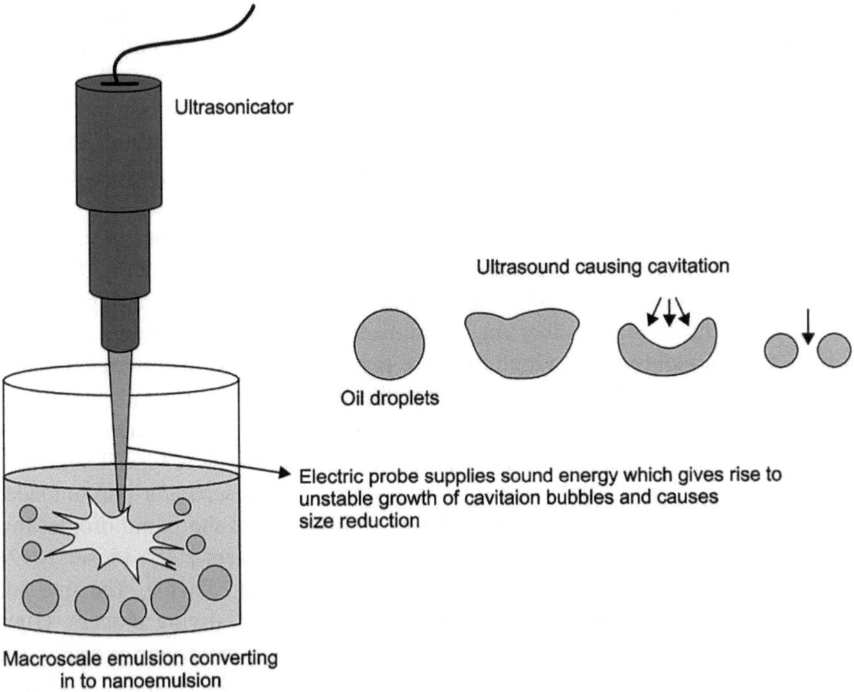

Fig. 20.2 Ultrasonic technique used to create small-size particles

20.2.2 Spray Drying

This technique is broadly utilized and is a highly economical method for encapsulating bioactive volatiles (Saifullah et al. 2016; Freiberger et al. 2015). The spray drying technique is applied for the generation of essence fine powders. This provides a boosted production rate with the lowest utilization prices. These advantages allow this technique to be utilized for the encapsulation of flavoring compounds on an industrial scale (Sanchez-Reinoso et al. 2017; Sultana et al. 2017). Moreover, spray-dried essence powder is applied to a broad range of food materials to elevate product influence and preferences in the consumer's eyes (Sultana et al. 2017). The preliminary principle on which the spray drying technique is based involves the atomization of a food sample fed into a drying chamber by using an atomizer (e.g., spray nozzle or stop cock), desiccating of liquid droplets inside the drying chamber, and powder restoration via a cyclone separator (Shishir and Chen 2017; Masters 2002). The scattered carrier materials should have satisfied water solubility because only water-based dispersals are beneficial for spray drying functioning. Moreover, the prerequisite condition for carrier materials is that they must have reduced viscosity at increased concentrations, better emulsification/film-forming features, and efficient drying properties (Shishir et al. 2017). The dissolution of the gist material and delivery compounds can be achieved by using high-speed blending or high-shear emulsification utilizing colloid mills to produce a coarse emulsion. After that, the emulsion can be processed further by utilizing diverse mechanical methods, for example, high-pressure homogenization, microfluidization, and ultrasonic emulsification. Prior to the beginning of the spray drying method, the emulsion formulated should be stable over a particular period of time (Liu et al. 2001). The carrier materials widely applied in spray drying technique are mostly lipophobic polysaccharides (e.g., maltodextrins, chitosan, alginate, and different types of gums) and proteins (e.g., whey protein), while the gist or core materials can be active lipophobic or lipophilic molecules (Zuidam and Shimoni 2010). The emulsion produced is utilized as the sample feed during spray drying. This is pumped into the drying chamber via the atomizer. There are two classes of atomizers which are commonly used: a high-pressure nozzle or a centrifugal wheel (Masters 2002; Phisut 2012). The atomizer fissures the liquid feed into small particles or droplets and sprays it into hot air. Three classes of airflow are applicable inside the drying chamber: co-current flow, counter-current, or mixed flow type. Ultimately, the encapsulated dry particles are separated via a cyclone separator and unloaded into the receiver. The total time needed for modification of the liquid droplet into powder form is a few milliseconds to a few seconds (Masters 2002; Phisut 2012), which is very appropriate for heat-labile compounds (e.g., flavor). The rate and viscosity of the feed have remarkable effects on the spray drying process. High-viscosity emulsions affect the atomization and, therefore, the drying process. The confinement of aroma compounds during spray drying is influenced by the conformation and properties of the emulsion and by the drying conditions (Gharsallaoui et al. 2007). The temperature of desiccating medium (hot air) is generally fixed at different levels depending upon the nature of the gist material. The abrupt drying of droplets maintains the core temperature below 100 °C despite the high temperatures (>150 °C) utilized in the process (Jafari et al. 2008). A crucial step is to sustain the temperature low for essence core materials as the essence may consist of different components with low boiling points, with high volatility. Hence, there is the probability of losing certain aromatics during the drying process (Apintanapong and Noomhorm 2003). A desired encapsulation displays the highest retention of gist material and the lowest accumulation of core material in the powder particle wall. The spray drying technique, however, exhibits restricted retention of volatile flavor compounds, especially those having a low boiling point, although ideal drying conditions are followed (Reineccius 2004). Two theories have been advanced, and the primary one is the selective diffusion theory. This theory describes that when the emulsion droplets desiccate to a moisture content of about 23 to 7% (aw <0.90), the dehydrated surface functions as semi-penetrable membrane permitting the steady evaporation (or diffusion) of water and the efficient confining of essence molecules (Reineccius 2001, 2004). Flavor diffusivity finally gets reduced as compared to water molecules after the formulation of dehydrated surface. Therefore, most loss of flavor generally occurs at the beginning of desiccating process. The second theory is the relative volatility of aroma compounds compared to water. The loss is equivalent to the proportionate volatility of flavor; hence, more loss of volatile flavors occurs at an increased rate in the primary drying stage (Reineccius 2001, 2004). In spray drying technique, there are three stages where flavor losses occur. Firstly, flavor losses occur during atomization because it creates increased surface area and extended turbulence in the emulsion system. Secondly, flavor losses occur when the droplets are disclosed to the dehydrating chamber and start desiccating. At this point, essence molecules spread out via the outer covering of the droplets with water molecules since the droplet membrane is not compact enough to capture the flavor molecules. Thirdly, terminal flavor losses occur when the boiling point of water droplets exceeds and forms froth inside the droplet that mostly explode to the surface along with volatile compounds. The losses during this third stage (i.e., during morphological development) are larger than during atomization and the initiation of surface drying (Hecht and King 2000). In a novel study, cocoa essence compounds were microencapsulated via spray drying techniques utilizing maltodextrin and Hi-cap 100. The study reported that the process yield reached up to 58.77%. The delivery material Hi-cap 100 displayed an increased retention of cocoa aroma (22.6–32.5%) and exceptional micro-structural properties in cocoa aroma microcapsules as compared to maltodextrin (Sanchez-Reinoso et al. 2017). Comparably, Sultana and coworkers revealed that the greatest retention of D-limonene and ethyl hexanoate by spray drying micro-encapsulation were 48% and 45%, respectively, with *Saccharomyces cerevisiae* utilized as a delivery material (Sultana et al. 2017). The final results reported that greater than 50% of flavor and aroma compounds were vaporized during spray drying microencapsulation. On the other hand, the highest microencapsulation efficacy by spray drying method was revealed up to 68.91%, 87.34%, and 80.2% for white champaca extract, sweet orange oil aroma, and oregano essence in skimmed milk powder, respectively (Samakradhamrongthai et al. 2016; Liu et al. 2012; Baranauskienė et al. 2006). In the case of nanoencapsulation, spray drying technique was only utilized to transform a suspension of colloidal nanoparticles into nanostructured powder form. Some workers reported the encapsulation of nanoemulsion particles of D-limonene by spray drying method (Jafari et al. 2007). Maltodextrin in amalgamation with a surface-active biopolymer, i.e., altered starch (Hi-Cap 100), whey protein fixate, or a surfactant (Tween 20), was utilized as the coat material. Hi-Cap exhibited greater retention of D-limonene by around 86.2% in spray-dried particles. Nevertheless, nano-spray drying technique has currently been applied for the production of nanoencapsulated bioactive compounds.

20.2.3 Electrospinning

Electrospinning, also known as electrospraying technique, is a typical way for encapsulation of aroma compounds. These techniques are easy, commercial, and scalable methods in addition to possessing functional utilization (Kayaci et al. 2014; Koo et al. 2014). In the electrospraying approach, nanofibers are formulated from a polymer solution in a spinneret by an enormous voltage potential, where particles are generated from a polymer solution in the stopcock via liquid atomization by electric power (Esfanjani and Jafari 2016). They are non-thermal operations, and there is no possibility of thermal degradation of heat-labile compounds, thus are very applicable for the encapsulation of aroma compounds. Apart from creating fibers and droplets with a confined size distribution, electrospinning methods impede particle cluster formation and agglomeration. Thus these techniques are very suitable for preparing both micro and nano-scale encapsulate systems encountered in food and medicine industries (Koo et al. 2014; Esfanjani and Jafari 2016; Tampau et al. 2017; Yao et al. 2016).

Electrospinning can also produce small droplet sizes when compared to nanospray drying. Beads and porous structures made by electrospraying can weaken the sustained release delivery system (Esfanjani and Jafari 2016). Nevertheless, the low throughput of electrospraying approach can hamper the large-scale production and commercial application (Yang et al. 2017). To date, several novel approaches have been introduced into electrospinning and electrospraying, e.g., coaxial and emulsion approaches. The successful applications of these novel approaches for the encapsulation of essence compounds have also been reported by several recent studies, e.g., rose hip seed oil by coaxial electrospraying (Yao et al. 2016), peppermint oil by coaxial electrospinning (Koo et al. 2014), carvacrol by emulsion electrospraying (Tampau et al. 2017), and D-limonene by emulsion electrospinning (Khoshakhlagh et al. 2017). This innovative approach provides one-step co-encapsulation of numerous compounds with varying solubility, elevated encapsulation efficacy, boosted stability, and elated controlled delivery profile of encapsulated compounds as compared to single stopcock electrospraying or electrospinning (Yao et al. 2016; Matsuura and Maruyama 2017). The emulsification approach permits the encapsulation of immiscible compounds with elated encapsulation efficacy, increased loading capacity, elevated thermal stability, and persistent release profile of aroma compounds (Tampau et al. 2017; Khoshakhlagh et al. 2017; Camerlo et al. 2013).

20.2.4 Freeze-Drying

The fundamental principle of freeze-drying or lyophilization technique involves the chilling of water content in a solution or suspension, accompanied by vaporization of the water molecules from the solution or suspension at a decreased temperature (Anandharamakrishnan et al. 2010). The entire dehydrating process takes place under conditions of low temperature and pressure. Therefore, it is an appropriate technique for heat-labile food compounds, including flavors, aromas, catalysts, etc. (Madene et al. 2006; Raja et al. 2019). Lyophilization provides appropriate characteristics to essence when compared with other drying and encapsulation techniques (Buffo and Reineccius 2001; Abdul-Fattah et al. 2007). The drying technique involves four main steps: freezing, primary drying, secondary drying, and final treatment (Abdul-Fattah et al. 2007). In freezing, the bioactive compound (raw material) is chilled to a solidified ice form, and the crystal's size is based on the chilling rate, the primary magnitude of chilling temperature, and the final temperature of chilling. The water molecules present in the food material form solidified ice, whereas food materials remain in an irregular and icy state, i.e., do not produce crystal (Abdul-Fattah et al. 2007). Primary drying includes the exclusion of ice by sublimation under a vacuum that is created, as the required energy for latent heat is provided to the chilled food compounds. Sublimation begins at the ice crystal surface and takes place at the junction between the chilled and dehydrated part of the compound. At the start of the event, the desiccating rate turns high as a result of the decreased resistance to transmission of heat and mass in the primary stages of desiccation. During the commencing of the desiccation process approaching the core of the material, the heat and mass transmission rate decreases due to the chilled portion of the material functioning as cushioning material, which in turn reduces the heat transfer to the crystal ice front and obstructing the mass transmission from the ice front (Abdul-Fattah et al. 2007). The stage of secondary drying starts when entire ice crystals are eliminated from the material and the water that is bound in the desiccating material. The conditions like temperature and pressure should be the same in the dehydrated chamber as in the primary dehydrating step (Abdul-Fattah et al. 2007). Various factors are responsible for producing better-featured frozen-dehydrated essence powders. Primarily, the complete dehydrating method is driven under vacuum conditions so that there is a workable deficiency of air that will prevent oxidative deterioration. Secondly, the dehydrating temperature is less than that of climatic temperature. Therefore, the materials, for example, flavors that are heat labile and very sensitive towards oxidation, can be dehydrated under vacuum with meager physical and chemical deterioration (Abdul-Fattah et al. 2007). The preparedness of the sample done for freeze desiccating is identical to spray drying technique. The flavor and aroma materials, stabilizer, and water are blended into the required particle size. In a notable study, aliphatic and aromatic alcohols were encapsulated by utilizing amylose molecular complex by fixed heating and freeze-drying method (Feng et al. 2018). At the beginning, a dispersed amylose suspension is formulated by heating and was cooled. The flavor compounds were then added to the amylose suspension and allowed to react. Then it was taken for ethanol washing and centrifugation. Consequently, coagulates formed were treated by freeze-drying method to get essence powder. The reports displayed that the encapsulation efficacy and outcome varied from 39% to 47% and 45–50%, respectively. All amylose-flavor inclusion complexes exhibited excellent thermal stability. Cardamom essence, i.e., D-limonene, 1,8-cineole, myrcene, terpineol, and linalool, were encapsulated utilizing whey protein isolate (WPI), guar gum (GG), and carrageenan (CG) through emulsification and freeze-drying method (Mehyar et al. 2014). The study reported that the microcapsules, consisting of only WPI, showed the elevated flavor entrapment (7.5%) and microencapsulation efficiency (98.5%). The 30% WPI and 30% WPI + GG formulations exhibited higher retention capacity of 1,8-cineole and D-limonene at the time storing process (Mehyar et al. 2014). Therefore, increased energy demands and prolonged processing time (more than 20 h) are the considerable limitations of freeze-drying (Wan et al. 2019). Furthermore, freeze-drying technique creates wide, porous network of coarse particles that are ineffective when a prolonged release of flavor materials is needed (Abdul-Fattah et al. 2007).

20.2.5 Extrusion

Extrusion is a novel advancement to encapsulate the flavor compounds when related to spray drying. In the method of extrusion, a delivery solution combined with bioactive compounds is run by using a nozzle into a jellifying environment. In laboratory-scale production, the combined solution consisting of a carrier and bioactive compounds is supplied into a syringe and run via a needle into jellifying conditions to formulate gelation (Khor et al. 2017). The pressure and temperature are mainly fixed below 100 psi and 118 °C in this procedure, respectively. Hence, the extrusion technique is very appropriate for thermo labile compounds, and it has demanding functions for the microencapsulation of aroma compounds (Khor et al. 2017; Rodríguez et al. 2016; Tackenberg et al. 2014). In the extrusion method, droplets are produced by various mechanisms based on the synergy of the gravitational, surface tension, impulse, and abrasion forces. Hence, the extrusion technique can be categorized into simple trickling, electrostatic extrusion, coaxial airflow, vibrating jet/nozzle, jet cutting, spinning disk atomization, and melt extrusion. When there is the formulation of droplets, they are directly solidified to capsules by physical (e.g., chilling or melting) or chemical process (e.g., jellifying) (Rodríguez et al. 2016). Calcium alginate/D-limonene beads were prepared by electrostatic extrusion (Lević et al. 2015). The study revealed that D-limonene was combined into Na-alginate solutions with forceful combination to form D-limonene/Na-alginate emulsions. Emulsions were then captured into an electrostatic immobilization unit fitted with a large voltage unit, magnetic stirrer, and shielding cage. The emulsions were modified into spherical droplets or calcium alginate beads via a blunt stainless-steel needle utilizing a syringe pump by the electrostatic extrusion system. For producing dried forms of beads, further air drying is accomplished at 25 °C for 48 h. The study displayed excellent immobilization of D-limonene within calcium alginate beads (immobilization efficacy 50–77%). Moreover, calcium alginate beads were able to confer protection to thermo labile D-limonene up to 200 °C (Lević et al. 2015). Encapsulation of orange terpenes in maltodextrin-sucrose carrier matrix was done by utilizing a counter-rotating twin screw extrusion process. Approximately 67% of aroma retention was reported from the loading capacity of 4.1%. The researchers elucidated that the cause of essence loss was because of inadequate blending during the procedure and essence vaporization after extrusion or swelling (Tackenberg et al. 2015). In another study, three ideal factorial designs were accomplished to explore the influence of barrel temperature, powdered feed rate, and vital medicinal ingredient content on the retention of flavor products. The final outcome revealed that a maximum of 6% loading of orange terpenes was acquired with an encapsulation efficiency of 86.7%. The orange terpene microcapsules were also largely stable during the storage process of 12 weeks (Tackenberg et al. 2015). Quercetin is a natural flavonoid exhibiting robustness in health-boosting and pharmaceutical advantages, but due to their very astringent and bitter taste, diminished the consumer's desirability to quercetin-enriched compounds. Moreover, it is very much prone to oxidative deterioration (Kawabata et al. 2015). Recently, the bitterness of quercetin was neutralized by encapsulation using carnauba wax, shellac, and zein by utilizing hot-melt extrusion. The quercetin-loaded microcapsules showed a remarkably decreased dispersal rate in the imitated salivary pH (6.8) in the order of zein > carnaubawax > shellac. In vitro bitterness investigation by electronic tongue revealed the quality taste concealing efficacy of the quercetin-loaded microcapsules (Khor et al. 2017). While the process of extrusion is two times economical than spray drying, the application of screw extruders at elevated pressure can lead to essence loss, and hence ideal conditions are required (Rodrı́guez et al. 2016).

20.2.6 Coacervation

Coacervation technique has been a promising encapsulation method for a long time, while currently it has been utilized for encapsulating food flavor compounds in food industries. Coacervation involves the liquid-liquid phase partition of colloid molecules from a solution and constitutes a new phase, which is known as coacervate (Yang et al. 2014). The phase partition can be categorized as aqueous and non-aqueous phase partition based on, if water is used as the solvent. For flavor encapsulation, the aqueous partition is the most accepted because aroma components are lipophilic in nature. The underlying mechanism included in this technique is the generation of an emulsion followed by the precipitation of the continuous phase encompassing the droplets of the discontinuous phase. This method includes a three-phase system, which involves a production vehicle (solvent), the active compound to be encapsulated, and the shell material. The coacervation procedure involves three major steps: creating the three immiscible phases, impeachment of coating material around the gist material(active compound), and solid microcapsule creation via condensation and coagulation (Mehyar et al. 2014). In the first step, phases are formulated during blending, and gist material, wall material, and continuous liquid phase are distributed in this step. In the case of food essence and aroma encapsulation, the choice of wall materials is stringently restricted by food additive management and gelatin is utilized in most cases. The second step includes interfacial adsorption of the lipophobic phase on the droplets of the gist material. The pH and temperature of the solution need to be adjusted for the progression of capsules which leads to the encapsulation separating out from the solution, precipitation at the top surface of the droplets, and creation of the wall coating. In the last step, condensation and hardening is done to form a condensed and solidified capsule. Heating, de-solvation, or cross-linking methods can be utilized to achieve this. In one study, β-pinene was encapsulated utilizing sodium caseinate, whey protein isolate, carboxy methylcellulose (CMC), and reticulating agents (i.e., glycerol and tannic acid) via complex coacervation (Koupantsis and Paraskevopoulou 2017). The terminal product exhibited that sodium caseinate -CMC microcapsules, with the addition of glycerol, boosted loading capacity and encapsulation efficacy reached up to ~55% and ~90%, respectively. Adding glycerol elevated encapsulation efficacy twofold. Additionally, glycerol addition revealed elated retention of volatile compounds in the microcapsules when stored at low relative humidity (RH) and temperature (0% RH, 25 °C) (Koupantsis and Paraskevopoulou 2017). Alternative studies have also revealed the fruitful encapsulation of vanilla oil and sweet orange oil by complex coacervation with elevated loading capacity, reinforced essence retention, and upgraded thermal stability (Yang et al. 2014; Jun-xia et al. 2011). As a result, the coacervation technique is regarded as more effective in essence retention than spray drying as it does not need elevated temperature, provides soaring core loading capacity, immense encapsulation efficacy, and an extreme restricted delivery of bioactive volatiles.

20.2.7 Inclusion Complexation

Cyclodextrins (i.e., α-, β-, γ-cyclodextrin) are modified starch molecules produced after enzymatic action. The most accepted kind of cyclodextrin is β-cyclodextrin (β-CD). Conventional utilization of cyclodextrin is to safeguard the unstable compounds and to contribute increased combined value, especially to essence materials by utilization of encapsulation. The atomic complex of cyclodextrin is identical to the empty curtailed cone. The inner cavity size is in nano order. The inner hydrophobic cavity of cyclodextrin captures the essence or aroma molecules of the appropriate size through hydrophobic synergy (called as molecular inclusion), while the external surface exhibits lipophobic features. Some workers formulated a β-CD atomic inclusion system for the encapsulation of sweet orange essence (Liang et al. 2012). The study reported that β-CD is mixed in lukewarm water sustained at 35 °C temperature. Then the blend is constantly agitated for 3 h at 35 °C. When its temperature gets diminished automatically to ambient temperature, the solution is left to remain overnight at 5 °C. The chilly coagulated material is then retrieved by vacuum filtration. The formed coagulate is then washed with alcohol and dehydrated in a convection oven at 50 °C for 24 h. Hence, the sweet orange essence-loaded β-CD atomic inclusion systems are formulated. Numerous studies have revealed the ameliorated features of essence and aroma when encapsulated with diverse cyclodextrins, e.g., carvacrol in α-CD, β-CD, and HP-β-CD (Liang et al. 2012), cinnamaldehyde in α-CD and β- (Chun et al. 2015), thymol in β-CD (Del Toro-Sánchez et al. 2010), and ethyl benzoate in HP-β-CD (Giri et al. 2013). Cyclic oligosaccharide cyclo-dextrins are the mostly utilized encapsulating polymer for EOs through this technique of inclusion complexation. Essential oil encapsulation within cyclodextrin biopolymer displayed a convincing increase in water solubility up to 16 times and diminished the photo-deterioration rate up to 44 times, providing physical and chemical stability along with elevation in its bio-potency (Kfoury et al. 2019).

20.2.8 Ionic Gelation

In ionic gelation technique nanoparticles are produced when polyelectrolyte cross-links with counter ions of essential oils to form nanoparticles (Kfoury et al. 2019). Ionic gelation is also known as fluid bed coating or air suspension coating, in which a shelling or coating of particles suspended in air is done (Saifullah et al. 2016). There are two types of fluid bed coating methods, Continuous and Batch. Three basic states are included in this method: at the start, there is fluidization of the molecules which are to be coated into the coating chamber by utilizing an airstream. Then, the shelling material is sprinkled by a stopcock onto the particles. Lastly, the solvent of the coating material is vaporized by the heated air and hence the shelling material sticks to the nanoparticle (Madene et al. 2006; Mehyar et al. 2014).This method can be utilized to develop a subsequent shell to spray-dried compounds or for compounds with a sensitive core, such as oils, flavors, and aromas (Sultana et al. 2017). This method permits a considerable range of shelling materials compared to traditional spray drying and is cost-effective in terms of operation and time when compared to freeze (Pandit et al. 2016). Therefore, this method is not profoundly analyzed for the encapsulation of essence compounds when compared to spray drying and freeze-drying. Some workers recently reported the influence of diverse drying techniques, i.e., fluidized bed coating, spray drying, and freeze-drying on the fish oil microcapsules by using maltodextrin, soluble soybean polysaccharide (SSPS), hydroxypropyl beta cyclodextrin (HPBCD), and octenyl succinic anhydride starch (OSA-starch) (Yuan et al. 2014). Fluidized bed coating proved to be an appropriate technique with superlative ranges of encapsulation efficiency (96.10–99.39%) in all matrices. Among the delivery systems, the gelatin emulsion delivery agent was more potent and exhibited elevated restriction of menthol discharge and elevated encapsulation efficacy (~95%) (Sun et al. 2013). *Syzygium aromaticum* (clove) EO within chitosan biopolymer displayed elevated antifungal activity against *Aspergillus niger* and also exhibited a controlled release effect (Giri et al. 2013). *Schinus molle* EO encapsulation via this technique increased its anti-fungicidal action against *Aspergillus parasiticus* and also enhanced its anti-aflatoxigenic efficiency at reduced concentrations (Anwar and Kunz 2011).

20.3 Restraints of Encapsulation Methods

So far, diverse processes of encapsulation have been exploited. However, no one proved to be correspondingly satisfactory for larger bioactive volatiles. The underlying reason for this lack of success is the specific structure of the bioactive compounds. Furthermore, nanoencapsulated EOs may agglomerate if the storage period is longer, which causes a reduction in bio-efficiency. The utilization of surfactants may lead to the dissipation of volatile alcoholic compounds of EOs, causing an effective decrease in biocide activity and initiation of some toxicological effects like allergic reactions and skin inflammation in the mammalian system (Das et al. 2019). The effects of such nanoencapsulated EOs on unspecific organisms are also a concern.

20.4 Conclusions and Future Perspectives

Implementing nanotechnology in bioactive volatile encapsulation provides many benefits, from food preparation to loading, which involves upgradation in stability and bioaccessibility. There are many health benefits presented by this technique, like the shielding and controlled release of bioactive volatiles. The bio-potency of EOs and their bioactive components is upgraded by encapsulation methods, which are utilized as agricultural preservatives by boosting their stability, biocidal potentiality, and antioxidant capability. The constraints of nanoencapsulation processes can be subdued by further research and studies and refinement of the prevailing techniques, formulations, and encapsulation systems. Moreover, emphasis should be put on the application of nanocarriers of bioactive volatiles in food and biological systems, an inspection of their effects on the cell viability and their uptake, dissemination, anabolism, catabolism, and evacuation profiles in humans and other living systems (Yang et al. 2014). Eventually, it is crucial to record that before using this technology in the field, studies on ecotoxicity should be accomplished broadly to estimate the effect of essential oils, but also of encapsulation grid on non-target organisms. In conclusion, there have been a lofty number of studies regarding EO encapsulation and the desire to expand the activity life of EOs. Additional research on the methods and products that have practicable interest should be taken under cogitation.

References

Abdul-Fattah AM, Kalonia DS, Pikal MJ (2007) The challenge of drying method selection for protein pharmaceuticals: product quality implications. J Pharm Sci 96:1886–1916. https://doi.org/10.1002/jps.20842

Anandharamakrishnan C, Rielly CD, Stapley AGF (2010) Spray-freeze-drying of whey proteins at sub-atmospheric pressures. Dairy Sci Technol 90:321–334. https://doi.org/10.1051/dst/2010013

Anton N, Vandamme TF (2011) Nano-emulsions and micro-emulsions: clarifications of the critical differences. Pharm Res 28:978–985. https://doi.org/10.1007/s11095-010-0309-1

Anwar SH, Kunz B (2011) The influence of drying methods on the stabilization of fish oil microcapsules: comparison of spray granulation, spray drying, and freeze drying. J Food Eng 105:367–378. https://doi.org/10.1016/j.jfoodeng.2011.02.047

Apintanapong M, Noomhorm A (2003) The use of spray drying to microencapsulate 2-acetyl-1-pyrroline, a major flavour component of aromatic rice. Int J Food Sci Technol 38:95–102. https://doi.org/10.1046/j.1365-2621.2003.00649.x

Badea ML, Iconaru SL, Groza A et al (2019) Peppermint essential oil-doped hydroxyapatite nanoparticles with antimicrobial properties. Molecules 24:2169. https://doi.org/10.3390/molecules24112169

Bajpai VK, Kamle M, Shukla S et al (2018) Prospects of using nanotechnology for food preservation, safety, and security. J Food Drug Anal 26:1201–1214. https://doi.org/10.1016/j.jfda.2018.06.011

Bakkali F, Averbeck S, Averbeck D, Idaomar M (2008) Biological effects of essential oils–a review. Food Chem Toxicol 46:446–475. https://doi.org/10.1016/j.fct.2007.09.106

Baranauskienė R, Venskutonis PR, Dewettinck K, Verhé R (2006) Properties of oregano (Origanum vulgare L.), citronella (Cymbopogon nardus G.) and marjoram (Majorana hortensis L.)

flavors encapsulated into milk protein-based matrices. Food Res Int 39:413–425. https://doi.org/10.1016/j.foodres.2005.09.005

Baser KHC, Buchbauer G (2015) Handbook of essential oils: science, technology, and applications. CRC Press

Bilia AR, Guccione C, Isacchi B et al (2014) Essential oils loaded in nanosystems: a developing strategy for a successful therapeutic approach. Evid Based Complement Alternat Med 2014:1. https://doi.org/10.1155/2014/651593

Boh B, Knez E (2006) Microencapsulation of essential oils and phase change materials for applications in textile products. Indian J Fibre Text Res 31:72

Bourkhiss M, Hnach M, Bourkhiss B et al (2007) Composition chimique et propriétés antimicrobiennes de l'huile essentielle extraite des feuilles de Tetraclinis articulata (Vahl) du Maroc. Afr Sci Rev Int Sci Technol 3:232

Bouzouita N, Kachouri F, Ben Halima M, Chaabouni MM (2008) Composition chimique et activités antioxydante, antimicrobienne et insecticide de l'huile essentielle de Juniperus phoenicea. J la Société Chim Tunisie 10:119–125

Buffo R, Reineccius G (2001) Comparison among assorted drying processes for the encapsulation of flavors. Perfum Flavorist 26:58–67

Burt S (2004) Essential oils: their antibacterial properties and potential applications in foods—a review. Int J Food Microbiol 94:223–253. https://doi.org/10.1016/j.ijfoodmicro.2004.03.022

Camerlo A, Vebert-Nardin C, Rossi RM, Popa A-M (2013) Fragrance encapsulation in polymeric matrices by emulsion electrospinning. Eur Polym J 49:3806–3813. https://doi.org/10.1016/j.eurpolymj.2013.08.028

Chitprasert P, Sutaphanit P (2014) Holy basil (Ocimum sanctum Linn.) essential oil delivery to swine gastrointestinal tract using gelatin microcapsules coated with aluminum carboxymethyl cellulose and beeswax. J Agric Food Chem 62:12641–12648. https://doi.org/10.1021/jf5019438

Chun J-Y, Jo Y-J, Bjrapha P et al (2015) Antimicrobial effect of α-or β-cyclodextrin complexes with trans-cinnamaldehyde against Staphylococcus aureus and Escherichia coli. Dry Technol 33:377–383. https://doi.org/10.1080/07373937.2014.957388

Das S, Singh VK, Dwivedy AK et al (2019) Antimicrobial activity, antiaflatoxigenic potential and in situ efficacy of novel formulation comprising of Apium graveolens essential oil and its major component. Pestic Biochem Physiol 160:102–111. https://doi.org/10.1016/j.pestbp.2019.07.013

Del Toro-Sánchez CL, Ayala-Zavala JF, Machi L et al (2010) Controlled release of antifungal volatiles of thyme essential oil from β-cyclodextrin capsules. J Incl Phenom Macrocycl Chem 67:431–441. https://doi.org/10.1007/s10847-009-9726-3

El Asbahani A, Miladi K, Badri W et al (2015) Essential oils: from extraction to encapsulation. Int J Pharm 483:220–243. https://doi.org/10.1016/j.ijpharm.2014.12.069

Esfanjani AF, Jafari SM (2016) Biopolymer nano-particles and natural nano-carriers for nano-encapsulation of phenolic compounds. Colloids Surf B Biointerfaces 146:532–543. https://doi.org/10.1016/j.colsurfb.2016.06.053

Feng T, Wang H, Wang K et al (2018) Preparation and structural characterization of different amylose–flavor molecular inclusion complexes. Starch-Stärke 70:1700101. https://doi.org/10.1002/star.201700101

Freiberger EB, Kaufmann KC, Bona E et al (2015) Encapsulation of roasted coffee oil in biocompatible nanoparticles. LWT Food Sci Technol 64:381–389. https://doi.org/10.1016/j.lwt.2015.05.039

Friedman M (2014) Chemistry and multibeneficial bioactivities of carvacrol (4-isopropyl-2-methylphenol), a component of essential oils produced by aromatic plants and spices. J Agric Food Chem 62:7652–7670. https://doi.org/10.1021/jf5023862

Friedman M (2017) Chemistry, antimicrobial mechanisms, and antibiotic activities of cinnamaldehyde against pathogenic bacteria in animal feeds and human foods. J Agric Food Chem 65:10406–10423. https://doi.org/10.1021/acs.jafc.7b04344

Gharsallaoui A, Roudaut G, Chambin O et al (2007) Applications of spray-drying in microencapsulation of food ingredients: an overview. Food Res Int 40:1107–1121. https://doi.org/10.1016/j.foodres.2007.07.004

Giri TK, Verma S, Alexander A et al (2013) Crosslinked biodegradable alginate hydrogel floating beads for stomach site specific controlled delivery of metronidazole. Farmacia 61:533–550

Gupta S, Variyar PS (2016) Nanoencapsulation of essential oils for sustained release: application as therapeutics and antimicrobials. In: Encapsulations. Elsevier, pp 641–672. https://doi.org/10.1016/B978-0-12-804307-3.00015-6

Hecht JP, King CJ (2000) Spray drying: influence of developing drop morphology on drying rates and retention of volatile substances. 1. Single-drop experiments. Ind Eng Chem Res 39:1756–1765. https://doi.org/10.1021/ie9904652

Hosseini SF, Zandi M, Rezaei M, Farahmandghavi F (2013) Two-step method for encapsulation of oregano essential oil in chitosan nanoparticles: preparation, characterization and in vitro release study. Carbohydr Polym 95:50–56. https://doi.org/10.1016/j.carbpol.2013.02.031

Hussein J, El-Bana M, Refaat E, El-Naggar ME (2017) Synthesis of carvacrol-based nanoemulsion for treating neurodegenerative disorders in experimental diabetes. J Funct Foods 37:441–448. https://doi.org/10.1016/j.jff.2017.08.011

Jafari SM, He Y, Bhandari B (2007) Production of sub-micron emulsions by ultrasound and microfluidization techniques. J Food Eng 82:478–488. https://doi.org/10.1016/j.jfoodeng.2007.03.007

Jafari SM, Assadpoor E, He Y, Bhandari B (2008) Encapsulation efficiency of food flavours and oils during spray drying. Dry Technol 26:816–835. https://doi.org/10.1080/07373930802135972

Jing L, Lei Z, Li L et al (2014) Antifungal activity of citrus essential oils. J Agric Food Chem 62:3011–3033. https://doi.org/10.1021/jf5006148

Jun-xia X, Hai-yan Y, Jian Y (2011) Microencapsulation of sweet orange oil by complex coacervation with soybean protein isolate/gum Arabic. Food Chem 125:1267–1272. https://doi.org/10.1016/j.foodchem.2010.10.063

Kawabata K, Mukai R, Ishisaka A (2015) Quercetin and related polyphenols: new insights and implications for their bioactivity and bioavailability. Food Funct 6:1399–1417. https://doi.org/10.1039/C4FO01178C

Kayaci F, Sen HS, Durgun E, Uyar T (2014) Functional electrospun polymeric nanofibers incorporating geraniol–cyclodextrin inclusion complexes: high thermal stability and enhanced durability of geraniol. Food Res Int 62:424–431. https://doi.org/10.1016/j.foodres.2014.03.033

Kerdudo A, Gonnot V, Njoh Ellong E et al (2015) Essential oil composition and biological activities of Petiveria alliacea L. from Martinique. J Essent Oil Res 27:186–196. https://doi.org/10.1080/10412905.2015.1014118

Kfoury M, Auezova L, Greige-Gerges H, Fourmentin S (2019) Encapsulation in cyclodextrins to widen the applications of essential oils. Environ Chem Lett 17:129–143. https://doi.org/10.1007/s10311-018-0783-y

Khor CM, Ng WK, Kanaujia P et al (2017) Hot-melt extrusion microencapsulation of quercetin for taste-masking. J Microencapsul 34:29–37. https://doi.org/10.1080/02652048.2017.1280095

Khoshakhlagh K, Koocheki A, Mohebbi M, Allafchian A (2017) Development and characterization of electrosprayed Alyssum homolocarpum seed gum nanoparticles for encapsulation of d-limonene. J Colloid Interface Sci 490:562–575. https://doi.org/10.1016/j.jcis.2016.11.067

Koo SY, Cha KH, Song D et al (2014) Microencapsulation of peppermint oil in an alginate–pectin matrix using a coaxial electrospray system. Int J Food Sci Technol 49:733–739. https://doi.org/10.1111/ijfs.12358

Koupantsis T, Paraskevopoulou A (2017) Flavour retention in sodium caseinate–carboxymethylcellulose complex coavervates as a function of storage conditions. Food Hydrocoll 69:459–465. https://doi.org/10.1016/j.foodhyd.2017.03.006

Kumar M, Bishnoi RS, Shukla AK, Jain CP (2019) Techniques for formulation of nanoemulsion drug delivery system: a review. Prev Nutr Food Sci 24:225. https://doi.org/10.3746/pnf.2019.24.3.225

Laokuldilok N, Thakeow P, Kopermsub P, Utama-Ang N (2016) Optimisation of microencapsulation of turmeric extract for masking flavour. Food Chem 194:695–704. https://doi.org/10.1016/j.foodchem.2015.07.150

Lardry J-M (2007) Les autres indications des huiles essentielles. Kinésithérapie, la Rev 7:35–42. https://doi.org/10.1016/S1779-0123(07)70312-1

Leong TSH, Wooster TJ, Kentish SE, Ashokkumar M (2009) Minimising oil droplet size using ultrasonic emulsification. Ultrason Sonochem 16:721–727. https://doi.org/10.1016/j.ultsonch.2009.02.008

Lević S, Lijaković IP, Đorđević V et al (2015) Characterization of sodium alginate/d-limonene emulsions and respective calcium alginate/d-limonene beads produced by electrostatic extrusion. Food Hydrocoll 45:111–123. https://doi.org/10.1016/j.foodhyd.2014.10.001

Liang H, Yuan Q, Vriesekoop F, Lv F (2012) Effects of cyclodextrins on the antimicrobial activity of plant-derived essential oil compounds. Food Chem 135:1020–1027. https://doi.org/10.1016/j.foodchem.2012.05.054

Liu X-D, Atarashi T, Furuta T et al (2001) Microencapsulation of emulsified hydrophobic flavors by spray drying. Dry Technol 19:1361–1374. https://doi.org/10.1081/DRT-100105293

Liu K, Xu Y, Wang X (2012) Microencapsulation of sweet orange oil terpeneless using the orifice method. J Food Eng 110:390–394. https://doi.org/10.1016/j.jfoodeng.2011.12.033

Lopez MD, Maudhuit A, Pascual-Villalobos MJ, Poncelet D (2012) Development of formulations to improve the controlled-release of linalool to be applied as an insecticide. J Agric Food Chem 60:1187–1192. https://doi.org/10.1021/jf204242x

Lou Z, Chen J, Yu F et al (2017) The antioxidant, antibacterial, antibiofilm activity of essential oil from Citrus medica L. var. sarcodactylis and its nanoemulsion. LWT 80:371–377. https://doi.org/10.1016/j.lwt.2017.02.037

Madene A, Jacquot M, Scher J, Desobry S (2006) Flavour encapsulation and controlled release–a review. Int J Food Sci Technol 41:1–21. https://doi.org/10.1111/j.1365-2621.2005.00980.x

Masters K (2002) Spray drying in practice. Spray Dry Consult International, Denmark

Matsuura T, Maruyama T (2017) Calcium phosphate-polymer hybrid microparticles having functionalized surfaces prepared by a coaxially electrospray technique. Colloids Surf A Physicochem Eng Asp 526:64–69. https://doi.org/10.1016/j.colsurfa.2016.10.036

Mehyar GF, Al-Isamil KM, Al-Ghizzawi HM, Holley RA (2014) Stability of cardamom (Elettaria Cardamomum) essential oil in microcapsules made of whey protein isolate, guar gum, and carrageenan. J Food Sci 79:C1939–C1949. https://doi.org/10.1111/1750-3841.12652

Neethirajan S, Jayas DS (2011) Nanotechnology for the food and bioprocessing industries. Food Bioprocess Technol 4:39–47. https://doi.org/10.1007/s11947-010-0328-

Nikolić M, Marković T, Mojović M et al (2013) Chemical composition and biological activity of Gaultheria procumbens L. essential oil. Ind Crop Prod 49:561–567. https://doi.org/10.1016/j.indcrop.2013.06.002

Obolskiy D, Pischel I, Feistel B et al (2011) Artemisia dracunculus L.(tarragon): a critical review of its traditional use, chemical composition, pharmacology, and safety. J Agric Food Chem 59:11367–11384. https://doi.org/10.1021/jf202277w

Pandit PS, Ranade DR, Dhakephalkar PK, Rahalkar MC (2016) A pmoA-based study reveals dominance of yet uncultured Type I methanotrophs in rhizospheres of an organically fertilized rice field in India. 3 Biotech 6:135. https://doi.org/10.1016/j.ijbiomac.2015.12.029

Phisut N (2012) Spray drying technique of fruit juice powder: some factors influencing the properties of product. Int Food Res J 19:1297

Raja KS, Taip FS, Azmi MMZ, Shishir MRI (2019) Effect of pre-treatment and different drying methods on the physicochemical properties of Carica papaya L. leaf powder. J Saudi Soc Agric Sci 18:150–156. https://doi.org/10.1016/j.jssas.2017.04.001

Reineccius GA (2001) The spray drying of food ingredients. In: The encapsulation of food ingredients. Leatherhead Publ, pp 151–184
Reineccius GA (2004) The spray drying of food flavors. Dry Technol 22:1289–1324. https://doi.org/10.1081/DRT-120038731
Rodríguez J, Martín MJ, Ruiz MA, Clares B (2016) Current encapsulation strategies for bioactive oils: from alimentary to pharmaceutical perspectives. Food Res Int 83:41–59. https://doi.org/10.1016/j.foodres.2016.01.032
Saifullah M, Ahsan A, Shishir MRI (2016) Production, stability and application of micro-and nanoemulsion in food production and the food processing industry. In: Emulsions. Elsevier, pp 405–442. https://doi.org/10.1016/B978-0-12-804306-6.00012-X
Samakradhamrongthai R, Thakeow P, Kopermsub P, Utama-ang N (2016) Microencapsulation of white champaca (Michelia alba DC) extract using octenyl succinic anhydride (OSA) starch for controlled release aroma. J Microencapsul 33:773–784. https://doi.org/10.1080/02652048.2016.1264493
Sanchez-Reinoso Z, Osorio C, Herrera A (2017) Effect of microencapsulation by spray drying on cocoa aroma compounds and physicochemical characterisation of microencapsulates. Powder Technol 318:110–119. https://doi.org/10.1016/j.powtec.2017.05.040
Shishir MRI, Chen W (2017) Trends of spray drying: a critical review on drying of fruit and vegetable juices. Trends Food Sci Technol 65:49–67. https://doi.org/10.1016/j.tifs.2017.05.006
Shishir MRI, Taip FS, Saifullah M et al (2017) Effect of packaging materials and storage temperature on the retention of physicochemical properties of vacuum packed pink guava powder. Food Packag Shelf Life 12:83–90. https://doi.org/10.1016/j.fpsl.2017.04.003
Sultana A, Miyamoto A, Hy QL et al (2017) Microencapsulation of flavors by spray drying using Saccharomyces cerevisiae. J Food Eng 199:36–41. https://doi.org/10.1016/j.jfoodeng.2016.12.002
Sumalan RM, Alexa E, Popescu I et al (2019) Exploring ecological alternatives for crop protection using Coriandrum sativum essential oil. Molecules 24:2040. https://doi.org/10.3390/molecules24112040
Sun P, Zeng M, He Z et al (2013) Controlled release of fluidized bed-coated menthol powder with a gelatin coating. Dry Technol 31:1619–1626. https://doi.org/10.1080/07373937.2013.798331
Tackenberg MW, Marmann A, Thommes M et al (2014) Orange terpenes, carvacrol and α-tocopherol encapsulated in maltodextrin and sucrose matrices via batch mixing. J Food Eng 135:44–52. https://doi.org/10.1016/j.jfoodeng.2014.03.010
Tackenberg MW, Krauss R, Schuchmann HP, Kleinebudde P (2015) Encapsulation of orange terpenes investigating a plasticisation extrusion process. J Microencapsul 32:408–417. https://doi.org/10.3109/02652048.2015.1035686
Tampau A, González-Martinez C, Chiralt A (2017) Carvacrol encapsulation in starch or PCL based matrices by electrospinning. J Food Eng 214:245–256. https://doi.org/10.1016/j.jfoodeng.2017.07.005
Tchoumbougnang F, Dongmo PMJ, Sameza ML et al (2009) Activité larvicide sur Anopheles gambiae Giles et composition chimique des huiles essentielles extraites de quatre plantes cultivées au Cameroun. Biotechnol Agron Soc Environ 13:77
Timung R, Barik CR, Purohit S, Goud VV (2016) Composition and anti-bacterial activity analysis of citronella oil obtained by hydrodistillation: process optimization study. Ind Crop Prod 94: 178–188. https://doi.org/10.1016/j.indcrop.2016.08.021
Tiwari S, Singh BK, Dubey NK (2020) Encapsulation of essential oils-a booster to enhance their bio-efficacy as botanical preservatives. J Sci Res 64:175. https://doi.org/10.1016/j.foodchem.2015.07.150
Wan J, Zhong S, Schwarz P et al (2019) Physical properties, antifungal and mycotoxin inhibitory activities of five essential oil nanoemulsions: impact of oil compositions and processing parameters. Food Chem 291:199–206. https://doi.org/10.1016/j.foodchem.2019.04.032

Yang Z, Peng Z, Li J et al (2014) Development and evaluation of novel flavour microcapsules containing vanilla oil using complex coacervation approach. Food Chem 145:272–277. https://doi.org/10.1016/j.foodchem.2013.08.074

Yang G-Z, Li J-J, Yu D-G et al (2017) Nanosized sustained-release drug depots fabricated using modified tri-axial electrospinning. Acta Biomater 53:233–241. https://doi.org/10.1016/j.actbio.2017.01.069

Yao Z-C, Chang M-W, Ahmad Z, Li J-S (2016) Encapsulation of rose hip seed oil into fibrous zein films for ambient and on demand food preservation via coaxial electrospinning. J Food Eng 191:115–123. https://doi.org/10.1016/j.jfoodeng.2016.07.012

Yuan Y, Gao Y, Zhao J, Mao L (2008) Characterization and stability evaluation of β-carotene nanoemulsions prepared by high pressure homogenization under various emulsifying conditions. Food Res Int 41:61–68. https://doi.org/10.1016/j.foodres.2007.09.006

Yuan C, Lu Z, Jin Z (2014) Characterization of an inclusion complex of ethyl benzoate with hydroxypropyl-β-cyclodextrin. Food Chem 152:140–145. https://doi.org/10.1016/j.foodchem.2013.11.139

Yuan C, Wang Y, Liu Y, Cui B (2019) Physicochemical characterization and antibacterial activity assessment of lavender essential oil encapsulated in hydroxypropyl-beta-cyclodextrin. Ind Crop Prod 130:104–110. https://doi.org/10.1007/s11947-019-02291-1

Zahi MR, Liang H, Yuan Q (2015) Improving the antimicrobial activity of D-limonene using a novel organogel-based nanoemulsion. Food Control 50:554–559. https://doi.org/10.1016/j.foodcont.2014.10.001

Zuidam NJ, Shimoni E (2010) Overview of microencapsulates for use in food products or processes and methods to make them. In: Encapsulation technologies for active food ingredients and food processing. Springer, pp 3–29. https://doi.org/10.1007/978-1-4419-1008-0_2

Aasiya Majeed is currently working as a PhD scholar in the Division of Biochemistry, SKUAST-Jammu, India. The title of her research is “Studies on Polymer Based Nanostructures for Controlled Release of Essential Oils, their Characterization and Bioactive Properties” (2019–2023). Asiya did her Masters in Biochemistry from the Department of Biochemistry, University of Kashmir (2013–2015).

Saima Masood is an Assistant Professor at Sher-e-Kashmir University of Agricultural Sciences and Technology. She has completed her MSc and PhD in Biochemistry and is actively involved in teaching and research.

Sadaf Nazir currently works at the Department of Nanotechnology, University of Kashmir, India, and did her Masters in Chemistry from the University of Kashmir, India, in 2019. Sadaf is presently working in the Laboratory for Multifunctional Nanomaterials at the University of Kashmir, where she works on the synthesis, characterization, optical and magnetic properties of nanomaterials prepared through ionic liquid-mediated microwave-assisted synthetic routes. She is interested in enhanced multimodal imaging modalities for biomedical and healthcare applications.

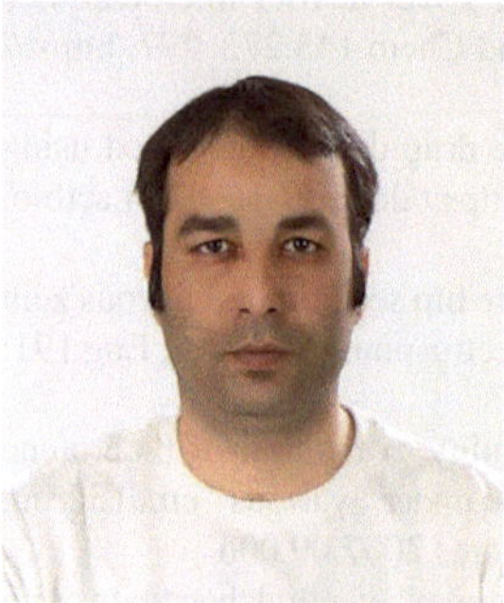

Faheem A. Sheikh is an Assistant Professor at the Department of Nanotechnology, University of Kashmir, India (2015). He served as an Assistant Professor at the Department of Biotechnology at the Central University of Kashmir, India (2015–2016); a Research Professor at Myongji University, South Korea (2014–2015); Assistant Professor at Hallym University, South Korea (2012–2014); Post-Doc/Research fellow at the University of Texas Rio Grande Valley, Texas, United States of America (2010–2012); and Research Professor at Myongji University, South Korea (2010). His research mainly focuses on fabricating nanomaterials used in tissue engineering.

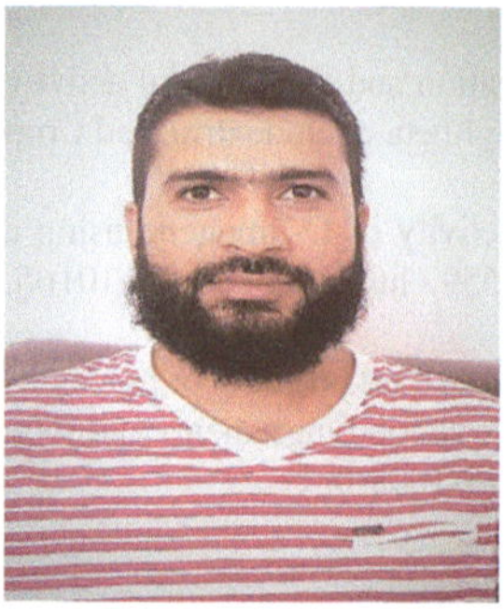

Mushtaq A. Beigh is working as an Assistant Professor at the Department of Nanotechnology, University of Kashmir, India. Dr. Beigh has postdoctoral research experience in cellular signalling and disease modelling (2013–2016). He had completed his Ph.D. from the University of Kashmir, India (2008–2012). He heads a laboratory at the University of Kashmir, where they work on a bio-nano interface to understand the molecular basis of nanoconjugate internalization. They also try to understand the growth factor signalling inputs directed at multiple cellular receptors like NRP1 and integrins in order to work out the mechanistic details of receptor-based internalization.

Shafquat Majeed is currently working as an Assistant Professor in the Department of Nanotechnology, University of Kashmir, India. Dr. Majeed did his Masters in Chemistry from the University of Kashmir. He did his Ph.D. from the prestigious Indian Institute of Science, Bangalore (2008–2015). He has a postdoctoral research experience in rare-earth-based nanomaterials for biomedical applications and fabrication of perovskite-based solar cells from the Centre for Nanoscience and Engineering, IISc Bangalore (2015–2016). He currently heads a Multifunctional Nanomaterials Laboratory at the University of Kashmir, where they work on synthesis, characterization, and optical and magnetic properties of nanomaterials prepared through various solution-based routes. More specifically, they are working on developing new and enhanced multimodal imaging modalities for biomedical and healthcare applications.

Review on Green Synthesis, Modification, Characterization, Properties, and Applications of Palladium Nanoparticles in Biomedical Applications

21

Rakesh Kumar Pradhan, Pooja Mohapatra, Lipsa Shubhadarshinee, Bigyan Ranjan Jali, Aruna Kumar Barick, and Priyaranjan Mohapatra

Abstract

This book chapter deals with the study of the green synthesis, modification, characterization, properties, and application of palladium (Pd) nanoparticles (NPs). We discuss that these NPs can be synthesized from plant extracts, bacteria, algae, pine needles, glucose, and honey. Although these NPs have been identified as effective materials for their catalytic, electrical, optical, hydrogen sensing, and magnetic properties, however, these NPs are tremendous biomedical applications. The Pd NPs are synthesized by various synthetic methods such as the Suzuki coupling reaction, sol-gel method, laser ablation, and green synthesis from various plant extracts and microorganisms. The Pd NPs are characterized by using ultraviolet-visible (UV-Vis) spectroscopy, FTIR (Fourier transform infrared) spectroscopy, X-ray diffraction (XRD) analysis, energy-dispersive X-ray (EDX) analysis, scanning electron microscopy (SEM), transmission electron microscopy (TEM), etc. In this chapter, we will understand the different applications of Pd NPs, such as their catalytic activity, antibacterial properties, drug delivery potential, biomedical application, and vehicle for gene therapy and for fuel cell preparation.

Keywords

Palladium nanoparticles · Green synthesis methods · Modifications · Characterizations · Properties · Applications

R. K. Pradhan · P. Mohapatra · L. Shubhadarshinee · B. R. Jali · A. K. Barick (✉) · P. Mohapatra
Department of Chemistry, Veer Surendra Sai University of Technology, Burla, Odisha, India
e-mail: akbarick_chem@vssut.ac.in; priya_chem@vssut.ac.in

F. A. Sheikh et al. (eds.), *Interaction of Nanomaterials With Living Cells*,
https://doi.org/10.1007/978-981-99-2119-5_21

21.1 Introduction

Nanotechnology is a branch of science concerned with the fabrication, characterization, and applications of nanostructured materials ranging from 1 to 100 nm. It is a nanoscale science that provides multiple facets to many branches of research, such as dentistry, pharmacology, and bioengineering (Gour and Jain 2019). The nanotechnologies involve the synthesis and use of materials at the nanometer range, either by scaling from a single atom cluster or by downsizing of bulk materials (National Nanotechnology Initiative (NNI) n.d.). Due to the vast range of applications in domains such as medicine, catalysis, and the fabrication of nanocomposites, using metal nanoparticles (NPs) has become an active area of research. These materials have unique qualities, including high electrical and thermal conductivity, tensile strength, and exceptional stiffness and hardness. Besides, having erosion resistance, it may be utilized to make satellite components, airplane spare parts, machinery parts, and electronic segments like microchip processors (Pradeep 2008). The NPs are often characterized as a cluster of atoms or molecules that have a variety of size-dependent features. Based on energy states, NPs in the nano range bridge the gap between smaller molecules and composite materials (Johnston and Wilcoxon 2012). The NPs are categorized based on dimensionality, morphology, composition, homogeneity, and aggregation of NPs (The Royal Society 2004; Buzea et al. 2007). On the basis of dimensionality, the NPs are classified as three types such as one dimension (1D), two dimensions (2D), and three dimensions (3D):

1D nanomaterials: These nanomaterials with a 1D structure. Thin films and surface coatings with 1D in the nanometer range are common. Thin films have been utilized in various disciplines for decades, including optical systems (Seshan 2002).

2D nanomaterials: At the nanometer scale, 2D NPs have 2D. Nanotubes, dendrimers, nanowires, fiber, and fibrils can be applied in information storage systems, electronics, and chemical and biological sensor systems. 2D nanomaterials are defined as free particles with a high aspect ratio and diameters in the nanoscale range. 2D systems have fewer well-understood characteristics and less sophisticated production capabilities.

3D nanomaterials: 3D nanomaterials are materials that are microscopic in all 3D. Quantum dots, fullerenes, particles, precipitates, and colloids are all examples of these materials.

In the form of topsoil, water, volcano, dust, and minerals, the NPs are thought to have been present on earth since their inception. Aside from their natural origins, humans have begun to synthesize NPs using a variety of ways (Maurer-Jones et al. 2013). The formulation of NPs may be done in two ways: (1) top-down and (2) bottom-up methods (Thakkar et al. 2010). In addition, three distinct tactics for synthesis of NPs are used such as chemical, biological, and physical approaches. Figure 21.1 depicts a schematic representation of the various procedures used to synthesize NPs and their uses. For the synthesis of NPs, physical approaches follow a top-down strategy, whereas chemical and biological approaches follow a bottom-up approach (Iravani et al. 2014).

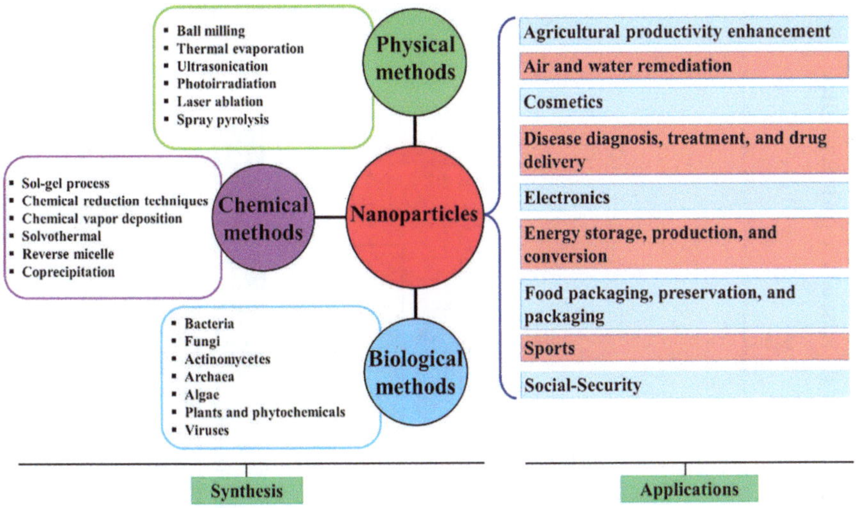

Fig. 21.1 Diagrammatic presentation of different techniques for production of NPs and their applications (Dikshit et al. 2021)

One of the most extensively utilized methods for large-scale NPs synthesis is the chemicals of NPs because of their corresponding salt by exploiting specific reducing agents. Chemical approaches are used for large-scale synthesis of metal NPs. Environmental damage is caused by the use of hazardous chemicals in these processes as well as the formation of dangerous by-products and restricting their clinical benefits (Gupta and Xie 2018). As a result, there is a great temptation for nontoxic, high-yielding, and environmentally sustainable procedures for the creation of nanostructured materials that can substitute traditional procedures. Hence, methods of biological synthesis offer an alternative solution to the techniques of physicochemical synthesis. Plants and plant components, microalgae, molds, yeast, bacterium, and viruses are among the many resources that may be used to make NPs biologically. The first step in the production of NPs is to combine the inorganic metal salt solution with biomaterials (Sriramulu et al. 2020). Plant-mediated NPs production may be accomplished in three ways: intracellular, extracellular, or individual using a specific phytoconstituent. Flavonoids, polyphenols, proteins, alkaloids, reducing sugars, and other chemicals found in biomaterials function as significantly reducing and stabilizing agents to make NPs from initial precursors of metal salts (Kuppusamy et al. 2016). Figure 21.2 displays the search results for the number of studies on the biosynthesis of NPs that have been published in the previous 10 years. With each passing year, the number of studies has increased, and from an estimation, 468 publications are published in 2020. These findings support the fact that interest in biologically synthesized NPs obtained from plant extract is growing each year (Dikshit et al. 2021).

The Pd NPs are among many known nanomaterials that are hot topics due to their capacity to act as a potential catalyst as well as their high aspect ratio and surface

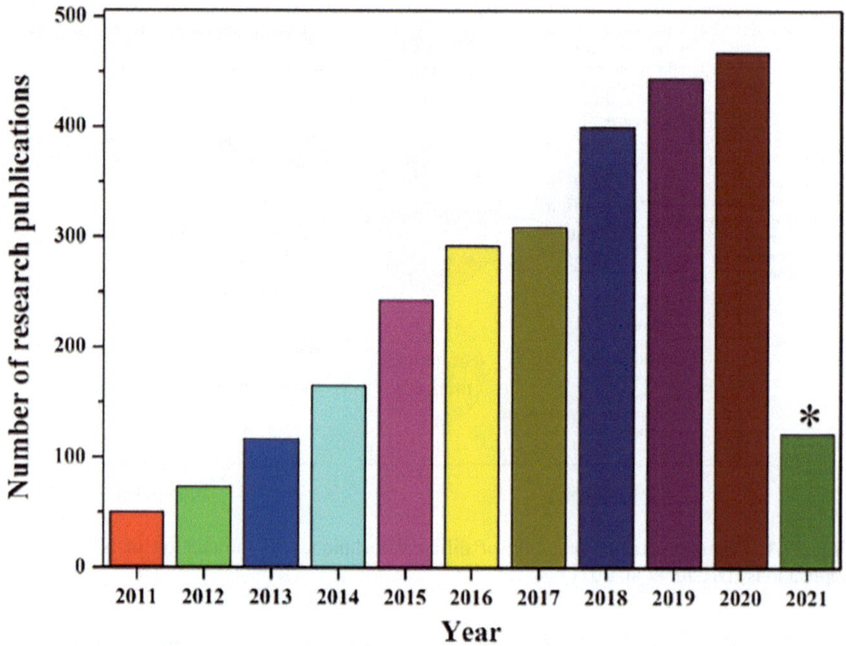

Fig. 21.2 Data of the Scopus search results indicating the total articles published in the previous 10 years on the production of NPs using green methods (Dikshit et al. 2021)

energy (Narayanan and El-Sayed 2005). The Pd NPs have also been utilized as catalysts in a number of coupling processes, including Suzuki coupling, Heck coupling, and allyl alcohol hydrogenation (Klingensmith and Leadbeater 2003; Karimi and Enders 2006; Wilson et al. 2006). The Pd NPs play an important role in catalysis and a variety of other applications, including hydrogen sensing and storage (Im et al. 2006). The Pd is one of the most effective metals as catalysts and has been extensively researched in catalytic applications such as oxidation, production of carbon-carbon bonds, and electrochemical processes in fuel cells (Astruc 2007; Bell 2003; Dimitratos et al. 2005; Beller et al. 1996; Cheong et al. 2010). The Pd NPs were reduced using microorganisms (MOs), plant extracts, and algae (Yenumula and Nagadesi 2018). The biogenesis of Pd NPs seems to be a secure, safe, and eco-friendly process to make NPs with a wide range of sizes, shape, and physicochemical and biological characteristics (Phan et al. 2020). Methods and circumstances for growing Pd NPs, such as temperature, light, pH, nutrients, and inoculation, are all important factors that should be considered. Scanning electron microscopy (SEM), transmission electron microscopy (TEM), Fourier transform infrared (FTIR) spectroscopy, X-ray diffraction (XRD) analysis, energy-dispersive X-ray (EDX) spectroscopy, etc. were used to analyze the produced Pd NPs. The main identification and characterization techniques of Pd NPs depend heavily on spectroscopic and diffractographic methods (Shahverdi et al. 2011).

21.2 Literature Review

Literature review indicates very few research articles are published regarding bio-synthesis of Pd NPs using plant extracts and biological entities, as tabulated in Table 21.1. In this vein, Sathiskumar et al. developed a process that is both

Table 21.1 Synthesis of Pd NPs using different green sources for the synthesis

Sl. no.	Name of plant/ biological entity	Part of plant	Average size (nm)	Shape	Ref. no.
1	*Cinnamomum camphora*	Bark	15–20	Hexagonal	Sathiskumar et al. (2009)
2	*Euphorbia granulate*	Leaf	25–35	Spherical	Nasrollahzadeh and Mohammad (2016)
3	*Glycine max*	Leaf	15	Spherical	Petla et al. (2012)
4	*Curcuma longa*	Tuber	10–15	Spherical	Sathishkumar et al. (2009)
5	*Musa paradisiaca*	Fruit peel	50	Irregular	Bankar et al. (2010)
6	*Anogeissus latifolia*	Gum	4.8	Spherical	Kora and Rastogi (2018)
7	*Gardenia jasminoides*	Leaf	3–5	–	Jia et al. (2009)
8	*Melia azedarach*	Leaf	10–20	Spherical	Bhakyaraj et al. (2017)
9	*Dioscorea bulbifera*	Tuber	10–25	Spherical	Ghosh et al. (2015)
10	*Sapium sebiferum*	Leaf	2–5	Spherical	Tahir et al. (2016)
11	Sodium alginate	–	13–33	Spherical	Ameri et al. (2020)
12	*Camellia sinensis*	Leaf	6–18	Spherical	Azizi et al. (2017)
13	Indian Khilar cow urine	–	–	Cylindrical	Prasad et al. (2020)
14	*Sargassum bovinum*	–	5	Octahedral	Momeni and Nabipour (2015)

Bankar et al. prepared spherical Pd NPs using *Musa paradisiaca* fruit peel extract as a reducing agent and reported an average size of 50 nm with irregular crystalline morphology (Bankar et al. 2010). Kora and Rostogi employed gum extract of *Anogeissus latifolia* to generate an average size spherical Pd NPs of dimension 4.8 nm (Kora and Rastogi 2018). Zhang et al. reported an eco-friendly greener approach for the synthesis of Pd NPs by utilizing *Gardenia jasminoides* leaves having an average size of 3–5 nm (Jia et al. 2009). Bhakyaraj et al. devised a simple and efficient way for the green synthesis of Pd NPs from *Melia azedarach* leaf extract with a spherical shape with size ranges from 10 to 20 nm (Bhakyaraj et al. 2017). Nitnavare et al. developed a facile route of biosynthesis of Pd NPs using *Dioscorea bulbifera* tuber extract with average size of 10–25 nm having spherical morphology (Ghosh et al. 2015). Tahir et al. employed the green synthesis of Pd NPs from *Sapium sebiferum* leaf extract and showed average size of 2–5 nm (Tahir et al. 2016). Ameri et al. studied the biosynthesis of spherical Pd NPs by using a polysaccharide (sodium alginate) and produced NPs with average size of 13–33 nm (Ameri et al. 2020). Azizi et al. reported a greener approach for synthesis of spherical Pd NPs with a normal size of 6.36 nm by using *Camellia sinensis* extract of tea powder (Azizi et al. 2017). Prasad et al. demonstrated new eco-friendly synthesis process of cylindrical Pd NPs employing Indian Khilar cow urine (Prasad et al. 2020). Momeni and Nabipour stated the biosynthesis of octahedral Pd NPs, including a size distribution of 5 nm using *Sargassum bovinum* bacterium (Momeni and Nabipour 2015).

eco-friendly and fast for synthesis of stable crystalline Pd NPs of standard size 15–20 nm by using *Cinnamomum camphora* (Sathiskumar et al. 2009). Moreover, Mohammad and Nasrollahzadeh revealed on the photosynthesis of Pd NPs having a typical size 25–35 nm by taking leaves extract from *Euphorbia granulate* (Nasrollahzadeh and Mohammad 2016). Petal et al. demonstrated the ecological synthesis of such spherical Pd NPs having size 15 nm using *Glycine max* fresh leaves (Petla et al. 2012). Sneha et al. reported rapid biosynthesis of Pd NPs by employing *Curcuma longa tuber* extracts with typical dimensions of 10–15 nm and spherical shape nanoparticles (Sathishkumar et al. 2009).

21.3 Synthesis of Pd NPs

For the successful production of NPs, a variety of physical and chemical approaches are available. All of these methodologies can be divided into two categories that can be applied to all nanoscale scientific research.

Figure 21.3a–c shows two approaches used for synthesizing Pd-based nanoparticles. Each of these has its own application. In top-bottom techniques, the majority of materials are subdivided into nanostructured materials, as shown in Fig. 21.3b (Ahmed et al. 2016). Mechanical milling/alloying and sputtering are two ways for lowering particle size (Hodaei et al. 2015). The procedure for synthesis in bottom-up techniques starts in conjunction with self-assembly of atoms or molecules into nuclei, subsequently producing the nanoscale particles as displayed in Fig. 21.3c (Ahmed et al. 2016). This method includes coprecipitation, sol-gel, and atomic condensation (Lesani et al. 2016). Chemical and biological techniques of production are used mostly in bottom-up synthesis.

The preparation of NPs relies on the chemical and physical processes in both top-bottom and bottom-up approaches. These approaches include employing

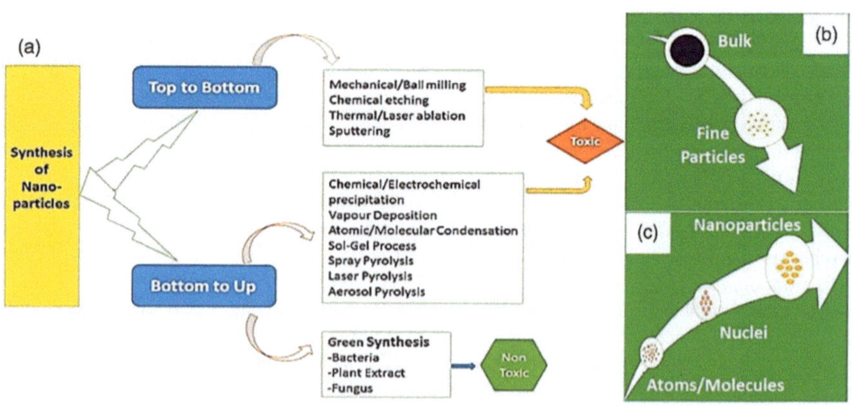

Fig. 21.3 (**a**) Different strategies to synthesize NPs: (**b**) top-to-bottom and (**c**) bottom-to-up approaches to synthesize NPs (Yadi et al. 2018)

poisonous and hazardous chemicals that are accountable for various biological issues and are costly and more likely to be environmentally hazardous (Mittal et al. 2013). As the production of plant-mediated NPs does not rely on physical or chemical techniques, so it is ecologically safe, biocompatible, and highly stable. As a result, researchers worldwide are interested in learning more about this method. For the environmentally sustainable production of Pd NPs from bacterial strains, extracts from plants, fungal/algae biomasses, and bio-based materials/chemicals, many types of biogenic processes are used (Kasthuri et al. 2009).

MOs (especially fungi) can bioreduce ions of Pd via enzyme activity and transform them into particles of Pd while keeping sufficient command over the dimensions of the NPs by biopatterning (Macaskie et al. 2017). Alternatively, the application of MOs in synthesizing of green materials is a time-consuming multistep process and may be expansive. Recently, microalgae have been linked to the production of Pd NPs. These are rich in widespread bioactive compounds and reducing agents that are naturally occurring, producing them ideal for the biogenic synthesis of Pd NPs. The Pd NPs were biosynthesized and tailored using a variety of algae, including red, brown, and green algae (Khanna et al. 2019). Green production of Pd NPs with phytoconstituents is a single-step process that eliminates the need for prolonged confinement, particular culture enhancement, and cleaning stages (Puja and Kumar 2019).

Plant extracts are also less expensive, safer, faster, and easier to produce. In comparison to MOs, the Pd NPs formed by photosynthesis are more stable and produced at a faster pace. Plants are rich in reducing and stabilizing compounds, e.g., phenols, flavonoids, alkaloids, saponins, vitamins, aldehydes, and steroids. The usage of herbs and plant-based products are suitable for the eco-friendly production of diversified Pd NPs. Antioxidant phytochemicals in the plant extract, such as phenolics and flavonoids, may be found liable for the bioreduction of Pd(II) ions to Pd(0). Plant-mediated biological manufacture of metallic NPs was used to make Pd NPs with controlled sizes and shapes under optimized conditions, so this method acts as an alternative to physicochemical approaches. Metal NPs (MNPs) have recently been synthesized using a variety of plant parts, including seeds, fruits, bark, roots, leaves, gums, and flowers (Shamaila et al. 2016; Devi and Ahmaruzzaman 2016). The biosynthesis of Pd NPs by using different plant extracts is shown in Fig. 21.4.

$$\text{Plant source} + \text{Pd salt} \rightarrow \text{biocompatible by} - \text{product of Pd NPs} + \text{biocompatible Pd NPs}$$

21.3.1 Mechanism of Synthesis of Pd NPs

The precise process for biogenesis of Pd NPs is unknown, and additional research is needed to provide more detailed information. Figure 21.5 shows a basic bottom-up system with four important phases (Durán et al. 2011). In the first phase of

Fig. 21.4 This figure shows different plant parts (via green methods) used to synthesize the Pd-based NPs (Nasrollahzadeh et al. 2020)

stimulation, the process for biosynthetic pathway of Pd NPs involves nucleation and bioreduction of the Pd ions, which converts Pd metallic ions to zero oxidation states. The minute Pd NPs are produced and aggregated into more significant and thermodynamically more stable Pd NPs in the next stage. Following the way, a termination stage happens in which a range of Pd NPs forms, such as rods, triangles, wires, spheres, pentagons, or hexagons, are formed, depending on the optimal circumstances. The plant extract, which includes several functional groups such as aldehydes, alcohols, amines, carboxylic acid, and ketones, might work as capping agents to stabilize the Pd NPs. Finally, centrifugation is used to purify and wash the biosynthesized Pd NPs (Malik et al. 2014).

21.3.2 Plant-Mediated Synthesis of Pd NPs

Manikandan et al. synthesized Pd NPs using *Prunus x yedoensis* leaves extracts. The leaf (50 g) was chopped into little parts and cooked in 200 mL clean nanopure water

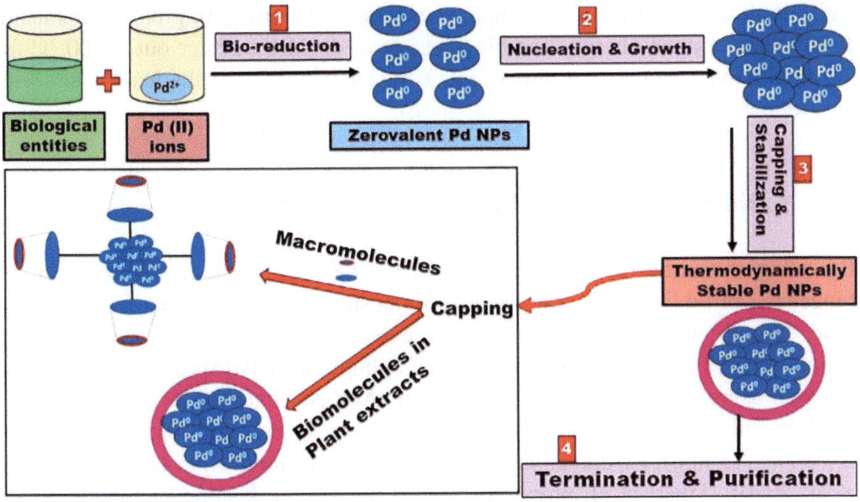

Fig. 21.5 Schematic representation of four steps for biosynthesis of Pd NPs (Fahmy et al. 2020)

and then filtered and stored at 4 °C until further experimentation. The Pd NPs were made by mixing 45 mL of 0.1 M Pd(II) chloride ($PdCl_2$) with 5 mL of pyle in a combination of reactions at 80 °C for 30 min with constant stirring. After the synthesis of Pd NPs was finished, it was centrifuged at 12000 rpm for 15 min before being rinsed with nanopure water and ethanol. To form a powder, the Pd NPs were freeze-dried (Manikandan et al. 2016). Nkosiet al. synthesized Pd NPs from *Moringa oleifera* flowers. Flowers were dried in a dark room for a few days. The floral solution was made in a 250 mL Erlenmeyer flask by weighing 10 g of fine flower powder into it, with 100 mL double distilled water. Then boil the entire combination for 20 min, cool it down, and store it at room temperature for up to a week. *M. oleifera* water extracts were put to conical flasks containing a 1 mM Pd acetate solution. The Pd NPs synthesis was monitored by the improvement of bright yellow color solution (Nkosi et al. 2016). Sharmila et al. reported synthesis of Pd NPs using Santalum album leaves. The leaves were washed and dried for a week. The powdered dried leaves were utilized to make the extract. 5 g of dried leaves was included to 50 mL of purified water and soaked in water at 60 °C for 10 min. The solution underwent filtering and Whatman filter paper was used to store the items for future purposes. At room temperature, 10 mL of aqueous *S. album* extract was combined with 90 mL of $PdCl_2$ solutions and incubated for 4 days. After that, centrifugation and drying take place to get Pd NPs (Sharmila et al. 2017a). The production of Pd NPs was reported by Vaghela et al. by utilizing *Bauhinia variegata* bark extract. After 10–15 min of at 6000 rpm, cylindrical Pd NPs with just a typical size of 2–9 nm were detected. The photo-synthesized Pd NPs have been demonstrated to exhibit substantial antibacterial and antifungal action against *Bacillus subtilis*, a gram-positive bacterium, as well as *Candida albicans*. The NPs also had a strong cytotoxic effect on MCF-7 breast cancer cells (Vaghela et al. 2018).

Veisi et al. commented on the simple eco-friendly manufacturing of Pd NPs using *Rosa canina* extract of fruit. The biogenesis was decided to carry out at 100 °C for 2 h employing acetone as an antisolvent, using biosynthesized spherical Pd NPs with an average size of 10 nm (Veisi et al. 2016).

21.3.3 Biological Systems for Biogenesis of Pd NPs

By using the urine of an Indian Khilar cow, Prasad et al. mentioned the fast biogenesis of Pd NPs with customizable dimensions, shape, and improved response time. Drop-by-drop introduction of cow excrement to Pd(II) ions with constant agitation at 80 °C produced the NPs. The cylindrical polydispersed Pd NPs were reported using the above mentioned strategy. At a concentrated level of 200 g/mL, the biosynthesized NPs were discovered to have strong antibacterial effects for *B. cereus* (15 mm), *Pseudomonas* (16 mm), and *Salmonella typhi* (16 mm) (Prasad et al. 2020). Ameri et al. proposed a basic microwave heating methodology for manufacturing of Pd NPs by utilizing vitamin C and sodium alginate. These were combined with the Pd(II) ions and cooked for 3 min in a microwave (850 W). The change of a light yellow color to a color dark brown indicated the fabrication of Pd NPs qualitatively. When compared to Pd(II) acetate, biologically synthesized Pd NPs show substantial antitumor activity on carcinoma cells in 48-h therapy (Ameri et al. 2020). Azizi et al. described the biogenesis of spherical Pd NPs with average sizes ranging from 6 to 18 nm, which was induced using *Camellia sinensis* powder extract. The p^H significantly reduced from 7.5 to 5.6 after 2 h of uninterrupted stirring of palladium (II) ions with white tea extract (1:1) at 40 °C. After the reduction process, the peak position of Pd(II) ions at 410 nm vanished, characterized by the emergence of a broadband spectrum suggesting the synthesis of Pd NPs (Azizi et al. 2017).

21.4 Modification of Pd NPs

Due to the nanoscale dispersion of NPs on the matrix (Zhang et al. 2002), the NPs have synergized characteristics between phases. Because of their very efficient catalytic and electrical properties (Zhang et al. 2014; Yin et al. 2014), Pd NPs have recently attracted a lot of attention. However, the high surface area of the Pd NPs accelerated the agglomeration and their use is severely limited. Pd catalysts have been anchored on various supports to prevent particle aggregation and mobility in order to improve performance. The majority of studies have focused on NPs that anchor on single support or many supports with naturally diverse morphologies but comparable chemical compositions. The qualities of NPs are closely linked to their size, morphology, phase, and structure. The surface charge distributions of NPs are determined by their morphologies (Hu and Yang 2013; Kim et al. 2020). The selective surface functionalization of Pd NPs changes size, which impacts catalytic performance.

21.4.1 Surface Modification by Metal Oxide

Unburned CH_4 has a huge global warming contributor than CO_2, causing environmental concerns if discharged directly into the atmosphere (Shao et al. 2017; Chen et al. 2011). The exhaust system of cars fueled by compressed natural gas (CNG) emits a significant quantity of CH_4, which is normally discharged at insufficient quantities (Zakaria and Kamarudin 2016). As a result, many countries have set emission of CH_4 restrictions for cars that use catalytic combustion of CH_4 in order to effectively eradicate CH_4 traces (Dai et al. 2019; Hong et al. 2018). The Pd NPs were effectively put down on the surface as modified metal oxides (mod-MOx) and used as catalyst materials in the combustion of lean CH_4. PdOx catalysts, with assistance, have been among the most productive materials for clean CH_4 combustion. The Langmuir-Hinshelwood mechanism was used for catalytic combustion of CH_4 over 1.0Pd/HfO_2 and 1.0Pd/mod-HfO_2 catalysts, and particular stages in the reactions suggested for 1.0Pd/mod-HfO_2 are shown in Fig. 21.6.

The chemical shift in Hf related to 4*f* of X-ray photoelectron spectroscopy (XPS) spectra indicates that the reaction cycle included electron transport from PdO to the PdO support contact first. In step 1, electrons from a silicon modifier were used to activate gas-phase oxygen, resulting in active O* (O, O_2^{2-}, O^{2-}) species. In step 2, the resultant O* sites fracture the CH bonds, facilitating the dehydrogenization of CH_4 into CH_3 and OH. Above 1.0Pd/mod-HfO_2, the conversion of CH_3 as $HCOO^-$ might be aided, resulting in either a quick transformation of CH_2O to $HCOO^-$ or no CH_2O being produced at the third step. During step 3, the coupled CH_3 and OH have to transit through additional dehydrogenization to create H^+ and CH_2O; after that the interactions between H^+ and O* while CH_2O was steadily transformed into oxidized form $HCOO^-$ over 1.0Pd/HfO_2. The $HCOO^-$ would then decompose, producing carbonate species with a single dentate, which were then bound to Pd by a single O_2 in the fourth step. In step 5, CO_2 was produced to complete the catalytic pathway.

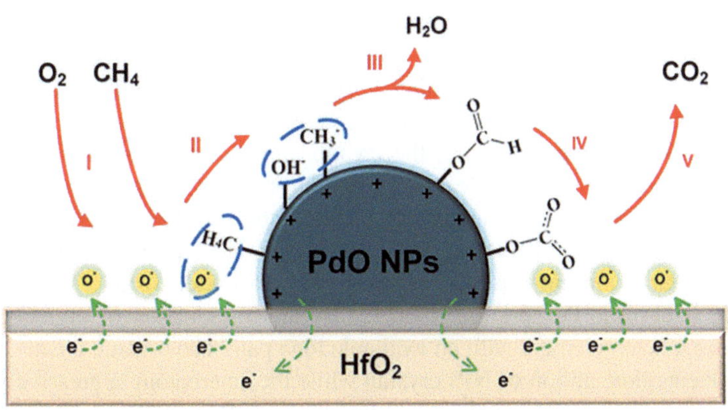

Fig. 21.6 Different stages for oxidation of CH_4 over 1.0Pd/mod-HfO_2 (Li et al. 2020)

After 60 h at 600 °C, 1.0Pd/mod-HfO_2 maintained a strong catalytic stability (Li et al. 2020).

21.4.2 Interstitial Doping of Boron Metal

The Lindlar catalysts made of Pd modified with lead acetate as well as quinoline are frequently utilized in the catalytic hydrogenization of alkynes in industry. However, due to the exceedingly nature of lead, their usage is restricted, especially in food, cosmetic, and medication manufacturing. Furthermore, the catalysts have low activity and selectivity in a number of situations (Nishimura 2001; Crespo-Quesada et al. 2011). The surface of unsupported Pd metal is very unselective, resulting over hydrogenation to undesirable alkanes. Modifying Pd metal with light elements by inserting them into the underlying locations of interstitial sites present inside the frame made of metal lattice. The modifications in the crystallite size and electrostatic interactions between the hosts and the guests may also have an impact upon that band structure; as a result the catalytic capabilities of the Pd metal supported on a surface increase (Teschner et al. 2008). As a sustainable alternative to the Lindlar for selective hydrogenation processes, it is essential to create novel Pd NPs catalysts doped with interstitial atoms. Chan et al. were curious in the production of boron interstitial in order to affect the electronic characteristics of metals and the catalytic efficiency of Pd nanocrystals. Boron atoms have been discovered to occur in Pd intermittent lattice positions with high thermal and chemical persistence. This is because of the intense electrical interactions between the host and the guest when it is supported Pd NPs are allowed to treat by using a borane in tetrahydrofuran solutions. The subterranean boron atom consequently modifies its selective and adsorbing characteristics of Pd (Theriot et al. 2014).

21.4.3 Sulfur-Based Ligand

The Pd NPs are bulk materials that have been finely split. In terms of agglomeration, they are thermodynamically unstable. As a result, they must be kinetically stabilized, which is normally accomplished with the use of a protective stabilizer. Electrostatic, steric, or a combination of the following forces is used to stabilize the system. The stabilizer is often supplied during the creation process of NPs, which is accomplished by means of a chemical reduction or thermal degradation of metallic precursors. The ensuing connection in between the stabilizer and the surface of NPs is very dynamic. This connection is in the form of a powerful covalent interaction, an atom that has been chemisorbed or an electrostatic attraction with an anion stack. The Pd NPs are stabilized by the introduction of organic ligands containing a heteroelement with an available lone pair. The organic chain of ligand inhibits the agglomeration of NPs crystal, while the heteroatom aggressively bonds to the surface of metal. The specific bonding between Pd metallic materials and soft

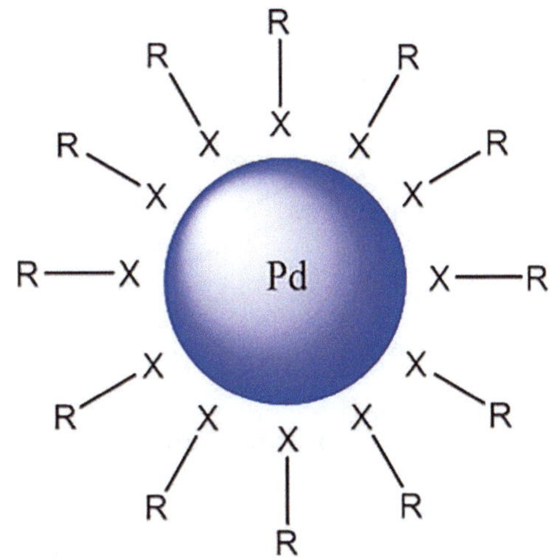

Fig. 21.7 Diagram depicting the stability of Pd NPs using ligand molecules (Cookson 2012)

sulfur-based donor ligands are highly efficient for stabilization of NPs (Cookson 2012) (Fig. 21.7).

21.5 Characterization of Pd NPs

21.5.1 UV-Vis Analysis

Amrutham et al. examined the UV-Vis spectra of Pd NPs. They devised a simple and very effective process for manufacturing Pd NPs with neem gum without the use of any hazardous chemicals in this study (Amrutham et al. 2020). When neem gum was added into the H_2PdCl_4 solution, it turned yellow color and was heated at 320 W in a microwave for the time interval of 10, 15, 20, and 25 min. Figure 21.8 illustrates the yellow color of H_2PdCl_4 in neem gum solution as well as the absorption peak centered at 425 nm for Pd(II) that transformed to Pd(III) that vanished due to the formation of Pd(0) NPs.

21.5.2 FTIR Analysis

Hekmati et al. studied the probable biomolecules accountable for the reduction of Pd NPs and capping of bioreduced NPs originated by means of leaves extract using FTIR (Hekmati et al. 2017). Figure 21.9 represents the FTIR spectra of Pd NPs. The peaks at 3410 and 1620 cm^{-1} imply the availability of phenolic and flavonoid substances in vegetable leaf extract, which act as a reducing agent to change Pd ion to Pd NPs.

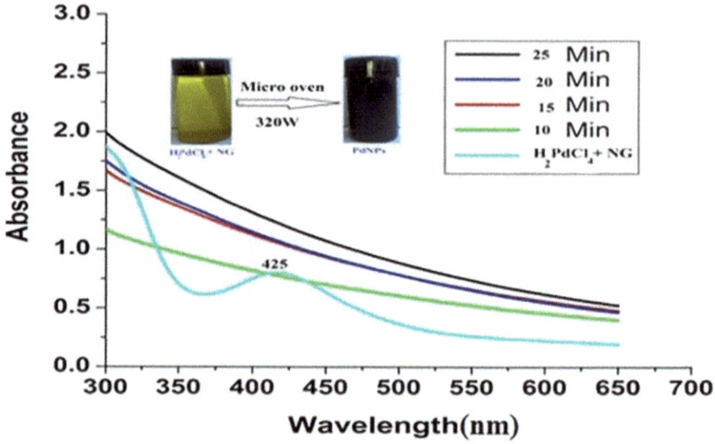

Fig. 21.8 UV-Vis spectra of Pd NPs synthesized using *Azadirachta indica* (Amrutham et al. 2020)

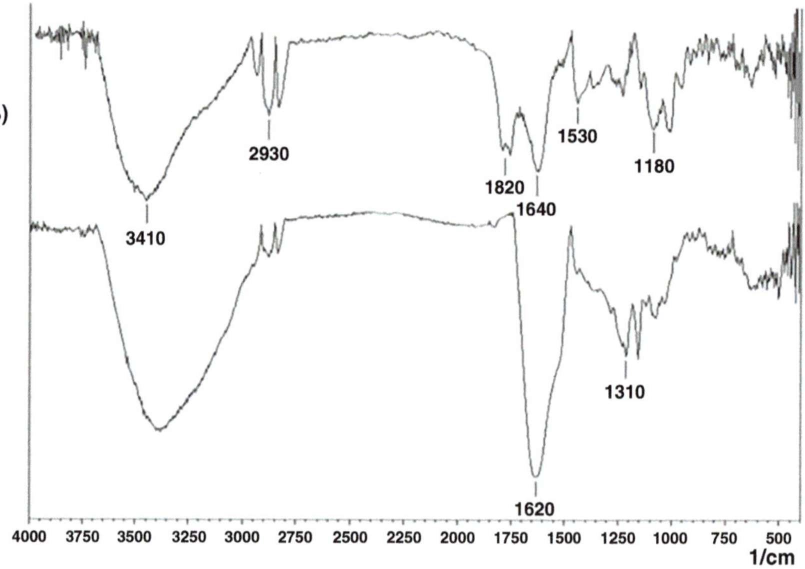

Fig. 21.9 FTIR spectra of Pd NPs synthesized using *Hibiscus sabdariffa* L. (Hekmati et al. 2017)

21.5.3 XRD Analysis

Shaik et al. used XRD analysis to check the crystallinity of the Pd NPs (Shaik et al. 2017). As seen in Fig. 21.10, the diffractogram shows five unique diffraction peaks centered at 2θ value of 40.10°, 46.49°, 68.12°, 81.60°, and 86.19° corresponding to (111), (200), (220), (311), and (222) crystal planes, which confirms the face-centered cubic crystal structure of the Pd NPs. In contradiction of the other four diffraction

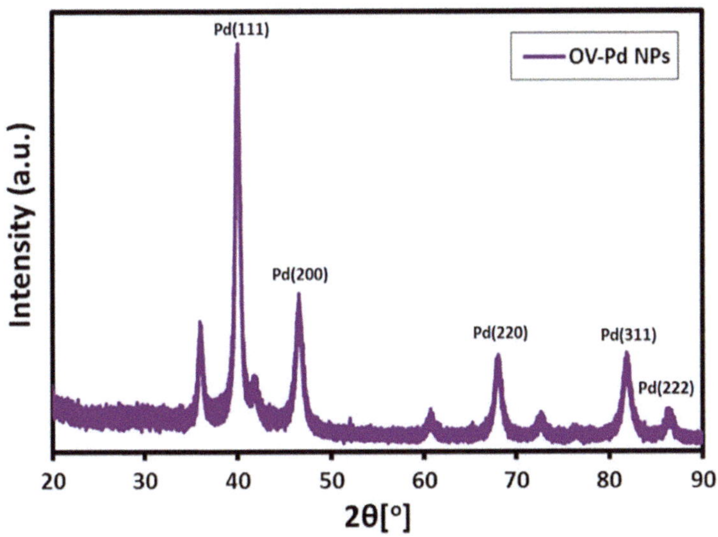

Fig. 21.10 XRD pattern of Pd NPs synthesized using *Origanum vulgare* L. (Shaik et al. 2017)

peaks, the strong diffraction at (111) plane can be described as the desired plane for the growth of Pd nanocrystals. The dimension of a typical crystalline material is the Pd NPs, i.e., 10 nm, which is calculated by applying the Debye-Scherrer equation based on the full width at half maximum of the (111) plane. Additionally, the characteristic diffraction of the Pd NPs synthesized using *Origanum vulgare* L. extract includes various other peaks.

21.5.4 EDX Analysis

Sharmila et al. used EDX to determine the elemental composition of the Pd NPs as represented in Fig. 21.11 (Sharmila et al. 2017b). The Pd peaks confirm the Pd NPs. The C and Cu peaks are attributed to the copper grid used to carry samples for TEM analysis, O peak corresponds to the aerial oxygen, and Cl peak is due to the precursor material $PdCl_2$.

21.5.5 SEM Analysis

Figure 21.12 represents the SEM images of the Pd NPs as reported by Siddiqi and Husen. The SEM image shows individual NPs and aggregates (Siddiqi and Husen 2016). On an aggregate, the particle size of the Pd NPs is close to 50 nm. Due to the accumulation of the Pd NPs, dendrites are formed matching with flower stems. Furthermore, dendrites are consisting of microcubes, which are arranged in a controlled pattern.

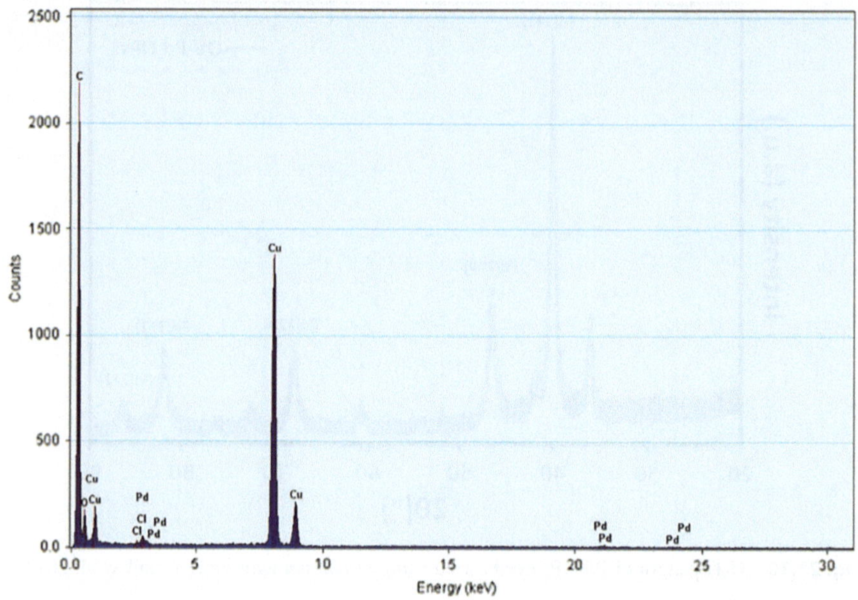

Fig. 21.11 EDX spectrum of Pd NPs synthesized using *Filicium decipiens* (Sharmila et al. 2017b)

21.5.6 TEM Analysis

Kumari et al. investigated the morphology of Pd NPs using TEM (Kumari et al. 2013). The Pd NPs show a spherical shape and narrow size distribution as shown in Fig. 21.13a–c. The average particle size of the Pd NPs was 11.3 nm. Moreover, the fringe spacing of 0.22 nm confirms the (111) diffraction plane of FCC crystal structure of the Pd NPs.

21.6 Properties of Pd NPs

21.6.1 Catalytic Properties

The Pd NPs are a type of catalyst that may be used in a wide variety of chemical processes. Their abilities are mostly because of the nanoscale particle diameter, which exhibits novel features and a high reactivity when compared to larger metals. The development of uniformly sized nanoscale catalysts has become a major topic in chemistry. The Pd NPs are used as an effective catalyst due to the large surface-to-volume ratio (Ngnie et al. 2016). The Pd NPs are well known for their ability to catalyze reduction processes and organometallic cross-coupling reactions like Heck and Suzuki-Miyaura. The Heck reaction is the arylation of alkynes by cross-coupling reaction in the vicinity of a 1:1 mixture of Pd(0) catalysts and a base. This reaction is commonly used to make substituted olefins, dienes, and a variety of other

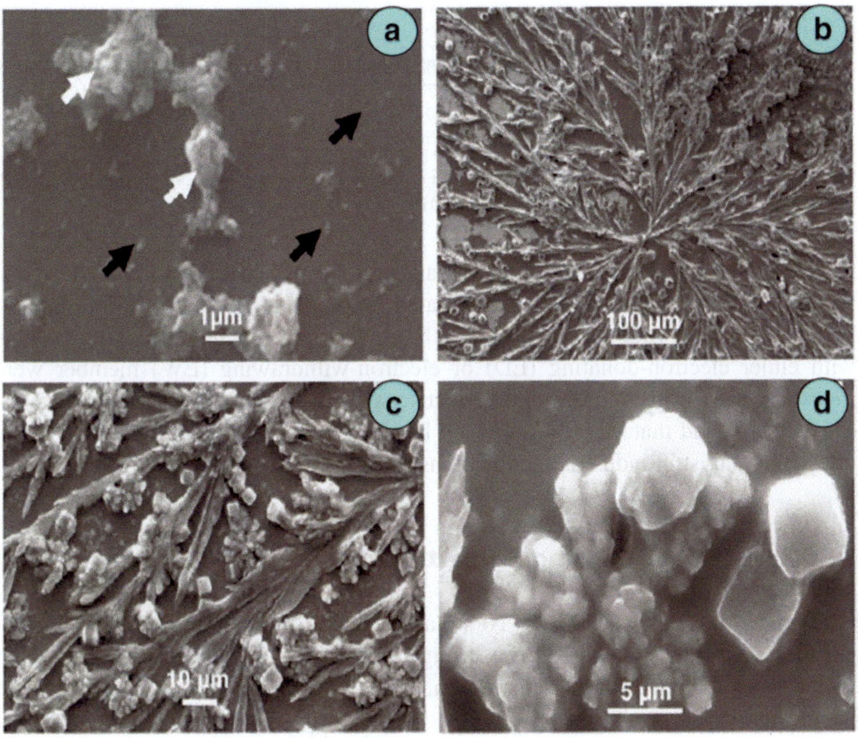

Fig. 21.12 SEM images of (**a**) Pd NPs synthesized using *Musa Paradisiaca* (**b–d**). Microwire networks at the periphery due to coffee ring effect (Siddiqi and Husen 2016)

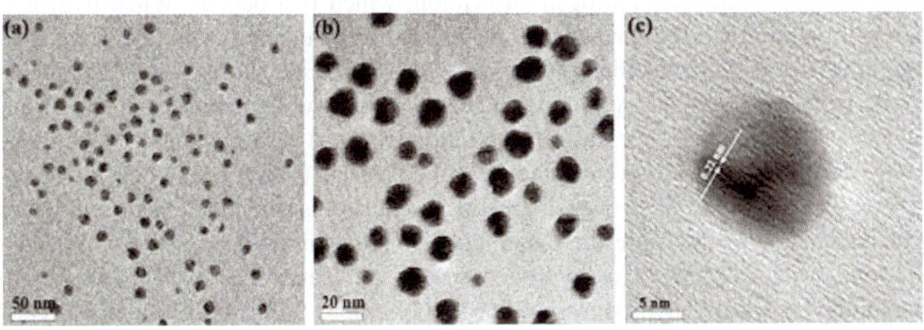

Fig. 21.13 TEM micrographs at different magnifications of Pd NPs synthesized using tannic acid (Kumari et al. 2013)

unsaturated compounds (Ohtaka et al. 2011). Bonds between aryls and aryls are prevalent in pharmaceuticals, polymeric materials, agricultural chemicals, and natural substances, and this coupling is employed to produce them. The Pd NPs were produced utilizing watermelon rind extract like a catalytic converter for such Suzuki coupling process by Lakshmipathy et al. as schematically depicted in Fig. 21.14. The

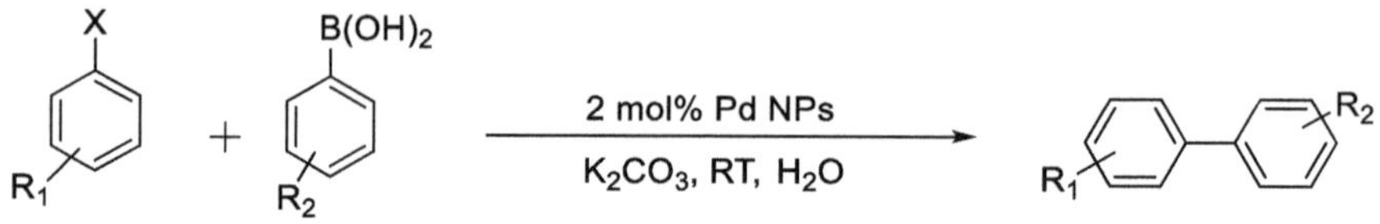

Fig. 21.14 Schematic illustration of the Suzuki coupling process using Pd NPs produced through watermelon rind extract (Lakshmipathy et al. 2015)

reaction was completed at room temperature in the presence of 2 mol.% Pd NPs, producing acceptable to excellent separated ratios of the product offering a short response time. A wide range of benzyl bromides with functionalized substituents with either electron-donating (ED) or electron-withdrawing (EW) member were investigated. The reactions went perfectly, with approaching quantitative yields. This study found that Pd NPs generated by rind extract have the potential to be an effective catalyst in industrial applications (Lakshmipathy et al. 2015).

21.6.2 Hydrogen Sensing Properties

Catalytic components such as Pd NPs play an important role in quantifying hydrogen. The response is predicted to be stronger if the catalytic surface area is amplified within a fixed device shape. This notion, when combined with recent advances in nanotechnology, produces promising outcomes in investigation of hydrogen sensors. Therefore, the simplicity of construction of the instrument and manufacturing still is a topic for additional research. The Pd NPs for sensor applications have been synthesized using a variety of approaches. The Pd NPs have a high affinity for hydrogen adsorption, which is a benefit. When hydrogen adsorbs on thin Pd films, the lattice expands, and when the hydrogen is removed, the original lattices are immediately restored. Pd hydride (PdH_x) is formed spontaneously in the Pd-hydrogen system. Depending on the hydrogen content, PdH_x occurs in two phases: α and β. For low concentrations below 1%, the α-phase is dominant, while over 1%, a transition from the α- to the β-phase (0.015×0.7) occurs. The β-phase, on the other hand, is undesirable because it causes a volume expansion, which can cause significant tensile stress and lead to device failure. Lower detection limits of the device are critical for hydrogen sensor applications, as well as the sensor including all feasible hydrogen gas sensors preferable if this limit values are really minimal. As a result, including all feasible hydrogen gas sensors, the greatest hydrogen percentage should not be more than 1% (Gupta et al. 2014). The ability of a hydrogen gas sensor with reversible sensing behavior has been shown using Pd NPs-supported carbon nanotubes (CNTs). Chemical reactions can take place on both side and inner walls of CNTs. During the first cycle of gas exposure, these processes cause irreversibility in sensory capabilities. In the meanwhile, NH_3, NO_2, and O_2 gas have shown irreversible behavior (Ong et al. 2002). Ghasempour et al. analyzed gas interaction as illustrated schematically in Fig. 21.15 for the sample in the presence of hydrogen to better explain the irreversibility behavior. The atomized hydrogen

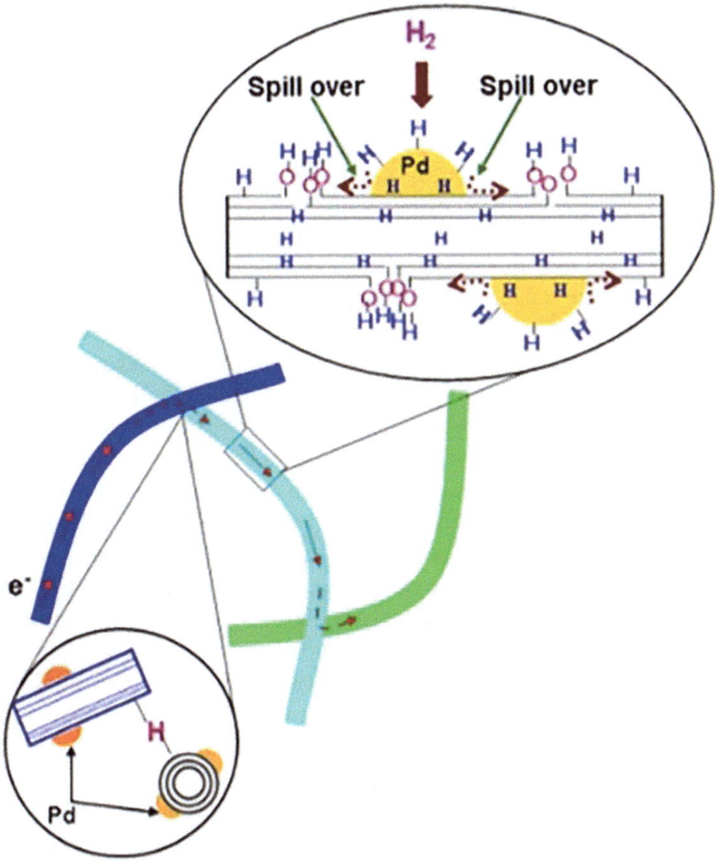

Fig. 21.15 Schematic illustration of hydrogen gas interaction with the surface of Pd/MWCNTs (Ghasempour and Mortazavi 2010)

produced by Pd particles can flow onto CNT walls and defects, as well as soak inside the Pd crystalline structure, causing a shift in phase between α and β (Ghasempour and Mortazavi 2010).

The hydrogen atoms that have been adsorbed can give electrons to the *p*-type semiconducting CNTs. As a result, the number of hole carriers is depleted, and electrical conductivity declines. However, there are a variety of reasons for increased conductivity. Small elements like hydrogen may enter into the CNT structures (Reddy and Ramaprabhu 2008). Rahimiet et al. anticipated them to chemically link to carbons and provide novel charge transfer channels in multiwalled carbon nanotubes (MWCNTs) structures and junctions (Ghasempour and Mortazavi 2010). All hydrogen atoms are desorbed when the hydrogen flow is turned off, and save those that have chemically adsorbed. Consequently, when the hydrogen flow is stopped off, the sample conductivity does not return to its original level (irreversibility behavior). Hydrogen atoms can physically adsorb on the empty space in subsequent adsorption-desorption cycles that attributed to the reversible gas sensing mechanism.

21.6.3 Magnetic Properties

Nanoscale Pd and 4*d* transition MNPs have been shown to have unique magnetic characteristics. Due to the fact that an unbound Pd atom has a $[Kr]4d^{10}$ configuration and as per the Hund rule, this arrangement is nonmagnetic. When a crystal pattern is developed by a cluster of Pd particles, the configuration changes, with just a couple of electrons shifting from 4d to 5 s. Bulk Pd improved the Pauli paramagnetism and large susceptibility with a structure of $4d^{(10-a)}5s^{a}$ ($a \geq 0$). Due to the superior paramagnetic susceptibility of bulk Pd, it accomplishing the Stoner criterion, which is recommended for 4d transition elements, $N(E_f)\ I > 1$, where $N(E_f)$ is the density of states just below the Fermi level (E_f), and I denotes the Stoner parameter (I= 0.71 eV for Pd). Taniyama et al. discovered a magnetic moment in Pd ensembles with a dimension little less than 7 nm (Taniyama et al. 1997). Because of the significant variation of surface energy, these Pd aggregates have a noncrystallographic ellipsoid shape rather the conventional massive FCC symmetry, which is thought to be the source of this ferromagnetism. Shinoharaet al. also discovered that lowering the size of the particles ranging from 9.9 to 5.9 nm, the ferromagnetic susceptibility increases (Shinohara et al. 2003). Cox et al. tested on 13–150 atom Pd clusters, on the other hand, found no indication of ferromagnetism. In addition, *hexagonal* closed packing (HCP) Pd has just been discovered to be ferromagnetic (Cox et al. 1994).

Litrán et al. investigated the Pd NPs nanostructures and magnetic behavior produced with distinct capping circumstances in a comparative manner (Litrán et al. 2006). On the other hand, weakly interacting dipole molecules such as tetraalkylammonium halides ($R_4N^+X^-$) salts have already been examined, with the entire length of hydrocarbon chains being varied to create variable degrees of surface oxidation of NPs. Due to the charge transfer mechanisms, alkane-thiol molecules covalently linked to Pd atoms in clusters usually produce localization of d-holes in every metalcore band. Two distinct particle sizes of Pd NPs were investigated in this scenario. Permanent magnetism was discovered at room temperature. The superconducting quantum interference device (SQUID) was used to measure the magnetic properties of Pd nanorods at various temperatures (Xiao et al. 2009). For the measurement, a specimen including an aspect ratio of 2.5 was being used to reduce the impact of size deviation as much as feasible. The hysteresis loops shown in Fig. 21.16 clearly indicate the ferromagnetic behavior of Pd nanorods at room temperature. With increasing measurement temperature, the coercive field (Hc) of Pd nanorods decreases: 112 Oe at 5 K, 80 Oe at 150 K, and 69 Oe at 300 K. Furthermore, the magnetic flux doesn't get affected at any temp. This might be because the particles contain atoms that are ferromagnetic and superparamagnetic. The presence of excess surfactants or surface oxidation might be answerable for the observed behavior. The M-H curve of polyvinylpyrrolidone (PVP) is shown in

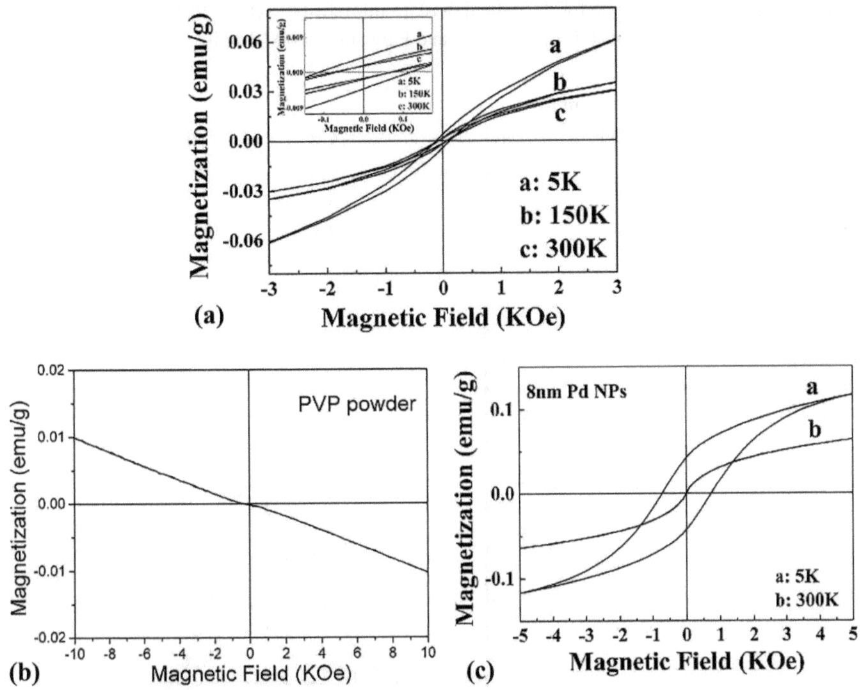

Fig. 21.16 (**a**) Temperature-dependent M-H curves of Pd nanorods and spherical NPs at 5, 150, and 300 K, (**b**) PVP M-H curve at ambient temperature, and (**c**) 8 nm M-H curves of spherical Pd NPs at 5, 150, and 300 K (Xiao et al. 2009)

Fig. 21.16b. On the other hand, it exhibits the diamagnetic character of the polymeric surfactant. When in association with the electron energy loss spectroscopy (EELS) study, which ruled out the presence of oxygen, it is clear that the permanent magnetic moment of Pd nanorods is an inherent characteristic. The hysteresis loop graph of produced 8 nm spherical PVP-capped Pd NPs using a compatible polyol procedure is shown in Fig. 21.16c. At low temperatures (5 K), the spherical Pd NPs are ferromagnetic, and at normal temperatures, they are superparamagnetic.

21.7 Application

21.7.1 Catalytic Activity

Since the production of Pd NPs from the surface of walnut shell produces composites with high catalytic activity. After confirming the specific qualities of catalysts, the Pd/walnut shell nanocomposite of catalytic efficiency was analyzed by checking the reduction of 4-nintro (4-NP), methyl orange (MO), Congo red (CR), rhodamine B (RhB), and methylene blue (MB) at ambient temperature in an aqueous

Table 21.2 Completion time for the reduction of 25 mL of 4-NP, MO, CR, RhB, and MB dyes in 25 mL $NaBH_4$ using 5 mg of Pd, walnut shell, and Pd/walnut shell (Bordbar and Mortazavimanesh 2017)

	Time (min)				
Catalyst	4-NP	MO	CR	RhB	MB
Pd/walnut shell	1	Immediately	0.41	0.21	Immediately
Pd NPs	4	5	4	5	3
Walnut shell	30[a]	30[a]	30[a]	30[a]	30[a]

[a]Not completed

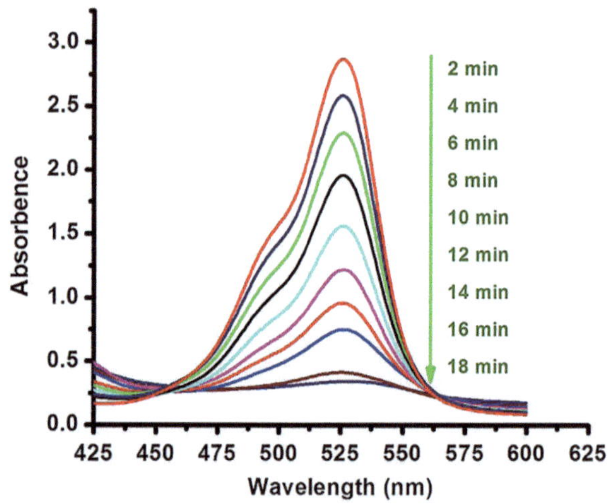

Fig. 21.17 Catalytic reduction of rhodamine 6G using Pd NPs (Amrutham et al. 2020)

environment as presented in Table 21.2. The absorption spectra were utilized to monitor the progress of the catalytic performance of the species as a protocol of reaction time. Initially, the reduction was carried out in the presence of the catalysts. The conversion rate was very slow while in the vicinity of only walnut shell. The Pd NPs-reinforced walnut shell nanocomposites exhibit effective catalytic activity than that of Pd NPs synthesized using *Equisetum arvense* L. leaf extract, which is owing to the Pd NPs' enormous surface area, tiny size, and homogeneous distribution over the walnut shell. The porous surface of the walnut shell provides effective catalytic actions. Furthermore, more amount of 4-NP is adsorbed on the surface of the Pd/walnut shell nanocomposites as compared to other organic dyes, which makes reduction process easier (Bordbar and Mortazavimanesh 2017).

The reduction of rhodamine 6G (Rh6G) dye was utilized to assess the catalyst behavior of Pd NPs stabilized in NG in the presence of $NaBH_4$ after incorporation of the Pd NPs into the reaction mixture. The magnitude of a maximum absorption centered at 527 nm steadily diminishes when response time rises as represented in Fig. 21.17. The complete reduction of the dye in the presence of Pd NPs requires 18 min (Amrutham et al. 2020).

21.7.2 Antibacterial Activity

The antibacterial activity of the Pd NPs synthesized from leaf extracts of *Filicium decipiens* was assessed using diffusion method. Gram-negative (*Escherichia coli* and *Pseudomonas aeruginosa*) and gram-positive (*Bacillus subtilis* and *Staphylococcus aureus*) bacteria were utilized as test microbe strains. The development of clear zones after incubation of 1 day in all the cases which confirms the usefulness of the Pd NPs dopes antibacterial drugs (Sharmila et al. 2016). The antibacterial activity of the Pd NPs is attributed to the internalization and affinity of the cell membranes for NPs that interrupts the transportation process by interacting with enzymes and deoxyribonucleic acid (DNA) and inhibition of respiratory system (Sharmila et al. 2017b).

The gram-negative *E. coli* was pushed to the limits to see if it has bactericidal characteristics of the glucosamine-stabilized Pd NPs as represented in Fig. 21.18. The Pd NPs are very effective against *E. coli* as compared to the tobramycin drug as reported from the agar plate experiment. The NPs are smoothly inserted into the *E. coli* cell wall having narrow covering of peptidoglycan due to the small size of NPs and interaction of the *E. coli* with NPs. The mechanism of the antibacterial activity of the NPs is carried out by various approaches such as annihilation of cell membranes and intracellular generation of reactive oxygen species (ROS) that inhibits this vital function. The ROS is an efficient mechanism for retardation of growth of bacteria (Ullah et al. 2018).

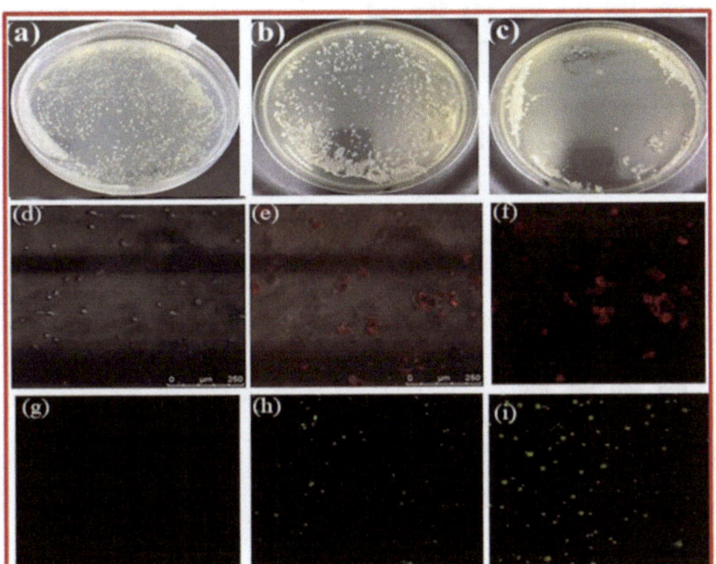

Fig. 21.18 Agar plate antibacterial analysis of (**a**) *E. coli* PBS control, (**b**) tobramycin, (**c**) glucosamine-mediated Pd NPs, (**d–f**) cell membrane analysis, and (**g–i**) intracellular ROS investigation in *E. coli* (Ullah et al. 2018)

21.7.3 Anticancer Activity

Both trichostatin A (*TSA*) and *Pd NPs* were tested for their ability for cytotoxicity in breast and cervical cancer cells. Vigushin et al. first reported the effectiveness of Pd NPs to arrest the increase in number of the MCF-7 cells in vitro of breast carcinoma (Vigushin et al. 2001). The cell viability was determined using the WST-8 (5-(2,4-disulfophenyl)-3-(2-methoxy-disulfophenyl)-3-(2-methoxy-disulfophenyl)-3-(2-methoxy-4-nitrophenyl)-2-(4-nitrophenyl)-2H-tetrazolium) TSA, which was noticed to have a significant inhibitory potential. The viability of MCF-7 cells is shown in Fig. 21.19a. The cell viability of both TSA and Pd NPs decreases with increase in concentration. The cell viability of MCF-7 cells was decreased by 50%,

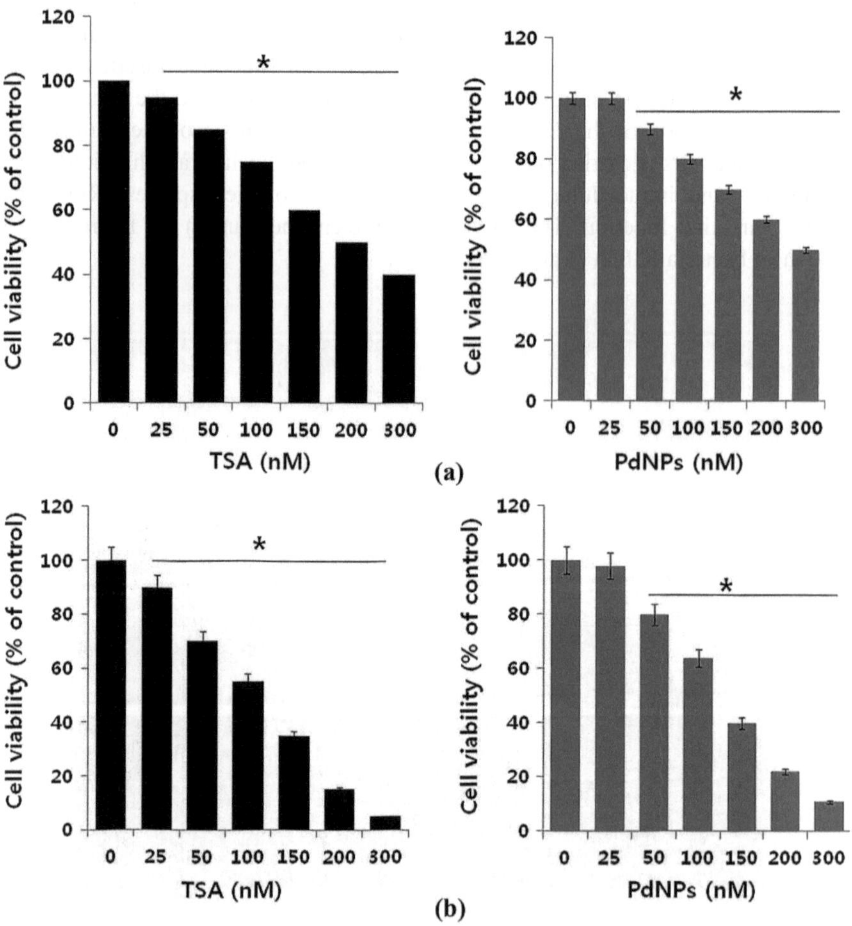

Fig. 21.19 Dose-dependent effect of TSA and Pd NPs on cell viability in human. (**a**) Breast and (**b**) cervical cancer cells (Zhang et al. 2016)

i.e., half-maximal inhibitory concentration (IC50) of 200 and 300 nM TSA and Pd NPs, respectively (Zhang et al. 2016).

According to Wu et al., the effect of TSA or Pd NPs on the cervical cancer cells was studied in a daily dosage basis as shown in Fig. 21.19b. The survival of cervical cancer cells decreases with increase in concentration of the TSA and Pd NPs. The cell viability of the cervical cancer cells attains IC50 values at 100 and 125 nM concentrations for TSA and Pd NPs, respectively (Wu et al. 2005).

21.7.4 Antioxidant Activity

The ability of Pd NPs to total antioxidant was studied by using 2,2-diphenyl-1-picrylhydrazyl (DPPH) free radical to assess its antioxidant activity. The purple color of the DPPH solution changed to pale yellow color with time in the presence of Pd NPs. The free radical scavenging of DPPH linearly increases with increase in Pd NPs (Gąsecka et al. 2018). The maximum radical inhibition of 77% was observed at 50 μg/ml concentration of Pd NPs, whereas ascorbic acid shows 64% inhibition at that concentration. However, Mohana and Sumathi reported the highest inhibition rate against the DPPH radical (Fig. 21.20).

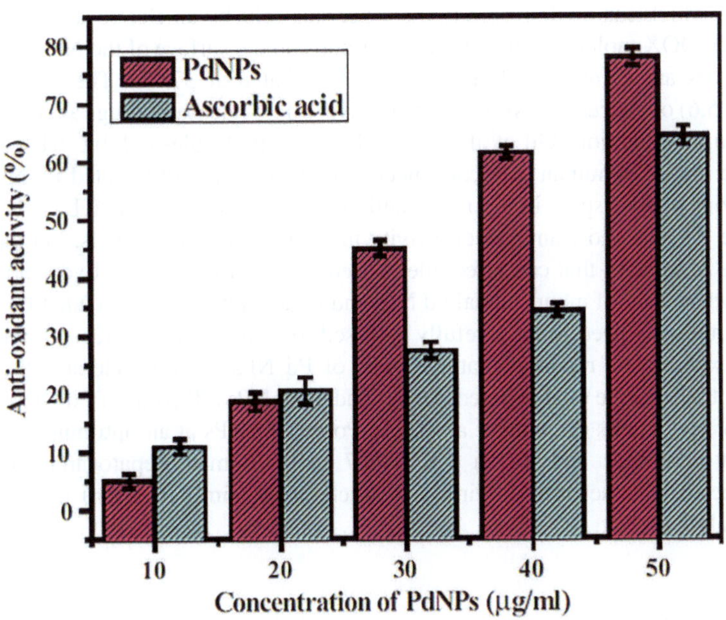

Fig. 21.20 Antioxidant activity of Pd NPs synthesized using *Agaricus bisporus* (Mohana and Sumathi 2020)

21.7.5 Biosensor

Pd NPs can be used in microelectronic devices. Recently, the use of Pd NPs in biosensing is getting lot of attention due to its amperometric measurement of generated electroactive species. Owing to its vast range of uses in the textile, paper, cleaning product, and food industries and H_2O_2 detection, Pd NPs are extremely significant. Pd NPs can be used as an electrode because it is catalytically active, relatively affordable, and difficult to oxidize. The dendritic Pd NPs-modified electrode can be employed to monitor H_2O_2 reduction at a negative potential to eliminate interference from other electroactive species (Zhou et al. 2007).

21.7.6 Gene and Drug Delivery

The technique of targeted medication distribution is rapidly growing to get out of the disadvantages of the conventional medicinal therapy, which include low selectivity, quick excretion, and high toxicity (Tiwari et al. 2012). The NPs can be designed to transport medicine and genes, which can be targeted and released at a controlled manner. The medicinal substances can be injected directly into the holes of NPs or completely attached with them through different chemical modifications. The Pd NPs can be employed for loading medicine and genes due to high porosity of the surface of the NPs (Kang et al. 2018). The cancer drugs like doxorubicin (DOX) can be inserted into the porous Pd NPs synthesized from aqueous *Inonotus obliquus* extract, which can be delivered to the cancerous cells by an electrostatic interaction between DOX molecules and absorbing groups on the surface of the NPs. The DOX molecules are completely filled after 6 h incubated at 37 °C. The acidic nature ($pH = 5.6$) of cancerous tissues can permit very high release of drugs as compared to the neutral condition (Gil et al. 2018). The DOX molecules exhibit pH sensitivity release profile in human cervical cancer cells (HeLa), and PEGylated Pd NPs via a hydrazine bond display in vivo high antitumor efficiency against HeLa cell.

The Pd NPs show anticancer activity in a variety of cancer cells, but there is a chance of toxicity that can affect the consequences of the drug delivery processes. The cytotoxicity of mono-modal Pd NPs and bimodal tumor drug-loaded Pd NPs on the tumor cells need to be carefully assessed to estimate the degree of toxicity. In vitro experiments reported that 20 g/mL of Pd NPs destroys HeLa cells, which account 20% of the total, whereas DOX-loaded Pd NPs (20 g/mL of Pd NPs) destroy 30% of HeLa cells (Shanthi et al. 2015). Porous Pd NPs at an optimum proportion exhibit antitumor activity in NS3-Huh7 cells (human hepatocarcinoma cells transfected with hepatitis C virus nonstructural protein 3 replicon) in an in vitro cell viability assay.

21.7.7 Lithium-Oxygen Battery

The Swagelok™ kind of cells was developed to evaluate the catalyst action of the Pd/MnO_2 nanorods on the air cathode of Li-air battery. Maria et al. evaluated the catalytic activity of three electrodes with and without catalysts, such as Pd/MnO_2 and MnO_2 along with the performance of the Pd/MnO_2-catalyzed electrode in a comparison with other two electrodes. The cycling performance of the catalyst-free KB-air cathode is compared with the lithium metal anode (Li/Li+) as shown in Fig. 21.21 (McCloskey et al. 2011). The batteries were investigated at ambient temperature and in an O_2 environment with a constant density of current 0.1 $mAcm^{-2}$ in the possible range of 2.0–4.3 V. The Pd/MnO_2 nanorod catalyst demonstrated a very high specific capacity in the first cycle as shown in Fig. 21.10a. With a discharging slope that is straight at 2.7 V, the initial discharge curve of Pd/MnO_2 nanorods resulted in a maximum capacity of 8526 $mAhg^{-1}$. With maximum reversibility, a maximum capacity of 8526 $mAhg^{-1}$ is attained when charging at 4.3 V (Lim et al. 2012). Furthermore, the difference between the possibilities of discharging and charging, or causing additional potential of 1.0 V, is very low and favors the reversibility of the rechargeable batteries. The oxygen reduction reaction (ORR) and oxygen evolution reaction (OER) performance of the Pd/MnO_2 catalyst in the nonaqueous phase significantly increases than that of the aqueous phase. The majority of the catalysts utilized for ORR/OER in aqueous phase have been proven to be effective catalysts in nonaqueous $Li\text{-}O_2$ batteries as shown in Fig. 21.21 (Zahoor et al. 2015).

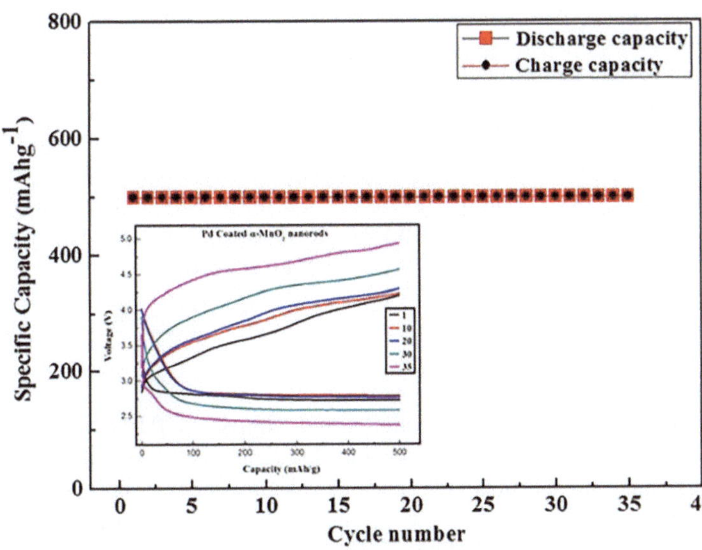

Fig. 21.21 Cycling properties of Pd/MnO_2 nanorod-catalyzed $Li\text{-}O_2$ battery at 500 $mAhg^{-1}$ capacity and variation of potential with applied capacity (insert) (Zahoor et al. 2015)

21.8 Future Scope of Pd NPs

In recent years, research on NPs and their potential uses are rapidly increasing. Numerous papers have been published and describe the phytoremediation of metallic NPs utilizing variety of biological materials, such as plants, plant part, bacteria, yeast, and fungus. However, significant obstacles remain, limiting its large-scale manufacture and subsequent uses. Some of the significant difficulties encountered during the processing are listed below:

- To regulate the size and morphology of the NPs, precise optimization studies on process parameters and reactants must be required.
- Research should also focus on improving a variety of physicochemical properties of NPs having particular applications.
- The role of each plant extract metabolite and microorganism cellular component in the creation of NPs should be thoroughly investigated.
- Improving NPs production rates and stability while reducing reaction time requires to adjust several reaction parameters.
- Many researchers are working on the green production of NPs in the future since there is no clear idea about it.
- Synthesis method of NPs from microorganisms is time-consuming, and the real optimal conditions are unknown.
- Many researchers are investigating the hazardous effects of NPs exposure interactions with biological systems and the implications of which are yet undiscovered.

By overcoming these obstacles, green synthesis methods might become more economical and equivalent to traditional approaches for generating NPs on a big scale. The possible therapeutic qualities of Pd NPs have lately been investigated in biological applications. Future study is devoted to understanding the anticancer and antibacterial processes of Pd NPs at the molecular level, as well as metabolic routes. Some potential harmful effects of Pd NPs on biological systems and their surroundings have recently been identified. However, our knowledge of the toxicities of Pd NPs is currently inadequate and requires further research. Furthermore, basic knowledge on the bioavailability of Pd NPs has been investigated.

21.9 Conclusions

This chapter focuses toward biosynthesis and the use of Pd NPs obtained from biomass and microbes and plant extracts. In comparison with other traditional procedures such as physical and chemical approaches, a fair, nontoxic, and ecologically sustainable method for producing Pd NPs has been developed. A variety of plant, products including extracts of leaves, fruit, seed, gum, bark, and so on, as well as MOs such as bacteria, fungus, filamentous fungi, and so on, have shown promise for the genesis of Pd NPs. The size and form of Pd NPs, as well as the reaction rate, are heavily influenced by the experimental factors as response rate, substrate concentrations, pH, temperatures, reaction time, washing, the amount of salt, etc. The Pd NPs have lately received a lot of interest due to their very excellent catalytic and electrical capabilities. The Pd NPs are quickly agglomerated due to their large surface area, and their application is highly restricted. The Pd NPs catalysts have been functionalized on the surface to avoid particle agglomeration and mobility and so improve performance. The size, morphology, and structure of biosynthesized Pd NPs have been determined using several characterizations, including UV-Vis spectroscopy, EDX, FTIR, XRD, SEM, and TEM. The Pd NPs have applications in catalysis, biological sensors, bioimaging, antimicrobial materials, hydrogen generation, optical devices, CH_4 combustion, and medicine. Because of their high surface-to-volume ratio, the Pd NPs can operate as a possible catalyst in a variety of reactions. The most essential uses of these clean Pd NPs structures created by plant extracts are related to hazardous chemicals removal. As a result, green chemistry has significant promise in the commercial and economic manufacture of Pd NPs. Plant-mediated manufacture of Pd NPs appears to be the most practical and effective for cheap and sustainable production among green chemistry techniques.

References

Ahmed S, Ahmad M, Swami BL (2016) A review on plants extract mediated synthesis of silver nanoparticles for antimicrobial applications: a green expertise. J Adv Res 7(1):17–28

Ameri A, Shakibaie M, Rahimi H-R, Adeli-Sardou M, Raeisi M, Najafifi A, Forootanfar H (2020) Rapid and facile microwave-assisted synthesis of palladium nanoparticles and evaluation of their antioxidant properties and cytotoxic effects against fibroblast-like (HSkMC) and human lung carcinoma (A549) cell lines. Biol Trace Elem Res 197:132–140

Amrutham S, Maragoni V, Guttena V (2020) One-step green synthesis of palladium nanoparticles using neem gum (*Azadirachta Indica*): Characterization, reduction of Rhodamine 6G dye and free radical scavenging activity. Appl Nanosci 10:4505–4511

Astruc D (2007) Palladium nanoparticles as efficient green homogeneous and heterogeneous carbon–carbon coupling precatalysts: a unifying view. Inorg Chem 46(6):1884–1894

Azizi S, Shahri MM, Rahman HS, Rahim RA, Rasedee A, Mohamad R (2017) Green synthesis palladium nanoparticles mediated by white tea (*Camellia sinensis*) extract with antioxidant, antibacterial, and antiproliferative activities toward the human leukemia (MOLT-4) cell line. Int J Nanomedicine 12:8841

Bankar A, Joshi B, Kumar AR, Zinjarde S (2010) Banana peel extract mediated novel route for the synthesis of palladium nanoparticles. Mater Lett 64(18):1951–1953

Bell AT (2003) The impact of nanoscience on heterogeneous catalysis. Science 299(5613):1688–1691

Beller M, Fischer H, Kühlein K, Reisinger C-P, Herrmann WA (1996) First palladium-catalyzed Heck reactions with efficient colloidal catalyst systems. J Organomet Chem 520:257

Bhakyaraj K, Kumaraguru S, Gopinath K, Sabitha V, Kaleeswarran PR, Karthika V, Arumugam A (2017) Eco-friendly synthesis of palladium nanoparticles using *Melia azedarach* leaf extract and their evaluation for antimicrobial and larvicidal activities. J Clust Sci 28:463–476

Bordbar M, Mortazavimanesh N (2017) Green synthesis of Pd/walnut shell nanocomposite using *Equisetum arvense L.* leaf extract and its application for the reduction of 4-nitrophenol and organic dyes in a very short time. Environ Sci Pollut Res 24:4093–4104

Buzea C, Pacheco II, Robbie K (2007) Nanomaterials and nanoparticles: sources and toxicity. Biointerphases 2(4):MR17–MR71

Chen J, Shi W, Zhang X, Arandiyan H, Li D, Li J (2011) Roles of Li^{+} and Zr^{4+} Cations in the catalytic performances of $Co_{1-x}M_XCr_2O_4$ (M= Li, Zr; x= 0–0.2) for methane combustion. Environ Sci Technol 45(19):8491–8497

Cheong S, Watt JD, Tilley RD (2010) Shape control of platinum and palladium nanoparticles for catalysis. Nanoscale 2(10):2045–2053

Cookson J (2012) The preparation of palladium nanoparticles. Platin Met Rev 56(2):83–98

Cox AJ, Louderback JG, Apsel SE, Bloomfield LA (1994) Magnetism in 4d-transition metal clusters. Phys Rev B 49(17):12295

Crespo-Quesada M, Yarulin A, Jin M, Xia Y, Kiwi-Minsker L (2011) Structure sensitivity of alkynol hydrogenation on shape-and size-controlled palladium nanocrystals: Which sites are most active and selective? J Am Chem Soc 133(32):12787–12794

Dai Y, Kumar VP, Zhu C, Wang H, Smith KJ, Wolf MO, MacLachlan MJ (2019) Bowtie-shaped $NiCo_2O_4$ catalysts for low-temperature methane combustion. Adv Funct Mater 29(8):1807519

Devi TB, Ahmaruzzaman M (2016) Bio-inspired sustainable and green synthesis of plasmonic Ag/AgCl nanoparticles for enhanced degradation of organic compound from aqueous phase. Environ Sci Pollut Res 23:17702–17714

Dikshit PK, Kumar J, Das AK, Sadhu S, Sharma S, Singh S, Gupta PK, Kim BS (2021) Green synthesis of metallic nanoparticles: applications and limitations. Catalysts 11(8):902

Dimitratos N, Porta F, Prati L, Villa A (2005) Synergetic effect of platinum or palladium on gold catalyst in the selective oxidation of D-sorbitol. Catal lett 99:181–185

Durán N, Marcato PD, Durán M, Yadav A, Gade A, Rai M (2011) Mechanistic aspects in the biogenic synthesis of extracellular metal nanoparticles by peptides, bacteria, fungi, and plants. Appl Microbiol Biotechnol 90:1609–1624

Fahmy SA, Preis E, Bakowsky U, Azzazy HMES (2020) Palladium nanoparticles fabricated by green chemistry: Promising chemotherapeutic, antioxidant and antimicrobial agents. Materials 13(17):3661

Gąsecka M, Magdziak Z, Siwulski M, Mleczek M (2018) Profile of phenolic and organic acids, antioxidant properties and ergosterol content in cultivated and wild growing species of *Agaricus*. Eur Food Res Technol 244:259–268

Ghasempour R, Mortazavi SZ, Rahimi F (2010) Hydrogen sensing properties of multi-walled carbon nanotube films sputtered by Pd. Int J Hydrog Energy 35(9):4445–4449

Ghosh S, Nitnavare R, Dewle A, Tomar GB, Chippalkatti R, More P, Chopade BA (2015) Novel platinum–palladium bimetallic nanoparticles synthesized by *Dioscorea bulbifera*: Anticancer and antioxidant activities. Int J Nanomedicine 10:7477

Gil YG, Kang S, Chae A, Kim YK, Min DH, Jang H (2018) Synthesis of porous Pd nanoparticles by therapeutic C*haga* extract for highly efficient tri-modal cancer treatment. Nanoscale 10 (42):19810–19817

Gour A, Jain NK (2019) Advances in green synthesis of nanoparticles. Artif Cells Nanomed Biotechnol 47(1):844–851

Gupta R, Xie H (2018) Nanoparticles in daily life: Applications, toxicity and regulations. J Environ Pathol Toxicol Oncol 37(3):209–230

Gupta D, Dutta D, Kumar M, Barman PB, Sarkar CK, Basu S, Hazra SK (2014) A low temperature hydrogen sensor based on palladium nanoparticles. Sens Actuators B Chem 196:215–222

Hekmati M, Bonyasi F, Javaheri H, Hemmati S (2017) Green synthesis of palladium nanoparticles using *Hibiscus sabdariffa L.* flower extract: Heterogeneous and reusable Nanocatalyst in Suzuki coupling reactions. Appl Organomet Chem 31(11):e3757

Hodaei A, Ataie A, Mostafavi E (2015) Intermediate milling energy optimization to enhance the characteristics of barium hexaferrite magnetic nanoparticles. J Alloys Comp 640:162–168

Hong E, Kim C, Lim DH, Cho HJ, Shin CH (2018) Catalytic methane combustion over Pd/ZrO_2 catalysts: effects of crystalline structure and textural properties. Appl Catal B Environ 232:544–552

Hu P, Yang H (2013) Insight into the physicochemical aspects of kaolins with different morphologies. Appl Clay Sci 74:58–65

Im Y, Lee C, Vasquez RP, Bangar MA, Myung NV, Menke EJ, Yun M (2006) Investigation of a single Pd nanowire for use as a hydrogen sensor. Small 2(3):356–358

Iravani S, Korbekandi H, Mirmohammadi SV, Zolfaghari B (2014) Synthesis of silver nanoparticles: Chemical, physical and biological methods. Res Pharm Sci 9(6):385–406

Jia L, Zhang Q, Li Q, Song H (2009) The biosynthesis of palladium nanoparticles by antioxidants in *Gardenia jasminoides Ellis*: long lifetime Nanocatalysts for P-nitrotoluene hydrogenation. Nanotechnology 20(38):385601

Johnston RL, Wilcoxon JP (eds) (2012) Metal Nanoparticles and Nanoalloys, Elsevier, Oxford, UK 3:2–302

Kang S, Shin W, Kang K, Choi MH, Kim YJ, Kim YK, Jang H (2018) Revisiting of Pd nanoparticles in cancer treatment: all-round excellence of porous pd nanoplates in gene-thermo combinational therapy. ACS Appl Mater Interface 10(16):13819–13828

Karimi B, Enders D (2006) New N-heterocyclic carbene palladium complex/ionic liquid matrix immobilized on silica: Application as recoverable catalyst for the Heck reaction. Org Lett 8 (6):1237–1240

Kasthuri J, Kathiravan K, Rajendiran NJNR (2009) Phyllanthin-assisted biosynthesis of silver and gold nanoparticles: A novel biological approach. J Nanopart Res 11:1075–1085

Khanna P, Kaur A, Goyal D (2019) Algae-based metallic nanoparticles: Synthesis, characterization and applications. J Microbiol Methods 163:105656

Kim YH, Yang X, Shi L, Lanham SA, Hilborn J, Oreffo RO, Dawson JI (2020) Bisphosphonate nanoclay edge-site interactions facilitate hydrogel self-assembly and sustained growth factor localization. Nat Commun 11(1):1365

Klingensmith LM, Leadbeater NE (2003) Ligand-free palladium catalysis of aryl coupling reactions facilitated by grinding. Tetrahedron lett 44(4):765–768

Kora AJ, Rastogi L (2018) Green synthesis of palladium nanoparticles using gum ghatti (*Anogeissus latifolia*) and its application as an antioxidant and catalyst. Arab J Chem 11 (7):1097–1106

Kumari MM, Aromal SA, Philip D (2013) Synthesis of monodispersed palladium nanoparticles using tannic acid and its optical non-linearity. Spectrochim Acta A Mol Biomol. Spectroscopy 103:130–133

Kuppusamy P, Yusoff MM, Maniam GP, Govindan N (2016) Biosynthesis of metallic nanoparticles using plant derivatives and their new avenues in pharmacological applications–An updated report. Saudi Pharm J 24(4):473–484

Lakshmipathy R, Reddy BP, Sarada NC, Chidambaram K, Pasha SK (2015) Watermelon rind-mediated green synthesis of noble palladium nanoparticles: Catalytic application. Appl Nanosci 5:223–228

Lesani P, Babaei A, Ataie A, Mostafavi E (2016) Nanostructured $MnCo_2O_4$ synthesized via co-precipitation method for SOFC interconnect application. Int J Hydrogen Ener 41(45):20640–20649

Li C, Li W, Chen K, Ogunbiyi AT, Zhou Z, Xue F, Yuan L (2020) Palladium nanoparticles supported on surface-modified metal oxides for catalytic oxidation of lean methane. ACS Appl Nano Mater 3(12):12130–12138

Lim HD, Park KY, Gwon H, Hong J, Kim H, Kang K (2012) The potential for long-term operation of a lithium–oxygen battery using a non-carbonate-based electrolyte. Chem Commun 48 (67):8374–8376

Litrán R, Sampedro B, Rojas TC, Multigner M, Sánchez-López JC, Crespo P, Fernández A (2006) Magnetic and microstructural analysis of palladium nanoparticles with different capping systems. Phys Rev B 73(5):054404

Macaskie LE, Mikheenko IP, Omajai JB, Stephen AJ, Wood J (2017) Metallic bionanocatalysts: potential applications as green catalysts and energy materials. Microb Biotechnol 10(5):1171–1180

Malik P, Shankar R, Malik V, Sharma N, Mukherjee TK (2014) Green chemistry based benign routes for Nanoparticle synthesis. J Nanopart:2014

Manikandan V, Velmurugan P, Park JH, Lovanh N, Seo SK, Jayanthi P, Oh BT (2016) Synthesis and antimicrobial activity of palladium nanoparticles from *Prunus* × *yedoensis* leaf extract. Mater Lett 185:335–338

Maurer-Jones MA, Gunsolus IL, Murphy CJ, Haynes CL (2013) Toxicity of engineered nanoparticles in the environment. Anal Chem 85(6):3036–3049

McCloskey BD, Scheffler R, Speidel A, Bethune DS, Shelby RM, Luntz AC (2011) On the efficacy of electrocatalysis in nonaqueous Li–O_2 batteries. J Am Chem Soc 133(45):18038–18041

Mittal AK, Chisti Y, Banerjee UC (2013) Synthesis of metallic nanoparticles using plant extracts. Biotechnol Adv 31(2):346–356

Mohana S, Sumathi S (2020) Multi-functional biological effects of palladium nanoparticles synthesized using *Agaricus bisporus*. J Clust Sci 31:391–400

Momeni S, Nabipour I (2015) A simple green synthesis of palladium nanoparticles with *Sargassum alga* and their electrocatalytic activities towards hydrogen peroxide. Appl Biochem Biotechnol 176:1937–1949

Narayanan R, El-Sayed MA (2005) Catalysis with transition metal nanoparticles in colloidal solution: Nanoparticle shape dependence and stability. J Phys Chem B 109(26):12663–12676

Nasrollahzadeh M, Sajadi SM (2016) Pd nanoparticles synthesized in situ with the use of *Euphorbia granulate* leaf extract: Catalytic properties of the resulting particles. J Colloid Interface Sci 462:243–251

Nasrollahzadeh N, Sajjadi M, Dadashi J, Ghafuri H (2020) Pd-based nanoparticles: Plant-assisted biosynthesis, characterization, mechanism, stability, catalytic and antimicrobial activities. Adv Colloid Interf Sci 276:102103

National Nanotechnology Initiative (NNI). www.nano.gov

Ngnie G, Dedzo GK, Detellier C (2016) Synthesis and catalytic application of palladium nanoparticles supported on kaolinite-based nanohybrid materials. Dalton Trans 45(22):9065–9072

Nishimura S (2001) Handbook of heterogeneous catalytic hydrogenation for organic synthesis. Wiley, New York, pp 213–215

Nkosi SM, Anand K, Anandakumar S, Singh S, Chuturgoon AA, Gengan RM (2016) Design, synthesis, anticancer, antimicrobial activities and molecular docking studies of novel quinoline bearing dihydropyridines. J Photochem Photobiol B Biol 165:266–276

Ohtaka A, Yamaguchi T, Teratani T, Shimomura O, Nomura R (2011) Linear polystyrene-stabilized PdO nanoparticle-catalyzed Mizoroki-Heck reactions in water. Molecules 16 (11):9067–9076

Ong KG, Zeng K, Grimes CA (2002) A wireless, passive carbon nanotube-based gas sensor. IEEE Sensors J 2(2):82–88

Petla RK, Vivekanandhan S, Mishra M, Mohanty AK, Satyanarayana N (2012) Soybean (*Glycine max*) leaf extract based green synthesis of palladium nanoparticles. J Biomater Nanobiotechnol 3:14–19

Phan TTV, Huynh T-C, Manivasagan P, Mondal S, Oh J (2020) An up-to-date review on biomedical applications of palladium nanoparticles. Nano 10(1):66

Pradeep T (ed) (2008) NANO: The essentials: understanding nanoscience and nanotechnology, 1st edn. McGraw-Hill Education, New Delhi, India

Prasad SR, Padvi MN, Suryawanshi SS, Shaikh YI, Chaudhary LS, Samant AP, Prasad NR (2020) Bio-inspired synthesis of catalytically and biologically active palladium nanoparticles using Bos taurus urine. SN Appl Sci 2:754

Puja P, Kumar P (2019) A perspective on biogenic synthesis of platinum nanoparticles and their biomedical applications. Spectrochim Acta A Mol Biomol Spectroscopy 211:94–99

Reddy ALM, Ramaprabhu S (2008) Hydrogen adsorption properties of single-walled carbon nanotube—Nanocrystalline platinum composites. Int J Hydrog Energy 33(3):1028–1034

Sathishkumar M, Sneha K, Yun YS (2009) Palladium nanocrystal synthesis using *Curcuma longa tuber* extract. Int J Mater Sci 4(1):11–17

Sathiskumar M, Sneha K, Kwak IS, Mao J, Tripathy SJ, Yun Y-S (2009) Phyto-crystallization of palladium through reduction process using *Cinnamom zeylanicum bark* extract. J Hazard Mater 171(1–3):400–404

Seshan K (ed) (2002) Handbook of thin film deposition techniques principles, methods, equipment and applications, 2nd edn. Noyes Publications, New York

Shahverdi A-R, Shakibaie M, Nazari P (2011) In: Rai M, Duran N (eds) Metal nanoparticles in microbiology. German, Springer, Berlin, Heidelberg, pp 177–195

Shaik MR, Ali ZJQ, Khan M, Kuniyil M, Assal ME, Alkhathlan HZ, Al-Warthan A, Siddiqui MRH, Khan M, Adil SF (2017) Green synthesis and characterization of palladium Nanoparticles using *Origanum vulgare L.* extract and their catalytic activity. Molecules 22 (1):165

Shamaila S, Sajjad AKL, N-u-L R, Farooqi SA, Jabeen N, Majeed S, Farooq I (2016) Advancements in nanoparticle fabrication by hazard free eco-friendly green routes. Appl Mater Today 5:150–199

Shanthi K, Vimala K, Gopi D, Kannan S (2015) Fabrication of a pH responsive DOX conjugated PEGylated palladium nanoparticle mediated drug delivery system: an in vitro and in vivo evaluation. RSC Adv 5(56):44998–45014

Shao C, Li W, Lin Q, Huang Q, Pi D (2017) Low temperature complete combustion of lean methane over cobalt–nickel mixed-oxide catalysts. Energ Technol 5(4):604–610

Sharmila G, Thirumarimurugan M, Sivakumar VM (2016) Optical, catalytic and antibacterial properties of phytofabricated CuO nanoparticles using *Tecoma castanifolia* leaf extract. Optik 127(19):7822–7828

Sharmila G, Haries S, Fathima MF, Geetha S, Kumar NM, Muthukumaran C (2017a) Enhanced catalytic and antibacterial activities of phytosynthesized palladium nanoparticles using *Santalum album* leaf extract. Powder Technol 320:22–26

Sharmila G, Fathima MF, Haries S, Geetha S, Kumar NM, Muthukumaran C (2017b) Green synthesis, characterization and antibacterial efficacy of palladium nanoparticles synthesized using *Filicium decipiens* leaf extract. J Mol Struct 1138:35–40

Shinohara T, Sato T, Taniyama T (2003) Surface ferromagnetism of Pd fine particles. Phys Rev Lett 91(19):197201

Siddiqi KS, Husen A (2016) Green synthesis, characterization and uses of palladium/platinum nanoparticles. Nanoscale Res Lett 11:482

Sriramulu M, Shanmugam S, Ponnusamy VK (2020) *Agaricus bisporus* mediated biosynthesis of copper nanoparticles and its biological effects: an in-vitro study. Colloid Interface Sci Commun 35:100254

Tahir K, Nazir S, Li B, Ahmad A, Nasir T, Khan AU, Shah SAA, Khan ZUH, Yasin G, Hameed MU (2016) *Sapium sebiferum* leaf extract mediated synthesis of palladium nanoparticles and in vitro investigation of their bacterial and photocatalytic activities. J Photochem Photobiol B Biol 164:164–173

Taniyama T, Ohta E, Sato T (1997) Observation of 4d ferromagnetism in free-standing Pd fine particles. Europhys Lett 38(3):195

Teschner D, Borsodi J, Wootsch A, Révay Z, Havecker M, Knop-Gericke A, Jackson SD, Schlogl R (2008) The roles of subsurface carbon and hydrogen in palladium-catalyzed alkyne hydrogenation. Science 320(5872):86–89

Thakkar KN, Mhatre SS, Parikh RY (2010) Biological synthesis of metallic nanoparticles. Nanomed Nanotechnol Bio Med 6:257–262

The Royal Society (2004) Nanoscience and nanotechnologies: opportunities and uncertainties. The Royal Society and The Royal Academy of Engineering, London, UK

Theriot CM, Koenigsknecht MJ, Carlson PE, Hatton GE, Nelson AM, Li B, Huffnagle GB, Li ZJ, Young VB (2014) Antibiotic-induced shifts in the mouse gut microbiome and metabolome increase susceptibility to Clostridium difficile infection. Nat Commun 5:3114

Tiwari G, Tiwari R, Sriwastawa B, Bhati L, Pandey S, Pandey P, Bannerjee SK (2012) Drug delivery systems: An updated review. Int J Pharm Investig 2(1):2–11

Ullah S, Ahmad A, Khan A, Zhang J, Raza M, Ur Rahman A, Tariq M, Zada S, Yuan Q (2018) Palladium nanoparticles synthesis, characterization using *glucosamine* as the reductant and stabilizing agent to explore their antibacterial & catalytic applications. Microb Pathog 125:150–157

Vaghela H, Shah R, Pathan A (2018) Palladium nanoparticles mediated through *bauhinia variegata*: Potent in vitro anticancer activity against mcf-7 cell lines and antimicrobial assay. Curr Nanomater 3(3):168–177

Veisi H, Rashtiani A, Barjasteh V (2016) Biosynthesis of palladium nanoparticles using *Rosa canina* fruit extract and their use as a heterogeneous and recyclable catalyst for Suzuki–Miyaura coupling reactions in water. Appl Organomet Chem 30(4):231–235

Vigushin DM, Ali S, Pace PE, Mirsaidi N, Ito K, Adcock I, Coombes RC (2001) Trichostatin A is a histone deacetylase inhibitor with potent antitumor activity against breast cancer in vivo. Clin Cancer Res 7(4):971–976

Wilson OM, Knecht MR, Garcia-Martinez JC, Crooks RM (2006) Effect of Pd nanoparticle size on the catalytic hydrogenation of allyl alcohol. J Am Chem Soc 128:4510–4511

Wu P, Meng L, Wang H, Zhou J, Xu G, Wang S, Xi L, Chen G, Wang B, Zhu T, Lu Y (2005) Role of hTERT in apoptosis of cervical cancer induced by histone deacetylase inhibitor. Biochem Biophys Res Commun 335(1):36–44

Xiao C, Ding H, Shen C, Yang T, Hui C, Gao HJ (2009) Shape-controlled synthesis of palladium nanorods and their magnetic properties. J Phys Chem C 113(31):13466–13469

Yadi M, Mostafavi E, Saleh B, Davaran S, Aliyeva I, Khalilov R, Milani M (2018) Current developments in green synthesis of metallic nanoparticles using plant extracts. Artif Cells Nanomed Biotechnol 46(sup3):S336–S343

Yenumula VR, Nagadesi PK (2018) Biogenic synthesis of engineered platinum nanomaterial: A review. Int J Sci Eng Dev Res 3(4):216–220

Yin H, Liu S, Zhang C, Bao J, Zheng Y, Han M, Dai Z (2014) Well-coupled graphene and Pd-based bimetallic nanocrystals nanocomposites for electrocatalytic oxygen reduction reaction. ACS Appl Mater Interfaces 6(3):2086–2094

Zahoor A, Christy M, Jeon JS, Lee YS, Nahm KS (2015) Improved lithium oxygen battery performance by addition of palladium nanoparticles on manganese oxide nanorod catalysts. J Solid State Electrochem 19:1501–1509

Zakaria Z, Kamarudin SK (2016) Direct conversion technologies of methane to methanol: an overview. Renew Sust Energ Rev 65:250–261

Zhang QM, Li H, Poh M, Xia F, Cheng ZY, Xu H, Huang C (2002) An all-organic composite actuator material with a high dielectric constant. Nature 419(6904):284–287

Zhang Z, Xiao F, Xi J, Sun T, Xiao S, Wang H, Liu Y (2014) Encapsulating Pd nanoparticles in double-shelled graphene@carbon hollow spheres for excellent chemical catalytic property. Sci Rep 4(1):4053

Zhang XF, Yan Q, Shen W, Gurunathan S (2016) Trichostatin A enhances the apoptotic potential of palladium nanoparticles in human cervical cancer cells. Int J of mol sci 17(8):1354

Zhou P, Dai Z, Fang M, Huang X, Bao J, Gong J (2007) Novel dendritic palladium nanostructure and its application in biosensing. J Phy Chem C 111(34):12609–12616

Rakesh Kumar Pradhan received his Master of Science in Chemistry with a specialization in Industrial Chemistry in 2022 from Veer Surendra Sai University of Technology, Burla, Sambalpur, Odisha, India. He is currently pursuing his Bachelor of Education at Nabakrushna Choudhury College of Teacher Education, Angul, Odisha, India.

Pooja Mohapatra received her Master of Science in Applied Chemistry from Centurion University of Technology and Management, Bhubaneswar, Odisha, India. She is currently pursuing her Doctor of Philosophy at Veer Surendra Sai University of Technology, Burla, Sambalpur, Odisha, India.

Lipsa Shubhadarshinee received her Master of Science in Applied Chemistry from Centurion University of Technology and Management, Bhubaneswar, Odisha, India. She is currently pursuing her Doctor of Philosophy at Veer Surendra Sai University of Technology, Burla, Sambalpur, Odisha, India.

Bigyan Ranjan Jali received his Master of Science in Chemistry in 2009 from Utkal University, Bhubaneswar, Odisha, India, and Doctor of Philosophy in 2014 from the Indian Institute of Technology Guwahati, Assam, India. Subsequently, he worked as a Lecturer at Government Polytechnic College, Gajapati, Odisha, India. In 2016, he joined as an Assistant Professor in the Department of Chemistry, Veer Surendra Sai University of Technology, Burla, Sambalpur, Odisha, India. His research interest includes design and synthesis of biological molecular probes and study of their fluorescence chemosensors and also study of their antibacterial, antimicrobial, cytotoxicity, and protein-binding properties.

Aruna Kumar Barick received his Master of Science in Chemistry in 2005 from Utkal University, Bhubaneswar, Odisha, India; Master of Technology in Plastic Engineering in 2007 from Central Institute of Petrochemicals Engineering and Technology, Bhubaneswar, Odisha, India; and Doctor of Philosophy in 2011 from Indian Institute of Technology Kharagpur, West Bengal, India, and also worked as a Postdoctoral Fellow at Hanyang University, Ansan, South Korea. Subsequently, he worked as an Assistant Professor at Kalinga Institute of Industrial Technology, Odisha. In 2014, he joined as an Assistant Professor in the Department of Chemistry, Veer Surendra Sai University of Technology,

India. His research interest includes the synthesis and characterization of nanomaterials and their polymer nanocomposites for various applications.

Priyaranjan Mohapatra received his Master of Science in Chemistry in 1997 and Master of Philosophy in Chemistry in 2001 from Ravenshaw University, Cuttack, Odisha, India, and Doctor of Philosophy in 2007 from Utkal University, Bhubaneswar, Odisha, India, and also worked as a Visiting Research Student at Gwangju Institute of Science and Technology, Gwangju, South Korea, and a Postdoctoral Fellow at Chonnam National University, Gwangju, South Korea. Subsequently, he worked as a Lecturer, Assistant Professor, and Associate Professor at Silicon Institute of Technology, Bhubaneswar, Odisha, India. In 2013, he joined as Reader in the Department of Chemistry, Veer Surendra Sai University of Technology, Burla, Sambalpur, Odisha, India, and continued as a Professor. His research interest includes the synthesis and characterization of quantum dots, nanocomposites, and its application as fluorescent chemosensors.

Innovative Nanomaterials with Profound Antibacterial Action Applied in Biomedical Sciences

22

Anwesha Kanungo and Sarbari Acharya

Abstract

Bacteria are omnipresent and can intensify their growth by colonization leading to the formation of biofilms. Biofilms are the primary cause of bacterial infection and mortality worldwide. For the prevention of these bacterial infections, numerous antibacterial agents like antibiotics are coming to light. These antibacterial agents are abundantly found in nature and referred to as natural antimicrobials, which can be obtained from plants, animals, and microorganisms as well. Regardless of numerous potent antimicrobial agents and antibiotics, infections caused by bacteria remain a prime threat to animals and humans. With the growing concern about the increased mortality due to bacterial infections, the development of highly potent antibacterial agents possessing unique combating properties is the need of the moment. Consequently, emerging nanoparticle-based antimicrobial materials are grabbing the attention of researchers for developing an antimicrobial-resistant regime for fighting against various infections. Moreover, it is also paramount to overcome the menace of bacterial drug resistance. In this context, this chapter highlights the recent insights concerning the antimicrobial actions of natural products, nanoparticles, and their mechanism of action featuring effective antimicrobial activity.

Keywords

Bacteria · Biofilms · Antimicrobial agents · Antibiotics · Natural products · Nanoparticle

---

Anwesha Kanungo and Sarbari Acharya contributed equally to this work.

A. Kanungo · S. Acharya (✉)
Department of Biology, School of Applied Sciences, Kalinga Institute of Industrial Technology, Bhubaneswar, Odisha, India

F. A. Sheikh et al. (eds.), *Interaction of Nanomaterials With Living Cells*,
https://doi.org/10.1007/978-981-99-2119-5_22

22.1 Introduction

Antibacterial agents are an invaluable weapon for aggressively fighting against various infectious and noninfectious diseases. Being helpful in medicines, food industries, pharmaceuticals, textile industries, disinfectant products, and water sterilization, antibacterial agents have emerged as the first strategy to fight against infection by inhibiting bacterial growth leading to bacterial cell death (Langeveld et al. 2014). One of the primary examples of antibacterial agents is antibiotics. Antibiotics are derived from various antibacterial agents and target the translational machinery, cell wall synthesis, and DNA replication machinery of the bacteria. However, excessive use of antibacterial agents through direct consumption by humans as well as used in animal food indicates bacterial resistance to antibiotics. Extensive usage of antibacterial agents leads to the formation of super-bacteria, which are approximately resistant to all antibiotics, making them life-threatening. Bacteria can gain resistance over their targets by modifying the enzymes and degrading the antibiotics, for instance, aminoglycosides and β-lactamases (Poole 2002), alteration of components of cells such as ribosomes in tetracycline resistance and cell wall in vancomycin resistance (Wang et al. 2017), and efflux pumps expression (Knetsch and Koole 2011). In 2017, WHO published a record of antibiotic-resistant pathogens that demanded the urgent need for antibiotics and were categorized into three groups, namely, critical, high, and medium (Fig. 22.1), which guided the development of new and effective antibiotics (Breijyeh et al. 2020).

Similarly, another hindrance observed during bacterial infection treatment includes the formation of biofilm. Biofilm is the medium for the interaction of

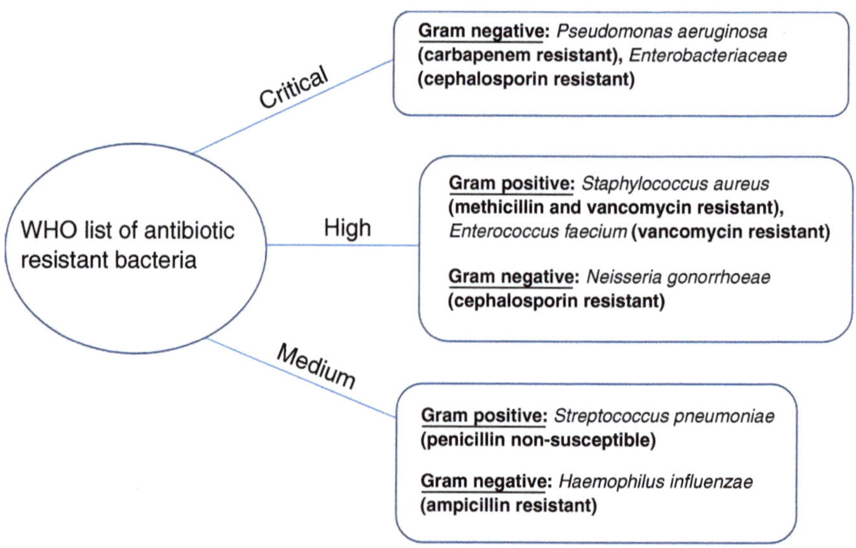

Fig. 22.1 WHO categorized pathogens according to their antibiotic resistance

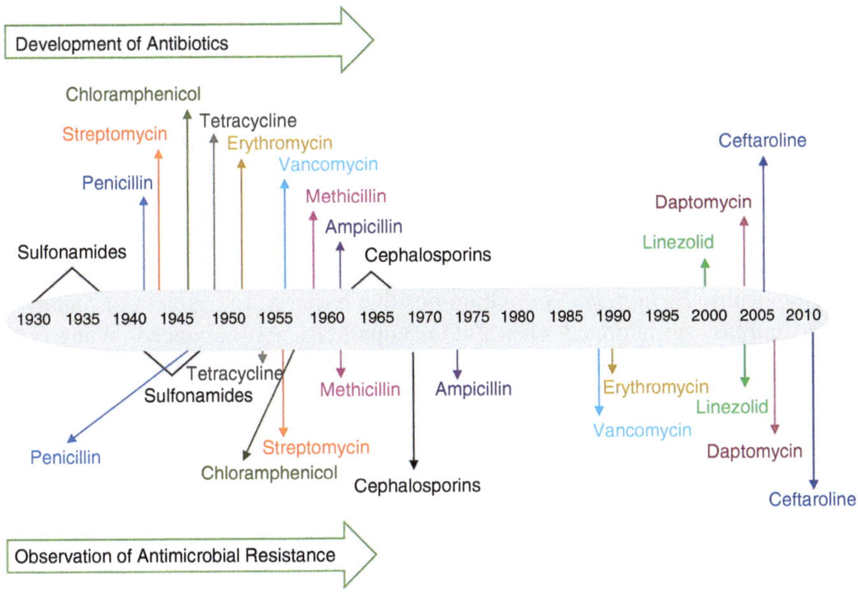

Fig. 22.2 Schematic representation of development in antibiotics and observation of antimicrobial resistance from 1930 to 2010

diverse species of bacteria with each other and their ecosystem, and they require a solid surface, for instance, extracellular polymeric substances (EPSs), which provide structural and functional integrity to the biofilms. Once the bacteria are properly attached to the surface, it multiplies aggressively to form a mature biofilm. At this point, bacteria being attached form a barricade that possesses the ability to resist antibiotics and cause chronic infections making biofilms hazardous for health (Huh and Kwon 2011; Hajipour et al. 2012). Additionally, superantigens are produced by the bacteria surrounded by biofilms which carry the ability to invade the immune system (Wang et al. 2017). Going by the phrase "Prevention is better than cure" if the adhesion of bacteria to healthy cells is avoided, then the proliferation of bacteria can be prevented (Kamaruzzaman et al. 2019; Olmos and González-Benito 2021). This suggests the requirement for extensive research about different antibacterial materials.

Natural products have been adapted to treat numerous bacterial diseases, for example, penicillin for infectious diseases and quinine from the cinchona tree for malaria treatment. Penicillin was discovered by Alexander Fleming in 1929; it possesses the ability to fight against various bacterial infections and left a tremendous impact on human health and lowered mortality rates all over the world (Breijyeh et al. 2020). Over the years, new and effective antibiotics have been developed and their antimicrobial resistance is studied (Fig. 22.2). With the emerging mutating power of the bacteria, the failing antibiotics can show desirable effects by combing themselves with other compounds such as repurposed drugs

(Brown 2015) and natural products, which bring out intensified efficacy of the drug. Furthermore, phytochemicals exerted antibacterial ability against resistant as well as sensitive pathogens through various mechanisms (Rossiter et al. 2017; Khameneh et al. 2019) and proved to be helpful in the disruption of biofilm (Guglielmi et al. 2020).

Nanomaterials can be defined as nano-sized materials present in the 3D space range with dimensions ranging from 1 to 100 nm formed through nanotechnology. Existing in nano-form, the nanomaterials show a wide range of antibacterial properties in the case of gram-negative, i.e., Ag nanoparticles show antimicrobial activity against *Escherichia coli*. Gram-positive bacteria, i.e., growth of *Staphylococcus aureus*, are inhibited when ZnO nanoparticles are introduced (Wang et al. 2017). Nanoparticles exhibit antimicrobial action by attaching themselves to one of the following models: metal ion release, oxidative stress induction, or non-oxidative mechanism. Some of the widely accepted nanoparticles used to prevent diseases are silver, gold, titanium, zinc, iron, copper, magnesium, nitric oxide, aluminum, and graphene. Usually, the activity of antibacterial materials depends on the action of polymers (chitosan) or various particles (ZnO, Cu, Ag), giving rise to the formation of reactive oxygen species (ROS), which leads to modification in the metabolism of the cell resulting in cell death (Olmos and González-Benito 2021). The principal objective of this chapter is to overview the current knowledge on the natural and antimicrobial nanoparticle's activity and their mode of action on various infectious agents and to highlight their contribution made to the research area. It is anticipated that the findings represented in this chapter will assist in developing new antibacterial agents in the coming future.

22.2 Factors Affecting the Antimicrobial Activity

An ideal antibacterial material should possess the following properties:

(a) Tolerance
Tolerance can be defined as the ability of the bacteria to withstand exposure to a raised concentration of antibiotics with no modification in MIC. Minimum inhibitory concentration (MIC) can be defined as the minimum concentration of antibiotics at which bacterial growth is inhibited. Exposure to higher concentrations of antibiotics is necessary to develop a tolerant strain of bacteria. These tolerant and non-tolerant bacteria do not show any difference in their MIC values which can be challenging further. The minimum duration of killing (MDK) can be expressed as the time required by the antibiotic to kill a known segment of bacteria. Hence, to evaluate the level of resistance between the bacterial strains, MIC is used, and for the tolerance between bacterial strains, MDK is used (Li et al. 2017).
(b) Susceptibility and Resistance
Susceptibility can be termed as no bacterial growth in the presence of antibiotics, whereas resistance means bacterial growth is observed in the

presence of antibiotics. Both vulnerability and resistance are calculated by MIC. To check the MIC values, to a chain of increasing antibiotic concentrations, the bacterial population was exposed for about 24 h (Wiegand et al. 2008). The susceptibility and resistance of the tested isolates can be determined by the epidemiological cutoff (ECOFF) value. Clinical resistance can be termed as a condition that separates clinically susceptible from clinically resistant bacteria (Turnidge and Paterson 2007). The relationship between antibiotic concentration at the infection site and in vivo antimicrobial activity is dependent on pharmacokinetic/pharmacodynamic (PK/PD) factors.

(c) Biofilm

Biofilms can be defined as the assemblage of bacterial cells surrounded by DNA, proteins, and polysaccharides (Hall-Stoodley et al. 2012), which cannot respond to any antibiotics. Due to the layers of cell, accumulation in a biofilm increases the antibiotic resistance of resistant bacteria, and the horizontal gene transfer of resistance determinants is facilitated by the extracellular DNA of the biofilm (Cheng et al. 2016). The antimicrobial resistance of a biofilm can appear by three distinctive mechanisms (Langeveld et al. 2014): occurrence of persistent bacterial cells (Poole 2002), damage done by the diffusion of antibiotics into the bacterial cells by extracellular matrix (Wang et al. 2017), and wounds in the DNA oxidative repair system or in the mismatch repair system which results in hypermutator (Penesyan et al. 2015). In vitro studies suggest that the biofilm formed by *Helicobacter pylori* is responsible for reducing antibiotic sensitivity and generating resistance mutations in a biofilm when compared to planktonic cells.

(d) Persistence

Persistence can be defined as when the bacterial population is not killed by antibiotics and when exposed to the same antibiotic, the heterogenous response is produced (Lewis 2007). Time-dependent persistence can be described as the particular circle of bacteria which has a slower growth rate or long lag time (Balaban et al. 2004). This time-dependent persistence expresses molecular mechanisms by slowing down the killings by antibiotics which are associated with tolerance (Adams et al. 2011). The bacterial population which survives the effect of antibiotics shows enhanced resistance.

Any material which possesses these properties are considered ideal antimicrobial activity.

22.3 Antibacterial Materials

Antibacterial materials are priceless compounds that are used to inhibit the growth of various microorganisms. They come in handy in preventing diseases by showing antibacterial abilities toward microorganisms. These antibacterial materials are abundantly found in nature as natural products and in the form of nanomaterials.

22.4 Natural Products

Natural antimicrobial materials are abundantly found in nature, possessing both biological and pharmacologic actions which are produced by living organisms (Koehn and Carter 2005). These living organisms give rise to diversified primary and secondary metabolites, where the primary metabolites carry out the chief function within the organism. On the contrary, the secondary metabolites might carry essential functions for their producers or result in waste materials. Conversely, these secondary metabolites have been proven to hold antibacterial activity and show constructive properties toward humans by combating numerous diseases like cancer, inflammation, and bacterial infections. These secondary metabolites acquiring antimicrobial activity are found in (a) plant sources like vegetables, spices, herbs, fruits, and seeds; (b) animal sources like eggs, milk, and tissues; and (c) microorganisms like bacteria, viruses, and fungi. Though approved for human consumption, chemical preservatives still carry a threat to our health, for which natural antimicrobials have been proven to be promising in decreasing food-borne illness rates (Mathew et al. 2007; Hayek 2014).

Natural antimicrobial peptides (AMPs) are naturally occurring biopolymers that can be both anionic and cationic. Some AMPs are histone-derived compounds, aromatic dipeptides, and oxygen-binding proteins like bacteriocins. Bacteriocins can be described as odorless, nontoxic, tasteless, and colorless (Farkas-Himsley 1980). Bacteriocins are widely accepted as a natural product due to their mode of action, where they can target a narrow range of bacterial species when compared to traditional antibiotics, and it also gets inactivated by proteolytic enzymes present in the digestive system of humans. Similarly, lactoferrin belonging to the transferrin family is a multifunctional globular glycoprotein found in human secretions like breast milk and tear. Lactoferrin was identified in 1939 in bovine milk by Sorensen and Sorensen. It inhibits the growth and multiplication of a wide range of gram-positive and gram-negative bacteria, fungi, viruses, and protozoa (Zhao 2011).

22.5 Antibacterial Nanomaterials

Currently, the excellency in nanotechnology has made it grab a superior position in the field of nanoscale antibacterial properties. Reportedly, antibacterial nanomaterials comprise silver nanoparticles (Ag) (Morones et al. 2005; Chen and Schluesener 2008), chitosan-based nano-biopolymer, nanosilver-based nanocomposites (Thomas et al. 2007), and carbon nanotubes (CNTs) (Fig. 22.3). Presently, the silver nanoparticle has grabbed more attention due to its antibacterial properties. To improve the efficacy of the chemical reaction, a one-pot synthesis was carried out between silver nanoparticle-polymer composites (Ag-PNCs) in water comprising polycondensation of methoxybenzyl chloride to produce highly emulsifiable microbial nanocomposites (Zhao 2011). Nevertheless, numerous metals and metal oxides are extensively researched to understand their antimicrobial actions (Loomba and Scarabelli 2013). Some of the highly potent metal oxide nanoparticles

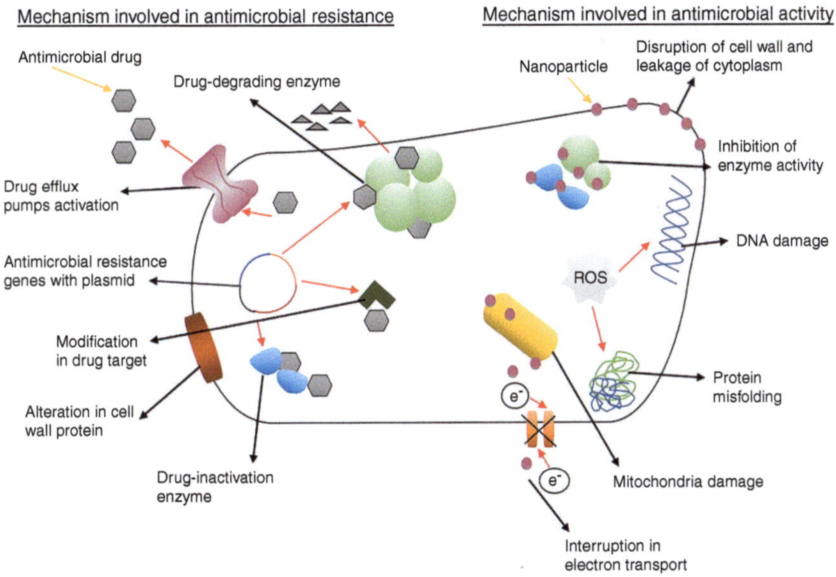

Fig. 22.3 Illustration of the mechanism involved in antimicrobial resistance (left) and mechanism involved in antimicrobial activity of NPs (right)

are silver (Ag), titanium oxide (TiO_2), zinc oxide (ZnO), gold (Au), iron oxide (Fe_3O_4), copper oxide (CuO), magnesium oxide (MgO), nitric oxide (NO), aluminum oxide (Al_2O_3), and graphene. Bactericidal properties of metal oxide nanoparticles are exhibited through reactive oxygen species (ROS) generation, metal ion release, and their physical structure (Beyth et al. 2015).

(a) **Silver Nanoparticles**

Silver nanoparticles have emerged as one of the leading and widely used antimicrobial agents which is effective against a broad range of bacteria, viruses, and fungi (Rai et al. 2009). Ag compounds are extensively used to treat a range of infectious diseases, wounds, and burns (Elliott 2010). Being an effective metal oxide nanoparticle, the antimicrobial effect of Ag is size-dependent; therefore, silver with a smaller diameter exhibits tremendous antimicrobial effect irrespective of the larger diameter (Morones et al. 2005; Beyth et al. 2015). Ag NPs are one of the extensively researched metal and metal oxide NPs with a great variety of applications in the area of medicine, microbiology, food technology, water purification, cell biology, pharmacology, and house appliances (Spirescu et al. 2021). These Ag NPs can be created by hydrothermal method, chemical vapor deposition, sol-gel method, thermal decomposition, and biogenic synthesis methods (Spirescu et al. 2021).

Even though Ag NPs demonstrated their efficiency against a broad range of microorganisms comprising gram-negative and gram-positive bacteria, viruses, and fungi, the accurate mechanism of silver nanoparticles is still to be interpreted (Ciriminna et al. 2020). The main mechanism of the positively charged Ag NPs is to interact with the plasma cell membrane, which is negatively charged, resulting in the deposition inside the membrane followed by modification in the structure of the bacteria and permeabilization due to cis-trans isomerization of unsaturated membrane fatty acids (Ciriminna et al. 2020; Hamad et al. 2020). This results in the release of Ag^+ from the NP and communicates with the proteins and nucleic acids, leading to the production of reactive oxygen species (ROS), for instance, singlet oxygen, superoxide anion, hypochlorous acid, and hydrogen peroxide (Loomba and Scarabelli 2013; Spirescu et al. 2021).

(b) **Gold NPs**

Gold has emerged as a potent nanoparticle that effectively inhibits the growth of fungi and gram-positive and gram-negative bacteria. The efficacy of Au NPs can be observed in their application in the field of medicine. For instance, Au NPs are used in the production of antibiotics like streptomycin, neomycin, and gentamycin, used for the production of soaps and cosmetics and biomedical applications (Loomba and Scarabelli 2013; Saidin et al. 2021). Principally, they are applied in the treatment of rheumatic diseases comprising of psoriasis, rheumatism, and juvenile arthritis. Au NPs show biocompatibility with the body and in association with photosensitizers for photodynamic antimicrobial chemotherapy (PACT) (Saidin et al. 2021). Au NPs show antimicrobial effects by interacting with the microbial cell wall regulated by protein binding, electrostatic forces, lipids, and carbohydrates resulting in the microbial cell membrane and cell wall damage, impeding the thiol groups within the bacterial cells and impairment of mitochondria and ribosome (Rónavári et al. 2021). Additionally, the antimicrobial actions lead to an increase in ROS concentration in *S. aureus*, bacterial cell lysis in *S. pneumoniae*, and inhibition of transcription in *E. coli* and *S. aureus* (Spirescu et al. 2021; Ortiz-Benítez et al. 2019).

(c) **Titanium**

Titanium dioxide (TiO_2) has shown its efficacy as a metal oxide antimicrobial agent by producing free radical oxides and peroxides, which possess high antimicrobial reactivity, and is broadly used to deal with gram-negative and gram-positive bacteria along with various parasites and viruses (Brady-Estévez et al. 2008). TiO_2 NPs, when exposed to UV light, lead to the production of high-energy electron-hole pairs (Azizi-Lalabadi et al. 2019). This UV light development gives rise to ROS (Nguyen et al. 2019). The production of ROS leads to the destruction of DNA, RNA, proteins, and the membrane of the bacterial cell, and it also hinders the functions inside the cell leading to cell death (Saidin et al. 2021). These TiO_2 NPs are produced by hydrothermal method, precipitation, sol-gel method, and electrochemical processes using titanium isopropoxide and titanium chloride as precursors (Spirescu et al. 2021; Waghmode et al. 2019).

(d) **Zinc Oxide**

The antimicrobial action of ZnO nanoparticles is dependent on the particles size and concentration. ZnO NPs are widely used to prevent the growth of methicillin-resistant *S. aureus* (MRSA), methicillin-sensitive *S. aureus* (MSSA), and methicillin-resistant *S. epidermidis* (MRSE) strains (Ansari et al. 2012). Also, due to ZnO NPs UV blocking ability and white color, they possess the capability to inhibit the formation of biofilm for which it is effectively used as a coating substance in medical and other devices of glass and fabric industries (Beyth et al. 2015), and Zn is used as a food additive by FDA (Blecher et al. 2011). Hence, the ZnO NPs lead to the production of ROS, disrupting membrane potential and integrity (Beyth et al. 2015; Pati et al. 2014). ZnO NPs are positively charged and specifically react with the microbial cell wall which is negatively charged via electrostatic forces, thus leading to disruption in membrane structure, the outflow of components of the cell, and loss in cell integrity which ultimately leads to cell death (Spirescu et al. 2021; Sirelkhatim et al. 2015).

(e) **Iron Oxide**

Iron oxide nanoparticles are tailored to utilize antimicrobial materials in healthcare sectors. Fe_3O_4 possesses anti-adherent characteristics leading to the inhibition of the growth and development of gram-positive and gram-negative bacteria. The Fe_2O_3 NPs acquire the ability of cell penetration and production of ROS (Saidin et al. 2021).

(f) **Copper Oxide**

Copper oxide (CuO) NPs exert antibacterial efficacy against numerous bacterial species and show antibacterial action by producing ROS and membrane disruption (Pelgrift and Friedman 2013; Hans et al. 2013). For example, the cell wall of *Bacillus subtilis* is abounding with carboxyl and amine groups which deeply bind to the CuO, making the organism more sensitive toward the NPs (Huh and Kwon 2011; Pelgrift and Friedman 2013). CuO belonging to the simplest member of the copper family possess a variety of applications like can withstand high temperature, superconductivity, electron correlation effects, and spin dynamics, making CuO NPs the most suitable NPs and having the capability to replace the use of silver NPs (Saidin et al. 2021). They show enhanced antimicrobial action toward *Escherichia coli* and *Bacillus subtilis*, including silver-resistant species like *Morganella morganii* and *Mycobacterium psychrotolerans* (Das and Patra 2017). Also, Cu NPs have emerged as an effective antifungal agent for plants by inhibiting the growth of *Curvularia lunata*, *Fusarium oxysporum*, and *Alternaria alternata* (Chen et al. 2019). However, in the field of medicine, Cu NPs show cyto- and genotoxic effects (Vincent et al. 2018).

(g) **Magnesium Oxide**

MgO NPs demonstrate antimicrobial action against both gram-negative and gram-positive bacteria, viruses, and spores. When compared to other metal NPs, the MgO NPs are readily available and used in the form of MgO or MgX_2 (Pelgrift and Friedman 2013; Lellouche et al. 2011). Besides the availability of Mg NPs, these NPs contain the capability to prevent biofilm formation in

Staphylococcus aureus and *Escherichia coli* (Beyth et al. 2015), inhibiting essential enzymes of the bacteria and inducing ROS (Blecher et al. 2011), causing phospholipid damage leading to membrane disruption and causing cell death (Maji et al. 2020). Synthesis of MgO NPs is done through various methods, namely, sol-gel, combustion, hydrothermal method, calcination, wet precipitation, and co-oxidation (Elliott 2010; Maji et al. 2020).

(h) **Nitric Oxide**

Nitric oxide (NO) NPs appear to be an assuring antimicrobial compound whose antimicrobial activity is dependent on shape and size; therefore, smaller particles are more efficacious than larger ones (Schairer et al. 2012). One of the major differences between other metal oxides and NO NPs is that it exclusively affects the reactive nitrogen species (RNS) than ROS. NO NPs effectually kill methicillin-resistant *Staphylococcus aureus* (MRSA) (Beyth et al. 2015) present in skin infections (Martinez et al. 2009) and improve the wound healing process in both normal and diabetic mice (Weller and Finnen 2006). Nitric oxide NPs facilitate the biofilm eradication of numerous bacterial species. Additionally, NO can possess as an accurate drug for topical treatment in subcutaneous and cutaneous wounds and facilitates a faster wound healing process (Saidin et al. 2021).

(i) **Aluminum Oxide**

The structure of aluminum is thermally stable and can withstand an extensive range of temperatures and carry a neutral charge making aluminum neutral in pH. Aluminum nanoparticles cover a broad range of antimicrobial applications starting from industrial use to personal care products. The antimicrobial effect of Al NPs can be observed in *Escherichia coli* as it leads to membrane disorganization, accumulation inside the cell, pit formation, and perforation, which results in cell death. Also, the formation of ROS creates cell wall disruption causing cell death (Saidin et al. 2021; Ansari et al. 2014). The adhesion rate of the aluminum NPs is dependent on the concentration of the nanoparticles, which results in growth inhibition. Hydrophobic interaction, electrostatic interaction, and polymer bridging guide in the adhesion of bacteria onto the nanoparticles (Saidin et al. 2021). Aluminum, when combined with silver, shows exceptional antimicrobial results by inhibiting the growth of various microbes like *Escherichia coli*.

(j) **Graphene**

Graphene holds special applications like a potential supercapacitor; exhibits thermal, electronic, and mechanical properties; and is applied in the field of nanosensors, nanoelectronics, and nanomedicine. Graphene is successfully utilized for drug delivery techniques and in the area of nanomedicine for diagnostics and biosensors (Saidin et al. 2021). Graphene oxide (GO), after surface modification, is exercised as a drug carrier. When compared for antimicrobial activity graphene oxide (GO), graphite (Gt), reduced graphene oxide (rGO), and graphite oxide (GtO) against *Escherichia coli*, the effectiveness of the materials was found to be decreasing in order GO > rGO > Gt > GtO under similar incubation conditions and concentrations (Saidin et al. 2021).

22.6 Applications of Antibacterial Materials

Antibacterial materials find their use in various sectors as enumerated below:

1. **Health Sector**

   Despite acquiring numerous prevention methods, aseptic measures, and strict sterilization techniques, bacterial infection continues to be a major threat. In dwelling devices like urinary catheters, endotracheal tubes, and enteral feeding tubes, bacterial infections serve as the major cause of healthcare-associated diseases worldwide, with an estimated one million cases every year in the USA for the year 2004. Moreover, the adhesion of the bacteria and the development of biofilm at the site of implantation also result in device-associated infections (Hetrick and Schoenfisch 2006). Thus, the use of antimicrobial agents to tackle this situation is the need of the hour.

   The antibacterial agents work in three different modes on the devices: (a) on the external surface, (b) on the active surface, and (c) in the time-release mode. The external mode symbolizes the use of disinfectants on the surface, which results in structural instability of the microorganisms leading to their death. The surface-active mode symbolizes the transfer of antimicrobial agents into the microorganism leading to the accumulation of toxins which ends in the disruption of the membrane causing cellular leakage. Lastly, the time-release mode contains antimicrobial agents that get discharged when there is a variation in their environments, such as a change in temperature, pH, pressure, and moisture, which is activated when the antimicrobials get attached to the surface. While the first mode of action is used in general as surface and container disinfectant, the last two modes are mostly used in clinical applications (Zhao 2011). Clinical applications of antimicrobial materials are common hospital appliances (hospital beds), urological equipment (catheters), medical packaging, intravenous (IV) devices, wound care products, and vascular grafts. Incorporating antimicrobial agents into these abovementioned products is necessary because it helps in controlling the growth of bacteria which is the major concern in hospital-acquired infections (Zhao 2011).

   Some of the clinical sectors where antimicrobial agents are applied are as follows:

   (a) **Wound Care**

   Combating wound infection has emerged as a major concern in the healthcare sector, for which antimicrobial materials are used. Wound care in the healthcare sector is ruling at an estimated value of US$10 billion in 2007 and is expected to rise to US$12.5 billion in 2012 (Zhao 2011). Discovered in the 1960s, silver sulfadiazine (AgSD) cream uses silver to treat burns (George et al. 1997). Lately, silver is effectively used as a cover for wound dressing (Atiyeh et al. 2007), and now, wound dressings that are impregnated with silver are used as gauze and are widely available. Some of the widely accepted wound dressings impregnated with antimicrobial materials are polyhexamethylene biguanide (PHMB) and iodine.

Microorganisms that cause wound infections are sorted into two groups (Langeveld et al. 2014): gram-positive bacteria, namely, *Clostridium*, *Bacillus*, *Streptococcus*, and *staphylococcus* spp. (Poole 2002), and gram-negative bacteria, namely, *Escherichia*, *Salmonella*, *Pseudomonas*, and *Klebsiella* spp. Infection caused by multiple bacterial species leads to chronic infections, which can be controlled by nanoparticles used as antimicrobial materials which can inhibit bacterial reproduction and growth. Currently, a fiber mat has been discovered for wound healing with the combination of nanosilver and polyvinyl alcohol with chitosan (Li et al. 2013).

(b) **Bone Cement**

Bone cement is a self-curing biomaterial made up of polymethyl methacrylate (PMMA) or methyl methacrylate (MMA) or modified PMMA and is popularly used as prostheses to fill the space between the bone and implant during replacement surgery of the knee, joint, or hip (Wang et al. 2017). As the cases of infectious diseases leading to infection in the wound are increasing, the situation can be life-threatening. In such a situation, the administration of nanoparticles is emerging as the research hotspot due to its ability to kill resistant bacteria (Pelgrift and Friedman 2013). According to the time-kill and Kirby-Bauer method, a reduction in surface biofilm formation is observed when PMMA bone cement was mixed with Ag NPs, where the chief function of PMMA-Ag NPs was to prevent the colonization of bacteria on the surface (Miola et al. 2015). This confirmed the use of Ag NPs as a potential antibacterial bone cement by replacing antibiotics (Prokopovich et al. 2015). Nanosilver at a concentration of 0.05% can successfully prevent surgery-related infections caused by *Acinetobacter baumannii*, *Staphylococcus epidermidis*, *Staphylococcus aureus*, and methicillin-resistant *Staphylococcus aureus* (MRSA) (Kose et al. 2016).

(c) **Cardiovascular**

Implants used for the arterial occlusive disease are vascular prosthetics that cause infection at the site of implantation in approximately 1–5% of the patients (Zhao 2011; Herrera et al. 2009). To prevent this infection, rifampicin-bonded prostheses have been used in some patients, and it confirmed potential results. The latest technique uses antibiotic-impregnated prosthesis for in situ graft replacement that is appropriate for inhibiting *Staphylococcus epidermis* infection (Wilson 2001). Similarly, for the treatment of vascular surgical site (VSS) infection, a combination of antibiotics like daptomycin, tobramycin, and vancomycin was filled with poly(methyl methacrylate) (PMMA) beads (Stone et al. 2006) which emerged as a promising approach. One more approach is the application of a rifampicin-bonded Gelsoft graft for the in situ replacement of methicillin-resistant *Staphylococcus aureus* (MRSA) (Zhao 2011; Vicaretti et al. 2000). Stents used for the treatment of arterial stenosis cause significant infections. To prevent methicillin-resistant *Staphylococcus aureus* in both animals and man, the stent was filled with gelatin and then immersed in rifampicin (Zhao 2011).

2. **Antimicrobial Materials in the Food Industry**
   Natural antimicrobials from plants are used in the process of cooking or directly as food. Some examples of such antimicrobial agents are:
   (a) **Herbs and Spices**
   Herbs and spices show variation in antimicrobial actions against food-borne microorganisms as they possess different structural configurations (Tajkarimi et al. 2010; Negi 2012), i.e., the hydroxyl (–OH) groups. This hydroxyl group interacts with the bacterial cell membrane resulting in the disruption of its structure and leakage of its components, leading to cell death (Quinto et al. 2019). Some of the antimicrobial materials from plants possessing both antimicrobial and antioxidant properties which bring out the more effective version of the compound are herbs and plants (lemongrass, garlic, rosemary, oregano), oils, spices (clove and cinnamon), and organic compounds (Gutierrez et al. 2008; Pisoschi et al. 2018). These compounds, alone or in combination, show efficacy in preserving food (Gavarić et al. 2015).

   One of the effective combinations of oils that were found to be effective against different microbes is oregano, lemon, basil, rosemary, thyme, and sage which inhibited the growth of *Escherichia coli*, *Pseudomonas aeruginosa*, and *Bacillus cereus* (Gutierrez et al. 2008). Furthermore, it was also proven that oregano essential oil showed more ability toward antibacterial properties for *Escherichia coli* and *Bacillus subtilis*. In addition, the main component in essential oil (from oregano) was found to be carvacrol which was about 80.5% which was sensitive to *Salmonella typhimurium* (Arshad and Batool 2017).
   (b) **Fruits and Vegetables**
   With the tremendous success in the antimicrobial activity of herbs and spices, fresh fruits and vegetables have been discovered to be containing organic acids and phenols that possess antimicrobial effects against different spoilage and pathogenic microorganisms. For instance, 3-hydroxycinnamic acid (coumaric acid) and phenol present in *Capsicum* and orange peel were responsible for its antimicrobial action (Hayek 2014; Dorantes et al. 2000). Bergamot peel is the by-product of a citrus fruit that produces flavonoids that were found to be inhibiting the growth of gram-negative bacteria like *Escherichia coli* and *Salmonella enterica*, and the effectiveness of the antimicrobial activity is enhanced after enzymatic de-glycosylation (Hayek 2014). Citrus extracts like lemongrass and lime peel have been effectively showing antimicrobial activity against meat products for *Salmonella typhimurium* and *Bacillus cereus*. Citrus fruit peels, juices, and seeds undergoing hot water extraction were revealed to be successfully fighting against bacteria such as *Escherichia coli*, *Staphylococcus aureus*, and *Pseudomonas aeruginosa* (Arshad and Batool 2017; Scallan et al. 2011). Pomegranate extracts show high potency against gram-positive bacteria like *Bacillus cereus* and *Staphylococcus aureus* at a concentration of 0.01% due to the presence of diverse flavonoids and phenols. Pomegranates have also shown great inhibition toward *Escherichia coli* (Arshad and Batool 2017).

Garlic extract is composed of organic sulfur, i.e., allicin which is responsible for inhibiting the growth of both gram-positive and gram-negative bacteria involving *Escherichia coli*, *Streptococcus*, *Salmonella*, *Helicobacter pylori*, and *Klebsiella* (Arshad and Batool 2017).

(c) **Phenolic Compounds**

Phenols are used for the preservation of food and enhancing its flavors such as hydroquinone and cresol. The phenolic compounds are of three types, namely, monophenols, diphenols, and triphenols (Quinto et al. 2019). Plants produce some of the effective phenolic acids which stall the growth of bacteria like *Escherichia coli* and *Aeromonas hydrophila* (Payne et al. 1989).

(d) **Hops**

Hops are the flower of the hops plant *Humulus lupulus*, used in the production of beer as it generates a bitter flavor. Hop comprises xanthohumol and prenylated acylphloroglucinols, which results in gram-positive bacteria growth inhibition (Kramer et al. 2015; Simpson and Smith 1992).

Natural antimicrobials are also derived from animals:

(a) **Peptides**

Antimicrobial peptides from animal sources actively destroy the cell membrane of both gram-positive and gram-negative bacteria (Pisoschi et al. 2018; Tiwari et al. 2009), proving their antibacterial activities and, additionally, exhibiting a broad range of antiviral activities (Jenssen et al. 2006). *Pleuronectes americanus* reveal to show antimicrobial activity against both gram-positive and gram-negative bacteria with the peptide, i.e., pleurocidin found in the secretions of its skin (Cole et al. 1997), such as *Saccharomyces cerevisiae*, *Vibrio parahaemolyticus*, *Escherichia coli*, and *Penicillium expansum* (Quinto et al. 2019). Protamine, defensins, magainin, and casocidin are peptides that show antimicrobial defenses against various microorganisms (Pisoschi et al. 2018). Protamine is a protein found in the sperm cells of vertebrates (Kim et al. 2015) and shows electrostatic affinity toward the negatively charged bacteria, whereas magainin is present on the skin of *Xenopus laevis* a frog; both the peptides actively fight against bacteria as well as fungi (Tiwari et al. 2009; Potter et al. 2005). Defensins are small proteins showing antimicrobial activities against bacteria, fungi, and viruses that are produced by phagocytic cells as well as epithelial cell lining of genitourinary and gastrointestinal tracts of mammals (Tiwari et al. 2009). Casocidin peptide expresses antibacterial activity against *Staphylococcus carnosus* and *Escherichia coli* is achieved from bovine milk (Quinto et al. 2019).

(b) **Polysaccharides**

Chitosan is a linear polysaccharide produced by treating chitin exoskeletons of arthropods and crustaceans (Pisoschi et al. 2018). Chitosan is capable of inhibiting the growth of bacteria like *Staphylococcus aureus*, *Escherichia coli*, *Lactobacillus fructivorans*, and *Yersinia enterocolitica*, as well as yeasts and molds, namely, *Aspergillus flavus*, *Rhizopus stolonifera*, *Saccharomyces cerevisiae*, and *Mucor racemosus* from food components (Oh et al. 2001; Rhoades and Roller 2000).

(c) **Lipids**

Lipids detected in milk are found to be inhibiting the growth of fungi and gram-positive and gram-negative bacteria by hindering the proliferation of microorganisms (Pisoschi et al. 2018; Sprong et al. 2001). The activity of *Escherichia coli*, *Campylobacter jejuni*, *Clostridium perfringens*, *Listeria monocytogenes*, and *Salmonella enteritidis* was observed to be affected due to the administration of lipids and triglycerides from bovine milk (Sprong et al. 2001).

(d) **Food Storage**

The food industry is a multifaceted sector, with the biggest concern being the quality, storage, security, and packaging of the food. Thus, food packaging materials play a vital role in both the food and health sectors. The food packaging industry can be divided into three types (Langeveld et al. 2014): conventional packaging, in which the content in the package is separated from the environment by a passive barrier created by the container (Poole 2002); active packaging, where the packaging material itself increased the shelf life of the food by improving its quality and safety (Wang et al. 2017); and smart packaging along with protecting the content from the environment, where it also delivers information regarding the changes in food, its environment, and the package. Antibacterial materials used to prevent disease also play a vital role in food packaging and food sectors (Azeredo 2009).

In the case of conventional packaging, the separating barrier between food in the package and the environment used is polyethylene film. Similarly, for active packaging, the materials used in the container synchronize with the food prolonging its shelf life by improving food quality and safety, for example, polyethylene film packed with TiO_2 and NPs, while smart or intelligent packaging protects the food content from the environment and in addition reacts to any change occurring in the food or its environment while delivering information regarding the product as well as the package, polyethylene film with TiO_2, with methylene blue and glycerol. Moreover, nanosensors present in smart packaging react to various factors such as concentration of oxygen, changes in the environment like variation in humidity and temperature, and occurrence of degradation products (Azeredo 2009; Mills et al. 2017). For instance, when polyethylene film is prepared with TiO_2 (titania dioxide) nanoparticle (anatase) in the presence of methylene blue and glycerol, it changes its color when O_2 is present.

3. **Textile Modification**

For the production of textiles with high performance, organic and inorganic antibacterial materials play a crucial role. Possessing unique properties, these inorganic and metallic nanostructured materials have grabbed a place in textile fabric modifications. Researchers have modified a procedure that focuses on bulk modification of filament yarns where different concentrations of nanocomposite fillers are used by mixing various silver-based fillers like Ag/TiO_2 and Ag/Zn with three different compositions of polymer powder to demonstrate its potential for mass production on a pilot plant scale. This industrial modification technique was found to be easily adaptable, eco-friendly, and high in quality and mutation frequencies (Saidin et al. 2021).

4. **Cosmetics**
   Antimicrobial agents are abruptly used in the preservation and quality maintenance and increase the action of antimicrobial compounds against the targeted microorganism. In deodorants, to inhibit the growth of *staphylococci* and diphtheroid due to the formation of sweat aluminum chlorohydrate, alcohol and chlorohexidine are used. In anti-dandruff shampoo, to reduce the *Malassezia* species and inhibit yeast growth, zinc and salicylic acids are used as antimicrobial agents. In antibacterial soap bars and wipes, to reduce the growth and accumulation of *Staphylococcus* and *Streptococcus*, triclocarban, alcohol, and glycerine are used as antibacterial agents. In toothpaste and mouthwash, to prevent bacterial growth and reduce plaque formation and prevent the development of *Proteobacteria* and *Actinobacteria*, some antimicrobial agents are used, such as alcohol, chlorohexidine, clove, and herbs (Halla et al. 2018).

22.7 Limitations of Antibacterial Materials

Even though antibacterial materials are very useful and are applied in various sectors, there are some disadvantages of these antibacterial agents. One of the major limitations is the choice of antibacterial materials which carry the ability for bacterial resistance. For instance, the overuse of silver as silver dressings has now become a major concern in the field of wound care as it gets exposed to more probable pathogens (Zhao 2011). Another crucial drawback is the lack of knowledge and research regarding the complex structure of the bacterial cell membrane for in vitro analysis. Because the in vitro analysis cannot replicate the in vivo environments, hence, it becomes difficult to understand the antibacterial mechanism of nanoparticles with the help of in vitro analysis (Wang et al. 2017). There is still a desperate need to understand the nanoneurotoxicity and to understand the crossing of nanoparticles through the cell membrane of bacteria (Wang et al. 2017). An additional shortcoming is the effect of antibacterial component durability. Essential oils are highly potent in preserving food. However, it exhibits cytotoxicity activity in humans when ingested, which can lead to several serious health hazards (Lucera et al. 2012). Hence, despite exceptional discoveries of various antibacterial materials, in the field of medicine, health, and food, it still holds numerous unavailability and lacunas which need further research and improvisation.

22.8 Conclusion and Future Trends

Antibacterial materials are essential for reducing the toxic effects of different harmful microorganisms, and their discovery brought a revolution in the medical sector. Moreover, it is projected that there will be a definite increase in the evolution of the antimicrobial material industry in the coming years (D'Arcy 2001). For instance, increasing demands for implants and other medical devices that possess anti-infection effects, escalating demand for the need of antimicrobial-treated surfaces to decrease disease spread, and growing use of plastics and antimicrobials to maintain hygiene are some of the areas where antibacterial agents are very much required. Asia has grown to be a high-end industry demanding an upgraded antibacterial product that meets with industrial hygiene conditions and medical and commercial consumer favor. Moreover, recent advancement in clinical research technologies has put light on the usage of nanoparticles as antibacterial materials for body repair as tissue engineering materials (Zhao 2011). On the top, progressive evolution in antibacterial materials exhibits the discovery of a surface with exceptional biocompatibility and indicates resistance toward bacterial adherence or dealing with controlled release of antibacterial materials to reduce the effect of cytotoxicity (Pearson and Abrutyn 1997). The increasing population demands more food, which requires a protective environment where it can be stored. So, now the researchers and food authorities are in desperate need of developing various natural products which enhance the quality of the food and ensured its nutritional safety without causing harm to human health. Hence, after numerous sustainability experiments of food on different natural products like enzymes, oils, organic acids, and chitosan are revealed as suitable food preservatives. Most importantly, plants hold a successful place for delivering the most natural drugs (Lucera et al. 2012). Essential oils portray great potential in preserving food without manipulating the balance of the food, but these oils possess strong aroma and toxicity complications. Keeping count of its remarkable antimicrobial benefits, it cannot be denied that essential oils are expensive, and the risk of toxicity and the discovery of effective doses for preservation techniques remains to be uncovered in the future (Lucera et al. 2012).

For enhanced functioning of antibacterial agents, detailed information on the natural compound should be known. But as of now, information regarding various natural compounds is still inadequate. Hence, for the advancement of new upgraded engineering techniques, further research is a must, and validation of the combination of active agents does not show amplification in antimicrobial effects (Lucera et al. 2012). Through various research and developments, polymer-based materials have emerged as potential antibacterial substances. But due to its complex nature, the problem arises in its study, which can be controlled by different procedures like developing a nanostructured and nanopatterned surface that possesses controlled surface factors and chemical composition to decrease bacterial adhesion (Olmos and González-Benito 2021). These compounds are tested mostly on animals, which does not determine their effectiveness toward normal microbiota and understanding the toxicity production and its effect on humans. Hence, researchers need to experiment on humans to get beneficial outcomes. Furthermore, in vitro testing and regulating the extraction method gives more methodological and clarified results (Cowan 1999). Therefore, further extensive study is required to understand the mechanism of action of the compound and know its appropriate use against specific microorganisms (Doyle and Stephens 2019). The day is not far when we will find suitable antibacterial material which possesses higher effectiveness against microorganisms; shows less toxicity toward humans; can withstand adverse environmental conditions like variation in temperature, pH, and pressure; is highly biodegradable; and is cost-effective.

References

Adams KN, Takaki K, Connolly LE, Wiedenhoft H, Winglee K, Humbert O et al (2011) Drug tolerance in replicating mycobacteria mediated by a macrophage-induced efflux mechanism. Cell 145(1):39–53

Ansari MA, Khan HM, Khan AA, Sultan A, Azam A (2012) Characterization of clinical strains of MSSA, MRSA and MRSE isolated from skin and soft tissue infections and the antibacterial activity of ZnO nanoparticles. World J Microbiol Biotechnol 28(4):1605–1613

Ansari MA, Khan HM, Khan AA, Cameotra SS, Saquib Q, Musarrat J (2014) Interaction of Al(2)O (3) nanoparticles with Escherichia coli and their cell envelope biomolecules. J Appl Microbiol 116(4):772–783

Arshad MS, Batool A (2017) Natural Antimicrobials, their Sources and Food Safety

Atiyeh BS, Costagliola M, Hayek SN, Dibo SA (2007) Effect of silver on burn wound infection control and healing: review of the literature. Burns 33(2):139–148

Azeredo HMC (2009) Nanocomposites for food packaging applications. Food Res Int 42(9): 1240–1253

Azizi-Lalabadi M, Ehsani A, Divband B, Alizadeh-Sani M (2019) Antimicrobial activity of Titanium dioxide and Zinc oxide nanoparticles supported in 4A zeolite and evaluation the morphological characteristic. Sci Rep 9(1):17439

Balaban NQ, Merrin J, Chait R, Kowalik L, Leibler S (2004) Bacterial persistence as a phenotypic switch. Science (New York, NY) 305(5690):1622–1625

Beyth N, Houri-Haddad Y, Domb A, Khan W, Hazan R (2015) Alternative antimicrobial approach: nano-antimicrobial materials. Evid Based Complement Alternat Med 2015:246012

Blecher K, Nasir A, Friedman A (2011) The growing role of nanotechnology in combating infectious disease. Virulence 2:395–401

Brady-Estévez AS, Kang S, Elimelech M (2008) A single-walled-carbon-nanotube filter for removal of viral and bacterial pathogens. Small (Weinheim an der Bergstrasse, Germany) 4(4):481–484

Breijyeh Z, Jubeh B, Karaman R (2020) Resistance of gram-negative bacteria to current antibacterial agents and approaches to resolve it. Molecules (Basel, Switzerland) 25(6):1340

Brown D (2015) Antibiotic resistance breakers: can repurposed drugs fill the antibiotic discovery void? Nat Rev Drug Discov 14(12):821–832

Chen X, Schluesener HJ (2008) Nanosilver: a nanoproduct in medical application. Toxicol Lett 176(1):1–12

Chen J, Mao S, Xu Z, Ding W (2019) Various antibacterial mechanisms of biosynthesized copper oxide nanoparticles against soilborne Ralstonia solanacearum. RSC Adv 9(7):3788–3799

Cheng G, Dai M, Ahmed S, Hao H, Wang X, Yuan Z (2016) Antimicrobial drugs in fighting against antimicrobial resistance. Front Microbiol 7:470

Ciriminna R, Albo Y, Pagliaro M (2020) New antivirals and antibacterials based on silver nanoparticles. ChemMedChem 15(17):1619–1623

Cole AM, Weis P, Diamond G (1997) Isolation and characterization of pleurocidin, an antimicrobial peptide in the skin secretions of winter flounder. J Biol Chem 272(18):12008–12013
Cowan MM (1999) Plant products as antimicrobial agents. Clin Microbiol Rev 12(4):564–582
D'Arcy N (2001) Antimicrobials in plastics: a global review. Plast Addit Compound 3(12):12–15
Das B, Patra S (eds) (2017) Antimicrobials: meeting the challenges of antibiotic resistance through nanotechnology
Dorantes L, Colmenero RMV, Hernández HJ, Mota L, Jaramillo ME, Fernández E et al (2000) Inhibition of growth of some foodborne pathogenic bacteria by Capsicum annum extracts. Int J Food Microbiol 57:125–128
Doyle AA, Stephens JC (2019) A review of cinnamaldehyde and its derivatives as antibacterial agents. Fitoterapia 139:104405
Elliott C (2010) The effects of silver dressings on chronic and burns wound healing. Br J Nurs (Mark Allen Publishing) 19(15):S32–S36
Farkas-Himsley H (1980) Bacteriocins—are they broad-spectrum antibiotics? J Antimicrob Chemother 6(4):424–426
Gavarić N, Kovac J, Kretschmer N, Kladar N, Smole Možina S, Bucar F et al (2015) Natural products as antibacterial agents—antibacterial potential and safety of post-distillation and waste material from Thymus vulgaris L., Lamiaceae. IntechOpen, pp 123–151
George N, Faoagali J, Muller M (1997) Silvazine (silver sulfadiazine and chlorhexidine) activity against 200 clinical isolates. Burns 23(6):493–495
Guglielmi P, Pontecorvi V, Rotondi G (2020) Natural compounds and extracts as novel antimicrobial agents. Expert Opin Ther Pat 30(12):949–962
Gutierrez J, Barry-Ryan C, Bourke P (2008) The antimicrobial efficacy of plant essential oil combinations and interactions with food ingredients. Int J Food Microbiol 124(1):91–97
Hajipour MJ, Fromm KM, Ashkarran AA, Jimenez de Aberasturi D, de Larramendi IR, Rojo T et al (2012) Antibacterial properties of nanoparticles. Trends Biotechnol 30(10):499–511
Halla N, Fernandes IP, Heleno SA, Costa P, Boucherit-Otmani Z, Boucherit K et al (2018) Cosmetics preservation: a review on present strategies. Molecules (Basel, Switzerland) 23(7): 1571
Hall-Stoodley L, Stoodley P, Kathju S, Høiby N, Moser C, Costerton JW et al (2012) Towards diagnostic guidelines for biofilm-associated infections. FEMS Immunol Med Microbiol 65(2): 127–145
Hamad AH, Khashan K, Hadi AA (2020) Silver nanoparticles and silver ions as potential antibacterial agents. J Inorg Organomet Polym Mater 30:4811–4828
Hans M, Erbe A, Mathews S, Chen Y, Solioz M, Mücklich F (2013) Role of copper oxides in contact killing of bacteria. Langmuir 29(52):16160–16166
Hayek S. Antimicrobial Natural Products. 2014
Herrera FA, Kohanzadeh S, Nasseri Y, Kansal N, Owens EL, Bodor R (2009) Management of vascular graft infections with soft tissue flap coverage: improving limb salvage rates—a veterans affairs experience. Am Surg 75(10):877–881
Hetrick EM, Schoenfisch MH (2006) Reducing implant-related infections: active release strategies. Chem Soc Rev 35(9):780–789
Huh AJ, Kwon YJ (2011) "Nanoantibiotics": a new paradigm for treating infectious diseases using nanomaterials in the antibiotics resistant era. J Control Release 156(2):128–145
Jenssen H, Hamill P, Hancock RE (2006) Peptide antimicrobial agents. Clin Microbiol Rev 19(3): 491–511
Kamaruzzaman NF, Tan LP, Hamdan RH, Choong SS, Wong WK, Gibson AJ et al (2019) Antimicrobial polymers: the potential replacement of existing antibiotics? Int J Mol Sci 20(11)
Khameneh B, Iranshahy M, Soheili V, Fazly Bazzaz BS (2019) Review on plant antimicrobials: a mechanistic viewpoint. Antimicrob Resist Infect Control 8:118
Kim YH, Kim SM, Lee SY (2015) Antimicrobial activity of protamine against oral microorganisms. Biocontrol Sci 20(4):275–280

Knetsch M, Koole LH (2011) New strategies in the development of antimicrobial coatings: the example of increasing usage of silver and silver nanoparticles. Polymers:3

Koehn FE, Carter GT (2005) The evolving role of natural products in drug discovery. Nat Rev Drug Discov 4(3):206–220

Kose N, Çaylak R, Pekşen C, Kiremitçi A, Burukoglu D, Koparal S et al (2016) Silver ion doped ceramic nano-powder coated nails prevent infection in open fractures: in vivo study. Injury 47(2):320–324

Kramer B, Thielmann J, Hickisch A, Muranyi P, Wunderlich J, Hauser C (2015) Antimicrobial activity of hop extracts against foodborne pathogens for meat applications. J Appl Microbiol 118(3):648–657

Langeveld WT, Veldhuizen EJ, Burt SA (2014) Synergy between essential oil components and antibiotics: a review. Crit Rev Microbiol 40(1):76–94

Lellouche J, Friedman A, Lellouche J-P, Gedanken A, Banin E (2011) Improved antibacterial and antibiofilm activity of magnesium fluoride nanoparticles obtained by water-based ultrasound chemistry. Nanomedicine 8:702–711

Lewis K (2007) Persister cells, dormancy and infectious disease. Nat Rev Microbiol 5(1):48–56

Li C, Fu R, Yu C, Li Z, Guan H, Hu D et al (2013) Silver nanoparticle/chitosan oligosaccharide/poly(vinyl alcohol) nanofibers as wound dressings: a preclinical study. Int J Nanomedicine 8: 4131–4145

Li J, Xie S, Ahmed S, Wang F, Gu Y, Zhang C et al (2017) Antimicrobial activity and resistance: influencing factors. Front Pharmacol 8:364

Loomba L, Scarabelli T (2013) Metallic nanoparticles and their medicinal potential. Part I: gold and silver colloids. Ther Deliv 4:859–873

Lucera A, Costa C, Conte A, Del Nobile MA (2012) Food applications of natural antimicrobial compounds. Front Microbiol 3:287

Maji J, Basu S, Pandey S (2020) Synthesis and evaluation of antibacterial properties of magnesium oxide nanoparticles. Bull Mater Sci 10

Martinez L, Han G, Chacko M, Mihu M, Jacobson M, Gialanella P et al (2009) Antimicrobial and healing efficacy of sustained release nitric oxide nanoparticles against staphylococcus aureus skin infection. J Invest Dermatol 129:2463–2469

Mathew AG, Cissell R, Liamthong S (2007) Antibiotic resistance in bacteria associated with food animals: a United States perspective of livestock production. Foodborne Pathog Dis 4(2): 115–133

Mills A, Hawthorne DG, Burns LD, Hazafy D (2017) Novel temperature-activated humidity-sensitive optical sensor. Sens Actuators B Chem 240:1009–1015

Miola M, Fucale G, Maina G, Verné E (2015) Antibacterial and bioactive composite bone cements containing surface silver-doped glass particles. Biomed Mater (Bristol, England) 10(5):055014

Morones JR, Elechiguerra JL, Camacho A, Holt K, Kouri JB, Ramírez JT et al (2005) The bactericidal effect of silver nanoparticles. Nanotechnology 16(10):2346–2353

Negi PS (2012) Plant extracts for the control of bacterial growth: efficacy, stability and safety issues for food application. Int J Food Microbiol 156(1):7–17

Nguyen VT, Vu VT, Nguyen TH, Nguyen TA, Tran VK, Nguyen-Tri P (2019) Antibacterial activity of TiO2- and ZnO-decorated with silver nanoparticles. J Compos Sci 3(2):61

Oh HI, Kim YJ, Chang EJ, Kim JY (2001) Antimicrobial characteristics of chitosans against food spoilage microorganisms in liquid media and mayonnaise. Biosci Biotechnol Biochem 65(11): 2378–2383

Olmos D, González-Benito J (2021) Polymeric materials with antibacterial activity: a review. Polymers 13(4)

Ortiz-Benítez EA, Velázquez-Guadarrama N, Durán Figueroa NV, Quezada H, Olivares-Trejo JJ (2019) Antibacterial mechanism of gold nanoparticles on Streptococcus pneumoniae. Metallomics 11(7):1265–1276

Pati R, Mehta RK, Mohanty S, Padhi A, Sengupta M, Vaseeharan B et al (2014) Topical application of zinc oxide nanoparticles reduces bacterial skin infection in mice and exhibits antibacterial

activity by inducing oxidative stress response and cell membrane disintegration in macrophages. Nanomedicine 10(6):1195–1208
Payne KD, Rico-Munoz E, Davidson PM (1989) The antimicrobial activity of phenolic compounds against listeria monocytogenes and their effectiveness in a model milk system. J Food Prot 52(3):151–153
Pearson ML, Abrutyn E (1997) Reducing the risk for catheter-related infections: a new strategy. Ann Intern Med 127:304–306
Pelgrift RY, Friedman AJ (2013) Nanotechnology as a therapeutic tool to combat microbial resistance. Adv Drug Deliv Rev 65(13–14):1803–1815
Penesyan A, Gillings M, Paulsen IT (2015) Antibiotic discovery: combatting bacterial resistance in cells and in biofilm communities. Molecules (Basel, Switzerland). 20(4):5286–5298
Pisoschi AM, Pop A, Georgescu C, Turcuş V, Olah NK, Mathe E (2018) An overview of natural antimicrobials role in food. Eur J Med Chem 143:922–935
Poole K (2002) Mechanisms of bacterial biocide and antibiotic resistance. J Appl Microbiol 92 (Suppl):55s–64s
Potter R, Truelstrup Hansen L, Gill TA (2005) Inhibition of foodborne bacteria by native and modified protamine: importance of electrostatic interactions. Int J Food Microbiol 103(1):23–34
Prokopovich P, Köbrick M, Brousseau E, Perni S (2015) Potent antimicrobial activity of bone cement encapsulating silver nanoparticles capped with oleic acid. J Biomed Mater Res B Appl Biomater 103(2):273–281
Quinto EJ, Caro I, Villalobos-Delgado LH, Mateo J, De-Mateo-Silleras B, Redondo-Del-Río MP (2019) Food safety through natural antimicrobials. Antibiotics (Basel, Switzerland) 8(4):208
Rai M, Yadav A, Gade A (2009) Silver nanoparticles as a new generation of antimicrobials. Biotechnol Adv 27(1):76–83
Rhoades J, Roller S (2000) Antimicrobial actions of degraded and native chitosan against spoilage organisms in laboratory media and foods. Appl Environ Microbiol 66(1):80–86
Rónavári A, Igaz N, Adamecz DI, Szerencsés B, Molnar C, Kónya Z et al (2021) Green silver and gold nanoparticles: biological synthesis approaches and potentials for biomedical applications. Molecules (Basel, Switzerland) 26(4):844
Rossiter SE, Fletcher MH, Wuest WM (2017) Natural products as platforms to overcome antibiotic resistance. Chem Rev 117(19):12415–12474
Saidin S, Jumat MA, Mohd Amin NAA, Saleh Al-Hammadi AS (2021) Organic and inorganic antibacterial approaches in combating bacterial infection for biomedical application. Mater Sci Eng C Mater Biol Appl 118:111382
Scallan E, Hoekstra RM, Angulo FJ, Tauxe RV, Widdowson MA, Roy SL et al (2011) Foodborne illness acquired in the United States—major pathogens. Emerg Infect Dis 17(1):7–15
Schairer DO, Chouake JS, Nosanchuk JD, Friedman AJ (2012) The potential of nitric oxide releasing therapies as antimicrobial agents. Virulence 3(3):271–279
Simpson WJ, Smith AR (1992) Factors affecting antibacterial activity of hop compounds and their derivatives. J Appl Bacteriol 72(4):327–334
Sirelkhatim A, Mahmud S, Seeni A, Kaus NHM, Ann LC, Bakhori SKM et al (2015) Review on zinc oxide nanoparticles: antibacterial activity and toxicity mechanism. Nanomicro Lett 7(3): 219–242
Spirescu VA, Chircov C, Grumezescu AM, Vasile B, Andronescu E (2021) Inorganic nanoparticles and composite films for antimicrobial therapies. Int J Mol Sci 22(9)
Sprong RC, Hulstein MF, Van der Meer R (2001) Bactericidal activities of milk lipids. Antimicrob Agents Chemother 45(4):1298–1301
Stone PA, Armstrong PA, Bandyk DF, Brumberg RS, Flaherty SK, Back MR et al (2006) Use of antibiotic-loaded polymethylmethacrylate beads for the treatment of extracavitary prosthetic vascular graft infections. J Vasc Surg 44(4):757–761
Tajkarimi MM, Ibrahim SA, Cliver DO (2010) Antimicrobial herb and spice compounds in food. Food Control 21(9):1199–1218

Thomas V, Yallapu MM, Sreedhar B, Bajpai SK (2007) A versatile strategy to fabricate hydrogel-silver nanocomposites and investigation of their antimicrobial activity. J Colloid Interface Sci 315(1):389–395
Tiwari BK, Valdramidis VP, O'Donnell CP, Muthukumarappan K, Bourke P, Cullen PJ (2009) Application of natural antimicrobials for food preservation. J Agric Food Chem 57(14): 5987–6000
Turnidge J, Paterson DL (2007) Setting and revising antibacterial susceptibility breakpoints. Clin Microbiol Rev 20(3):391–408, table of contents
Vicaretti M, Hawthorne W, Ao PY, Fletcher JP (2000) Does in situ replacement of a staphylococcal infected vascular graft with a rifampicin impregnated gelatin sealed Dacron graft reduce the incidence of subsequent infection? Int Angiol 19(2):158–165
Vincent M, Duval RE, Hartemann P, Engels-Deutsch M (2018) Contact killing and antimicrobial properties of copper. J Appl Microbiol 124(5):1032–1046
Waghmode MS, Gunjal AB, Mulla JA, Patil NN, Nawani NN (2019) Studies on the titanium dioxide nanoparticles: biosynthesis, applications and remediation. SN Appl Sci 1(4):310
Wang L, Hu C, Shao L (2017) The antimicrobial activity of nanoparticles: present situation and prospects for the future. Int J Nanomedicine 12:1227–1249
Weller R, Finnen MJ (2006) The effects of topical treatment with acidified nitrite on wound healing in normal and diabetic mice. Nitric Oxide 15(4):395–399
Wiegand I, Hilpert K, Hancock RE (2008) Agar and broth dilution methods to determine the minimal inhibitory concentration (MIC) of antimicrobial substances. Nat Protoc 3(2):163–175
Wilson SE (2001) New alternatives in management of the infected vascular prosthesis. Surg Infect 2(2):171–175; discussion 5-7
Zhao X (2011) Antibacterial bioactive materials. Bioact Mater Med:97–123

Anwesha Kanungo is currently a Research Scholar at the KIIT School of Applied Sciences under the guidance of Dr. Sarbari Acharya. Her area of interest is cancer research; besides, she has experience in the field of virology, including COVID-19. She has completed her MSc in Microbiology (Honours) at GITAM, Vizag, and her BSc in Microbiology (Honours) at the College of Basic Science and Humanities, OUAT, Bhubaneswar.

Dr. Sarbari Acharya has been part of the prestigious KIIT-DU since 2018. She is working as an Assistant Professor in the School of Applied Sciences. She completed her Ph.D. from the Institute of Life Sciences, Bhubaneswar, in 2017. Her area of research includes nanotechnology and drug delivery for cancer therapy. Besides academics, she is actively engaged in research in the field of cancer therapy. She basically works in formulating different nanoparticles and studying their efficacy and mode of action in a wide spectrum of cancer. She has many peer-reviewed journals and book chapters to her name. Recently she has been studying the effect of these formulations in targeting the tumor microenvironment and cancer stem cells which are some of the major reasons for cancer aggressiveness, drug resistance, and relapse. She has recently received a project grant from UGC.

Musculoskeletal Pains and its Common Diseases: Novel Insights in Treatments Using Biomaterials

23

Azadeh Izadyari Aghmiuni, Saeed Heidari Keshel, and Mohammad Zahraei

Abstract

Nowadays, tissue engineering and regenerative medicine are two critical and interrelated topics in health sciences, particularly for treating diseases, which need to promote several cell types, such as musculoskeletal pains and diseases like rheumatoid arthritis and osteoarthritis. In this field, several methods exist to decrease pain and/or treat disease. The use of stem cells, their extracellular vesicles, growth factors, and biomaterials to deliver or modulate these cells and signaling molecules, differentiate cells, proliferate and migrate to targeted tissue, etc. are potential solutions that can be led to tissue repair. Hence, this chapter aims to review these methods and study effective polymers/biomaterials in the induction of signaling pathways of osteogenesis, anti-inflammation, pain reduction, and chondrogenesis, as well as highlight potential advantages. These materials can improve chronic joint inflammatory diseases such as osteoarthritis, tendinitis, rheumatoid arthritis, bone fracture, and neck and low back pains. Active herbal ingredients and their combination with the mentioned materials have been assessed as attractive candidates to increase cellular interactions, regenerate tissue, and create novel insights into the treatment of musculoskeletal disorders/diseases.

A. I. Aghmiuni (✉)
Department of Nanobiotechnology, Pasteur Institute of Iran, Tehran, Iran

S. H. Keshel
Department of Tissue Engineering and Applied Cell Science, School of Advanced Technologies in Medicine, Shahid Beheshti University of Medical Sciences, Tehran, Iran

M. Zahraei
Department of Orthopedics, Nikan Education & Research center (NERC), West Nikan Hospital, Tehran, Iran

F. A. Sheikh et al. (eds.), *Interaction of Nanomaterials With Living Cells*,
https://doi.org/10.1007/978-981-99-2119-5_23

Keywords

Musculoskeletal · Tissue engineering · Regenerative medicine · Biomaterials · Herbal active ingredients

23.1 Introduction

Musculoskeletal diseases are known as the set of conditions that limit the movement of patients via physical injuries or inflammatory pains in the musculoskeletal sites, such as tendons, joints, bones, ligaments, and muscles, caused by lifestyle, occupation, family histories, aging, autoimmune disorders, etc. (Cieza et al. 2020). Studies have shown that these tissues possess poor recovery ability, so their unsatisfactory regeneration is led to the disability of patients and pain in the injured areas (Li et al. 2021). In this field, rheumatoid arthritis (RA) and osteoarthritis (OA) are the most common musculoskeletal diseases and cause permanent disability among some elderly individuals. In particular, RA, defined as an autoimmune inflammatory disorder, is the main factor in synovial joint swelling and inflammations (Guo et al. 2018). Likewise, OA, a chronic inflammatory disease of joints, can affect the joint tissues of the musculoskeletal system, such as articulations of the hips, hands, and knees (Hunter et al. 2020; Zeng et al. 2020; Englund et al. 2012).

There is no cure for these diseases, and most methods are intended to slow down the disease process or pain of patients (Abramoff and Caldera 2020; Messina et al. 2019; Kuang et al. 2021; Bullock et al. 2018). However, surgery to replace damaged joints as the graft of autologous, allograft, or xenograft tissue is considered one of the pain reduction methods, despite limitations of reproducibility and availability, poor plasticity (in the structure or shape), donor scarcity, morbidities of donor sites, risk of transmitting infectious diseases, immune rejection, and the possibility of graft being rejected (Campana et al. 2014). Moreover, using nonsteroidal anti-inflammatory drugs (NSAIDs) is another strategy in this field that can be administered to manage the pains caused by musculoskeletal disorders/diseases. However, they are frequently ineffective in restoring functions/movements of tissues (Abramoff and Caldera 2020). Hence, the current task for researchers is to develop novel techniques to create effective therapeutic methods for the regeneration of the musculoskeletal system. In this field, tissue engineering is known as a promising strategy in regenerative medicine for restoring structure/function of musculoskeletal system.

Indeed, these methods can act as growth strategies that lead to the regeneration of damaged tissue and the treatment of diseases such as OA and RA. In this regard, applying multifactorial approaches such as the design of biomaterial-based substrates and tissue substitutes combined with autologous cells, stem cells, and/or bioactive molecules or active herbal ingredients not only can be led to the imitation of the functions of the native extracellular matrix (ECM) but also result in the stimulation of cell interactions and improvement of the target tissue function. According to the importance of the subject, this chapter has been aimed at studying the current advances and research regarding musculoskeletal tissue engineering and regenerative medicine to treat its diseases, focusing on the role of biomaterials, herbal active ingredients, and bioactive materials to deliver micro-/biomolecules such as growth factors, stem cells, and extracellular vesicles (EVs: byproducts of stem cells), as future strategies for repairing the musculoskeletal system.

23.2 Musculoskeletal System Tissue Engineering

Tissue engineering provides effective approaches to assist the body in imitating organogenesis and regeneration of diseased/damaged tissues (Mooney and Mikos 1999). However, to obtain an efficient regeneration, especially for tissues that need to promote several cell types, such as the musculoskeletal system, there are two key components, which include the stem cells/cells as regenerative factors and the polymer substrates for delivering these cells. Given that the structures/functions of bones, cartilages, skeletal muscle tissue, tendons, and ligaments play an essential role in the selection of these two elements, in this section, we focus on sources of cells, biomaterials, and active ingredients that can play a crucial role in supporting a correct regeneration process musculoskeletal system.

23.3 Cellular Sources for Musculoskeletal Tissue Engineering

The creation of engineered musculoskeletal tissues in vitro and then in vivo needs to use the cells for populating designed polymer substrates to mimic the function and structure of native tissue matrix. Although these cells alone have limitations due to their invasive nature, they can be successful when used in tissue engineering protocols. In this field, we can refer to embryonic stem cells (ESCs), bone marrow-derived mesenchymal stem cells (MSCs), and umbilical cord-derived MSCs (UCMSCs). The characteristics of these cells have been listed in Table 23.1.

23.4 Polymers and Biomaterials

Tissue engineering and regenerative medicine develop alternative solutions to reduce the risks of disease transfer of allografts and the lack of availability of autografts (Delloye et al. 2007; Seiler 3rd and Johnson 2000). The strategy of these two methods includes the regeneration of the guided tissues via the natural regeneration ability of the tissue, cell therapy by allograft/autograft cells, and cell therapy supported by substrates/scaffolds. In this regard, natural and synthetic polymers as scaffolding materials can play a central role in the mentioned approaches. Nowadays, enormous progress in biomaterial research has been achieved, which led to the use of biomaterials or polymers to design novel substrates or scaffolds. This section provides the types of biomaterial for applications in musculoskeletal tissue engineering and analyzes their functions for developing new scaffolds.

Table 23.1 Sources of the cells for musculoskeletal tissue engineering

Sources of the cells	Characteristics	Ref.
ESCs	Derived from the inner blastocyst mass, the maintenance/storage for long culture periods that can provide large amounts of these cells; true pluripotent nature; highest differentiation potential than other stem cells; the ability to differentiate into multiple cell types via asymmetric division, as well as produce adult tissues from all three germ layers (ectoderm, mesoderm, and endoderm); unchecked growth of tissues; teratomas formation; sharing genetic program with cancer stem cells; have limitation due to the ethical problems	Law et al. (1993); Thomson et al. (1998); Wong et al. (2008); Barberi et al. (2007); Hewitt et al. (2007)
MSCs	The cells isolated from bone marrow (stromal cells); have the benefits of primary cells; multi-lineage differentiation along with the high propensity to divide cells; the ability to differentiate into the chondrogenic and osteogenic; suitable for regenerating bone, stabilizing artificial joints of hips; the reduction in differentiating potential after age 30	Jaiswal et al. (1997); Johnstone et al. (1998); Bruder et al. (1998); Bolland et al. (2007)
UCMSCs	The multi-lineage differentiation, the gene expression profile similar to bone marrow-derived MSCs, ability to differentiate into osteoblasts and chondrocytes cells; these cells can improve the matched tissue availability and act as the classical bone marrow MSC	Jeong et al. (2005); Lee et al. (2004)
Primary cells	These cells are known as autologous cells and can be used to regenerate musculoskeletal tissue by myoblast and tendon tissue by tenocyte; they can be removed, ideal from the immunological stance; requires surgical intervention to remove tissues for allowing cell production (similar to the grafting process) that led to pain at the graft sites; some autologous primary cells possess a low propensity for division and their expansion is complex; suitable for treating cartilage defects	Beier et al. (2006); Cao et al. (2006)

Generally, biomaterials are known as the materials that can treat/augment tissues or their functions via intracorporeal implantation. Previously, metallic biomaterials and silicone gels were considered popular implants (Ratner et al. 2004), but today, the ideal biomaterials encompass the concepts of biodegradability or bioabsorbability. It means that biodegradable material-based implants/scaffolds/substrates can be degraded/disintegrated into smaller fragments and cleared from the body after tissue regeneration. This degradation can occur through two dominant mechanisms, i.e., hydrolytic and enzymatic degradations.

Polyesters are a sample of this group that can disintegrate into lactic acid and/or glycolic acid (Gilding and Reed 1979). However, only the materials with a degradation rate similar to targeted tissue are considered bioabsorbable or biodegradable biomaterials. Notably, the existence of hydrophilic groups and molecular weights of the used materials plays a prominent role in controlling the rate of biomaterial degradation (Barrows 1986; Middleton and Tipton 2000; Pietrzak et al. 1997). Accordingly, the researches on biomaterials continue for better substitutes; however, biomaterial-based approaches illustrate their potential to develop engineered grafts (such as substrates or scaffolds) to repair or regenerate musculoskeletal defect. Given that the musculoskeletal system includes the bones, cartilages, tendons, ligaments, muscles, and meniscus tissue, as well as the interfaces of bone-cartilage, muscle-tendon, etc. that support body shape and structure, and its locomotions, nowadays, various types of biomaterials with physical properties similar to targeted tissue such as elasticity, stiffness, flexibility, etc. are being developed to mimic mechanical characteristics of musculoskeletal tissues. Some of these biomaterials and their characteristics have been listed in Table 23.2.

The biomechanical and biophysical properties of the biomaterials/polymers are paramount to ensure the improvement of their function as replacements for the native tissues. Here, some of the common biomaterials, with the classification of hard, soft, and flexible/elastic biomaterials, that can be used for musculoskeletal tissue engineering applications, along with their biomechanical properties, have been highlighted (Fig. 23.1).

1. **Hard Biomaterials**
   Bio-ceramics and bio-glasses are samples of hard biomaterials that can be used for bone repair by orthopedic procedures (Hench 1991). Likewise, there is a wide range of biomaterials with associated advantages to targeted tissues. Nowadays, combinations of phosphate, sulfate, and calcium as bone cement are widely used for regenerating bone tissue due to compatibility and bio-resorbability (Bajammal et al. 2008; Goedhart et al. 2014). In this regard, hydroxyapatite is vital due to its similarity to the mineral structure of hard tissues such as bones and teeth. Hence, the use of this biomaterial and its carbonated forms, alone or in combination with other natural and synthetic materials, have been developed for the fabrication/design of composite scaffolds and regeneration of hard tissues such as bones and osteochondral.

Table 23.2 Biomaterials or polymers of musculoskeletal tissue engineering

Biomaterials/ polymers	Groups	Characteristics	Ref.
Natural	Natural protein materials such as collagen, fibrin, and silk Polysaccharides such as alginate, hyaluronic acid, and starch, chitosan, cellulose	Capable of enzymatic degradation; biocompatible; contain functional groups to conjugate with other molecules like growth factors (as both chemical and enzymatic); uncontrolled enzymatic degradation rate; involve batch-to-batch variability in purity and molecular weight	Gunatillake et al. (2003); Moroni et al. (2014); Van Blitterswijk et al. (2008); Shelke et al. (2014); Lee and Mooney (2012); Andersen et al. (2015); Venkatesan et al. (2015); Machida-Sano et al. (2009); Augst et al. (2006); Racine et al. (2017); Collins and Birkinshaw (2013); Fakhari and Berkland (2013); Tohidi et al. (2022)
Synthetic	Simplest linear and aliphatic polymers such as polyglycolic acid (PGA) An aliphatic polymer such as polylactic acid (PLA) Polyesters such as polycaprolactone (PCL) Charged polyelectrolytes such as polystyrene sulfonate (PSS), polyethylene imine (PEI), polyvinyl alcohol (PVA), polydimethyldiallylammonium chloride (PDADMAC), polyallylamine hydrochloride (PAH), and poly-acrylic acid (PAA) Thermo-responsive polymers such as poly (N-isopropylacrylamide) (PNIPAM)	Attractive group to apply in musculoskeletal tissue engineering; capable of being designed based on the specific need of tissues; ability to manipulate degradation kinetics and mechanical properties of engineered scaffolds for musculoskeletal system (like bones, tendons, ligaments, etc.); suitable for fastening bone tissue as screw and pin; promotion of flexibility for adding cell-binding peptides and consequently the increase of biomimetic properties of the polymer Disadvantages: unfavorable byproduct,	Middleton and Tipton (2000); Gunatillake et al. (2003); Nair and Laurencin (2005); Boland et al. (2001); Kobayashi et al. (2013); Casalini et al. (2019); Malikmammadov et al. (2018); Mochane et al. (2019); Schmidt et al. (2010)

(continued)

Table 23.2 (continued)

Biomaterials/ polymers	Groups	Characteristics	Ref.
		bio-incompatibility, poor cell adhesion	
Hybrid materials	PGA/hyaluronic acid for restoring cartilage, PGA/collagen for stimulation of vascularization, and hydroxyapatite-coated chitosan/PLA nanofibers for the acceleration of bone tissue regeneration	These groups are created based on synthetic and natural polymers and targeted tissue characterizations; via mixing the mentioned polymers to develop biomaterials with biological and mechanical properties similar to tissue of bones, tendons, ligaments, etc.	Lee et al. (2011); Li et al. (2006a); Deng et al. (2008); Patrascu et al. (2013); Lin et al. (2014); Li et al. (2009)
Copolymers	The combination of two or more monomeric species, such as poly(lactide-co-glycolide) (PLGA) consists of poly (lactide) and poly(glycolide); polyester-polyamide copolymer consists of poly(ester amide) and electroactive tetraaniline (PEA-g-TA)	Combining the beneficial/positive properties of monomers; possessing benefits of hybrid materials; fine-tuning glass transition temperature (Tg), durability, and degradation rate; suitable for musculoskeletal tissue scaffolds; tunable biochemical and biophysical properties	Astete and Sabliov (2006); Cui et al. (2012)
Polymer-ceramic composites	These groups have ceramic materials and biodegradable polymers to mimic structures and functions of the musculoskeletal system, such as bones (inorganic mineral compounds and organic collagen fibers). Indeed, they are the result of combining natural and synthetic polymers like elastin, gelatin, collagen, PCL, PLGA, and poly-l-lactide (PLLA) with bone mineral substitutes such as bio-ceramic	These biomaterials are biomimetic and can stimulate the calcium phosphate formation/ deposition/ precipitation from simulated body fluid (SBF) that increases bone-matrix interface strength. They can promote the mechanical properties, chemical	Albrektsson and Johansson (2001); Bose and Tarafder (2012); Li et al. (2017a); Li et al. (2017b); Cui et al. (2019); Xavier et al. (2015); Ho et al. (2006); Santin et al. (2004); Marra et al. (1999)

(continued)

Table 23.2 (continued)

Biomaterials/ polymers	Groups	Characteristics	Ref.
	like 2D nano-silicates and calcium phosphate (CaP), tricalcium phosphate (TCP), hydroxyapatite (HA), etc. to create bone tissue-liked composite scaffolds	stability, biocompatibility, and bioactivity of the biopolymer and improve levels of osteoconductive and osteoinductive	

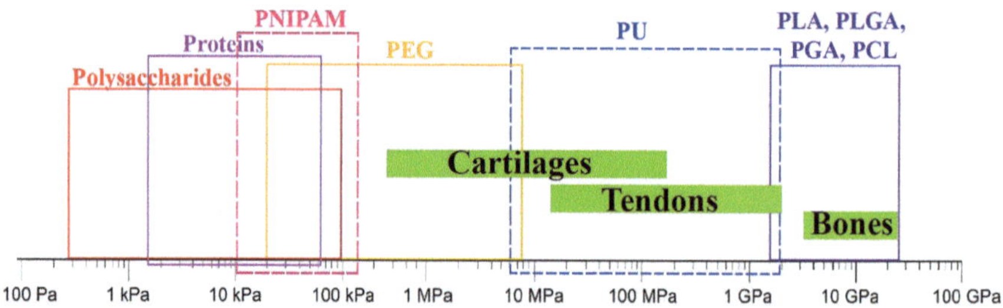

Fig. 23.1 Mechanical properties of some materials and musculoskeletal sites (Abalymov et al. 2020)

2. **Soft Biomaterials**
   In musculoskeletal tissue engineering, natural and/or synthetic hydrogels and porous sponges can be used to regenerate articular cartilage tissue (Wang et al. 2005; Fritz et al. 2009). In this field, alginate, agarose, collagen (I), fibrin gels, hyaluronan, and PLA and PGA sponges play an essential role in the reconstruction of cartilage (Johnstone et al. 2013). Hydrogels and sponges obtained from these soft biomaterials as cartilage substrates/substitutes/scaffolds are biocompatible and can maintain the innately hydrated structure, increasing chondrogenesis of articular chondrocytes, and incorporate chemical cues (Johnstone et al. 2013; Yang et al. 2018; Slaughter et al. 2009).
3. **Flexible/Elastic Biomaterials**
   Several flexible tissues in the musculoskeletal system are known as elastic tissues, such as ligaments, tendons, and meniscus. These tissues possess poor vascular designs and low repair capacities. The reports have indicated that some of the elastic biomaterials can mimic the mechanical properties of such tissues. These biomimetic biomaterials involve elastin, collagen, PCL, polyurethane (PU), and PLLA and can be used in the structure of polymeric scaffolds or substitutes for elastic tissues (Boys et al. 2017; Patel et al. 2018; Chen et al. 2013a; Coenen et al. 2018; Thayer et al. 2016).

23.5 Herbal Active Ingredients or Bioactive Herbal Extracts

Currently, the use of active herbal ingredients or bioactive herbal extracts, such as flavonoids, polyphenolic biomolecules, alkaloids, tannin, etc., in musculoskeletal tissue engineering is booming. In this field, reports have demonstrated that some herbs can be beneficial for bone health and lead to signaling pathways of osteogenesis and chondrogenesis, as well as a decrease in joint inflammation.

Based on the study of Qi et al., Du-Zhong cortex extract (DZCE) can reduce lead acetate (PbAc)-induced bone loss (Qi et al. 2019). Their study on female Sprague-Dawley rats and the analysis of the bone mineral density, bone marrow adipocyte parameters, and bone histomorphology demonstrated that drinking water containing DZCE for 60 days could decrease the PbAc-induced mineral density loss of lumbar spine and femur bones. Moreover, the traditional intake of DZCE was not only led to an increase in the serum phosphorus and serum calcium but also decreased the level of alkaline phosphatase (ALP), osteocalcin, and RANKL. Likewise, the analyses of bone histomorphometric illustrated that the volume of bones and trabecular thickness in the femoral trabecular bones, as well as mean adipocyte diameter, the percent adipocyte volume per tissue volume (AV/TV), and the number of bone marrow adipocytes, have been restored in the group treated with DZCE. Hence, Qi et al. stated that the DZCE possesses the potential to treat and/or prevent PbAc-induced osteoporosis. *A. bidentata* is also one of the herbs in this field. Based on the reports, polysaccharides extracted from the roots of this herb can stimulate bone formation. In this regard, Zhang et al. showed that *Achyranthes bidentata* polysaccharides by intragastric administration were led to the inhibition of glucocorticoid-induced osteoporosis (GIOP), as well as the increase in mineral contents and density of bones and trabecular thickness (Zhang et al. 2018). They stated that such polysaccharides possess a high potential to be applied in anti-osteoporosis medicines.

The evaluation of the herb Radix Dipsaci-isolated asperosaponin-VI (ASA-VI) and its effects and action mechanism on the osteogenic differentiation and proliferation of bone marrow stromal cells in another study in this field was performed by Ke et al. (Ke et al. 2016). Based on the results, the ASA-VI can promote the proliferation of the mentioned cells, ALP activities, and nodule formation. Likewise, it can increase the expression levels of osteocalcin, runt-related transcription factor 2 (RUNX2), and collagen 1. This study shows that ASA-VI promotes osteoblast formation by acting on the signaling pathways of phosphatidylinositol 3-kinase/AKT.

The research of Liang et al. also indicates that plastrum testudinis extracts (PTEs) can reverse spinal osteoporosis induced by the glucocorticoid by targeting osteoblastic and osteoclastic markers such as by targeting the levels of osteoprotegerin, RUNX2, and CTSK in mRNA and protein (Liang et al. 2016). This herb can also be led to an increase in osteoblastic functions. The results of histomorphology approve that oral administration of the extracts of this herb can increase the quantity and quality of bones, such as their thickness and density. Song et al. also demonstrated that Drynariae Rhizoma herb-extracted total flavonoids could be applied in clinical practice to treat joint diseases and bone fractures (Song et al. 2017). Such that the administration for 4 weeks led to the inhibition of bone loss induced by hindlimb unloading in rats. There are many herbs in this field; the epimedium plant is one of them that can be used to treat bone diseases like osteoporosis and/or bone fracture due to having factors like icariin that resulted in osteogenic effects and bone regeneration (Zhang et al. 2014; Zhao et al. 2008; Fan et al. 2011). Buguzhi (psoralea fruit)-isolated psoralen is another active component that can promote osteoblastic differentiation (Xiong et al. 2003). Based on the reports, plants-extracted kaempferol (Kaem) flavonoid can also decrease glucocorticoid-induced bone loss and be led to the promotion of osteoblast differentiation (Prouillet et al. 2004; Adhikary et al. 2018). The herbs that can be applied as attractive osteoinductive candidates in bone tissue engineering have been listed in Table 23.3.

23.6 Effective Parameters in the Design of Tissue Engineering Scaffolds

Pathologically, the design of ideal scaffolds for tissue engineering applications plays a crucial role in regenerating damaged tissues. These temporary matrices that support the formation of new tissues should be able to facilitate cell attachment, induce cell proliferation and differentiation, and form extracellular matrix (ECM) deposition (Hu and Ma 2011). Hence, providing mechanical stability, mimicking the function and structure of targeted tissue, and delivering and localizing the bio-/micro-molecules (such as cells, growth factors, and bioactives into damaged tissue) are considered the prominent roles of the designed substitutes. Generally, the design criteria of musculoskeletal tissue engineering scaffolds involve the following points:

1. Biocompatibility.
2. Ability to adhere, differentiate, proliferate, and deliver cells.
3. Osteoconductivity, i.e., encouraging osteoconduction to create strong bonds between the designed scaffolds and host bones.
4. Biodegradability, i.e., the degradation rate similar to the rates of tissue repair.
5. Mechanical properties depended on the biomaterial properties and porosity of scaffolds.
6. Fabrication capability, i.e., easy formation of materials.
7. Porosity, i.e., the creation of porous structures (>90%) with a pore size of 300–500 μm to penetrate and grow cells and deliver nutrients.

In this field, studies have shown that some scaffolds with a minimum pore diameter of 100 μm can lead to an increase in cell attachment and enhance their proliferation and differentiation into osteoprogenitor or osteoblasts cells (bone-forming cells) (Karageorgiou and Kaplan 2005). Likewise, it is illustrated that microsphere scaffolds with a certain percentage of pores being more than 300 μm can promote large-area bone formation (Amini et al. 2012). Table 23.4 shows some

Table 23.3 Use of herbal active ingredients in musculoskeletal tissue engineering

Active ingredients or bioactive factors/ plants	Classification	Musculoskeletal tissue engineering applications	Ref.
Ginsenosides such as Rb1, Rb2, Rc, Rd, Re, and Rg1/ginseng	Steroid glycosides and triterpene saponins	Osteogenic differentiation of human BM-MSCs	Wang et al. (2020a)
		Fish scale collagen/chitosan composite scaffolds containing hydroxyapatite, β-tricalcium phosphate, and ginseng compound K The reduction of inflammation and increase of bone morphogenetic protein-2 production and the growth of MG-63 cells; suitable for approaches to bone regeneration	Muthukumar et al. (2016)
Naringin/gu sui-bu herb and citrus fruits	Flavanone-7-O-glycoside	Naringin can increase bone mineral densities at the trabecular-rich bone, stress-strain index at the lumbar spine and distal femur, and biomechanical strength at tibia diaphysis in ovariectomized mice; improvement of estrogen-like activities in rat osteoblast-like cell (UMR-106); and protection of against ovariectomized-induced bone loss via ligand-independent activation of estrogen receptor in osteoblastic cells	Pang et al. (2010)
		Genipin-cross-linked gelatin/β-tricalcium phosphate composites containing naringin The increase of osteoblast proliferation, promotion of osteoclast activities, and enhancement of nodule formation without affecting the ALP activities of osteoblasts and mitochondrial activities of mixed bone cells; new bone formation	Chen et al. (2013b)
		The increase in proliferation, differentiation, and osteogenesis induction of the human periodontal ligament stem cells (hPDLSCs) The promotion of effectiveness and expression of bone-related genes (RUNX2, COL1A2, osteopontin, and osteocalcin) at a concentration of 1 μM; creation of	Yin et al. (2015)

(continued)

Table 23.3 (continued)

Active ingredients or bioactive factors/plants	Classification	Musculoskeletal tissue engineering applications	Ref.
		typical trabecular structure surrounded by osteoblasts	
		Electrospun poly(ε-caprolactone) (PCL) and poly(ethylene glycol)-block-poly (ε-caprolactone) (PEG-b-PCL) nano-scaffold containing naringin The improvement of MC3T3-E1 osteoblast adhesion, proliferation, differentiation, and mineralization on the scaffold, as well as suppression of osteoclast formation via controlled release of naringin	Ji et al. (2014)
Icariin/epimedium pubescens herb	Prenyl flavonoid glycoside	The formation of new bone formation, promotion of their thickness, and bone regeneration via transplanting icariin-calcium phosphate cement tablet in a mouse calvarial defect model Mixing icariin and helioxanthin-derived compounds can be led to the induction of osteogenic differentiation of MC3T3-E1 cells. Likewise, combining icariin and medium enriched with calcium can increase mineralization	Zhao et al. (2010)
		β-Tricalcium phosphate ceramic disks loaded with icariin The increase in the ability of calcium phosphate-based scaffolds to repair bone defects; promotion of the osteogenic differentiation (osteogenesis) of stem cells	Park et al. (2014)
		The use of small intestine submucosa as the scaffold for controlled delivery of icariin and improvement of osteoinductivity The increase in the level of osteogenic differentiation marker expression (i.e., osteocalcin, bone sialoprotein, alkaline phosphatase); bone regeneration at 8 weeks; formation of new bones at both 4 and 8 weeks	Li et al. (2017c)

(continued)

Table 23.3 (continued)

Active ingredients or bioactive factors/ plants	Classification	Musculoskeletal tissue engineering applications	Ref.
		Porous β-tricalcium phosphate ceramic disks loaded with icariin for repairing bone defects The ability to promote osteoblast differentiation, improving proliferation and differentiation of Ros17/28 cells on the designed disks, forming bone-like apatite on the disk surface, and increasing new bone formation Based on the reports, although icariin does not change the biocompatibility of the ceramic disk, however it promotes ceramic bioactivity in vivo	Zhang et al. (2011)
		Chitosan/nanohydroxyapatite system to control the release of loaded icariin, simultaneously with the degradation of mentioned system, stimulate alkaline phosphatase activity of bone marrow-derived stroma cells, form mineralized nodules, and increase bone regeneration. Suitable for the design of bone tissue engineering scaffolds	Fan et al. (2012)
		Icariin-loaded calcium phosphate cement scaffold to repair bone defects and treat osteoporosis The increase in osteogenic differentiation of BMSCs, promotion of angiogenic gene expression in the mentioned cells, and inhibition of osteoclast formation	Wu et al. (2017)
		The calcinated antler cancellous bone composite scaffolds containing icariin, velvet antler polypeptide (VAP), or recombinant human bone morphogenetic protein-2 (rhBMP-2) The scaffolds containing icariin led to enhanced cell attachment and proliferation, expression of osteogenic genes, alkaline phosphatase activities, and mineralization of rat bone marrow	Zhang et al. (2013)

(continued)

Table 23.3 (continued)

Active ingredients or bioactive factors/plants	Classification	Musculoskeletal tissue engineering applications	Ref.
		mesenchymal stem cells (BMSCs) than scaffolds containing VAP and rhBMP-2	
		Creation of chondrogenic medium via adding icariin (1×10^{-6} M) to the chondrogenic medium containing transforming growth factor-$\beta 3$ to express genes of chondrocytes in rat BMSCs The increase in chondrogenic differentiation markers such as aggrecan, collagen II, and SRY (sex-determining region Y)-box 9 [SOX9], an effective accelerant of growth factor for applications of cartilage tissue engineering via increasing effect of chondrogenic differentiation and decreasing hypertrophic differentiating effect	Wang et al. (2014)
		Icariin-conditioned serum along with hyaluronic acid to repair osteochondral tissue in rabbit knees Icariin (0.94 g/kg) can promote the proliferation of chondrocytes and increases the secretion of GAG; icariin serum containing hyaluronic acid can be led to an increase in regeneration of the cartilage defects and neoformation of cartilages	Zhang et al. (2019)
Psoralen/dried fruits of *Psoralea corylifolia* L.	Linear furanocoumarins	Collagen matrix mixed with psoralen to produce bone cells and new bones The 454% production of new bones and increase of bone-forming osteoblasts in psoralen-collagen matrix-grafted bone defects compared to the collagen matrix	Wong and Rabie (2011)
Kaempferol/rhizome of *Kaempferia galangal*	Flavonoid	58S bioactive glass-based scaffolds containing zein coating (7%wt) and kaempferol The improvement of mechanical properties of scaffold; controlled delivery and sustained release of kaempferol; promotion of cell attachment; improvement of	Ranjbar et al. (2021)

(continued)

Table 23.3 (continued)

Active ingredients or bioactive factors/ plants	Classification	Musculoskeletal tissue engineering applications	Ref.
		skeletal diseases and bone defects; suitable for bone regeneration applications	
		Kaempferol immobilized on TiO_2 implants to regenerate bones around the dental implant The increase of ALP activities, calcium deposition, and osteogenic differentiation of rat bone marrow stromal cells; stimulation of new bone formation around the dental implant	Tsuchiya et al. (2018)
Ursolic acid	Triterpenoids	Stimulation of osteoblast differentiation; the increase of osteoblast mineralization; induction of expression of osteoblast-specific genes and new bone formation via activating activator protein-1, protein kinases activated with mitogen, and nuclear factor-kappaB (NF-kB); promotion of bone mass and improvement of bone architecture	Lee et al. (2008)
		Meso-porous bio-glass/chitosan scaffolds loaded with ursolic acid to regenerate bones and apply orthopedics The increase of ALP activities; collagen I-related osteogenic differentiation; the expression of runt-associated transcription factor-2 and osteoblast-related protein; promotion of new bone formation	Ge et al. (2019)
		Anti-inflammatory properties regulate inflammation pathways of NF-κB/NLRP3; the protection of chondrocytes and prevention of cartilage degeneration; improvement of osteoarthritis	Wang et al. (2020b)
Curcumin/rhizome of turmeric	Polyphenolic pigments	Fish collagen nanohydroxyapatite composite scaffolds loaded with curcumin The sustained curcumin release, increase in repairing bones, and	Li and Zhang (2018)

(continued)

Table 23.3 (continued)

Active ingredients or bioactive factors/ plants	Classification	Musculoskeletal tissue engineering applications	Ref.
		inhibition of the reactive oxygen species production	
		Poly(ε-caprolactone) nanofibers containing curcumin for promoting osteogenesis and growth of bones via ALP expression	Jain et al. (2016)
		3D calcium phosphate scaffolds loaded with liposome-encapsulated curcumin Creation of cytotoxicity in osteosarcoma (bone cancer) cells; increased cell viability and proliferation of osteoblasts (healthy bone cell); treatment of bone defects	Sarkar and Bose (2019)
Thymol/oregano leaves	Monoterpenoid	Titanium modified with plasma-sprayed hydroxyapatite coats to apply in cemented implants and regenerate bones Inhibiting *Staphylococcus epidermidis*; no cytotoxic effects on osteoblast proliferation; reduction of osteoclastic bone desorption; suitable for dental and orthopedic applications	Vu and Bose (2020)

of these parameters that are effective in the design of musculoskeletal tissue engineering scaffolds, along with their fabrication methods.

23.7 Summary and Future Perspectives

Nowadays, tissue engineering and regenerative medicine have excitingly led to excellent results in repairing and treating musculoskeletal diseases/disorders, so that it can overcome the limitations of some techniques such as grafts of autologous, allograft, or xenograft tissue, surgeries, and cell transplantation. Indeed, musculoskeletal tissue engineering has been developed to meet mechanical demands, reduce pains, and avoid or minimize the degradation of tissues. In this regard, understanding the functions of the musculoskeletal system and using design criteria/parameters combined with fabrication techniques of the scaffold can be led to ideal substrates/ substitutes based on targeted tissue properties. Likewise, the use of new biomaterials such as bioactive materials or herbal ingredients and their combination with other polymer materials can provide biochemical, biomechanical, and biophysical

Table 23.4 Main parameters in design of musculoskeletal tissue engineering scaffolds

Scaffold materials	Methods of fabrication	Channel/pore size, porosity	Mechanical properties	Morphology	Ref.
PLLA and PLGA	The polymer solution was found to dissolve in the solvent and then cast onto the paraffin	Smaller pore size: 100–200 μm Larger pore sizes: 250–350 or 420–500 μm Porosity: ~96%	The decrease in compressive modulus with increasing porosity; the increase in modulus with increasing foam density Modulus: ~300 kPa	Pores: circular in shape and uniform in size	Ma and Choi (2001)
PLGA (75/25)	The frozen solution immersion in H_2O to extract particles from the precipitated polymers	The average pore size: 1.44 mm with interconnections of 0.3–0.8 mm Porosity: 92%	The scaffolds provided mechanical support to regenerate bone tissue	3D morphology similar to trabecular bone	Holy et al. (2003)
PLLA with a polydopamine (PDA) layer and quercetin (Qu)	3D-printed	The mesh with square pores and interconnected pores Thread diameter: 350 μm Pore sizes: 460 μm Porosity: 56.9%	The dry and wet compression strengths: 14.52 MPa and 10.81 MPa, respectively	3D-printed scaffolds with rough surfaces	Chen et al. (2019)
Poly(*L*-lactide) and chitosan containing PDA layer and Qu	Combination of 3D printing and the technology of thermally induced phase separation	Fiber diameter: 410 μm Theoretical porosity: 60–65% Transverse distance between fibers: 390 μm Longitudinal distance: 380 μm Irregular chitosan	Compressive strength and modulus (under dry conditions): 13.09 MPa and 0.112 GPa, respectively Compressive strength and modulus (under wet conditions): 12.42 MPa and 0.105 GPa, respectively	A micro-/nanofiber hierarchical structure with pores of square-like	Zhu et al. (2020)

(continued)

Table 23.4 (continued)

Scaffold materials	Methods of fabrication	Channel/pore size, porosity	Mechanical properties	Morphology	Ref.
		nanofibers with a diameter of 80–600 nm			
PLGA microspheres and poorly crystalline calcium phosphate	Fusion between microspheres and crystallites	Pore size: >100 μm Porosity: 75%	Modulus: 65 MPa	The porous 3D scaffold	Khan et al. (2004)
Chitosan–gelatin/β-tricalcium phosphate	Co-cross-linking with glutaraldehyde	Pore size: 323 to 355 μm Porosity: 92–98%	Yield compressive strength: 0.3–0.88 MPa Compressive modulus: 3.94–10.88 MPa	Macroporous scaffold	Yin et al. (2003)
Nanohydroxyapatite and polyamide	Thermally induced phase inversion processing technique	Pore size: 50–500 μm Porosity: 52–70%	Compressive strength: 13.20–33.90 MPa Modulus: 0.29–0.85 GPa	Composite macropores scaffold	Wang et al. (2007)
PCL and hydroxyapatite	The precision extrusion deposition (PED) process	Group 1: Pore size: 750 μm Porosity: 70% Group 2: Pore size: 450 μm Porosity:60%	Group 1: Compressive modulus: 76 MPa Group 2: Compressive modulus: 84 MPa	3D composite scaffold with internal pore structure of the 0°/90° pattern and pore interconnectivity to regenerate bones	Shor et al. (2007)
PLAGA and hydroxyapatite	Solvent evaporation technique	Hydroxyapatite particle size: 53–150 μm Volume-based median pore diameter of scaffold: ~152–183 μm	Compressive strength: up to 120 MPa Elastic modulus: up to 300 MPa	3D composite microsphere matric (cylindrical and tubular)	Li et al. (2006b)

Gelatin-PCL blend treated with nanohydroxyapatite (nHA)	Electrospinning technique	Average fiber diameter: ~615 nm Average pore size: ~4.7 μm	The improvement of mechanical properties with adding PCL and nHA to gelatin	Fibrous scaffold	Gautam et al. (2021)
PCL and octa-calcium phosphate (OCP)	Coprecipitation method and electrospinning technique	Fiber diameter distribution: 0.1–1.7 μm Average fiber diameter: ~0.52 μm	Average ultimate tensile strength: ~4.34 MPa Maximum tensile strain: ~182.01% Young's modulus: ~5.59 MPa	Nanofibrous scaffold	Heydari et al. (2017)
PCL and nanoclay (NC: as an inorganic filler)	Electrospinning technique	Fiber diameter distribution: 300–700 nm	Tensile strength: ~1–3.5 MPa Young's modulus: ~5–7 MPa	Fibrous scaffold	Nitya et al. (2012)
Plasma-treated PCL	Electrospinning technique	Average fiber diameter: 1.76 μm	–	Nanofiber scaffold	Ko et al. (2015)
PCL and PLGA, along with biphasic tricalcium phosphate (BCP)	Solvent evaporation method	Pore size: 50–100 μm	–	Porous scaffold	Thi Hiep et al. (2017)
Hydroxyapatite (HA) containing PCL and polydiisopropyl fumarate (PDIPF)	Blending method	HA average size: ~225 nm	Ultimate tensile stress: 7.4 MPa Elastic modulus: 182 MPa Elongation at breaking point: 20%	Composite scaffold with narrow size distribution and homogeneous distribution of pores	Fomby et al. (2010)
PCL and cuttlefish bone-derived hydroxyapatite	Solvent casting and particulate leaching method	Average pore dimension: 200–300 μm Porosity: ~85%	Compressive modulus: 0.46 MPa	Composite porous scaffold	Kim et al. (2014)

(continued)

Table 23.4 (continued)

Scaffold materials	Methods of fabrication	Channel/pore size, porosity	Mechanical properties	Morphology	Ref.
Nanohydroxyapatite with PCL and PCL-polyethylene glycol (PEG)-PCL	Electrospinning technique	Average pore size: 20 μm Average fiber diameter: 0.63 μm Porosity: 92%	Tensile strength: 12 MPa (75% in 90 days) Elongation at breaking point: ~269% Young's modulus: ~17.2 MPa Storage modulus at 37 °C: 10.41 MPa	Composite scaffold	Remya et al. (2013)

properties similar to targeted tissues, such that future successes in regulating biological pathways and regenerating musculoskeletal tissues will likely depend more on these issues/factors, and provide new promises for improving diseases or congenital defects.

References

Abalymov A, Parakhonskiy B, Skirtach AG (2020) Polymer- and hybrid-based biomaterials for interstitial, connective, vascular, nerve, visceral and musculoskeletal tissue engineering. Polymers 12:620. https://doi.org/10.3390/polym12030620

Abramoff B, Caldera FE (2020) Osteoarthritis: pathology, diagnosis, and treatment options. Med Clin 104:293–311

Adhikary S, Choudhary D, Ahmad N, Karvande A, Kumar A, Banala VT, Mishra PR, Trivedi R (2018) Dietary flavonoid kaempferol inhibits glucocorticoid-induced bone loss by promoting osteoblast survival. Nutrition 53:64–76. https://doi.org/10.1016/j.nut.2017.12.003

Albrektsson T, Johansson C (2001) Osteoinduction, osteoconduction and osseointegration. Eur Spine J 10:S96–S101. https://doi.org/10.1007/s005860100282

Amini AR, Laurencin CT, Nukavarapu SP (2012) Bone tissue engineering: recent advances and challenges. Crit Rev Biomed Eng 40:363–408

Andersen T, Auk-Emblem P, Dornish M (2015) 3D cell culture in alginate hydrogels. Microarrays 4:133–161

Astete CE, Sabliov CM (2006) Synthesis and characterization of PLGA nanoparticles. J Biomater Sci Polym Ed 17:247–289

Augst AD, Kong HJ, Mooney DJ (2006) Alginate hydrogels as biomaterials. Macromol Biosci 6: 623–633

Bajammal SS, Zlowodzki M, Lelwica A, Tornetta P III, Einhorn TA, Buckley R, Leighton R, Russell TA, Larsson S, Bhandari M (2008) The use of calcium phosphate bone cement in fracture treatment: a meta-analysis of randomized trials. JBJS 90:1186–1196

Barberi T, Bradbury M, Dincer Z, Panagiotakos G, Socci ND, Studer L (2007) Derivation of Engraftable skeletal myoblasts from human embryonic stem cells. Nat Med 13:642–648

Barrows TH (1986) Degradable implant materials: a review of synthetic absorbable polymers and their applications. Clin Mater 1:233–257

Beier JP, Stern-Straeter J, Foerster VT, Kneser U, Stark GB, Bach AD (2006) Tissue engineering of injectable muscle: three-dimensional myoblast-fibrin injection in the syngeneic rat animal model. Plast Reconstr Surg 118:1113–1121

Boland ED, Wnek GE, Simpson DG, Pawlowski KJ, Bowlin GL (2001) Tailoring tissue engineering scaffolds using electrostatic processing techniques: a study of poly (glycolic acid) electrospinning. J Macromol Sci A 38:1231–1243

Bolland B, New AMR, Madabhushi SPG, Oreffo ROC, Dunlop DG (2007) Vibration-assisted bone-graft compaction in impaction bone grafting of the femur. The journal of bone and joint surgery. British 89:686–692

Bose S, Tarafder S (2012) Calcium phosphate ceramic systems in growth factor and drug delivery for bone tissue engineering: a review. Acta Biomater 8:1401–1421

Boys AJ, McCorry MC, Rodeo S, Bonassar LJ, Estroff LA (2017) Next generation tissue engineering of orthopedic soft tissue-to-bone interfaces. MRS communications 7:289–308

Bruder SP, Jaiswal N, Ricalton NS, Mosca JD, Kraus KH, Kadiyala S (1998) Mesenchymal stem cells in osteobiology and applied bone regeneration. Clin Orthop Relat Res 1976-2007(355): S247–S256

Bullock J, Rizvi SAA, Saleh AM, Ahmed SS, Do DP, Ansari RA, Ahmed J (2018) Rheumatoid arthritis: a brief overview of the treatment. Med Princ Pract 27:501–507

Campana V, Milano G, Pagano E, Barba M, Cicione C, Salonna G, Lattanzi W, Logroscino G (2014) Bone substitutes in Orthopaedic surgery: from basic science to clinical practice. J Mater Sci Mater Med 25:2445–2461. https://doi.org/10.1007/s10856-014-5240-2

Cao D, Liu W, Wei X, Xu F, Cui L, Cao Y (2006) In vitro tendon engineering with avian tenocytes and polyglycolic acids: a preliminary report. Tissue Eng 12:1369–1377

Casalini T, Rossi F, Castrovinci A, Perale G (2019) A perspective on polylactic acid-based polymers use for nanoparticles synthesis and applications. Front Bioeng Biotechnol 7:259

Chen Q, Liang S, Thouas GA (2013a) Elastomeric biomaterials for tissue engineering. Prog Polym Sci 38:584–671

Chen KY, Lin KC, Chen YS, Yao CH (2013b) A novel porous gelatin composite containing naringin for bone repair. Evid Based Complement Alternat Med 2013:283941. https://doi.org/10.1155/2013/283941

Chen S, Zhu L, Wen W, Lu L, Zhou C, Luo B (2019) Fabrication and evaluation of 3D printed poly (l-Lactide) scaffold functionalized with quercetin-polydopamine for bone tissue engineering. ACS Biomater Sci Eng 5:2506–2518. https://doi.org/10.1021/acsbiomaterials.9b00254

Cieza A, Causey K, Kamenov K, Hanson SW, Chatterji S, Vos T (2020) Global estimates of the need for rehabilitation based on the global burden of disease study 2019: a systematic analysis for the global burden of disease study 2019. Lancet 396:2006–2017

Coenen AMJ, Bernaerts KV, Harings JAW, Jockenhoevel S, Ghazanfari S (2018) Elastic materials for tissue engineering applications: natural, synthetic, and hybrid polymers. Acta Biomater 79: 60–82

Collins MN, Birkinshaw C (2013) Hyaluronic acid based scaffolds for tissue engineering—a review. Carbohydr Polym 92:1262–1279

Cui H, Liu Y, Deng M, Pang X, Zhang P, Wang X, Chen X, Wei Y (2012) Synthesis of biodegradable and electroactive tetraaniline grafted poly (ester amide) copolymers for bone tissue engineering. Biomacromolecules 13:2881–2889

Cui Z-K, Kim S, Baljon JJ, Wu BM, Aghaloo T, Lee M (2019) Microporous methacrylated glycol chitosan-montmorillonite nanocomposite hydrogel for bone tissue engineering. Nat Commun 10:3523. https://doi.org/10.1038/s41467-019-11511-3

Delloye C, Cornu O, Druez V, Barbier O (2007) Bone allografts: what they can offer and what they cannot. J bone Joint Surg 89:574–580

Deng M, Nair LS, Nukavarapu SP, Kumbar SG, Jiang T, Krogman NR, Singh A, Allcock HR, Laurencin CT (2008) Miscibility and in vitro osteocompatibility of biodegradable blends of poly [(ethyl alanato)(p-phenyl phenoxy) phosphazene] and poly (lactic acid-glycolic acid). Biomaterials 29:337–349

Englund M, Roemer FW, Hayashi D, Crema MD, Guermazi A (2012) Meniscus pathology, osteoarthritis and the treatment controversy. Nat Rev Rheumatol 8:412–419

Fakhari A, Berkland C (2013) Applications and emerging trends of hyaluronic acid in tissue engineering, as a dermal filler and in osteoarthritis treatment. Acta Biomater 9:7081–7092

Fan J-J, Cao L-G, Wu T, Wang D-X, Jin D, Jiang S, Zhang Z-Y, Bi L, Pei G-X (2011) The dose-effect of icariin on the proliferation and osteogenic differentiation of human bone mesenchymal stem cells. Molecules 16:10123–10133. https://doi.org/10.3390/molecules161210123

Fan J, Bi L, Wu T, Cao L, Wang D, Nan K, Chen J, Jin D, Jiang S, Pei G (2012) A combined chitosan/nano-size hydroxyapatite system for the controlled release of icariin. J Mater Sci Mater Med 23:399–407. https://doi.org/10.1007/s10856-011-4491-4

Fomby P, Cherlin AJ, Hadjizadeh A, Doillon CJ, Sueblinvong V, Weiss DJ, Bates JHT, Gilbert T, Liles WC, Lutzko C, Rajagopal J, Prockop DJ, Chambers D, Giangreco A, Keating A, Kotton D, Lelkes PI, Wagner DE, Prockop DJ (2010) Stem cells and cell therapies in lung biology and diseases: conference report. Ann Am Thorac Soc 12:181–204. https://doi.org/10.1002/term

Fritz JR, Pelaez D, Cheung HS (2009) Current challenges in cartilage tissue engineering: a review of current cellular-based therapies. Curr Rheumatol Rev 5:8–14

Gautam S, Sharma C, Purohit SD, Singh H, Dinda AK, Potdar PD, Chou CF, Mishra NC (2021) Gelatin-polycaprolactone-nanohydroxyapatite electrospun nanocomposite scaffold for bone tissue engineering. Mater Sci Eng C 119:111588. https://doi.org/10.1016/j.msec.2020.111588

Ge YW, Lu JW, Sun ZY, Liu ZQ, Zhou J, Ke QF, Mao YQ, Guo YP, Zhu ZA (2019) Ursolic acid loaded-mesoporous bioglass/chitosan porous scaffolds as drug delivery system for bone regeneration. Nanomedicine 18:336–346. https://doi.org/10.1016/j.nano.2018.10.010

Gilding DK, Reed AM (1979) Biodegradable polymers for use in surgery—polyglycolic/poly (actic acid) homo-and copolymers: 1. Polymer 20:1459–1464

Goedhart LM, Ploegmakers JJW, Kroon HM, Zwartkruis ECH, Jutte PC (2014) The presentation, treatment and outcome of periosteal chondrosarcoma in The Netherlands. Bone Joint J 96:823–828

Gunatillake PA, Adhikari R, Gadegaard N (2003) Biodegradable synthetic polymers for tissue engineering. Eur Cell Mater 5:1–16

Guo Q, Wang Y, Xu D, Nossent J, Pavlos NJ, Xu J (2018) Rheumatoid arthritis: pathological mechanisms and modern pharmacologic therapies. Bone research 6:1–14

Hench LL (1991) Bioceramics: from concept to clinic. J Am Ceram Soc 74:1487–1510

Hewitt Z, Priddle H, Thomson AJ, Wojtacha D, McWhir J (2007) Ablation of undifferentiated human embryonic stem cells: exploiting innate immunity against the gal α1-3Galβ1-4GlcNAc-R (α-gal) epitope. Stem Cells 25:10–18

Heydari Z, Mohebbi-Kalhori D, Afarani MS (2017) Engineered electrospun polycaprolactone (PCL)/octacalcium phosphate (OCP) scaffold for bone tissue engineering. Mater Sci Eng C 81:127–132. https://doi.org/10.1016/j.msec.2017.07.041

Ho E, Lowman A, Marcolongo M (2006) Synthesis and characterization of an injectable hydrogel with tunable mechanical properties for soft tissue repair. Biomacromolecules 7:3223–3228

Holy CE, Fialkov JA, Davies JE, Shoichet MS (2003) Use of a biomimetic strategy to engineer bone. J Biomed Mater Res A 65:447–453. https://doi.org/10.1002/jbm.a.10453

Hu J, Ma PX (2011) Nano-fibrous tissue engineering scaffolds capable of growth factor delivery. Pharm Res 28:1273–1281

Hunter DJ, March L, Chew M (2020) Osteoarthritis in 2020 and beyond: a lancet commission. Lancet 396:1711–1712

Jain S, Krishna Meka SR, Chatterjee K (2016) Curcumin eluting nanofibers augment osteogenesis toward phytochemical based bone tissue engineering. Biomedical Materials (Bristol) 11:55007. https://doi.org/10.1088/1748-6041/11/5/055007

Jaiswal N, Haynesworth SE, Caplan AI, Bruder SP (1997) Osteogenic differentiation of purified, culture-expanded human mesenchymal stem cells in vitro. J Cell Biochem 64:295–312

Jeong JA, Hong SH, Gang EJ, Ahn C, Hwang SH, Yang IH, Han H, Kim H (2005) Differential gene expression profiling of human umbilical cord blood–derived mesenchymal stem cells by DNA microarray. Stem Cells 23:584–593

Ji Y, Wang L, Watts DC, Qiu H, You T, Deng F, Wu X (2014) Controlled-release Naringin Nanoscaffold for osteoporotic bone healing. Dent Mater 30:1263–1273. https://doi.org/10.1016/j.dental.2014.08.381

Johnstone B, Hering TM, Caplan AI, Goldberg VM, Yoo JU (1998) In vitro chondrogenesis of bone marrow-derived mesenchymal progenitor cells. Exp Cell Res 238:265–272

Johnstone B, Alini M, Cucchiarini M, Dodge GR, Eglin D, Guilak F, Madry H, Mata A, Mauck RL, Semino CE (2013) Tissue engineering for articular cartilage repair—the state of the art. Eur Cell Mater 25:e67

Karageorgiou V, Kaplan D (2005) Porosity of 3D biomaterial scaffolds and osteogenesis. Biomaterials 26:5474–5491

Ke K, Li Q, Yang X, Xie Z, Wang Y, Shi J, Chi L, Xu W, Hu L, Shi H (2016) Asperosaponin VI promotes bone marrow stromal cell osteogenic differentiation through the PI3K/AKT signaling pathway in an osteoporosis model. Sci Rep 6:2–10. https://doi.org/10.1038/srep35233

Khan YM, Katti DS, Laurencin CT (2004) Novel polymer-synthesized ceramic composite-based system for bone repair: an in vitro evaluation. J Biomed Mater Res A 69:728–737. https://doi.org/10.1002/jbm.a.30051

Kim BS, Yang SS, Lee J (2014) A Polycaprolactone/cuttlefish bone-derived hydroxyapatite composite porous scaffold for bone tissue engineering. J Biomed Mater Res B Appl Biomater 102:943–951. https://doi.org/10.1002/jbm.b.33075

Ko YM, Choi DY, Jung SC, Kim BH (2015) Characteristics of plasma treated electrospun polycaprolactone (PCL) nanofiber scaffold for bone tissue engineering. J Nanosci Nanotechnol 15:192–195. https://doi.org/10.1166/jnn.2015.8372

Kobayashi H, Terada D, Yokoyama Y, Moon DW, Yasuda Y, Koyama H, Takato T (2013) Vascular-inducing poly (glycolic acid)-collagen nanocomposite-fiber scaffold. J Biomed Nanotechnol 9:1318–1326

Kuang X, Chiou J, Lo K, Chunyi WEN (2021) Magnesium in joint health and osteoarthritis. Nutr Res 90:24–35

Law PK, Goodwin TG, Fang Q, Deering MB, Duggirala V, Larkin C, Florendo JA, Kirby DS, Li HJ, Chen M (1993) Cell transplantation as an experimental treatment for Duchenne muscular dystrophy. Cell Transplant 2:485–505

Lee KY, Mooney DJ (2012) Alginate: properties and biomedical applications. Prog Polym Sci 37: 106–126

Lee OK, Kuo TK, Chen W-M, Lee K-D, Hsieh S-L, Chen T-H (2004) Isolation of multipotent mesenchymal stem cells from umbilical cord blood. Blood 103:1669–1675

Lee SU, Park SJ, Kwak HB, Oh J, Min YK, Kim SH (2008) Anabolic activity of ursolic acid in bone: stimulating osteoblast differentiation in vitro and inducing New bone formation in vivo. Pharmacol Res 58:290–296. https://doi.org/10.1016/j.phrs.2008.08.008

Lee K, Silva EA, Mooney DJ (2011) Growth factor delivery-based tissue engineering: general approaches and a review of recent developments. J R Soc Interface 8:153–170

Li Y, Zhang ZZ (2018) Sustained curcumin release from PLGA microspheres improves bone formation under diabetic conditions by inhibiting the reactive oxygen species production. Drug Des Devel Ther 12:1453–1466. https://doi.org/10.2147/DDDT.S154334

Li M, Mondrinos MJ, Chen X, Gandhi MR, Ko FK, Lelkes PI (2006a) Co-electrospun poly (lactide-co-glycolide), gelatin, and elastin blends for tissue engineering scaffolds. J Biomed Mater Res A 79:963–973

Li M, Mondrinos MJ, Chen X, Gandhi MR, Ko FK, Lelkes PI (2006b) Elastin blends for tissue engineering scaffolds. J Biomed Mater Res A 79:963–973. https://doi.org/10.1002/jbm.a

Li X, Xie J, Lipner J, Yuan X, Thomopoulos S, Xia Y (2009) Nanofiber scaffolds with gradations in mineral content for mimicking the tendon-to-bone insertion site. Nano Lett 9:2763–2768

Li Z, Bi S, Thompson BC, Li R, Khor KA (2017a) Multifunctional bioceramic-based composites reinforced with silica-coated carbon nanotube Core-Shell structures. Ceram Int 43:16084–16093. https://doi.org/10.1016/j.ceramint.2017.08.125

Li Z, Khun NW, Tang X-Z, Liu E, Khor KA (2017b) Mechanical, Tribological and biological properties of novel 45S5 bioglass® composites reinforced with in situ reduced graphene oxide. J Mech Behav Biomed Mater 65:77–89. https://doi.org/10.1016/j.jmbbm.2016.08.007

Li M, Gu Q, Chen M, Zhang C, Chen S, Zhao J (2017c) Controlled delivery of icariin on small intestine submucosa for bone tissue engineering. Mater Sci Eng C 71:260–267. https://doi.org/10.1016/j.msec.2016.10.016

Li Z, Xiang S, Li EN, Fritch MR, Alexander PG, Lin H, Tuan RS (2021) Tissue engineering for musculoskeletal regeneration and disease modeling. Handb Exp Pharmacol 265:235–268. https://doi.org/10.1007/164_2020_377

Liang D, Ren H, Qiu T, Shen G, Xie B, Wei Q, Yao Z, Tang J, Zhang Z, Jiang X (2016) Extracts from plastrum testudinis reverse glucocorticoid-induced spinal osteoporosis of rats via targeting osteoblastic and osteoclastic markers. Biomed Pharmacother 82:151–160. https://doi.org/10.1016/j.biopha.2016.04.068

Lin C-C, Fu S-J, Lin Y-C, Yang I-K, Gu Y (2014) Chitosan-coated electrospun PLA fibers for rapid mineralization of calcium phosphate. Int J Biol Macromol 68:39–47

Ma PX, Choi J-W (2001) Biodegradable polymer scaffolds with well-defined interconnected spherical pore network. Tissue Eng 7:23–39

Machida-Sano I, Matsuda Y, Namiki H (2009) In vitro adhesion of human dermal fibroblasts on iron cross-linked alginate films. Biomed Mater 4:25008

Malikmammadov E, Tanir TE, Kiziltay A, Hasirci V, Hasirci N (2018) PCL and PCL-based materials in biomedical applications. J Biomater Sci Polym Ed 29:863–893

Marra KG, Szem JW, Kumta PN, DiMilla PA, Weiss LE (1999) In vitro analysis of biodegradable polymer blend/hydroxyapatite composites for bone tissue engineering. J Biomed Mater Res 47: 324–335

Messina OD, Vidal Wilman M, Vidal Neira LF (2019) Nutrition, osteoarthritis and cartilage metabolism. Aging Clin Exp Res 31:807–813

Middleton JC, Tipton AJ (2000) Synthetic biodegradable polymers as orthopedic devices. Biomaterials 21:2335–2346

Mochane MJ, Motsoeneng TS, Sadiku ER, Mokhena TC, Sefadi JS (2019) Morphology and properties of electrospun PCL and its composites for medical applications: a mini review. Appl Sci 9:2205

Mooney DJ, Mikos AG (1999) Growing New organs. Sci Am 280:60–65

Moroni L, Schrooten J, Truckenmüller R, Rouwkema J, Sohier J, van Blitterswijk CA (2014) Tissue engineering, 2nd edn. Elsevier, pp 1–21. https://doi.org/10.1016/B978-0-12-420145-3.00001-8

Muthukumar T, Aravinthan A, Sharmila J, Kim NS, Kim JH (2016) Collagen/chitosan porous bone tissue engineering composite scaffold incorporated with ginseng compound K. Carbohydr Polym 152:566–574. https://doi.org/10.1016/j.carbpol.2016.07.003

Nair LS, Laurencin CT (2005) Polymers as biomaterials for tissue engineering and controlled drug delivery. Adv Biochem Eng Biotechnol 102:47–90

Nitya G, Nair GT, Mony U, Chennazhi KP, Nair SV (2012) In vitro evaluation of electrospun PCL/nanoclay composite scaffold for bone tissue engineering. J Mater Sci Mater Med 23:1749–1761. https://doi.org/10.1007/s10856-012-4647-x

Pang WY, Wang XL, Mok SK, Lai WP, Chow HK, Leung PC, Yao XS, Wong MS (2010) Naringin improves bone properties in ovariectomized mice and exerts oestrogen-like activities in rat osteoblast-like (UMR-106) cells. Br J Pharmacol 159:1693–1703. https://doi.org/10.1111/j.1476-5381.2010.00664.x

Park JW, Kim JM, Lee HJ, Jeong SH, Suh JY, Hanawa T (2014) Bone healing with oxytocin-loaded microporous β-TCP bone substitute in ectopic bone formation model and critical-sized osseous defect of rat. J Clin Periodontol 41:181–190. https://doi.org/10.1111/jcpe.12198

Patel S, Caldwell J, Doty SB, Levine WN, Rodeo S, Soslowsky LJ, Thomopoulos S, Lu HH (2018) Integrating soft and hard tissues via Interface tissue engineering. J Orthop Res 36:1069–1077

Patrascu JM, Krüger JP, Böss HG, Ketzmar A, Freymann U, Sittinger M, Notter M, Endres M, Kaps C (2013) Polyglycolic acid-hyaluronan scaffolds loaded with bone marrow-derived mesenchymal stem cells show Chondrogenic differentiation in vitro and cartilage repair in the rabbit model. J Biomed Mater Res B Appl Biomater 101:1310–1320

Pietrzak WS, Sarver DR, Verstynen ML (1997) Bioabsorbable polymer science for the practicing surgeon. J Craniofac Surg 8:87–91

Prouillet C, Mazière J-C, Mazière C, Wattel A, Brazier M, Kamel S (2004) Stimulatory effect of naturally occurring flavonols quercetin and kaempferol on alkaline phosphatase activity in MG-63 human osteoblasts through ERK and estrogen receptor pathway. Biochem Pharmacol 67:1307–1313

Qi S, Zheng H, Chen C, Jiang H (2019) Du-Zhong (Eucommia Ulmoides Oliv.) cortex extract alleviates lead acetate-induced bone loss in rats. Biol Trace Elem Res 187:172–180. https://doi.org/10.1007/s12011-018-1362-6

Racine L, Texier I, Auzély-Velty R (2017) Chitosan-based hydrogels: recent design concepts to tailor properties and functions. Polym Int 66:981–998

Ranjbar FE, Foroutan F, Hajian M, Ai J, Farsinejad A, Ebrahimi-Barough S, Dehghan MM, Azami M (2021) Preparation and characterization of 58S bioactive glass based scaffold with kaempferol-containing zein coating for bone tissue engineering. J Biomed Mater Res B Appl Biomater 109:1259–1270. https://doi.org/10.1002/jbm.b.34786

Ratner BD, Hoffman D, Schoen FJ, Lemons JE (2004) Biomaterial science, An introduction to materials in medicine. Academic press, San Diego

Remya KR, Joseph J, Mani S, John A, Varma HK, Ramesh P (2013) Nanohydroxyapatite incorporated electrospun polycaprolactone/polycaprolactone-polyethyleneglycol-polycaprolactone blend scaffold for bone tissue engineering applications. J Biomed Nanotechnol 9:1483–1494. https://doi.org/10.1166/jbn.2013.1640

Santin M, Motta A, Borzachiello A, Nicolais L, Ambrosio L (2004) Effect of PMMA cement radical polymerisation on the inflammatory response. J Mater Sci Mater Med 15:1175–1180

Sarkar N, Bose S (2019) Liposome-encapsulated curcumin-loaded 3D printed scaffold for bone tissue engineering. ACS Appl Mater Interfaces 11:17184–17192. https://doi.org/10.1021/acsami.9b01218

Schmidt S, Zeiser M, Hellweg T, Duschl C, Fery A, Möhwald H (2010) Adhesion and mechanical properties of PNIPAM microgel films and their potential use as switchable cell culture substrates. Adv Funct Mater 20:3235–3243

Seiler JG 3rd, Johnson J (2000) Iliac crest autogenous bone grafting: donor site complications. J South Orthop Assoc 9:91–97

Shelke NB, James R, Laurencin CT, Kumbar SG (2014) Polysaccharide biomaterials for drug delivery and regenerative engineering. Polym Adv Technol 25:448–460

Shor L, Güçeri S, Wen X, Gandhi M, Sun W (2007) Fabrication of three-dimensional polycaprolactone/hydroxyapatite tissue scaffolds and osteoblast-scaffold interactions in vitro. Biomaterials 28:5291–5297. https://doi.org/10.1016/j.biomaterials.2007.08.018

Slaughter BV, Khurshid SS, Fisher OZ, Khademhosseini A, Peppas NA (2009) Hydrogels in regenerative medicine. Adv Mater 21:3307–3329

Song S, Gao Z, Lei X, Niu Y, Zhang Y, Li C, Lu Y, Wang Z, Shang P (2017) Total flavonoids of Drynariae Rhizoma prevent bone loss induced by hindlimb unloading in rats. Molecules 22:1–17. https://doi.org/10.3390/molecules22071033

Thayer PS, Verbridge SS, Dahlgren LA, Kakar S, Guelcher SA, Goldstein AS (2016) Fiber/collagen composites for ligament tissue engineering: influence of elastic moduli of sparse aligned fibers on mesenchymal stem cells. J Biomed Mater Res A 104:1894–1901

Thi Hiep N, Chan Khon H, Dai Hai N, Byong-Taek L, Van Toi V, Thanh Hung L (2017) Biocompatibility of PCL/PLGA-BCP porous scaffold for bone tissue engineering applications. J Biomater Sci Polym Ed 28:864–878. https://doi.org/10.1080/09205063.2017.1311821

Thomson JA, Itskovitz-Eldor J, Shapiro SS, Waknitz MA, Swiergiel JJ, Marshall VS, Jones JM (1998) Embryonic stem cell lines derived from human blastocysts. Science 282:1145–1147

Tohidi H, Maleki-Jirsaraei N, Simchi A, Mohandes F, Emami Z, Fassina L, Naro F, Conti B, Barbagallo F (2022) An electroconductive, thermosensitive, and injectable chitosan/pluronic/gold-decorated cellulose nanofiber hydrogel as an efficient carrier for regeneration of cardiac tissue. Materials 15:5122

Tsuchiya S, Sugimoto K, Kamio H, Okabe K, Kuroda K, Okido M, Hibi H (2018) Kaempferol-immobilized titanium dioxide promotes formation of New bone: effects of loading methods on bone marrow stromal cell differentiation in vivo and in vitro. Int J Nanomedicine 13:1665–1676. https://doi.org/10.2147/IJN.S150786

Van Blitterswijk C, De Boer J, Thomsen P, Hubbell J, Cancedda R, De Bruijn JD, Lindahl A, Sohier J, Williams DF (2008) Tissue engineering. Elsevier

Venkatesan J, Bhatnagar I, Manivasagan P, Kang K-H, Kim S-K (2015) Alginate composites for bone tissue engineering: a review. Int J Biol Macromol 72:269–281

Vu AA, Bose S (2020) Natural antibiotic oregano in hydroxyapatite-coated titanium reduces osteoclastic bone resorption for orthopedic and dental applications. ACS Appl Mater Interfaces 12:52383–52392. https://doi.org/10.1021/acsami.0c14993

Wang Y, Kim U-J, Blasioli DJ, Kim H-J, Kaplan DL (2005) In vitro cartilage tissue engineering with 3D porous aqueous-derived silk scaffolds and mesenchymal stem cells. Biomaterials 26: 7082–7094. https://doi.org/10.1016/j.biomaterials.2005.05.022

Wang H, Li Y, Zuo Y, Li J, Ma S, Cheng L (2007) Biocompatibility and osteogenesis of biomimetic nano-hydroxyapatite/polyamide composite scaffolds for bone tissue engineering. Biomaterials 28:3338–3348. https://doi.org/10.1016/j.biomaterials.2007.04.014

Wang ZC, Sun HJ, Liu MZ, Fu C (2014) Icariin promotes directed chondrogenic differentiation of bone marrow mesenchymal stem cells but not hypertrophy in vitro. Exp Ther Med 8:1528–1534. https://doi.org/10.3892/etm.2014.1950

Wang Z, Jiang R, Wang L, Chen X, Xiang Y, Chen L, Xiao M, Ling L, Wang Y (2020a) Ginsenoside Rg1 improves differentiation by inhibiting senescence of human bone marrow mesenchymal stem cell via GSK-3 β and β-catenin. Stem Cells Int 2020:2365814. https://doi.org/10.1155/2020/2365814

Wang C, Gao Y, Zhang Z, Chen C, Chi Q, Xu K, Yang L (2020b) Ursolic acid protects chondrocytes, exhibits anti-inflammatory properties via regulation of the NF-κB/NLRP3 inflammasome pathway and ameliorates osteoarthritis. Biomed Pharmacother 130:110568. https://doi.org/10.1016/j.biopha.2020.110568

Wong RWK, Rabie ABM (2011) Effect of psoralen on bone formation. J Orthop Res 29:158–164. https://doi.org/10.1002/jor.21124

Wong DJ, Liu H, Ridky TW, Cassarino D, Segal E, Chang HY (2008) Module map of stem cell genes guides creation of epithelial cancer stem cells. Cell Stem Cell 2:333–344

Wu Y, Cao L, Xia L, Wu Q, Wang J, Wang X, Xu L, Zhou Y, Xu Y, Jiang X (2017) Evaluation of osteogenesis and angiogenesis of icariin in local controlled release and systemic delivery for calvarial defect in ovariectomized rats. Sci Rep 7:1–14. https://doi.org/10.1038/s41598-017-05392-z

Xavier JR, Thakur T, Desai P, Jaiswal MK, Sears N, Cosgriff-Hernandez E, Kaunas R, Gaharwar AK (2015) Bioactive nanoengineered hydrogels for bone tissue engineering: a growth-factor-free approach. ACS Nano 9:3109–3118. https://doi.org/10.1021/nn507488s

Xiong Z, Wang D, Xu Y, Li F (2003) Osteoblastic differentiation bioassay and its application to investigating the activity of fractions and compounds from Psoralea Corylifolia L. Pharmazie 58:925–928

Yang Y, Lin H, Shen H, Wang B, Lei G, Tuan RS (2018) Mesenchymal stem cell-derived extracellular matrix enhances chondrogenic phenotype of and cartilage formation by encapsulated chondrocytes in vitro and in vivo. Acta Biomater 69:71–82

Yin Y, Ye F, Cui J, Zhang F, Li X, Yao K (2003) Preparation and characterization of macroporous chitosan-gelatin/β-tricalcium phosphate composite scaffolds for bone tissue engineering. J Biomed Mater Res A 67:844–855. https://doi.org/10.1002/jbm.a.10153

Yin L, Cheng W, Qin Z, Yu H, Yu Z, Zhong M, Sun K, Zhang W (2015) Effects of naringin on proliferation and osteogenic differentiation of human periodontal ligament stem cells in vitro and in vivo. Stem Cells Int 2015:758706. https://doi.org/10.1155/2015/758706

Zeng N, Yan Z-P, Chen X-Y, Ni G-X (2020) Infrapatellar fat pad and knee osteoarthritis. Aging Dis 11:1317

Zhang X, Guo Y, Li DX, Wang R, Fan HS, Xiao YM, Zhang L, Zhang XD (2011) The effect of loading icariin on biocompatibility and bioactivity of porous β-TCP ceramic. J Mater Sci Mater Med 22:371–379. https://doi.org/10.1007/s10856-010-4198-y

Zhang X, Xu M, Song L, Wei Y, Lin Y, Liu W, Heng BC, Peng H, Wang Y, Deng X (2013) Effects of compatibility of deproteinized antler cancellous bone with various bioactive factors on their osteogenic potential. Biomaterials 34:9103–9114. https://doi.org/10.1016/j.biomaterials.2013.08.024

Zhang X, Liu T, Huang Y, Wismeijer D, Liu Y (2014) Icariin: does it have an osteoinductive potential for bone tissue engineering? Phytother Res 28:498–509. https://doi.org/10.1002/ptr.5027

Zhang S, Zhang Q, Zhang D, Wang C, Yan C (2018) Anti-osteoporosis activity of a novel achyranthes bidentata polysaccharide via stimulating bone formation. Carbohydr Polym 184: 288–298. https://doi.org/10.1016/j.carbpol.2017.12.070

Zhang J, Zhang D, Wu C, Liu A, Zhang C, Jiao J, Shang M (2019) Icariin-conditioned serum engineered with hyaluronic acid promote repair of articular cartilage defects in rabbit knees. BMC Complement Altern Med 19:1–9. https://doi.org/10.1186/s12906-019-2570-0

Zhao J, Ohba S, Shinkai M, Chung U, Nagamune T (2008) Icariin induces osteogenic differentiation in vitro in a BMP- and Runx2-dependent manner. Biochem Biophys Res Commun 369: 444–448. https://doi.org/10.1016/j.bbrc.2008.02.054

Zhao J, Ohba S, Komiyama Y, Shinkai M, Chung UI, Nagamune T (2010) Icariin: a potential osteoinductive compound for bone tissue engineering. Tissue Eng A 16:233–243. https://doi.org/10.1089/ten.tea.2009.0165

Zhu L, Chen S, Liu K, Wen W, Lu L, Ding S, Zhou C, Luo B (2020) 3D poly (L-lactide)/chitosan micro/nano fibrous scaffolds functionalized with quercetin-polydopamine for enhanced osteogenic and anti-inflammatory activities. Chem Eng J 391:123524. https://doi.org/10.1016/j.cej.2019.123524

Azadeh Izadyari Aghmiuni holds a Ph.D. degree in Chemical Engineering-Tissue Engineering. Currently, she is working on stem cells, biological macromolecules, and natural/synthetic polymers to construct engineered scaffolds and smart biological substrates to regenerate hard and soft tissues. In this field, she has one patent entitled “Hybrid nano-substrate containing collagen to regenerate skin wounds, IR101120.” Moreover, she investigates the application of the active herbal ingredient and targeted drug delivery (TDD) method in tissue engineering and regenerative medicine to treat diseases.

Saeed Heidari Keshel has a Ph.D. in Applied Proteomics. He is working on stem cells and their differentiation into corneal epithelial cells, keratinocytes, ameloblastin, heart muscle, osteoblast, and hepatocytes. He has one patent entitled “Method of co-culturing human endometrial stem cells and rat embryonic tooth bud cells to obtain Ameloblast cells, US 258275975989 A8.” Currently, he is working in regenerative medicine research for eye-related diseases.

Mohammad Zahraei is an orthopedic specialist in bone and joint surgery. Currently, he is working in Iran Pars Polyclinic, as well as the Department of Orthopedics, Nikan Education & Research Center (NERC), West Nikan Hospital, Tehran, Iran.

Electrospun Cellulose- and Derivatives-Based Nanofibers Loaded with Bioactive Agents for Wound Dressing Applications

24

Sibusiso Alven, Sindi P. Ndlovu, and Blessing A. Aderibigbe

Abstract

The delayed wound recovery process in chronic wounds is challenging in wound treatment. Factors commonly leading to delayed wound recovery include infections, diseases, poor nutrition, smoking, etc. Some presently utilized wound dressing products suffer from various drawbacks, such as poor antioxidant and antibacterial activity, inability to absorb exudate and provide suitable moisture for wound recovery, failure to induce cell migration and proliferation, and weak mechanical performance. A series of preclinical studies on cellulose-based nanofibrous scaffolds have demonstrated promising outcomes in wound dressing applications. The advantages of nanofibrous materials from cellulose and its derivatives that make them ideal wound dressings include good biodegradability and biocompatibility, nonallergic, non-toxicity, good porosity, exudate absorption, and their capability to provide moisture to accelerate the process of wound healing. These materials can be further loaded with bioactive agents (e.g., growth factors, antibiotics, vitamins, and nanoparticles) to improve their biological activities. This book chapter reports the outcomes of electrospun nanofibers based on cellulose and its derivatives incorporated with therapeutic agents together with the in vivo and in vitro experiments, making them potential wound dressing materials.

Keywords

Nanofibers · Electrospinning · Cellulose · Cellulose acetate · Wound healing · Antibacterial activity

S. Alven (✉) · S. P. Ndlovu · B. A. Aderibigbe
Department of Chemistry, University of Fort Hare, Alice, South Africa
e-mail: 201214199@ufh.ac.za; 201304407@ufh.ac.za; baderibigbe@ufh.ac.za

F. A. Sheikh et al. (eds.), *Interaction of Nanomaterials With Living Cells*,
https://doi.org/10.1007/978-981-99-2119-5_24

24.1 Introduction

Chronic wounds are a critical challenge in biomedicine and tissue engineering because of their delayed healing and complications during treatment. Numerous factors contributing to the retarded healing mechanism of chronic wounds include microbial infections, underlying conditions (such as cancer and diabetes mellitus), smoking, infections, obesity, and poor nutrition (Mutlu et al. 2018). The treatment and management of wounds result in negative socioeconomic impacts worldwide. In 2015 and 2018, approximately 6.1 million and 8.2 million people suffered from injuries, respectively (Martinengo et al. 2019). The treatments of wounds cost between \$28.1 billion and \$96.8 billion annually. With a frequency of 1–2% in the universal population, chronic injuries are reported to be the highest case of skin lesions, especially surgical wounds and foot/leg ulcers (diabetic foot ulcer and pressure ulcer) (Walker et al. 2014). Some of the products that are presently utilized for the treatment of wounds in clinical applications suffer from various shortcomings, such as poor biological efficacies (antioxidant and antimicrobial activity), weak mechanical features, and they do not provide the appropriate moisture for accelerated wound healing mechanism (Dhivya et al. 2015; Alven et al. 2022a). There is a pressing necessity to design ideal wound dressing scaffolds that meet the standard for clinical application. The expected properties of perfect wound dressings include excellent biocompatibility and biodegradability, non-toxic and nonallergic, protection against pathogens, optimum local moisture, proper gaseous exchange, absorption of wound exudate, mechanical safety, easy to apply and remove, and affordability (Marin et al. 2018; Trevisol et al. 2019; Patil et al. 2020).

Polymers are biomaterials that are commonly employed for the preparation of wound dressing scaffolds. These biomaterials are typically categorized into biopolymers (natural polymers) and synthetic polymers. Biopolymers are mainly found in natural sources, such as plants, animals, and microorganisms. The properties of biopolymers include excellent biological activity, biodegradability, and biocompatibility (Hussain et al. 2017). The natural origin of biopolymers results in them being appropriate for replacing natural extracellular matrix (ECM) constituents, making them suitable materials for wound treatment (Malafaya et al. 2007). Examples of biopolymers include cellulose, alginate, chitosan, gelatin, silk fibroin, etc. However, wound dressings that are fabricated from biopolymers only suffer from weak mechanical performance (Huang and Fu 2010). This shortcoming can be simply overcome by combining them with synthetic polymers. Various synthetic biodegradable/nonbiodegradable, biocompatible polymers or copolymers have been utilized to fabricate wound dressing materials (Baldwin and Kiick 2010; Zhong et al. 2010). Examples of synthetic polymers include poly(lactic-co-glycolic acid) (PLGA), poly(vinyl pyrrolidone) (PVP), poly(ethylene oxide) (PEO), poly (ethylene glycol) (PEG), poly(ε-caprolactone) (PCL), poly(hydroxyethyl methacrylate) (PHEMA), polylactide (PLA), and poloxamer (Alven et al. 2020).

The polymer-based scaffolds designed for wound dressing applications are frequently fabricated in various forms (e.g., nanofibers, films, hydrogels, membranes, sponges, composites, wafers, and foams). Nanofibers that are prepared using the electrospinning method have attracted the fantastic attention of biomedical researchers due to their attractive properties. Distinct features of electrospun nanofibers include nanoscale structure, ability to imitate ECM, high porosity, and large surface area (Mbese et al. 2021). These features are vital in promoting cell adhesion and proliferation during wound recovery. Furthermore, nanofibers can also be utilized as therapeutic delivery systems to improve their biological outcomes (Mbese et al. 2021; Heydari et al. 2018). The bioactive agents that can be loaded into the nanofibers for the treatment of wounds include antibiotics, vitamins, growth factors, and nanoparticles. This chapter discusses the efficacy of electrospun cellulose- and derivatives-based nanofibers encapsulated with bioactive agents in wound treatment.

24.2 Classification of Wound Dressings

Wound dressing materials are crucial in the clinical management and treatment of acute or chronic wounds. The primary purpose of wound dressings is to keep injuries away from bacterial invasion and accelerate the mechanism of injury recovery (Zhang and Zhao 2020). Wound dressings are generally categorized into five classes: passive (traditional) dressings, skin substitutes, dermal grafts, interactive wound dressings, and bioactive wound dressings (Table 24.1) (Alven et al. 2022b). Traditional wound dressings are designed to protect the injury from foreign attack or contamination, control bleeding, cover the wound, absorb wound exudates, provide a dry environment, and cushion the wound (Alven et al. 2022b). Traditional dressings include bandages, gauze, plaster, and wool dressing. The shortcomings of some traditional dressings are the dripping of wound exudate, which usually results in microbial infections and pain, and skin damage during removal (Felgueiras and Amorim 2017).

Skin substitutes are biomaterials composed of tissue-engineered constituents classically from cell-cultured materials, and they are very suitable for skin regeneration. Nevertheless, they can transmit diseases, cause infections, are rejected by the host body, are high-priced, and have a short shelf life. Examples of skin substitutes are OrCel, Apligraf, and TransCyte (Mir et al. 2018). Dermal grafts are the most critical biomaterials in the arena of dermatology and plastic surgery. Their examples are acellular xenografts, autografts, and allografts. These grafts are frequently used in traumatic wounds, burn reconstruction, post-oncologic resection defects, scar contracture release, hair restoration, and vitiligo. However, they are unsuitable for the management of complicated injuries (i.e., circumstances where bones are exposed and there are deep scars) (Alven et al. 2022b; Shimizu and Kishi 2012).

Interactive dressings such as films, composites, foams, gels, and sprays act as a defensive barrier against microbial invasions, and they accelerate the process of wound healing by offering moisture to the wound site, improving reepithelialization and granulation, and improving water transmission rate with good mechanical performance (Ambekar and Kandasubramanian 2019). These wound dressing materials can also be incorporated with bioactive agents to form bioactive wound

Table 24.1 Summary of classification of wound dressings

Classes of wound dressings	Examples	Advantages	Disadvantages
Passive/ traditional dressings	Gauze, plaster, bandages, and cotton wool	They shield the wound from foreign attacks, absorb wound exudates, terminate bleeding, offer a dry environment, and cushion the wound	Dripping of wound exudate from wound dressing can cause microbial infections and cause pain and further skin damage during removal
Skin substitutes	OrCel, Apligraf, and TransCyte	They are very appropriate for skin regeneration	They can transmit diseases, cause wound infections, are expensive, are rejected by the host body, and have a short shelf life
Dermal grafts	Allografts, autografts, and acellular xenografts	They are instrumental in burn reconstruction, traumatic injuries, congenital skin deficiencies, etc.	They are not appropriate for the treatment of complicated wounds
Interactive dressings	Films, composites, foams, gels, and sprays	Ability to offer suitable moisture for wound healing, enhance granulation and reepithelialization, and excellent mechanical properties	Most of them suffer from poor antibacterial activity
Bioactive dressings	Nanofibers, sponges, hydrocolloids, hydrogels, foams, wafers, collagens, and films	Drug delivery systems of various therapeutic agents and improved process of wound healing	No obvious limitations

dressings. Bioactive dressings such as nanofibers, hydrocolloids, sponges, hydrogels, wafers, foams, collagens, films, and composites are biodegradable and biocompatible. They can be used as drug delivery systems for various bioactive agents, such as growth factors (GFs), antibiotics, vitamins, and nanoparticles, with the improved process of wound recovery (Fahimirad and Ajalloueian 2019; Naseri-nosar and Maria 2018).

24.3 Physiological Process of Wound Healing

It is crucial to study and understand the mechanism of wound healing to choose the appropriate wound dressing material for a specific phase of the healing process. The wound recovery process is a well-coordinated and multifaceted process that comprises a diversity of biochemical and cellular reactions, which require a dynamic cascade of biological processes for the restoration of skin layers (i.e., epidermis, dermis, and subcutaneous layer), anatomical continuity, tissue regeneration, and functions of the skin (Patel et al. 2018; Massee et al. 2015).

24.3.1 Phases of Wound Healing Process

The process of wound healing comprises four different consecutive phases that typically overlap: hemostasis, inflammation, proliferation, and maturation (remodeling) phases (Fig. 24.1) (Rezvanian et al. 2017). The hemostasis phase is the first and shortest phase (5–10 min) of wound recovery that happens instantly after the lesion. The fundamental purpose of this stage is to terminate bleeding via vasoconstriction, primary hemostasis (thrombocyte accumulation with the development of thrombocyte plugs), and secondary hemostasis (formation of fibrin clot) (Iacob et al. 2020). The inflammation phase that typically coincides with the hemostasis involves the recruitment of macrophages and neutrophils, secretion of cytokines, destruction of microorganisms (e.g., bacteria), and development of wound bed. Inflammation stimulates leukocyte build-up in the wound, promoting various chemotactic factors and mediators 1–2 days after an injury (Kanikireddy et al. 2020; Cañedo-Dorantes and Cañedo-Ayala 2019).

The proliferation phase involves a complex mechanism that includes the formation of granulation tissue via collagen deposition and fibroblast proliferation, neoangiogenesis, reepithelization, production of ECM, and wound retirement, and all these occur concurrently. This phase of the wound recovery process starts after 3 days of tissue disruption and can occur for over 2 weeks (Pawar et al. 2013). The last stage of the injury recovery process is the maturation phase (remodeling), the most extended stage of all four steps, and can take weeks to 1–2 years, even more. The critical event of this stage is the remodeling of granulation tissue, where collagen type I substitutes collagen type III because type I is more stable. This also involves developing cellular connective tissue and hardening the new epithelium, creating the ending scar (Yang et al. 2022).

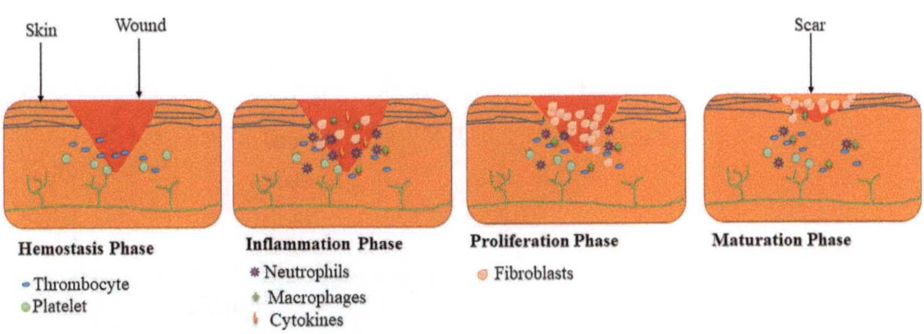

Fig. 24.1 Sequential phases of wound healing process

24.3.2 Factors that Delay the Process of Wound Healing

Numerous factors can delay the process of wound healing and subsequently result in chronic wounds. These factors can be classified into two groups: systematic and local factors (Fig. 24.2). Systematic factors comprise the health status or conditions of the patients. The systematic factors include immunocompromised diseases, conditions, smoking or alcohol, obesity, stress, medications, nutrition, age, gender, and sex hormones. Old age is one of the significant factors that retard wound healing because of the numerous comorbidities (Powers et al. 2016). Acute wounds have a prolonged duration of the wound repair process for elderly females than elderly males, and this can be elucidated through sex hormones that play an essential role in injury healing. Stress also induces the reduction of chemoattractant expression and decreases the level of pro-inflammatory cytokines, which are involved in the inflammation phase of wound repair (Tudoroiu et al. 2021).

Health condition such as diabetes mellitus negatively affects the wound healing process by causing a reduction of fibroblast proliferation and inhibiting the expansion of neutrophils and macrophages and hypoxia (Okonkwo and DiPietro 2017).

Fig. 24.2 Factors that affect the process of wound healing

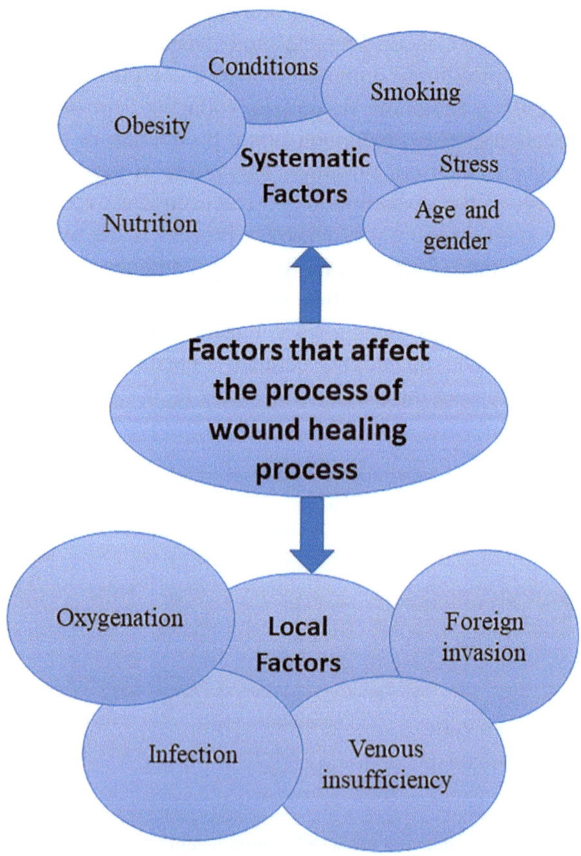

Obesity is another major factor that impedes the wound recovery mechanism because it results in an augmented workload of the heart that offers oxygenated blood to the skin tissues; it cannot perfuse them and lead to ischemia and the development of infections that retard wound recovery. The medications such as anticancer drugs and steroids can also cause delayed wound healing by disturbing the proliferation phase, decreasing the deposition of fibrin, inhibiting the contraction of the lesion, proliferation of skin cells, and production of collagen (Serra et al. 2017). The consumption of alcohol and smoking delay wound repair by causing injury dehiscence and infection, reduction of neutrophils, angiogenesis lessening, and necrosis of tissues (McDaniel and Browning 2014). Lastly, malnutrition can result in an extension of inflammation, declining angiogenesis, affecting fibroblast functions, and reducing biosynthesis and deposition of collagen, ultimately resulting in delayed wound repair (Sen 2019).

Examples of local factors that can lead to a slow wound recovery process are oxygenation, infections, foreign invasions, and venous insufficiency. Vascular damage creates oxygen depletion in injured tissues, leading to hypoxia and impaired lesion healing (Desmet et al. 2018). Oxygen plays several roles in the damaged tissues, including activation of angiogenesis, prevention of infections, improvement of reepithelialization, differentiation of keratinocytes, increased proliferation of fibroblasts, collagen biosynthesis, and inducing contraction of wound (Gueldner et al. 2017). The physical invasion of skin when tissues are injured results in local infections of the wounds. Furthermore, endotoxins and bacteria can stimulate prolonged elevation of pro-inflammatory cytokines (IL-1 and TNF-α) and matrix metalloproteinases, leading to an extended inflammation phase (El-Ashram et al. 2021).

24.4 Properties of Cellulose and Derivatives in Wound Healing Applications

Cellulose is a biopolymer and an organic polysaccharide from a plant source. It is non-toxic with a structural role, being the universe's most abundant and renewable natural polymer. It was discovered in the nineteenth century by Anselme Payen. Cellulose is a linear biopolymer comprised of β-1,4 linked D-glucose units linked to produce cellobiose repeating units (Fig. 24.3) (Mano et al. 2007). Cellulose is

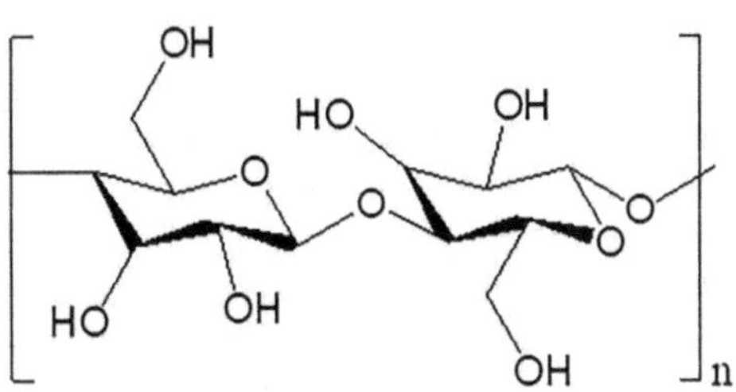

Fig. 24.3 Molecular structure of cellulose

extensively utilized in biomedical applications, including wound dressings. Some research investigations have demonstrated that cellulose and its derivatives possess excellent biocompatibility because of its smaller inflammatory response. Furthermore, the resorption of this biopolymer in a biological environment does not take place due to the inability of cells to produce cellulases (Helenius et al. 2006).

Several studies have reported the wound repair efficacies of cellulose, which indicated that cellulose promotes the acceleration of the wound recovery process via the release and maintenance of many growth factors (GFs) in the wound environment. Examples of GFs released by cellulose include the epidermal growth factor (EGF), basic fibroblast growth factor (bFGF), and phosphodiesterase growth factor. The GFs induce the movement and growth of skin cells and hinder the invasion of microorganisms in the wound site (Barrientos et al. 2014). GFs are biologically active polypeptides that play an essential role in cell differentiation, growth, migration, proliferation, and metabolism. A wide variety of GFs and cytokines controlled all phases of the wound recovery process (Barrientos et al. 2014). Bioengineered cellulose, such as microbial cellulose, is mainly used as a healing matrix for chronic wound dressings, reducing healing time and reducing pain. For partial- and full-thickness injuries, cellulose induces granulation and epithelialization (Alven and Aderibigbe 2020; De Sousa Moraes et al. 2016). Wound dressings with modified cellulose can be loaded with various bioactive molecules such as enzymes, antioxidants, antimicrobial drugs, plant extracts, vitamins, and hormones. Combining cellulose-based wound dressing materials with bioactive molecules significantly results in a synergistic effect in the wound healing mechanism (Zheng et al. 2020). The examples of cellulose derivatives that are regularly used for the fabrication of wound dressings include carboxymethyl cellulose (CMC), methylcellulose (MC), hydroxypropyl cellulose (HPC), hydroxypropylmethylcellulose (HPMC), hydroxyethyl cellulose (HEC), and hydroxyethylmethylcellulose (HEMC).

24.5 Electrospinning Technique and Properties of Electrospun Nanofibers

Several techniques are used to fabricate nanofibers or nanofibrous scaffolds (e.g., nanofibrous films, nanofibrous mats, nanofibrous sheets, and nanofibrous hydrogels). These techniques include melt blowing, phase separation, template synthesis, self-assembly, etc. (Gupta et al. 2014). Nevertheless, electrospinning is the most advantageous technique. The electrospinning procedure has been used since the 1890s (Last et al. 2002). The preparation of nanofibers utilizing the electrospinning method is presented in Fig. 24.4. This method employs electrostatic force to pull the fibers from the droplet formed at the spinneret tip. Various studies have been conducted to study appropriate parameters for electrospinning nanofibers with excellent physicochemical features for biomedical applications (Kenawy et al. 2003). These three parameters influence the formation of nanofibers during the electrospinning method: ambient parameters (e.g., temperature and humidity), process parameters (e.g., applied voltage, electric field induced by the collector, flow

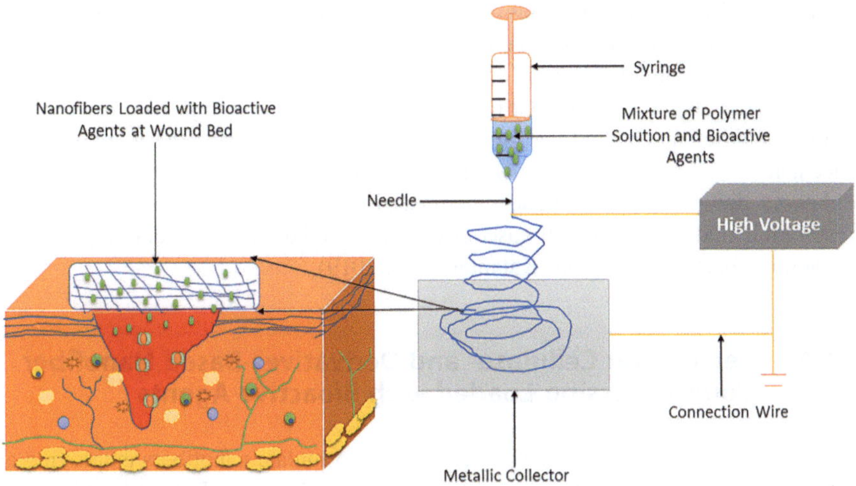

Fig. 24.4 A schematic diagram of the electrospinning method

rate, and the distance between tip and collector), and solution parameters (e.g., surface tension, solution viscosity, and conductivity) (Alven et al. 2021a).

The electrospinning procedure can form continuous nanofibers using a diversity of materials, such as polymers and their composites, with a mean diameter that ranges from micron to nanosized diameter (Wu et al. 2012). The biomedical applications of electrospun nanofibrous materials include wound dressing, tissue regeneration, drug delivery systems, etc. (Uppal et al. 2011). Polymers are frequently used for the fabrication of nanofibers, resulting in a significant role in the arena of wound recovery. The selection of suitable polymers for the formulation of nanofibers is substantial so that they would be eligible for the anticipated properties of ideal wound dressing. Biopolymers such as cellulose, chitosan, collagen, silk fibroin, alginate, hyaluronic acid (HA), and gelatin have been employed for the preparation of electrospun nanofibers for drug delivery and wound healing applications (Ranjith et al. 2019). Recent research experiments on biopolymers' physical and biological properties have shown that these polymers possess excellent biocompatibility and play a vital role when fabricating nanofibrous materials for wound dressing (Liu et al. 2017).

The electrospun nanofibrous scaffolds formulated from synthetic polymers possess excellent mechanical performance than those fabricated from biopolymers. Moreover, synthetic polymers are soluble in various solvents, which promotes their application in the formulation of nanofibers using an electrospinning procedure (Zhang et al. 2005). Biopolymers are commonly combined with synthetic polymers during electrospinning to improve the morphology, control the degradation rate, and enhance the mechanical properties of the nanofibers. The electrospun nanofibrous scaffolds have been employed in wound dressing applications to accelerate the process of wound recovery and to avoid postsurgical infections with sustained and controlled drug-release kinetics (Xu et al. 2011). The other advantages of electrospun nanofibrous materials include a high surface-area-to-volume ratio, high porosity, and small pore size (Alven et al. 2021b). Nanofibrous wound dressings imitate the extracellular matrix (ECM), resulting in improved cell proliferation of epithelial cells and the formation of new tissues during the wound healing mechanism (Abrigo et al. 2014). Their nanometer fiber diameter significantly enhances fluid absorption, induces hemostasis of disrupted tissues, improves cell respiration, and promotes high-gaseous exchange and dermal drug delivery, thereby preventing bacterial invasion (Kalantari et al. 2019).

24.6 Electrospun Cellulose- and Derivatives-Based Nanofiber Wound Dressing Loaded with Bioactive Agents

Cellulose and its derivatives have been used to formulate nanofibers employing an electrospinning technique for application in wound dressing. These nanofibers can be loaded with bioactive agents (e.g., essential oils, antibiotics, plant extracts, vitamins, antioxidants, growth factors, and nanoparticles) to improve their biological activities during wound dressing.

24.6.1 Cellulose-Based Nanofibers

Due to the various exciting properties of cellulose, some preclinical studies report cellulose-based nanofibers loaded with bioactive agents formulated using the electrospinning technique for wound healing applications. Haider et al. prepared lignin-mediated cellulose nanofibers loaded with CuO nanoparticles via an electrospinning method for wound care (Haider et al. 2021). The Fourier transform infrared (FTIR) and X-ray diffraction (XRD) analysis confirm the successful formulation of CuO nanoparticles-loaded cellulose nanofibers. The in vitro antibacterial studies employing the disk diffusion procedure exhibited that the nanofibers containing CuO nanoparticles possess a higher zone of inhibition (ZOI) of 2.46 ± 0.39 and 2.81 ± 0.29 mm on *Escherichia coli* (*E. coli*) and *Staphylococcus aureus* (*S. aureus*), respectively. In contrast, pristine nanofibers did not display ZOI on both bacteria strains. These results indicated that the good antibacterial activity of nanofibers was due to the loading of CuO nanoparticles, suggesting that CuO nanoparticle-encapsulated nanofibers are potential scaffolds for managing microbial-infected wounds. Moreover, the in vitro cytotoxicity analysis showed more than 80% cell viability of NIH3T3 fibroblasts when cultured with CuO nanoparticles containing cellulose nanofibers revealing good cytocompatibility and non-toxicity of nanofibrous scaffolds (Haider et al. 2021). Phan et al. formulated electrospun cellulose nanofibers loaded with either Cu or Ag nanoparticles that exhibited superior antibacterial activity against *Bacillus subtilis* (*B. subtilis*) and *E. coli* than the pristine cellulose nanofibers (Phan et al. 2019).

Sofi and co-workers formulated cellulose nanofibers incorporated with hydroxyapatite and Ag nanoparticles. These nanofibers suppressed the microbial growth of *S. aureus* and *E. coli* effectively, indicating superior bactericidal activity against bacteria strains that clinically infect injuries. The in vitro cytotoxicity analysis exhibited that these nanofibers significantly stimulated the cell growth and proliferation of the chicken embryo fibroblasts, demonstrating their potential to induce wound healing acceleration (Sofi et al. 2021). Abdul et al. formulated electrospun cellulose-based nanofibers incorporated with Ag nanoparticles for wound dressing application. The in vitro antimicrobial study showed superior antibacterial efficacy (area of inhibition zone of 347.1305 mm^2) against *S. aureus* in comparison to *E. coli* (area of inhibition zone of 269.9637 mm^2) (Wahab et al. 2019). The cellulose/chitosan nanofibers enriched with cinnamon extract fabricated by Kefayat demonstrated significantly higher bactericidal effects against *E. coli* and *S. aureus* than the plain nanofibers with good biocompatibility on L929 normal skin fibroblasts, suggesting the properties of ideal wound dressing (Kefayat et al. 2022).

Yazdanbakhsh and co-workers successfully prepared electrospun cellulose-based nanofibers incorporated with ciprofloxacin hydrochloride confirmed by FTIR for bacterial-infected wounds. The antibacterial studies of ciprofloxacin-loaded nanofibers showed excellent antibacterial activity against *S. aureus*. The in vivo experiments utilizing created lesions on male Wistar rats showed that the ciprofloxacin-loaded cellulose nanofibers significantly accelerated the wound healing process as they are permeable to oxygen and humidity (Yazdanbakhsh et al. 2018). Meamar et al. reported cellulose nanofibers co-loaded with venlafaxine and doxycycline for the treatment of diabetic foot ulcers (DFU) (Meamar et al. 2021). These nanofibers exhibited a high loading efficiency of about 37.8 $\pm$ 1.6% and 48 $\pm$ 1.9% for doxycycline and venlafaxine, respectively. The ex vivo permeation experiments of dual drug-loaded nanofibers on rat skin showed permeation effectiveness of 40% for doxycycline after 7–29 h and 83% for venlafaxine during the 105 h. They also conducted clinical studies on dual drug-loaded cellulose nanofibers in DFU patients. A quicker reduction after 3 months for patients treated with cellulose nanofibers containing venlafaxine and doxycycline than the control group (ordinary wound treatment system) indicated that these nanofibers are potential candidates for diabetic wound care. Moreover, the abnormal responses in the Michigan Neuropathy Screening Instrument (MNSI) survey showed that reduced and pain-free walking distance significantly increased in the patients treated with nanofibers compared to the control group (Meamar et al. 2021) (Table 24.2).

24.6.2 Electrospun Cellulose Acetate-Based Nanofibers

Cellulose acetate (CA) is the acetate ester of cellulose that has been recently evaluated and considered an excellent biomaterial upon producing electrospun nanofibrous wound dressing candidates due to its capacity to improve the cellular interaction between the candidates and skin fibroblast cells (Fischer et al. 2008). The other exciting properties of CA nanofibers that are advantageous in wound care

Table 24.2 Summary of electrospun cellulose nanofibers

Polymers combined with cellulose	Loaded bioactive agents	Main findings	Ref.
Lignin	CuO nanoparticles	Good antibacterial activity with excellent cytocompatibility and non-toxicity	Haider et al. (2021)
–	Cu or Ag nanoparticles	Superior antibacterial activity	Phan et al. (2019)
–	Hydroxyapatite and Ag nanoparticles	Improved cell growth and proliferation	Sofi et al. (2021)
–	Ag nanoparticles	Superior antibacterial efficacy	Wahab et al. (2019)
Chitosan	Cinnamon extract	Good antibacterial activity and biocompatibility	Kefayat et al. (2022)
–	Ciprofloxacin hydrochloride	Acceleration of wound healing process	Yazdanbakhsh et al. (2018)
–	Venlafaxine and doxycycline	Reduced and pain-free walking distance in clinical trials	Meamar et al. (2021)

include biocompatibility, biodegradability, chemical resistance, good hydrolytic stability, high entrapment efficiency, and combined delivery of diverse bioactive agents (Khoshnevisan et al. 2018). In the last decade, numerous preclinical experiments have demonstrated that the electrospun CA-based nanofibers loaded with various therapeutic agents are potential scaffolds for treating and managing injuries. Gomaa et al. fabricated electrospun CA/PLA nanofibers incorporated with thymoquinone (an antimicrobial drug) for wound dressing application (Gomaa et al. 2017). The scanning electron microscope (SEM) analysis of nanofibers displayed round-shaped and bead-free with uniform diameters that ranged between 390 ± 45 and 830 ± 82 nm, mimicking the ECM structure. The in vitro cytotoxicity experiments of thymoquinone-loaded nanofibers displayed more than 97% cell viability of mouse fibroblast cells, indicating excellent cytocompatibility and non-toxicity of CA/PLA nanofibers. The in vivo wound recovery studies employing bulb c male mice wound model showed a faster rate of wound closure for lesions covered with thymoquinone-loaded nanofibers than those treated with reference (gauze) (Gomaa et al. 2017).

The propolis-loaded CA/PCL nanofiber mats prepared by Khoshnevisan and co-workers showed superior antioxidant and antibacterial activity than the free propolis against *S. aureus*, *S. epidermidis*, *P. aeruginosa*, and *E. coli*, suggesting their ability to target prolonged inflammatory phase of the wound recovery process (Khoshnevisan et al. 2019). Liu et al. fabricated CA/zein nanofiber membranes loaded with sesamol to treat diabetic wounds. The hydrophilicity analysis displayed that the water contact angles of nanofibers decreased from 134.2° to 118.1° after incorporating sesamol, suggesting that the hydrophilicity of electrospun nanofibers was improved after the incorporation. The in vivo experiments demonstrated that the nanofiber membranes loaded with high-dose sesamol significantly induced myofibroblast cell development by enhancing the TGF-β signaling pathway transduction and promoting keratinocyte growth cells and inhibiting chronic inflammation in a lesion, consequently improving the process of the wound healing in diabetic mice (Liu et al. 2020). Suteris et al. fabricated electrospun curcumin-loaded cellulose acetate/PCL nanofibers that significantly promoted a higher proliferation and expression of actin in 3T3 mouse fibroblast cells than the plain nanofibers for wound healing applications (Suteris et al. 2022). The electrospun nanofiber mats co-enriched with curcumin and honey were fabricated by Gaydhane et al. for wound recovery application. They demonstrated good antibacterial activity toward *E. coli* with approximately 90% antioxidant efficacy against diphenyl-picrylhydrazyl (DPPH)-free radical, revealing their potential to target the inflammation stage of wound recovery mechanism prolonged in chronic clinical-infected wounds (Gaydhane et al. 2020).

Ullah et al. formulated electrospun CA nanofibrous mats enriched with Manuka honey that demonstrated significantly high porosity that ranged between 85 and 90%, water vapor transmission rate (WVTR) values of 2600 to 1950 g/m^2/day, and excellent cytocompatibility on NIH3T3 mouse fibroblast cells, revealing them as an ideal wound dressing that can lead to suitable wound breathability/moisture and non-toxicity to subsequently result in the accelerated wound healing process (Ullah et al. 2020). Elsayed and co-workers designed electrospun cellulose acetate/PLA mats encapsulated sulfonamide analog for wound dressing application (Elsayed et al. 2020). The antimicrobial experiments of sulfonamide-loaded nanofibers exhibited a superior antibacterial efficacy against *S. aureus* and *E. coli* up to 85%, indicating that sulfonamide-loaded nanofibers are potential scaffolds for microbial-infected injuries. Moreover, the in vivo studies on mice wound model exhibited a significant wound area reduction on the fifth and seventh day of injuries dressed with nanofiber mats, suggesting that the sulfonamide analog has significantly improved the wound recovery process without toxic effect on the developing tissue (Elsayed et al. 2020). The sambong oil-enriched CA nanofibers fabricated by Ullah and co-workers using an electrospinning technique demonstrated good antioxidant activity against the DPPH solution and high bactericidal effects against *E. coli* and *S. aureus* with high cell viability of approximately 92% of NIH3T3 cells in prolonged time of incubation, demonstrating suitable biological activities and cytocompatibility that are suitable for the acceleration of wound recovery process (Ullah et al. 2021).

The antimicrobial analysis of ceftriaxone-incorporated CA/PVA nanofibers reported by Youdhestar et al. demonstrated effective inhibitory activity against *S. aureus* and *E. coli* with ZOI of about 20–25 mm, indicating their suitability for the management of infected injuries (Youdhestar et al. 2022). Aly and Ahmed fabricated CA nanofibers loaded with ZnO nanoparticles and graphene oxide for wound dressing applications. XRD and FTIR analysis confirmed electrospun nanofibers' physicochemical properties and successful fabrication. The in vitro experiments of dual drug-incorporated nanofibers exhibited a viability of about 97.38 ± 3.9% for human fibroblasts with good cell migration and proliferation that can significantly promote wound healing (Aly and Ahmed 2021). Wang et al.

prepared CA/PCL nanofibers co-loaded with Ag nanoparticles and lavender oil for wound management. The in vitro antimicrobial experiments of co-enriched nanofibers using disk diffusion assay showed superior inhibition zone of 20.08 ± 0.63 and 19.75 ± 0.96 mm for *S. aureus* and *E. coli*, respectively, demonstrating synergistic antibacterial activity (Wang et al. 2022). Al-Saeedi and co-workers formulated CA nanofibrous materials containing CuO nanoparticles and showed excellent antibacterial efficacy against *E. coli* and *S. aureus* with good cytocompatibility on HFB4 fibroblasts (Al-saeedi et al. 2021).

The graphene oxide/curcumin/TiO_2-integrated CA nanofibers developed by Prakash and co-workers using electrospinning showed sustained drug release with antibacterial activity against *E. coli*, *S. aureus*, *E. faecalis*, and *P. aeruginosa* that can significantly improve wound recovery and skin regeneration. The in vitro studies of nanofibers exhibited superior cell viability and migration of NIH3T3 fibroblast cells, confirming excellent biocompatibility and higher healing properties (Prakash et al. 2021). Samadian et al. formulated electrospun CA/gelatin nanofibers encapsulated with berberine for diabetic wound management. The berberine-loaded nanofibers showed high % porosity of 78.17 ± 1.04, which can promote cell migration and proliferation. The in vivo studies on the streptozotocin-induced diabetic Wistar rats demonstrated that the berberine-loaded CA/gelatin nanofibers improved the wound healing process (Samadian et al. 2020).

Sharaf et al. formulated CA-based nanofibers incorporated with bioactive glass nanoparticles for wound treatment. The in vivo analysis of CA nanofibers containing bioactive glass nanoparticles employing a diabetic rat model induced with wounds demonstrated efficient acceleration of wound closure by the tenth day (Sharaf et al. 2022). The gallic acid-loaded CA nanofibers reported by Wutticharoenmongkol and co-workers using the electrospinning showed superior scavenging activity against DPPH and good antimicrobial efficacy against *S. aureus* revealing their potential use as wound dressings (Wutticharoenmongkol et al. 2019). The SEM micrographs of gentamicin-loaded CA nanofibers formulated by Bie and co-workers displayed beadless morphology with a fiber diameter of around 320 ± 2 nm, mimicking natural ECM with excellent antibacterial efficacy against *E. coli* and *B. subtilis* (Bie et al. 2020). Antimicrobial efficacy of tetracycline hydrochloride-incorporated CA nanofibers formulated by Gouda et al. exhibited effectiveness of 77–88% and 83–85% bacteria reduction for *E. coli* and *S. aureus*, demonstrating superior antibacterial efficacy (Gouda et al. 2014). The Li and Yang fabricated rutin-loaded CA/PEO nanofibers showed higher antioxidant efficacy of 98.3% and antimicrobial properties of 95.0% and 93.5% against *S. aureus* and *E. coli*, respectively (Bachelor and Yang 2020).

Khan et al. fabricated Ag sulfadiazine-loaded CA nanofibers for bacterial-infected wound treatment. XRD and FTIR analysis confirmed the successful fabrication and physicochemical properties of nanofibers. The in vitro antimicrobial studies of Ag sulfadiazine-loaded CA nanofibers displayed superior antimicrobial efficacy against *B. subtilis* and *E. coli* bacteria with significant sustainability for repetitive use, indicating their suitability for treating wound pathogens that are common in chronic injuries (Khan et al. 2019). The core-shell Au@Se nanoparticles-incorporated CA/polyvinylidene fluoride nanofibers formulated by Aldalbahi and co-workers demonstrated high cell viability of 88.2 ± 4.3% and 91.1 ± 3.4% for HFB4 cells at the both highest and lowest concentrations of nanoparticles, respectively, showing excellent biocompatibility (Aldalbahi et al. 2020). Lei et al. prepared electrospun CA nanofibers containing tirilazad mesylate for wound healing applications (Lei et al. 2022). The in vitro drug-release profile exhibited initial burst release of tirilazad mesylate at physiological conditions followed by sustained drug release. The cytotoxicity analysis utilizing MTT assay exhibited high cell viability of L929 cells when cultured with tirilazad mesylate-encapsulated CA nanofibers, indicating superior cytocompatibility. The results of wound size closure analysis using a diabetic rat model showed that the tirilazad mesylate-loaded CA nanofibers closed the injuries up to 75.19 ± 3.28% and 94.79 ± 2.61% on days 7 and 14, respectively, while the plain CA nanofibers showed wound closure of 58.59 ± 4.19 and 75.08 ± 3.05% on days 7 and 14, respectively, demonstrating that loading of tirilazad mesylate into nanofibers effectively promoted wound healing process (Lei et al. 2022) (Table 24.3).

24.6.3 Carboxymethyl Cellulose-Based Nanofibers Loaded with Bioactive Agents

Carboxymethyl cellulose (CMC) is a cellulose derivative made up of carboxymethyl groups (–CH_2-COOH) bound to a backbone of cellulose. This derivative is highly water-soluble with a diversity of applications because of its biodegradable and biocompatible nature (Elshazly et al. 2019). Fabrication of nanofibers using the electrospinning method from plain CMC is difficult due to its rigid structure that does not permit chain entanglement (El-Newehy et al. 2016). Therefore, combining CMC with other polymers, especially biocompatible and non-toxic synthetic polymers, is a way out of this issue. Electrospun CMC-based nanofibers are gaining popularity because of their unique features, such as hydrophilicity, bioadhesive, biocompatibility, good biodegradability, and non-toxicity (Zhang et al. 2020). Some research reports revealed the potential of CMC-based nanofibers containing bioactive agents for treating injuries.

Alipour et al. prepared electrospun CMC/PVA composite nanofibers incorporated with Ag sulfadiazine for wound management (Alipour et al. 2019). The successful preparation and physicochemical properties of nanofibers were confirmed by FTIR, XRD, atomic absorption spectroscopy (AAS), and thermogravimetric analysis (TGA). The in vitro drug-release kinetics at physiological conditions exhibited that Ag sulfadiazine's release was mainly controlled by its diffusion from the polymer surface. The cytotoxicity analysis using MTT assay showed that composite nanofibers containing Ag sulfadiazine possessed more than 85% cell viability of HSF-PI 18 fibroblast cells after 48 h of incubation, suggesting excellent cytocompatibility and non-toxicity of nanofibers. The in vitro antimicrobial studies of Ag sulfadiazine-loaded CMC/PVA nanofibers exhibited significant antibacterial efficacy against both *P. aeruginosa* and *S. aureus* than pristine

Table 24.3 Summary of electrospun CA nanofibers

Polymers combined with CA	Loaded bioactive agents	Main findings	Ref.
PLA	Thymoquinone	An accelerated rate of wound recovery	Gomaa et al. (2017)
PCL	Propolis	Superior antioxidant activity and antibacterial activity	Khoshnevisan et al. (2019)
Zein	Sesamol	Induced development and growth of skin cells	Liu et al. (2020)
PCL	Curcumin	Induced proliferation and expression of actin in fibroblast cells	Suteris et al. (2022)
–	Curcumin and honey	Good antibacterial and antioxidant activity	Gaydhane et al. (2020)
–	Manuka honey	High porosity, moderate WVTR, and excellent cytocompatibility	Ullah et al. (2020)
PLA	Sulfonamide analog	The accelerated wound healing process and superior antibacterial activity	Elsayed et al. (2020)
–	Sambong oil	Good antibacterial and antioxidant activity with excellent cytocompatibility	Ullah et al. (2021)
PVA	Ceftriaxone	Effective inhibitory efficacy against bacterial strains	Youdhestar et al. (2022)
–	ZnO nanoparticles and graphene oxide	High cell viability, proliferation, and migration	Aly and Ahmed (2021)
PCL	Ag nanoparticles and lavender oil	Synergistic antibacterial activity	Wang et al. (2022)
–	CuO nanoparticles	Excellent antibacterial activity and cytocompatibility	Al-saeedi et al. (2021)
	Graphene oxide, curcumin, and TiO2	Good antibacterial efficacy and improved wound recovery and skin regeneration	Prakash et al. (2021)
Gelatin	Berberine	Promoted cell migration and proliferation with the enhanced process of diabetic wound healing	Samadian et al. (2020)
–	Bioactive glass nanoparticles	Acceleration of diabetic wound closure	Sharaf et al. (2022)
–	Gallic acid	Superior antioxidant activity and good antimicrobial efficacy	Wutticharoenmongkol et al. (2019)

(continued)

Table 24.3 (continued)

Polymers combined with CA	Loaded bioactive agents	Main findings	Ref.
–	Gentamicin	Beadless morphology mimicking natural ECM and excellent antibacterial efficacy	Bie et al. (2020)
–	Tetracycline hydrochloride	Superior antibacterial efficacy	Gouda et al. (2014)
PEO	Rutin	Higher antioxidant efficacy and good antimicrobial properties	Bachelor and Yang (2020)
–	Ag sulfadiazine	Superior antimicrobial efficacy	Khan et al. (2019)
Polyvinylidene fluoride	Core-shell Au@Se nanoparticles	Excellent biocompatibility	Aldalbahi et al. (2020)
–	Tirilazad mesylate	Superior cytocompatibility and induced wound recovery	Lei et al. (2022)

nanofibers, revealing that Ag sulfadiazine-loaded CMC/PVA nanofibers are effective antibacterial wound dressing materials to protect injuries from known pathogens. The in vivo experiments showed that wounds on rabbits treated with Ag sulfadiazine-loaded CMC/PVA nanofibers were faster when compared to those treated with control (plain CMC/PVA nanofibers) (Alipour et al. 2019).

The CMC/PVA nanofibers formulated by Allafchian and co-workers showed an initial burst release of flufenamic acid (an antioxidant and analgesic drug) followed by sustained release, revealing their potential application for treating injury pain and inflammation (Allafchian et al. 2020). Kazeminava et al. reported CMC/PVA nanofibers incorporated with colistin as the potential antibacterial wound dressing materials. The in vitro cytocompatibility experiments showed more than 80% cell viability of human skin fibroblast cell line (HFF-1 cells) when incubated with colistin-loaded nanofibers, revealing that these nanofibers could potentially be regarded as safe wound dressing candidates. Moreover, the in vitro antibacterial study employing disk diffusion procedures demonstrated that blank nanofibers (reference) revealed no inhibitory efficacy. In contrast, the colistin-loaded nanofibers displayed superior antibacterial activity against various bacteria strains (*S. aureus*, *K. pneumoniae*, *P. aeruginosa*, and *E. coli*) than chitosan films used as the control sample, demonstrating that colistin-loaded nanofibers can be utilized as effective antibacterial wound dressings (Kazeminava et al. 2022) (Table 24.4).

24.6.4 Ethyl Cellulose-Based Nanofibers Loaded with Bioactive Agents

Ethyl cellulose (EC) is an inert water-insoluble hydrophobic cellulose derivative used to develop sustained drug-release scaffolds (Park et al. 2015). Among the

Table 24.4 Summary of electrospun CMC nanofibers

Polymers combined with CMC	Loaded bioactive agents	Main findings	Ref.
PVA	Ag sulfadiazine	Excellent cytocompatibility, non-toxicity, and faster recovering	Alipour et al. (2019)
PVA	Flufenamic acid	Initial rapid drug-release mechanism followed by sustained release	Allafchian et al. (2020)
PVA	Colistin	Good antibacterial activity and cytocompatibility	Kazeminava et al. (2022)

numerous biomaterials electrospun into nanofibers, EC is a perfect candidate as an amalgamation material due to its film-forming capability, high flexibility, non-toxicity, hydrophobicity, and thermoplasticity for application in pharmaceutics, food, and microencapsulation (Wasilewska 2019). This cellulose derivative can be used for the formulation of electrospun nanofibers for the treatment of wounds due to the properties mentioned above. Ahmadian et al. formulated EC/PLA/collagen hybrid nanofiber mats loaded with Ag sulfadiazine for infected wound care (Ahmadian et al. 2020). The mechanical characterization of nanofibers displayed a tensile strength that ranges between 5.90 ± 0.11 MPa and 10.32 ± 0.65 MPa, depending on the increasing weight of Ag sulfadiazine, indicating their compatibility to the human skin and easy application during wound dressing. The in vitro experiments exhibited enhanced cell proliferation of NIH 3T3 cells when incubated with Ag sulfadiazine-loaded nanofibers, suggesting that Ag sulfadiazine did not cause any toxicity on the cells. The antibacterial studies showed that nanofibers incorporated with higher content of Ag sulfadiazine possessed superior ZOI than nanofibers containing less Ag sulfadiazine for both *E. coli* and *Bacillus* (Ahmadian et al. 2020).

Godakanda et al. prepared EC/PVP nanofibers encapsulated with naproxen (an anti-inflammatory agent) using electrospinning for wound treatment. The SEM micrographs of nanofibers displayed smooth and cylindrical morphologies with average diameters ranging from 647 to 802 nm, imitating the ECM. The in vitro drug-release profile exhibited tunable release of naproxen from nanofibers with zero-order drug release over 20 and 80 h, showing that these nanofibers possess great potential in managing pain and inflammation (Godakanda et al. 2019). The electrospun EC/PVP nanofibers fabricated by Li and co-workers for drug delivery of ciprofloxacin (a broad-spectrum antibiotic) displayed high cell growth and proliferation of human dermal fibroblasts (HDFs), indicating excellent cytocompatibility. The antimicrobial studies demonstrated that the development of *E. coli* and *S. aureus* could be effectively hindered due to the presence of ciprofloxacin in the EC/PVP nanofibers. These results revealed that ciprofloxacin-loaded nanofibers are potential antibacterial wound dressing scaffolds with excellent biocompatibility and reduced toxicity, which are the properties of ideal wound dressings (Li et al. 2019).

Ghorbani and co-workers fabricated EC/gum tragacanth nanofibers loaded honey via electrospinning technique for wound management (Ghorbani et al. 2021). The

Table 24.5 Summary of electrospun EC nanofibers

Polymers combined with EC	Loaded bioactive agents	Main findings	Ref.
PLA and collagen	Ag sulfadiazine	Excellent mechanical properties, enhanced cell proliferation, and good antibacterial activity	Ahmadian et al. (2020)
PVP	Naproxen	Tunable drug release with zero-order drug release over a prolonged time	Godakanda et al. (2019)
PVP	Ciprofloxacin	High cell growth and proliferation and good antibacterial activity	Li et al. (2019)
Gum tragacanth	Honey	Suitable biological activities and improved cell proliferation and adhesion	Ghorbani et al. (2021)

in vitro antioxidant experiments exhibited that the honey-enriched nanofibers significantly demonstrated strong radical scavenging activity against DPPH, which might be due to the plentiful phenolic antioxidants of honey, revealing their potential to control the overproduction of ROSs that is strongly associated with chronic injuries and extended inflammation. The cell proliferation analysis displayed that NIH3T3 fibroblasts adhered well to the honey-enriched nanofibers' surface, revealing good interaction between the skin cells and nanofibers. The incorporation of honey into the electrospun EC/gum tragacanth nanofibers resulted in superior antibacterial activity against *E. coli* and *S. aureus*. In contrast, the unmodified nanofibers did not show any inhibition effect against these strains. This study showed that the fabricated honey-enriched nanofibers could be excellent candidates for wound treatment and tissue regeneration due to their non-toxicity, good biocompatibility, and promising biological activities (antibacterial and antioxidant properties) (Ghorbani et al. 2021) (Table 24.5).

24.6.5 Other Electrospun Cellulose Derivatives-Based Nanofibers Loaded with Bioactive Agents

The electrospun nanofibers containing bioactive agents can be fabricated from the other cellulose derivatives for wound dressing applications. Those derivatives include NaCMC, MC, hydroxypropyl cellulose (HPC), hydroxyethyl cellulose (HEC), ethyl hydroxyethyl cellulose (EHEC), HPMC, and HEMC. HEC is a non-ionic moderately substituted hydroxyethyl ether of cellulose (Rošic et al. 2011). This water-soluble cellulose derivative is used as a binder, thickener, stabilizer, coating agent, dispersing agent, stabilizer, and emulsifier in the pharmaceutical, cosmetic, food, and biotechnological industries because of its reduced toxicity, splendid biocompatibility, and non-susceptibility to immunogenic responses (El Fawal et al. 2018). Ullah et al. formulated electrospun HEC/soy protein isolate nanofibers co-loaded with diclofenac sodium and halloysite nanotubes to manage microbial-infected wounds. The in vitro drug-release studies at physiological conditions exhibited sustained release of diclofenac sodium from the nanofibers for over 2 weeks, showing their effectiveness in protecting the wound from bacterial invasion. The antimicrobial analysis of dual drug-loaded nanofibers using agar disk diffusion procedure showed higher inhibition against *B. subtilis* and *E. coli* than the plain nanofibers with excellent cytocompatibility on COS-7 cells. These nanofibers exhibited the properties of an ideal wound dressing (Ullah et al. 2022).

HPC is also a nonionic derivative of cellulose that can be produced by substituting with hydroxypropyl ether groups. HPC is widely utilized in wound dressing and tissue regeneration for drug delivery applications. Nevertheless, as far as we know, there are few research studies in the literature associated with the electrospinning of HPC, especially the loading of bioactive agents into the electrospun HPC nanofibers (Aytac et al. 2015). Aytac et al. fabricated HPC/hydroxypropyl-beta-cyclodextrin nanofibers incorporated with sulfisoxazole. The SEM images of sulfisoxazole-loaded nanofibers showed bead-free and uniform nanofibers with a diameter of about 60 ± 25 nm, imitating the natural ECM that supports the growth of skin cells and restores injured tissues. The in vitro drug-release kinetics showed an initial burst release of sulfisoxazole from nanofibers followed by sustained release, demonstrating that these nanofiber scaffolds can significantly destroy microorganisms at the wound bed and protect the wound from further microbial invasions (Aytac et al. 2015).

EHEC is a nonionic cellulose ether used in various applications, such as rheology control agents in detergents, paints, cosmetics, and oil recovery. This cellulose derivative is also commonly employed in biomedical fields because of its biodegradability, biocompatibility, and non-toxicity. The electrospinning of pristine EHEC solution is problematic because it displays a complex solution behavior with the inherent feature of temperature-stimulated phase separation. Hence, using synthetic polymers as co-spinning biomaterials in the electrospinning procedure can solve this shortcoming (Wali et al. 2019). Wali and co-workers prepared electrospun EHEC/PVA nanofibers incorporated with gentamicin sulfate and halloysite clay for an accelerated wound recovery process. The XRD and FTIR spectroscopy confirmed the successful formulation of electrospun EHEC/PVA nanofibers containing gentamicin sulfate and halloysite clay. The mechanical analysis showed that the incorporation of halloysite clay into the nanofibers significantly improved the mechanical properties of nanofiber wound dressings. The in vitro cytotoxicity studies showed more than 100% cell viability of L929 fibroblast cells over 5 days, revealing the excellent cytocompatibility and cell adhesion, migration, and proliferation on the surface of dual drug-loaded EHEC/PVA nanofibers. The in vivo wound healing studies of dual drug-loaded EHEC/PVA nanofibers using a rat wound model showed wound closure of 100% on the 14th day. In comparison, injuries treated with marketed ointment (povidone-iodine used as a control) demonstrated only 84% wound closure, revealing the capability of EHEC/PVA nanofibers to accelerate the wound healing process (Wali et al. 2019) (Table 24.6).

Table 24.6 Summary of electrospun nanofibers from other cellulose derivatives

Polymers used	Loaded bioactive agents	Main findings	Ref.
HEC and soy protein	Diclofenac sodium and halloysite nanotubes	Sustained drug release and higher inhibition against bacteria strains	Ullah et al. (2022)
HPC and hydroxypropyl-beta-cyclodextrin	Sulfisoxazole	Initial burst release of sulfisoxazole followed by the sustained drug release	Aytac et al. (2015)
EHEC and PVA	Gentamicin sulfate and halloysite clay	Excellent cytocompatibility, cell attachment, migration, and proliferation enhance wound healing properties	Wali et al. (2019)

24.7 Commercially Available Cellulose-Based Wound Dressing Products

Very few wound dressing products are fabricated from cellulose and its derivatives, which are available on the market. An example of a commercial cellulose wound dressing product is Aquacel® Hydrofiber dressing, an antimicrobial carboxymethyl derivative of cellulose. On the wound bed, it absorbs exudate and forms a gel-like network that sustains moisture in the wound environment and subsequently accelerates the mechanism of wound recovery (Hussain et al. 2017). Another commercially available cellulose wound dressing product is Hcel® NaT. This product possesses excellent cell proliferation and adhesion characteristics on HDF cells using fibrin. Some studies demonstrated that the porous forms of Hcel® NaT have better cell proliferation and adhesion properties when compared to the homogeneous Hcel® NaT forms (Bacakova et al. 2018).

24.8 Conclusion and Future Perspective

The electrospun scaffolds containing bioactive agents formulated from cellulose and its derivative are promising candidates for wound treatment. These scaffolds have demonstrated the properties of ideal wound dressing, such as non-toxicity and excellent biocompatibility, by showing high cell viability of various skin cells, enhanced drug delivery, and good antibacterial and antioxidant properties. The combination of cellulose with other polymers significantly improved the mechanical performance of nanofiber dressings, suggesting their compatibility with human skin and easy application during wound dressing. These preclinical outcomes demonstrate electrospun cellulose-based nanofiber scaffolds loaded with bioactive agents possess the potential to solve several wound-associated issues in clinical applications and enhance wound management in the future. Undoubtedly, the fabrication and exploration of cellulose and its derivatives-based nanofibers and their extended advances remain crucial emphases of future research.

Acknowledgments The financial support of the South African Medical Research Council (SAMRC) and National Research Foundation (NRF), South Africa, toward this research is hereby acknowledged.

References

Abrigo M, McArthur S, Kingshott P (2014) Electrospun nanofibers as dressings for chronic wound care: advances, challenges, and future prospects. Macromol Biosci 14:772–792

Ahmadian S, Ghorbani M, Mahmoodzadeh F (2020) Silver sulfadiazine-loaded electrospun ethyl cellulose/polylactic acid/collagen nanofibrous mats with antibacterial properties for wound healing. Int J Biol Macromol 162:1555–1565. https://doi.org/10.1016/j.ijbiomac.2020.08.059

Aldalbahi A, El-naggar ME, Ahmed MK, Periyasami G, Rahaman M, Menazea AA (2020) Core-shell Au@Se nanoparticles embedded in cellulose acetate/polyvinylidene fluoride scaffold for wound healing. J Mater Res Technol 9(6):15045–15056. https://doi.org/10.1016/j.jmrt.2020.10.079

Alipour R, Khorshidi A, Shojaei AF, Mashayekhi F (2019) Silver sulfadiazine-loaded PVA/CMC nanofibers for the treatment of wounds caused by excision. Fibers Polym 20(12):2461–2469

Allafchian A, Hosseini H, Allafchian MS (2020) Electrospinning of PVA-carboxymethyl cellulose nanofibers for flufenamic acid drug delivery. Int J Biol Macromol 163:1780–1786

Al-saeedi SI, Al-kadhi NS, Al-senani GM, Almaghrabi OA, Nafady A (2021) Antibacterial potency, cell viability and morphological implications of copper oxide nanoparticles encapsulated into cellulose acetate nanofibrous scaffolds. Int J Biol Macromol 182:464–471. https://doi.org/10.1016/j.ijbiomac.2021.04.013

Alven S, Aderibigbe BA (2020) Chitosan and cellulose-based hydrogels for wound management. Internatinal J Mol Sci 21:9656

Alven S, Nqoro X, Aderibigbe B (2020) Polymer-based materials loaded with curcumin for wound healing application. Polymers (Basel) 12:2286

Alven S, Buyana B, Feketshane Z, Aderibigbe BA (2021a) Electrospun nanofibers/nanofibrous scaffolds loaded with silver nanoparticles as effective antibacterial wound dressing materials. Pharmaceutics 13:964

Alven S, Khwaza V, Oyedeji O, Aderibigbe B (2021b) Polymer-based scaffolds loaded with aloe vera extract for the treatment of wounds. Pharmaceutics 13:961

Alven S, Peter S, Aderibigbe B (2022a) Polymer-based hydrogels enriched with essential oils: a promising approach for the treatment of infected wounds. Polymers (Basel). 14:3772

Alven S, Peter S, Mbese Z, Aderibigbe BA (2022b) Polymer-based wound dressing materials loaded with bioactive agents: potential materials for the treatment of diabetic wounds. Polymers (Basel). 14:724

Aly AA, Ahmed MK (2021) Nanofibers of cellulose acetate containing ZnO nanoparticles/graphene oxide for wound healing applications. Int J Pharm 598:120325. https://doi.org/10.1016/j.ijpharm.2021.120325

Ambekar RS, Kandasubramanian B (2019) Advancements in nanofibers for wound dressing: a review. Eur Polym J 117:304–336

Aytac Z, Sener H, Durgun E, Uyar T (2015) Sulfisoxazole/cyclodextrin inclusion complex incorporated in electrospun hydroxypropyl cellulose nanofibers as drug delivery system. Colloids Surfaces B Biointerfaces 128:331–338. https://doi.org/10.1016/j.colsurfb.2015.02.019

Bacakova M, Pajorova J, Sopuch T, Bacakova L (2018) Fibrin-modified cellulose as a promising dressing for accelerated wound healing. Materials (Basel) 11:2314. https://doi.org/10.3390/ma11112314

Bachelor BL, Yang X (2020) Rutin-loaded cellulose acetate/poly (ethylene oxide) fiber membrane fabricated by electrospinning: a bioactive material. Mater Sci Eng C 109:110601. https://doi.org/10.1016/j.msec.2019.110601

Baldwin A, Kiick K (2010) Polysaccharide-modified synthetic polymeric biomaterials. Biopolymers 94:128–140

Barrientos S, Brem H, Stojadinovic O, Tomic-Canic M (2014) Clinical application of growth factors and cytokines in wound healing. Wound Repair Regen 22(5):569–578

Bie X, Khan MQ, Ullah A, Ullah S, Kharaghani D, Phan D-N et al (2020) Fabrication and characterization of wound dressings containing gentamicin/silver for wounds in diabetes mellitus patients fabrication and characterization of wound dressings containing gentamicin/silver for wounds in diabetes mellitus patients. Mater Res Express 7:045004

Cañedo-Dorantes L, Cañedo-Ayala M (2019) Skin acute wound healing: a comprehensive review. Int J Inflamm 2019:3706315

De Sousa Moraes PRF, Saska S, Barud H, De Lima LR, Da Conceicao Amaro Martins V, De Guzzi Plepis AM et al (2016) Bacterial cellulose/collagen hydrogel for wound healing. Mater Res 19(1):106–116

Desmet C, Preat V, Gallez B (2018) Nanomedicines and gene therapy for the delivery of growth factors to improve perfusion and oxygenation in wound healing. Adv Drug Deliv Rev 129:262–284

Dhivya S, Padma V, Santhini E (2015) Wound dressings—a review. Biomedicine 5:24–28

El Fawal G, Abu-Serie M, Hassan M, Elnouby M (2018) Hydroxyethyl cellulose hydrogel for wound dressing: fabrication, characterization, and in vitro evaluation. Int J Biol Macromol 111:649–659. https://doi.org/10.1016/j.ijbiomac.2018.0

El-Ashram S, El-Samad L, Basha A, El Wakil A (2021) Naturally-derived targeted therapy for wound healing: beyond classical strategies. Pharmacol Res 170:105749

El-Newehy M, El-Naggar M, Alotaiby S, El-Hamshary H, Moydeen M, Al-Deyab S (2016) Preparation of biocompatible system based on electrospun CMC/PVA nanofibers as controlled release carrier of diclofenac sodium. J Macromol Sci Part A 53:566–573. https://doi.org/10.1080/10601325.2016.120

Elsayed RE, Madkour TM, Azzam RA (2020) Tailored-design of electrospun nanofiber cellulose acetate/poly (lactic acid) dressing mats loaded with a newly synthesized sulfonamide analog exhibiting superior wound healing. Int J Biol Macromol 164:1984–1999. https://doi.org/10.1016/j.ijbiomac.2020.07.316

Elshazly E, Yu L, Zhang Y, Wang H, Chen K, Zhang S et al (2019) Fabrication of folate-phytosterol-carboxymethyl cellulose nanoparticles derived from plant material as carrier of anticancer drug. Micro Nano Lett 14:1111–1116. https://doi.org/10.1049/mnl.2019.0080

Fahimirad S, Ajalloueian F (2019) Naturally-derived electrospun wound dressings for target delivery of bioactive agents. Int J Pharm 566:307–328

Felgueiras H, Amorim M (2017) Functionalization of electrospun polymeric wound dressings with antimicrobial peptides. Coll Surf B Biointer 156:133–148

Fischer S, Thümmler K, Volkert B, Hettrich K, Schmidt I, Fischer K (2008) Properties and applications of cellulose acetate. Macromol Symp 262:89–96

Gaydhane MK, Kanuganti JS, Sharma CS (2020) Honey and curcumin loaded multilayered polyvinylalcohol/cellulose acetate electrospun nanofibrous mat for wound healing. J Mater Res 35:600–609. https://doi.org/10.1557/jmr.2020.52

Ghorbani M, Ramezani S, Rashidi M (2021) Fabrication of honey-loaded ethylcellulose/gum tragacanth nanofibers as an effective antibacterial wound dressing. Colloids Surfaces A Physicochem Eng Asp 621:126615. https://doi.org/10.1016/j.colsurfa.2021.126615

Godakanda VU, Li H, Alquezar L, Zhao L, Zhu L, De Silva R et al (2019) Tunable drug release from blend poly(vinyl pyrrolidone)-ethyl cellulose nanofibers. Int J Pharm 562:1272–1179. https://doi.org/10.1016/j.ijpharm.2019.03.035

Gomaa SF, Madkour TM, Moghannem S, El-sherbiny IM (2017) New polylactic acid/cellulose acetate-based antimicrobial interactive single dose nanofibrous wound dressing mats. Int J Biol Macromol 105:1148–1160. https://doi.org/10.1016/j.ijbiomac.2017.07.145

Gouda M, Hebeish AA, Aljafari AI (2014) Synthesis and characterization of novel drug delivery system based on cellulose acetate electrospun nanofiber mats. J Ind Text 43(3):319–329

Gueldner J, Zhang F, Zechmann B, Bruce E (2017) Evaluating a novel oxygenating therapeutic for its potential use in the advancement of wound healing. Toxicol Vitr 43:62–68

Gupta K, Haider A, Choi Y, Kang I (2014) Nanofibrous scaffolds in biomedical applications. Biomater Res 18:5

Haider MK, Ullah A, Nauman M, Saito Y, Sun L, Park S et al (2021) Lignin-mediated in-situ synthesis of CuO nanoparticles on cellulose nanofibers: a potential wound dressing material. Int J Biol Macromol 173:315–326. https://doi.org/10.1016/j.ijbiomac.2021.01.050

Helenius G, Backdahl H, Bodin A, Nannmark U, Gatenholm P, Risberg B (2006) In vivo biocompatibility of bacterial cellulose. J Biomed Mater Res Part A 76A:431–438

Heydari P, Varshosaz J, Kharazi A, Karbasi S (2018) Preparation and evaluation of poly glycerol sebacate/poly hydroxy butyrate core—shell electrospun nanofibers with sequentially release of ciprofloxacin and simvastatin in wound dressings. Polym Advaced Technol 2018:1–9

Huang S, Fu X (2010) Naturally derived materials-based cell and drug delivery systems in skin regeneration. J Control Release 142:149–159

Hussain Z, Thu HE, Shuid AN, Katas H, Hussain F (2017) Recent advances in polymer-based wound dressings for the treatment of diabetic foot ulcer: an overview of state-of-the-art. Curr Drug Targets 18:527–550

Iacob A, Dragan M, Ionescu O, Profire L, Ficai A, Andronescu E, Confederat LG, Lupascu D (2020) An overview of biopolymeric electrospun nanofibers based on polysaccharides for wound healing management. Pharmaceutics 12:983

Kalantari K, Afifi A, Jahangirian H, Webster T (2019) Biomedical applications of chitosan electrospun nanofibers as a green polymer—review. Carbohydr Polym 207:588–600

Kanikireddy V, Varaprasad K, Jayaramudu T, Karthikeyan C, Sadiku R (2020) Carboxymethyl cellulose-based materials for infection control and wound healing: a review. Int J Biol Macromol 164:963–975

Kazeminava F, Javanbakht S, Nouri M, Adibkia K, Ganbarov K, Yousefi M et al (2022) Electrospun nanofibers based on carboxymethyl cellulose/polyvinyl alcohol as a potential antimicrobial wound dressing. Int J Biol Macromol 214:111–119. https://doi.org/10.1016/j.ijbiomac.2022.05.175

Kefayat A, Farahani RH, Farahani M, Hezrati E, Yekta NH (2022) Synthesis and characterization of cellulose nanofibers/chitosan/cinnamon extract wound dressing with significant antibacterial and wound healing properties. J Iran Chem Soc 19:1191–1202. https://doi.org/10.1007/s13738-021-02374-x

Kenawy E, Layman J, Wartkins G, Bowlin J, Matthews A, Simpson A et al (2003) Electrospinning of poly(ethylene-covinyl alcohol) fibers. Biomaterials 24:907

Khan MQ, Kharaghani D, Sanaullah, Shahzad A, Saito Y, Yamamoto T et al (2019) Fabrication of antibacterial electrospun cellulose acetate/silver-sulfadiazine nanofibers composites for wound dressings applications. Polym Test 74:39–44. https://doi.org/10.1016/j.polymertesting.2018.12.015

Khoshnevisan K, Maleki H, Samadian H, Shahsavari S, Sarrafzadeh M, Larijani B et al (2018) Cellulose acetate electrospun nanofibers for drug delivery systems: applications and recent advances. Carbohydr Polym 198:131–141

Khoshnevisan K, Maleki H, Samadian H, Doostan M, Khorramizadeh MR (2019) Antibacterial and antioxidant assessment of cellulose acetate/polycaprolactone nanofibrous mats impregnated with propolis. Int J Biol Macromol 140:1260–1268. https://doi.org/10.1016/j.ijbiomac.2019.08.207

Last I, Levy Y, Jortner J (2002) Beyond the Rayleigh instability limit for multicharged finite systems: from fission to coulomb explosion. Proc Natl Acad Sci U S A 99:9107

Lei L, Huang W, Liu K, Liu X, Dai M, Liu Z et al (2022) Trilazad mesylate-loaded electrospun cellulose acetate nanofibrous wound dressings promote diabetic wound healing by modulation of immune response and protection against oxidative damage. J Drug Deliv Sci Technol. 69: 102863. https://doi.org/10.1016/j.jddst.2021.102863

Li H, Zhang Z, Godakanda VU, Chiu Y, Angkawinitwong U, Patel K et al (2019) The effect of collection substrate on electrospun ciprofloxacin-loaded poly (vinylpyrrolidone) and ethyl cellulose nanofibers as potential wound dressing materials. Mater Sci Eng C 104:109917. https://doi.org/10.1016/j.msec.2019.109917

Liu M, Duan X, Li Y, Yang D, Long Y (2017) Electrospun nanofibers for wound healing. Mater Sci Eng C 76:1413–1423

Liu F, Li X, Wang L, Yan X, Ma D, Liu Z et al (2020) Sesamol incorporated cellulose acetate-zein composite nanofiber membrane: an efficient strategy to accelerate diabetic wound healing. Int J Biol Macromol 149:627–638. https://doi.org/10.1016/j.ijbiomac.2020.01.277

Malafaya PB, Silva GA, Reis R (2007) Natural-origin polymers as carriers and scaffolds for biomolecules and cell delivery in tissue engineering applications. Adv Drug Deliv Rev 59: 207–233

Mano J, Silva G, Azevedo H et al (2007) Natural origin biodegradable systems in tissue engineering and regenerative medicine: present status and some moving trends. J R Soc Interface 4:999–1030

Marin S, Albu Kaya M, Ghica M, Dinu-Pirvu C, Popa L, Udeanu D et al (2018) Collagen-polyvinyl alcohol-indomethacin biohybrid matrices as wound dressings. Pharmaceutics. 10:224

Martinengo L, Olsson M, Bajpai R, Soljak M, Upton Z, Schmidtchen A, Car J, Jarbrink K (2019) Prevalence of chronic wounds in the general population: systematic review and meta-analysis of observational studies. Ann Epidemiol 29:8–15

Massee M, Chinn K, Lim J, Godwin L, Young C, Koob T (2015) Type I and II diabetic adipose-derived stem cells respond in vitro to dehydrated human amnion/chorion membrane allograft treatment by increasing proliferation, migration, and altering cytokine secretion. Adv Wound Care 2015:5

Mbese Z, Alven S, Aderibigbe B (2021) Collagen-based nanofibers for skin regeneration and wound. Polymers (Basel). 13:4368

McDaniel J, Browning K (2014) Smoking, chronic wound healing, and implications for evidence-based practice. J Wound Ostomy Cont Nurs 41:415–423

Meamar R, Chegini S, Varshosaz J, Aminorroaya A, Amini M, Siavosh M (2021) Alleviating neuropathy of diabetic foot ulcer by codelivery of venlafaxine and matrix metalloproteinase drug—loaded cellulose nanofiber sheets: production, in vitro characterization and clinical trial. Pharmacol Rep 73:806–819. https://doi.org/10.1007/s43440-021-00220-8

Mir M, Ali M, Barakullah A, Gulzar A, Arshad M, Fatima S et al (2018) Synthetic polymeric biomaterials for wound healing: a review. Prog Biomater 1:1–21

Mutlu G, Calamak S, Ulubayram K, Guven E (2018) Curcumin-loaded electrospun PHBV nanofibers as potential wound-dressing material. J Drug Deliv Sci Technol. 43:185–193

Naseri-nosar M, Maria Z (2018) Wound dressings from naturally-occurring polymers: a review on homopolysaccharide-based composites. Carbohydr Polym 189:379–398

Okonkwo U, DiPietro L (2017) Diabetes and wound angiogenesis. Int J Mol Sci 18:1419

Park J, Kim J, Lee I (2015) Fabrication and characterization of antimicrobial ethyl cellulose nanofibers using electrospinning techniques. J Nanosci Nanotechnol 15(8):5672–5675

Patel S, Srivastava S, Rawat M, Singh D (2018) Preparation and optimization of chitosan-gelatin films for sustained delivery of lupeol for wound healing. Int J Biol Macromol 107:1888–1897

Patil P, Reagan M, Bohara R (2020) Silk fibroin and silk-based biomaterial derivatives for ideal wound dressings. Int J Biol Macromol 164:4613–4627

Pawar H, Tetteh J, Boateng J (2013) Preparation, optimisation and characterisation of novel wound healing film dressings loaded with streptomycin and diclofenac. Colloid Surf B Biointerfaces 102:102–110

Phan D-N, Dorjjugder N, Khan MQ, Saito Y, Taguchi G, Mukai Y et al (2019) Synthesis and attachment of silver and copper nanoparticles on cellulose nanofibers and comparative antibacterial study. Cellulose 2019(26):6629–6640. https://doi.org/10.1007/s10570-019-02542-6

Powers J, Higham C, Broussard K, Phillips T (2016) Wound healing and treating wounds chronic wound care and management. J Am Acad Dermatol 74:607–625

Prakash J, Venkataprasanna KS, Bharath G, Banat F, Niranjan R, Venkatasubbu DG (2021) In-vitro evaluation of electrospun cellulose acetate nanofiber containing graphene oxide/TiO 2/curcumin for wound healing application. Colloids Surfaces A Physicochem Eng Asp 627:127166. https://doi.org/10.1016/j.colsurfa.2021.127166

Ranjith R, Balraj S, Ganesh J, Milton M (2019) Therapeutic agents loaded chitosan-based nanofibrous mats as potential wound dressings: a review. Mater Today Chem 12:386–395

Rezvanian M, Ahmad N, Cairul M, Mohd I, Ng S (2017) Optimization, characterization, and in vitro assessment of alginate-pectin ionic cross-linked hydrogel film for wound dressing applications. Int J Biol Macromol 97:131–140

Rošic R, Kocbek P, Baumgartner S, Kristl J (2011) Electro-spun hydroxyethyl cellulose nanofibers: the relationship between structure and process. J Drug Deliv Sci Technol 21:229–236. https://doi.org/10.1016/S1773-2247(11)500

Samadian H, Zamiri S, Ehterami A, Farzamfar S, Vaez A, Khaster H et al (2020) Electrospun cellulose acetate/gelatin nanofibrous wound dressing containing berberine for diabetic foot ulcer healing: in vitro and in vivo studies. Sci Rep 10:8312

Sen C (2019) Human wounds and its burden: an updated compendium of estimates. Adv Wound Care 8:39–48

Serra M, Barroso W, Silva N, Silva S, Borges A, Abreu I et al (2017) From inflammation to current and alternative therapies involved in wound healing. Int J Inflamm 2017:3406215

Sharaf SS, El-Shafei AM, Refaie R, Gabriel A, Abdel-Sattar R (2022) Antibacterial and wound healing properties of cellulose acetate electrospun nanofibers loaded with bioactive glass nanoparticles; in-vivo study. Cellulose 29:4565–4577. https://doi.org/10.1007/s10570-022-04570-1

Shimizu R, Kishi K (2012) Skin graft. Plast Surg Int 2012:563493

Sofi HS, Akram T, Shabir N, Vasita R, Jadhav AH, Sheikh FA (2021) Regenerated cellulose nanofibers from cellulose acetate: incorporating hydroxyapatite (HAp) and silver (Ag) nanoparticles (NPs), as a scaffold for tissue engineering applications. Mater Sci Eng C 2021(118):111547. https://doi.org/10.1016/j.msec.2020.111547

Suteris NN, Yasin A, Misnon II, Roslan R, Sulkifli FH, Rahim MHA et al (2022) Curcumin loaded waste biomass resourced cellulosic nanofiber cloth as a potential scaffold for regenerative medicine: an in-vitro assessment. Int J Biol Macromol 198:147–156. https://doi.org/10.1016/j.ijbiomac.2021.12.006

Trevisol T, Fritz A, de Souza S, Bierhalz A, Valle J (2019) Alginate and carboxymethyl cellulose in monolayer and bilayer films as wound dressings: effect of the polymer ratio. J Appl Polym Sci 136:46941

Tudoroiu EE, Dinu-Pîrvu CE, Kaya MGA, Popa L, Anuța V, Prisada RM et al (2021) An overview of cellulose derivatives-based dressings for wound-healing management. Pharmaceuticals 14(12):1215

Ullah A, Ullah S, Khan MQ, Hashmi M, Nam PD, Kato Y et al (2020) Manuka honey incorporated cellulose acetate nanofibrous mats: fabrication and in vitro evaluation as a potential wound dressing. Int J Biol Macromol 155:479–489. https://doi.org/10.1016/j.ijbiomac.2020.03.237

Ullah A, Saito Y, Ullah S, Haider K, Nawaz H, Duy-nam P et al (2021) Bioactive sambong oil-loaded electrospun cellulose acetate nanofibers: preparation, characterization, and in-vitro biocompatibility. Int J Biol Macromol 166:1009–1021. https://doi.org/10.1016/j.ijbiomac.2020.10.257

Ullah A, Sarwar M, Wang F, Kharaghani D, Zhu L, Yoshiko Y et al (2022) In vitro biocompatibility, antibacterial activity, and release behavior of electrospun soy protein isolate/hydroxyethyl

cellulose nanofibers. Curr Res Biotechnol 4:445–458. https://doi.org/10.1016/j.crbiot.2022.09.008
Uppal R, Ramaswamy G, Arnold C, Goodband R, Wang Y (2011) Hyaluronic acid nanofiber wound dressing-production, characterization, and in vivo behaviour. J Biomed Mater Res B Appl Biomater 97B:20
Wahab JA, Kim IS, Ni QQ (2019) A comparative study on synthesis of AgNPs on cellulose nanofibers by thermal treatment and DMF for antibacterial activities. Mater Sci Eng C 2019(98): 1179–1195
Wali A, Gorain M, Inamdar S, Kundu G, Badiger M (2019) In vivo wound healing performance of halloysite clay and gentamicin-incorporated cellulose ether-PVA electrospun nanofiber mats. ACS Appl Bio Mater 2:4324–4334
Walker J, Cullen M, Chambers H, Mitchell E, Steers N, Khalil H (2014) Identifying wound prevalence using the mobile wound care program. Int Wound J 11:319–325
Wang M, Yu D, Williams GR (2022) Co-loading of inorganic nanoparticles and natural oil in the electrospun janus nanofibers for a synergetic antibacterial effect. Pharmaceutics 14:1208
Wasilewska KW (2019) Ethylcellulose-a pharmaceutical excipient with multidirectional application in drug dosage forms development. Materials (Basel) 12:3386
Wu H, Pan W, Lin D, Li H (2012) Electrospinning of ceramic nanofibers: fabrication, assembly and applications. J Adv Ceram 1:2
Wutticharoenmongkol P, Hannirojram P, Nuthong P (2019) Gallic acid—loaded electrospun cellulose acetate nanofibers as potential wound dressing materials. Polym Adv Technol 2019: 1–13
Xu C, Xu F, Wang B, Lu T (2011) Electrospinning of poly(ethylene-co-vinyl alcohol) nanofibres encapsulated with ag nanoparticles for skin wound healing. J Nanomater 2011:201834
Yang L, Liang F, Zhang X, Jiang Y, Duan F, Li L et al (2022) Remodeling microenvironment based on MOFs-hydrogel hybrid system for improving diabetic wound healing. Chem Eng J 427: 131506
Yazdanbakhsh MF, Rashidi A, Rahimi MK, Khajavi R, Shafaroodi H (2018) The effect of impregnated alpha-cellulose nanofibers with ciprofloxacin hydrochloride on Staphylococcus aureus in vitro and healing process of wound in rat. Regen Eng Transl Med 4:247–256
Youdhestar, Mahar FK, Das G, Tajammul A, Ahmed F, Khatri M et al (2022) Fabrication of ceftriaxone-loaded cellulose acetate and polyvinyl alcohol nanofibers and their antibacterial evaluation. Antibiotics 11:352
Zhang M, Zhao X (2020) Alginate hydrogel dressings for advanced wound management. Int J Biol Macromol 162:1414–1428
Zhang Y, Ouyang H, Lim C, Ramakrishna S, Huang Z (2005) Electrospinning of gelatin fibers and gelatin/PCL composite fibrous scaffolds. J Biomed Mater Res Part B 72:156
Zhang C, Yang X, Li Y, Qiao C, Wang S, Wang X et al (2020) Enhancement of a zwitterionic chitosan derivative on mechanical properties and antibacterial activity of carboxymethyl cellulose-based films. Int J Biol Macromol 159:1197–1205
Zheng L, Li S, Luo J, Wang X (2020) Latest advances on bacterial cellulose-based antibacterial materials as wound dressings. Front Bioeng Biotechnol 8:1–15
Zhong S, Zhang Y, Lim C (2010) Tissue scaffolds for skin wound healing and dermal reconstruction. Nanomed Nanobiotechnol 2:510–525

Sibusiso Alven is a qualified candidate for the degree of Doctor of Philosophy in Chemistry at the University of Fort Hare (UFH), and the degree will be conferred in May 2023. His research focus is based on polymer-based materials (such as hydrogels, topical gels, nanofibers, sponges, polymer-drug conjugates, etc.) loaded with various therapeutic agents for treating malaria, cancer, and chronic wounds. He has published 19 articles in various refereed journals (*Polymer*, *Nanomaterials*, and *Journal of Drug Delivery Science and Technology*) and 6 book chapters. He was considered a top publisher by Sasol Bursary Committee among his fellow bursars in 2021.

Sindi P. Ndlovu is a second-year Ph.D. candidate in the Department of Chemistry at the University of Fort Hare. Her research focuses on polymer-based wound dressings encapsulated with various selected bioactive agents to treat microbial-infected wounds. She published three articles in various journals. She orally presented her research results findings at the STEM conference (International conference, 2020) that was held in East London, Eastern Cape, South Africa, and also presented at a postgraduate conference (GMRDC, 2020) held at her institution, University of Fort Hare.

Blessing A. Aderibigbe is a professor of organic chemistry at the University of Fort Hare and an NRF C2-rated researcher. She was a postdoctoral fellow between December 2010 and February 2014 at different universities: the University of Johannesburg (Auckland Park and Doornfontein campuses) and Tshwane University of Technology (Polymer division, which was previously based at CSIR, Pretoria). She obtained her Ph.D. in Polymer/Medicinal Chemistry from the University of Witwatersrand, South Africa, under the supervision of late Emeritus Prof. E.W. Neuse. Her Ph.D. research was focused on the design of polymer-drug conjugates with promising antimalarial activity. Her master's degree was in organic chemistry, also from the University of Witwatersrand, South Africa, under the supervision of Prof. W. van Otterlo and Prof. J. P. Michael. Her MSc research was focused on the synthesis of 6-, 7-, 8-, and 9-membered benzo-fused heterocyclic compounds using Grubbs II catalyst. She joined the University of Fort Hare as a Lecturer on the 1st of March 2014. She was promoted to the position of Senior Lecturer in January 2016, to Associate Professor in January 2018, and to Full Professor in January 2020. Her research themes include designing and developing polymer-based wound dressings, polymer-based drug delivery systems, and organic molecules, specifically hybrid molecules via a combination of two molecules for enhanced efficacy.

Co-delivery of Anticancer Drugs Using Polymer-Based Nanomedicines for Lung and Prostate Cancer Therapy

25

Sijongesonke Peter, Tobeka Naki, Sibusiso Alven, and Blessing A. Aderibigbe

Abstract

Although there are many treatment advances in oncology, cancer remains the cause of morbidity and mortality worldwide. The most common cancer types diagnosed in men are lung and prostate cancer. The use of anticancer drugs is known as the best strategy for the treatment of these cancer types. However, various limitations exist, including multidrug resistance, poor tumor tissue drug accumulation, and nonspecific cytotoxicity or off-target effects. On the other hand, combination chemotherapy is the most promising one, with numerous advantages among the approaches employed to solve the drawbacks. On the other hand, polymer-based nanomedicines are drug delivery systems that can further enhance the therapeutic outcomes of combination chemotherapy. This book chapter reports the preclinical results of polymer-based nanomedicines co-encapsulated with anticancer agents for efficient lung and prostate cancer therapy. This chapter discusses the challenges and advantages of nanomedicine; especially the delivery routes (e.g., by using nanoparticles, dendrimers, micelle, drug conjugate, nanocapsules, nanoliposomes, exosome, noisome, nanogel, etc.) are elaborated. Furthermore, a brief scenario of clinical trials and future perspectives using these approaches are discussed in this chapter.

Keywords

Lung cancer · Prostate cancer · Anticancer drugs · Combination chemotherapy · Nanomedicines · Nanoparticles · Micelles

S. Peter · T. Naki · S. Alven (✉) · B. A. Aderibigbe
Department of Chemistry, University of Fort Hare, Alice, Eastern Cape, South Africa
e-mail: 201414787@ufh.ac.za; tnaki@ufh.ac.za; 201214199@ufh.ac.za; baderibigbe@ufh.ac.za

F. A. Sheikh et al. (eds.), *Interaction of Nanomaterials With Living Cells*,
https://doi.org/10.1007/978-981-99-2119-5_25

25.1 Introduction

Globally, in developed and developing countries, public health systems are disrupted because of the growing number of cases and deaths caused by noncommunicable diseases such as cancer, ischemic heart disease, strokes, and low respiratory tract infections, with cancer regarded as the fatal one among them all (Rana et al. 2015; Mahdi et al. 2022; Wende et al. 2021; Matthews et al. 2022; Peter and Aderibigbe 2019). Cancer is caused by an unusual growth (uncontrolled cell replication and proliferation) of body cells resulting in tumor cells that can spread to other organs leading to metastatic cancer (advanced cancer). Generally, cancer can be caused by a gene mutation (5–10%), and 90–95% are attributed to aging, poor dietary habits, and individual lifestyles, such as lack of physical activity, high consumption of alcohol, obesity, and smoking. Cancer types are also named according to the body organs they affect, including lung, prostate, breast, liver, brain, colorectal, and stomach cancer (Peter and Aderibigbe 2019; Peter et al. 2021, 2022).

Statistically, before the high transmission of COVID-19, reports from the World Health Organization (WHO) and GLOBOCAN estimated an increasing number of cancer incidents and deaths. There were more than 19 million cancer cases and more than 9 million deaths in 2020 globally, with developing countries in Africa and Asia contributing more to these numbers than developed countries (Peter et al. 2022; Gao et al. 2020; Sung et al. 2021). Additionally, further predictions indicate that the numbers could double by 2040 due to their drastic increase since the year 2000 (Peter et al. 2022). Among the 30 types of cancer reported internationally, lung and prostate cancers are cancer types that are lethal in both men and women (Xia et al. 2022; Barta et al. 2019; Culp et al. 2020). Globally, lung carcinoma is reported as a leading cause of cancer-related deaths in men and women, the second leading diagnosed cancer type among women after breast cancer, and the first leading diagnosed cancer type among men. Additionally, prostate cancer is reported as the sixth leading cause of cancer deaths and the second diagnosed cancer among men. Thus, lung cancer is responsible for 1 in 5 cancer-associated deaths, and prostate cancer affects 1 in 25 men worldwide (Culp et al. 2020; Mirzaei et al. 2022). Due to the limitations of the present procedures, including chemotherapy (anticancer drugs), that are used for the treatment of lung and prostate cancer, there is an urgent need to develop enhanced therapeutics.

The issue with the currently used anticancer drugs for lung and prostate cancer is that they suffer from myriad shortcomings, such as drug toxicity and multidrug resistance (MDR), lack of specificity, poor solubility, poor delivery system, and poor bioavailability (Peter et al. 2021; Gao et al. 2020). Therefore, the global desperation to discover a cure for cancer since the number of cancer incidents is still increasing has resulted in several medicinal researchers developing novel and effective anti-cancer therapeutic agents (Dadashpour and Emami 2018; Khalifa et al. 2019; Beik et al. 2019). Moreover, despite the use of different strategies such as chemotherapy, radiotherapy, and surgery, nanotechnology has caught the attention of various researchers, and this strategy is a potent approach for combating cancer because it improves drug transportation resulting in enhanced therapeutic effects with reduced side effects (Khalifa et al. 2019; Zheng et al. 2021; Bhatia 2016). Thus, reports indicate that more than 250 nanomedicine therapies or technologies with effective anticancer effects are in the final stages of clinical practice. Some have been approved by the Food and Drug Administration (FDA) recently (Tang et al. 2018).

Nanotechnology is the application of small-size (1–1000 nm) materials to develop nanomedicine, including polymers, for chemotherapeutic purposes (Tang et al. 2018). This approach can overcome the shortcomings associated with therapeutic agents that are used for lung and prostate cancer therapy by loading the selected anticancer drugs, and they can be easily modified for improved specificity. Furthermore, applying nanotechnology in cancer management and treatment is a promising approach. This strategy can be used to tailor the therapeutic response, drug delivery, and diagnosis (Qin et al. 2021; Kirtane et al. 2021; Ali et al. 2020). Moreover, combination therapy of anticancer agents using nanomedicines can result in various advantages, including synergistic anticancer effects against lung and prostate cancer cells. Polymers that are frequently utilized for the preparation of nanomedicines are categorized into two classes, i.e., natural and synthetic polymers. Natural polymers include proteins, chitosan, alginate, hyaluronic acid (HA), nucleic acids, peptides, etc. Examples of synthetic polymers include poly(ε-caprolactone) (PCL), poly(L-aspartate), poly(ethylene glycol) (PEG), etc. (Ali et al. 2020; Zhao et al. 2018). Several publications on polymer therapeutics have reported improved therapeutic effects (biocompatibility, biodegradability, specificity, etc.) with limited shortcomings in vitro and in vivo. However, natural polymers show some shortfalls, such as uncontrollable drug release rate and short half-lives (Ali et al. 2020; Rao et al. 2018). Therefore, this chapter will review the co-delivery of antitumor agents using polymer-based nanomedicines for lung and prostate cancer therapy, an update, and future perspectives about this approach in treating and managing these cancer types.

25.2 Classification of Anticancer Drugs

Anticancer drugs are classified according to their mechanism of action into four groups: anti-tubulin agents, alkylating agents, topoisomerase inhibitors, and antimetabolites (Figs. 25.1 and 25.2) (Alven et al. 2020; Fathi et al. 2021). Alkylating agents (**A**) are anticancer drugs that covalently bind to the deoxyribonucleic acid (DNA) and cross-link them, destroying the DNA (Fathi et al. 2021; Huang et al. 2017). Some examples include cyclophosphamide, melphalan, carboplatin, cisplatin, and oxaliplatin. Anti-tubulin agents (**B**) terminate mitotic spindles and stop mitosis. Their examples include vinblastine, vincristine, docetaxel, and paclitaxel (the most active chemotherapeutic that is used for the treatment of various cancer types) (Alven et al. 2020; Fathi et al. 2021).

Antimetabolites (**C**) hinder the biosynthesis of nucleic acid, and their examples include bortezomib, bevacizumab, methotrexate, and 5-fluorouracil (Fathi et al. 2021; Kaye 1998). Topoisomerase inhibitors (**D**) impede DNA replication and are responsible for binding to the active site of topoisomerase, hindering the binding of

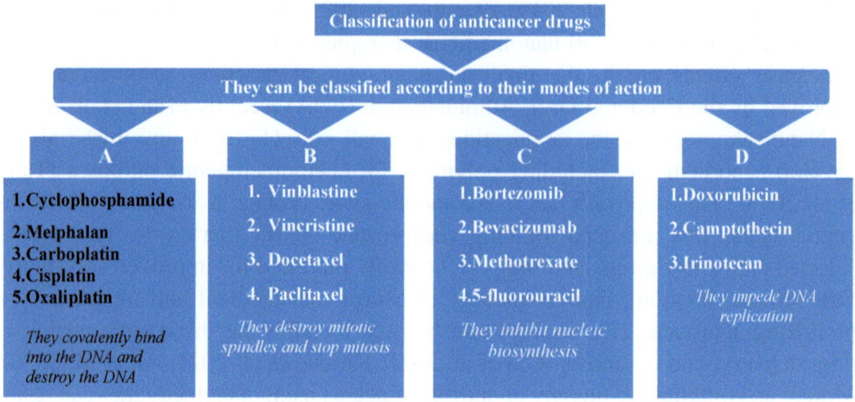

Fig. 25.1 Classification of anticancer drugs according to their mode of action

the substrate of DNA (Fathi et al. 2021; Goftar et al. 2014). Furthermore, topoisomerase inhibitors result in a cleavage complex that avoids enzyme turnover and the build-up of excessive amounts of the cytotoxic cleavage complex within the tumor cell (Fathi et al. 2021; Hevener et al. 2018). Their examples include doxorubicin, camptothecin, and irinotecan (Fathi et al. 2021; Khadka and Cho 2013).

25.3 Challenges in Lung and Prostate Cancer Treatment

The increasing numbers of cancer incidents are due to the management and treatment of cancer being hampered by various limitations. Several medicinal scientists are still investigating and developing effective anticancer therapies to control cancer. For instance, clinical therapies, including chemotherapy, radiotherapy, surgery, immunotherapy, and hormone-targeting drugs, are still being investigated. Moreover, nanotechnology (nanomedicines) is also used to combat cancer (van der Spek et al. 2020; Navya et al. 2019; Kanno et al. 2021; Hegde and Chen 2020). Contrarily, some of these therapies or drugs used in these therapies are limited by drug resistance leading to a reoccurrence of the disease, drug toxicity mostly leading to severe drug side effects, poor bioavailability, lack of specificity, poor solubility, poor stability, and disrupted metabolism resulting to poor pharmacokinetics which sometimes leads to some anticancer drugs not reaching clinical trials, let alone the market due to adverse side effects (van der Spek et al. 2020; Navya et al. 2019; Kanno et al. 2021; Hegde and Chen 2020).

Specifically, in most countries, although the increasing aging population is a drawback to the treatment of lung cancer due to lack of compliance, high consumption of tobacco is another major drawback in the treatment of lung cancer as it contributes to the increasing number of cases (Didkowska et al. 2016; To et al. 2021). Furthermore, an early lung cancer diagnosis is challenging. Therefore, the treatment of metastatic lung cancer has some challenges as the currently used

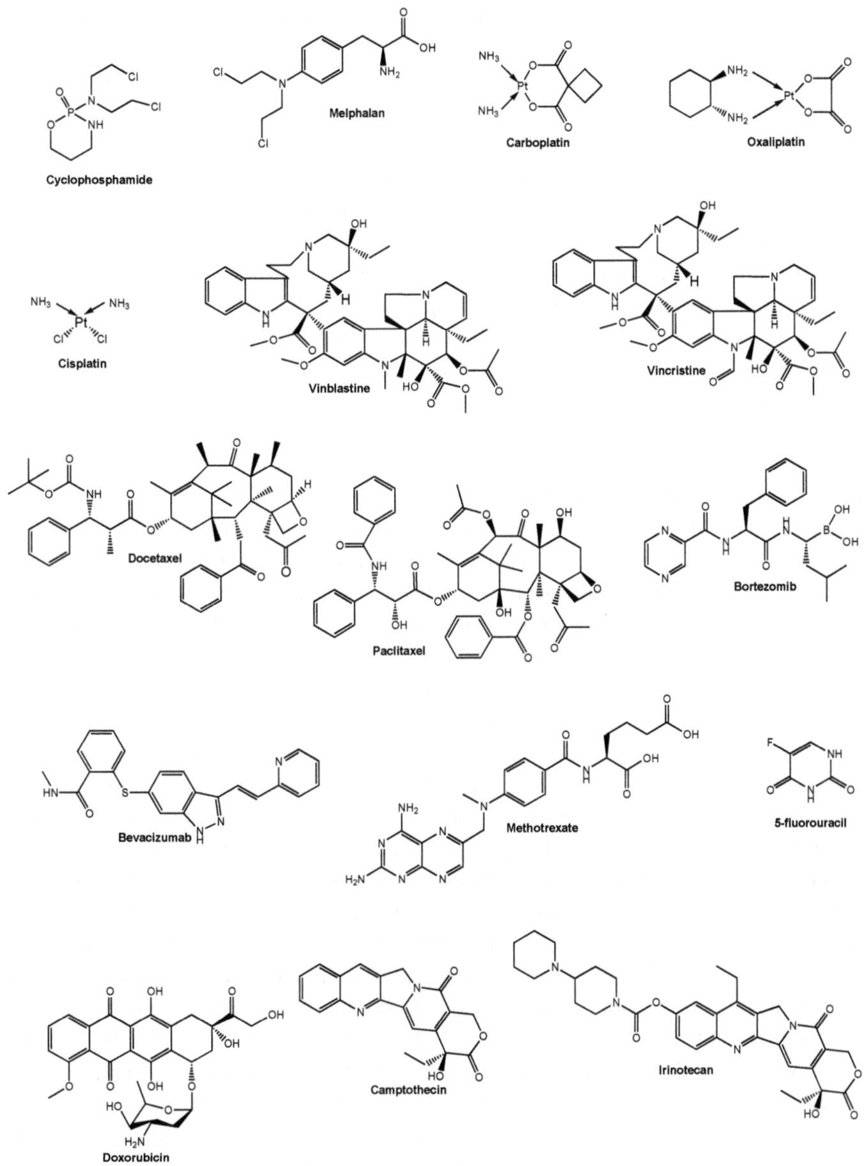

Fig. 25.2 Chemical structures of anticancer drugs used to treat cancers

therapies, including chemotherapy to treat it, lack specificity and they depend on the tumor type histology resulting in relapse cases. Thus, only 1–2% of patients treated with these therapies exceed a 5-year survival rate (Ilie et al. 2016). Moreover, the classical locations for metastatic lung cancer are the bone, liver, nervous system, adrenal glands, and respiratory system. The first treatment in these locations is surgical removal. However, this approach is not convenient; as a result, it is used in 10–20% of patients as it depends on the respiratory status of patients and the site of lesions (Mangal et al. 2017). Furthermore, anticancer drugs' ability to pass the blood-brain barrier (BBB) is another drawback to the treatment of metastatic lung cancer in the nervous system (Zhong et al. 2019a).

Additionally, the delivery system for the treatment of metastatic lung cancer is also a challenge. For instance, intravenous administration results in drugs being distributed to various organs, leading to low drug concentration on tumor site, which leads to dose increase to keep the therapeutic effect leading to drug toxicity and severe adverse side effects. Another example is the use of inhalation chemotherapy as a delivery system where it also leads to drugs distribution to several organs and compromises drug concentration to the tumor site resulting to the use of high doses leading to drug toxicity and side effects such as coughing, respiration-related complications, and damaging of healthy lung cells. Consequently, these scenarios compromise the compliance of patients and the efficacy of the anticancer drugs for the treatment of metastatic lung cancer (Mangal et al. 2017). Other respiratory diseases, such as COVID-19, are also a drawback to treating lung cancer patients as there are several complications to consider (Calabrò et al. 2020). In essence, early diagnosis of lung cancer is a significant challenge in the treatment of lung cancer.

Prostate cancer management experiences similar challenges as lung cancer, such as a lack of specific and highly effective therapies, drug resistance, etc. The most common challenge with treating this type of cancer is its detection. For instance, the use of biopsy needles guided by transrectal ultrasound is sometimes not accurate. They can detect low-grade prostate cancer and miss higher-grade prostate cancer. Furthermore, the use of prostate-specific antigen screening can result in false detection (Barqawi et al. 2012; Barrett and Haider 2017; Crocetto et al. 2022). Thus, these false detections of prostate cancer stages may result in overdiagnosis leading to overtreatment. Consequently, overtreatment results to some serious side effects (Crocetto et al. 2022). Nevertheless, another issue is the therapy resistance due to inaccurate detection, which sometimes leads to undertreatment of prostate cancer as it may be detected in its advanced stages. For instance, the use of typical hormone therapy can display resistance over time, resulting in castrate-resistant prostate cancer (Flores-Téllez and Baena 2022; Haberkorn et al. 2016; Silberstein et al. 2013). Basically, screening for prostate cancer is a significant drawback to its management. Despite these limitations hampering lung and prostate cancer treatment and management, it is noted that global pandemics such as COVID-19, Ebola, severe acute respiratory syndrome coronavirus 2, etc. sometimes increase cancer cases and deaths. During the pandemic, the focus is on managing the pandemic crisis resulting in reduced screening for cancer, whereby cancer is diagnosed at the advanced stages.

25.4 Advantages of Combination Chemotherapy Using Nanomedicines

Combination chemotherapy is one potent approach commonly utilized to solve the drawbacks of anticancer agents. The advantages that are exhibited by combination therapy using polymer-based nanomedicines include synergistic anticancer effects, co-delivery of bioactive molecules with different physicochemical and pharmacological properties, ratiometric drug loading and controlled drug release mechanism, and stimuli-responsive.

25.4.1 Synergistic Anticancer Effects

Combination chemotherapy nanomedicines have been widely studied and have shown the capability to promote synergistic cytotoxic effects against various cancer cell lines leading to superior anticancer activity than monotherapy with reduced side effects (Pan et al. 2019). Monotherapy which is based on the application of a single bioactive agent is not enough to suppress tumor regression. But with the combination chemotherapy via nanomedicines, the synergistic effect of two or more drugs targeting various cell-cycle checkpoints, pathways, or genes in the cancer process possesses the chance to eradicate cancer (Xu et al. 2015). Henceforth, combination therapy nanomedicines are a suitable and promising strategy that can be utilized to improve the activity of chemotherapy. The synergistic antitumor effect of co-delivered anticancer drugs employing nanomedicines is not easily achieved using conventional therapeutics administration because of several factors such as injection schedule, various pharmacokinetics features, metabolism, and nonuniform distribution. The anticancer drugs co-loaded-based nanomedicines are delivered to the targeted tumor site sustainably (Mahira et al. 2019; Unsoy and Gunduz 2018).

25.4.2 Co-delivery of Bioactive Molecules with Different Physicochemical and Pharmacological Properties

Nanomedicines-mediated combination chemotherapy can deliver more anticancer drugs with various physicochemical and pharmacological properties in one nanomedicine (Gurunathan et al. 2018). Since most polymer-based nanomedicines have been employed for the co-delivery of therapeutic agents that have similar properties, biomedical researchers focused on formulating novel nanocarriers that can multi-deliver anticancer molecules with different pharmacological and physicochemical properties such as the mechanism of action, charge, size, solubility, weight, etc. Moreover, nanomedicines formulated for the co-delivery of therapeutics can deliver both hydrophobic and hydrophilic drugs (Alven and Aderibigbe 2020; Kanai et al. 2011). This can be done by chemical conjugation or modification procedures whereby the drugs are modified, making them more compatible with the polymeric nanomedicines (Kolishetti et al. 2010).

25.4.3 Ratiometric Drug Loading and Drug Release Mechanism

Ratiometric drug loading and release mechanism of the drugs is another significant aspect that must be considered in designing multidrug delivery nanomedicines because the drug-to-drug ratio strongly affects the effectiveness of combination chemotherapies. The research studies that were recently conducted have demonstrated that the degree of antagonism and synergism between the therapeutic agents in combinations is mainly dependent on their respective concentrations (Raitanen et al. 2002). The nanomedicines that are used for co-delivery can stimulate overlap between the pharmacological features of the loaded drugs, also resulting in the opportunity of delivering many drugs at a prearranged ratio to accomplish the good activity of the combination chemotherapy. The examples in this aspect are liposomes CPX-1 (1:1 molar ratio of irinotecan and floxuridine) and CPX-351 (5:1 molar ratio of cytarabine and daunorubicin), which are loaded with therapeutic agents at fixed drug molar ratios, exhibit potential in vivo activity, and are presently in clinical trials (Lancet et al. 2014). Moreover, improvements in polymer chemistry led to the designing of the systems capable of ratiometric drug loading and controlled release of more than two anticancer drugs in a single nanomedicine (Liao et al. 2014).

25.4.4 Stimuli-Responsive

Nanomedicines that are stimuli-responsive can alter drug release and physicochemical properties when they encounter specific environmental conditions, which can significantly result in high amount of the loaded drugs to be delivered to the targeted diseased cells/tissues. These stimuli can be the application of adjustments in extracellular conditions such as near-infrared light, heating, or ultrasound to generate an effect or intracellular conditions such as enzymatic reactions, introduction of reducing agents, or lower pH. Some stimuli-responsive nanomedicines have been developed successfully, and they are in clinical trials for the treatment of various cancer types (Gadde 2015). Nevertheless, these nanomedicines are mainly appropriate for the delivery of single therapeutic agents and might not be ideal for co-delivery. Biomedical researchers are focusing on designing novel multidrug delivery nanomedicines that possess stimuli-responsive abilities, as they can lead to an advantage in controlling the release of loaded drugs, which is vital in achieving maximum synergy (Gadde 2015; Jia et al. 2021).

The aforementioned advantages of drug co-delivery nanomedicines can further result in improved biological efficacy of the encapsulated/incorporated drugs, overcome MDR, increased drug stability, reduced drug toxicity, improved drug internalization and accumulation in the cancer cells, and extended the plasma circulation time of the encapsulated molecular agents. Although there are many exciting properties of nanomedicines for the co-delivery of anticancer drugs, some challenges need to be considered in their formulation, including those associated with the entrapment of drugs with various physicochemical features and solubilities, controlling their sequential drug release, and increasing drug build-up in the cancer site. Nanomedicines evaluated for co-delivery of anticancer drugs include nanoparticles, dendrimers, micelles, polymer-drug conjugates, nanocapsules, and nanoliposomes. The polymers utilized to fabricate the nanomedicines are classified into six categories: thermo-sensitive polymers, block co-polymer conjugates, MMP-sensitive polymers, magnetic-responsive polymers, pH-sensitive polymers, and redox-sensitive polymers (Kanai et al. 2011).

25.5 Co-delivery of Anticancer Drugs Using Polymer-Based Nanomedicines

25.5.1 Polymeric Nanoparticles

Polymeric nanoparticles are colloidal particle nanomedicines or nanocarriers with a mean diameter of 1–1000 nm (Sailaja et al. 2011). An example of the polymeric nanoparticles is shown in Fig. 25.3. These nanomedicines are frequently prepared from synthetic, natural, and semisynthetic polymers, whereby the therapeutic drugs can be loaded, chemically incorporated, entrapped, encapsulated, absorbed, or dissolved. Nanoparticles possess fascinating properties, such as excellent biodegradability, biocompatibility, and versatility in their use. Many biodegradable polymers are usually used for the formulation of polymeric nanoparticles, such as poly(D, L-lactic acid) (PLA), polyalkylcyanoacrylates, polyethylene glycol (PEG), poly (e-caprolactone) (PCL), poly(D,L-lactic-co-glycolic acid) (PLGA), etc. (Dennis et al. 2015). Nanoparticles, predominantly stimuli-sensitive nanoparticles, exhibit a

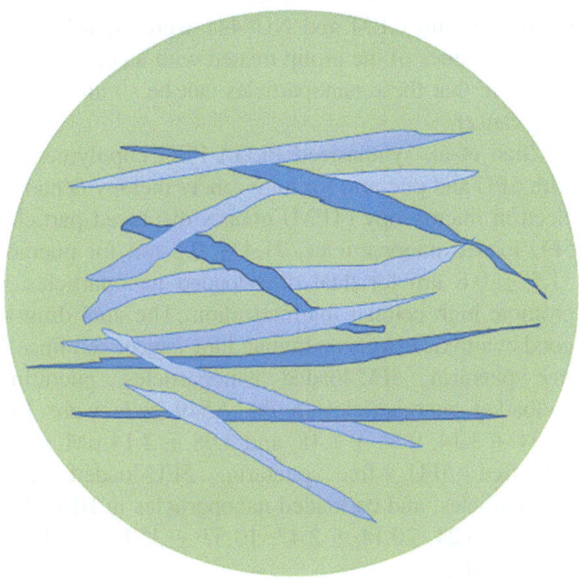

Fig. 25.3 Schematic diagram of a polymeric nanoparticle

high drug-encapsulating ability and controlled drug release kinetics with enhanced in vivo stability for co-distribution of several antitumor drugs. The benefits of polymeric nanoparticles include targeted drug delivery, sustained and controlled drug release mechanisms, the improved therapeutic activity of the loaded bioactive agent, enhanced drug hydrophilicity, reduced drug toxicity, and appropriateness for co-delivery of drugs (Jawahar et al. 2019). Several polymeric nanoparticles have been reported for combination chemotherapy to treat lung and prostate cancer (Table 25.1).

25.5.1.1 Nanoparticles for Lung Cancer Therapy

Many research studies have reported polymeric nanoparticles co-encapsulated with anticancer agents for lung cancer therapy. Zhou et al. prepared PLGA-PEG nanoparticles co-loaded with gemcitabine (GEM) and 5-fluorouracil (5FU) for lung cancer therapy (Zhou et al. 2021). The drug release at physiological conditions (pH 7.2 and 37°) was controlled and sustained for both loaded anticancer drugs. The in vitro cytotoxicity analysis showed half-maximal inhibitory concentrations (IC_{50}) of 4.62 ± 0.97 and 5.16 ± 2.80 μM for dual drug-loaded nanoparticles against lung cancer cell lines, A549 and HEL-299, respectively, while the IC_{50} values of the nanoparticles loaded with one drug ranged between 9.05 ± 2.11 and 11.20 ± 0.98. These results suggested that polymeric nanoparticles co-loaded with GEM and 5FU possessed superior anticancer efficacy against lung carcinoma than nanoparticles loaded with only one drug, which could be due to synergistic anticancer efficacy (Zhou et al. 2021). The PLGA-based nanoparticles containing GEM and NU7441 (potent radiosensitizer) developed by Menon et al. demonstrated high drug loading efficiencies for GEM and NU7441 (88% and 52%), respectively (Menon et al. 2017). The in vivo studies using H460 (lung cancer cell line)-tumor nude mice showed that the volumes of tumors for the group administered with nanoparticles co-loaded with GEM and NU7441 were significantly smaller when compared to tumor volumes of the group treated with a combination of free GEM and NU7441, revealing that these nanoparticles can be simultaneously used for the treatment of lung cancer.

Zhao et al. synthesized PLGA-PEG copolymer nanoparticles co-encapsulated with 5FU and puerarin for lung cancer therapy (Zhao et al. 2021). The transmission electron microscopy (TEM) images displayed particle sizes of 72.4 ± 0.7 nm for 5FU-loaded nanoparticles, 71.4 ± 0.3 nm for puerarin-loaded nanoparticles, and 81.5 ± 0.6 nm for dual drug-loaded nanoparticles, indicating their potential to promote high cellular internalization. The dual drug-loaded nanoparticles showed good cytotoxic effects on human lung cancer cell lines. The IC_{50} values of free 5FU, free puerarin, 5FU-loaded nanoparticles, puerarin-loaded nanoparticles, and co-loaded nanoparticles in the A549 cells were 13.04 ± 2.54, 14.45 ± 2.47, 10.24 ± 3.14, 9.54 ± 1.48, and 3.98 ± 2.14 μM, respectively. The IC_{50} values of the free 5FU, free puerarin, 5FU-loaded nanoparticles, puerarin-loaded nanoparticles, and co-loaded nanoparticles in HEL-299 lung cancer cell lines were 19.47 ± 3.24, 20.14 ± 2.47, 10.57 ± 3.28, 11.24 ± 3.24, and 5.12 ± 2.45 μM, respectively.

Table 25.1 Polymeric nanoparticles co-loaded with anticancer drugs for lung and prostate cancer therapy

Polymers used	Loaded anticancer drugs	Targeted cancer type	Therapeutic outcomes	Factors that made nanoparticles suitable for drug co-delivery	Ref
PLGA and PEG	GEM and 5FU	Lung cancer (A549 and HEL-299 cells)	Controlled drug release kinetics with high cytotoxicity against cancer cells	The oil/water solvent evaporation technique used for nanoparticle preparation resulted in loading both hydrophobic and hydrophilic drugs	Zhou et al. (2021)
PLGA	GEM and NU7441	Lung cancer (H460 cells)	Reduced lung tumor volume	High drug loading capacity	Menon et al. (2017)
PLGA and PEG	5FU and puerarin	Lung cancer (A549 cells)	High cellular internalization	High loading and encapsulation efficiencies	Zhao et al. (2021)
PLGA and PEG	Metformin and silibinin	Lung cancer (A549 cells)	Superior anticancer activity	Good nanoencapsulation efficiency of the drugs	Mogheri et al. (2021)
PEG and PCL	Crizotinib and sorafenib	Lung cancer (4T1 cells)	Reduce tumor development, reduce side effects, and longer survival rate	High encapsulation efficiency of drugs	Zhong et al. (2021)
PLA and PEG	Cisplatin and docetaxel	Lung cancer (A549 cells)	Higher cytotoxicity against cancer cells	Larger particle size of more than 200 nm	Wu et al. (2020)
PEG	Cisplatin and DOX	Lung cancer (A549 cells)	Synergistic anticancer effects	Good combination index of 0.57	Nan (2019)
mPEG and PLG	Cisplatin and DOX	Lung cancer (C57BL/6 cells)	High cellular uptake and accumulation of loaded drugs	High drug loading content and drug loading efficiency	Xu et al. (2019)
PLGA and PEG	Cisplatin and etoposide	Lung cancer (344SQ allograft and H460 xenograft)	Good anticancer efficacy	Large particle size	Zhang et al. (2021)
PEG	PTX and siRNA	Lung cancer (A549 cells)	Inhibition of tumor growth with prolonged survival rate	High drug loading capacity and good dispersibility	Jin et al. (2018)

(continued)

Table 25.1 (continued)

Polymers used	Loaded anticancer drugs	Targeted cancer type	Therapeutic outcomes	Factors that made nanoparticles suitable for drug co-delivery	Ref
PLGA	PTX and cisplatin	Lung cancer (A549 cells)	Higher antitumor activity	The larger particle size of approximately 190 nm	Wang et al. (2018)
PLGA	DTX and siCCAT2	Lung cancer	Excellent tumor targeting ability with improved anticancer effect	High drug loading efficiency	Xiao et al. (2022)
HA	Pt (IV) drug and survivin siRNA	Lung cancer (A549/DDP cells)	Reduced drug efflux and improved cancer cell apoptosis	Large hydrodynamic diameters of more than 180 nm and high encapsulation efficiency	Ma et al. (2021a)
Polydopamine	Indocyanine green and dimethylcurcumin	Prostate cancer (22Rv1 cells)	Enhanced cellular uptake with superior antitumor effect	Large hydrodynamic diameters range between 200 and 300 nm. And high drug encapsulation efficiencies	Hong et al. (2022)
PEG	Methylene blue and indocyanine green	Prostate cancer (DU145 and PC3 cells)	Improved cancer cell apoptosis	Excellent dispersive index and a particle size of more than 103.12 nm	Ma et al. (2022)
Chitosan and PCL	Bicalutamide and hesperetin	Prostate cancer (PC3 cells)	Improved cellular uptake with superior anticancer effect	The big average particle size of 281 $\pm$ 13.7 nm and positive zeta potential of 12.10 $\pm$ 1.11 mV	Arya et al. (2016)
PLGA	Indocyanine green and resiquimod	Prostate cancer (PC3 cells)	Good anticancer activity	The average particle diameter of about 157.7 nm	Lin et al. (2021)
PLGA	Cabazitaxel and orlistat	Prostate cancer (PCa cells)	Higher inhibition of tumor volume	Minimum combination index of about 0.57	Qu et al. (2021)
PEG	DTX and as enzalutamide	Prostate cancer (PCa cells)	Higher cell uptake with improved anticancer activity	The particle size of more than 140 nm with an excellent dispersive index	Chen et al. (2022)

Mogheri et al. developed PLGA-PEG nanoparticles for the co-delivery of silibinin and metformin. The dynamic light scattering (DLS) data of dual drug-loaded polymeric nanoparticles revealed a mean particle size of 35 ± 8.38 nm, polydispersity index (PDI) of 0.109, and surface charges of −2.9 ± 1.2 mV, demonstrating their suitability for effective internalization in cancerous cells. The cytotoxicity studies employing MTT assays indicated superior anticancer efficacy against the A549 lung cell line for polymeric nanoparticles co-encapsulated with metformin and silibinin than the free drugs and blank nanoparticles (Mogheri et al. 2021). Zhong et al. prepared PEG-PCL-PEG nanoparticles co-loaded with crizotinib and sorafenib, demonstrating a significant reduction of tumor development, reduced side effects, and longer survival rate than the free drugs when intravenously administered to mice xenograft model with 4T1 lung cancer cells (Zhong et al. 2021). The aptamer-decorated PLA-PEG nanoparticles developed by Wu for the co-delivery of cisplatin and docetaxel showed significantly low cell viability of A549 lung cancer cells than the free drugs and non-aptamer-decorated nanoparticles, especially in lower concentrations, suggesting good anticancer activity against lung cancer (Wu et al. 2020). Nan reported epidermal growth factor receptor (EGFR)-targeted lipid PEG nanoparticles co-encapsulated with doxorubicin (DOX) and cisplatin for lung cancer treatment. The synergistic effects studies and in vivo antitumor studies using mice bearing the lung cancer model showed that dual drug-loaded PEG nanoparticles induced synergistic anticancer effects with a combination index (CI) of 0.57 and reduced tumor volume with inhibition of tumor growth ratio to 74.5%, respectively, indicating significant anticancer activity useful for lung carcinoma therapy (Nan 2019).

Xu et al. formulated mPEG-OEI-PLG nanoparticles co-loaded with cisplatin and DOX for lung carcinoma therapy (Xu et al. 2019). The in vitro drug release studies at acidic conditions (pH 5.0 and 6.8) showed initial burst release of cisplatin and DOX from the nanoparticles followed by slow and sustained release, demonstrating that these dual drug-loaded nanoparticles will be advantageous for therapeutic activity in anticancer application because of rapid release that would take place in the acidic tumor microenvironment. The in vivo cellular uptake studies showed that the lungs administered with co-loaded nanoparticles by pulmonary administration demonstrated very solid fluorescence intensity of DOX in the cancer environment indicating high cellular uptake and accumulation of DOX but no prominent DOX fluorescence in normal lung tissues. Furthermore, the in vivo anticancer evaluation using metastatic lung cancer on C57BL/6 mice showed that the mice treated with co-loaded nanoparticles were more effective in hindering tumor growth than DOX nanoparticles and cisplatin nanoparticles, indicating that the co-loading of cisplatin and DOX in polymeric nanoparticles is a potential antitumor approach when compared to DOX nanoparticles or cisplatin nanoparticles (Xu et al. 2019). The PLGA-PEG nanoparticles formulated by Zhang et al. for the co-delivery of cisplatin and etoposide showed good anticancer efficacy in murine models utilizing both the H460 xenograft and 344SQ allograft (Zhang et al. 2021).

Jin et al. designed PEG nanoparticles co-incorporated with survivin small interfering RNA (siRNA) and paclitaxel (PTX) for lung cancer treatment. The

DLS results of co-loaded nanoparticles displayed a particle size and positive surface charge of about 82.4 nm and 4.1 mV, respectively. The in vitro cytotoxicity analysis utilizing cell counting kit-8 assay showed that the nanoparticles possessed an enhanced anti-proliferation effect on A549 lung carcinoma cells. Furthermore, the in vivo studies using mice showed that the accumulated nanoparticles in the tumor site could significantly inhibit tumor development and prolong the survival rate of the animals (Jin et al. 2018). The in vitro studies of PLGA nanoparticles co-loaded with PTX and cisplatin by Wang et al. demonstrated an IC_{50} value of 26.7 μg/mL with higher cellular uptake. In comparison, the free drug combination showed an IC_{50} value of 75.3 μg/mL, revealing a significantly higher antitumor activity of the nanoparticles co-encapsulated with PTX and cisplatin when compared to individual drugs against A549 lung carcinoma cells. Moreover, the DLS results displayed an average particle size of approximately 190 nm and a zeta potential of −35 mV, which could be suitable for high cellular internalization (Wang et al. 2018).

The PLGA nanoparticles developed by Xiao et al. for co-delivery of docetaxel (DTX) and si-colon cancer-associated transcript-2 (siCCAT2) exhibited exceptional tumor targeting potential. They led to an improved anticancer effect when compared to the DTX-loaded nanoparticles due to the silencing of CCAT2 levels in the lung tumor, resulting in downregulated expressions of multidrug resistance-associated proteins 1 (MRP1) and P-glycoprotein (P-GP) (Xiao et al. 2022). Ma et al. formulated HA-based nanoparticles to co-deliver Pt (IV) and survivin siRNA to lung cancer cells. The drug release was sustained for Pt (IV) that gradually reduced to toxic Pt (II) species, which reduced drug efflux and induced apoptosis of cancer cells. The in vivo anticancer studies showed that the co-loaded nanoparticles possessed a prominently enhanced therapeutic effect (TIR, 82.46%) in nude mice induced with A549/DDP lung tumor model, with no noticeable side effects (Ma et al. 2021a).

25.5.1.2 Nanoparticles for Prostate Cancer

Several biomedical researchers have reported polymeric nanoparticles co-encapsulated with anticancer drugs for the treatment of prostate cancer. Hong et al. designed PEG2000- and PEG5000-modified polydopamine nanoparticles co-encapsulated with indocyanine green and dimethylcurcumin for castration-resistant prostate cancer therapy (Hong et al. 2022). The Fourier-transform infrared spectroscopy (FTIR) analysis confirmed the successful formulation of the modified co-loaded nanoparticles. These polymeric nanoparticles displayed drug encapsulation efficiencies of 62.1 ± 3.1% and 73.2 ± 2.8% for indocyanine green and dimethylcurcumin, respectively. The drug release kinetics was a sustained release of the loaded bioactive agents from nanoparticles at the pH of 5.0 and 7.4. The cellular uptake in vitro displayed that these two PEG-modified nanoparticles possessed enhanced cellular uptake than the free indocyanine green and dimethylcurcumin. The in vitro cytotoxicity analysis on 22Rv1 prostate cancer cells using MTT assay showed that dual drug-loaded PEG2000- and PEG5000-modified nanoparticles have superior in vitro anticancer effect than the free drugs upon 808 nm laser irradiation, which could be due to their improved cellular uptake (Hong et al. 2022).

Ma et al. formulated PEG-based nanoparticles co-loaded with indocyanine green and methylene blue for prostate cancer therapy. The in vitro anticancer studies exhibited dose-dependent cytotoxicity against PC3 and DU145 prostate tumor cells than the plain nanoparticles, with a significantly higher cell apoptotic rate of 80% (Ma et al. 2022). Arya et al. prepared chitosan-PCL nanoparticles for the co-delivery of an anti-androgen drug, bicalutamide and hesperetin. The DLS investigation of co-loaded nanoparticles exhibited a mean particle size of 281 ± 13.7 nm and a surface charge of $+12.10 \pm 1.11$ mV, which can improve cellular internalization. The in vitro cytotoxicity studies of chitosan-PCL nanoparticles co-incorporated bicalutamide and hesperetin showed cell cytotoxicity (CC_{50}) of 22.27 μM against PC3 cells which were threefold lower than the combination of free bicalutamide and hesperetin (66.38 μM), indicating good antitumor efficacy of dual drug-encapsulated nanoparticles with improved cellular uptake (Arya et al. 2016).

Lin et al. synthesized the PLGA-based nanoparticles co-loaded with resiquimod and indocyanine green, which showed the anticancer activity for prostate cancer therapy by combining PTT with immunotherapy (Lin et al. 2021). The PLGA nanoparticles formulated by Qu et al. for co-delivery of cabazitaxel-hyaluronic acid prodrug and orlistat demonstrated sustained drug release behaviors with significantly higher inhibition of tumor volume in PCa xenograft mice than the free drug and plain nanoparticles, suggesting good anticancer activity of the dual drug-loaded nanoparticles against prostate cancer in vivo (Qu et al. 2021). Chen et al. synthesized PEG-based nanoparticles co-loaded with DTX and enzalutamide for prostate cancer treatment. These nanoparticles possessed an average particle size of 143.7 ± 4.1 nm, a surface charge of $+29.1 \pm 2.4$ mV, and PDI of 0.162 ± 0.037 with high cell uptake by PCa cells of approximately 70%, which can significantly improve anticancer activity against prostate cancer (Chen et al. 2022).

25.5.2 Dendrimer

Dendrimers are nanomedicines with three-dimensional, hyperbranched, and spherical structures (Fig. 25.4) (Narmani et al. 2019). They are regularly used in biomedical applications for drug delivery of different therapeutic molecules, including anticancer, antimalarial, antiviral drugs, antitubercular, etc. Their globular nanosize (1–10 nm) and well-defined surface functional groups make them appropriate for drug delivery. Polymer-based dendrimers such as polyamidoamine (PAMAM) dendrimers have attracted significant attention from biomedical researchers because of their low cytotoxicity (Uram et al. 2018). Other dendrimers utilized in biomedical applications, particularly in oncology, are phosphorus, poly-lysine, carbosilane, and PPI dendrimers (Qiu and Bae 2006; Caminade 2020). These nanomedicines have been demonstrated to be able to successfully deliver anticancer agents for the treatment of various cancer types. The advantages of dendrimers in cancer treatment include improved drug activity, reduced drug toxicity, reduced drug resistance,

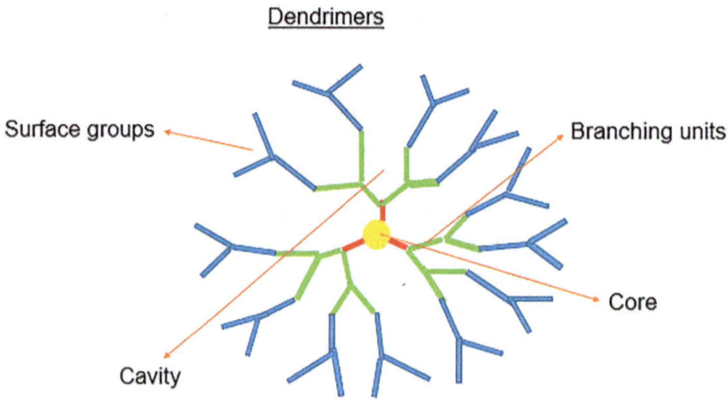

Fig. 25.4 Schematic diagram of a dendrimer showing the core, cavity, branching units, and surface groups. Here various drugs can be entrapped inside the dendrimer

Table 25.2 Dendrimers co-loaded with anticancer agents for lung cancer therapy

Polymers used	Loaded anticancer drugs	Targeted cancer type	Therapeutic outcomes	Factors that made dendrimers suitable for drug co-delivery	Ref
PEG and PAMAM	siRNA and paclitaxel	Lung cancer (HT-1080 and A375 cells)	The suppressed growth rate of the tumor and superior anti-relapse effect	Uniform diameter of about 160 nm and sustained drug release	Li et al. (2018)
PAMAM	Erlotinib and survivin-shRNA	Lung cancer (H1975 cells)	Improved cellular uptake	The particle size of about 200 nm	Lv et al. (2018)
PAMAM and PEG	DOX and siRNA	Lung cancer (A549 cells)	Improved serum stability and enhanced transfection effectiveness	Triblock polymeric G4-PAMAM-PEG dendrimers with various functional groups	Biswas et al. (2013)
PAMAM	siRNA and cis-diamine platinum	Lung cancer (H1299 cells)	Cell growth inhibition of cancer cells	High drug encapsulation efficiency	Amreddy et al. (2018)

enhanced drug biocompatibility and bioavailability, and sustained and controlled drug release mechanisms (Aqil et al. 2013). Various anticancer agents have been co-encapsulated into polymeric dendrimers for lung cancer therapy (Table 25.2).

25.5.2.1 Dendrimers for Lung Cancer

Li et al. prepared PAMAM-PEG dendrimers co-loaded with siRNA and paclitaxel for lung cancer treatment (Li et al. 2018). The particle size studies of co-loaded dendrimers utilizing DLS showed a uniform diameter of about 160 nm, indicating their suitability for therapeutic distribution to cancer cells. The drug release experiments displayed an initial rapid release of siRNA and paclitaxel from the dendrimers, followed by a slow and sustained drug release mechanism after 12 h. The cytotoxicity results showed that the dual drug-loaded dendrimers possessed stronger growth inhibition effects on HT-1080 and A375 lung tumor cell lines than the single drug-loaded dendrimers after 48 h of incubation, demonstrating good anticancer activity for combination therapy of siRNA and paclitaxel using dendrimers. The in vivo antitumor studies using a tumorigenesis-relapse model of A375 cell-xenografted mice revealed that co-loaded polymeric dendrimers displayed the lowest growth rate of the tumor and superior anti-relapse effect with the highest survival rate in comparison with the control (Taxol) and free drugs (Li et al. 2018).

Lv et al. reported aptamer-modified PAMAM dendrimers for survivin-shRNA co-delivery and erlotinib (Lv et al. 2018). Chloroquine was incorporated with dendrimers to regularize tumor vessels for necessary gene/drug delivery to overcome drug resistance in non-small cell lung cancer (NSCLC). TEM results of nanoparticles displayed spherically shaped particles with a particle size of approximately 200 nm that can significantly improve cellular uptake. The in vitro drug release of the PAMAM dendrimers at pH 7.4 or 5.4 was a biphasic pattern that was characterized by a rapid release followed by a sustained release under acidic conditions. This research showed that the normalization of tumor vessels promoted the intracellular survivin-shRNA/erlotinib delivery, and the downregulation of survivin acted synergistically with erlotinib to reverse erlotinib resistance in EGFR mutation-positive NSCLC.

Biswas et al. developed triblock polymeric generation 4 (G4)-PAMAM-PEG dendrimers to co-deliver DOX and siRNA. These nanomedicines demonstrated a high DOX and siRNA loading rate, improved serum stability due to the extended PEG chains, and improved transfection effectiveness due to lipid modification. Moreover, PAMAM dendrimers can also be used in inhaled nanocarriers for lung cancer therapy (Biswas et al. 2013). Amreddy et al. formulated folic-modified PAMAM dendrimer nanocarriers for combined cis-diamine platinum and siRNA delivery for lung carcinoma therapy. These dual drug-loaded dendrimers showed significant cell growth inhibition of more than 50% on H1299 lung cancer cells, indicating potential anticancer efficacy against lung cancer (Amreddy et al. 2018). To the best of our knowledge, no polymer-based dendrimers were reported for the co-delivery of anticancer drugs to prostate cancer cells.

25.5.3 Micelles

Micelles are self-aggregated or colloidal nanomedicines with a mean particle size of 5–100 nm (Fig. 25.5) (Valenzuela-oses et al. 2017). They are composed of surfactants or amphiphiles and comprise two regions: hydrophilic heads and hydrophobic tails (Zhong et al. 2019b). Critical micelle concentration (CMC) is a concentration whereby the micelles are formed (Chen et al. 2020). Various factors affect the development of these nanomedicines, such as the solvent employed during their formulation, the temperature, the concentration of amphiphiles, and the size of the hydrophobic domain in the amphiphilic molecule (Suroshe et al. 2019). The benefits of micelles include high drug encapsulation efficiency, high drug loading capacity, high cellular uptake, enhanced drug stability, and protection of normal body cells from drug toxicity, and they are simply removed from the system after biodegradation; they improve pharmacokinetic parameters of loaded anticancer drugs and are suitable for combination chemotherapy (Senevirathne et al. 2016). Various anticancer agents have been co-encapsulated into the micelles leading to an improved anticancer efficacy against lung and prostate cancer lines in the preclinical trials (Table 25.3).

25.5.3.1 Micelles for Lung Cancer

Ma et al. formulated HA-vitamin E succinate copolymer micelles co-loaded with curcumin and DOX for synergistic antitumor efficacy against lung carcinoma (Ma et al. 2017). The DLS analysis exhibited a mean particle size of about 223.8 nm and a surface charge of -10 ± 0.12 mV due to the carboxyl groups on the HA surface. The encapsulation rates of micelles for DOX and curcumin were 94.8% and 72.5%, respectively. The drug release studies of the drug-co-loaded micelles displayed sustained release at 7.4 (simulating physiological pH), 6.5 (simulating tumor extracellular pH), 5.5 (simulating endosomal pH), and pH 4.5 (simulating lysosomal pH). The in vitro antitumor experiments of the dual

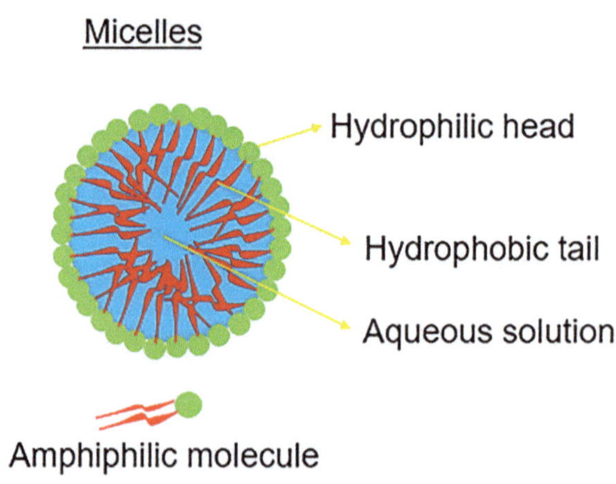

Fig. 25.5 Schematic diagram of a micelle

Table 25.3 Polymeric micelles co-loaded with anticancer agents for lung and prostate cancer therapy

Polymers used	Loaded anticancer drugs	Targeted cancer type	Therapeutic outcomes	Factors that made micelles suitable for drug co-delivery	Ref
HA	Curcumin and DOX	Lung cancer (MCF-7/Adr cells)	Sustained drug release with superior cytotoxicity on cancer cells	High encapsulation rates of more than 70% for both therapeutic agents	Ma et al. (2017)
PEG-pp-PEI-PE	PTX and siRNA	Lung cancer (A549 cells)	Increased cellular uptake with increased cytotoxicity on cancer cells	Multiple functional groups that can promote the co-loading of drugs	Zhu et al. (2014a)
mPEsG-b-PLG-b-PLL	PTX and DOX	Lung cancer (A549 cells)	Highly effective targeting and build-up at the implanted site of xenograft tumor	Multiple functional groups that can promote the co-delivery of drugs	Lv et al. (2014)
PEG	PTX and parthenolide	Lung cancer (A549-T24 and A549 cells)	Higher intracellular delivery of paclitaxel	High encapsulation efficiency of more than 90%	Gill et al. (2016)
PEG	Cisplatin and etoposide	Lung cancer (A549 cells)	Superior anticancer efficacy	Wide range of drug combination ratios and high two-drug loading of more than 50% wt	Wan et al. (2018)
mPEG	DOX and gene p53	Lung cancer (H1299 and A549 cells)	Effectively deliver and release co-loaded bioactive agents into cancer cells	Multiple functional groups	Liu et al. (2013)
PEG	PTX with shRNA	Lung cancer (A549 cells)	Superior antitumor activity	High drug loading capacity	Shen et al. (2012)

(continued)

Table 25.3 (continued)

Polymers used	Loaded anticancer drugs	Targeted cancer type	Therapeutic outcomes	Factors that made micelles suitable for drug co-delivery	Ref
Polymetformin	Cisplatin prodrug and IL-12 cytokine gene	Lung cancer (A549 cells)	Hindering of the carcinoma growth and prolonged overall survival of tumor-bearing mice	High encapsulation rates	Sun et al. (2021a)
PEG	PTX and siRNA	Lung cancer (A549 cells)	Improved cell apoptosis of lung cancer cell lines	A sustained drug release mechanism	Zhu et al. (2014b)
PEG	PTX and embelin	Prostate cancer (PC3 cells)	Excellent anticancer activity	The high drug loading efficiency of more than 60%	Lu et al. (2013)
PEG	DOX and dasatinib	Prostate cancer (PC3 cells)	Sustained drug release with significant cytotoxic effect on prostate cells	The high drug loading efficiency of more than 90%	Zhang et al. (2015)
Hydrazone-caproyl-maleimide	DOX and si-Beclin1	Prostate cancer (PC3 cells)	Higher cytotoxicity and apoptosis in cancer cells	Multiple available functional groups for drug incorporation	Hu et al. (2021)
PEG	DTX and siRNA	Prostate cancer (PC3 cells)	Prolonged blood circulation, selective targeting, and improved anticancer effects	High drug loading and encapsulation efficiency	Zhang et al. (2022a)
PEG	DTX and retinoic acid	Prostate cancer (22Rv1 and C4-2 cells)	Synergistic anticancer effect with improved cytotoxicity	High drug loading capacity	Hu and Xu (2021)
Polylactide and polyethylene	Marizomib and crizotinib	Prostate cancer (LNCAP and PC3pip cells)	Dose-dependent cytotoxic effects	Multiple functional groups	Ma and Dong (2021)
PEG	PTX and lapatinib	Prostate cancer (DU145-TXR cells)	Good antitumor activity	Multiple functional groups	Li et al. (2011)

drug-loaded micelles employing MTT analyses demonstrated superior cytotoxicity on MCF-7/Adr lung cancer cells when compared to the individual drug solutions of either doxorubicin-curcumin or doxorubicin (Ma et al. 2017).

The PEG-pp-PEI-PE micelles reported by Zhu et al. for the co-delivery of siRNA and PTX to lung cancer cells demonstrated significant increased cellular uptake from 400% to 650% after the cleavage of matrix metalloproteinase 2 when incubated with A549 lung cancer cells (Zhu et al. 2014a). These results revealed polymeric micelles' capability to induce cellular internalization and subsequently result in good anticancer activity. The in vitro synergistic experiments of micelles, when cultured with A549 T24 or A549 cells, revealed increased cytotoxicity of PTX than the free PTX, resulting from an enhanced cellular uptake of the drug-loaded micelles (Zhu et al. 2014a).

Lv et al. formulated deoxycholate-decorated mPEsG-b-PLG-b-PLL micelles co-encapsulated with PTX and DOX for lung carcinoma therapy. The in vitro cytotoxicity experiments of co-delivered nanoparticles showed a synergistic anti-cancer effect on A549 lung cells by promoting apoptosis of cancer cells. Furthermore, the ex vivo DOX fluorescence imaging showed that these micelles resulted in highly effective targeting and accumulation at the in vivo implanted site of the A549 xenograft tumor (Lv et al. 2014). Gill et al. fabricated EGFR-functionalized PEG micelles for the combined delivery of PTX and parthenolide. The DLS analysis of micelle nanomedicines exhibited a particle size of 15 ± 3 nm with a negative zeta potential of -34 mV. The cellular uptake studies showed that the EGF-targeted micelles resulted in superior intracellular delivery of paclitaxel than the non-targeted micelles in both A549-T24 and A549 lung cancer cell lines (Gill et al. 2016). The amphiphilic block copolymer micelles formulated by Wan et al. for combination chemotherapy of cisplatin and etoposide displayed superior anticancer efficacy than the individual drug-loaded micelles or free drugs against NSCLC, which could be due to synergetic antitumor effects (Wan et al. 2018).

Liu et al. synthesized mPEG-based micelles co-encapsulated with DOX and tumor suppressor gene p53 that showed superior antitumor activity on H1299 and A549 lung carcinoma cells with effective delivering of the co-loaded bioactive agents in A549 cells than the single drug or gene therapy (Liu et al. 2013). The vitamin E-coated (d-α-tocopherol PEG glycol 1000 succinate) micelles fabricated by Shen et al. for simultaneous delivery of PTX with shRNA showed an IC_{50} value lower than that of the free drug by 360-fold, indicating excellent anticancer activity. In addition, the in vivo experiments employing A549 tumor-induced mice models exhibited superior antitumor activity for the co-loaded micelles (Shen et al. 2012). Sun et al. formulated HC/pIL-12/polymetformin micelles co-loaded with cisplatin prodrug and IL-12 cytokine gene for lung cancer treatment. The prolonged circulating micelles promoted the accumulation of bioactive in the tumor, which led to significant inhibition of the lung carcinoma growth and extended the survival rate of the cancer-bearing mice (Sun et al. 2021a). Zhu et al. designed self-assembled block copolymer micelles for combination therapy of PTX and siRNA. These polymeric micelles showed several significant properties, e.g., passive tumor targeting, enhanced stability, and improved cell apoptosis of lung cancer cell lines (Zhu et al. 2014b).

25.5.3.2 Micelles for Prostate Cancer

Lu et al. formulated PEG-based micelles for co-distribution of embelin (a natural product with anticancer efficacy) and PTX with a low CMC value of 0.002 μM. The particle size analysis of micelles displayed a particle size that ranges between 20 and 30 nm with or without encapsulated drugs. The in vitro cell viability experiments of the micelles containing PTX and embelin showed more potent anticancer efficacy than Taxol in PC3 and DU145, which are the two androgen-independent human prostate cancer cell lines (Lu et al. 2013). Zhang et al. reported PEG-lysyl-(-α-Fmoc-ε-Cbz-lysine)$_2$ for potential combination chemotherapy of dasatinib (oncogenic tyrosine kinases inhibitor) and DOX in the treatment of various cancer types. The in vitro drug release kinetics in physiological environments exhibited a significantly sustained release of both drugs from polymeric micelles than the free drugs. The cytotoxicity assay showed that the drug-free micelles did not induce a cytotoxic effect on PC3 cells. In contrast, dual drug-loaded micelle nanocarriers improved the inhibition of cell proliferation which could be due to a strong synergistic effect of DOX and dasatinib (Zhang et al. 2015). The hydrazone-caproyl-maleimide micelles formulated by Hu et al. for the co-delivery of DOX and si-Beclin1 demonstrated 2.7-fold more significant cytotoxicity and cell apoptosis in PC-3 cancer cells when compared to the free DOX, indicating that the si-Beclin1 hindered the autophagy functioning of prostate cancer cells by targeting the type III PI3K pathway and enhanced the sensitivity to the DOX. The in vivo studies showed that the micelles co-loaded with DOX and si-Beclin1 significantly resulted in a 3.4-fold more considerable tumor inhibition, revealing a synergistic anti-proliferative effect (Hu et al. 2021).

Zhang et al. designed micelles to deliver DTX and siRNA for castration-resistant prostate cancer therapy. These micelles showed selective targeting, extended blood circulation, and improved anticancer effects (Zhang et al. 2022a). Hu and Xu synthesized PEG-based micelles co-loaded with DTX and retinoic acid for synergistic prostate cancer therapy. These co-delivered micelles displayed enhanced cellular uptake than the free drugs. The cytotoxicity analysis using 22Rv1 and C4-2 cells showed that DTX in combination with retinoic acid showed a synergistic anticancer effect, and the dual drug-loaded micelles showed improved cytotoxicity against 22Rv1 and C4-2 prostate cancer cells than the free drugs, indicating a significant potential for prostate cancer therapy (Hu and Xu 2021). Ma and Dong developed polylactide-polyethylene succinate glycol micelle nanomedicines for the co-delivery of marizomib and crizotinib to prostate cancer cell lines. The cytotoxicity tests of micelles showed significant dose-dependent cytotoxic effects in both LNCAP and PC3pip prostate cancer cells (Ma and Dong 2021). Li et al. formulated the PEG-b-poly(2-methyl-2-carboxylpropylene carbonate-graft-dodecanol) micelles for combination therapy of PTX and lapatinib destroyed 70% and 80% of DU145-TXR prostate cancer cells, indicating good antitumor activity. At the same time, monotherapy did not show any cytotoxic effect (Li et al. 2011).

25.5.4 Polymer-Drug Conjugates

Polymer-drug conjugates, also known as polymeric prodrugs, are nanocarriers fabricated for chemically incorporating bioactive molecules into the polymers employing chosen functionalities (e.g., amides, alcohols, esters, hydrazine, etc.). These nanomedicines comprise three components: therapeutic agent, targeting moiety, and solubilizing unit. These components are combined via covalent bonds into the backbone of the polymer. This model was firstly proposed by Helmut Ringsdorf in 1975 (Fig. 25.6) (Elvira et al. 2005; Pang and Yang 2014). The solubilizing unit improves the water solubility of polymer prodrugs, while the targeting moiety enhances its delivery to the targeted biological site (Larson and Ghandehari 2012). Examples of targeting moieties that are typically used include engineered antibodies, peptides, folic acid, and sugars (Kumar 2012). The linkers or functionalities play a vital role in releasing the incorporated therapeutic agent under certain conditions, such as the change in pH, the presence of enzymes, and overexpressed groups of diseased tissue (Patil and Mahajan 2015). The polymers that can be employed for the preparation of polymer-drug conjugates include polyamidoamine carriers, polyglutamic acid, PEG, polyaspartamides, etc. (Marasini et al. 2017). The advantages of polymer-drug conjugates include enhanced drug solubility, reduced drug toxicity, improved drug bioavailability, enhanced pharmacokinetics, pharmacodynamics, and pharmacological properties. Polymer-drug conjugates can be used for combination therapy, which is very useful for the treatment of various diseases including cancer (Sanchis et al. 2010). Moreover, these nanomedicines preserve and protect loaded drugs during circulation from attacks of enzymes (Alven et al. 2019). There are very few polymeric prodrugs that have been reported for combination chemotherapy to treat lung and prostate cancer (Table 25.4).

25.5.4.1 Polymer-Drug Conjugates for Lung Cancer

Hong et al. formulated U11 peptide-functionalized pH-sensitive polymeric prodrugs co-incorporated with DOX and curcumin through an amide linkage to lung cancer cells (Hong et al. 2019). The TEM results of the nanomedicines exhibited a spherically shaped morphology with a core-shell structure. The DLS analysis employing Zetasizer Nano average displayed an average particle size of about 121.3 nm and zeta potential of −33.5 mV, with a DOX and curcumin encapsulation rate of 81.7

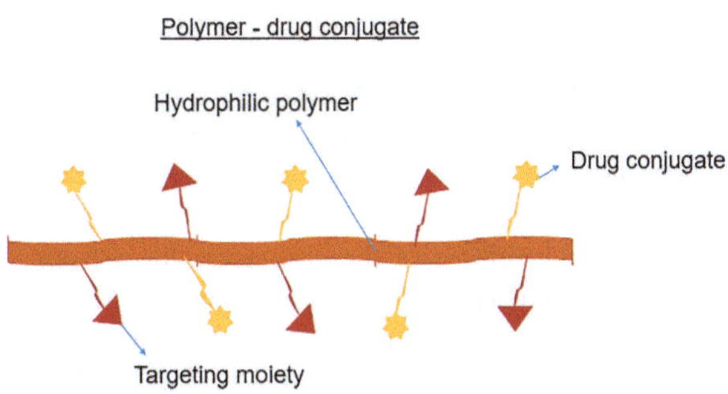

Fig. 25.6 Structure of a polymer-drug conjugate

Table 25.4 Polymeric prodrugs co-loaded with anticancer agents for lung and prostate cancer therapy

Polymers used	Loaded anticancer drugs	Targeted cancer type	Therapeutic outcomes	Factors that made polymeric prodrugs suitable for drug co-delivery	Ref
U11 peptide	DOX and curcumin	Lung cancer (A549 cells)	pH-dependent release kinetics, higher cellular uptake, and good anticancer activity	The particle size of about 121.3 nm and the high encapsulation rate	Hong et al. (2019)
Glutathione	DOX and cisplatin	Lung cancer (A549 cells)	Superior tumor inhibition efficiency	The particle size of 128.6 nm and dispersive index of 0.196 ± 0.021	Jin et al. (2020)
Poly(oligo (ethylene glycol) methacrylate)	DOX and dasatinib	Prostate cancer (PC3 cells)	Slow sustained drug release with good anticancer efficacy	The high drug loading efficiency of 90% and above	Sun et al. (2017)

and 90.5%, respectively. The in vitro drug release analysis performed at pH 5.0, 6.0, and 7.4 exhibited the pH-dependent release kinetics for both drugs. U11 peptide-functionalized conjugates showed notably higher cellular uptake than the non-U11 peptide-functionalized conjugates. The cytotoxicity studies of the co-loaded conjugates showed better antitumor efficacy against A549 cells compared to the plain drugs. The in vivo studies showed that the decorated nanocarriers possessed enhanced tumor growth inhibition than the non-decorated nanocarriers and free drugs. The sustained release kinetics of the dual drug-loaded polymeric prodrugs significantly reduced systematic toxicity and enhanced anticancer efficacy (Hong et al. 2019). The DLS analysis of polymeric prodrugs designed by Jin et al. for simultaneous delivery of DOX and cisplatin displayed particle size of 128.6 ± 3.2 nm, a zeta potential of 15.7 ± 1.7 mV, and PDI of 0.196 ± 0.021. These polymer-drug conjugates showed superior tumor inhibition efficiency of about 79.9% in A549 lung cancer-induced BABL/c mice with slight body weight loss, demonstrating synergistic anticancer effects against lung cancer (Jin et al. 2020).

25.5.4.2 Polymer-Drug Conjugates for Prostate Cancer

Sun et al. developed poly(oligo(ethylene glycol) methacrylate)-based polymeric prodrugs co-loaded with DOX and dasatinib to treat various cancer types. The in vitro drug release kinetics at physiological conditions exhibited slow and sustained drug release of the co-loaded drugs from the polymeric prodrugs. The in vitro cytotoxicity analysis revealed that the dual drug-loaded polymeric prodrugs possessed lower IC_{50} in the PC-3 prostate cancer cell line than the free drugs and single drug-loaded prodrugs, demonstrating good anticancer efficacy of co-loaded prodrugs that might be caused by a synergistic effect of dasatinib and the co-loaded DOX (Sun et al. 2017).

25.5.5 Polymeric Nanocapsules

Nanocapsules are polymer-based nanomedicines made up of a core and protective shell where bioactive agents are loaded (Fig. 25.7) (Kothamasu et al. 2012). Different techniques can be used to prepare these nanomedicines, such as the solidification of droplet shells, self-assembly of block copolymers, nanoemulsion polymerization, etc. (Ariga et al. 2011). The average particle size of nanocapsules ranged between 10 and 1000 nm. Nanocapsules exhibit several advantages in drug delivery and biomedicines, including reduced drug toxicity, high drug encapsulation efficiency, controlled and sustained drug release kinetics, and enhanced drug bioavailability and biodegradability (Alven and Aderibigbe 2020). Several polymeric nanocapsules have been reported for combination chemotherapy to treat lung and prostate cancer (Table 25.5).

25.5.5.1 Nanocapsules for Lung Cancer

Rudnik et al. formulated PEG-PCL nanocapsules to co-deliver curcumin and methotrexate to lung cancer cells. The DLS results of polymeric nanocapsules displayed an average particle size that ranged between 287.83 and 325.16, PDI varying from 0.290 to 0.351, and negative zeta potential between −33 and −41 mV with a high encapsulation rate of 82.4–99.4%. These nanocapsules showed prolonged drug release kinetics with no rapid effect when compared to the free drugs. The in vitro cytotoxicity experiments of dual drug-loaded nanocapsules using MTT assay exhibited an enhanced reduction in Calu-3 lung cancer cell viability when 4.94:

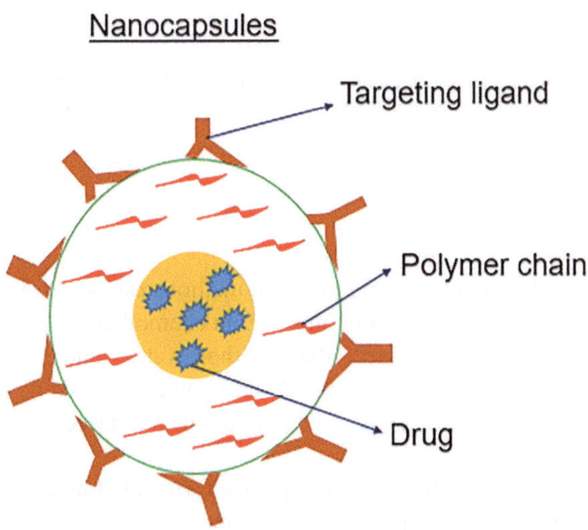

Fig. 25.7 Structure of nanocapsule

Table 25.5 Polymeric nanocapsules co-loaded with anticancer agents for lung and prostate cancer therapy

Polymers used	Loaded anticancer drugs	Targeted cancer type	Therapeutic outcomes	Factors that made nanocapsules suitable for drug co-delivery	Ref
PEG and PCL	Curcumin and methotrexate	Lung cancer (Calu-3 cells)	Higher cytotoxicity against cancer cells	High encapsulation efficiency of 90%	Rudnik et al. (2020)
PEG and PCL	Methotrexate and cisplatin	Lung cancer (A549 cells)	Reduced cell viability of cancer cells	Narrow particle size distribution with an average particle diameter of about 100 nm	Akbari et al. (2022)
HA and chitosan	Indocyanine green and gold nanoparticles	Lung cancer (A549 cells)	Inhibition of tumor volume growth	Regular round sphere with a particle size of approximately 250 nm	Zhu et al. (2022)
mPEG	PTX and erlotinib	Lung cancer (Calu-3 cells)	Time- and concentration-dependent cellular uptake with higher antitumor activity	High drug loading and loading efficiency	Gupta et al. (2018)
Albumin	Etoposide and berberine	Lung cancer (A549 cells)	Improved internalization and cytotoxicity into cancer cells	Particle size diameter of about 148 nm	Elgohary et al. (2018)
PCL	DOX and siRNA	Prostate cancer (PC3 cells)	Higher antitumor activity	High encapsulation efficiency	Chen et al. (2014)
PLA and PEG	Carboplatin and NaHCO3	Prostate cancer (RM-1 and PC3 cells)	Potential cellular uptake ability with improved anticancer efficacy	Particle sizes that range between 88 and 144 nm	Fu et al. (2020)

2 and 9.88:4 ratios of curcumin/methotrexate were evaluated, indicating a synergistic anticancer effect of combination chemotherapy against lung cancer (Rudnik et al. 2020). Furthermore, the methotrexate- and cisplatin-containing PEG-PCL nanocapsules reported by Akbari showed reduced cell viability of A549 non-small cell lung cancer cells, revealing anticancer effects against lung carcinoma (Akbari et al. 2022).

The HA/chitosan nanocapsules formulated by Zhu et al. for concurrent delivery of indocyanine green and gold nanoparticles demonstrated that lung tumor-bearing mice without treatments showed sustained growth of tumor volume. Still, the tumors in mice treated with nanocapsules showed significant inhibition of the tumor (Zhu et al. 2022). Gupta et al. designed mPEG-based nanocapsules co-loaded with PTX and erlotinib for lung cancer therapy. These nanocapsules showed time- and concentration-dependent cellular uptake by NCI-H23 lung cancer cells with a significantly higher antitumor activity than the free drugs and PTX/erlotinib (Gupta et al. 2018). Elgohary et al. prepared boronic acid-decorated albumin nanocapsules for the combined delivery of etoposide and berberine to treat cancer. TEM images showed spherically shaped morphology and smooth surface non-decorated and decorated nanoparticles with diameters of 148.3 and 201.21 nm, respectively. Furthermore, boronic acid-decorated albumin nanocapsules showed significantly improved internalization and cytotoxicity in the A549 lung cancer cells through binding to sialic acid residues overexpressed by cancer cells than the non-decorated nanocapsules and free drugs (Elgohary et al. 2018).

25.5.5.2 Nanocapsules for Prostate Cancer

Chen et al. formulated allyl-functionalized PCL nanocapsules for combination chemotherapy of DOX and siRNA. TEM micrographs of nanocapsules confirmed their capsule-like morphology with surface dimensions of about 50 nm. The in vitro cytotoxicity assays showed significantly higher antitumor efficacy of the dual drug-loaded nanocapsules to the PC3 prostate cancer cells than the free DOX (Chen et al. 2014). Fu et al. designed PLA-PEG nanocapsules co-loaded with carboplatin and $NaHCO_3$ for prostate cancer therapy. The in vitro release experiment showed a significantly rapid release of bioactive agents from nanocapsules in a pH of 5.0 (simulating a cancer cell environment) compared to the neutral pH (pH 7.4). The cellular uptake analysis showed potential cellular uptake ability into RM-1 cells, indicating the capacity to improve the anticancer efficacy of loaded drugs. The in vitro cytotoxicity studies showed superior antitumor activity for dual drug-loaded polymeric nanocapsules when compared to the carboplatin solution on RM-1 and PC3 prostate cancer cell lines, suggesting synergistic anticancer effects of carboplatin and $NaHCO_3$ loaded in nanocapsules (Fu et al. 2020).

25.5.6 Nanoliposomes

Nanoliposomes are smaller and artificial spherical-shaped nanomedicines fabricated from cholesterol and phospholipid (Fig. 25.8). Nanoliposomes were discovered in the 1960s by Bangham and co-workers. These nanomedicines exhibit hydrophilic and hydrophobic properties with a size range of 30 nm to a few micrometers (Danaei et al. 2018). The features of nanoliposomes are different depending on the particle size, surface charge, formulation method, and composition of the lipids. Nanoliposomes are comprised of lipid bilayers encircled by aqueous constituents, whereby the polar head groups are oriented in the pathway of the exterior and interior aqueous phases. The form of nanoliposomes' bilayers, such as rigidity,

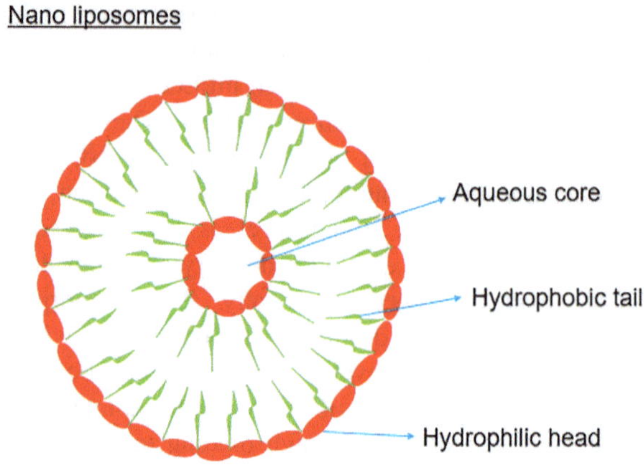

Fig. 25.8 A schematic diagram of a nanoliposome

bilayer charge, and permeability, is influenced by the components employed for their formulation. These nanomedicines are appropriate for drug delivery of hydrophobic and hydrophilic bioactive agents (Khosravi-Darani and Mozafari 2010). The advantages of nanoliposomes include good biocompatibility and biodegradability, reduced drug toxicity to normal body cells, site-specific drug delivery, usefulness for the co-delivery of therapeutic agents, etc. (Shade 2016). All these features of nanoliposomes make them appropriate for the co-delivery of anticancer drugs to treat lung and prostate cancer (Table 25.6).

25.5.6.1 Nanoliposomes for Lung Cancer

Ghosh et al. formulated PEG-based nanoliposomes for the co-delivery of DOX and vincristine to lung and breast cancer cell lines (Ghosh et al. 2021). These nanoliposomes displayed significantly higher encapsulation efficiency of more than 95% for both antitumor drugs. The DLS analysis of co-loaded nanoliposomes displayed particle size of about 95 nm, PDI of less than 0.1, and adverse surface charge of -9.17 ± 1.19 mV. The in vitro experiments showed higher cellular uptake of nanoliposomes by A549 lung cancer cells that were visualized through confocal microscopy. At the same time, the MTT assay demonstrated IC_{50} values for the dual-drug nanoliposome, which showed an eightfold reduction in cellular viability in A549 cells as compared to DOX-loaded liposomal nanomedicines, indicating superior anticancer activity of co-loaded nanoliposomes than the single drug-loaded nanoliposomes that is due to synergistic antitumor activity. Furthermore, the in vivo anticancer experiments using A549 tumor xenograft models exhibited a more significant reduction in tumor volume than the DOX-encapsulated liposomes, vincristine-encapsulated liposomes, and a mixture of liposomes.

The methoxy-PEG-2000 nanoliposomes prepared by Yang et al. for the co-delivery of DOX and lapatinib demonstrated efficient delivery of DOX into GLC-82 lung adenocarcinoma cells because the cancer cells effectively internalize liposome nanomedicines through the hydrophobic and electrostatic interactions between their membranes (Yang et al. 2019). Li et al. formulated

Table 25.6 Polymeric nanoliposomes co-loaded with anticancer agents for lung and prostate cancer therapy

Polymers used	Loaded anticancer drugs	Targeted cancer type	Therapeutic outcomes	Factors that made nanoliposomes suitable for drug co-delivery	Ref
PEG	DOX and vincristine	Lung cancer (A549 cells)	High cellular uptake and superior anticancer activity	High encapsulation efficiency of about 95% for both drugs with a particle size of 95 nm	Ghosh et al. (2021)
PEG	DOX and lapatinib	Lung cancer (GLC-82 cells)	Efficient delivery of loaded drugs into cancer cells	High loading rate without drug leakage	Yang et al. (2019)
PEG	DTX and baicalein	Lung cancer (A549 cells)	Good anticancer efficacy	High drug loading and loading efficiency	Li et al. (2017)
PEG	GEM and cisplatin	Lung cancer (A549 cells)	Better inhibition of cellular proliferation and lower cell viability of cancer cells	Various functional groups for drug incorporation	Liu et al. (2021)
PEG	Cabazitaxel and β-elemene	Lung cancer (A549/T cells)	Superior anticancer activity	Encapsulation efficiencies of more than 95%	Zeng et al. (2019)
PEG	DXT and shRNA	Lung cancer (A549 cells)	Higher reduction in tumor volume	High entrapment efficiencies that range between 43.6 and 98.9%	Chowdhury et al. (2017)
PEG and PLA	Cisplatin and siRNA	Lung cancer (A549 cells)	Excellent pharmacokinetic properties and potential antitumor effects	High entrapment efficiencies of more than 70% for both drugs	Patel et al. (2021)
PEG	Curcumin and SN38 prodrug	Lung cancer (A549 cells)	Most potent inhibitory effects and more significant cell apoptosis in cancer cells	Mean particle size of about 171.21 ± 1.10 nm	Gao et al. (2022)

(continued)

Table 25.6 (continued)

Polymers used	Loaded anticancer drugs	Targeted cancer type	Therapeutic outcomes	Factors that made nanoliposomes suitable for drug co-delivery	Ref
PEG	DTX and resveratrol	Prostate cancer (PC3 cells)	Sustained drug release kinetics and potential antitumor efficiency	Spherically shaped morphology with an average particle size of about 99.67 nm	Zhang et al. (2022b)
PEG	DOX and tanshinones	Prostate cancer (LNCaP cells)	Higher cellular uptake with inhibition effects on tumor development	High encapsulation efficiencies	Sun et al. (2021b)
PEG	Curcumin and genistein	Prostate cancer (PC3 cells)	Superior cell growth inhibition in cancer cells	High drug entrapment efficiency of more than 75%	Aditya et al. (2013)

transferrin-decorated PEG nanoliposomes for combination chemotherapy of DTX and baicalein. The antitumor studies demonstrated better anticancer efficacy for dual drug-loaded transferrin-decorated PEG nanoliposomes in A549 cells than the non-decorated nanoliposomes and single drug-loaded nanoliposomes with the significant synergistic anticancer effects (Li et al. 2017). The micelle-containing PEGylated hybrid nanoliposomes reported by Liu et al. for simultaneous delivery of GEM and cisplatin exhibited better inhibition of cellular proliferation and lower cell viability of A549 lung cancer cells at all evaluated concentrations when compared to GEM/cisplatin mixture, demonstrating good anticancer activity of hybrid nanoliposomes co-encapsulated with GEM and cisplatin (Liu et al. 2021). Zeng et al. formulated nanoliposomes co-loaded with cabazitaxel and β-elemene for paclitaxel-resistant lung adenocarcinoma treatment. The in vivo anticancer studies showed that dual drug-loaded nanoliposomes had a superior anticancer effect on nude mice utilizing the human paclitaxel-resistant A549/T cells when compared to the free drugs and blank nanoliposomes (Zeng et al. 2019).

Chowdhury et al. formulated DXT and shRNA plasmid-containing nanoliposomes that showed a 3.8-fold reduction in tumor volume in the A549 lung cancer xenograft model, which was significantly higher when compared to the free drugs (Chowdhury et al. 2017). The PEG-PLA di-block copolymer nanoliposomes that Patel et al. reported for combination chemotherapy of cisplatin and siRNA demonstrated significantly increased half-life and tumor regression in an A549 xenograft model in BALB/c nude mice, suggesting excellent pharmacokinetic properties and potential antitumor effects for the treatment of lung cancer (Patel et al. 2021). Gao et al. formulated PEG-based nanoliposomes co-encapsulated with curcumin and SN38 prodrug for lung cancer therapy (Gao et al. 2022). The DLS results of co-encapsulated nanoliposomes displayed a mean particle size of 171.21 ± 1.10 nm and negative zeta potential of 5.96 ± 0.32, revealing that the dual drug-loaded nanoliposomes possess the good potential to improve chemotherapy activity because of their suitable particle size that is within the range of enabled endocytosis. The in vitro cytotoxicity studies of co-loaded nanoliposomes using CCK-8 assay exhibited the most potent inhibitory effects in A549 cell lines than the free drugs. The dual drug-loaded nanoliposomes showed significantly greater cell apoptosis of 49.79% when compared to the free curcumin (14.61%) and SN38 prodrug (36.39%).

25.5.6.2 Nanoliposomes for Prostate Cancer

Zhang et al. formulated PEGylated nanoliposomes co-loaded with DTX and resveratrol for prostate cancer therapy. The nanoliposome particle size analysis displayed a mean diameter of about 99.67 nm with a spherically shaped morphology. The in vitro drug release studies at physiological conditions showed a sustained release of DTX and resveratrol. The in vivo anticancer experiments using PC3-bearing male Balb/c nude mice model exhibited potential antitumor efficiency for dual drug-loaded nanoliposomes, while the saline-treated groups demonstrated rapidly increased tumor weight (Zhang et al. 2022b). Sun et al. reported PEGylated nanoliposomes for co-delivery of DOX and tanshinones to prostate cancer cells. The DLS analysis of co-encapsulated nanoliposomes exhibited an average particle size and positive surface charge of 139.7 ± 4.1 nm and 11.2 ± 1.6 mV, respectively. The dual drug-loaded nanoliposomes presented higher cellular uptake by the LNCaP prostate cancer cells of over 58.9 ± 1.9%, with more in vivo inhibition effect on tumor growth than DOX and tanshinones which could be due to cell internalization and synergistic anticancer effect (Sun et al. 2021b). The curcumin- and genistein-containing nanoliposomes formulated by Aditya et al. exhibited significantly higher drug entrapment efficiency of more than 75% with superior cell growth inhibition in PC3 prostate cancer cells, indicating good effective antitumor effectiveness of liposomes for the treatment of prostate cancer (Aditya et al. 2013).

25.5.7 Other Nanomedicines Co-loaded with Anticancer Drugs for Lung and Prostate Cancer Therapy

Several other nanomedicines can be used as potential delivery systems for treating lung and prostate tumors. Those nanomedicines include nanogels/hydrogels, exosomes, niosomes, etc. Figure 25.9 represents the diagram of some nanomedicines. Some polymeric nanoparticles have been reported for combination chemotherapy to treat lung and prostate cancer (Table 25.7).

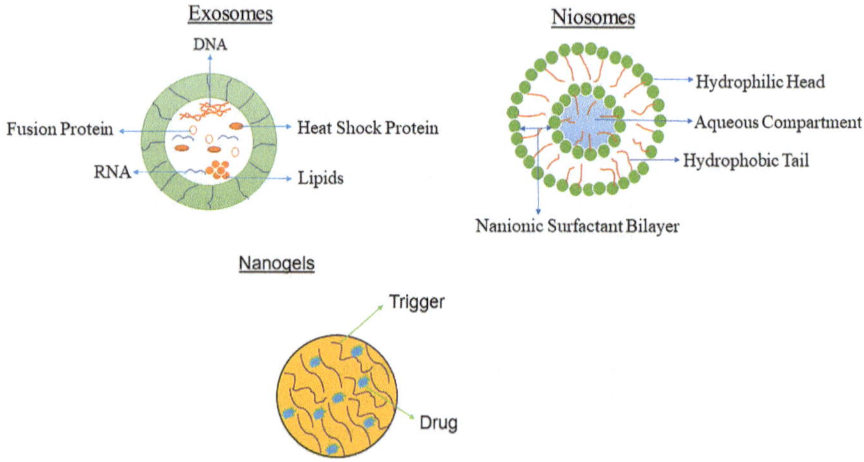

Fig. 25.9 Schematic diagram of exosome, noisome, and nanogel

25.5.7.1 Other Nanomedicines for Co-delivery of Anticancer Drugs to Lung Cancer Cells

Lv et al. formulated polypeptide hydrogels co-loaded with DOX and cisplatin for lung cancer therapy (Lv et al. 2020). The in vitro cytotoxicity assays showed that the combination chemotherapy of DOX and cisplatin using hydrogels significantly resulted in synergistic anticancer effects by inhibiting both A549 and A549/CDDP lung cancer cells with increased inhibition of the expression of the anti-apoptosis B-cell lymphoma 2 and multidrug-resistant (MDR) genes. Moreover, the in vivo studies of DOX/cisplatin-co-encapsulated polymeric hydrogels using nude mouse bearing A549 or A549/CDDP xenografts exhibited remarkably high antitumor activity than a single drug-loaded hydrogel or dual-drug solution. These preclinical results revealed that dual drug-loaded hydrogels are potential candidates for the treatment of lung carcinoma.

The 192.3 nm chondroitin-lactoferrin nanocomposites reported by Elwakil et al. for combined delivery of ellagic acid and DOX to lung cancer showed higher cellular internalization and superior cytotoxicity on A549 lung cancer cells mediated through Tf and CD44 receptors (Elwakil et al. 2018). Said-Elbahr et al. formulated nebulization-friendly nanoemulsions co-loaded with naringin and celecoxib for lung cancer treatment (Said-Elbahr et al. 2022). The DLS analysis of dual drug-loaded nanocomposites displayed particle sizes that ranged between 75 and 106 nm and negative zeta potential that ranged from −3.42 to −4.86 mV, suggesting the ability to promote superior internalization by cancer cells. The nanoemulsions showed good stability properties and better anticancer on A549 lung cancer cells.

Table 25.7 Polymeric nanomedicines co-loaded with anticancer agents for lung and prostate cancer therapy

Types of nanomedicines	Polymers used	Loaded anticancer drugs	Targeted cancer type	Therapeutic outcomes	Factors that made nanomedicines suitable for drug co-delivery	Ref
Hydrogels	Polypeptide	DOX and cisplatin	Lung cancer (A549 and A549/CDDP cells)	Synergistic anticancer effects	High drug loading capacity	Lv et al. (2020)
Nanocomposites	Chondroitin	Ellagic acid and DOX	Lung cancer (A549 cells)	Higher cellular internalization and superior cytotoxicity against cancer cells	The particle size of about 192.3 nm	Elwakil et al. (2018)
Nanoemulsions	–	Naringin and celecoxib	Lung cancer (A549 cells)	Superior anticancer activity on cancer cells	Particle sizes that ranged between 75 and 106 nm	Said-Elbahr et al. (2022)
Niosomes	PEG	miRNA-16-1 and miRNA-15a	Prostate cancer (PC3 cells)	Higher cytotoxicity against cancer cells	High drug loading efficiency	Ghaffari et al. (2021)
Nanovesicle	–	PSMA and PSA	Prostate cancer (C4-2B cells)	High cellular uptake with reduced tumor growth	High encapsulation efficiency	Ma et al. (2021b)

25.5.7.2 Other Nanomedicines for Co-delivery of Anticancer Drugs to Treat Prostate Cancer

Ghaffari et al. formulated PEGylated niosomes for the combined delivery of miRNA-16-1 and miRNA-15a to lung cancer cells (Ghaffari et al. 2021). The DLS analysis exhibited a particle size of 69.7 nm and a surface charge of +14.83 mV. The anticancer experiments showed that noisome-loaded miRNAs possessed superior cytotoxicity against PC3 prostate cancer cells when compared to the free miRNAs, suggesting synergistic anticancer activity of co-loaded miRNAs in niosomes. The bioinspired nanovesicle designed by Ma et al. for dual targeting of prostate-specific membrane antigen (PSMA) and prostate-specific antigen (PSA) demonstrated high cellular uptake with significantly reduced C4-2B prostate tumor growth in vivo than the free PSA-DOX and nontargeted nanovesicles (Ma et al. 2021b).

25.6 Nanomedicines in Clinical Trials for Lung and Prostate Cancer Treatment

Although a variety of nanomedicines have been reported for the treatment of lung and prostate cancer, only a few nanotherapeutics have reached clinical trials to date. Doxil-loaded liposomal nanomedicine is the first FDA-approved cancer nanomedicine sold under the tradename “Abraxane,” with improved circulation half-life and reduced toxic side effects. Other nanomedicines, such as Myocet, Marqibo, and DaunoXome, were later approved (Anselmo and Mitragotri 2016). The lipid-based nanomedicine (DOTAP/Chol TUSC2) is under phase I clinical trial which is being investigated for the treatment of metastatic lung cancer. In the nanoformulation, solid lipid nanoparticle (SLN)-carrier p53 was compared with Lipofectin, in which SLN-carrier p53 was found to be efficient and a potential therapeutic in treating the transfected p53-null H1299 lung cancer cells. This study demonstrated this approach’s effectiveness and encouraged a preclinical and clinical investigation for its application in gene/chemotherapy (Choi et al. 2008). Furthermore, one of the polymeric nanomedicines, “Genexol-PM,” is presently being evaluated for phase II clinical trials in NSCLC patients (Ahn et al. 2014).

In prostate cancer clinical trials, nanomicelles loaded with curcumin were administrated with radiotherapy in a phase II clinical trial (NCT02724618), while albumin nanoparticles co-encapsulated with PTX and lapatinib have accomplished a phase I trial (NCT00313599). Cyclodextrin for the delivery siRNA is another nanomedicine that reached a phase I trial ((NCT00689065) for prostate cancer therapy. Nevertheless, the study was stopped in 2013. A phase II trial (NCT02341404) was accomplished with calcium sulfate gel loaded with 2-hydroxyl flutamide in 2015. The adenoviral nanoformulations also reached clinical trials for prostate cancer therapy. The administration of IL-12 (NCT00406939) gene therapy has accomplished a phase I clinical trial. Likewise, another phase I trial is the administration of TLR5 receptor, and its agonist protein 502 s Mobilan (M-VM3) (NCT02654938) was accomplished in 2017. Besides, many natural compounds are verified to be potential therapeutics, including genistein, currently under phase II clinical trials as it presented multifactorial inhibition of cell invasion and proliferation (Lazarevic et al. 2011; Johnson et al. 2022). Another compound silibinin, acting efficiently against prostate cancer cell lines (DU145, LNCaP, and PC-3 cells), is in phase I/II clinical trials (NCT01871662) (Johnson et al. 2022). Although numerous articles have reported the potential of nanomedicines to treat lung and prostate cancer in the past decade, very few have reached clinical trials.

25.7 Conclusion and Future Perspective

Polymer-based nanomedicines reported in preclinical studies for the co-delivery of anticancer drugs have demonstrated promising and interesting therapeutic outcomes in treating lung and prostate cancer. These nanomedicines showed selective cytotoxicity because they were toxic against various cancer cell lines. At the same time, they possessed high cell viability on normal cell lines, suggesting reduced toxicity of co-loaded drugs. The superior anticancer activity of nanomedicines containing two or more anticancer agents than the free drugs and plain nanocarriers was promoted by various factors, such as initial burst drug release followed by a sustained release at tumor microenvironment, high cellular uptake and internalization by the cancer cells due to nanosized particle range, synergistic anticancer effects which significantly overcome MDR, and modification of nanocarriers by ligands. However, biomedical researchers must consider some drawbacks of nanomedicines, such as unclear anticancer mode of mechanism, toxicological profiles over a long-term biomedical application, high cost, etc. Moreover, there is a pressing need for more in vitro and in vivo studies of the dual drug-loaded nanomedicines that are presently reported for lung and prostate cancer therapy. Nevertheless, due to the significant therapeutic outcomes revealed by nanomedicines, there is no doubt that they will reach clinical application for the treatment of various cancer types in the near future.

Acknowledgments The financial support of the South African Medical Research Council (SAMRC) and National Research Foundation (NRF), South Africa, toward this research is acknowledged.

References

Aditya N, Shim M, Lee I, Lee Y, Im M-H, Ko S (2013) Curcumin and genistein coloaded nanostructured lipid carriers: in vitro digestion and antiprostate cancer activity. J Agric Food Chem 61:1878–1883

Ahn H, Jung M, Sym S, Shin D, Kang S, Kyung S et al (2014) A phase II trial of cremophor EL-free paclitaxel (Genexol PM) and gemcitabine in patients with advanced non-small cell lung cancer. Cancer Chemother Pharmacol 74(2):277–282

Akbari E, Mousazadeh H, Hanifehpour Y, Mostafavi E, Gorabi AM, Najati K et al (2022) Co-loading of cisplatin and methotrexate in nanoparticle-based PCL-PEG system enhances lung cancer chemotherapy effects. J Clust Sci 2022(33):1751–1762. https://doi.org/10.1007/s10876-021-02101-9

Ali I, Alsehli M, Scotti L, Tullius Scotti M, Tsai ST, Yu RS, Hsieh MF, Chen JC (2020) Progress in polymeric nano-medicines for theranostic cancer treatment. Polymers 12(3):598. https://doi.org/10.3390/polym12030598

Alven S, Aderibigbe BA (2020) Efficacy of polymer-based nanocarriers for co-delivery of curcumin and selected anticancer drugs. Nano 10:1556

Alven S, Aderibigbe BA, Balogun MO, Matshe WMR, Ray SS (2019) Polymer-drug conjugates containing antimalarial drugs and antibiotics. J Drug Deliv Sci Technol 53:101171. https://doi.org/10.1016/j.jddst.2019.101171

Alven S, Nqoro X, Buyana B, Aderibigbe BA (2020) Polymer-drug conjugate, a potential therapeutic to combat breast and lung cancer. Pharmaceutics. 12(5):406. https://doi.org/10.3390/pharmaceutics12050406

Amreddy N, Babu A, Panneerselvam J, Srivastava A, Ms RM, Ms AC et al (2018) Chemo-biologic combinatorial drug delivery using folate receptor-targeted dendrimer nanoparticles for lung cancer treatment. Nanomedicine 2018(14):373–384. https://doi.org/10.1016/j.nano.2017.11.010

Anselmo A, Mitragotri S (2016) Nanoparticles in the clinic. Bioeng Transl Med 1(1):10–29

Aqil F, Munagala R, Jeyabalan J, Vadhanam MV (2013) Bioavailability of phytochemicals and its enhancement by drug delivery systems. Cancer Lett 334:133–141. https://doi.org/10.1016/j.canlet.2013.02.032

Ariga K, Lvov Y, Kawakami K, Ji Q, Hill J (2011) Layer-by-layer self-assembled shells for drug delivery. Adv Drug Deliv Rev 63:762–771

Arya A, Khandelwal K, Ahmad H, Laxman S, Sharma K, Mittapelly N et al (2016) Co-delivery of hesperetin enhanced bicalutamide induced apoptosis by exploiting mitochondrial membrane potential via polymeric nanoparticles in aPC-3 cell line. RSC Adv 2016(6):5925–5935

Barqawi AB, Krughoff KJ, Eid K (2012) Current challenges in prostate cancer management and the rationale behind targeted focal therapy. Adv Urol 2012:862639. https://doi.org/10.1155/2012/862639

Barrett T, Haider MA (2017) The emerging role of MRI in prostate cancer active surveillance and ongoing challenges. Am J Roentgenol 208(1):131–139. https://doi.org/10.2214/AJR.16.16355

Barta JA, Powell CA, Wisnivesky JP (2019) Global epidemiology of lung cancer. Ann Glob Heal 85(1):1–16. https://doi.org/10.5334/aogh.2419

Beik J, Khateri M, Khosravi Z, Kamrava SK, Kooranifar S, Ghaznavi H et al (2019) Gold nanoparticles in combinatorial cancer therapy strategies. Coord Chem Rev 387:299–324. https://doi.org/10.1016/j.ccr.2019.02.025

Bhatia S (2016) Natural polymer drug delivery systems nanoparticles: nanoparticles, mammals and microbes, vol 1. Springer, pp 1–225. https://doi.org/10.1007/978-3-319-41129-3

Biswas B, Deshpande P, Navarro G, Dodwadkar N, Torchilin V (2013) Lipid modified triblock PAMAM-based nanocarriers for siRNA drug co-delivery. Biomaterials 34(4):1289–1301

Calabrò L, Peters S, Soria JC, Di Giacomo AM, Barlesi F, Covre A, Altomonte M, Vegni V, Gridelli C, Reck M, Rizvi N (2020) Challenges in lung cancer therapy during the COVID-19 pandemic. Lancet Respir Med 8(6):542–544. https://doi.org/10.1016/S2213-2600(20)30170-3

Caminade A-M (2020) Phosphorus dendrimers as nanotools against cancers. Molecules 25:3333

Chen C-K, Law W-C, Aalinkeel R, Yu Y, Nair B, Wu J et al (2014) Biodegradable cationic polymeric nanocapsules for overcoming multidrug resistance and enabling drug–gene co-delivery to cancer cells†. Nanoscale 6:1567–1572

Chen T, Tu L, Wang G, Qi N, Wu W, Zhang W et al (2020) Multi-functional chitosan polymeric micelles as oral paclitaxel delivery systems for enhanced bioavailability and anti-tumor efficacy. Int J Pharm 578:119105

Chen Y, Xu Z, Lu T, Luo J, Xue H (2022) Prostate-specific membrane antigen targeted, glutathione-sensitive nanoparticles loaded with docetaxel and enzalutamide for the delivery to prostate cancer. Drug Deliv 29(1):2705–2712. https://doi.org/10.1080/10717544.2022.2110998

Choi S, Jin S, Lee M, Lim S, Park J, Kim B et al (2008) Novel cationic solid lipid nanoparticles enhanced p53 gene transfer to lung cancer cells. Eur J Pharm Biopharm 68(3):545–554

Chowdhury N, Vhora I, Patel K, Doddapaneni R, Mondal A, Singh M (2017) Liposomes co-loaded with 6-phosphofructo-2-kinase/Fructose-2, 6-biphosphatase 3 (PFKFB3) shRNA plasmid and docetaxel for the treatment of non-small cell lung cancer. Pharm Res 3:2371–2384

Crocetto F, Russo G, Di Zazzo E, Pisapia P, Mirto BF, Palmieri A, Pepe F, Bellevicine C, Russo A, La Civita E, Terracciano D (2022) Liquid biopsy in prostate cancer management—current challenges and future perspectives. Cancers 14(13):3272. https://doi.org/10.3390/cancers14133272

Culp MBB, Soerjomataram I, Efstathiou JA, Bray F, Jemal A (2020) Recent global patterns in prostate cancer incidence and mortality rates. Eur Urol 77(1):38–52. https://doi.org/10.1016/j.eururo.2019.08.005

Dadashpour S, Emami S (2018) Indole in the target-based design of anticancer agents: a versatile scaffold with diverse mechanisms. Eur J Med Chem 150:9–29. https://doi.org/10.1016/j.ejmech.2018.02.065

Danaei M, Kalantari M, Raji M, Samareh F, Saber R, Asnani G et al (2018) Probing nanoliposomes using single particle analytical techniques: effect of excipients, solvents, phase transition and zeta potential. Heliyon 4:e01088

Dennis E, Peoples VA, Johnson F, Bibbs RK, Topps D, Coats MT (2015) Utilizing nanotechnology to combat malaria. Infect Dis Ther 3(4):1–6

Didkowska J, Wojciechowska U, Mańczuk M, Łobaszewski J (2016) Lung cancer epidemiology: contemporary and future challenges worldwide. Ann Transl Med 4(8):1–11. https://doi.org/10.21037/atm.2016.03.11

Elgohary MM, Helmy MW, Abdelfattah EA, Ragab DM, Mortada SM, Fang J-Y et al (2018) Targeting sialic acid residues on lung cancer cells by inhalable boronic acid-decorated albumin nanocomposites for combined chemo/herbal therapy. J Control Release 2018(285):230–243. https://doi.org/10.1016/j.jconrel.2018.07.014

Elvira C, Gallardo A, San Roman J, Cifuentes A (2005) Covalent polymer-drug conjugates. Molecules 10:114–125

Elwakil MMA, Mabrouk MT, Helmy MW, Abdelfattah E-ZA, Khiste KA, Elzoghby AO (2018) Inhalable lactoferrin—chondroitin nanocomposites for combined delivery of doxorubicin and ellagic acid to lung carcinoma. Nanomedicine 13(16):2015–2035

Fathi M, Vakili K, Jazi K, Sadeghi MA, Hajiesmaeili M, Mohamadkhani A et al (2021) Challenges of cancer immunotherapy and chemotherapy during the COVID-19 pandemic. Tumori 108(5): 407–419. https://doi.org/10.1177/03008916211063939

Flores-Téllez TDNJ, Baena E (2022) Experimental challenges to modeling prostate cancer heterogeneity. Cancer Lett 524:194–205. https://doi.org/10.1016/j.canlet.2021.10.012

Fu S, Liang S, Jiang D, Yang R, Zhang Z, Chang L et al (2020) Gas-blasting nanocapsules to accelerate carboplatin lysosome release and nucleus delivery for prostate cancer treatment. Asian J Pharm Sci 16(2):192–202. https://doi.org/10.1016/j.ajps.2020.05.002

Gadde S (2015) Multi-drug delivery nanocarriers for combination therapy. MedChemComm 6(11): 1916–1929. https://doi.org/10.1039/c5md00365b

Gao F, Sun Z, Kong F, Xiao J (2020) Artemisinin-derived hybrids and their anticancer activity. Eur J Med Chem 188:112044. https://doi.org/10.1016/j.ejmech.2020.112044

Gao C, Zhang L, Xu M, Luo Y, Wang B, Kuang M et al (2022) Pulmonary delivery of liposomes co-loaded with SN38 prodrug and curcumin for the treatment of lung cancer. Eur J Pharm Biopharm 2022(179):156–165. https://doi.org/10.1016/j.ejpb.2022.08.021

Ghaffari M, Kalantar SM, Mahdie K, Ali H, Firoozabadi AD, Asri A et al (2021) Co-delivery of miRNA-15a and miRNA-16-1 using cationic PEGylated niosomes downregulates Bcl-2 and induces apoptosis in prostate cancer cells. Biotechnol Lett 2021(43):981–994. https://doi.org/10.1007/s10529-021-03085-2

Ghosh S, Lalani R, Maiti K, Bunerjee S, Bhatt H, Bobde YS et al (2021) Synergistic co-loading of vincristine improved chemotherapeutic potential of pegylated liposomal doxorubicin against triple negative breast cancer and non-small cell lung cancer. Nanomedicine 2021(31):102320. https://doi.org/10.1016/j.nano.2020.102320

Gill KK, Kamal MM, Kaddoumi A, Nazzal S (2016) EGFR targeted delivery of paclitaxel and parthenolide co-loaded in PEG-phospholipid micelles enhance cytotoxicity and cellular uptake in non-small cell lung cancer cells. J Drug Deliv Sci Technol. 2016(36):150–155. https://doi.org/10.1016/j.jddst.2016.10.005

Goftar MK, Rayeni NA, Rasouli S (2014) Topoisomerase inhibitors and types of them. Int J Adv Biol Biom Res 2(8):2431–2436

Gupta B, Poudel BK, Regmi S, Pathak S, Ruttala HB, Gautam M et al (2018) Paclitaxel and erlotinib-co-loaded solid lipid core nanocapsules: assessment of physicochemical characteristics and cytotoxicity in non-small cell lung cancer. Pharm Res 2018(35):96

Gurunathan S, Kang MH, Qasim M, Kim JH (2018) Nanoparticle-mediated combination therapy: two-in-one approach for cancer. Int J Mol Sci 19(10):3264. https://doi.org/10.3390/ijms1910326

Haberkorn U, Eder M, Kopka K, Babich JW, Eisenhut M (2016) New strategies in prostate cancer: prostate-specific membrane antigen (PSMA) ligands for diagnosis and therapy. Clin Cancer Res 22(1):9–15. https://doi.org/10.1158/1078-0432.CCR-15-0820

Hegde PS, Chen DS (2020) Top 10 challenges in cancer immunotherapy. Immunity 52(1):17–35. https://doi.org/10.1016/j.immuni.2019.12.011

Hevener K, Verstak TA, Lutat KE, Riggsbee DL, Mooney JW (2018) Recent developments in topoisomerase-targeted cancer chemotherapy. Acta Pharm Sin B 8(6):844–861. https://doi.org/10.1016/j.apsb.2018.07.008

Hong Y, Che S, Hui B, Yang Y, Wang X, Zhang X et al (2019) Lung cancer therapy using doxorubicin and curcumin combination: targeted prodrug based, pH-sensitive nanomedicine. Biomed Pharmacother 112:108614

Hong J, Tang Y, Zhou M, Deng J, Hu H, Xu D (2022) Polyethylene glycol-modified mesoporous polydopamine nanoparticles co-loaded with dimethylcurcumin and indocyanine green for combination therapy of castration-resistant prostate cancer. J Drug Deliv Sci Technol. 2022(69):103158. https://doi.org/10.1016/j.jddst.2022.103158

Hu H, Xu D (2021) Co-delivery of docetaxel and retinoic acid by poly (ethylene glycol)-retinoic acid conjugates based micelles for synergistic. Micro Nano Lett 2021(16):336–343

Hu C, Gu F, Gong C, Xia Q, Gao Y, Gao S (2021) Co-delivery of the autophagy inhibitor si-Beclin1 and the doxorubicin nano-delivery system for advanced prostate cancer treatment. J Biomater Appl 36(7):1317–1331

Huang CY, Ju DT, Chang CF, Muralidhar Reddy P, Velmurugan BK (2017) A review on the effects of current chemotherapy drugs and natural agents in treating non-small cell lung cancer. Biomedicine 7(4):23. https://doi.org/10.1051/bmdcn/2017070423

Ilie M, Hofman V, Dietel M, Soria JC, Hofman P (2016) Assessment of the PD-L1 status by immunohistochemistry: challenges and perspectives for therapeutic strategies in lung cancer patients. Virchows Arch 468(5):511–525. https://doi.org/10.1007/s00428-016-1910-4

Jawahar N, Baruah UK, Singh V (2019) Co-delivery of chloroquine phosphate and azithromycin nanoparticles to overcome drug resistance in malaria through intracellular targeting. J Pharm Sci Res 11(1):33–40

Jia R, Teng L, Gao L, Su T, Fu L, Qiu Z, Bi Y (2021) Advances in multiple stimuli-responsive drug-delivery systems for cancer therapy. Int J Nanomedicine 16:1525. https://doi.org/10.2147/IJN.S293427

Jin M, Jin G, Kang L, Chen L, Gao Z, Huang W (2018) Smart polymeric nanoparticles with pH-responsive and PEG-detachable properties for co-delivering paclitaxel and survivin siRNA to enhance antitumor outcomes. Int J Nanomedicine 2018(13):2405–2426

Jin Y, Zhou J, Liu X, Jing Z, Wang X, Feng H et al (2020) Synergistic combination chemotherapy of lung cancer: cisplatin and doxorubicin conjugated prodrug loaded, glutathione and pH-sensitive nanocarriers. Drug Des Devel Ther 14:5205–5215

Johnson RP, Ratnacaram CK, Kumar L, Jose J (2022) Combinatorial approaches of nanotherapeutics for inflammatory pathway targeted therapy of prostate cancer. Drug Resist Updat 64:100865. https://doi.org/10.1016/j.drup.2022.100865

Kanai M, Yoshimura K, Asada M, Imaizumi A, Suzuki C, Matsumoto S et al (2011) A phase I/II study of gemcitabine-based chemotherapy plus curcumin for patients with gemcitabine-resistant pancreatic cancer. Cancer Chemother Pharmacol 68:157–164

Kanno Y, Chen CY, Lee HL, Chiou JF, Chen YJ (2021) Molecular mechanisms of chemotherapy resistance in head and neck cancers. Front Oncol 11:640392. https://doi.org/10.3389/fonc.2021.640392

Kaye SB (1998) New antimetabolites in cancer chemotherapy and their clinical impact. Br J Cancer 78 Suppl 3(Suppl 3):1–7. https://doi.org/10.1038/bjc.1998.747

Khadka DB, Cho WJ (2013) Topoisomerase inhibitors as anticancer agents: a patent update. Expert Opin Ther Pat 23(8):1033–1056. https://doi.org/10.1517/13543776.2013.790958

Khalifa SAM, Elias N, Farag MA, Chen L, Saeed A, Hegazy MF, Moustafa MS, Abd El-Wahed A, Al-Mousawi SM, Musharraf SG, Chang FR, Iwasaki A, Suenaga K, Alajlani M, Göransson U, El-Seedi HR (2019) Marine natural products: a source of novel anticancer drugs. Mar Drugs 17(9):491. https://doi.org/10.3390/md17090491

Khosravi-Darani K, Mozafari M (2010) Nanoliposome potentials in nanotherapy: a concise overview. Int J Nanosci Nanotechnol 6:3–13

Kirtane AR, Verma M, Karandikar P, Furin J, Langer R, Traverso G (2021) Nanotechnology approaches for global infectious diseases. Nat Nanotechnol 16(4):369–384. https://doi.org/10.1038/s41565-021-00866-8

Kolishetti N, Dhar S, Valencia PM, Lin LQ, Karnik R, Lippard SJ, Langer R, Farokhzad OC (2010) Engineering of self-assembled nanoparticle platform for precisely controlled combination drug therapy. Proc Natl Acad Sci 107(42):17939–17944. https://doi.org/10.1073/pnas.1011368107

Kothamasu P, Kanumur H, Ravur N, Maddu C, Parasuramrajam R, Thangavel S (2012) Nanocapsules: the weapons for novel drug delivery systems. Bioimpacts 2:71–81

Kumar K (2012) Targeted delivery of nanomedicines. ISRN Pharmacol 2012:1–9

Lancet JE, Cortes JE, Hogge DE, Tallman MS, Kovacsovics TJ, Damon LE, Komrokji R, Solomon SR, Kolitz JE, Cooper M, Yeager AM (2014) Phase 2 trial of CPX-351, a fixed 5:1 molar ratio of cytarabine/daunorubicin, vs cytarabine/daunorubicin in older adults with untreated AML. Blood 123(21):3239–3246. https://doi.org/10.1182/blood-2013-12-540971

Larson N, Ghandehari H (2012) Polymeric conjugates for drug delivery. Chem Mater 13(24): 840–853

Lazarevic B, Boezelijn G, Diep L, Kvernrod K, Ogren O, Ramberg H et al (2011) Efficacy and safety of short-term genistein intervention in patients with localized prostate cancer prior to radical prostatectomy: a randomized, placebo-controlled, double-blind phase 2 clinical trial. Nutr Cancer 63:889–898

Li F, Danquah M, Singh S, Wu H, Mahato RI (2011) Paclitaxel- and lapatinib-loaded lipopolymer micelles overcome multidrug resistance in prostate cancer. Drug Deliv Transl Res 1:420–428

Li S, Wang L, Li N, Liu Y, Su H (2017) Combination lung cancer chemotherapy: design of a pH-sensitive transferrin-PEG-Hz-lipid conjugate for the co-delivery of docetaxel and baicalin. Biomed Pharmacother 2017(95):548–555

Li X, Sun A, Liu Y, Zhang W, Pang N, Cheng S et al (2018) Amphiphilic dendrimer engineered nanocarrier systems for co-delivery of siRNA and paclitaxel to matrix metalloproteinase-rich tumors for synergistic therapy. NPG Asia Mater 10:238–254. https://doi.org/10.1038/s41427-018-0027-4

Liao L, Liu J, Dreaden EC, Morton SW, Shopsowitz KE, Hammond PT, Johnson JA (2014) A convergent synthetic platform for single-nanoparticle combination cancer therapy: ratiometric loading and controlled release of cisplatin, doxorubicin, and camptothecin. J Am Chem Soc 136(16):5896–5899. https://doi.org/10.1021/ja502011g

Lin W, Li C, Xu N, Watanabe M, Xue R, Xu A et al (2021) Dual-functional PLGA nanoparticles co-loaded with indocyanine green and resiquimod for prostate cancer treatment. Int J Nanomedicine 2021(16):2775–2787

Liu J, He J, Zhang M, Xu G, Ni P (2013) A synergistic polyphosphoester-based co-delivery system of an anticancer drug doxorubicin and the tumor suppressor gene P53 for lung cancer therapy. J Mater Chem B 00:1–3

Liu Z, Chu W, Sun Q, Zhao L, Tan X, Zhang Y et al (2021) Micelle-contained and PEGylated hybrid liposomes of combined gemcitabine and cisplatin delivery for enhancing antitumor activity. Int J Pharm 2021(602):120619. https://doi.org/10.1016/j.ijpharm.2021.120619

Lu J, Huang Y, Zhao W, Marquez RT, Meng X, Li J et al (2013) PEG-derivatized embelin as a nanomicellar carrier for delivery of paclitaxel to breast and prostate cancers. Biomaterials 2013(34):1591–1600. https://doi.org/10.1016/j.biomaterials.2012.10.073

Lv S, Tang Z, Li M, Lin J, Song W, Liu H et al (2014) Co-delivery of doxorubicin and paclitaxel by PEG-polypeptide nanovehicle for the treatment of non-small cell lung cancer. Biomaterials 2014(35):6118–6129. https://doi.org/10.1016/j.biomaterials.2014.04.034

Lv T, Li Z, Xu L, Zhang Y, Chen H, Gao Y (2018) Chloroquine in combination with aptamer-modified nanocomplexes for tumor vessel normalization and efficient erlotinib/Survivin shRNA co-delivery to overcome drug resistance in EGFR-mutated non-small cell lung cancer. Acta Biomater 2018(76):257–274. https://doi.org/10.1016/j.actbio.2018.06.034

Lv Q, Yu S, Quan F, He C, Chen X (2020) Thermosensitive polypeptide hydrogels co-loaded with two anti-tumor agents to reduce multi-drug resistance and enhance local tumor treatment. Adv Ther 1900165:1–10

Ma Z, Dong Z (2021) Dual anticancer drug-loaded self-assembled nanomaterials delivery system for the treatment of prostate cancer. J Biomater Sci Polym Ed 32(16):2103–2117. https://doi.org/10.1080/09205063.2021.1958449

Ma W, Guo Q, Li Y, Wang X, Wang J, Tu P (2017) Co-assembly of doxorubicin and curcumin targeted micelles for synergistic delivery and improving anti-tumor efficacy. Eur J Pharm Biopharm 2017(112):209–223

Ma S, Li X, Ran M, Ji M, Gou J, Yin T et al (2021a) Fabricating nanoparticles co-loaded with survivin siRNA and Pt(IV) prodrug for the treatment of platinum-resistant lung cancer. Int J Pharm 2021(601):120577. https://doi.org/10.1016/j.ijpharm.2021.120577

Ma G, Severic M, Barker M, Pereira S, Ruiz A, Cheung CCL et al (2021b) Dually targeted bioinspired nanovesicle delays advanced prostate cancer tumor growth in vivo. Acta Biomater 2021(134):559–575

Ma Z, Ji T, Ji G, Niu Q, Han W (2022) Facile construction of dual-drug loaded nanoparticles for improvement synergistic chemotherapy in prostate cancer. Int J Polym Mater Polym Biomater:1–11. https://doi.org/10.1080/00914037.2022.2066667

Mahdi H, Mula-Hussain L, Ramzi ZS, Tolba M, Abdel-Rahman O, Abu-Gheida I, Khorshid O, Al Sukhun S, Siddiqi NP, Al Mandhari Z, Al Hussaini M (2022) Cancer Burden Among Arab-World Females in 2020: working toward improving outcomes. JCO Glob Oncol 2022: e2100415. https://doi.org/10.1200/GO.21.00415

Mahira S, Kommineni N, Husain GM, Khan W (2019) Cabazitaxel and silibinin co-encapsulated cationic liposomes for CD44 targeted delivery: a new insight into nanomedicine based combinational chemotherapy for prostate cancer. Biomed Pharmacother 110:803–817. https://doi.org/10.1016/j.biopha.2018.11.145

Mangal S, Gao W, Li T, Zhou QT (2017) Pulmonary delivery of nanoparticle chemotherapy for the treatment of lung cancers: challenges and opportunities. Acta Pharmacol Sin 38(6):782–797. https://doi.org/10.1038/aps.2017.34

Marasini N, Haque S, Kaminskas L (2017) Polymer-drug conjugates as inhalable drug delivery systems: a review. Curr Opin Colloid Interface Sci 31:8–29

Matthews HK, Bertoli C, de Bruin RAM (2022) Cell cycle control in cancer. Nat Rev Mol Cell Biol 23(1):74–88. https://doi.org/10.1038/s41580-021-00404-3

Menon JU, Kuriakose A, Iyer R, Hernandez E, Gandee L, Zhang S et al (2017) Dual-drug containing Core-Shell nanoparticles for lung cancer therapy. Sci Rep 7:13249. https://doi.org/10.1038/s41598-017-13320-4

Mirzaei S, Paskeh MDA, Okina E, Gholami MH, Hushmandi K, Hashemi M, Kalu A, Zarrabi A, Nabavi N, Rabiee N, Sharifi E, Karimi-Maleh H, Ashrafizadeh M, Kumar AP, Wang Y (2022) Molecular landscape of LncRNAs in prostate cancer: a focus on pathways and therapeutic targets for intervention. J Exp Clin Cancer Res 41(1):214. https://doi.org/10.1186/s13046-022-02406-1

Mogheri F, Jokar E, Afshin R, Akbari AA, Dadashpour M, Firouzi-amandi A et al (2021) Co-delivery of metformin and silibinin in dual-drug loaded nanoparticles synergistically improves chemotherapy in human non-small cell lung cancer A549 cells. J Drug Deliv Sci Technol. 2021(66):102752. https://doi.org/10.1016/j.jddst.2021.102752

Nan Y (2019) Lung carcinoma therapy using epidermal growth factor receptor-targeted lipid polymeric nanoparticles co-loaded with cisplatin and doxorubicin. Oncol Rep 42:2087–2096

Narmani A, Mohammadnejad J, Yavari K (2019) Synthesis and evaluation of polyethylene glycol- and folic acid-conjugated polyamidoamine G4 dendrimer as nanocarrier. J Drug Deliv Sci Technol. 50:278–286. https://doi.org/10.1016/j.jddst.2019.01.037

Navya PN, Kaphle A, Srinivas SP, Bhargava SK, Rotello VM, Daima HK (2019) Current trends and challenges in cancer management and therapy using designer nanomaterials. Nano Converg 6(1):23. https://doi.org/10.1186/s40580-019-0193-2

Pan J, Rostamizadeh K, Filipczak N, Torchilin V (2019) Polymeric co-delivery systems in cancer treatment: an overview on component drugs' dosage ratio effect. Molecules 24(6):1035

Pang X, Yang XZG (2014) Polymer-drug conjugates: recent progress on administration routes. Expert Opin Drug Deliv 11:75–86

Patel V, Lalani R, Vhora I, Bardoliwala D, Patel A, Ghosh S et al (2021) Co-delivery of cisplatin and siRNA through hybrid nanocarrier platform for masking resistance to chemotherapy in lung cancer. Drug Deliv Transl Res 2021(11):2052–2071. https://doi.org/10.1007/s13346-020-00867-5

Patil J, Mahajan H (2015) A review on polymer drug conjugate-what, why and how? Int J Pharm Sci Res 6:4611–4621

Peter S, Aderibigbe BA (2019) Ferrocene-based compounds with antimalaria/anticancer activity. Molecules 24(19):3604. https://doi.org/10.3390/molecules24193604

Peter S, Jama S, Alven S, Aderibigbe BA (2021) Artemisinin and derivatives-based hybrid compounds: promising therapeutics for the treatment of cancer and malaria. Molecules 26(24):7521. https://doi.org/10.3390/molecules26247521

Peter S, Alven S, Maseko RB, Aderibigbe BA (2022) Doxorubicin-based hybrid compounds as potential anticancer agents: a review. Molecules 27(14):4478. https://doi.org/10.3390/molecules27144478

Qin J, Gong N, Liao Z, Zhang S, Timashev P, Huo S, Liang XJ (2021) Recent progress in mitochondria-targeting-based nanotechnology for cancer treatment. Nanoscale 13(15): 7108–7118. https://doi.org/10.1039/d1nr01068a

Qiu LY, Bae YH (2006) Polymer architecture and drug delivery. Pharm Res 23:1–30. https://doi.org/10.1007/s11095-005-9046-2

Qu Z, Ren Y, Shen H, Shi H, Shi L, Tong D (2021) Combination therapy of metastatic castration-recurrent prostate cancer: hyaluronic acid decorated, cabazitaxel-prodrug and orlistat co-loaded nano-system. Drug Des Devel Ther 2021(15):3605–3616

Raitanen M, Rantanen V, Kulmala J, Helenius H, Grénman R, Grénman S (2002) Supra-additive effect with concurrent paclitaxel and cisplatin in vulvar squamous cell carcinoma in vitro. Int J Cancer 100(2):238–243. https://doi.org/10.1002/ijc.10472

Rana A, Alex JM, Chauhan M, Joshi G, Kumar R (2015) A review on pharmacophoric designs of antiproliferative agents. Med Chem Res 24(3):903–920. https://doi.org/10.1007/s00044-014-1196-5

Rao NV, Ko H, Lee J, Park JH (2018) Recent progress and advances in stimuli-responsive polymers for cancer therapy. Front Bioeng Biotechnol 6:110. https://doi.org/10.3389/fbioe.2018.00110

Rudnik LAC, Farago PV, Budel JM, Lyra A, Barboza FM, Klein T et al (2020) Co-loaded curcumin and methotrexate nanocapsules enhance cytotoxicity against non-small-cell lung cancer cells. Molecules 25:1913

Said-Elbahr R, Nasr M, Alhnan MA (2022) Simultaneous pulmonary administration of celecoxib and naringin using a nebulization-friendly nanoemulsion: a device-targeted delivery for treatment of lung cancer. Expert Opin Drug Deliv 19(5):611–622

Sailaja AK, Amareshwar P, Chakravarty P (2011) Different technique used for the preparation of nanoparticles using natural polymers and their application. Int J Pharm Pharm Sci 3:45–50

Sanchis M, Canal J, Lucas F, Vicent R (2010) Polymer-drug conjugates for novel molecular targets. Nanomedicine 5:915–935

Senevirathne S, Washington K, Biewer M, Stefan M (2016) PEG based anti-cancer drug conjugated prodrug micelles for the delivery of anti-cancer agents. J Mater Chem B 4:360–370

Shade C (2016) Liposomes as advanced delivery systems for nutraceuticals. Integr Med 15:33–36

Shen J, Yin Q, Chen L, Zhang Z, Li Y (2012) Co-delivery of paclitaxel and survivin shRNA by pluronic P85-PEI/TPGS complex nanoparticles to overcome drug resistance in lung cancer. Biomaterials 2012(33):8613–8624

Silberstein JL, Pal SK, Lewis B, Sartor O (2013) Current clinical challenges in prostate cancer. Transl Androl Urol 2(3):122. https://doi.org/10.3978/j.issn.2223-4683.2013.09.03

Sun J, Liu Y, Chen Y, Zhao W, Zhai Q, Rathod S et al (2017) Doxorubicin delivered by a redox-responsive dasatinib-containing polymeric prodrug carrier for combination therapy. J Control Release 2017(258):43–55. https://doi.org/10.1016/j.jconrel.2017.05.006

Sun Y, Yang J, Yang T, Li Y, Zhu R, Hou Y et al (2021a) Co-delivery of IL-12 cytokine gene and cisplatin prodrug by a polymetformin-conjugated nanosystem for lung cancer chemo-gene treatment through chemotherapy sensitization and tumor microenvironment modulation. Acta Biomater 2021(128):447–461. https://doi.org/10.1016/j.actbio.2021.04.034

Sun G, Sun K, Sun J (2021b) Combination prostate cancer therapy: prostate-specific membranes antigen targeted, pH-sensitive nanoparticles loaded with doxorubicin and tanshinone targeted, pH-sensitive nanoparticles loaded with doxorubicin and tanshinone. Drug Deliv 28(1): 1132–1140. https://doi.org/10.1080/10717544.2021.1931559

Sung H, Ferlay J, Siegel RL, Laversanne M, Soerjomataram I, Jemal A, Bray F (2021) Global cancer statistics 2020: GLOBOCAN estimates of incidence and mortality worldwide for 36 cancers in 185 countries. CA Cancer J Clin 71(3):209–249. https://doi.org/10.3322/caac.21660

Suroshe S, Nerkar P, Patil K, Chalikwar S (2019) Breast cancer: recent review on micelles as nano-carriers for treatment Indo Am J Pharm Res 9

Tang H, Zhao W, Yu J, Li Y, Zhao C (2018) Recent development of pH-responsive polymers for cancer nanomedicine. Molecules 24(1):4. https://doi.org/10.3390/molecules24010004

To KK, Fong W, Cho WC (2021) Immunotherapy in treating EGFR-mutant lung cancer: current challenges and new strategies. Front Oncol 11:635007. https://doi.org/10.3389/fonc.2021.635007

Unsoy G, Gunduz U (2018) Smart drug delivery systems in cancer therapy. Curr Drug Targets 19: 202–212

Uram Ł, Filipowicz A, Misiorek M, Pienkowska N, Markowicz J, Walajtys-Rode E et al (2018) Biotinylated PAMAM G3 dendrimer conjugated with celecoxib and/or Fmoc-L-leucine and its cytotoxicity for normal and cancer human cell lines. Eur J Pharm Sci 124:1–9

Valenzuela-oses JK, García MC, Feitosa VA, Pachioni-vasconcelos JA, Gomes-filho SM, Lourenço FR et al (2017) Development and characterization of miltefosine-loaded polymeric micelles for cancer treatment. Mater Sci Eng C 81:327–333. https://doi.org/10.1016/j.msec.2017.07.040

van der Spek YM, Kroep JR, Tollenaar RAEM, Mesker WE (2020) Chemotherapy resistance and stromal targets in breast cancer treatment: a review. Mol Biol Rep 47(10):8169–8177. https://doi.org/10.1007/s11033-020-05853-1

Wan X, Min Y, Bludau H, Keith A, Sheiko SS, Jordan R et al (2018) Drug combination synergy in worm-like polymeric micelles improves treatment outcome for small cell and non-small cell lung cancer. ACS Nano:0.1021/acsnano.7b07878

Wang G, Wang Z, Li C, Duan G, Wang K, Li Q et al (2018) RGD peptide-modified, paclitaxel prodrug-based, dual-drugs loaded, and redox-sensitive lipid-polymer nanoparticles for the enhanced lung cancer therapy. Biomed Pharmacother 2018(106):275–284. https://doi.org/10.1016/j.biopha.2018.06.137

Wende M, Sithole S, Chi GF, Stevens MY, Mukanganyama S (2021) The effects of combining cancer drugs with compounds isolated from Combretum zeyheri Sond. and Combretum platypetalum Welw. Ex. M.A. Lawson (Combretaceae) on the viability of Jurkat T cells and HL-60 cells. Biomed Res Int 2021:6049728. https://doi.org/10.1155/2021/6049728

Wu R, Zhang Z, Wang B, Chen G, Mao J, Wang L (2020) Combination chemotherapy of lung cancer—co-delivery of docetaxel prodrug and cisplatin using aptamer-decorated lipid—polymer hybrid nanoparticles. Drug Des Devel Ther 2020(14):2249–2261

Xia C, Dong X, Li H, Cao M, Sun D, He S et al (2022) Cancer statistics in China and United States, 2022: profiles, trends, and determinants. Chin Med J 135(5):584–590. https://doi.org/10.1097/CM9.0000000000002108

Xiao D, Xinlei H, Juan Z (2022) Tumour targeted polymer nanoparticles co-loaded with docetaxel and siCCAT2 for combination therapy of lung cancer. J Drug Target 30(5):534–543

Xu JL, Jin B, Ren ZH, Lou YQ, Zhou ZR, Yang QZ, Han BH (2015) Chemotherapy plus erlotinib versus chemotherapy alone for treating advanced non-small cell lung cancer: a meta-analysis. PLoS One 10(7):e0131278. https://doi.org/10.1371/journal.pone.0131278

Xu C, Wang Y, Guo Z, Chen J, Lin L, Wu J et al (2019) Pulmonary delivery by exploiting doxorubicin and cisplatin co-loaded nanoparticles for metastatic lung cancer therapy. J Control Release 2019(295):153–163. https://doi.org/10.1016/j.jconrel.2018.12.013

Yang J, Wen C, Pan C, Guo H, Zhao W, Zhang J et al (2019) Nanoliposomal multi-drug delivery system with reduced toxicity and multi-drug resistance. J Mater Sci 2019(54):9718–9728. https://doi.org/10.1007/s10853-019-03573-x

Zeng Y-Y, Zeng Y-J, Zhang N-N, Li C-X, Xie T, Zeng Z-W (2019) The preparation, determination of a flexible complex liposome co-loaded with cabazitaxel and β-elemene, and animal pharmacodynamics on paclitaxel-resistant lung adenocarcinoma. Molecules 24:1697

Zhang P, Li J, Ghazwani M, Zhao W, Huang Y, Zhang X et al (2015) Effective co-delivery of doxorubicin and dasatinib using a PEG-Fmoc nanocarrier for combination cancer chemotherapy. Biomaterials 2015(67):104–114. https://doi.org/10.1016/j.biomaterials.2015.07.027

Zhang M, Hagan CT, Foley H, Tian X, Yang F, Man K et al (2021) Co-delivery of etoposide and cisplatin in dual-drug loaded nanoparticles synergistically improves chemoradiotherapy in non-small cell lung cancer models. Acta Biomater 2021(124):327–335

Zhang Y, Duan H, Zhao H, Qi L, Liu Y, Zhang Z et al (2022a) Development and evaluation of a PSMA-targeted Nanosystem co-packaging docetaxel and androgen receptor siRNA for castration-resistant prostate cancer treatment. Pharmaceutics 2022(14):964

Zhang L, Lin Z, Chen Y, Gao D, Wang P, Lin Y et al (2022b) Co-delivery of docetaxel and resveratrol by liposomes synergistically boosts antitumor efficiency against prostate cancer. Eur J Pharm Sci 2022(174):106199. https://doi.org/10.1016/j.ejps.2022.106199

Zhao CY, Cheng R, Yang Z, Tian ZM (2018) Nanotechnology for cancer therapy based on chemotherapy. Molecules 23(4):826. https://doi.org/10.3390/molecules23040826

Zhao Y, Liu K, Li J, Liao J, Ma L (2021) Engineering of hybrid anticancer drug-loaded polymeric nanoparticles delivery system for the treatment and care of lung cancer therapy. Drug Deliv 28(1):1539–1547

Zheng W, Zhou Q, Yuan C (2021) Nanoparticles for oral cancer diagnosis and therapy. Bioinorg Chem Appl 2021:9977131. https://doi.org/10.1155/2021/9977131

Zhong G, Yang C, Liu S, Zheng Y, Lou W, Teo JY, Bao C, Cheng W, Tan JP, Gao S, Park N (2019a) Polymers with distinctive anticancer mechanism that kills MDR cancer cells and

inhibits tumor metastasis. Biomaterials 199:76–87. https://doi.org/10.1016/j.biomaterials.2019.01.036
Zhong G, Yang C, Liu S, Zheng Y, Lou W, Yng J et al (2019b) Polymers with distinctive anticancer mechanism that kills MDR cancer cells and inhibits tumor metastasis. Biomaterials 199:76–87. https://doi.org/10.1016/j.biomaterials.2019.01.036
Zhong T, Liu X, Li H, Zhang J (2021) Co-delivery of sorafenib and crizotinib encapsulated with polymeric nanoparticles for the treatment of in vivo lung cancer animal model. Drug Deliv 28(1):2108–2118. https://doi.org/10.1080/10717544.2021.1979129
Zhou N, Wang W, Li H, Jiang D, Zhong X (2021) Development and investigation of dual potent anticancer drug-loaded nanoparticles for the treatment of lung cancer therapy. Process Biochem 2021(106):42–49. https://doi.org/10.1016/j.procbio.2021.03.018
Zhu L, Perche F, Wang T, Torchilin VP (2014a) Matrix metalloproteinase 2-sensitive multifunctional polymeric micelles for tumor-specific co-delivery of siRNA and hydrophobic drugs. Biomaterials 35(13):4213–4222. https://doi.org/10.1016/j.biomaterials.2014.01.060
Zhu L, Perche F, Wang T, Torchili V (2014b) Matrix metalloproteinase 2-sensitive multifunctional polymeric micelles for tumor-specific co-delivery of siRNA and hydrophobic drugs. Biomaterials 35(13):4213–4222
Zhu L, Yang Y, Li X, Zheng Y, Li Z, Chen H et al (2022) Facile preparation of indocyanine green and tiny gold nanoclusters co-loaded nanocapsules for targeted synergistic sono-/photo-therapy. J Colloid Interface Sci 627(2022):596–609. https://doi.org/10.1016/j.jcis.2022.07.084

Sijongesonkespiepr146 Peter is a second year PhD student at the University of Fort Hare and his research is based on organic synthesis (drug discovery). His research focus is on the synthesis of hybrid molecules through drug repurposing/repositioning to enhance efficacy. He has published six papers in refereed journals, with one conference proceedings (H3D Symposium), and two book chapters. He co-supervised two honors students in the University of Fort Hare since he started his PhD studies. He was awarded as a best presenter during 2022 Sasol bursar's presentation.

Tobeka Naki is a lecturer at the Department of Chemistry, University of Fort Hare, South Africa. She is currently a PhD student. Her research is focused on the design and in vitro cytotoxicity studies of nanocarriers for the treatment of cancer. She is currently developing nanocarriers for brain targeting, and the emphasis is on evaluating the toxic side effects of drug-loaded nanocarriers to the brain.

Sibusiso Alven is a qualified candidate for the degree of Doctor of Philosophy in Chemistry at the University of Fort Hare (UFH), and the degree will be conferred in May 2023. His research focus is based on polymer-based materials (such as hydrogels, topical gels, nanofibers, sponges, polymer-drug conjugates, etc.) loaded with various therapeutic agents for the treatment of malaria, cancer, and chronic wounds. He has published 19 articles in various refereed journals (*Polymer*, *Nanomaterials*, and *Journal of Drug Delivery Science and Technology*) and 6 book chapters. He was considered as a top publisher by Sasol Bursary Committee among his fellow bursars in 2021. He reviewed more than ten manuscripts for various refereed journals including cytokines, pharmaceutics, and polymers. He has orally presented his research results in the international conference (ICTERM, 2017) that was held in Vanderbijlpark, South Africa, and a regional conference (SAMRC Symposium) that was held in East London (2019). In 2022, he presented poster at H3D Symposium that was held at Stellenbosch, South Africa. He also presented twice at faculty conferences in his institution (UFH). Sibusiso will start the position of postdoctoral research fellowship in the Department of Chemistry at Nelson Mandela University (NMU) in 2023, and his postdoctoral research will be focused on polymer-based membranes for metal recovery and water treatment that will be funded by the National Research Foundation (NRF).

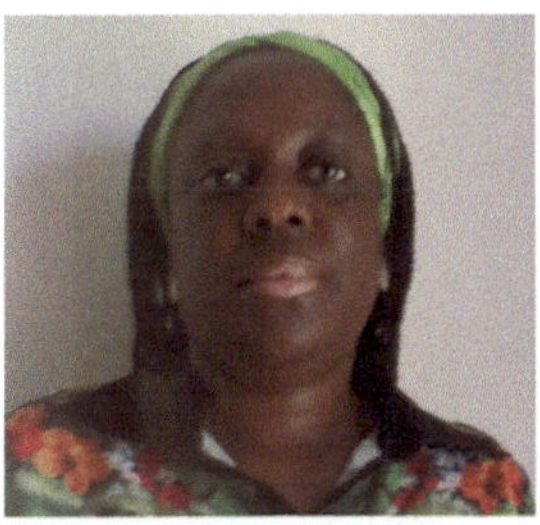

Blessing A. Aderibigbe is a professor of organic chemistry at the University of Fort Hare and an NRF C2 rated researcher. She was a postdoctoral fellow between December 2010 and February 2014 at different universities: the University of Johannesburg (Auckland Park and Doornfontein campuses) and Tshwane University of Technology (Polymer division, which was previously based at CSIR, Pretoria). She obtained her PhD in Polymer/Medicinal Chemistry from the University of Witwatersrand, South Africa, under the supervision of late Emeritus Prof. E.W. Neuse. Her PhD research was focused on the design of polymer-drug conjugates with promising antimalarial activity. Her master's degree was in organic chemistry also from the University of Witwatersrand, South Africa, under the supervision of Prof. W. van Otterlo and Prof. J. P. Michael. Her MSc research was focused on the synthesis of 6-, 7-, 8-, and 9-membered benzo-fused heterocyclic compounds using Grubbs II catalyst. She joined the University of Fort Hare as a lecturer on the 1st of March 2014. She was promoted to the position of a senior lecturer in January 2016, to an associate professor in January 2018, and to full professor in January 2020. Her research themes include the design and development of polymer-based wound dressings, polymer-based drug delivery systems, and organic molecules, specifically hybrid molecules via a combination of two molecules for enhanced efficacy.

Silver Nanoparticle-Incorporated Textile Substrate for Antimicrobial Applications

26

T. Hemamalini, Pon G. K. Gokul, R. S. Mahalakshmi, R. Rajesh, and V. R. Giri Dev

Abstract

One of the significant challenges humanity faces is attack by microbes. Silver (Ag) nanoparticles (NPs) have been used from time immemorial to treat infections. Ag NPs are effective against many bacteria, fungi, and viruses and work by binding to and inactivating the enzymes that bacteria need to grow and reproduce. The incorporation of Ag into textiles has a broad scope ranging from health care to household applications, but the major limitation is to ensure durability during laundering. The durability of Ag on the fabrics depends on the form of Ag used, the type of substrate, and the method of incorporation. The present chapter gives a holistic picture of incorporating silver nanoparticles into textile substrates. A few examples of cotton, protein, polyester, nylon, polyolefin, and high-performance-based textile recently impregnated with Ag NPs are discussed here. Moreover, the potential of biodegradable and nonbiodegradable is highlighted briefly regarding antimicrobial activities.

Keywords

Fabric · Fiber · Nanofibers · Silver nanoparticles

26.1 Introduction

Silver (Ag) nanoparticles (NPs) offer bactericidal activity against a wide range of multidrug-resistant bacteria and viral strains due to the cationic nature of Ag ions (Ag^+) as it interacts with the anionic cell wall of bacteria/microbes, thereby causing

T. Hemamalini · P. G. K. Gokul · R. S. Mahalakshmi · R. Rajesh · V. R. G. Dev (✉)
Department of Textile Technology, Anna University, Chennai, Tamil Nadu, India
e-mail: vrgiridev@annauniv.edu

F. A. Sheikh et al. (eds.), *Interaction of Nanomaterials With Living Cells*,
https://doi.org/10.1007/978-981-99-2119-5_26

cell lysis. The Ag NPs were applied on the textile substrate as it provides structural stability and integrity upon usage. The Ag ions can be introduced into the textile substrate either through coating or incorporation into a polymeric solution followed by spinning into fibers. The mechanism of cell lysis of Ag NPs involves the attachment of Ag ions within the bacterial cell wall as it gradually penetrates through microbial cells resulting in structural changes causing the death of the cell. Ag is nonreactive in its metal form (Ag). For bactericidal properties against a wide range of microorganisms, Ag in the ionic state (Ag^+) offers higher advantages compared to other metal salts and metal nanoparticles, namely, gold (Au), titanium oxide (TiO_2), copper oxide (CuO), zinc oxide (ZnO), and magnesium oxide (MgO). Ag NPs can be synthesized in biological, physical, and chemical ways. The biological method includes the utilization of enzymes present in bacteria, fungi, algae, and plant extract. In contrast, physical synthesis includes irradiation, utilization of silica gel and activated carbon, and electrokinetic coagulation. The chemical synthesis includes oxidation and ozonation processes, sodium hypochlorite, and electrochemical and photochemical destruction processes (Hemamalini et al. 2021). The chapter summarizes the treatment of Ag NPs on the textile substrate and their potential in antimicrobial applications.

26.2 Incorporation of Ag NPs on the Surface of the Textile Substrate

Ag NPs can be treated on the textile substrate in various forms, namely, fiber, yarn, and fabric. The coating of Ag NPs on the fabric produced through the woven, knitted, and nonwoven structures can be carried out using padding or exhaustion. The padding process involves coating the fabric with NPs followed by passing through the pair of positively loaded rollers, which in turn apply pressure on the substrate as it decides the final add-on value of the Ag NPs. The latter can be increased by altering the processing parameters, such as the number of nips and dips, the concentration of liquor, and pressure acting between the nip of padding rollers. The exhaustion technique involves treating the fabric with Ag nanoparticles containing liquor on the machine. The technique is suitable for the substrate, which offers an affinity for the Ag components, thereby offering the chemical interaction between the substrate and NPs in order to increase the serviceability and durability of the fabric (Cay et al. 2009).

26.2.1 Cotton-Based Textile Substrate

Ag NPs were prepared by adding $AgNO_3$ and *Fusarium oxysporum* biomass in distilled water. $AgNO_3$ reacts with fungal filtrate solution resulting in the formation of NPs confirmed by forming a peak around 420 nm confirmed by plasmon resonance. The cotton fabrics were impregnated in Ag NPs solution followed by a padding and drying process. Cytotoxicity assay of Ag NPs was studied by culturing fibroblast cells for 48 h, and the semiconfluent culture was exposed to different concentrations of Ag NPs ranging from 0 to 22 μM. The neutral red uptake (NRU) and methylthiazol tetrazolium (MTT) reduction assays confirmed that the Ag NPs showed no cytotoxicity till 16 μM. The spherical-shaped Ag NPs have a diameter of 7.30 nm, examined using transmission electron microscope (TEM) images. Using the disc diffusion method, the antibacterial performance of Ag-loaded cotton fabrics showed better antibacterial activity with the inhibition zone of 35 mm, 18 mm, 20 mm, 22 mm, and 21 mm against *S. aureus*, *S. epidermis*, *E. coli*, and *Salmonella enterica*, respectively. It was reported that the adhesion of biogenic Ag NPs on cotton fabrics showed good antibacterial activity even after 30 washing cycles (Marcato et al. 2012). Ag offers strong biocidal activity on pathogenic bacteria, thereby functionalizing the cotton fibers results in the functionalization of fibers for technical applications. The cotton fibers were surface modified by treatment with ammonium cerium (IV) nitrate/nitric acid (Ce^{4+}/HNO_3) followed by GMA-IDA solution (glycidyl methacrylate-iminodiacetic acid), which resulted in the formation of grafted cotton fibers. The polymer radicals were formed between cotton and Ce^{4+}, which in turn react with GMA-IDA solution resulting in grafted fibers that react with $AgNO_3$ to form NPs upon UV irradiation. It was reported that increasing the concentration of the grafting agent increases the weight of Ag^+ thereby increasing the diameter of Ag NPs. The antibacterial effect of Ag-grafted cotton fibers and bulk Ag metal against *E. coli* was studied by colony-counting method. The fibers incorporated with Ag NPs of 75 nm offered high antibacterial activity, thereby offering the potential for antimicrobial applications (Chen and Chiang 2008). Colloidal Ag NPs were coated onto cotton fabrics with the aid of 3-mercaptopropyltrimethoxysilane (3-MPTMS). The colloidal solution was prepared by the addition of 3-MPTMS and Ag in isopropanol. The cotton fabrics were dipped in the colloidal solution, then rinsed in propanol, and then dried. It was reported that no traces of Ag were detected in the cotton fabric treated with a colloid solution without 3-MPTMS, thereby indicating that 3-MPTMS aids in trapping Ag within the structure. It was reported that an increase in temperature resulted in an increase in the loading of Ag onto the fabric. Cotton treated with Ag NPs/3-MPTMS showed 100% reduction against both the microbes, whereas cotton treated with 3-MPTMS alone showed inferior antibacterial activity. It showed considerable wash fastness than the pad-dry-cure process by retaining 66% of Ag after five washing cycles (Kim et al. 2011). Cotton fabric was coated with a colloidal Ag solution prepared by microwaving $AgNO_3$, polyvinylpyrrolidone, and ethylene glycol to yield a yellow-colored colloidal solution. The fabric was padded with different concentrations of the Ag colloidal solution followed by squeezing, drying, and washing in order to assess the antibacterial property of the fabric according to the colony-counting method. It was reported that increasing the concentration of Ag resulted in higher bacterial reduction, and upon washing, antibacterial efficiency of the developed fabric decreased due to weak physical interaction between Ag NPs and cotton fabric (Ke Thanh and Phuong Phong 2009). Ag NPs were synthesized using green synthesis with a reducing agent of hydroxypropyl starch (HPS). The cotton fabric was padded with 50 and 100 ppm of Ag NPs followed by drying and curing at 70 and 100 °C, respectively. It was reported that increasing the concentration of Ag treatment resulted in increased deposition of nanoparticles on the substrate, and Ag-treated cotton fabric offered bacterial growth reduction of 95% against *E. coli* and *S. aureus* prior to washing. Upon washing the fabric from 5 to 20 washing cycles, bacterial growth reduction was decremented to be 56%, which may be due to poor attachment of nanoparticles on the substrate. It was reported that the fabric treated with Ag NPs in the presence of a binder showed a higher bacterial reduction of above 90% on the selected bacteria (Hebeish et al. 2011). Synthesis of dodecanethiol-capped Ag NP sol-gel solution was prepared by the robust stirring of $AgNO_3$, chloroform and tetraoctylammonium bromide, a phase transfer catalyst. Migrated Ag ions from the organic phase after removal of dodecanethiol were added to aqueous sodium borohydride followed by stirring. Ag NPs were added in a dropwise manner of desired concentration to the sol containing tetraethyl orthosilicate, alcohol, and water in a ratio of 1:1:4, respectively. The cotton fabric samples were padded with the sol doped with dodecanethiol-capped Ag NPs, followed by drying and curing at 60 and 150 °C, respectively. Antibacterial activity of the cotton fabrics treated with sol dopped with 15 mL of dodecanethiol-capped Ag NPs showed better inhibitory effect and bactericidal performance up to 40% against *E. coli* (Tarimala et al. 2006).

26.2.2 Protein-Based Textile Substrate

The deposition of the fatty acid on the surface of the pristine wool fiber causes a rough surface aiding in microbial growth in cuticle structure and generating a static charge. The removal of rough hydrophobic surface and coating with hydrophilic material decreases the mechanical property of the fibers and also lowers the electrical resistivity and creates charge dissipation. The surface of the wool fibers was treated with different concentrations of protease enzyme and loaded with Ag NPs by varying the pH level from 4 to 7. The scales on the wool surface were entirely degraded by the action of pneumonia, aiding in forming a hydrophilic layer. Increasing the concentration of protease enzyme resulted in a smooth surface resulting from scales stripping, thereby increasing the friction between fibers. It was reported that lowering the pH resulted in an increase in the absorption of Ag NPs due to the protonation of internal amino acids and resulted in a decrease in the specific resistance of the fibers. The antibacterial efficiency of developed fibers was higher at 6% protease enzyme and Ag NPs at pH 7 against *S. aureus* and *K. pneumoniae* due to the interaction between thiol groups of bacteria and inhibited their growth. The half-life period of wool decreased at 3% protease concentration due to the accumulation of charges created by Ag NPs (Memon et al. 2018). Wool fibers (Merino type) were treated with SNSE (sulfur, nano-Ag, ethanol-based colloid) solution using the pad-dry-cure method. The antibacterial properties of the samples were assessed qualitatively against *Staphylococcus* and *Klebsiella pneumoniae* through AATCC 100-1999 method. The bacteriostatic activity was evaluated through the percent reduction of bacteria. Mothproofing test was conducted according to International Organization for Standardization 3998-1177 with *Tinea pellionella* as larvae. The antibacterial efficiency of *S. aureus* was consistent irrespective of Ag ppm, whereas, in *K. pneumoniae*, lower ppm showed lower bacterial reduction (99.70%) than higher ppm. It was reported that mothproofing test showed no pupation of larvae on the developed samples (Ki et al. 2007).

The antibacterial activity of Ag-coated silk fabric was studied against *E. coli* and *S. aureus* by subjecting the fabric to plasma treatment in the presence of non-polymerizing gases such as oxygen and argon. The samples were immersed in the solution containing stannous chloride and hydrochloric acid to obtain surface sensitization and rinsed with deionized water. Then, the fabric was subjected to electroless plating of Ag utilizing the solution containing $AgNO_3$, NaOH, and NH_4^+ solution and glucose, followed by curing at 150 °C. It was seen that oxygen plasma-treated fabric showed severe damage on its surface due to a faster rate of etching compared to argon plasma-treated fabric, which was confirmed using a surface morphology study. The Ag deposition on the oxygen-treated fabric was higher due to the presence of higher hydrophilic groups compared to argon-treated samples and offered higher antibacterial activity against bacteria with the inhibitory zone of 11 mm and 10 mm, respectively, and also provided better antistatic and ultraviolet radiation properties compared to untreated samples (Jiang et al. 2009). In order to utilize the amphoteric nature of silk, degumming of silk was carried out in the presence of plant extract (Keliab), sodium chloride, enzyme (Alcalase), and sodium carbonate (Na_2CO_3). Uniform surface without fibrillation was seen on raw silk upon degumming; the fiber became soft and shiny due to the removal of sericin which was confirmed using a morphology study. The fibers were treated with Ag NPs to impart antibacterial activity against *Escherichia coli* and *Staphylococcus aureus*. The treated fibers offered higher antibacterial activity with a Ag concentration above 50 ppm confirmed using the bacterial-counting method. Due to the degumming process, silk fiber attains more positive charges at acidic pH, thereby increasing the interaction with negatively charged Ag NPs resulting in higher uptake. It was reported that the antibacterial activity of degummed silk was lower than raw silk at the same pH due to an increase in water absorption. Sericin showed poor solubility at an isoelectric point at pH of 4 and also possessed a low positive charge and limited interaction with Ag NPs resulting in lower antibacterial activity. It was reported that increasing the concentration of sodium chloride and sonication time resulted in an increase in nanoparticle uptake as it increases the surface tension and ionic strength, thereby decreasing the vapor pressure. It was seen that the sonication process increased the dispersion and uniform absorption of nanoparticles on the fibrous surfaces due to the collision of Na^+. It was reported that the deposition of Ag NPs was lower in the fabric treated with Na_2CO_3 as it interacts with water resulting in decreased antibacterial activity (Moazami et al. 2010).

26.2.3 Polyester and Nylon-Based Textile Substrate

The antibacterial property of nonwoven polyester (PET) fabrics was studied by incorporating different nanoparticles such as Ag, Cu, and ZnO NPs. The thin film of organosilicon of 70 nm thickness was deposited using a plasma jet deposition system followed by immersion in an ethanol solution containing a suspension of nanoparticles. The second layer of organosilicon was deposited on the nanoparticle-treated nonwoven as the barrier layer. The antibacterial test was carried out against *Staphylococcus aureus* and *Escherichia coli*; it was reported that the polyester fabrics incorporated with Ag, Cu, and ZnO NPs exhibit effective antibacterial activity against the selected microorganism (Deng et al. 2015a). PET nonwoven fabrics with antimicrobial properties were produced via a double layer of the plasma-deposited organic film. The two separate layers of organosilicon film were deposited with an intermediate step of immersing the fabric into Ag NPs/ethanol suspension. X-ray photoelectron spectroscopy (XPS) showed that the oxygen and silicon concentrations in the fibrous mat decreased upon the incorporation of Ag NPs. Ag concentration was reduced with the deposition of a barrier layer, and it was reported that the cracks and pores were formed on the treated nonwoven web due to thermal stress containing minimum Ag concentration. The release of the Ag from the web was controlled by increasing the thickness of the barrier layer. The test microbes taken for measuring the antibacterial activity were *S. aureus*, *P. aeruginosa*, and *C. albicans*. The samples without a barrier layer offered the highest reduction of microbes, about 90%, but in the presence of a barrier layer, the decline was around 50% which was due to the slow release of Ag NPs from the fibrous web. The mechanism of microbial reduction was due to the destruction of cell membranes by Ag ions by inactivating enzymes by reacting with -SH groups. The cytotoxic effect of Ag NPs was mediated by the generation of oxidative stress. It was reported that for the samples with a barrier layer, Ag concentration had almost negligible fluctuations for all washing cycles, but upon increasing the washing cycle to five times, the antimicrobial activity reduced significantly due to loss of Ag NPs (Deng et al. 2015b). Desized and bleached polyester (PET) samples were subjected to low-temperature plasma treatment followed by immersion in Ag NP colloid that was prepared from $AgNO_3$ and $NaBH_4$. The antibacterial efficiency of the developed samples was quantitatively assessed using *E. coli* and *S. aureus*. Scanning electron microscopy (SEM) images showed that untreated PET (UPET) fabrics offered a smooth surface, whereas plasma-treated PET (PPET) showed a change in morphology with uneven cracks, pits, and striations with an increase in fiber surface roughness. It was reported that the plasma treatment resulted in the uniform deposition of Ag NPs by imparting hydrophilicity and increased surface roughness. The total amount of Ag deposited onto PPET fabrics was approximately two times higher than that of UPET fabrics. It was concluded that the release of Ag during washing might be due to the simple detachment of Ag NPs from fiber to the washing bath or due to the oxidation of Ag, thereby reducing the activity against selected bacteria (Ilić et al. 2010).

Modified silica sols were coated on the polyamide (PA) fabric by pad-dry-cure technique for antimicrobial application. Silica sols were prepared by hydrolysis of tetraethoxysilane (TEOS) and 3-glycil-oxypropyltriethoxysilane (GLYEO) to which $AgNO_3$ and hexadecyltrimethylammonium-p-toluene sulfonate (HTAT) were added. Upon application of sol-gel on the fabric, the appearance of the fabric was changed, and a change in property was reported as the weight of the fabric increased and air permeability and surface resistance were decreased. The release of Ag ions from the substrate was slower, which may be related to the reducing $AgNO_3$ to Ag NPs; thus, the treated fabric offered better antimicrobial activity against *E. coli* (Mahltig and Textor 2010). Ag NPs were prepared from $AgNO_3$ using hydrazine as a reducing agent and polyvinylpyrrolidone as a stabilizer and applied on degummed mulberry silk fabric through the exhaustion method. The usage of a reducing agent results in the conversion of $AgNO_3$ to Ag NPs resulting in the agglomeration of particles, but increasing the dispersion speed results in a decreased particle size. The time for reducing to Ag NPs was less as hydrazine was a powerful reducing agent compared to glucose. At lower temperatures, the interaction between the reducing agent and $AgNO_3$ was low, resulting in the formation of small nanoparticles. It was reported that the interaction between silk fabric and nanoparticles was higher at acidic pH, and the developed fabric offered higher antibacterial activity against *S. aureus* (Gulrajani et al. 2008). Exposure to Ag beyond a particular limit (0.01 mg/m^3) creates adverse effects, namely, argyria, in humans. To detect the amount of release of Ag, fabric samples were coated with different concentrations of Ag ranging from 0, 0.5, 1, 5, to 10 respectively. The fabrics were then soaked in four different formulations of artificial sweat with pH ranging between 4.3, 5.5, 6.5, and 8.0, respectively. The initial Ag content and release of Ag from the fabric upon soaking in the artificial sweat were detected through GFAAS (graphite furnace atomic absorption spectroscopy) and ICP (inductively coupled plasma) spectrometer. It was reported that the amount of Ag released from the fabric was based on the initial concentration of Ag present in the fabric as the higher release was seen upon initial contact with sweat and gradually decreased. The sweat formulation of 5.5 pH released a very low concentration of Ag which was similar to the pH of human skin. Thus the developed Ag NP-coated fabric showed higher antibacterial activity against *S. aureus* than against *E. coli* (Kulthong et al. 2010). To increase the stability of the Ag NPs on the surface of the fabric, a cross-linking agent was used. Ag NPs were applied to the nylon-knitted fabric by pad-dry technique with the aid of butanetetracarboxylic acid (BTCA). Ag NPs of 100 and 200 ppm were dispersed in deionized water in the presence of BTCA and sodium hypophosphate as the catalyst. The cross-linking agents react with the amine end group of the polyamide chain in the nylon backbone, resulting in cross-linking between nanoparticles and fabric. The antimicrobial activity of cross-linked fabric offered higher activity for more than 30 washes than uncross-linked fabric. It was reported that fabric loaded with 100 ppm of Ag NPs offered low bacterial reduction against *S. aureus* after 30 washes due to the thicker cell wall of bacteria (Montazer et al. 2012).

Surface modification of PET and PA was carried out using the corona discharge technique by applying a low frequency or pulsed, high voltage across fabrics placed between the two electrodes. Ag NPs were synthesized by dissolving $AgNO_3$ in argon-purged water with the reducing agent $NaBH_4$. The corona-treated PET and PA fabrics were immersed in the colloidal suspension, followed by curing at 100 °C, drying, rinsing, and again drying at room temperature. Atomic absorption spectrometer (AAS) confirmed that the presence of Ag NPs on corona-treated PET and PA fabrics was 8.61 and 4.46 mg/g, respectively, and the nanoparticles of 10 nm were formed confirmed using a transmission electron microscope (TEM). The corona-treated fabrics showed a formation of polar functional groups between Ag NPs and PET/PA fabrics. It was reported that corona-treated Ag-loaded PA fabrics showed better antibacterial activity of 99.8% and 99.9% reduction for *E. coli* and *S. aureus* than the corona-treated Ag-loaded PET fabrics, and corona-treated Ag-loaded PET fabrics showed excellent laundering durability compared to corona-treated Ag-loaded PA fabrics (Radetić et al. 2008).

26.2.4 Polyolefin-Based Textile Substrate

The antibacterial efficiency of Ag-coated polypropylene (PP) fibers and Ag-incorporated composite fibers was studied by spinning the polymer chips to fibers using the lab-scale melt spinning technique. The polymer was powdered and spun into composite fibers with Ag concentration at a weight ratio of 0.3 and 0.5, respectively. Ag NPs of 20 nm and 50 nm were sputter coated on the surface of PP fibers. It was reported that Ag on sputter-coated PP fibers was higher than that of Ag/PP nanocomposite fibers which was confirmed using atomic force microscopy (AFM) and energy-dispersive X-ray (EDX) spectrum. The antibacterial activity of PP, Ag/PP-spun, and Ag-coated PP fibers was analyzed against *S. aureus* using the shake flask test method. It was reported that PP fibers offered no antibacterial activity; sputter coating of Ag on PP fibers offered better antibacterial activity, thereby reducing the growth of bacteria compared to Ag/PP-spun composite fibers (Wei et al. 2009). Polyethylene/polypropylene-based nonwovens act as a carrier for infectious disease-causing bacteria. The Ag colloidal solutions were prepared by dispersing in water and ethanol and dispersing nanosized Ag/sulfur composites in ethanol. The resulting solutions were padded using padding mangle, and the antibacterial activity of the developed fabric was studied using AATCC 100. It was reported that the stability of Ag NPs decreased with an increase in time, and the antibacterial activity of the fabric increased with an increase in Ag concentration. Uniform deposition of Ag NPs was seen on the nonwovens confirmed using surface morphology studies (Jeong et al. 2005). The outer layer of the surgical mask was meant to filter the microbes and gaseous substances from the environment or patients, thereby offering protection to the wearer. Oleophobol C aids in the homogenous dispersion of nanoparticles within the solution. The antibacterial activity of the developed surgical mask was tested according to AATCC 100-1999 against *E. coli* and *S. aureus*. The minimum inhibitory concentration was 1/128 and 1/512, and the minimal bacterial concentration was 1/128 and 1/64 against *E. coli* and *S. aureus*. It was reported that upon coating with Ag-based nanoparticles, surgical masks offered 100% bacterial reduction compared to an uncoated mask. It was reported that upon treatment, the surgical mask offered no signs of skin inflammation or itchiness on the skin (Li et al. 2006).

26.2.5 Blend of Natural and Synthetic Fabric-Based Textile Substrate

Ultrasonication is a technique adopted for the synthesis and uniform application of nanoparticles onto the textile substrate. The Ag NPs were incorporated onto the substrate by the addition of $AgNO_3$ in the solution containing water and ethylene glycol, which acts as a reducing agent in the presence of argon gas to remove the traces of oxygen on the surface of the fabric, namely, nylon, polyester, and cotton. Then the material was sonicated utilizing an ultrasonic horn. Aqueous ammonia solution was added as it slowed down the reduction process resulting in the formation of Ag NPs. The fixation of NPs on the textile substrate was confirmed by the color change from clear to gray fabric, and the fabric was washed with aqueous ethanol. The amount of Ag deposited on the surface of nylon, polyester, and cotton fabric were 1.4, 1.2, and 1.1 wt%, respectively, and no chemical interaction was imparted by the sonication process between nanoparticles and fabric. Rather a physical adsorption of nanoparticles was reported. The mechanical properties of the fabric tend to decrease by 30% compared to the pristine fabric as the ultrasonication process results in faster migration of Ag NPs on the fabric with an average diameter of Ag NPs of 80 nm. It was reported that sonochemically treated Ag-incorporated fabric offered higher antibacterial activity against *E. coli* and *S. aureus* (Perelshtein et al. 2008). Ag NPs were reduced and applied to the cotton and SMS (spunbond meltblown spunbond) fabric by pad-dry-cure technique. The reducing agents used were tulsi leaf extract, neem extract, and quaternary ammonium compound, and citric acid was used as a binder. The treated samples were tested for antibacterial activity against *S. aureus*, *E. coli*, and *P. aeruginosa*. It was reported that spherical Ag NPs were formed at 58–82 nm in neem extract as a reducing agent. Percentage reduction test (AATCC 100) showed maximum reduction in cotton fabrics against *P. aeruginosa* followed by a drug-resistant strain of *P. aeruginosa* and *S. aureus*, whereas maximum reduction in SMS nonwoven fabrics was observed against the drug-resistant strain of *P. aeruginosa*. Wash durability test of cotton fabrics treated with Ag NPs synthesized from neem leaf extract offered superior durability to laundering. Still, on the other hand, SMS nonwoven fabrics offered poor resistance due to the low absorption capacity of Ag NPs as the latter cannot penetrate through the fabric surface reducing the binding ability between the nanoparticles and the fiber surface (Desai 2015).

Pristine wool, polyester, and a blend of wool/polyester fabric were treated with $AgNO_3$ and NaCl utilizing the exhaustion process. The antibacterial activity of the treated fabrics was analyzed according to ASTM E 2149-01, and the antifungal property was assessed against *Aspergillus niger*. Uniform deposition of spherical Ag

NPs of a diameter of 30–200 nm was reported. It was reported that increasing the concentration of $AgNO_3$ and NaCl resulted in an increase in Ag nanoparticles. The amount of Ag NPs adsorbed on the surface of the fabric was in the increasing order of polyester, blend of polyester/wool, and wool fabric. It was reported that bacterial reduction was dependent on adsorbed Ag NPs and fiber properties as wool fabric offered poor bacterial reduction as it hinders the release of Ag cations due to chemical bond interaction but polyester fabric offered higher antibacterial activity as Ag was physically adsorbed on the surface of the fabric due to crystallinity and hydrophobic nature. The antifungal property of the developed fabric was found to be similar to antibacterial activity with a concentration of 0.5 mM of $AgNO_3$ and NaCl solutions. It was reported that increasing the washing cycle decreased the antibacterial activity of the developed fabrics (Klemenčič et al. 2013). Cotton, polyester, blend of polyester/cotton, and polyester/spandex were coated with nano-Ag colloids before dyeing and after dyeing in order to study the antibacterial activity evaluated against *Staphylococcus aureus* and *Klebsiella pneumoniae*. It was reported that TEM (transmission electron microscope) confirmed the presence of Ag NPs that were spherical in shape with an average diameter of 2–5 nm. The bacterial reduction was more effective in the samples treated with Ag colloids after dyeing compared to before-dyed samples. Fabrics with higher concentrations of Ag colloidal solution offered better bacteriostasis. ICP-MS (inductively coupled plasma mass spectrometry) measurements showed that the concentration of Ag NPs decreased rapidly after five times of laundering. Ag NPs offered better bacteriostasis even at smaller concentrations because of the high particle size-to-volume ratio (Lee et al. 2003).

26.2.6 High-Performance Fiber-Based Textile Substrate

Ag nanoparticles was applied to the m-aramid fibers using the pad-dry-cure technique. The padding technique involves immersion of m-aramid fibers in glycidyltrimethylammonium chloride (GTAC, a quaternary ammonium salt) solution followed by drying at 80 °C and curing at 180 ° C, respectively, and the unreacted GTAC on the fibers was removed by washing in distilled water. The GTAC-treated and GTAC-untreated m-aramid fibers were coated with Ag colloids and dried. It was reported that increasing the concentration of Ag increased the surface roughness of the fibers. The antimicrobial (GTAC/Ag NPs)-treated m-aramid fibers showed 75.63% retention of Ag after five wash cycles using an aqueous detergent solution and 3% lower tensile strength and poor thermal stability compared to GTAC-treated and GTAC-untreated m-aramid fibers. Using the disc diffusion method, the efficacy of antimicrobial activity against *P. aeruginosa* was higher in the samples treated with GTAC/Ag NPs compared to GTAC and Ag-treated fibers (Kang et al. 2017). Radical mediated dispersion polymerization technique was used to produce Ag/PMMA (polymethyl methacrylate) nanofibers by injecting 2-2-azobis(isobutyronitrile) and methyl methacrylate into a polymeric aqueous solution containing polyvinyl alcohol and $AgNO_3$ followed by precipitation and washing. The formation of Ag was confirmed using an organic reagent test, namely, rhodamine, into $AgNO_3$ and Ag NP solution. The developed fibers were tested against two bacteria such as *E. coli* and *S. aureus* using the colony-counting method. It was reported that the organic reagent confirmed the formation of Ag NPs as it forms a fade solution due to the formation of Ag-rhodamine complex present on the surface, whereas in the case of $AgNO_3$, red-yellow precipitates that gradually changed to black-brown precipitates indicate the presence of Ag in the solution. The developed nanofibers offered higher antibacterial activity against the selected bacteria, which can be attributed to ion reduction or salt formation between the Ag ions and bacterial cell walls (Kong and Jang 2008).

26.3 Incorporation of Ag NPs onto the Structure of the Textile Substrate

The Ag NPs can be incorporated within the fibrous matrix utilizing the electrospinning technique. Electrospinning is an electrohydrodynamic process composed of three major components as voltage supplier, syringe pump, and collector for spinning the polymeric solution into nanofibers. The principle involves loading polymeric solution on the syringe and charging the solution with a positive charge by application of an electric field through the metal needle in the syringe. The polymeric solution undergoes instability caused by the application of charges, and upon increasing the electric field, the charged polymeric solution undergoes repulsive force within the polymeric chain in the solution. On increasing the voltage beyond the threshold or critical value, the polymeric solution undergoes deformation from spherical to the conical shape called a Taylor cone as the repulsive force overcomes the surface tension of the polymeric solution followed by whipping instability resulting in nanofiber formation by the evaporation of the solvent before reaching the grounded or negatively charged metal conductor (Hemamalini and Giri Dev 2018; Thillaipandian and Rengaswami 2021).

26.3.1 Biodegradable Polymers

Electrospun membrane comprises of silk fibroin and polyethylene oxide was prepared for antibacterial applications by the photoreduction method. Aqueous silk fibroin solution was prepared along with polyethylene oxide and electrospun. The fibrous membrane was crystallized in methanol, and Ag NPs were coated by dipping the fibrous membrane in different concentrations of $AgNO_3$ and then irradiated with UV light, which resulted in the photoreduction of $AgNO_3$ to Ag NPs. To study the interaction of Ag NPs and silk, polyethylene oxide was washed from the fibrous membrane with the aid of a solvent. It was reported that the morphology of the fiber changed to irregular due to crystallization in methanol, and an increase in Ag NP size was seen with an increase in the concentration of $AgNO_3$. It was reported that Ag NPs and silk nanofibers offered interaction between nitrogen and oxygen fibroin groups with nanoparticles which was confirmed using XPS (X-ray photoelectron spectroscopy) (Kang et al. 2007). Chitosan, a second abundant polyelectrolyte natural polymer with 1,4-glycosidic linkage, is widely used for biomedical applications. The chitosan-Ag NP complex was prepared by dissolving chitosan in NaOH to which Ag NPs were added. The spinning solution for electrospinning was prepared by dissolving the complex in acetic acid, to which gelatin was added and electrospun. The fibrous web comprises nanoparticles in the range of 2–10 nm and a fibrous structure of 220–400 nm and finds application in the antibacterial field (Zhuang et al. 2010). Lignin, a renewable biopolymer, interacts with other polymers to form stable polymers; electrospinning of lignin was demanded to increase the spinnability; it was blended with synthetic polymers to form smooth fibers. Lignin was extracted from acacia and blended with polyvinyl alcohol, to which Ag NPs were added to produce polyvinyl alcohol-lignin-Ag blend nanofibers (PLS). Antimicrobial activity of the developed nanofibers was tested against *Bacillus circulans* and *Escherichia coli* using the agar diffusion method, and the inhibition zone was around 11 mm against selected bacteria. It was reported that lignin offers antimicrobial agents because of the presence of the phenolic group, which can damage the bacterial cell membrane (Aadil et al. 2018).

The ideal wound dressing should provide a moist environment for enhancing the wound healing properties, thereby hindering the growth of microorganisms. Polyvinyl alcohol was electrospun into nanofibers comprising $AgNO_3$, followed either by heat treatment or UV irradiation to enhance the antibacterial property. To increase the stability of polyvinyl alcohol fibrous matrix in an aqueous medium, heat treatment was carried out as it increases the crystallinity of polyvinyl alcohol by removal of water molecules rather than chemical cross-linking to avoid toxicity to the newly formed cells. The fibrous web was prepared by altering the sequence of UV irradiation techniques; as such the solution was irradiated before electrospinning. The fibrous web was then illuminated and pristine polyvinyl alcohol web without irradiation to study the deposition of Ag NPs. In the case of solution irradiated with UV, larger Ag NPs were seen compared to the electrospun and pristine fibrous web, which can be due to the earlier reduction of $AgNO_3$ to Ag NPs resulting in agglomeration of particles. The rate of release of Ag^+ was in a burst manner initially followed by the gradual emancipation in the case of non-UV-treated electrospun samples. The fibrous web offered higher antibacterial activity of 99.9% against *S. aureus*, *K. pneumoniae*, and *Pseudomonas aeruginosa* confirmed using the plate-counting method. It was reported that the heat-treated electrospun samples were an ideal candidate for wound dressing applications compared to UV-treated samples (Hong 2007). Electrospinning of polyvinyl alcohol/$AgNO_3$ was carried out by utilizing a plastic tip and graphite electrode in place of a stainless-steel needle and copper or iron electrode. As the metallic needle results in the conversion of metallic Ag or Ag oxide, resulting in blockage of the spinneret or needle, a plastic syringe was used. In the presence of a copper or iron electrode, Ag^+ can be electrochemically reduced to its metallic form; thus graphite electrode was used to avoid the reduction of Ag^+ ions. The spinning solution was maintained at constant pH of 4 as organic materials in polyvinyl alcohol solution tend to reduce Ag ions. The fibrous web was cross-linked with methanol to increase the physical stability, and upon heat treatment, Ag^+ ions were reduced and aggregated to Ag NPs. With the aid of UV irradiation, the number of Ag NPs increased and also offered higher antibacterial activity against *S. aureus* and *K. pneumoniae.* It was reported that strong interaction was found between Ag^+ and polyvinyl alcohol polymer and also the release of Ag ions from the fibrous web was faster in heat-treated samples compared to UV-treated nanofibrous web (Hong et al. 2006). Electrospun polyvinyl alcohol nanofibers comprise Ag NPs by dissolving polyvinyl alcohol in distilled water to which Ag NPs and sodium citrate were added. The solution was electrospun using a Nanospider electrospinning machine. It was confirmed that no new peaks were formed in polyvinyl alcohol/Ag nanofibrous web, which was similar to the pure polyvinyl alcohol web implying no chemical interaction between polyvinyl alcohol and Ag nanoparticles confirmed through Fourier transform infrared (FTIR) spectroscopy. In an agar medium, the bacterial spore suspension, fungal suspension, and bacterial non-spore suspension were prepared from the respective cultures to assess the antibacterial activity of the developed nanofibers using the agar diffusion method. It was inferred that an increase in the concentration of Ag NPs showed a higher reduction rate against gram-negative bacteria *Escherichia coli* and *Pseudomonas aeruginosa*, spore-forming bacteria *Bacillus cereus*, and fungi *Candida albicans* (Pupkevičiūtė et al. 2015).

Polyvinylpyrrolidone (PVP) nanofibers comprising Ag NPs were electrospun for antimicrobial application. Ag NPs were synthesized using ethanol as a reducing and polyvinylpyrrolidone as a stabilizing agent. The resulting solution was irradiated in a microwave oven, and the color change from yellow to brown indicates the formation of Ag NPs. It was reported that the diameter of the Ag NPs was influenced by the concentration of $AgNO_3$, polyvinylpyrrolidone, and microwave energy. The spinning solution for the technique was prepared by dissolving low and high molecular weight of polyvinylpyrrolidone and synthesized Ag NPs and electrospun. It was reported that increasing the concentration of polymer resulted in lengthy, bead-free, uniform, and higher-diameter fibers, while a low concentration of polymer tends to form beads. Ag NP-incorporated polyvinylpyrrolidone nanofibers offered better thermal stability than pure polyvinylpyrrolidone nanofibers and also better antimicrobial activity against bacteria but poor antifungal activity (Çavuşoğlu Vatansever and Meriçboyu 2019). Ag NPs were biosynthesized using *Piper nigrum* (black pepper leaves) and Ag NPs in the presence of ethanol. Poly(ε-caprolactone) of different concentrations was dissolved in acetone, to which biosynthesized Ag NPs were added and electrospun into nanofibers. Incorporation of Ag NPs (0.5 wt%) in the poly(ε-caprolactone) matrix showed higher tensile strength, tensile modulus, and elasticity properties, and the fibrous web offered higher antibacterial activity with inhibiting zone of 11.6 mm against *S. aureus* than *E. coli* of 7.9 mm by using agar diffusion method which may be due to the unstable outer cell wall of *S. aureus* (Augustine et al. 2016).

Poly(L-lactide) (PLA) was synthesized from L-lactide by ring-opening polymerization with tri-octanoate as a catalyst. The polymer was dissolved in dimethylformamide and dichloromethane, to which various percentages of $AgNO_3$

were added and electrospun into nanofibers. The fibrous web was treated in a tubular furnace in a hydrogen atmosphere to convert $AgNO_3$ to Ag NPs. It was reported that increasing the concentration of $AgNO_3$ resulted in an increase in the diameter of the fibers due to phase separation of PLA. Atomic absorption spectroscopy showed that the release of Ag ions was higher at the beginning, followed by a gradual decrease concerning the accumulative release of Ag ions. It was reported that the antibacterial efficiency of the developed fibrous web was 98.5% and 94.2% against *S. aureus* and *E. coli* (Xu et al. 2006). The electrospinning solution was prepared by dissolving PLA in dichloromethane and N,N-dimethylacetamide, to which different concentration of $AgNO_3$ was added by means of ultrasonic treatment. It was reported that the electrical conductivity of the polymeric solution increased with an increase in the concentration of $AgNO_3$. The particle size of the Ag NPs was 4.9 and 8.8 nm on increasing the concentration of $AgNO_3$ due to agglomeration of Ag NPs during electrospinning technique. It was reported that the antibacterial activity of the neat PLA nanofibers was around 60% against *S. aureus* and *E. coli* upon the incorporation of $AgNO_3$, resulting in a 99% growth reduction against the selected bacteria (Kim et al. 2010).

Poly(L-lactide-co-glycolide) (PLGA) is an aliphatic biodegradable polymer that can degrade within the body via the citric acid cycle. PLGA with L:G ratio of 50:50 (PLGA 50/50) and 75:25 (PLGA 75/25) was electrospun by incorporating Ag NPs for antimicrobial application. The polymer was dissolved in chloroform and dimethylformamide mixture, and then Ag NPs were added at different concentrations and electrospun to nanofiber. It was reported that an increase in the concentration of Ag NPs resulted in an increase in electrical conductivity, thereby decreasing the fiber diameter of the fabric. It was also concluded that increasing the concentration of lactide resulted in an increase in fiber diameter and tensile properties of the fabric. To study the cytotoxicity of developed samples, human dermal fibroblast (HDF) was cultured, and it was inferred that increasing the concentration beyond 6 wt% $AgNO_3$ resulted in toxicity to fibroblast cells, thereby hindering the cell growth and also reducing the mechanical property of the nanofibrous mat. It was concluded that Ag with 3 wt% offered higher antibacterial activity and was also biocompatible with fibroblast cells, thereby offering good mechanical properties similar to human skin tissue (Gõra et al. 2015).

26.3.2 Nonbiodegradable Polymers

Nylon 6, along with Ag NP-incorporated nanofibers, was spun by dissolving different concentration of nylon 6 in formic acid to which Ag NPs were added and electrospun. It was reported that increasing nylon resulted in an increase in the diameter of the fibers and increasing the concentration from 15 to 25 wt % resulted in uniform bead-free fibers. The addition of Ag NPs decreased the fiber diameter due to an increase in the conductivity of the polymeric solution. The size of Ag NPs was 5–10 nm which was confirmed using TEM analysis, and nanofibers with 1000 ppm of Ag NPs offered 99.9% antibacterial activity against *S. aureus* and *K. pneumoniae* by cell-counting method (Park et al. 2009). In situ Ag NP-incorporated polyamide 6 nanofibers were electrospun by dissolving a solvent mixture of formic acid/acetic acid in the ratio of 1:4, to which $AgNO_3$ was added and reduced using sodium borohydride. It was reported that an increase in Ag concentration resulted in a decrease in fiber diameter due to the rise in the conductivity of the polymeric solution. The conversion of $AgNO_3$ to Ag metal was less in formic acid, which was confirmed by the formation of a peak around 275 nm in the UV-visible spectrophotometer, whereas pristine Ag NPs formed a peak at 400 nm, which was indicated by the appearance of yellow color from colorless solution. It was confirmed that nylon-Ag NP nanofibers showed almost 98–100% growth reduction against *E. coli* and *S. aureus* (Montazer and Malekzadeh 2012). Nylon 6-based fibrous web was prepared by dissolution of polymer in formic acid and acetic acid solvent mixture to which $AgNO_3$ and methoxy polyethylene glycol as reducing agent were added. The polymeric solution was allowed to stand for 5 and 12 h for the reduction of $AgNO_3$ to Ag NPs before electrospinning, and it has an influence on the fiber diameter as complete reduction of $AgNO_3$ resulted in a decrease in fiber diameter and also reduction time resulted in degradation of the polymer due to acetic acid and methoxy polyethylene glycol. Antibacterial activity of Ag/nylon 6 mats was measured against *S. aureus* and *E. coli*, and it was reported that the mat with a short reaction time showed better antibacterial activity against both microorganisms (Pant et al. 2012).

Cellulose acetate (CA) nanofibers incorporated with $AgNO_3$ were electrospun in a solvent mixture, namely, acetone and water, to avoid the clogging of the needles. To the homogenous solution, $AgNO_3$ was added and then UV radiated to reduce to Ag NPs. It was reported that an increase in the concentration of Ag resulted in a decrease in fiber diameter due to the rise in the conductivity of CA solution, which may be due to an increase in charge density increasing the elongation force. It was reported that the viscosity and surface tension of the solution tends to unalter upon the addition of $AgNO_3$. The developed cellulose acetate-incorporated Ag NPs offered maximum inhibition zone against the selected bacteria such as *S. aureus*, *K. pneumoniae*, *E. coli*, and *P. aeruginosa* confirmed using zone of inhibition study (Son et al. 2004). Composite nanofibers comprising polyether amide (PEA) and Ag were prepared for antibacterial applications using the electrospinning technique. The spinning solution was designed by dissolving polyether amide in isopropanol to yield a homogenous solution in which a different mass ratio of $AgNO_3$ was added and electrospun followed by UV irradiation for the conversion of $AgNO_3$ to Ag NPs. It was reported that the incorporation of Ag in the polymeric solution resulted in a decrease in the regularity of the internal structure of PEA and increased the flexibility of the polymeric chain, which resulted in a drop in crystallinity and glass transition temperature, confirmed using DSC (differential scanning calorimetry). The thermal stability of the nanofibrous web was decreased due to an increase in the expansion of the nanocrystalline region by the addition of $AgNO_3$ resulting in the expansion of molecular arrangement. It was reported that the fibrous membrane offered no bacterial growth and offered an inhibition rate of 99.9% against *E. coli* and *S. aureus* with a minimum Ag concentration of 0.15% (Liang et al. 2014).

Ag NP-incorporated polyurethane (PU) nanofiber-based wound dressing was prepared by blending the synthesized Ag NPs into the polymeric solution and electrospinning them into nanofibers. $AgNO_3$ was synthesized using the organic method using dimethylformamide and the aqueous method using polyetherimide and tetrahydrofuran, respectively. The solution was electrospun, and the resulting web was analyzed for antibacterial activity against *Klebsiella* bacteria and cytotoxicity study. It was reported that increasing the concentration of Ag resulted in a decrease in fiber diameter, which may be due to an increase in the conductivity of the polymeric solution. The developed Ag-PU web's water absorption ability was higher than the pristine PU web due to its high porosity and surface area. The developed Ag-incorporated web offered high resistance to the selected bacteria by inhibiting their growth. It was reported that the organic method of nanofiber production was cytotoxic to cells due to the presence of dimethylformamide, which was not vaporized completely. In contrast, the aqueous method was nontoxic as tetrahydrofuran evaporated completely as it offered a low boiling point (Lakshman et al. 2010). Electrospinning of rubber and Ag NPs involves a complex mechanism as the elasticity of the rubber material hinders the formation of nanofibers due to the movement of macromolecules in the polymer structure, thereby disintegrating the fibrous structure. Rubber-based nanofibers were electrospun using an in situ chemical cross-linking technique to yield uniform bead-free fibers. Rubber composite fibers, namely, butadiene rubber, isobutylene-isoprene rubber, and silicon rubber, were prepared by dispersing $AgNO_3$ and polyvinylpyrrolidone in dimethylformamide to which tetrahydrofuran was added. Butadiene rubber was added to the solution and irradiated with UV irradiation to form Ag NPs. Camphorquinone (CQ) was added for in situ cross-linking prior to the electrospinning. A similar technique was adopted for the preparation of Isobutylene-isoprene rubber (IIR)-based solution, but for spinning Silicon rubber (SiR)-based nanofibers, platinum was used as a curing agent in the absence of CQ as a cross-linking agent. It was reported that polyvinylpyrrolidone as stabilizing agent prevented the agglomeration of Ag NPs and an in situ chemical cross-linking technique produced uniform bead-free fibers. It was reported that a limited concentration of Ag NPs offered higher optical absorbance and higher antibacterial activity against *E. coli* (Hu et al. 2012).

Polyacrylonitrile (PAN) nanofibers comprising a different percentage of TiO_2 and $AgNO_3$ were prepared by dissolving the polymer in dimethylformamide and spinning it into nanofibers using the electrospinning technique. The viscosity of the polymeric solution depended on the concentration of TiO_2 as the viscosity increased with the concentration of TiO_2. It was reported that increasing the concentration of TiO_2 increased the diameter of the nanofibers from 400 to 600 nm resulting in the formation of a beaded structure. The composite polyacrylonitrile/TiO_2 nanofibers were functionalized by immersing the fibrous mat in an aqueous $AgNO_3$ solution as Ag^+ ions bind onto the negatively charged titania incorporated in the fibers. Upon the photoreduction process, pure PAN/titania nanofibers possessed no color change. Still, on the treatment with $AgNO_3$, the color change was seen to brownish gray, indicating the reduction of Ag^+ to Ag NPs. It was also reported that increasing titania concentration resulted in an increase in the uptake of

Ag NPs in the polyacrylonitrile nanofibers, confirmed using UV-visible spectroscopy (Lim et al. 2006). Coaxial electrospinning was carried out to spin nanofibers comprising $AgNO_3$ on the sheath to assess the antibacterial activity. The core solution includes polyacrylonitrile in dimethylacetamide, whereas the sheath solution of $AgNO_3$ in dimethylacetamide and electrospinning was carried out using a homemade concentric spinneret with 1 mm intended capillary gap between the core and sheath. The collected electrospun nanofibers were treated at 254 nm under ultraviolet light to reduce Ag ions to Ag NPs. During the bending and whipping process of the electrospinning technique, the sheath fluid of $AgNO_3$ with the flow rate of 0.1 was unable to catch the flow rate of core fluid PAN, which resulted in an irregular distribution of $AgNO_3$ on the surface of the nanofibers, but upon increasing the flow rate to 0.2 mL/h of the sheath fluid resulted in better distribution. The antibacterial activity of the coaxial electrospun nanofibers showed strong antibacterial activity against *B. subtilis* and *E. coli* with a Ag content of 4% on the surface of nanofibers compared to the traditional electrospun nanofibers which require 30% of Ag content as the nanoparticles were buried on the core of the nanofibers (Yu et al. 2012).

26.4 Conclusion

As discussed in the chapter, Ag has been known as an antimicrobial agent for a long time. But the major challenge in the incorporation of Ag onto textiles is durability. It depends on the type of fiber used and the method of application. A majority of industries work on the exhaust method of application; the loading of Ag can be scaled at the industry level if the limitations in the processing route are addressed. The type of fiber used in industry and the challenges in bringing Ag to the substrate have been dealt with in detail. The potential applications of Ag in various medical fields have also been explored. The major limitation in applying Ag NPs to biodegradable and nondegradable polymers by exhaust method or by padding process knows the exact quantity of loading onto the fabric and its distribution on the substrate. Moreover, the mechanism of binding Ag to the fabric is to be researched further to provide insights into the release mechanism. Aging studies of Ag in the fabric will provide excellent knowledge of the product's shelf life and durability on exposure to various environments. In addition, studies of the interaction of Ag and colorant used in textiles will help us understand the challenge to be faced while scaling up the process. The research on the above points would result in a durable and practical product against bacteria.

References

Aadil KR, Mussatto SI, Jha H (2018) Synthesis and characterization of silver nanoparticles loaded poly(vinyl alcohol)-lignin electrospun nanofibers and their antimicrobial activity. Int J Biol Macromol 120:763–767. https://doi.org/10.1016/j.ijbiomac.2018.08.109

Augustine R, Kalarikkal N, Thomas S (2016) Electrospun PCL membranes incorporated with biosynthesized silver nanoparticles as antibacterial wound dressings. Appl Nanosci 6:337–344. https://doi.org/10.1007/s13204-015-0439-1

Çavuşoğlu Vatansever H, Meriçboyu AE (2019) Production of antibacterial polyvinylpyrrolidone nanofibers containing silver nanoparticles via electrospinning method. Manas J Eng 7:13–23

Cay A, Tarakçioğlu I, Hepbasli A (2009) Assessment of finishing processes by exhaustion principle for textile fabrics: an exergetic approach. Appl Therm Eng 29:2554–2561. https://doi.org/10.1016/j.applthermaleng.2008.12.032

Chen CY, Chiang CL (2008) Preparation of cotton fibers with antibacterial silver nanoparticles. Mater Lett 62:3607–3609. https://doi.org/10.1016/j.matlet.2008.04.008

Deng X, Nikiforov A, Vujosevic D, Vuksanovic V, Leys C (2015a) Nanoparticle incorporated nonwoven fabric prepared by atmospheric pressure plasma process for antibacterial property. pp 22–24

Deng X, Yu Nikiforov A, Coenye T, Cools P, Aziz G, Morent R, De Geyter N, Leys C (2015b) Antimicrobial nano-silver nonwoven polyethylene terephthalate fabric via an atmospheric pressure plasma deposition process. Sci Rep 5:1–10. https://doi.org/10.1038/srep10138

Desai MHPPB (2015) Antimicrobial effect of herbal nanosilver finished fabrics on drug resistant pathogens. Int J Sci Res 4:427–432

Gõra A, Prabhakaran MP, Eunice GTL, Lakshminarayanan R, Ramakrishna S (2015) Silver nanoparticle incorporated poly(l-lactide-co-glycolide) nanofibers: evaluation of their biocompatibility and antibacterial properties. J Appl Polym Sci 132:1–11. https://doi.org/10.1002/app.42686

Gulrajani ML, Gupta D, Periyasamy S, Muthu SG (2008) Preparation and application of silver nanoparticles on silk for imparting antimicrobial properties. J Appl Polym Sci 108:614–623. https://doi.org/10.1002/app.27584

Hebeish A, El-Naggar ME, Fouda MMG, Ramadan MA, Al-Deyab SS, El-Rafie MH (2011) Highly effective antibacterial textiles containing green synthesized silver nanoparticles. Carbohydr Polym 86:936–940. https://doi.org/10.1016/j.carbpol.2011.05.048

Hemamalini T, Giri Dev VR (2018) Comprehensive review on electrospinning of starch polymer for biomedical applications. Int J Biol Macromol 106:712–718. https://doi.org/10.1016/j.ijbiomac.2017.08.079

Hemamalini T, Vrishni Ritvic J, Premitha R, Divya Dharshini AK, Giri Dev VR (2021) Facile and cost-effective development of silver based cellulosic wound dressing using electrospraying process. J Nat Fibers 19:7436–7444. https://doi.org/10.1080/15440478.2021.1946887

Hong KH (2007) Preparation and properties of electrospun poly(vinyl alcohol)/silver fiber web as wound dressings. Polym Eng Sci 47:43–49. https://doi.org/10.1002/pen.20660

Hong KH, Park JL, Sul IH, Youk JH, Kang TJ (2006) Preparation of antimicrobial poly(vinyl alcohol) nanofibers containing silver nanoparticles. J Polym Sci Part B Polym Phys 44:2468–2474. https://doi.org/10.1002/polb.20913

Hu Q, Wu H, Zhang L, Fong H, Tian M (2012) Rubber composite fibers containing silver nanoparticles prepared by electrospinning and in-situ chemical crosslinking. Express Polym Lett 6:258–265. https://doi.org/10.3144/expresspolymlett.2012.29

Ilić V, Šaponjić Z, Vodnik V, Lazović S, Dimitrijević S, Jovančić P, Nedeljković JM, Radetić M (2010) Bactericidal efficiency of silver nanoparticles deposited onto radio frequency plasma pretreated polyester fabrics. Ind Eng Chem Res 49:7287–7293. https://doi.org/10.1021/ie1001313

Jeong SH, Hwang YH, Yi SC (2005) Antibacterial properties of padded PP/PE nonwovens incorporating nano-sized silver colloids. J Mater Sci 40:5413–5418. https://doi.org/10.1007/s10853-005-4340-2

Jiang SX, Yuen CWM, Zhang L, Guo RH, Choi PSR (2009) Characteristics of silver-plated silk fabric with plasma pre-treatment. Fibers Polym 10:791–796. https://doi.org/10.1007/s12221-009-0791-4

Kang M, Jung R, Kim HS, Youk JH, Jin HJ (2007) Silver nanoparticles incorporated electrospun silk fibers. J Nanosci Nanotechnol 7:3888–3891. https://doi.org/10.1166/jnn.2007.056

Kang C, Kim SS, Ahn D, Kim SJ, Lee J (2017) Effective surface attachment of ag nanoparticles on fibers using glycidyltrimethylammonium chloride and improvement of antimicrobial properties. RSC Adv 7:23407–23414. https://doi.org/10.1039/c7ra01636k

Ke Thanh NV, Phuong Phong NT (2009) Investigation of antibacterial activity of cotton fabric incorporating nano silver colloid. J Phys Conf Ser 187:012072. https://doi.org/10.1088/1742-6596/187/1/012072

Ki HY, Kim JH, Kwon SC, Jeong SH (2007) A study on multifunctional wool textiles treated with nano-sized silver. J Mater Sci 42:8020–8024. https://doi.org/10.1007/s10853-007-1572-3

Kim ES, Kim SH, Lee CH (2010) Electrospinning of polylactide fibers containing silver nanoparticles. Macromol Res 18:215–221. https://doi.org/10.1007/s13233-010-0316-4

Kim SS, Park JE, Lee J (2011) Properties and antimicrobial efficacy of cellulose fiber coated with silver nanoparticles and 3-mercaptopropyltrimethoxysilane (3-MPTMS). J Appl Polym Sci 119: 2261–2267. https://doi.org/10.1002/app.32975

Klemenčič D, Tomšič B, Kovač F, Žerjav M, Simončič A, Simončič B (2013) Antimicrobial wool, polyester and a wool/polyester blend created by silver particles embedded in a silica matrix. Colloids Surfaces B Biointerfaces 111:517–522. https://doi.org/10.1016/j.colsurfb.2013.06.044

Kong H, Jang J (2008) Antibacterial properties of novel poly(methyl methacrylate) nanofiber containing silver nanoparticles. Langmuir 24:2051–2056. https://doi.org/10.1021/la703085e

Kulthong K, Srisung S, Boonpavanitchakul K, Kangwansupamonkon W, Maniratanachote R (2010) Determination of silver nanoparticle release from antibacterial fabrics into artificial sweat. Part Fiber Toxicol 7:2–10. https://doi.org/10.1186/1743-8977-7-8

Lakshman LR, Shalumon KT, Nair SV, Jayakumar R, Nair SV (2010) Preparation of silver nanoparticles incorporated electrospun polyurethane nano-fibrous mat for wound dressing. J Macromol Sci Part A Pure Appl Chem 47:1012–1018. https://doi.org/10.1080/10601325.2010.508001

Lee HJ, Yeo SY, Jeong SH (2003) Antibacterial effect of nanosized silver colloidal solution on textile fabrics. J Mater Sci 38:2199–2204. https://doi.org/10.1023/A:1023736416361

Li Y, Leung P, Yao L, Song QW, Newton E (2006) Antimicrobial effect of surgical masks coated with nanoparticles. J Hosp Infect 62:58–63. https://doi.org/10.1016/j.jhin.2005.04.015

Liang S, Zhang G, Min J, Ding J, Jiang X (2014) Synthesis and antibacterial testing of silver/poly (ether amide) composite nanofibers with ultralow silver content. J Nanomater 2014:684251. https://doi.org/10.1155/2014/684251

Lim SK, Lee SK, Hwang SH, Kim H (2006) Photocatalytic deposition of silver nanoparticles onto organic/inorganic composite nanofibers. Macromol Mater Eng 291:1265–1270. https://doi.org/10.1002/mame.200600264

Mahltig B, Textor T (2010) Silver containing sol-gel coatings on polyamide fabrics as antimicrobial finish-description of a technical application process for wash permanent antimicrobial effect. Fibers Polym 11:1152–1158. https://doi.org/10.1007/s12221-010-1152-z

Marcato PD, Nakasato G, Brocchi M, Melo PS, Huber SC, Ferreira IR, Alves OL, Durán N (2012) Biogenic silver nanoparticles: antibacterial and cytotoxicity applied to textile fabrics. J Nano Res 20:69–76. https://doi.org/10.4028/www.scientific.net/JNanoR.20.69

Memon H, Wang H, Yasin S, Halepoto A (2018) Influence of incorporating silver nanoparticles in protease treatment on fiber friction, antistatic, and antibacterial properties of wool fibers. J Chem 2018:4845687. https://doi.org/10.1155/2018/4845687

Moazami A, Montazer M, Rashidi A, Rahimi MK (2010) Antibacterial properties of raw and degummed silk with nanosilver in various conditions. J Appl Polym Sci 118:253–258. https://doi.org/10.1002/app.32289

Montazer M, Malekzadeh SB (2012) Electrospun antibacterial nylon nanofibers through in situ synthesis of nanosilver: preparation and characteristics. J Polym Res 19:9980. https://doi.org/10.1007/s10965-012-9980-8

Montazer M, Shamei A, Alimohammadi F (2012) Stabilized nanosilver loaded nylon knitted fabric using BTCA without yellowing. Prog Org Coatings 74:270–276. https://doi.org/10.1016/j.porgcoat.2012.01.003

Pant B, Pant HR, Pandeya DR, Panthi G, Nam KT, Hong ST, Kim CS, Kim HY (2012) Characterization and antibacterial properties of Ag NPs loaded nylon-6 nanocomposite prepared by one-step electrospinning process. Colloids Surfaces A Physicochem Eng Asp 395:94–99. https://doi.org/10.1016/j.colsurfa.2011.12.011

Park S-W, Bae H-S, Xing Z-C, Kwon OH, Huh M-W, Kang I-K (2009) Preparation and properties of silver-containing nylon 6 nanofibers formed by electrospinning. J Appl Polym Sci 112:2320–2326. https://doi.org/10.1002/app.29520

Perelshtein I, Applerot G, Perkas N, Guibert G, Mikhailov S, Gedanken A (2008) Sonochemical coating of silver nanoparticles on textile fabrics (nylon, polyester and cotton) and their antibacterial activity. Nanotechnology 19:245705. https://doi.org/10.1088/0957-4484/19/24/245705

Pupkevičiūtė S, Adomavičiūtė E, Pavilonis A, Stanys S, Prosyčevas I (2015) Formation and antibacterial property analysis of electrospun PVA nonwoven material with a small amount of silver nanoparticles. Fibers Text East Eur 23:48–54. https://doi.org/10.5604/12303666.1167419

Radetić M, Ilić V, Vodnik V, Dimitrijević S, Jovančić P, Šaponjić Z, Nedeljković JM (2008) Antibacterial effect of silver nanoparticles deposited on corona-treated polyester and polyamide fabrics. Polym Adv Technol 19:1816–1821. https://doi.org/10.1002/pat.1205

Son WK, Youk JH, Lee TS, Park WH (2004) Preparation of antimicrobial ultrafine cellulose acetate fibers with silver nanoparticles. Macromol Rapid Commun 25:1632–1637. https://doi.org/10.1002/marc.200400323

Tarimala S, Kothari N, Abidi N, Hequet E, Fralick J, Dai LL (2006) New approach to antibacterial treatment of cotton fabric with silver nanoparticle-doped silica using sol-gel process. J Appl Polym Sci 101:2938–2943. https://doi.org/10.1002/app.23443

Thillaipandian H, Rengaswami GDV (2021) Biofunctional textile fibers and their applications. In: Fundamentals of natural fibers and textiles. Elsevier, pp 263–302

Wei Q, Tao D, Deng B, Huang F (2009) Comparative studies of silver nanocomposite fibers. J Ind Text 38:309–316. https://doi.org/10.1177/1528083708092013

Xu X, Yang Q, Wang Y, Yu H, Chen X, Jing X (2006) Biodegradable electrospun poly(l-lactide) fibers containing antibacterial silver nanoparticles. Eur Polym J 42:2081–2087. https://doi.org/10.1016/j.eurpolymj.2006.03.032

Yu DG, Zhou J, Chatterton NP, Li Y, Huang J, Wang X (2012) Polyacrylonitrile nanofibers coated with silver nanoparticles using a modified coaxial electrospinning process. Int J Nanomedicine 7:5725–5732. https://doi.org/10.2147/IJN.S37455

Zhuang X, Cheng B, Kang W, Xu X (2010) Electrospun chitosan/gelatin nanofibers containing silver nanoparticles. Carbohydr Polym 82:524–527. https://doi.org/10.1016/j.carbpol.2010.04.085

T. Hemamalini, teaching fellow and part-time research scholar in the Department of Textile Technology, Anna University, Chennai, with a degree in master of textile technology. With an area of interest in medical textiles, my research works have been carried out in wound dressing and blood clot applications to explore and gain knowledge in the mentioned field, and I successfully published peer-reviewed articles in reputed journals.

Pon G. K. Gokul, completed BTech in Textile Technology in Alagappa College of Technology, Anna University, Chennai, and working as Six Sigma coordinator and R&D in Lush Hair, Nigeria. I have an excellent opportunity to enhance my writing skills by contributing a literature review regarding the antimicrobial activities of silver nanoparticle-incorporated textile substrates.

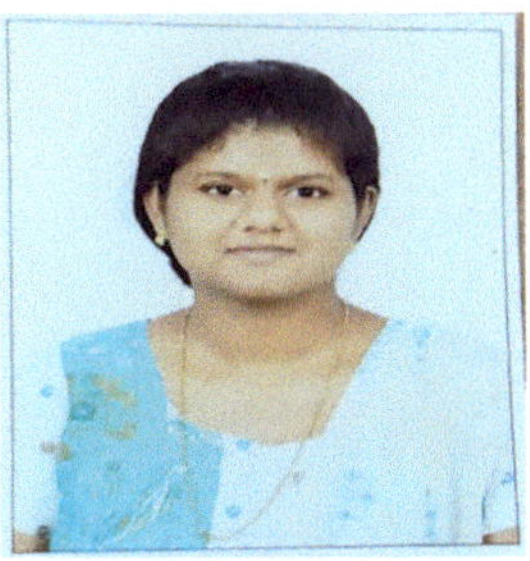

R. S. Mahalakshmi, completed BTech in Textile Technology in Anna University, Chennai, and working in Zen Linen. I was involved in data analysis and research. In addition to my research training and experience, I have gained coordination and administrative skills, which enhance my effectiveness in further research work.

R. Rajesh, Bachelor of Technology in Textile Technology, Anna University, Chennai, and working in Voltas. I have an interest in reading and writing articles which aided in writing a review article on the incorporation of silver nanoparticles in a textile substrate which provides better antimicrobial properties.

V. R. Giri Dev, a textile technologist who has a passion in fields of fiber science, chemical processing, wound dressing, and composites. I am an inquisitive learner who constantly feels that the research should yield symbiotic benefits to both industry and academia, aiming to establish a good industry and academic network to achieve the goals. I am well-trained in DOE and its analysis and successful in executing various government-funded research projects in the field of fiber science leading to conclusive findings; have good networking between various students, academics, and industry persons; and manage alumni and textile association in the department. At present, I am working in the field of sustainable chemical processing, medical textile, and biocomposites.

Recent Advancement of Gelatin for Tissue Engineering Applications

27

Faisal Raza, Jing Su, Jian Zhong, and Mingfeng Qiu

Abstract

Traumatic injuries of bones, cartilage, ligaments, and tendons require novel tissue engineering technology using more biocompatible materials. For this purpose, many biomaterials are contributing successfully toward developing regenerative medicine for healing and repairing in thousands of clinical cases yearly. Today, biomaterials have shifted toward more advancements with the integration of other fields such as pharmaceutics, biology, biomedical engineering, and much more. Gelatin, in this regard, has played a pivotal role in novel tissue engineering technology. The remarkable properties of gelatin, such as biocompatibility, biodegradability, and its favorable physicochemical characteristics, have made using this biomaterial more convenient. This book chapter presents a thorough overview of gelatin for its significant contribution to the field of tissue engineering. The source, composition, and general characteristics of gelatin are explained, followed by various biomaterials based on gelatin. Furthermore, the significant recent advances in gelatin for tissue engineering are discussed.

Keywords

Gelatin · Bone · Gelatin nanofibers · Electrospinning

F. Raza · J. Su · M. Qiu (✉)
School of Pharmacy, Shanghai Jiao Tong University, Shanghai, China
e-mail: faisalraza@sjtu.edu.cn; jingsu@sjtu.edu.cn; mfqiu@sjtu.edu.cn

J. Zhong
Xinhua Hospital, Shanghai Institute for Pediatric Research, Shanghai Key Laboratory of Gastroenterology and Nutrition, Shanghai Jiao Tong University School of Medicine, Shanghai, China
e-mail: jzhong@shsmu.edu.cn

F. A. Sheikh et al. (eds.), *Interaction of Nanomaterials With Living Cells*,
https://doi.org/10.1007/978-981-99-2119-5_27

27.1 Introduction

There is increasing demand to devise new and biomaterial-based technologies and fabricate the existing ones with the latest technologies for the healing, repair, or regeneration of diseased or injured tissues (Echave et al. 2017; Raza et al. 2022). All types of tissues require these technologies, but acute and chronic musculoskeletal conditions such as bone (Laurent et al. 2020), cartilage (Carstairs and Genever 2014), tendon (Gallagher and Schall Jr 2017), ligament trauma and other injuries (Rodrigues et al. 2013), degenerative disc disease, and osteoarthritis enormously need the development of such techniques (Padilla et al. 2017). These conditions and others are treated in the United States, with 34 million surgical procedures yearly (Lee and Mooney 2001). It is well-known that scaffolds made of biomaterials and biomaterial-based technologies have critical roles in medicine for regenerative purposes as they optimize and drive cellular fate and function (Lutolf and Hubbell 2005). In recent years, a remarkable shift in the designing and fabrication criterion for advanced biomaterials has been observed, completely integrating the principles from pharmaceutics, biology, and biomedical engineering (Kamal et al. 2022; Hoque et al. 2015). Gelatin is a naturally occurring macromolecule, amphiphilic in nature, usually obtained from the collagen of mammals (Joyce et al. 2021), fish (Huang et al. 2019), and insects (Chen et al. 2019) by hydrolysis. Gelatins have been immensely explored in various other fields, including food (Zhang et al. 2020), the pharmaceutical industry (Wasswa et al. 2007), and tissue engineering (Hoque et al. 2015). As gelatin has been used as a plasma expander for a long time, it probably has the opportunity to be considered as a biopolymer for nanomaterials along with other naturally occurring polymers (Aggarwal et al. 2021). About 13% of positively charged amino acids constitute gelatin protein (arginine and lysine) (Gupta et al. 2021). Polymer backbone comprises hydrophobic amino acids, about 11%, and negatively charged amino acids, such as aspartic acid and glutamic acid, which make up 12% of the protein (Zohuriaan-Mehr et al. 2009). The rest of the chain is comprised of lysine, proline, and hydroxyproline (Koshti et al. 2021). There are no toxic side products as the chain is enzymatically transformed to corresponding amino acid residues (Wu et al. 2004). Commercially, there are two forms of gelatin, i.e., A and B, having distinct isoelectric points (IEPs) (Masuelli and Sansone 2012; Baydin et al. 2022). A form is known as cationic with isoelectric points 7–9 making acid hydrolysis part of collagen. Anionic B form of gelatin has an IEP of 4.7–5.5 and is acquired from alkaline collagen treatment (Madkhali et al. 2019). Gelatin also contains amino acid (Arg-Gly-Asp series) of targeting ligand that has importance in tissue engineering and nanomedicine in cancer studies and treatment (Dong et al. 2015). This book chapter is a state-of-the-art overview presenting the use of gelatin and its composites used in tissue engineering as well as its use as micro- and nanoparticles and scaffolds for the said purpose in current studies of gelatin-based formulations (Fig. 27.1). The nuclear role of gelatin as a biomaterial in regeneration, tissue repair, and engineering is highlighted in this chapter.

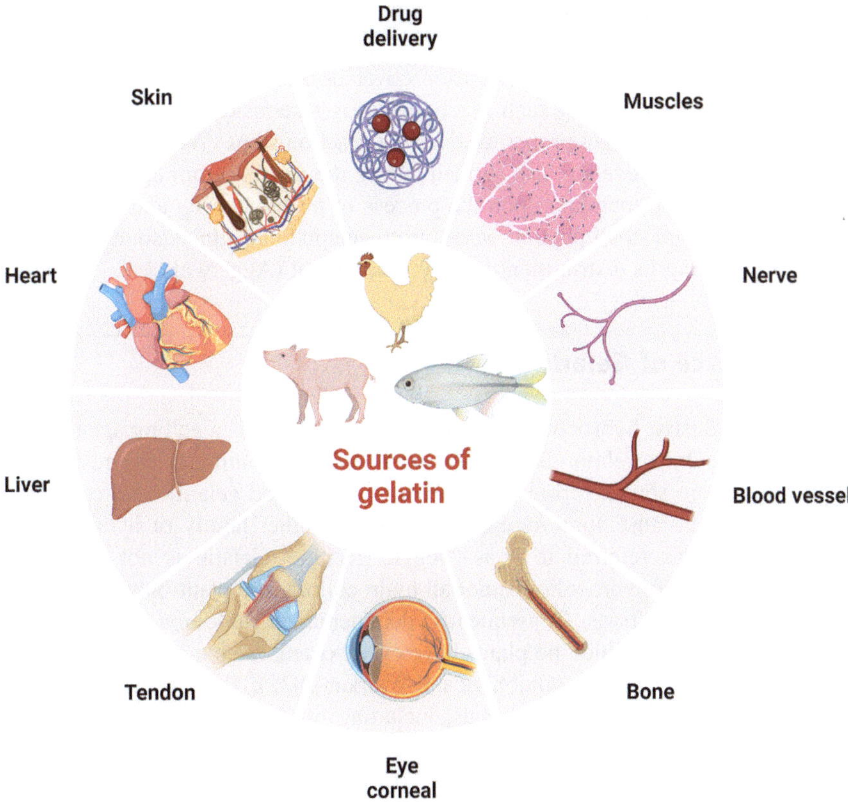

Fig. 27.1 Schematic representation of sources of gelatin and its application in tissue engineering and drug delivery

27.2 Composition of Gelatin

The composition of gelatin is similar to the collagen molecules in that it has most of the denatured collagen. There are variations in native collagen and gelatin because of differences in gelatin's synthesis and molecular organization. However, collagen can be transformed into gelatin by changing the amino acid molecular composition (Joyce et al. 2021). Deamination of glutamine into glutamic acid and asparagine into aspartic acid in the alkaline process leads to a higher proportion of aspartic acid and glutamic acid in type B gelatin than type A (Hafidz et al. 2011). Gelatin amino acid composition is not well-defined. However, the proportion of hydroxyproline and proline in mammalian gelatins is 30% of the total amino acid residues. Cysteine was not reported to be present in pig skin, according to Farris et al. (Farris et al. 2010). Bailey and Light found it in type III collagen (Stanton and Light 1990). There is no evidence in the literature about the abundance of this collagen in pig skin. However, it is an essential part of the human skin, constituting 15% of the total collagen. Despite careful purification, gelatin contains salt, lipid, and sugar moieties, as bone and skin contain sugars, lipids, small molecules, and ions. Upon interaction with protein fibers, these molecules form covalent bonding with gelatin. Maillard reaction is an example of a such reaction that is responsible for brown color of gelatin and changes its gel properties during extraction steps (Wasswa et al. 2007). Peptides of variable size are also formed during the process from collagen and are present in gelatin. Depending upon the process of manufacturing and raw material used, the number of small peptides varies from gelatin to gelatin. Also, it changes its function following its distribution of molecular weight (Aggarwal et al. 2021).

27.3 Source of Gelatin

In the food industry, hydrocolloids are commonly utilized as a gelling agent (Shyni et al. 2014). Only gelatin is usually extracted from animals; the rest of the hydrocolloids are sourced from plants. Hydrocolloids and gelatin have commonly been confused because some researchers favor the other family of hydrocolloids, particularly plants, referred to it as "veggie gelatin." Gelatin is not a naturally occurring family of hydrocolloids; not all hydrocolloids are plant-based; it must be derived from animal parts. Consequently, demineralized cattle bones (ossein), pig, and bovine skins, for which no plants have ever existed, are the only commercially feasible sources of collagen (Gajbhiye and Wairkar 2022). The most easily accessible resources of gelatin are mammals, including pig and cow hides, bones, and hooves, which together make up 46% of the world's gelatin source. Fish and other marine forms, including bones and hooves, represent the remaining 1% (Rakhmanova et al. 2018). In Europe, porcine and bovine hides account for 95% of gelatin production and their bones for 5% (Alipal et al. 2021). Mammalian gelatin is thermoreversible and also has high boiling and gelling points. Traditional mammalian sources of gelatin also include bones, cattle hides, and skin from pigs (Alipal et al. 2021). For industrial uses, high-quality gelatin derived from cow bone is recommended. Pork gelatin and gelatin obtained from other animals not slaughtered in accordance with Islamic law are not utilized due to aesthetic and religious concerns (Rakhmanova et al. 2018). These factors expanded the halal meals and additive market and attracted academics and entrepreneurs (Ab Talib et al. 2016). As a result, scientists are looking for new, different, halal sources of gelatin. The commercial potential for fish and poultry byproduct gelatin has grown recently. However, because of low yields, commercial synthesis of gelatin from poultry wastes is currently restricted (Abedinia et al. 2020). Compared to gelatin produced from mammals, much research has been conducted on how to extract gelatin from fish skin and poultry manure. From the perspective of packing, the physical, chemical, and functional characteristics of the films which can be generated from a specific source of gelatin are significantly influenced by the source of gelatin (Alfaro et al. 2015).

Conversely, Lestari et al.'s term "veggie gelatin" refers to a substitute for animal-based gelatins like Konjac (a type of gelatin for use in Japanese cuisine) and yam that can be made from plant hydrocolloids. Agar, carrageenan, pectin, xanthan gum, modified corn starch, and celluloid could all be sources of this vegetable gelatin (Lestari et al. 2019). The recent effort to produce “veggie gelatin” from sago starch combined with carrageenan for the hard capsule industries was a study undertaken by Oladzadabbasabadi et al. (Oladzadabbasabadi et al. 2017). The dual-modified sago starch used to create the capsule composite film shows that the composite had properties comparable to those of bovine gelatin film, where it was more resilient at greater relative humidity (Oladzadabbasabadi et al. 2017). There are currently few options that can match mammalian-based gelatin’s quick and affordable manufacturing, particularly when it comes from quickly reproducing animals like pigs (Kamatchi and Leela 2016). For this purpose, fish-based gelatin was more advantageous in the marketplace than plant-based gelatin due to the widespread mass production of fish byproducts (Ismail and Abdullah 2017). Fish-based gelatin remains one of the most promising replacements for mammalian gelatin in modern times, with “vegetarian gelatin” applications remaining in the gray area (Lin et al. 2017). Fish-based gelatin is the ideal substitute for mammalian gelatin due to its rapid dissolution rate and low melting point (Karim and Bhat 2009). The rheological and thermostability characteristics of fish-based gelatin are similar to those of human gelatin, but they also rely on the species, the sort of raw material, and the process conditions (Gómez-Guillén et al. 2011).

27.4 Chemical Structure of Gelatin

Gelatin is a polyampholyte containing cations and anions, and hydrophobic groups are also present in the approximate ratio 1:1:1, making it a more particular polypeptide (Mohanty et al. 2005). Gelatin molecule has 13% positive charge in the form of lysine and arginine and about 12% negatively charged amino acid, i.e., glutamic and aspartic acid, and approximately 11% of the chain is hydrophobic as leucine, isoleucine, methionine and valine are present in it (Hashizume 2014). The rest of the chain is constituted by glycine, proline, and hydroxyproline. The triple helical structure of gelatin is presented by $(\text{Gly-X-Pro})_n$ where X is the amino acid including lysine, arginine, methionine, and valine, about 6% (Hathout and Omran 2016). Glycine constitutes one-third of the polypeptide chain (33%), and the other 33% of the polypeptide chain is composed of hydroxyproline (Gupta et al. 2021). The rest of the gelatin is composed of other residues. Both cationic and anionic forms are commercially available. The cationic form includes type I collagen with an isoelectric point (pI) 7–9 and prepared by acid hydrolysis of type I collagen from pig skin, while the anionic form includes type B of gelatin with pI 4.8–5. The type is prepared from bovine collagen by alkaline hydrolysis (Mohanty et al. 2005; Kommareddy et al. 2007). Figure 27.2 shows the composition of gelatin in terms of amino acids.

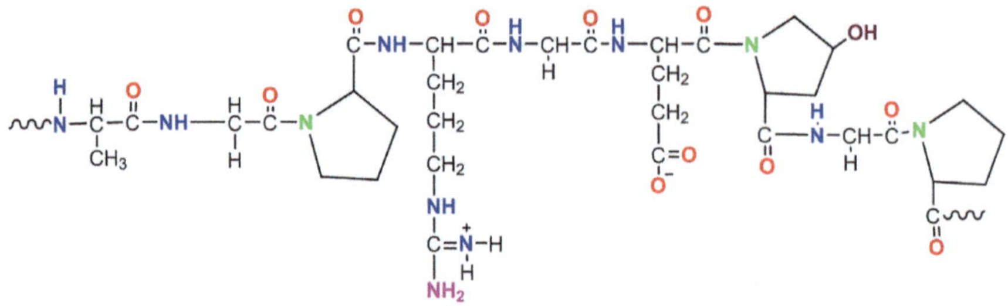

Fig. 27.2 Chemical structure of gelatin. Reproduced with permission from Thakur et al. 2017

27.5 General Characteristics of Gelatin

Gelatin has a long history of safe use in medicinal, cosmetic, and food products. It is one of the most adaptable naturally occurring biopolymers. It is a generally recognized as safe (GRAS) biomaterial by the US Food and Drug Administration (FDA) and is nontoxic, noncarcinogenic, biodegradable, and nonirritating. The gelatin polymer's functional groups are available for chemical reaction. Additionally, the amino acid combination Arg-Gly-Asp (arginine, glycine, and aspartic acid, RGD) is present in gelatin and influences cell adhesion (Vickers 2017). As gelatin can be degraded to its constituent amino acids, it does not form hazardous side products (Elzoghby 2013). Gelatin is extracted from animal collagen via partial acid or alkaline hydrolysis (Chan et al. 2017). Collagen is one of the vital structural proteins, constituting approximately 25% to 35% of the whole-body protein content in which the majority are fibrous tissues such as skin, tendons, and ligaments (about one-half of total body collagen) and also in corneas, bones, cartilages, blood vessels, the gut, and intervertebral discs (Gorgieva and Kokol 2011). The distinctive triple-helix structure, which consists of three polypeptide chains twisted around one another in a superhelical form, is a common characteristic of collagens and, consequently, gelatin polypeptides. This structure is because of the conserved amino acid sequence shown by $(Gly–X–Y)_n$, where X represents the amino acid including lysine, arginine, methionine, and valine about 6% (Hathout and Omran 2016). Glycine constitutes one-third of the polypeptide chain (33%), and the other 33% of the polypeptide chain is composed of hydroxyproline (Pal et al. 2018). Interchain hydrogen bonds stabilize collagen molecules. Acidic or basic hydrolysis of collagen destabilizes collagen's tertiary structure, leading to the loss of the triple-helix conformation. Commercially available forms are prepared by partial acid hydrolysis of collagen (type A gelatin, cationic, positively charge compound at neutral pH, and type B gelatin, synthesized via an alkaline treatment of collagen, negatively charged compound at neutral pH) (Madkhali et al. 2019). In the alkaline process, the amide groups of glutamine and asparagine are targeted hydrolyzing them into carboxyl groups producing gelatin with a high concentration of carboxyl groups, making it negatively charged (Samal et al. 2012).

27.6 Biomaterials Based on Gelatin

Because of its cell stimulatory characteristics, gelatin has been extensively used in regenerative medicine and tissue engineering (Echave et al. 2019). Gelatin is a naturally occurring biopolymer with many advantages over its precursor, collagen (Sell et al. 2009). One of the primary restrictions for biomedical uses is the low water solubility of collagen in neutral conditions (Farris et al. 2010). With the use of gelatin extraction, this limitation in collagen can be fixed. Another noteworthy characteristic of gelatin is its capacity to assemble charged medicinal substances such as proteins, growth hormones, nucleotides, and polysaccharides into poly-ionic complexes (Larsen et al. 2006), thus making it an ideal delivery vehicle for many biomolecules. Gelatin has an isoelectric point at pH 7.4 that is either net positive (IEP = 9, type A gelatin) or net negative (IEP = 5, type B gelatin), depending on the conditions used for extraction, allowing oppositely charged proteins to be sequestered while maintaining bioactivity. Gelatin is a critical component in producing microcapsules and microspheres for medication delivery (Diba et al. 2017). Moreover, the source and the conditions are decisive factors for obtaining gelatins with a diversity of physicochemical properties (melting temperature, gel modulus, or viscosity) due to the variations in the proportions of amino acids and the molecular weights of the materials being produced (Karim and Bhat 2009). Furthermore, gelatin is readily functionalized to create materials with specific properties, bringing up new therapeutic possibilities. Gelatin-based 3D microgels, for instance, can be utilized to promote cell division and differentiation of different encapsulated cells, such as stem cells, and can enhance the regenerative impact of cell-loaded microbeads injected in lesion areas (Li et al. 2017). Moreover, these microgels can protect the cells from shear-force-associated mortality during injection and equip them with an environment that promotes cell retention at the intended location (Zhao et al. 2021). Scaffolds made of cross-linked gelatin can be utilized for 3D cell culture. However, thermal gelatinization often yields thin, weak gels. Covalently cross-linked gelatin-based hydrogels are created using traditional chemical techniques to remedy this (Petros et al. 2020). Enzymatic cross-linking of gelatin systems utilizing different enzymes, like horseradish peroxidase or microbial transglutaminase, has been one of the intriguing methods suggested by multiple studies (Echave et al. 2019). Due to the successful development of diverse hydrogels with adjustable gelation rates and final physical strengths, these systems have been examined for various tissue engineering applicability (Kuo and Ma 2001). Injectable enzymatically cross-linked gelatin-hydroxyphenyl propionic acid composite hydrogel, for instance, has been studied for tissue repair in both cartilage and the brain. Additionally, photo-cross-linkable gelatin methacrylate (GelMA) hydrogels have attracted much interest in various therapeutic applications, including creating cartilage constructs and engineering corneal tissue and peripheral nerve regeneration (Zhao et al. 2022). Furthermore, pendant tetrazine or norbornene click chemistry pairs in modified polymers have recently been used to create injectable covalently cross-linked gelatin hydrogels. When combined, these gelatin polymers quickly cross-link, and when injected in vivo, they break down. In addition, they encourage cell survival and can induce 3D elongated morphologies in encapsulated cells (Echave et al. 2019). The continued development of 3D bioprinting offers previously unheard-of control over the spatial distribution of components, cells, and biomolecules, ultimately making it possible to create more realistic 3D tissue structures. Jia W and colleagues developed a direct 3D model comprising many layers of coaxial extrusion printing methods to create highly perfusable and well-organized vessels as one notable example. Gelatin methacrylate, sodium alginate, and 4-arm poly(ethylene glycol)-tetraacrylate were the main components of the composite bioink (PEGTA). In the 3D-printed scaffolds, this doubly cross-linked (covalently and ionically) technology allowed encapsulated endothelium and mesenchymal stem cells to multiply and spread, ultimately building native-like perfusable vessels (Jia et al. 2016).

27.7 Recent Advances in Gelatin for Tissue Engineering

Creating artificial systems or structures that mimic some of the most crucial functionalities of genuine tissues is the overriding goal of gelatin for tissue engineering. The body's complex tissue topologies and the numerous synergistic biological elements included in natural tissues make it difficult to achieve this goal. The most recent research in this area is focused on fusing various biomaterials to provide native-like synergistic effects and to supply critical native-like tissue microarchitectures. Over the past 10 years, several possible gelatin-based systems have been created using this method.

Finding a new method for the treatment of heart disease is one of the key subjects researched by researchers in recent decades due to the rising number of patients with heart disease and the relative inefficiency of present treatments. Esmaeili et al. created an electrospun cardiovascular tissue engineering scaffold out of polyurethane, chitosan, *Vicia ervilia* protein, gelatin, and heparin. In this study, heparin, an anticoagulant drug, will be placed onto nanofibers. As a result, by electrospinning using a two-phase voltage of 16.5 kW and a distance needle up to 18 cm with a feed rate of 0.2, nanofibers of chitosan complex (2%)/polyurethane (14.5%)/cow protein (1%)/gelatin (20%) with *A. gossypinus* coating (3%)/gelatin were created. The prepared nanofibers' morphology was evaluated. Additionally, there was no hazy separation and the resultant fibers are miscible. Additionally, the heparin and toxicity of the generated nanofibers are released using the MTT dialysis bag. In 120 h, this medicine can eliminate roughly 82%. This system's lack of cytotoxicity was examined, and these nanofibers containing the medication demonstrated 91% cell survival after 7 days. This study demonstrated improved endothelial cell adhesion and growth in the vessel's inner wall (Heydari and Esmaeili 2022). A bacterial infection is one of the main causes of tissue engineering scaffold failure. Therefore, for tissue engineering applications, it is crucial to create multifunctional scaffolds that can direct tissue regeneration and prevent bacterial colonization. The study team of Handa et al. created a scaffold based on silk fibroin (SF) and gelatin methacryloyl (GelMA) that is highly antimicrobial, biocompatible, and biodegradable for this purpose.

Cross-linking GelMA and SF created interpenetrating network (IPN) hydrogels with a porous structure under UV irradiation and methanol treatment, respectively. In addition, NO-releasing scaffolds with potent antibacterial capabilities were created by impregnating the hydrogels with the nitric oxide (NO) donor molecule S-nitroso-N-acetylpenicillamine (SNAP). The results showed that adding SF to GelMA hydrogels improved their mechanical characteristics and NO-release kinetics, while preventing quick enzymatic degradation in aqueous conditions. Moreover, a bacterium reduction efficacy of >99.9% against Gram-positive (*Staphylococcus aureus*) and Gram-negative (*Escherichia coli*) bacteria was achieved by inflating the GelMA-SF scaffolds with SNAP. By promoting the proliferation and assisting the adherence of 3T3 murine fibroblast cells, the scaffolds also demonstrated excellent cytocompatibility in vitro. Overall, the GelMA-SF-SNAP scaffold demonstrated significant promise for usage in tissue engineering and wound healing applications (Ghalei et al. 2021). Chen et al. created a spiral wound scaffold based on gelatin/PCL/heparin (GPH) nanofiber membranes for tendon tissue engineering. They created a high-resilience scaffold for the high-loading environment that tendons face by inserting sutures in dual layers of aligned GPH nanofiber membranes made from mixed electrospinning of gelatin and PCL/heparin solutions, as shown in Fig. 27.3. In order to keep the tenogenic phenotype, the necessary fibroblast growth factor (bFGF) was anchored to the GPH scaffold through an affinity between heparin and bFGF. Additionally, it is anticipated that the aligned nanofiber shape will offer physical indications toward seeded tenocytes. GPH-bFGF can increase proliferation, upregulate tenogenic gene expression, and increase the synthesis of tendon-specific proteins by tenocytes in vitro with the prolonged release of bFGF. Furthermore, GPH-bFGF/tenocyte constructions surpassed GPH-bFGF in terms of mechanical qualities by correctly retaining tendon phenotypes. Histological staining and biomechanical analysis from an in vivo investigation using GPH-bFGF/tenocytes to repair rabbit Achilles tendon lesions revealed the development of new tendon tissue. These results show that the newly developed GPH-bFGF scaffold could offer a niche for stimulating tendon tissue regeneration by successfully repairing the structure and function of the tendon tissue (Darshan et al. 2022).

Being a biodegradable material, gelatin has weak mechanical properties. For efficient use in biomedicine as nontoxic material, it is necessary to improve its mechanical strength. Various methods exist to increase its mechanical strength using ceramics and bioglass as external fillers. Boron nitride (BN)-reinforced gelatin was introduced by Bechelany et al. as a new bidimensional and biocompatible nanomaterial category (Fig. 27.4). The nanofiller effect on mechanical strength is checked. Fourier transform infrared (FTIR) spectroscopy and X-ray diffraction (XRD) show that the BN is efficiently exfoliated. The gelatin electrospun fibers are strengthened by the exfoliated BN, raising Young's modulus. After glutaraldehyde cross-linking, the electrospun mats (ESM) are stable, and the fiber morphology is retained. Scanning electron microscopy has demonstrated that the cross-linked gelatin/BN ESM is highly bioactive in forming hydroxyapatite that resembles bone. As they have immense mineralization properties, cross-linked ESM is tested on human bone cells, especially cell lines of HOS osteosarcoma. ESM are nontoxic due

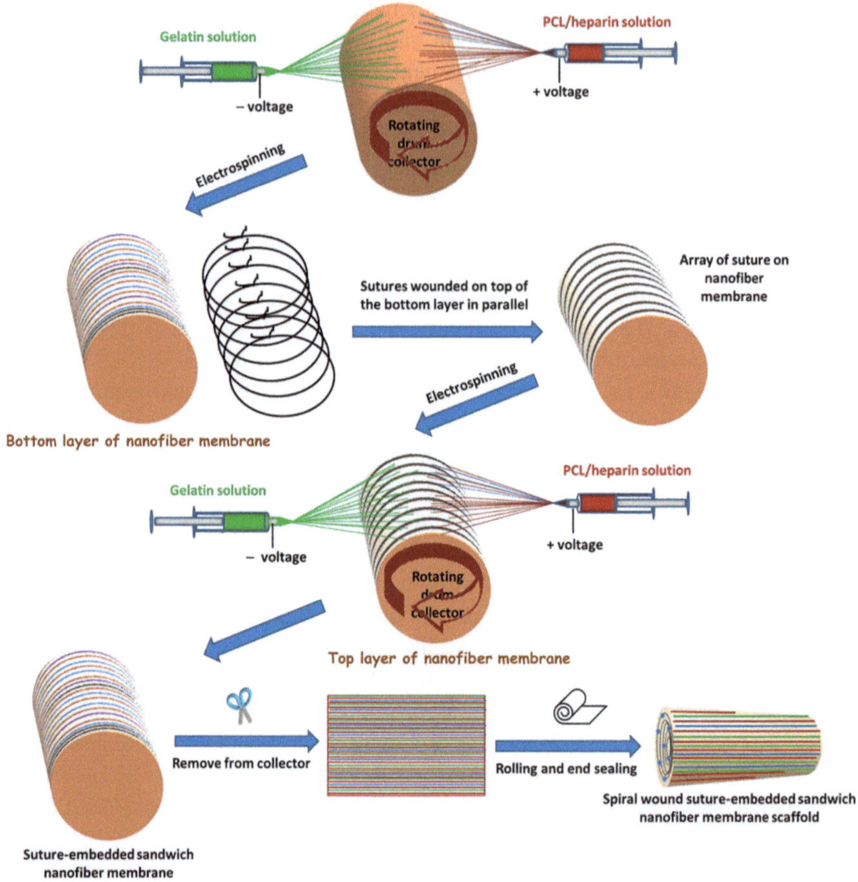

Fig. 27.3 Electrospinning prepares gelatin/PCL/heparin (GPH) suture-embedded sandwiched nanofiber membrane and scaffold. Reproduced with permission from Darshan et al. 2022

to enhanced cell attachment, proliferation, and biocompatibility. Gene expression in osteoblast and alkaline phosphatase activity measurement confirmed their suitability for tissue and bone engineering (Nagarajan et al. 2017).

As compared to the chemical cross-linking technique, the potential of a radiation cross-linking technology for creating gelatin scaffolds for tissue engineering was assessed. Gelatin that had been exposed to radiation and cross-linked had a more excellent visible light transmittance than gelatin that had been chemically cross-linked. The efficiency of radiation cross-linking gelatin was about 91%; hydroxyl (OH) radicals produced by the radiolysis of the aqueous gelatin solution in the presence of nitrous oxide or nitrogen saturation had an impact on it. With an increase in absorbed dose, the quantities of phenylalanine, tyrosine, and histidine in the gelatin dramatically decreased. The quantity of the other 12 amino acids remained the same and did not alter after exposure to γ-ray radiations. These amino acids did

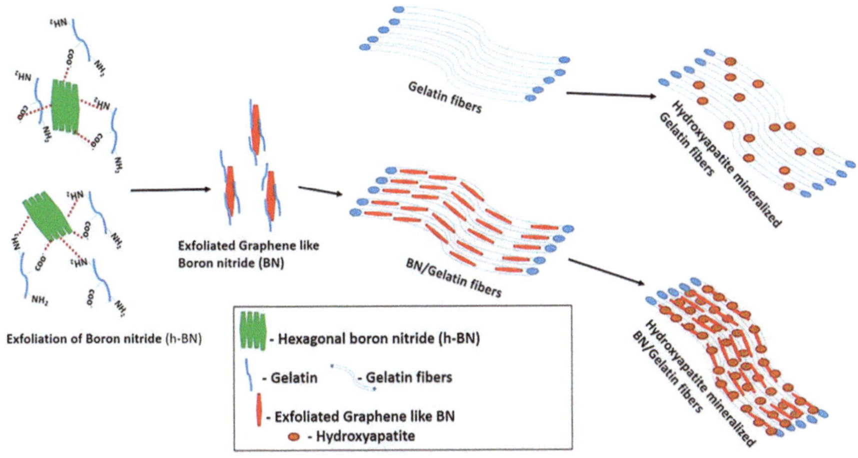

Fig. 27.4 Schematic illustration of boron nitride/gelatin electrospun nanofibers for bone tissue engineering. Reproduced with permission from Nagarajan et al. 2017

not take part in the cross-linking reactions. For radiation cross-linking reactions, the essential contents were tyrosine, phenylalanine, and histidine in gelatin. Cell adhesion would be maintained of the radiation-cross-linked gelatin before and after γ-ray exposure as the reduction in the proportion of the cell adhesion was not reported in active sequence (arginine-glycine-aspartic acid, RGD motif). The dityrosine formation marked the point at which cross-linking occurred in the gelatin during radiation cross-linking. Contrary to chemical cross-linking approaches, the radiation cross-linking process can change the gelatin while preserving high transparency since the cross-linked gelatin and degradation products have no absorption bands in the visible spectrum (Kimura et al. 2021). The dominant trends to address are the complex need for tissue engineering scaffolds in multipotent bone tissues for bone repair. A porous composite scaffold was created using nano-hydroxyapatite (nHAp) and polymers (natural and synthetic) to replicate the microstructure and components of real bone by Wu et al. research group, as shown in Fig. 27.5. The extracellular matrix was stimulated by the assembly of chitosan (CS), polyvinyl alcohol (PVA), and gelatin (Gel). Properties of the simulated matrix include swelling, mechanical strength, pH, porosity, and tunable pore size. The composite scaffolds showed increased compression resistance, pH adaptability, surface bioactivity, and biomimetic structure after adding nHAp. The composite scaffolds can also efficiently encourage cell proliferation and adhesion; scaffolds containing 12.5% nHAp were shown to have a good capacity for osteogenic differentiation. These findings, therefore, indicate that Gel/CS/PVA/nHAp composite scaffolds may be a potential biomimetic scaffold for bone tissue engineering (Ma et al. 2021).

Gelatin-based adhesives are used in various biomedical fields because of their immense biocompatibility, easy processability, transparency, nontoxicity, and good mechanical properties to induce the extracellular matrix (ECM). Modifying gelatin adhesives to acquire various viscoelastic and mechanical qualities that make ocular

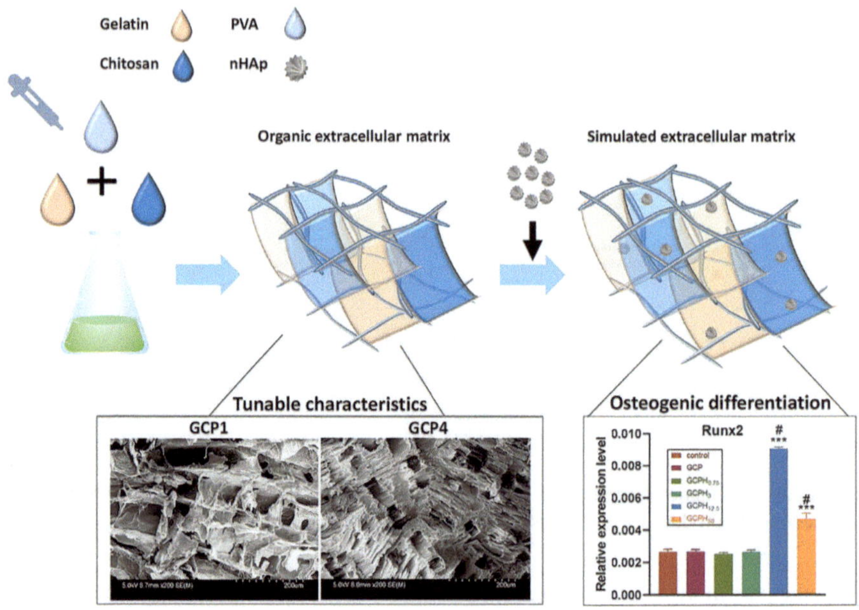

Fig. 27.5 Fabrication, tunable characteristics, and osteogenic differentiation of gelatin/chitosan/PVA/nHAp (GCPH) composite scaffolds. Reproduced with permission from Ma et al. 2021

application easier is simple. Gonzalez-Andrades et al. grafted glycidyl methacrylate onto the gelatin backbone using a straightforward chemical modification of the precursor. They did this using visible light cross-linking and epoxide ring-opening processes. An elastic protein-based hydrogel (GELGYM) with outstanding biomimetic properties that are close to those of genuine tissue can be created thanks to this chemical alteration. In contrast to gelatin methacryloyl (GelMA), the most researched gelatin derivative utilized as a bioadhesive, GELGYM may be adjusted to be stretched up to four times its initial length and resist high tensile stresses up to 1.95 MPa with compressive strains as high as 80%. Additionally, GELGYM stimulates cellular adhesion, proliferation, and migration in both two- and three-dimensional cell cultures and is highly biocompatible. These properties, as well as its super adhesion property to biological tissues including cornea, aorta, heart, muscle, kidney, liver, and spleen, depict its broader applications of hydrogel in various biomedical fields, for instance, in organ transplants, tissue engineering as adhesive, dressing of the wound, bioprinting, and in drug delivery (Sharifi et al. 2021).

27.8 Conclusion and Future Perspective

Gelatin has excellent potential to be used in regenerative medicine due to the inherent properties of biocompatibility and novel approaches in the biomedical field. As an example, cell-laden gelatin-based 3D tissue models could be created in the future to support drug development, assessment, and understanding of diseases. These 3D physiological models are expected to replace contemporary in vivo studies. Quoting an example of gelatin microparticles that have a role in clarifying complicated stoma cancer interrelationship, in combination with pancreatic cancer cells and fibroblasts (Brancato et al. 2017). In another unique approach, the cardiac microtissues of humans have been bioengineered via cardiomyocytes derived from human pluripotent stem cells, encapsulated in gelatin hydrogels with tunable degradation rates and stiffness (Lee et al. 2017). In accordance with the description, Agrawal and colleagues developed a 3D skeletal muscle-on-a-chip platform with cell-laden gelatin hydrogel being used as a screening platform for drugs and toxicities such as cardiotoxin (Agrawal et al. 2017), which can be used for preclinical discovery and development of drugs. Gelatin holds a wide range of applications and possibilities in tissue repair and regeneration, targeted drug delivery, and the development of nanoparticle-based drugs by using 3D biomimetic scaffolds. This area is expected to enhance the future translation of gelatin-based biomaterials into clinics and treatment based on ongoing research and progression.

Conflict of Interest The authors declare no conflict of interest.

References

Ab Talib MS, Sawari SSM, Hamid ABA, Chin TA (2016) Emerging halal food market: an institutional theory of halal certificate implementation. Manag Res Rev 39(9):987–997

Abedinia A, Nafchi AM, Sharifi M, Ghalambor P, Oladzadabbasabadi N, Ariffin F et al (2020) Poultry gelatin: characteristics, developments, challenges, and future outlooks as a sustainable alternative for mammalian gelatin. Trends Food Sci Technol 104:14–26

Aggarwal S, Chakravarty A, Ikram S (2021) A comprehensive review on incredible renewable carriers as promising platforms for enzyme immobilization & thereof strategies. Int J Biol Macromol 167:962–986

Agrawal G, Aung A, Varghese S (2017) Skeletal muscle-on-a-chip: an in vitro model to evaluate tissue formation and injury. Lab Chip 17(20):3447–3461

Alfaro AT, Balbinot E, Weber CI, Tonial IB, Machado-Lunkes A (2015) Fish gelatin: characteristics, functional properties, applications and future potentials. Food Eng Rev 7(1): 33–44

Alipal J, Pu'ad NM, Lee T, Nayan N, Sahari N, Basri H et al (2021) A review of gelatin: properties, sources, process, applications, and commercialisation. Mater Proceed 42:240–250

Baydin T, Aarstad OA, Dille MJ, Hattrem MN, Draget KI (2022) Long-term storage stability of type A and type B gelatin gels: the effect of Bloom strength and co-solutes. Food Hydrocoll 127: 107535

Brancato V, Comunanza V, Imparato G, Corà D, Urciuolo F, Noghero A et al (2017) Bioengineered tumoral microtissues recapitulate desmoplastic reaction of pancreatic cancer. Acta Biomater 49: 152–166

Carstairs A, Genever P (2014) Stem cell treatment for musculoskeletal disease. Curr Opin Pharmacol 16:1–6

Chan GG, Koch CM, Connors LH (2017) Blood proteomic profiling in inherited (ATTRm) and acquired (ATTRwt) forms of transthyretin-associated cardiac amyloidosis. J Proteome Res 16(4):1659–1668

Chen C, Li D, Yano H, Abe K (2019) Insect cuticle-mimetic hydrogels with high mechanical properties achieved via the combination of chitin nanofiber and gelatin. J Agric Food Chem 67(19):5571–5578
Darshan T, Chen C-H, Kuo C-Y, Shalumon K, Chien Y-M, Kao H-H et al (2022) Development of high resilience spiral wound suture-embedded gelatin/PCL/heparin nanofiber membrane scaffolds for tendon tissue engineering. Int J Biol Macromol 221:314–333
Diba M, Pape B, Klymov A, Zhang Y, Song J, Löwik DW et al (2017) Nanostructured raspberry-like gelatin microspheres for local delivery of multiple biomolecules. Acta Biomater 58:67–79
Dong R, Zhou Y, Huang X, Zhu X, Lu Y, Shen J (2015) Functional supramolecular polymers for biomedical applications. Adv Mater 27(3):498–526
Echave MC, Burgo LS, Pedraz JL, Orive G (2017) Gelatin as biomaterial for tissue engineering. Curr Pharm Des 23(24):3567–3584
Echave MC, Hernáez-Moya R, Iturriaga L, Pedraz JL, Lakshminarayanan R, Dolatshahi-Pirouz A et al (2019) Recent advances in gelatin-based therapeutics. Expert Opin Biol Ther 19(8): 773–779
Elzoghby AO (2013) Gelatin-based nanoparticles as drug and gene delivery systems: reviewing three decades of research. J Control Release 172(3):1075–1091
Farris S, Song J, Huang Q (2010) Alternative reaction mechanism for the cross-linking of gelatin with glutaraldehyde. J Agric Food Chem 58(2):998–1003
Gajbhiye S, Wairkar S (2022) Collagen fabricated delivery systems for wound healing: a new roadmap. Biomater Adv 2022:213152
Gallagher S, Schall MC Jr (2017) Musculoskeletal disorders as a fatigue failure process: evidence, implications and research needs. Ergonomics 60(2):255–269
Ghalei S, Douglass M, Handa H (2021) Nitric oxide-releasing gelatin methacryloyl/silk fibroin interpenetrating polymer network hydrogels for tissue engineering applications. ACS Biomater Sci Eng 8(1):273–283
Gómez-Guillén M, Giménez B, López-Caballero MA, Montero M (2011) Functional and bioactive properties of collagen and gelatin from alternative sources: a review. Food Hydrocoll 25(8): 1813–1827
Gorgieva S, Kokol VJN (2011) Biomaterials and their biocompatibility: review and perspectives. Biomater Appl:17–52
Gupta SS, Mishra V, Mukherjee MD, Saini P, Ranjan KR (2021) Amino acid derived biopolymers: recent advances and biomedical applications. Int J Biol Macromol 188:542–567
Hafidz R, Yaakob C, Amin I, Noorfaizan A (2011) Chemical and functional properties of bovine and porcine skin gelatin. Int Food Res J 18(2):787–791
Hashizume H (2014) Adsorption of twenty bio-amino acids by natural allophane and imogolite. Viva Origino 42(3):18–22
Hathout RM, Omran MK (2016) Gelatin-based particulate systems in ocular drug delivery. Pharm Dev Technol 21(3):379–386
Heydari S, Esmaeili A (2022) Synthesize of polyurethane/chitosan/Vicia ervilia protein/gelatin/heparin-coated Astragalus gossypinus scaffold for cardiovascular tissue engineering. Process Biochem 122:205
Hoque ME, Nuge T, Yeow TK, Nordin N, Prasad RJPRJ (2015) Gelatin based scaffolds for tissue engineering—a review. Poly Res J 9(1):15
Huang T, Tu Z-c, Shangguan X, Sha X, Wang H, Zhang L et al (2019) Fish gelatin modifications: a comprehensive review. Trends Food Sci Technol 86:260–269
Ismail N, Abdullah HZ (eds) (2017) Influence of extraction process (Pre-treatment time) on the characteristic of black tilapia fish skins gelatin. Materials science forum; Trans Tech Publ
Jia W, Gungor-Ozkerim PS, Zhang YS, Yue K, Zhu K, Liu W et al (2016) Direct 3D bioprinting of perfusable vascular constructs using a blend bioink. Biomaterials 106:58–68
Joyce K, Fabra GT, Bozkurt Y, Pandit AJST, Therapy T (2021) Bioactive potential of natural biomaterials: identification, retention and assessment of biological properties. Signal Transduct Target Ther 6(1):1–28

Kamal Z, Su J, Yuan W, Raza F, Jiang L, Li Y et al (2022) Red blood cell membrane-camouflaged vancomycin and chlorogenic acid-loaded gelatin nanoparticles against multi-drug resistance infection mice model. J Drug Deliv Sci Technol 76:103706

Kamatchi P, Leela K (2016) Extraction, characterization and application of gelatin from Carcharhinus amblyrhyncho and Sphyraena barracuda. IOSR J Biotechnol Biochem 2:40–49

Karim AA, Bhat R (2009) Fish gelatin: properties, challenges, and prospects as an alternative to mammalian gelatins. Food Hydrocoll 23(3):563–576

Kimura A, Yoshida F, Ueno M, Taguchi M (2021) Application of radiation crosslinking technique to development of gelatin scaffold for tissue engineering. Radiat Phys Chem 180:109287

Kommareddy S, Shenoy DB, Amiji MM (2007) Gelatin nanoparticles and their biofunctionalization

Koshti B, Kshtriya V, Singh R, Walia S, Bhatia D, Joshi KB et al (2021) Unusual aggregates formed by the self-assembly of proline, hydroxyproline, and lysine. ACS Chem Neurosci 12(17):3237–3249

Kuo CK, Ma PX (2001) Ionically crosslinked alginate hydrogels as scaffolds for tissue engineering: part 1. Structure, gelation rate and mechanical properties. Biomaterials 22(6):511–521

Larsen M, Artym VV, Green JA, Yamada KM (2006) The matrix reorganized: extracellular matrix remodeling and integrin signaling. Curr Opin Cell Biol 18(5):463–471

Laurent A, Darwiche SE, Hirt-Burri N, Scaletta C, Michetti M, Laurent P et al (2020) Banking progenitor cells for hippiatric regenerative medicine: optimized establishment of safe and consistent cell sources for standardized veterinary therapeutic protocols. Am J Biomed Sci Res 8(4):252–271

Lee KY, Mooney DJ (2001) Hydrogels for tissue engineering. Chem Rev 101(7):1869–1880

Lee S, Serpooshan V, Tong X, Venkatraman S, Lee M, Lee J et al (2017) Contractile force generation by 3D hiPSC-derived cardiac tissues is enhanced by rapid establishment of cellular interconnection in matrix with muscle-mimicking stiffness. Biomaterials 131:111–120

Lestari W, Octavianti F, Jaswir I, Hendri R (2019) Plant-based substitutes for gelatin. In: Hassan F, Osman I, Kassim E, Haris B, Hassan R (eds) Contemporary management and science issues in the halal industry. Springer, pp 319–322

Li F, Truong VX, Thissen H, Frith JE, Forsythe JS (2017) Microfluidic encapsulation of human mesenchymal stem cells for articular cartilage tissue regeneration. ACS Appl Mater Interfaces 9(10):8589–8601

Lin L, Regenstein JM, Lv S, Lu J, Jiang S (2017) An overview of gelatin derived from aquatic animals: properties and modification. Trends Food Sci Technol 68:102–112

Lutolf M, Hubbell JA (2005) Synthetic biomaterials as instructive extracellular microenvironments for morphogenesis in tissue engineering. Nat Biotechnol 23(1):47–55

Ma P, Wu W, Wei Y, Ren L, Lin S, Wu J et al (2021) Biomimetic gelatin/chitosan/polyvinyl alcohol/nano-hydroxyapatite scaffolds for bone tissue engineering. Mater Des 207:109865

Madkhali O, Mekhail G, Wettig SD (2019) Modified gelatin nanoparticles for gene delivery. Int J Pharm 554:224–234

Masuelli MA, Sansone MG (2012) Hydrodynamic properties of gelatin-studies from intrinsic viscosity measurements. In: Verbeek J (ed) Products and applications of biopolymers. InTech, pp 85–116

Mohanty B, Aswal V, Kohlbrecher J, Bohidar H (2005) Synthesis of gelatin nanoparticles via simple coacervation. J Surf Sci Technol 21(3/4):149

Nagarajan S, Belaid H, Pochat-Bohatier C, Teyssier C, Iatsunskyi I, Coy E et al (2017) Design of boron nitride/gelatin electrospun nanofibers for bone tissue engineering. ACS Appl Mater Interfaces 9(39):33695–33706

Oladzadabbasabadi N, Ebadi S, Nafchi AM, Karim A, Kiahosseini SR (2017) Functional properties of dually modified sago starch/κ-carrageenan films: an alternative to gelatin in pharmaceutical capsules. Carbohydr Polym 160:43–51

Padilla S, Sánchez M, Orive G, Anitua E (2017) Human-based biological and biomimetic autologous therapies for musculoskeletal tissue regeneration. Trends Biotechnol 35(3):192–202

Pal A, Bajpai J, Bajpai A (2018) Easy fabrication and characterization of gelatin nanocarriers and in vitro investigation of swelling controlled release dynamics of paclitaxel. Polym Bull 75(10): 4691–4711

Petros S, Tesfaye T, Ayele MJ (2020) A review on gelatin based hydrogels for medical textile applications. J Eng 2020:8866582

Rakhmanova A, Khan Z, Sharif R, Lu X (2018) Meeting the requirements of halal gelatin: a mini review. MOJ Food Process Technol 6:477–482

Raza F, Siyu L, Zafar H, Kamal Z, Zheng B, Su J et al (2022) Recent advances in gelatin-based nanomedicine for targeted delivery of anti-cancer drugs. Curr Pharm Des 28(5):380–394

Rodrigues MT, Reis RL, Gomes ME (2013) Engineering tendon and ligament tissues: present developments towards successful clinical products. J Tissue Eng Regen Med 7(9):673–686

Samal SKJCSR, Dash M, Van Vlierberghe S, Kaplan DL, Chiellini E, van Blitterswijik C, Moroni L, Dubruel P (2012) Cationic polymers and their therapeutic potential. Chem Soc Rev 41(21):7147–7194

Sell SA, McClure MJ, Garg K, Wolfe PS, Bowlin GL (2009) Electrospinning of collagen/ biopolymers for regenerative medicine and cardiovascular tissue engineering. Adv Drug Deliv Rev 61(12):1007–1019

Sharifi S, Islam MM, Sharifi H, Islam R, Koza D, Reyes-Ortega F et al (2021) Tuning gelatin-based hydrogel towards bioadhesive ocular tissue engineering applications. Bioactive Materials 6(11): 3947–3961

Shyni K, Hema G, Ninan G, Mathew S, Joshy C, Lakshmanan P (2014) Isolation and characterization of gelatin from the skins of skipjack tuna (Katsuwonus pelamis), dog shark (Scoliodon sorrakowah), and rohu (Labeo rohita). Food Hydrocoll 39:68–76

Stanton C, Light N (1990) The effects of conditioning on meat collagen: part 3—Evidence for proteolytic damage to endomysial collagen after conditioning. Meat Sci 27(1):41–54

Thakur S, Govender PP, Mamo MA, Tamulevicius S, Thakur VK (2017) Recent progress in gelatin hydrogel nanocomposites for water purification and beyond. Vacuum 146:396–408

Vickers NJ (2017) Animal communication: when i'm calling you, will you answer too? Curr Biol 27(14):R713–R7R5

Wasswa J, Tang J, Gu X (2007) Utilization of fish processing by-products in the gelatin industry. Food Rev Int 23(2):159–174

Wu N, Deiters A, Cropp TA, King D, Schultz PG (2004) A genetically encoded photocaged amino acid. J Am Chem Soc 126(44):14306–14307

Zhang T, Xu J, Zhang Y, Wang X, Lorenzo JM, Zhong J et al (2020) Gelatins as emulsifiers for oil-in-water emulsions: extraction, chemical composition, molecular structure, and molecular modification. Trends Food Sci Technol 106:113–131

Zhao X, Hu DA, Wu D, He F, Wang H, Huang L et al (2021) Applications of biocompatible scaffold materials in stem cell-based cartilage tissue engineering. Front Bioeng Biotechnol 9: 603444

Zhao Y, Song S, Ren X, Zhang J, Lin Q, Zhao Y (2022) Supramolecular adhesive hydrogels for tissue engineering applications. Chem Rev 122(6):5604–5640

Zohuriaan-Mehr M, Pourjavadi A, Salimi H, Kurdtabar M (2009) Protein-and homo poly (amino acid)-based hydrogels with super-swelling properties. Polym Adv Technol 20(8):655–671

Faisal Raza is currently a PhD student at the School of Pharmacy, Shanghai Jiao Tong University (SJTU), China. He has broad research interests in the areas of pharmaceutics and drug delivery. His specific field of research focused on the design and synthesis of novel peptides and polymers for drug delivery, analyzing therapeutic efficacy and toxicity studies in cancer mouse models, formulating and developing advanced drug delivery systems for poorly water-soluble anticancer drugs, and developing stimuli-responsive nanomedicine and biomimetic nanotechnology for cancer therapy.

Jing Su is an associate professor at the School of Pharmacy, Shanghai Jiao Tong University (SJTU). He obtained his PhD in medicinal chemistry at Shenyang Pharmaceutical University under Prof. Maosheng Cheng and Prof. Song Li. He joined SJTU in 2008 as a faculty of the School of Pharmacy. He once worked as a visiting scholar with Prof. Liangfang Zhang at UC San Diego (2012.04–2013.03). His research now focuses on developing novel drug delivery systems based on biomaterials and biomimetic membranes.

Jian Zhong is an associate professor at Xinhua Hospital, Shanghai Institute for Pediatric Research, Shanghai Key Laboratory of Pediatric Gastroenterology and Nutrition, Shanghai Jiao Tong University School of Medicine (SJTUSM). He obtained his PhD in biomechanics and medical engineering at Peking University with the supervision of Profs. Jing Fang and Yinlin Sha. He was a postdoc at University College Dublin (2007–2011), an associate prof. at National Engineering Research Center for Nanotechnology (2011–2015), and an associate prof. at Shanghai Ocean University. He joined SJTUSM in 2022 as a faculty. His research now focuses on the processing and nutrition of pediatric marine foods.

Mingfeng Qiu is a professor and vice-president at the School of Pharmacy, Shanghai Jiao Tong University (SJTU). He obtained his PhD in pharmaceutics of traditional Chinese medicine at Chengdu University of Traditional Chinese Medicine with the supervision of Prof. Xiuqiong Xie. He joined SJTU in 2003 as a faculty of the School of Pharmacy. His research now focuses on the development of sustained-release and quick-release preparations of traditional Chinese medicine and its evaluation methods.

Biomedical Applications of the Fused Filament Fabrication (FFF) Technology

28

Irene Buj-Corral, Carolina Herranz-Diez, Aitor Tejo-Otero, and Jordi Otero

Abstract

This chapter deals with the main biomedical applications of the fused filament fabrication (FFF) 3D printing technology. It begins with an overview of the main groups of 3D printing technologies. After, a detailed explanation of the FFF technique is presented, focusing on the main process parameters, the characteristics of the 3D printers, and the different materials that can be used, both polymeric materials and composite materials. Then, the biomedical applications of the FFF technology are addressed, which include 3D scaffolds for cell culture, surgical models, and prostheses. Fabrication of complex porous structures of the extracellular matrix, surgical guides used to help surgeons during medical intervention, and prostheses mimicking bone mechanical properties are discussed. The chapter closes with future perspectives of FFF in medicine, paying particular attention to the development of new materials, and the main conclusions that can be extracted from the use of FFF in biomedical applications.

I. Buj-Corral (✉) · A. Tejo-Otero
Department of Mechanical Engineering, Barcelona School of Industrial Engineering (ETSEIB), Universitat Politècnica de Catalunya (UPC), Barcelona, Spain
e-mail: Irene.buj@upc.edu; aitor.tejo@upc.edu

C. Herranz-Diez
Unit of Biophysics and Bioengineering, Faculty of Medicine and Health Sciences, Universitat de Barcelona (UB), Barcelona, Spain
e-mail: carolinaherranz@ub.edu

J. Otero
Unit of Biophysics and Bioengineering, Faculty of Medicine and Health Sciences, Universitat de Barcelona (UB), Barcelona, Spain

CIBER de Enfermedades Respiratorias, Madrid, Spain
e-mail: jorge.otero@ub.edu

F. A. Sheikh et al. (eds.), *Interaction of Nanomaterials With Living Cells*,
https://doi.org/10.1007/978-981-99-2119-5_28

Keywords

Fused filament fabrication · Porous structures · Scaffolds · Surgical models · Prostheses

28.1 Introduction

The industry is nowadays submerged in the fourth industrial revolution, which involves a lot of different fields such as nanotechnology, the Internet of Things (IoT), and Artificial Intelligence (AI), among others. In this group, 3D printing is also included, which was initially called rapid prototyping and nowadays is known as additive manufacturing (AM). It is defined as the process of joining materials to make objects from a 3D model data, normally layer upon layer (each one has a thickness of 0.001–0.1 inch (Wohlers 2016)) through a series of cross-sectional slices, as opposed to subtractive manufacturing technologies (ASTM International 2015) (see Fig. 28.1). ISO/ASTM 52900 Standard (ASTM International 2015) classifies all technologies into seven categories: binder jetting (BJ), directed energy deposition (DED), material extrusion (MEX) (including fused filament fabrication, FFF, and direct ink writing, DIW), material jetting (MJ), powder bed fusion (PBF) (includes selective laser sintering, SLS, and selective laser melting, SLM), sheet lamination, and vat photopolymerization (VPP) (includes stereolithography, SLA, and digital light processing, DLP, as well as volumetric 3D printing, V3DP).

3D printing is a technological field that began its path many decades ago with different patents. During the early 1980s, Hideo Kodama invented two methods for manufacturing 3D plastic models with photo-hardening thermoset polymers (Kodama 1981), although he did not patent them. Then, in 1984, Jean-Claude André, Alain La Mehauté, and Olivier de Witte made the patent FR 2567668

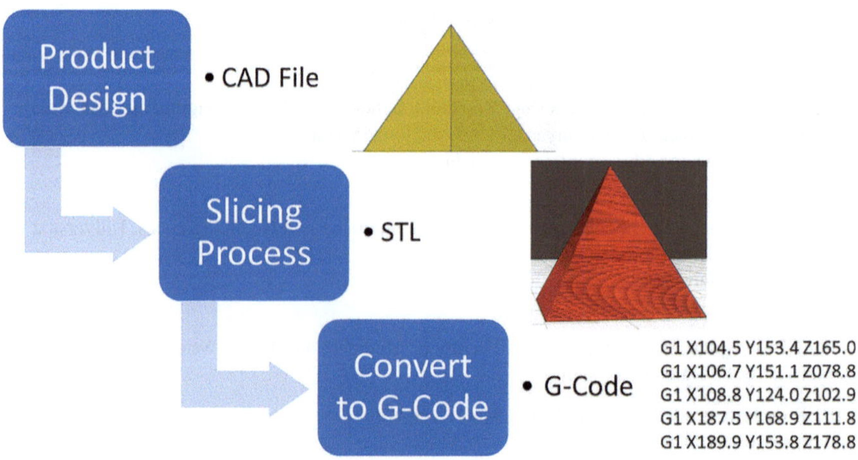

Fig. 28.1 Basic steps involved the 3D printing process

(Andre et al. 1984), a device whose aim was to produce a model of an industrial part. However, it was rejected due to the "lack of marketing perspectives". It is important to remark that, in those times, subtractive technologies were the most common approach used in factories; therefore, introducing these newly developed techniques was complicated. In 1986, Charles W. Hull, founder of 3D Systems, patented stereolithography (SLA), USA Patent 4,575,330 (Hull 1986). Additionally, he also made another contribution to the field, which was introducing the STL ("standard triangle language") format. It is a computer-aided design (CAD) computer file format that defines the geometry of 3D objects, excluding characteristics such as color, textures, or physical properties. The following year, Carl Deckard developed the SLS (Deckard 1989), which would not be marketed until 1992. During the late 1980s, Scott Crump, co-founder of Stratasys, developed the FDM technology (Crump 1992), USA Patent 5,121,329. Due to this, this technology did not grow steadily as an open-source one.

Additive manufacturing technologies offer a lot of benefits over subtractive techniques: (1) a wide range of materials (polymers including hydrogels and resins, ceramics, or metals), (2) complex architectures can be manufactured, (3) high accuracy, (4) less material waste, (5) more cost-effective, and (6) some processes are faster. Thanks to 3D printing technology, different types of objects and devices can be manufactured: (1) parts of an airplane, (2) daily-life gadgets such as a phone case, (3) implants, etc.

3D printers are easy-to-use "personal" 3D printing machines that can be used in either home or work environment, since they are both low-cost and straightforward to use. Others, instead, are usually more suitable for use in industry. They are larger machines capable of meeting different user requirements. Consequently, although more expertise is required to operate them, they offer various possible results and effects. As a result of this, AM technologies started to bloom in different fields such as aeronautics, industry, or health. The latter mentioned has undergone an important transformation since it is currently used in different applications: surgical planning prototypes, surgical guides, scaffolds, medical devices, etc. 3D printing allows the fabrication of parts of almost any geometrical complexity in relatively low time, with reduced cost (if desk machines are employed) and without significant requirements in technical expertise. These unique features changed the industry's manufacturing rules, making it possible to develop prototypes in a very short time to test the product's final characteristics before being released to the market, thus reducing costs. The fact that 3D printing was able to reproduce very complex geometries was a key factor in moving from industry to medical applications. More than 2000 years ago, humans tried to replace damaged body parts with substitutes made of gold, silver, iron, wood, etc. (Herrerín et al. 2014). Unfortunately, not all the approaches were successful, because of a lack of biocompatibility or a lower resemblance with the part to be replaced. In the present, different materials have been proven to be biocompatible, and many others have been developed for such purposes: metals, ceramics, polymers, glasses, and composite materials (Table 28.1).

Fused deposition modeling (FDM) or fused filament fabrication (FFF) is a widespread 3D printing technique because it is cheap, it provides smooth operation,

Table 28.1 Main materials used in biomedicine and its field of application (Park and Lakes 2007)

Material	Strong points	Weak points	Medical application
Ceramics	Highly compatible	Brittle, not resilient, weak in tension	Dental and orthopedic implants
Metals	Strong, though, ductile	Potential corrosion, dense and difficult to manufacture	Dental and orthopedic implants, pacer and suture wire, and bone plates and screws
Polymers	Resilience, easy manufacturing	Weakness, potential degradation over time, and deformation overuse	Hip socket, ear and nose prostheses, blood vessel substitutes, contact lenses, and sutures
Composites	Strong and tailor-made	Difficult to manufacture	Bone cement and dental resin

and the raw material needed is easy to handle (Sathies et al. 2020). Despite the sound characteristics of FFF to be used in medical applications, the choice of the material is a limiting factor: FFF does not allow working with high melting point materials (Salem Bala et al. 2016). Besides the limitations of the FFF technique, there is another constraint that is intrinsic to biomedical applications: the material to be used must be biocompatible.

When a device is implanted in the body, the body reacts to neutralize the potentially harmful effect it may cause by what is called the foreign body reaction. Basically, this reaction is expressed as an inflammation process and fibrous encapsulation (Williams et al. 1950; Anderson 2001; Anderson et al. 2008). From the injured tissue, cells and plasma release different chemicals that mediate the inflammatory process. Neutrophils and monocytes, cells from the immune system, produce a provisional cell matrix that activates platelets, inflammatory cells, and endothelial cells that restore the damaged tissue. In the case of FFF, the need for a material that can be transformed into a filament, heated, and extruded limits the biocompatible materials that can be used mainly for polymers and composites (Pugliese et al. 2021; Tappa and Jammalamadaka 2018). In this book chapter, the introduction of the FFF technology is given, with the different parameters that need to be considered for a successful outcome of the 3D printing process. Additionally, different 3D printers are described. Next, the possible filaments used for biological purposes are described. And, finally, the biomedical applications of the FFF technology are highlighted.

28.2 Fused Filament Fabrication

28.2.1 Definition

Fused filament fabrication (FFF), commonly known as fused deposition modeling (FDM), is the process of manufacturing 3D objects using a continuous filament, usually based on a thermoplastic material (see Fig. 28.2). The process is as follows:

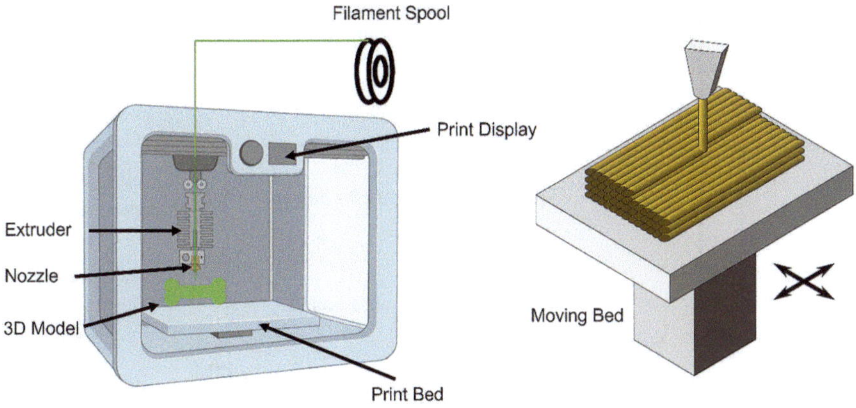

Fig. 28.2 Fused filament fabrication (FFF) scheme

(1) the filament is led to the extruder from the filament spool; (2) the extruder uses torque and pinch systems to feed and retract the filament in precise amounts; (3) the heater melts the filament to a useable temperature (this depends on the material used); (4) the heated filament is forced out of the heated nozzle, and laid down on the heated print bed or build plate, following a certain printing pattern; and, finally, (5) this process is repeated layer by layer until the process is finished. Once it is completed, the 3D object is detached from the print bed, and the required post-processes are carried out. For example, in case printing supports are used they need to be removed from the part.

This technology offers a wide range of possibilities depending on the target application. The parameters can be divided into two different groups (Prabhakar et al. 2021): (1) material parameters and (2) machine parameters. The latter group is classified into geometry-based, process-based, and structural-based (Prabhakar et al. 2021) (see Fig. 28.3).

On the one hand, the material parameters need to be considered depending on the target application. For example, if the desired 3D objects need more elasticity, in this way, a material with a more elastic behavior should be chosen. On the other hand, the machine parameters are those modified by the user, depending on the requirements. Firstly, regarding geometry-based, the nozzle size is the element that extrudes the filament to create the 3D objects. There are different ranges, starting from 10 to 100 μm in diameter. However, the most usual nozzle size is 0.4 mm, because in this case the print speed and precision are well-balanced. Additionally, the filament size is the thermoplastic feedstock for 3D printing. There are typically two diameters: 1.75 and 2.85 mm, depending on the machine used. Secondly, in terms of the process, the melting temperature is defined as the extruder exit temperature of the molten material. Then, heated beds are essential for 3D printing, with usual values ranging between 55 and 70 °C. Print speed is related to the displacement of the printing head. And finally, in the structural-based parameters, the layer thickness is the vertical resolution of the Z-axis and is one of the important technical features of

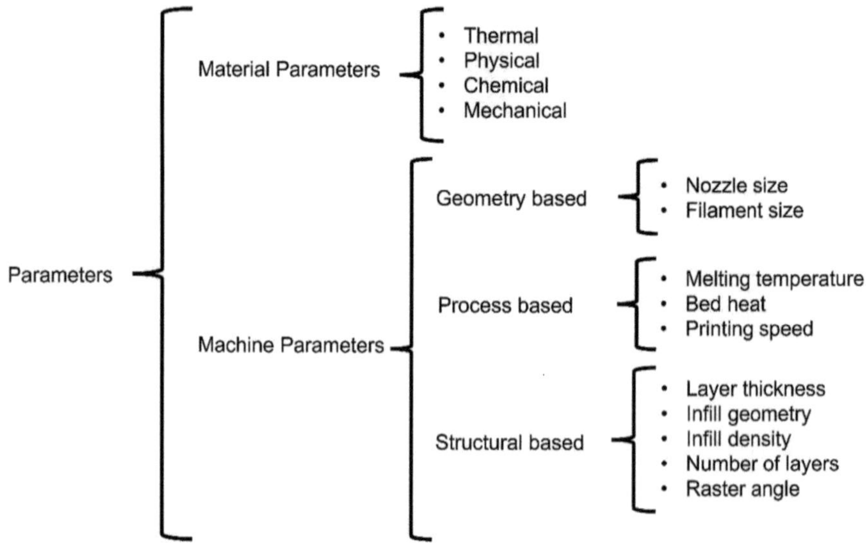

Fig. 28.3 Parameters for the fused deposition modeling process (Prabhakar et al. 2021)

a 3D printer. The higher the layer thickness, the faster the printing process, but the resolution is lower. Then, infill geometry, commonly known as infill pattern, influences printing not only time and speed but also the mechanical properties of the 3D printed object. Overall, there are different types of infill patterns: triangular, rectangular, hexagonal, or honeycomb and wiggle. Then, the infill density is the proportion of material within a certain infill structure. There may be from parts (0%) to solid 3D parts (100%). And last but not least, the raster angle is defined as the angle between the nozzle path and the X-axis of the 3D printer board and can directly affect not only the shape precision but also the mechanical properties of the parts. For example, the raster angle between two consecutive layers can vary by 90°.

28.2.2 Desktop FFF 3D Printers

Most of the 3D printers that use FFF are easy to employ and low-cost. They can be operated both at home and in a work environment. Different examples of desktop FFF 3D printers are described in terms of several important parameters (see Table 28.2). Most of this information was obtained from the websites of the manufacturers. Figure 28.4 shows pictures of some of the described machines.

28.2.3 Industrial FFF 3D Printers

However, FFF 3D printers used in industry are different, since they are larger machines and they are also capable of meeting additional user requirements. As

Table 28.2 Characteristics of different desktop FFF 3D printers

Parameters	Ultimaker S3	BCN3D SIGMA D25	Creality 3D Ender	Prusa i3 MK3S+
Build volume [mm]	230 × 190 × 200	420 × 300 × 200	220 × 220 × 250	250 × 210 × 210
Print speed [mm/s]	<24 [mm^3/s]	–	<180	200
Minimum layer height [μm]	20	50	100	50
Number of extruders	2	2	1	1
Filament diameter [mm]	2.85	2.85	1.75	1.75
Nozzle diameters [mm]	0.25–0.8	0.4–0.8	0.2–0.4	0.4
Operating temperature [°C]	15–32	15–30	≈21	20–30
Extruder temperature [°C]	180–280	<300	<280	<300
Heated bed temperature [°C]	20–140	<80	<100	<120
Multi-material option	Yes	Yes	No	Need a kit
Price [€]	4500	3500	210	800–1000

described previously, in the present section, different industrial FDM 3d printers are introduced (see Table 28.3).

Some 3D printer specifications cannot be found easily on either an additive manufacturing website or the official website.

28.2.4 Filaments for the FFF Technology

First, FFF printers were prepared to 3D print plastic materials. Different materials employed are summarized next (3D Natives n.d.; Kechagias et al. 2022):

Polylactic acid (PLA). PLA is made from renewable raw materials such as cornstarch or sugar cane. This material is among the easiest to be printed. However, it tends to shrink after printing. The printing temperature is lower than that of ABS, between 190 and 230 °C. Its Tg ranges between 50 and 70 °C. PLA is more difficult to manipulate due to its high speed of cooling and solidification. It is also important to mention that the models can deteriorate when in contact with water. Despite this,

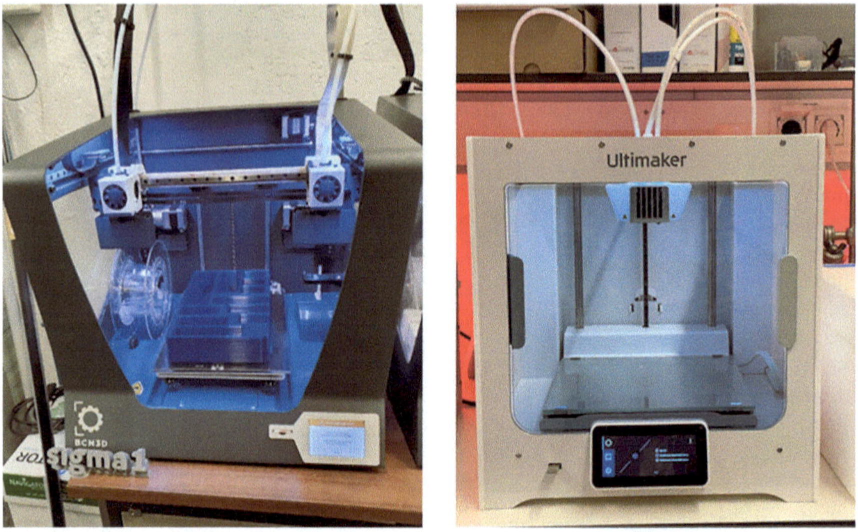

(a) (b)

Fig. 28.4 (**a**) Sigma R19 from BCN3D technologies, (**b**) Ultimaker S3

Table 28.3 Characteristics of different industrial FFF 3D printers

Parameters	F900 Stratasys	Fortus FDM 400 mc	Zortrax Endureal
Build volume [mm]	914 × 609 × 914	406 × 355 × 406	400 × 300 × 300
Printing speed [mm/s]	–	–	<180
Minimum layer height [μm]	127	127	200
Number of extruders	–	–	2
Filament diameter [mm]	–	–	1.75
Nozzle diameters [mm]	–	–	0.4
Operating temperature [°C]	–	–	17–30
Extruder temperature [°C]	–	<450	<480
Heated bed temperature [°C]	–	<350	<100
Multi-material option	Yes	Yes	Yes
Price [€]	100,000–250,000	150,000	210

the material is consistent and simple to use and possesses many colors. It is biodegradable under certain conditions.

Acrylonitrile butadiene styrene (ABS). ABS is a very commonly used filament. Its applications include car bodies, household appliances, cell phone covers, etc. It contains a polybutadiene-based elastomer base that makes it flexible and shock resistant. Its printing temperature ranges between 230 and 260 °C. Its Tg is around 105 °C. It requires the printing bed to be heated to minimize deformation. In addition, it is recommended to use a closed chamber when printing since ABS produces significant emissions when printing. It is not biodegradable.

Polyethylene terephthalate (PET). PET is the ideal filament for any part intended for contact with food. In addition, the material is very rigid and has good chemical resistance. The ideal temperatures for printing PET are 75–90 °C. PET is marketed with variants such as PETG (polyethylene terephthalate glycol), EPET (extrusion grade variant of PET), or OPET (oriented PET). An advantage of using PET is that no odors are released during printing and that it is 100% recyclable.

Nylon or polyamide (PA). PA corresponds to a synthetic polymer with good strength and certain flexibility, moderate chemical resistance, and high fatigue strength. Its printing temperature ranges between 220 and 270 °C. Because of their biocompatibility, polyamides can be used to create parts that come into contact with food (except food containing alcohol). Polyamides consist of semicrystalline structures and have a good balance in chemical and mechanical characteristics to provide good stability, stiffness, and impact strength. Because of their high quality, polyamides are used in the manufacture of gears, parts for the aerospace market, automation, robotics, medical prosthetics, injection molds, and essentially in applications involving press-fit plugs, tools with press-fit inserts, components fitting inserts, components subjected to high vibrations, and parts containing threaded inserts.

Polyvinyl alcohol (PVA). PVA is a synthetic polymer that is usually employed to print supports because of its water solubility. It is biocompatible. Its printing temperature lies between 190 and 220 °C.

Polyether ether ketone (PEEK). PEEK provides high mechanical strength. It requires high printing temperatures between 350 and 400 °C.

Thermoplastic polyurethane (TPU). TPU is more flexible than other materials used in FFF printing. Its printing temperature is between 195 and 230 °C.

28.3 Composite Filaments

Composite filaments are composed of one or more materials and have different properties than the original materials. Thermoplastic bases can be reinforced with either particles or fibers.

Regarding particles, recently, several metal-filled and ceramic-filled materials have been developed. For example, filaments with low metal content are supplied by ColorFabb: steelFill, copperFill, and bronzeFill. With a post-processing technique such as sanding or polishing, the parts achieve a shiny appearance. Although the main purpose of these materials is to provide metallic appearance to the parts, steelFill has been proven to enhance cell growth (Buj-Corral et al. 2022). In contrast, copperFill and bronzeFill did not provide satisfactory results, given their low biocompatibility. SteelFill also shows a weak magnetic property. Filaments with higher metallic content allow obtaining parts with higher mechanical strength than the low content ones, provided that a sintering process is carried out. For example, Zetamix by Nanoe provides a filament with high AISI316 content. As for ceramic filaments, the same company supplies zirconia and alumina filaments with high zirconia and alumina content. After a chemical debonding and a sintering process, the parts achieve high mechanical strength values, similar to those of the solid parts. It is also possible to use a filament with silicon carbide. When fibers are considered, for instance, Ultrafuse from BASF is an example of a polypropylene base (PP) with glass fiber. It is also possible to add carbon fiber to the polymeric base.

28.4 Biomedical Applications of FFF

As it was stated before, the capability of FFF to reproduce complex shapes, the ease of cost, and the choice of materials available, which include biocompatible polymers, have made FFF a suitable technique to be used in medical applications. Because of all these characteristics, the main areas where FFF is used are as follows:

- The development of 3D scaffolds for 3D culture
- Models
- Prostheses

28.4.1 Scaffolds for 3D Culture

FFF 3D printing allows replicating the intricated structure fibers of the extracellular matrix. With this technique, a 3D environment for cell culture with more physiological features than conventional 2D cell culture is obtained and will help to replicate cell behavior to some in vivo conditions, being translated into a better understanding of cell and tissue behavior (Duval et al. 2017; Li et al. 2003).

PLA, ABS, and HIPS (high impact polystyrene) scaffolds made by FFF have been successfully used as a tool to characterize cancer stem cell behavior and drug testing (Sathies et al. 2020; Polonio-Alcalá et al. 2018; Herreros-Pomares et al. 2021). As well as the choice of material, when designing 3D-printed scaffolds it is important to take into account that the scaffold should be able to diffuse oxygen and nutrients to the cells. Pore sizes between 200 to 500 μm are recommended to allow bone regeneration and vascularization of the scaffolds (Thavornyutikarn et al. 2014) (Fig. 28.5). The minimum pore size that supports oxygen and nutrient diffusion for cell survival when cultured on the scaffolds is 100 μm (Rouwkema et al. 2008). Thus, special attention should be paid to the scaffold design and the FFF parameters to ensure that the resulting biomedical scaffold possesses interconnected networks to allow oxygen and nutrient diffusion (Buj-Corral et al. 2018; Velasco et al. 2015).

28.4.2 Surgical Models

Surgeons have rapidly adopted the use of 3D printing techniques to plan surgeries, for operation rehearsal, or for teaching purposes (Malik et al. 2015). This rapid adoption of the FFF technique was because it is used in a very similar way it is used in industry: to make prototypes. The main difference in how FFF is used in industry

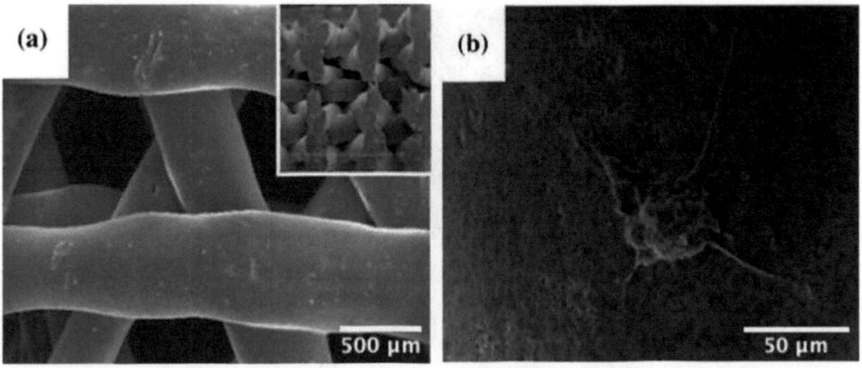

Fig. 28.5 (**a**) Top view of a PCL/TCP composite scaffold printed by FDM. The insert shows a cross-sectional view of the scaffold, (**b**) image of an osteoblast cell attached on the scaffold surface evidencing its biocompatibility (Thavornyutikarn et al. 2014). Images distributed under the terms of the Creative Commons CC BY license

and medicine is that the model to print is made with the help of MRI and CT image files. These files can be processed by CAD/CAM technology to reproduce in deep detail the structure to be printed, obtaining a model that resembles the original area in great detail.

In dentistry, models made by FFF are used to reproduce surgical guides to help the surgeons during the operation (Lüchtenborg et al. 2021; Pieralli et al. 2020). Models produced by FFF are used in other areas where the morphology of the body part is to be replaced by the characteristics of each patient. One of these cases is reverse shoulder arthroplasty, where the size and positioning of the prosthesis depend to a large extent of the degree of degradation and the glenoid morphology of each patient. With the help of FFF, it is possible to make guides for better positioning during operation, providing the best mobility and anchorage of the prosthesis (Piles et al. 2019). In complex congenital heart defects, the use of 3D-printed models of the cardiovascular anatomy of the patient helps to give a more accurate diagnosis and approach to the surgery from different perspectives before the operation (Valverde et al. 2017). When treating total joint arthroplasty, bone tumors, and bone deformities, it has also been proven the usefulness of the 3D models to improve the resection area and to match the patient's anatomy (Wong et al. 2017). FFF is also a proper 3D printing technique to represent anatomical models, which brings the students the opportunity to get in contact with the structures before operation (Fig. 28.6) (Lee et al. 2018; Damon et al. 2020).

Another way 3D printing is used in medicine is to make improved models for the casting of hip prostheses. Singh et al. (Singh et al. 2018) demonstrated that FFF might aid other manufacturing techniques to produce hip implants with better chemical and biological characteristics. In this work, they made an ABS mold by FFF with two different orientations and three different densities. The mold was smoothed with vapor at other exposure times and cycles. Then the mold was used to

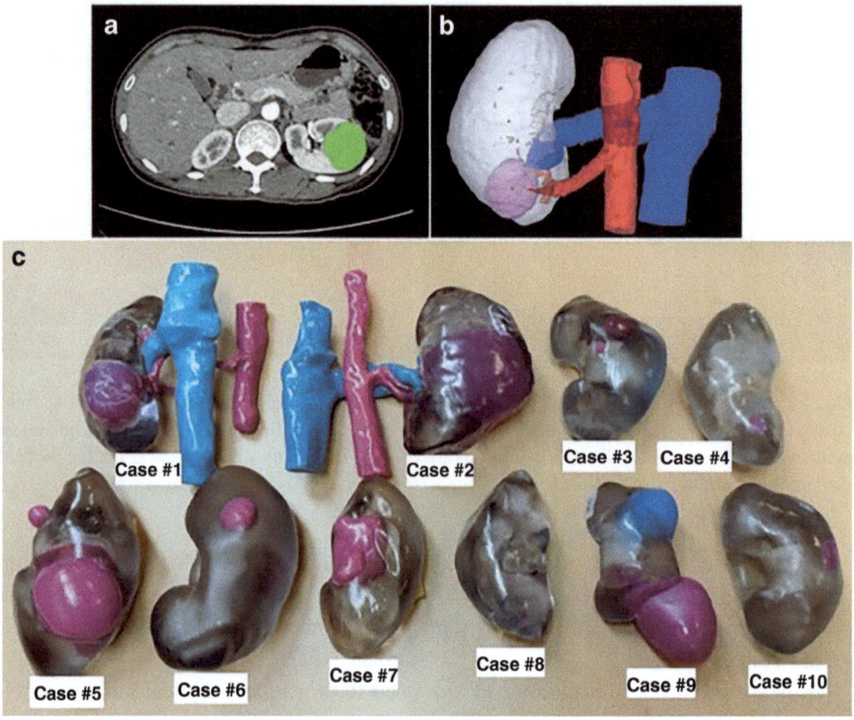

Fig. 28.6 (**a**) Computed tomography (CT) image of a kidney tumor, (**b**) the 3D-rendered image of the CT image, (**c**) renal 3D-printed models from different subjects (Lee et al. 2018). Images distributed under the terms of the Creative Commons CC BY license

cast an SS 316 L hip prosthesis. SS 316 L has mainly been used for fracture plates, screws, and hip nails as its corrosive resistance is compromised by high stress and/or hypoxia in the body environment (M. 2002; Park and Fung 1980). The use of FFF to make the casting mold of the implant increased its corrosion resistance and improved cell proliferation on the surface and broadened the SS 316 L area of application.

28.4.3 Prostheses

When fabricating prostheses, it is essential to consider that bone is a living tissue that can remodel itself due to loading and stress forces (Wolff 2010; Ruff et al. 2006). Although it is still not clear how the process happens (Van Oers et al. 2011, 2014), it has been demonstrated that prostheses made from materials with an elastic modulus higher than that of bone (Rho et al. 1993) lead to what is called the stress shielding effect. The consequence of the stress shielding effect is a reduction of bone density around the implant, which leads to implant loosening (Geetha et al. 2009). Traditional bone prostheses are made mainly from metal alloys (AISI 316 L, Ti6Al4V,

CrCo, TiNi) because they present an elastic modulus closer to that of bone. But they are not perfect. Besides the mechanical properties of the materials to be used, biocompatibility is also a crucial question. The ions released from metal prostheses have been related to skin sensitization (M. 2002), carcinogenicity (Uo et al. 1999), Alzheimer, neuropathy, and osteomalacia (Geetha et al. 2009; Swierenga et al. 1987).

There is another critical factor that takes part in the success of the prosthesis implantation: osseointegration. Osseointegration is defined as "the formation of a direct interface between an implant and bone, without intervening soft tissue" (Miller, 2005). If the prosthesis' material possesses low osseointegration potential, the implant is not well integrated with the surrounding bone, which results in the loosening of the implant with the consequent revision surgery (Ridzwan et al. 2007).

All this being said, FFF is a potential fabrication technique for a bone prosthesis as it allows the manufacture of bespoke prosthesis, saving time and money (Boland et al. 2007), and there are several materials that can overcome the problems derived from the use of metal alloys regardless of the material's specific needs such as low melting point, specific fiber size to feed the extrusion nozzle, and the need to recalibrate and optimize the feeding parameters for each new material (El-Katatny et al. 2010).

Polyether ether ketone (PEEK) is a polymer with a stiffness comparable to that of bone, does not elicit a response from the host, and also fulfils the mechanical characteristics needed to be used as an FFF material (Salem Bala et al. 2016). Because of its good properties, PEEK has been used in several 3D-printed bone prostheses. PEEK has been proven an excellent material to 3D print rib prostheses. The rib cage plays an essential role in preserving vital organs and also in respiratory function. If a rib cage replacement is made from titanium, the stress distribution on the ribs will change, and thus, the stress shielding effect may debilitate the cage, putting at risk its organ protective role as well as potentially affecting respiratory mechanics. Combining PEEK and FFF technology has improved the performance of a rib prosthesis compared with the titanium one. Thanks to the 3D printing technique, the rib cage shape was studied in detail, and the strain forces were replicated in a FFF model. This allowed printing a bespoke section of the rib cage with the exact shape of the section to be replaced. Using PEEK as a manufacturing material provided a rib replacement with mechanical characteristics similar to the original rib cage and good clinical performance (Kang et al. 2018) (Fig. 28.7).

PEEK has been also used to make a condylar prosthesis (Guo et al. 2021) (Fig. 28.8). In maxillofacial uses, it is very important to replicate the shape of the part to be replaced, as the fit is of paramount importance. With FFF and the aid of CAD/CAM, it is possible to design the prosthesis in great detail for a perfect match. Additionally, the possibility to use polymers or composites instead of metals is an advantage, as they avoid the artifacts produced by metals that affect the quality of the MRI or CT images that are conventional tests to assess the implant after the surgery (Wang et al. 2019; Do et al. 2018).

Other materials, such as carbon fiber polyamide 12 composites, have been used to 3D print hip prostheses by FFF (Nyiranzeyimana et al. 2022). The strengths of the

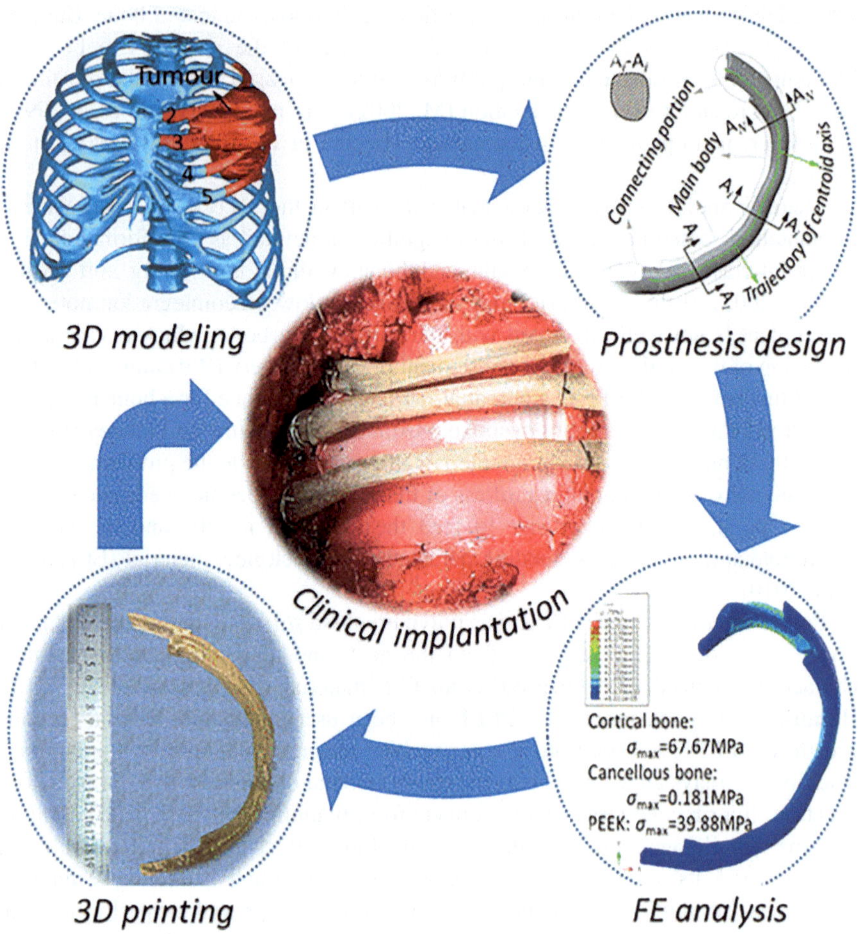

Fig. 28.7 Illustration of the different steps needed to 3D print a rib prosthesis, from 3D modeling to clinical implantation (Kang et al. 2018). Images distributed under the terms of the Creative Commons CC BY license

printed prosthesis matched those of the femur's cortical bone during regular bone load and showed good fatigue resistance. Polymers like ultra-high-molecular-weight polyethylene (UHMWPE) present good properties but, because of their characteristics, they are not able to be used in FFF. To overcome this problem, a blend of ultra-high-molecular-weight polyethylene - high density polyethylene - polyethylene-glycol (UHMWPE-HDPE-PEG) has been developed, improving their melting properties and showing good mechanical properties to be used as an implant (Ramli et al. 2016).

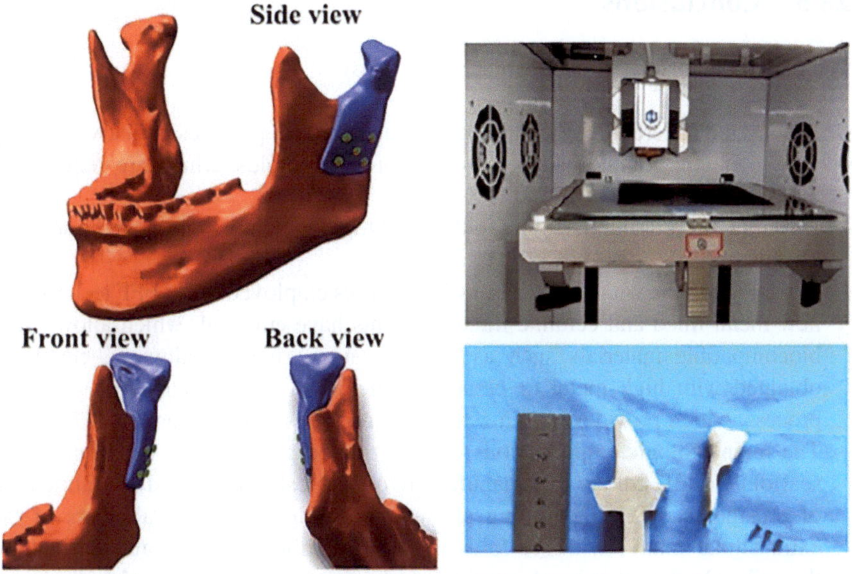

Fig. 28.8 On the left the CAD model for condylar prosthesis and its fixation to the lower mandible of the patient is shown. On the right, the 3D printing machine as well as the 3D-printed prosthesis are shown, demonstrating the low weight it has and, thus, its small size (Guo et al. 2021). Images distributed under the terms of the Creative Commons CC BY license

28.5 Future Perspective of FFF in Medicine

3D printing techniques and FFF have increased their use in treatment since they were first used in 1993 (Pugliese et al. 2021). The ranges of materials and applications include polymers, ceramics, and different areas of medicine spanning from 3D in vitro models to prostheses, but there is still room for improvement. Nowadays, there are a bunch of materials that can be used in FFF for biomedical purposes, but they present some limitations. To improve the performance of FFF in biomedical applications, new materials are being studied for biomedical use.

Even though PLA has been extensively used to 3D print scaffolds, researchers have worked toward foamed PLA-based resins to reduce the problems faced with bare PLA (Choi et al. 2020). The developed filaments showed better melting stability, rheological properties, and foamability than raw PLA. Also, foamed PLA scaffolds gave a better cell proliferation rate than not foamed PLA, demonstrating it as a good candidate material to be used when designing biological scaffolds.

28.6 Conclusions

The main conclusions of the chapter are as follows:

- Among the different AM technologies available, FFF provides a relatively cheap way to obtain complex shapes and/or porous scaffolds, which are required in biological applications.
- Different 3D printing machines, either desktop or industrial ones, can be selected for this purpose.
- In addition to the wide range of plastic materials employed in the FFF technology, new metal-filled and ceramic-filled filaments have emerged, which allow using biocompatible materials such as stainless steel, zirconia, alumina, etc. Parts obtained with high metal or ceramic content can be subjected to a sintering process, which provides high mechanical properties to these parts.
- The main biological applications of FFF printed parts are the development of scaffolds for 3D culture, the manufacture of surgical models, and the fabrication of prostheses. FFF 3D printing allows replicating the structural fibers of the extracellular matrix with scaffolds that can distribute oxygen and nutrients to the cells. On the other hand, surgical models reproduced the complex shapes of organs with the purpose of helping surgeons to rehearse before surgical operations. Finally, prostheses have been 3D printed with the FFF technique, for example, with carbon fiber-reinforced plastic filaments.
- Future trends include both the research in new metal- and ceramic-filled materials and the use of resins mixed with PLA in order to increase its melting stability, rheological properties, and foamability.

References

3D Natives (n.d.) 3D Printer Materials Guide. https://www.3dnatives.com/en/plastics-used-3d-printing110420174/. Accessed 11 Oct 2022

Anderson JM (2001) Biological responses to materials. Annu Rev Mater Sci 31:81–110

Anderson JM, Rodriguez A, Chang DT (2008) Foreign body reaction to biomaterials. Semin Immunol 20:86–100

Andre JC, Le Mehaute A, De Witte O (1984) Dispositif pour realiser un module de piece industrielle. Brevet François 84:11241

ASTM International (2015) ASTM52900–15 standard terminology for additive manufacturing—general principles—terminology. ASTM International, West Conshohocken, PA

Boland T, Ovsianikov A, Chickov BN, Doraiswamy A, Narayan RJ, Yeong WY, Leong KF, Chua CK (2007) Rapid prototyping of artificial tissues and medical devices. Adv Mater Process 165: 51–53

Buj-Corral I, Bagheri A, Petit-Rojo O (2018) 3D printing of porous scaffolds with controlled porosity and pore size values. Materials (Basel) 11:1532

Buj-Corral I, Sanz-Fraile H, Ulldemolins A, Tejo-Otero A, Domínguez-Fernández A, Almendros I, Otero J (2022) Characterization of 3D printed metal-PLA composite scaffolds for biomedical applications. Polymers (Basel) 14:2754

Choi WJ, Hwang KS, Kwon HJ, Lee C, Kim CH, Kim TH, Heo SW, Kim JH, Lee JY (2020) Rapid development of dual porous poly(lactic acid) foam using fused deposition modeling (FDM) 3D printing for medical scaffold application. Mater Sci Eng C 110:110693

Crump SS (1992). U.S. Patent No. 5,121,329

Damon A, Clifton W, Valero-Moreno F, Nottmeier E (2020) Orientation planning in the fused deposition modeling 3d printing of anatomical spine models. Cureus 12(2):e7081

Deckard CR (1989). U.S. Patent No. 4,863,538

Do TD, Sutter R, Skornitzke S, Weber MA (2018) CT and MRI techniques for imaging around orthopedic hardware. RoFo 190:31–41

Duval K, Grover H, Han LH, Mou Y, Pegoraro AF, Fredberg J, Chen Z (2017) Modeling physiological events in 2D vs. 3D cell culture. Physiology 32:266–277

El-Katatny I, Masood SH, Morsi YS (2010) Error analysis of FDM fabricated medical replicas. Rapid Prototyp J 16:36–43

Geetha M, Singh AK, Asokamani R, Gogia AK (2009) Ti based biomaterials, the ultimate choice for orthopaedic implants—a review. Prog Mater Sci 54:397–425

Guo F, Huang S, Hu M, Yang C, Li D, Liu C (2021) Biomechanical evaluation of a customized 3D-printed polyetheretherketone condylar prosthesis. Exp Ther Med:21

Herrerín J, Sánchez MA, Onstine S, Reckard V, Warkentin E, Redman T (2014) Prosthesis for the afterlife in TT16, Luxor, Egypt. J Am Res Cent Egypt 50:127–145

Herreros-Pomares A, Zhou X, Calabuig-Fariñas S, Lee SJ, Torres S, Esworthy T, Hann SY, Jantus-Lewintre E, Camps C, Zhang LG (2021) 3D printing novel in vitro cancer cell culture model systems for lung cancer stem cell study. Mater Sci Eng C 122:111914

Hull CW (1986). US Patent No 4,575,330

Kang J, Wang L, Yang C, Wang L, Yi C, He J, Li D (2018) Custom design and biomechanical analysis of 3D-printed PEEK rib prostheses. Biomech Model Mechanobiol 17:1083–1092

Kechagias J, Chaidas D, Vidakis N, Salonitis K, Vaxevanidis NM (2022) Key parameters controlling surface quality and dimensional accuracy: a critical review of FFF process. Mater Manuf Process 37:963–984

Kodama H (1981) Automatic method for fabricating a three-dimensional plastic model with photo-hardening polymer. Rev Sci Instrum 60:677–688

Lee H, Nguyen NH, Hwang S, Lee HJ, Hong SK, Byun SS (2018) Personalized 3D kidney model produced by rapid prototyping method and its usefulness in clinical applications. Int Braz J Urol 44:952–957

Li S, Lao J, Chen BPC, Li Y, Shuan, Zhao Y, Chu J, Den Chen K, Tsou TC, Peck K, Chien S (2003) Genomic analysis of smooth muscle cells in 3-dimensional collagen matrix. FASEB J 17:97–99

Lüchtenborg J, Burkhardt F, Nold J, Rothlauf S, Wesemann C, Pieralli S, Wemken G, Witkowski S, Spies BC (2021) Implementation of fused filament fabrication in dentistry. Appl Sci 11:6444

M. (2002) Niinomi recent metallic materials for biomedical applications. Metall Mater Trans A 33: 477

Malik HH, Darwood ARJ, Shaunak S, Kulatilake P, Abdulrahman A, Mulki O, Baskaradas A (2015) Three-dimensional printing in surgery: a review of current surgical applications. J Surg Res 199:512–522

Nyiranzeyimana G, Mutua JM, Mose BR, Mbuya TO (2022) A grey-based Taguchi method to optimize fused deposition modelling process parameters for manufacture of a hip joint implant. Materwiss Werksttech 53:89–108

Park JB, Fung YC (1980) Biomaterials, an introduction. J Biomech Eng 102:161–161

Park JB, Lakes RS (2007) Structure-property relationships of biological materials. In: Biomaterials. Springer, Boston, pp 225–263

Pieralli S, Spies BC, Hromadnik V, Nicic R, Beuer F, Wesemann C (2020) How accurate is oral implant installation using surgical guides printed from a degradable and steam-sterilized biopolymer? J Clin Med 9:1–12

Piles L, Reig MJ, Seguí VJ, Pla R (2019) Reverse shoulder arthroplasty: methodology improvement through personalized modelling techniques and FDM technology. Procedia Manuf 41:755–762

Polonio-Alcalá E, Rabionet M, Guerra AJ, Yeste M, Ciurana J, Puig T (2018) Screening of additive manufactured scaffolds designs for triple negative breast cancer 3D cell culture and stem-like expansion. Int J Mol Sci 19(10):3148

Prabhakar MM, Saravanan AK, Lenin AH, Jerin I, Mayandi K, Ramalingam PS (2021) A short review on 3D printing methods, process parameters and materials. Mater Today Proc 45:6108–6114

Pugliese R, Beltrami B, Regondi S, Lunetta C (2021) Polymeric biomaterials for 3D printing in medicine: an overview. Ann 3D Print Med 2:100011

Ramli MS, Wahab MS, Ahmad M, Bala AS (2016) FDM preparation of bio-compatible UHMWPE polymer for artificial implant. ARPN J Eng Appl Sci 11:5474–5480

Rho JY, Ashman RB, Turner CH (1993) Young's modulus of trabecular and cortical bone material: ultrasonic and microtensile measurements. J Biomech 26:111–119

Ridzwan MIZ, Shuib S, Hassan AY, Shokri AA, Mohammad Ibrahim MN (2007) Problem of stress shielding and improvement to the hip implant designs: a review. J Med Sci 7:460–467

Rouwkema J, Rivron NC, van Bliterswijk CA (2008) Vascularization in tissue engineering. Trends Biotechnol 26:434–441

Ruff C, Holt B, Trinkaus E (2006) Who's afraid of the big bad Wolff?: "Wolff's law" and bone functional adaptation. Am J Phys Anthropol 129:484–498

Salem Bala A, Bin Wahab S, Binti Ahmad M (2016) Elements and materials improve the FDM products: a review. Adv Eng Forum 16:33–51

Sathies T, Senthil P, Anoop MS (2020) A review on advancements in applications of fused deposition modelling process. Rapid Prototyp J 26:669–687

Singh D, Singh R, Boparai KS, Farina I, Feo L, Verma AK (2018) In-vitro studies of SS 316 L biomedical implants prepared by FDM, vapor smoothing and investment casting. Compos Part B Eng 132:107–114

Swierenga SHH, Gilman JPW, McLean JR (1987) Cancer risk from inorganics. Cancer Metastasis Rev 6:113–154

Tappa K, Jammalamadaka U (2018) Novel biomaterials used in medical 3D printing techniques. J Funct Biomater 9:17

Thavornyutikarn B, Chantarapanich N, Sitthiseripratip K, Thouas GA, Chen Q (2014) Bone tissue engineering scaffolding: computer-aided scaffolding techniques. Prog Biomater 3:61–102

Uo M, Watari F, Yokoyama A, Matsuno H, Kawasaki T (1999) Dissolution of nickel and tissue response observed by X-ray scanning analytical microscopy. Biomaterials 20:747–755

Valverde I, Gomez-Ciriza G, Hussain T, Suarez-Mejias C, Velasco-Forte MN, Byrne N, Ordoñez A, Gonzalez-Calle A, Anderson D, Hazekamp MG et al (2017) Three-dimensional printed models for surgical planning of complex congenital heart defects: an international multicentre study. Eur J Cardio-thoracic Surg 52:1139–1148

Van Oers RFM, Van Rietbergen B, Ito K, Huiskes R, Hilbers PAJ (2011) Simulations of trabecular remodeling and fatigue: is remodeling helpful or harmful? Bone 48:1210–1215

Van Oers RFM, Klein-Nulend J, Bacabac RG (2014) The osteocyte as an orchestrator of bone remodeling: an engineer's perspective. Clin Rev Bone Miner Metab 12:2–13

Velasco MA, Narváez-Tovar CA, Garzón-Alvarado DA (2015) Design, materials, and mechanobiology of biodegradable scaffolds for bone tissue engineering. Biomed Res Int 2015:729076

Wang Z, Vandersteen C, Demarcy T, Gnansia D, Raffaelli C, Guevara N, Delingette H (2019) Deep learning based metal artifacts reduction in post-operative Cochlear implant CT imaging. Lect Notes Comput Sci 11769:121–129

Williams RW, Mason LB, Bradshaw HH (1950) Factors affecting wound healing. Surg. Forum:410–417

Wohlers T (2016) Rapid prototyping & tooling state of the industry: 1998 worldwide progress report. Mater Technol 13:174–176

Wolff J (2010) The classic: on the inner architecture of bones and its importance for bone growth. 1870. Clin Orthop Relat Res 468:1056–1065

Wong TM, Jin J, Lau TW, Fang C, Yan CH, Yeung K, To, M, Leung F (2017) The use of three-dimensional printing technology in orthopaedic surgery: a review. J Orthop Surg 25: 2309499016684077

Irene Buj-Corral is an industrial engineer and holds a PhD in environmental engineering from the Universitat Politècnica de Catalunya (UPC), Spain, where she currently is an associate professor. She is the coordinator of the TECNOFAB Research Group in Manufacturing Technologies, and she is the director of the Common Laboratory of Mechanical Engineering of UPC. She worked for automotive, railway, and aeronautic companies. One of her main research areas is the study and analysis of dimensional accuracy and surface finish in 3D printing processes. She is the principal investigator of the project PROMIM3D (obtaining prostheses for bone tissue replacement utilizing extrusion 3D printing and subsequent sintering).

Carolina Herranz-Diez holds a BEng in mechanical engineering, MEng in biomedical engineering, and a PhD in biomedical engineering from the Universitat Politècnica de Catalunya (UPC), Spain. She has taught at the University of Sheffield and the Universitat Politècnica de Catalunya (UPC), collaborated with different research institutions in the United States, Germany, and Australia, and also worked for medical devices companies and nonprofit organizations related to health. She is currently a post-doctoral researcher at the Universitat de Barcelona (UB), Spain. Her research is focused on studying new biomaterials and their effect on cell behavior.

Aitor Tejo-Otero did his PhD research at CIM UPC and received his PhD in mechanical engineering at the Universitat Politècnica de Catalunya (UPC), Spain. Before that, he did his bachelor's degree in biomedical engineering at the University of Navarra, Spain, and an MSc in biomaterials and biomechanics at the University of Nottingham, UK. His research career has been focused on the application of AM technologies to the medical field in different applications such as surgical planning prototypes, medical devices, or scaffolds. During his PhD, he also did a research stay at 3D FAB (Lyon, France) using the FRESH 3D printing technique.

Jordi Otero, currently assistant professor at the Unit of Biophysics and Biomedicine of the medical school at the University of Barcelona, is a computer scientist and electronic and biomedical engineer and holds a PhD in biomedicine. He has published in top journals ranging from engineering and physics to biology. Dr. Otero has been active in national and European research projects during his predoctoral and postdoctoral formation. He is currently working on biomedical engineering at the school of medicine, where he is the principal investigator of the project HYBRID (high-yield fabrication of biomechatronic hydrogel devices for respiratory injury and disease modeling).

Role of Stem Cells in the Delivery of Essential Pharmaceuticals

29

Asma Rafiq, Iqra Jan, Mehvish Ali, Sabreena Aashaq, Bilal A. Reshi, Zaffar Amin Shah, Faheem A. Sheikh, Shafquat Majeed, and Mushtaq A. Beigh

A. Rafiq · I. Jan · M. Ali
Cellular signalling laboratory, Department of Nanotechnology, University of Kashmir Hazratbal, Srinagar, Jammu and Kashmir, India
e-mail: asmarafiq.scholar@kashmiruniversity.net; iqrajan.scholar@kashmiruniversity.net

S. Aashaq
Cellular signalling laboratory, Department of Nanotechnology, University of Kashmir Hazratbal, Srinagar, Jammu and Kashmir, India

Department of Immunology and Molecular Medicine, Sher-i-Kashmir Institute of Medical Sciences, Srinagar, Jammu and Kashmir, India

B. A. Reshi
Department of Biotechnology, University of Kashmir Hazratbal, Srinagar, Jammu and Kashmir, India
e-mail: bareshi@uok.eu.in

Z. A. Shah
Department of Immunology and Molecular Medicine, Sher-i-Kashmir Institute of Medical Sciences, Srinagar, Jammu and Kashmir, India

F. A. Sheikh
Nanostructured and Biomimetic Lab, Department of Nanotechnology, University of Kashmir Hazratbal, Srinagar, Jammu and Kashmir, India
e-mail: faheemnt@uok.edu.in

S. Majeed
Laboratory for Multifunctional Nanomaterials, Department of Nanotechnology, University of Kashmir Hazratbal, Srinagar, Jammu and Kashmir, India
e-mail: smshah@uok.edu.in

M. A. Beigh (✉)
Cellular Signalling and Nanotherapeutics Laboratory, Department of Nanotechnology, University of Kashmir Hazratbal, Srinagar, Jammu and Kashmir, India
e-mail: beighm@uok.edu.in

F. A. Sheikh et al. (eds.), *Interaction of Nanomaterials With Living Cells*,
https://doi.org/10.1007/978-981-99-2119-5_29

Abstract

Stem cell-based therapies hold great promise in treating different pathologies. Stem cells are able to self-renew and differentiate into multiple cell lineages, and this broad plasticity defines the therapeutic potential of stem cells. The risk of tissue rejection, inflammatory response, and bioavailability of drugs at the target site is still the underlying issue in treating many pathological conditions. The use of stem cells and stem cell-based drug delivery systems have overcome these issues. Stem cells have homing ability to migrate toward the specific target site. They are being engineered to express particular bioactive molecules loaded with drugs for targeted drug delivery. They also offer cell-free therapies where stem cell secretome can be used for the treatment. It included a broad range of bioactive molecules, including growth factors, cytokines, and angiogenic factors. They are being recognized as key regulators of physiological processes with their paracrine and autocrine roles. The chapter will focus on the properties of stem cells and the application of stem cells in regenerative medicine.

Keywords

Stem cells · Drug delivery systems · Inflammation · Hematopoietic

29.1 Introduction

Stem cells comprise a unique population of cells that are defined by their broad plasticity and self-renewal properties. They can indefinitely self-renew, differentiate into different types of cell lineages, and form a clonal cell population from a single cell. There are various sources of stem cells, and they vary in their potencies. After the fertilization of an oocyte, the totipotent zygote divides and gives rise to a multicellular organism. After the embryo develops, cells differentiate into specific fates to perform a particular function. In a fully grown organism, the specified cells, in addition to their particular function, also maintain homeostasis. They replenish the cell pool after a disease or an injury. Being tissue-specific stem cells, they are oligopotent and terminally differentiate into specific tissue. Their known properties of repairing damaged tissue provide a setup for understanding the disease mechanism. Stem cells in modern medicine are of great importance as they provide opportunities for both research and development of stem cell-based therapies. Stem cell-based therapies have been investigated in many diseases, with positive results obtained from preclinical and clinical studies. Among stem cells, mesenchymal stem cells (MSCs) are widely studied and investigated for regenerative medicine. MSCs are derived from different tissue resources and have multidirectional differentiation potential. The source of tissue affects the differentiation and proliferation tendencies of MSCs. They are also preferred over other stem cells as they can be cultured for longer times, have larger-scale expansion, and have higher anti-inflammatory effects. In addition to the already existing properties, MSCs are also engineered and modified to carry out a particular function. MSCs are used for gene therapies, where they are limited to secreting bioactive compounds at the target site.

MSCs have the ability to home and migrate toward a specific point. This homing results from substances released at sites of tissue injury, thus making them a potential candidate for targeted drug delivery. MSCs are also loaded with drugs, proteins, and therapeutic agents where they act as a vehicle to carry them to specific sites. This in turn helps in increasing the bioavailability of the drugs and other bioactive molecules at the target sites. Although MSCs are engrafted to the disease sites, engineered to produce specific substances, their therapeutic effect also accounts for their secretome. MSCs as cell have a genome that is encoded by many bioactive molecules, including proteins, enzymes, growth factor, angiogenic factors, apoptotic proteins, cytokines, interleukin, etc. These molecules are secreted by the cell into its extracellular space and are known as secretome. This secretome is potentially more active in the repair and regeneration process as it is responsible for carrying the cell-cell communication, which is responsible for regulating many biochemical processes. This chapter will discuss stem cells in regenerative medicine and how they are used as drug delivery systems.

29.2 Stem Cells

Stem cells are defined as primary undifferentiated cells having the ability to maintain themselves through self-renewal and differentiate into different cell types. This ability allows them to replenish the cell pool and act as a repair system for the body. At the cell division, the first progeny retains the stem properties giving rise to new stem cells, and the second progeny may lose the self-renewal property and generate mature cells giving rise to a particular tissue and finally yielding an organ or a population of functional blood cells. Stem cells are present both in embryos and adults. However, with each step of development, there is a decrease in the stem cell potency, which means as the cell develops, it may not be able to differentiate into many cell types as pluripotent (PSC) one. Pluripotent stem cells are derived from the inner cell mass of the blastocyst and are able to give rise to all germ layers except for the extraembryonic structures, such as the placenta. Embryonic stem cells (ESC) are an example of pluripotent stem cells. Their potency, not being constant, starts from the completely pluripotent cells and ends at the less potent ones, including multi-, oligo-, and unipotent cells. Multipotent stem cells differentiate into specific cell lineages as they have a narrower spectrum of differentiation. Hematopoietic stem cell is an example of multipotent stem cells that develop into different types of blood cells as shown in Fig. 29.1. On differentiation, the multipotent stem cells are restricted to a specific cell lineage, giving rise to the oligopotent stem cells. These cells further narrow down in the differentiation. A myeloid stem cell divides into white blood cells but not red blood cells and is an example of oligopotent stem cells. Further down the tree are the unipotent cells, which can only generate one specific cell type. These cells have the narrowest differentiation capability, but they possess the special property of repeated division. These properties make unipotent stem cells a promising candidate for regenerative medicine. Hepatocytes with long-term repopulating ability and the epidermal stem cells of the basal layer which only produce keratinized squamous cells are the examples of unipotent stem cells. Unipotent stem cells ensure an organism's long-term survival as they primarily act

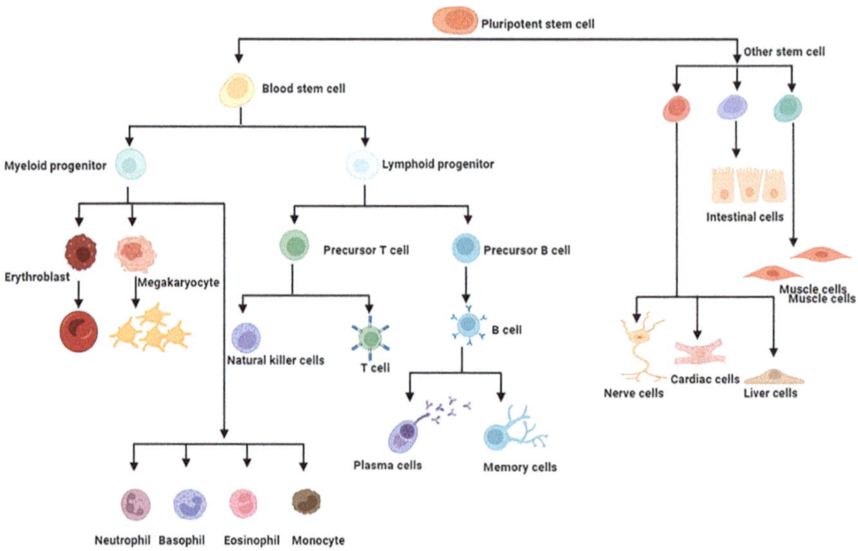

Fig. 29.1 Differentiation of pluripotent stem cell into different cell lineages

in tissue renewal. Stem cells further have been classified into five types on the basis of their origin: embryonic, induced pluripotent stem (IPS) cells, perinatal, fetal, and adult. Embryonic and IPS cells are pluripotent. The fetal and perinatal cells are multipotent, whereas the adult stem cells are generally oligopotent or unipotent.

29.3 Stem Cells in Regenerative Medicine

The progress made in science and medicine has significantly improved healthcare. However, there are still many medical conditions that have limited capacity for repair. Also, there are patients with diseased organs which can be treated with organ transplants, but the patients surpass the organ supply. Regenerative medicine is an emerging field in therapeutics and research and holds the promise of repairing the damaged tissues and organs of the body. As stated by Daar and Greenwood, regenerative medicine aims at repair, replacement, or regeneration of cell tissues and organs to restore impaired functions. With the properties like self-renewal and differentiation into various cell lineages, stem cells represent an important building block of regenerative medicine. The regenerative potential of stem cells can be used to replace lost or defective tissue. Thus, with broad therapeutic potential, stem cells can be employed to treat many pathological conditions.

29.3.1 Strategies for Regenerative Medicine

The stem cell field is opening new avenues for therapeutics, for example, cellular reprogramming and therapeutic cloning paving the way for tissue engineering.

Tissue engineering exploits the aspects of living cells, biomedical engineering, and material science in ways that can restore damaged tissues and organs. The technique utilizes the implantation of the functional cells into the damaged tissue or organ to facilitate the regeneration process or functional cells, along with biocompatible materials that can also be used, which can guide their differentiation and assembly into a functional tissue. Thus, regenerative medicine strategies fall into three categories:

1. Cell-based therapy
2. Use of biomaterials
3. Biomaterials seeded with cells

In this chapter, we will be discussing cell-based therapy. In one of the cell-based therapies, cells for tissue engineering are obtained and further cultured in vitro. The cell source can be allogenic (from the donor) or autologous (from the host). The autologous source is preferred as it doesn't elicit the immune response, and the use of immunosuppressant drugs is also avoided. In another cell-based therapy, a cell-based targeted drug delivery system (TDDS) has developed as a promising strategy. In targeted delivery system, low immunogenicity, longtime circulation, low intrinsic mutation, innate targeting capability, and low toxicity and tumorigenicity can be achieved. Various body cell types, including red blood cells, neutrophils, B lymphocytes, leukocytes, tumor cells, bacteria, viruses, etc., are found to be effective carriers as can be seen in Fig. 29.2. However, there are also some clinical limitations associated with using their cell types; some cannot be obtained in enough quantities, have issues with the modifications, and cannot be proliferated in vitro.

On the other hand, adult stem cells, mesenchymal stem cells (MSCs), can be used as the carriers as they are easy to isolate and can be increased in vitro. Also, they are free from both ethical concerns and teratoma formation.

29.4 Mesenchymal Stem Cells and Drug Delivery

The regenerative potential of stem cells can be attributed to three properties: migration toward the affected site via chemical gradients, differentiation into the particular cell type to function or replace the damaged tissue, and release of the bioactive substances with the potential for affecting both local and systemic physiological processes. Out of different type of stem cells, Mesenchymal stem cells comprise a unique cell population and are well studied and understood. MSCs are multipotent cells that possess the differentiation potential for lineages of mesenchymal tissues, self-renew, and form clonal cell populations. The regenerative potential of MSCs with drug loading and genetic modifications has found its application in drug delivery and antitumor therapy. Earlier studies primarily focused on how to exploit the plasticity of the stem cells and use them locally to engraft and differentiate into particular tissue types to restore the local injury. In recent years, bioactive substances produced by stem cells have also gained attention for their role in numerous physiological processes.

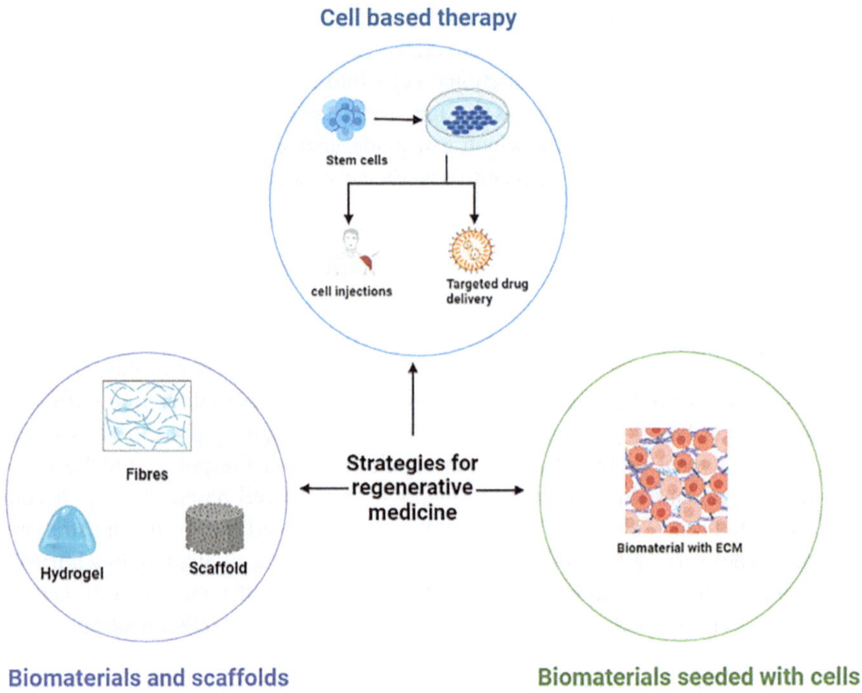

Fig. 29.2 Regenerative medicine strategies to restore damaged tissues or organs

The diverse assortment of bioactive molecules, such as cytokines, growth factors, apoptotic factors, angiogenic factors, chemokines, etc., known as secretomes have several advantages over the traditional use of stem cells in regenerative medicine. Not only do the stem cell transplant and drugs carried by the stem cells initiate the regeneration process, but the stem cell secretome with its paracrine/autocrine roles also mediates regenerative effects and has been well observed after the therapeutic stem cell administration. Earlier, the main benefit of stem cell therapy was restricted to the stem cell's ability to restore the function of locally damaged tissue, but now with the stem cells being drug delivery systems exploiting the stem cell secretome, genetically modifying stem cells and using them as the carriers for targeted drug delivery have expanded the stem cell-based therapies. Hence, stem cells can be used now in diverse ways to cure many pathological conditions.

29.4.1 Mesenchymal Cells as Gene Carriers

MSCs have participated in many therapeutic treatments. MSC-based gene therapy has become a plausible strategy alternative to viral and nonviral gene therapies. In many studies, MSCs' homing has been demonstrated. MSCs have been seen to selectively home the site of injury, inflammation, and tumor. Karp et al. has

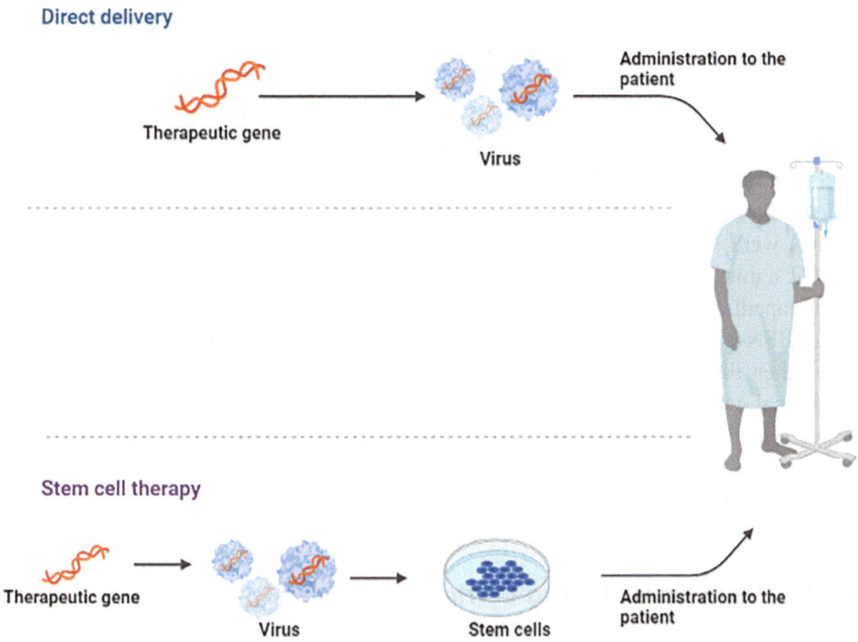

Fig. 29.3 Stem cells as gene delivery system

described MSCs' homing as the arrest of MSCs within the vasculature of a tissue followed by transmigration across the endothelium. Furthermore, MSCs have been observed to migrate to lymphoid organs, the lungs, bone marrow, and tumor sites. During the wound-healing process, many bioactive substances, including cytokines, hormones, and growth factors, regulate the requisition of MSCs at the wound site. Similarly, at the inflammation sites, extracellular matrix, cytokines, chemokines, etc. play an important role in the homing and recruitment of MSCs to the repair sites. In the case of cancer treatment, the cancer microenvironment can act as inflammation or the injury site and can be used to direct the MSC migration toward it. Genetically modified MSCs producing bioactive substances can be used for a particular pathological treatment. Inducible nitric oxide gene therapy has been identified as an antitumor strategy. In induced nitric oxide gene therapy, the nitric oxide synthase gene has been used, which generates high levels of nitric oxide, leading to cell apoptosis. The nitric oxide gene therapy has also shown increased vascular dilation, increasing the blood flow, which can be used to sensitize the area for other treatments, including radiotherapy and chemotherapy.

For transferring genes into MSCs, various methods, including viral vectors, nonviral vectors, and transfection systems, are being used (Fig. 29.3). However, viral systems often cause toxicity, immunogenicity, and even carcinogenicity. Therefore, nonviral methods draw attention to be used as gene transfer systems with no detrimental effects and no limitation on the plasmid size. Nonviral vectors, including DNA/polymer complexes, liposomes, and magnetic gene complexes, have been used to deliver genes into the MSCs. In a study carried out by Hong-HuiWu et al., bone marrow stem cells (BMSCs) were transfected to a suicide gene cytomegalovirus thymidine kinase (CMV-TK) using the spermine pullulan transfection method. After the transfection, the TK-BMCs activated the ganciclovir (GCV) to its toxic form (GCVTP) and suicide. The study was conducted on mice pulmonary melanoma models with the coadministration of liposomal formulation of CGV and suicide gene-expressing cells. An increase in survival rate and better penetration of BMCs were seen in these models. In other studies with genetic modifications of MSCs, a number of genes were transfected, and the cells were used for the delivery of therapeutic proteins, including TRAIL (TNF-related apoptosis-inducing ligand), IL-12, IFN-α and β, and FL-TRAIL. In TRAIL-modified MSCs, high tumor tropism and other therapeutic effects were seen. Also sodium iodide symporter (NIS), epidermal growth factor receptor (EGFR), and transforming growth factor (TGF)-β were transfected in MSCs and administered to the tumor site where these cells secreted the cytokines, prodrug converting enzymes, growth factor antagonists, and proapoptotic protein and acted as antitumor therapy.

29.4.2 Stem Cells as Drug Carriers

The homing ability of MSCs to migrate to the site of injury, allergy, and other microenvironments has led to them being rendered for use as targeted delivery vehicles. There are still many diseases where the potential drugs are not fully able to treat the condition because of their short half-life and their limited delivery to the target site. Also, in the case of cancer, drugs have detrimental effects on the vital organs and tissues of the body. MSCs exhibit potent pathotropic migratory properties with high specificity rendering them attractive for targeted drug delivery. MSCs can not only be used to deliver drugs at specific site but can also be engineered to continuously release the drugs with short half-lives. Furthermore, their easy availability and expansion procedures produce requisite cells for the treatment, making them a convincing approach for drug and gene delivery.

The pathotropic potential of MSCs is not fully understood. However, extensive studies have shown that the release of cytokines from the affected site regulates MSC migration. In many cancer studies, MSCs have been used to deliver anti-angiogenic drugs, interferons, and interleukins to the tumor site. In many studies, interleukins have been used as payloads in MSCs acting as tumor-suppressing agents. They are seen to exert their effects either directly working as tumoricidal or positively stimulating the immune response, activating cytotoxic lymphocytes and natural killer cells. MSCs with interleukin (IL)-12, interferon (IFN)-γ, and interleukin (IL)-2 have been used in tumor studies. MSCs with IL-12 have shown antitumor effect where they prevented metastasis and also increased tumor apoptosis. Similarly, umbilical cord MSCs with IL-18 have demonstrated a strong tumoricidal effect by inhibiting proliferation and metastasis in mice with breast cancer. Also, many interferons which can suppress tumor growth have been studied in which Interferon β (IFN-β) loaded MSCs have shown tumor suppressive effects and have also induced apoptosis. There are also many prodrugs and prodrug activators that have been studied for cancer therapies. A prodrug is an inactive form of the drug which after the administration is metabolized and converted to an active drug. Using prodrugs, in MSCs, increases their bioavailability at the target sites and also reduces their adverse effects as can be seen with the chemotherapeutic drugs. Also, prodrug activators which convert a nontoxic prodrug into toxic metabolites have been used to kill tumor cells. Cytosine deaminase (CD) is a chemotherapeutic agent and is used in MSC drug delivery. It is rapidly released in the tumor microenvironment and converts a nontoxic prodrug 5-fluorocytosine to active form 5-fluorouracil, selectively killing dividing cells.

Besides drugs and proteins, microRNAs have also gained an interest in disease treatment. MicroRNAs are known for the posttranscriptional regulation of gene expression and can be used as therapeutic agents to cure diseases. MSCs can be engineered to express a variety of microRNAs and deliver them to the specific pathological site. MSCs have also been loaded with oncolytic viruses and have been reported as an effective antitumor therapy. Natural or engineered oncolytic viruses upon infection, selectively replicate in the neoplastic tissues and kill them. The systemic administration of oncolytic viruses can illicit immune response and, therefore, can be cleared by the immune system and hence cannot reach the target site. Cell-based delivery of these oncolytic viruses can shield them from the immune attack and direct them toward the target site.

In contrast to MSCs loaded with drugs, their therapeutic effects can also be achieved via paracrine effects. MSCs are known to secrete extracellular vesicles, including microvesicles and exosomes as mentioned in Fig. 29.4. The exosomes are also found to be potentially a good candidate for therapeutic use. MSC-derived exosomes are found to regulate cell-cell communication, immune response, and tissue regeneration. Their administration in many animal models have revealed their beneficial effects and may represent a new therapeutic tool in the future. Exosomes were once known to be protein trash bags, but now studies have shown them as an important regulator of cell-cell communication that therefore can affect the neighboring cell environment or a distant target. Exosomes originate from the inward budding of the cell membrane and can be taken up by the adjacent cells or carried to a distant tissue or organ. MSC-derived exosomes are reported to carry a cargo of many proteins and RNAs, which elicit many biochemical processes and interactions with other cell types. During tissue injury in any disease, the homeostasis of the tissue environment gets disrupted and affects tissue functioning. The exosomes derived from the MSCs act as stromal cell support, which in turn helps maintain homeostasis with the tissue microenvironment. The proteins and enzymes present in these vesicles are activated by the change in the environment (change in pH or substrate concentration) of the tissue and make them well equipped for this role. These exosomes are also good for carrying RNAs and proteins to a distant target as they protect them from degradation and help in their cellular uptake.

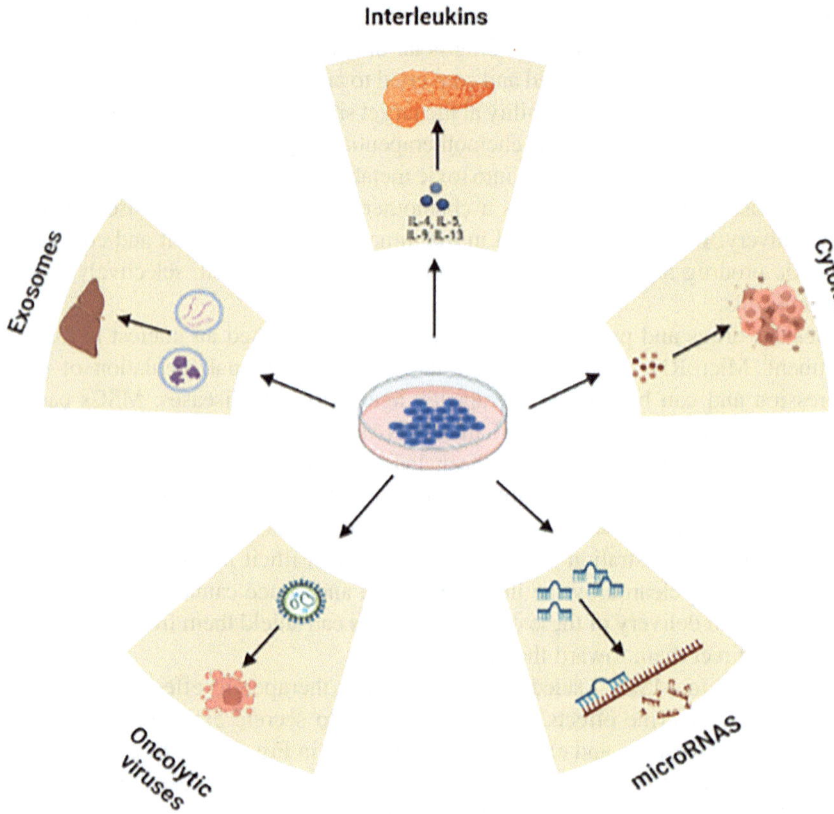

Fig. 29.4 Stem cells as drug carriers for targeted drug delivery

29.4.3 Application of Stem Cell Secretome for Regeneration

Apart from the roles of where stem cells are engineered or modified for repair and regeneration, stem cells in themselves produce certain bioactive molecules, which are found in the regulation of numerous physiological processes. Earlier stem cell studies were concerned in exploiting stem cell plasticity for its therapeutic use to locally engraft and differentiate into multiple cell types. However, the bioactive molecules secreted by stem cell have also gained attention as they are found to mediate angiogenesis, tissue scarring, and apoptosis at the sites of tissue injury. This milieu of bioactive molecules secreted by the stem cells is known as secretome and has become a subject of growing interest. Along with the known paracrine/autocrine effects of stem cells, stem cell secretome has found a potential for therapeutic use. The secretome of stem cell is thought to be encoded by 10% of the cell genome. These genes encode a milieu of proteins, including hormones, angiogenic factors, growth factors, serum proteins, extracellular matrix proteins, etc. This secretome is essential for replication, cell-cell communication, apoptosis, differentiation, angiogenesis, and adhesion. The stem cells secrete these molecules through both classical and nonclassical mechanisms. Biomolecules that are secreted through classical route are taken care by the endomembrane system, where they undergo proper folding, peptide signal cleavage, and carbohydrate addition. Following these posttranslational modifications, they are then secreted via exocytosis. However, the ones which are processed through a nonclassical pathway do not require an endomembrane system and are secreted via membrane translocation, or protein-coated vesicles are used. The use of stem cell secretome in regenerative medicine significantly reduces the time associated with the expansion of cells and can also prevent the transmission of infection related to cell therapies.

As stem cells differentiate and regenerate the damaged tissue, studies have shown that stem cell therapy alone cannot carry out the repair process; the cell secretome has a part to play. Also, studies have shown that cell secretome is adequate to account for beneficial effects and promote tissue recovery. In a study carried out in mice with myocardial infraction, stem cells were injected into the heart ventricle of mice. After a few administrations, it was found that the repaired tissue had very low number of stem cells but the therapeutic effects can still be seen. The authors suggested the role of the secretome in the therapeutic process.

As the therapeutic potential of stem cell secretome has been implicated in many studies, the secretome is found to play important roles in many physiological processes, including angiogenesis, immune modulation, anti-apoptosis, and tissue repair and healing. During the innate response of the body toward the infection or the injury, severe immune inflammation is also reported sometimes. MSCs found to possess immunomodulatory effects can be used to eliminate the severe inflammatory effects. MSCs also lack co-stimulatory cell surface markers which are responsible for eliciting the inflammatory response. As they lack co-stimulatory markers, MSCs play an important role in immune suppression and can be used to prevent the rejection of tissue or organ grafts. It is also seen in a study where patients with myocardial infraction were treated with the MSCs, and it was found that those patients had a better recovery rate than those who received a normal placebo. Also, no signs of rejection were observed.

MSCs also play a role in angiogenesis, and a number of pro-angiogenic stimulators and inhibitors are reported in MSC secretome. The normal angiogenesis is carried out by a number of biomolecules including growth factors, matrix proteins, enzymes, etc. MSCs are found to include fibroblast growth factors, interleukin-16, vascular endothelial growth factor, etc., and their secretion can also be modified by changing the media condition. In contrast to the apoptotic role of MSCs in tumor therapy, MSCs also prevent cell death in the injured tissues and organs. To avoid cell death, MSCs secrete the proteins which are known to have anti-apoptotic effects and also decrease the expression of apoptotic proteins. A study on cardiac injury also showed that the tissue treated with MSCs had decreased apoptotic proteins (caspase 3 and BAX) and increased pro-angiogenic factors, including fibroblast growth factors and vascular endothelial growth factors.

29.5 Conclusion

Since the discovery of stem cells, they have been studied and potentially used for many clinical disorders. The ability of stem cells to self-renew and differentiate into different lineages, specific migration toward the injury or disease, and non-immunogenic nature have made them an attractive candidate for cell-based therapies. They act as a tool for understanding organogenesis and regeneration. Stem cell-based therapies have brought new hope for the treatment of many life-threatening diseases. Among stem cells, MSCs have been extensively studied and used in preclinical and clinical studies, but many mechanisms still need to be clarified. Also, the long-term toxicological effects of these therapies need to be defined, and new methods for the efficient delivery of stem cells also need to be worked. Clearing the underlying issues could help to develop proper strategies leading to novel therapeutic opportunities.

Bibliography

Abe Y, Oshika Y, Ohnishi Y et al (1997) A xenograft line of human teratocarcinoma established by serial transplantation in severe combined immunodeficient (SCID) mice. APMIS 105:283–289

Akiyama Y, Radtke C, Honmou O et al (2002) Remyelination of the spinal cord following intravenous delivery of bone marrow cells. Glia 39:229–236

Alam T, Sollinger HW (2002) Glucose-regulated insulin production in hepatocytes. Transplantation 74(12):1781–1787

Ali NN, Edgar AJ, Samadikuchaksaraei A, Timson CM, Romanska HM, Polak JM, Bishop AE (2002) Derivation of type II alveolar epithelial cells from murine embryonic stem cells. Tissue Eng 8:541

Allen TD (1981) Haemopoietic microenvironments in vitro: ultrastructural aspects

Ben-David U, Benvenisty N (2012) High prevalence of evolutionarily conserved and species-specific genomic aberrations in mouse pluripotent stem cells. Stem Cells 30:612–622

Buttery LDK, Bourne S, Xynos JD, Wood H, Hughes FJ, Hughes SPF, Episkopou V, Polak JM (2001) Differentiation of osteoblasts and in vitro bone formation from murine embryonic stem cells. Tissue Eng 7:89

Campbell KH, McWhir J, Ritchie WA et al (1996) Sheep cloned by nuclear transfer from a cultured cell line. Nature 380:64–66

Cao Y, Vacanti JP, Ma X et al (1994) Generation of neo-tendon using synthetic polymers seeded with tenocytes. Transpl Proc 26(6):3390–3392

Carrel A, Lindbergh CA (1935) The culture of whole organs. Science 81:621–623

Cavallaro JF, Kemp PD, Kraus KH (1994) Collagen fabrics as biomaterials. Biotechnol Bioeng 43: 781–791

Cen L, Liu W, Cui L et al (2008) Collagen tissue engineering: development of novel biomaterials and applications. Pediatr Res 63:492–496

Chamberland DL, Agarwal A, Kotov N et al (2008) Photoacoustic tomography of joints aided by an etanercept-conjugated gold nanoparticle contrast agent: an ex vivo preliminary rat study. Nanotechnology 19:095101

Chun SY, Lim GJ, Kwon TG et al (2007) Identification and characterization of bioactive factors in bladder submucosa matrix. Biomaterials 28:4251–4256

Cibelli JB, Kiessling AA, Cunniff K, Richards C, Lanza RP, West MD (2001) Somatic cell nuclear transfer in humans: pronuclear and early embryonic development. Ebiomed J Regen Med 2:25

Compte M, Cuesta AM, Sanchez-Martin D, Alonso-Camino V, Vicario J, Sanz L, Alvarez-Vallina L (2009) Tumor immunotherapy using gene modified human mesenchymal stem cells loaded into synthetic extracellular matrix scaffolds. Stem Cells 27:753–760

Daar AS, Greenwood HL (2007) A proposed definition of regenerative medicine. J Tissue Eng Regen Med 1:179–184

Dahms SE, Piechota HJ, Dahiya R et al (1998) Composition and biomechanical properties of the bladder acellular matrix graft: comparative analysis in rat, pig and human. Br J Urol 82:411–419

de Almeida PE, Meyer EH, Kooreman NG et al (2014) Transplanted terminally differentiated induced pluripotent stem cells are accepted by immune mechanisms similar to self-tolerance. Nat Commun 5:3903

European Commission (2003). http://ec.europa.eu/research/press/2003/pdf/sec2003–441reporten.pdf. Accessed 2 Feb 2007

Friedenstein D, Kulagina A, Panasuk N, Rudakowa A, Luriá S, Ruadkow E (1974) Precursors for fibroblasts in different populations of hematopoietic cells as detected by the in vitro colony assay method. Exp Hematol 1974:83–92

Friedenstein AJ, Chailakhyan RK, Gerasimov UV (1987) Bone marrow osteogenic stem cells: in vitro cultivation and transplantation in diffusion chambers. Cell Tissue Kinet 20(3):263–272

Gambardella L, Schneider-Maunoury S, Voiculescu O, Charnay P, Barrandon Y (2000) Pattern of expression of the transcription factor Krox-20 in mouse hair follicle. Mech Dev 96:215–218

Gonczy P, Matunis E, DiNardo S (1997) Bag-of-marbles and benign gonial cell neoplasm act in the germline to restrict proliferation during drosophila spermatogenesis. Development 124:4361–4371

Goshima J, Goldberg VM, Caplan AI (1991) The osteogenic potential of culture-expanded rat marrow mesenchymal cells assayed in vivo in calcium phosphate ceramic blocks. Clin Orthop 262:298–311

Graves KH, Moreadith RW (1993) Derivation and characterization of putative pluripotential embryonic stem cells from preimplantation rabbit embryos. Mol Reprod Dev 36:424

Guild WR, Harrison JH, Merrill JP et al (1955) Successful homotransplantation of the kidney in an identical twin. Trans Am Clin Climatol Assoc 67:167–173

Haseltine W (1999) A brave new medicine. A conversation with William Haseltine. Interview by joe flower. Health Forum J 42:28–30

Hebert JM, Rosenquist T, Gotz J, Martin GR (1994) FGF5 as a regulator of the hair growth cycle: evidence from targeted and spontaneous mutations. Cell 78:1017–1025

Horwitz EM, Prockop DJ, Gordon PL et al (2001) Clinical responses to bone marrow transplantation in children with severe osteogenesis imperfecta. Blood 97:1227–1231

Hovatta O, Stojkovic M, Nogueira M, Varela-Nieto I (2010) European scientific, ethical, and legal issues on human stem cell research and regenerative medicine. Stem Cells 28:1005–1007

Hu YL, Fu YH, Tabata Y, Gao JQ (2010) Mesenchymal stem cells: a promising targeted-delivery vehicle in cancer gene therapy. Control Rel 147:154–162

Huang B, Tabata Y, Gao JQ (2012) Mesenchymal stem cells as therapeutic agents and potential targeted gene delivery vehicle for brain diseases. J Control Release 162:464–473

Huang B, Jiang X, Zhang T, Hu Y, Tabata Y, Chen Z, Pluchino S, Gao JQ (2017) Peptide modified mesenchymal stem cells as targeting delivery system transfected with miR-133b for the treatment of cerebral ischemia. Int J Pharm 531:90–10

Iams RL, Hilton DJ, Pease S, Willson TA, Stewart CI, Gearing DP, Wagner EF, Metcalf D, Nicola NA, Gough NM (1988) Myeloid leukaemia inhibitory factor maintains the developmental potential of embryonic stem cells. Nature 36:684

Jaiswal S, Jamieson CH, Pang WW et al (2009) CD47 is upregulated on circulating hematopoietic stem cells and leukemia cells to avoid phagocytosis. Cell 138:271–285

Kang WJ, Kang HJ, Kim HS et al (2006) Tissue distribution of 18F-FDG-labeled peripheral hematopoietic stem cells after intracoronary administration in patients with myocardial infarction. J Nuclear Med 47:1295–1301

Karanu FN, Murdoch B, Gallacher L et al (2000) The notch ligand jagged-1 represents a novel growth factor of human hematopoietic stem cells. J Exp Med 192:1365–1372
Kim BS, Baez CE, Atala A (2000) Biomaterials for tissue engineering. World J Urol 18:2–9
Klug MG, Soonpaa MH, Koh GY, Field LJ (1996) Genetically selected cardiomyocytes from differentiating embryonic stem cells form stable intracardiac grafts. J Clin Invest 98:216
Kuroda T, Yasuda S, Kusakawa S et al (2012) Highly sensitive in vitro methods for detection of residual undifferentiated cells in retinal pigment epithelial cells derived from human iPS cells. PLoS One 7:e37342
Le Blanc K, Tammik L, Sundberg B et al (2003) Mesenchymal stem cells inhibit and stimulate mixed lymphocyte cultures and mitogenic responses independently of the major histocompatibility complex. Scand J Immunol 57:11–20
Le Blanc K, Rasmusson I, Sundberg B et al (2004) Treatment of severe acute graft-versus-host disease with third party haploidentical mesenchymal stem cells. Lancet 363:1439–1441
Lee AS, Tang C, Rao MS et al (2013) Tumorigenicity as a clinical hurdle for pluripotent stem cell therapies. Nat Med 19:998–1004
Li L, Jiang Q, Ding G et al (2010) Effects of administration route on migration and distribution of neural progenitor cells transplanted into rats with focal cerebral ischemia, an MRI study. J Cereb Blood Flow Metab 30:653–662
Li M, Zhang D, Hou Y, Jiao L, Zheng X, Wang WH (2003) Isolation and culture of embryonic stem cells from porcine blastocysts. Mol Reprod Dev 65:429
Lin H, Spradling AC (1993) Germline stem cell division and egg chamber development in transplanted drosophila germaria. Dev Biol 159:140–152
Marchesi C, Pluderi M, Colleoni F et al (2007) Skin-derived stem cells transplanted into resorbable guides provide functional nerve regeneration after sciatic nerve resection. Glia 55(4):425–438
Mummery C, Oostwaard DW, Doevendans P, Spijker R, Van Den Brink S, Hassink R, Van Der Heyden M, Opthof T, Pera M, De La Riviere AB, Passier R, Tertoolen L (2003) Differentiation of human embryonic stem cells to cardiomyocytes: role of coculture with visceral endoderm-like cells. Circulation 107:2733
Nguyen PK, Nag D, Wu JC (2010) Methods to assess stem cell lineage, fate and function. Adv Drug Deliv Rev 62:1175–1186
Nishimura EK, Jordan SA, Oshima H et al (2002) Dominant role of the niche in melanocyte stem-cell fate determination. Nature 416:854–860
Pacioni S, D'Alessandris QG, Giannetti S, Morgante L, De Pascalis I, Coccè V et al (2015) Mesenchymal stromal cells loaded with paclitaxel induce cytotoxic damage in glioblastoma brain xenografts. Stem Cell Res Ther 6(1):1
Parson AB (2004) The proteus effect: stem cells and their promise for medicine. Joseph Henry, Washington
Piechota HJ, Dahms SE, Nunes LS et al (1998) In vitro functional properties of the rat bladder regenerated by the bladder acellular matrix graft. J Urol 159:1717–1724
Sabetkish S, Kajbafzadeh AM, Sabetkish N et al (2014) Whole-organ tissue engineering: decellularization and recellularization of three-dimensional matrix liver scaffolds. J Biomed Mater Res A 103(4):1498–1508
Salehi H, Calas-Bennasar I, Durand JC, Middendorp E, Valcarcel J, Larroque C et al (2014) Confocal Raman spectroscopy to monitor intracellular penetration of TiO2 nanoparticles. J Raman Spectrosc 45(9):807–813
Schulz TC, Young HY, Agulnick AD et al (2012) A scalable system for production of functional pancreatic progenitors from human embryonic stem cells. PLoS One 7:e37004
Sharon JL, Puleo DA (2008) The use of N-terminal immobilization of PTH(1-34) on PLGA to enhance bioactivity. Biomaterials 29:3137–3142
Shegokar S, Shaal L (2016) Applications of cell-based drug delivery systems: use of single cell assay. In: Tseng FG, Santra T (eds) Essentials of single-cell analysis. Springer, New York, pp 325–346

Shih IM, Wang TL, Traverso G et al (2001) Top-down morphogenesis of colorectal tumors. Proc Natl Acad Sci U S A 98:2640–2645

Slack JMW (2000) Stem cells in epithelial tissues. Science 287:1431–1433

Sotiropoulou P, Perez S, Salagianni M, Baxevanis C, Papamichail M (2006) Characterization of the optimal culture conditions for clinical scale production of human mesenchymal stem cells. Stem Cells 24:462–471

Spradling A, Drummond-Barbosa D, Kai T (2001) Stem cells find their niche. Nature 414:98–104

Streilein JW (2003) Ocular immune privilege: therapeutic opportunities from an experiment of nature. Nat Rev Immunol 3:879–889

Tabata Y (2008) Current status of regenerative medical therapy based on drug delivery technology. Reprod Biomed Online 16:70–80

Thanuja M, Anupama C, Ranganath S (2018) Bioengineered cellular and cell membrane-derived vehicles for actively targeted drug delivery: so near and yet so far. Adv Drug Deliv Rev 132:57–80

Thomson JA, Kalishman J, Golos TG, Durning M, Harris CP, Becker RA, Hearn JP (1995) Isolation of a primate embryonic stem cell line. Proc Natl Acad Sci U S A 92:7844

Thomson JA, Itskovitz-Eldor J, Shapiro SS et al (1998) Embryonic stem cell lines derived from human blastocysts. Science 282:1145–1147

Till JE, McCulloch EA (1961) A direct measurement of the radiation sensitivity of normal mouse bone marrow cells. Radiat Res 14:1419–1430

Twombly V et al (1996) The TGF-b signalling pathway is essential for drosophila oogenesis. Development 122:1555–1565

Vacanti CA, Bonassir LJ, Vacanti MP et al (2001) Replacement of an avulsed phalanx with tissue-engineered bone. N Engl J Med 344:1511–1514

Warren L, Manos PD, Ahfeldt T et al (2010) Highly efficient reprogramming to pluripotency and directed differentiation of human cells with synthetic modified mRNA. Cell Stem Cell 7:618–630

Watt FM, Hogan BLM (2000) Out of Eden: stem cells and their niches. Science 287:1427–1430

Wei X, Yang X, Han Z, Qu F, Shao L, Shi Y (2013) Mesenchymal stem cells: a new trend for cell therapy. Acta Pharmaco Sin 34:747 754

Wong MH, Saam JR, Stappenbeck TS, Rexer CH, Gordon JI (2000) Genetic mosaic analysis based on Cre recombinase and navigated laser capture microdissection. Proc Natl Acad Sci U S A 97: 12601–12606

Wu JC, Spin JM, Cao F et al (2006) Transcriptional profiling of reporter genes used for molecular imaging of embryonic stem cell transplantation. Physiol Genomics 25:29–38

Wurmser AE, Nakashima K, Summers RG et al (2004) Cell fusion-independent differentiation of neural stem cells to the endothelial lineage. Nature 430(6997):350–356

Xu C, Inokumi MS, Denham J, Golds K, Kundu P, Gold JD, Carpenter MK (2001) Feeder-free growth of undifferentiated human embryonic stem cells. Nat Biotechnol 19:971

Yannas IV, Burke JF (1980) Design of an artificial skin. I. Basic design principles. J Biomed Mater Res 14:65–81

Yannas IV, Burke JF, Gordon PL et al (1980) Design of an artificial skin. II. Control of chemical composition. J Biomed Mater Res 14:107–132

Ying Y, Liu XM, Marble A, Lawson KA, Zhao GQ (2000) Requirement of Bmp8b for the generation of primordial germ cells in the mouse. Mol Endocrinol 14:1053–1063

Yoo JJ, Meng J, Oberpenning F et al (1998) Bladder augmentation using allogenic bladder submucosa seeded with cells. Urology 51:221–225

Asma Rafiq received her BSc in 2016 from the University of Kashmir, India, and later joined MSc in biochemistry. She completed her master's program in biochemistry at the University of Kashmir, India, in 2018. She qualified for GATE in 2020 and is currently pursuing her PhD in the Department of Nanotechnology at the University of Kashmir, India. She has authored a couple of publications appearing in reputed journals. Her research interests include understanding the regulatory pathways in cancer and how these pathways help cancer cells avoid destruction by the immune attack, so that these pathways can be exploited for therapeutic intervention in cancer treatment.

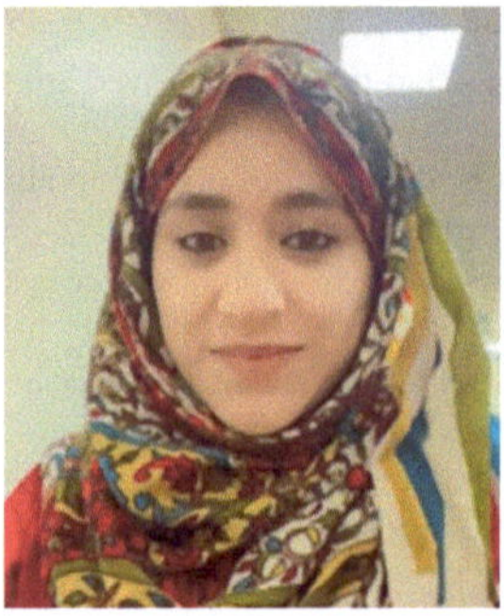

Iqra Jan was born in J & K, India. She completed her BSc in 2016 from the University of Kashmir, India, and later joined her MSc in biotechnology and completed the degree in 2018 from the University of Kashmir, India. She qualified for the GATE in 2020. She has been enrolled in a doctoral program in the Department of Biotechnology, University of Kashmir, India, since 2020 under Dr. Mushtaq A. Beigh. Her research interests include studying the role of posttranslational modifications in cancer metastasis and exploiting these pathways in drug therapeutics.

Mehvish Ali completed her MSc in nanotechnology from the University of Kashmir in 2021, and she is actively involved in understanding the nanotherapeutics for the delivery of phytochemicals from endangered plants.

Dr. Sabreena Aashaq was born in J & K, India. She completed her BSc in 2011 from the University of Kashmir, India, and later joined MSc in biotechnology and completed the degree in 2013 from the University of Kashmir, India. She completed her doctoral program in biotechnology from the University of Kashmir, India, in 2018 and is currently working as a research associate in the Department of Immunology and Molecular Medicine, Sher-i-Kashmir Institute of Medical Sciences, India. She has qualified for various national-level exams, including CSIR UGC NET and JRF, GATE, and ASRB-NET, and has been the recipient of various national-level fellowships. She has contributed extensively to world literature, with publications appearing in reputed journals. Her research interests include the regulation of various signal transduction pathways in cells, how dysregulation of these pathways results in different cancers, and the exploitation of these pathways in cancer therapeutics.

Bilal Ahmed Reshi has been working as an assistant professor in the Department of Biotechnology, University of Kashmir, since 2005. He has done his master's (2001) and MPhil (2005) in biochemistry from the University of Kashmir. He has qualified for UGC NET (2001) and is currently pursuing PhD (part-time) in the Department of Biotechnology in 2019. His area of specialization is plant science (exploring the potential of plant compounds as therapeutics). He has 17 years of teaching experience in the field.

Dr. Zaffar Amin Shah is a professor and head of the Department of Immunology and Molecular Medicine at Sher-i-Kashmir Institute of Medical Sciences, Soura, Srinagar. He holds BVSc, MVSc, and PhD degrees and is actively involved in teaching and research. He has around 170 publications to his credit and has a total citation of 1500.

Dr. Faheem A. Sheikh is an assistant professor at the Department of Nanotechnology, University of Kashmir, India (2015). He served as an assistant professor at the Department of Biotechnology at the Central University of Kashmir, India (2015–2016); a research professor at Myongji University, South Korea (2014–2015); assistant professor at Hallym University, South Korea (2012–2014); postdoc/research fellow at the University of Texas Rio Grande Valley, Texas, United States of America (2010–2012); and research professor at Myongji University, South Korea (2010). His research mainly focuses on fabricating nanomaterials used in tissue engineering.

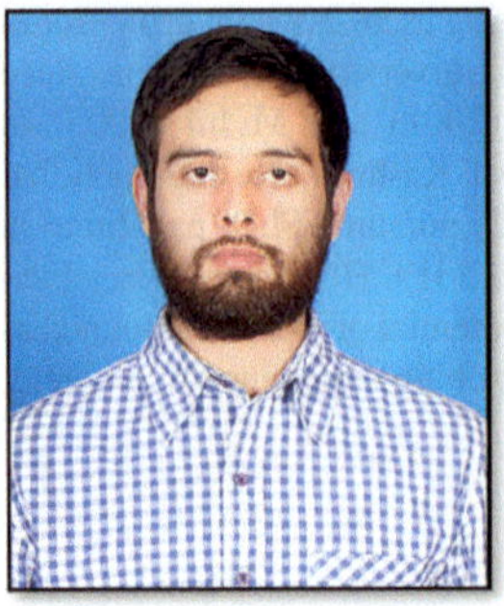

Dr. Shafquat Majeed is currently working as an assistant professor in the Department of Nanotechnology, University of Kashmir, India. Dr. Majeed did his master's in chemistry from the Department of Chemistry, University of Kashmir. He did his PhD at the prestigious Indian Institute of Science Bangalore (2008–2015). He has a postdoctoral research experience in rare-earth-based nanomaterials for biomedical applications and fabrication of perovskite-based solar cells from the Centre for Nano Science and Engineering, IISc Bangalore (2015–2016). He currently heads a multifunctional nanomaterial laboratory at the University of Kashmir, where they work on synthesis, characterization, and optical and magnetic properties of nanomaterials prepared through various solution-based routes. More specifically, they are working on developing new and enhanced multimodal imaging modalities for biomedical and healthcare applications.

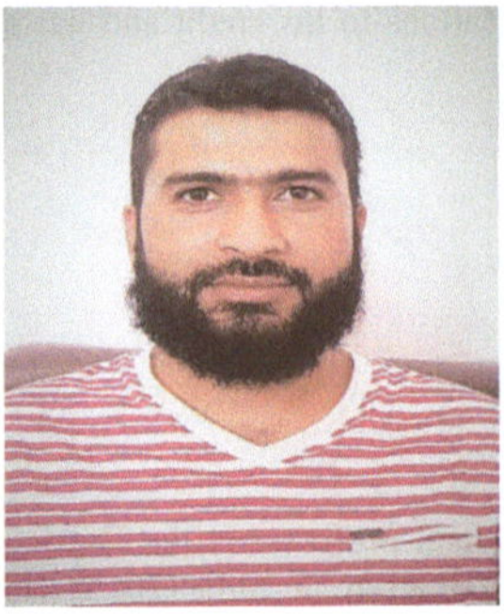

Dr. Mushtaq A. Beigh is working as an assistant professor at the Department of Nanotechnology, University of Kashmir, India. Dr. Beigh has postdoctoral research experience in cellular signaling and disease modeling (2013–2016). He had completed his PhD at the University of Kashmir, India (2008–2012). He heads a laboratory at the University of Kashmir, where they work on a bio-nano interface to understand the molecular basis of nanoconjugate internalization. They also try to understand the growth factor signaling inputs directed at multiple cellular receptors like NRP1 and integrins in order to work out the mechanistic details of receptor-based internalization.

Biomaterials in Autoimmune Diseases 30

Iqra Jan, Asma Rafiq, Mehvish Ali, Sabreena Aashaq, Bilal A. Reshi, Faheem A. Sheikh, Shafquat Majeed, and Mushtaq A. Beigh

Abstract

For a safe and efficient study of disease pathogenesis, biomaterials are utilized to treat tissue engineering and regeneration applications. Scientists have made an

I. Jan · A. Rafiq · M. Ali
Cellular Signalling Laboratory, Department of Nanotechnology, University of Kashmir Hazratbal, Srinagar, Jammu and Kashmir, India
e-mail: iqrajan.scholar@kashmiruniversity.net; asmarafiq.scholar@kashmiruniversity.net

S. Aashaq
Cellular Signalling Laboratory, Department of Nanotechnology, University of Kashmir Hazratbal, Srinagar, Jammu and Kashmir, India

Department of Immunology and Molecular Medicine, Sher-i-Kashmir Institute of Medical Sciences, Srinagar, Jammu and Kashmir, India

B. A. Reshi
Department of Biotechnology, University of Kashmir Hazratbal, Srinagar, Jammu and Kashmir, India
e-mail: bareshi@uok.eu.in

F. A. Sheikh
Nanostructured and Biomimetic Lab, Department of Nanotechnology, University of Kashmir Hazratbal, Srinagar, Jammu and Kashmir, India
e-mail: faheemnt@uok.edu.in

S. Majeed
Laboratory for Multifunctional Nanomaterials, Department of Nanotechnology, University of Kashmir Hazratbal, Srinagar, Jammu and Kashmir, India
e-mail: smshah@uok.edu.in

M. A. Beigh (✉)
Cellular Signalling and Nanotherapeutics Laboratory, Department of Nanotechnology, University of Kashmir Hazratbal, Srinagar, Jammu and Kashmir, India
e-mail: beighm@uok.edu.in

F. A. Sheikh et al. (eds.), *Interaction of Nanomaterials With Living Cells*,
https://doi.org/10.1007/978-981-99-2119-5_30

immense contribution to the stimulation of antigen-specific responses and the development of immunotherapeutics. One such disorder, e.g., autoimmune disease, needs to be controlled at molecular and cellular levels. Biomaterials have also been considered as means of delivering autoantigens or drugs for treating autoimmune disorders in a sustained and specific-release manner. However, their use has certain drawbacks, like the development of chronic inflammation, compatibility issues, etc. Despite these concerns, using biomaterials to engineer the immune system for various applications in the autoimmunity field is primarily considered beneficial. This chapter discusses how biomaterials have been planned for drug delivery and disease detection of autoimmune diseases. Biomaterials have also been engineered to treat autoimmune diseases, such as rheumatoid arthritis, multiple sclerosis, and type 1 diabetes, which are also discussed.

Keywords

Autoimmune · Immunotherapeutics · Biomaterial-based immunotherapy

30.1 Introduction

30.1.1 Immune Tolerance

To safeguard a host from materials referred to as pathogens, the immune system of mammals, which happens to be composed of a highly synchronized network of cells and organs, comes into play (Banchereau et al. 2000; Lewis et al. 2014). The immune system comprises two subtypes: innate immunity and adaptive immunity. Innate immunity involves instantaneous and nonspecific responses by conferring physical barriers to pathogens, stimulating complement system activation, prompting cytokine production causing local inflammation, and uptake of pathogens by antigen-presenting cells (APCs) like macrophages, neutrophils, and dendritic cells (DCs). Adaptive immunity is related to T-cell and B-cell activation, antibody production, and immunological memory and is very specific to the antigenic epitope. Adaptive immunity also bears the ability to bring about an immune response against self-antigens, which in turn may lead to a loss of immune tolerance.

Immune tolerance may be explained as an incompetent approach of a cell to elicit an inflammatory response against a self-antigen. This inability is crucial for safeguarding immune homeostasis and is achievable by two means: central tolerance and peripheral tolerance. Central tolerance occurs when adaptive cells develop from stem cells, and peripheral tolerance refers to the recurrence of immune regulation throughout the host (von Boehmer 1990). Central tolerance refers to the recognition of self-antigens, with the aid of T-cells and B-cells, through their membrane-bound receptors (Nossal 1994). This recognition in T-cells has been found to occur in the thymus, and the recognition in B-cells has been found to take place in the bone marrow. Also, peripheral tolerance encompasses induction of anergy, ignorance, receptor editing outside the primary lymphoid organs, and apoptosis. The regions not occupied by antigen receptors are continuously expressed by an anergic cell. However, they remain unresponsive against antigen stimulation (Andrews and Wilson 2010; Yarkoni et al. 2010). The process in which a cell undergoes programmed death is often stimulated by Fas/Fas ligand interactions commonly found between T-cells and thymic epithelial cells (Castro et al. 1996). A cell becomes ignorant when adaptive lymphocytes do not get exposed to their respective cognate antigen because of being an immunologically privileged site (e.g., testis). Ignorant B-cells may also undergo apoptosis when they become self-reactive to sequestered antigens in tissues, which are never stimulated because of inaccessibility. But, they can be autoimmune if antigen gets released from an immunologically privileged site (Forrester et al. 2008). So, central tolerance and peripheral tolerance are vital for the identification, differentiation, and targeting of self- and non-self-antigens for the immune system.

30.1.2 A Collapse in Immune Tolerance: Autoimmune Disease

Whenever a breakdown occurs, intolerance, it gives birth to autoimmunity. Terminologically, "autoimmune disease" refers to the condition where the immune system causes impairment of its/host's tissues by responding destructively (Agusti et al. 2003). In these circumstances, the self-reactive lymphocytes resort to an inflammatory response against self-antigens and hence refrain from mechanisms that cause activation of central tolerance and peripheral tolerance. Self-antigens act explicitly on organs or can be ubiquitous in the body. They are recognized by autoreactive lymphocytes and may vary in nature. The damage inflicted by autoimmune disorders ranges from causing damage to tissues to abnormal growth and unstable organ functioning. Moreover, self-antigens get released in case the tissue gets damaged. These self-antigens are released into the periphery leading to aggravation of the disease. This mainly happens when the antigen is concealed in an immunologically privileged site, e.g., the eye (Forrester et al. 2008).

30.2 Biomaterial-Based Immunotherapy

The foreign body response (FBR) elicited by the host immune system essentially degrades the versatility of the materials or devices (e.g., insulin pump) that get implanted in the host. However, biomaterials can manipulate such immune reactions. Biomaterials have been known to be capable of being engineered to exploit the immune system for therapeutic and diagnostic applications. Some of the properties of biomaterials that have been brought under control to generate immune responses include biomaterial surface chemistry, surface topography, microscale architecture, and other physiochemical properties (Fig. 30.1). The main aim of researchers remains to focus on the strategies that can be employed to diminish self-generated immunity with primary stress upon particulate engineering, tolerogenic drug delivery, and lymph node conditioning.

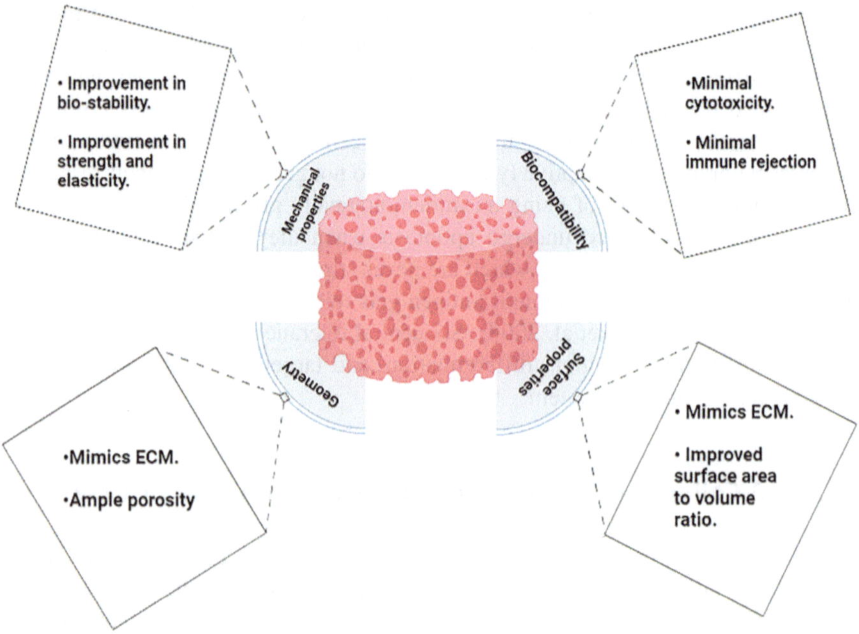

Fig. 30.1 Summary of the desired scaffold properties when designing biocompatible scaffolds. Created with Biorender

30.2.1 Adjuvant-Induced Autoimmune Syndrome

The concept of the adjuvant-induced autoimmune syndrome (ASIA) was first recognized in 2011, although patients having diverse symptoms after treatment with silicone or paraffin fillers were reported by Miyoshi in 1964 (Alijotas-Reig 2015). According to the ASIA concept, chronic systemic inflammation and autoimmune disorders can develop by repeated exposure to biomaterials, especially to the materials used in plastic surgeries, such as silicone implants used for breast augmentation (Hajdu et al. 2011). Silicone is a polymer of dimethylsiloxane that has been extensively explored in biomedical applications due to its variable viscosity, which depends on the length of the polymer used. Silicone mainly gained popularity in breast implants despite its wide usage in implantable products like ventriculoperitoneal and heart valves (Hajdu et al. 2011). Today breast, buttock, and face augmentation by the use of liquid silicone is the most commonly adopted plastic surgery method in the United States and other countries. However, capsular fibrosis and contracture, scleroderma, local adverse reactions, and other autoimmune disorders are associated with silicone implants despite their acceptable biocompatibility nature (Hajdu et al. 2011; Gabriel et al. 1994; Sanchez-Guerrero et al. 1995; França et al. 2013). Association of silicone with autoimmune disorders was first reported in 1984, when a spectrum of autoimmune diseases like rheumatoid arthritis, systemic lupus erythematosus (SLE), polymyositis, and sclerosis were reported in 24 patients who had received silicone injections (Kumagai et al. 1984). In another study, one out of three silicone-implanted patients developed RA symptoms, the second patient developed symptoms like SLE, and the third acquired connective tissue abnormality (Van Nunen et al. 1982). During the last few years, many cases of connective tissue disorders have been documented with silicone implants (Spiera et al. 1994; Hölmich et al. 2007; Nesher et al. 2015; Singh et al. 2016). Besides silicone, other biomaterials like alum and hyaluronic acid may also be responsible for ASIA (Alijotas-Reig 2015; Alijotas-Reig et al. 2012). Causative factors and symptoms identified for ASIA syndrome include repeated biomaterial injection, development of local chronic inflammation, and systemic symptoms such as joint pain, fever, fatigue, and vasculitis. Risk factors for developing ASIA include a family history of autoimmunity, an autoimmune reaction against implants, and an allergic reaction against the injected materials (Goren et al. 2015). Very recently, a case of the adjuvant-induced autoimmune syndrome was reported when dextranomer/hyaluronic acid copolymer (Deflux) was used for the injection to treat vesicoureteral reflux (VUR). Deflux injections cure about 85% of primary VUR abnormalities in females; however, 0.6–0.7% of patients are refractory to the treatment. Suda et al., from Japan, reported occurrence of autoimmune disease in a female patient having ureteral obstruction, who has received repeated injections of Deflux (Suda et al. 2016). Besides the polymeric biomaterials, ASIA has also been reported with metallic implants. The most common types of metals reported for induction of autoimmunity are nickel and titanium. In a case study, a 23-year-old female had nickel-titanium chain implant for cosmetic purpose. After 1 year postimplantation, she developed autoimmune symptoms like thrombocytopenia, anemia, and neutropenia (Loyo et al. 2013).

30.2.2 Biomaterials for Drug Delivery and Disease Detection in Autoimmune Diseases

Biomaterials meant for medical use are often met with the objection of being able to acquire the requisite functionality, while exhibiting biocompatibility and minimal off-target effects. With the aid of material designing, the responses outside the desired target can be limited. Stimuli-responsive materials provide an opportunity to specifically target receptors, cells, or organs. Physiological changes like changes in local pH or temperature, exposure to the magnetic field or a light source, and biomolecular interactions such as enzymatic degradation can be used to bring about relevant stimuli for biomedical devices (Fig. 30.2). A wide range of natural as well as synthetic materials have been investigated, which include polymers and hydrogels, metal nanoparticles and quantum dots (QD), and biomolecules like antibodies and enzymes, to ensure the maintenance of biocompatibility for in vivo devices or the high selectivity critical for external diagnostics.

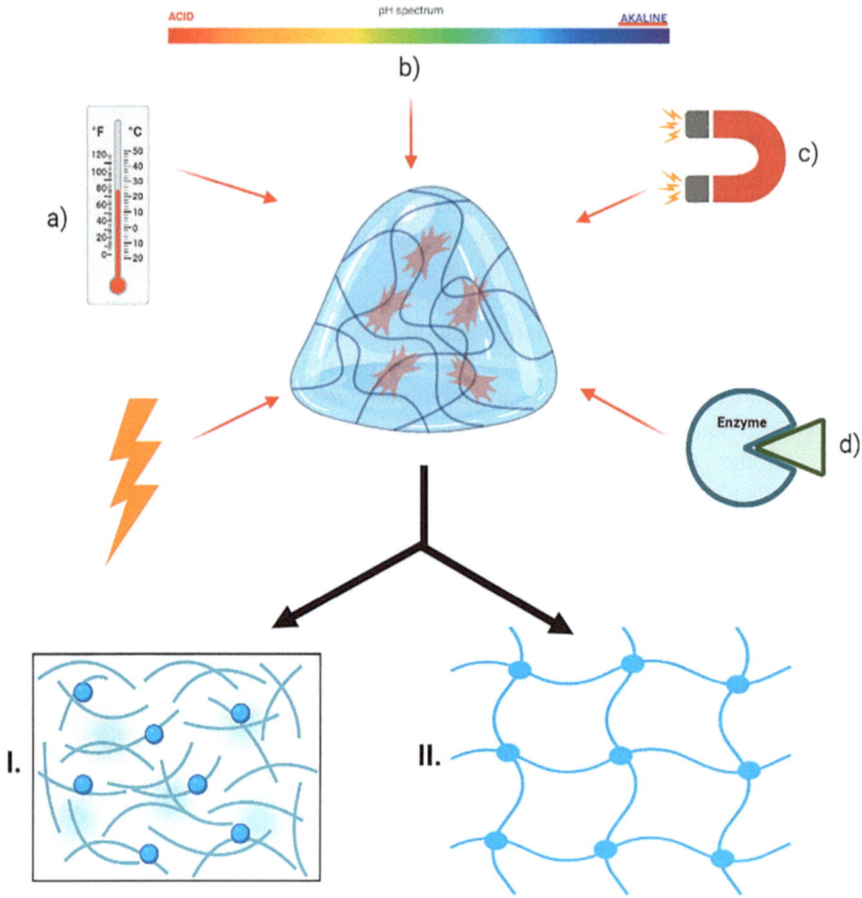

Fig. 30.2 Schematic of intelligent biomaterial response to external stimuli such that the biomaterial loaded with cells or drugs to be delivered is exposed to external stimuli changes in (**a**) temperature, (**b**) pH, (**c**) magnetic fields, (**d**) enzymatic activity, and (**e**) light that triggers (I) swelling and/or (II) degradation of the biomaterial

30.2.3 Polymers

Polymers are the most frequently used materials meant for biomedical applications. Polymers owe their extensive usage to their characteristics like high turnability, making them more flexible in modifying their properties to be potent enough to meet the requirements of strength, biocompatibility, or degradation, which happen to be requirements for their application. The most frequently studied polymers are polymeric hydrogels, known for advancement in drug delivery and tissue engineering. Polymeric hydrogels are generally developed either by polymerization of desirable functional monomers to synthesize highly tunable synthetic polymer systems or through cross-linking of naturally derived polymers which are potentially biocompatible and easily degradable (Clegg et al. 2017). In order to safeguard and support encapsulated cells, using hydrogels as an artificial extracellular matrix can be put into practice (Truong et al. 2015; Chan and Neufeld 2010). Also, for the protection of bioactive agents meant for drug delivery applications, nano- and micro-gel systems can be exploited (Xiong et al. 2011). Since these systems offer a variable range of responsive behaviors through the use of responsive monomers, these systems have been extensively studied. The most often used systems are pH-responsive systems comprising of monomers such as methacrylic acid (MAA) or diethylaminoethyl methacrylate, which, when prepared as a cross-linked hydrogel network, manifests pH-dependent swelling behavior.

The polymers, when in a charged state, face electrostatic repulsion from each other, which leads to the expansion of hydrogel. The expansion can be reversed by removing the charge on the interacting groups, which can be achieved by changing the pH conditions. This behavior can be altered by incorporating co-monomers containing hydrophobic groups, which are able to modify pKa of the ionizable groups (Liechty et al. 2013), or by changing the ionic strength of the solution, which in turn can change the strength of electrostatic interactions (Convertine et al. 2009; Sonawane et al. 2003). Thermo-responsive behavior can be exploited through temperature-dependent solubility of polymers such as poly (N-vinylcaprolactam) and poly(N-isopropylacrylamide) (PNIPAAm), which experience a hydrophilic-to-hydrophobic transition near body temperature that can be adjusted to lower or higher values by making use of co-monomers (Zhang et al. 2015). Such systems have been used for drug delivery applications, for increased temperature areas such as in tumors, etc. The backbone of pure synthetic polymers is usually comprised of polyvinyl, polyacrylate, or polyacrylamide and is essentially nondegradable under biological conditions. In order to incorporate the desired degradation properties to the hydrogel, cross-linking agents containing disulfide bonds or degradable polymers such as poly(lactic-co-glycolic acid) (PLGA) are often used (Tang et al. 2009; Lv et al. 2014).

The polymers derived from natural sources such as alginates, chitosan, and collagen have also been studied by researchers for their application in systems where biodegradability and nontoxicity are of utmost importance. Since they tend to be highly biocompatible and biodegradable simultaneously, they are used in tissue engineering applications (Sionkowska 2011). But manipulating their physical properties is not easily achieved as they are synthesized using natural biomaterials. Elasticity, strength, and degradation rate are some essential characteristics to be considered while designing a scaffold. Despite the disadvantages of natural biomaterials having unfavorable properties when used alone (Nicodemus and Bryant 2008), natural polymers have been considerably used for cell encapsulation (Chou and Nicoll 2009; Utech et al. 2015) and drug delivery. In order to overcome this deficiency and better control polymer properties, natural biomaterials are being used in conjunction with synthetic materials as polymers to exploit the properties of both materials. Natural biomaterials provide biocompatibility and synthetic polymers provide tunability (Patenaude and Hoare 2012).

30.2.4 Inorganic Materials

The issues like toxicity and bioaccumulation have always been a matter of concern, and in order to combat this, the focus has been laid on responsive inorganic biomaterials, mainly for disease detection and biomarker quantification applications. More recently, theranostic and drug delivery systems have also been used. The most commonly used inorganic materials for biomedical applications include gold nanomaterials and semiconductor quantum dots, mostly known for their unique optical properties as well as iron oxide nanoparticles (IONPs), which are used for magnetic interactions. These materials can easily be conjugated to bioactive molecules in order to be potent stimuli-responsive drug carriers for precision medicine as well as for sensing and targeting strategies.

Gold nanoparticles (GNPs) offer various advantages due to many of their flexible properties, such as optical properties, biocompatibility, and localization, which can be modified by changing particle shape and surface coating. Gold nanospheres and nanorods are the most common structures used for biomedical applications, owing their predominant use to their well-characterized synthesis. Some other structures like nano-shells, nano-stars, and nano-pyramids have also been studied. Gold nanoparticles are predominantly used because of their localized surface plasmon resonance (LSPR), which is caused by the collective oscillation of the electrons when exposed to light at a wavelength greater than the particle size. This leads to the absorption of light at plasmon resonance wavelength, characteristic of the size and shape of nanomaterials, and can be modified by altering the local refractive index, like mass changes on the surface of nanoparticles (Su et al. 2003; Shodeinde et al. 2020). Other metal nanoparticles and alloys, like silver and copper, offer similar effects as that of gold nanoparticles; they have been found to be of good use in LSPR (Sekhon and Verma 2011; Bansal et al. 2014). The optical properties of gold nanoparticles have made them potent for photothermal therapy within infrared wavelength windows and have been used for targeted destruction of tumors (Van de Broek et al. 2011; Dickerson et al. 2008; Owens III et al. 2007). Gold nanoparticles have also been used for sensitive biomarker detection and conjugating particular probes to their surface by altering LSPR wavelength upon direct antigen binding or upon subsequent aggregation (Zeng et al. 2011; Petryayeva and Krull 2011). Gold nanoparticles have also been found to be potent to be used for performing immunoassays (Kim et al. 2009; Cao et al. 2011) and used as drug carriers and are still believed to be a leading technology for medical devices in the future (Lasagna-Reeves et al. 2010; Chen et al. 2009; Alkilany and Murphy 2010; Alkilany et al. 2012).

30.2.5 Bioactive Molecules

Mostly stimuli-responsive materials are made synthetically; natural biomolecules have been found to be a basis for targeted drug delivery and biomarker recognition. These materials have been used to attain selective recognition or activation of biological pathways, subject to the particular interaction and functions of antibodies, peptides, enzymes, or small molecules (Venkatesh et al. 2005; Peppas and Kim 2006). It is difficult for artificial systems to achieve targeted interaction within the biological milieu. So, in order to accomplish this, biomolecule conjugates are dependent on the exploitation of the natural function of the extracted or genetically modified biomolecules.

In order to achieve selective and high-affinity binding, proteins are most commonly explored. The proteins are specific to bind to their specific antigen or substrate. The proteins owe their specificity to the combinatorial contribution of shape, charge, and degree of hydrophilicity of the binding site, in addition to changes in the conformation of proteins, enzymes, and antibodies. The high affinity of antibodies and enzymes to their specific targets has paved the way for them to be used for applications in biomedical engineering (Hu et al. 2014). To avoid the occurrence of false positives, biosensors demand for high selectivity for the marker of interest. This requirement of the biosensors is met by antibodies or enzymes which are frequently used as recognition elements to minimize nonspecific binding. Enzymatically active systems are being explored in both drug delivery through the addition of enzyme-labile groups like disulfides (Choh et al. 2011; McRae Page et al. 2013) and peptides in drug carriers (Foster et al. 2017; Rodell et al. 2015) and biosensing through glucose oxidase-induced pH changes from reaction with glucose (Unnikrishnan et al. 2013; Periasamy et al. 2011).

Bioactive molecules can also be used for drug delivery applications by conjugating them with a molecule, which is able to recognize specific ligands, or by conjugating them to a surface molecule so that they become easily detectable in the targeted biological milieu. Antibodies are believed to be versatile in order to achieve targeted delivery of therapeutic agents by an efficient delivery system where there is elevated antigen expression (Ulbrich et al. 2016). The targeted drug delivery has successfully been demonstrated by using antibodies against cell surface antigens such as vascular endothelial growth factor A (Jain et al. 2006; Ferrara et al. 2005), transferrin receptor (Tacchini et al. 2008; Tirosh et al. 2009), intercellular adhesion molecule-1 (Mane and Muro 2012; Higa et al. 2017), HER2/Neu, (Mahato et al. 2011; Tai et al. 2010), etc. The sugars, peptides, and other small molecules have been explored to facilitate cellular interactions at a greater level. For example, cell surface integrins like αυ-β3 are profusely expressed at the sites of rapid angiogenesis, which can be targeted by conjugating the Arg-Gly-Asp (RGD) peptide sequence to the surface of nanocarrier (Desgrosellier and Cheresh 2010). The receptors of transferrin and folate are targeted by their respective ligands, transferrin and folate (Assaraf et al. 2014; Daniels et al. 2006). Additionally, other molecules like biotin (vitamin B) or hormones like melanocyte-stimulating hormone have also been studied to be used as alternatives for site-specific drug delivery (Le Droumaguet et al. 2012; Vannucci et al. 2012).

30.2.6 Biomaterials for the Treatment of Autoimmune Diseases

Around the time a disease begins to develop, treatments are often initiated in order to prevent its further development. However, if a disease has already been established, intervention therapy evades its worsening. Even though prior to clinical diagnosis most of the common autoimmune disorders start many years or maybe decades before, vaccination becomes the best means to establish tolerance toward self-antigens, and efforts are in place to develop vaccines for autoimmune diseases (Correale et al. 2008; Nicholas et al. 2011; Rosenthal et al. 2015). In the next section, biodegradable materials which are used as delivery vehicles for drugs and vaccines in order to achieve their targeted delivery have been discussed (Hunter et al. 2014).

30.2.6.1 Rheumatoid Arthritis

Autoantigens often administered via the oral route is a routine practice to attain peripheral immune tolerance against autoimmune diseases such as rheumatoid arthritis (RA), diabetes, and experimental uveitis. The efficacy of oral tolerance relies upon various factors like nature, dose, frequency of administered antigen, the genetic background of the recipient, and the degree of antigen uptake by antigen-presenting cells of the gastrointestinal tract (GI). The quantity of antigen delivered to the mucosa is minimal due to the harsh environment of the GI tract. So, a highly governed release of antigens at the mucosal sites is required, and at the same time, protection of antigens from GI environment is to be ensured. The optimal level of peripheral immune tolerance is believed to be achieved by nano- and micron-sized particle systems. For instance, C-II was used to encapsulate poly(lactide-co-glycolide) (PLGA) nanoparticles which were synthesized by water in oil in water emulsion (w/o/w) by the solvent evaporation method. It was found that when mice were fed with a single dose of C-II-encapsulated PLGA particles, which contained 40 μg of C-II, they were found to have reduced clinical symptoms of arthritis and low level of anti-C-II antibodies in the serum (Kim et al. 2002; Lee et al. 2005). For rheumatoid arthritis, there are not many developed systems that have the specificity to deliver the drug at the site of interest, so a significant amount of drug is required, which may, later on, prove to be toxic. In order to find a balance between efficacy and minimal side effects, systems like liposomes, soft gels, microemulsions, nano-dispersions, topical formulations, and pellets have been developed. The liposome-based delivery system is the most explored, biocompatible, degradable, and immunologically inert and can encapsulate a significant amount of drugs (Kapoor et al. 2014) (Fig. 30.3).

Another strategy where polyethylene glycol (PEG) polymer is attached to drug/antigen has been explored (Fig. 30.4). This strategy enhances the therapeutic potential of antirheumatic drugs like TNF-related apoptosis-inducing ligands (TRAIL). Also, to improve the bioactivity of TRAIL, negatively charged hyaluronic acid (HA) was used to synthesize nano-sized complexes of PEGylated TRAIL protein and was found to show immense therapeutic potential in treating arthritis in mice, owing to the continuous release of TRAIL or good stability of proteins

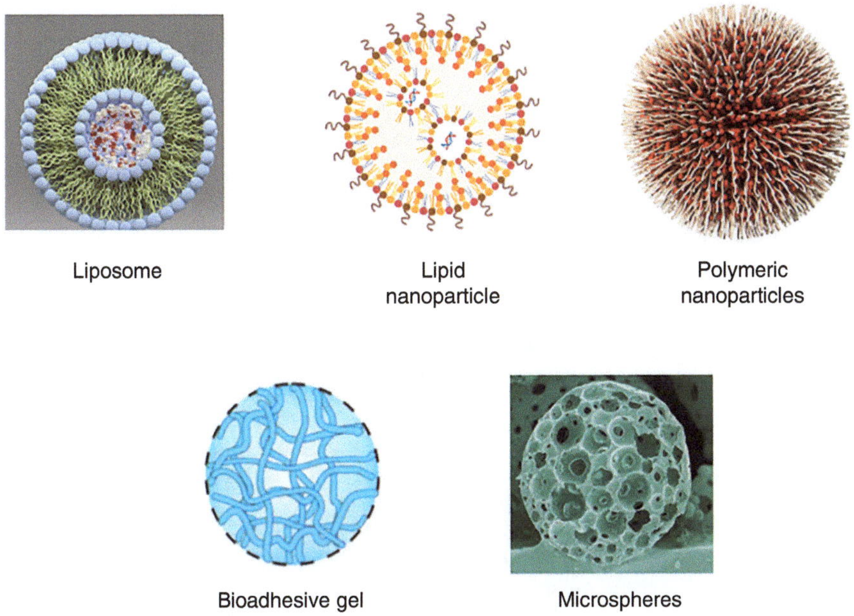

Fig. 30.3 Different vehicle systems used for delivery of antigens or drugs used in immunotherapy

under in vivo system (Kim et al. 2011; Jiang et al. 2011). PEGylation also improves the shelf life of proteins in vivo by minimizing their renal clearance. It also enhances drug solubility and stability and reduces immune reactions against therapeutic proteins by guarding proteins against water molecules.

30.2.6.2 Multiple Sclerosis

Demyelinating diseases like multiple sclerosis (MS) have always been challenging conditions to be treated. Transplantation of cells is believed to be efficient enough as a therapeutic approach to treat such diseases. Promising results have been shown by neuronal precursor cells in remyelination, which possess the ability to differentiate into neural cells. Also, various delivery systems for the efficient delivery of these precursor cells into the central nervous system have been explored. For example, three-dimensional nanofiber-based poly(glycolic-co-lactide)/chitosan (PLGA/CS) scaffold was able to induce the precursor cells (PC12) into neural cell differentiation in the presence of nerve growth factor (NGF) and basic fibroblast growth factor (bFGF). The cells of PC12 seeded on the PLGA/CS scaffold were found to show enhanced expression of nestin, β tubulin, and microtubule-associated proteins. The PC12-seeded scaffolds upon transplantation into lateral ventricles of EAE-affected mice led to a significant decrease in clinical signs of the disease (Hoveizi et al. 2015).

In another system, a colloidal gel containing an immune-modulating peptide was explored, with its release assisted via vaccines, for the treatment of EAE (experimental allergic encephalomyelitis). The colloidal gel from alginate-chitosan and

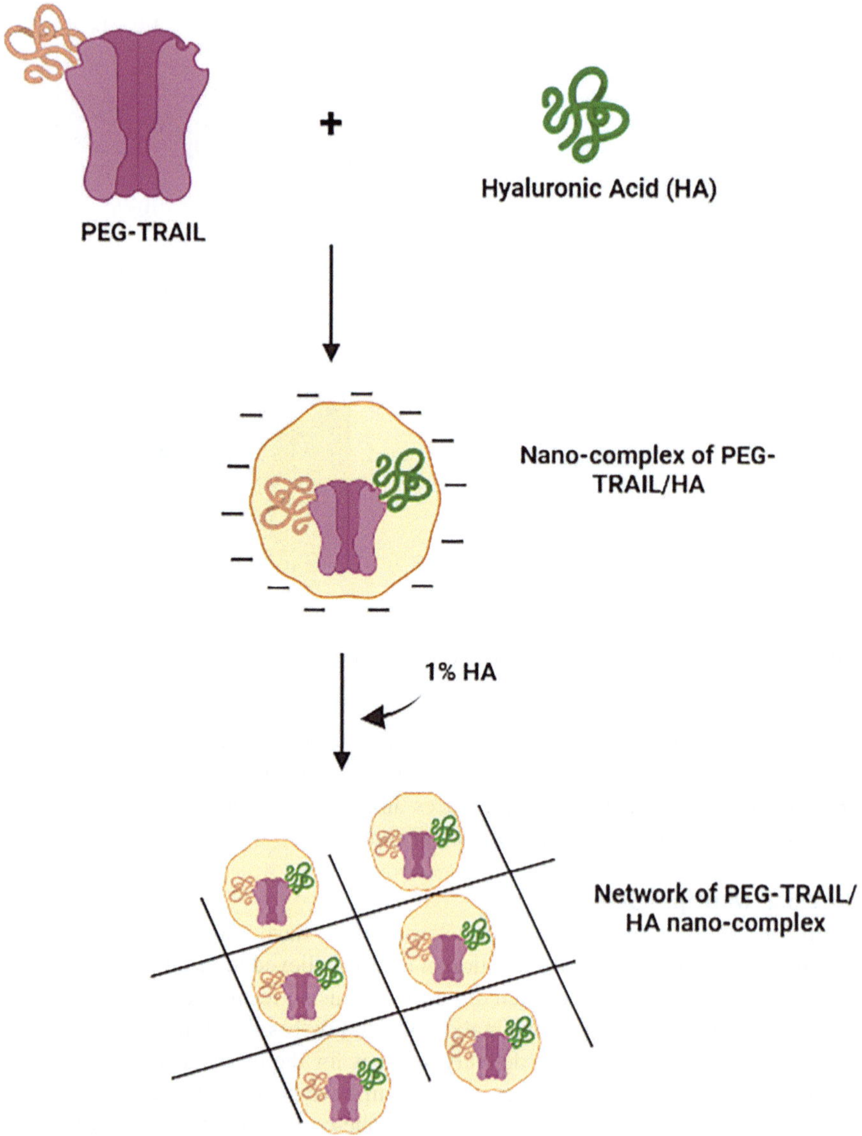

Fig. 30.4 Schematic representation of nano-complex release system formation through ionic interaction between PEG-TRAIL molecules and hyaluronic acid (HA). HA exists in anionic form in biological fluids which formed the complex with cationic charged PEG-TRAIL molecules. Created with Biorender

PLGA polymers delivered the Ac-PLP-BPI-NH2-2 peptide in a controlled manner. The same peptide was designed in such a way that it binds to the MHC II and intercellular adhesion molecule-I (ICAM-I) molecules simultaneously. The mice

Table 30.1 Delivery systems and administrative routes of antirheumatic drugs/biomolecules

Antirheumatic drug/ molecule	Delivery systems	Mode of delivery	References
Ketoprofen	Bioadhesive gels	Transdermal delivery	Singh et al. (2009)
	Eudragit-based microparticles	Oral delivery	El-Kamel et al. (2001)
Methotrexate	Polylactic acid-based microspheres	Intra-articular delivery	Liang et al. (2004)
	Solid-in-oil nanocarrier	Transdermal delivery	Yang et al. (2012)
Prednisolone	Cyclodextrins	Intravenous	Hwang et al. (2008)
Diclofenac	Solid lipid nanoparticles	Transdermal delivery	Liu et al. (2010)
Indomethacin	Liposomes	Oral delivery	Soehngen et al. (1988)
Prostaglandin E1	Lipid microspheres	Intravenous delivery	Moriuchi-Murakami et al. (2000)
Collagen type II	PLGA nanoparticles	Oral delivery	Kim et al. (2002)

having encephalomyelitis, when subjected to a colloidal gel-based vaccine injected subcutaneously, showed suppression and delay in the severity of EAE symptoms in comparison to the control group (Table 30.1).

Also, it was seen that there was downregulation in the expression of IL-6 and IL-7 in mice that were subjected to vaccination. This reinstates the fact that there occurs long-term suppression of the disease (Büyüktimkin et al. 2012; Wang et al. 2011). It has also been studied that micron-sized PLGA particles, when loaded with antigen, were found to deliver encephalitogenic peptides through intravenous infusion to circumvent initiation of the disease, hence, turning the clinical symptoms of EAE to a good side. Peptide-loaded micron-sized particles (MPs) were taken up by marginal zone macrophages expressing the MARCO receptors that modulate the activity of T-regulatory cell (Tregs) (Getts et al. 2012; Maldonado et al. 2015). Another study was carried out where it was demonstrated that PLGA nanoparticles were able to treat EAE in mice. Established EAE has been found to have been prevented as well as treated by nanoparticles coupled with myelin antigen (Yeste et al. 2012). In another study, it was found that a single dose of nanoparticles was enough to improve established EAE by activating Tregs (Hunter et al. 2014; Carambia et al. 2015).

30.3 Type 1 Diabetes

Type 1 diabetes (T1D) is a well-known autoimmune disorder wherein there occurs destruction of β-cells which are insulin-secreting cells, by autoreactive T-cells (Fig. 30.5). The condition where there is a deficiency of insulin secretion by β-cells causes hyperglycemia and overall abnormal metabolism of glucose (Nicholas

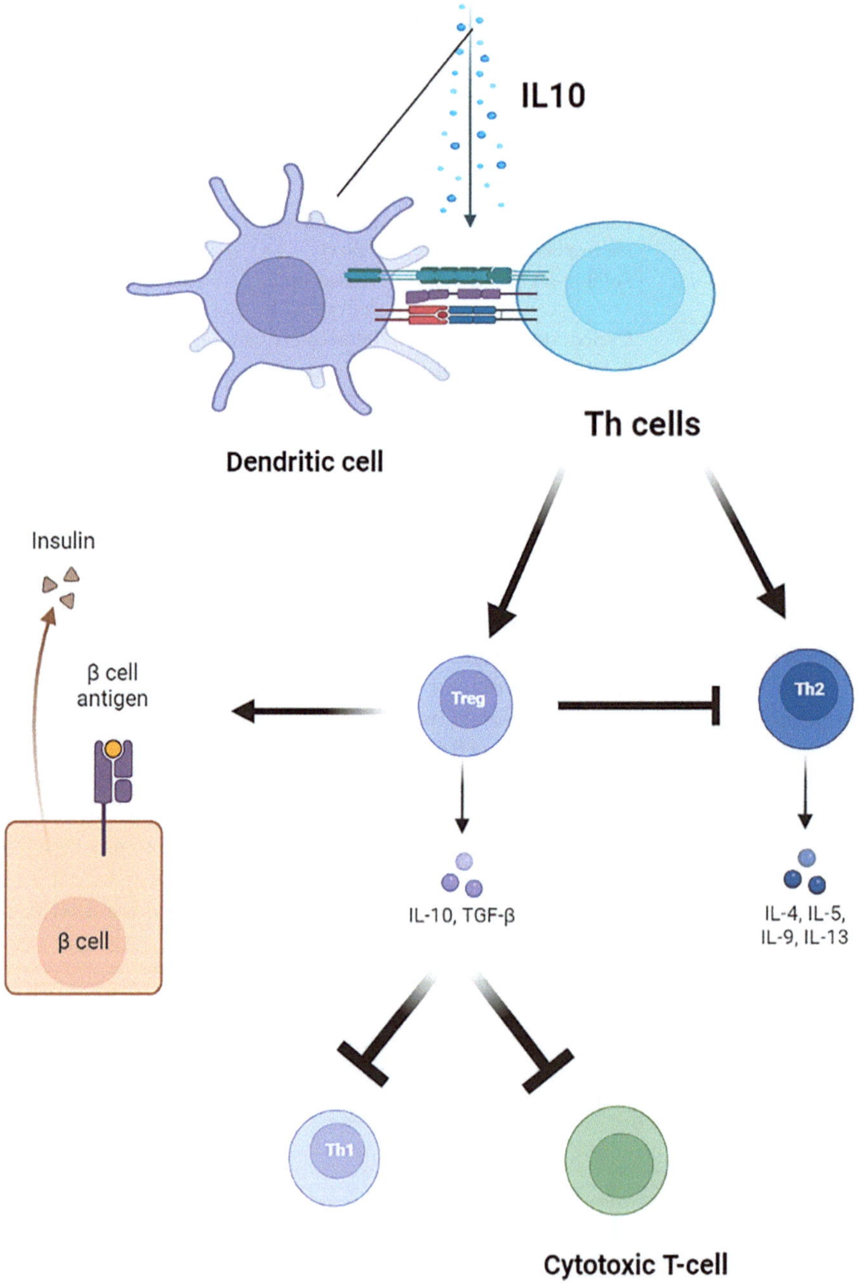

Fig. 30.5 Mechanism of autoantigen-specific immunotherapy. During antigen-specific immunotherapy, DCs process the antigen and present to naïve Th cells through secretion of IL-10 cytokine. In this context, activated Th cells activate the Tregs and Th2 cells, which block the proliferation of autoreactive Th1 and CTL cells through secretion of IL-10. Activation of Tregs leads to the prevention of beta cell death, and this way immunotolerance is achieved against the autoantigen. Created with Biorender

et al. 2011). The development of type 1 diabetes is mostly governed by both genetic and environmental factors, with autoreactive T-cells being the key player in the establishment of this disease (Van Belle et al. 2011; Bluestone et al. 2010; Lehuen et al. 2010). So, one of the various ways to treat type 1 diabetes is the immune modulation of autoreactive T-cell response. For the effective delivery of a drug/vaccine against T1D, a number of biomaterials have been explored. For example, PLGA polymeric nanoparticles were employed to encapsulate insulin as an autoantigen and CpG oligodeoxynucleotides as an adjuvant. To ensure the prolonged release of the antigen, nanoparticles loaded with antigen were embedded in PuraMatrix peptide hydrogel. PuraMatrix is a biodegradable, biocompatible, and self-assembling peptide-based material, which can be used to form a 3D matrix in the presence of physiologic medium with 5–200 nm pore size. The nonobese diabetic (NOD) mice, when treated with antigen-loaded matrix-based hydrogel subcutaneously, induced the formation of temporary granuloma, which had huge filtration of lymphocytes and macrophages. Therefore, this approach demonstrates a favorable immunotolerance model for treating type 1 diabetes (Yoon et al. 2015; Verbeke et al. 2017).

30.4 Conclusion

The rapidly evolving use of biomaterials to alter and control immune responses can potentially create therapeutic applications for autoimmune disorders. Nanoparticles have the potential to present antigens to dendritic cells and facilitate antigen-specific responses from the immune system with the aid of pathogen mimicking. The various parameters, such as selection of the material, geometry, structural properties, and surface topology, are necessary factors to be considered to optimize the immune response toward an antigen of interest. Hydrogels and scaffolds have been found to augment this immune response. The response of T-regulatory cells can be modulated, paving the way for its use as a treatment against autoimmune diseases. Despite the potential, the sphere of immunoengineering is unexplored mainly as the question of how the particles interact with immune cells is mechanistically unknown.

References

Agusti A et al (2003) Hypothesis: does COPD have an autoimmune component? BMJ Publishing Group Ltd.

Alijotas-Reig J (2015) Human adjuvant-related syndrome or autoimmune/inflammatory syndrome induced by adjuvants. Where have we come from? Where are we going? A proposal for new diagnostic criteria. Lupus 24(10):1012–1018

Alijotas-Reig J et al (2012) Autoimmune/inflammatory syndrome (ASIA) induced by biomaterials injection other than silicone medical grade. Lupus 21(12):1326–1334

Alkilany AM, Murphy CJ (2010) Toxicity and cellular uptake of gold nanoparticles: what we have learned so far? J Nanopart Res 12(7):2313–2333

Alkilany AM et al (2012) Gold nanorods: their potential for photothermal therapeutics and drug delivery, tempered by the complexity of their biological interactions. Adv Drug Deliv Rev 64(2):190–199

Andrews SF, Wilson PC (2010) The anergic B cell. Blood 115(24):4976–4978

Assaraf YG, Leamon CP, Reddy JA (2014) The folate receptor as a rational therapeutic target for personalized cancer treatment. Drug Resist Updat 17(4–6):89–95

Banchereau J et al (2000) Immunobiology of dendritic cells. Annu Rev Immunol 18(1):767–811

Bansal A, Sekhon JS, Verma S (2014) Scattering efficiency and LSPR tunability of bimetallic Ag, Au, and Cu nanoparticles. Plasmonics 9(1):143–150

Bluestone JA, Herold K, Eisenbarth G (2010) Genetics, pathogenesis and clinical interventions in type 1 diabetes. Nature 464(7293):1293–1300

Büyüktimkin B et al (2012) Vaccine-like controlled-release delivery of an immunomodulating peptide to treat experimental autoimmune encephalomyelitis. Mol Pharm 9(4):979–985

Cao X, Ye Y, Liu S (2011) Gold nanoparticle-based signal amplification for biosensing. Anal Biochem 417(1):1–16

Carambia A et al (2015) Nanoparticle-based autoantigen delivery to Treg-inducing liver sinusoidal endothelial cells enables control of autoimmunity in mice. J Hepatol 62(6):1349–1356

Castro JE et al (1996) Fas modulation of apoptosis during negative selection of thymocytes. Immunity 5(6):617–627

Chan AW, Neufeld RJ (2010) Tuneable semi-synthetic network alginate for absorptive encapsulation and controlled release of protein therapeutics. Biomaterials 31(34):9040–9047

Chen Y-S et al (2009) Assessment of the in vivo toxicity of gold nanoparticles. Nanoscale Res Lett 4(8):858–864

Choh S-Y, Cross D, Wang C (2011) Facile synthesis and characterization of disulfide-cross-linked hyaluronic acid hydrogels for protein delivery and cell encapsulation. Biomacromolecules 12(4):1126–1136

Chou AI, Nicoll SB (2009) Characterization of photocrosslinked alginate hydrogels for nucleus pulposus cell encapsulation. J Biomed Mater Res A 91(1):187–194

Clegg JR, Wechsler ME, Peppas NA (2017) Vision for functionally decorated and molecularly imprinted polymers in regenerative engineering. Regen Eng Transl Med 3(3):166–175

Convertine AJ et al (2009) Development of a novel endosomolytic diblock copolymer for siRNA delivery. J Control Release 133(3):221–229

Correale J, Farez M, Gilmore W (2008) Vaccines for Multiple sclerosis. CNS Drugs 22(3):175–198

Daniels TR et al (2006) The transferrin receptor part I: biology and targeting with cytotoxic antibodies for the treatment of cancer. Clin Immunol 121(2):144–158

Desgrosellier JS, Cheresh DA (2010) Integrins in cancer: biological implications and therapeutic opportunities. Nat Rev Cancer 10(1):9–22

Dickerson EB et al (2008) Gold nanorod assisted near-infrared plasmonic photothermal therapy (PPTT) of squamous cell carcinoma in mice. Cancer Lett 269(1):57–66

El-Kamel A et al (2001) Preparation and evaluation of ketoprofen floating oral delivery system. Int J Pharm 220(1–2):13–21

Ferrara N, Hillan KJ, Novotny W (2005) Bevacizumab (Avastin), a humanized anti-VEGF monoclonal antibody for cancer therapy. Biochem Biophys Res Commun 333(2):328–335

Forrester JV et al (2008) Immune privilege or privileged immunity? Mucosal Immunol 1(5): 372–381

Foster GA et al (2017) Protease-degradable microgels for protein delivery for vascularization. Biomaterials 113:170–175

França DCC, de Castro AL, Soubhia AMP (2013) Evaluation of the biocompatibility of silicone gel implants–histomorphometric study. Acta Inform Med 21(2):93

Gabriel SE et al (1994) Risk of connective-tissue diseases and other disorders after breast implantation. N Engl J Med 330(24):1697–1702

Getts DR et al (2012) Microparticles bearing encephalitogenic peptides induce T-cell tolerance and ameliorate experimental autoimmune encephalomyelitis. Nat Biotechnol 30(12):1217–1224

Goren I, Segal G, Shoenfeld Y (2015) Autoimmune/inflammatory syndrome induced by adjuvant (ASIA) evolution after silicone implants. Who is at risk? Clin Rheumatol 34(10):1661–1666
Hajdu SD, Agmon-Levin N, Shoenfeld Y (2011) Silicone and autoimmunity. Eur J Clin Investig 41(2):203–211
Higa LH et al (2017) Ultra-small solid archaeolipid nanoparticles for active targeting to macrophages of the inflamed mucosa. Nanomedicine 12(10):1165–1175
Hölmich LR et al (2007) Breast implant rupture and connective tissue disease: a review of the literature. Plast Reconstr Surg 120(7):62S–69S
Hoveizi E, Tavakol S, Ebrahimi-Barough S (2015) Neuroprotective effect of transplanted neural precursors embedded on PLA/CS scaffold in an animal model of multiple sclerosis. Mol Neurobiol 51(3):1334–1342
Hu Q, Katti PS, Gu Z (2014) Enzyme-responsive nanomaterials for controlled drug delivery. Nanoscale 6(21):12273–12286
Hunter Z et al (2014) A biodegradable nanoparticle platform for the induction of antigen-specific immune tolerance for treatment of autoimmune disease. ACS Nano 8(3):2148–2160
Hwang J et al (2008) α-Methylprednisolone conjugated cyclodextrin polymer-based nanoparticles for rheumatoid arthritis therapy. Int J Nanomedicine 3(3):359
Jain RK et al (2006) Lessons from phase III clinical trials on anti-VEGF therapy for cancer. Nat Clin Pract Oncol 3(1):24–40
Jiang HH et al (2011) PEGylated TNF-related apoptosis-inducing ligand (TRAIL) for effective tumor combination therapy. Biomaterials 32(33):8529–8537
Kapoor B et al (2014) Application of liposomes in treatment of rheumatoid arthritis: quo vadis. Sci World J 2014:978351
Kim WU et al (2002) Suppression of collagen-induced arthritis by single administration of poly (lactic-co-glycolic acid) nanoparticles entrapping type II collagen: a novel treatment strategy for induction of oral tolerance. Arthritis Rheum 46(4):1109–1120
Kim D, Daniel WL, Mirkin CA (2009) Microarray-based multiplexed scanometric immunoassay for protein cancer markers using gold nanoparticle probes. Anal Chem 81(21):9183–9187
Kim TH et al (2011) PEGylated TNF-related apoptosis-inducing ligand (TRAIL)-loaded sustained release PLGA microspheres for enhanced stability and antitumor activity. J Control Release 150(1):63–69
Kumagai Y et al (1984) Clinical spectrum of connective tissue disease after cosmetic surgery. Arthritis Rheum 27(1):1–12
Lasagna-Reeves C et al (2010) Bioaccumulation and toxicity of gold nanoparticles after repeated administration in mice. Biochem Biophys Res Commun 393(4):649–655
Le Droumaguet B et al (2012) Versatile and efficient targeting using a single nanoparticulate platform: application to cancer and Alzheimer's disease. ACS Nano 6(7):5866–5879
Lee W-K et al (2005) Preparation and characterization of biodegradable nanoparticles entrapping immunodominant peptide conjugated with PEG for oral tolerance induction. J Control Release 105(1–2):77–88
Lehuen A et al (2010) Immune cell crosstalk in type 1 diabetes. Nat Rev Immunol 10(7):501–513
Lewis JS, Roy K, Keselowsky BG (2014) Materials that harness and modulate the immune system. MRS Bull 39(1):25–34
Liang LS et al (2004) Methotrexate loaded poly (L-lactic acid) microspheres for intra-articular delivery of methotrexate to the joint. J Pharm Sci 93(4):943–956
Liechty WB, Scheuerle RL, Peppas NA (2013) Tunable, responsive nanogels containing t-butyl methacrylate and 2-(t-butylamino) ethyl methacrylate. Polymer 54(15):3784–3795
Liu D et al (2010) Solid lipid nanoparticles for transdermal delivery of diclofenac sodium: preparation, characterization and in vitro studies. J Microencapsul 27(8):726–734
Loyo E et al (2013) Autoimmunity in connection with a metal implant: a case of autoimmune/autoinflammatory syndrome induced by adjuvants. Autoimmun Highlights 4(1):33–38
Lv S et al (2014) Co-delivery of doxorubicin and paclitaxel by PEG-polypeptide nanovehicle for the treatment of non-small cell lung cancer. Biomaterials 35(23):6118–6129

Mahato R, Tai W, Cheng K (2011) Prodrugs for improving tumor targetability and efficiency. Adv Drug Deliv Rev 63(8):659–670
Maldonado RA et al (2015) Polymeric synthetic nanoparticles for the induction of antigen-specific immunological tolerance. Proc Natl Acad Sci 112(2):E156–E165
Mane V, Muro S (2012) Biodistribution and endocytosis of ICAM-1-targeting antibodies versus nanocarriers in the gastrointestinal tract in mice. Int J Nanomedicine 7:4223
McRae Page S et al (2013) Disulfide cross-linked phosphorylcholine micelles for triggered release of camptothecin. Mol Pharm 10(7):2684–2692
Moriuchi-Murakami E et al (2000) Treatment of established collagen induced arthritis with prostaglandin E1 incorporated in lipid microspheres. J Rheumatol 27(10):2389–2396
Nesher G et al (2015) Severe ASIA syndrome associated with lymph node, thoracic, and pulmonary silicone infiltration following breast implant rupture: experience with four cases. Lupus 24(4–5): 463–468
Nicholas D, Odumosu O, Langridge WH (2011) Autoantigen based vaccines for type 1 diabetes. Discov Med 11(59):293
Nicodemus GD, Bryant SJ (2008) Cell encapsulation in biodegradable hydrogels for tissue engineering applications. Tissue Eng Part B Rev 14(2):149–165
Nossal G (1994) Negative selection of lymphocytes. Cell 76(2):229–239
Owens DE III et al (2007) Temperature-responsive polymer–gold nanocomposites as intelligent therapeutic systems. J Biomed Mater Res A 83(3):692–695
Patenaude M, Hoare T (2012) Injectable, mixed natural-synthetic polymer hydrogels with modular properties. Biomacromolecules 13(2):369–378
Peppas N, Kim B (2006) Stimuli-sensitive protein delivery systems. J Drug Deliv Sci Technol 16(1):11–18
Periasamy AP, Chang Y-J, Chen S-M (2011) Amperometric glucose sensor based on glucose oxidase immobilized on gelatin-multiwalled carbon nanotube modified glassy carbon electrode. Bioelectrochemistry 80(2):114–120
Petryayeva E, Krull UJ (2011) Localized surface plasmon resonance: nanostructures, bioassays and biosensing—a review. Anal Chim Acta 706(1):8–24
Rodell CB et al (2015) Selective proteolytic degradation of guest–host assembled, injectable hyaluronic acid hydrogels. ACS Biomater Sci Eng 1(4):277–286
Rosenthal KS et al (2015) Rheumatoid arthritis vaccine therapies: perspectives and lessons from therapeutic ligand epitope antigen presentation system vaccines for models of rheumatoid arthritis. Expert Rev Vaccines 14(6):891–908
Sanchez-Guerrero J et al (1995) Silicone breast implants and the risk of connective-tissue diseases and symptoms. N Engl J Med 332(25):1666–1670
Sekhon JS, Verma S (2011) Refractive index sensitivity analysis of Ag, Au, and Cu nanoparticles. Plasmonics 6(2):311–317
Shodeinde AB et al (2020) Recent advances in smart biomaterials for the detection and treatment of autoimmune diseases. Adv Funct Mater 30(37):1909556
Singh S et al (2009) Enhanced transdermal delivery of ketoprofen from bioadhesive gels. Pak J Pharm Sci 22(2)
Singh M et al (2016) Silicone-induced granuloma after buttock augmentation. Plast Reconstr Surg Glob Open 4(2)
Sionkowska A (2011) Current research on the blends of natural and synthetic polymers as new biomaterials. Prog Polym Sci 36(9):1254–1276
Soehngen EC et al (1988) Encapsulation of indomethacin in liposomes provides protection against both gastric and intestinal ulceration when orally administered to rats. Arthritis Rheum 31(3): 414–422
Sonawane ND, Szoka FC Jr, Verkman A (2003) Chloride accumulation and swelling in endosomes enhances DNA transfer by polyamine-DNA polyplexes. J Biol Chem 278(45):44826–44831
Spiera R, Gibofsky A, Spiera H (1994) Silicone gel filled breast implants and connective tissue disease: an overview. J Rheumatol 21(2):239–245

Su K-H et al (2003) Interparticle coupling effects on plasmon resonances of nanogold particles. Nano Lett 3(8):1087–1090

Suda K et al (2016) Deterioration of autoimmune condition associated with repeated injection of dextranomer/hyaluronic acid copolymer: a case report. J Pediatr Surg Case Rep 4:10–12

Tacchini L et al (2008) Role of HIF-1 and NF-κB transcription factors in the modulation of transferrin receptor by inflammatory and anti-inflammatory signals. J Biol Chem 283(30): 20674–20686

Tai W, Mahato R, Cheng K (2010) The role of HER2 in cancer therapy and targeted drug delivery. J Control Release 146(3):264–275

Tang L-Y et al (2009) Shell-detachable micelles based on disulfide-linked block copolymer as potential carrier for intracellular drug delivery. Bioconjug Chem 20(6):1095–1099

Tirosh B et al (2009) Transferrin as a luminal target for negatively charged liposomes in the inflamed colonic mucosa. Mol Pharm 6(4):1083–1091

Truong VX et al (2015) Simultaneous orthogonal dual-click approach to tough, in-situ-forming hydrogels for cell encapsulation. J Am Chem Soc 137(4):1618–1622

Ulbrich K et al (2016) Targeted drug delivery with polymers and magnetic nanoparticles: covalent and noncovalent approaches, release control, and clinical studies. Chem Rev 116(9):5338–5431

Unnikrishnan B, Palanisamy S, Chen S-M (2013) A simple electrochemical approach to fabricate a glucose biosensor based on graphene–glucose oxidase biocomposite. Biosens Bioelectron 39(1):70–75

Utech S et al (2015) Supplementary data for article: Utech, S.; Prodanovic, R.; Mao, AS; Ostafe, R.; Mooney, DJ; Weitz, DA Microfluidic Generation of Monodisperse, Structurally Homogeneous Alginate Microgels for Cell Encapsulation and 3D Cell Culture. Adv Healthc Mater 4(11): 1628–1633. https://doi.org/10.1002/adhm.201500021. Advanced Healthcare Materials, 2015

Van Belle TL, Coppieters KT, Von Herrath MG (2011) Type 1 diabetes: etiology, immunology, and therapeutic strategies. Physiol Rev 91(1):79–118

Van de Broek B et al (2011) Specific cell targeting with nanobody conjugated branched gold nanoparticles for photothermal therapy. ACS Nano 5(6):4319–4328

Van Nunen SA, Gatenby PA, Basten A (1982) Post-mammoplasty connective tissue disease. Arthritis Rheum 25(6):694–697

Vannucci L et al (2012) Selective targeting of melanoma by PEG-masked protein-based multifunctional nanoparticles. Int J Nanomedicine 7:1489

Venkatesh S et al (2005) Applications of biomimetic systems in drug delivery. Expert Opin Drug Deliv 2(6):1085–1096

Verbeke CS et al (2017) Multicomponent Injectable Hydrogels for Antigen-Specific Tolerogenic Immune Modulation. Adv Healthc Mater 6(6):1600773

von Boehmer H (1990) Developmental biology of T cells in T cell-receptor transgenic mice. Annu Rev Immunol 8(1):531–556

Wang Q et al (2011) PLGA-chitosan/PLGA-alginate nanoparticle blends as biodegradable colloidal gels for seeding human umbilical cord mesenchymal stem cells. J Biomed Mater Res A 96(3): 520–527

Xiong W et al (2011) Dual temperature/pH-sensitive drug delivery of poly (N-isopropylacrylamide-co-acrylic acid) nanogels conjugated with doxorubicin for potential application in tumor hyperthermia therapy. Colloids Surf B: Biointerfaces 84(2):447–453

Yang F, Kamiya N, Goto M (2012) Transdermal delivery of the anti-rheumatic agent methotrexate using a solid-in-oil nanocarrier. Eur J Pharm Biopharm 82(1):158–163

Yarkoni Y, Getahun A, Cambier JC (2010) Molecular underpinning of B-cell anergy. Immunol Rev 237(1):249–263

Yeste A et al (2012) Nanoparticle-mediated codelivery of myelin antigen and a tolerogenic small molecule suppresses experimental autoimmune encephalomyelitis. Proc Natl Acad Sci 109(28): 11270–11275

Yoon YM et al (2015) A combination hydrogel microparticle-based vaccine prevents type 1 diabetes in non-obese diabetic mice. Sci Rep 5(1):1–13

Zeng S et al (2011) A review on functionalized gold nanoparticles for biosensing applications. Plasmonics 6(3):491–506

Zhang H et al (2015) A new comonomer design for enhancing the pH-triggered LCST shift of thermosensitive polymers. Polym Chem 6(37):6644–6650

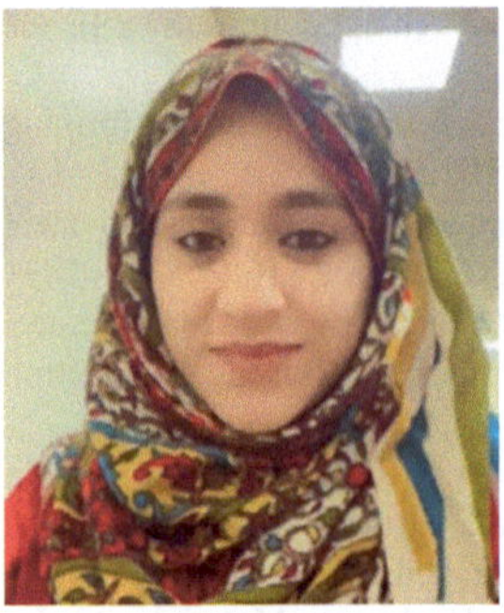

Iqra Jan was born in J & K, India. She completed her B.Sc. in 2016 from the University of Kashmir, India, and later joined her M.Sc. Biotechnology and completed the degree in 2018 from the University of Kashmir, India. She qualified for the GATE in 2020. She is currently enrolled in a doctoral program in the Department of Biotechnology, University of Kashmir, India, since 2020 under the supervision of Dr. Mushtaq A. Beigh. Her research interests include studying the role of post-translational modifications in cancer metastasis and exploiting these pathways in drug therapeutics.

Asma Rafiq received her B.Sc. in 2016 from the University of Kashmir, India, and later joined M.Sc. Biochemistry. She completed her master's program in Biochemistry at the University of Kashmir, India, in 2018. She qualified for GATE in 2020 and is currently pursuing her PhD. in the Department of Nanotechnology at the University of Kashmir, India. She has authored a couple of publications appearing in reputed journals. Her research interests include understanding the regulatory pathways in cancer and how these pathways help cancer cells avoid destruction by the immune attack, so these pathways can be exploited for therapeutic intervention in cancer treatment.

Mehvish Ali completed her M.Sc. in Nanotechnology from the University of Kashmir in 2021, and her reserach interests include targeted therapeutics and cancer metastasis.

Dr. Sabreena Aashaq was born in J & K, India. She completed her B.Sc. in 2011 from the University of Kashmir, India, and later joined M.Sc. Biotechnology and completed the degree in 2013 from the University of Kashmir, India. She completed her doctoral program in Biotechnology from University of Kashmir, India, in 2018 and is currently working as a Research associate in the Department of Immunology and Molecular Medicine, Sher-i-Kashmir Institute of Medical Sciences, India. She has qualified for various national-level exams, including CSIR UGC NET and JRF, GATE, and ASRB-NET, and has been the recipient of various national-level fellowships. She has contributed extensively to world of literature, with publications appearing in reputed journals. Her research interests include the regulation of various signal transduction pathways in cells, how dysregulation of these pathways result in different cancers, and the exploitation of these pathways in cancer therapeutics.

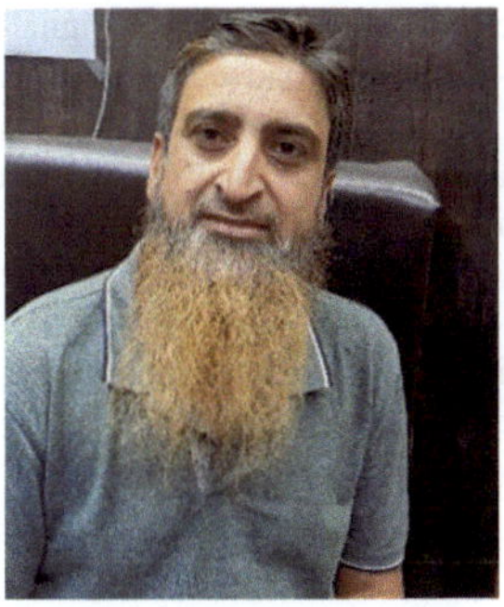

Bilal Ahmed Reshi has been working as an Assistant Professor in the Department of Biotechnology, University of Kashmir, since 2005. He has done his Mastes (2001) and M.Phil. (2005) in Biochemistry from University of Kashmir. He has qualified for UGC NET (2001) and is currently pursuing PhD. (part-time) in the Department of Biotechnology in 2019. His area of specialization is plant science (exploring the potential of plant compounds as therapeutics). He has 17 years of teaching experience in the field.

Dr. Faheem A. Sheikh is an assistant professor at the Department of Nanotechnology, University of Kashmir, India (2015). He served as an assistant professor at the Department of Biotechnology at the Central University of Kashmir, India (2015–2016); a research professor at Myongji University, South Korea (2014–2015); assistant professor at Hallym University, South Korea (2012–2014); post-doctoral/research fellow at the University of Texas Rio Grande Valley, Texas, United States of America (2010–2012); and research professor at Myongji University, South Korea (2010). His research mainly focuses on fabricating nanomaterials used in tissue engineering.

Dr. Shafquat Majeed is currently working as an assistant professor in the Department of Nanotechnology, University of Kashmir, India. Dr. Majeed did his masters in Chemistry from the Department of Chemistry, University of Kashmir. He did his PhD at the prestigious Indian Institute of Science, Bangalore (2008–2015). He has a post-doctoral research experience in rare-earth-based nanomaterials for biomedical applications and fabrication of perovskite-based solar cells from the Centre for NanoScience and Engineering, IISc, Bangalore (2015–2016). He heads the laboratory, where they work on the synthesis, characterization, and optical and magnetic properties of nanomaterials prepared through various solution-based routes. More specifically, they are working on developing new and enhanced multi-modal imaging modalities for biomedical and healthcare applications.

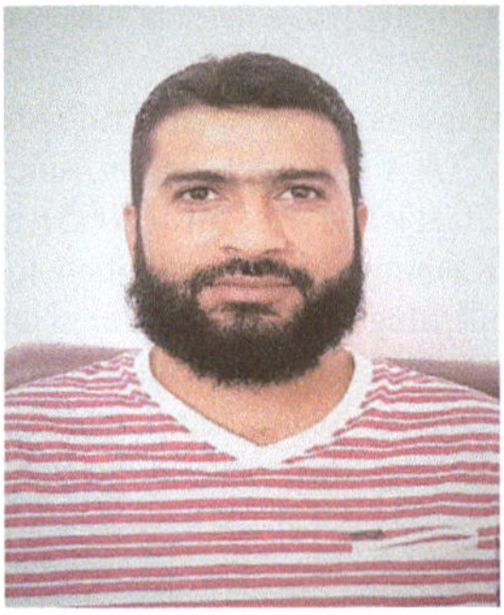

Dr. Mushtaq A. Beigh is working as an assistant professor at the Department of Nanotechnology, University of Kashmir, India. Dr. Beigh has post-doctoral research experience in cellular signaling and disease modelling (2013–2016). He had completed his PhD at the University of Kashmir, India (2008–2012). He heads a laboratory at the University of Kashmir, where they work on a bio-nano interface to understand the molecular basis of nanoconjugate internalization. They also try to understand the growth factor signaling inputs directed at multiple cellular receptors like NRP1 and integrins in order to work out the mechanistic details of receptor-based internalization.

31 Regulatory and Ethical Issues Raised by the Utilization of Nanomaterials

Faamiya Shajar, Seerat Saleem, Naveed Ul Mushtaq, Wasifa Hafiz Shah, Aadil Rasool, Sajad Ahmed Padder, Inayatullah Tahir, and Reiaz Ul Rehman

Abstract

Nanotechnology has emerged as an area constantly expanding, affecting practically every sphere of science and daily life. It offers new product possibilities and provides promising solutions to various problems in areas including pharmaceutics, agri-nanotechnology, electronics, the food industry, environmental science, etc. Nanotechnology is believed to be a double-edged sword because it brings in new possibilities for development. Still, at the same time, it poses some severe concerns to human health and the environment. Since this field is still in its infancy, an accurate assessment of risks posed by nanotechnology is complex. Currently, the major risk associated with using nanomaterials is the lack of control over it; there is no adequate legislation to regulate nanomaterials. Some countries have established standards and regulations; however, they are not free of loopholes. This chapter discusses significant application areas of nanomaterials, their potential risks, current law employed in different countries, main problems encountered in regulation, and some aspects of ethics.

Keywords

Nanotechnology · Nanomaterials · Nanoregulation · Ethical issues

F. Shajar · S. Saleem · N. U. Mushtaq · W. H. Shah · A. Rasool · S. A. Padder · I. Tahir · R. U. Rehman (✉)
Department of Bioresources, School of Biological Sciences, University of Kashmir, Srinagar, India

F. A. Sheikh et al. (eds.), *Interaction of Nanomaterials With Living Cells*,
https://doi.org/10.1007/978-981-99-2119-5_31

31.1 Introduction

31.1.1 Nanomaterials: An Overview

As we know, nanomaterials (NMs) are a group of materials having at least one dimension in the range of 1–100 nm (Laurent et al. 2008). Even though nanomaterials were used centuries ago, major concepts, advances, and understandings only emerged during the middle of the twentieth century (Vajtai 2013). Nano-sized particles exist in nature and can also be created from various products, such as carbon or minerals like silver. Depending on their properties, morphology, size, and origin, nanomaterials can be classified in a variety of categories (Fig. 31.1). Below are some of the most well-known nanoparticle (NP) classes based on their physical and chemical characteristics.

31.1.2 Carbon-Based NPs

Various applications have been developed to employ carbon-based nanomaterials due to their exceptional properties, such as electrical conductivity and thermal conductivity, high strength, structure, electron affinity, and versatility (Astefanei et al. 2015). Fullerenes, carbon quantum dots, and carbon nanotubes are some of the critical examples belonging to this class.

Fullerenes. Fullerenes are a class of carbon allotropes consisting of 60 C atoms connected by single and double bonds to form a closed or partially closed mesh forming a hollow sphere, ellipsoid, or a tube. Because of their excellent biocompatibility, fullerenes can be potentially used as drug carriers for targeted drug delivery. The use of fullerenes as possible adsorbents for organic compounds has also been investigated (Nimibofa et al. 2018).

Graphene. Graphene is a two-dimensional monolayer of carbon atoms tightly arranged in a hexagonal network of honeycomb lattices (Salem et al. 2022). Due to its extraordinarily high mechanical strength, outstanding electrical conductivity, high optical transparency, and favorable biocompatibility, graphene has been widely regarded as a promising candidate for industrial and biomedical applications (Liao et al. 2018).

Carbon nanotubes (CNTs). The internal diameter of CNTs is in the range of nanometers, while their lengths are in micrometers. There are two types of CNTs, i.e., single-walled and multiwalled. Single-walled CNTs (SWCNTs) are formed by rolling up a single graphene sheet into a tube, whereas multiwalled CNTs (MWCNTs) are formed by rolling multiple concentric graphene sheets into cylindrical shapes held together via van der Waals force. These sheets are spaced at a distance of 0.34 nm (Sajid 2021).

CNTs have emerged as a promising class of field emitters, mechanical reinforcements in high-performance composites, nanoprobes in metrology and biological and chemical investigations, and templates for creating other nanostructures.

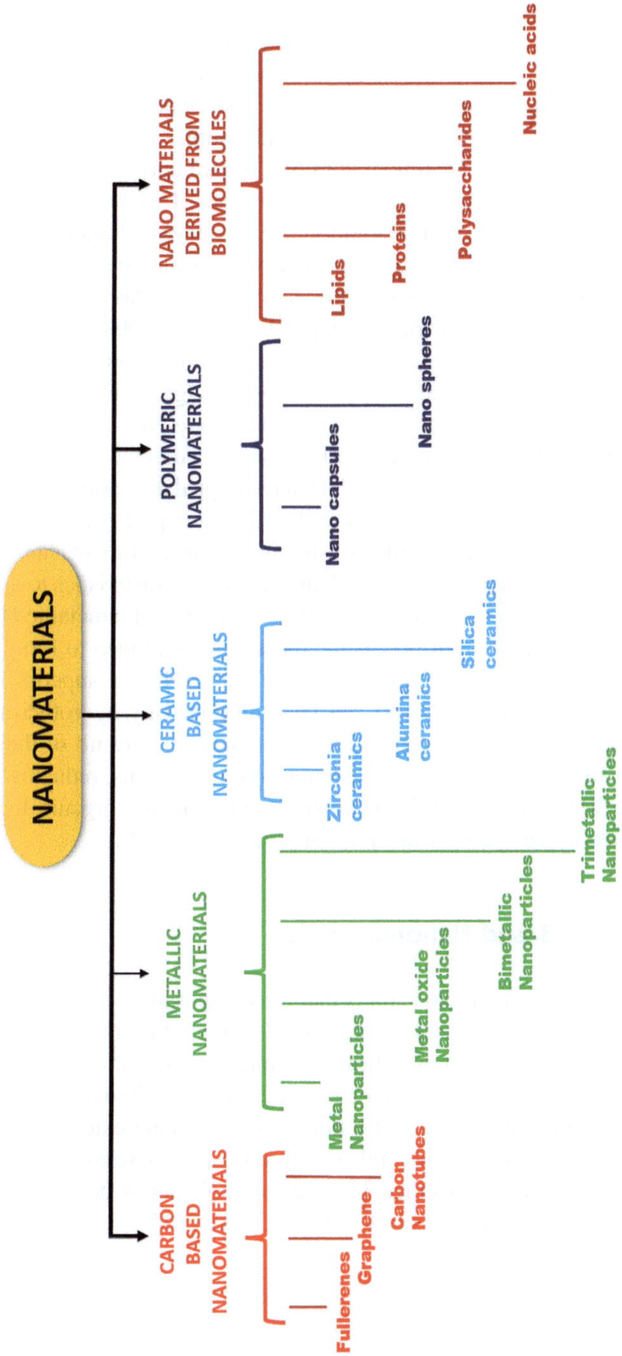

Fig. 31.1 Types of nanomaterials

31.1.3 Metallic Nanomaterials

These nanomaterials are derived from metal precursors. In general, metallic NPs can be classified as metal NPs, metal oxide NPs, bimetallic NPs, and trimetallic NPs. Metal NPs, specifically gold NPs, have vast importance in optical sensors for imaging purposes (Sajid 2021). Gold NPs find another use in scanning electron microscopy (SEM), where a coating of gold NPs is used during the sampling stage to create a 1–2 nm conductive layer on the sample surface. This enhances the electronic stream, thereby producing high-resolution SEM images. Various metallic NPs exhibit antibacterial properties and are currently being investigated as alternatives for antibiotics, given the increasing antibiotic resistance (Slavin et al. 2017). Gold NPs also find use in targeted cancer therapy (Patra et al. 2008). Metal oxides like Fe_2O_3, Al_2O_3, TiO_2, SiO_2, CuO, and SnO_2 are frequently used in several research fields due to their small size and high density of edge, which gives them their distinctive physicochemical characteristics. Metal oxide NPs are particularly interesting for optoelectronic applications, antimicrobial/medicinal, sensing, and energy storage (Niederberger 2007). Bimetallic NPs have piqued the curiosity of researchers worldwide. As the name suggests, these NPs are made up of two metals, and their characteristics are determined by the constituent metals and their nanometric size. The bimetallization of NPs can enhance their catalytic characteristics to a significant extent than monometallic catalysts (Sajid 2021). Gold-based bimetallic NPs can act as efficient catalysts and biosensors. Three metals are combined to form trimetallic NPs. The electrochemical catalytic activity of these NPs is superior to that of monometallic or bimetallic NPs and can be synthesized by hydrothermal, microwave, microemulsion, and coprecipitation techniques. As a result of their unstable surface area, trimetallic NPs readily precipitate from solutions, reducing their catalytic activity. Some of the stabilizers of these NPs include organic ligand block copolymers, dendrimers, and surfactants (Sharma et al. 2017).

31.1.4 Ceramic-Based Nanomaterials

Ceramic-based NPs are inorganic solids that are composed of carbonates of metals/metalloids, carbides, oxides, and phosphates. Ceramic NPs made from silica and alumina have a wide range of applications due to their remarkable properties, such as high heat resistance and chemical inertness. Since ceramic NPs have controlled sizes and are biocompatible with cells and tissues, they are being extensively studied as possible carriers of drugs, proteins, and enzymes. Despite these advantages, there are a few concerns with respect to their biomedical applications because of their low biodegradability and potential toxicity (Thomas et al. 2015).

31.1.5 Polymeric Nanomaterials

Polymeric nanomaterials are organic, having a particle size of 1–1000 nm in one dimension. Like ceramic-based nanomaterials, polymeric nanomaterials have also shown an excellent potential for the targeted delivery of drugs. Due to their surface area-to-volume ratio, shelf life, solubility, and nontoxic and stable nature, they are outstanding drug carriers.

Polymeric NPs as carriers can carry vaccination antigens, proteins, and medicines to the targeted site of action (Han et al. 2018). Besides drug delivery, separation, catalysis, nanoreactors, and surface coatings are some of the key applications. It has been suggested that polymer nanocomposite materials could be used in environmental remediation and contaminant treatment (Khin et al. 2012). Polymeric nanomaterials used for drug delivery can be classified into nanocapsules and nanospheres. The drug of interest is put into a cavity enclosed by a polymer membrane in nanocapsules, whereas the drug is physically and evenly incorporated in the matrix in nanospheres (Sajid 2021).

31.1.6 Biomolecule-Derived Nanomaterials

Biomolecules have been successfully used in the preparation of nanomaterials. Biomolecule-derived nanomaterials have gathered significant attention over the past few decades because of their potential biomedical applications in tissue engineering and analytical science (Wang et al. 2019b). These materials can also form hybrid nanomaterials, which combine the properties of both inorganic and organic components (Taylor-Pashow et al. 2010).

31.2 Applications

Because of their extraordinary properties, nanomaterials offer a wide range of applications that have the potential to not only improve but also revolutionize a multitude of sectors, including pharmaceutics, food technology, and environmental science (Chrimes et al. 2012). Some of the important applications have been discussed below.

31.3 Nanomaterials and the Environment

- **Environmental monitoring and remediation**
  One of the major challenges the world is currently dealing with is environmental pollution. It is becoming worse each year and causing irreparable damage. Thus, there is a need to develop sensitive, reliable, and affordable monitoring programs that can effectively characterize the quality of the environment and quantify the impact of various anthropogenic activities on our ecosystem. Environmental

monitoring is crucial to protecting human health and the environment in general. It helps us determine if the quality of the environment is improving or deteriorating (Hashem et al. 2021).

Significant progress has been made in the development of effective biosensor platforms based on nanomaterials such as metal NPs, metal oxide nanomaterials, carbon nanomaterials, and biomaterials (Maduraiveeran and Jin 2017). Biosensors that can be used for the detection and monitoring of various environmental pollutants, including pesticides, have been devised (Zhang et al. 2014).

In addition to environmental monitoring, environmental remediation must also be prioritized. The limitations of conventional methods used for environmental remediation have prompted the researchers to explore other alternatives to combat environmental challenges. Because of their novel physicochemical properties, NPs are constantly being investigated as potential remediating tools.

Environmental remediation technologies utilize physical, chemical, and biological processes for the removal of contaminants (e.g., heavy metals, dyes, herbicides, volatile organic compounds, etc.) from the environment. Nanomaterials are particularly suitable for such processes due to their increased reactivity, high surface area-to-volume ratio, porosity, etc. (Guerra et al. 2018). Besides, nanomaterials can make use of unique surface chemistry, which enables them to be functionalized or grafted with functional groups that can target and recognize specific pollutants from a mixture for effective remediation. Furthermore, the effectiveness of the nanomaterials for remediation can be significantly improved by deliberately tuning the physical properties of the nanomaterials (such as size, shape, porosity, and chemical composition). A variety of materials in their nano form, like iron, titanium dioxide, silica, zinc oxide, carbon nanotube, dendrimers, polymers, etc., are increasingly being used to purify air and water and to decontaminate the soil. After they have been used, it is crucial to monitor that the materials used for remediation do not become new sources of pollution themselves.

- **Agriculture**

  Currently, global agricultural systems are facing several unprecedented challenges. The ever-increasing population has led to an increase in demand for food and other resources around the globe. To deal with issues related to food security, a lot of chemical fertilizers have been used in agricultural fields. However, the unreasonable application of agrochemicals has severely affected the environment and human health. Therefore, there is a need to adopt alternate, sustainable approaches that can improve agricultural productivity, while ensuring environmental safety and stability (Pande and Arora 2019). Nanotechnology presents enormous opportunities to revolutionize modern agriculture by significantly improving plant nutrition, soil health, water management, and abiotic stress tolerance (Islam 2019). Several nanomaterials have been identified for use in agriculture in order to help reduce the consumption of agrochemicals by use of intelligent delivery systems, minimize nutrient losses, and boost crop yield through optimized water and nutrient management (Dwivedi et al. 2016).

Extensive research in the field of agricultural nanotechnology is being conducted in an effort to develop agriproducts like nanobiosensors, nano-fertilizers, nano-pesticides, nano-fertigation products, etc. (Dwivedi et al. 2016). Nanobiosensors are comparatively more sensitive than conventional biosensors and can efficiently sense soil pH, moisture, disease-causing pathogens, and residues of agrochemicals (Ghaffar et al. 2020). In addition, the use of a NP-based smart delivery system holds the potential for site-targeted administration of different agrochemicals required for improved disease resistance and effective nutrient utilization in an environmentally benign manner (Panpatte et al. 2016). Additionally, some studies found that certain carbon-based nanomaterials when used at a lower concentration can increase the vegetative growth and yield of fruit/seed (Verma et al. 2019). Despite such promising benefits, agri-nanotechnology is still in the research and development stage, and the prospective applications of nanotechnology in agriculture have not yet been fully realized. This is due to possible threats to consumer health and the lack of proper guidelines on the risk assessment of nanotechnology.

31.4 Nanotechnology in Biomedical Sciences

Due to their minimal size, extremely high surface area-to-volume ratio, tunable optical emission, improved mechanical qualities, and superparamagnetic properties, NPs have grown in popularity in biomedical applications. Currently, cutting-edge research is being carried out in an attempt to exploit these remarkable properties of nanomaterials in order to apply them in the field of biomedical sciences (Das et al. 2013). The unique features of NPs can be exploited for the detection, prevention, and treatment of diseases, drug delivery, and gene therapy of cancer (Ahmed et al. 2021). Most research in the field of nanomedicine is mainly focused on the applications of nanomaterials in drug delivery systems (DDS). Nanomaterials have been incorporated into a drug formulation, such as active constituents (nanocrystals), excipients (drug-metal complexes), drug carriers (liposomes), or complexes/conjugates (drug-protein) (D'Mello et al. 2017). Liposomal drug formulations have been successfully used for the treatment of breast and ovarian cancers and Kaposi's sarcoma. When combined with anticancer medications like amphotericin and hamycin, liposomal drug formulations demonstrated significantly greater efficacy and safety than traditional formulations (Tyner et al. 2017). Besides, some NPs have been employed for the identification and diagnosis of cancer cells due to their high sensitivity and multi-contrast imaging capabilities (Morrow Jr et al. 2007). Advances in nanotechnology have hinted at their potential use in surgical oncology (Singhal et al. 2010) for minimally invasive, precision, and intelligent oncological surgery (Wang et al. 2019a). Additionally, the development of surgical instruments, suture materials, targeted drug therapy, visualization methods, and wound healing techniques is one of the clinically relevant applications of nanomaterials (Mariappan 2019). Currently, nanotechnology in biomedical sciences still has quite a few roadblocks to overcome. Further research is required to get an insight into the long-term effects of nanotechnology on the environment. Authorities need to establish more precise regulations on the potential health risks of nanotech-based devices.

31.5 Nanomaterials and Food Science

Nanotechnology has the potential to impact various aspects of the food system, viz., processing, packaging, storage, transportation, and shelf life (Sharma and Hussain 2020). With nanotechnological interventions, competent delivery of nutrients, nanoencapsulation of nutraceuticals, bioseparation of proteins, rapid sampling of biological and chemical contaminants, solubilization, delivery, and color in food systems can be significantly enhanced (Ravichandran 2010). The development of interactive or functional foods that adapt to the needs of the consumer can be aided by nanotechnology. These days, researchers are also focused on developing new on-demand foods which allow consumers to alter the food as per their nutritional requirements. The idea is that the nanocapsules containing flavor enhancers or nutrients like vitamins would remain dormant in the food and will be released only when triggered by the consumer (Ameta et al. 2020). NPs can also be incorporated into existing foods for increased nutrient absorption.

In recent years, workers have been keen on combining nanotechnology with biosensing techniques to develop nanosensors. These nanosensors are diagnostic tools that check the safety and efficacy of food production operations. They can detect even minute quantities of pathogenic bacteria, viruses, toxins, or chemical pollutants in food systems (Otles and Yalcin 2010). Food degradation is one of the most frequently debated topics concerning the production, processing, transportation, and storage of food. Some NPs exhibit antimicrobial properties, which make them capable of preventing food deterioration (Mitura and Zarzycki 2018). It has been found that a variety of metal and metal oxide NPs work well as antimicrobials. These NPs are thought to form reactive oxygen species (ROS), which causes oxidative stress resulting in cell damage. Food is a perishable item and can be easily contaminated or degraded at any stage of the food chain. Therefore, it is necessary to use high-quality packaging that is safe, nontoxic, and economical. Nanomaterial-based packaging controls the pH, temperature, moisture, and freshness of the food material contained inside the package. Along with managing the atmosphere to increase the shelf life, it also includes consumer information (Ameta et al. 2020). The applicability of nanomaterials in the field of food science is expected to increase in the coming years. However, its success will depend on consumer acceptance and a uniform international regulatory framework.

31.6 Potential Risks and Hazards

The remarkable achievements of nanotechnology cannot be denied. Owing to their unique properties, nanomaterials have diverse applications in various sectors like pharmaceutics, agriculture, food sciences, etc. Nanomaterials have the potential to lower waste generation by reducing the use of nonrenewable resources of energy. In the sphere of medical research, tremendous progress has been made; small sensors or diagnostic probes, as well as entire testing systems, are now being built and implanted for diagnostic purposes. Nanotechnology is thus emerging as a multidisciplinary science and many more benefits are anticipated from ongoing nanotechnology research. However, serious concerns have been expressed about the potential hazards of nanomaterials that can pose a severe threat to the environment, ecosystems, and human health (Hegde et al. 2015). According to studies, most nanomaterials have a dual character; their usage is both appealing and risky due to the toxicity of these materials (Fig. 31.2) (Baran 2016).

31.7 Health Risks

31.7.1 Pulmonary Toxicity

Due to their small size, NPs can gain easy access to airway cells and even intracellular components. Studies on the inhalation toxicity of ultrafine particles by laboratory animals at very high particle concentrations have been reported. Data from both animal and human studies indicate that NPs can have both immediate and long-term effects on the lung, including inflammation, asthmatic flare-ups, genotoxicity, and carcinogenesis (Borm et al. 2006). Certain studies suggest that inhaled ultrafine particles evade alveolar macrophage surveillance after depositing in the lung and gain better access to the pulmonary interstitium via translocation from alveolar spaces through epithelium (Donaldson et al. 2001). NPs have high deposition rates in the lungs of healthy individuals and even higher in individuals with asthma or chronic obstructive pulmonary disease (COPD) (Oberdörster et al. 2005).

31.7.2 Neurotoxicology

In studies of NPs interaction with neurons, both positive and negative effects have been found. The majority of research on the interaction of CNS with NPs has focused on metals or metal oxides (including Cu, CuO, Zn, and Ag) with specific neuronal cell lines (PC-12, CA1, and CA3) (Yang et al. 2010). NPs have demonstrated favorable biological roles, such as eliminating dangerous bacteria and viruses (e.g., flu). Still, research has also shown that NPs may have negative consequences in human cells. NP has the capacity to penetrate through biological membranes and may have significant functional and harmful impacts on human brain cells, which is one of the most serious problems in research today (Brooking et al. 2001). There has been no detailed investigation of the effects of metal NP accumulation in the brain or through the blood-brain barrier (BBB). The migration of NPs over the BBB is expected to occur via two mechanisms: passive diffusion or carrier-mediated endocytosis (Hoet et al. 2004). After inhalation through the olfactory epithelium, NPs may enter the brain via transsynaptic transport (Oberdörster

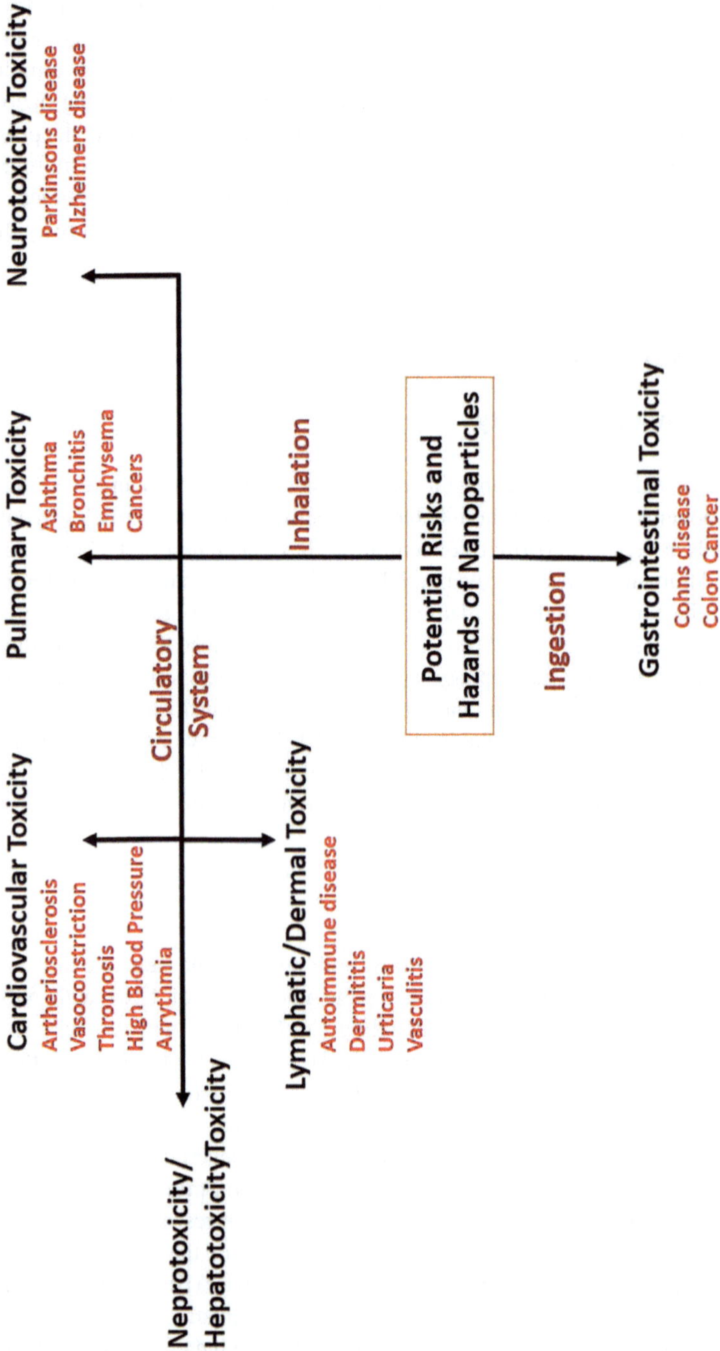

Fig. 31.2 Potential risks and hazards of nanomaterials on the human health

et al. 2005). For example, Ag NPs may penetrate the BBB and aggregate in various parts of the brain, which may aid in medication delivery but may possibly harm the patient (Yang et al. 2010). NP exposure may induce oxidative stress, and this has been linked to the development of neurodegenerative diseases such as Parkinson's and Alzheimer's. It is quite possible that the long-term effects of NP exposure might include a decrease in cognitive function (Borm et al. 2006).

31.7.3 Dermal Toxicity

Particles with sizes ranging from 50 to 500 nm are frequently used in cosmetic products to increase the uniformity of the formulations or to serve as UV filters against solar radiation (Borm et al. 2006). The main subject of discussion is whether these particles can penetrate the skin. Dermal exposure to NPs of smaller sizes (10 nm) is more harmful than exposure to those of larger sizes (>30 nm). NPs of smaller sizes may penetrate more readily than those of bigger sizes. According to reports, exposure to NPs smaller than 10 nm resulted in eschar development, persistent erythema, and edema. After exposing rabbits to NPs of smaller size, hyperkeratosis and papillomatosis in the irregular epidermis, as well as fibrosis, hyperemia, erythema, intracellular edema, and hyalinization, were observed. Once NPs infiltrate the skin, they may impact the integrity of the nuclear DNA because of their ability to induce the production of ROS and elicit oxidative stress within the cell (Gautam et al. 2011). Since NPs used in the cosmetic industry are usually greater in size than 10 nm, the health risk from dermal exposure to NPs is relatively low. However, further exploration and investigation are required to expand the current data set.

31.8 Environmental Toxicity of Nanomaterials

NPs can be produced naturally through processes like combustion or nucleation. It is predicted that around 50,000 kg of nano-sized materials is created each year through unintentional human causes. NPs are also being produced, and depending on the processes utilized, NPs might be discharged into the air and water and eventually contaminate soil and food products (Reijnders 2006). Intentionally or unintentionally produced, the occurrence of NPs in the environment is continuously rising. As a result, all stakeholders are becoming concerned. The benefits of NPs, such as their small size, high reactivity, and large capacity, might become deadly factors by triggering cellular toxicity (Khan et al. 2019). There is a scarcity of data on the effects of synthetic nanomaterials on ecologically important species, but some studies have indicated that fullerenes and their derivatives may cause toxicity in fish, daphnia, and bacteria (Borm et al. 2006). Reportedly, metal nanomaterials, metal oxide nanomaterials, quantum dots, fullerenes, and fibrous nanomaterials have the ability to damage or interact with DNA. Many of the man-made nanomaterials examined were found to elicit genotoxic effects, such as chromosomal fragmentation, DNA strand breakages, point mutations, oxidative DNA adducts, and abnormalities in gene expression profiles (Singh et al. 2009).

The environmental impact of NPs is predicted to worsen in the future. One of the hazardous properties of NPs is their ability to organize around protein concentrations, which is determined by particle size, curvature, shape, and surface characteristics such as charge, functionalized groups, and free energy. Because of this interaction, some particles cause adverse biological effects such as protein unfolding, fibrillation, thiol cross-linking, and loss of enzyme activity (Khan et al. 2019). With the increasing number of newly emerging nanomaterials, there is a need for continued and extensive research to understand the properties and possible toxicity of these materials (Bratovcic 2019).

31.9 Regulatory Landscape of Nanomaterials

Establishing a unified worldwide regulation of nanotechnology activities is critical for addressing the safety issues caused by the usage of nanomaterials. However, due to the use of nanotechnology in numerous areas of the economy, this is not a simple task. There is currently no clear international rule governing nanotechnology or nano-derived goods. A few government bodies or groups, however, have set rules and laws to define and control the usage of nanoproducts (Jain et al. 2018). In the following section, a general overview of nanoregulation in different countries has been discussed.

31.9.1 United States of America

Like many other developed countries, the United States has regulations that can help to reduce and control the threats that nanomaterials pose to public health and environmental concerns. For instance, the US Food and Drug Administration (FDA) is legally empowered to control nanomaterials used in foods, medications, medical devices, and cosmetics in order to safeguard the general population from their hazardous effects (Resnik 2019). The FDA approaches the regulation of nanotechnology by acknowledging that all developing technologies come with a promise, danger, and uncertainty. No new rules and regulations specific to nanomaterials have been introduced because of the assumption that the existing framework is adequate to regulate nanomaterials. The FDA now relies on horizon scanning to stay up to date on new discoveries and possible hazards posed by nano-derived goods. Through the National Nanotechnology Initiative (NNI), the FDA has been working with various government departments and agencies on developing nanotoxicity studies. NNI is a joint endeavor of more than 20 US agencies and departments and is supervised by the National Science and Technology Council of the White House Office of Science and Technology. The NNI priority is to focus on topics like infrastructure for nanomaterial measurement, risk assessment and management, exposure assessment, etc. (Allan et al. 2021).

31.9.2 European Union

Nanomaterials are recognized as a type of chemical substance in the European Union and are subjected to legislative control like other chemicals. The criteria in this respect are set in two major regulations, i.e., Registration, Evaluation, Authorisation and Restriction of Chemicals (REACH) ((EC) No 1907/2006) and Convention on the Limitation of Chemicals/Classification, Labelling and Packaging (CLP) ((EC) No 1272/2008). REACH and CLP is primarily concerned with chemical substances, and while no special provisions have been made with regard to nanomaterials, they are mainly covered by these regulations (Schwirn et al. 2014; Alessandrelli and Polci 2011). Moreover, REACH was modified in 2018 to include nanotechnology-related information, as well as additional rules for chemical safety assessment and downstream user duties (Allan et al. 2021).

An independent European executive body, the European Food Safety Authority (EFSA), evaluates the food and feed chain risks. It covers biological risks, animal welfare and health, feed and food additives, pesticides, plant protection, and health. The guidance on risk assessment of the application of nanoscience and nanotechnologies in the food and feed chain and human and animal health has been updated by EFSA. The updated guidance has taken into account the important scientific findings that offer insight into the physicochemical properties, exposure assessment, and hazard characterization of nanomaterials and areas of applicability (EFSA Scientific Committee 2021).

31.9.3 Canada

Several Canadian departments and agencies are relatively active in nanoregulation, like Health Canada, Environment and Climate Change Canada, Agriculture and Agri-Food Canada, and the Canadian Food Inspection Agency. These organizations focus on risk assessment and mitigation and the creation of a register of nanomaterials and nano-derived products in the market. According to the Working Definition of Nanomaterials adopted by Health Canada, nanomaterials include manufactured substances, products, components, ingredients, devices, and structures that are either within or near the nanoscale (1–100 nm) in at least one spatial dimension or are smaller or larger than the nanoscale in all spatial dimensions and exhibit nanoscale properties and phenomena. By using this working definition as a tool, regulated products and substances containing nanomaterials can be identified. Canada relies on current legal frameworks for the approval of nanotechnology products. In order to identify and evaluate the product's risks and characteristics during the early development phase, Health Canada urges manufacturers to consult with the relevant regulatory body (Health Canada 2011). Furthermore, Canadian regulatory authorities are working closely with stakeholders in the United States and Europe to create a uniform framework for prioritizing actions and analyzing hazards connected with NPs.

31.9.4 Asia

Regulation in this region is being overseen by the Japanese Ministry of Health, Labour, and Welfare (MHLW), the Pharmaceutical and Medical Devices Agency (PMDA), the National Institute of Health Sciences (NIHS), and the Agency for Medical Research and Development (AMED). In 2018, the PMDA launched a regulatory science center to stimulate innovative approaches to sophisticated therapies and technology, such as nanotechnologies, with a heavy emphasis on horizon scanning to assist regulators in keeping up with emerging breakthroughs. (Allan et al. 2021). In other Asian countries like Thailand, China, and India, the regulatory scenario to comprehensively deal with risks posed by nanotechnology are yet to evolve.

31.10 Regulatory Challenges

Although nanotechnology has received the necessary funding for research and product development, the need for establishing a unifying regulatory framework to address environmental and health risks associated with nanotechnology has more or less been ignored. This is critical given the scientific uncertainty surrounding the effects of specific nanomaterials on human health and the environment. Despite being deemed competent, the current fragmented regulatory frameworks worldwide have certain loopholes that preclude comprehensive regulation of nanotechnology. There is a need for an internationally integrated regulatory framework that can address the ethical, legal, and social aspects of nanotechnology (Devasahayam 2019). Most of the challenges faced by the regulators at present mainly arise from uncertainties surrounding nanomaterials and huge gaps in knowledge. The following major challenges with respect to the regulation of nanomaterials have been identified (Fig. 31.3):

1. Lack of globally acknowledged definition of nanomaterials
2. Insufficient ability to track the origins and pathways of nanomaterials
3. Difficulty in nanomaterial exposure assessment
4. Lack of understanding of toxic mechanisms of nanomaterials
5. Difficulty in evaluating the bioavailability of nanomaterials

A detailed explanation of these regulatory challenges is given in the following sections.

31.11 Lack of a Globally Acknowledged Definition

Definitions are crucial for establishing rules and regulations for material testing. Although some regulatory bodies have attempted to create standards, the term "nanomaterial" still lacks a globally acknowledged definition. In the absence of an

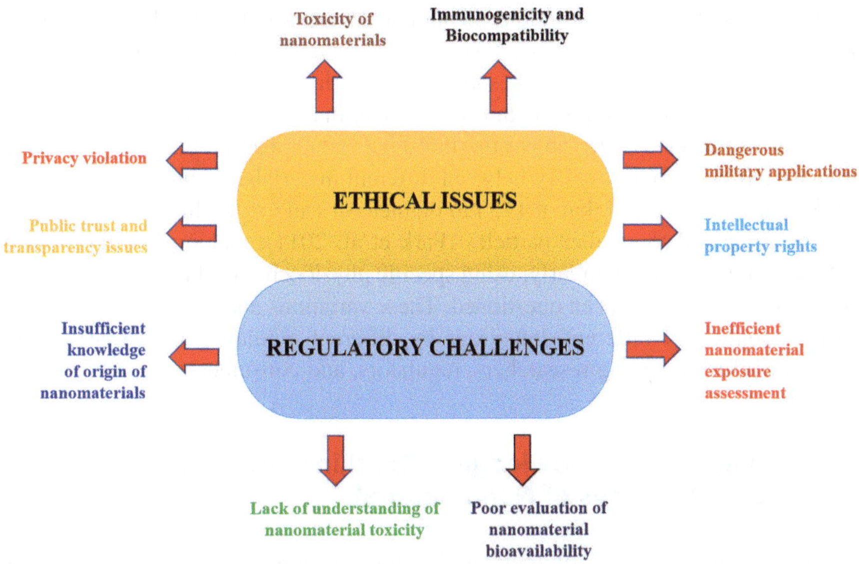

Fig. 31.3 Ethical issues and regulatory challenges surrounding the nanomaterials

internationally agreed terminology, the effective regulation of nanomaterials at a global level seems impossible. Nanomaterials can be defined based on a size threshold, their nano-specific properties, or a combination of size and properties (Lai et al. 2018). International Organization for Standardization (ISO) defines nanomaterial as a "material with any external dimension in the nanoscale or having an internal structure or surface structure in the nanoscale." Working Party on Manufactured Materials (WPMM), established by the Organization for Economic Co-operation and Development (OECD), defines nanomaterial as a "Material which is either a nano-object or is nanostructured" (Lövestam et al. 2010). Health Canada considers any manufactured product, material, substance, ingredient, device, system, or structure to be nanomaterial if:

(a) It is at or within the nanoscale (1–100 nm) in at least one spatial dimension
(b) It is smaller or larger than the nanoscale in all spatial dimensions and exhibits one or more nanoscale phenomena

Nanomaterials are defined by the Danish Ministry of the Environment as "materials that are fewer than 100 nanometres in length along the shortest side or have structures that have such tiny dimensions but are incorporated into bigger materials" (i.e., nanostructured surfaces). Nanomaterials can be manufactured from existing chemical substances and entirely novel chemical compounds, and they can be formed from one or more components. The materials' unique properties are due to their tiny size (Lövestam et al. 2010).

Today, most existing definitions mention a 100 nm size threshold for a substance to qualify as a nanomaterial. However, none of these definitions are workable. This is because studies have shown that some nanomaterials exhibit unique physicochemical properties at sizes below the threshold value. In contrast, other materials exhibit unusual properties at sizes above the threshold value. The particle-size-dependent toxicity has been found to vary between different materials. A study on the toxicity of Ag particles revealed that at 113 nm diameter it can exhibit higher toxicological effects compared to smaller particles (Park et al. 2011), making the threshold of 100 nm unreasonable. Similarly, using specific physicochemical properties to define nanomaterials has also been questioned. These variations and uncertainties have led to the adoption of different definitions by different regulating bodies, which has created confusion among researchers, regulators, and consumers.

31.12 Inadequate Knowledge of Tracking the Origins and Pathways of Nanomaterials

Because of the diversity of their applications, nanomaterials may infiltrate the environment through many pathways. This infiltration results in the deposition of nanomaterials in soils and aquatic ecosystems (Tourinho et al. 2012). It is imperative to trace and remove the nanomaterials from the system in order to safeguard the environment. In some cases, there is a possibility to track and remove NMs, but in other cases, this task is rather difficult. For example, nanomaterials in commercial products like pharmaceuticals, paints, etc. that may be released into the sewage system can be collected and eventually dumped into landfills (Lai et al. 2018). On the other hand, NMs that may be released into the environment through accidental spills can easily escape the tracking, collection, and removal systems. Further, the transport and fate of nanomaterials in the environment are decided by their physicochemical properties and environmental factors. Some nanomaterials, after entering the environment, undergo a transformation like agglomeration and aggregation (Tourinho et al. 2012). In order to prevent agglomeration and settling of NMs in aqueous environments, manufacturers coat the surfaces of nanomaterials, which selectively alters their properties. In the absence of surface coatings, some metal-based nanomaterials may undergo dissolution. This dissolution results in the release of ionic species that may be toxic. To understand the potential long-term effects of nanomaterials, it is imperative to consider the degree of dissolution and the toxicities of nanoparticulate and dissolved forms.

With so many different sources, transit modes, and potential influencing factors, it is challenging to predict nanomaterial depositions once they enter the environment. This creates a lack of understanding regarding the transit and fate of NMs, which restricts the regulators' ability to measure the exposure of nanomaterials during risk assessment. Thus, regulators might not be able to identify vulnerable areas and organisms for regulation and protection.

31.13 Difficulty in Nanomaterial Exposure Assessment

Assessment of nanomaterial exposure is critical for accurate hazard assessment and for creating a practical regulatory framework. Nanomaterial sampling and exposure require different techniques from those employed for their bulk counterparts. There are several strategies for characterizing NPs for exposure, each with advantages and limitations for the ecotoxicologist (Handy et al. 2008). Currently, novel techniques are being developed in addition to conventional labor-intensive procedures (filtration and electron microscopy). However, all of these techniques need to be refined before being applied to environmental investigations (Lai et al. 2018). In addition to the lack of reliable methods for quantifying nanomaterial exposure, there are no established metrics for airborne NP exposure assessment. Furthermore, there are no accepted occupational exposure levels (OELs) of manufactured nanomaterials (Lee et al. 2010) which further complicates the issue. Without effective quantifying techniques and protocols, regulators find it difficult to determine the potential risks of nanomaterials. This may prevent vulnerable areas and groups from receiving protection.

31.14 Lack of Understanding of Toxic Mechanisms of Nanomaterials

Toxicity assessment of nanomaterials is essential for an in-depth understanding of the potential adverse effects and their fate in the environment. However, assessing and validating the nanotoxicity in biological systems are hard. The ongoing discussion in this area is whether or not nanomaterials should be treated differently from their bulk counterparts. The toxicity of nanomaterials depends on their size and dose (Kim et al. 2015). However, some studies suggest that size-dependent toxicity varies among different NMs and species. For instance, a study on the toxicity of oxide NPs revealed that ZnO NPs are more toxic to *Chlorella* sp. than their bulk counterparts (Ji et al. 2011). On the contrary, other studies have suggested that ZnO NPs show similar or less toxicity to *P. subcapitata* (Aruoja et al. 2009).

Besides, a major challenge in the toxicity assessment of nanomaterials is the presence of degraded or transformed nanomaterials in the environment. Nanomaterials after transformation may have altered biological influences and toxicities than their pristine forms, which is why it is crucial that these transformed nanomaterials must be evaluated and assessed for their potential toxicity. However, at present, most of the studies are conducted on pristine nanomaterials (Gupta et al. 2019).

Studies have shown that nanomaterials can exhibit toxicity via various mechanisms—by triggering signaling pathways that damage the cell or by releasing toxic metal ions that impair essential enzyme functions or by inducing the production of reactive oxygen species (ROS) (Buchman et al. 2019). Different categories of NPs adopt other mechanisms to manifest their toxicity. In some cases, different species evoke different responses after being exposed to the same nanomaterial

(Neal et al. 2012; Su et al. 2015). Thus, there are still a lot of questions and uncertainties about the toxicity and mechanism of action of nanomaterials, particularly in light of their diverse physicochemical properties, various environmental conditions, and the lack of reliable toxicity test protocols. Without a proper understanding of the toxicity mechanisms of nanomaterials, it is difficult for regulators to formulate a risk and hazard assessment framework.

31.15 Difficulty in Evaluating the Bioavailability of Nanomaterials

Since bioavailable forms of nanomaterials can have direct biological effects on organisms and can accumulate along the food chain, determining the bioavailability of nanomaterials is crucial. As discussed earlier, nanomaterials have a tendency to transform once they enter the environment. These transformations may determine the bioavailability of NMs to biota. Agglomeration results in decreased mobility of nanomaterials, which causes them to sediment and become less accessible to organisms in the water column (Behra et al. 2013). However, it is believed that large aggregated nanomaterials might still be bioavailable to organisms like filter feeders and suspension feeders. Furthermore, the intake of large nanomaterials does not guarantee that organisms are safe from NMs because agglomerations of nanomaterials connected by weak interactions may dissociate to smaller particles when exposed to changes in ambient conditions (Lai et al. 2018). Surface coatings and stabilizing surfactants prevent the agglomeration of NMs, which allows for the retention of NMs in the upper layers (Darlington et al. 2009). While this reduces the availability of NMs to sediment-dwellers, it exposes them to pelagic species (Ankley et al. 2010).

Moreover, the NMs used in environmental remediation interact with various pollutants, which alters their bioavailability. Because of their strong affinities to aromatic compounds, carbon nanotubes (CNTs) are considered to be excellent remediating tools. Their strong affinity reduces their bioavailability to organisms. On the other hand, the interaction of CNTs to compounds like pyrene is weak and can be reversed. This results in the desorption of these compounds, thereby making both the CNTs and the pollutant (pyrene) bioavailable to biota. Since bioavailable forms of nanomaterials can have direct biological effects on organisms and can accumulate along the food chain, determining the bioavailability of these nanomaterials is the first step in evaluating their potential dangers. But before their potential dangers can be accurately assessed for regulators, further research is necessary given the complex interplay between nanomaterials and their physicochemical properties, environmental variables, and interaction with other pollutants.

31.16 Ethical Issues

The development of nanotechnology has raised many ethical questions related to public trust, potential risks, environmental impact issues, and information transparency. Nanotechnology must not only be safe and valuable, but it must also earn public acceptance. There are numerous speculations among the masses regarding nanotechnology, which have resulted in exaggerated public hopes and fears, so much so that nanotechnology is discussed mainly in terms of the societal and ethical implications of these speculations. The prevalence of such speculations hampers the identification of actual and relevant ethical issues. The public must be made aware of the benefits, potential risks, and necessary measures related to the use of nanotechnology so that they can make informed decisions and form independent opinions. Here, researchers could play a vital role in informing the public about the potential benefits and hazards of nanomaterials (Baran 2016).

Identification of ethical issues that arise due to the use of nanomaterials is crucial as it helps with proper decision-making. It is, therefore, imperative for policymakers, employers, workers, investors, and healthcare authorities to be aware of these issues (Schulte and Salamanca-Buentello 2007). Some of the main ethical issues have been discussed below (Fig. 31.3).

31.16.1 Privacy Violation

Nanotechnology has the potential to dramatically improve surveillance devices. These devices are essential law enforcement tools that are mostly aimed at safeguarding the public. However, on the flip side, these devices can result in intrusive surveillance—the biggest privacy concern surrounding nanotechnology (Ferreira and Filipe 2022). Moreover, the use of devices like microchips for customized drug delivery has been a topic of discussion among the scientific community. These devices could also threaten individual security and privacy. With nanotechnology bringing new types of surveillance devices that have the potential to invade individual privacy, legislation should be passed not necessarily to prohibit surveillance but to ensure that information received from such surveillance is fair, transparent, and subject to the law.

31.16.2 Issues Arising from Military Applications

Nanotechnology research can be used to develop deadly weapons, remarkably miniaturized guns and explosives, that might pose new threats and spark a new arms race among nations. Additionally, there is a serious risk to public security in case these weapons end up in the hands of terrorists. It is also believed that nanotechnology can be used to augment the military performance of soldiers by enhancing their physical and mental strength. Such applications can prove to be dangerous since they might result in long-term psychological and physiological effects. However, many of the worries are based on conjecture because most military research is classified (Schummer 2007).

31.16.3 Public Trust and Transparency Issues

Government organizations are abusing the danger of terrorism to classify research and disregard scientific findings that conflict with their political objectives (Unesco 2006). As a result, important information is kept away from the public domain. This adds to the fears and speculation among people, which is why authorities should encourage open access to research findings. This is essential for maintaining transparency and gaining public trust.

31.16.4 Intellectual Property Rights

Companies and innovators can protect their intellectual property rights with patents and trademarks. However, in recent times, few disputes have plagued the use of intellectual property in science and science-based commerce, such as new database laws, which effectively give single corporations rights over facts, and overly liberal patent granting, which can result in increased litigation costs and overlapping patent claims (Unesco 2006). Further, academic knowledge that was once published in academic journals and was a part of the public domain is now increasingly being protected by patents and licensed on the market. While this might have helped universities transfer technology to local businesses and provided these universities with new revenue streams, it has made industrial development much more complex and expensive because every piece of fundamental knowledge must now be purchased. This will significantly widen the technology gap between developed and developing countries since the latter might not be able to pay license fees (Schummer 2007).

31.16.5 The "Gray Goo" Myth

There is a dystopian narrative that at some point in nanotechnology development, self-replicating nanobots or devices will be created that could potentially escape, violate the biosphere, and become uncontrollable, leading to a catastrophe called "gray goo." However, this narrative is purely hypothetical and is based on mere speculation.

The "gray goo" scenario at the moment is scientifically implausible and is a distraction as it steers the conversation away from the real scientific and ethical issues that exist today (Schummer 2007).

31.17 Conclusion

Although nanotechnology-derived products have the potential to improve and dramatically revolutionize a multitude of sectors, their use may pose risks that are still difficult to identify, monitor, and quantify. Even though commendable progress has been made in the quantification, risk assessment, and regulation of nanomaterials, there still are some knowledge gaps that hamper the process of regulation. Filling these knowledge gaps is extremely crucial for establishing more effective and stringent regulatory guidelines. A formal, international regulatory framework for nanotechnology is the need of the hour, as most of the existing regulatory guidelines are inconsistent. Like every new technology, nanotechnology is not free from its concomitant ethical issues. The main issues that need to be addressed are privacy, security, public trust, and transparency of information. If the study of these issues lags behind the rate of scientific development, nanotechnology may have to face a premature end.

References

Ahmed HM, Roy A, Wahab M, Ahmed M, Othman-Qadir G, Elesawy BH et al (2021) Applications of nanomaterials in agrifood and pharmaceutical industry. J Nanomater 2021:1472096

Allan J, Belz S, Hoeveler A, Hugas M, Okuda H, Patri A et al (2021) Regulatory landscape of nanotechnology and nanoplastics from a global perspective. Regul Toxicol Pharmacol 122: 104885

Alessandrelli M, Polci ML (2011) CLP application to nanomaterials: a specific aspect. Ann Ist Super Sanita 47:146–152

Ameta SK, Rai AK, Hiran D, Ameta R, Ameta SC (2020) Use of nanomaterials in food science. In: Biogenic nano-particles and their use in agro-ecosystems. Springer, Singapore, pp 457–488

Ankley GT, Bennett RS, Erickson RJ, Hoff DJ, Hornung MW, Johnson RD et al (2010) Adverse outcome pathways: a conceptual framework to support ecotoxicology research and risk assessment. Environ Toxicol Chem Int J 29(3):730–741

Aruoja V, Dubourguier HC, Kasemets K, Kahru A (2009) Toxicity of nanoparticles of CuO, ZnO and TiO2 to microalgae Pseudokirchneriella subcapitata. Sci Total Environ 407(4):1461–1468

Astefanei A, Núñez O, Galceran MT (2015) Characterisation and determination of fullerenes: a critical review. Anal Chim Acta 882:1–21

Baran A (2016) Nanotechnology: legal and ethical issues. Ekonomia i Zarządzanie 8(1):47–54

Behra R, Sigg L, Clift MJ, Herzog F, Minghetti M, Johnston B et al (2013) Bioavailability of silver nanoparticles and ions: from a chemical and biochemical perspective. J R Soc Interface 10(87): 20130396

Borm PJ, Robbins D, Haubold S, Kuhlbusch T, Fissan H, Donaldson K et al (2006) The potential risks of nanomaterials: a review carried out for ECETOC. Part Fibre Toxicol 3(1):1–35

Bratovcic A (2019) Different applications of nanomaterials and their impact on the environment. Int J Mater Sci Eng 5(1):1–7

Brooking J, Davis SS, Illum L (2001) Transport of nanoparticles across the rat nasal mucosa. J Drug Target 9(4):267–279

Buchman JT, Hudson-Smith NV, Landy KM, Haynes CL (2019) Understanding nanoparticle toxicity mechanisms to inform redesign strategies to reduce environmental impact. Acc Chem Res 52(6):1632–1642

Thomas SC, Kumar Mishra P, Talegaonkar S (2015) Ceramic nanoparticles: fabrication methods and applications in drug delivery. Curr Pharm Des 21(42):6165–6188

Chrimes AF, Khoshmanesh K, Stoddart PR, Kayani AA, Mitchell A, Daima H et al (2012) Active control of silver nanoparticles spacing using dielectrophoresis for surface-enhanced Raman scattering. Anal Chem 84(9):4029–4035

Darlington TK, Neigh AM, Spencer MT, Guyen OT, Oldenburg SJ (2009) Nanoparticle characteristics affecting environmental fate and transport through soil. Environ Toxicol Chem Int J 28(6):1191–1199

Das S, Mitra S, Khurana SP, Debnath N (2013) Nanomaterials for biomedical applications. Front Life Sci 7(3–4):90–98

Devasahayam S (2019) Nanotechnology and nanomedicine in market: a global perspective on regulatory issues. In: Characterization and biology of nanomaterials for drug delivery. Elsevier, pp 477–522

D'Mello SR, Cruz CN, Chen ML, Kapoor M, Lee SL, Tyner KM (2017) The evolving landscape of drug products containing nanomaterials in the United States. Nat Nanotechnol 12(6):523–529

Donaldson K, Stone V, Clouter A, Renwick L, MacNee W (2001) Ultrafine particles. Occup Environ Med 58(3):211–216

Dwivedi S, Saquib Q, Al-Khedhairy AA, Musarrat J (2016) Understanding the role of nanomaterials in agriculture. In: Microbial inoculants in sustainable agricultural productivity. Springer, New Delhi, pp 271–288

EFSA Scientific Committee (2021) Guidance on risk assessment of nanomaterials to be applied in the food and feed chain: human and animal health

Ferreira MA, Filipe JA (2022) Ethical considerations on nanotechnology. arXiv preprint arXiv:2202.01063

Gautam A, Singh D, Vijayaraghavan R (2011) Dermal exposure of nanoparticles: an understanding. J Cell Tissue Res 11(1):2703–2708

Ghaffar N, Farrukh MA, Naz S (2020) Applications of nanobiosensors in agriculture. In: Nanoagronomy. Springer, Cham, pp 179–196

Guerra FD, Attia MF, Whitehead DC, Alexis F (2018) Nanotechnology for environmental remediation: materials and applications. Molecules 23(7):1760

Gupta A, Kumar S, Kumar V (2019) Challenges for assessing toxicity of nanomaterials. Biochemical toxicology-heavy metals and nanomaterials

Han J, Zhao D, Li D, Wang X, Jin Z, Zhao K (2018) Polymer-based nanomaterials and applications for vaccines and drugs. Polymers 10(1):31

Handy RD, Von der Kammer F, Lead JR, Hassellöv M, Owen R, Crane M (2008) The ecotoxicology and chemistry of manufactured nanoparticles. Ecotoxicology 17(4):287–314

Hashem A, Hossain MM, Al Mamun M, Simarani K, Johan MR (2021) Nanomaterials based electrochemical nucleic acid biosensors for environmental monitoring: a review. Appl Surface Sci Adv 4:100064

Health Canada (2011) Policy Statement on Health Canada's Working Definition for Nanomaterial

Hegde K, Goswami R, Sarma SJ, Veeranki VD, Brar SK, Surampalli RY (2015) Environmental hazards and risks of nanomaterials. In: Nanomaterials in the environment, 1st edn. American Society of Civil Engineers, Reston, pp 357–382

Hoet PH, Brüske-Hohlfeld I, Salata OV (2004) Nanoparticles–known and unknown health risks. J Nanobiotechnol 2:1–15

Islam MT (2019) Applications of nanomaterials for future food security: challenges and prospects. Malays J Halal Res 2(1):6–9

Jain A, Ranjan S, Dasgupta N, Ramalingam C (2018) Nanomaterials in food and agriculture: an overview on their safety concerns and regulatory issues. Crit Rev Food Sci Nutr 58(2):297–317

Ji J, Long Z, Lin D (2011) Toxicity of oxide nanoparticles to the green algae Chlorella sp. Chem Eng J 170(2–3):525–530

Khan I, Saeed K, Khan I (2019) Nanoparticles: properties, applications and toxicities. Arab J Chem 12(7):908–931

Khin MM, Nair AS, Babu VJ, Murugan R, Ramakrishna S (2012) A review on nanomaterials for environmental remediation. Energy Environ Sci 5(8):8075–8109

Kim IY, Joachim E, Choi H, Kim K (2015) Toxicity of silica nanoparticles depends on size, dose, and cell type. Nanomedicine 11(6):1407–1416
Taylor-Pashow KM, Della Rocca J, Huxford RC, Lin W (2010) Hybrid nanomaterials for biomedical applications. Chem Commun 46(32):5832–5849
Lai RW, Yeung KW, Yung M, Djurišić AB, Giesy JP, Leung KM (2018) Regulation of engineered nanomaterials: current challenges, insights and future directions. Environ Sci Pollut Res 25(4): 3060–3077
Laurent S, Forge D, Port M, Roch A, Robic C, Vander Elst L et al (2008) Magnetic iron oxide nanoparticles: synthesis, stabilization, vectorization, physicochemical characterizations, and biological applications. Chem Rev 108(6):2064–2110
Lee JH, Moon MC, Lee JY, Yu IJ (2010) Challenges and perspectives of nanoparticle exposure assessment. Toxicol Res 26(2):95–100
Liao C, Li Y, Tjong SC (2018) Graphene nanomaterials: synthesis, biocompatibility, and cytotoxicity. Int J Mol Sci 19(11):3564
Lövestam G, Rauscher H, Roebben G, Klüttgen BS, Gibson N, Putaud JP et al (2010) Considerations on a definition of nanomaterial for regulatory purposes. Joint Res Centre Ref Rep 80:00–41
Maduraiveeran G, Jin W (2017) Nanomaterials based electrochemical sensor and biosensor platforms for environmental applications. Trends Environ Analyt Chem 13:10–23
Mariappan N (2019) Recent trends in nanotechnology applications in surgical specialties and orthopedic surgery. Biomed Pharmacol J 12(3):1095–1127
Mitura KA, Zarzycki PK (2018) Biocompatibility and toxicity of allotropic forms of carbon in food packaging. Role Mater Sci Food Bioeng:73–107
Morrow KJ Jr, Bawa R, Wei C (2007) Recent advances in basic and clinical nanomedicine. Med Clin N Am 91(5):805–843
Neal AL, Kabengi N, Grider A, Bertsch PM (2012) Can the soil bacterium Cupriavidus necator sense ZnO nanomaterials and aqueous Zn2+ differentially? Nanotoxicology 6(4):371–380
Niederberger M (2007) Nonaqueous sol–gel routes to metal oxide nanoparticles. Acc Chem Res 40(9):793–800
Nimibofa A, Newton EA, Cyprain AY, Donbebe W (2018) Fullerenes: synthesis and applications. J Mater Sci 7:22–33
Oberdörster G, Maynard A, Donaldson K, Castranova V, Fitzpatrick J, Ausman K et al (2005) Principles for characterizing the potential human health effects from exposure to nanomaterials: elements of a screening strategy. Part Fibre Toxicol 2(1):1–35
Otles S, Yalcin B (2010) Nano-biosensors as new tool for detection of food quality and safety. Log Forum 6(4):7
Pande A, Arora S (2019) Nanotechnological interventions for improving plant health and productivity. In: Plant biotechnology: progress in genomic era. Springer, Singapore, pp 375–395
Panpatte DG, Jhala YK, Shelat HN, Vyas RV (2016) Nanoparticles: the next generation technology for sustainable agriculture. In: Microbial inoculants in sustainable agricultural productivity. Springer, New Delhi, pp 289–300
Park MV, Neigh AM, Vermeulen JP, de la Fonteyne LJ, Verharen HW, Briedé JJ et al (2011) The effect of particle size on the cytotoxicity, inflammation, developmental toxicity and genotoxicity of silver nanoparticles. Biomaterials 32(36):9810–9817
Patra CR, Bhattacharya R, Mukhopadhyay D, Mukherjee P (2008) Application of gold nanoparticles for targeted therapy in cancer. J Biomed Nanotechnol 4(2):99–132
Ravichandran R (2010) Nanotechnology applications in food and food processing: innovative green approaches, opportunities and uncertainties for global market. Int J Green Nanotechnol Phys Chem 1(2):72–96
Reijnders L (2006) Cleaner nanotechnology and hazard reduction of manufactured nanoparticles. J Clean Prod 14(2):124–133

Resnik DB (2019) How should engineered nanomaterials be regulated for public and environmental health? AMA J Ethics 21(4):363–369
Sajid M (2021) Nanomaterials: types, properties, synthesis, emerging materials, and toxicity concerns. Curr Opin Environ Sci Health 14:100319
Salem SS, Hammad EN, Mohamed AA, El-Dougdoug W (2022) A comprehensive review of nanomaterials: types, synthesis, characterization, and applications. Biointerface Res Appl Chem 13(1):41
Schulte PA, Salamanca-Buentello F (2007) Ethical and scientific issues of nanotechnology in the workplace. Environ Health Perspect 115(1):5–12
Schummer J (2007) Identifying ethical issues of nanotechnologies. Nanotechnol Ethics Polit:79–98
Schwirn K, Tietjen L, Beer I (2014) Why are nanomaterials different and how can they be appropriately regulated under REACH? Environ Sci Eur 26(1):1–9
Sharma D, Hussain CM (2020) Smart nanomaterials in pharmaceutical analysis. Arab J Chem 13(1):3319–3343
Sharma G, Kumar D, Kumar A, Ala'a H, Pathania D, Naushad M et al (2017) Revolution from monometallic to trimetallic nanoparticle composites, various synthesis methods and their applications: a review. Mater Sci Eng C 71:1216–1230
Singh N, Manshian B, Jenkins GJ, Griffiths SM, Williams PM, Maffeis TG et al (2009) NanoGenotoxicology: the DNA damaging potential of engineered nanomaterials. Biomaterials 30(23–24):3891–3914
Singhal S, Nie S, Wang MD (2010) Nanotechnology applications in surgical oncology. Annu Rev Med 61:359
Slavin YN, Asnis J, Häfeli UO, Bach H (2017) Metal nanoparticles: understanding the mechanisms behind antibacterial activity. J Nanobiotechnol 15(1):1–20
Su G, Zhang X, Giesy JP, Musarrat J, Saquib Q, Alkhedhairy AA et al (2015) Comparison on the molecular response profiles between nano zinc oxide (ZnO) particles and free zinc ion using a genome-wide toxicogenomics approach. Environ Sci Pollut Res 22(22):17434–17442
Tourinho PS, Van Gestel CA, Lofts S, Svendsen C, Soares AM, Loureiro S (2012) Metal-based nanoparticles in soil: fate, behavior, and effects on soil invertebrates. Environ Toxicol Chem 31(8):1679–1692
Tyner KM, Zheng N, Choi S, Xu X, Zou P, Jiang W et al (2017) How has CDER prepared for the nano revolution? A review of risk assessment, regulatory research, and guidance activities. AAPS J 19(4):1071–1083
Unesco (2006) The ethics and politics of nanotechnology. Unesco
Vajtai R (ed) (2013) Springer handbook of nanomaterials. Springer Science & Business Media
Verma SK, Das AK, Gantait S, Kumar V, Gurel E (2019) Applications of carbon nanomaterials in the plant system: a perspective view on the pros and cons. Sci Total Environ 667:485–499
Wang C, Fan W, Zhang Z, Wen Y, Xiong L, Chen X (2019a) Advanced nanotechnology leading the way to multimodal imaging-guided precision surgical therapy. Adv Mater 31(49):1904329
Wang L, Gong C, Yuan X, Wei G (2019b) Controlling the self-assembly of biomolecules into functional nanomaterials through internal interactions and external stimulations: a review. Nano 9(2):285
Yang Z, Liu ZW, Allaker RP, Reip P, Oxford J, Ahmad Z et al (2010) A review of nanoparticle functionality and toxicity on the central nervous system. J R Soc Interface 7(suppl_4):S411–S422
Zhang W, Asiri AM, Liu D, Du D, Lin Y (2014) Nanomaterial-based biosensors for environmental and biological monitoring of organophosphorus pesticides and nerve agents. TrAC Trends Anal Chem 54:1–10

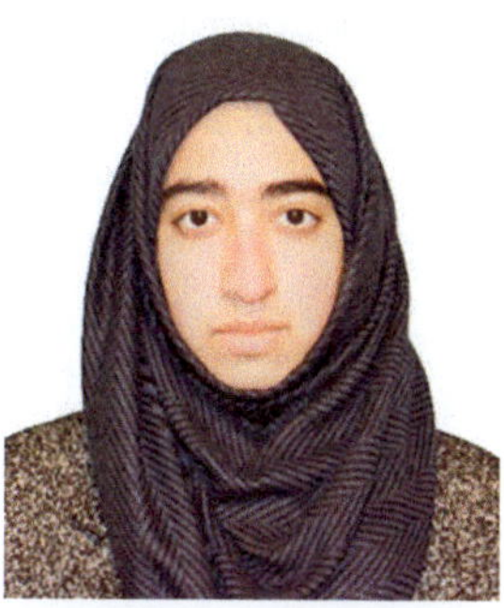

Faamiya Shajar is currently pursuing PhD in plant biology from the Department of Bioresources, University of Kashmir. Her study mainly focusses on elucidating the role of lipids in salt stress responses in proso millet.

Seerat Saleem is currently pursuing PhD in plant biology from the Department of Bioresources, University of Kashmir. She is working on mineral nutrition and abiotic stress amelioration in foxtail millet.

Naveed Ul Mushtaq is currently pursuing PhD in plant biology from the Department of Bioresources, University of Kashmir. He is working on mineral nutrition and abiotic stress amelioration in proso millet.

Dr. Wasifa Hafiz Shah is currently working as research associate (ICMR) in the Department of Clinical Biochemistry, University of Kashmir. She is presently working on phytochemical modulation of drug-induced type 2 diabetes.

Dr. Aadil Rasool is currently working as research associate (ICMR) in the Department of Clinical Biochemistry, University of Kashmir. He is presently working on identification of bioactive molecules with glycation inhibition potential from medicinal plants.

Sajad Ahmed Padder is a graduate student and is working as an ICMR research fellow in the field of human fungal pathogenesis at the Department of Bioresources, University of Kashmir.

Dr. Inayatullah Tahir is currently working as a professor, Department of Botany, University of Kashmir, Srinagar, India. He is working on mineral nutrition and abiotic stress amelioration in various underutilized crops. He has also been doing research on flower senescence and postharvest physiology of the ornamentals.

Dr. Reiaz Ul Rehman is currently working as an assistant professor, Department of Bioresources, University of Kashmir, Srinagar, India. He is working on mineral nutrition and abiotic stress amelioration in various underutilized crops.

MIX
Papier aus verantwortungsvollen Quellen
Paper from responsible sources
FSC® C105338

If you have any concerns about our products,
you can contact us on
ProductSafety@springernature.com

In case Publisher is established outside the EU,
the EU authorized representative is:
Springer Nature Customer Service Center GmbH
Europaplatz 3, 69115 Heidelberg, Germany

Printed by Libri Plureos GmbH
in Hamburg, Germany